The
ENCYCLOPEDIA
of
APPLIED GEOLOGY

ENCYCLOPEDIA OF EARTH SCIENCES SERIES

Series Editor: Rhodes W. Fairbridge

ENCYCLOPEDIA OF EARTH SCIENCES, VOLUME XIII

The
ENCYCLOPEDIA
of
APPLIED GEOLOGY

EDITED BY

Charles W. Finkl, Jnr.
Florida Atlantic University

VNR VAN NOSTRAND REINHOLD COMPANY
NEW YORK   CINCINNATI   TORONTO   LONDON   MELBOURNE

Manufactured in the United States of America.

Published by Van Nostrand Reinhold Company Inc.
135 West 50th Street
New York, New York 10020

Van Nostrand Reinhold Company Limited
Molly Millars Lane
Wokingham, Berkshire RG11 2PY, England

Van Nostrand Reinhold
480 Latrobe Street
Melbourne, Victoria 3000, Australia

Macmillan of Canada
Division of Gage Publishing Limited
164 Commander Boulevard
Agincourt, Ontario MIS 3C7, Canada

15   14   13   12   11   10 9 8 7 6 5 4 3 2 1

Library of Congress Cataloging in Publication Data
Main entry under title:
The Encyclopedia of applied geology.
   (Encyclopedia of earth sciences; v. 13)
   Includes bibliography and index.
   1. Geology—Dictionaries. I. Finkl, Charles W., 1941- . II. Series.
QE5.E5   1983   624.1′51′0321   83-12765
ISBN 0-442-22537-7

PREFACE

Applied geology is a diffuse field of endeavor that embraces diverse disciplines in the geosciences. Unlike cognate sciences, which are based on rather definitive points of view, applied geology involves the application of various aspects of geology to economic, engineering, water-supply, and environmental problems. According to the *Glossary of Geology* (American Geological Institute, 1980), applied geology is "geology related to human activity." This perspective distinguishes applied geology from other branches of geological science that do *not* focus on immediate problems of primary environmental concern. Many geologists are, in a sense, "applied geologists" because they study the earth as the home of man: the conservation and use of earth materials in construction, man's adaptation to geological-geomorphic processes in places where he lives or works, and man's perception and use of environmental resources. The phrase *applied geology,* however, is used to designate those geological efforts that bear direct or indirect application to practical problems related, for example, to construction in potentially hazardous materials such as swelling clays or in geologically active environments such as flood plains, along fault zones, or coastal areas.

Geologists working under the banner of applied geology are professional geoscientists with specializations in engineering geology, environmental geology, economic geology, geochemistry, geomorphology, or geophysics. Applied geology also encompasses some interesting and diverse subfields that incorporate unrelated disciplines, for example, earthquake engineering, forensic geology, geostatistics, hydromechanics, medical geology, military geoscience, mineral economics, river engineering, soil mechanics, and urban geology, among others. Some colleges and universities offer courses in some of these subfields: "Applied Geophysics for Geologists" (University of Houston, Geology Foundation, Fifth Annual Short Course, 1976), "Applied Geodesy" (George Washington University, School of Engineering and Applied Science, 1982), "Applied Well Log Interpretation" (GeoQuest International, Houston, 1982), and so on.

Coverage and Scope

The Encyclopedia of Applied Geology, Volume XIII in the Encyclopedia of Earth Science Series, contains an agglomeration of topics linked by the common thread of geology in the service of man. Economy of space has necessitated an eclectic approach to the topic. Since its inception and initial preparation in 1974, the table of contents has been continually pared to manageable proportions, so much so that other complete volumes "grew" out from under the egis of applied geology, namely, *The Encyclopedia of Soil Science,* Parts 1 and 2 (Volume XII), *The Encyclopedia of Petroleum Geology* (in preparation), and *The Encyclopedia of Mining and Mineral Resources* (in preparation). The topic was still too encompassing for a single volume in this series, however, so another volume was recently spun off from the manuscript. This new volume, *Field and General Geology* (Volume XIV), now in press, is viewed as a companion to *Applied Geology* because it deals with topics related to exploration surveys, field techniques, mapping, prospecting, mining, and popular geology. These two volumes thus closely complement one another.

Topics in this volume largely center around the field of engineering geology and deal with landscapes, earth materials, or the "management" of geological processes. The engineering geology of alluvial plains, arid lands, beaches and coasts, deltaic plains, cold regions, glacial landscapes, and urban environments receives major consideration, as do the geotechnical properties of caliche, clay, duricrust, soil, laterite, marine sediments, and rocks. Entries are included to discuss the structural response of dams, building foundations, trenches, tunnels, and shafts to earth processes such as earthquakes, slope stability, soil conditions, and erosion; in addition, some entries discuss engineering geology applications to channelization, bank stabilization, and coastal erosion control. Still other examples of applied geoscience topics are drawn from socioeconomics and focus on computerized information banks, conferences and congresses, earth science information sources, the communication of geological information, economic geology, and mineral economics. This volume, then, is essentially a potpourri of miscellany that follow an applied theme. In this geological smorgasbord, each topic might be considered an apophhoreta, something to be savored for the information it contains. Because it is simply not possible to consider every aspect of applied geology in a reference work of this sort and with a mandate so broad, this book represents an attempt to provide an overview of diverse subjects through a sample of a wide range of interrelated topics. Each entry is an introduction to some aspect of applied geology. For further details, the reader is directed to other works cited in the references following each entry.

Literature

The scope of applied geology is great, but the combined literature of its cognate fields is even greater. For those working only within relatively narrowly defined applied fields such as coastal engineering, foundation engineering, soil mechanics, or urban hydrology, it may be possible to keep abreast of current literature, especially when professional association dues include subscriptions to society publications. Consultants, academicians, and other professionals dealing with broad environmental issues or a range of applied topics are no doubt challenged by the need to consult periodicals from multiple fields. There are, in addition, abstracts and computerized databanks that support literature searches in a number of specialized fields. Some useful texts and reference works are listed to introduce sources of information for many applied topics.

Abstracts and Abstracting Services. Many abstracting services typically group reference materials by subject and direct the searcher to appropriate sources. The range of topics embraced by applied geology necessitates familiarity with a rather large number of abstracts in fields as diverse as civil engineering, chemical engineering, geomorphology, geotechnology, hydraulics, information science and documentation, mineral engineering, remote sensing, rheology, and soil science. The following list contains a selection of abstracts that are currently published or available through computer services. Most listings give the date when the service was initiated and the address of the professional society, government agency, or publishing house responsible for production. Although the list concentrates on compilations prepared in English, it still provides a representative sample of the kind of abstracting services available to geoscientists dealing with practical problems.

Abstracts of Geochronology and Isotope Geology, Belgian Centre for Geochronology, c/o Museum, Steenweg Op Leuven, B-1980 Tervuran, Belgium.

American Society for Information Science, Journal, 1950-, 1155 16th Street N.W., Washington, D.C. 20036.

Applied Mechanics Reviews, 1948-, American Society of Mechanical Engineers, 345 East 47th Street, New York, N.Y. 10017.

Applied Science and Technology Index, 1958-, H. W. Wilson Co., 950 University Avenue, Bronx, New York 10452.

ASCE Publications Abstract, 1970-, American Society of Civil Engineers, 345 East 47th Street, New York, N.Y. 10017.

Asian Geotechnical Engineering Abstracts, 1971-, UNESCO, Regional Office for Education in Asia, Darakarn Bldg., Box 1425, Bangkok 11, Thailand.

Bibliography and Index of Geology, 1974-, Geological Society of America, 3300 Penrose Place, Boulder, Colorado 80301. (Formerly *Bibliography and Index of Geology Exclusive of North America,* 1933-1974).

Bibliography on Petroleum and Applied Literature, 1970-, Oil and Natural Gas Commission, Dehradun, Institute of Petroleum Exploration, Kaulagarh, Dehradun, India. (Supersedes *Bulletin of Current References on Petroleum and Geological Sciences.*)

Bitten Dokumentacije, Rudarstvo I Geologija, 1950-, Jugoslovenski Centar za Technicku i Naucnu Dokumentaciju, Sl. Penezica, Krucuna 29-31, Box 724, 11000 Belgrade, Yugoslovia.

British Geological Literature, 1972-, Bibliographic Press Ltd., 160 North Gower Street, London NW1 2ND, England.

Bulletin Analytique Petrolier, 1955-, Comité Professionnel du Petrole, 51 Blvd. de Courcelles, 75008 Paris, France.

Bulletin Signaletique, Part 101: Sciences de l'Information, 1970-, Centre National de la Recherche Scientifique, Service des Abonnements, 26 rue Boyer, 75971 Paris 20, France.

Bulletin Signaletique—Bibliographie des Sciences de la Terre, Section 223, Cahier D, Roches Sédimentaires, Géologie Marine; Section 224, Cahier E, Stratigraphie, Géologie Régionale et Générale; Section 225, Cahier F, Tectonique; Section 226, Cahier G, Hydrologie, Géologie de l'engenieur, Formations Superficielles, Centre National de la Recherche Scientifique, Service des Abonnements, 26 rue Boyer, 75791 Paris 20, France.

Chemical Abstracts, Applied Chemistry and Chemical Engineering Sections, 963-, Chemical Abstracts Service, Box 3012, Columbus, Ohio 43210.

Civil Engineering Hydraulics, 1968-, British Hydromechanics Research Association, Cranfield, Bedford MK43 0AJ, England.

Current Bibliography on Sciences and Technology: Earth Science, Mining and Metallurgy, 1958-, Information Center of Science and Technology, Nihon Kagaku Gijustsu Joho Senta 2-5-2, Nagata-cho, Chiyoda-Ku, Tokyo 100, Japan.

Fluid Flow Measurement Abstracts, 1974-, British Hydromechanics Research Association, Cranfield, Bedford MK43 0AJ, England.

Geo Abstracts A (Landforms and the Quaternary), 1960-; *B* (Climatology and Hydrology), 1966-; *G* (Remote Sensing and Cartography), Geo Abstracts Ltd., University of East Anglia, Norwich NR4 7TJ, England.

Geodex Retrieval System for Geotechnical Abstracts, 1970- (International Association for Soil Mechanics and Foundation Engineering), Geodex International, 669 Broadway, P.O. Box 279, Sonoma, California 95476.

Geoscience Documentation, 1969-, Geosystems, Box 1024, Westminster, London SW1, England.

Geotechnical Abstracts, 1970- (International Society for Soil Mechanics and Foundation Engineering), Deutsche Gesellschaft fur Erd- und Grundbau, Kronprinzenstrasse 35a, 4300 Essen, West Germany.

Geotitles Weekly, 1969-, Geosystems, Box 1024, Westminster, London SW1, England.

I.C.E. Abstracts, 1974-, (Institution of Civil Engineers), Thomas Telford Ltd., 26-34 Old Street, London EC1V 9AD, England.

I.M.M. Abstracts, 1950-, Institution of Mining and Metallurgy, 44 Portland Place, London W1N 4BR, England.

Index to Scientific and Technical Proceedings, 1978-, Institute for Scientific Information, 325 Chestnut Street, Philadelphia, Pennsylvania 19106.

Index to U.S. Government Periodicals, 1974-, Infordata

International, Suite 4602, 175 East Delaware Place, Chicago, Illinois 60611.

Information Service Abstracts, 1966- (American Society for Information Science), Documentation Abstracts, Box 8510, Philadelphia, Pennsylvania 19101.

Mineralogical Abstracts, 1959- (Mineralogical Society of America), Mineralogical Society, 41 Queens Gate, London SW7, England.

Offshore Abstracts, 1974-, Offshore Information Literature, 30 Baker Street, London WIM 20S, England.

Quarterly Review of Remote Sensing, 1974-, University of New Mexico, Technology Application Center, Albuquerque, New Mexico 87131.

Referativnyi Zhurnal—Geologiya, 1956-; Geodeziya Aeros Emka, 1963-; Geophizika, 1963-, Usesoyuynyi Institut Nauchno-Technicheskoi Informatsii (VINITI), Baltiiskaya ul. 14, Moscow A-219, USSR.

Rheology Abstracts, 1958- (British Society of Rheology), Pergamon Press, Maxwell House, Fairview Park, Elmsford, New York 10523.

Science and Technology Information Sources, 1969-, Trent Polytechnic, National Centre for School Technology, Burton Street, Nottingham N51 4BU, England.

Weekly Government Abstracts—Civil Engineering: Ocean Technology and Engineering, National Technical Information Service, 5285 Port Royal Road, Springfield, Virginia 22151.

Zentralablatt für Mineralogie, Teil 2, Petrographie, Technische Mineralogia, Geochemica und Lagerstättenkunde, 1807-, Schweizerbart'sche Verlagsbuchhandlung, Johannesstrasse 3a, 7000 Stuttgart, West Germany.

Computerized Databanks. In addition to bound and printed abstracts held in libraries and other depositories, many geoscientists now have access to computer databanks via proprietary or personal computer terminals. Several services provide "online only" or "online/print" retrieval formats. Depending on the databank, retrospective searches may be based on author, cited author, title words, institutional affiliations, cited references, journal, year of publication, document type, and language for specific time frames. Most databanks are accessible via Telenet, Tymnet, and direct dial in the United States, and search services (e.g., Lockheed Information Systems, Dialog, Orbit) normally provide an online customer service representative during business hours. The following examples of databanks offer a wide range of services of interest to geoscientists:

AGRICOLA, United States Department of Agriculture, National Agricultural Library, Beltsville, Maryland 20705.

AQUALINE, Water Research Centre, Medmenham Laboratory, P.O. Box 16, Buckinghamshire SL7 2HD, England.

ENVIROLINE, Environmental Information Center, Inc., 292 Madison Avenue, New York, N.Y. 10017.

GEOARCHIVE, Geosystems, P.O. Box 1024, Westminster, London SW1, England.

GEOFILE, GTS Corporation, 4321 Directors Row, Houston, Texas 77092.

GEOREF, American Geological Institute, One Skyline Place, 5205 Leesburg Pike, Falls Church, Virginia 22041.

ISI/GeoSciTech, Institute for Scientific Information, 3501 Market Street, University City Science Center, Philadelphia, Pennsylvania 19104.

Periodicals. The following list of periodicals includes journals that currently publish or have recently published information dealing with applied geoscience topics. They range from journals that regularly emphasize engineering geology to those that incidentally publish material relevant to some aspect of applied geology. The decision to include or exclude journals at the latter end of the spectrum was somewhat arbitrary because many journals publish articles of applied interest only occasionally. Most listings contain the title of the periodical, date of first publication, sponsoring organization, and publisher. Publications of special interest and additional sources of information are given in *Earth Science, Information and Sources* and *Information Centers.*

Acta Geodaetica, Geophysica et Montanistica, 1966-, Akademiai Kiado, P.O. Box 24, H-1363 Budapest, Hungary.

Acta Geologica, 1952-, Magyar Tudomanyos Akademia, H-1363 Budapest, Hungary.

Acta Geológica Hispanica, 1966-, Instituto Nacional de Geología, Avenida Jose Autonia 585, Barcelona, Spain.

Acta Geologica Polonica, 1950-, Polska Akademia Nauk, Komitet Geologiczny, Warsaw, Poland.

Acta Geologica Sinica, 1964-, Plenum Publishing Corporation, 227 West 17th Street, New York, N.Y. 10011.

Acta Geologica Taiwanica, 1947-, Science Reports of the National Taiwan University, Taipei, Taiwan.

Advances in Tunneling Technology and Subsurface Use, 1980-, International Tunneling Association, Pergamon Press, Maxwell House, Fairview Park, Elmsford, New York 10523.

Agua, 1971-, Diaccion General de Recursos Hidraulicos Secretaria Technica, Biblioteca F., Caracas 101, Venezuela.

Ambio, 1972-, Royal Swedish Academy of Sciences, Universitetsforlaget, Box 307, Blindern, Oslo 3, Norway.

American Society of Civil Engineers, Journal (Geotechnical Engineering), 1956-, Geotechnical Engineering Division, 345 East 47th Street, New York, N.Y. 10017.

American Society of Civil Engineers, Journal (Hydraulics), 1956-, 345 East 47th Street, New York, N.Y. 10017.

American Water Works Association, Journal, 1914-, 6666 West Quincy Avenue, Denver, Colorado 80235.

Asian Geotechnical Engineering in Progress, 1974-, Asian Institute of Technology, Box 2754, Bangkok, Thailand.

Australia-New Zealand Conference on Geomechanics, Proceedings, 1971-, Institution of Engineers (Australia), 157 Gloucester Street, Sydney, N.S.W. 2000, Australia.

Australian Mineral Development Laboratories, Bulletin, 1966-, Flemington Street, Fernville, South Australia 5063, Australia.

Australian Road Research, 1962-, Australian Road Research Board, 500 Burwood Road, Vermont, Victoria 3133, Australia.

BMR Journal of Australian Geology and Geophysics, 1976-, Bureau of Mineral Resources, Geology and Geophysics, Box 378, Canberra City, A.C.T. 2601, Australia.

Boden und Gesundheit, 1953-, Gesellschaft Boden und Gesundheit, 7183 Langenburg, West Germany.

Bolletino di Geofisica, Teorica ed Applicata, 1959-, Osservatorio Geofisico Sperimentale, 34123 Trieste, Italy.

Bulletin of the Czechoslovak Seismological Stations, 1940-, Publishing House of the Czechoslovak Academy of Sciences, Vodickova 40, 112 29 Prague 1, Czechoslovakia.

Bureau de Recherche Géologiques et Miniers, Bulletin, 1963-, Département de Documentation, B.P. 6009, 45018 Orléans, France.

California Geology, 1948-, Division of Mines and Geology, 1416 Ninth Street, Sacramento, California 95814.

Canadian Geotechnical Journal, 1963-, National Research Council of Canada and Canadian Geotechnical Society, Ottawa, Ontario K1A OR6, Canada.

Canadian Journal of Civil Engineering, 1974-, National Research Council of Canada, Ottawa, Ontario K1A OR6, Canada.

Canadian Journal of Earth Sciences, 1964-, National Research Council of Canada and the Geological Association of Canada, Ottawa, Ontario K1A OR6, Canada.

Canadian Mining Journal, 1879-, National Business Publications Ltd., 310 Victoria Avenue, Westmount, Quebec H32 2M9, Canada.

Civil Engineering and Public Works Review, 1906-, Morgan-Grampian (Construction Press) Ltd., 30 Calderwood Street, Woolwich, London SE18 6QH, England.

Civil Engineering Transactions, 1959-, Institution of Engineers (Australia), Science House, 157 Gloucester Street, Syndey, N.S.W. 2000, Australia.

Coastal Engineering, 1977-, Elsevier Scientific Publishing Company, Box 211, Amsterdam, The Netherlands.

Coastal Zone Management Journal, 1973-, Crane, Russak and Company, 347 Madison Avenue, New York, N.Y. 10017.

Colorado School of Mines, Quarterly, 1905-, Golden, Colorado 80401.

Commonwealth Scientific and Industrial Research Organization, Division of Applied Geomechanics Abstracts of Published Papers, 1972-, C.S.I.R.O., Division of Applied Geomechanics, P.O. Box 54, Mount Waverly, Victoria 3149, Australia.

Construction, 1933-, Construction Publishing Company, 2420 Wilson Boulevard, Arlington, Virginia 22201.

Cuadernos Geologia, 1970-, Universidad de Granada, Secretariado de Publicaçiones, Granada, Spain.

Doklady—Earth Science Sections, 1959-, Scripta Technica, Inc., 1511 K Street, N.W., Washington, D.C. 20005.

Dokumentation für Bodenmechanik—Grundbau-Felsmechanik-Ingenieurgeologie, 1970-, Gesellschaft für Dokumentation Bodenmechanik und Grundbau e.V., Kronzprinzenstrasse 35a, 4300 Essen, West Germany.

Earth and Mineral Sciences, 1931-, College of Earth and Mineral Sciences, Pennsylvania State University, University Park, Pennsylvania 16802.

Earthmover and Civil Contractor, 1963-, Earthmovers and Road Contractors of Australia, Ltd., Box 75, Woolahara, N.S.W. 20025, Australia.

Earthquake Engineering and Structural Dynamics, 1972-, International Association for Earthquake Engineering, John Wiley & Sons Ltd., Baffins Lane, Chichester, Sussex, England. (Expediters of the Printed World Ltd., 527 Madison Avenue, New York, N.Y. 10022).

Earthquake Information Bulletin, U.S. Geological Survey, 12201 Sunrise Valley Drive, Reston, Virginia 22092.

Earthquake Notes, 1929-, Seismological Society of America, U.S. Geological Survey, Denver Federal Center, Denver, Colorado 80225.

Earthquake Prediction Research, 1982-, D. Reidel Publishing Company, 190 Old Derby Street, Hingham, Massachusetts 02043.

Eclogae Geologicae Helvetiae, 1888-, Birkhäuser Verlag, Elissabethenstrasse 19, CH-4010 Basel 1, Switzerland.

Engineering and Mining Journal, 1866-, McGraw-Hill, 1221 Avenue of the Americas, New York, N.Y. 10020.

Erdbau, 1964-, Firma Baumachinen GmbH., Industriezentrum NOe-Sved, Strasse Nr. 1, Objekt 27, A-2351 Naudorf, Austria.

Excavating Contractor, 1905-, Cummins Publishing Co., 21590 Greenfield Road, Oak Park, Michigan 48237.

Excavator, 1936-, Association Graphique et Artistique, 225 Avenue Molière, 1060 Brussels, Belgium.

Foundation Facts, 1965-, International Society for Soil Mechanics and Foundation Engineering, Raymond International Inc., Box 22718, Houston, Texas 77027.

Geodesy, Mapping and Photogrammetry, 1962-, American Geophysical Union, 1909 K Street N.W., Washington, D.C., 20006.

Geoforum, Pergamon Press, Maxwell House, Fairview Park, Elmsford, New York 10523.

Geophysical Surveys, 1973-, D. Reidel Publishing Company, Box 17, 3300 AA Dordrecht, The Netherlands.

Geotechnica, 1954-, Associazione Geotechnica Italiana, Instituto Propaganda Internazionale, Via Friuli 32, Milan, Italy.

Géotechnique, 1948-, Thomas Telford Ltd., Publications Division, 26-34 Old Street, London EC14 9AD, England.

Gerlands Beiträge zur Geophysik, 1887-, Akademische Verlagsgesellschaft Geest und Portig K.G., Sternwartenstrasse 8, 701 Leipzig, East Germany.

Ground Engineering, 1968-, Foundation Publications Ltd., 7 Ongar Road, Brentwood, Essex, England.

Ground Water Newsletter, 1972-, Water Information Center, 7 High Street, Huntington, New York 11743.

Highways and Road Construction International, 1934-, Embankment Press Ltd., Building 59, G.E.C. Estate, East Lane, Wembley, Middlesex, England.

Hydrotechnical Construction, 1967-, American Society of Civil Engineers, 345 East 47th Street, New York, N.Y. 10017.

Indian Geophysical Union, Journal, Hyderabad 7, India.

Indian Geotechnical Society, Journal, 1971-, Institution of Engineers (India), Delhi Centre, Bahadur Shah Zafar Marg, New Delhi, 110001, India.

Indian Journal of Earth Sciences, 1974-, Indian Society of Earth Sciences, Department of Geology, Presidency College, Calcutta 700073, India.

Indian Mining and Engineering Journal, 1962-, Mining Engineers Association, Colaba Causeway, Bombay 1, India.

Indian Society of Earthquake Technology, Bulletin, 1964-, Roorkee, Uttar Pradesh, India.

International Association of Engineering Geology, Bulletin, 1970-, Geologisches Landesamt N.W., Dr.-Grieff-Strasse 195, Postfach 1080, 4150 Krefeld, West Germany.

International Journal of Mining Engineering, 1983-, Chapman & Hall, 11 New Fetter Lane, London EC4P 4EE, England.

International Seismological Centre, Bulletin, 6 South Oswald Road, Edinburgh EH9 2HX, Scotland.

International Union of Geological Sciences, Geological Newsletter, 1967-, Box 379, Haarlem, The Netherlands.

Israel Journal of Earth Sciences, 1951-, Weigmann Science Press of Israel, Box 801, Jerusalem 91000, Israel.

Japanese Association of Groundwater Hydrology, Journal, Geological Survey of Japan, 135 Hisamoto, Takatasu-ku, Kawasaki 213, Japan.

Japanese Society of Soil Mechanics and Foundation Engineering, Journal, 1972-, Doshitsu Kogakkai, 1-13-5 Nishi Shinbashi, Minato-ku, Tokyo 105, Japan.

Journal of Environmental Management, 1973-, Academic Press, 111 Fifth Avenue, New York, N.Y. 10003.

Journal of Environmental Quality, 1972-, American Society of Agronomy, 677 South Segoe Road, Madison, Wisconsin 53711.

Journal of Environmental Sciences, 1958-, Institute of Environmental Sciences, 940 East Northwest Highway, Mt. Prospect, Illinois 60056.

Journal of Geophysical Research, 1896-, American Geophysical Union, 1909 K Street N.W., Washington, D.C. 20006.

Journal of Geophysics, 1924-, Springer-Verlag, 175 Fifth Avenue, New York, N.Y. 10010.

Journal of Hydrology, 1963-, Elsevier Scientific Publishing Company, Box 211, Amsterdam, The Netherlands.

Journal of Mining and Geology, 1964-, Nigerian Mining, Geological and Metallurgical Society, Geology Department, University of Ife, Ile-Ife, Nigeria.

Landslide, 1973-, Box 1347, Eureka, California 95501.

Large Dams, 1956-, Japanese Committee on Large Dams, Toden-kyukan Building, 1-1-13 Shinbashi, Minato-ku, Tokyo 105, Japan.

Mine and Quarry, 1924-, Ashire Publishing Ltd., 42 Grays Inn Road, London WC1X 8LR, England.

Mineral Industry Quarterly, 1976-, Department of Mines, Box 151, Eastwood, South Australia 5063, Australia.

Mines, 1910-, Colorado School of Mines, Alumni Association, Golden, Colorado 80401.

Mining Congress Journal, 1915-, American Mining Congress, 1100 Ring Building, Washington, D.C. 20036.

Mining Engineer, 1960-, Institution of Mining Engineers (Great Britain), Hobart House, Grosvenor Place, London SW1X 7AE, England.

Mining Engineering, 1949-, American Institute of Mining, Metallurgical and Petroleum Engineers, Society of Mining Engineers, 540 Arapeen Drive, Box 8300, Salt Lake City, Utah 84108.

Mining Geology, 1951-, Society of Mining Geologists of Japan, Nihon Kogyo Kaikan Building, 8-5-4 Ginza-ku, Tokyo 104, Japan.

Mining Journal, 1835-, Mining Journal Ltd., 15 Wilson Street, London EC2M 2TR, England.

New Zealand Journal of Geology and Geophysics, 1958-, Department of Scientific and Industrial Research, Box 9741, Wellington, New Zealand.

Norges Geologiske Undersökelse, Bulletin, 1972-, Universitetsforlaget, Oslo, Norway.

Office de la Recherche Scientifique et Technique Outre-mer de M'Bour, Bulletin Seismique, 1974-, Centre de Geophysique, B.P. 50, M'Bour, Senegal.

Pollution Engineering, 1969-, Technical Publishing Company, 1301 South Grove Avenue, Barrington, Illinois 60010.

Ports and Harbors, 1969-, International Association of Ports and Harbors, Kokusai Kowan Kyokai, 1 Shiba Kotahira-cho, Minato-ku, Tokyo 105, Japan.

Pure and Applied Geophysics, 1939-, Birkhäuser Verlag, Elisabethenstrasse 19, Ch-4010 Basel 1, Switzerland.

Quarterly Journal of Engineering Geology, 1968-, The Geological Society of London, Scottish Academic Press Ltd., 33 Montgomery Street, Edinburgh EH7 55X, Scotland.

Revista Brasileira de Geosciencias, 1971-, Conselho Nacional de Pesquisas, Sociedade Brasileira de Geológia, Rua Peixoto Gomide 1400, C.P. 5450, São Paulo, Brazil.

Revue de Géographie Physique et de Géologie Dynamique, 1928-, Masson et Cie, 120 Blvd. Saint-Germain, 75280 Paris Cedex 06, France.

Revue des Ingenieurs, 1948-, Associations des Anciens Elèves des Ecoles des Mines, 19 rue du Grand Moulin, 42029 Saint-Etienne, France.

Rock Mechanics, 1963-, Springer-Verlag, 175 Fifth Avenue, New York, N.Y. 10010.

Sciences Géologiques Bulletin, 1920-, Institut de Géologie, Université Louis Pasteur de Strasbourg, 1 rue Blessig, 67084 Strasbourg Cedex, France.

Scripta Geologia, 1971-, Universita J. E. Purkyne, Brno, Czechoslovakia.

Sedimentary Geology, 1967-, Elsevier Scientific Publishing Company, Box 211, Amsterdam, The Netherlands.

Sedimentology, 1962-, Blackwell Scientific Publications Ltd., Osney Mead, Oxford OX2 0EL, England.

Seismological Society of America, Bulletin, 1911-, Box 826, Berkeley, California 94701.

Société Géologique de Belgique, Annales, 1877-, Université de Liège, 7 Place du Vingt-aout, 4000 Liège, Belgium.

Society of Mining Engineers of AIME, Transactions, 1962-, American Institute of Mining, Metallurgical and Petroleum Engineers, 540 Arapeen Drive, Salt Lake City, Utah 84108.

Soil Mechanics and Foundation Engineering, 1964-, Consultants Bureau, Plenum Publishing Corporation, 227 West 17th Street, New York, N.Y. 10011.

Soils and Foundations, 1960-, Doshitsu Kogabbai, 1-13-5 Nishi Shinbashi, Minato-ku, Tokyo 105, Japan.

Sols-Soils, 1962-, Editions Sols-Soils, 54 Avenue de la Motte-Picquet, 75015 Paris, France.

South African Tunnelling, Pithead Press, Box 9002, Johannesburg, South Africa.

Soviet Mining Science, Consultants Bureau, Plenum Publishing Corporation, 227 West 17th Street, New York, N.Y. 10011.

Tsukumo Earth Science, 1966-, Institute of Earth Science, Kyoto Daigaku Kyoyabu, Chigaku Kyoshitsu, Yoshida-Honmachi, Sakyo-ku, Kyota 606, Japan.

Tunneling Technology Newsletter, 1973-, National Academy of Sciences, Committee on Tunneling Technology, 2101 Constitution Avenue N.W., Washington, D.C. 20418.

Tunnels and Tunnelling, 1969-, Morgan-Grampian (Construction Press) Ltd., 30 Calderwood Street, Woolwich, London SE18 6QH, England.

Underground Engineering Contractors' Association, UECA Publication (UECA), 8615 Florence Avenue, Suite 205, Downey, California 90240.

Underground Space, 1976-, American Underground-Space Association, Pergamon Press, Maxwell House, Fairview Park, Elmsford, New York 10523.

United States Geological Survey, Professional Papers, 1902-, *Bulletin*, 1883-, Washington, D.C. 20402.

World Environment Report, 1975-, Center for International Environment Information, 300 East 42nd Street, New York, N.Y. 10017.

Zeitschrift für Angewandte Geologie, 1955-, Akademie-Verlag GmbH, Leipzigerstrasse 3/4, 108 Berlin, East Germany.

Texts and Reference Works. The growth of applied geology in recent decades is perhaps most dramatically emphasized by the publication of textbooks and specialized volumes. Because this encyclopedia gives an overview of the field, the reader is directed to relevant publications in specific branches of the applied geosciences. A selection of a few recent texts and some older reference works that complement this volume are listed here.

American Society of Civil Engineers Staff, 1958. Glossary of terms and definitions in soil mechanics, *A.S.C.E. Proceedings* **84**(SM4), Paper 1826, 1-43.

Amstutz, G. C., 1971. *Glossary of Mining Geology.* Amsterdam: Elsevier, 196p.

Anonymous (n.d.). *Elsevier's Oilfield Dictionary.* Amsterdam: Elsevier, 162p.

Anonymous 1956 et seq. *AGI Data Sheets.* Falls Church, Va.: American Geological Institute, var. pag.

Anstey, N.A., 1982. *Simple Seismics.* Boston, Mass.: IHRDC Publications, 168p.

Armstrong, A. T., 1975. *Handbook on Quarrying.* Adelaide, South Australia: Department of Mines, 238p.

Bergman, S. M., 1981. *Subsurface Space.* New York: Pergamon Press, 1500p.

Brantly, J. E., 1961. *Rotary Drilling Handbook.* New York: Palmer Publications, 825p.

British Standards Institution (London), 1964. *Glossary of Mining Terms.* Section 5, Geology, British Standard 3618, sec. 5, 18p. and Section 6, Drilling and Blasting, British Standard 3618, sec. 6, 20p.

Brown, E. T. (ed.), 1981. *Rock Characterization, Testing and Monitoring.* New York: Pergamon Press, 200p.

Carson, A. B., 1961. *General Excavation Methods.* New York: McGraw-Hill, 392p.

Carson, A. B., 1965. *Foundation Construction.* New York: McGraw-Hill, 424p.

Chapman, R. E., 1973. *Petroleum Geology.* Amsterdam: Elsevier, 304p.

Coates, D. R., 1976. *Geomorphology and Engineering.* Stroudsburg, Pa.: Dowden, Hutchinson and Ross, 360p.

Cochran, W.; Fenner, P.; and Hill, M., 1979. *Geowriting.* Falls Church, Va.: American Geological Institute, 80p.

Craig, R. G., and Craft, J. L., 1982. *Applied Geomorphology.* Winchester, Mass.: Allen & Unwin, 350p.

Cummins, A. B., and Given, I. A. (eds.), 1973. *SME Mining Engineering Handbook,* 2 vols. New York:

Society of Mining Engineers of the American Institute of Mining, Metallurgical, and Petroleum Engineers.

Davis, J. C., 1973. *Statistics and Data Analysis in Geology.* New York: Wiley, 550p.

Dennis, J. G. (ed.), 1967. *International Tectonic Dictionary.* Tulsa, Okla.: American Association of Petroleum Geologists, 196p.

Dohr, G., 1981. *Applied Geophysics.* New York: Wiley, 256p.

Du Pont de Nemours & Company, 1966. *Blaster's Handbook: A Manual Describing Explosives and Practical Methods of Use.* Wilmington, Del.: Du Pont, 524p.

Forrester, J. D., 1946. *Principles of Field and Mining Geology.* New York: Wiley, 647p.

Fraenkel, K. H. (ed.), 1953-1961 with supplements. *Manual on Rock Blasting,* 3 vols. Stockholm: Atlas Copco Abteibolag.

Glazov, N. V., and Glazov, A. N., 1959. *Engineering Geology.* New York: Consultants Bureau, 91p.

Golze, A., 1977. *Handbook of Dam Engineering.* New York: Van Nostrand Reinhold, 816p.

Green, N., 1981. *Earthquake Resistant Building Design and Construction.* New York: Van Nostrand Reinhold, 192p.

Hamilton, E. I., 1965. *Applied Geochronology.* London: Academic Press, 267p.

Harvey, A. P., and Diment, J. A., 1979. *Geoscience Information.* Boston, Mass.: IHRDC Publications, 287p.

Hatheway, A. W., and McClure, C. R., 1979. *Geology in the Siting of Nuclear Power Plants.* Boulder, Colo.: Geological Society of America.

Holman, W. W.; McCormack, R. K.; Minard, J. P.; and Jumikis, A. R., 1957. *Practical Applications of Engineering Soil Maps.* New Brunswick, N. I.: Rutgers University Engineering Research Bull. No. 36, 114p.

Huntington, W. C., 1957. *Earth Pressure and Retaining Walls.* New York: Wiley, 534p.

Journal, A. G., and Huijbregts, C. J., 1981. *Mining Geostatistics.* New York: Academic Press, 610p.

Knill, J. L., 1978. *Industrial Geology.* Oxford, England: Oxford University Press, 344p.

Lapedes, D. M. (ed.), *McGraw-Hill Encyclopedia of the Geological Sciences.* New York: McGraw-Hill, 915p.

Lapinski, M., 1978. *Road and Bridge Construction Handbook.* New York: Van Nostrand Reinhold, 168p.

Lee, I. K., 1968. *Soil Mechanics.* Sydney, Australia: Butterworths, 625p.

Lee, I. K.; Ingles, O. O.; and White, W., 1982. *Geotechnical Engineering.* London: Pitman, 432p.

Legget, R. F., 1967. *Geology and Engineering.* New York: McGraw-Hill, 884p.

Lewis, R. S., and Clark, G. B., 1964. *Elements of Mining.* New York: Wiley, 768p.

Lillesand, T. M., and Kiefer, R. W., 1979. *Remote Sensing and Image Interpretation.* New York: Wiley, 612p.

Loudon, T. V., 1979. *Computer Methods in Geology.* New York: Academic Press, 278p.

Mandel, S., and Shifton, Z. L., 1981. *Groundwater Resources.* New York: Academic Press, 288p.

Mathess, G., 1981. *The Properties of Groundwater.* New York: Wiley, 432p.

Mayer-Gurr, A., 1976. *Petroleum Engineering.* New York: Wiley (Halsted Press), 208p.

Megaw, T. M., and Bartlett, J., 1981. *Tunnels: Planning, Design, Construction,* 2 vols. New York: Wiley.

Miller, D. H., 1982. *Water at the Surface of the Earth.* New York: Academic Press, 576p.

Nelson, A., and Nelson, K. D., 1967. *Dictionary of Applied Geology.* London: George Newnes, 421p.

Nichols, H. L., Jr., 1962. *Mining the Earth: The Workbook of Excavation.* Greenwich, Conn.: North Castle Books, var. pag.

Parasnis, D. S., 1962. *Principles of Applied Geophysics.* London: Methuen, 176p.

Pavoni, J., 1978. *Handbook of Water Quality Management Planning.* New York: Van Nostrand Reinhold, 448p.

Peck, R. B.; Hansen, W. E.; and Thornburn, T. H., 1974. *Foundation Engineering.* New York: Wiley, 514p.

Pfannkuch, H.-O., 1969. *Elsevier's Dictionary of Hydrogeology.* Amsterdam: Elsevier, 168p.

Roberts, A., 1981. *Applied Geotechnology.* Oxford, England: Pergamon Press, 352p.

Rodda, J. C.; Downing, R. A.; and Law, F. M., 1976. *Systematic Hydrology.* London: Newnes-Butterworths.

Schach, R.; Garshol, K.; and Heltzen, A. M., 1979. *Rock Bolting.* New York: Pergamon Press, 96p.

Siegal, B. S. and Gillaspie, A. R., 1980. *Remote Sensing in Geology.* New York: Wiley, 702p.

Siegel, F. R., 1974. *Applied Geochemistry.* New York: Wiley-Interscience, 353p.

Sinclair, J., 1958. *Geological Aspects of Mining.* London: Pitman, 343p.

Stewart, G. A. (ed.), 1968. *Land Evaluation.* Melbourne, Australia: Macmillan, 392p.

Stiegler, S. E., 1976. *A Dictionary of Earth Sciences.* London: Macmillan, 301p.

Stout, K. S., 1980. *Mining Methods and Equipment.* New York: McGraw-Hill, 218p.

Thrush, P. W., 1968. *A Dictionary of Mining, Mineral, and Related Terms.* Washington, D.C.: Superintendent of Documents, 1,269p.

Todd, D. K., 1959. *Groundwater Hydrology.* New York: Wiley, 535p.

Tomlinson, M. J., 1980. *Foundation Design and Construction.* London: Pitman, 818p.

Ven et Chow (ed.), 1981. *Advances in Hydroscience,* Vol. 12. New York: Academic Press, 456p.

Visser, A. D. (ed.), 1965. *Elsevier's Dictionary of Soil Mechanics.* Amsterdam: Elsevier, 359p.

Visser, W. A. (ed.), 1980. *Geological Nomenclature.* The Hague: Martinus Nijhoff, 540p.

Vollmer, E. (ed.), 1967. *Encyclopedia of Hydraulics, Soil and Foundation Engineering.* Amsterdam: Elsevier, 398p.

Walton, W. C., 1970. *Groundwater Resource Evaluation.* New York: McGraw-Hill.

Ward, D. C.; Wheeler, M. W.; and Pangborn, M. W., Jr., 1972. *Geologic Reference Sources.* Metuchen, N. J.: Scarecrow Press, 453p.

Winter, J., 1978. *Power Plant Siting.* New York: Van Nostrand Reinhold, 224p.

Zeevaert, L., 1983. *Foundation Engineering for Difficult Subsoil Conditions.* New York: Van Nostrand Reinhold, 704p.

Zilly, R. G., 1977. *Handbook of Environmental Civil Engineering.* New York: Van Nostrand Reinhold, 1,000p.

How to Use This Encyclopedia

As in most encyclopedic works, topics follow alphabetical order. This organization of our three categories of articles—reviews of major disciplines, component building blocks, and elements and phenomena—permits direct access to primary subject matter. The wide range of subjects, however, often necessitates grouping specific concepts or applications under general headings or according to key words or phrases. To aid the reader, entries are cross-referenced both in the body of the text and at the very end of each article. (Coastal engineering, for example, appears under that title, listed in the C's, and is also cross-referenced at the end of the "Engineering Geology" entry.) The abbreviation q.v. *(quod vide)*—as in "rock mechanics (q.v.)"—means that an entry with that title appears elsewhere in this volume. Cross-references to related subjects in other volumes in this series are also included. For a list containing published and planned volumes in The Encyclopedia of Earth Science Series see Table 1.

Should a subject not appear as a main entry or cross-reference or as an in-text cross-reference, the comprehensive subject index should be consulted. An author citation index is also provided, as some researchers find it useful to locate specific subjects by author or want to look up original publications by the experts quoted.

References to cited works follow each entry and are designed to lead the interested reader to additional, more detailed information or to other summaries that contain extensive bibliographies. Every effort was made to include complete bibliographic information in each citation so that materials could be easily retrieved in libraries, through reference services, or from computerized databanks. Journal titles are generally abbreviated according to the style recommended by *Serial Publications Commonly Cited in Technical Bibliographies of the United States Geological Survey.* Citations for books follow an abbreviated format that gives main titles and a key word for the publishing house; thus Hutchinson Ross Publishing Company, Inc. becomes Hutchinson Ross, and John Wiley & Sons, Inc., becomes Wiley.

Abbreviations, Units, and Symbols

Standard abbreviations are used for measures of area, length, volume, mass, pressure, and temperature. The Système International d'Unites (S.I.), as established by the International Organization for Standardization in 1960, is generally used throughout this volume. Equivalent units are, however, usually given in parentheses because some tables, figures, and maps retain units of the English system (Customary Units), or because it is convenient to retain older units for practical purposes. Precious metal grades, for example, are commonly reported in grams per tonne but converted back to ounces troy prior to sale on the London market; the pascal is the S.I. unit of pressure but the millibar is still found in many practices; kinematic viscosity should be expressed as square meters per second (m^2/s) or square millimeters per second (mm^2/s), but the use of stokes (St) will probably continue for some time.

With such a wide range of topics in applied geology, it is impossible to list here all the units used in this volume. Most units are defined locally as they are used in each article, but for those readers wishing to brush up on refinements of the International System of Units (S.I.), the following tables may be useful. Table 2 gives values and symbols for basic S.I. units. Table 3 provides prefixes, symbols, and multiplying factors that can be applied to basic units. These prefixes, which indicate fractions or multiples of basic or derived S.I. units, are useful because some primary S.I. units have been found to be of inconvenient size for practical applications. Some units that are not part of S.I. but widely used by specialists are listed in Table 4. Even though most nations use S.I. units

in scientific and industrial applications, the United States, Brunei, Burma, and North and South Yemen remain the last to officially make the transition from Customary Units to S.I. Because momentum of the national conversion process has been temporarily stifled in the United States, both kinds of units continue to be used by geoscientists working in applied situations. Table 5 is provided to facilitate working in both systems. It lists a number of traditional units that can be converted from Customary to metric units using a multiplying factor. Other specialized units used to advantage by geoscientists are given in Table 6 for drilling, cementing, and formation testing; general geology; reservoir geology and engineering; pipeline operations; seismological investigations; gravity and magnetic survey; and applications in rock mechanics. A more detailed discussion of abbreviations, units, and symbolization may be found in the entry entitled "Units, Numbers, Constants and Symbols" in Volume II in this series.

TABLE 1. Published and Planned Volumes in The Encyclopedia of Earth Science Series

Volume

I   THE ENCYCLOPEDIA OF OCEANOGRAPHY/*Rhodes W. Fairbridge*
II   THE ENCYCLOPEDIA OF ATMOSPHERIC SCIENCES AND ASTROGEOLOGY/*Rhodes W. Fairbridge*
III THE ENCYCLOPEDIA OF GEOMORPHOLOGY/*Rhodes W. Fairbridge*
IVA THE ENCYCLOPEDIA OF GEOCHEMISTRY AND ENVIRONMENTAL SCIENCES/
      Rhodes W. Fairbridge
IVB THE ENCYCLOPEDIA OF MINERALOGY/*Keith Frye*
VI   THE ENCYCLOPEDIA OF SEDIMENTOLOGY/*Rhodes W. Fairbridge and Joanne Bourgeois*
VII THE ENCYCLOPEDIA OF PALEONTOLOGY/*Rhodes W. Fairbridge and David Jablonski*
VIII THE ENCYCLOPEDIA OF WORLD REGIONAL GEOLOGY, Part 1: Western Hemisphere (Including
      Antarctica and Australia/*Rhodes W. Fairbridge*
XII THE ENCYCLOPEDIA OF SOIL SCIENCE, Part 1: Physics, Chemistry, Biology, Fertility, and Technology/
      Rhodes W. Fairbridge and Charles W. Finkl, Jnr.
XIII THE ENCYCLOPEDIA OF APPLIED GEOLOGY/*Charles W. Finkl, Jnr.*
XV   THE ENCYCLOPEDIA OF BEACHES AND COASTAL ENVIRONMENTS/*Maurice L. Schwartz*

To Be Published

THE ENCYCLOPEDIA OF WORLD REGIONAL GEOLOGY, Part 2: Europe and Asia
THE ENCYCLOPEDIA OF STRUCTURAL GEOLOGY
THE ENCYCLOPEDIA OF FIELD AND GENERAL GEOLOGY
THE ENCYCLOPEDIA OF IGNEOUS AND METAMORPHIC PETROLOGY, VOLCANOLOGY, AND
   GEOTHERMAL RESOURCES
THE ENCYCLOPEDIA OF SOIL SCIENCE, Part 2: Morphology, Genesis, Classification, and Geography
THE ENCYCLOPEDIA OF WORLD REGIONAL GEOLOGY, Part 3: Africa and the Middle East
THE ENCYCLOPEDIA OF GEO-ARCHEOLOGY
THE ENCYCLOPEDIA OF SNOW, ICE, AND GLACIOLOGY
THE ENCYCLOPEDIA OF CLIMATOLOGY
THE ENCYCLOPEDIA OF GEOPHYSICS
THE ENCYCLOPEDIA OF STRATIGRAPHY

TABLE 2. The International System of Units (S.I.).

Physical Quantity	Name of Unit	Value	Symbol
Length	meter	base unit	m
	millimeter	0.001 m	mm
	centimeter	0.01 m	cm
	kilometer	1,000 m	km
	international nautical mile (navigation)	1,852 m	n mi
Mass (commonly called "weight")	kilogram	base unit (1,000 g)	kg
	gram	0.001 kg	g
	tonne	1,000 kg	t
Time interval	second	base unit	s
Area	square meter	S.I. unit	m^2
	square millimeter	$0.000001\ m^2$	mm^2
	square centimeter	$0.0001\ m^2$	cm^2
	hectare	$10,000\ m^2$	ha
Volume	cubic meter	S.I. unit	m^3
	cubic millimeter	$10^{-9}\ m^3$	mm^3
	cubic centimeter	$0.000001\ m^3$	cm^3
	cubic decimeter	$0.001\ m^3$	dm^3
Volume (fluids only)	liter	$0.001\ m^3$	l
	milliliter	$0.001\ l$	ml
	kiloliter	$1,000\ l\ (1\ m^3)$	kl
Velocity and speed	meter per second	S.I. unit	m/s or m s^{-1}
	kilometer per hour	0.27 m/s	km/h or km h^{-1}
	knot (navigation)	1 n mi/h or 0.514 m/s	kn
Force	newton[a]	S.I. unit	N
Energy	joule[a]	S.I. unit	J
Power	watt[a]	S.I. unit	W
Density	kilogram per cubic meter	S.I. unit	kg/m^3 or kg m^{-3}
	tonne per cubic meter	$1,000\ kg/m^3$	t/m^3 or tm^{-3}
	gram per cubic centimeter	$1,000\ kg/m^3$	g/cm^3 or g cm^{-3}
Density (fluids only)	kilogram per liter	$1,000\ kg/m^3$	kg/l or kgl^{-1}
	gram per milliliter	$1,000\ kg/m^3$	g/ml or g ml^{-1}
Pressure	pascal	S.I. unit (N/m^2)	Pa
Pressure (meteorology)	bar	100,000 Pa	bar
	millibar	100 Pa	mbar
Electric current	ampere[b]	base unit	A
Potential difference or electromotive force	volt [a, b]	S.I. unit	V
Electrical resistance	ohm[a, b]	S.I. unit	
Frequency	hertz[a]	S.I. unit	Hz
	revolution per minute	$\frac{1}{60}$ Hz	rpm or rev/min
Temperature	kelvin	base unit	K
	degree Celsius[c]	K	°C
Plane angle	radian	S.I. unit	rad
	milliradian	0.001 rad	mrad
	degree	$\pi/180$ rad	°
	minute	$\frac{1°}{60}$	'
	second	$\frac{1'}{60}$	''
Amount of substance	mole	base unit	mol

[a]Decimal multiples commonly associated with this unit are *kilo* ($\times$ 1,000), *mega* ($\times$ 1,000,000), and *giga* ($\times$ 1,000,000,000).

[b]Decimal submultiples associated with this unit are *milli* ($\times$ 0.001) and *micro* ($\times$ 0.0000001).

[c]The units of temperature on the Celsius scale (°C) and the thermodynamic scale (K) are equal. A temperature t on the Celsius scale is related to a temperature T on the thermodynamic scale by the relationship $t = T - 273.15$.

Source: After Berkman, D. A., 1976. *Field Geologists' Manual.* Parkville, Victoria: Australasian Institute of Mining and Metallurgy, pp. 275–276.

TABLE 3. Prefixes, Symbols, and Multiplying Factors

Multiplying Factor		Prefix	Symbol
1,000,000,000,000 =	10^{12}	tera	T
1,000,000,000 =	10^{9}	giga	G
1,000,000 =	10^{6}	mega	M
1,000 =	10^{3}	kilo	k
100 =	10^{2}	hecto	h
10 =	10^{1}	deca	da
0.1 =	10^{-1}	deci	d
0.01 =	10^{-2}	centi	c
0.001 =	10^{-3}	milli	m
0.000001 =	10^{-6}	micro	μ
0.000000001 =	10^{-9}	nano	n
0.000000000001 =	10^{-12}	pico	p
0.000000000000001 =	10^{-15}	femto	f
0.000000000000000001 =	10^{-18}	atto	a

TABLE 4. Some Non-S.I. Units with S.I.

Condition of Use	Unit	Symbol	Value in S.I. Units
Permissable Universally with S.I.	minute	min	1 min = 60 s
	hour	h	1 h = 3,600 s
	day	d	1 d = 86,400 s
	year	a	1 a = 3.1536×10^{7} s
	degree (of arc)	°	$1° = (\pi/180)$ rad
	minute (of arc)	′	$1' = (\pi/10,800)$ rad
	second (of arc)	″	$1'' = (\pi/648,000)$ rad
	liter[a]	l	1 l = 1dm³
	tonne[b]	t	1 t = 10^{3} kg
	degree Celius	°C	[f]
	revolution	r	1 r = 2πrad
Permissible in Specialized fields	electronvolt	eV	1 eV = 1.60219×10^{-19} J
	unit of atomic mass	u	1 u = 1.66053×10^{-27} kg
	astronomical unit[c]		1 AU = 149.600 Gm
	parsec	pc	1 pc = .30857 Tm
Permissible for a Limited Time	nautical mile		1 nautical mile = 1,852 m
	knot[d]		1 nautical mile per hour = (1,852/3,600) m/s
	angstrom	Å	1 Å = 0.1 nm = 10^{-10} m
	hectare[e]	ha	1 ha = 10^{4}m²
	bar	bar	1 bar = 100kPa
	standard atmosphere	atm	1 atm = 101.325 kPa

[a]The word liter standing alone must be typed in full unless the typewriter is equipped with a special looped "ell."
[b]Care must be taken in the interpretation of this word when it occurs in French text of Canadian origin where the implication may be a "ton of 2,000 lb."
[c]This unit does not have an international symbol. The abbreviations used are AU in French and UA in French.
[d]There is no internationally recognized symbol for knot (1 nautical mile per hour), but kn is frequently used.
[e]The "acre" and "hectare," by international agreement, can be used for a limited period of time as needed in the agriculture and surveying sectors.
[f]The S.I. unit of temperature is the kelvin. The Celsius temperature scale (previously called Centigrade) is the commonly used scale for temperature measurements, except for some scientific work where a thermodynamic scale is used.

Source: After Kenting Limited, n. d. *The International System of Units (S.I.) for the Petroleum Industry in Canada.* Calgary, Alberta, Canada.

TABLE 5. Alphabetical List of Traditional Units Showing the Conversion Factor to Selected Metric (S.I.) Units

Traditional Unit	Multiply Value in Traditional Units by Factor × 10E to Obtain Value in Metric Units[a]	Selected Metric (S.I.) Unit	
	Factor E	Name	Symbol
acre	4.046856 E + 03	square meter	m^2
acre	4.046856 E − 01	hectare	ha
acre-foot	1.233482 E + 03	cubic meter	m^3
ampere hour	3.6* E + 00	kilocoulomb	kC
API gravity at 60°F	ASTM D1250 Table 3 × 1000	kilogram per cubic meter at 15°C	kg/m^3
angstrom unit	1.0* E − 01	nanometer	nm
arpent	3.418894 E − 01	hectare	ha
atmosphere (atm)	1.01325* E + 02	kilopascal	kPa
atmosphere technical (at or atm)	9.80665* E + 01	kilopascal	kPa
bar	1.0* E + 02	kilopascal	kPa
barrel (42 U.S. gal)	1.589873 E − 01	cubic meter	m^3
barrel per foot	5.216118 E − 01	cubic meter per meter	m^3/m
barrel per inch	6.25934 E − 01	cubic meter per centimeter	m^3/cm
barrel per acre-foot	1.28893 E − 04	cubic meter per cubic meter	m^3/m^3
barrel per cubic mile	3.81431 E − 02	cubic meter per cubic kilometer	m^3/km^3
barrel per long ton (U.K.)	1.564763 E − 01	cubic meter per tonne	m^3/t
barrel per short ton (U.S.)	1.752535 E − 01	cubic meter per tonne	m^3/t
barrel per day	6.624471 E − 03	cubic meter per hour	m^3/h
barrel per day psi	2.305916 E − 02	cubic meter per day kilopascal	$m^3/(d \cdot kPa)$
BCF (billion cubic feet: 60°F 1 atm)	2.826231 E + 07	cubic meter (API)	m^3(API)
	1.195307 E + 00	gigamole	Gmol
Btu (International Table)	1.055056 E + 00	kilojoule	kJ
Btu per barrel	6.636102 E + 00	kilojoule per cubic meter	kJ/m^3
Btu per brake horsepower hour	3.930148 E − 01	watt per kilowatt	W/kW
Btu per square foot second	1.135653 E + 01	kilowatt per square meter	kW/m^2
Btu inch per square foot second °F	5.192204 E − 01	kilowatt per meter degree Celsius	$kW/(M \cdot °C)$
Btu per square foot hour °F	5.678263 E − 03	kilowatt per square meter degree Celsius	$kW/(m^2 \cdot °C)$
Btu per square foot second °F	2.044175 E +01	kilowatt per square meter degree Celsius	$kW/(m^2 \cdot °C)$
Btu per cubic foot	3.725895 E + 01	kilojoule per cubic meter	kJ/m^3
Btu per standard cubic foot (60°F − 1 atm)	8.826705 E − 01	kilojoule per mole	kJ/mol
Btu per gallon	2.320808 E − 01	kilojoule per liter	kJ/l
Btu per gallon (U.S.)	2.787163 E − 01	kilojoule per liter	kJ/l
Btu per hour	2.930711 E − 04	kilowatt	kW
Btu per minute	1.758427 E − 02	kilowatt	kW
Btu per pound	2.326010 E + 00	kilojoule per kilogram	kJ/kg
Btu per pound °F	4.1868 E + 00	kilojoule per kilogram degree Celsius	$kJ/(kg \cdot °C)$
Btu per pound mole	2.326000 E + 00	joule per mole	J/mol
Btu per second	1.055056 E + 00	kilowatt	kW
Btu foot per square foot hour °F	1.73073 E + 00	watt per meter degree Celsius	$W/(m \cdot °C)$
calorie (International Table)	4.1868* E + 00	joule	J
calorie (thermochemical)	4.184* E + 00	joule	J

*The conversion factor is exact.
[a]To go from metric to customary units, divide the conversion factor instead of multiplying.

continued

TABLE 5. continued

Traditional Unit	Multiply Value in Traditional Units by Factor × 10E to Obtain Value in Metric Units[a]	Selected Metric (S.I.) Unit	
	Factor E	Name	Symbol
calorie per centimeter °C second	4.1868* E + 02	watt per meter degree Celsius	W/m · °C)
calorie (IT) per cubic centimeter second	4.1868* E + 12	microwatt per cubic meter	$\mu W/m^3$
calorie (IT) per square centimeter second	4.1868* E + 07	milliwatt per square meter	mW/m^2
calorie (IT) per gram °C	4.1868* E + 00	kilojoule per kilogram degree celsius	$kJ/(kg \cdot °C)$
calorie (thermochemical) per pound	9.224141 E + 00	joule per kilogram	J/kg
centigrade (= Celsius)			
centipoise	1.0* E + 00	millipascal second	mPa · s
centistoke	1.0* E + 00	square millimeter per second	mm^2/s
chain	2.01168* E + 01	meter	m
cubem (cubic mile)	4.168182 E + 00	cubic kilometer	km^3
cubic inch	1.638706 E − 02	cubic decimeter	dm^3
cubic foot (see standard cu. ft.)	2.831685 E − 02	cubic meter	m^3
cubic foot gas (60°F 1 atm) per acre foot	9.690510 E − 04 2.291262 E − 05	mole per cubic meter cubic meter API per cubic meter	mol/m^3 m^3API/m^3
cubic foot of gas (60°F − 1 atm) per barrel	1.777646 E − 01	cubic meter API per cubic meter	m^3API/m^3
cubic foot of gas per barrel of oil	7.518255 E + 00	mole per cubic meter	mol/m^3
cubic foot per foot	9.290304* E − 02	cubic meter per meter	m^3/m
cubic foot per pound	6.242797 E + 01	cubic decimeter per kilogram	dm^3/kg
cubic yard	7.645549 E − 01	cubic meter	m^3
cycle per second	1.0* E + 00	hertz	Hz
"cc"	1.0* E + 00	cubic centimeter	cm^3
CV (cheval vapeur)	7.354990 E − 01	kilowatt	kW
darcy	9.869233 E − 01	square micrometer	μm^2
decibel	0.1 × $\log_{10}$ (ratio of two intensities, e.g., watts)		dB
degree (angle) per foot	5.726145 E − 02	radian per meter	rad/m
degree (angle)	1.745329 E − 02	radian	rad
degree Centigrade (= Celsius)			
degree Fahrenheit	(°F − 32) 5/9* E + 00	degree Celsius	°C
degree Fahrenheit as interval	5/9* E + 00	degree Celsius	°C
degree Rankine	5/9* E + 00	kelvin	K
degree F per hundred foot	1.822689 E − 02	degree Celsius per meter	°C/m
dyne	1.0* E − 05	newton	N
dyne per centimeter	1.0* E + 00	millinewton per meter	mN/m
dyne per square centimeter	1.0* E − 01	pascal	Pa
erg	1.0* E − 07	joule	J
erg per square centimeter	1.0* E + 00	millijoule per square meter	mJ/m^2
erg per year	3.170979 E − 15	watt	W
fathom	1.8288* E + 00	meter	m
foot	3.048* E − 01	meter	m
foot-candle	1.076391 E + 01	lux	lx

*The conversion factor is exact.
[a]To go from metric to customary units, divide the conversion factor instead of multiplying.

continued

TABLE 5. continued

Traditional Unit	Multiply Value in Traditional Units by Factor × 10E to Obtain Value in Metric Units[a] — Factor E	Selected Metric (S.I.) Unit — Name	Symbol
foot per barrel	1.917134 E + 00	meter per cubic meter	m/m^3
foot per cubic foot	1.07639 E − 01	meter per cubic meter	m/m^3
foot per °F	5.4864* E − 01	meter per kelvin	m/K
foot per gallon (U.S.)	8.051964 E + 01	meter per cubic meter	m/m^3
foot per gallon	6.7046 E + 01	meter per cubic meter	m/m^3
foot per mile	1.893939 E − 01	meter per kilometer	m/km
foot pound-force	1.355818 E + 00	joule	J
foot pound-force per minute	2.259697 E − 02	watt	W
gal (see milligal)			
gallon (Cdn. & new U.K.)	4.54609* E + 00	liter	l
gallon (old U.K.)	4.546092 E + 00	liter	l
gallon (U.S.)	3.785412 E + 00	liter	l
gallon (U.S.) per foot	1.241933 E − 02	cubic meter per meter	m^3/m
gallon per foot	1.4914 E − 02	cubic meter per meter	m^3/m
gallon per horsepower hour	1.693466 E + 00	liter per megajoule	l/MJ
gallon per mile	2.824809 E + 02	liter per 100 kilometer	l/100km
gallon (U.S.) per mile	2.352146 E + 02	liter per 100 kilometer	l/100km
gallon per pound	1.002241 E + 01	liter per kilogram	l/kg
gallon (U.S.) per pound	8.345405 E + 00	liter per kilogram	l/kg
gallon (U.S.) per short ton	4.172702 E + 00	liter per tonne	l/t
gallon (U.S.) per long ton	3.725627 E + 00	liter per tonne	l/t
gamma (magnetic flux density)	1.0* E + 00	nanotesla	nT
gas constant: value	8.31432 E + 00	joule per mole kelvin	J/(mol · K)
gas gravity (density relative to air)	2.896 E + 01	gram per mole	g/mol
grain	6.479891* E + 01	milligram	mg
grain per 100 SCF	2.292768 E + 01	milligram per cubic meter API	mg/m^3
	5.421110 E − 01	milligram per mole	mg/mol
grain per gallon	1.42538 E − 02	gram per liter	g/l
gram mole	1.0* E + 00	mole	mol
horsepower (boiler)	9.80950 E + 00	kilowatt	kW
horsepower (550 ft-lb/s)	7.456999 E − 01	kilowatt	kW
horsepower (electric)	7.46* E − 01	kilowatt	kW
horsepower (hydraulic)	7.46043 E − 01	kilowatt	kW
horsepower (metric)	7.35499 E − 01	kilowatt	kW
horsepower (U.K.) & "indicated" or "brake"	7.457 E − 01	kilowatt	kW
hundredweight	4.535924 E + 01	kilogram	kg
inch	2.54* E + 00	centimeter	cm
inch to the fourth power	4.162314 E + 05	millimeter to the fourth power	mm^4
inch of mercury (Hg) at 0°C	3.386389 E + 00	kilopascal	kPa
inch of mercury (Hg) at 60°C	3.37685 E + 00	kilopascal	kPa
inch of water (H_2O) at 60°F	2.48843 E − 01	kilopascal	kPa
kilogram-force (kgf)	9.80665* E + 00	newton	N
kilogram-force per square centimeter	9.80665* E + 01	kilopascal	kPa
kilogram-force per square millimeter	9.80665* E + 00	megapascal	MPa

*The conversion factor is exact.
[a]To go from metric to customary units, divide the conversion factor instead of multiplying.

continued

TABLE 5. continued

Traditional Unit	Multiply Value in Traditional Units by Factor × 10E to Obtain Value in Metric Units[a] Factor E	Selected Metric (S.I.) Unit Name	Symbol
kilopond (kp)	9.80665* E + 00	newton	N
kilowatt hour	3.6* E + 02	kilojoule	kJ
kip	4.448222 E + 00	kilonewton	kN
knot (international)	5.144444 E − 01	meter per second	m/s
link	2.01168 E − 01	meter	m
magnetic permeability (cgs e.m.u.)	1.256637 E + 00	microhenry per meter	μH/m
magnetic susceptibility (cgs e.m.u.)	1.579137 E + 01	microhenry per meter	μH/m
MCF (thousand cubic foot 60°F−1 atm)	2.826231 E + 01	cubic meter (API)	m^3API
	1.195307 E + 00	kilomole	kmol
MCF per acre foot (60°F−1 atm)	9.690510 E − 01	mole per cubic meter	mol/m^3
	2.291262 E − 02	cubic meter API per cubic meter	m^3API/m^3
microcalorie per square centimeter second	4.1868* E + 01	milliwatt per square meter	mW/m^2
micron	1.0* E + 00	micrometer	μm
microsecond per foot	3.280840 E + 00	microsecond per meter	μs/m
mil	2.54* E + 01	micrometer	μm
mile (U.S. and Canada)	1.609344* E + 00	kilometer	km
mile per gallon	3.540060 E − 01	kilometer per liter	km/l
mile per U.S. gallon	4.251437 E − 01	kilometer per liter	km/l
mile (international nautical)	1.852* E + 00	kilometer	km
millicalorie per second centimeter °C	4.1868* E + 02	milliwatt per meter degree Celsius	mW/(m · °C)
millidarcy	9.869233 E − 04	square micrometer	μm^2
milligal	1.0* E + 01	micrometer per second squared	μm/s^2
millimeter of mercury (Hg) 0°C	1.333222 E − 01	kilopascal	kPa
millimho	1.0* E + 00	millisiemens	mS
millimicron	1.0* E + 00	nanometer	nm
millimicrosecond	1.0* E + 00	nanosecond	ns
MMCF (million cubic foot 60°F − 1 atm	2.826231 E + 04	cubic meter (API)	m^3API
	1.195307 E + 00	megamole	Mmol
million years	1.0* E + 00	megayear	Ma
millisecond per foot	3.289474 E + 00	millisecond per meter	ms/m
neper per foot	3.777207 E − 01	decibel per meter	dB/m
oersted	7.957747 E + 01	ampere per meter	A/m
ounce (avdp)	2.834952 E + 01	gram	g
ounce (fluid UK)	2.841308 E + 01	cubic centimeter	cm^3
parts per billion (mass basis)	1.0* E + 00	microgram per kilogram	μg/kg
parts per million (ppm) (mass basis)	1.0* E + 00	milligram per kilogram	mg/kg
parts per million (ppm) (by volume)	1.0* E + 00	cubic meter per liter	m^3/l
	multiply parts by density in kg/m^3	milligram per cubic meter	mg/m^3
parts per thousand (0/00) (mass basis)	1.0 E + 00	gram per kilogram	g/kg
parts per thousand (0/00) (by volume)	1.0* E + 00	cubic centimeter per liter	cm^3/l
pint	5.68261 E − 01	liter	l
pound-force	4.448222 E + 00	newton	N
pound-force foot (see foot pound-force)			

*The conversion factor is exact.

[a]To go from metric to customary units, divide the conversion factor instead of multiplying.

continued

TABLE 5. continued

Traditional Unit	Multiply Value in Traditional Units by Factor × 10E to Obtain Value in Metric Units[a] Factor E	Selected Metric (S.I.) Unit Name	Symbol
poundal	1.382550 E − 01	newton	N
pound-force per 100 square foot	4.788026 E − 01	pascal	Pa
pound-force per square foot	4.788026 E + 01	pascal	Pa
pound-force per square inch (psi)	6.894757 E + 00	kilopascal	kPa
pound-force second per square foot	4.788026 E + 01	pascal second	Pa · s
pound-mass (avdp)	4.535924 E − 01	kilogram	kg
pound-mass per horsepower hour	1.689659 E + 02	milligram per kilojoule	mg/kJ
pound-mass per barrel	2.853010 E + 00	kilogram per cubic meter	kg/m^3
pound-mass per foot	1.488164 E + 00	kilogram per meter	kg/m
pound-mass foot per second	1.352549 E − 01	kilogram meter per second	kg · m/s
pound-mass per cubic foot	1.601846 E + 01	kilogram per cubic meter	kg/m^3
pound-mass per gallon	9.97763 E + 01	kilogram per cubic meter	kg/m^3
pound-mass per gallon (U.S.)	1.198264 E + 02	kilogram per cubic meter	kg/m^3
pound-mass per cubic inch	2.767990 E + 04	kilogram per cubic meter	kg/m^3
pound-mass per thousand cubic foot	1.601846 E − 02	kilogram per cubic meter	kg/m^3
pound-mass per square foot	4.882428 E + 00	kilogram per square meter	kg/m^2
pound mole	4.535924 E − 01	mole	mol
psi (pound-force per square inch)	6.894757 E + 00	kilopascal	kPa
psi per foot	2.262059 E + 01	kilopascal per meter	kPa/m
quart	1.136522 E + 00	liter	l
quart (U.S.)	9.463529 E − 01	liter	l
quarter section (160 acres)	6.474970 E + 01	hectare	ha
RPM	1.0* E + 00	revolution per minute	r/min
second per quart (U.S.)	1.056882 E + 00	second per liter	s/l
section (640 acres)	2.589988 E + 02	hectare	ha
square inch	6.4516* E + 00	square centimeter	cm^2
square foot	9.290304* E − 02	square meter	m^2
square mile	2.589988 E + 00	square kilometer	km^2
square yard	8.361274 E − 01	square meter	m^2
standard cubic foot (60°F 1 atm − ideal gas)	2.826231 E − 02	cubic meter API	m^{3}API
	1.195307 E + 00	mole	mol
TCF (trillion cubic foot 60°F 1 atm)	1.195307 E + 00	teramole	Tmol
	2.826231 E + 10	cubic meter (API)	m^{3}API
"thou"	2.54* E + 00	micrometer	µm
thirty-second of an inch	7.93750 E − 01	millimeter	mm
ton (U.S. short − 2,000 lb)	9.071847 E − 01	tonne	t
ton (U.K. long − 2,240 lb)	1.016047 E + 00	tonne	t
ton-mile	1.431744 E + 01	megajoule	MJ
ton-mile per foot	4.697322 E + 01	megajoule per meter	MJ/m
ton (metric)	1.0* E + 00	tonne	t
yard	9.144* E − 01	meter	m

Note: Although the conversion factors have been calculated for the S.I. unit judged to be the most frequently required, for any particular application reference should be made to other tables of recommended units.

*The conversion factor is exact.

[a]To go from metric to customary units, divide the conversion factor instead of multiplying.

Source: From Kenting Limited, n. d. *The International System of Units (S.I.) for the Petroleum Industry in Canada.* Calgary, Alberta, Canada.

TABLE 6. S.I. Units Commonly Used

Item	Name	Symbol	Printer
Drilling, Cementing, and Formation Testing			
Linear	meter	m	M
(tool dimensions—always)	millimeter	mm	MILLIM
Area	square meter	m^2	M2
	hectare	ha	HECTARE
Volume and capacity	cubic meter	m^3	M3
	liter	l	LITRE
Mass	kilogram	kg	KG
	tonne	t	TONNE
Other			
Time	second	s	S
	minute	min	MIN
	hour	h	HR
	day	day	D
	week	wk	WEEK
	month	mo	MONTH
	year	yr	ANN
General Geology, Geophysics, and Reservoir Engineering			
Dip, gradient	meter per kilometer	m/km	M/KILOM
	degree	°	DEG
Geographical coordinates	degree or decimal degree	°	DEG
	minute	′	MNT
	second	″	S
Universal transverse mercator coordinates	meter	m	M
Distance	kilometer	km	KILOM
Elevation	meter	m	M
Depth	meter	m	M
Thickness of formations	meter	m	M
Area	square meter	m^2	M2
	hectare	ha	HECTARE
Volume of sediment in a basin	cubic kilometer	km^3	KILOM3
Geological age	megayear	Ma	MEGAANN
Reservoir Geology and Engineering			
Volume of reservoir or fluid	cubic meter	m^3	M3
Volume of pore space or fluid per volume of sediment	cubic meter per cubic meter	m^3/m^3	M3/M3
Permeability	square micrometer	μm^2	MICROM2
Formation pressure	megapascal	MPa	MEGAPA
Capillary pressure	pascal	Pa	PA
Head	meter	m	M
Pressure gradient	kilopascal per meter	kPa/m	KILOPA/M
Gas-oil ratio	cubic meter API per cubic	m^{3}API/m^3	M3/M3
	meter kilomole per cubic meter	kmol/m^3	KILOMOL/M3
Productivity index	cubic meter per day kilopascal	m^3/(d · kPa)	M3/(D.KILOPA)
Pipeline Operations			
Flow	liter per second	l/s	LITRE/S
	cubic meter per second	m^3/s	M3/S
	cubic meter per hour	m^3/hr	M3/HR
Pressure	pascal	Pa	PA
Force	newton	N	N
Energy, work			
—quantity of heat	joule	J	J
Compressor rating:			
—compressor heads	kilopascal	kPa	KILOPA
—flow	cubic meter per second	m^3/s	M3/S
Pumping rating:			
—dynamic head	meter	m	M
—flow	liter per second	l/s	LITRE/S
—conversion factor	kilopascal per meter	kPa/m	KILOPA/M

TABLE 6. continued

Item	Name	Symbol	Printer
Gradient:			
—pressure	kilopascal	kPa/km	KILOPA/KILOM
—slope	meter per kilometer	m/km	M/KILOM
Viscosity:			
—dynamic (gas)	micropascal second	μPa $\cdot$ s	MICROPA.S
—kinematic	square millimeter per second	mm^2/s	MILLIM2/S
Rotational frequency	revolution per minute	r/min	R/MIN
	radian per second	rad/s	RAD/S
Concentration	mole per cubic meter	mol/m^3	MOL/M3
	gram per cubic meter	g/m^3	G/M3
	milligram per kilogram	mg/kg	MILLIG/KG
	cubic centimeter per cubic meter	cm^3/m^3	CENTIM3/M3
Density—gas	kilogram per cubic meter	kg/m^3	KG/M3
	gram per mole	g/mol	G/MOL
Gravity—liquid density	kilogram per cubic meter	kg/m^3	KG/M3
Velocity	meter per second	m/s	M/S
	kilometer per hour	km/hr	KILOM/HR
Sound intensity	watt per square meter	W/m^2	W/M2
	decibel	dB	DECIBEL
	Seismological Investigations and Survey		
Amount of explosive	kilogram	kg	KG
Attenuation	decibel	dB	DECIBEL
Energy of source	megajoule	MJ	MEGAJ
Frequency	hertz	Hz	HZ
Pressure of shock wave	gigapascal	GPa	GIGAPA
Travel time	second	s	S
Velocity	meter per second	m/s	M/S
Wavelength	meter	m	M
	Gravity and Magnetic Surveys		
Gravitational variation	micrometer per second squared	μm/s^2	MICROM/S2
Density	kilogram per cubic meter	kg/m^3	KG/M3
Magnetic flux density (intensity)	nanotesia	nT	NANOT
Magnetic permeability	microhenry per meter	μH/m	MICROH/M
Magnetic susceptibility	microhenry per meter	μH/m	MICROH/M
	Rock Mechanics		
Strength, stress,	megapascal	MPa	MEGAPA
bulk modulus, elastic constant	gigapascal	GPa	GIGAPA
Elastic modulus	gigapascal	GPa	GIGAPA
Viscosity	pascal second	Pa $\cdot$ s	PA.S

Source: From Kenting Limited, n. d. *The International System of Units (S.I.) for the Petroleum Industry in Canada.* Calgary, Alberta, Canada.

Acknowledgments

The preparation of a work of this sort involves the generous cooperation of contributors (acknowledged in the section headed "Contributors") and many others who worked behind the scenes to get the job done. I am especially indebted to our contributors for their patience and perseverance during the preparation of this volume. I am also most grateful to my colleagues who critically read papers submitted for inclusion and who made suggestions for improvement both in content and style.

To those contributors who volunteered to write additional articles, to other correspondents who indicated topics or concepts that should be incorporated into the volume, and to those who suggested names of other experts as potential contributors, a special word of thanks is due: G. D. Aitchison (Australia), Frederick Betz, Jr. (USA), Julie Bichteler (USA), Regina Brown (USA), George V. Chilingarian (USA), Donald R. Coates (USA), Robert G. Font (USA), Keith Frye (USA), A. A. Geodekyan (USSR), Keith Grant (Australia), Alfred R. Jumikis (USA), Charles R. Kolb (USA), J. Lag (Norway), Peter F. Lagasse (USA), I. K. Lee (Australia), James T. Kirkland (USA), Henry N. McCarl (USA), Christopher C. Mathewson (USA), John L. Mero (USA), M. Hamid Metwali (Poland), Kurt L. Othberg (USA), Richard H. Pearl (USA), Nikola Prokopovich (USA), John C. Reed, Jr. (USA), Ronald J. Tanenbaum (USA), Robert L.

Schuster (USA), V. A. Stroganov (USSR), and William White (USA).

To my series editor, Rhodes W. Fairbridge, goes a special word of thanks for encouragement and inspiration. His interest and cooperation in the development of this volume is warmly appreciated and especially deserved as he proved to be a real *fons et origo* of information for many subjects.

The efforts of Pamela M. Matlack my graduate assistant, are appreciated for her attention to details in the filing of manuscripts, assistance in the preparation of main cross-references, and redrafting of figures.

Last but not least, I thank the publisher for continued support and interest throughout this lengthy project. Of special note are the heroic efforts of Shirley End, Executive Editor, and Mary Dorian, my production editor. Both deserve a special word of thanks for enhancing the production of this book beyond the call of duty.

CHARLES W. FINKL, JNR.

MAIN ENTRIES

CONTRIBUTORS

DUWAYNE M. ANDERSON, Faculty of Natural Sciences and Mathematics, State University of New York, Buffalo, New York 14260. *Permafrost, Engineering Geology.*

HENRY C. BARKSDALE, Geological Survey of Alabama, P.O. Drawer O, University, Alabama 35486. *Urban Hydrology.*

JOHN P. BARA, U.S. Bureau of Reclamation, Mid-Pacific Region, 2800 Cottage Way, Sacramento, California 95825. *Soil Classification System, Unified.*

DAVID J. BARR, Geological Engineering Department, 125 Mining Building, University of Missouri, Rolla, Missouri 65401. *Remote Sensing, Engineering Geology.*

ERIC J. BEST, Canberra College of Advanced Education, P.O. Box 1, Belconnen, A.C.T., Australia 2616. *Dams, Engineering Geology.*

FREDERICK BETZ, JR., Mar de Alboran, Apt. 3B, Avenida Antonio Belon, 28, Marbella, Málaga, Spain. *Geological Communication; Geology, Applied; Military Geoscience.*

JULIE BICHTELER, Graduate School of Library and Information Science, University of Texas, University Station Box 7576, Austin, Texas 78712. *Conferences, Congresses, and Symposia.*

ROBERT F. BLACK, Dept. of Geology and Geophysics, University of Connecticut, Groton, Connecticut 06340. *Geocryology.*

W. R. P. BOUCAUT, South Australia Department of Mines, P.O. Box 151, Eastwood, South Australia, Australia 5063. *Pumping Stations and Pipelines, Engineering Geology.*

R. BOWEN, Dept. of Geology, University of Zambia, P.O. Box 2379, Lusaka, Zambia. *Grout, Grouting.*

JERRY BROWN, U.S. Army Cold Regions Research and Engineering Laboratory, Hanover, New Hampshire 03755. *Permafrost, Engineering Geology.*

REGINA BROWN, Orton Memorial Library of Geology, Ohio State University, 155 South Oval Drive, Columbus, Ohio 43210. *Earth Science, Information and Sources.*

JAMES A. CALKINS, Office of Resource Analysis, U.S. Geological Survey, National Center, Reston, Virginia 22092. *Computerized Resources Information Bank.*

GEORGE V. CHILINGARIAN, Petroleum Engineering Department, University of Southern California, Los Angeles, California 90007. *Electrokinetics; Formation Pressures, Abnormal.*

DONALD R. COATES, Dept. of Geological Sciences, State University of New York at Binghamton, Binghamton, New York 13901. *Glacial Landscapes, Engineering Geology; Medical Geology; Urban Geomorphology.*

GEORGE V. COHEE, U.S. Geological Survey, National Center, Reston, Virginia 22092. *International Geochronological Time Scale.*

EDWARD L. DILLON, Occidental of Libya, Inc., P.O. Box 2134, Tripoli, Libya. *Well Data Systems.*

MOHAMED T. EL-ASHRY, Office of Natural Resources, Tennessee Valley Authority, Knoxville, Tennessee 37902. *Magnetic Susceptibility, Earth Materials.*

I. ENGELSTEIN, International Institute for Aerial Survey and Earth Sciences, 144 Boulevard 1945, P.O. Box 6, Enschede, The Netherlands. *Photogrammetry.*

CHARLES W. FINKL, JNR., Dept. of Geology, Florida Atlantic University, Boca Raton, Florida 33431. *Field Geology; Hydrogeology and Geohydrology; Information Centers.*

JOSEPH A. FISCHER, Geoscience Services, 50 Division Avenue, Millington, New Jersey 07946. *Earthquake Engineering; Nuclear Plant Siting, Offshore.*

ROBERT G. FONT, Conoco, P.O. Box 1959, Midland, Texas 79702. *Atterberg Limits and Indices; Urban Engineering Geology.*

PETER G. FOOKES, Winters Wood, 47 Crescent Road, Caterham, Surrey, England CR3 6LH. *Arid Lands, Engineering Geology.*

JOHN A. FRANKLIN, Franklin Geotechnical Engineering Limited, The Stream R.R. #1, Orangeville, Ontario, Canada L9W 2Y8. *Rock Structure Monitoring.*

GORDON S. FRASER, BP Alaska Exploration, Inc., 100 Pine Street, San Francisco, California 94111. *Beach Replenishment, Artificial.*

R. A. FRASER, C.S.I.R.O. Division of Applied Geomechanics, P.O. Box 54, Mount Waverly, Victoria, Australia 3149. *Reinforced Earth.*

M. F. DE FREITAS, Dept. of Geology, Royal School of Mines, Imperial College of Science and Technology, Prince Consort Road, South Kensington, London, England SW7 2BP. *Rock Slope Engineering.*

CHARLES P. GIAMMONA, Civil Engineering Dept., Texas A & M University, College Station, Texas 77843. *Oceanography, Applied.*

J. E. GILLOTT, Dept. of Civil Engineering, University of Calgary, Calgary, Alberta, Canada T2N 1N4. *Clay, Engineering Geology.*

SAMUEL S. GOLDICH, Geology Dept., Colorado School of Mines, Golden Colorado 80401. *Geochronology.*

KEITH GRANT, C.S.I.R.O. Division of Land Use Research, P.O. Box 1666, Canberra City, A.C.T., Australia 2601. *Duricrust, Engineering Geology.*

THOMAS E. HOWARD, Division of Advanced Mining Systems, Joy Manufacturing Company, 600 Broadway, Denver, Colorado 80217. *Tunnels, Tunneling.*

ROY E. HUNT, Av. Atlantica 290/Ap. 65, 22010 Rio de Janeiro, R. J., Brazil. *Geotechnical Engineering.*

ALFREDS R. JUMIKIS, Dept. of Civil Engineering, Rutgers State University, New Brunswick, New Jersey 08903. *Foundation Engineering; Soil Mechanics, History of; Wells, Aerial.*

GEORGE H. KELLER, School of Oceanography, Oregon State University, Corvallis, Oregon 97331. *Marine Sediments, Geotechnical Properties.*

CHARLES R. KOLB, deceased. *Alluvial Plains, Engineering Geology; Alluvial Valley Engineering; Deltaic Plains, Engineering Geology.*

R. J. KRIZEK, Dept. of Civil Engineering, Northwestern University, Evanston, Illinois 60201. *Rheology, Soil and Rock.*

PETER F. LAGASSE, Simons, Li and Associates, Inc., P.O. Box 1816, Fort Collins, Colorado 80522. *River Engineering.*

PHILIP LAMOREAUX, Alabama Geological Survey, P.O. Drawer O, University, Alabama 35486. *Urban Hydrology.*

P. F. F. LANCASTER-JONES, 17 Century Drive, Northam, Bideford, North Devon, England EX39 1B6. *Shaft Sinking.*

LAURENCE H. LATTMAN, College of Mines and Mineral Industries, University of Utah, Salt Lake City, Utah 84112. *Caliche, Engineering Geology.*

FITZHUGH T. LEE, Engineering Geology Branch, U.S. Geological Survey, MS904K AE, Denver Federal Center, Denver, Colorado 80225. *Rocks, Engineering Properties.*

I. K. LEE, Dept. of Civil Engineering Materials, University of New South Wales, P.O. Box 1, Kensington, New South Wales, Australia 2033. *Soil Mechanics.*

ROBERT F. LEGGET, 531 Echo Drive, Ottawa, Ontario, Canada K1S 1N7. *Urban Geology.*

HENRY N. McCARL, Dept. of Economics, University of Alabama, Birmingham, Alabama 35294. *Mineral Economics.*

MICHAEL E. McCORMICK, Dept. of Ocean Engineering, U.S. Naval Academy, Annapolis, Maryland 21402. *Hydromechanics.*

ROBERT M. McHAM, Dept. of Geology, Texas A & M University, College Station, Texas 77843. *Coastal Inlets, Engineering Geology.*

JULES A. MACKALLOR, U. S. Geological Survey, National Center, Reston, Virginia 22092. *Economic Geology.*

JOHN R. McWILLIAMS, Division of Mining Research, U.S. Bureau of Mines, Washington, D.C. 20240. *Tunnels, Tunneling.*

DAVID P. MANIAGO, Offshore Power Systems, P.O. Box 8000, Jacksonville, Florida 32211. *Nuclear Plant Siting, Offshore.*

CHRISTOPHER C. MATHEWSON, Dept. of Geology, Texas A & M University, College Station, Texas 77843. *Coastal Inlets, Engineering Geology; Geological Information, Marketing.*

BERNARD Le MEHAUTE, Rosenstiel School of Marine and Atmospheric Science, University of Miami, Miami, Florida 33149.

DAVID G. MOORE, Scripps Institution of Oceanography, University of California, La Jolla, California 92093. *Marine Sediments, Geotechnical Properties.*

PAUL H. MOSER, Geological Survey of Alabama, P.O. Drawer O, University, Alabama 35486. *Urban Hydrology.*

M. P. MOSLEY, Water and Soil Science Centre, Ministry of Works and Development, P.O. Box 1479, Christchurch, New Zealand. *Channelization and Bank Stabilization.*

RAYMOND C. MURRAY, Dept. of Geology, Rutgers University, New Brunswick, New Jersey 08903. *Forensic Geology.*

U. R. NEJIB, Engineering Dept., Wilkes College, Wilkes-Barre, Pennsylvania 18703. *Magnetic Susceptibility, Earth Materials.*

THOMAS C. NICHOLS, JR., Engineering Geology Branch, U.S. Geological Survey, Denver Federal Center MS903KAE, Denver, Colorado 80225. *Residual Stress, Rocks.*

HEROLD W. OLSEN, U.S. Geological Survey, Denver Federal Center, P.O. Box 25046, Denver, Colorado 80225. *Consolidation, Soil.*

J. T. PARRY, Dept. of Geography, McGill University, Montreal, Quebec, Canada H3A 2K6. *Terrain Evaluation, Military Purposes.*

BENJAMIN S. PERSONS, Dames & Moore Consulting Engineers, 455 East Paces Ferry Road, Atlanta, Georgia 30327. *Laterite, Engineering Geology.*

RICHARD J. PROCTOR, 327 Fairview Avenue, Arcadia, California 91006. *Rapid Excavation and Tunneling; Urban Tunnels and Subways.*

NIKOLA PROKOPOVICH, U.S. Bureau of Reclamation, Mid-Pacific Region, 2800 Cottage Way, Sacramento, California 95825. *Engineering Geology Reports; Lime Stabilization; Soil Classification System, Unified.*

HERMAN H. RIECKE, III, Dept. of Petroleum Engineering, University of Southern California, Los Angeles, California 90007. *Electrokinetics.*

ROBERT E. RIECKER, U.S. Air Force Cambridge Research Laboratories, Hanscom Air Force Base, Maine 01730. *Rock Mechanics.*

JOHN ROCKAWAY, Dept. of Mining, Petroleum and Geological Engineering, University of Missouri, Rolla, Missouri 65401. *Maps, Engineering Purposes.*

I. TH. ROSENQVIST, Institute for Geology, Blindern, Oslo, Norway. *Clays, Strength of.*

SYLVIA H. ROSS, Aermotor Division, Braden Industries, Rt. 4—All Seasons, Great Bend, Kansas 67530. *Wells, Water.*

ANTHONY J. SABATINO, Technosolo S.A., Rua Pedro Alves 15, Santo Cristo, Rio de Janeiro, RJ, Brazil. *Seismological Methods.*

ADRIAN SCHEIDEGGER, Institut für Geophysik, Technische Universität Wien, Gusshausstrasse 27-29, Vienna, Austria A-1040. *Hydrodynamics, Porous Media.*

ROBERT L. SCHUSTER, Engineering Geology Branch, U.S. Geological Survey, Denver Federal Center, Denver, Colorado 80225. *Consolidation, Soil.*

FREDERICK R. SIEGEL, Dept. of Geology, The George Washington University, Washington, D.C. 20052. *Geochemistry, Applied.*

KEITH SMITH, Dept. of Geography, Livingstone Tower, University of Strathclyde, Glasgow, Scotland G1 1XH2. *Hydrology.*

TRUMAN STAUFFER, JR., Center for Underground Space Studies, University of Missouri, Kansas City, Missouri 64110. *Cavity Utilization.*

RONALD J. TANENBAUM, College of Engineering and Technology, South Academic Center, North Arizona University, P.O. Box 15600, Flagstaff, Arizona 86001. *Pipeline Corridor Evaluation.*

DAVID J. VARNES, U.S. Geological Survey, Federal Center, Box 25046, Mail Stop 903 KAE, Denver, Colorado 80225. *Residual Stress, Rocks.*

IAN WATSON, Dept. of Geology, Florida Atlantic University, Boca Raton, Florida 33431. *Nuclear Plant Siting, Offshore.*

A. W. G. WHITTLE, P.O. Box 102, Stansbury, South Australia, Australia 5582. *Mineragraphy.*

JOHN B. WILSON, National Institute of Ocean Sciences, Wormley, Godalming, Surrey, England GU8 5VB. *Submersibles.*

HANS F. WINTERKORN, deceased. *Soil Mineralogy.*

LEONARD E. WOOD, School of Civil Engineering, Purdue University, West Lafayette, Indiana 47907. *Shale Materials, Engineering Classification.*

ROBERT L. WRIGHT, Dept. of Geography, The University, Sheffield, England S1O 2TN. *Geomorphology, Applied.*

A

**AERIAL PHOTOGRAPHY—See
PHOTOGRAMMETRY; REMOTE
SENSING, ENGINEERING GEOLOGY.
Vol. XIV: AERIAL SURVEYS, GENERAL;
PHOTOGEOLOGY; PHOTO
INTERPRETATION.**

**AIRBORNE GEOPHYSICS—See Vol. XIV:
EXPLORATION GEOPHYSICS.**

ALLUVIAL PLAINS, ENGINEERING GEOLOGY

The term *alluvial plain* is defined as a plain consisting of alluvium; it is used in both a restricted and a general sense. The AGI *Glossary* (Bates and Jackson, 1980) defines it restrictively as "a level or a gently sloping tract or a slightly undulating land surface produced by extensive deposition of alluvium, usually adjacent to a river that periodically overflows its banks; it may be situated on a flood plain, a delta, or an alluvial fan." Others consider the term in its broader sense, such as "the alluvial plain of the Lower Mississippi." Fisk (1944), in his classic volume on the Lower Mississippi, uses it principally in this broader sense. He considers the flood plain, the deltaic plain, the alluvial valley, and older braided stream surfaces that form low terraces within the valley walls as portions of the more inclusive landform alluvial plain. Thus the alluvial plain of the Lower Mississippi River applies to a vast flatland some 965 km long and 60 km wide that includes active and abandoned deltas, alluvial fans, low terraces, and the flood plains of a number of rivers.

Flood plains are often equated with alluvial plains, particularly on smaller streams; in the larger river systems, however, the term *flood plain* usually has a more restrictive connotation. A common definition of a flood plain is that portion of a river valley, adjacent to the river channel, which is periodically inundated when the river overflows its banks. Legal definitions of the term sometimes involve the frequency of flooding. Hydrophytic plants, drift lines, scarps, and so on are used to delineate its extent. In a restricted sense, sev-eral flood plains can occur within a given alluvial plain (see the discussion by Schmudde, 1968).

The interrelationship among the terms *alluvial plain, alluvial valley,* and *deltaic plain* is poorly defined. Should an alluvial plain include its deltaic plain, and should the term *alluvial valley* be restricted to that portion of the alluvial plain upstream from the deltaic plain usually marked by well-defined valley walls? Fisk (1944) uses these terms in this sense. Useful concepts involving distinction among these terms are advanced by Coleman and Wright (1971), Wolman and Leopold (1957), Schumm (1977), and others. Consider a river system that enters the sea. The drainage basin of the system is defined as a region drained by this system, an area that can be delineated with considerable accuracy. The entire drainage basin functions as a supplier of water and sediment, and consists of upland areas, alluvial valleys, and a deltaic plain. Upland areas are the principal areas subjected to erosion, the principal suppliers of sediment. According to Coleman and Wright (1971):

> The system is essentially "closed" at its upper limit. The alluvial valley, on the other hand, is a more or less balanced system, "open" at both ends, in which the river flows over and through its own deposits. Typically, there is neither net accumulation nor net erosion of sediment.... The volume of material deposited within the alluvial valley tends to be about equal to the volume eroded. Material eroded from the drainage basin is only temporarily stored in the alluvial valley....The deltaic plain, on the other hand, is that portion of a river system in which the sediment load is distributed and deposited rather than supplied, traded, or carried in transit.

It is further defined (see *Deltaic Plains, Engineering Geology*) as that area which merges with the alluvial valley where the main stream channel branches, or has branched in the past, into multiple distributaries.

Thus, as shown in Figure 1, a drainage basin can be conveniently divided into (a) upland areas, (b) alluvial valleys of the master stream and its tributaries, and (c) its deltaic plain. Alluvial valleys and deltaic plains are characterized by distinctive suites of landforms. An understanding of the distribution, thickness, and lateral extent of the soils that constitute these landforms is useful for the engineering geologist.

To summarize, an alluvial plain, in its broader and more commonly used sense, is defined as

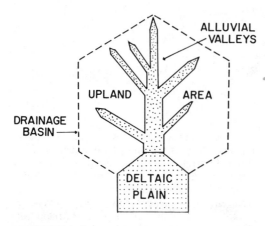

FIGURE 1. Components of a river system. The alluvial plain consists of the alluvial valleys and the deltaic plain of a given system. Based on Schumm's (1977) zones I, II, and III.

consisting of both the alluvial valley and the deltaic plain. In a restricted sense, the term is applied to any plain consisting of alluvium.

CHARLES R. KOLB*

*Deceased

References

Bates, R. L., and Jackson, J. A., 1980. *Glossary of Geology.* Falls Church, Va.: American Geological Institute, 749p.

Coleman, J. M., and Wright, L. D., 1971. *Analysis of Major River Systems and Their Deltas: Procedures and Rationale.* Tech. Rept. No. 95. Baton Rouge, La.: Coastal Studies Institute, Louisiana State University.

Fisk, H. N., 1944. *Geological Investigation of the Alluvial Valley of the Lower Mississippi River.* Vicksburg, Miss.: U.S. Army Corps of Engineers, Mississippi River Commission.

Schmudde, T. H., 1968. Flood plain, *in* R. W. Fairbridge, ed., *The Encyclopedia of Geomorphology.* Stroudsburg, Pa.: Dowden, Hutchinson & Ross, 359-362.

Schumm, S. A., 1977. *The Fluvial System,* New York: Wiley, 388p.

Wolman, M. G., and Leopold, L. B., 1957. River flood plains: some observations on their formation, *U.S. Geol. Survey Prof. Paper 282-C.*

Cross-references: *Channelization and Bank Stabilization; Coastal Engineering; Dams, Engineering Geology; Deltaic Plains, Engineering Geology; River Engineering.* Vol. XIV: *Alluvial Systems Modeling; Canals and Waterways, Sediment Control; Land Drainage; Photogeology.*

ALLUVIAL VALLEY ENGINEERING

An *alluvial valley* is a gently sloping plain consisting of alluvium. It is normally delimited by uplands on either side that rise above the level of the valley to varying heights. More specifically, the alluvial valley of a given stream is that portion of its *alluvial plain* upstream from its *deltaic plain* (see *Alluvial Plains, Engineering Geology; Deltaic Plains, Engineering Geology).* An alluvial valley is considered to be a more or less balanced system (see Vol. XIV: *Alluvial Systems Modeling).* Sediment supplied to it from the surrounding upland and its tributary alluvial valleys is gradually traded downstream until it is eventually deposited in the deltaic plain. Typically, there is neither net accumulation nor net removal of sediment from the alluvial valley. The volume of material deposited within the valley tends to about equal the amount of material removed from it and carried to the delta.

Alluvial valleys vary widely in width, and the thickness of alluvium within a given valley ranges from the depth of maximum scour during floods to many times the depth of scour. Regional subsidence, changes in base level caused by local obstructions, faulting, and related factors significantly affect the thickness of alluvium. The most prominent base-level change affecting thickness of alluvium was the Late Wisconsinan drop in sea level. Nearly all the larger river valleys of the world were entrenched during the last drop in sea level. Where data are available, they almost always indicate that the thickness of the alluvium through which the larger rivers flow is considerably greater than the present depth of scour of the river.

Sediments within valleys almost always grade upward from coarse at the base to finer materials at the valley surface. Deviations from this generality can usually be attributed to remnants of older materials left within the valley from previous periods of alluviation, the entrance of tributaries into the valley, ponding within tributaries caused by rapid aggradation of the master stream, or tectonic influences such as faulting.

In temperate and humid areas of the world, alluvial valley fill consists exclusively of alluvium. In arid regions, however, significant thicknesses of eolian sands and silts are often intercalated with the valley fill (Glennie, 1970). Stream beds are often dry for long periods in such areas, and wind tends to rework previously deposited alluvium significantly. Because of this, landforms and depositional processes common to most river valleys vary to some extent from those in more humid areas. The same is true to a somewhat lesser extent in arctic regions, where permafrost and extended periods during which streams are frozen interject conditions that are geomorphically and depositionally unique to the arctic (see *Permafrost, Engineering Geology).*

The great majority of the world's river valleys are characterized by landforms produced by braided streams, meandering streams, or a combination of the two. The alluvial valley of the lower Mississippi River belongs to the last category and is one of the world's more intensively studied river valleys. Most of the environments of deposition within the Mississippi Valley were formed by meandering streams; however, within the valley large areas of an older, slightly higher surface exist which reflect a time when the stream was braided. The classification of Mississippi Valley sediments used here is essentially that followed by Fisk (1944) and Kolb et al. (1968). For useful variations to this classification, particularly as such classification might apply to streams considerably smaller than the Mississippi, see Schumm (1977) and Happ (1971).

An understanding of the geological processes involved in the formation of the various environments of deposition within an alluvial valley, of the lithological and physical characteristics of these environments, and of their distribution in plan and profile are of importance in engineering, petroleum geology, environmental science, law, and many other disciplines. The reconstruction of ancient petroliferous sedimentary environments based on a knowledge of the landforms and lithological characteristics found within present-day alluvial valleys has become of ever-increasing importance in oil exploration. The development of agriculture, housing, and industry within valleys is often predicated on geological factors, as are environmental or flood control considerations, which just as often tend to discourage or modify such developments (see *Urban Engineering Geology; Urban Geology*). Problems involving property rights and boundaries along rivers more and more frequently involve the expertise of the geologist, the fluvial morphologist.

Of particular importance is the application of geological parameters or techniques in the prediction of river behavior (see *River Engineering*), in the search for aggregate in areas where aggregate sources are sometimes scarce, and in resolving foundation problems (see *Foundation Engineering*). The value of geological studies cannot be overemphasized in locating and designing engineering structures within river valleys and in economizing on soil borings and laboratory testing programs. The major types of depositional environments within river valleys produce predictable sequences of soils, each having a fairly well-defined range of engineering properties.

The Alluvial Valley of the Lower Mississippi River

The alluvial valley of the Mississippi was profoundly affected by sea-level fluctuations during the Pleistocene. The latest drop in sea level is generally believed to have reached its maximum about 17,000 years ago. This drop caused the scouring of an entrenched valley beneath the present flood-plain surface. There is controversy about the effect of this relatively short-lived drop in sea level on the sediments forming the alluvial valley. Some contend that sediments of the previous interglacial deposits were removed entirely and that the valley was entrenched to new depths. Others believe that large remnants of the mid-Wisconsinan interglacial deposits remain at depths beneath more recent Holocene deposits. Still others hypothesize that entrenchment of the mid-Wisconsinan interglacial deposits extended no farther upstream than the latitude of Baton Rouge, Louisiana.

For our purposes, it is sufficient to characterize the present alluvial valley as consisting of a substratum of sand and gravel—whatever its age—overlain by a topstratum of finer-grained materials. The substratum, the massive sand and gravel sequence that underlies the topstratum and overlies older Tertiary and Cretaceous strata, is treated as a single unit.

Figure 1 is a generalized map showing the distribution of the major depositional types that form the lower Mississippi alluvial plain. The alluvial valley, by definition, does not include the deltaic plain. Figure 2 shows the environments of deposition in plan and profile within a small portion of the valley in the vicinity of Greenville, Mississippi.

The Substratum. The substratum consists of a wedge of coarse-grained material laid down during the earlier stages of the filling of the entrenched valley of the Mississippi River. Figure 2 illustrates the thickness and general nature of the substratum deposits at a centrally located site in the valley. The unit is composed predominantly of clean sand and gravel, with the material normally becoming coarser with depth. Cobbles up to 10 cm in diameter are sometimes encountered near the base of the unit. Occasional lenses of clay, sandy silt, or silty sand are also found, but they are rare and discontinuous. Thicknesses of the substratum and the depth to the top of the substratum generally increase in a down-valley direction. The substratum is often encountered at depths as shallow as 3 m in the northern part of the valley and may average only 15 m in thickness. The depth to the substratum near Baton Rouge, Louisiana, on the other hand, is as much as 40 m and the thickness of the substratum may be more than 100 m.

Although a shallow depth to firm substratum sands may be desirable from a foundation standpoint, the relatively high permeability of the substratum may often result in an expensive dewatering operation if deep excavations are involved. Pressure-relief wells are often installed to minimize undesirable uplift pressures in cases in which the bottom of the excavation is in clays overlying substratum sands at shallow depths. Where the excavation bottom is in the substratum, the problem of keeping the excavation free of water may be sizable. Permeabilities of the substratum generally range from 400×10^{-4} cm/sec to $2,000 \times 10^{-4}$ cm/sec, with the larger values associated with the deeper and coarser part of the deposit.

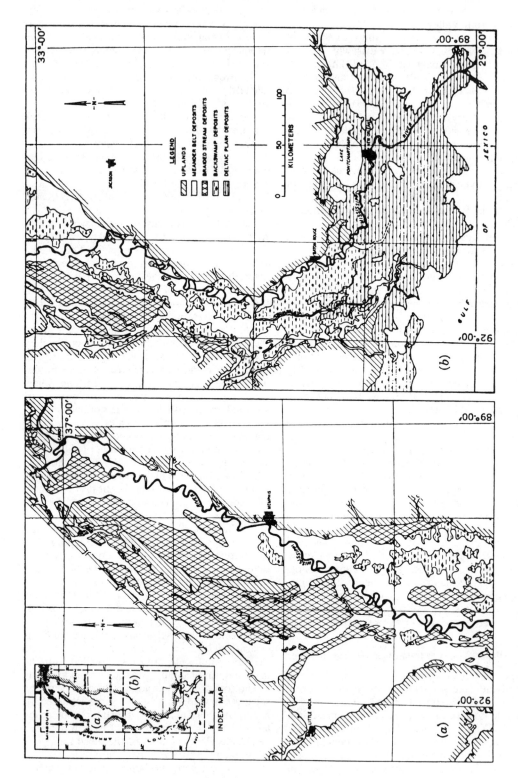

FIGURE 1. Distribution of depositional types in the lower Mississippi Valley (from Kolb and Shockley, 1959).

4

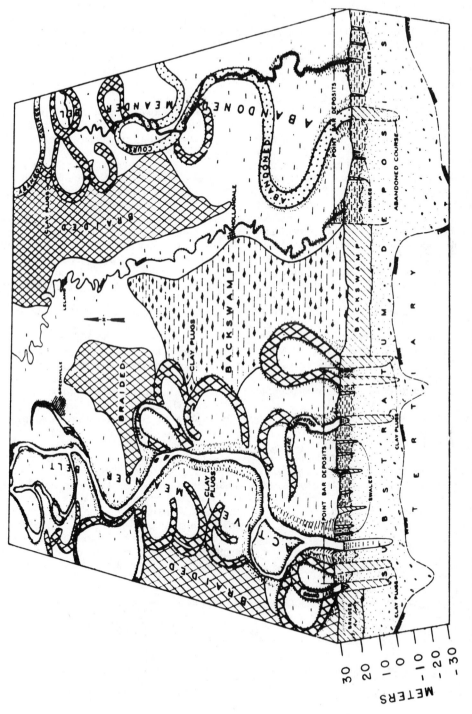

FIGURE 2. Major environments of deposition in the vicinity of Greenville, Mississippi (from Kolb and Shockley, 1959).

Water tables within the valley are near the surface; thus the highly permeable substratum provides a convenient and important aquifer for agricultural, industrial, municipal, and individual use. The volume of the alluvial valley substratum is enormous. Kolb (1961) estimated that the amount of storage in this aquifer equals the total flow of the Mississippi for two and one-half years. Wells 70 m deep yield 15,000 to 18,000 l/min (see *Wells, Water*).

The substratum is also an important source of concrete aggregate and base-course fill. Dredging from gravel bars is common along the river, and numerous pits within the flood plain exploit substratum sands and gravels that occur at shallow depths. Even more important as an aggregate source are substratum sands and gravels in the alluvial terraces that border the valley.

Natural Levees. Natural levees (Fig. 3) are low ridges that flank both sides of streams that periodically overflow their banks. Since the coarsest and greatest quantities of sediment are deposited closest to the stream channels, the natural levees are highest and thickest in these areas and gradually thin away from the channels. In general, the greater the distance from the stream, the greater the percentage of the finer-grained sediments. Minute drainage channels trending at right angles to the parent stream (down the backslope of the levees) are rather common and tend to fill with fine sand. Major crevasses through the levee sometimes form during floods and often carve surprisingly deep channels, which tend to fill with some of the coarsest material carried in suspension by the river.

Natural levees attain crest heights of 3 to 4 m above adjacent backswamp areas and may be 4 km wide. They typically consist of stiff to very stiff, brown to grayish brown silts, silty clays, and clays that exhibit moderate to high degrees of oxidation. Natural water contents of the deposits are typically low, and organic matter is seldom present except in the form of roots. Early roads through the Mississippi Valley typically were sited on these higher, well-drained features. As highway technology advanced, however, roads were built as often as not without regard to these natural ridges. As a result, many of the older roads built on natural levees have performed well over the years, while more modern highways have posed serious problems particularly where they cross low-strength, highly compressible clays (see *Consolidation, Soil)* characteristic of some of the other alluvial valley environments.

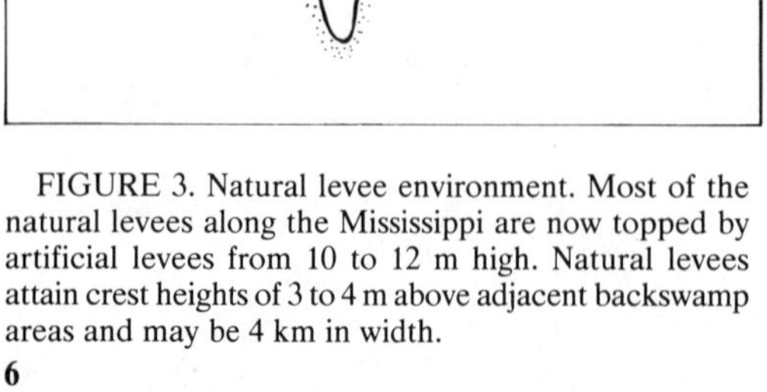

Alluvial Aprons. Alluvial aprons (Fig. 4) are combinations of alluvial and colluvial deposits that overlie flood-plain deposits along the valley walls and along the sides of upland remnants within the valley. Typically, symmetrical alluvial fans are present at the mouths of streams that drain the uplands. When these streams are closely spaced, the fans coalesce to form alluvial aprons. When the streams are more widely spaced, the fans are separated and the intervening portions of the aprons are composed mainly of sediments that have washed down from the uplands or that have been moved downslope by soil creep.

Alluvial aprons are common, particularly along the high eastern valley walls of the Mississippi alluvial valley. They occur but are less pronounced along the lower western valley walls. They are best developed near the mouths of the small streams that enter from the uplands and particularly where they overlie backswamp deposits and thus have not been affected by migration of the river. Alluvial apron widths of more than 3 km are common in the Yazoo Basin, for example, and elevations of 4 to 6 m above the flood-plain level occur near the valley edge. Borings made on alluvial aprons encounter soils that reflect the composition of the materials in the uplands. Along the eastern valley wall of the Mississippi Valley, the thick loess of the uplands is the predominant source material. Thus, clayey silt, silt, and fine sand are its typical constituents. Of interest to the engineering geologist and the environmental geologist is the accelerated rate at which streams in the loessial uplands are being entrenched. The result is an accelerated rate of growth of the alluvial aprons that these

CREVASSE
CHANNEL

FIGURE 3. Natural levee environment. Most of the natural levees along the Mississippi are now topped by artificial levees from 10 to 12 m high. Natural levees attain crest heights of 3 to 4 m above adjacent backswamp areas and may be 4 km in width.

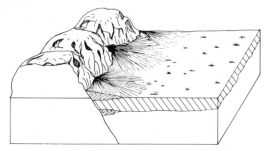

FIGURE 4. Alluvial apron environment. Aprons are particularly well developed along the loess-covered eastern valley wall of the lower Mississippi Valley.

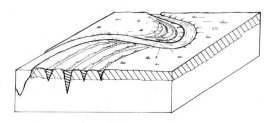

streams build as they drop their loads and build alluvial aprons at the valley borders. Exploitation of timber and land use practices in the uplands are causing significant problems not only in the uplands but where alluvial aprons flank the valley borders.

Braided Stream Deposits. Braided stream deposits (Fig. 5) consist of the sediments that were laid down by rapidly shifting, aggrading steams during the earlier stages of valley alluviation. The braided stream deposits were formed by shallow, anastomosing, ancestral streams of the Ohio, the Mississippi, the Arkansas and smaller streams emerging from the uplands adjacent to the entrenched valley. The great majority of the thick substratum deposits that underlie the valley and extend to depths of 100 or more meters were laid down by braided streams. By the time alluviation within the valley has reached its present level, however, fines were being deposited and braided channels were gradually consolidating into single meandering streams, particularly in the southern portions of the valley. As meander belts became established and were extended up-valley, valley gradients were reduced and valley surfaces were built to levels slightly lower than the older braided stream surfaces. Consequently, these surfaces in the Mississippi Valley often stand as low terraces above the level of the meander belt portions of the valley, as shown on Figure 5.

Braided stream topstratum covers a fairly significant area within the lower Mississippi Valley (Fig. 1). It consists of light gray to tan clays, silts, and well-graded fine sands on the order of 3 to 10 m thick. Both water and organic contents of the sediments are low. The average grain size of the sediments increases toward the northern end of the basin.

Point Bar Deposits. Point bar deposits (Fig. 6) consist of sediments laid down on the insides of river bends at a result of the meandering of the stream. Although the deposits extend to a depth equal to the deepest portion, or *thalweg,* of the parent stream, only the uppermost, fine-grained portion is considered part of the topstratum. Within the point bar topstratum, there are two types of

FIGURE 6. Point bar environment. Sediments laid down on the insides of the river bend as the stream meanders. They consist of an alternating series of swales (clay bodies) and ridges (silty sands and sandy silts) overlying substratum at shallow depths.

deposits: silty and sandy elongate bar deposits, or *ridges,* which are laid down during high stages of the stream, and silty and clayey deposits in arcuate depressions, or *swales,* which were laid down during falling river stages. Characteristically, the ridges and swales form an alternating series, the configuration of which conforms to the curvature of the migrating channel and indicates the direction and extent of meandering.

Point bar deposits are most widespread along the present course of the Mississippi River and along the abandoned courses of the river. Because of successive occupations of certain meander belts by streams of different sizes, complex patterns of ridge-and-swale topography are common. Point bar deposits consist of tan to gray clays, clayey silts, silts, and fine sands in the ridges, and soft, gray clays and silty clays in the swales. Excluding the larger swales, which occasionally may be filled with clays over 20 m thick, the topstratum varies from 6 to 12 m thick. Both water and organic contents are high in the swale deposits, whereas they are both commonly low in the ridge deposits.

A common and potentially hazardous phenomenon associated with flooding of the Mississippi River is seepage beneath the levees and the formation of *sand boils,* particularly in the thin topstratum deposits characteristic of point bar ridges. Sand boils consist of sand carried by seepage forces to the surface on the landward side of levees. These features often form conical mounds with water—sometimes muddy water—issuing from the top of the mound. Although limited underseepage and through-seepage of the levees are generally acceptable, seepage beneath levees in the form of sand boils indicates active piping and poses a threat to the safety of the levee. Comprehensive studies of these features (Mansur et al., 1956, and Kolb, 1976) have shown that the disposition of the various environments of deposition, the juxtaposition of pervious versus impervious flood-plain deposits beneath the levee, and the angle at which such bodies are crossed by the overlying levees are controlling factors in localizing sand boils. Figure 7 shows the effect of elongate swales and channel fill deposits where these clayey deposits pass beneath

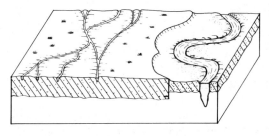

FIGURE 5. Braided stream environment. Sediments laid down by a shallow, aggrading network of streams during earlier phases of Mississippi Valley development. These now stand as low terraces within the valley above the level of those environments of deposition associated with meandering streams.

7

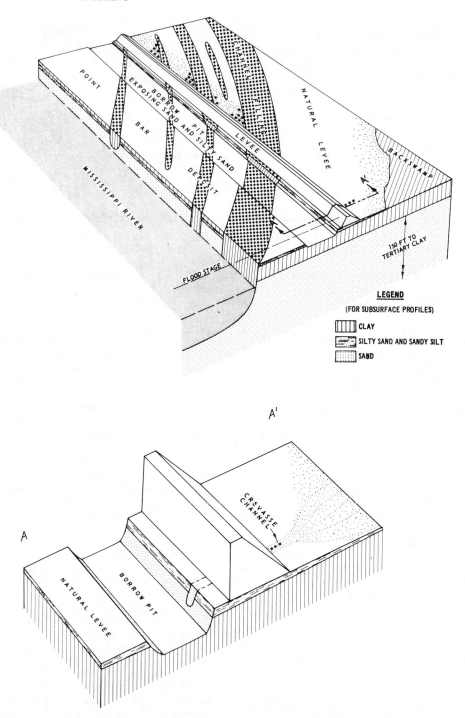

FIGURE 7. Sand boils (shown with asterisks) and seepage (shown with a dot pattern) that formed during the 1937 Mississippi flood. This is an example of only one of thousands of such phenomena that occurred on the landward side of the levees during this flood. A special case is illustrated in the expanded section shown along A-A'. Here a well-developed, semi-pervious natural levee deposit lies between backswamp clays and the artificial levee. In such instances seepage may occur in the extreme landward portions of the natural levee and boils may form in old natural levee crevasses backfilled with sand (from Kolb, 1976).

a levee at an angle. Seepage is often heaviest and boil formation most marked within the acute angle. The clay body tends to concentrate seepage in the pervious ridge areas where the geometry of the levees vis-à-vis the trend of the swales resembles that shown in Figure 7. Note that boils also tend to form adjacent to the swale within the obtuse angle formed by the swale and the levee. Such seepage is generally less pronounced, however. Also of interest in this figure is the development of boils where a sand-filled crevasse channel lies beneath the levee; in this instance, a crevasse channel developed through natural levee deposits overlying backswamp clay.

Abandoned Channels. Abandoned channels (Fig. 8)—or clay plugs, as filled abandoned channels are commonly called—are segments of stream channels formed when the stream shortens its course. The abandoned segment may consist of an entire meander loop when the river cuts directly across the narrow neck between two converging arms of a loop (a *neck cutoff*), or it may be a portion of a loop formed when a stream occupies a large point bar swale during a flood stage and abandons the outer portion of the loop (a *chute cutoff*).

Deposits filling abandoned channels are predominantly clay and silty clay. Shortly after cutoff, a wedge of sand fills both arms of the abandoned channel at the point of cutoff, and soon thereafter an oxbow lake is formed. The only material deposited within the oxbow lake is that from periodic overbank flow from the river. As the channel migrates away from the point of cutoff, only the finest materials find their way to the oxbow lake. Consequently, clays may fill these abandoned channels to the depth of the former active channel; clay deposits from 30 to 50 m thick are not uncommon. Abandoned channels of the Mississippi are usually 8 to 15 km long (following the loop) and from 500 to 1500 m wide (channel width).

Clays settling within deep oxbow lakes are characteristically high in water content. Because plants grow within the lake area only when it is filled or nearly filled with sediment, organic content is low

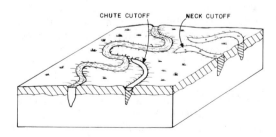

FIGURE 8. Abandoned channel environment. Abandoned segments of a loop formed when a stream shortens its course. These form lunate or horseshoe-shaped lakes that eventually fill with clays—clay plugs.

except in the upper 3 m of the clay body. Clay plugs are among the deepest clay bodies found in the alluvial valley. They are relatively unconsolidated and tend to exhibit high compressibility and low strength. Special design and construction measures are usually necessary where levees, highways, and other engineering structures cross abandoned channel deposits. Their distribution often vitally affects the location of drainage and navigation channels. In addition, these channel deposits have a significant effect on river migration. Wherever these deep clay bodies occur along the river, they act as "hard" points that radically alter the normal rate and direction of stream migration and greatly influence revetment placement and levee location (Fisk, 1947). Valid predictions of the direction in which the river may migrate during the life of an installation is of prime importance to the engineering geologist entrusted with site location along the banks of the Mississippi. Bridges, pipelines, overbank flow structures, docks, levees, and factories anxious to utilize cheap transportation facilities provided by the river are examples of the numerous installations that are affected by the stability of the river.

Of equal importance has been the application of fluvial morphology to the ubiquitous problem of boundary lines between states, counties, and private properties that border on the Mississippi and its tributaries within the alluvial valley. Because the boundaries between states often follow the thalweg of the river, changes in the thalweg radically affect these boundaries. Simplistically speaking, rapid changes *(avulsions)* in the thalweg caused by a chute or a neck cutoff fix the boundary along the thalweg of the abandoned channel. Thus, enclaves of Louisiana within Mississippi, for example, and of Mississippi within Arkansas are found along the Mississippi and its tributaries. Here, a portion of one state may be entirely surrounded by another. Conversely, state boundaries change as the thalweg shifts, if the shift is *gradual*. Huge portions of states adjoining the river are carved away each year as the river shifts and are added to the state across the river. Fluvial morphologists are thus often called on to testify in court cases involving riparian boundaries.

Abandoned channels formed by meandering streams of all sizes are numerous in the lower Mississippi Valley. Figure 9 shows the major clay plugs mapped in that portion of the alluvial valley known as the Yazoo Basin, a football-shaped flatland with Vicksburg, Mississippi, at its southern end and Memphis, Tennessee, at its northern end. The figure includes abandoned courses (discussed below), but most of the units delineated are abandoned channels or clay plugs.

Abandoned Courses. Abandoned courses (Fig. 10) are lengthy segments of river abandoned when the stream forms a new course across the flood

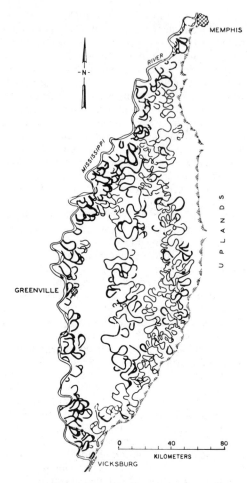

FIGURE 9. Abandoned channel and abandoned course deposits in the Yazoo Basin. Literally thousands of these massive clay bodies are found within the Mississippi alluvial valley (from Kolb et al., 1968).

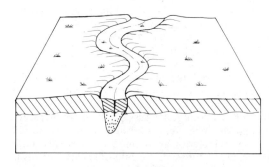

FIGURE 10. Abandoned course environment. Lengthy segments of the river are abandoned when the stream chooses a new course across the floodplain. These features tend to fill with a wedge of sand thickest at the point of diversion and thinning downstream.

plain. The abandoned course, varying from a few kilometers (but always more than one meander loop) up to hundreds of kilometers in length, gradually fills with sediment and is often occupied by a smaller or underfit stream. Studies of these features suggest that the old course fills with a wedge of sand, thickest where the new course diverges from the old, gradually thinning downstream. In many cases, the smaller stream meanders within the confines of the larger meander belt and destroys surficial segments of the original abandoned course. In other cases, the smaller stream delineates the extent of the abandoned course where where are not other indications of its presence.

Many of the surface expressions of Mississippi River abandoned courses have been destroyed by the meandering of smaller streams. The short, isolated segments that are recognized are similar in size and shape to abandoned channels. Abandoned courses of smaller streams are numerous throughout the valley. Unlike abandoned channels or clay plugs, topstratum deposits in the former are relatively thin, particularly in the immediate area of the point of diversion and from some distance downstream from this point. (Otherwise, the topstratum of the abandoned courses resembles those of the abandoned channels, i.e., highly compressible clays and silty clays.

Based on available data (Fisk et al., 1952), it appears that the abandonment of one Mississippi River course and the development of another takes place over a fairly lengthy period (up to a century) and also, that such a traumatic change in the river happens only once in 500 years or so. As such, it might seem that such changes are so rare as to be of little practical consequence during the life of a project. It becomes very significant, however, if this change is about to occur. This is true of the Mississippi. The Atchafalaya River (Fig. 1) leaves the Mississippi north of Baton Rouge, Louisiana, and diverts water through a direct, markedly shorter route to the Gulf than the present channel, which flows past Baton Rouge and New Orleans. The Mississippi is on the verge of abandoning the latter course for the former and undoubtedly would have done so by now had not extensive and expensive control structures been built in the early 1950s to prevent it. Consequences of such a major diverstion of the river are monumental. Some (Kolb, 1980) argue that the future of the southern part of the alluvial valley would be better served if the river were not shackled with artificial and costly engineering restraints and were permitted to choose the shorter path to the sea it would have chosen under natural conditions. Comprehensive studies might well conclude that a gradual, controlled diversion down the Atchafalaya, with necessary safeguards to ensure an adequate amount of fresh water down the present channel south of the point of diversion, might be a viable option.

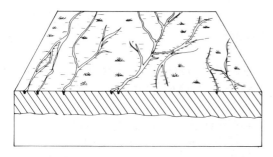

FIGURE 11. Backswamp environment. Fine-grained sediments are laid down in shallow ponded areas during floods. Because backswamp clays are exposed to desiccation and oxidation after each flood, they tend to develop a preconsolidated strength considerably higher than similarly fine materials in the abandoned channel, abandoned course, and swale environments.

Backswamp Deposits. Backswamp deposits (Fig. 11) consist of fine-grained sediments laid down in shallow ponded areas during periods of stream flooding. The coarser material from overbank flow is dropped near the stream to form natural levees. The finer material settles slowly in low-lying areas as the ponded water gradually drains off, seeps into the ground, or evaporates. Backswamp areas typically have very low relief, and a distinctive, complex drainage pattern develops in which the channels alternately serve as tributaries and distributaries at different times of the annual flood cycle.

Backswamp deposits are widespread within the alluvial valley and increase in areal extent in the wider portions of the valley, particularly in the southern portion. Deposits are continuous over areas as large as 500 km², and their thickness generally increases toward the south, where deposits as thick as 30 m are found. Varying thicknesses of natural levee or alluvial apron deposits often overlie backswamp deposits. Three to 4 m of backswamp deposits, in turn, often overlie other valley deposits such as point bars and abandoned channels.

Soft to stiff, gray to dark gray-brown clays and silty clays are typical of backswamp deposits, and occasional thin layers of silt or sand may be found, but more than 80 percent of the deposit normally consists of clay. Organic matter in the form of disseminated particles, peat layers, and large wood fragments are common. Because backswamp clays are exposed to desiccation and oxidation after each flood, they often develop strength properties considerably higher than similarly fine materials that settle out in an aqueous environment, such as the oxbow lakes that are the sites for abandoned channel deposits. This preconsolidated strength is apparent not only in laboratory measurements of shear strengths but is also reflected in the lower water content of these sediments when compared with abandoned-channel and abandoned-course topstratum.

Clays are generally added to the backswamp in increments ranging from paper-thin to several centimeters thick. On drying between floods and after local precipitation, the clay shrinks and thousands of small cubical fragments often form, fragments that can be scooped up by the handful from the dried backswamp surface. The size and hardness of these dry clay pellets give rise to the term "buckshot" clay. When wet, the extremely fine clay becomes maddeningly viscous and sticky and is referred to locally as "gumbo."

Because the backswamp deposits typically contain variable quantities of organic materials, they are sometimes used as a source for lightweight aggregate. On firing, organic particles in the clay form gases that expand and result in a light, porous, durable clinker that is much in demand in the southern part of the valley where aggregate is scarce and expensive and where lightweight aggregate is often used in high-rise buildings to impose as small a load as possible on soils that are characteristically low in bearing strength.

Summary

Alluvial valleys are gently sloping plains consisting of alluvium delimited on either side by uplands rising to varying heights. The principal depositional types are those associated with (1) braided streams, and (2) meandering streams. The major alluvial valleys of the world fall in the latter category. The alluvial valley of the Mississippi River is characterized chiefly by depositional landforms associated with meandering channels; however, it also contains remnants of surfaces left by an older, braided Mississippi River.

Principal depositional types forming surficial deposits in the Mississippi Valley are natural levees, alluvial aprons, braided streams, point bars, abandoned channels, abandoned courses, and backswamps. Each is characterized by a predictable sequence of soils, with predictable thicknessess and lateral limits, and having fairly well-defined ranges of engineering properties. A knowledge of the geological processes forming each depositional type and the distribution of these types has proved invaluable in locating and designing engineering structures in the valley and in economizing on soil borings and laboratory testing programs.

An understanding of the fluvial processes in alluvial valleys and the depositional types that result have numerous applications. Among them: (1) predictions of future meandering and migration of the river based on past migratory history; (2) age determinations for archaeological sites on abandoned streams; (3) reconstruction of ancient petroliferous sedimentary environments based on present alluvial valley stratigraphy; (4) siting industrial, housing, and agricultural developments having the least environmental impact; (5) resolving boundary disputes along rivers among states, counties, and individuals; (6) locating sources and assessing resources for sand and gravel supplies; and (7) solving a multitude of problems associated with foundation stability beneath locks, dams, drainage facilities, and similar engineering structures associated with flood control and general valley development.

CHARLES R. KOLB*

*Deceased

References

Fisk, H. N., 1944. *Geological Investigation of the Alluvial Valley of the Lower Mississippi River.* Vicksburg, Miss.: Mississippi River Commission, 78p.

Fisk, H. N., 1947. *Fine-grained Alluvial Deposits and Their Effects on Mississippi River Activity,* 2 vols. Vicksburg, Miss.: U.S. Army Corps of Engineers Waterways Experiment Station, 82p.

Fisk, H. N.; Kolb, C. R.; and Wilbert, L. J., 1952. *Geological Investigation of the Atchafalaya Basin and the Problem of Mississippi River Diversion,* 2 vols. Vicksburg, Miss.: U.S. Army Corps of Engineers Waterways Experiment Station.

Glennie, K. W., 1970. *Desert Sedimentary Environments.* Developments in Sedimentology 14. Amsterdam: Elsevier, 222p.

Happ, S. C., 1971. Genetic classification of valley sediment deposits, *Am. Soc. Civil Engineers Proc., Jour. Hydraulics Div.,* **97,** 43-53.

Kolb, C. R., 1961. Geology of the Lower Mississippi River Valley and its economic aspects, in James E. Noblin, Jr. (ed.), *Our Nuclear Future.* Jackson, Miss.: Mississippi Industrial and Technological Research Commission, 129-146.

Kolb, C. R., 1976. Geologic control of sand boils along Mississippi River levees, in D. R. Coates (ed.), *Geomorphology and Engineering.* Stroudsburg, Pa.: Dowden, Hutchinson and Ross, 99-113.

Kolb, C. R., 1980. Should we permit Mississippi-Atchafalaya diversion? *Gulf Coast Assoc. Geol. Soc. Trans.,* **30,** 145-150.

Kolb, C. R., and Shockley, X. X., 1959. Engineering geology of the Mississippi Valley, *Am. Soc. Civil Engineers Trans.,* **124,** 633.

Kolb, C. R.; Steinriede, W. B., Jr.; Krinitzsky, E. L.; Saucier, R. T.; Mabrey, P. R.; Smith, F. L.; and Fleetwood, A. R., 1968. Geological Investigation of the Yazoo Basin, Lower Mississippi Valley, *U.S. Army Corps Engineers, Waterways Expt. Sta. Tech. Rept. 3-480,* 160p.

Mansur, C. I.; Kaufmann, R. I.; and Schultz, J. R., 1956. Investigation of Underseepage and Its Control: Lower Miss. River Levees, *U.S. Army Corps Engineers Waterways Expt. Sta. Tech. Memo 3-242,* 421p., appendixes, and 241 plates.

Schumm, S. A., 1977. *The Fluvial System.* New York: Wiley, 338p.

Cross-references: *Alluvial Plains, Engineering Geology; Channelization and Bank Stabilization; Coastal Engineering; Dams, Engineering Geology; Deltaic Plains, Engineering Geology; Geomorphology, Applied; River Engineering.* Vol. III: *Flood Plain; Terraces, Fluvial.*

Vol. XIV: *Alluvial Systems Modeling; Canals and Waterways, Sediment Control.*

ANTHROPOLOGICAL GEOLOGY—See Vol. XIV: GEOANTHROPOLOGY.

APPLIED MINERALOGY—See MINERAGRAPHY; MINERAL ECONOMICS.

APPLIED OCEANOGRAPHY—See COASTAL ENGINEERING; COASTAL INLETS, ENGINEERING GEOLOGY; SUBMERSIBLES. Vol. XIV: COASTAL ZONE MANAGEMENT; OCEAN, OCEANOGRAPHIC ENGINEERING.

ARCHAEOLOGICAL GEOLOGY—See Vol. XIV: GEOARCHAEOLOGY.

ARID LANDS, ENGINEERING GEOLOGY

Deserts are dry areas of sparse or nonexistent vegetation that comprise more than one-third of the Earth's land surface if semi-arid regions are included. The general term *desert* usually refers to the hot, dry regions of the world. About 5 percent of the earth's land surface can be classified as hot and extremely arid, and about 15 percent as hot and arid. These areas owe their existence largely to meteorological causes, being located along the earth's two great subtropical belts of minimal rainfall or far away from centers of rainfall. Because the development and form of their ground conditions arise from past and present climates, a definition of a hot desert area from an engineering viewpoint is essentially related to climate.

Attempts to delineate boundaries of arid zones have resulted in the development of a number of indices of aridity, including the widely used classification of Köppen (1931). For example, Köppen suggested that boundaries could be established in terms of mean annual temperature and mean annual precipitation. Indices of aridity, however, generally suffer from several disadvantages. In particular, it is questionable whether mean annual values of precipitation indicate sufficient similarities between two regions to let them be placed in the same category. For example, if each of two regions of similar temperatures had 250 mm of rainfall annually but one experienced about 20 mm of rain per month and the other only three large storms per year, their vegetation, drainage, soil, and other characteristics would be significantly different. Even though aridity is a measure of dryness, the best index is probably one based on the water balance of the area, that is, the difference between the moisture received and the moisture lost. Moisture received is chiefly in the form of precipitation, whereas losses are from evaporation, runoff, and seepage. There are several difficulties in any water balance evaluation, and this is particularly true of the vast arid areas with limited data and vegetation cover. Thornthwaite (1948) devised various indices that are closely related to water balance. His aridity index, for example, is related to the potential total evaporation from a continually damp area covered in vegetation, to the length of the day, and to the temperature. Thornthwaite's index is easy to apply, with the use of nomograms: Meigs (1953) chose to use it in the preparation of his small-scale maps depicting the world's arid regions. Figure 1 shows Meigs's subdivision of the world's arid lands by climatic zones. Hot deserts, which occupy the great majority of the area, comprise arid and extremely arid areas.

Ground Conditions of Significance to Engineering

For areas as large as those depicted by Figure 1, it is extremely difficult, if not actually misleading, to try to rationalize ground conditions for engineering purposes. Yet in some significant respects it is reasonable to do so because of a certain unity of conditions imposed by the overall climatic regime. The arid climate tends to produce particular forms of erosion with a dominance of mechanical weathering on high land which supplies coarse debris, which is subsequently transported by stream- or sheetflood to low land. There is also a transport of fine sediments by wind, in addition to a general upward leaching (q.v. in Vol. XI) and surface precipitation of salts by evaporation. The ground is therefore commonly saline and covered with granular alluvial sediments, usually without humus (see Vol. VI: *Desert Sedimentary Environments*).

The principal engineering problems associated with these desert conditions include (a) unstable terrain, such as wind-blown silt (loess) and sand (sheets, drifts, and dunes); (b) aggressive salty ground, such as sabkhas, salinas, salt playas, and some duricrusts; (c) unsuitable construction materials, such as some silts, sands, weak carbonate sediments, and some duricrusts; (d) rapid erosion and deposition, by wind and floods—especially flash floods—and debris flows. Other problems occur but are generally similar to those found in temperate regions. For background discussions, see Geological Society of London (1978).

As an aid to engineering feasibility evaluation of hot desert ground conditions, and Middle East conditions in particular, Fookes and Knill (1969) proposed a simple model based on "mountain and piedmont plain" terrain and natural desert processes. Four zones were recognized, each with different desert surface characteristics and with different engineering behavior. Figure 2 outlines the characteristics of these zones in cross-section. An important part of the concept is that the width of the mountain (or hill) and the plain can vary from only a few hundred meters to tens or even hundreds of kilometers, the width of the individual zones stretching or shortening commensurately (Cooke and Warren, 1973) (see Vol. III: *Deserts and Desert Landforms*).

The overwhelming majority of engineering soils comprising the zones are granular (see *Soil Classification System, Unified*). Their grading is related to the zone by reference to the distance from the mountains supplying the erosion debris—the farther from the mountains, the finer the deposits. Gravity and water transport the particles from the mountains into the plains where wind and water move the finer fractions around. In general, the grading of materials indicates their geotechnical engineering characteristics and their Casagrande group symbol classification (see *Soil Classification System, Unified*). The base-leveled central plains are mainly composed of sands and silts. The general level of the water table is such that the capillary rise often reaches ground level; thus, a high water table is the single most significant factor affecting ground engineering problems (see Tables 1 and 2 for examples of index properties). Further discussion of the ground engineering conditions outlined here are also given in Fookes (1976), Fookes and Higginbottom (1980), Epps (1980), and Oweiss and Bowman (1981).

Zone I—Mountains. Even though mechanical weathering—which involves the splitting, exfoliation, and crumbling of rocks—is dominant here, chemical weathering in the presence of moisture from dew and occasional rain or snow, although slow, also plays an important role. By decomposition and solution, rocks that would otherwise successfully resist the stresses set up by temperature changes are gradually weakened until they are shattered. Minute quantities of dissolved matter are brought to the surface by evaporation. Commonly, the loose salts are blown away, but oxides of iron, accompanied by traces of manganese and other oxides, form a red, brown, or black surface film that is known as desert varnish (see Vol. VI: *Weathering in Sediments*).

General problems associated with engineering works in Zone I include major landslips (Wiseman et al., 1970) and karst topography (limestone surface solution features and subsurface cavities) while lesser slips and slope failures occur in any situation with unfavorable combinations of discontinuities, such as joints and faults (Hoek and Bray, 1981; Fookes and Sweeney, 1976; Schuster and Krizek, 1978). Within the mountains, the size

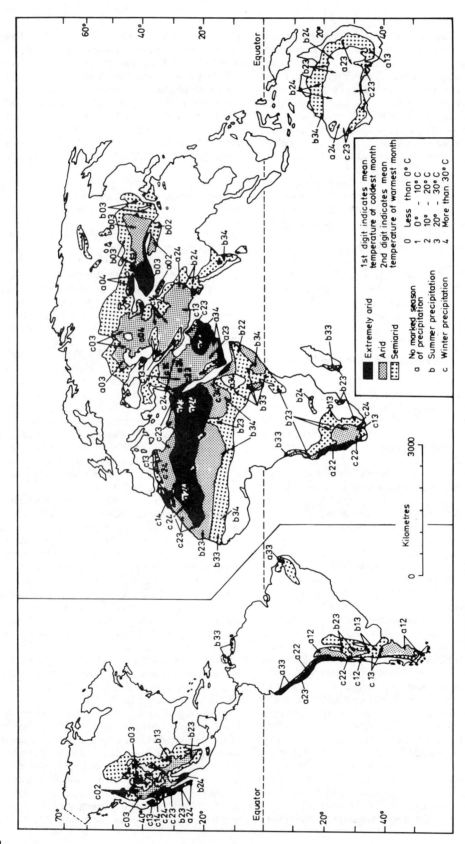

FIGURE 1. Extremely arid, arid, and semi-arid climatic zones of the world (after Meigs, 1953).

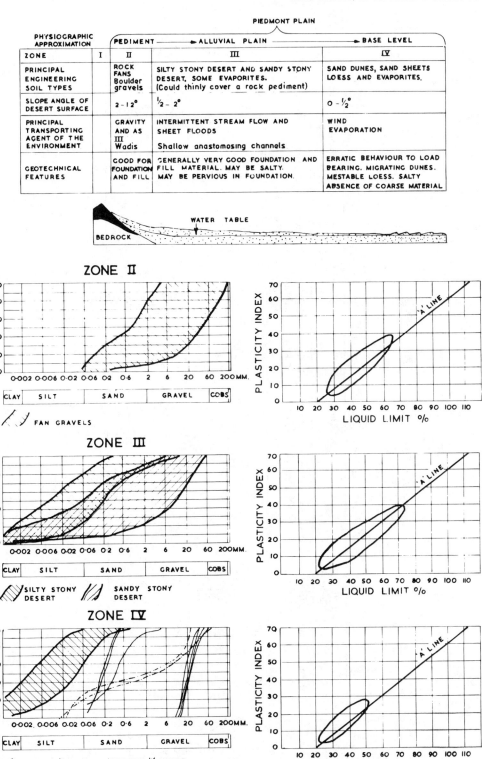

FIGURE 2. Idealized profile across mountain and desert plain showing engineering zones I-IV and grading envelopes, grading curves, and plasticity chart data from zones II, III, and IV.

TABLE 1. Granular Soils—Some Classification Test Results

Zone	Sample Type	Particle Size (% Passing)				Atterberg Limits			Density/moisture Relationship		California Bearing Ratio (%)		Water-Soluble Matter
		20 mm	2 mm	0.06 mm	0.002 mm	Plastic Limit	Liquid Limit	Plasticity Index	Max. Dry Density (mg/m³)	Optimum Moisture Content	Unsoaked	Soaked	
II	Fan	87	38	—	—	—	—	—	2.12	8.4	53.2	38.7	10.8
II	Fan	63	27	—	—	22	27	6	2.24	7.0	46.3	39.3	0.1
III	Sandy desert	93	54	21	8	18	36	18	—	—	—	—	2.25
III	Sandy desert	83	30	—	—	—	—	—	2.11	9.5	59.2	67.10	1.63
III	Silty desert	97	77	31	—	58	112	50	1.60	20.3	15.7	2.8	41.90
III	Silty desert	—	99	76	35	19	39	18	1.75	18.2	21.8	4.8	1.00
III	Silty desert	—	96	63	20	15	29	17	1.94	12.0	24.0	14.0	0.70
IV	Silty desert	—	100	57	15	26	39	23.5	1.93	13.0	29.8	14.4	0.50
IV	Loess	—	—	83	21	26	64	38	1.57	22.80	11.1	6.5	0.50
IV	Loess	—	—	89	18	17	33	6	1.81	11.59	9.8	4.3	0.60
IV	Loess	—	—	90	20	19	36	17	1.57	23.0	5.1	6.7	0.45
IV	Sandy desert	—	96	14	—	—	—	—	1.96	12.00	15.0	16.5	0.65
IV	Dune sand	—	99	83	9	22	39	17	1.74	18.3	3.9	3.2	1.75

TABLE 2. Outline Summary of Runoff and Soil Characteristics of Desert Zones

Desert Zone	Typical Soils (Casagrande Symbol)	Runoff Hazard	Notes on Road Design
I	Scarce GP GU GW	Stormflow down hillsides and in mountain canyons. High runoff coefficients, say, › 0.55	Conventional mountain road designs. Do not underestimate potential flood conditions because of "arid" nature of terrain. Generally good subgrade conditions.
II	Boulder gravels GP GU GW	Storm wadi flow, possibly some sheetflow. Low to moderate runoff coefficients, say, 0.3-0.55	Volume of dumped water-transported debris during storm flow may be large. Scour also a hazard. For roads parallel to streamflows, low embankments with strengthened stream crossing areas may suffice. For roads transverse to stream flow, high embankments, numerous wide culverts, and bridges. Scour protection for abutments by gabions or similar. Generally good subgrade conditions.
III	GF SW SP	Storm sheetflow and deep wadi flow. Low to moderate runoff coefficients, say, 0.2-0.5	Scour may be a major hazard. Dumped water-transported debris may also be a hazard. For roads parallel to streamflow, low embankments with strengthened stream-crossing areas. For roads transverse to streamflow, moderate embankments, numerous wide culverts and bridges, training bunds, and scour protection. Upstream sides of embankment may require armoring or the whole construction by rockfill. Generally poor to moderately good subgrade conditions.
IV	SU SF ML	Storm sheetflow and shallow wadi flow. Moderate to high runoff coefficients—ground may quickly get saturated (especially ground with high water table), say, 0.25-0.7	Scour may be a hazard. Generally low embankments, armored in potential stream flow areas. Training bunds and many small culverts may be necessary in some areas. Generally poor to moderate subgrade conditions especially where ground water table is high.

of erosional debris ranges from poorly sorted medium angular gravel to very large boulders. Hazards occur from streamflooding and from talus slope movements, especially in semi-arid areas (Fig. 1). Duricrusts occur extensively in some areas of limestone terrain, such as the Near East, the Middle East, and Australia (Fig. 1). Pre-Tertiary hardrocks in semi-arid mountain areas—for example, the Mediterranean and the Ethiopian highlands—may have residual soils developed on them. In areas marginal to wetter climates, such as the eastern Mediterranean, terra rossa soils may have developed on limestones (Chapman, 1971, 1974; Goudie, 1973).

Zone II—Aprons and Sediments. When the apron consists of rock, it may be covered by a thin mantle of sand and gravel deposits, which sometimes hide an irregularly eroded, or even a terraced, rock surface. The engineering performance of the bedrock is directly related to its rock type, which may have a duricrusted surface with a leached or softer underlying zone (see Vol. VI: *Duricrust*).

Gravel fans are, however, generally the most extensive form of apron, especially in parts of the Middle East and southwestern United States. They are almost entirely composed of angular particles, boulders, and cobbles on the upper slopes grading to fine gravels downslope (Fig. 3). The gravel fans occur in rough layers, which reflect deposition in

FIGURE 3. Gravel fan; zone II, Iran.

times of flood—and, being reasonably compact, they usually have good load-bearing characteristics, except where occassional silt or clay layers or debris flow materials occur. These materials have a fairly high permeability and sometimes form aquifers. They generally have a good borrow potential, but one should check their chemistry in addition to conducting the usual mechanical-physical properties tests when considering them for use in concrete. Such terrains are subject to flash floods, which is a particularly important consideration in road design.

Zone III—Alluvial Plains. Alluvial plains mainly consist of extensive splays of fine gravels and sands (Fig. 4). Two common types of alluvial plain (q.v.) include the sandy-stony and silty-stony deserts, which respectively represent deposition from stream flow and overbank sheet flow. Zone III grading envelopes are shown in Figure 2, the two envelopes being derived from bundles of a large number of grading curves taken from adjacent coarse and fine layers found in exploratory pits (Fig. 5). Such deposits generally provide good foundation conditions because the sands and fine gravels are water-laid and reasonably dense. They also make good borrow for fill and aggregates, depending on their grading. Pavements composed of a single layer of single-sized stone may also occur extensively in this zone (and zones II and IV). These pavements can be scraped to supply aggregate, but their removal may expose underlying finer material that is susceptible to erosion by wind and water. Alluvial plains are subject to sheet- and stream-flood hazards.

Zone IV—Base-Leveled Plains. These surfaces are composed of wind-blown silts and sands, which are frequently modified by subsequent flooding or marine action. Perhaps the most extensive of all the zones, base-leveled plains are very common in arid areas (Fig. 1), especially the Near and Middle East and parts of Australia. In areas where there is

FIGURE 5. Wadi wall in Iran showing fine and coarse layers in zone III. See also grading envelopes in Figure 2.

a high water table and where aggressive salty conditions occur, load bearing and other engineering performance is reduced, requiring dewatering in excavations and tanking of foundations. In areas of uniform or thin-bedded, uncemented sands, permeabilities are similar to those generally associated with grading, typically ranging from $k = 10^{-1}$ to 10^{-3} or 10^{-4} cm/s. Layered deposits are often cemented by carbonate and therefore their overall permeabilities tend toward 10^{-4} to 10^{-6} cm/s. Until more experience is gained in high water table locations, foundations for major works should be inspected by deep pits (see *Foundation Engineering*). Successful large dewatering operations can usually be conducted by well point systems. In layered deposits, difficulties may be expected with dredging or pile driving through thin, cemented layers.

Because many of the unconsolidated deposits have a large, finely granular component that is often not bound by clay or cement, erosion by wind or water is common and therefore protection may be required. Similarly, filter protection against migration of fines may be necessary in underground and surface drainage systems and also in dewatering systems. Because metal filter screens readily corrode in certain ground waters, filter media constructed of man-made fibers are widely used.

Most potential engineering problems are associated with this zone, namely clays, silts, wind-blown sands, salty soils, and duricrusts.

Clays These fine materials are uncommon, except near the coast and in areas of clay plains in Australia. The clays are usually calcareous and normally fall in the CI-CH range in the Casagrande classification. Some may have marked shrink-swell characteristics (see *Clay, Engineering Geology*), which can cause problems for shallow foundations. Coastal clays are usually consolidated,

FIGURE 4. Alluvial plain in foreground (zone III) with zones I and II in background; Jordan.

but many have a desiccated crust; their strength ranges from soft to stiff (Tomlinson, 1957) (see Vol. XV: *Coastal Soils*). Some residual clays occur in the marginal highlands, especially in the semi-arid areas, where chemical weathering probably occurred during pluvial periods or may even be occurring today. Special engineering problems are associated with estuarine clays, such as the plastic muds in the Nile delta (see *Deltaic Plains, Engineering Geology*).

Silts. Hot desert silts may have been blown in from higher latitudes or produced, at least in part, within the desert system. Some silts may be loesses, which are potentially metastable, that is, they collapse on wetting when under load because the individual grains are not packed in a dense configuration. This condition also exists in some fine sands. When a higher load than any previous loading by nature is applied, the normal consolidation curve is followed. If the loess is wetted under its new load, however, a sudden collapse may occur (Fig. 6). Loess areas are also suscepti-

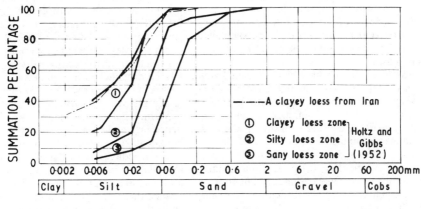

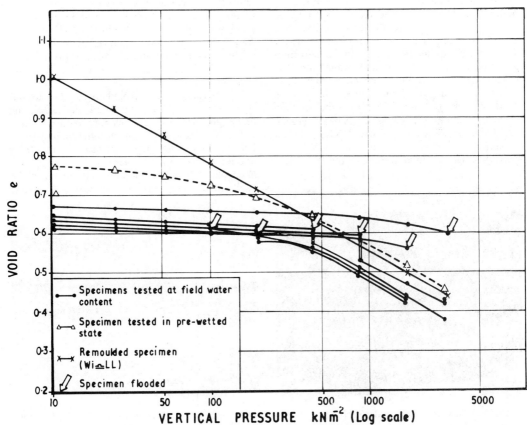

FIGURE 6. Some consolidation test data on clayey loess from Iran.

ble to piping erosion and underground drainage channels (Holtz and Gibbs, 1951; Fookes and Best, 1969). When loess is reworked and deposited from water, it will generally have the properties of conventional water-laid alluvial silt deposits.

Windblown Sands. An eolian dune is a mound of windblown sand; the smallest may be only a meter or so high and may cover only 10 m², whereas the largest may be over 40 m high and may cover several square kilometers. Sand sheets, in distinction, are large areas of gently undulating sandy surfaces with low relief. True dunes cover extensive areas mainly in the big sand seas, but they also occur in isolated areas on hard surfaces. Partly vegetated dunes occur in semi-arid regions on the margins of deserts and in moist coastal areas.

Dune sands commonly have median grain diameters between 0.2 and 0.4 mm and range between extremes of 0.1 and 0.7 mm. It is of note that most dunes are well sorted and samples of sand from one dune usually have particles of similar size and rounded shape. Sand sheets, on the other hand, are poorly sorted and bimodal, comprising both coarse (0.6 mm) and fine (0.1 mm) grains. Dunes frequently have poor load-bearing capacities (loose to medium-dense) and can be difficult to compact as fill. Sheet sands (Figs. 7 and 8) have slightly better bearing capacity and exhibit compaction characteristics similar to those that have been reworked, transported, and deposited by water. The latter are quite common around desert margins, in coastal areas, and in semi-humid climates.

Mobile dunes may blow over roads or buildings moving at a rate of several meters per month. Methods for controlling drifting sand include removal of dunes, realigning routes or structures, and stabilization by oiling (see Vol. XI, Pt. 1: *Soil Conditioners*), fencing, planting, or paving.

Salty Soils Ground waters in the Near and Middle East, parts of the southwestern United States, and Africa are frequently saline due to the presence of salts dissolved from the local bedrock or from seawater. Salts may occur at the soil surface in the form of salty crusts and elsewhere as wind-blown particles of sand or silt. Figure 9 shows common relationships between ground water, capillary rise, and the ground surface (see, e.g., Cooke, 1981). Where the capillary water does not reach the ground surface, its depth can be inferred from characteristic desiccation ground patterns (Neal, 1969). Salty capillary moisture reaching the ground surface usually produces dry, caked surfaces (Fig. 10), whereas ground water near the surface produces a damp, puffy surface. All these conditions depend on local circumstances, but in general an intuitive feel for salty ground can usually be obtained rather quickly. Figure 11 gives ranges of capillary rises for different soil gradings under laboratory conditions.

Natural salt surfaces have a variety of names, depending on location and country (Neal, 1969). Fookes and Collis (1975a) have, with much simplification, reduced these for engineering purposes to *sabkha* (coastal salt marsh), *playa* (an ephemeral lake flat), *salt playa* (a playa with a salty surface due to evaporation of salty lake waters), and *salina* (local depression with high salt water table and attendant formation of salt crusts). The local salt regime and chemical conditions are frequently complex and often vary with the season so that each engineering site must be investigated separately (Table 3) (see Vol. XII, Pt. 1: *Management of Soils*).

Salts may actually help bind unsealed roads, but they can severely damage sealed roads. Building and engineering foundations may require tanking because of salt attack on masonry or concrete. Salts may also contaminate aggregate sources (Ellis and Russell, 1973; Netterberg et al., 1974; Fookes and Collis, 1975b, 1976; Fookes and French, 1977). Reclaimed land adjacent to coasts, or in any

FIGURE 7. Layered sheet sands exposed in walls of a wadi; zone IV, Jordan.

FIGURE 8. Shallow wadi in sheet sand terrain; zone IV, Libya.

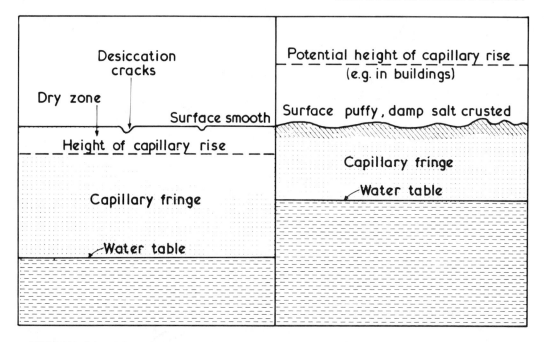

FIGURE 9. Schematic section showing low and high water tables, capillary rise, and related surface features in zone IV sands (after Brunsden et al., 1976).

FIGURE 10. Sabkha; zone IV, United Arab Emirates.

high-water-table situation, may become aggressively saline where capillary moisture reaches the surface and evaporation takes place (see Vol. XII, Pt. 1: *Evaporation*).

Figure 12 shows some natural forms of salt weathering occurring on a Tertiary limestone sequence in Bahrain. Salt weathering is perhaps the most common form of weathering in salty lowland areas. The damaging effects of salt weathering are evident on the lower part of the new block wall shown in Figure 13.

Duricrusts. Duricrusts occur in a variety of forms, commonly with a hardened surface, ranging from millimeters to tens and hundreds of centimeters thick, often with a leached, cavernous, porous, or friable zone underneath. The term *duricrust* (q.v. in Vol. VI) was first applied in Australia to denote a surface or near-surface hardened accumulation of silica, alumina, or iron oxides in varying proportions. Admixtures of other substances also occur, and the term is now applied by extension to encrusting layers of calcium or magnesium carbonate, gypsum, and salt, the latter two being common products of desert climates. For engineering purposes, those duricrusts ending in -*crete*—such as calcrete ($CaCO_3$) and silicrete (SiO_2)—indicate hardened surfaces. Those ending in -*crust*—gypcrust ($CaSO_4 2H_2O$) and salcrust (NaCl), for example—are softer accumulations usually occurring in areas of centrally draining deserts (salt playas or salinas) or coastal sabkhas. Sands may be locally cemented with carbonate to form cap rock or miliolite, especially near the coast. Mixtures of nodular calcrete, calcrete fragments, and drifted sand form the so-called desert fill (Fig. 2, zone IV grading curve). Massive development of calcrete on limestone rocks, common in the semi-arid highland areas of the Middle East, probably represents a pedogenic relic.

Surface inspection of the different types of cretes can be quite misleading because the thickness of surface hardening will vary depending on the local circumstances, as will the leached or otherwise altered zone underneath. Each site should therefore be investigated by drilling if a knowledge

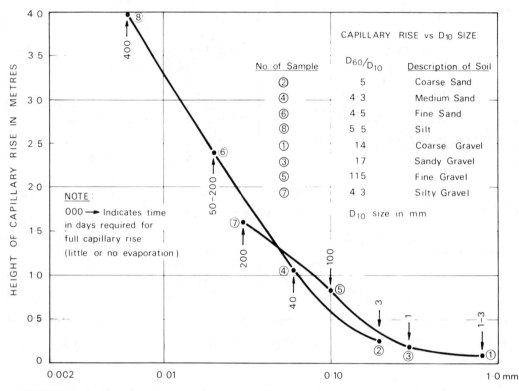

FIGURE 11. Capillary rise of moisture in granular soils under laboratory conditions (from data in Lane and Washburn, 1946).

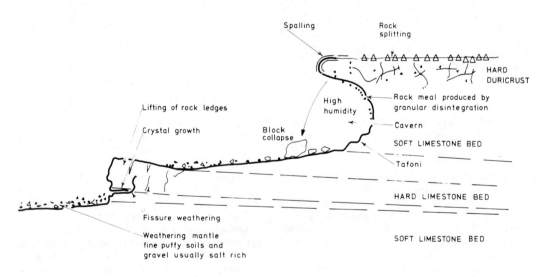

FIGURE 12. Schematic cross-section of near-surface Tertiary limestone bedrock showing some forms of weathering in Bahrain (after Brunsden et al., 1976).

TABLE 3. Some Specific Salty Soil Types in Zone IV and Their Engineering Significance

Name	Terrain	Ground Water Table	Salts	Special Significance	Construction Technique
Sabkha	Coastal flat, inundated by sea water either tidally or during exceptional floods.	Very near the surface.	Thick surface salt crusts from evaporating sea brines. Salts usually include carbonates, sulfates, chlorides, and others.	Generally aggressive to all types of foundations by salt weathering of stone and concrete and/or sulfate attack on cement-bound materials. Evaluate bearing capability.	Carefully investigate. Consider tanking concrete foundations; using SR cement. For surfaced roads consider using inert aggregate, capillary break layer, or positive cutoff below sub-bases. Use as fill suspect. May not be deleterious to unsurfaced roads.
Playa	Inland, shallow, centrally draining basin—of any size.	Too deep for capillary moisture zone to reach ground surface, but area will be temporary lake during floods.	None if temporary lake is of salt-free water.	Nonspecial. Ground surface may be silt/clay or covered by wind-blown sands. Evaluate bearing capability.	Nonspecial.
Salt playa	As playa, but often smaller.	As above, but lake of salty water.	Surface salt deposits from evaporating temporary salty lake water. Salts usually include chlorides and sometimes nitrates, sulfates, and carbonates.	Can be slightly to moderately aggressive to all types of foundations by salt weathering and sulfate attack. More severe near water table.	As sabkha.
Salina	As playa.	Near surface; capillary moisture zone from salty ground water can reach surface.	Surface crusts from evaporating salty groundwater. Salts include carbonates, sulfates, chlorides, and many others.	Can be slightly to exceptionally aggressive to all types of foundations by salt weathering and sulfate attack.	As sabkha.

Source: After Fookes and Collis (1975a).

FIGURE 13. Deleterious effects of salt attack on lower part of a new retaining wall constructed in zone IV.

subsurface characteristics is required for engineering purposes. Indurated crusts make suitable aggregate borrow, but the chemical properties must also be evaluated. A depth profile of chemical and physical properties for a surface-altered limestone bedrock may show, for example, an increase of gypsum and halite toward the surface, making it unacceptable for crushing as concrete aggregate; a decrease in porosity and an increase in unconfined compressive strength, however, make it otherwise attractive (Fookes and Collis 1975b; Fookes and Higginbottom, 1980).

Site Investigation

Where there is adequate exposure of the terrain, walkover surveys, inspection of air photos, engineering geology, and geomorphological mapping are economical and successful techniques. In addition to basic mapping and air photo interpretation for feasibility and site investigation planning, maps can be prepared for specific requirements (*Quart. Jour. Eng. Geology,* 1972), such as sand dune migration, potential salt hazard, potential flood hazard, borrow locations, and route and urban planning (see *Engineering Geology Reports*). Various forms of color air photography and satellite imagery are also valuable aids (see Doornkamp, et al., 1980).

Site investigation contracts for desert conditions should generally include pitting as a main tool of the investigation because many boring techniques tend to lose fines and break up large stones. There is also a need to develop improved techniques of boring (see Vol. XIV: *Borehole Drilling*), sampling, and evaluating the predominantly granular desert soils (Fookes and Higginbottom, 1975). The durability of natural and man-made materials in salty or wet environments must be estimated if appropriate to the works. The nature

of the ground chemistry and seasonal changes in the water table should be established, as they are important parameters in the planning stages of engineering works in arid environments.

Descriptions and classification systems (e.g., *Quart. Jour. Eng. Geology,* 1972) generally do not adequately cover the range of calcareous soils and rocks found in many desert areas, especially the Near and Middle East. Accordingly, a pilot carbonate soil and rock classification system has been proposed by Fookes and Higginbottom (1975).

PETER GEORGE FOOKES

References

Brunsden, D.; Doornkamp, J. C.; and Jones, D. K. C. (eds.), 1976. Geomorphology and superficial materials, *Bahrain Surface Materials Resources Survey.* **4**, 1-124.

Chapman, R. W., 1971. Climatic changes and the evolution of landforms in the Eastern Province of Saudi Arabia, *Geol. Soc. America Bull.,* **82**, 2713-2728.

Chapman, R. W., 1974. Calcareous duricrust in Al-Hasa, Saudi Arabia, *Geol. Soc. America Bull.,* **85**, 119-130.

Cooke, R. U., 1981. Salt weathering in deserts, *Geol. Assoc. Proc.,* **92**, 1-16.

Cooke, R. U., and Warren, A., 1973. *Geomorphology in Deserts.* London: Batsford, 374p.

Doornkamp, J. C.; Brunsden, D.; and Jones, D. K. C. (eds.), 1980. *Geology, Geomorphology and Pedology of Bahrain,* Norwich, England: Geo Abstracts Ltd., 443p.

Ellis, C. I., and Russell, R. B. C., 1973. The use of salt-laden soils (sabkha) for low cost roads, *Department of Environment TRRL Paper PA 78/74,* 1-26.

Epps, R. J., 1980. Geotechnical practice and ground conditions in coastal regions of the United Arab Emirates, *Ground Eng.,* **13**, 12-25.

Fookes, P. G., 1976. Road geotechnics in hot deserts, *Highway Engineer,* **23**, 11-29.

Fookes, P. G., and Best, R., 1969. Consolidation characteristics of some late Pleistocene, periglacial metastable soils in East Kent, *Quart. Jour. Eng. Geology,* **2**, 103-128.

Fookes, P. G., and Collis, L., 1975a. Problems in the Middle East, *Concrete,* **9**, 12-17.

Fookes, P. G., and Collis, L., 1975b. Aggregates and the Middle East, *Concrete* **9**, 14-19.

Fookes, P. G., and Collis, L., 1976. Cracking and the Middle East, *Concrete,* **10**, 14-19.

Fookes, P. G., and French, W. J., 1977. Soluble salt damage to surfaced roads in the Middle East. *Highway Engineer,* **24**, 10-20.

Fookes, P. G., and Higginbottom, I. E., 1975. The classification and description of near-shore carbonate sediments for engineering purposes, *Geotechnique,* **25**, 406-411.

Fookes, P. G., and Higginbottom, I. E., 1980. Some problems of construction aggregates in desert areas with particular reference to the Arabian Peninsula: 1) Occurrence and special characteristics 2) Investigation, production and quality control, *Proc. Inst. Civil Engineers* **68**, pt. 1, 39-90.

Fookes, P. G., and Knill, J. L., 1969. The application of engineering geology to the regional development of

Northern and Central Iran, *Eng. Geology—Internat. Jour.* **3,** 81-120.

Fookes, P. G., and Sweeney, M., 1976. Stabilization and control of local rock falls and degrading rock slopes, *Quart. Jour. Eng. Geology* **9,** 37-55.

Geological Society of London, 1978. Proceedings of the conference on engineering problems associated with ground conditions in the Middle East, *Quart. Jour. Eng. Geology* **11,** 1-112.

Goudie, A., 1973. *Duricrusts in Tropical and Subtropical Landscapes.* Oxford: Clarendon Press, 174p.

Hoek, E., and Bray, J. W., 1981. *Rock Slope Engineering,* 3rd ed. London: Institution of Mining and Metallurgy, 360p.

Holtz, W. B., and Gibbs, J. H., 1951. Consolidation and related properties of loessial soils, *Am. Soc. Testing and Materials, Spec. Tech. Pub.* **126,** 9-33.

Köppen, W., 1931. *Grundriss der Klimakunde.* Berlin: Walter de Gruyter, 388p.

Lane, K. S., and Washburn, D. E., 1946. Capillary tests by capillarimeter and by soil filled tubes, *Highway Research Board Proc.,* **26,** 460-473.

Meigs, P., 1953. World distribution of arid and semi-arid homoclimates, in *Reviews of Research of Arid Zone Hydrology,* Paris: UNESCO, 203-209.

Neal, J. T., 1969. Playa variation, in W. G. McGinnies and B. J. Goldman eds., *Arid Lands in Perspective.* Tucson: University of Arizona Press, and Washington, D.C.: American Association for the Advancement of Science, 13-44.

Netterberg, F.; Blight, G. C.; Theron, P. F.; and Marais, G. P., 1974. Salt damage to roads with bases of crusher-run Witwatersrand quartzite, *Proc. 2nd Conf. Asphalt Pavements S. Africa, Durban,* Session 7, 134-153.

Oweis, I., and Bowman, J., 1981. Geotechnical considerations for construction in Saudi Arabia, *Am. Soc. Civil Engineers Proc., Jour. Geotech. Eng. Div.,* **107,** 319-338.

Quart. Jour. Eng. Geology, (1972), The preparation of maps and plans in terms of engineering geology, **5,** 295-382.

Schuster, R. L., and Krizek, R. J. (eds.), 1978. *Landslides Analysis and Control,* special report 176. Washington, D.C.: National Academy of Sciences, 234p.

Thornthwaite, C. W., 1948. An approach toward a rational classification of climate, *Geog. Rev.,* **38,** 55-94.

Tomlinson, M. J., 1957. Saline calcareous soils, *Inst. Civil Engineers Proc.,* **8** (Nov.), 232-246.

Wiseman, G.; Hayati, G.; Frydman, S.; Aisenstein, R.; David, D.; and Flexer, A., 1970. A study of a landslide in Galilee, Israel, *Proc. 1st Internat Congr. Internat. Assoc. Eng. Geology Paris* **1,** 50-61.

Cross-references: *Alluvial Plains, Engineering Geology; Deltaic Plains, Engineering Geology; Engineering Geology Reports; Foundation Engineering; Laterite, Engineering Geology; Lime Stabilization; Rock Slope Engineering; Soil Mechanics; Soil Classification System, Unified.* Vol. III: *Deserts and Desert Landforms.* Vol. VI: *Desert Sedimentary Environments; Duricrust; Sensitive Clays; Weathering in Sediments.* Vol. XII, Pt. 1: *Evaporation; Management of Soils; Soil Conditioners.* Vol. XII, Pt. 2: *Duricrust.* Vol. XIV: *Borehole Drilling; Dispersive Clays; Expansive Soils.*

ATTERBERG LIMITS AND INDICES

In 1911 Atterberg defined the states of consistency in which a soil can exist. In passing from a "wet" to a "dry" condition, a soil goes from a liquid state through semiliquid, plastic, semisolid, and finally solid states. In the liquid state the soil has no shear strength. In the semiliquid state the soil resembles a viscous fluid and has practically no shear strength. In the plastic state the soil can be deformed without cracking and retains its deformed shape. As it passes into the semisolid state, the soil begins to crack and crumble with deformation. Finally, in the solid state the soil has all the characteristics of a solid. Atterberg referred to the transitional boundary between the semiliquid and plastic states as the *upper plastic limit,* and defined it as the level of moisture content where shear strength is just observed. The boundary between the plastic and semisolid states he called the *lower plastic limit,* defining it as the level of moisture content at which the soil just starts to crack and crumble with deformation. He defined the *shrinkage limit* as the level of moisture content at which the soil ceases to shrink with further drying, and called it the boundary between the semisolid and solid states (Atterberg, 1911).

In 1932 Casagrande introduced the use of these limits in *Soil Mechanics* (q.v.). Casagrande (1932) called Atterberg's upper plastic limit the *liquid limit* (LL), and defined it as the level of moisture content at which a 2-mm groove cut into a pat of soil closes along a distance of 0.5 in. at 25 blows of a disc falling from a height of 1 cm (using the standard liquid limit device he designed). Casagrande named Atterberg's lower plastic limit the *plastic limit* (PL), and defined it as the level of moisture content at which soil threads $\frac{1}{8}$ in. in diameter begin to crack and crumble with deformation. He defined the *shrinkage limit* (SL) as the level of moisture content at which the soil ceases to shrink with drying, or as the level of minimum *moisture content* (w) at 100-percent saturation. The shrinkage limit (w) is determined by oven-drying a pat of soil and weighing it. A small dish is then placed inside a bigger one, and the small dish is filled with mercury. The soil pat is placed into the small dish, and the displaced mercury accumulates in the bigger dish. The weight of the displaced mercury is equal to (W_m). The volume of the solids (V_s) is

$$V_s = \frac{W_m}{\delta_m}$$

where $\delta_m = 13.55$ g/cm^3, the unit weight of mercury. Once V_s is determined, the shrinkage limit (SL) can be found:

$$SL = \frac{\delta_w V_s}{W_s} - \frac{G_t}{G_s}$$

where δ_w = the unit weight of water, W_s = the weight of the oven-dried soil pat, G_t = the specific gravity of water at the temperature of the test, and G_s = the specific gravity of the solids.

Casagrande referred to the difference LL - PL as the plasticity index (PI). Thus, PI = LL - PL. Similarly, the shrinkage index (SI) is defined as SI = PL - SL. The *liquidity index* (LI), sometimes called the water-plasticity ratio, is defined as

$$LI = \frac{w_n - PL}{PI}$$

where w_n = the natural or *in situ* water (moisture) content. The *activity ratio* (A_r), defined by Skempton (1953), is closely related to the plasticity index. Skempton discovered that the plasticity index (PI) varies directly with the percentage of clay minerals, that is, the percentage of particles by weight finer than 2 microns (Fig. 1). This relationship can be expressed as

$$A_r = \frac{PI}{\%2\mu}$$

Uses and Importance

The Atterberg limits and indices are useful in soil identification and classification. In general, the Atterberg limits increase as (a) the grain size decreases, (b) the percentage of clay and disc-shaped particles increases, and (c) the organic content increases. Montmorillonite-rich clays, and clays with absorbed monovalent cations, tend to have high values of liquid limit.

In general, as the Atterberg limits increase, the soil's shear strength decreases. The potential volume change (shrink-swell characteristics) of a soil can be estimated from the liquid limit; any LL greater than 40 indicates severe potential volume change. The optimum moisture content of a soil may be estimated from the plastic limit; in many cases it is approximately 3.6% less than the plastic limit.

Finally, the Atterberg limits are closely related to the Unified Soil System of Classification (q.v.). This system, developed by Casagrande, makes use of the liquid limit and the plasticity index. The relationship is best illustrated by the Casagrande plasticity chart (Casagrande, 1948). (See Fig. 2.)

ROBERT G. FONT

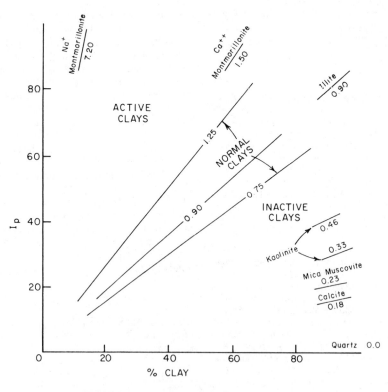

FIGURE 1. Activity of clays (after Skempton, 1953).

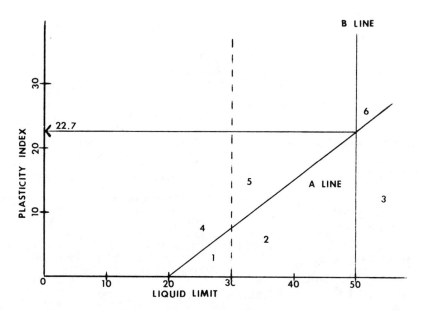

FIGURE 2. Plasticity chart (after Casagrande, 1948). 1. Inorganic silts of low compressibility. 2. Inorganic silts of medium compressibility. 3. Inorganic silts of high compressibility and organic clay. 4. Inorganic clays of low plasticity. 5. Inorganic clays of medium plasticity. 6. Inorganic clays of high plasticity.

References

Atterberg, A., 1911. On the investigation of the physical properties of soil and on the plasticity of clays, *Int. Mitt. für Bodenkunde,* **1,** 10-43.

Casagrande, A., 1932. The structure of clay and its importance in foundation engineering, contributions to soil mechanics, *Boston Soc. Civil Engineers Jour.,* April, 72-112.

Casagrande, A., 1948. Classification and identification of soils, *Am. Soc. Civil Engineers Trans.,* **113,** 901-991.

Skempton, A. W., 1953. The colloidal activity of clays, *3rd Internat. Conf. Soil Mechanics Found. Eng. Proc. (Switzerland),* **1,** 57-61.

Cross-references: *Clay, Engineering Geology.; Clays, Strength of; Foundation Engineering; Soil Mechanics.* Vol. XII, Pt. 1: *Activity Ratio; Soil Mechanics.*

B

BANK STABILIZATION—See
CHANNELIZATION AND BANK
STABILIZATION.

BATHYMETRIC SURVEYS—See
COASTAL ENGINEERING; NUCLEAR
PLANT SITING, OFFSHORE. Vol. XIV:
ACOUSTIC SURVEYS, MARINE;
HARBOR SURVEYS; MARINE
MAGNETIC SURVEYS; SEA SURVEYS.

BEACH ENGINEERING—See BEACH
REPLENISHMENT, ARTIFICIAL;
COASTAL ENGINEERING.

BEACH REPLENISHMENT, ARTIFICIAL

Artificial beach replenishment is a method of constructing and maintaining beaches by the artificial emplacement of sand. In recent decades the method has received greater acceptance because it has been recognized that protective structures such as groins and seawalls frequently serve merely to shift the erosional problem to another area, whereas artificial nourishment remedies the basic cause of most erosion, often benefits adjoining shores, and is more aesthetically pleasing (Finkl, 1981). In addition, artificial nourishment is often less expensive than the construction and maintenance of artificial structures.

Design Requirements

Before an artificial replenishment project can be undertaken, the net loss of material must be determined, the physical processes affecting the beach must be known, and the material requirements for the project must be calculated.

Determining Net Deficit. The net deficit is most accurately determined by making a detailed study of the sand budget along the coast (Table 1). Rarely, however, have all the necessary data for an area been collected for a period of time long enough to determine the rate of net loss of beach sediment. Reasonably accurate estimates may be made by comparison of detailed maps and hydro-

TABLE 1. Data Needed to Calculate the Sand Budget for an Area[a]

Sediment Added to Area	Sediment Lost from Area
Longshore transport	Longshore transport
Onshore transport	Offshore transport
Wind transport	Wind transport
River transport	Solution
Biogenous deposition	Mining and dredging
Hydrogenous deposition	Washover
Erosion of cliffs	Inlets
	Submarine canyons

[a]Based on information in U.S. Army Coastal Engineering Research Center (1973) *Shore Protection Manual,* and Bowen and Inman (1966).

graphic charts drawn up at frequent intervals over decades. Where such surveys are unavailable, volumetric analyses of sediment movement in the area can be made by noting the amount of sediment accumulation on the updrift side of beach barriers and sediment loss on the downdrift side, or by comparing the amounts of material removed from harbors, channels, and inlets along the coast by dredging operations (see Vol. VI: *Sedimentation*). A rough estimate can also be made when the rate of shoreline recession can be measured. A loss of 1 ft^2 of shore area is, for example, considered equivalent to the loss of 1 yd^3 of material from an exposed beach (Hall, 1952).

Determining Sediment Transport Direction. The dominant direction of longshore transport is determined by studying the shape of natural shoreline features such as spits, shoals, and beach ridges, or by noting where sand accumulates around beach obstructions. Other methods include the use of sand tracer studies, and the determination of long-term flow directions of wave energy from wave-recording stations or from hindcast wave data derived from synoptic meteorological charts.

Determining Material Requirements. The material requirements needed to restore the beach depend on the desired width of the restored beach, the design height of the beach berm, and the slope of the beach from the berm to the toe of the fill. The material required to maintain the restored beach depends on the rate of net loss of beach material. The actual amount of fill required is the sum of the amount of fill necessary to meet design specifications plus the amount of fill expected to be removed from the area by the sorting action of shore processes. In the case where the fill material and the natural beach sediments have identical grain-size distributions, little of the fill will be lost and the initial fill requirements can be determined directly from the design specifications. This is not often the case, however, and estimates of initial material requirements must take into consideration an estimate of the proportion of the fill that will be stable on the equilibrium beach.

Krumbein and James (1965) have proposed a formula that considers the effect that differences in grain-size characteristics will have on the ratio (R_{crit}) between the volume of fill required and the volume of material that will be stable on the beach. In this formula,

$$R_{\text{crit}} = \frac{\delta_b}{\delta_n} e^{-\left[\frac{(M_b - M_n)^2}{2(\delta_n^2 - \delta_b^2)}\right]}$$

where δ_b = standard deviation (sorting) of borrow (fill) material
δ_n = standard deviation of natural beach material
M_b = mean grain size of borrow material
M_n = mean grain size of native beach material

in ϕ units

On the basis of this equation, four cases can be outlined (Table 2). In the first two cases, the borrow material is more poorly sorted than the natural beach material. In case 1 the borrow material is finer-grained than the natural beach sediment; R_{crit} defines the ratio between the amount of fill placed on the beach and the amount remaining after equilibrium is attained and the fine-grained material has been winnowed from the shore zone. Much of this fine material will be transported to the offshore zone, where it will serve to decrease the beach gradient.

In case 2, R_{crit} probably represents a near-maximum value because of the initial coarseness of the borrow material relative to that on the natural beach. The small amount of fine material in the fill will be winnowed out, but the excess coarse material will remain on the beach. After equilibrium is attained, therefore, the sediment of the restored beach will be coarser-grained than that of the natural beach.

R_{crit} cannot be defined in either case 3 or case 4. The borrow material in these cases is better sorted than the beach sediment and cannot be transformed into a sediment matching that on the original beach. When the material is coarser-grained, the borrow is deficient in the finer grain sizes and much of the fill will be stable on the beach. When the borrow is finer-grained, however, the equation implies that none of the material will be stable. While large initial losses should be expected in such a case, some of the borrow will remain—how much depends on the difference in mean grain size between the borrow and the natural beach sediment.

The method has not been adequately tested in the field. The equation was applied to a restoration project at Carolina Beach, North Carolina (Valianos, 1970), but only after the project had begun. The initial beach sediment had not been sampled, and R_{crit} was defined on the assumption that the sediment on the equilibrium beach, formed after the project was completed, was representative of the native beach material. Initial losses exceeded calculations; apparently much of the material moved offshore, where it reduced the nearshore—foreshore gradient.

Source of Borrow Material. To be economically feasible, the borrow material should be roughly similar to that of the natural beach and the source should be near the problem area. If the borrow

TABLE 2. Applicability of R_{crit} Calculations for Various Combinations of the Graphic ϕ Moments of Borrow and Native Material Grain-Size Distributions [a]

Case	Relationship of ϕ Means	Relationship of ϕ Standard Deviations	Response to Sorting Action
1	$M_b > M_n$ Borrow material is finer than native material.	$\delta_b > \delta_n$ Borrow material is more poorly sorted than native material.	Best estimate of overfill ratio is given by R_{crit}.
2	$M_b < M_n$ Borrow material is coarser than native material.		Required overfill ratio is probably less than that computed for R_{crit}.
3	$M_b < M_n$ Borrow material is coarser than native material.	$\delta_b > \delta_n$ Borrow material is better sorted than native material.	Distributions cannot be matched, but all fill material should be stable.
4	$M_b > M_n$ Borrow material is finer than native material.		Distributions cannot be matched. Fill loss cannot be predicted but will be large.

[a]Based on information from U.S. Army Coastal Engineering Research Center (1973) *Shore Protection Manual.*

is too fine, much of it will be winnowed from the shore and redeposited in the offshore zone so that high-volume maintenance feeding will be necessary. If the material is too coarse, however, the equilibrium beach may be unsuitable for recreational purposes.

Sources in offshore areas, back-barrier lagoons, inlets, harbors and coastal waterways, and the updrift side of beach obstructions have been used successfully. Nearby land sources may also be considered, especially along constructional coasts, where suitable borrow material is often available within coastal sand bodies. Borrow material can be mined from previously deposited beaches within these sand bodies; this material should be very nearly identical to the natural beach sediment (Fraser and Hester, 1974).

Borrow sources in bays, lagoons, and offshore areas are particularly subject to environmental degradation caused by dredging operations. The impact of dredging on the stability of the bottom sediments, on the exchange and recycling of nutrients and decay products in the sediments, alteration of wave and current patterns, and the effect on the flora and fauna must be carefully considered (Thompson, 1973).

Methods

Three general methods of artificial replenishment have been used: (1) offshore dumping; (2) stockpiling; and (3) direct placement. Once the design factors have been established, the choice of which method to use will be based on the direction and magnitude of the forces that affect the beach, and the availability and cost of supplying the borrow material, as well as the position of the source of the borrow relative to the beach area.

Offshore Deposit. The borrow material may be placed directly offshore of the project area so that the sand may be transported onshore by shoaling waves. Offshore stockpiles can be economically emplaced by hopper dredges; this method may be useful where sand is available in offshore areas or from dredging operations in harbors and coastal waterways. Initial efforts using this method were unsuccessful because the hopper dredges could not be operated in water shallow enough for onshore transport of the borrow material to take place (Hall, 1952; Harris, 1954).

Stockpiling. Where strong, consistent longshore currents operate, borrow material may be stockpiled on the updrift shore of the problem area, to be distributed along the coast by the currents. This method is relatively inexpensive where sand can be dredged and piped as a slurry to the beaches (Olmstead and Lynde, 1958), but where the borrow is very coarse the use of land-based vehicles may be necessary (Muir Wood, 1970). The method is commonly used in conjunction with sand *bypassing operations* across inlets and

around shore obstructions (see U.S. Army Coastal Engineering Research Center, 1973, for a review of major bypassing projects). The main drawback to the method is that erosion downdrift of the stockpile will continue until the currents have distributed the borrow material along the coast. This may be alleviated by the use of multiple stockpiles.

Direct Placement. The replenishment project may be completed at one time, except for maintenance feeding, when the borrow material is distributed along the whole length of the eroding shore. Although the method may be expensive, it is often required in cases where erosion of the shore is extensive and ongoing. Most projects have utilized pipelines fed by dredging operations in lagoons and inlets (U.S. Army Corps of Engineers, 1950; Watts, 1956, 1959; Valianos, 1970) or from offshore areas where material may be pumped onto the shore from floating dredges or from anchored barges supplied by hopper dredges with pump-out capability (Watts, 1958; Mauriello, 1967; Fisher, 1969). Feeder stockpiles may be used in combination with direct-placement methods (Vesper, 1967), and since the replenishment project usually benefits adjoining shores, the project area itself may be considered a feeder beach for downdrift shores.

GORDEN S. FRASER

References

Bowen, A. J., and Inman, D. L., 1966. Budget of littoral sands in the vicinity of Point Arguello, California, *U.S. Army, Corps Engineers Coastal Eng. Research Center Tech. Mem. 19*, 41p.

Finkl, C. W., 1981. Beach nourishment: a practical method of erosion control, *Geo-Marine Letters*, 1 (2), 155-161.

Fisher, C. H., 1969. Mining the oceans for beach sand, *Proceedings of the Conference on Civil Engineering in the Oceans, II*. New York: American Society of Civil Engineers, 717-723.

Fraser, G. S., and Hester, N. S., 1974. Sediment distribution in a beach-ridge complex and its application to artificial beach replenishment, *Illinois Geol. Survey Environmental Geology Note 67*, 26p.

Hall, J. V., 1952. Artificially nourished and constructed beaches, *U.S. Army Corps Engineers, Beach Erosion Bd. Tech. Mem. 29*, 25p.

Harris, R. L., 1954. Restudy of test-shore nourishment by offshore deposition of sand, Long Branch, New Jersey *U.S. Army Corps Engineers, Beach Erosion Bd., Tech. Mem. 77*, 33p.

Krumbein, W. C., and James, W. R., 1965. A lognormal size distribution model for estimating stability of beach fill material, *U. S. Army Corps Engineers Coastal Eng. Research Center Tech. Mem. 16*, 17p.

Mauriello, L. J., 1967. Experimental use of self-unloading hopper dredge for rehabilitation of an ocean beach, *Proceedings of the World Dredging Conference*, 367-396.

Muir Wood, A. M., 1970. Characteristics of shingle beaches: the solution to some practical problems, *Proceedings of the Twelfth Conference on Coastal Engineering*, Vol. 2. New York: American Society of Civil Engineers, 1059-1075.

Olmstead, L. W., and G. A. Lynde, 1958. Feeder beaches and groins restore Presque Isle Peninsula, *Civil Engineering* **42,** 172-175.

Thompson, J. R., 1973. Ecological effects of offshore dredging and beach nourishment, *U.S. Army Corps Engineers Coastal Eng. Research Center Misc. Paper 1-73,* 39p.

U.S. Army Coastal Engineering Research Center, 1973. *Shore Protection Manual,* 3 vols. Washington, D.C.: U.S. Government Printing Office.

U.S. Army Corps of Engineers, 1950. *Beach Erosion at Santa Barbara, California.* House of Representatives, Doc. 552, 75th Cong., 3rd Sess.

Valianos, L., 1970. Recent history of erosion at Carolina Beach, North Carolina. *Proceedings of the Twelfth Conference on Coastal Engineering,* Vol. 2. New York: American Society of Civil Engineers, 1223-1242.

Vesper, W. H., 1967. Behavior of beach fill and borrow area at Sherwood Island State Park, Westport, Connecticut, *U.S. Army Corps. Engineers Coastal Eng. Research Center Tech. Mem. 20,* 11p.

Watts, G. M., 1956. Behavior of beach fill at Ocean City, New Jersey, *U.S. Army Corps Engineers, Beach Erosion Bd. Tech. Mem. 77,* 33p.

Watts, G. M., 1958. Behavior of beach fill and borrow area at Harrison County, Mississippi, *U.S. Army Corps of Engineers, Beach Erosion Bd. Tech. Mem. 107,* 17p.

Watts, G. M., 1959. Behavior of beach fill at Virginia Beach, Virginia, *U.S. Army Corps of Engineers, Beach Erosion Bd. Tech. Mem. 113,* 13p.

Cross-references: *Coastal Engineering; Coastal Inlets, Engineering Geology; Geomorphology, Applied.* Vol. VI: *Lateral and Vertical Accretion.* Vol. XIV: *Coastal Zone Management; Environmental Engineering.*

BIOGEOCHEMICAL PROSPECTING — See Vol XIV: BIOGEOCHEMISTRY.

BIOGEOLOGY — See Vol XIV: GEOMICROBIOLOGY.

BLASTING, EXPLOSIVES — See Vol. XIV: BLASTING AND RELATED TECHNOLOGY; CRATERING, MAN-MADE.

BORE LOG DATA PROCESSING — See ELECTROKINETICS; WELL DATA SYSTEMS. Vol. XIV: WELL LOGGING.

BORING AND DRILLING — See RAPID EXCAVATION AND TUNNELING; SHAFT SINKING; TUNNELS, TUNNELING. Vol. XIV: AUGERS, AUGERING; BOREHOLE DRILLING; BOREHOLE MINING.

C

CALICHE, ENGINEERING GEOLOGY

Caliche is a general term given to deposits of secondary calcium carbonate widespread throughout arid and semi-arid regions. Although all such deposits are chemically similar—dominantly calcite—the physical properties, distribution, and degree of development vary due to different modes of formation. Generally, caliche is deposited as part of the soil-forming process in arid and semi-arid regions (Gile et al., 1966). Such pedogenic caliche is formed below the topographic surface. Some caliche is, however, of geogenetic origin and forms by deposition on the surface and as surface coatings (see Vol. VI: *Caliche, Calcrete*). Details of the origin of the various forms of caliche are beyond the scope of this article; refer to Gile et al. (1966), Lattman (1973), and Goudie (1973) for further information.

From the engineering viewpoint, the detailed classification of Aridisols (which include calichified soils) used by the U.S. Department of Agriculture (Soil Survey Staff, 1975) is not needed; thus a simpler classification is used here. In general, the engineering and hydrologic properties of caliche depend on the type of caliche, and also on the age of the caliche that affects degree of development.

Pedogenetic Caliche

Pedogenetic caliche forms at various depths below the topographic surface. It begins as surface coatings on pebbles and cobbles and as stringers and nodules within the unconsolidated sediments. With time, the calcium carbonate fills the pore spaces within the sediment and becomes a continuous layer. The term *calcic horizon* (Fig. 1) is given to soil horizons in which caliche has or is accumulating. With further calcium carbonate deposition, the horizon may become very hard and not slake in water; such a horizon is termed a *petrocalcic horizon* (Fig. 1).

Of special significance is the *laminar layer*, which is frequently found on top of a petrocalcic horizon but is also found alone. Such a layer is made of thin laminae of calcite and perhaps clay, each generally about 1 mm thick (Fig. 2). The laminar layer contains very little material coarser than silt and is continuous over local areas, but not usually continuous over areas of more than about 1,000 m^2—although it may be discontinuously present over large areas. There appear to be two possible origins of such laminar layers. One theory (Gile et al., 1966) holds that the formation occurs on a "plugged" calcic horizon due to the perching of downward-percolating water on this horizon. The overlying horizons are raised to make space for the new laminar horizon. Another possible origin (Lattman, 1973) is from evaporation of surface water trapped in broad, shallow topographic lows. The second origin does not require an underlying plugged horizon.

Calcic horizons vary in thickness from a few centimeters to 10 or more meters. Very thick horizons may be of polygenetic origin (Bretz and Horberg, 1949) or accretional (Brown, 1956). Calcic horizons may be modified by surface hardening after exposure, as discussed below.

Geogenetic Caliche and Modification

Calcium carbonate deposits developed on the surface rather than within the soil are of particular engineering and hydrologic importance. The deposits consist of new material, or are the result of

FIGURE 1. An exposure of caliche in the McCollough fan, Las Vegas, Nevada. Angular caliche rubble litters the surface. Beneath the rubble is a boulder-free petrocalcic horizon about 33 cm thick. Under the petrocalcic horizon is a bouldery, well-cemented calcic horizon. Hammer at lower right shows scale.

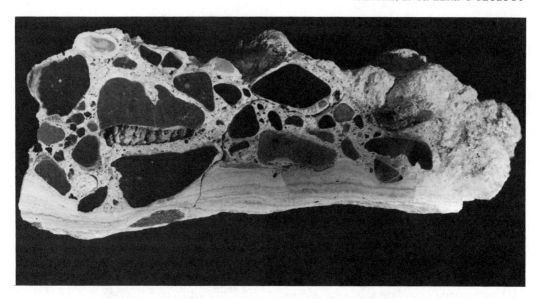

FIGURE 2. An example of a laminar layer overlying a petrocalcic horizon.

modification of pedogenetic caliche by surface processes. These deposits tend to be impermeable and hard.

Case-hardening (Lattman and Simonberg, 1971) develops on steep to vertical slopes cut into carbonate sediment (Fig. 3). It is very weak, or absent, on noncarbonate sediments. Case-hardening is apparently due to solution and redeposition of fine-grained calcitic material on the surface and decreases in development inward. It forms very rapidly, and within a few months it may cement a fresh exposure so that cobbles cannot be removed by hand. Case-hardening does not form uniformly on heterogeneous sediments. In alluvium, for example, poorly sorted lenses undergo more rapid and stronger cementation than adjacent, better sorted layers.

FIGURE 3. Case-hardening of vertical gully sides cut into an alluvial fan near Las Vegas, Nevada.

Case-hardening increases in thickness and strength with time, and on old exposures (10^3 years old) it forms an extremely hard and impermeable surface on calcareous sediments. After about 35 years, road cuts in southern Nevada have cemented to a degree that the surface breaks through limestone cobbles rather than between the cobble and the cement.

Surface-hardening is generally similar to case-hardening. It occurs on calcic horizons exposed at the surface and is most common in flat areas and on low and gentle slopes. It is also a rapid process and can cause a soft calcic horizon to become a "petrocalcic horizon," which is hard and non-slaking. Calcic horizons that have been surface-hardened may still be soft and punky just below the exposed surface.

Geogenetic laminar layers are present in shallow topographic lows and on gentle slopes. Such layers may not have a petrocalcic horizon underneath, but rather a thin silt or clay layer that has retained water until evaporation and precipitation of the calcite.

Distribution of Caliche

Caliche, which is developed in unconsolidated sediments and soils, is widely distributed in arid and semi-arid regions. The thickness and degree of development is a function primarily of the availability of calcium carbonate, the rock types of which the sediments are made, age, and perhaps climatic history.

Obviously calcareous sediments can supply the calcium carbonate for caliche directly; however, noncalcareous sediments may also be strongly cemented. Additional sources of calcium carbonate may be wind-blown material (Brown, 1956), atmospheric sources (Gardner, 1972), and perhaps bacterial reduction of gypsum under former wetter conditions (Lattman and Lauffenberger, 1974). In any case, well-developed caliche deposits commonly occur on, and downwind from, sources of eolian calcite and gypsum. Additionally, calcium carbonate may be derived from calcium liberated by the weathering of feldspars in basic igneous rocks. Generally, in the southwestern United States, basic igneous rock detritus exhibits well-developed caliche, as do calcareous sediments. Mixed sedimentary rock detritus in a particular area possesses caliche in rough proportion to the amount of calcareous detritus. Acid igneous rock detritus is generally poorly cemented throughout the southwestern United States.

Natural Destruction of Caliche

Throughout arid areas of the southwestern United States, caliche is being destroyed today. Two processes appear active in the destruction: solution and mechanical breakup. The hard caliche types—such as laminar layers, petrocalcic horizons, and surface-hardened layers—generally break up by mechanical means. The result is sharp-edged caliche rubble littering the surface (see Fig. 1). This rubble shows little solution effect. Lattman (1973) has suggested that freeze and thaw of water is the mechanical process involved. Softer caliche types—such as calcic layers and locally case- and surface-hardened caliche—undergo destruction by solution. Surface caliche attacked by solution becomes porous. The porosity commonly is due to the loss of pebbles and cobbles loosened by solution. Additionally, solution-attacked caliche may be crumbly. Calcite, of course, has a hardness of 3 on the Mohs scale. However, the common inclusion of fine-grained siliceous material in the calcareous cement frequently causes the caliche to be harder than calcite. A calichified layer of coarse detritus has a variable hardness controlled by the detritus and cement; thus each individual case must be studied.

Hydrologic Effects of Caliche

Caliche causes unconsolidated material to lose its vertical permeability. In many calichified alluvial fans, the caliche covers the interfluves and sides and bottoms of washes, resulting in a severe total reduction in infiltration.

Cooley and coworkers (1973) have studied infiltration characteristics of two calichified fans in the Las Vegas, Nevada, area. In both cases caliche significantly reduced infiltration, but for different reasons on the Red Rock Canyon and McCollough fans. The Red Rock Canyon fan is composed of limestone detritus. Surface-hardening and laminar layers were the two features that caused the major reduction in infiltration. The fan exhibited permeabilities "equivalent to or, in the case of the maximum value, approaching those for either stratified or unweathered clay, for all practical purposes impervious" (Cooley et al., 1973, p. 25). The McCollough fan (Fig. 1) is composed of calichified andesite and basalt detritus. A boulder-free petrocalcic horizon overlies a calichified bouldery horizon. Laminar layers are absent. The petrocalcic horizon is fractured and the fractures are widened by solution. The underlying bouldery caliche is unfractured. The fractured caliche has high sustained rates of infiltration, while the underlying unfractured bouldery caliche markedly reduced infiltration. Both alluvial fans have a history of flash-flooding. In the Las Vegas area, the calichified fans show a more frequent and stronger flash-flooding history than adjacent noncalichified fans.

In general, caliche reduces infiltration on alluvial fans and increases flood hazard. The particular effect of the caliche on an alluvial fan varies; hence each fan must be studied. Cooley et al. (1973) describe a device for measuring infiltration characteristics of different caliches in the field and for analyzing the results.

Caliche Properties of Possible Construction Significance

Caliche locally is a valuable construction material in arid and semi-arid areas (see *Arid Lands, Engineering Geology*). In unconsolidated sediments, the presence of a thick petrocalcic horizon serves as a base on which heavy structures may be placed. Care must be taken to drill such horizons because a soft calcic horizon may have a thin surface-hardened layer and may superficially resemble a petrocalcic horizon. Removal of the surface layer will expose the soft, crumbly interior.

When exposed—and thus subjected to traffic—caliche is very dusty. A true petrocalcic horizon or a calcic horizon with a well-developed laminar layer will not be dusty. Secondary roads cut on calcic horizons are generally dusty because traffic continually removes any newly formed surface-hardened layer. A calcic horizon exposed in a secondary road infrequently used, or cut and not used for some time, may develop a surface-hardened layer that will greatly reduce dusting. Slopes in road cuts made in calcareous sediments are stabilized by case-hardening very quickly, generally within one year. Once case-hardening begins, it progresses rapidly, and even vertical slopes require no attention or maintenance in such cases.

Case-hardening has another useful effect. It is noticed that blocks of caliche cut from a soft calcic horizon, if allowed to "weather" for some time, develop case-hardening quickly and become useful construction material.

LAURENCE H. LATTMAN

References

Bretz, J. H. and Horberg, L., 1949. Caliche in southeastern New Mexico, *Jour. Geology,* **57,** 491-511.

Brown, C. N., 1956. The Origin of caliche on the northeastern Llano Estacado, Texas, *Jour. Geology,* **64,** 1-5.

Cooley, R. L.; Fiero, G. W.; Lattman, L. H.; and Mindling, A. L., 1973. *Influence of surface and near-surface caliche distribution on infiltration characteristics, Las Vegas, Nevada.* Reno, Nev.: Center for Water Resources Research, Desert Research Institute, Project Report No. 21, 41p.

Gardner, R. G., 1972. Origin of the Mormon Mesa Caliche, Clark County, Nevada, *Geol. Soc. America Bull.,* **83,** 143-156.

Gile, L. H.; Peterson, F. F.; and Grossman, R. B., 1966. Morphological and genetic sequences of carbonate accumulation in desert soils, *Soil Sci.,* **101,** 347-360.

Goudie, A., 1973. *Duricrusts in Tropical and Subtropical Landscapes.* Oxford: Clarendon, 174p.

Lattman, L. H., 1973. Calcium carbonate cementation of alluvial fans in southern Nevada, *Geol. Soc. America Bull.,* **84,** 3013-3028.

Lattman, L. H., and Lauffenberger, S. K., 1974. Proposed role of gypsum in the formation of caliche, *Zeitschr. Geomorphologie Suppl.* Band **20,** 140-149.

Lattman, L. H., and Simonberg, E. M., 1971. Case-hardening of carbonate alluvium and colluvium, Spring Mountains, Nevada, *Jour. Sed. Petrology* **41,** 274-281.

Soil Survey Staff, 1975. Soil taxonomy, a basic system of soil classification for making and interpreting soil surveys, *U. S. Dept. Agriculture, Agriculture Handb. 436,* 754p.

Cross-references: *Arid Lands, Engineering Geology; Duricrust, Engineering Geology; Rocks, Engineering Properties.* Vol. IVA: *Calcium Carbonate: Geochemistry.* Vol. XII, Pt. 1: *Soil Mineralogy.* Vol. XII, Pt. 2: *Calcrete; Duricrust.*

CAPILLARY, CAPILLARY WAVE—See Vol. II.

CAVITY UTILIZATION

Biblical references and archaeological evidence seem to show that the use of natural caves and caverns for habitation, storage, and religious rites is as old as man himself. Abraham, for example, purchased burial caves, the Indian Brahmans used caverns for religious instruction, and the American Indians used caves for storage of grain and roots (Whiting, 1935).

More recently, the Maori of New Zealand stored sweet potatoes in pits, and early Middle East cultures stored water in underground reservoirs, qanats, and countless cisterns. Cold storage had its beginnings in spring houses, root cellars, and ice caves. The cliffdwellers of the U.S. Southwest used the lower level of the semi-underground space in their multilevel dwellings for storage. The use of underground vaults for storing vital records is exemplified by the Dead Sea Scrolls, which were found preserved after having been stored almost 2,000 years in a rock cave. The Germans stored art treasures during World Wars I and II in salt mines just north of Frankfurt.

The Goreme Valley of Cappodocia in Turkey contains multistoried living space and churches, which third-century Christians carved into tufa rock columns, eroded remnants of a volcanic era. The ruins of Joktheel, an ancient city of the Near East, indicate that its inhabitants dwelt in spaces hewn from stone. More recent examples of underground space use are religious shrines and a hospital that have been built in salt mines in Poland.

First-century Romans and eighteenth-century Parisians seem to have originated the use of man-made underground space in urban areas. The tufa under Rome and the limestone under Paris were quarried for construction of the cities, leaving vast subterranean rooms of mined-out space. The space under Rome became a place for sepulture, for storage, and even for housing. The space under Paris became a refuge and shelter for the poor and a place where thieves stored their loot; subsequently, it became the storage place for the bones of some 6 million bodies removed from their original sites of interment (Knox, 1873; Utudjian, 1964).

Although each of these uses relates in its own way to some aspect of underground usage, but none even approaches the current extensive and organized use of underground space in the Kansas City, Missouri, area. Other urban areas in the United States (and elsewhere, but on a less extensive scale) are beginning to view the underground as a resource to be developed. Because man-made caves offer savings in space and energy in an increasing congested world faced with depleting resources they merit public attention (Stauffer, 1972*a*, 1982*b*).

In greater Kansas City, approximately 2,000 people are employed daily in factories, warehouses, and offices located from 7 to 30 m below the ground. Because the development of the underground space by private industry has not been attended by the publicity that usually accompanies such projects supported by public funds, few people outside the city are aware of what has been done. Extensive uses have been developed for the subsurface, and mining techniques have been altered in anticipation of secondary use of the mined space. Today there is general public acceptance in Kansas City of the use and development of underground space; residents see it as an extra dimension of the city (Fig. 1).

Kansas City has worked to achieve urban and

FIGURE 1. One of the older mines in the Kansas City area, now successfully converted to a warehouse of dry and frozen foods.

mining compatibility and has also recovered abandoned mines for second and continuing uses. Because the insular qualities of the rock conserve energy, a full complement of cold-storage and freezer-storage warehouses have been developed. Urban space has been extended by expansion into a subsurface dimension. In Kansas City, use of subsurface space is no longer theory; it has been successfully practiced since the early 1950s (Stauffer, 1972a).

The Rock Matrix

The rock in which underground development occurs in the Kansas City area is a Pennsylvanian massive limestone consistently 3 to 3.5 m thick, dipping imperceptibly. It underlies the northwestern part of Missouri and extends into Kansas to the Oklahoma border. Locally known as the Bethany Falls limestone, it is continuous except where dissected by valleys of the Missouri and Kansas rivers and their tributaries. The bluffs created by these dissections have exposed the Bethany Falls limestone layer so that room and pillar mining from these bluffs has been the general mode of operation. Over 11,960,000 m² of mined-out space has resulted. The limestone is of commercial value mainly as aggregate for concrete and asphalt mix. The overlying shale and sealant clay have eliminated vadose ground-water erosion, so that karst features, such as caves, do not occur in this part of Missouri, leaving the shape and size of these rooms entirely to the discretion of man. None of the underground development in Kansas City mentioned here is of the basement type. All space is geologically separated from the surface by an undisturbed overburden of 7 to 30 m of additional limestone and shale (Parizek, 1975).

Physiography

The natural accessibility from valley into bluff has also been an important factor in the secondary use of the space, because industry and transportation are usually located within the valleys between the mined-out bluffs. Today, we rationalize the use of the subsurface by citing such factors as the preservation of the surface as aesthetically valuable, being nondisruptive to existing neighborhoods, preserving a tax base, and offering many other reasonable benefits. In Kansas City, however, subsurface use began simply because the space is economical. Industry needed to expand, and the hills, with vast areas of mined-out space, were available. In the early 1950s, when an overstock of cars was stored in an abandoned mine, the concept of underground storage was born. Today about 1 million m² of underground space is used for warehousing in the greater Kansas City area, with an additional 200,000 m² used for light and heavy industry and approximately 100,000 m² for retail sales offices, lounges, display areas, and so on.

A national survey conducted by Stauffer in 1971 revealed that Missouri led the contiguous United States in the secondary use of space mined out of a limestone matrix, with the greatest concentration occurring in the Kansas City area (Stauffer, 1978). Regardless of rationalization in behalf of the use of underground space, the Kansas City experience seems to indicate that it must be made economically feasible to gain wide acceptance.

Modification of Mining Methods

Optimal land use requirements in urbanized settings underscore the need for correct mining practices (see Vol. XIV: *Mining Preplanning*). Urban land can no longer be sacrificed to a single economic venture such as the removal of rock. The rock must be removed in such a manner as to leave the surface area available for future and continued usage. Where a mined-out subsurface exists, as in the Kansas City setting, both the surface and subsurface can be preserved for the community's economic benefit. A city is much more than a surficial adjunct superimposed on the topography. Its development and growth are deeply rooted in the resources, benefits, and problems of its underlying geology.

Earlier mining methods were focused on recovery of the limestone rock, via the "mine-and-abandon" method. The robbing of pillars and ceiling was common, and an aftermath of wasted land was accepted as inevitable. Large tracts of land were abandoned, and their mine-scarred surface, punctuated by subsidence pits, became a common sight in the rural-urban fringe of many cities (see Vol. XIV: *Mine Subsidence Control*). These abandoned wastelands often stood in the path of orderly urban growth, forcing streets to terminate and traffic to be permanently detoured. Residential and industrial development was forced to leapfrog these areas because their unstable surfaces make them unusable—and thus unproductive in terms of taxable activities.

The secondary use of mined-out space brings a reorientation of purpose in limestone mining. A new economic resource—the value of the mined space—now competes with the value of the rock, and attention is directed toward the dual objective of mining for the rock and for the later use of the created space. This places increased emphasis on the stability of the overburden as a surface for potential development above ground and as a ceiling to the development below ground. Some sites are impossible to convert to secondary usage due to the hazardous condition of the overburden, which may have been encumbered through mining methods not compatible to later use of the mined space.

Current mining practice in the Kansas City area is to remove 4 to 5 m of the lower part of the limestone layer and leave 3 to 4 m overhead, coinciding with a prominent bedding plane in the limestone layer. This supports the roof of the mine, reduces scale to a minimum, and leaves a level ceiling amenable to many secondary uses. Roof bolting is unnecessary, although it is sometimes used in the renovation of abandoned mines that were not mined for secondary use.

A new blasting technique, *smooth blasting* (see Vol. XIV: *Blasting and Related Technology*), is the widely accepted method for controlling overbreak. All mining in the Kansas City area is done by the means of drill-and-blast method. No machine mining is currently being undertaken.

The second major change in mining methods involves the distribution of pillars. Irregularly spaced pillars, relics of the day when no thought was given to use of the mined space, are not adaptable to secondary-use layout. Companies requiring long assembly lines are handicapped by a lack of linear patterns. Storage facilities requiring lengthy rail and truck lines have difficulty bending their routes through and around an irregular pillar system. Quarry operators have therefore altered their methods of pillar spacing. Pillars are now 7 to 8 m square and regularly spaced 17 to 20 m apart on the center. The grid pattern facilitates incorporation of pillars into partitioning walls, descriptive location of subdivided areas, designation of subdivision for recovery of stored items, arrangement of assembly lines, and linear office patterns. Pillars of the older mine, although irregularly spaced, have been incorporated into partitioning walls, business entrances, and office decor (Fig. 2).

Although this modification of mining practice in the Kansas City area has reduced income from rock extraction only slightly, it has stabilized the surface area by virtually freeing it from danger of collapse. It has at the same time created a stable ceiling for the mined-out rooms, leaving them

FIGURE 2. A mine pillar incorporated into office decor.

available for secondary use. Where more rock is left for pillar support and additional ceiling support, the recovery is reduced from about 90 percent, attempted in the days before secondary use, to about 80 percent, creating a loss of approximately 11 percent of the mineral recovery when compared to earlier mine-and-abandon methods. Drilling and blasting makes up about 40 percent of the cost of mining the rock; removal and crushing, about 60 percent. With smooth blasting requiring additional labor costs, and better pillar and ceiling support reducing recovery, the owner and operator are somewhat penalized by the new methods required to mine for secondary use of the mined space. Although the owner and operator are penalized, when the rock is crushed and retailed or processed into asphalt paving, the profit from the rock is still rewarding, and the space is an added bonus (Armstrong, 1975). Uses for the limestone in the Kansas City area normally run as follows: crushed stone, 20%; concrete, 40%; asphalt, 30%; agriculture, 5%; cement, 5%; and mineral filler, negligible.

Secondary Uses

Where limestone mining was once the primary use and the space left by mining only a by-product, new uses for this space have proved a second and continuing role for mined areas. In Kansas City, major secondary uses are warehousing, factories, and offices, in that order. The low cost of underground space has enabled the city's warehousing capability to expand: one-seventh of its warehousing is now underground. As a major railroad hub, Kansas City is an ideal place to store food and other items that will be shipped across the United States. Eighty freight cars, each capable of holding 45,000 kg of food, can be accommodated at one time on Inland Storage Distribution Center's two underground rail spurs, and many jobs have been created by the receiving, handling, and redistribution of goods (Figs. 3 and 4) (Vineyard, 1975).

The geographic location of Kansas City in the "bread basket" of the United States and midway between the western area of the United States—which produces about 50 percent of the nation's processed and frozen foods—and the eastern region—which buys two-thirds of all foods produced—is an ideal storage-in-transit point. The first underground freezer-storage room was developed in 1953, and Inland Storage Distribution Center leads as the world's largest refrigerated warehouse, handling 4 million kg daily. Over ½ kg of food for each person in the United States can be stored in this facility at any given time. The Department of Agriculture reports that Kansas City has 995-m³ capacity for frozen-food storage, most of which is underground, about one-tenth of the total such capacity of the entire nation.

Amber Brunson was the first to quarry rock as a secondary process, his primary objective being to obtain the underground space for a factory. The underground factory was occupied in 1960, and

FIGURE 3. Freight cars accommodating an underground storage facility in a once-abandoned mine.

FIGURE 4. Underground warehousing.

his facilities have since been an object of national and international interest. Brunson is recognized as the father of the planned use of underground space because the previous uses of underground space had been an afterthought. Brunson led in preplanned mining and arrangement of pillars, which he aligned to serve the purpose of secondary occupancy. Tunneled into one of the numerous limestone bluffs that characterize the terrain around Kansas City, the Brunson Instrument Company manufactured surveying and optical instruments that were used on the moon. The number of people in the 13,000-m² factory, which is located 25 m below the ground, ranges as high as 435. Precision settings can be made at any hour in this vibration-free environment; only the low-traffic hours from 2:00 to 4:00 a.m. could be used for this purpose in the company's former above-surface location. The economy of construction—the plant was built at a third of a comparable surface cost—and the vibration-free environment were Brunson's reasons for locating underground (Fig. 5).

One operation has an extensive two-tier development, with 110 hectares of industrial park on the surface and a choice of either an elevator or a ramp entrance to an additional 70 hectares of offices, industry, and warehousing over 35 m below ground in a former limestone mine. Additional mall shopping space is being created about 50 m below the surface at a cost slightly over half of comparable surface costs. Heating, air conditioning, maintenance, and security can be provided at a cost 60 to 70 percent lower than at a similar surface location. At a time when national attention is focused on energy conservation—any energy savings of 6 to 7 percent gains public attention—it would seem that underground utilization that approaches an energy savings of 60 to 70 percent would merit national acclaim. The amount of energy conserved varies with the type of function

FIGURE 5. Brunson Instrument Company, located 25 m below the surface.

one compares. An underground factory or a general warehouse requiring only worker comfort would need less than 10 percent of comparable surface heating to raise the constant natural subsurface temperature of 7-12°C to 15-18°C. However, one cannot assume this economy for a special function such as refrigerated and frozen-food space. A general figure of 50-percent reduction in use of energy through use of the subsurface for refrigerated space is an ideal: daytime lighting also requires energy, as do the electric forklifts and carts needed to control atmospheric pollution. Capital outlay for extra equipment to handle temperature extremes is not required, however, because the temperature fluctuation in an underground site is less than 6°C over the whole year.

Energy must also be expended for air conditioning. Air is pumped into the underground space through all open air space, and is exhausted to the surface through prepared vents or open tunnels. The air is usually treated by heating or cooling to accommodate the particular needs of the various users of the underground facility. This treatment occurs immediately after the air enters the intake, and the treated air is then forced through the facility.

Security is more easily effected in a subsurface site, because there are fewer entrances to guard or control. During the riots and burnings in Kansas City in the late 1960s, valuable goods, supplies, and technical equipment were hastily stored in the vacant subsurface areas for protection. Additional civil defense uses of the underground are evident in the storage of a 400-bed hospital and survival supplies in mined-out space.

Contrary to what may be expected, the underground rooms are easily kept dry with a minimum amount of dehumidification. Outside surface air does not rush into the subsurface, so that air once dried is easily and economically kept at whatever amount of humidity one desires. Metal machinery and factory equipment do not rust, and metal parts may be stored without damage. Allis Chalmers Farm Equipment Company makes use of a large underground storage area for a parts warehouse, and the Ford Motor Company also uses a large subsurface area for storage.

The easily controlled humidity and the ease of establishing adequate security have made the underground facilities ideal for the storage of film and records. Company standby records for use in case of flood, storm, or fire, as well as bank and university microfilm, are increasingly being stored in underground vaults.

In the greater Kansas City area 28 sites have usable subsurface space; 13 of these have developed some of their underground space for secondary usage, and there are 50 to 60 users of subsurface space occupying approximately 130,000 m². Analysis of the types of use made of underground

FIGURE 6. Retail sales room of drapery firm, accessible by elevator in a mall about 40 m below the street surface.

space shows warehousing and storage as the principal use, making up 89 percent of the total; manufacturing accounts for an additional 7 percent, and offices and retail sales make up 4 percent (Fig. 6).

Conclusions

Certain physiographic and geological features are necessary to the development of underground space. In the Kansas City area these features were found to be a massively bedded limestone, with sufficient thickness to be a matrix for rock removal without endangering roof support, an overlying clay and protective shale with sealant quality, nearly level stratigraphy, and a competent overburden.

Otherwise once-mined and wasted space has been converted into an employment and tax base for the community. Roughly 400 people are employed in the stone production, with an annual payroll of approximately $4 million for the primary rock production. Some 2,000 employees earn approximately $16 million in the secondary use of the mined-out space. Secondary usage has not exceeded stone production as an economic wage factor in the Kansas City area by some 5 times in numbers employed and 4 times in annual wages.

Locational analysis has clearly demonstrated several distinct advantages in use of the underground environment. For example: (1) mined-out space can be purchased or leased at a fraction of the cost of comparable surface facilities; (2) roof and foundation problems are reduced or eliminated; (3) floors are capable of supporting very high storage weight and equipment weight; (4) complete noise and vibration control is possible; (5) the areas themselves are fireproof and command the lowest insurance rates in the area; (6) greatly reduced costs in heating, air conditioning, or freezing result from the insulation properties of the rock (savings are as much as 90 percent for simple warehousing, with approximately 50 percent savings in refrigerated areas); and (7) security for equipment, records, and personal protection can be vastly improved. The main advantage of an underground location is the savings in rental costs, which average one-third of the cost of surface facilities of the same quality.

Systematic monitoring of stresses, strains, and seismic movement near these developments should be maintained. Societal, economic, and environmental factors should be mapped and studied on a regular basis to analyze the socioeconomic effect of these developments on the neighborhood, community, and region. An atlas of the subsurface and all its components should be made, and international cooperation should be sought to standardize symbols and pool data in identifying this late frontier.

TRUMAN STAUFFER, SR.

References

Armstrong, E. L., 1975. Underground space as a resource, *Proceedings of the Symposium on the Development and Utilization of Underground Space.* Kansas City, Mo.: University of Missouri-Kansas City/National Science Foundation, 135-142.

Knot, T., 1873. *Underground.* Hartford, Conn.: J. B. Burr & Hyde, 942p.

Parizek, E. J., 1975. Geologic setting of greater Kansas City, *Proceedings of the Symposium on the Development and Utilization of Underground Space.* Kansas City, Mo.: University of Missouri-Kansas City/National Science Foundation, 9-23.

Stauffer, T. P., 1972a. *Guidebook to the Occupance and Use of Underground Space in the Greater Kansas City Area,* Geographic Publication No. 1. Kansas City, Mo.: Department of Geology-Geography, University of Missouri-Kansas City, 56p.

Stauffer, T. P., 1972b. Use of mined-out space in greater Kansas City, in W. P. Adams and F. M. Helleiner, eds., *International Geography,* Vol. 1, 22nd International Geographic Congress. Toronto: University of Toronto Press, 672-673.

Stauffer, T. P., 1978. *Underground Utilization: A Reference Manual of Selected Works,* 8 vols. Kansas City, Mo.: Center for Underground Space Studies, University of Missouri-Kansas City, 1160p.

Utudjian, E., 1964. *L'Urbanisme Souterrain,* Que Sais-Je Series No. 533. Paris: Presses Universitaires de France, 125p.

Vineyard, J., 1975. Kansas City's limestone mines do double duty, *Missouri Environment,* **1**(4), 6.

Whiting, J. D., 1935. Petra, ancient caravan stronghold, Natl. Geog. Mag. **67,** 129-165.

Cross-references: *Urban Engineering Geology; Urban Tunnels and Subways.* Vol. XIV: *Mine Subsidence Control; Mining Preplanning; Open Space.*

CHANNEL EXCAVATION—See CHANNELIZATION AND BANK STABILIZATION.

CHANNELIZATION AND BANK STABILIZATION

Channelization involves human modification and control of natural, existing waterways, usually to permit or promote economic development or to protect already established urban, agricultural, and industrial developments. A specific channelization project may be undertaken for one or more of a number of reasons: (1) for flood control; (2) to drain wetlands; (3) to improve navigation; and (4) to prevent bank erosion and channel migration, and thus to protect neighboring property. There has been much controversy about whether channelization, especially over the long term, is effective, and whether harmful effects may exceed the benefits.

Aims of Channelization

Flood Damage. Flood damage to structures, crops, and so on amounts to about $1 billion annually in the United States alone. In addition, flooding causes widespread disruption of human activity, and may result in great loss of life. Floods can be controlled and abated by a number of means, including flood control dams, levees, and floodways, as well as channelization. Frequently, a comprehensive program will utilize all available measures to maximize protection. Channelization of a waterway aims to increase the hydraulic efficiently of the channel so that flood waters from upstream may pass as rapidly as possible through the reach. The boundary roughness of the channel is reduced by smoothing the channel perimeter and removing obstacles to flow such as trees, and the form roughness may also be reduced by realigning the channel to produce a straight or smoothly sinuous course (Acheson, 1968). By making the channel more hydraulically efficient, the cross-sectional area of the water is reduced, and hence water depths and the chance of overbank flow are decreased.

Drainage. While channelization for flood control aims to deal with excess water from upstream of the affected reach, channelization for drainage is undertaken to remove excess water from the land in the immediate vicinity of the waterway. It is often extremely difficult, however, to distinguish between the flood control and land drainage (q.v.) effects of a particular project, and both benefits are frequently realized. Channelization for drainage has the same immediate aim as for flood control—to increase the hydraulic efficiency of a waterway, facilitate the evacuation of excess water, and lower water levels. The water table in adjacent lands will thus be lowered, permitting agricultural development of or construction on former wetlands.

Navigation Improvement. Channelization projects designed to improve navigation facilities aim to provide a navigable, shoal-free channel that will be easily negotiated by ships. The waterway must hence be trained to follow a desirable course, aligned to produce smoothly sinuous bends—which, to obviate the need for frequent dredging—should be self-maintaining (see Vol. XIV: *Alluvial Systems Modeling*). By constricting the channel width, flow velocities are increased and bed scour promoted. This helps maintain the desired channel depth throughout the channelized reach, although dredging may still be necessary.

Bank Protection. A frequent aim of channel modification is to prevent bank caving and channel migration, which may destroy valuable farmland and threaten buildings, levees, and other structures. On a more limited scale, channelization of a stretch of river may be undertaken in conjunction with, for example, construction or relocation of a highway, in which the desired alignment encroaches on the channel. Channelization may also improve wildlife habitat, by preventing destructive bank erosion and channel migration (reviewed by Stevens, Simons, and Richardson, 1975, pp. 558-561). For example, Mifkovic and Petersen (1975) have described methods under test along the Sacramento River to improve riverine environments, by preventing erosion of the basins between channel and levee, and White (1975) described a project on a Wisconsin stream undertaken specifically to improve fish habitat.

A given channelization project may well provide benefits in all the preceding categories, and projects are increasingly undertaken for multipurpose water resource development. In turn, channelization may be only one aspect of river basin regulation, and may be associated with the construction of levees and flood control dams, land treatment for erosion control, and so on.

Channelization is usually undertaken on river flood plains, valley bottomlands, and coastal plains, except where urban flood control or protection of structures against erosion is desired. The English fenlands or the Rhein and Scheldte deltas in the Low Countries—all highly productive and populous agricultural regions—are good examples of low-lying, gently sloping areas in which channelization has provided substantial drainage, floor control, and navigation benefits. Channelization is largely restricted to the highly developed, economically advanced nations of the world, and especially to the North American and European continents.

Although many channelization projects have been on a large scale, data summarized by Little (1973), and Acheson (1968) suggest that projects in the United States and New Zealand, for example, tend to be quite small. Apart from small-scale channel modification carried out by individual landowners for drainage purposes, 65 percent of the projects completed by the U.S. Army Corps of Engineers before 1972 were less than 8 km long, and 10 percent of the approved Soil Conservation Service projects averaged 6.6 km in modified channel length (Little, 1973). Nevertheless, between 1940 and 1970, the Corps of Engineers and the Soil Conservation Service modified over 55,000 km of river.

Channelization Techniques

Channel modification may be accomplished in a number of ways, several of which may be used in a given project. Two basic references to channelization techniques, in New Zealand and the United States respectively, are Acheson (1968) and Winkley (1972).

Clearing and Snagging. This procedure consists of removing large obstacles such as trees and rocks from the channel. Such obstacles retard flow by increasing turbulence and energy dissipation, and catch floating debris that exacerbates the problem. Removal of obstacles achieves a modest improvement in hydraulic efficiency and channel capacity, speeds evacuation of water, and lowers water levels. Clearing and snagging is thus of particular importance in flood control or drainage projects, especially in small upstream channels. With bank grading, it is also a prerequisite to bank stabilization and protection, discussed below.

Channel Excavation. Greater improvement in the flow capacity of a waterway than is provided by clearing and snagging may be obtained by channel excavation. Two broad types of excavation may be identified: conventional and dredging.

Conventional. Equipment such as draglines, power shovels, and bulldozers work from the bank of the waterway to widen and deepen the channel, and to form a more hydraulically efficient cross-section. To minimize costs and problems due to sedimentation downstream, this type of work is done as far as possible "in the dry," during periods of low flow. Conventional excavation is especially suitable for relatively small waterways, or for wide, braided rivers, in which shallow depths permit excavation of both bed and banks. It may also be used for smoothing and grading the upper banks of larger rivers, usually to prepare for bank stabilization with revetment.

The aim of conventional channel excavation is to increase flow capacity and to develop a more efficient channel shape. The trapezoid is the most common shape for unlined channels; although it is hydraulically efficient, it is an unstable shape for alluvial channels. Side slopes are dictated by the

TABLE 1. Suitable side slopes for channels built in various materials (from Chow, 1959).

Material	Side Slope
Rock	nearly vertical
Muck and peat soils	$\frac{1}{4}$:1
Stiff clay or earth with concrete lining	$\frac{1}{2}$:1 to 1:1
Earth with stone lining, or earth for large channels	1:1
Firm clay or earth for small ditches	1$\frac{1}{2}$:1
Loose, sandy earth	2:1
Sandy loam or porous clay	3:1

stability of the material through which the channel is cut; permissible side slopes for stability in various materials are listed in Table 1. The rectangle, with vertical side slopes, is a special case of the trapezoid, and is commonly used for channels built of stable materials, such as masonry.

Dredging. There are three broad types of dredge—dipper, ladder, and suction (Huston, 1967). A *dipper dredge* is merely a floating power shovel, and would be restricted to work in shallow water. *Ladder dredges* have an endless chain of buckets, which bring the bottom material to the surface and discharge it onto a conveyor. *Suction dredges* pick up the bottom material and water in suction pipes, and discharge the slurry via a spoil pipe supported by floats to the desired spoil area. Rather than being merely dumped, the spoil should be used to supplement other channelization operations, such as the closing of chutes as the heads of point bars or islands.

Channel Cuts. A logical extension to enlarging existing channels is to excavate completely new sections of channel (Fig. 1). Natural rivers have meandering courses, which have been regarded as undesirable for flood control and navigation purposes. Removing and bypassing bends by excavating channel cuts reduces resistance to flow

due to form roughness and reduces the distance water must travel. For example, the cutoffs on the Mississippi River between Memphis and Baton Rouge described by Matthes (1948) shortened the river by 270 km from an original length of 1,095 km. This shortening resulted in an appreciable lowering of flood stages on the river (Fig. 2). Cutoffs also bring benefits to navigation and bank protection. Elimination of sharp meander bends simplifies negotiation of the waterway by ships, and reduces journey time. The caving banks in the cutoff bends are bypassed, and properly designed cutoffs may substantially reduce the lengths of eroding banks that must be maintained.

Realignment. An existing channel can be realigned by training the flow into the desired alignment with structures such as *dikes* and *jetties*. Whether for navigation or flood control purposes, the function of river-training structures is to persuade the flow into a smoothly sinuous alignment that is hydraulically more efficient than the natural channel. The channel is constructed so that widths are reduced and depths increased, thus promoting bed scour, which prevents deposition and makes the channel more or less self-maintaining. In other situations the training structures are intended to move the main flow away from the bank lines to prevent bank caving and channel migration.

A common type of realignment structure is the *stone-fill dike*. Its principal function is to direct the flow away from the bank; angling the dike upstream, downstream, or normal to the bank accomplishes this purpose in any given location. Two configurations using stone dikes that deserve specific mention are the vane dike and the L-head dike system. *Vane dikes* are angled about 10° to the bank in a downstream direction, and generate less eddying an scouring than dikes constructed normal to the bank. The L-head dike includes a section of dike extending downstream from the main dike and parallel to the flow. When the L-heads close about half the gap between dikes in

FIGURE 1. An artificial cutoff on the Missouri River. Note that the excavated channel is smoothly curved, not straight.

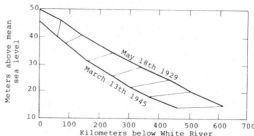

FIGURE 2. Effect of cutoffs on lower Mississippi River surface profiles. Both curves are on rising stage at 1.5 million f³ sec; upper is before and lower is after cutoffs. Lines between curves connect same locations on river before and after cutoffs (after Carey, 1966).

the dike field, they promote deposition between the dikes, decrease scour at the ends of the dikes, and provide bank protection. Both dike configurations may be included in navigation channel or bank-protection projects.

Other types of river-training structures are retards and jetty fields. *Retards* are permeable devices, such as timber piles, which are placed parallel to river banks to decrease flow velocities and prevent erosion. *Jetty fields* are intended to train the main stream into a selected alignment, reduce flow velocities along the banks, and eliminate erosion. They may be constructed of jacks, typified by those used in the middle Rio Grande to stabilize the braided channel and protect levees and adjacent areas. The unit consists of three 5-m-long steel angles placed at right angles to each other, bolted together at the center, and reinforced with wire. The jacks are then connected with cables to form a jetty line. Two types of jetty lines are used in a jetty field: *diversion lines* are placed along the desired location of the channel, and *retard lines* are placed at an angle to the diversion lines and spaced between 40 and 80 m apart. Other types of structures are available, and many variations and combinations have been used.

Bank Protection and Stabilization. The river-training structures discussed in the preceding section provide one means of protecting banks against erosion, by moving high-velocity flows away from the bank line. They may be said to provide intermittent protection; continuous protection is provided by revetment, in which the entire bank is covered with some type of erosion-resistant material. High flow velocities are permitted against the bank line, but the revetment increases the resistance of the bank to erosion.

The simplest form of bank protection, at least of the upper banks, is *vegetation*. The foliage reduces flow velocities at the soil surface and the roots bind the bank material, thus reducing or eliminating erosion. Vegetation needs periodic maintenance to prevent deterioration of the channel; bank slopes must therefore be sufficiently gentle to permit the use of mowing machinery.

The standard practice in New Zealand has been to place willow poles along the bankline to be protected, wire them together and anchor them with cable, then bulldoze river gravel over the toe of the slope. The willows rapidly sprout and, after repeated layering (cutting of branches that then resprout), provide a dense line of trees along the bank. Because willows may spread so readily beyond the location they are needed and because of difficulties with supply of appropriate poles, this type of work is being partly supplanted by the use of rock riprap (Acheson, 1968).

Riprap is perhaps the most common type of bank protection; it consists of a layer of rock fragments, preferably with a smooth size grada-

tion and with an angular shape. The specifications for riprap on the upper banks of the Mississippi require a 25 cm ($\pm$ 5 cm) layer of rock with individual particles weighing between 6 and 25 lb, with an approximate gradation as follows (1 lb = 0.45 kg):

> 75-125 lb: 10% max
> 25-74 lb: 40-60%
> 6-24 lb: 20-40%
> ‹ 6 lb: 15% max

With a good gradation of sizes, the interstices between the larger rocks are filled with the smaller sizes; interlocking is enhanced when the rock fragments are angular. Riprap may be used for the whole bank on smaller rivers, and for the upper banks only on rivers such as the Mississippi. When adequate riprap sizes are not available, rocks of cobble size may be placed in wire mesh mats or baskets and laid along the bank to provide protection.

Continuous bank protection that is fabricated in large sections and sunk against the underwater bank is termed *mattress.* The most effective yet designed is the *articulated concrete mattress,* formed of 20 concrete blocks spaced on a continuous wire mesh reinforcing. The sections are assembled on a barge, and lowered to the bottom as the barge moves out from the bank. Because of the scale of operation necessary, articulated concrete mattresses have been used solely on the lower Mississippi, on which over 1,200 km have been laid. Other types of revetment in use are woven willow brush and woven *lumber mattresses,* which have been used on various rivers, such as the Missouri, Arkansas, and Red rivers. The lumber mattress consists of a mat of 10-cm × 2.5-cm boards woven together and sunk to the bottom with the aid of stone-filled cribs. Several other types of fabricated bank protection have been used; they have in common, apart from their ability to protect the bank from erosion, a degree of flexibility that permits them to conform to bank contours and to adjust to any undercutting and caving that may occur at the toe of the bank.

The ultimate in bank protection is complete lining of the channel with masonry or concrete, as has been done with the Los Angeles River, for example. *Lining* substantially reduces channel roughness, improves hydraulic efficiency and reduces flow stages, and completely eliminates bank erosion. It also permits vertical sidewalls, which substantially reduce the area covered by the channel, but which also represent a potential safety hazard. Only in urban areas where land values are high and structures are in close proximity to the channel can complete lining be economically justified.

Effects of Channelization

Channelization has brought major benefits, both direct and indirect, to agriculture, transportation, and other sectors of the economy. Little's report to the Council on Environmental Quality (1973) concluded that the direct benefits of the channelization projects that it studied were generally somewhat conservatively stated; the report has been severely criticized, however. Jahn and Trefethen (1972) considered that, for many projects, benefits are overstated and costs are understated. Since the late 1960s, there has been recognition that there are significant environmental and other costs that must be considered in project evaluations.

In addition, it appears that the benefits of channelization may be only temporary, unless continued maintenance is undertaken. Thus, for example, bank vegetation must be cut back on channelized reaches to prevent deterioration of the channel and loss of capacity, while on the Mississippi River continual dredging is necessary to maintain the navigation channel. Moreover, there is uncertainty about the long-term effects of channelization. For example, there is some concern that the beneficial floor control effects of middle Mississippi River channelization are being lost, and that the situation may in fact be exacerbated in the future (e.g., Stevens, Simons, and Schumm, 1975; Belt, 1975).

Several costs of channelization have been identified:

1. Channelization causes the loss of large numbers of different, and increasingly rare, habitats for plant, animal, and fish species. Drainage of wetlands and bottomlands represent an obvious loss of habitat, and an unavoidable loss given that drainage is a major justification for a large proportion of projects. In addition, excavated channels—with their wide, shallow flows, bare or eroding banks, and high sediment loads—are less ecologically productive than are natural rivers, while the excavation of cutoffs frequently leaves the bypassed bends stagnant and silted, and of limited value for fish and wildlife. Duvet et al. (1976) concluded that channelization had no long-term deleterious effects on forage fish species and benthic macroinvertebrates, but reduced trout populations by removing overhead cover (overhanging banks and vegetation) and deep pools. Examples of "ecological disasters" in which the natural stream ecosystem has been almost completely disrupted—such as the attempt at channelization for flood mitigation in Crow Creek, Tennessee—are numerous, but measures are available to mitigate the environmental effects of channel modification. For example, Keller (1975) suggested that the bed of a channelized reach be excavated to leave a winding low-water channel with the characteristic of a natural

river. The Tennessee Valley Authority modified Bear Creek in such a way that the original channel was relatively untouched (Jahn and Trefethen, 1972). Meanders were cut off by a shallow grassed channel to take flood flows, but the natural channel was permitted to carry a normal flow at other times, so that fishery values were maintained. Mifkovic and Petersen (1975) listed a number of bank-protection techniques, tested on the Sacramento River, that protect levees against bank erosion with minimum disruption of riparian habitat and aesthetic values. They concluded that bank protection can protect and preserve environmental values, given proper planning.

2. Channelization frequently results in severe erosion and downstream sedimentation. For example, a section of the Blackwater River in Missouri was shortened in 1910 from 53.6 km to 29 km, with an increase in gradient from 1.67 m/km to 3.1 m/km (Emerson, 1971). The area of the original excavated cross-section was 38 m², but because of severe erosion induced by the increased slope and flow velocities, cross-sectional areas now range between 160 and 484 m². Channelization enabled the utilization of new floodplain land, but also caused erosional loss of farmland, and necessitated expensive bridge renewal and repair. Bird's (1979) case study of Lang Lang River, Australia, described unintended impacts of channelization that are, if anything, even more disastrous. Koloseus (1972) suggested that these effects may be predicted, using analytical procedures developed by hydraulic engineers. To prevent such effects, measures such as installation of drop sills may be adopted, as is already done in many cases by the U.S. Soil Conservation Service.

3. There may also be effects on reaches and water bodies downstream of a channelized stretch of river. On the Blackwater River, for example, channelization was followed by sedimentation and increased flooding downstream. If flood waters or drainage water is rapidly passed through a hydraulically efficient, channelized section of waterway, flood peaks in unchannelized reaches downstream will be increased. Thus, additional channelization may be required to deal with the negative effects of the original work. Channelization may also cause a deterioration in water quality, by shortening the time period in which the water can be purified. Channelization on the Kissimmee River, Florida, has led to severe deterioration of the quality of the waters of Lake Okeechobee, because flood waters flow straight to the lake, rather than spilling onto, and being filtered by, the floodplain.

4. Many people regard channelized waterways as aesthetically objectionable, especially when maintained to limit vegetation regrowth.

Because of these and other environmentally undesirable effects, channelization is an extremely controversial subject in the United States, although it is less so in other countries. The main targets for criticism are projects designed for flood control and drainage; projects for erosion control and navigation do not seem to have so many unintended side effects. Many engineers now believe that channelization is in many cases the least desirable course of action to take to achieve a specific aim (Schoof, 1980). For flood control purposes, for example, it is pointed out that natural river flood plains provide ready-made reservoirs to reduce flood peaks, and that channelization to prevent inundation induces more problems downstream than it solves. Alternative measures such as flood plain zoning or floodway construction are being considered; the flood plain must obviously be maintained under a land-use system, such as pasture, that is not adversely affected by periodic inundation. There seems less chance of resolving to everyone's satisfaction the conflicts over channelization for drainage of wetlands. Both the U.S. Sol Conservation Service and the U.S. Army Corps of Engineers have severely cut back their rural channelization programs for flood mitigation and drainage, but in many situations channelization is justified. Given adequate planning, design, execution, and maintenance, as required under Public Law 566, many effects of channelization, such as induced erosion and loss of visual amenity, may be mitigated or avoided. It is necessary that channelization be viewed as just one means of achieving a desired end, and that it be used when other methods are not appropriate.

M. P. MOSLEY

References

Acheson, A. R., 1968. *River Control and Drainage in New Zealand.* Wellington, N.Z.: New Zealand Ministry of Works, 296p.

Belt, C. B., Jr., 1975. The 1973 flood and man's constriction of the Mississippi River, *Science,* **189,** 681-684.

Bird, J. F., 1979. Geomorphological implications of flood control measures, Lang Lang River, Victoria, *Australian Geog. Studies,* **17,** 169-183.

Carey, W. C., 1966. Comprehensive river stabilization, *Am. Soc. Civil Engineers Proc., Jour. Waterways and Harbors Div.,* **92** (WW1), 87-108.

Chow, V. T., 1959. *Open Channel Hydraulics.* New York: McGraw-Hill, 680p.

Duvel, W. A.; Volkmar, R. D.; Specht, W. L.; and Johnson, F. W, 1976. Environmental impact of stream channelization, *Water Resources Bull.,* **12,** 799-812.

Emerson, J. W., 1971. Channelization: a case study, *Science,* **173,** 325-326.

Huston, J., 1967. Dredging fundamentals, *Am. Soc. Civil Engineers Proc., Jour. Waterways and Harbors Div.,* **93** (WW1), 45-69.

Jahn, L. R., and Trefethen, J. B., 1972. Placing channel modification in perspective in S. C. Csallany, T. G. McLaughlin, and W. D. Striffer (eds.), *Watersheds in Transition, Symposium Proceedings.* Urbana, Ill.: American Water Resources Association, 15-21.

Keller, E. A., 1975. Channelization: a search for a better way, *Geology,* **3,** 246-248.

Koloseus, H. J., 1972. Channel changes, *Civil Engineering,* **42** (2), 46.

Little, A. D., 1973. *Report on Channel Modifications: Report to the Council on Environmental Quality,* Vol. 1. Washington, D.C.: U.S. Government Printing Office, 394p.

Matthes, G. H., 1948. Mississippi River cutoffs, *Am. Soc. Civil Engineers Trans.,* **113,** 1-15.

Mifkovic, C. S., and Petersen, M. S., 1975. Environmental aspects—Sacramento bank protection: *Am. Soc. Civil Engineers Proc., Jour. Hydraulics Div.,* **101** (HY5), 543-555.

Schoof, R., 1980. Environmental impact of channel modification, *Water Resources Bull.,* **16,** 697-701.

Stevens, M. A.; Simons, D. B.; and Schumm, S. A., 1975. Man-induced changes of middle Mississippi River, *Am. Soc. Civil Engineers Proc., Jour. Waterways, Harbors and Coastal Engineering Div.,* **101** (WW2), 119-133.

Stevens, M. A.; Simons, D. B.; and Richardson, E. V., 1975. Non-equilibrium river form, *Am. Soc. Civil Engineers Proc., Jour. Hydraulics Div.,* **101** (HY5), 557-566.

White, R. J., 1975. Trout population responses to streamflow fluctuations and habitat management in Big Roche-a-Cri Creek, Wisconsin, *Verh. Internat. Verein. Limnologie,* **19,** 2469-2477.

Winkley, B. R., 1972. Practical aspects of river regulation and control, in H. W. Shen (ed.), *River Mechanics,* Vol. 1. Fort Collins, Colo.: H. W. Shen, 1-79.

Cross-references: *Alluvial Plains, Engineering Geology; Hydromechanics; River Engineering; Urban Engineering Geology; Urban Hydrology. Vol. VI: Alluvium; Flow Regimes; Fluvial Sediment Transport. Vol. XIV: Alluvial Systems Modeling; Canals and Waterways, Sediment Control; Environmental Engineering; Environmental Geology; Slope Stability Analysis.*

CHEMICAL MINING—See Vol. XIV: BOREHOLE MINING

CLAY, ENGINEERING GEOLOGY

The engineering geologist is primarily concerned that the effects of geological factors on the "location, planning, design, construction, operation and maintenance of engineering structures and the development of ground-water resources" are adequately provided for and recognized (Bates and Jackson, 1980, p. 204). The history, origins, and composition of earth materials are of importance insofar as they affect the applied objectives (Quigley, 1980) (see *Geotechnical Engineering*). It is well known that clay significantly affects the behavior of earth materials including the important property of cohesion in soils.

The term clay as used today carries with it three implications: (1) a natural material with plastic properties; (2) an essential composition of particles of very fine size grades; and (3) an essential composition of crystalline fragments of minerals that are essentially hydrous aluminum silicates or occasionally hydrous magnesium silicates. (Howell, 1966, p. 52)

The upper limit of grain size for minerals in the clay-size range is commonly taken as 2 μm. While clay is composed of the smallest mineral particles, its role is large in engineering geology (Gillott, 1963, 1968); the part it and other microstructural features play in determining behavior is receiving ever greater recognition.

In engineering, *soil* generally refers to unconsolidated earth material (see Vol. XII, Pt. 1: *Soil*). Clay soils often contain organic compounds and noncrystalline materials that sometimes have important effects on engineering behavior. The properties of clays are important in geotechnical and materials engineering, as well as in industry.

Classification

Confusion has sometimes resulted from the numerous differences in terminology between geology and engineering. The problem has recently been discussed by Morgenstern and Eigenbrod (1974), who also propose a classification of argillaceous soils and rocks based on a simple slaking test. In engineering there are several systems of classification in use, such as those of the American Association of State Highway and Transportation Officials (1978, 1982) for highway construction the Unified Soil Classification System (q.v.) (U.S. Waterways Experiment Station, 1953; Wagner, 1957) and the method used in Britain (Dumbleton, 1968). In geology the proportions of sand, silt, and clay have been used as the basis for the classification of intermediate- and fine-grained sediments (Shepard, 1954). Clays are also subdivided into two categories: residual and transported. Residual clays form by weathering and by hydrothermal action. Gravity, water, ice, and wind are the agents of transportation.

Genesis, Composition, and Structure

Clay mineral formation (q.v. in Vol. XII, Pt. 1) involves physical and chemical processes; in addition, the role of biological agencies is being increasingly recognized. Physical break up of rocks and minerals results from all varieties of transportation—from unloading, thermal cycling, crystal growth and frost action, wetting and drying cycling, and volcanic action. Chemical reactions commonly take place via true or colloidal solution, and it seems probable that organic stabilizers influence the course of reactions involving amorphous materials. Surface reactions and solid-state transformations are also important in the genesis of clay

minerals (see Vol. VI: *Clay as a Sediment*).

The largest proportion of clay minerals in sediments are of detrital origin; types found reflect on a broad scale the weathering conditions in the source areas. Glacial weathering leads to significant proportions of primary minerals in the clay-size range, together with illites and chlorite. Chemical processes become more important under conditions of temperate, warm or tropical weathering, and kaolinites and smectites such as montmorillonite are more abundant. Poorly ordered minerals such as allophane are more common than was once believed.

Clay minerals form authigenically in the weathering zone and under diagenetic conditions (Fig. 1). Borst (1972) described formation of kaolinite from gels, by silicification of gibbsite and from solution under conditions of low pH in the microenvironment of foraminiferal chambers containing decomposing organic matter. Similarly, there is evidence for authigenic formation of other clay minerals under a variety of conditions (Millot, 1970; Wilson and Pittman, 1977). Montmorillonites form by weathering of 2:1 and 2:2 layer structure silicates and under alkaline conditions in soils and from volcanic material.

The origin and history of a clay soil determine its composition and structure. The inorganic minerals generally make up the largest proportion of the solids, although organics are locally predominant in deposits such as peat. Liquids and gases occupy the void spaces. In general, the solids are mainly crystalline, but inorganic and organic noncrystalline components sometimes have an important effect on properties. The inorganic components are principally silicates, together with oxides, hydrous oxides and hydroxides, and carbonates and phosphates. Secondary minerals are most typical, but primary minerals of clay size range are locally important in deposits of glacial erosion. The clay minerals are hydrous aluminosilicates of platy or fibrous morphology and high specific surface area. Many carry a net negative charge, which is balanced by cations that are exchangeable to an extent that varies between different minerals and between different samples of the same mineral (Fig. 2). The clay minerals are responsible for most of the qualities that characterize clay (see Vol. XII, Pt. 1: *Clay Minerals, Silicates*).

Structure can be considered both on a large scale and at the microscopic level. Gross features include changes due to stratigraphy and other causes, fissures, joints, cracks, shears and other failure surfaces and zones, ice wedges, sand dikes and pipes, and bedding disturbances. These features result from conditions at the time the soil formed and from subsequent changes such as those that accompany consolidation, syneresis, slides, tectonic processes, stress relief, weathering, stratigraphic relations (as when a stiff layer overlies a soft deposit), and forces due to ice

FIGURE 1. Scanning electron micrographs of authigenic clay minerals: *A.* kaolinite. Bluesky formation, cretaceous, N.W. Alberta; *B.* chlorite, Falher formation, cretaceous, N.W. Alberta.

MINERAL	STRUCTURE*	CATION EXCHANGE CAPACITY IN MILLIEQUIVS. PER 100 gms	LAYER CHARGE	BASAL SPACING
KAOLINITE	1:1 (T / O,T)	~ 10	~ 0	~ 7 Å
ILLITE	2:1 (T / K+ / T / O / T / K+ / T)	10 – 40	≲ 2	~ 10 Å
MONTMORILLONITE	2:1 (T / T / O / T / T)	80 – 150	0.5 – 1.0	VARIABLE DEPENDING ON RELATIVE HUMIDITY

* $O = Mg(OH)_2$ OR $A\ell(OH)_3$, OCTAHEDRAL UNIT, $T = SiO_4$, TETRAHEDRAL UNIT

FIGURE 2. Structure and properties of principal clay minerals.

movement. Classification has been considered by Chandler (1973). Features such as fissures and joints may be mapped as in structural geology (Skempton et al., 1969; Fookes, 1969) and have been found to have a systematic orientation that can sometimes be related to causes such as the direction of ice movement (McGown, 1973). Such structural features have an important bearing on the engineering behavior of soils. For example, cracks and joints allow access by water, which may cause softening that leads to accelerated soil movements; thin seams of montmorillonite have frequently acted as planes of weakness along which slides have developed.

Fine structure of soils has been investigated by optical (Lafeber, 1965, 1967) and electron microscopy (Smart, 1966; Gillott, 1969, 1970; Tovey, 1973), by x-ray diffraction (Borodkina and Osipov, 1973), by gas adsorption, by sound propagation, by thermal conduction, and by measurement of permeability and electrical and magnetic properties (Osipov and Sokolov, 1973). The electron microscopic methods are the only ones that make possible the direct observation of the minerals themselves, their mutual contacts, and the very fine pore system. In all other techniques averaging is involved. This is sometimes an advantage, but visual observation of soil fabric elements should contribute greatly to the proper interpretation of behavior. The size of the sample varies a good deal among the different methods; the amount of sample pretreatment required also varies. In microscopic methods, for example, it is an important consideration (Gillott, 1975).

Early suggestions concerning the fabric of soils (see Vol. XIV: *Soil Fabric*) and sediments were made by Terzaghi (1925) and Casagrande (1940). Somewhat later the state of flocculation of clay mineral platelets was taken into consideration by Lambe (1953, 1958), who proposed an edge-to-edge or edge-to-face arrangement for flocculated clay minerals and a close-packed parallel type of disposition for deflocculated clay minerals. More recently it has become popular to believe that clays are composed of aggregates, sometimes termed *peds,* which act as more or less coherent units and which are joined to neighboring units by bridges of particles (see Vol. IVB: *Clays and Clay Minerals;* Vol. XII, Pt. 1, *Clay Minerals, Silicates*). Other terms, such as block and microblock, have been used by Russian workers (Shibakova, 1965; Bochko, 1973); other schemes of classification have also been discussed (Yong and Sheeran, 1973). On the basis of electron micrographs, Pusch (1973) considers that larger and denser aggregates exist in marine clays than in fresh-water clays. Electron micrographs of sensitive soils in eastern Canada suggest that agglomerations do exist (Fig. 3*A*), but some samples are composed of very open arrangements of platelets, which are relatively extensive on a microscopic scale (Fig. 3*B*).

Hence, fabric analysis may require consideration of the arrangement of the clay minerals within aggregations, the size distribution and disposition of agregations with respect to one another, the nature of the interconnecting particle bridges, and the characteristics of voids between and within aggregations. The term *microstructure* is sometimes used when consideration is also given to the forces and bonds between the fabric elements of the soil.

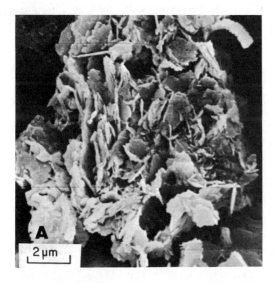

FIGURE 3. Scanning electron micrographs of sensitive soils: *A*. agglomeration and interagglomerate and intraagglomerate pores, Oakville, Ontario; *B*. open fabric — Kars Bridge, Ontario.

Strength and Rheology

The strength and rheology (q.v.) of clay soils are influenced by composition, structure, and fabric (Sergeyev et al., 1973). For kaolinite, illite, and montmorillonite, shear strength is reportedly influenced mainly by physical effects controlled by the size and shape of the individual particles; large equidimensional particles are associated with high strengths while small thin particles lead to low strengths (Olson, 1974). Soil strength is generally calculated in terms of the shear strength by use of an equation known as *Coulomb's law*. Shear strength is expressed as the sum of two terms, one of which is called the cohesion and the other the internal friction. Because the values of these two parameters are probably affected by many variables, the wisdom of the terminology has been questioned. Allowance is commonly made in calculations for the pressure of pore solutions that develops due to shearing. Plasticity and other rheological properties are important in the ceramic formation of clay bodies and in the deformation behavior of soil foundations. The clay minerals make a large contribution to these properties.

Consolidation

The rate and degree of settlement of structures is calculated on the basis of *consolidation theory* (see *Consolidation, Soil*), the principles of which were developed by Terzaghi (1923) and are described in standard texts (Lambe and Whitman, 1969). Due to discrepancies between observation and theory, numerous modifications have been proposed. *Rheological models* have been used, one of the earliest of which consisted of a *Hookean spring* in series with a *Kelvin body* (Taylor and Merchant, 1940). In a later simplification, the top spring was removed (Fig. 4*A*). Elastic properties of the soil are simulated by the spring. Dissipation of the pressure, carried initially by the pore fluids when the load is applied, is represented by the dashpot. The difference between the load and the hydrostatic pressure is termed the *effective pressure*. The rate at which the load is transferred from the pore water to the soil fabric depends on the time required for the water to escape, which itself depends on the permeability. This is sometimes called the *hydrodynamic lag*. Terzaghi regarded *Darcy's law* as valid, assumed constant permeability, and considered pore fluid and minerals to have negligible compressibility. He thus obtained a linear relationship that simplified computation. However, many of these propositions are known to be invalid. In addition to the hydrodynamic lag considered by Terzaghi are other effects sometimes called *plastic lag*. These effects result from particle interactions at and near points of contact and from interactions between particles and pore solutions. Under load the most highly

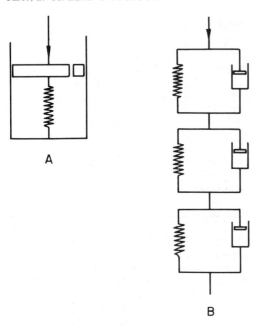

FIGURE. 4. Rheological models of soil behavior.

stressed bonds fail and particles slide relative to one another while the load is transferred to less stressed bonds. Water adsorbed on the surfaces of the clay minerals is thought to be more viscous than normal water, so it may retard the rate of particle sliding. In a natural soil it has been suggested that particles in bridges between peds are rearranged before the fabric of the peds themselves is altered. It is assumed that the peds take up a denser, more closely packed configuration. If hydrodynamic lag is disregarded, soil behavior can be simulated by Kelvin bodies arranged in series (Fig. 4B). Microstructural models and the mathematical theory of computational methods are still in an active state of development (Sandhu and Wilson, 1969; Poskitt, 1971; Mesri and Rokhsar, 1974; Lowe, 1974; Bazant et al., 1975). The behavior of unsaturated soils on consolidation is more complicated, and a general model that considers the swelling-collapse effect, variations in fluid permeability coefficients, and other problems has been discussed by Lloret and Alonso (1980).

Heave

Heave causes damage worldwide to roads, aircraft runways, building foundations, underground service lines, channel and reservoir linings, and retaining walls. Differential movements are particularly destructive. Several methods are used to combat these problems. Footings, piers, or foundations may be placed at depths below those affected by seasonal changes in moisture content, or anchored in nonswelling strata. The structure

may be designed to accommodate movements without failure by making it sufficiently strong or flexible (Lytton and Meyer, 1971; Williams and Donaldson, 1980). The soil may be removed and replaced by more stable backfill, moisture movements may be restricted, or stabilizers (see Vol. XII, Pt. 1: *Soil Conditioners*) may be employed.

Moisture interaction with a soil is considered in terms of the *capillary potential* or soil water potential. Thermodynamic reasoning has led to expressions in which moisture retention and migration are described by the sum of the hydraulic pressure, the osmotic potential, the capillary potential, and temperature.

Heave results from moisture uptake and migration that occurs in response to gradients resulting from physical and chemical processes. Natural causes or construction may lead to an increase in the supply of water or may cause the soil to dry. Clays are *hydrophilic;* when dry, clay soils have a capacity for moisture uptake that varies with the nature of the clay minerals, the stress history, the load, and so on. Rate of moisture movement is affected by the gradient, the soil fabric, the permeability, and other factors. Montmorillonites (smectites) have the greatest potential for expansion; soils that contain appreciable amounts of this group of minerals may exert high swelling pressures on moisture uptake. The response of this type of clay to moisture depends on physicochemical and mechanical factors (Sankaran and Rao, 1974).

Frost action is another physical factor that causes heave. In silts and lean clays, freezing leads to growth of ice lenses accompanied by moisture movements, and considerable forces are developed (Penner, 1970; Jessberger, 1979).

Heave of lightly loaded structures founded on black pyritic shales has been attributed to mineralogical transformation (see Vol. XIV: *Expansive Soils*). Gypsum and jarosite are alteration products formed by biogeochemical oxidation of pyrite and by reactions involving other minerals (Figs. 5 and 6) (Quigley et al., 1973; Gillott et al., 1974).

Fabric affects rate of heave, rate of consolidation, and rate of yield of water or oil from porous beds. Small amounts of clay may migrate and block pore spaces. State of flocculation (q.v. in Vol. XI) and type of clay mineral are factors in this behavior. Recent work on soil fabric has been used to account for permeability anomalies in clays in terms of migrating particles of mineral and organic matter (Hansbo, 1973). Such particle movements clog pores and lead to decreased permeability with time, under conditions of constant direction of flow.

There are numerous reported discrepancies between observed and predicted swelling behavior of clays. These have been variously attributed to shortcomings in experimental technique and

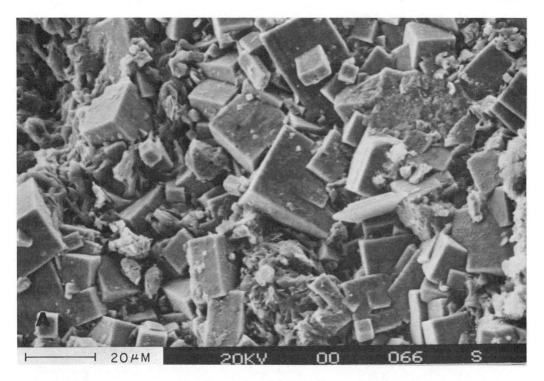

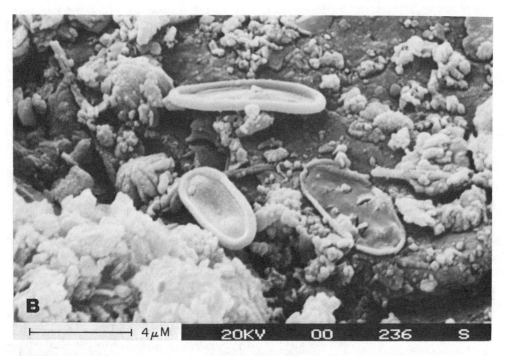

FIGURE 5. Scanning electron micrographs of Billings shale, Ottawa, Ontario: *A.* pyrite; *B.* micro-organisms or spores.

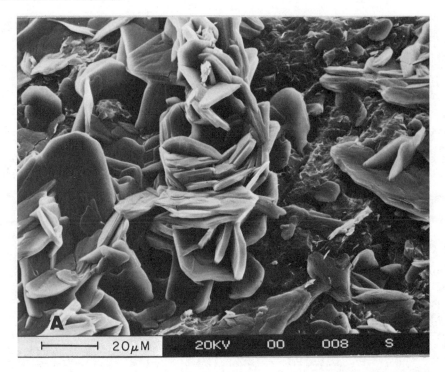

FIGURE 6. Scanning electron micrographs of alteration products on Billings shale, Ottawa, Ontario: *A.* radiating gypsum; *B.* jarosite.

limitations of present theory. Attempts to predict heave potential have been based on index properties, consolidation tests, evaluation of effective stress, and climatic rating systems (Gromko, 1974).

Collapsing Soils

In certain soils moisture uptake under load leads to a decrease in bulk volume rather than normal heave. The effect has been referred to as surface subsidence, hydroconsolidation, and collapse when settlement is rapid. The phenomenon is distinct from consolidation (q.v.) since moisture is being absorbed, not expelled. Soils of this sort are known to exist in many parts of the world. Loose eolian deposits such as loess are the best known examples, but residual and alluvial soils and man-made fills that show collapse have also been described. Such soils characteristically are only partially saturated, have a large void ratio, and are composed of granules of sand, silt, or clay aggregations (peds). Fine particles of silt or clay are thought to form a connecting neck or bridge at the points of contact between grains (Fig. 7A, B, C). Clay may form an "onion-skin" coating on the

primary minerals due to either authigenesis or deposition, or it may, like silt particles, be deposited from suspension by percolating pore solutions or during drying to function as a buttress in the cusps between the larger grains. The open-packed arrangement of bulky grains is maintained by bonds between the clay minerals at the junctions, by surface tension forces in water filling capillary spaces, and by the cementitious action of chemical precipitates (Fig. 7D). Under sufficient load the structure becomes unstable, particularly when capillary forces are diminished by water uptake. The bulk volume decreases because the grains assume a more closely packed configuration. The subject has been discussed in more detail by Dudley (1970) and by Barden et al. (1973).

Soil Stabilization

Soil stabilization (q.v. in Vol. XII) is the term applied when the properties of a soil are modified so that its engineering performance is improved. An alternative approach is to take account of unfavorable soil properties in engineering design (this will not be described here, however, since it is

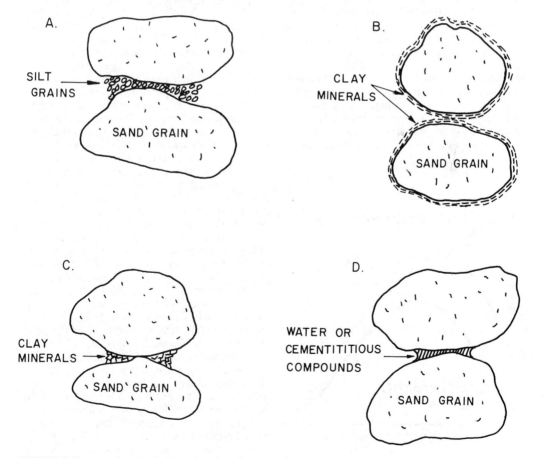

FIGURE 7. Buttressed or bonded junctions in collapsing soils (adapted from Dudley, 1970; Barden et al., 1973).

outside the scope of engineering geology). Soil stabilization is important in geotechnical engineering in the control of soil movements resulting from compressibility, settlement, uplift, creep, and slides. Likewise in the use of soils as materials of construction, the objectives are improved strength, durability, and impermeability. When soils are used as foundations or as materials of construction at the site, there are economic benefits since high costs are incurred by transportation or removal of such heavy, bulky materials.

Soil stabilization has been used in road and aircraft runway construction; in the building of earth dams, canal walls, and low-cost housing; in erosion control; and so on. It may be achieved by mechanical, chemical, and physical means.

Mechanical Stabilization. In general, compaction increases soil strength and decreases compressibility, volume change, permeability, and susceptibility to frost action. It also modifies the resilience of the soil. When there are just sufficient fines to fill the void spaces between the coarse particles, maximum bulk density can be achieved. Grain-size distribution may be found by mechanical analysis (see Vol. XII, Pt. 1: *Particle-Size Analysis*) and the distribution adjusted to give an optimum grading. At a given particle-size distribution, there is an optimum moisture content at which the maximum dry density is achieved for a given compactive effort (Fig. 8) (American Society for Testing and Materials, 1982 *a, b*). In cohesive soils maximum strength is found at a water content slightly below optimum. This is because compression of the soil generates lower pore pressures than arise when soils containing optimum water content are loaded; high pore pressures

cause a marked reduction in strength. The possibilities of heave due to moisture uptake or shrinkage due to drying also require consideration in the case of clay soils. Soils compacted wetter than optimum are believed to induce a parallel orientation of the phyllosilicates (Lambe, 1958); macropores are constricted, and water permeability is low. Soils compacted drier than optimum are believed to be composed of packets of clay particles that aggregate into pellets in the presence of water; macropores and channels are continuous, and in the saturated state water permeability is high (Barden, 1974). Fabric is also believed to be affected by the type of compaction, but nonetheless compaction has been found to have very little effect on the pore size distribution of laboratory-compacted illitic clay (Ahmed et al., 1974). Kneading, impact, vibratory, and static methods have been used (Hall, 1968; Krizek and Fernandez, 1971). The magnitude of the shear component probably also varies with the moisture content relative to the wet or dry sides of optimum.

Chemical Stabilization. Because the most commonly used additives must be mixed with the soil, traveling plant mixers and stationary plants are used (see Vol. XII, Pt. 1: *Soil Conditioners*). Tillers, pugmills, mullers, kneaders, and the like are used to break the soil into small lumps and blend it with the stabilizer. Alternative procedures are under investigation (Suh and Lee, 1974).

When soil is mixed with Portland cement and compacted at an appropriate water content, the mixture forms a hard, durable material known as *soil cement,* which has been used in the construction of roads since 1915 (Highway Research Board, 1961). There are laboratory tests to determine the moisture content at which maximum compacted density of soil-cement mixtures can be achieved (Catton, 1937, 1940; American Society for Testing and Materials, 1982c). Strength is commonly estimated by the unconfined compression test (Symons, 1970) and durability may be assessed by cyclical wetting and drying and freezing and thawing tests (British Standards Institution, 1982; American Society for Testing and Materials, 1982 *d, e*). The properties and mode of failure of soil cement depend on the nature of the soil (Croft, 1967), the type of cement, and the manner of loading (Alhashimi and Chaplin, 1973). In clay soils an increase in Ca^{2+} ion concentration of the pore water promotes flocculation and reduces the liquid limit and plasticity index; the clay minerals tend to aggregate into silt-sized domains. Chemical attack on the clay minerals forms compounds that contribute to the cementitious action of the hydrating Portland cement (Herzog and Mitchell, 1963). Sulfates and organic matter in soils have a detrimental effect on the properties of soil cement.

The use of lime as a stabilizing agent in road construction dates back to the early 1920s (McCaustland, 1925). Presently lime is generally

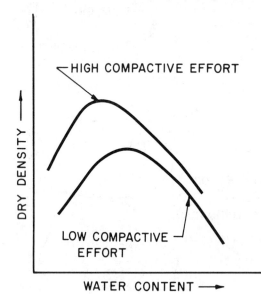

FIGURE 8. Effect of moisture content and compactive effort on density.

applied by mechanical means (Fig. 9), which often involves mixing or slurry injection. Quicklime, hydrated high-calcium lime, and dolomitic limes have all been employed. Lime is generally added in the proportion of about 5 percent of the dry weight of the soil, but laboratory tests are required to establish the optimum proportion. The method is found to be most effective in soils that contain a high content of fines. Addition of lime to a clay soil generally causes an increase in unconfined compressive strength (Fig. 10), a decrease in liquid limit, and an increase in plastic limit so the plasticity index falls to an extent that varies with the clay mineralogy and other factors (see *Lime Stabilization*). Ion exchange, attack on the clay minerals with formation of calcium silicate hydrates, and carbonation of unreacted calcium hydroxide are involved in the mechanism of lime stabilization. Addition of *Pozzolans* together with the lime has been reported to give favorable results. A manual on *Lime Stabilization Construction* is available from the National Lime Association (1982) and the topic has been discussed in *Transportation Research Circular* (1976) and by Ballantine and Rossouw (1972).

Sodium and calcium chloride reduce plasticity and may have other effects, and various other chemicals have been used as stabilizers (Arora and Scott, 1974). Phosphates and other acids and alkalis have been investigated experimentally (Ingles, 1970) but have had limited use in practice.

Physical Stabilization. Methods that depend on control or purposeful variation of the soil moisture content fall into this category. Use is made of inert fillers intended to plug voids, waterproofers, moisture barriers designed to maintain constant water content, drainage systems, prewetting, and electrokinetic techniques. Change of soil fabric by control of flocculation state of the clay minerals depends on surface chemical and physical effects and is borderline between chemical and physical methods insofar as classification.

Bitumens are used and function principally as waterproofers (see Vol. XII, Pt. 1: *Soil Conditioners*). Various chemicals such as acrylamides have been used as fillers in applications such as injection grouting. Similar use has been made of pure bentonite suspensions and more complex bentonite-based suspensions (Boyes, 1972). Surface active agents have been investigated with the object of decreasing the hydrophilic character of the clay minerals and thereby lowering the capacity of the soil for water uptake and giving it greater volumetric stability. Dewatering has been accomplished by electrical methods, and more complex changes due to ionic migrations under the potential gradient have been shown to take place (Titkov et al., 1965; Bjerrum et al., 1967; Gray, 1970; Robinson, 1975).

Frost action causes problems in construction due to the generation of uplift pressures (Penner, 1974). Frost heave can be reduced or prevented through the control of temperature regime, for example, by using thermal insulation beneath roads, by using an additive to modify soil properties (Lambe et al., 1971), by removing of susceptible soil, and by controlling ground-water movements.

FIGURE 9. Bulk pneumatic spreader truck applying 5-percent hydrated lime on Wisconsin subgrade project (photo: National Lime Association)

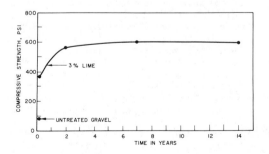

FIGURE 10. Increase in strength of soil on treatment with lime (data supplied by National Lime Association).

Ground Movements and Slides

Soil undergoes a slow downhill movement under gravity. It results from daily and seasonal heating and cooling, freezing and thawing, and wetting and drying. Its rate depends on the steepness of slope, type of soil, stress history, intensity of weathering, vegetation cover, and so on. Structural damage to tunnels and other structures has been attributed to this cause.

Slides are of particular importance in engineering because of their destructive potential. They are sometimes triggered by engineering operations, by seismic shocks or other vibrations, by effects of water such as change in position of the ground-water table, dissolution of intergranular cement, leaching, and the like (see Vol. XIV: *Landslide Control*). A common diagrammatic cross-section of a slide shows a rotational movement on a curved shear plane; a widely used method of analysis is based on this "slip circle." While slides of this sort occur in thick, normally consolidated clays, in reality many slides do not have this cross-section. This is because of local inhomogeneities caused by planes of weakness or layers of soft strata such as clay that has picked up percolating water. Slides are sometimes progressive. Ruptures develop at increasing distances from the source of instability, at places where shear strength is exceeded. An excavation at the toe of a slope in overconsolidated clays or construction of an embankment at the top of an old slide on soft clays is sometimes followed by this kind of progressive failure. The surface along which displacement takes place is often a plane of weakness, such as a bentonite seam, as in a number of slides in the Bearpaw and Pierre shales that underlie the central United States and Canada (Wilson, 1970).

In sensitive soils, landslides have been classified as rotational slips, retrogressive rotational flowslides, or earthflows (Kenney and Drury, 1973; Mitchell and Markell, 1974). (*Sensitivity* is the ratio of the strength of the undisturbed material to its strength after remolding—see Vol. XII, Pt. 1: *Thixotropy, Thixotropism*.) Slides of this sort are common in postglacial marine clays in eastern Canada and Scandinavia. The sensitivity of the Bootlegger Cover clay was a factor in the landslide at Anchorage, Alaska, in 1964. These soils commonly contain a significant proportion of primary minerals in the less-than-2-μm size range, together with the clay minerals illite and chlorite (Gillott, 1971). Apart from the content of primary minerals, sensitivity does not correlate with mineralogy. It is believed that the fabric of these soils is metastable due to changes in geological conditions between the time of deposition and the present. Many studies have been published and a variety of theories advanced to account for their properties; these proposals include salt leaching, action of dispersing agents, rupture of cementitious bonds between particles, weathering, nature of cations and ion exchange, shape of fine quartz particles, thixotropy, and planes of weakness (Penner, 1965; Rosenquist, 1966; Söderblom, 1966; Mitchel and Houston, 1969; Eden and Mitchell, 1970; Gillott, 1970; Pusch, 1970; Moum et al., 1971; Sangrey, 1972; Hammond et al., 1973; Smalley, 1979).

Clays as Materials

Some of the earliest man-made structures were made of dried clay; clays are still often used today to form the impermeable cores of earth and earth rock dams. Most clays are mixtures, but some deposits are composed of essentially one type of clay mineral. Examples are *china clay*, composed mainly of kaolin; bentonites, in which montmorillonite (smectite) is the principal clay mineral; and attapulgite clays. These have important industrial uses as in the production of paper, ceramics, and pharmaceuticals and other chemicals. Newly produced clay-organic complexes are being marketed increasingly. Bentonites are used as bonding and plasticizing agents, in pelletizing and well drilling; in civil engineering bentonites are used in a method of trench excavation, in constructing diaphragm walls, in grouting, and in bentonite-sand backfills (Boyes, 1972).

Clay products can be grouped into four major categories: (1) heavy clay products, which include building bricks, structural tile, lightweight clay block, sewer pipe, and so on; (2) cement; (3) lightweight aggregate; and (4) refractories. The strength of clay products has been investigated intensively by materials scientists, and its origins are better understood than ever before (see *Clays, Strength of*). Durability is known to be adversely affected by moisture expansion, chemical attack, and exposure to the weather. Moisture migration sometimes produces white efflorescence on brick masonry due to leaching soluble salts such as sulfates from the interior (Cole, 1961).

Argillaceous rocks are unfavorable as sources of concrete aggregates, as they break up easily on handling and may cause deterioration of the concrete due to their capacity for moisture imbibition (Dolar-Mantuani and Laakso, 1974). The clay content has been held responsible for poor durability of concrete caused by the alkali-carbonate rock reaction (Gillott and Swenson, 1969). Some varieties of alkali-aggregate reaction may be controlled by the use of Pozzolans, which also sometimes improve other properties of concrete. Some Pozzolans are produced by calcination of clay.

J. E. GILLOTT

References

Ahmed, S.; Lovell, C. W.; and Diamond, S., 1974. Pore sizes and strength of compacted clay, *Am. Soc. Civil Engineers Proc., Jour. Geotech. Eng. Div.*, **100**, 407-425.

Alhashimi, K., and Chaplin, T. K., 1973. An experimental study of deformation and fracture of soil cement, Géotechnique, **23**, 541-550.

American Association of State Highway and Transportation Officials, 1978. Appendix B, Soil Classification, *Manual on Foundation Investigations*, 146p.

American Association of State Highway and Transportation Officials, 1982. Classification of soils and soil aggregate mixtures for highway construction purposes, *Standard Specification for Transportation Materials and Methods of Sampling and Testing, Standard Spec.*, M145, Vol. 1.

American Society for Testing and Materials, 1982a. Standard test methods for moisture-density relations of soils and soil-aggregate mixtures using 5.5-lb (2.49-kg) rammer and 12-in. (305-mm) drop, *Ann. Book ASTM Standards, ASTM D698-78*, 202-208.

American Society for Testing and Materials, 1982*b*. Standard test methods for moisture-density relations of soils and soil-aggregate mixtures using 10-lb (4.54 kg) rammer and 18-in. (457-mm) drop, *Ann. Book ASTM Standards, ASTM D698-78*, 278-284.

American Society for Testing and Materials, 1982*c*. Standard test methods for moisture-density relations of soil-cement mixtures, *Ann. Book ASTM Standards, ASTM D558-57*, 146-151.

American Society for Testing and Materials, 1982*d*. Standard methods for wetting-and-drying tests of compacted soil-cement mixtures, *Ann. Book ASTM Standards ASTM D559-57*, 152-157.

American Society for Testing and Materials, 1982*e*. Standard methods for freezing-and-thawing tests of compacted soil-cement mixtures, *Ann. Book ASTM Standards, ASTM D560-57*, 158-163.

Arora, H. S., and Scott, J. B., 1974. Landslides by ion exchange, *California Geology*, **27**, 99-107.

Ballantine, R. W., and Rossouw, A. J., 1972. *Lime Stabilization of Soils*. Johannesburg, South Africa: Northern Lime Co. Ltd., 109p.

Barden, L., 1974. Consolidation of clays compacted "dry" and "wet" of optimum water content, *Géotechnique*, **24**, 605-625.

Barden, L.; McGown, A.; and Collins, K., 1973. The collapse mechanism in partly saturated soil, *Eng. Geology*, **7**, 49-60.

Bates, R. L. and Jackson, J. A. eds. 1980. *Glossary of Geology*, 2nd ed. Falls Church, Virginia: American Geological Institute, 751p.

Bazant, X. P.; Ozaydin, I. K.; and Krizek, R. J., 1975. Micromechanics model for creep of anisotropic clay, *Am. Soc. Civil Engineers Proc., Jour. Eng. Mechanics Div.*, **101**, 57-78.

Bjerrum, L.; Moum, J.; and Eide, O., 1967. Application of electro-osmosis to a foundation problem in a Norwegian quick clay, *Géotechnique*, **17**, 214-235.

Bochko, R., 1973. Types of microtextural elements and microporosity in clays, *International Symposium on Soil Structure Proceedings*. Gothenburg: Swedish Geotechnical Society, 97-101.

Borodkina, M. M., and Osipov, V. I., 1973. Automatic X-ray analysis of clay microfabrics, *International Symposium on Soil Structure Proceedings*. Gothenburg: Swedish Geotechnical Society, 15-20.

Borst, R. L., 1972. Authigenic kaolinite crystals within microfossils of the Danian Limestone, North Sea, *International Clay Conference, Kaolin Symposium, Madrid*, 41-48.

Boyes, R. G. H., 1972. Uses of bentonite in civil engineering, *Inst. Civil Engineers Proc.*, **52**, 25-37.

British Standards Institution, 1982. *Methods of Test for Stabilized Soils*. London, 96p.

Casagrande, A., 1940. The structure of clay and its importance in foundation engineering, *Contributions to Soil Mechanics, 1925 to 1940*. Boston, Mass.: Boston Society of Civil Engineers, 72-125.

Catton, M. D., 1937. Basic principles of soil-cement mixtures and exploratory laboratory results, *Highway Research Board Proc.* **17**, 7-31.

Catton, M. D., 1940. Research on the physical relations of soil and soil-cement mixtures, *Highway Research Board Proc.*, **20**, 821-855.

Chandler, R. J., 1973. A study of structural discontinuities in stiff clays using a polarizing microscope, *International Symposium on Soil Structure Proceedings*.

Gothenburg: Swedish Geotechnical Society, 78-85.

Cole, W. F., 1961. Terracota roofing tile deterioration in Australia, *British Clay-worker*, **70**, 249-259.

Croft, J. B., 1967. The influence of soil mineralogical composition on cement stabilization, *Géotechnique*, **17**, 119-135.

Dolar-Mantuani, L. L. M., and Laakso, R., 1974. Results of ethylene glycol swelling test on argillaceous limestone, *Canadian Jour. Earth Sci.*, **11** (3), 430-436.

Dudley, J. H., 1970. Review of collapsing soils, *Am. Soc. Civil Engineers Proc., Jour. Soil Mechanics and Found. Div.*, **96**, 925-947.

Dumbleton, M. J., 1968. *The Classification and Description of Soils for Engineering Purposes: A Suggested Revision of the British System*. Great Britain Road Research Laboratory, RRL Rept. LR182, 41p.

Eden, W. J., and Mitchell, R. J., 1970. The mechanics of landslides in Leda clay, *Canadian Geotech. Jour.*, **7**, 285-296.

Fookes, P. G., 1969. Geotechnical mapping of soils and sedimentary rock for engineering purposes with examples of practice from the Mangla dam project, *Géotechnique*, **19**, 52-74.

Gillott, J. E., 1963. Clay mineralogy in building research, *Clays and Clay Minerals*, **11**, 296.

Gillott, J. E., 1968. *Clay in Engineering Geology*. Amsterdam: Elsevier, 296p.

Gillott, J. E., 1969. Study of the fabric of fine-grained sediments with the scanning electron microscope, *Jour. Sed. Petrology*, **39**, 90-105.

Gillott, J. E., 1970. Fabric of Leda clay investigated by optical, electron-optical and X-ray diffraction methods, *Jour. Eng. Geology*, **4**, 133-153.

Gillott, J. E., 1971. Mineralogy of a Leda clay, *Canadian Mineralogist*, **10**, 797-811.

Gillott, J. E., 1976. Importance of specimen preparation in microscopy, *Soil Specimen Preparation STP 599*. American Society for Testing and Materials, 289-307.

Gillott, J. E., and Swenson, E. G., 1969. Mechanism of the alkali-carbonate rock reaction, *Quart. Jour. Eng. Geology*, **2**, 7-23.

Gillott, J. E.; Penner, E.; and Eden, W. J., 1974. Microstructure of billings, Shale and biochemical alteration products, Ottawa, Ontario, *Canadian Geotech. Jour.*, **11**, 484-489.

Gray, D. H., 1970. Electrochemical hardening of clay soils, *Géotechnique*, **20**, 81-93.

Gromko, G. J., 1974. Review of expansive soils, *Am. Soc. Civil Engineers Proc., Jour. Geotech. Eng. Div.* **100**, 667-687.

Hall, J. W., 1968. Soil compaction investigation: evaluation of vibratory rollers on three types of soils, *U.S. Army Corps Engineers Waterways Expt. Sta. (Vicksburg, Miss.)* Tech. Mem. 3-271, Report 10, 32p.

Hammond, C.; Moon, C. F.; and Smalley, I. J., 1973. High voltage electron microscopy of quartz particles from post-glacial clay soils, *Jour. Materials Sci.*, **8**, 509-513.

Hansbo, S., 1973. Influence of mobile particles in soft clay on permeability, *International Symposium on Soil Structure Proceedings*. Gothenburg: Swedish Geotechnical Society: 132-135.

Herzog, A., and Mitchell, J. K., 1963. Reactions accompanying stabilization of clay with cement, *Highway Research Board Proc.*, **36**, 146-171.

Highway Research Board, 1961. Soil stabilization with Portland cement, *Highway Research Board Bull.*, **292**.

Howell, J. V. (ed.), 1966. *Glossary of Geology and Related Sciences,* Washington, D.C.: American Geological Institute, 72p.

Ingles, O. G., 1970. Mechanisms of clay stabilization with inorganic acids and alkalis, *Australian Jour. Soil Research,* **8,** 81-95.

Jessberger, H. L. (ed.), 1979. *Ground Freezing— Developments in Geotechnical Engineering,* vol. 26. Amsterdam: Elsevier, 558p.

Kenney, T. C., and Drury, P., 1973. Case record of the slope failure that initiated the retrogressive quick-clay landslide at Ullensaker, Norway, *Géotechnique,* **23,** 33-47.

Krizek, R. J., and Fernandez, J. I., 1971. Vibratory densification of damp clayey sands, *Am. Soc. Civil Engineers Proc., Jour. Soil Mechanics and Found. Div.,* **97,** 1069-1079.

Lafeber, D., 1965. The graphical representation of planar pore patterns in soils, *Australian Jour. Soil Research,* **3,** 143-164.

Lafeber, D., 1967. The optical determination of the spatial orientation of platy clay minerals in soil thin sections, *Geoderma,* **1,** 359-369.

Lambe, T. W., 1953. The structure of inorganic soil, *Am. Soc. Civil Engineers Proc., Jour. Soil Mechanics and Found. Div.,* **79,** 1-49.

Lambe, T. W., 1958. The structure of compacted clay, *Am. Soc. Civil Engineers Proc., Jour. Soil Mechanics and Found. Div.,* **84,** 1654-1-1654-34.

Lambe, T. W. and Whitman, R. V., 1969. *Soil Mechanics.* New York: Wiley, 553p.

Lambe, T. W.; Kaplan, C. W.; Lambie, T. J., 1971. Additives for modifying the frost susceptibility of soils, *U.S. Army Corps. Engineers Tech. Rep. 123,* Parts 1 and 2, 80p.

Lloret, A., and Alonso, E. E., 1980. Consolidation of unsaturated soils including swelling and collapse behavior, *Géotechnique,* **30,** 449-477.

Lowe, J., 1974. New concepts in consolidation and settlement analysis, *Am. Soc. Civil Engineers Proc., Jour. Geotech. Eng. Div.,* **100,** 574-612.

Lytton, R. L., and Meyer, K. T., 1971. Stiffened mats on expansive clay, *Am. Soc. Civil Engineers Proc., Jour. Soil Mech. Found. Div.,* **97,** 999-1019.

McCaustland, D. E. J., 1925. Lime dirt in roads, *Natl. Lime Assoc. Proc.,* **7,** 12-17.

McGown, A., 1973. The nature of the matrix in glacial ablation tills, *International Symposium on Soil Structure Proceedings.* Gothenburg: Swedish Geotechnical Society, 87-95.

Mesri, G., and Rokhsar, A., 1974. Theory of consolidation for clays, *Am. Soc. Civil Engineers Proc., Jour. Geotech. Eng. Div.,* **100,** 889-904.

Millot, G. M., 1970. Geology of Clays. New York: Springer-Verlag, 429p.

Mitchell, J. K., and Houston, W. N., 1969. Causes of clay sensitivity, *Am. Soc. Civil Engineers Proc., Jour. Soil Mechanics and Found. Div.,* **95** (SM3), 845-871.

Mitchell, R. J., and Markell, A. R., 1974. Flowsliding in sensitive soils, *Canadian Geotech. Jour.* **11,** 11-31.

Morgenstern, N. R., and Eigenbrod, K. D., 1974. Classification of argillaceous soils and rocks, *Am. Soc. Civil Engineers Proc., Jour. Geotech. Eng. Div.,* **100,** 1137-1156.

Moum, J.; Løken, T.; and Torrance, J. K., 1971. A geochemical investigation of the sensitivity of a normally consolidated clay from Drammen, Norway, *Géotechnique,* **21,** (4), 329-340.

National Lime Association, 1982. *Lime Stabilization Construction Manual. Bull. 326.* Arlington, Va., 48p.

Olson, R. E., 1974. Shearing strengths of kaolinite, illite, and montmorillonite, *Am. Soc. Civil Engineers Proc., Jour. Geotech. Eng. Div.,* **100,** 1215-1229.

Osipov, J. B., and Sokolov, B. A., 1973. On the texture of clay soils of different genesis investigated by magnetic anisotropy method, *International Symposium on Soil Structure Proceedings.* Gothenburg: Swedish Geotechnical Society. 21-28.

Penner, E., 1965. A study of sensitivity of Leda clay, *Canadian Jour. Earth Sci.,* **2,** 425-441.

Penner, E., 1970. Frost-heaving forces in Leda clay, *Canadian Geotech. Jour.,* **7,** 8-16.

Penner, E., 1974. Uplift forces on foundations in frost-heaving soils, *Canadian Geotech. Jour.* **11** (3), 323-328.

Poskitt, T. J., 1971. Consolidation of clay and peat with variable properties, *Am. Soc. Civil Engineers Proc., Jour. Soil Mechanics and Found. Div.,* **97,** 841-879.

Pusch, R., 1970. Microstructural changes in soft quick clay at failure, *Canadian Geotech. Jour.,* **7,** 1-7.

Pusch, R., 1973. Structural variations in boulder clay, *International Symposium on Soil Structure Proceedings.* Gothenburg: Swedish Geotechnical Society, 113-121.

Quigley, R. M., 1980. Geology, mineralogy, and geochemistry of Canadian soft soils: a geotechnical perspective, *Canadian Geotech. Jour.* **17,** 261-285.

Quigley, R. M.; Zajic, J. E.; McKyes, E.; and Yong, R. N., 1973. Biochemical alteration and heave of black shale; detailed observations and interpretations, *Canadian Jour. Earth Sci.,* **10,** 1005-1015.

Robinson, D. E., 1975. Successful electroshock therapy for deteriorated bridges, *Transportation Research News,* **58,** 3-4.

Rosenquict, I. T., 1966. Norwegian research into the properties of Quick clay—a review, *Eng. Geology,* **1,** 445-450.

Sandhu, R. S., and Wilson, E. L., 1969. Finite element analysis of land subsidence, *Land Subsidence, I.A.S.H.—Unesco, Publ. No.* **89,** 2, 393-400.

Sangrey, D. A., 1972. Naturally cemented sensitive soils, *Géotechnique,* **22,** 139-152.

Sankaran, K. S., and Rao, D. V., 1974. Mechanistic response of expansive clays, *Soil Sci.,* **118,** 289-298.

Sergeyev, Y. M.; Budin, D. Y.; Osipov, V. I.; and Shibakova, V. S., 1973. The importance of the fabric of clays in estimating their engineering-geological properties, *International Symposium on Soil Structure Proceedings.* Gothenburg: Swedish Geotechnical Society, 243-251.

Shepard, F. P., 1954. Nomenclature based on sand-silt-clay ratios, *Jour. Sed. Petrology,* **24,** 151-158.

Shibakova, V. S., 1965. *Textural Changes in Argillaceous Rocks by Hydrostatic Compression.* Moscow State University Geology Series, No. 2.

Skempton, A. W.; Schuster, R. L.; and Petley, D. J., 1969. Joints and fissures in the London clay at Wraysbury and Edgeware, *Géotechnique,* **19,** 205-217.

Smalley, I. J. (ed.), 1979. Sensitive soils and quick clays,

Eng. Geology, **14,** 81-217.

Smart, P., 1966. Particle arrangements in kaolin, *Clays and Clay Minerals, Natl. Conference, Pittsburgh, Proc.,* **15,** 241-254.

Söderblom, R., 1966. Chemical aspects of quick-clay formation, *Eng. Geology,* **1,** 415-431.

Suh, N. P., and Lee, R. S., 1974. Centrifugal method of mixing soil with stabilizer, *Am. Soc. Civil Engineers Proc., Jour. Geotech. Eng. Div.,* **100,** 295-307.

Symons, I. F., 1970. The effect of size and shape of specimen upon the unconfined compressive strength of cement stabilized materials, *Mag. of Concrete Research,* **22,** 45-50.

Taylor, D. W., and Merchant, W., 1940. A theory of clay consolidation accounting for secondary compressions, *Jour. Math. and Phys.,* **19.**

Terzaghi, K., 1925, *Erdbaumechanik auf bodenphysikalischer Grundlage.* Vienna: Deuticke, 399p.

Terzaghi, K., 1927. Soil classification for foundation purposes, *Internat. Congr. Soil Sci. Trans.* **4,** 127-157.

Titkov, N. I.; Petrov, V. P.; and Neretina, A. Y., 1965. *Mineral Formation and Structure in the Electrochemical Induration of Weak Rocks.* New York: Consultants Bureau, 74p.

Tovey, N. K., 1973, Quantitative analysis of electron micrographs of soil structure, *International Symposium on Soil Structure Proceedings.* Gothenburg: Swedish Geotechnical Society, 50-57.

Transportation Research Circular, 1976. State of the art: lime stabilization, reactions, properties, design, construction, *Transportation Res. Circ.,* **180,** 31p.

U.S. Waterways Experiment Station, 1953. The unified soil classification system, *U.S. Army Corps Engineers Waterways Expt. Sta. (Vicksburg, Miss.) Tech. Mem. 3-357,* 50p.

Wagner, A. A., 1957. The use of the unified soil classification system by the Bureau of Reclamation, *Proceedings of the Fourth International Conference of the Soil Mechanics Foundation of Engineers,* 125-134.

Williams, A. A. B., and Donaldson, G. W., 1980. Developments relating to building on expansive soils in South Africa: 1973-1980, *Fourth Internat. Conf. on Expansive Soils, Denver.*

Wilson, M. D., and Pittman, E. D., 1977. Authigenic clays in sandstones: recognition and influence on reservoir properties and paleoenvironmental analysis, *Jour. Sed. Petrology,* **47,** 3-31.

Wilson, S. D., 1970. Observational data on ground movements related to slope instability, *Am. Soc. Civil Engineers Proc., Jour. Soil Mechanics and Found. Div.,* 1521-1544.

Yong, R. N., and Sheeran, D. E., 1973. Fabric unit interaction and soil behavior, *International Symposium on Soil Structure Proceedings.* Gothenburg: Swedish Geotechnical Society, 176-183.

Cross-references: *Clays, Strength of; Consolidation, Soil; Foundation Engineering; Geotechnical Engineering; Hydrodynamics, Porous Media; Lime Stabilization; Marine Sediments, Geotechnical Properties; Rheology, Soil and Rock.* Vol. XII, Pt. 1: *Clay Minerals, Silicates; Soil Conditioners.* Vol. XIV: *Cat Clays; Dispersive Clays; Engineering Soil Science; Expansive Soils; Landslide Control; Soil Fabric.*

CLAY MINERALOGY—See CLAY, ENGINEERING GEOLOGY. Vol. IVB: SOIL MINERALOGY. Vol. XII, Pt. 1: CLAY MINERALOGY; CLAY MINERALS, SILICATES. Vol. XIV: DISPERSIVE CLAYS; EXPANSIVE SOILS.

CLAYS—See Vol. XIV: CAT CLAYS; DISPERSIVE CLAYS; EXPANSIVE SOILS.

CLAYS, STRENGTH OF

The English word *clay* refers to natural, inorganic materials consisting of fine-grained minerals and water. In dictionaries we may find the word *clay* translated to German *Ton,* French *argile,* Russian *glina,* Scandinavian *lere,* and so on. This suggests that clays of different countries are identical. Clay, however, does not represent any definite chemical or mineralogical compound and may vary considerably even within a limited area (see Vol. XII, Pt. 1: *Clay Minerals, Silicates*). Different clays may have very different mechanical properties.

Until x-ray studies, combined with various other methods of micromineralogy, made it clear how the mineral phases varied from clay to clay, there was little accurate idea of their nature. The shear strength properties of all soils have, however, some features in common. These had been recognized by Coulomb in 1776, when he introduced the shear strength equation

$$s = c + \sigma \tan \phi$$

which states that the total shear resistance of a soil(s) can be considered as the sum of cohesive resistance (c) and the frictional term ($\sigma \tan \phi$). The normal concept until the end of the 1920's was that clays were characterized by cohesion and normally little or no friction, whereas sands, silt, and gravel had friction and no cohesion. In 1925, Terzaghi established the fundamental basis for soil mechanics (q.v.). Using Coulomb's equation, he introduced a concept of effective stresses, showing that the total stresses of Coulomb's equation gave misleading results. Terzaghi's equation may be given as follows:

$$s = c + (\sigma - \overline{u}) \tan \phi$$

where $\overline{u}$ is the pore water pressure, that is, the hydrostatic pressure that corresponds to the pressure of the voids in the soil. In natural soil deposits this pressure will be higher or lower than what can be evaluated from the position of the sample relative to the ground-water level. Excess pore water pressure is normally called *artesian pressure*.

The next major step forward was taken by M. J. Hvorslev in 1937. He confirmed in principle Coulomb's law, but as cohesion was found to be a function of water content, he extended Coulomb's equation:

$$s = \bar{c} + \bar{\sigma} \tan \bar{\phi}$$

where c depends on the water content. This dependence can be expressed by an introduction of the equivalent consolidation pressure σ_e:

$$s = H\sigma_e + \bar{\sigma} \tan \bar{\phi}$$

The significance of the result of Hvorslev's investigations is very profound. The two main results—dependence of the cohesion on water content alone and recognition of the angle of internal friction as a soil characteristic that determines the strength of even rather fat clays—were most important steps in the advancement of knowledge of the fundamental strength properties of soils.

Hvorslev's investigation was carried out on remolded soils, and his results are valid only for such material. This has often been forgotten by many research workers dealing with soil properties. As seen from an electron micrograph of a normally consolidated clay, these consist of aggregates of platy minerals touching each other corner-to-plane in more or less dense cardhouse structures. Any deformation of such a body will necessarily involve a mutual movement of the particles relative to each other and a corresponding flow of the water in interstitial voids. As chemical and electrostatic forces will act between the individual minerals at the point of contact, the deformation process must necessarily be complex; indeed it may be stated that *the strength of clay material is not fully understood.*

Present Concept

Terzaghi's (1925, 1943) introduction of the concept of effective stresses is still fully valid. If we consider a soil sample, we conclude from a general mathematical principle on invariance that any scientifically correct failure criterion must include all three effective principle stresses $-\bar{\sigma}_1$, $\bar{\sigma}_2$, and $\bar{\sigma}_3$. Several attempts have been made to solve the general failure problem, but because of lack of experimental data and insufficient theoretical understanding, we are unable to treat the complete problem, even for isotropic soils, especially as these turn anisotropic as a function of deformation. Using original anisotropic bodies with changing anisotropy, a complete solution of the frictional term has been tried, but in vain. That cohesion is a function of the stress history and not

only of the actual set of stresses adds to the complexity of the problem.

In practice, however, it has been proved that $\bar{\sigma}_1$ and $\bar{\sigma}_3$ are of the greatest importance for the strength, and the normal practice is to work with failure criteria where only these two principal stresses are used. We choose to put $\sigma_2 = \sigma_3$. This is done by all tests of the triaxial compression type. Thus we ignore the influence of the intermediate principal stress. As the void ratio or water content of a saturated soil is a function of the initial water content and the consolidation pressure σ_e, it is possible to prepare clay samples of constant water content at various stress conditions by starting from initial materials of the same minerals, but of different water contents. Under such conditions, the envelope of several *Mohr circles* for a given soil of constant porosity may be given, each of the Mohr circles representing a set of values for σ_1 and σ_3 at conditions of failure. In practice we assume that the envelope curve is a straight line representing the *real conditions of failure* or *the Hvorslev conditions of failure*. The *Hvorslev line* is characterized by the c_r, which is the intercept value of the line and the ordinate, further by the angle ϕ_r, which is the angle to the abscissa; c_r is called true or real cohesion, and ϕ is the true or real angle of friction.

According to the Hvorslev concept of failure, a clay body exposed to a stress ellipsoid of rotational symmetry will have a plane of failure with the angle $45° + \frac{1}{2}\phi^r$ to the σ_3 direction. As already said, the Hvorslev true or real conditions of failure are based on effective stresses in a given soil material having constant pore ratio under variable external conditions. If, on the other hand, the pore ratio varies as a function of external stresses, we may determine a set of Mohr circles for the effective stresses at failure. In this way, the envelope curve will determine the effective failure conditions, and we obtain the values $\bar{c}$ and $\bar{\phi}$, called the *effective cohesion* and the *effective angle of friction*.

To determine c_r, ϕ_r, $\bar{c}$, and $\bar{\phi}$, respectively, we have to know the pore water pressure at failure. In certain cases it is possible to analyze the problem of failure approximately only by total stresses. This gives us Coulomb's conditions of failure c and ϕ, called the *apparent cohesion* and the *apparent angle of friction*.

Several hypotheses have been advanced as to the nature of cohesion and friction in soils. The fact that clays are colloidal materials indicates that electrochemical forces are involved, at least in the cohesional term, probably also in the frictional term. As all the failure equations from Coulomb to Hvorslev deal only with the bulk properties, the macroscopic stress ellipsoid, and the pore water pressure, it is obvious that none of these equations can predict the attractive and repulsive forces acting between the charred mineral particles, nor do they evaluate the cementing action of precipitated mineral substances at the points and areas of contact between the minerals. This stresses the importance of the title of Hvorslev's paper. His equation was given for completely remolded material where any cementing action could be ignored. In all natural geological deposits, we have to assume that secondary cementation may act.

Among other authors, Denisov and Reltov (1961) have demonstrated the influence of natural cementing bonds. Such a cementation may be included as a nonreversible cohesion term, whereas some of the electrochemical forces may be included in the stress conditions as a set of internal stresses adding to the set of external effective stresses. The last-mentioned term, especially, will depend on chemical changes in the pore water and mineralogical changes of the minerals. Thus, for example, the subpolar *quick clays* have obtained their peculiar properties from secondary chemical changes in flocculated or slightly overconsolidated clays (see *Consolidation, Soil;* Vol. XII, Pt. 1: *Flocculation*), these chemical changes being of a nature that increased the *zeta potential* of the minerals.

The simplest chemical change is represented by the leaching (see Vol. XII, Pt. 1: *Leaching*) of salt from marine deposits due to the seepage of rain water. Another reaction that will cause a similar effect on the zeta potential is represented by the action of organic substance from decaying humus. In both cases, soils of high sensitivity may develop. By *sensitivity* we understand a permanent loss of strength by remolding (see *Soil Mechanics*), observed in many clays. If a clay sample of a given natural strength is thoroughly remolded and the strength immediately measured, we mostly observe a considerable drop in the shear strength value. By storage, the strength gradually increases, but in many soils, it will never arrive at the initial value unless the water content drops, either by desiccation or by syneresis. Such a stored clay will again by a subsequent remolding arrive at the same low shear strength as the original sample (Fig. 1). Thus there is reversible loss of strength in remolded samples (thixotropy) and nonreversible loss of strength in natural soils (see Vol. XII, Pt. 1: *Thixotropy, Thixotropism*). This loss of strength is called *sensitivity;* the ratio between the initial strength and the remolded value is called *sensitivity value*. In typical quick clays, the sensitivity may have values of more than 100, whereas the thixotropy value may be only 2 to 5. Thus these clays will lose 99 percent of their strength by remolding, and only 2 to 5 percent of the initial strength will be regained by storage.

These actions seem to be restricted to clays of nonswelling minerals (illites, chlorites, kaolinite). In soils of swelling minerals such as montmoril-

FIGURE 1. Illustrating the loss of strength by remolding in a typical quick clay. The clay cylinder to the left is able to carry 11 kg. After remolding, it turns liquid and remains liquid (courtesy of Carl Crawford).

lonites, the chemical changes that cause the high sensitivity of quick clays may cause a decrease in the sensitivity. This is probably due to the volume change of the mineral phase by swelling (see Vol. XIV: *Expansive Soils*).

Ever since the early days of modern soil mechanics (q.v.), much emphasis has been given to the problem of the water in clay soils. The typical adsorption phenomena (q.v. in Vol. XII, Pt. 1) and hygroscopicity of clay minerals made it reasonable to assume that the aqueous phase of clays had properties fundamentally different from those of free water. Only recently, modern techniques, using diffusion of isotope tracers combined with nuclear magnetic resonance and infrared radiation, have shown that the greater part of the aqueous phase of clay soils has properties that are not fundamentally different from those of water. However, through a transition zone of more or less gradually increasing rigidity, this water of the voids becomes solidly bound water molecules on the surface of the clay minerals. Thermodynamically, and probably also mechanically, these innermost one to three water layers seem to have properties like those of a solid body. As any failure in a clay involved a shearing in this innermost water, it is necessary to have a better knowledge of the innermost water in order to understand the physical basis for any rheological model of clay.

Johnson (1970) developed a rheological model describing the flow behavior of remolded clay material at a finite rate of deformation. This model is called a *Coulomb-viscous model*. It has the following form:

$$T = c + \sigma_\eta \tan \phi + \eta \dot{\epsilon}, \qquad T > c + \sigma_\eta \tan \phi$$

where T is internal shear stress, σ_η is internal normal stress, ϕ is the angle of internal friction, η is viscosity, and $\dot{\epsilon}$ is the rate of shear strain (velocity gradient). In this case, it is well to be aware that the cohesional term is not identical to the cohesion at rest.

I. T. ROSENQVIST

References

Coulomb, C. A., 1776. Essai sur une application des règles de maximis et minimis à quelques problemes de statique, relatifs à l'architecture, *Mem. Div. Sav. Academie des Sciences.*

Denisov, N. Y., and Reltov, B. F., 1961. The influence of certain processes on the strength of soils, *Proc. 5th Internat. Conf. Soil Mechanics Foundation Eng.,* Paris, 75-78.

Hvorslev, M. J., 1937. Über die Festigkeitseigenschaften gestörter bindiger Böden, *Ingeniorvidenskab. Skrifter A,* No. 45.

Johnson, A. M., 1970. *Physical Processes in Geology.* San Francisco: Freeman, Cooper & Co., 577p.

Terzaghi, K., 1925. *Erdbaumechanik auf bodenphysikalischer Grundlage.* Leipzig: Frantz Deuticke.

Terzaghi, K., 1943. *Theoretical Soil Mechanics.* New York: Wiley, 510p.

Cross-references: *Clay, Engineering Geology; Consolidation, Soil; Marine Sediments, Geotechnical Properties; Soil Mechanics.* Vol. XII, Pt. 1: *Clays, Physical Chemistry.* Vol. XIV: *Expansive Soils.*

COASTAL ENGINEERING

Coastal engineering deals with the conservation, development, and exploitation of the coastline and coastal resources. It is a branch of *civil engineering,* which requires a specialized knowledge of the oceanographic environment and hydrodynamics (q.v.). It is also considered as a branch of oceanographical engineering applied to the "triple point", that is, where the air, sea, and land interact (Wiegel, 1964; Horikawa, 1978).

In a broad sense, coastal engineering extends to investigations, design, and civil construction from the deep tidal estuaries to the limit of the continental shelf (≈ 200 m). A limited classification of coastal engineering problems is presented in Figure 1.

The scope of coastal engineering includes the design and construction of structures and civil works (Fig. 2) for (1) shoreline protection against erosion by waves and currents; (2) ports, harbor, and marina development and control of navigation channels in estuaries; (3) waste disposal and control of shoreline water quality, (4) offshore installation for under-sea drilling and mining operations; and (5) any coastal problems related to nearshore installations, such as thermal, nuclear, and desalination plants, tanker terminals, extensions of airport runways, installations for mariculture, (Silvester, 1974).

Coastal engineering is based on a great number of basic scientific disciplines, the most important being hydrodynamics (q.v.), structural dynamics, soil mechanics (q.v.), material science, and branches of geophysics, including physical oceanography, coastal morphology, geology, and sedimentology. Finally, like all engineering projects, coastal installations are constrained by economic and sociological requirements that are the determining factors of any development. Furthermore, aesthetic and architectural development of the coastline and the ecological balance of the nearshore environment are of increasing importance to coastal engineers because of growing public interest in such matters (Steers, 1971; Bascom, 1980). They may have to compromise among political forces to determine whether the area should be conserved for recreational use of be devoted to industrial development. *Coastal Zone Management* (Vol. XIV) is an attempt at rationalizing the development of the coastline, and at establishing a large master plan by means of compromise among conflicting interests.

Since most coastal engineering developments occur over decades, if not centuries, depending on local conditions, a rational master plan of coastal development is very difficult to achieve. Instead, coastal engineering consists mainly of solving problems as they occur, rather than working toward an established goal for the optimum exploitation of the marine resources for conservation, recreational, commercial, and industrial uses.

Coastal Oceanographic Environment

The *shoreline* is sometimes called the "triple point" as it characterizes the line where land, sea, and air meet. The coastal environment is characterized by the interaction of these three elements, as exemplified by wind-generated waves breaking on beaches, causing sand transportation (littoral drift) and giving rise to shoreline evolution.

For any coastal engineering installation, the following environmental information is always required: (1) bottom topography or bathymetry and local hydrography; (2) sediment transport (which can actually be determined from wave statistics by use of a *littoral drift formula*), and observations of shoreline evolution, erosion, and

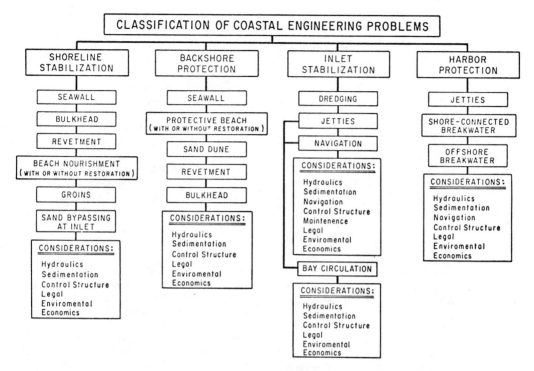

FIGURE 1. Schematic diagram illustrating the scope of coastal engineering (courtesy of the U.S. Army Corps of Engineers).

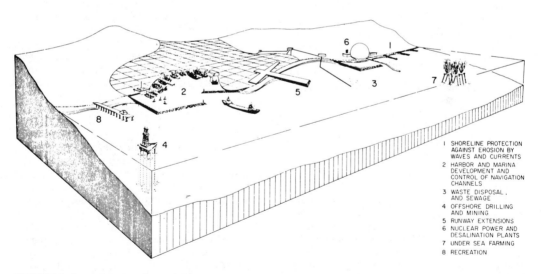

1 SHORELINE PROTECTION AGAINST EROSION BY WAVES AND CURRENTS
2 HARBOR AND MARINA DEVELOPMENT AND CONTROL OF NAVIGATION CHANNELS
3 WASTE DISPOSAL, AND SEWAGE
4 OFFSHORE DRILLING AND MINING
5 RUNWAY EXTENSIONS
6 NUCLEAR POWER AND DESALINATION PLANTS
7 UNDER SEA FARMING
8 RECREATION

FIGURE 2. Coastal engineering activities.

soil characteristics; (3) the ecological population pattern; (4) tidal amplitude and currents (usually obtained from tidal tables); and (5) wave statistics. The sea-state probability distribution (wave height, period, and direction) is of paramount importance for determining wave design criteria during and after construction of any structure at sea. This can be estimated either by long-term analysis of wave records and ship observations or by "wave-hindcasting" methods. The latter consists of calculating the wave characteristics and their evolution from the wind velocity field (weather maps giving the wind field at 6-hour intervals). The sea state (Kinsman, 1965) is conveniently characterized by an average or significant wave height and wave period, or by directional energy spectrum, giving the root mean square energy density as a function of frequency and direction.

Wave spectrum evolve over the continental shelf as a result of the combined effects of local wind, energy dissipation by bottom friction, white caps, wave refraction, and "wave shoaling" due to the change of water depth over complex bottom topography. The wave pattern tends to be influenced by the bottom as soon as the water depth becomes shallower than one-half the wave length. In the case of a storm wave, that is, at the limit of the continental shelf. The wave phase velocity being a decreasing function of depth, individual wave crests tend to become parallel to the shore, and their heights initially decrease then ultimately increase up to the inception of breaking. Breaking occurs when the wave height is of the same order of magnitude as the water depth.

The momentum of waves breaking at an angle to the shore causes a *longshore current,* which, added to the high rate of turbulence due to the breakers, is the main cause of littoral transport of sand and shoreline evolution. Long waves (swell) generated by a distant storm tend to push the sand forward as a result of the transport of mass toward the shore near the bottom (Le Méhauté, 1968). Locally generated waves (sea) tend to erode beaches as a result of a transport of mass toward the shore near the free surface, causing a return flow near the bottom. During storms, masses of water pushed toward the coast by atmospheric pressure and wind stress (storm surge) may also cause extensive flooding in shallow coastal areas, requiring dike protection (Coastal Engineering Research Center Staff, 1977). In some areas, coastal construction modify sediment erosion and deposition patterns in areas such as estuaries, navigation channels, and harbor entrances. Such changes must be predicted by engineers before construction of coastal works (see Vol. XIV: *Harbor Surveys*).

Wave Forces

One of the determining factors in the design of coastal structures is the selection of a design wave as a basis for assessment of maximum probable forces compatible with an acceptable economical risk. The cost of offshore structures is highly affected by the selection of the design structures. Wave climatological information, particularly concerning extreme events that the structures must stand, are generally insufficient, making this choice the most difficult task in offshore engineering.

In the case of a floating structure (see Vol. XIV: *Floating Structures*), one of the principal problems is how to estimate maximum mooring forces. A moored ship at quay can be considered a (nonlinear) spring-mass system excited by wave action. Resonance effects are possible under a long-period wave oscillation of the harbor basin (seiche),

causing the breaking of mooring lines; that is, mooring force may exceed 200 tons at quays. At sea, the problem has been solved for cargo liquid bulk carriers by the single point mooring system, which is fixed to the bow of the ship while loading or unloading. Open sea berths are also constructed in relatively calm areas.

Wave forces on fixed structures are estimated with various degrees of accuracy, depending on the relative complexity of the structure. The case of wave reflection (standing wave) from a vertical wall is well known analytically. For example, the amplitude pressure variation at the base of a vertical wall is given to a good approximation by the formula

$$P = \overline{\omega} \left(z + \frac{H \cos \sigma t}{\cos h\, 2\pi d/L} \right)$$

where $\overline{\omega}$ is the specific weight of sea water 1030 kg/m^3 (64 lb/ft^3); z is the elevation of the SWL (still-water level); H is the incident wave height; d is the water depth; L is the wave length, $\sigma = 2\pi/T$; T is the wave period; and t is the time. Due to the nonlinearity of the free-surface boundary condition and convective inertia, more complex formulas are recommended as being more exact.

Wave forces due to breaking waves are best taken from laboratory or field measurement, because there are no reliable theories for flow velocities at breaking. The calculation of wave forces on a cylindrical pile is similarly empirical. From a theoretical viewpoint, the problem is three-dimensional, nonlinear, and viscous (alternate wake), that is, beyond the state of the art of deterministic approach. Engineers then use the empirical equation

$$F = \rho C_D D \frac{u^2}{2} + \rho C_M \frac{\pi D^2}{4} \frac{\partial u}{\partial t}$$

where F is the force per unit pile length, D the pile diameter, u the horizontal particle velocity of the undisturbed velocity field due to incident waves at the location of the pile, and C_D and C_M coefficients that are functions of Reynolds number (uD/ν) and $D^2/T\nu$ ($C_d \simeq 1$, and $C_M \approx 2$), where ν is the kinematic viscosity and T the wave period. C_D is the drag coefficient due to wake development and C_M is the inertia coefficient due to the local inertia force induced by the presence of the pile in a time-dependent (alternate) flow. In fact, the pile modifies the velocity field so that C_D and C_M are not constant but complex time-dependent functions that still remain to be determined. Large safety coefficients are recommended due to these uncertainties. The dynamic responses of elastic steel structures subjected to a wave spectrum are analyzed, based both on these basic principles and on the probabilistic distribution of wave characteristics (Carneiro and Brebbia, 1978).

Coastal Engineering Structures

Coastal engineering structures are characteristically rugged and functional with few architectural niceties. Rocks, sand fill, concrete, timbers, and steel are the raw materials used for breakwaters, dikes, levees, seawalls, bulkheads, piers, groins, jetties, quays, wharves, floating platforms, Texas towers, and so on.

Rubble-mound Breakwaters. Due to the variety of characteristics of the coastal environment, standard solutions do not exist. *Rubble-mound breakwaters* are made of a core of quarry run covered by layers of larger stone for stability (Coastal Engineering Research Center Staff, 1977). The equilibrium profile of a rubble mound under wave action during phases of construction is shown in Figure 3. The slope α is a function of wave height, size, and density of the upper layers as given by the nomograph (Fig. 4). The following formula, devel-

oped by the U.S. Army Corps of Engineers Waterways Experiment Station, is commonly used:

$$W = \frac{\gamma_r H^3}{K(S_r - 1)^3 \cot \alpha}$$

where W is the rock weight in tons, γ_r the specific gravity of the stone (≈ 150 lb/ft^3), H is the wave height in feet, S_r is the ratio of specific gravity of the stone to water (≈ 2.6), and K is a coefficient that varies from 2.0 for rounded stones piling up "pell-mell" and subjected to breaking shallow-water waves to 8.0 for sharp, keyed, and fitted stones subjected to nonbreaking deep-water waves.

When, as often happens, no stones of sufficient weight are available in nearby quarries, artificial concrete blocks are used to achieve a more economical and a more stable breakwater (Fig. 5). Typical of these blocks are *tetrapods* (Fig. 6) thrown pell-mell with a three-quarters slope. Such a structure is very porous, absorbs energy well, limits wave runup and overtopping, and is structurally sound (even without steel reinforcement). It is also stable to wave action, and has been very widely used, particularly in Japan. *Tribars*, developed in the United States, and *akmon blocks*, developed in Holland, are other typical shapes (Fig. 6). All three types of concrete blocks are more stable against shifting than the natural stones of the same weight. The *dolos* shape, developed more recently, appears to be the most stable of all, and tends to be the most commonly recommended for future construction of small, shallow breakwa-

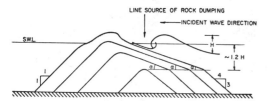

FIGURE 3. Natural equilibrium profiles of rubble mound under wave action.

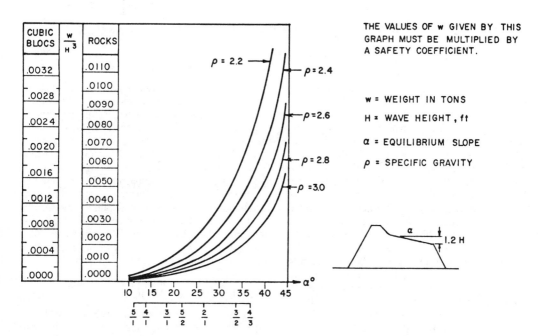

FIGURE 4. Equilibrium slope of rubble mound and cubic blocks versus weight and density.

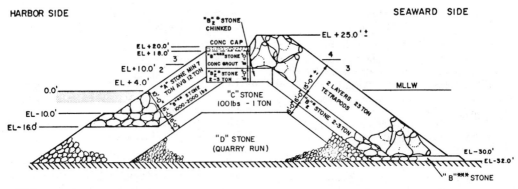

FIGURE 5. Cross-section of rockfill breakwater with tetrapods (Crescent City, California).

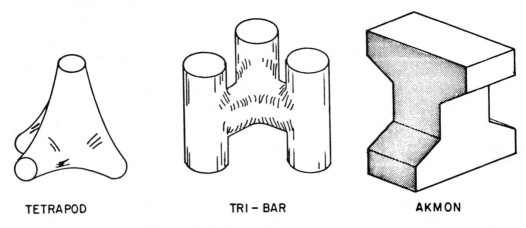

TETRAPOD   TRI – BAR   AKMON

FIGURE 6. Artificial concrete blocks for breakwater protection.

ter (Fig. 7). Dolosse are fragile and not recommended for deep breakwaters.

Rubble-mound breakwaters are easily maintained, because failures due to storms can be repaired by simply dumping more stones or artificial blocks.

Pre-cast Vertical Breakwaters. Although conceptually very strong, pre-cast vertical breakwaters (Fig. 8) have been largely abandoned because of their higher initial cost, and also because very high peak pressures are generated by breaking waves striking the vertical face, such that undermining and pressure-tilting may cause total failure.

Crib Breakwaters. *Crib breakwaters* comprise a vertical crib of timber, sheet-pile, or concrete that has been filled by large stones. They are particularly economical in shallow water ($\simeq 20$ ft).

Groins. Timbers, rocks pell-mell, or crib are also used for the construction of *groins*, structures built perpendicularly to the shoreline, extending from the berm of the beach to the beginning of the surf zone. Their main purpose is to stop or limit the littoral drift (sand transport), to trap the sand and to increase the beach width, or to limit erosion. Groins also tend to divert longshore currents seaward, creating rip currents; therefore, they may actually contribute to beach depletion. T-shaped groins or constructions parallel to the coast are now preferred. Plastic bags filled with sand offer good protection against erosion in case of emergency. The use of groins in one section of a beach, however, may result in depletion and severe erosion of beaches on the downdrift side. Shorelines are otherwise protected by seawalls, bulkheads (Fig. 9), and revetments of various types, including vegetation.

Coastal Engineering Methods

Coastal engineering is an art based on past local experience (past mistakes), and an increasing use of theoretical and experimental tools including computers. On one hand, theorists, and in particu-

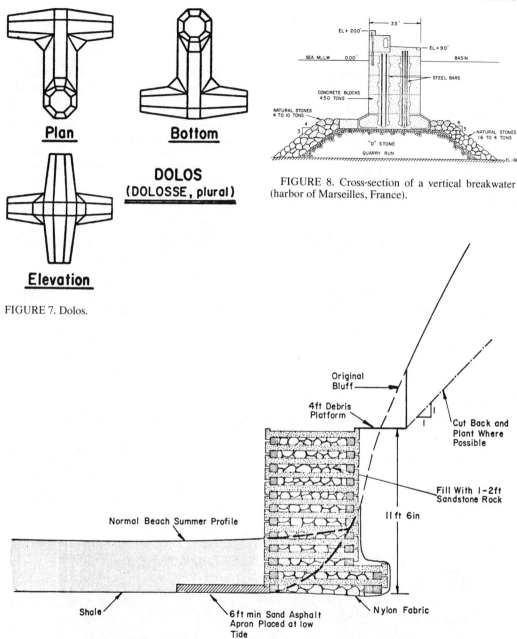

Plan

Bottom

DOLOS
(DOLOSSE, plural)

Elevation

FIGURE 7. Dolos.

FIGURE 8. Cross-section of a vertical breakwater (harbor of Marseilles, France).

FIGURE 9. Cross-section of concrete crib protection against bluff erosion.

lar, water-wave hydrodynamicists, have developed advanced sophisticated theories in which the mathematical intricacies are not necessarily rewarded by a better fit with experimental facts (Le Méhauté, 1976). On the other hand, practical engineers have often ignored the tremendous resources of the theoretical tool for better design, safety, and economy. Empirical methods have prevailed in the past.

Numerical techniques and computers now permit theoretical solutions to complex cases that are of interest to design engineers. For example, long-wave agitation in a harbor of arbitrary shape can now be investigated economically by computer with engineering accuracy. But in many cases, it is impossible to predict the behavior of coastal structures by analytical means owing to the complexity of the phenomena or the boundary conditions.

The gap between the theorist and the design engineer has been largely overcome by the widely accepted use of scale-model technology. It is common to investigate coastal engineering problems by scale model in a hydraulic laboratory before finalizing the design of coastal structures. The following are typical instances when the use of scale models is particularly useful.

1. The study of breakwater stability and determination of height to avoid overtopping, including wave energy dissipation and stability during various construction phases. These studies can be performed relatively inexpensively with the aid of a two-dimensional wave tank at a scale of 1/20 to 1/50 (Fig. 10).
2. Ship dynamics, ship resistance, and mooring forces at quays. These factors can be advantageously determined by scale-model studies (scales of 1/40 to 1/100). Towing tanks are expensive installations that require sophisticated monitoring systems (this activity actually borders on naval hydrodynamics).
3. Harbor agitation. The determination of orientation and length of a harbor entrance, location of wave absorbers, and design of the best general layout for ensuring calm water are aided by 1/100 to 1/150 scale models.
4. Shoreline evolution and harbor silting. This type of model requires a long and delicate adjustment, during which it is attempted to reproduce an observed field-bottom evolution under wave action and currents. These movable-bed scale models are widely used in Europe, but less frequently used in the United States, for these models are generally distorted. For example, the vertical scale is larger than the horizontal scale.

The scale model is just a tool which, like the theoretical tool, needs to be interpreted intelligently from a sound knowledge of similitude law and "scale effects." Some practical problems, such as the determination of wave force on cylindrical piles, cannot realistically be studied by scale model unless the inertia force (approximately in similitude) overcomes the drag force due to wake (subject to scale effects). This is the case of caissons and of piles of very large diameter.

Because coastal engineering is still largely empirical, past experience remains the most valuable guide to dealing with the difficult marine environment. However, the need for further research is certainly apparent, in view of the high initial cost of civil engineering structures at sea, and considering the amount of money spent every year in dredging harbors as a result of inadequate breakwater layout, the frequent flooding due to storm surge, and generally speaking, the enormous potential future of oceanographic engineering projects that will arise in the quest for human survival and improvement of living conditions.

BERNARD LE MÉHAUTÉ

References

Bascom, W. B., 1980. *Waves and Beaches.* New York: Anchor/Doubleday, 366p.

Carneiro, F., and C. Brebbia, 1978. *Offshore Structures Engineering.* Gulf Publishing Company.

Coastal Engineering Research Center Staff, 1977. *Shore Protection Manual,* 3 vols. Washington, D.C.: U.S. Government Printing Office.

Horikawa, K., 1978. *Coastal Engineering.* Tokyo: University of Tokyo Press.

Kinsman, B., 1965. *Wind Waves.* Englewood Cliffs, N.J.: Prentice-Hall, 676p.

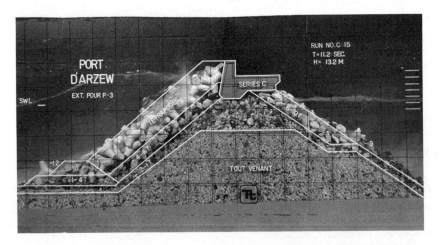

FIGURE 10. Two-dimensional scale-model breakwater stability study (courtesy of Tetra Tech, Inc., Pasadena, California).

Le Méhauté, B., 1968. Littoral processes quantitative treatment, in R. W. Fairbridge (ed.), *Encyclopedia of Geomorphology.* New York: Reinhold Book Corporation, 667-672.

Le Méhauté, B., 1976. *An Introduction to Hydrodynamics and Water Waves.* New York: Springer-Verlag, 320p.

Silvester, R., 1974. *Coastal Engineering,* 2 vols. New York: Elsevier, 795p.

Steers, J. A., 1971. *Applied Coastal Geomorphology.* Cambridge, Mass.: MIT Press, 227p.

Wiegel, R. L., 1964. *Oceanographical Engineering.* Englewood Cliffs, N.J.: Prentice-Hall, 532p.

Cross-references: *Beach Replenishment, Artificial; Coastal Inlets, Engineering Geology; Oceanography, Applied.* Vol. I: *Oceanography, Nearshore; Sea State; Tides; Tsunamis.* Vol. XIV: *Floating Structures; Harbor Surveys; Nuclear Plant Protection, Offshore; Ocean, Oceanographic Engineering.*

COASTAL INLETS, ENGINEERING GEOLOGY

Formation of Natural Inlets

Inlets result from a variety of processes that include the action of rivers, storms, longshore drift, and other geological phenomenan. Rivers entering the ocean form natural inlets, although the flow regimes differ from tidal inlets. Basic characteristics of both types may be similar (Schmeltz and Sorensen, 1973).

Perhaps the most predominant agent of inlet formation along barrier island coasts is storm activity. Pierce (1970) noted that such inlets are most likely to form where the island is narrow and the adjacent lagoonal depths are relatively deep. Inlets created by storms tend to be ephemeral, however, and they normally close relatively rapidly if the breach of the island is at a relatively wide point. Brown Cedar Cut on the Texas Gulf coast is an example of such a *storm-created inlet* (Mason and Sorensen, 1971).

Lateral deposition and growth of a bar across the mouth of a bay may also form an inlet. Longshore currents may transport sedimentary material and thus create a laterally accreting bar. Material is transported along shore until it reaches deeper, quieter water on the bay side, where it is eventually deposited. This process of lateral deposition and growth across the bay continues until the resulting inlet reaches a state of dynamic equilibrium with the physical and hydrologic parameters that act on it.

Glacial erosion during the Pleistocene formed extensive valleys. When the sea level rose in the Holocene, many natural inlets were formed as rising water inundated the lower portions of the valleys. The *fiords* of Scandanavia and Alaska are inlets of this type. Because fiords typically have rocky gorges, Bruun and Gerritsen (1960) note

that normal laws for inlets in alluvial material are not applicable to this type.

Inlet Physiography and Characteristics

Figure 1 shows the general physiographic features of a typical *alluvial inlet.* The offshore bar, a typical feature along many sandy coasts, is bowed on the seaward side of the inlet as a result of the jetting action of the ebb tide. The gorge occurs at a point within the inlet where current velocity is at a maximum, and thus the deepest portion. A tidal delta, or bay shoal area, results from material deposited by flood tides. Where the ebb tide is a dominant tidal force, such as along the Pacific coast of the United States, extensive offshore shoals may form at the expense of bay shoals.

The predominant source of material for the shoal area, in most cases, is sediment supplied by littoral drift. At flood tide, some of the material supplied to the mouth of the inlet is drawn in and discharged into the bay. Due to several factors, the amount of material supplied to the bay will not equal that carried back out on the ebb tide, even if the two tidal currents are of the same magnitude.

Figure 2 shows the relationship between current velocity and sediment grain size as a function of deposition, transportation, and erosion. As can be seen, the velocity of flow required to erode, and thus make sediment available for transport, is higher than that required for transport. The difference between the minimum velocity required

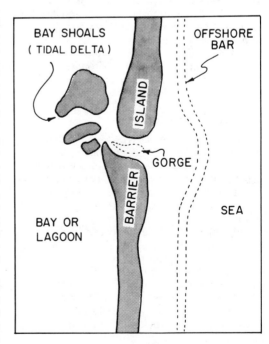

FIGURE 1. Diagrammatic sketch of the typical physiographic features of a tidal inlet.

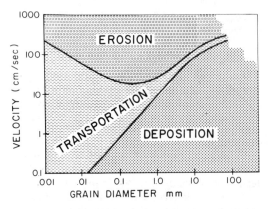

FIGURE 2. Relationship between grain size and flow velocity as it affects sediment deposition, transport, and erosion (after Hjülstrom, 1935).

to erode and that to transport increases with decreasing grain size because the finer material, once deposited, is more cohesive and presents a smoother hydrodynamic surface to the flow.

As material is jetted out into the relatively quiet waters of the bay, deposition occurs. The average grain size decreases as the energy regime of the flood current decreases away from the inlet channel. The distribution of current energy regime of the ebb tide is generally the same as that of the flood tide, with flow velocities decreasing away from the inlet channel. Because the velocity required to lift finer materials into suspension is higher than that required to transport it, some material will be left behind; that is, chances of transport decrease with decreases in energy regime and grain size. The other major contribution to this process is related to the wave energy environment, which also puts material into suspension, but it is of lesser intensity on the bay side.

Littoral Drift and Inlet Stability

Littoral drift, the process by which sedimentary material is moved along a coastline (see vol. VI: *Littoral Sedimentation*), occurs when waves intercept the shoreline at an angle. Agitation of bottom sediments by incoming waves helps transport material along the beachface and within the surf zone. A longshore current is capable of transporting vast quantities of sediment because the net movement of material is parallel to the coastline in the direction of wave propagation.

Coastal inlets trap littoral material by drawing it into the inlet during flood tides. Although much of the material is captured by the inlet and stored in shoal areas, some of it bypasses the inlet by way of the offshore bar, particularly if the bar is in relatively shallow water.

Not only does littoral drift supply material for shoaling; it is also responsible for *inlet migration*.

In areas of high net littoral drift rates, lateral deposition along the updrift side of the inlet mouth forces the maximum velocity against the opposite side, causing erosion. This causes the inlet to migrate, which in turn results in an overall lengthening of the channel. As the channel becomes more elongated, increased frictional resistance causes a loss in the competence of the tidal currents to erode and transport material. This results in shoaling and eventual closure of the inlet.

Inlet Stabilization

Due to the need for safe, navigable entrances to the sea, inlets are often stabilized by *jetties*. These structures are designed to halt migration, to restrict the movement of littoral material into the inlet, to maintain a safe channel depth free from shoaling, and to provide protection from storms. The most important factor considered in jetty system design is the channel configuration created by the structures. If the channel is too wide, low velocity will result in shoaling; if it is too narrow, erosive velocities may undermine jetty foundations. Jetty construction is treated in detail by the U.S. Army Corps of Engineers (1966).

The need to create access to the sea and/or to improve bay or lagoonal circulation in some areas has led to the creation of artificial inlets. As with jettied natural inlets, proper channel configuration is an important factor. Figure 3 relates flow velocity to hydraulic radius with respect to inlet stability. If the value of the hydraulic radius lies on the curve between A and B, siltation and closure of the inlet occurs. If it lies between B and C, erosion will ensue until stabilization is reached (point C). Between C and D, the inlet will continue to silt until it reaches the stable point at C. Failure to take this relationship into account has led to excessive inlet erosion or to inlet failure due to inadequate hydraulic flushing of sediments—and thus to expensive coastal property damage.

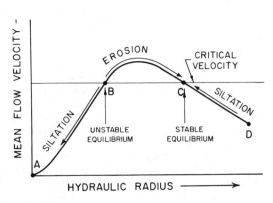

FIGURE 3. Relationship between flow velocity and hydraulic radius as it affects the stability of natural and man-made inlets (after Escoffier, 1940).

Rollover Fish Pass on the Texas Gulf coast, an artificially created inlet completed in 1955, experienced severe erosion problems within its channel and along adjacent Gulf shores (Prather and Sorensen, 1972). Original plans for the project visualized a channel depth of 2.4 m (8 ft) and a width of 24 m (80 ft). As the project neared completion, however, the resulting channel eroded to a depth of 9.1 m (30 ft) and the gulfward mouth of the inlet eroded to a width of 152 m (500 ft). The original hydraulic radius of this inlet evidently fell between unstable equilibrium and stable equilibrium (points *B* and *C* on Fig. 3).

In 1965, an attempt was made to reopen Brown Cedar Cut on the Texas Gulf coast after it had been closed the previous year by hurricane activity (Mason and Sorenson, 1971). A long, narrow cut was excavated connecting the bay to the gulf. The resulting inlet remained open for about one week before it silted up completely. This attempt apparently resulted in a hydraulic radius of the channel that placed it between points *A* and *B* on Figure 3.

Artificial Bypassing of Littoral Material

Jetties, while serving to stabilize an inlet, often create problems, especially in areas with high rates of littoral drift. Material that once was carried downshore by the dominant longshore current becomes impounded by the updrift jetty wall. Since the material can no longer bypass the inlet, undernourishment of the downdrift shoreline occurs and beach erosion ensues (Fig. 4). The only apparent solution to this problem is a method of artificially bypassing the littoral material to the downdrift side of the inlet.

Because physical conditions along a coastal zone can differ greatly, artificial bypass systems must be individually tailored to the particular inlet involved. Watts (1962) lists some important factors to be considered in designing a bypass system: coastal geomorphology; subsurface conditions; water-level variations; winds and waves; allowable wave runup and overtopping; shoreline and offshore depth changes; direction, rate, character, and quantity of littoral drift to be bypassed; effects of nearby inlets or other structures; effects of prior constructive works; zone of collection and placement of littoral drift; availability of materials for construction and maintenance; effective time of operation of the bypass system; and comparative costs.

In an attempt to improve circulation within South Lake Worth, Florida, a jettied inlet was constructed in 1927. Because of the interruption to littoral drift (301,300 m³/yr, or approximately 230,000 yd³/yr), serious erosion occured on the downdrift side of the inlet. In 1937 a fixed-bypass plant was constructed on the updrift jetty which

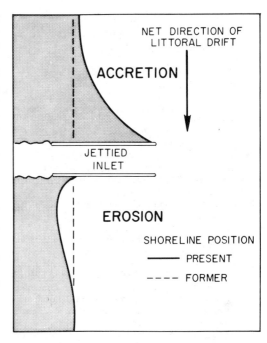

FIGURE 4. Diagrammatic sketch of the effect of artificial stabilization of an inlet on beach erosion and deposition.

transfers between 98,250 and 131,000 m³ (between 75,000 and 100,000 yd³) of littoral material per year to the downdrift shore. Although the amount transferred is less than the net drift rate, it has apparently been sufficient for maintaining the downdrift beach.

A successful bypass system involving two inlets has been employed in Ventura County, California. Jettying of the entrance to Port Hueneme Harbor in 1938 caused the diversion of an estimated 1,048,000 to 2,096,000 m³ (800,000 to 1,600,000 yd³) of littoral material per year into the Hueneme Submarine Canyon (Watts, 1962). Soon after construction, shoreline erosion approximately equal to the net drift rate occurred downdrift of the port. In 1961, jetties and an offshore breakwater were constructed at the entrance to the Channel Islands Harbor (formerly Ventura County Harbor) approximately 1.6 km (1 mi) updrift of Port Hueneme. The breakwater was designed to protect the inlet by creating a trap for littoral drift. Material deposited at the breakwater is removed by dredging and is transferred by submerged discharge lines to the area downdrift of Port Hueneme. In addition, material dredged from within the Channel Islands inlet is added to that transferred from the breakwater. The system has been successful in halting the shoreline recession downdrift of Port Hueneme.

ROBERT M. MCHAM
CHRISTOPHER C. MATHEWSON

References

Bruun, P., and Gerritsen, F., 1960. *Stability of Coastal Inlets.* Amsterdam: North-Holland Publishing Company, 123p.

Escoffier, F. F., 1940. The stability of tidal inlets, *Shore and Beach,* **8,** 28.

Hjülstrom, F., 1935. The morphological activity of rivers as illustrated by River Fyris, *Bull. Geol. Institute Uppsala,* **25,** 221.

King, C. A. M., 1972. *Beaches and Coasts.* London: Edward Arnold Ltd., 570p.

Mason, C., and Sorensen, R. M., 1971. *Properties and Stability of a Texas Barrier Beach Inlet,* College Station: Texas A&M University, Sea Grant Publication No. TAMU-SG-71-217, 166p.

Pierce, J. W., 1970. Tidal inlets and washover fans, *Jour. Geology,* **78,** 230-234.

Prather, S. H., and Sorensen, R. M., 1972. *A Field Investigation of Rollover Fish Pass, Bolivar Peninsula, Texas,* College Station: Texas A&M University, Sea Grant Publication No. TAMU-SG-72-202, 116p.

Price, W. A., 1951. Reduction of maintenance by property orientation of ship channels through tidal inlets, *Proceedings, Second Conference on Coastal Engineering, Council on Wave Research.* Berkeley: University of California, 243-255.

Schmeltz, E. J., and Sorensen, R. M., 1973. *A Review of the Characteristics, Behavior and Design Requirements of Texas Gulf Coast Tidal Inlets,* College Station: Texas A&M University, Sea Grant Publication No. TAMU-SG-73-202, 88p.

Silvester, R., 1974. *Coastal Engineering I and II.* New York: Elsevier, 795p.

U.S. Army Corps of Engineers, 1966. *Shore Protection Manual,* 3 vols. *U.S. Army Corps Engineers Coastal Eng. Research Center Tech. Report No. 4,* 401p.

Watts, G. M., 1962. Mechanical by-passing of littoral drift at inlets, *Am. Soc. Civil Engineers Proc., Jour. Waterways and Harbors Div.* **88,** 83-99.

Cross-references: *Beach Replenishment, Artificial; Coastal Engineering.* Vol. VI: *Tidal-Inlet and Tidal-Delta Facies.* Vol. XIV: *Coastal Zone Management; Ocean, Oceanographic Engineering.* Vol. XV: *Inlets and Inlet Migration; Jetties; Protection of Coasts.*

COLD REGIONS, ENGINEERING GEOLOGY—See GEOCRYOLOGY; PERMAFROST, ENGINEERING GEOLOGY.

COMPUTERIZED RESOURCES INFORMATION BANK

A burgeoning activity in the geological sciences is the collection and storage of descriptive and numeric information in the form of computerized data files.[1] These computerized files are commonly

[1]The terms *Data* and *information* are synonyms in most of their senses and are used interchangeably here.

called *storage and retrieval files* because they store information and provide for quick access to that information by means of a simple retrieval language. The formalized organization, collection, and storage of information in a computer file has now become a separate subspecialty for a growing group of geologists. This trend toward computerized data files has developed because: (1) computers and do-it-yourself file-management program packages have become generally available and (2) we have learned that the computer can help organize information.

Many computer files now exist—although most are incomplete as yet—and many more are under development. One of the earliest files dealing with mineral resources information is CRIB (*Computerized Resources Information Bank*), which is discussed here, together with some preliminary remarks on manual versus computerized filing methods and the nature of mineral resources information.

Manual Files Versus Computer Files

The purpose of generating any file, manual or computer, is to organize information. Organization is required because order is a precondition to successful activity; the human brain cannot cope with randomized data but is highly effective with organized data. The standard goal in organizing information is to reduce the given subject to a set of key attributes that characterize the subject. When a subject is incompletely known, as in geology and mineral resources, manual files are advantageous because they can be loosely organized yet interpreted, and they can carry a wealth of information, expressed in free text, without the requirement of rigorous organization. Some aspects of the subject of geology and mineral resources are, however, capable of being organized into computer files, which then can be manipulated. The present status in this subject, therefore, is a mixture of manual files and computer files, which supplement each other. Computer files provide a quick look, mass storage, and the ability to manipulate a predefined set of data items. The manual file provides detail not available in a computer file. Most computer information files (including CRIB) in geology and mineral resources, however, are still in the buildup stage—they are technically sound and functional but not yet complete enough to provide their maximum utility.

The advantages of computerized methods stem from the fact that a given subject file of information can be disassembled into its component parts (fields and subfields), that operations can be performed on those parts (read, sort, lookup, match, compare, move, arithmetic, etc.), and that the file or part of it can then be reassembled into any other arrangement desired.

Manual files provide for *limited lookup* only, as they are based on those two or three main categories (fields) within the records that have been sorted (alphabetized) in advance. In some situations, limited lookup is all that is required, in which case the manual file (e.g., the telephone book) is highly effective. Once built, a manual file is frozen; a new arrangement is not possible, except via a new compilation. Only the two or three presorted fields can be manipulated. All other fields in the record are nonaddressable because the information contained in these fields is not in a predictable order.

On the other hand, the computer file must be rigorously and logically organized because the computer is unable to read between the lines. A high degree of consistency and standardization is required. These and other restraints are built-in disadvantages associated with the computer file, and they limit our freedom of expression.

The manual file can be loosely organized because the human brain can resolve inconsistencies and ambiguities. A high degree of formalization is not required. Therefore, the manual file allows for a high degree of freedom of expression.

Table 1 summarizes the different characteristics of manual files and computer files. In general, the manual file functions at its optimum when it contains few basic records (is a small file), has few fields (data items) per record, and/or is designed for a single purpose, mainly spot lookup. The computer file, on the other hand, functions optimally when it contains many basic records (is a large file), has many fields (data items) per record, and/ or involved manipulations are required.

When the intended use of a large and complex file of information involves frequent comparisons, correlations, computations, or other operations across a large number of fields or across a large number of records, computer filing methods are potentially superior to manual filing methods.

Mineral Resources Data

The subject of geology and mineral resources is broad in scope, the nomenclature varied and ambiguous, and the general subject matter loosely organized and awkward to deal with from the standpoint of a computer operation. The breadth of the subject results in a bewildering array of data that

involve geology, mining engineering, economics, environment, geopolitics, and other related subjects. Individuals working with mineral resources information have differing use patterns and approaches. They have varying needs, requirements, and preferences, and consequently they have differing views on what information should be included in the description of a mineral deposit and how this information should be presented.

The diversity of mineral resources data is exemplified by individual mineral commodities, of which there are about 100 common ones. These commodities can be combined into various groups to suit a given purpose. A few of the common groups are listed in Table 2.

These groups are only a selected sampling of an endless variety of combinations and permutations. In view of the wide variety of the basic raw materials and wide range of uses of these materials, it is understandable why data on minerals resources are difficult to organize and use effectively.

The CRIB File

CRIB (Computerized Resources Information Bank) is designed for the storage of information on metallic and nonmetallic mines and mineral occurences found in the United States and, to a certain extent, in other countries. It is an *attribute,* or *properties,* file (in contrast to a *bibliographic* file) (see *Earth Science, Information and Sources*), and as such the basic record contains descriptive and numerical information on a set of attributes (variables) that characterize a particular mine or mineral occurrence.

Developed by the U.S. Geological Survey to augment its manual filing methods, the CRIB system became operational in 1972 (Calkins et al., 1973). The long-term process of data collection and file buildup has been underway since that time. As of May 1981, the CRIB file contained 51,408 records in the master file. Concurrent with file buildup, many changes have been made and continue to be made to the file structure, and several enhancements have been added to the system, such as the implementation of interactive access and interactive map-plotting capabilities (see Vol. XIV: *Cartography, Automated*).

The CRIB system operates on the Survey's IBM 370-155 computer at the National Center in Reston, Virginia and it provides both batch and time-share remote terminal access to government and other "official users" of the system. Various types of cooperative arrangements are or have been in effect with the Tennessee Valley Authority (TVA), the Forest Service, the Bureau of Land Management (BLM), the State Department, the U.S. Bureau of Mines, and with 12 different states. These "participating users" contribute data to the CRIB file and in return can access the file themselves via

TABLE 1. Characteristics of manual and computer files.

	Manual File	Computer File
Information content	Maximum	Minimum
Ability to manipulate	Minimum	Maximum
Freedom of expression	Maximum	Minimum
Addressibility	Minimum	Maximum

TABLE 2. Diversity of mineral resource data as shown by a selection of individual mineral commodities.

Applied Mineral Groups	Commodity
Natural mineral groups	Metals/nonmetals
	Ferrous metals
	Clays
Chemical groups	Sulfides
	Oxides
Genetically related groups	Copper/molybdenum
	Lead/zinc/silver
	Cobalt/nickel
End-use groups	Mineral pigments
	Fertilizer minerals
	Abrasives
	Refractories
Resources groups	Reserves/potential resources
	Identified/undiscovered resources
	Recoverable/marginal resources
Strategic minerals	Chromium
	Manganese
Mineral fuels	Petroleum
	Oil shale
Geological deposit types (natural mineral concentrations)	Placer
	Vein
	Massive sulfide

remote terminal, or they can request data retrievals from the CRIB staff. The CRIB staff also provides special services to "participating users" and to Survey geologists, including training, user assistance, and specialized products such as map plots.

Coverage is still spotty and incomplete, however, even in terms of the larger mines of the world. Under the present system of data collection by voluntary contributions, it is estimated that complete coverage will not be available until late 1980s.

In 1976 the CRIB file was made available to the general public through the University of Oklahoma and the General Electric Company's international computer services network (Calkins et al., 1978). Telecommunications via satellites, under-sea cables, and land lines provide service by local telephone to 500 cities in the world.

In 1978 CRIB became part of the Mineral Data System (MDS), which was created as a means for accommodating additional separate computer files on mineral and energy resources. The MDS provides for the storage of any number of resources-related computer files—all under a common system and common addressing protocol. The MDS is a single-file (as opposed to a multifile) access system. In other words, the user has access to only one file at a time. The MDS currently contains CRIB (information on mines and occurrences), GEOTHERM (information on geothermal energy and resources), NATLAS (information on mining districts), and SARI (information on annual mineral production, consumption, exports, imports and reserves for all commodities produced or imported by South Africa).

The CRIB file uses the *GIPSY program* (General Information Processing System) to process the data through the computer. GIPSY (Oklahoma University, 1975) is a *file-management system* that performs all tasks associated with a storage and retrieval operation. The CRIB file and the GIPSY program are operating on the Survey's IBM 370-155 computer at the National Center in Reston, Virgina.

File Structure and Content. CRIB consists of a set of variable-length records on metallic and nonmetallic mines and mineral occurrences of the United States and other countries. The information is in the form of descriptive text, numerical data, mixed descriptive text and numerical data, codes, and certain keywords or phrases. The record accommodates about 400 individual data items under the following main topics: Name and Location, Commodity Information, Selected Analytical Data and Mineral Economics Factors, Description of Deposit, Geology and Mineralogy, Production, Reserves, Resources, and References. The file contains no secret or confidential information. All information in the file is "open" and publicly available.

As of May 1981, the CRIB file contained 51,408 records in the master file. Other blocks of records are in several "holding" areas; these are being moved into the master file batch by batch as editing and updating are completed.

Most records refer to a specific location (mine,

prospect, or mineral occurrence) and to the mineral commodity or commodities associated with that location. A few records are summary records — that is, they summarize information on a given country or state or on a given mineral commodity or mineral district in some way, for example, zinc production country by country. A few records refer to some aspect of a mineral commodity, for example, the origin of bauxite.

A general indication of the content of the CRIB file by area is given in Figure 1, and by mineral commodity in Figure 2.

Reporting Form. The reporting form (also called "input form" or "source document") serves as a means for organizing the raw data in a logical manner suitable for computer processing. The forms are completed by geologists or technicians, who compile the information for the most part on a voluntary basis. The CRIB reporting form (Fig. 3 on page 79) contains the main topics accommodated and most of the individual data items available within each main topic. The data items are arranged into logical elements, called *fields*. Information supplied on these forms is "keyed" to a magnetic disk at the University of Oklahoma and entered into the computer file. Most fields on the reporting form contain a field name, a label, a space for information, and a set of delimiters (⟨ ⟩) that mark the beginning and end of the field. The label identifies the field to the GIPSY program, whereas the field name identifies the field to the reporter. Thus, the label *A10* is equivalent to the field name *Deposit Name.* Fields containing descriptive text are not formatted and may be of any length, to a maximum of 32,000 characters per record. Other fields on the form are rigidly formatted into fixed-length fields, or are partially formatted. These are primarily fields that contain numbers, coded information, or keywords. A few fields consist of a label only. No data are entered into these fields; the label itself entered into the record is equivalent to the data.

Quality and Consistency. CRIB records vary greatly in the amount, quality, and consistency the information they contain — due to like variations in the original data; the absence of a single, centralized data source; and the voluntary nature of the contributions. Contributors (reporters) to the file differ in their emphasis, preferences, and approaches to the subject, depending on their interests, needs, and specialties. It follows, therefore, that the resulting records reflect these differences.

The very complexity of the subject of geology is itself a prime reason for computerizing it: to reduce the subject to manageable and addressable components so that data on the subject can be better utilized by interested parties. An example of a detailed record — of the Foskor mine in South Africa — is shown in Figure 4 (on page 85). The arrangement is in the *standard output record format,* which prints the entire content of the record in a predefined arrangement.

GIPSY (General Information Processing System). The usefulness of a computer file depends not only on its content and logical organization, but also on the computer program used to run it. Three general-purpose *file-management program systems* were operating on the U.S. Geological Survey's computer system in 1970 when the CRIB file was being considered. One of these — the General Information Processing System (GIPSY) — was chosen as the processing system for CRIB because it accommodates variable-length records, has a powerful retrieval mechanism, and was already operational. The variable-length record format provides great flexibility relating to file design and subsequent changes to a file, because the fields are independent of start position, position in sequence, and declared field length.

GIPSY is a file-management system that performs the computer processing tasks needed to build, operate, and maintain a storage and retrieval file. It operates on IBM 360-370 computers and disk-storage devices. It was designed to handle variable-length records and to provide the non-computer specialist with a means for making highly selective retrievals. Retrievals can be made in batch or time-sharing modes, both of which are operational on the Survey's IBM 370-155 computer in Reston. An indexing feature also is available, by which one or more fields in the record can be indexed to create an indexed (inverted) file structure.

GIPSY functions as four separate but interrelated files: (1) a dictionary file, (2) a records file, (3) a selected records file (SRF), and (4) an optional index file.

The *dictionary file* consists of a list of labels and other information that identifys the data items (fields) contained in the records file; it also controls the format of the printed record. Any number of dictionaries can be constructed for special purposes, such as for printing the records in different ways. All labels are of equal rank; that is, there is no hierarchical structure. The *records file* contains the records themselves, which are stored in random order on a disk. The maximum length for a given record is 32,000 characters. The *selected records file* (SRF) is a preallocated disk-storage space used to store the track addresses of records selected as a result of a retrieval. In the usual case the SRF is only a transient file, occupied only during a retrieval and then automatically freed. The primary function of the SRF is to reduce search time, in that subsequent search steps of a given retrieval are directed against the relatively small SRF rather than against the entire records file. The *index file* is an auxiliary indexed sequential search file of the parent records file, that is, a set of ordered lookup tables. It consists of a

RECORD INVENTORY BY COUNTRY
MASTER CRIB FILE....... 51,348

CODE	NAME	# OF RECS	CODE	NAME	# OF RECS	CODE	NAME	# OF RECS
	UNKNOWN	20	CS	COSTA RICA	0	GP	GUADELOUPE	0
AF	AFGHANISTAN	0	CU	CUBA	11	GQ	GUAM	0
AL	ALBANIA	6	CY	CYPRUS	0	GT	GUATEMALA	6
AG	ALGERIA	4	CZ	CZECHOSLOVAKIA	1	GV	GUINEA	3
AQ	AMERICAN SAMOA	0	YQ	RYUKYU IS, SOUTHERN	0	PU	GUINEA BISSAU (PORT G)	0
AN	ANDORRA	0	SS	SP SAHARA (W SAHARA)	1	GY	GUYANA	15
AO	ANGOLA	3	SM	SAN MARINO	0	HA	HAITI	1
AC	ANTIGUA	0	TP	SAO TOME AND PRINCIPE	2	HM	HEARD AND MC DONALD IS	0
AY	ANTARCTICA	0	SA	SAUDI ARABIA	2	HO	HONDURAS	3
IY	ARAB NEUT ZN	0	SG	SENEGAL	0	HK	HONG KONG	2
AR	ARGENTINA	64	SE	SEYCHELLES	0	HU	HUNGARY	10
AT	ASHMORE AND CARTIER IS	0	SL	SIERRA LEONE	9	IC	ICELAND	0
AU	AUSTRIA	6	SK	SIKKIM	0	IN	INDIA	38
AS	AUSTRALIA	342	SN	SINGAPORE	0	ID	INDONESIA	19
BF	BAHAMAS	1	SO	SOMALIA	2	EI	IRELAND	6
BA	BAHRAIN	0	SF	SOUTH AFRICA	714	IR	IRAN	7
BB	BARBADOS	0	RH	SOUTHERN RHODESIA	15	IZ	IRAQ	0
BE	BELGIUM	0	YS	SOUTHERN YEMEN	0	IT	ITALY	16
WB	BERLIN, WEST	0	UR	SOVIET UNION	542	IS	ISRAEL	0
BD	BERMUDA	0	SP	SPAIN	32	IW	ISRAEL-JORDAN DEMIL ZN	0
BT	BHUTAN	0	PG	SPRATLY ISLAND	0	IU	ISRAEL-SYRIA DEMIL ZN	0
BL	BOLIVIA	48	SC	ST. CHRIS-NEVIS-ANGUILA	0	IV	IVORY COAST	0
BC	BOTSWANA	2	SH	ST. HELENA	0	JA	JAPAN	81
BV	BOUVET ISLAND	0	ST	ST. LUCIA	0	JM	JAMAICA	14
BH	BR HONDURAS (BELIZE)	305	SB	ST. PIERRE AND MIQUELON	1	JN	JAN MAYEN	0
BR	BRAZIL	0	DM	DAHOMEY (BENIN)	0	JQ	JOHNSTON ATOLL	0
IO	BRIT INDIAN OCN TERR	2	DA	DENMARK	0	JO	JORDAN	0
BP	BRITISH SOLOMON ISLANDS	0	DO	DOMINICA	1	KE	KENYA	6
VI	BRITISH VIRGIN ISLANDS	0	DR	DOMINICAN REPUBLIC	5	KN	NORTH KOREA	0
BX	BRUNEI	0	FT	DJIBOUTI	0	VC	ST. VINCENT	0
BU	BULGARIA	1	EC	ECUADOR	10	SU	SUDAN	1
BM	BURMA	15	ES	EL SALVADOR	0	SV	SVALBARD	0
BY	BURUNDI	4	EK	EQUATORIAL GUINEA	0	SQ	SWAN ISLANDS	0
CB	CAMBODIA	47	ET	ETHIOPIA	4	WZ	SWAZILAND	1
CM	CAMEROON	5	FD	FAERDE ISLANDS	0	SW	SWEDEN	17
CA	CANADA	1,484	FA	FALKLAND ISLANDS	0	SZ	SWITZERLAND	0
PQ	CANAL ZONE	0	FJ	FIJI	1	NS	SURINAM	10
EQ	CANTON AND ENDERBURY IS	0	FI	FINLAND	15	SY	SYRIA	0
CV	CAPE VERDE	1	FR	FRANCE	39	TZ	TANZANIA	6
CJ	CAYMAN ISLANDS	0	FS	FR. STH. & ANTARCT LNDS	0	TH	THAILAND	82
CT	CENTRAL AFRICAN REP	1	FG	FRENCH GUIANA	0	TO	TOGO	0
CE	CEYLON	2	FP	FRENCH POLYNESIA	2	TL	TOKELAU ISLANDS	0
CD	CHAD	0	GB	GABON	2	TN	TONGA	0
CI	CHILE	173	GA	GAMBIA	0	TD	TRINIDAD AND TOBAGO	0
TW	CHINA, REPUBLIC OF	31	GZ	GAZA STRIP	0	TC	TRUCIAL STATES	0
CH	CHINA, M/INLAND	17	GC	GERMANY, EAST	8	TS	TUNISIA	12
KT	CHRIST IS (INDIAN OCN)	0	GE	GERMANY, WEST	27	TU	TURKEY	4
CL	CNTRL & S LINE ISLANDS	0	GH	GHANA	6	TK	TURKS AND CAICOS IS	0
CK	COCOS ISLANDS	0	GI	GIBRALTAR	0	UG	UGANDA	6
CO	COLOMBIA	16	GN	GILBERT & ELLICE IS	0	KS	KOREA, SOUTH	73
CN	COMORO ISLANDS	0	GR	GREECE	9	KU	KUWAIT	0
CF	CONGO REPUBLIC	4	GL	GREENLAND	40	LA	LAOS	27
CW	COOK ISLANDS	0	GJ	GRENADA	0	LE	LEBANON	0

RECORD INVENTORY BY COUNTRY

MASTER CRIB FILE........ 51,348

CODE	NAME	# OF RECS
LT	LESOTHO	0
LI	LIBERIA	1
LY	LIBYA	0
LS	LIECHTENSTEIN	0
LU	LUXEMBOURG	0
MC	MACAO	2
MI	MALAWI	6
MY	MALAYSIA	23
MA	MALAGASY REP (MADAG)	
MV	MALDIVES	2
ML	MALI	0
MT	MALTA	24
MB	MARTINIQUE	52
MR	MAURITANIA	5
MP	MAURITIUS	3
MX	MEXICO	68
MQ	MIDWAY ISLANDS	0
MN	MONACO	2
MG	MONGOLIA	0
MH	MONTSERRAT	3
MO	MOROCCO	0
MZ	MOZAMBIQUE	14
MU	MUSCAT AND OMAN	4
NM	NAMIBIA	7
NR	NAURU	0

CODE	NAME	# OF RECS
NP	NEPAL	0
NL	NETHERLANDS	0
NA	NETHERLANDS ANTILLES	0
NC	NEW CALEDONIA	18
NH	NEW HEBRIDES	0
NZ	NEW ZEALAND	45
NU	NICARAGUA	1
NI	NIGERIA	0
NG	NIGER	1
NE	NIUE	0
NF	NORFOLK ISLAND	1
NY	NORWAY	24
PP	PAPUA AND NEW GUINEA	52
PK	PAKISTAN	5
PN	PANAMA	3
PF	PARACEL ISLANDS	0
PA	PARAGUAY	2
PE	PERU	224
RP	PHILIPPINES	149
PC	PITCAIRN ISLAND	0
PL	POLAND	1
PO	PORTUGAL	2
PT	PORTUGUESE TIMOR	0
RQ	PUERTO RICO	3

CODE	NAME	# OF RECS
QA	QATAR	0
RE	REUNION	0
RO	ROMANIA	5
RW	RWANDA	0
EG	UNITED ARAB REPUBLIC	1
UK	UNITED KINGDOM	8
US	UNITED STATES	46,000
UV	UPPER VOLTA	5
UY	URUGUAY	5
BQ	US MISC CARIBBEAN IS	0
IQ	US MISC PACIFIC ISLANDS	0
TQ	US TRUST IS PACIFIC	0
VT	VATICAN CITY	0
VE	VENEZUELA	26
VN	VIET-NAM, NORTH	12
VS	VIET-NAM, SOUTH	9
VQ	VIRGIN ISLANDS (U.S)	0
WQ	WAKE ISLAND	0
WF	WALLIS AND FUTUNA	0
WS	WESTERN SAMOA	0
YE	YEMEN	1
YO	YUGOSLAVIA	20
ZR	ZAIRE	21
ZA	ZAMBIA	9

RECORD INVENTORY BY STATE

MASTER CRIB FILE........ 51,348

CODE	NAME	# OF RECS
	UNKNOWN	6
01	ALABAMA	187
02	ALASKA	4,317
04	ARIZONA	729
05	ARKANSAS	102
06	CALIFORNIA	10,629
08	COLORADO	6,063
09	CONNECTICUT	525
10	DELAWARE	15
11	DISTRICT OF COLUMBIA	25
12	FLORIDA	43
13	GEORGIA	64
15	HAWAII	7
16	IDAHO	4,312
17	ILLINOIS	37
18	INDIANA	5
19	IOWA	2
20	KANSAS	4

CODE	NAME	# OF RECS
21	KENTUCKY	27
22	LOUISIANA	4
23	MAINE	153
24	MARYLAND	270
25	MASSACHUSETTS	438
26	MICHIGAN	161
27	MINNESOTA	593
28	MISSISSIPPI	18
29	MISSOURI	73
30	MONTANA	984
31	NEBRASKA	3
32	NEVADA	1,100
33	NEW HAMPSHIRE	63
34	NEW JERSEY	163
35	NEW MEXICO	265
36	NEW YORK	687
37	NORTH CAROLINA	1,920

CODE	NAME	# OF RECS
38	NORTH DAKOTA	9
39	OHIO	560
40	OKLAHOMA	16
41	OREGON	2,855
42	PENNSYLVANIA	437
44	RHODE ISLAND	45
45	SOUTH CAROLINA	525
46	SOUTH DAKOTA	48
47	TENNESSEE	1,715
48	TEXAS	76
49	UTAH	2,791
50	VERMONT	131
51	VIRGINIA	761
53	WASHINGTON	1,373
54	WEST VIRGINIA	67
55	WISCONSIN	18
56	WYOMING	578

FIGURE 1. Inventory of CRIB records, by country and state.

RECORD INVENTORY BY COMMODITY
MASTER CRIB FILE........ 51,348

05/26/81

CODE	NAME	# OF RECS
	NO COMMODITY	10
ALM	ALUM	7
AL	ALUMINUM (GENERAL)	36
AL2	ALUMINUM (OTHER)	33
AL3	ALUNITE	1
AMB	AMBER	0
COA1	ANTHRACITE	33
SB	ANTIMONY	762
AS	ARSENIC	165
ASB	ASBESTOS	165
CLY4	BALL CLAY	48
BA	BARIUM, BARITE	515
CLY1	BENTONITE	17
BE	BERYLLIUM	287
BI	BISMUTH	187
COA2	BITUMINOUS COAL	786
BIT	BITUMENS (ASPHALT)	15
CLY6	BLOATING MATERIAL	33
B	BORON - BORATES	4
AL1	BAUXITE	248
BRI	BRINES/SALINES	37
CD	CADMIUM	87
CA	CALCIUM	96
C	CARBON	7
CAR	CARBONATES	17
CER	CEMENT ROCK (NATURAL)	3
CE	CERIUM	20
CS	CESIUM	2
CR	CHROMIUM	1,740
CLY	CLAY (GENERAL)	821
CLY7	COMMON BRICK CLAY	47
COA	COAL	855
CO	COBALT	403
CON	CONCENTRATE	0
CU	COPPER	7,131
COR	CORUNDUM	65
STN1	CRUSHED/BROKEN STN	470
CRY	CRYOLITE	0
DIA	DIAMOND	86
DIT	DIATOMITE	69
STN2	DIMENSION STONE	264
DOL	DOLOMITE	34
EMY	EMERY	16
EVA	EVAPORITES	6
FLD	FELDSPAR	438
CLY5	FIRE CLAY (REFRACTORY)	185
MIC3	FLAKE MICA	6
F	FLUORINE, FLUORITE	582
CLY2	FULLER'S EARTH	24
GA	GALLIUM	7
GAR	GARNET	144
GAS	GAS (NATURAL)	12
GEM	GEMSTONES	184
GE	GERMANIUM	30
GLA	GLAUCONITE	2
AU	GOLD	17,895
GRT	GRANITE	295
GRP	GRAPHITE	92
GYP	GYPSUM, ANHYDRITE	147
HF	HAFNIUM	0
HE	HELIUM	32
HAL	HALITE	1
LST2	HIGH CALCIUM LIMESTONE	31
DOL2	HIGH MAGNESIAN DOLOMITE	0
IN	INDIUM	1
IR	IRIDIUM	384
FE	IRON	3,466
CLY3	KAOLIN	90
KYN	KYANITE	108
LAT	LATERITE	0
PB	LEAD	5,841
COA4	LIGNITE	46
LST	LIMESTONE (GENERAL)	1,054
LI	LITHIUM	100
LWA	LIGHTWEIGHT AGGREGATE	0
MGS	MAGNESITE (BRUCITE)	12
MG	MAGNESIUM (BRUCITE)	62
MN	MANGANESE	1,018
MBL	MARBLE	131
BG	MERCURY	1,718
MIC	MICA	331
MPG	MINERAL PIGMENTS	16
MO	MOLYBDENUM	1,002
MON	MONAZITE	268
NI	NICKEL	854
NB	NIOBIUM (COLUMBIUM)	142
N	NITROGEN - NITRATES	1
OI	OSMIUM + IRIDIUM	31
SAO	OIL SANDS	9
SHO	OIL SHALE	176
OLV	OLIVINE	35
ORE	ORE	0
OS	OSMIUM	376
CVB	OVERBURDEN	0
OXD	OXIDES	0
FD	PALLADIUM	262
FEA	PEAT	69
FER	PERLITE	34
CIL	PETROLEUM (OIL)	63
P	PHOSPHORUS-PHOSPHATES	391
FT	PLATINUM	1,619
PGM	PLATINUM GROUP METALS	268
K	POTASSIUM	21
PUM	PUMICE	103
PYR	PYRITE	331
PYRI	PYRRHOTITE	30
PYF	PYROPHYLLITE	34
QTZ	QUARTZ	392
RA	RADIUM	9
RAE	RARE EARTHS	237
RAM	RADIO-ACTIVE MATERIALS	86
REF	REFRACTORY	2
RE	RHENIUM	40
RH	RHODIUM	124
RB	RUBIDIUM	1
RU	RUTHENIUM	70
SDG	SAND AND GRAVEL	4,107
SAM	SAND, MOLDING	25
SST	SANDSTONE	195
SAP	SAPROLITE	2
SC	SCANDIUM	2
MIC2	SCRAP MICA	41
SE	SELENIUM	23
SHL	SHALE	40
MIC1	SHEET MICA	175
SIL	SILICA	194
AG	SILVER	6,964
SLA	SLATE	78
NA	SODIUM	9
STN	STONE	366
SR	STRONTIUM	13
COA3	SUB-BITUMINOUS COAL	274
SUL	SULFIDES	31
S	SULFUR	352
SLF	SULFURIC ACID	57
TLC	TALC, SOAPSTONE	181
TA	TANTALUM	82
TE	TELLURIUM	29
TL	THALLIUM	1
TH	THORIUM	436
SN	TIN	505
TI	TITANIUM	895
W	TUNGSTEN	2,182
DOL1	ULTRA PURE DOLOMITE	1
LST1	ULTRA PURE LIMESTONE	10
UNF	UNIDENTIFIED COMMODITY	763
U	URANIUM	4,043
V	VANADIUM	1,280
VOL	VOLCANIC MATERIALS	63
VRM	VERMICULITE	75
WOL	WOLLASTONITE	1
YT	YTTRIUM	27
ZEO	ZEOLITES	10
ZN	ZINC	4,290
ZN1	ZINC OXIDE	1
ZR	ZIRCONIUM	323

FIGURE 2. Inventory of CRIB records, by occurrences of commodities.

FIGURE 3. Reporting form for the CRIB mineral resources file.

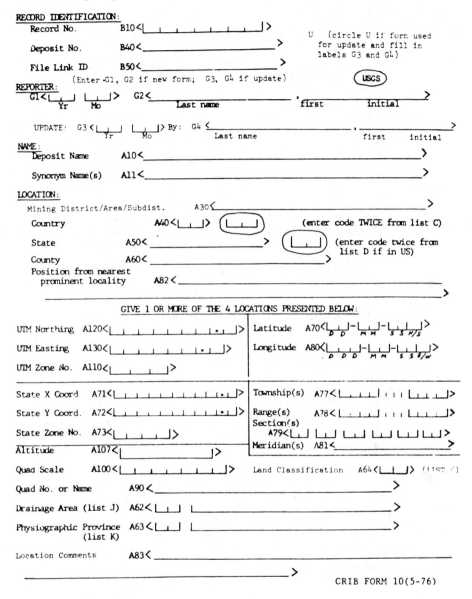

(Record No.   )   2.

COMMODITY INFORMATION

                                  1   2   3   4   5   6   7

Commodities Present   C10 < _____ >

Commodity Specialist   C20 < _____ >
   Information

Significance   MAJOR < _____ >

             MINOR < _____ >   (code from list E)

   COPROD < _____ >   POTEN < _____ >

   BYPROD < _____ >   OCCUR < _____ >

Ore Minerals, rocks, etc.   C30 < _____ >

Commodity Subtypes   C41 < _____ >
  or Use Categories
Commodity Comments   C50 < _____ >

ANALYTICAL DATA

  Reference   C44 < _____ >

  BTU's   BTU < _____ > >   Volatiles   C45 < _____ [%] >

  Sulfur   SUL < _____ [%] >   Moisture   C46 < _____ [%] >

  Ash   ASH < _____ [%] >   Thickness of Coal C47 < _____ |__| > (FT or M)

  Fixed Carbon CARB < _____ [%] >

  Analytical Data (General)   C43 < _____ >

MINERAL ECONOMICS FACTORS

  Exploration M$ C42A < _____ >   Mill   M$ C42D < _____ >

  Development M$ C42B < _____ >   Total Investments M$ C42E < _____ >

  Expansion M$ C42C < _____ >   Mill Capacity Per Yr C42F < _____ > C42G < ____ >
                                                        thous units   yr appl

  Economic Comments C42 < _____ >

EXPLORATION AND DEVELOPMENT

  Status of Exploration or Development   A20 < |__| >   (code from list B)

  Year of Discovery   L10 < _____ > By Whom L20 < _____ >

  Nature of Discovery   L30 < _____ > (List L) Present or Last Owner A12 < _____ >

  Year of First Production L40 < _____ > Present or Last Operator A13 < _____ >

Work Done by USGS   (earliest to present)
     Year   Type of work (List M)   Geologist and Results

  1) L41 < _____ | , _____ | , _____ >

  2) L42 < _____ | , _____ | , _____ >

  3) L43 < _____ | , _____ | , _____ >

Work Done by Other Organizations   (earliest to present)
     Year   Type of work (List M)   Organization and Results

  1) L50 < _____ | , _____ | , _____ >

  2) L60 < _____ | , _____ | , _____ >

  3) L70 < _____ | , _____ | , _____ >

                               CRIB FORM 10(5-76)

FIGURE 3 (continued)

EXPLORATION AND DEVELOPMENT (Contd.)   (Record No.   )   3.

Reports Available   L100<_____

_____

_____>

Comments   L110<_____

_____

_____>

DESCRIPTION OF DEPOSIT

Deposit Type(s)   (List F)   C40<_____>

Deposit Form/Shape (List N)   M10<_____>

Max Thickness   M60<_____>   M61<___(units)___>   Size   M15<_____>

Depth to Top   M20<_____>   M21<_____>   Strike   M70<_____>

Depth to Bottom   M30<_____>   M31<_____>   Dip   M80<_____>

Max Length   M40<_____>   M41<_____>   Plunge   M90<_____>

Max Width   M50<_____>   M51<_____>   Plunge dir. M100<_____>

Property is: (Active) A21   (Inactive) A22   (Circle One)

Comments   M110<_____

_____

_____>

DESCRIPTION OF WORKINGS   (Circle
                                                           Appropriate
Workings are: (Surface) M120   (Underground) M130   (Both) M140   Labels)

For Underground Workings: (units)

   Depth Below Surface   M160<_____>   M161<_____>

   Length of Workings   M170<_____>   M171<_____>

For Open Workings (surface and underground):   (units)

   Overall Length of Mined Area   M190<_____>   M191<_____>

   Overall Width of Mined Area   M200<_____>   M201<_____>

   Overall Area   M210<_____>   M211<_____>

Comments   M220<_____

_____

_____>

GENERAL REFERENCES

1)   F1<_____

_____>

2)   F2<_____

_____>

3)   F3<_____

_____>

4)   F4<_____

_____>

CRIB FORM 10(5-76)

FIGURE 3 (continued)

GEOLOGY AND MINERALOGY

(Record No.   )   4.

Age (List O)

Host Rocks and Age   K1 < |_____ |K|

Assoc. Igneous Rocks   K2 < |_____ |K|
  and age

Age of Mineralization   K3 < |_____ |>

Pertinent Mineralogy Other K4 <
  than Ore Minerals

Important Ore Control   K5 <
  or Locus

Major Regional Structures N5 <
  or Trends

Tectonic Setting   N15 <

Significant Local   N70 <
  Structures

Significant Alteration   N75 <

Process of Concentration N80 <
  or Enrichment

Age (List O)   Age, Names of Formations or Rock Types

N30 < |_____ |K| >
N35 < |_____ |K| >
N40 < |_____ |K| >
N45 < |_____ |K| >

Age (List O)   Age, Names of Igneous Units or Rock Types

N50 < |_____ |K| >
N55 < |_____ |K| >
N60 < |_____ |K| >
N65 < |_____ |K| >

Comments (Geology and Mineralogy) N85 <

GENERAL COMMENTS   GEN <

CRIB FORM 10(5-76)

FIGURE 3 (continued)

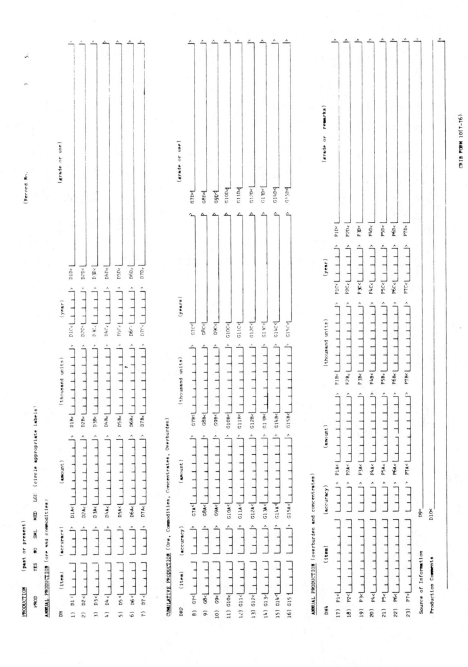

FIGURE 3 (continued)

83

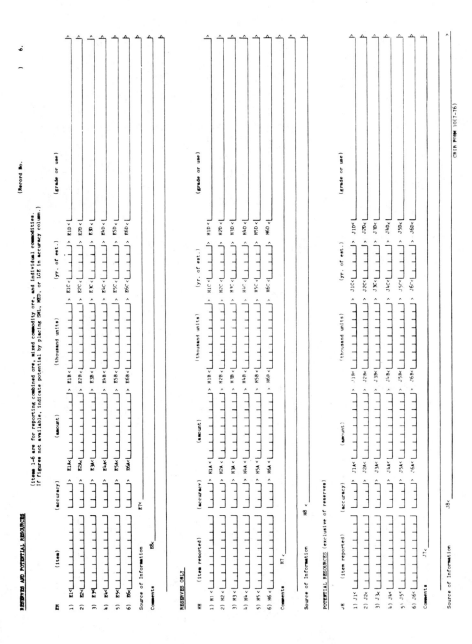

FIGURE 3 (continued)

84

FIGURE 4. An example of a detailed record in the CRIB file: the Foskor mine, South Africa.

```
                              RECORD IDENTIFICATION
                                 RECORD NO.............  W028277
                                 COUNTRY/ORGANIZATION.  USGS
                                 MAP CODE NO. OF REC..

                              REPORTER
                                 NAME .....................  SWEETWOOD, CHARLES W.
                                 DATE .....................  78 12
                                 UPDATED...................  80 05
                                 BY........................  SWEETWOOD, CHARLES W.

NAME AND LOCATION
    DEPOSIT NAME.................  PHALABORWA IGNEOUS COMPLEX
    SYNONYM NAME.................  FOSKOR MINE

    MINING DISTRICT/AREA/SUBDIST.  PHALABORWA AREA

    COUNTRY CODE.................  SF
    COUNTRY NAME:
       SOUTH AFRICA

    STATE CODE................  TRANSVAAL

    COUNTY...................  LETABA DISTRICT
    LAND CLASSIFICATION......  40

    LATITUDE          LONGITUDE
    24-00- S          031-08- E

    ALTITUDE.. 457 METERS

    POSITION FROM NEAREST PROMINENT LOCALITY:  NEAR WESTERN BOUNDARY OF KRUGER NATIONAL PARK, NORTH OF OLIFANT AND
       SELATI RIVERS.  6 KILOMETERS SOUTH OF PHALABORWA TOWNSITE.

    LOCATION COMMENTS: COPPER, GOLD, SILVER, URANIUM, VERMICULITE, NICKEL, IRON ORE (AS MAGNETITE) AND SULFURIC ACID
       (AS SULFIDE ORES) ARE MINED AND PROCESSED FROM THIS DEPOSIT SEPARATELY BY THE PALABORA MINING COMPANY LIMITED.
       SEE RECORD NO. W007520 .

COMMODITY INFORMATION
    COMMODITIES PRESENT..........  P    CU  FE  ZR

       PRODUCER(PAST OR PRESENT):
               MAJOR PRODUCTS.. P
               MINOR PRODUCTS.. CU  FE  ZR

    ORE MATERIALS (MINERALS,ROCKS,ETC.):
       MAGNETITE-OLIVINE-APATITE AND PYROXENE-VERMICULITE-OLIVINE ROCKS WITH ASSOCIATED CHALCOPYRITE, BORNITE,
       CHALCOCITE, AND VALLERITE IN CARBONATITE.

    COMMODITY COMMENTS:
       THE DEPOSIT, IN THE FORM OF A PIPE, GRADES INWARDS FROM PYROXENITE (FORMING 95 PERCENT OF THE OUTCROP AREA),
       THROUGH FOSKORITE ( 3 PERCENT OF THE OUTCROP AREA ), TO A CENTRAL CORE OF CARBONATITE ( 2 PERCENT OF THE AREA ).
       THE PYROXENITE BODY CONTAINS 17 PERCENT APATITE, THE FOSKORITE CONTAINS 25 PERCENT APATITE (PLUS 30 PERCENT
       MAGNETITE AND SOME COPPER AND ZIRCONIUM MINERALS), AND THE CARBONATITE CONTAINS COPPER AND IRON MINERALS.

    ANALYTICAL DATA(GENERAL)
       SEE PRODUCTION ENTRIES
MINERAL ECONOMICS FACTORS

    ECONOMIC COMMENTS:
       THIS OPERATION IS SOUTH AFRICA'S PRINCIPAL SUPPLIER OF PHOSPHATE ROCK FOR THE DOMESTIC MARKET.  6 - 9 PERCENT
       P2O5 ORE IS CONCENTRATED TO 36.5 PERCENT P2O5 ECONOMICALLY.  OPEN PIT MINING AND LARGE SCALE FLOTATION PROCESS
       HOLDS OPERATING COSTS WITHIN PROFITABLE LIMITS.  LONG RANGE OUTLOOK EXCELLENT.

EXPLORATION AND DEVELOPMENT
    STATUS OF EXPLOR. OR DEV.  4
                               PROPERTY IS ACTIVE
    YEAR OF DISCOVERY........  1906
    BY WHOM..................  DR. E. T. MELLOR
    NATURE OF DISCOVERY......  B
    YEAR OF FIRST PRODUCTION.  1955
    PRESENT/LAST OWNER.......  PHOSPHATE DEVELOPMENT CORPORATION LIMITED
    PRESENT/LAST OPERATOR....  PHOSPHATE DEVELOPMENT CORPORATION LIMITED

    REPORTS AVAILABLE:
       PHOSPHATE DEVELOPMENT CORPORATION LIMITED, ANNUAL REPORTS, 1972 - 1978 FOSKOR BROCHURE, CIRCA 1976 "FOSKOR IN A
       NUTSHELL"; INFORMATION PAMPHLET, CIRCA 1978

DESCRIPTION OF DEPOSIT

    DEPOSIT TYPES:
       METAMORPHIC; ALKALINE IGNEOUS; PEGMATITE; PIPE
    FORM/SHAPE OF DEPOSIT: RING STRUCTURE

    SIZE/DIRECTIONAL DATA
       SIZE OF DEPOSIT......  LARGE
       DEPTH TO TOP ........  3.0    M
       MAX LENGTH...........  6.5    KM
       MAX WIDTH............  2.5    KM
       MAX THICKNESS........  1000.0 M
       STRIKE OF OREBODY....  N-S
       DIP OF OREBODY.......  90 DEG.
    COMMENTS(DESCRIPTION OF DEPOSIT):
       AROUND A CENTRAL CORE OF CRYSTALLINE DOLOMITIC LIMESTONE (CARBONATITE) MEASURING 1,000x600 METERS AND CONTAINING
       VARYING AMOUNTS OF MAGNETITE, APATITE AND SERPENTINIZED OLIVINE, THERE OCCURS A PERIPHERAL ZONE OF COARSE
       SERPENTINE-MAGNETITE-APATITE ROCK (PHOSCORITE).   THE LATTER IS ENVELOPED BY A MORE EXTENSIVE BODY OF PYROXENITE
       ROUGHLY 1,500 HECTARES IN SURFACE AREA.  THE PYROXENITE IS LARGELY COMPOSED OF DIOPSIDE, PHLOGOPITIC MICA AND
       APATITE.  A PHLOGOPITIC SERPENTINITE, AS A PIPE-LIKE INTRUSION, OCCURS IN THE NORTHERN PART OF THE PYROXENITE
       BODY.  THE ENTIRE DEPOSIT IS ROUGHLY CIRCULAR.  MAX. THICKNESS REFERS TO DEEPEST DRILLING DONE TO DATE.  DEPTH TO
       BOTTOM NOT KNOWN.

DESCRIPTION OF WORKINGS
    SURFACE
```

FIGURE 4 (continued)

DESCRIP. OF UNDERGRND WORKINGS

DESCRIP. OF OPEN WORKINGS (SURFACE OR UNDERGRND)
    OVERALL LENGTH OF MINED AREA..... 500   M
    OVERALL WIDTH OF MINED AREA..... 500   M

COMMENTS(DESCRIP. OF WORKINGS):
    MECHANIZED OPEN PIT MINING PLAN. 12,90 - TON DIESEL-ELECTRIC TRUCKS, 2 ELECTRIC ROTARY DRILLS AND 4 ELECTRIC
    SHOVELS. CURRENT PIT DEPTH: 50 METERS.

PRODUCTION
    YES
        LARGE PRODUCTION

ANNUAL PRODUCTION (ORE AND COMMODITIES)

ITEM	ACC	AMOUNT	THOUS.UNITS	YEAR	GRADE ,REMARKS
1 ORE	ACC	11649.00	MET TONS	1977	8.0 % P205 (APPROXIMATELY)

CUMULATIVE PRODUCTION (ORE,COMMOD..CONC..OVERBUR.)

ITEM	ACC	AMOUNT	THOUS.UNITS	YEAR	GRADE ,REMARKS
8 ORE P	ACC	92041.00	MET TONS	1971 - 1979	6-9 % P205 (APPROXIMATELY)
9 CON P	ACC	13728.00	MET TONS	1971 - 1979	36.5 % P205 (APPROXIMATELY)
10 CON CU	ACC	0209.183	MET TONS	1972 - 1979	35.0 % CU (APPROXIMATELY)

OVERBURDEN AND CONCENTRATES (ANNUAL)

ITEM	ACC	AMOUNT	THOUS.UNITS	YEAR,GRADE ,REMARKS	
17 CON P	ACC	1630.000	MET TONS	1977	36.5 % P205 (APPROXIMATELY)
18 CON CU	ACC	0043.797	MET TONS	1977	35.0 % CU (APPROXIMATELY)

SOURCE OF INFORMATION (PRODUCTION).. PHOSPHATE DEVELOPMENT CORPORATION ANNUAL REPORTS 1978 AND 1979

PRODUCTION COMMENTS.... 1973 AND 1974 COPPER CONCENTRATE PRODUCTION ESTIMATED. ALL OTHER DATA ACCURATE. COMPANY
REPORTS DO NOT GIVE ANNUAL GRADES OF ORE OR GRADES OF CONCENTRATES PRODUCED. VOLUMES AND GRADES OF ZIRCONIUM
(BADDELEYITE) PRODUCED NOT AVAILABLE.

RESERVES ONLY

ITEM	ACC	AMOUNT	THOUS.UNITS	YEAR GRADE OR USE	
1 ORE P	EST	5140000.	MET TONS	1976	6 - 9 % P205

SOURCE OF INFORMATION (RESERVES).. MINERAL RESOURCES OF THE REPUBLIC OF SOUTH AFRICA, FIFTH EDITION, HANDBOOK 7 ,
DEPARTMENT OF MINES, GEOLOGICAL SURVEY: 1976

GEOLOGY AND MINERALOGY

AGE OF HOST ROCKS............ PREC
HOST ROCK TYPES.............. GRANITE-GNEISSIC COMPLEX OF THE TRANSVAAL SYSTEM

AGE OF ASSOC. IGNEOUS ROCKS.. PREC
IGNEOUS ROCK TYPES........... GRANITE-GNEISSIC COMPLEX

AGE OF MINERALIZATION........ PREC

PERTINENT MINERALOGY......... DOLERITE, BANDED CARBONATITES, MAGNETITE, SERPENTINIZED OLIVINE, DIOPSIDE,
PHLOGOPITIC MICA AND SYENITES.

IMPORTANT ORE CONTROL/LOCUS.. TRANSGRESSIVE CARBONATITE

GEOLOGY (SUPPLEMENTARY INFORMATION)
  REGIONAL GEOLOGY
    MAJOR REGIONAL STRUCTURES.. PYROXENITE, SYENITE AND CARBONATITE INTRUSIVES IN GRANITE-GNEISSIC HOST ROCKS
    TECTONIC SETTING........... SHIELD

  LOCAL GEOLOGY
    NAMES/AGE OF FORMATIONS,UNITS,OR ROCK TYPES
      AGE: QUAT   SURFACE LIMESTONES

    NAMES/AGE OF IGNEOUS UNITS OR IGNEOUS ROCK TYPES
      AGE: PREC   PHALABORWA IGNEOUS COMPLEX AND GRANITE-GNEISSIC COMPLEX

    SIGNIFICANT LOCAL STRUCTURES:
    "PHALABORWA IGNEOUS COMPLEX", AN INTRUSIVE NEAR-VERTICAL PIPE COMPOSED LARGELY OF PYROXENITE.

    GEOLOGICAL PROCESSES OF CONCENTRATION OR ENRICHMENT:
    INTENSE VOLCANIC ACTIVITY DEVELOPED, FIRST, A LARGE INTRUSION OF PYROXENITE FOLLOWED BY SYENITIC INTRUSIONS AND
    LATER STAGES OF PHOSCORITE (COPPER, IRON AND PHOSPHATE-BEARING), VERMICULITIC SERPENTINE AND CARBONATITE
    (COPPER AND IRON-BEARING). COPPER MINERALIZATION OCCURRED BETWEEN THE BANDED AND TRANSGRESSIVE CARBONATITE
    PERIODS AND AGAIN AFTER EMPLACEMENT OF THE TRANSGRESSIVE CARBONATITE.

GENERAL COMMENTS
    THE PHOSPHATE DEVELOPMENT CORPORATION AND PALABORA MINING COMPANY MINE FROM THE SAME (PHALABORWA COMPLEX)
    OCCURRENCE. (SEE RECORD NO. W007520) PMC HAS RIGHTS TO THE COPPER AND OTHER MINERALS CONTAINED IN THE CARBONATITE
    AND PART OF THE PHOSPHATE-BEARING ORE. FOSKOR HAS THE RIGHTS TO ALL THE PHOSPHATE-BEARING MINERALS IN THE PMC OFF,
    AS WELL AS TO ALL MINERALS WITHIN ITS OWN, SEPARATE, LEASE. PMC REMOVES THE COPPER AND OTHER MINERALS TO WHICH IT
    HAS RIGHTS AND PIPES ITS TAILINGS (CONTAINING PHOSCORITE) TO FOSKOR, WHICH THEN RECOVERS THE RESIDUAL APATITE.
    ALSO, TO MINE ITS OWN ORE EFFECTIVELY, PMC MUST MOVE PORTIONS OF THE PHOSCORITE TO WHICH FOSKOR HAS CLAIM; AND
    WHICH IS WASTE OVERBURDEN TO PMC. TO FOSKOR, HOWEVER, PMC'S PHOSCORITE IS "ORE" STOCKPILED AT THE FOSKOR PLANT FOR
    PROCESSING AS TIME, PLANT CAPACITY AND MARKETING CONDITIONS PERMIT. COPPER CONTAINED IN THE ORE MINED BY FOSKOR
    AND RECEIVED FROM PMC IS CONCENTRATED IN THE FOSKOR PLANT (FLOTATION) AND DELIVERED TO PMC FOR SMELTING, REFINING
    AND MARKETING ON BEHALF OF FOSKOR.

GENERAL REFERENCES
    1) MINERAL RESOURCES OF THE REPUBLIC OF SOUTH AFRICA, FIFTH EDITION, HANDBOOK 7 , DEPARTMENT OF MINES, GEOLOGICAL
    SURVEY: 1976
    2) MINERAL RESOURCES OF SOUTH-CENTRAL AFRICA: PELLETIER, R. A.: 1964

FORM

CRIB

SELECT

TERM

A. A40< US >
        COUNTRY CODE............

B. A50< 09 >
        STATE CODE..............

LOGIC A AND B
    SEARCH
15:52:46.7 SEARCH BEGINNING
15:52:49.2 SEARCH COMPLETED

SEARCHED   2067

SELECTED   37   SUBSET   1

VARIABLES SATISFIED

     A   1045

     B   3

INDEX

   SUBSET 1 HAS   403 RECORDS

ITERATE

A. A70<N >
       LATITUDE

B. A80<W >
       LONGITUDE

LOGIC A AND B
    SEARCH
15:52:59.3 SEARCH BEGINNING
15:53:01.8 SEARCH COMPLETED

SEARCHED   403

SELECTED   401   SUBSET   2

VARIABLES SATISFIED

     A   401

     B   401

ITERATE

A. A70< 41-30-   N > THRU < 42-00-00N >
        LATITUDE

B. A80< 072-22-30W > THRU < 073-00-00W >
        LONGITUDE

LOGIC A AND B
    SEARCH
15:53:02.0 SEARCH BEGINNING
15:53:04.1 SEARCH COMPLETED

SEARCHED   401

SELECTED   152   SUBSET   3

VARIABLES SATISFIED

     A   243

     B   193

SORT

A70 9
END OF SORT

COPY

  ' '

B10 8

A10 30

  ' '

A50 3

A70 10

A80 12

MAJOR 12

  ' '

C41 30

  ' '

YES 'YES' '

NO 'NO' ' '

W000484	FELDSPAR PEGMATITE QUARRY NO.	09	41-30-01N	072-31-41W	FLD	CERAMICS	YES
W000485	GRANITIC GNEISS ROCK QUARRY NO	09	41-30-06N	072-30-10W	STN	CONSTRUCTION STONE	YES
W000397	FELDSPAR PEGMATITE DIKE QUARRY	09	41-30-10N	072-33-39W	FLD	CERAMICS	YES
W000486	GRANITE GNEISS ROCK QUARRY NO	09	41-30-17N	072-30-45W	STN	CONSTRUCTION STONE	YES
W000485	ROCK LANDING FELDSPAR QUARRIES	09	41-30-31N	072-31-23W	FLD	CERAMICS	YES
W000494	GRANITIC GNEISS ROCK QUARRY NO	09	41-30-32N	072-32-58W	STN	CONSTRUCTION STONE	YES
W000493	GRANITIC GNEISS ROCK QUARRY NO	09	41-30-53N	072-32-40W	STN	CONSTRUCTION STONE	YES
W000404	ANDERSON MICA MINES AND SWANSO	09	41-30-56N	072-31-16W	LI MIC	LITHIUM AS DEOXIDIZER, MICA FO	YES
W000403	ENEGREN (POWER) MICA MINE	09	41-31-00N	072-31-17W	MIC	ELECTRICAL INSULATORS	YES
W000298	CRUSHED DIABASE ROCK QUARRY NO	09	41-31-01N	072-55-31W	STN1	CRUSHED ROCK	YES
W000402	FELDSPAR PEGMATITE PROSPECTS N	09	41-31-07N	072-31-53W		CERAMICS	NO
W000396	FELDSPAR PEGMATITE QUARRY NO.	09	41-31-08N	072-35-47W	FLD	CERAMICS	YES
W000304	CHESHIRE BARYTES CO. MINE	09	41-31-09N	072-54-44W	BA		YES
W000944	COPPER MINE NO. DJ-038	09	41-31-11N	072-49-27W	CU	METAL	YES
W000491	GRANITIC GNEISS ROCK QUARRY NO	09	41-31-14N	072-33-45W	STN	CONSTRUCTION STONE	YES
W000943	BROWNSTONE SANDSTONE ROCK QUAR	09	41-31-18N	072-53-25W	STN2	BROWNSTONE BUILDING STONE	YES
W000599	UPPER TRIASSIC COPPER PROSPECT	09	41-31-32N	072-32-05W			NO
W000492	GRANITIC GNEISS ROCK QUARRY NO	09	41-31-34N	072-31-11W	STN	CONSTRUCTION STONE	YES
W000395	FELDSPAR PEGMATITE QUARRY NO.	09	41-31-41N	072-35-54W	FLD	CERAMICS	YES
W000401	EAST SELDEN MICA, BERYL, FELDS	09	41-31-47N	072-31-44W	FLD MIC	FELDSPAR FOR FLOOR-CLEANING C	YES
W000297	CRUSHED DIABASE ROCK QUARRY NO	09	41-31-47N	072-54-56W	STN1	CRUSHED ROCK	YES
W000659	SLATER'S FELDSPAR PEGMATITE QU	09	41-31-50N	072-28-18W	FLD	CERAMICS	YES
W000394	TOLLGATE MICA-FELDSPAR MINE	09	41-31-56N	072-36-40W	FLD	CERAMICS, ELECTRICAL INSULATOR	YES
W000296	CRUSHED DIABASE ROCK QUARRY NO	09	41-31-58N	072-54-39W	STN1	CRUSHED ROCK	YES
W000303	NATHAN BOOTH AND WILLIAM MINMA	09	41-31-58N	072-54-29W	BA		YES
W000658	SLOCUM BERYL FELDSPAR MINE	09	41-32-16N	072-27-59W	FLD	CERAMICS	YES
W000600	MAROMUS GRANITE GNEISS ROCK QU	09	41-32-17N	072-33-59W	STN2	BUILDING STONE	YES
W000301	COPPER PROSPECT NO. DJ-012	09	41-32-35N	072-53-03W			NO
W000601	BENVENUE GRANITE GNEISS ROCK Q	09	41-32-38N	072-34-49W	STN2	BUILDING STONE	YES
W000302	CHESHIRE MINING AND MANUFACTUR	09	41-32-43N	072-54-17W	BA		YES
W000400	REEB MICA MINE	09	41-32-48N	072-32-10W	MIC	MICA FOR ELECTRICAL INSULATORS	YES
W000398	MARKHAM MICA MINE	09	41-32-58N	072-32-07W	MIC	ELECTRICAL INSULATORS	YES
W000300	NEW HAVEN MINERAL CO. BARITE M	09	41-33-01N	072-53-42W	BA		YES
W000628	GRANITE GNEISS ROCK QUARRIES N	09	41-33-02N	072-34-42W	STN	BUILDING STONE	YES
W000295	BARITE OCCURRENCE NO. DJ-010	09	41-33-05N	072-53-58W	BA		YES
W000490	METAMORPHIC ROCK QUARRY NO. CJ	09	41-33-08N	072-35-02W	STN	CONSTRUCTION MATERIAL	YES
W000393	WHITE ROCK PEGMATITE QUARRIES	09	41-33-17N	072-35-52W	FLD	CERAMICS	YES
W000489	ARKOSE ROCK QUARRY NO. CJ-037	09	41-33-19N	072-36-54W	STN2	BUILDING STONE	YES
W000293	CRUSHED BASALT ROCK QUARRY NO.	09	41-33-32N	072-47-54W	STN1	CRUSHED ROCK	YES
W000294	YORK HILL TRAPROCK CO. QUARRY	09	41-33-34N	072-45-14W	STN1	CRUSHED ROCK	YES
W000603	MIDDLETOWN LEAD-SILVER MINE	09	41-33-34N	072-36-44W	PB AG		YES
W000292	CRUSHED BASALT ROCK QUARRY NO.	09	41-33-40N	072-47-47W	STN1	CRUSHED ROCK	YES
W000392	BERYL FELDSPAR PEGMATITE PROSP	09	41-33-49N	072-35-56W		SPECIAL METAL ALLOYS, CERAMICS	NO
W000291	CRUSHED BASALT ROCK QUARRY NO.	09	41-33-51N	072-47-28W	STN1	CRUSHED ROCK	YES
W000317	NO NAME GIVEN	09	41-34- N	072-33- W			YES
W000488	GRANITE GNEISS ROCK QUARRY NO.	09	41-34-02N	072-34-18W	STN	BUILDING STONE	YES
W000391	FELDSPAR PEGMATITE QUARRY NO.	09	41-34-15N	072-34-17W	FLD	CERAMICS	YES
W000790	STATE FOREST NO. 2 MICA MINE	09	41-34-30N	072-33-09W	MIC	ELECTRICAL INSULATORS	YES
W000604	CHATHAM GREAT HILL COBALT NICK	09	41-34-38N	072-33-03W	AS FE CO		YES
W000389	HALE-WALKER BERYL PROSPECT AND	09	41-35-17N	072-35-17W	FLD	CERAMICS	YES
W000662	AMPHIBOLITE GNEISS ROCKQUARRY	09	41-35-21N	072-35-39W	STN	NATURAL BUILDING MATERIAL	YES
W000388	STRICKLAND-CRAMER FELDSPAR-MIC	09	41-35-32N	072-35-31W	FLD MIC	CERAMICS, ELECTRICAL INSULATOR	YES
W000406	GREAT HILL STATE FOREST NO. 1	09	41-35-45N	072-32-25W	MIC	BOOK MICA, SCRAP MICA FOR ELEC	YES
W000646	BASALT ROCK QUARRY NO. CJ-071	09	41-36-08N	072-41-16W	STN1	CRUSHED STONE	YES
W000387	FELDSPAR PEGMATITE QUARRY NO.	09	41-36-17N	072-34-21W	FLD	CERAMICS	YES
W000645	BARITE OCCURRENCES NO. CJ-070	09	41-36-53N	072-43-34W	BA	DRILLING MUDS	NO
W000290	CRUSHED BASALT ROCK QUARRY NO.	09	41-37-04N	072-47-40W	STN1	CRUSHED ROCK	YES
W000299	BERLIN MOORE'S MILL BARITE MIN	09	41-37-09N	072-47-40W	BA		YES
W000386	LIMESTONE QUARRY NO. DJ-002	09	41-37-09N	072-49-51W	LST	HYDRAULIC CEMENT	YES
W000386	GOTTA-WALDEN FELDSPAR PROSPECT	09	41-37-09N	072-35-39W	FLD	CERAMICS	YES

FIGURE 5. Conditions statements and printed output resulting from an area search using latitude and longitude. The search was for deposits located around Hartford, Connecticut. The retrieval was made in batch mode.

FIGURE 6. A search for coal mines in South Africa in which the volatile content in the coal is greater than 25.0 percent. Shown are the conditions statements and part of the printed output using the LIST output option. The retrieval was made on a time-share terminal.

```
   G I P S Y - UNIVERSITY OF OKLAHOMA                 8:28 A.M. THURSD
RUARY    3.1977

?
FORM
-------
CRIB

?
SELECT
-------
     FULL OR TERM SEARCH?
F
A. A40< SF >
        COUNTRY CODE.................
B.
LOGIC A
    SEARCH
LOGIC

  8:29:45.9 SEARCH BEGINNING

  8:31:45.9    22652 RECORDS SEARCHED        9 RECORDS SELECTED

  8:33:45.9    33039 RECORDS SEARCHED       69 RECORDS SELECTED


  8:34:16.2 SEARCH COMPLETED
  SEARCHED    37037
  SELECTED       79      SUBSET      1
  VARIABLES SATISFIED
     A           79

ITERATE?
Y
A. MAJOR< COA>
                MAJOR......
B. C45 GT 25.0
     VOLATILES........
C.
LOGIC A AND B
    SEARCH
LOGIC

  8:35:52.8 SEARCH BEGINNING


  8:35:56.6 SEARCH COMPLETED
  SEARCHED       79
  SELECTED       20      SUBSET      2
  VARIABLES SATISFIED
     A           30
     B           20

ITERATE?
N

?
SORT
-------
ASCENDING OR DESCENDING ORDER?
A
A10 10
/
END OF SORT
```

FIGURE 6 (continued)

```
?
LIST
-------
TERMINAL OR PRINTER?
T
ENTER LABEL(S)
B10
G2
A10
A40
A70
A30
MAJOR
BTU
ASH
CARB
C45
/

RECORD NO............ W012707
NAME:  SWEETWOOD, CHARLES W.
DEPOSIT NAME................     ERMELO - BELFAST COALFIELD
COUNTRY CODE................     SF
LATITUDE 26-15-  S
LONGITUDE  029-55-  E
            MAJOR...... COA2
BTU.............. 11524
ASH.............. 15.1 %
FIXED CARBON...... 52.8 %
VOLATILES......... 29.1 %

RECORD NO............ W012710
NAME:  SWEETWOOD, CHARLES W.
DEPOSIT NAME................     ERMELO-CAROLINA COALFIELD
COUNTRY CODE................     SF
LATITUDE 26-27-  S
LONGITUDE  029-58-  E
            MAJOR...... COA2
♦♦♦

BTU.............. 11732
ASH.............. 15.1 %
FIXED CARBON...... 54.3 %
VOLATILES......... 27.6 %

RECORD NO............ W012718
NAME:  SWEETWOOD, CHARLES W.
DEPOSIT NAME................     ERMELO-CAROLINA COALFIELD
COUNTRY CODE................     SF
LATITUDE 26-15-  S
LONGITUDE  030-00  E
            MAJOR...... COA3
BTU.............. 11868
ASH.............. 13.9 %
FIXED CARBON...... 52.2 %
VOLATILES......... 31.0 %

RECORD NO............ W012706
NAME:  SWEETWOOD, CHARLES W.
DEPOSIT NAME................     WITBANK - MIDDELBURG COALFIELD
♦♦♦

COUNTRY CODE................     SF
LATITUDE 26-10-  S
LONGITUDE  029-20-  E
            MAJOR...... COA2
BTU.............. 12040
ASH.............. 13.1 %
FIXED CARBON...... 57.4 %
VOLATILES......... 27.0 %
```

FIGURE 7. A count, by county, of CRIB records on Tennessee. The count was made using the COUNT command.

```
G I P S Y  -  UNIVERSITY OF OKLAHOMA  12:38 P.M. TUESDAY  SEPTEMBER 15,1976

FORM

CRIB

SELECT

A. A40< US >
        COUNTRY CODE.................

B. A50< 47 >
        STATE CODE...............

LOGIC A AND B
    SEARCH
12:38:40.3 SEARCH BEGINNING
12:42:13.8 SEARCH COMPLETED

SEARCHED   33400

SELECTED    1397       SUBSET    1

VARIABLES  SATISFIED

    A      29647

    B       1397

COUNT

A60 40
```

VALUE	FREQUENCY
	70
ANDERSON	33
ASHE	1
BEDFORD	1
BENTON	38
BLEDSOE	3
BLOUNT	14
BRADLEY	6
CAMPBELL	34
CANNON	13
CARROLL	12
CARTER	51
CLAIBORNE	91
CLAIBORNE AND UNION, TENNESSEE	1
CLAY	13
COCKE	21
COFFEE	9
CUMBERLAND	11
DAVIDSON	11
DE KALB	2
DECATUR	4
DEKALB	69
DICKSON	1
DYER	2
FAYETTE	7
FENTRESS	7
FRANKLIN	7
GIBSON	1
GILES	18
GRAINGER	11
GRAINGER, HANCOCK, HAWKINS, AND UNION	1
GREENE	24
GREENE AND WASHINGTON	1
GRUNDY	4
HAMBLEN	3
HAMILTON	7
HANCOCK	10
HARDEMAN	2
HARDIN	7
HAWKINS	9
HENDERSON	12
HENRY	26
HICKMAN	18
HUMPHREYS	2
JACKSON	17
JEFFERSON	7
JEFFERSON, KNOX, AND SEVIER	1
JOHNSON	8
KNOX	11
LAUDERDALE	9
LAWRENCE	1
LINCOLN	3
LOUDON	25
MACON	8
MADISON	3
MARION	13
MARSHALL	3

FIGURE 7 (continued)

VALUE	FREQUENCY
MAURY	68
MC MINN, MONROE, LOUDON	1
MCMINN	36
MCNAIRY	2
MEIGS	1
MITCHELL	1
MONROE	139
MONROE-MCMINN	1
MONTGOMERY	1
MOORE	5.
MORGAN	6
OBION	6
OVERTON	4
PERRY	1
PICKETT	5
POLK	8
PUTNAM	19
RHEA	1
ROANE	8
ROBERTSON	3
RUTHERFORD	8
SCOTT	17
SEQUATCHIE	8
SEVIER	10
SHELBY	23
SMITH	9
STEWART	2
SULLIVAN	18
SUMNER	11
TROUSDALE	4
UNICOI	39
UNICOI AND WASHINGTON	2
UNICOI	2
UNION	25
VAN BUREN	6
WARREN	3
WASHINGTON	19
WASHINGTON AND UNICOI	1
WAYNE	2
WEAKLEY	25
WHITE	2
WILLIAMSON	37
WILSON	11

TOTAL	1397

G I P S Y — UNIVERSITY OF OKLAHOMA   12:43 P.M. TUESDAY   SEPTEMBER 15,1976

FIGURE 8A. Interactive map plot of copper deposits in Peru. Retrieval and data preparation step.

```
OLOGON
IKJ56700A ENTER USERID -
████
ENTER PASSWORD FOR WG9195J-
███
ENTER PROCEDURE NAME -
GIPSY
WG9195J LOGON IN PROGRESS AT 08:37:48 ON FEBRUARY 4, 1977
WELCOME TO THE RESTON COMPUTER CENTER FE1 OS 21.8 TCAM 5E.
TO LIST ALL AVAILABLE TSO COMMANDS, KEY IN:    HELP  11/3/76
READY
TERM LINESIZE(80)
READY
EXEC CRIB 'GIPFILE(RIF.W0001.CRIB1) MEMBER(PERU)'
                    LINK CRIB FILE AND INTERACTIVE CAM
DATA SET S4 NOT IN CATALOG

  G I P S Y - UNIVERSITY OF OKLAHOMA                    8:40 A.M. FRIDAY    FEB

RUARY    4,1977


?
FORM
-------
CRII
++++ FORM NOT IN DICTIONARY, TRY AGAIN
CRIB


?
SELECT
-------
      FULL OR TERM SEARCH?
F
A. A40< PE >
         COUNTRY CODE.................
B.
LOGIC A
     SEARCH
LOGIC

 8:42:29.9 SEARCH BEGINNING

 8:44:29.9   27932 RECORDS SEARCHED        1 RECORDS SELECTED


 8:45:27.2 SEARCH COMPLETED
 SEARCHED    37037
 SELECTED      200      SUBSET      1
 VARIABLES SATISFIED
     A         200

ITERATE?
Y
A. MAJOR< CU >
               MAJOR......
B. A70
        LATITUDE
C.
LOGIC A AND B
     SEARCH
LOGIC

 8:46:22.6 SEARCH BEGINNING


 8:46:46.6 SEARCH COMPLETED
 SEARCHED     200
 SELECTED      80      SUBSET
 VARIABLES SATISFIED
     A          81
     B         183

ITERATE?
N
```

FIGURE 8*A* (continued)

```
?
SORT
-------
ASCENDING OR DESCENDING ORDER?
A
A70 10
/
END OF SORT

?
COPY
-------
TERMINAL OR WORKFILE?
W
'-1116'
A70 10
A80 10
/◆

?
END
-------
  G I P S Y - UNIVERSITY OF OKLAHOMA                         8:49 A.M. FRIDAY    FEB

RUARY    4,1977
UTILITY DATA SET NOT FREED, IS NOT ALLOCATED
READY
EDIT
ENTER DATA SET NAME -
'RIF.W0020.MWRK(PERU)'
ENTER DATA SET TYPE-
DATA
DATA SET 'RIF.W0020.MWRK(PERU)' NOT LINE NUMBERED, USING NONUM
EDIT
TOP
L
-1116                                                      064400S0784600W
-1116                                                      064600S0783600W
-1116                                                      065900S0781930W
-1116                                                      070030S0781700W
-1116                                                      070300S0781800W
-1116                                                      072300S0784200W
-1116                                                      072400S0783430W
-1116                                                      073630S0782900W
-1116                                                      073730S0781400W
-1116                                                      075000S0780130W
-1116                                                      080030S0782000W
-1116                                                      080100S0783700W
-1116                                                      081200S0774700W
-1116                                                      083500S0774430W
-1116                                                      083600S0774430W
-1116                                                      085000S0775900W
-1116                                                      093000S0773400W
-1116                                                      093400S0770500W
-1116                                                      095230S0695900W
-1116                                                      095300S0765730W
-1116                                                      095800S0771200W
-1116                                                      100000S0770400W
-1116                                                      102830S0761800W
-1116                                                      103200S0770700W
-1116                                                      103300S0772230W
-1116                                                      104000S0755800W
-1116                                                      104100S0770300W
-1116                                                      110700S075250!
EDIT
SAVE
SAVED
END
READY
LOGOFF
CCD008A   JOB 'WG9195J' TSO CHRG: DSK-IO TERM-IO    CPU  CNNCT     TOTAL
                                    8.53     .65   16.81  2.13    $28.12
```

FIGURE 8*B*. Interactive map plot of copper deposits in Peru. Resulting map plot. Azimuthal equal area projection; scale—1:10,000,000.

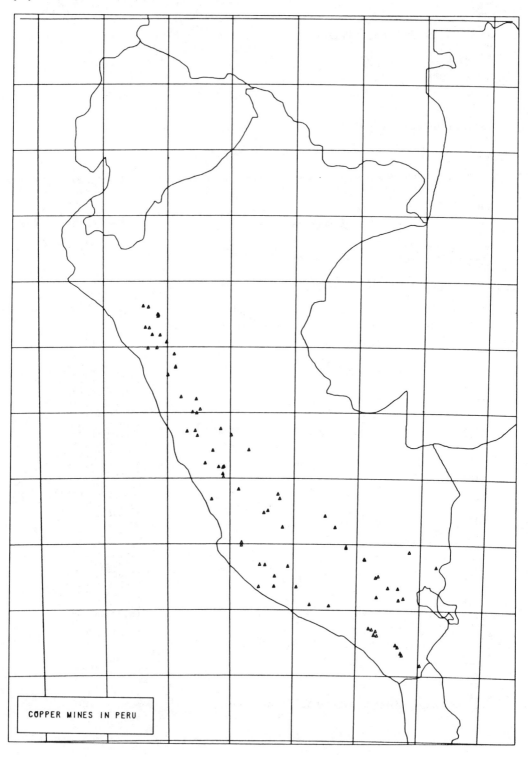

COPPER MINES IN PERU

directory of the data associated with one or more labels in the parent records file. Labels indexed in the CRIB file are A40 (country code), A50 (state code), and C10 (commodities present). In general, when an inquiry involves any of those labels, it is faster to make the initial search on the index file rather than on the parent records file.

The GIPSY system consists of separate programs that perform various tasks. The following are principal programs used by file managers to build and maintain a GIPSY file:

CREATE—preallocates disk space for the basic record, dictionary, and selected records files
RBUILD—builds files on disk from original input
TRBUILD—builds files on tape rather than disk
DUMP—copies disk files onto backup tapes
RESTORE—restores tape backups to disk
DICTLOOK—prints out the contents of a dictionary
FILELOOK—shows the disk space being used for a file
UPDATE—modifies existing records
QUESTRAN—searches and retrieves data

The average user is interested only in interrogating the file, not in building or maintaining it. The file is interrogated through a set of commands associated with the QUESTRAN search and retrieval program. The principal commands available to the user are as follows:

FORM—designates the dictionary to be used
SELECT—initiates (activates) the search and retrieval module
INDEX—initates an index search (in contrast to the usual sequential search)
ITERATE—returns the search to the previous subset (if the BACK command is used first, the ITERATE command returns the search to the subset designated under the BACK command)
BACK—returns the search to any previous subset
DELETE—deletes selected records (used only by file managers)
SORT, SORTD—sorts selected records into ascending or descending sequence
SUM—produces a record count and the maximum, minimum, and average values for any designated field
TOTAL—provides the same "within-field" information as the SUM command, but also provides a total across the fields specified
COUNT—produces a sorted list of the different entries contained under given label and the number of times each of the different entries occurs
PRINT—produces a preformatted printout of entire record
LIST—produces a line-by-line printout of user-selected fields

COPY—produces a user-arranged, fixed-field, fixed-record-length output (which is either printed or used as input to another program)
END—closes out the current retrieval session on a time-share terminal

Retrieval Procedures. The search and retrieval module, QUESTRAN (QUEstion TRANslater), provides the mechanism for retrieval, various kinds of intermediate processing, and printing. It is the vehicle by which the user communicates with the computer—without the need for an intermediate programmer between the user and his information. By means of the QUESTRAN commands, the user translates a question into a set of commands and operations that the computer can understand. During the first pass of a retrieval operation, the system reads a directory in front of each record, searching for the presence of a specified label. If the label is present in the directory, the system moves to the record itself and reads the data associated with that label, searching for the specified word, phrase, or other condition as defined by the user when he formulated his inquiry. The user may also search the file for the presence of a label only. Addresses of records retrieved on the first pass are placed in the selected records file. Subsequent passes (iterations) are directed against the selected records file.

A wide assortment of specific methods are available to the user for retrieving information. These methods fall into one of three modes, the *word mode, numeric mode,* or *label-only mode:*

1. *Word mode*—all entries, including numerical data, treated as character data.
  a. Word search—a search for a desired word or phrase. A *word* is defined as one or more alphanumeric characters with a leading and trailing blank. A *phrase* consists of two or more words.
  b. Prefix search.
  c. Suffix search.
  d. Existence search—search for the existence (presence) of some desired character string. A character string, in this instance, is defined as one or more alphanumeric characters without leading or ending blanks.
  e. Character range search—a search for all entries under a given label that lie between two specified end members.
2. *Numeric mode*—operates on numbers only. Character data, if present, are ignored. This allows for the presence in the record of mixed character and numeric data in a numeric field—for example, 1,500 ft.
  a. Equal to (EQ).
  b. Less than (LT).
  c. Greater than (GT).

d. Number range—a search for all numeric entries under a given label that lie between two specified end numbers.

e. Intrarecord comparison—the numeric value under another label.

3. *Label-only mode*—searches for the presence of a label only. No search operations are performed on the data entry (if any) under that label.

Complete information about retrieval procedures and the GIPSY program is contained in the GIPSY user's guide (Oklahoma University, 1975).

Uses of CRIB

At the general level the CRIB file is used as a storage medium and as a source of information for mineral resources inventory and evaluation studies, land use planning, and geological correlations and associations. In addition, it contributes in management decisions relating to the above.

At the functional level the CRIB file is used in three ways: (1) for primary (spot) retrievals; (2) for secondary applications; and (3) as a primary storage and inventory mechanism.

1. *Primary retrievals* (Figs. 5, 6, 7 on pages 87-91)— The user desires a particular subset of information from the file and, in most cases, a printed report (printout) of the results. Primary retrievals are handled entirely within the GIPSY program itself.

2. *Secondary applications* (Fig. 8 on pages 92-94)— Secondary applications involve a retrieval followed by additional processing to create a derivative product. The results of a CRIB retrieval are copied to a workfile by means of the COPY command. The workfile is then available for additional processing as a second job step. The following are some common secondary applications:

a. Map-plotting applications—using the CAM (Cartographic Automatic Mapping) program (Calkins et al., 1980), and the DPS (Data Presentation System) (Wachter et al., 1968) to generate plots of mineral resources data.

b. Specialized output reports—tabulated listings, group listings, summary totals, and other types of "report-generator" products using RPG, COBAL, Assembly, and other program languages and software packages.

c. Data transfer and exchange—the use of CRIB data to begin, or add to, other files or specialized data sets, and the acquisition, conversion, and entry of data from other files into the CRIB file.

3. *Primary storage and inventory mechanism*—The CRIB file can be used as an addressable storage medium for subsets of data on particular areas or mineral commodities. In this latter use the contributor (reporter) in most cases also is a user, using the CRIB file as if it were his own file drawer.

JAMES A. CALKINS

References

Calkins, J. A.; Kays, Olaf; and Keefer, E. K., 1973. CRIB—The mineral resources data bank of the U.S. Geological Survey, *U.S. Geol. Survey C. 681,* 39p.

Calkins, J. A.; Keefer, E. K.; Ofsharick, R.A.; Mason, G. T.; Tracy, P.; and Atkins M., 1978. Description of CRIB, the GIPSY retrieval mechanism, and the interface to the General Electric Mark III Service, *U.S. Geol. Survey Circ. 755-A,* 49p.

Calkins, J. A.; Crosby, A. S.; Huffman, T. E.; Clark, A. L.; Mason, G. T.; and Bascle, R. J., 1980. Interactive computer methods for generating mineral-resource maps, *U.S. Geol. Survey Circular 815,* 70p.

Oklahoma University, Office on Information Systems Programs, 1975. *GIPSY Documentation Series, Volume II—User's Guide.* Norman, Okla. var. pages.

Wachter, R. A., 1968. *Data Presentation System—User's Manual,* U.S. Dept. Commerce, Natl. Tech. Information Service AD-844 551, 236p.

Cross-references: *Earth Science, Information and Sources; Geological Communication; Geological Information, Marketing.* Vol. XIV: *Cartography, Automated; Punch Cards, Geologic Referencing.*

COMPUTER MAPPING—See Vol. XIV: CARTOGRAPHY, AUTOMATED.

CONFERENCES, CONGRESSES, AND SYMPOSIA

Entries in this volume of the *Encyclopedia of Earth Science* illustrate the very wide range of topics within the scope of *applied geology.* The following are criteria for inclusion in the lists: national organizations of the United States and Canada; international organizations with three or more countries represented in the membership; and selected United States, Canadian, and international meetings that have been held for succeeding years. Short courses and workshops are excluded, as are local, state, provincial, and regional organizations and those that are sections of national societies. Useful sources of information on these organizations and their publications are the *Union List of Geologic Field Trip Guidebooks of North America* (Geoscience Information Society, 1978) and the *American Association of Petroleum Geologists Bulletin.* Geological surveys and other governmental and intergovernmental agencies (see Vol. XIV: *Associations, Institutes, and Publications*) are omitted, as are associations of more general interest such as the American Association for the Advancement of Science.

Although the majority of the entries represent a scientific or technical orientation, the list does include selected major trade or commercial organizations. And some associations that are not themselves primarily geological in nature may appear on the list if their interests overlap with those of applied geologists or if they cosponsor joint meetings with geological organizations from time to time. In general, an organization must currently be holding meetings to be eligible.

Entries are arranged alphabetically within two groups. In the first group are associations, societies, institutes, committees, federations, and other bodies that sponsor conferences. The second group consists of named symposia, conferences, seminars, congresses, colloquia, and other types of meetings.

Because addresses and officers of associations are constantly changing, they are not included. Suggestions for locating recent information of this type do, however, appear in the annotated list of references. Obtaining publications resulting from conferences, congresses, and symposia may present problems; standing orders are often not accepted, and small printings are typical. Most associations will supply a list of their publications upon request. Kyed and Matarazzo (1981) and *Proceedings in Print* provide information about some available publications.

Sources used in compiling the present lists appear in the references. In addition, calendars and lists of meetings in the various geological journals were invaluable for current information on associations and their meetings. For example, see *Geotimes, Earth Science Reviews, EOS,* and *EPISODES.*

Associations, Societies, and Institutes

Air Pollution Control Association
American Academy of Environmental Engineers
American Association for Crystal Growth
American Association of Petroleum Geologists
American Association of Petroleum Landmen
American Association for Quaternary Research
American Association of Stratigraphic Palynologists
American Astronautical Society
American Astronomical Society
American Ceramic Society
American Coke and Coal Chemicals Institute
American Crystallographic Association
American Federation of Mineralogical Societies
American Gas Association
American Geographical Society
American Geological Institute
American Geophysical Union
American Institute of Aeronautics and Astronautics
American Institute of Mining, Metallurgical, and Petroleum Engineers
American Institute of Professional Geologists
American Meteorlogical Society
American Nuclear Society
American Oceanic Organization
American Petroleum Institute
American Polar Society

American Society of Agronomy
American Society of Cartographers
American Society of Civil Engineers
American Society of Limnology and Oceanography
American Society for Metals
American Society for Oceanography
American Society for Photogrammetry
American Society of Surveying and Mapping
American Society for Testing and Materials
American Water Resources Association
American Water Works Association
Arctic Circle
Arctic Institute of North America
Association of American Geographers
Association of American State Geologists
Association of Earth Science Editors
Association of Engineering Geologists
Association of Exploration Geochemists
Association of Geoscientists for International Development
Association for Mexican Cave Studies
Association of Professional Geological Scientists
Association of Soil and Foundation Engineers
Association of Southeast Asian Nations Council on Petroleum
Association of Teachers of Geology
Association of Women Geoscientists
Atlantic Estuarine Research Society
Benelux Society on Metallurgy
Bio-Energy Council
Bituminous Coal Research, Inc.
Botanical Society
Canada Centre for Remote Sensing
Canadian Association of Aerial Surveyors
Canadian Diamond Drilling Association
Canadian Exploration Geophysics Society
Canadian Geophysical Union
Canadian Geoscience Council
Canadian Geotechnical Society
Canadian Geothermal Resources Association
Canadian Hydrographers Association
Canadian Institute of Marine Engineers
Canadian Institute of Mining and Metallurgy
Canadian Institute of Surveyors
Canadian Meteorological and Oceanographic Society
Canadian Oil Scouts Association
Canadian Palaeontological Association
Canadian Petroleum Association
Canadian Society of Exploration Geophysicists
Canadian Society of Oceanology
Canadian Society of Petroleum Geologists
Canadian Society of Soil Science
Canadian Water Quality Association
Canadian Water Resources Association
Canadian Well Logging Society
Carpatho-Balkan Geological Association
Cartographic Information Society
Cave Research Foundation
Circum-Pacific Jurassic Research Group

Citizens and Scientists Concerned About Dangers in Environment
Clay Minerals Society
Coal Association of Canada
Coastal Engineering Research Council
Coastal Society
Commission on Atmospheric Chemistry and Global Pollution
Commission on Atmospheric Physics and Meterology
Commission on the Coastal Environment
Commission for the Geological Map of the World
Committee for Coordination of Joint Prospecting for Mineral Resources in Southwest Pacific Areas
Committee on Pacific Neogene Stratigraphy
Committee on Space Research
Committee on Storage, Automatic Processing, and Retrieval of Geological Data
Commonwealth Committee on Mineral Resources and Geology
Commonwealth Consultative Space Research Committee
Council of Active Independent Oil and Gas Producers
Council of Commonwealth Mining and Metallurgical Institutions
Council of Earth Science Societies in Canada
Deep Foundations Institute
Domestic Petroleum Council
Domestic Wildcatters Association
Earthquake Engineering Research Institute
East African Academy
Editeast
Editerra (European Association of Earth Science Editors)
Environmental Research Institute
Estuarine Research Federation
European Association for Earthquake Engineering
European Association of Exploration Geophysicists
European Center for Marine Environments' Problems
European Ceramic Association
European Geophysical Society
European Liquefied Petroleum Gas Association
European Oceanic Association
European Seismological Commission
Federation of Astronomical and Geophysical Services
Fine Particle Society
Forum on the Geology of Industrial Minerals
Gas Processors Association
Gas Research Institute
Gemological Institute of America
Geochemical Society
Geological Association of Canada
Geological Society of America
Geoscience Information Society
Geothermal Resources Council
Ground Water Council
Ground Water Institute

Gypsum Association
Independent Petroleum Association of America
Independent Petroleum Association of Canada
Institute of Environmental Sciences
Institute of Gas Technology
Institute on Lake Superior Geology
Institute of Polar Studies
Instrument Society of America
International Antarctic Glaciological Project
International Association for the Advancement of Earth and Environmental Sciences
International Association of Atmospheric Physics
International Association of Drilling Contractors
International Association for Earthquake Engineering
International Association on Engineering Geology
International Association on the Genesis of Ore Deposits
International Association of Geochemistry and Cosmochemistry
International Association of Geodesy
International Association of Geomagnetism and Aeronomy
International Association of Geophysical Contractors
International Association for Great Lakes Research
International Association for Hydraulics Research
International Association of Hydrogeologists
International Association for Hydrological Sciences
International Association for Mathematical Geology
International Association of Meteorology and Atmospheric Physics
International Association for the Physical Sciences of the Ocean
International Association of Planetology
International Association of Scientific Hydrology
International Association of Sedimentologists
International Association of Seismology and Physics of the Earth's Interior
International Association for the Study of Clays
International Association of Theoretical and Applied Limnology
International Association of Volcanology and Chemistry of the Earth's Interior
International Association on Water Pollution Research
International Atmospheric Ozone Commission
International Bryozoology Association
International Cartographic Association
International Centre for Geothermal Research
International Centre for Heat and Mass Transfer
International Commission of Atmospheric Chemistry and Global Pollution
International Commission on Atmospheric Electricity
International Commission on Cloud Physics
International Commission on Dynamic Meteorology
International Commission on the Meteorology of the Upper Atmosphere
International Commission on Polar Meteorology

International Commission for the Scientific Exploration of the Mediterranean Sea
International Commission on Snow and Ice
International Commission on Subsurface Water
International Committee for Palynology
International Council for the Exploration of the Sea
International Energy Agency
International Federation of Automatic Control
International Federation of Societies of Economic Geologists
International Gas Union
International Geological Correlation Programme
International Glaciological Society
International Hydrographic Organization
International Hydrological Programme
International Institute for Aerial Survey and Earth Sciences
International Metallographic Society
International Mineralogical Association
International Ocean Institute
International Oil Scouts Association
International Organization of Palaeobotany
International Ozone Association
International Palaeontological Association
International Precious Metals Institute
International Quaternary Association
International Seismological Centre
International Society for Geothermal Engineering
International Society for Hydrothermal Techniques
International Society for Mine Surveying
International Society for Photogrammetry
International Society for Rock Mechanics
International Society for Soil Mechanics and Foundation Engineering
International Society of Soil Science
International Solar Energy Society
International Speleological Union
International Tsunami Information Center
International Union of Air Pollution Prevention Associations
International Union of Crystallography
International Union of Geodesy and Geophysics
International Union of Geological Sciences
International Union for Quaternary Research
International Union of Radio Science
International Volcanological Institute
International Water Resources Association
Interstate Natural Gas Association of America
Inter-Union Commission on Geodynamics
Inter-Union Commission on Solar-Terrestrial Physics
Jesuit Seismological Association
Joint Committee on Atmospheric-Ocean Interactions
Latin American Association of Editors in the Earth Sciences
Latin American Energy Organization
Marine Technology Society
Mediterranean Association for Marine Biology and Oceanology
Metal Industries Association

Metallurgical Society of AIME
Meteoritical Society
Microbeam Analysis Society
Mid-Continent Oil and Gas Association
Mineralogical Association of America
Mineralogical Association of Canada
Minerals Professional Association
Mining Association of Canada
Mining and Metallurgical Society of America
National Association of Conservation Districts
National Association of Geology Teachers
National Association of State Groundwater Officials
National Ceramic Association
National Coal Association
National Crushed Stone Association
National Institute of Ceramic Engineers
National Lime Association
National Limestone Institute
National Ocean Industries Association
National Petroleum Council
National Speleological Society
National Water Resources Association
National Water Supply Improvement Association
National Water Well Association
Nordic Association of Applied Geophysics
Nordic Geodetic Commission
Nordic Geological Association
Nordic University Group on Physical Oceanography
North American Micropaleontological Society
North American Thermal Analysis Society
North Sea Hydrographic Commission
Open Pit Mining Association
Pacific Science Association
Paleontological Research Institution
Paleontological Society
Prospectors and Developers Association
Regional Center for Seismology for South America
Salt Institute
Scientific Committee on Antarctic Research
Scientific Committee on Oceanic Research
Scientific Committee on Problems of the Environment
Scientific Committee on Water Research
Seismological Society of America
Seismological Society of the South-West Pacific
Society to Adapt Building to the Environment Reasonably
Society for the Advancement of Material and Process Engineering
Society of Economic Geologists
Society of Economic Paleontologists and Mineralogists
Society for Environmental Geochemistry and Health
Society for Experimental Stress Analysis
Society of Exploration Geophysicists
Society of Explosives Engineers
Society of Independent Professional Earth Scientists
Society of Mining Engineers
Society of Petroleum Engineers of AIME
Society of Photo-Optical Instrumentation Engineers

Society of Professional Well Log Analysts
Society of Vertebrate Paleontology
Soil Conservation Society of America
Soil Science Society of America
Southeast Asian Society of Soil Engineering
Southeast Asia Petroleum Exploration Society
Southern African Regional Commission for the Conservation and Utilisation of the Soil
Underwater Mining Institute
United States Committee for the Gobal Atmospheric Research Program
United States Committee on Irrigation, Drainage and Flood Control
United States National Committee on Geology
United States National Committee on History of Geology
Universities Council on Water Resources
Water Forum
Water Pollution Control Federation
Water Quality Association
Water Systems Council
Western Oil and Gas Association
World Environment and Resources Council
World Meteorological Organization

Conferences, Congresses, and Symposia

Alaskan Science Conference
American Congress on Surveying and Mapping
American Mining Congress
Arab Petroleum Congress
Baltic Conference of Soil Mechanics and Foundation Engineering
Binghamton Geomorphology Symposium
Canadian Conference on Coal
Canadian Conference on Earthquake Engineering
Canadian Hydrology Symposium
Canadian Permafrost Conference
Canadian Symposium on Rock Mechanics
Caribbean Geological Conference
Chapman Conference [topics vary]
Circum-Pacific Energy and Mineral Resources Conference
Coal Convention
Coastal Engineering Conference
Coastal Zone Management Conference
Colloquium on African Geology
Colloquium on Planetary Water
Commonwealth Mining and Metallurgical Congress
Conference on African Geology
Conference on Coastal Engineering
Conference on Coastal Meteorology
Conference on Earth Resources Observation and Information Analysis Systems
Conference on Energy and Environment
Conference of Environmental Earth Sciences and Engineering
Conference and Exposition on Energy
Conference on Great Lakes Research

Conference on Hydrometeorology
Conference on Materials for Coal Conversion and Utilization
Conference on Natural Gas Research and Technology
Conference on Ocean-Atmosphere Interaction
Conference on Port and Ocean Engineering Under Arctic Conditions
Conference on Productivity in Mining
Conference on Sedimentary Tectonics
Conference on Solar Wind
Conference on Transfer of Water Resources Knowledge
Congress on Large Dams
Dissertation Symposium in Chemical Oceanography
Drillers' Seminar
Eastern Snow Conference
Energy Technology Conference
Environmental Engineering and Science Congress
Environmental Technology Seminar
ESLAB Symposium on Physics of Solar Variations
European Colloquium on Geochronology, Cosmochronology and Isotope Geology
European Conference on Microwaves
European Conference on Soil Mechanics and Foundation Engineering
European Conodont Symposium
European Symposium on Earthquake Engineering
European Symposium on Particle Size Measurement
Maurice Ewing Symposium
Exploration Update [year]
Geochautauqua [topics vary]
GEOCOME: Geological Congress on the Middle East
Geodesy/Solid Earth and Ocean Physics Research Conference
Geomorphology Symposium
Gondwana Symposium
Guelph Symposium on Geomorphology
Hanford Life Sciences Symposium on Coal Conversion and the Environment
Highway Geology Symposium
Hot Dry Rock Geothermal Energy Conference
Hydrographic Conference
Hydrotechnical Conference
Inter-American Conference on Materials Technology
International Archaean Symposium
International Clay Conference
International Cloud Physics Symposium
International Conference on Application of Statistics and Probability to Soil and Structural Engineering
International Conference on Asbestos
International Conference on Atmospheric Electricity
International Conference on Basement Tectonics
International Conference on Cartography
International Conference on Coastal Engineering

International Conference on Computational Methods in Nonlinear Mechanics

International Conference on Crystal Growth

International Conference on Engineering in the Ocean Environment

International Conference on Environmental Sensing and Assessment

International Conference and Exhibition for Marine Technology

International Conference on Finite Elements in Water Resources

International Conference on Fluvial Sediments

International Conference on Geochronology, Cosmochronology, and Isotope Geology

International Conference on Geological Information

International Conference on Geophysics of the Earth and Oceans

International Conference on Industrial Minerals

International Conference on Liquefied Natural Gas

International Conference on Marine Sciences and Ocean Engineering

International Conference on Mid-Cretaceous Events

International Conference on the New Basement Tectonics

International Conference on Numerical Methods in Geomechanics

International Conference on Offshore Site Investigation

International Conference on Permafrost

International Conference on Physicochemical Hydrodynamics

International Conference on the Physics and Chemistry of Asbestos Minerals

International Conference on Planetology

International Conference on Stable Isotopes

International Conference on Tektites and Impact Craters

International Conference on Urban Storm Drainage

International Conference on Water Planning

International Congress on Bentonites

International Congress of Carboniferous Stratigraphy and Geology

International Congress of Crystallography

International Congress of Ecology

International Congress of Engineering Geology

International Congress on the Exploitation of Stones and Non-Metallic Ores

International Congress on the History of Oceanography

International Congress on Mediterranean Neogene

International Congress for Photogrammetry

International Congress of Prehistorical and Protohistorical Sciences

International Congress on Rheology

International Congress on Sedimentology

International Congress on Siliceous Deposits of the Pacific Region

International Congress of Soil Mechanics and Foundation Engineering

International Congress of Soil Science

International Congress of Speleology

International Congress for the Study of Bauxites, Alumina and Aluminum

International Congress on Waves and Instabilities in Plasmas

International Cosmic Ray Conference

International Energy Exhibition, ENERGY [year]

International Estuarine Research Conference

International Gas Research Conference

International Geochemical Exploration Symposium

International Geodetic Symposium on Satellite Doppler Positioning

International Geological Congress

International Kimberlite Conference

International Liege Colloquium on Ocean Hydrodynamics

International Mine Planning and Development Symposium

International Meeting of European Quaternary Botanists

International Meeting on Organic Geochemistry

International Mineral Processing Congress

International Mining Congress

International Ocean Development Conference

International Palynological Conference

International Platinum Symposium

International Pollution Engineering Exposition and Congress

International Refuse to Energy Conference

International Sedimentological Congress

International Seminar on Education in Water Resources

International Seminar on Hypergraph-Based Data Structures

International Speleological Congresses

International Strata Control Conference

International Symposium on Antarctic Geology and Geophysics

International Symposium on Antarctic Glaciology

International Symposium on Arctic Geology

International Symposium on the Cambrian System

International Symposium on Coral Reefs

International Symposium on Crustal Movements in Africa

International Symposium on Earthquake Prediction

International Symposium on Earth Tides

International Symposium on Environmental Biogeochemistry

International Symposium on Equatorial Aeronomy

International Symposium on the Genesis of Ore Deposits

International Symposium on the Geochemistry of Natural Water

International Symposium on Geodesy and Physics of the Earth

International Symposium on Geothermometry and Geobarometry

International Symposium on Groundwater

International Symposium in Hydrology

International Symposium on Induced Seismicity
International Symposium on Inertial Technology for Surveying and Geodesy
International Symposium on Living and Fossil Diatoms
International Symposum on Management of Geodetic Data
International Symposium on Mathematical Geophysics
International Symposium on Mine Surveying
International Symposium on the Ordovician System
International Symposium on Ostracoda
International Symposium on Ozone for Water and Waste Water Treatment
International Symposium on Problems Related to the Redefinition of North American Vertical Geodetic Networks
International Symposium on Remote Sensing of the Environment
International Symposium on Salt
International Symposium on Soil Conditioning
International Symposium on Solar-Terrestrial Physics
International Symposium on Stratified Flows
International Symposium on Urban Hydrology
International Symposium on Urban Storm Runoff
International Symposium on Water-Rock Interaction
International Uranium Symposium
International Water Quality Symposium
International Weather Modification Conference
International Working Meeting on Soil Micromorphology
Interstate Conference on Water Problems
Joint Conference on Sensing of Environmental Pollutants
Joint Oceanographic Assembly
Lake Symposium
Latin American Geological Congress
Lunar and Planetary Science Conference
Midwest Ground-Water Conference
Mineral Waste Symposium
National Conference on Clays and Clay Minerals
National Conference on Complete Water Use
National Conference on Earthquake Engineering
National Ground Water Quality Symposium
National Oceanographic Symposium
National Western Mining Conference
Northern Resources Conference
Offshore Technology Conference
Oil Shale Symposium
Pan American Conference on Soil Mechanics and Foundation Engineering
William T. Pecora Memorial Symposium
Planetological Symposium
Planktonic Conference
Rapid Excavation and Tunneling Conference
Regional Conference on Geology and Mineral Resources of Southeast Asia

Regional Conference for Soil Mechanics
Research and Applied Technology Symposium on Mined-Land Reclamation
Rocky Mountain Ground-Water Conference
Salt Symposium
Seminar on the Cerro Prieto Geothermal Field
Southeast Asian Conference on Soil Engineering
Symposium on Application of Computers and Mathematics in the Mining Industry
Symposium on Arctic Geology
Symposium in Earthquake Engineering
Symposium on Engineering Geology and Soils Engineering
Symposium on Geochemical Exploration
Symposium of Geomorphology
Symposium of Global Tectonics
Symposium on Improved Oil Recovery
Symposium on Inland Waterways for Navigation, Flood Control, and Water Diversions
Symposium on Meteorological Observations
Symposium on New Concepts in Sedimentation
Symposium on the Petroleum Geology of the Georgia Coastal Plain
Symposium on Rehabilitation of Drastically Disturbed Surface Mined Lands
Symposium on Remote Sensing of the Environment
Symposium on Rock Mechanics
Symposium on Scanning Electron Microscopy
Tunneling [year]
Underground Operators Conference
Water Resources Congress
Water Resources Seminar
World Climate Conference
World Conference on Earthquake Engineering
World Congress of Underwater Activities
World Dredging Conference
World Energy Conference
World Mining Congress
World Oceanology International Conference
World Petroleum Congress

JULIE BICHTELER

References

Colgate, C., Jr. (ed.), 1982. *National Trade and Professional Associations of the U.S.*, 17th ed. Washington, D.C.: Columbia Books, 377p.
Alphabetical listing of 5,700 national associations with information on date founded, address and telephone, chief administrative officer, number of members and size of staff, budget, major serial publications, place and date of meetings. Several indexes. Revised and updated annually.

"A Directory of Societies in Earth Science," 1982. *Geotimes*, **27** (8), 19-27.
Alphabetical international listing of more than 425 societies with addresses. U.S. state surveys appear in a separate list. Revised annually (August issue).

Encyclopedia of Associations, 17th ed., 1983. Detroit: Gale Research. Vol. 1, 1562p.; Vol. 2, 993p.; Vol. 3, periodic supplement.

Major guide to national and international organizations of all types, including more than 1,170 scientific, engineering, and technical organizations. Each entry includes name, address and telephone; officers; date founded; number of members and staff; description of the association (membership, objectives, and activities); affiliates, divisions, etc.; committees; publications; conventions/meetings. Several indexes.

Geoscience Information Society, 1978. *Union List of Geologic Field Trip Guidebooks of North America,* 3rd ed. Falls Church, Va.: American Geological Institute, 253p.
A union list of guidebooks, incorporating monographic titles. Includes holdings of 108 libraries.

Kyed, J. M., and Matarazzo, J. M. (eds.), 1981. *Scientific, Engineering, and Medical Societies Publications in Print, 1980-1981,* 4th ed. New York: Bowker, 626p.
Bibliographic and sales information on publications of U.S. scientific and engineering societies and related organizations, including current addresses. Subject and author indexes.

Proceedings in Print, 1964-   . Bimonthly. Mattapan, Mass: Proceedings in Print.
Index of proceedings in all subject areas. Includes reports and proceedings of conferences, symposia, lecture series, congresses, hearings, institutes, colloquia, meetings, etc. Order information included. Entries arranged alphabetically by unique title with subject and sponsor index.

Scientific, Technical and Related Societies in the United States, 9th ed., 1971. Washington, D.C.: National Academy of Sciences, 213p.
Alphabetical listing of societies including information on officers, history and organization, purpose, membership, meetings, professional activities, and publications. Subject index.

Yearbook of International Organizations, 19th ed., 1981. Brussels: Union of International Associations, unpaged.
Classified listing of international organizations with information on addresses, officers, history, aims, structure, activities, publications, and members. Several indexes.

Zils, M., and Gorznyl, W. (eds.), 1982. *World Guide to Scientific Associations and Learned Societies,* 3rd ed. Munich: Internationalest Verzeichnis Wissenschaftlicher Verbände und Gesellschaften, 619p.
Current data on associations (from over 130 countries) concerned with all fields of research, medical schools, and organizations of physicians and educators. Associations are arranged alphabetically by country within continents. Names of officers, year founded, and addresses. Subject index.

Cross-references: *Earth Science, Information and Sources; Geological Information, Marketing.* Vol. XIV: *Abbreviations, Ciphers, and Mnemonicons; Associa-* *tions, Institutes, and Publications; Geological Surveys, State and Federal; Map and Chart Depositories; Professional Geologists' Associations; Punch Cards, Geologic Referencing; Remote Sensing and Photogrammetry, Societies and Periodicals.*

CONSOLIDATION, SOIL

The term *consolidation* describes the process by which a soil mass decreases in volume in response to either natural or man-made loadings. Natural loadings that induce consolidation result from geological processes such as sedimentation. Man-induced consolidation is a result of surface loadings associated with the construction of buildings or soil embankments and of subsurface load changes associated with underground excavations and the extraction of subsurface fluids (see Vol. XII, Pt. 1: *Microstructure, Manipulation*).

This meaning of the term *consolidation* has been used generally by soil and geotechnical engineers since Terzaghi (1925, 1943) developed the theory of consolidation in the 1920s. It is to be distinguished from the meaning of the term that geologists have used during the same period. They have commonly used the terms *consolidated* or *compacted sediments* to describe shales that formed from sedimentary clays (Krynine and Judd, 1957). In this usage the term *consolidation* is associated not only with the process of volume reduction in response to loadings but also with diagenetic chemical and mineralogical changes that take place during the formation of shales (see *Shale Materials, Engineering Classification*).

Geotechnical engineers are primarily concerned with the magnitudes and rates of consolidation associated with man-made loadings. These factors are generally small in cohesionless soils such as clean sands and gravels. In clays (q.v.), however, they are often critical to the successful design of engineered works. Dramatic examples of major consolidation include the few meters of subsidence in Mexico City and the continuing tilting of the Leaning Tower of Pisa, which has been settling slowly for the past 800 years.

The basic concepts and techniques that geotechnical engineers use to predict the magnitudes and rates of consolidation include the principle of effective stress, the mechanics of consolidation, the laboratory consolidation test, and the one-dimensional theory of consolidation (see *Foundation Engineering; Soil Mechanics*). This discussion will describe the above in terms that apply to saturated systems. Their quantitative description for partially saturated soils is more complex, owing to capillarity, and thus beyond the scope here. (See Bishop and Blight, 1963; Lee, 1968.)

Many authors distinguish between the magnitude and rate aspects of the consolidation process by using the terms *compression* and *settlement* to describe the total magnitude of soil volume change that occurs in response to a given loading, and by restricting the use of the term *consolidation* to describe the dynamic part of the process in which the volume of a soil mass changes with time.

Principle of Effective Stress

The *principle of effective stress* describes how a stress applied to a soil mass is distributed between the soil grains and soil pore fluid (Skempton, 1961). The effective stress—the portion of the applied stress that is carried by the soil grains—is equal to the difference between the total applied stress and the fluid pressure in the soil pores (see Vol. XII, Pt. 1: *Soil Mechanics*). For water-saturated soils, this relation may be written as follows:

$$\sigma' = \sigma - u \qquad (1)$$

where σ' is the effective stress in the soil grains, σ is the total stress applied to the soil mass, and u is the pore pressure.

Mechanics of Consolidation

A soil mass consolidates as the effective stress in the soil grains is increased. According to Figure 1, this will occur either when the external stress on a soil mass is increased or when its pore pressure is reduced. The former takes place when buildings and embankments are constructed on the ground surface. The latter occurs when fluids such as water or oil are extracted from the ground.

Consolidation occurs rapidly in a dry soil mass because the skeleton of soil grains is much less compressible than the air in the soil voids. An applied stress is carried immediately and directly by the soil grain skeleton. Conversely, consolidation occurs slowly in a saturated soil mass. In this case, the skeleton of soil grains is more compressible than the water in the soil voids. An applied stress, therefore, is initially carried by the soil pore water and is gradually transferred to the soil grains as water is expelled from the soil voids.

The mechanics of consolidation in a saturated soil mass and the interrelated principle of effec-

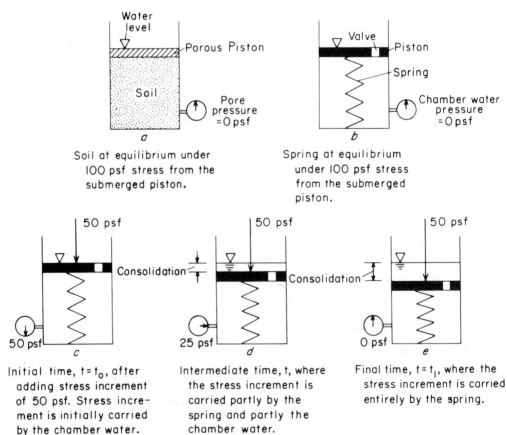

a — Soil at equilibrium under 100 psf stress from the submerged piston.

b — Spring at equilibrium under 100 psf stress from the submerged piston.

c — Initial time, $t = t_0$, after adding stress increment of 50 psf. Stress increment is initially carried by the chamber water.

d — Intermediate time, t, where the stress increment is carried partly by the spring and partly the chamber water.

e — Final time, $t = t_1$, where the stress increment is carried entirely by the spring.

FIGURE 1. Hydromechanical analogy of consolidation in a saturated soil mass.

tive stress are illustrated by the hydromechanical analogy shown in Figure 1. The soil model shown in Figure 1a consists of a soil mass whose cross-sectional area is 1 ft^2 that is loaded by a porous piston having a submerged weight of 100 lb. The voids in the soil mass and the porous piston are filled with water. This piston represents a layer of material that is highly permeable compared to the soil mass under study. For the purposes of this illustration, the piston provides a means for applying stress to the soil mass and allowing water to escape from the soil as it consolidates under the applied load.

The hydromechanical analogue shown in Figures 1b through 1e simulates the soil model. The spring represents an elastic soil grain skeleton and its resistance to compression. The water in the chamber represents the water in the soil voids. The impermeable piston, having a buoyant weight equal to that of the porous piston shown in Figure 1a, contains a valve that governs the rate at which water can discharge from the chamber. The resistance that the valve imposes on the discharge of water from the chamber represents the resistance that the pores in the soil grain skeleton impose on the discharge of water from the soil voids.

Figures 1a and 1b show the initial state, where the soil mass and the spring are in equilibrium with the submerged weights of the pistons. In both figures the applied stress equals the submerged weight of the piston (100 lb) divided by the cross-sectional area of the model (1 ft^2); that is, 100 lb/ft^2, or psf. Because the systems are at equilibrium, the pore pressure shown in Figure 1a and the chamber water pressure shown in Figure 1b equal zero. Hence, the effective stress equals the applied stress, 100 psf, because the applied stress is carried solely by the soil grain skeleton shown in Figure 1a and by the spring shown in Figure 1b.

Figures 1c through 1e illustrate the time-dependent changes that take place when the applied stress is increased by 50 psf. Immediately after loading (time, $t = t_0$), this applied stress increment is carried by the pore water, as shown in Figure 1c. The pore pressure induced thereby causes water to discharge from the chamber. As time elapses, the loss of water from the chamber (or soil voids) is accompanied by compression of the spring (or soil grain skeleton), which leads to the transfer of the applied stress increment from the chamber water (soil pore water) to the spring (soil grain skeleton). Figure 1d shows this case at time t, when the chamber water (soil pore water) and the spring (soil grain skeleton) each carry half the applied load increment, so that the effective stress is increased from 100 to 125 psf. Figure 1e shows the final case at time $t = t_1$, when sufficient water has been discharged to permit the spring (soil grain skeleton) to carry the applied stress increment by itself. The system is again in equilibrium when the

chamber (soil pore) water pressure equals zero and the effective stress in the spring (soil grain skeleton) equals 150 psf, the sum of the initial stress plus the increment of stress applied to the system.

Figure 2 further illustrates the consolidation process in a soil stratum that is bounded above and below by pervious layers. For simplicity the water table is initially assumed to be at the top of the soil stratum. The figure shows the variation with depth of the total stress, the pore pressure, and the effective stress at four stages of the consolidation process.

The initial conditions shown in Figure 2a include the stratum thickness, H_0; the total stress, σ_0; the pore pressure, u_0; and the effective stress, σ'_0. Figure 2b shows the immediate changes induced by the application of a total stress increment, $\Delta\sigma$. The thickness is unchanged because no time has elapsed for pore water to be discharged from the system. The total stress is increased to $\sigma_0 + \Delta\sigma$. The stress increment $\Delta\sigma$ is initially carried solely by the pore water; hence the pore pressure equals $u_0 + \Delta\sigma$ and the effective stress is unchanged.

Figures 2c and 2d show how the conditions change during the consolidation process. The decreasing stratum thickness is described by $U\Delta H$, where ΔH is the total thickness change induced by the stress increment $\Delta\sigma$ and the consolidation ratio, U, describes the fraction of ΔH that has occurred during the consolidation process. (The consolidation ratio is frequently expressed in percentage and called the *percent consolidation*.) The total stress, $\sigma + \Delta\sigma$, remains unchanged. As pore water is expelled from the system, the pore pressure decreases and the effective stress increases. Figure 2c illustrates that during the consolidation process the pore pressures nearest the drainage boundaries approach equilibrium most rapidly, resulting in parabolic distributions of pore pressure and effective stress during consolidation. Finally, as shown in Figure 2d, the pore pressure returns to its initial-state value, u_0, after the applied stress increment, $\Delta\sigma$, is fully transferred from the pore water to the soil grain skeleton as an increase in effective stress.

Laboratory Consolidation Test

The laboratory consolidation test simulates the consolidation process shown in Figure 2. Figure 3 illustrates how, for this test, a cylindrical soil specimen is encased in a ring that prevents lateral deformation and how it is sandwiched between porous plates. The test procedure involves the application of a series of load increments to the porous plates. During each load increment the specimen thickness is measured as it varies with time, until compression virtually ceases. At this point the excess pore pressure has been dissipated and the effective stress is equal to the total stress.

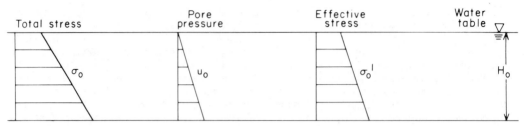

a. Initial (equilibrium) stresses in a saturated soil stratum that is bounded above and below by relatively pervious materials

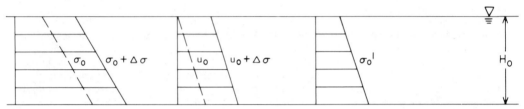

b. Stresses immediately after adding an increment of total stress, $\Delta\sigma$, to the stratum. Time, $t = t_0$.

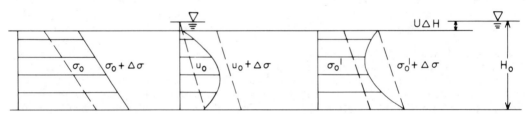

c. Stresses after a time, t when the soil stratum has partially consolidated under the total stress increment, $\Delta\sigma$, and some pore water has been extruded from the stratum.

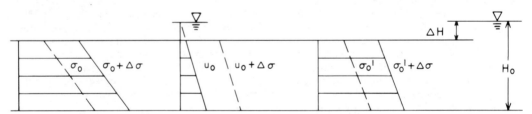

d. Stresses after complete consolidation at time $t = t_1$. The pore pressure has returned to its initial value and the total stress increment, $\Delta\sigma$, is supported entirely by the effective stress

FIGURE 2. Distribution of stresses in a saturated soil stratum during consolidation (modified from Sowers and Sowers, 1970).

The one-dimensional consolidation of a soil mass, either in a laboratory specimen or a horizontal soil stratum, is commonly expressed in terms of its vertical strain and its void ratio. These quantities and the relation between them are illustrated in Figure 4. The *vertical strain, $\Delta H H_0$*, is simply the ratio of the change in thickness of the soil mass, ΔH, to its initial thickness, H_0. The *void ratio, e,* is defined by the ratio of the volume of voids, V_v, to the volume of solids, V_s; that is, $e = V_v/V_s$. In Figure 4 the volume of solids is constant and for convenience has been defined as unity. It follows that $V_v/V_s = V_v = e_0$, $V/V_s = V = 1 + e_0$, $\Delta V = \Delta V_v = \Delta e$, and finally

$$\frac{\Delta H}{H_0} = \frac{\Delta e}{1 + e_0} \qquad (2)$$

This equation is commonly used to relate the void ratio changes in a consolidation test to the thickness changes in a horizontal soil stratum.

The consolidation test produces two experimental relationships from a sample: (1) The variation of the soil void ratio with effective stress, which provides a basis for estimating the magnitudes of consolidation in practical problems; and (2) the variation of the soil void ratio with time for each loading increment, which is used with the theory of consolidation to estimate rates of consolidation in practical problems.

The relationship between void ratio and effective stress obtained in a consolidation test is illustrated in Figure 5, which displays the data on arithmetic and semilogarithmic plots. This relationship is based on a sequence of loading, unloading, and reloading increments. The *virgin loading curve* shows the behavior that occurs in soil that

has not previously been consolidated. The *unloading curve* shows the expansion that takes place in soils that are unloaded by geological processes, during engineering activities, or by extraction of samples from the ground. The *reloading curve* shows how the behavior of preconsolidated soil differs from its behavior in the virgin state. In addition, Figure 5b shows the behavior of soil in the remolded state to illustrate the profound effect that sample disturbance can have on the results of a consolidation test.

The principal characteristics of the relationship void ratio versus effective stress are the following: (1) the *coefficient of compressibility, a_v,* is the slope of the relation on an arithmetic plot,

$$a_v = \frac{de}{d\sigma'} \qquad (3)$$

(b) the *compression index, C_c,* is the slope of the relation on a semilogarithmic plot,

$$C_c = -\frac{de}{d(\log \sigma')} \qquad (4)$$

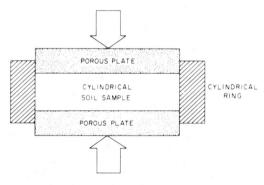

FIGURE 3. Cross section of the elements of the laboratory consolidation test apparatus. The cylindrical ring prevents lateral deformation of the soil sample. The porous plates permit drainage of water from the top and bottom surfaces of the soil sample.

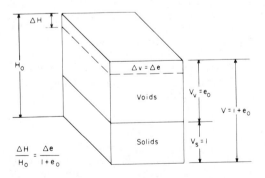

FIGURE 4. Relationship between thickness and void ratio changes associated with the one-dimensional consolidation of a soil mass. For simplicity, the model has a volume of solids equal to unity.

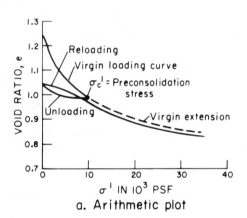

a. Arithmetic plot

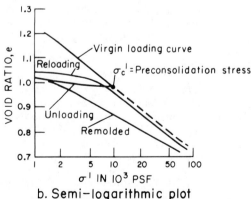

b. Semi-logarithmic plot

FIGURE 5. Variation of soil void ratio, *e*, with effective stress, *σ'*, in virgin loading, unloading, and reloading phases of consolidation (modified from Sowers and Sowers, 1970).

107

and (c) the *preconsolidation stress* is the maximum effective stress to which the soil has previously been consolidated.

The semilogarithmic plot of this relationship is particularly useful because the virgin curve is approximately a straight line. This provides a constant value of C_c for the virgin compression of a given soil. Also, because of the abrupt change in C_c between the recompression and virgin ranges, this plot provides a convenient basis for defining the magnitude of the preconsolidation stress, σ_c'.

The preconsolidation stress, σ_c', is used together with the effective overburden stress, σ' *(in situ)*, to define the initial consolidation state of a soil mass. The ratio of these stresses is known as the *overconsolidation ratio*, OCR, where

$$OCR = \frac{\sigma_c'}{\sigma' \ (in \ situ)} \qquad (5)$$

The OCR equals 1.0 in a soil that is normally consolidated. This term describes the case where a soil is fully consolidated under the weight of the overlying material. The OCR exceeds 1.0 in soils that have been consolidated in the past under effective stresses greater than are currently acting on them. This condition occurs where overburden materials have been removed by geological processes or engineering activities and also where soils have been consolidated by desiccation. The OCR is less than 1.0 where soils are not yet fully consolidated under the weight of the overlying materials.

The magnitudes of consolidation in compressible strata are commonly estimated from the relations described by equations (2), (4), and (5). Combining equations (2) and (4),

$$\Delta H = \frac{H_0 C_c}{1 + e_0} \ \Delta \ (log \ \sigma') \qquad (6)$$

where ΔH is the change in stratum thickness induced by the anticipated increase in effective stress in the material. The effective stresses and the OCR govern the selection of C_c from the e-versus-log σ' relationship. For a normally consolidated clay, the appropriate C_c is the slope of the virgin curve. In a heavily preconsolidated clay, the appropriate C_c is the slope of the reloading curve. In cases where the clay is initially preconsolidated and the effective stress is increased beyond the preconsolidation stress, the consolidation can be estimated in terms of its reloading and virgin loading components.

For each increment of load in the consolidation test, the variation of the soil void ratio with time is illustrated in Figure 6. The first stage of consolidation, which occurs instantaneously, is known as the *initial consolidation*. The second stage is known as the *primary,* or *hydrodynamic, consolidation;* this stage is accompanied and governed by the

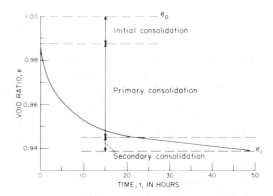

FIGURE 6. Variation of soil void ratio with time for an increment of load.

expulsion of water from the soil voids, illustrated in Figures 1 and 2. The final stage, *secondary consolidation,* is controlled by plastic deformation of the soil fabric, which is small for inorganic clays but can be substantial in organic clays. For inorganic clays, the magnitude of the *primary consolidation* is much the greatest of the three, and is the stage to which the following discussion mainly applies.

One-Dimensional Theory of Consolidation

The time rate of consolidation in a soil stratum can be estimated from laboratory data using Terzaghi's one-dimensional theory of consolidation (Terzaghi, 1943). This theory is based on the following assumptions: the soil is homogeneous; its voids are completely filled with water (i.e., saturated); the compressibilities of its water and solid constituents are negligible; the movement of water through the soil pores is governed by Darcy's law; the permeability and compressibility of the soil remain constant throughout the duration of a loading increment; the soil is laterally confined so that volume change occurs only in the vertical direction; and water movement within and out of the soil occurs only in the vertical direction.

Terzaghi's theory of consolidation describes the degree to which consolidation has progressed with time in a soil mass in terms of the consolidation ratio, U, and a dimensionless time factor, T, as illustrated in Figure 7 and expressed in the following equation:

$$U = f(T) \qquad (7)$$

This relationship is a mathematical function that is independent of the soil characteristics or the amount of consolidation. It does vary with initial loading and drainage conditions, however. Figure 7 applies to the case illustrated in Figure 2, in which the induced excess pore pressure is initially

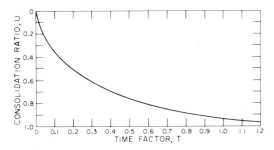

FIGURE 7. Variation of the consolidation ratio, *U*, with the time factor, *T*, for an increment of load. The relationship shown is for a soil stratum drained at both faces.

constant with depth and drainage is unrestricted above and below the stratum. For other loading and drainage conditions of practical importance, appropriate *U*-versus-*T* relationships are readily available in the literature (Lambe and Whitman, 1969; Leonards, 1962; Sowers and Sowers, 1970; Taylor, 1948; Terzaghi and Peck, 1968; and others).

The parameters in Figure 7 are related to those in Figure 6 as follows: the consolidation ratio equals the change in void ratio attained at any time, *t*, divided by the total change in void ratio that occurs during a given loading increment. It is expressed by

$$U = \frac{e_0 - e}{e_0 - e_1} \qquad (8)$$

where *e* is the void ratio at time *t*, and e_0 and e_1 are the initial and final void ratios, respectively. The time factor, *T*, varies in part with time, *t*, and also with the properties and geometry of a soil mass. For horizontal soil strata, the theory of consolidation defines the time factor by

$$T = \left(\frac{c_v}{(H/N)^2}\right) t \qquad (9)$$

where *H* is the soil stratum thickness, *N* is the number of drainage boundaries associated with the stratum (commonly one or two), *t* is the time elapsed since the application of a loading increment, and c_v is the coefficient of consolidation.

The *coefficient of consolidation* is defined by the basic differential equation used by Terzaghi (1943) to describe the dissipation of pore water pressure at any point in a soil layer with time,

$$c_v \cdot \frac{d^2u}{dx^2} = \frac{du}{dt} \qquad (10)$$

where the *x* coordinate is measured downward to the point in question from the top of the consolidating layer and c_v is a complex soil property that

is proportional to the permeability of the soil, *k*, and inversely proportional to the *soil compressibility*, a_v, as follows:

$$c_v = \frac{k(1 + e)}{a_v \gamma_w} \qquad (11)$$

In this relationship, γ_w is the unit weight of water and the other terms are as previously defined.

C_v values are obtained from laboratory measurements by comparing the experimental relationship between soil void ratio versus time, illustrated in Figure 6, with the theoretical relationship between the consolidation ratio and the dimensionless time factor illustrated in Figure 7. For this purpose, various curve-fitting techniques have been developed that are readily available in the literature (Lambe and Whitman, 1969; Sowers and Sowers, 1970; Taylor, 1948; Terzaghi and Peck, 1968; and others).

Finally, the time rate of consolidation can be estimated—for example, in a horizontal soil stratum—as follows: the variation of the time factor *T* with time *t* is calculated from equation (9) using the geometry of the stratum and a determination of the coefficient of consolidation as described above. This relation can be substituted for *T* on the abcissa in Figure 7, which leads to a relation between the consolidation ratio *U* and time *t* for a particular soil stratum.

H. W. OLSEN
R. L. SCHUSTER

References

Bishop, A. W., and Blight, G. E., 1963. Some aspects of effective stress in saturated and partially saturated soils, *Géotechnique*, **13**, 177-197.
Krynine, D. P., and Judd, W. R., 1957. *Principles of Engineering Geology and Geotechnics.* New York: McGraw-Hill, 730p.
Lambe, T. W., and Whitman, R. V., 1969. *Soil Mechanics.* New York, Wiley, 553p.
Lee, I. K. (ed.), 1968. *Soil Mechanics: Selected Topics.* New York: American Elsevier, 625p.
Leonards, G. A. (ed.), 1962. Engineering properties of soils, in *Foundation Engineering.* New York: McGraw-Hill, 66-240.
Skempton, A. W., 1961. Effective stress in soils, concrete, and rocks, in *Pore Pressure and Suction in Soils.* London: Butterworths, 4-16.
Sowers, G. B., and Sowers, G. F., 1970. *Introductory Soil Mechanics and Foundations,* 3rd ed. New York: Macmillan, 556p.
Taylor, D. W., 1948. *Fundamentals of Soil Mechanics.* New York: Wiley, 700p.
Terzaghi, Karl, 1925. *Erdbaumechanik auf bodenphysikalischer Grundlage.* Vienna: Franz Deuticke, 399p.
Terzaghi, Karl, 1943. *Theoretical Soil Mechanics.* New York: Wiley, 510p.
Terzaghi, Karl, and Peck, R. B., 1968. *Soil Mechanics in Engineering Practice,* 2nd ed. New York: Wiley, 729p.

Cross-references: *Atterberg Limits and Indices; Clays, Strength of; Geotechnical Engineering; Marine Sediments, Geotechnical Properties; Soil Mechanics.* Vol. VI: *Compaction in Sediments.* Vol. XII, Pt. 1: *Pore-Size Distribution; Soil Mechanics; Soil Pores* Vol. XIV: *Soil Fabric.*

CONSTRUCTION MATERIALS — See ARID LANDS, ENGINEERING GEOLOGY; CALICHE, ENGINEERING GEOLOGY; COASTAL ENGINEERING; DAMS, ENGINEERING GEOLOGY; LATERITE, ENGINEERING GEOLOGY.

D

DAMS, ENGINEERING GEOLOGY

A *dam* is an engineering structure constructed across a valley or natural depression to create a water storage reservoir. Such reservoirs are required for three main purposes: (1) provision of a dependable water supply for domestic and/or irrigation use; (2) flood mitigation; and (3) generation of electric power.

In providing a water supply, the reservoir storage is filled during the periods of above-average streamflow, thus ensuring a steady supply of water during periods of little or no streamflow. For flood mitigation, the storage reservoir is kept nearly empty during drought and periods of low rainfall, so that when the flood-generating rainstorms occur, the storage volume available in the reservoir provides a buffer against severe flooding in the river valley downstream of the dam. For power generation, the storage reservoir provides a head of water upstream of the dam, and the potential energy of this water is converted first to kinetic energy by passing the water through turbines, and then to electrical energy by generators.

A large dam has two essential requirements. First, it must be reasonably watertight. Therefore, either the dam is constructed of impermeable material (e.g., concrete), or it incorporates an impermeable membrane in its structure (e.g., an earth core). Also, the dam foundations must be made watertight by grouting (q.v.) or other means if necessary. Second, the dam must be stable. Movement and deformation of the dam and its foundations cannot be eliminated, but they must be predicted and allowed for in the design.

Because of these requirements, the location and design of dams are invariably influenced to some extent by geological features. In many cases, geological factors such as foundation conditions and the proximity of construction materials are of overriding importance in determining the type of dam constructed at a given site. It therefore follows that a detailed knowledge of the geology of a prospective dam site and its environs is necessary before an informed decision can be made on the most suitable dam design and the estimated cost of construction.

The most important basic geological data required are the distribution and nature of the various rock types present in the area, the weathering profile, and details of the structural geology. These data are obtained by a site investigation program that uses a wide range of data-gathering techniques, such as outcrop mapping (see *Field Geology*), bulldozed trenching to expose bedrock below overburden (see *Pipeline Corridor Evaluation*), diamond core drilling (see Vol. XIV: *Borehole Drilling*), water pressure testing, geophysical surveys (q.v.), joint surveys, and laboratory testing of rock samples. The geological and physical data so obtained are then integrated to form a geomechanical model of the site, which provides the design engineer with a reasonably realistic and quantitative basis on which to design the dam and its associated structures. This process of collecting relevant geological data and presenting them in a form useful to the engineer is the main function of the engineering geologist (Attewell and Farmer, 1976; Bell 1980; Leggett, 1962; Paige, 1950).

Types of Dams

In designing a dam for a particular site, the engineer has several basic types of dams to choose from (Thomas, 1976). Figure 1 summarizes the layout and significant characteristics of these basic dam types.

At some dam sites, the most economical design has been a composite of two or more basic dam types. One particular type of composite concrete dam is the multiple arch design, which consists of several cylindrical arches supported by buttresses. This design is well suited to sites with geologically variable foundations—the buttresses are located on strong parts of the foundations, while the arches are located so that they bridge weak zones in the foundations.

Dam Foundations

The foundations of a dam have to support the weight force of the dam, plus a significant component of the force that the reservoir of water exerts on the upstream side of the dam. For the dam types illustrated in Figure 1, there is a progressive decrease in the area of foundations, for a given dam height, from the earth dam (largest area) to the double curvature arch dam (smallest area). This means that the bearing pressure (i.e., force per unit area) that must be supported by the foundations progressively increases from a minimum for an earth dam to a maximum for the arch dam; in other words, the sequence of dam types

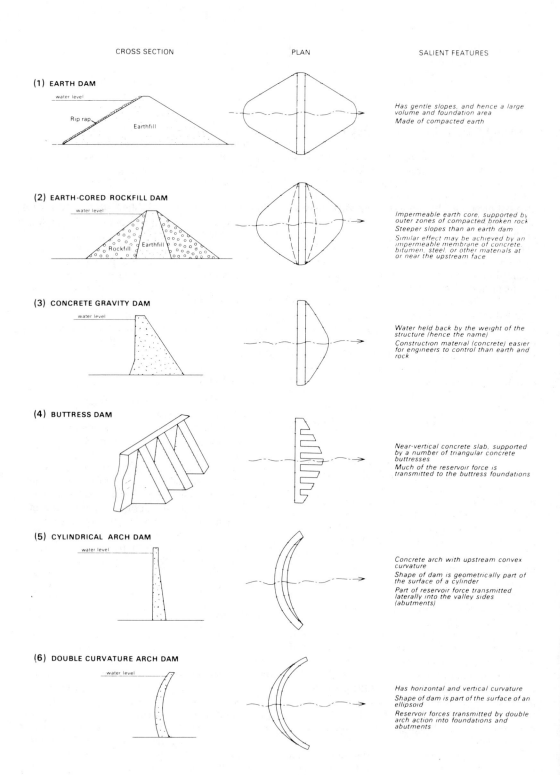

FIGURE 1. Basic types of dam design (from Best, 1981).

from (1) to (6) in Figure 1 requires progressively stronger foundations. The overall strength of dam foundations is determined by the detailed geological characteristics of the area; it therefore follows that foundation geology (see *Foundation Engineering*) at a proposed dam site is an important factor in deciding the most appropriate type of dam for the site (Walters, 1971; Wahlstrom, 1974).

Construction Materials

Dams are constructed from large volumes of naturally occurring earth materials. For example, an earth-cored rockfill dam requires regolith with certain physical properties for the earth core, sand and gravel for filter zones, and broken rock for rockfill; a concrete dam requires suitable broken rock or gravel for concrete aggregate, together with sand and cement. These construction materials must be obtained from as close to the dam site as possible if the dam is to be economically feasible. Therefore, the location and cost of extraction of construction materials constitute another important factor in determining the type of dam to be built. For instance, a site may have highly weathered bedrock (see Vol. XIV: *Rock Weathering Classification*) that appears to be suitable only for the foundations of an earth dam. If there is little suitable earth material close to the site, however, it may well be cheaper to excavate the foundations to a depth suitable for a concrete dam than to transport earth material over a long distance to the site.

The location of adequate quantities of suitable construction materials is determined very much by the bedrock geology and the geological history of the area. It is therefore fairly obvious that a search for construction materials should be based on detailed geological mapping (q.v.)—both bedrock and surficial—of the area around the dam site. However, the cost of extraction is based not only on the volume of the particular required material that is available, but also on the geometry of the deposit and its relationship to the ground surface. For instance, a quarry developed in a particular rock sequence to supply concrete aggregate will invariably involve the unavoidable extraction of a substantial volume of unsuitable material because of the geological variability (often caused by folding and faulting) of the quarry area. The expense of extracting and dumping such unusable material is unavoidable, and can be minimized only by a thorough geological investigation of alternative sources of construction materials so that those finally selected are the most economical for dam construction.

Some idea of the volumes of construction materials required for large dams can be obtained from some dam statistics. The largest dam (by volume) yet constructed is Tarbela Dam in Pakistan, which is an earth dam containing 122×10^6 m^3 of materi-

al. Several large dams have also been constructed in North America, such as those at Fort Peck, Montana (96×10^6 m^3), Oahe, South Dakota (70×10^6 m^3), Gardiner, Saskatchewan, Canada (66×10^6 m^3), and Oroville, California (60×10^6 m^3). The largest concrete dam in the world is Sayany in the USSR (9.1×10^6 m^3), followed by Grande Coulee Dam in the United States (8.5×10^6 m^3).

Choice of Dam Type

Once a site has been selected for a dam, consideration has to be given to deciding which type of dam is most suited to the site. Some dam types may be impracticable because of the topography of the site; for instance, the valley may be too wide for the construction of an arch dam, or the valley sides may be too steep for an earth dam. However, at any site several types of dam should be considered, and estimates of quantities of materials, cost of materials, amount of foundation excavation, amount and type of foundation treatment, costs of river diversion and spillway construction, and so on, should be prepared for each type. The final choice should be the dam with the lowest estimated construction cost. In general, three factors control this final decision: (1) the topography of the dam site and reservoir area; (2) the strength and variability of the foundations; and (3) the availability and suitability of construction materials. These factors are largely controlled by the geological structure and history of the site, and an informed decision requires a great deal of geological data analysis and interpretation, particularly for the second and third factors, presented in a manner that the engineer can use in design calculations.

The influence of topography and foundation geology on the selection of dam type is particularly evident where composite dams are constructed. For example, when studying the site for Aviemore Dam in New Zealand, investigators discovered a fault beneath the river valley, with good dam foundations on one side of the fault and poor foundations on the other. Accordingly, it was decided to construct a concrete gravity dam, incorporating the power station, on that part of the site with good foundations, and an earth-cored rockfill dam on the poor foundations, with the two dam types abutting against each other in the middle of the valley. Another example is the Rio Torto Dam in Italy, which has a concrete arch dam across the main river valley and a buttress dam and an earth dam section farther up the valley sides.

Site Investigation for Dams

The progress from an initial proposal to dam a river to the completion of construction of a dam is marked by a series of investigations of the site, using a wide range of techniques such as surveying and various engineering tests. However, the major component by far of a dam site investigation is the geological investigation of the site, the reservoir area, and prospective sources of construction materials. The importance of geology in dam construction is evident in just about all technical reports (see *Engineering Geology Reports*) prepared during the investigation of major dams; many case histories have been published in engineering and geological journals and conference proceedings.

The ultimate aim of a site investigation is the identification of a geomechanical model, with quantitative parameters, which represents the engineering properties of the site, and on which the design of the dam can be based. Intermediate stages in this process are the development of (1) a geological model, which describes the lithology, geological structure, weathering, and geological history of the site; followed by (2) an engineering geological model, in which the physical properties of the geological features are described in some semi-quantitative manner (e.g., rock materials properties; description of joints; orientation, spacing, and continuity analysis of joints). From the engineering geological model, quantification of physical properties using field and laboratory tests (often accompanied by some reasonable simplification of the geological model) produces the geomechanical model.

Phases of Investigation

The geological investigation of a large dam is usually carried out in several phases associated with the sequence of engineering decisions to be made. Following is a brief description of the general sequence of dam site investigations.

Preliminary or Reconnaissance Investigation. At this stage, alternative locations are being considered for the dam, and the objective is to provide sufficient data to select the best (cheapest) site for dam construction.

Feasibility Investigation. Once the preferred site has been selected, investigations are carried out to confirm the economic feasibility of the project at that site. Alternative types of structures are evaluated at this stage (e.g., concrete dam versus rockfill dam), and the investigation should provide sufficient data to allow comparative designs and cost estimates to be prepared.

Design Investigation. Once the general type of engineering structure has been determined, still more information is required for the detailed design of the structure. For instance, it may have been decided that a rockfill dam be constructed, but the spillway and diversion tunnel have yet to be specifically located. The data from this phase are used to complete the detailed design of the structure, to work out quantity estimates (such as volumes of rock excavation and construction materials) and to draw up the tender documents. This is a critical stage, as all geological conditions to be

encountered during construction should be predicted and addressed in the design documents.

Construction Investigation. In any major project, conditions actually encountered may necessitate changes in the proposed design—one hopes that these will be minor and result in little extra cost for the project. However, to ensure that variations from the anticipated geological conditions are detected as soon as possible, a systematic and progressive record of the geology exposed as construction progresses must be maintained. If conditions are very much as predicted, then little or no investigation will be required, but if a major unexpected geological problem arises, a proper investigation using a wide range of techniques may be necessary. The important feature of this stage is the maintenance of detailed, up-to-date records of the "as-constructed" geology of the entire site.

Post-Construction Investigation. Once the structure or project becomes operational, regular inspections should be undertaken to monitor the behavior of the structure and to detect any unexpected changes that could affect the stability or efficient operation of the structure (e.g., blocking of ground-water drains, instability of rock or soil slopes, deterioration of exposed rock). If a problem does occur, a properly planned site investigation may be required to evaluate the problem and provide a solution. The plans of the "as-constructed" site geology are invaluable in evaluating post-construction problems.

Site Investigation Techniques

During site investigations, information is required from the surface and near-surface bedrock, from the rock at depth, and on the nature and distribution of surficial materials. Following are brief details of the main techniques used to obtain this information. (For more detailed descriptions, see Attewell and Farmer, 1976; Bell, 1980; and Best and Hill, 1967.)

Geological Mapping. The main component of all site investigations is detailed large-scale mapping of natural and artificial exposures of bedrock (and of surficial materials where appropriate). The scale and detail of mapping depends to some extent on the phase of the investigation and the geological complexity of the site, but scales of between 1:100 and 1:500 are common. It is usually essential to use a detailed and accurate topographic map as a base for such mapping, to enable three-dimensional analysis of geological structures (see Vol. XIV: *Exploration Geology*). One very important point about this mapping is that factual data should be kept separate from interpretive data. The geological map should show as much information as it is possible to record from the field observations, and should include as much quantitative (e.g., orientation) data as possible. Geological inferences and interpretation must inevitably be made from these basic data, but should be presented on separate, clearly labeled, interpretive plans and sections. Special engineering geological maps and sections are often prepared to present the geological information in the most appropriate manner (UNESCO, 1976).

Trenching and Pitting. Dam sites rarely have anything like 100-percent exposure of bedrock; 10 to 25 percent is much more common. At such sites, there will be very large gaps in the geological map where soil and scree deposits occur. Once the outcrop geology map has been prepared, critical areas of covered bedrock can be selected for exposure by bulldozing away the overburden. Apart from extending the knowledge of bedrock geology, bulldozed trenches may also be used to provide information on soil and weathering profiles, and to assess the excavation characteristics of selected areas (see *Pipeline Corridor Evaluation*).

Excavation of pits with a mechanical backhoe is done extensively in the mapping and evaluation of prospective sources of construction materials such as core material, sand, and gravel. The information obtained by mapping the surficial deposits exposed in these pits enables a three-dimensional picture of their distribution to be drawn, which is essential for the economic evaluation of alternative sources of construction materials.

Sluicing. At geologically complex sites, narrow strips of exposed bedrock from trenching are sometimes not sufficient to provide the required information; larger areas of foundations need to be exposed. If the overburden thickness is not too great, most of the overburden can be bulldozed away and the bedrock can then be cleaned with high-pressure water jets. While this technique may appear costly, the large amounts of geological information obtained and the increased reliability of the geological interpretation may well justify the expense. Another important advantage of this technique is that it makes available large areas of the bedrock for inspection by prospective tenderers; consequently, the prices for foundation excavation and treatment will be more realistic than if the bedrock were visible only in drill cores and isolated pits and trenches.

Diamond Core Drilling. This is the most common way to obtain information and samples from depth at an engineering site. It is an expensive technique, so great care needs to be taken in planning drill holes to obtain the maximum amount of geological information with the minimum of drilling. A 50-mm-diameter core is commonly used for site investigations; it should be obtained with a special core barrel incorporating a split inner tube capable of hydraulic ejection at the end of each drilling run. The drilling must be of high quality, and 100-percent core recovery should be obtained in all but very bad ground. Care should be taken to recover all seams and crushed zones of rock, as

these are much more important than the sticks of core obtained from fresh rock. The driller's skill and motivation are the most important factors in diamond drilling for engineering site investigation (provided, of course, that the right equipment is used), as it is essential that the core recovered is as undisturbed as possible by the drilling process (see Vol. XIV: *Borehole Drilling*).

Water Pressure Testing (Lugeon Testing). Quantitative data on the leakage properties of rock can be obtained by conducting water pressure tests in drill holes. Packers are used to isolate a section of the drill hole, water is pumped into the section under a known pressure, and the rate of water leakage into the rock is measured. Leakage rates are measured for a range of water pressures, and after appropriate corrections are made to the field data, the results give a good indication of the water tightness and degree of open jointing in the rock at depth.

Exploratory Shafts or Adits. Where an evaluation of the rock foundations at some distance below the ground surface is necessary for engineering design criteria (e.g., the abutments of an arch dam), direct access to these areas must be furnished by shafts or adits (see *Shaft Sinking*). Careful excavation techniques should be used in the areas of interest, as excessive blasting will give an incorrect picture of the rock properties. Detailed geological mapping of the walls, floor, and roof of the excavations is carried out to provide a three-dimensional picture of the rock mass properties.

Geophysical Surveys. Seismic and resistivity surveys (see Vol. XIV: *Exploration Geophysics*) are frequently carried out during dam site investigations, as they provide very useful data that complement the information from diamond drilling. They are also useful in assessing "average" geological conditions, and can give information on rock mass properties that diamond drilling cannot provide. They cannot, however, be relied on to evaluate particular geological features (such as an individual clay seam or a narrow shear zone), and they are no replacement for a drill hole in identifying what the rock material is at a particular point some distance below ground surface.

The seismic refraction method (see *Seismological Methods*) is the most common geophysical technique used in engineering site investigations; special equipment has been specifically developed for such investigations. The information obtained enables useful extrapolation and interpolation of drill-hole data, such as depth of weathering and tightness of jointing, and also provides values for certain physical properties of the rock mass (such as moduli of elasticity) (see Vol. XIV: *Well Logging*).

Laboratory Testing. The quantification of rock material properties for the engineering geological and geomechanical models is provided by laboratory tests on drill-core specimens. Parameters commonly determined include density, porosity, compressive strength, shear strength, Young's modulus (both by direct tests and by geophysical tests), and Poisson's ratio (see *Rock Mechanics; Soil Mechanics*). Some physical properties of defects in the rock—such as the friction characteristics of joints—can also be determined on core samples.

Field testing. Some important rock foundation properties such as shear strength, friction characteristics, and module of elasticity can be determined by in situ field tests. They usually give much more reliable results than equivalent laboratory tests because the rock is tested in its natural environment. Also, the volume of the tested sample is much larger than laboratory test specimens, and so is much more likely to be representative of the rock mass. On the other hand, in situ field tests are very expensive, and even on major projects only a small number of tests can be afforded. It is therefore important that the test sites be selected with great care or unrepresentative data may be obtained.

Another category of field testing includes trial tests of engineering procedures that need to be evaluated before construction commences. For example, a trial blast (see Vol. XIV: *Blasting and Related Technology*) may be carried out at a proposed quarry site to confirm that suitable rock for construction materials is obtainable. Other such field tests include trial grouting, trial ripping, trial loading, trial rockbolting, and trial compaction tests.

Examples of the Influence of Geology on Dam Location, Design, and Construction

The importance of geology in dam construction is evident in just about all technical reports (see *Engineering Geology Reports*) prepared during the investigation of major dams; many case histories have been published in engineering and geological journals and conference proceedings. To illustrate the importance and influence of geology in dam design and construction, some details will now be given on certain aspects of the investigations for two dams that have been constructed to provide the water supply for Canberra, the capital city of Australia.

Bendora Dam. After federation of the six Australian states to form the Commonwealth of Australia in 1901, an area of about 2,400 km in southeastern Australia was set aside to form the Australian Capital Territory (A.C.T.). The construction of Canberra, the Federal Capital, commenced in 1910, and since then four dams have been constructed to provide all the water supply requirements for the city.

Bendora Dam was the second of these dams, and at the time of the preliminary site investigation (1953), four possible sites—designated A, B, C, and D—on the Cotter River had been selected

on the basis of topography. The regional geology near these sites is shown in Figure 2. The early geological mapping of the sites indicated that site B, located on granite, would be the best site, followed by site C, which had foundations of gently dipping quartzite. Site A was considered less suitable because it had foundations of both granite and quartzite and was very close to a major fault zone (the Cotter Fault), while site D was rejected because it straddled the Cotter Fault. Therefore, although geological investigations continued on sites A, B, and C, there was an initial concentration on site B. Later mapping raised doubts about this site, however, as a deep zone of weathered granite was suspected on the east bank. This suspicion was followed up by a geophysical survey, which indicated weathered rock to depths of 20 to 30 m; site B was therefore rejected.

Although at this stage site C appeared to be geologically preferable to site A, it had the disadvantage of having a smaller reservoir capacity than a dam of similar height at site A. In 1955, the

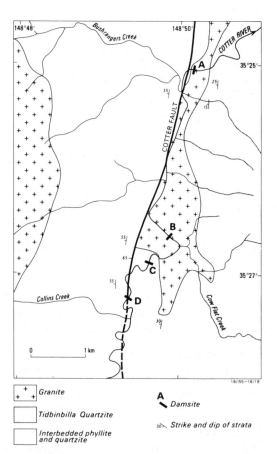

FIGURE 2. Regional geology of the Cotter valley near Bendora Dam, showing prospective dam sites (from Best, 1981).

Parliamentary Works Committee recommended that the next water supply dam be constructed at site A, and a program of diamond drilling and trenching was carried out to test the suitability of the site for a double curvature arch dam. These investigations showed that the foundations of the site were not strong enough for an arch dam but were suitable for a concrete gravity dam, which would be more expensive. A similar investigation was therefore carried out at site C, which showed that the foundations were adequate for a double curvature arch dam. Because of the significant difference in the amount of concrete necessary for these two dam types, a geological survey for suitable construction materials was done, which revealed a scarcity of suitable sand for concrete but more than adequate resources of rock for concrete aggregate.

Despite the significant advantage of site A with respect to the reservoir storage/dam height relationship, a costing by the engineers on the basis of the geological information available showed that an arch dam at site C would be cheaper to construct than the lower concrete gravity dam necessary at site A to give the same storage capacity. The final decision was made in 1958 to design and construct a 47-m-high, double curvature arch dam at site C to form Canberra's next water supply storage reservoir. The dam was constructed during 1960 and 1961, and although some foundation stability problems were encountered, they were successfully investigated and overcome (Best, 1981).

Corin Dam. Because of the rapid growth of Canberra, site investigations for a third dam were under way before the completion of Bendora Dam. The site eventually selected was on the Cotter River, 14 km upstream from Bendora Dam, and after extensive feasibility and design investigations, a 75-m-high, earth-cored rockfill dam was constructed at the site between 1966 and 1968.

The general arrangement of Corin Dam and its associated structures is shown in Figure 3, together with a nominal dam cross-section to show the main features of the dam design. (No attempt has been made to show the foundation geology because of the variability of the lithology and the complexity of folding and faulting at the site.) In summary, the foundations consist of a sequence of interbedded quartzite, silicified sandstone, and laminated siltstone, which has been tightly folded and considerably faulted. The regional strike is northwest, and bedding dips are generally steeper than 45°. The dam itself is made up of three zones: the impermeable core zone (zone 1), the filter zone (zone 2), and the rockfill zone (zone 3). The upstream and downstream rockfill zones are both subdivided into four zones, which increase in coarseness from the filter zone outward; zone 3a was placed and compacted in 45-cm layers, 3b in 90-cm layers, 3c in 1.8-m layers, and 3d is a protective layer of very

coarse rockfill placed on both faces of the dam. This progressive grading of the material in a rockfill dam is an important design feature to prevent migration of fine-grained material from the core zone, which could lead to the failure of the impermeable membrane.

The investigation and construction of Corin Dam offered many examples of the influence of geology on dam construction. Following are details of one example from each of the preliminary, design, and construction phases of the investigation.

Location of the Dam. The site for Corin Dam was originally selected because of the pronounced widening of the Cotter valley immediately upstream, which results in a large reservoir storage capacity; downstream of Corin Dam, the Cotter River flows in a narrow, youthful valley with overlapping spurs and steep valley sides (Fig. 4).

This sudden change in the physiography of the Cotter valley is a direct result of variations in the regional geology. Downstream of Corin Dam, the Cotter River flows over a sequence of folded, northwesterly striking metasediments. This sequence is terminated just west of the river by the Cotter Fault—a major structural feature that has controlled the overall course of the Cotter River—and to the west of the fault there is a sequence of westerly dipping phyllites and quartzites. Upstream of the damsite, however, faulted blocks of quartz porphyry occur to the east of the main Cotter Fault, and the preferential weathering and erosion of this zone of porphyry (which is up to 600 m wide) has resulted in the development of a comparatively broad and straight valley floor. The valley sides are generally as steep as those downstream from the dam site, but the extra width of the valley floor has effectively doubled the storage capacity of an equivalent dam located in the downstream tract of the Cotter River.

Location of Construction Materials. The faulted blocks of quartz porphyry have also provided suitable core material close to the dam site. Weathered granitic rocks usually form good core-zone material; the weathered feldspars provide the clay binder that makes the material impermeable, while the residual quartz grains give the weathered material useful strength and rigidity properties. A program of augering and trenching during the design investigation was carried out in the areas where the quartz porphyry was exposed, and a spur of deeply weathered (up to 40 m) porphyry was located 1,300 m upstream from the dam site (Figs. 4 and 5). Engineering tests proved its suitability for use as core material, and the bulk of the 262,000 m^3 required for the impermeable core was obtained from this site.

The presence of the quartz porphyry in the valley floor upstream of the dam site also provided a cheap source of filter material for the dam. The broad valley floor on the porphyry, together with a

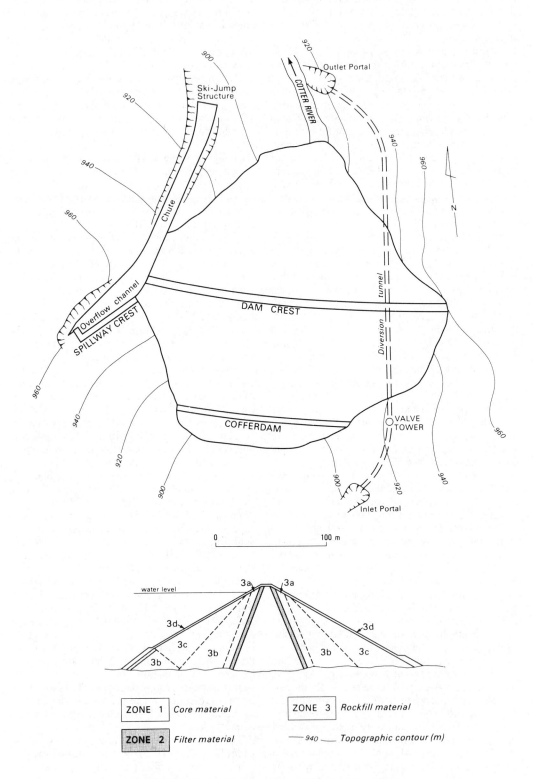

FIGURE 3. Design layout of Corin Dam (from Best, 1981).

FIGURE 4. Aerial view of the Cotter valley, looking downstream from the upper reaches of the Corin Dam reservoir (from Best, 1981).

slight but significant decrease in river gradient on the less resistant porphyry, resulted in the deposition of substantial volumes of alluvium in the valley within 1.5 km of the dam site (Fig. 5). A total of 89,000 m³ of filter material was placed in the dam, most of which was obtained from an extensive alluvial deposit 500 to 800 m upstream of the dam site.

Rockfill material for a dam must fulfill certain requirements of strength, size range, and shape. In particular, the material must not contain more than a specified proportion of fine-grained fragments, and must not disintegrate or slake during long-term saturation. In the context of Corin Dam, the rockfill had to be composed essentially of sandstone or quartzite, with only a small proportion of siltstone interbeds allowable.

During the feasibility investigation, a number of

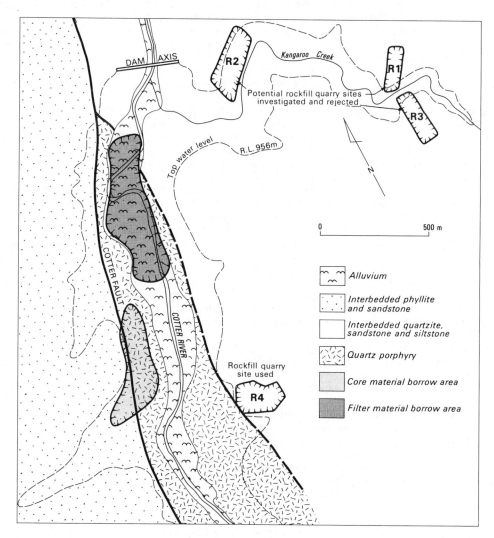

FIGURE 5. Geological map of the area immediately upstream of Corin Dam, showing sources of construction materials (from Best, 1981).

prominent outcrops of sandstone were noted close to the site, and one of these (R1 in Fig. 5) was investigated by a diamond drill hole. It was concluded that, should a rockfill dam be chosen for the site, a suitable source of rockfill would be present close to the site. This was also the opinion of outside consultants, who stated that the area labeled R2 in Figure 5 would probably be a suitable quarry site.

Once the decision had been made to construct a rockfill dam, R2 was investigated by diamond drilling, which revealed the presence of siltstone interbeds scattered throughout the sandstone sequence. It became evident that the siltstone/sandstone ratio was too high for suitable rockfill, so investigations moved to site R3. These showed that sites R1 and R3 combined could just provide sufficient rockfill material; however, as sandstone sequence dips at 80° and would have to be quarried to depths of up to 75 m, the waste rock/rockfill ratio would be high, giving a high cost per unit volume for the rockfill.

The search was then extended up the Cotter valley, and a promising site was located 1.5 km upstream. A prominent cliff of quartzite 45 m high was found exposed at the end of a westerly trending spur formed between two tributary creeks of the Cotter River. The quartzite dipped east at only 15°, and mapping along the spur to the east revealed a sudden change in bedding attitude to 75° west, suggesting a syncline that would bring the prospective quartzite back to the surface. Bulldozed trenching and diamond drilling confirmed the synclinal structure, and further detailed geological investigations showed that this would be an ideal location for the rockfill quarry.

The stratigraphic succession in the quarry area consists of four rock units. At the base is a sequence of well-bedded quartzite, occurring as beds ranging from 5 cm to 75 cm thick; this is overlain by the massive quartzite that formed the prominent cliff. On top of this quartzite is a sequence of sandstone with siltstone interbeds, which is overlain by laminated siltstone. Only the laminated siltstone sequence was unsuitable for rockfill, and this was removed from the site before quarrying for rockfill commenced. The proportion of siltstone in the upper-bed inter-bedded sequence was sufficiently high for excess fines to be formed, but mixing with good-quality rock reduced the volume of material rejected from the quarry to an insignificant amount.

As expected, the quarry produced rock of adequate durability and grading for use as rockfill. However, the engineering properties of the rockfill and the economics of quarrying were considerably enhanced by particular geological features of the quarry area:

1. The three rock sequences quarried had different ranges in spacing of bedding planes, which helped provide suitable rock for the four rockfill zones in the dam. The interbedded sandstone and siltstone provided material for zone 3b and, to a lesser extent, 3a; the massive quartzite provided rock for zones 3c and 3d; and well-bedded quartzite was used for most of zone 3a and some of zone 3b.
2. The folding and faulting of the rock sequence had formed well-developed joint sets, even at depth, and this resulted in very good and economical rock fragmentation during blasting. No secondary blasting was necessary during the entire quarrying operation.
3. The absence of a dominant joint set resulted in the formation of irregular, roughly equidimensional rock fragments, which are ideal for rockfill.
4. The folding of the geological succession was very convenient for quarrying operations, because benches could be developed so as to expose the different rock sequences in the same quarry face. This enabled working of the most suitable rock sequence to provide rockfill for any particular zone of the dam at any time. It also allowed easy mixing of good rock with the marginal-quality rock from the top of the interbedded sequence to produce suitable rockfill.

The total volume of solid rock quarried was 789,500 m³, which, when placed and compacted in the dam, produced 1,039,400 m³ of rockfill. This gives a bulking factor of 1.32, which indicates the high quality of the rockfill.

Stability of Spillway Crest. The spillway crest is a concrete structure that controls the overflow level of the reservoir. It is constructed on the lip of a concrete-lined excavation in rock, called the overflow channel, which directs the water down a concrete chute into the river valley downstream of the dam (Fig. 3). The force of water flowing over the crest exerts a considerable overturning moment on the crest structure; this force is resisted by a number of steel cables, anchored to the bedrock, which effectively tie the crest structure to solid rock at depth.

The overflow channel is a large excavation into bedrock, with a maximum vertical depth below original bedrock surface of 25 m on the uphill side. It was obviously desirable to design the walls of the excavation at as steep an angle as possible to minimize the amount of rock to be removed. A joint survey of the spillway area, carried out during the site investigation, revealed the presence of six joint sets in the rock; however, a stereographic analysis of joints indicated that none of these sets was unfavorably oriented with respect to a steep rock face on the uphill side of the proposed excavation. The overflow channel excavation was therefore designed with walls at 65° angles.

Excavation of the spillway channel proceeded according to the design, and no problems with rock slope stability were encountered. However, when the contractor blasted the rock on the downhill side of the excavation to form the spillway crest foundations, considerable overbreak of rock occurred at the lip of the excavation along two-thirds of the length of the crest area. The overbreak resulted from the unusual degree of development of one of the joint sets in this particular area of the spillway. Although the downhill wall of the excavation was only 5 to 7 m high, the well-developed joint set was adversely oriented with respect to the rock slope, and the forces imposed by the spillway crest structure would obviously aggravate the situation. A detailed joint survey of the crest area was therefore carried out to provide a basis for deciding how best to stabilize the crest and its foundations.

The joint set causing the problem dips at 30° in direction 340°: that is, it dips toward the channel excavation. It is present in bedrock over a large area of the western abutment, but in the spillway crest area the joints of this set are more closely spaced and much more continuous (have much larger joint surface areas) than elsewhere. This condition was not apparent during the original joint survey of weathered surface outcrops.

The original design for the spillway crest structure is shown in Figure 6, along with the modified design necessary to overcome the foundation instability. There were three stages in the stabilization of the spillway crest:

1. Installation of 137 grouted rock bolts, each 4 m long, along the wall of the excavation. These replaced the grouted anchor bars of the original design.
2. Placement of a concrete slab on the irregular rock surface of the crest foundation, followed by the installation of 46 grouted rock bolts, each 6 m long, on a 2 = m grid spacing (to tie the concrete slab to the bedrock and to consolidate the foundations).
3. Installation of 26 post-tensioned cables, each 11 m long and tensioned with a force of 498 MN (50 tons). In the original design, these cables were vertical and would therefore not contribute to the stability of the jointed rock. However, re-orienting the cables to an angle of 65° (i.e., roughly perpendicular to the adversely-oriented joint set), considerably enhanced the stability of the crest and its foundations.

This particular example is typical of the problems that arise during a major construction project as a result of the geological variability of rock, particularly at depth below the surface. While the remedial work resulted in extra cost (due mainly to extra materials and the three-stage construction of the concrete crest), the problem was recognized, investigated, and evaluated quickly enough to avoid costly delay in the overall construction schedule. The problem could not have been pre-

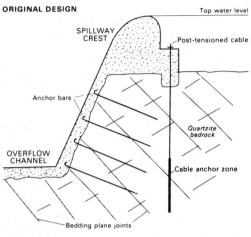

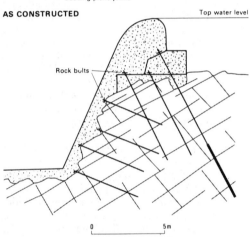

FIGURE 6. Sections through the spillway crest of Corin Dam, showing the original design and the design as constructed (from Best, 1981).

dicted from the information available before construction started.

The Importance of Geology in Dam Construction

When designing a large dam, the engineer has two prime goals: (1) the dam must be stable and successfully fulfill its function throughout its projected life; and (2) the dam and its associated structures must be constructed as economically as possible. Of course there is a dilemma here, because the two objectives act against each other; ensuring stability by overdesign increases the cost, while cost-cutting methods, if taken to extremes, will lead to an unsafe structure.

Stability of Dams

On a worldwide scale, it is clear that the objective of constructing stable dams is not always achieved. During the 1900–1965 period, for example, about 1 percent of the 9,000 large dams in service throughout the world have failed, and another 2 percent have suffered serious accidents; significantly, in more than half these incidents, the failure or damage could be related to geological causes (Stapledon, 1976). Brief details of some recent failures illustrate the magnitude of the problem.

Malpasset Dam, France—a 60-m-high double curvature arch dam that failed in 1959, causing 400 deaths. The cause of failure was uplift and sliding of a section of the rock foundations (Jaeger, 1963; Londe, 1967).

Vajont Dam, Italy—a 265-m high double curvature arch dam (at the time, the highest dam in the world). In 1963, some 260 million m^3 of rock slid into the reservoir just upstream of the dam, and over 2,000 lives were lost in the flood caused by the displaced water (Muller, 1963). Because the dam itself did not give way, this is considered a reservoir failure rather than a dam failure.

Baldwin Hills Storage Basin, U.S.A.—failed in 1963, after 12½ years of service, with the loss of five lives and $15 million in damages. The cause of failure was subsidence and erosion due to movement along fault planes below the structure (James, 1968).

Teton Dam, U.S.A.—a 92-m-high earth dam that failed in 1976, with the loss of 11 lives and damage estimated at about $400 million. The investigating committee concluded that failure was probably caused by a combination of inadequate sealing of bedrock in a critical part of the dam foundations, and inadequate protection of the core material against internal erosion (U.S. Department of the Interior, 1977).

It might be expected that progressive advances in dam design and construction techniques would result in a lower incidence of failures. This does not appear to be the case, however, for two main reasons.

First, with any technological advance, there are always likely to be unforeseen factors that can produce unexpected problems. For example, when Malpasset Dam was constructed, drainage of foundations to reduce hydrostatic pressures downstream of arch dams was not considered necessary (Malpasset was one of the first dams constructed in the double curvature thin arch design). The subsequent failure was caused by hydrostatic uplift in the foundations.

Second, most of the obvious and easy dam sites around the world have now been utilized. This means that future dam construction will be necessary at progressively more difficult and geologically complex dam sites, which increases the probability of foundation problems. It is therefore clear that if dam failures and accidents are to be minimized in the future, the role of geology must be maintained or enhanced during the investigation, design, and construction of dams.

Economic Construction of Dams

Hundreds of dams have cost much more than the original contract price; unexpected geological features, many of which could have been identified by a more thorough investigation, are often the cause of expensive problems during construction. Even today too many dam sites are underinvestigated, and the resultant extra cost of construction is far greater than the additional expense that would have been necessary to carry out a thorough investigation. The problem is, of course, that the assessment of what constitutes a reasonable investigation has to be done well before the scheduled commencement of construction, when there is little information available. This is compounded by the fact that an investigation that is perfectly adequate for a simple site could well be quite inadequate for a geologically complex site. It is extremely difficult to plan site investigations that are adequate without being wasteful of resources, but it is clear that, on a worldwide scale, the pendulum needs to swing toward more thorough pre-construction investigations.

Conclusions

Geological investigations play an important role in the four main phases of dam construction: (1) the selection of the most suitable site; (2) the selection of the type of dam best suited to the site; (3) the design of the most economical dam; and (4) the safe and economical construction of the dam.

The geological and physical data necessary to fulfill these four objectives are gathered by means of a wide range of techniques, and are then integrated to provide a geomechanical model of the area investigated. The translation from geological data to geomechanical data is a critical step; this is the interface between geology and engineering, and it is essential that there are no communication gaps across this interface. In his report on the Malpasset Dam inquiry Jaeger (1963) stated, "There is such a discrepancy between the approach of the geologist and the way of thinking of the engineer that an intermediary would be desirable." Engineering geology (q.v.) (or geological engineering) is an area of applied geology (q.v.) that has developed strongly in recent decades to provide this essential interface between the science of geology and the technology of engineering.

The figures and much of the material in this paper have previously been published in the *B. M. R. Journal of Australian Geology and Geophysics* (Best, 1981); the permission of the Director, Bureau of Mineral Resources, Geology and Geophysics, Canberra, Australia, to use this material is gratefully acknowledged.

ERIC BEST

References

Attewell, P. B., and Farmer, I. W., 1976. *Principles of Engineering Geology.* London: Chapman & Hall, 1045p.

Bell, F. G., 1980. *Engineering Geology and Geotechnics.* London: Newnes-Butterworths, 497p.

Best, E. J., 1981. The influence of geology on the location, design and construction of water supply dams in the Canberra area, *Bur. Mineral Resources Jour. Australian Geology and Geophysics,* **6** (2), 161-179.

Best, E. J., and Hill, J. K., 1967. Site investigation techniques used at Corin Dam site, Cotter River, A. C. T., *Proceedings of the 5th Australia-New Zealand Conference on Soil Mechanics and Foundation Engineering,* Australia: Institution of Engineers, 1-8.

Jaeger, C., 1963. The Malpasset report, *Water Power,* **15** (2), 55-61.

James, L. B., 1968. Failure of Baldwin Hills Reservoir, Los Angeles, California, *Geol. Soc. America Eng. Geology Case Histories* **6,** 1-11.

Legget, R. F., 1962. *Geology and Engineering.* New York: McGraw-Hill, 884p.

Londe, P., 1967. Panel discussion, *Proceedings of the 1st Congress of the International Society of Rock Mechanics,* Vol. 3, Madrid: Editorial Blume, 449-453.

Muller, L., 1963. The rock slide in the Vajont valley, *Rock Mechanics Eng. Geology,* **2,** 149-212.

Paige, S. (ed.), 1950. *Application of Geology to Engineering Practice.* Berkey Volume, New York: Geological Society of America, 327p.

Stapeldon, D. H., 1976. Geological hazards and water storage, *Bull. Internat. Assoc. Eng. Geology,* **14,** 249-262.

Thomas, H. H., 1976. *The Engineering of Large Dams,* 2 vols. London: Wiley, 777p.

UNESCO, 1976. *Engineering Geological Maps—A Guide to Their Preparation.* Paris: Unesco Press, 79p.

U.S. Department of Interior, 1977. *Failure of Teton Dam: A Report of Findings.* Washington, D.C.: U.S. Government Printing Office, 107p. plus 638p. appendixes.

Wahlstrom, E. E., 1974. *Dams, Dam Foundations and Reservoir Sites.* Amsterdam: Elsevier, 278p.

Walters, R. C. S., 1971. *Dam Geology.* London: Butterworths, 470p.

Cross-references: *Alluvial Plains, Engineering Geology; Engineering Geology Reports; Foundation Engineering; Grout, Grouting; Maps, Engineering Purposes; Pipeline Corridor Evaluation; Pumping Stations and Pipelines, Engineering Geology; Rapid Excavation and Tunneling; River Engineering; Shaft Sinking.* Vol. XIV: *Augers, Augering; Blasting and Related Technology; Borehole Drilling.*

DATA BANKS—See COMPUTERIZED RESOURCES INFORMATION BANK; WELL DATA SYSTEMS.

DATA CODING—See COMPUTERIZED RESOURCES INFORMATION BANK. Vol. XIV: PUNCH CARDS, GEOLOGIC REFERENCING.

DELTAIC PLAINS, ENGINEERING GEOLOGY

Definitions and Types

A *deltaic plain* consists of active or abandoned deltas, which are either overlapping or contiguous to one another. A *delta* is a relatively flat area at the mouth of a river or a river system in which sediment load is deposited and distributed (see Vol. VI: *Delta Sedimentation*). That portion of a drainage basin within which the sediment load is traded or carried in transit is referred to as an *alluvial valley* (see *Alluvial Plains, Engineering Geology*). The *alluvial valley* merges downstream with the *deltaic plain,* often where the main stream channel branches, or has branched in the past, into multiple distributaries. The *deltaic plain* does not necessarily begin downstream from its most upstream distributary or its most upstream abandoned distributary. In the case of the Mississippi River, the most upstream distributary at present is the Atchafalaya River (Fig. 1). However, much of the area through which the Atchafalaya now flows consists of fluviatile backswamp deposits. Moreover, individual abandoned Mississippi River deltas have been mapped in some detail, and only the stippled area shown on Figure 1 contains such deltaic masses. Thus landforms and environments of deposition characteristic of deltas identify a *deltaic plain (*see Vol. VI: *Deltaic Sediments*); fluviatile landforms and environments of deposition are characteristic of and identify an *alluvial valley.*

This concept is useful in describing those characteristics of the deltaic plain of interest to disciplines as disparate as fluvial morphology, environmental geology (q.v.), hydrology, and river engineering (q.v.). It is of particular importance to the engineering geologist and the soil mechanics engineer (see *Soil Mechanics)* concerned with the distribution and strength parameters of the notoriously weak soils that support the levees, docks, industrial structures, urban developments, harbors, and similar installations that are built to develop and make deltaic plains habitable. The problems associated with such development and the cost involved in resolving them are often the primary reasons why many of the world's larger deltaic plains are undeveloped or underdeveloped.

The deltaic plains of the world are as varied as the river systems that give them birth (see Vol. III: *Deltaic Evolution*). Table 1 lists 46 of the world's rivers in the order of their average discharge. Note that although the Amazon ranks first in average discharge, it ranks third in yearly suspended load; that the Hwang Ho, which ranks first in yearly suspended load, ranks twenty-ninth in average discharge. Obviously, the discharge of a given river and the amount of sediment it carries to the sea are important parameters in the distribution of environments of deposition and their associated

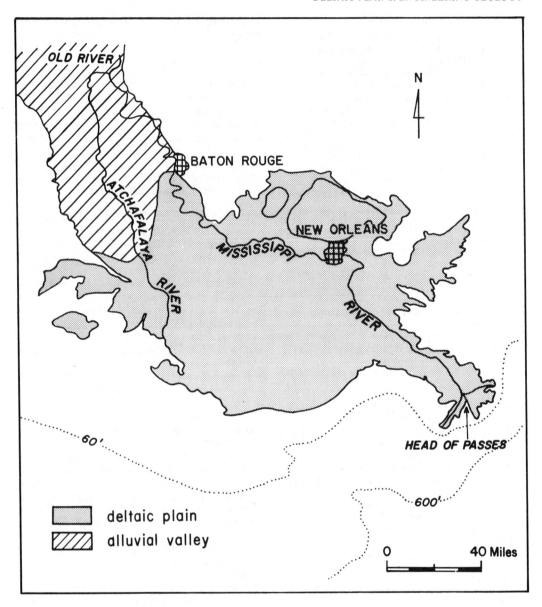

FIGURE 1. Mississippi River deltaic plain. The boundary between the deltaic plain and upstream fluviatile deposits of the alluvial valley is approximate.

sediment types within the growing deltaic plain. Other parameters of equal or greater importance include the depth of water into which the delta is being built (expressed in Table 1 in terms of the slope of the offshore areas), the magnitude of the tides, the climate, and the offshore wave action and currents. These factors, together with the tectonics and geometry of the receiving basin and interrelated parameters, are used by Coleman (1981) and Coleman and Wright (1975) to describe and classify world deltas. A somewhat similar group of factors is proposed by Morgan (1970), who divides

the four factors controlling and influencing delta formation into: (1) river regime, mainly particle size and quantity of material transported by a river to its delta and variations in these properties during seasonal fluctuations in flow; (2) coastal processes, essentially the influence of waves, tides, and currents on the seaward margins of the delta; (3) structural behavior and the relation of sea level to the depositional site; and (4) climatic factors, particularly those that affect vegetation within the delta.

Galloway (1975) proposes the more simplified

TABLE 1. Data on selected world deltas (after Inman and Nordstrom, 1971; Coleman, 1981; climate types after Trewartha).

RIVER (LOCATION)	AVERAGE (m³/sec) DISCHARGE	RANK YEARLY SUSPENDED LOAD	CLIMATE	OFFSHORE SLOPE (%)	TIDAL RANGE (m)	AVERAGE WAVE POWER (x10⁷ ergs/sec)
1. Amazon (Brazil)	149,736	3	Tropical Rainforest	0.5	4.9	0.193
2. Congo (Zaire)	40,441	21	Subtropical Steppe		1.70	0.586
3. Ganges-Brahmaputra (Bangladesh)	34,500	2	Tropical Rainforest	1.5	3.6	
4. Orinoco (Venezuela)	25,200	19	Tropical Rainforest	2.8	1.8	0.127
5. Yangtze-Kiang (China)	22,231	8	Humid Subtropical	0.013	3.7	
6. Yenisey (USSR)	17,190	34	Tundra		0.40	
7. Lena (USSR)	15,661		Tundra	0.8	0.21	0.034
8. Mississippi (USA)	15,631	6	Humid Subtropical	7.0	0.43	
9. Mekong (Vietnam)	14,168	4	Tropical Savanna	4.3	2.6	
10. St. Lawrence (Canada)	14,160	35	Humid Continental			
11. Parana (Argentina)	12,658	17	Tropical Rainforest	0.3	0.64	
12. Ob (USSR)	12,631	31	Subarctic		0.70	0.193
13. Irrawaddy (Burma)	12,558	10	Tropical Rainforest	1.4	2.7	
14. Tocantins (Brazil)	10,110		Tropical Rainforest		4.30	
15. Amur (USSR)	9,714	23	Subarctic	6.2	2.30	2.007
16. Niger (Nigeria)	8,769	22	Tropical Rainforest	1.9	1.4	
17. Mackenzie (Canada)	8,532	20	Tundra		0.34	
18. Volga (USSR)	7,736	27	Steppe	36.0	0	206.25
19. Magdalena (Colombia)	7,500	11	Tropical Savanna		1.1	
20. Columbia (USA)	7,278	25	Marine West Coast			
21. Zambezi (Mozambique)	7,164	16	Tropical Savanna	14.1	4.00	0.034
22. Danube (Romania)	6,250		Humid Continental		0	
23. Yukon (USA)	5,154	18	Subarctic		1.20	

(continued)

TABLE 1. *(continued)*

RIVER (LOCATION)	AVERAGE (m³/sec) DISCHARGE	RANK YEARLY SUSPENDED LOAD	CLIMATE	OFFSHORE SLOPE (%)	TIDAL RANGE (m)	AVERAGE WAVE POWER (x10⁷ ergs/sec)
24. Indus (W. Pakistan)	4,274	9	Tropical Desert	9.6	2.6	14.15
25. Red (N. Vietnam)	3,913	14	Humid Subtropical	3.7	1.9	30.415
26. Sao Francisco (Brazil)	3,420		Tropical Savanna	11.2	1.86	
27. Pechora (USSR)	3,362		Tundra	6.2	0.73	
28. Godavari (India)	3,180		Tropical Savanna	12.8	1.2	
29. Hwang Ho (China)	2,571	1	Humid Continental	1.5	1.13	0.218
30. Rhine (Netherlands)	2,223	36	Marine West Coast		1.7	
31. Rhone (France)	1,528	24	Mediterranean		0.20	
32. Po (Italy)	1,484		Humid Subtropical	5.6	0.73	
33. Nile (Egypt)	1,480	15	Desert	7.3	0.43	10.25
34. Dneiper (USSR)	1,370		Steppe	4.4	0	
35. Shatt al Arab (Iran and Iraq)	1,300	5	Desert	0.470	2.5	
36. Klang (Malaysia)	1,100		Trop. Rainforest	4.1	4.2	0.218
37. Senegal (Senegal)	867		Steppe	16.0	1.22	112.42
38. Chao Phraya (Thailand)	831	33	Topical Savanna	0.6	2.4	0.736
39. Colorado (USA)	595	12	Desert		8.2	
40. Ebro (Spain)	552		Mediterranean	36.0	0	0.155
41. Colville (Alaska)	491		Tundra	2.1	0.21	0.001
42. Burdekin (Australia)	475		Humid Subtropical	9.2	2.2	6.414
43. Murray (Australia)	400	25	Mediterranean		2.80	
44. Tana (Kenya)	172		Tropical Savanna	0.032	2.9	
45. Ord (Australia)	166		Tropical Savanna	3.9	5.8	1.062
46. Rio Grande (USA)	142	28	Steppe			

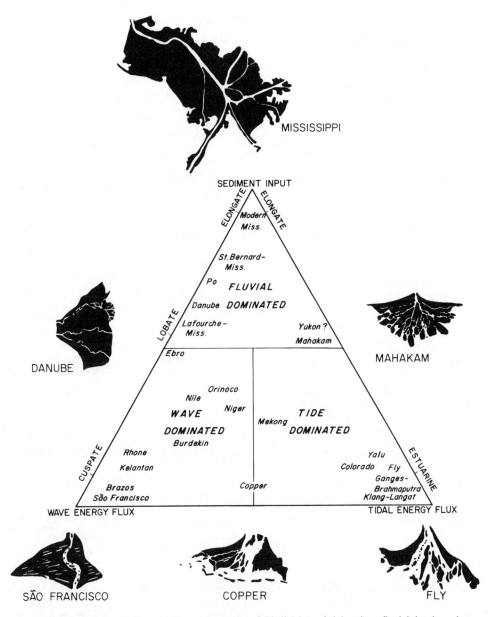

FIGURE 2. Schematic diagram illustrating a threefold division of deltas into fluvial-dominated, wave-dominated, and tide-dominated types (based on Galloway, 1975).

threefold classification shown in Figure 2, based on whether a delta is fluvially dominated, tidally dominated, or principally affected by wave energy. The modern Mississippi is fluvially dominated with characteristically high discharge and sediment load. Elongate land areas tend to form along a limited number of distributaries. *Lobate deltas* like the Ebro, the Danube, and abandoned deltas of the Mississippi are considered typical of deltas influenced chiefly by fluvial and wave processes. In tidally dominated rivers such as the Ganges-Brahmaputra and the Colorado, *estuarine deltas* are typical and distributaries with flaring mouths and many mid-channel bars are common. Where wind and wave processes are influenced by more moderate tidal action, barrier beaches and off-shore sandy tidal flats are typical, such as in the Copper and the Niger rivers. *Cuspate deltas* are thought to be characteristic of rivers, such as the Sao Francisco, where wave forces are predominant.

Despite the number and variety of factors that affect delta formation, some depositional processes and environments are common to all deltas. One of the more intensively studied of the world's

127

deltaic plains is that of the Mississippi. As indicated earlier, the present Mississippi delta is a good example of a fluvially dominated system, a system affected, moreover, by a negligible tidal range and by only moderate wave effects. It is commonly described as a *birdfoot* delta, for it has a few well-developed distributaries separated by wide, shallow bays. The distributaries extend seaward in elongate projections like a bird's foot. Most of the abandoned deltas of the Mississippi, on the other hand, formed lobate deltas, often described as *horsetail deltas,* where numerous distributaries were occupied and abandoned as the deltas built seaward as huge fans. No bays of consequence occurred between distributaries. This is attributed to the fact that the ancient Mississippi deltas were built rapidly across a shallow continental shelf with water depth ranging from a few meters to generally less than 30 m. The modern Mississippi, in contrast, is dumping its load at the very edge of the continental shelf into water that was 150 to 200 m deep in the recent past. Figure 3 shows the major Mississippi River deltas that were built and abandoned in the recent geological past. Figure 4B illustrates the types of sediments associated with the present delta and contrasts these in Figure 4A with the sedimentary units associated with past Mississippi deltas.

The soils that form the deltaic plain of the Mississippi have been extensively studied by geologists and engineers, particularly during the past half-century. Tens of thousands of shallow borings were made and comprehensive reports and papers published. (Some of the more notable of these are listed as references at the end of this entry.) Although differences exist between environments and associated soil units characteristic of the Mississippi deltaic plain and other deltas, the similarities between other deltas and the Mississippi are remarkable.

The following discussion emphasizes the situation found in the Mississippi deltaic plain. Occasional comments are included where landforms or their constituent soils are known to differ significantly from those of the Mississippi. Sources for data on the Mississippi are principally from Kolb and Van Lopik (1958), Fisk et al. (1954), and Bernard and LeBlanc (1965). Data contrasting the Mississippi with other world deltas were based on many references, chief among them Allen (1965) for the Niger; Andel (1967) for the Orinoco; Arnborg, Walker and Peippo (1962) for the Colville; Coleman (1981) for the Klang, the Ord, the Burdekin, the Sao Francisco, and the Senegal; Kolb and Dornbusch (1975) for the Mekong; Naidu and Mowatt (1975) for the Colville; Nelson (1970) for the Po; and Rodolfo (1975) for the Irrawaddy. The references cited at the end of this entry contain additional sources. For more extensive bibliographies on deltas, refer to LeBlanc (1975) and Coleman (1981).

Pre-Deltaic Plain Deposits

Deltaic plain deposits of world rivers are underlain by a variety of soil and rock types, the depth to and the nature of which are often extremely important parameters in areas where good foundations for heavy structures are at a premium. Deltaic plain soils of the Mekong, the Colorado, and the Burdekin, for example, overlie and are flanked in places by ancient rock strata. The deltaic plain of the Ebro is flanked and underlain by Mesozoic and Pliocene sedimentary units in some areas. Deltaic plain deposits of the Mississippi, probably in common with most of the world's deltaic plains, overlie Pleistocene deposits. These deposits normally consist of soils that are nonindurated but that have been subjected to tens of thousands of years of consolidation, desiccation, oxidation, and erosion during the last Wisconsinan drop in sea level. Multistoried buildings, major highways, harbors, and so on are usually founded on pilings driven to this horizon. As shown by the slanted lines in Figure 3, the Pleistocene flanks the deltaic plain on the west and north. The irregularly eroded surface slopes gradually seaward beneath Lake Pontchartrain and, beneath New Orleans, it averages about 20 m below the surface. It is readily recognized in borings. The colors of the overlying Holocene deltaic soils are generally gray to blue-gray. The upper portion of the Pleistocene, in contrast, is characteristically oxidized to a yellow or orange color. The Pleistocene is also marked by a distinctive stiffening in soil consistency and soil strength, a decrease in water content, and the occurrence of calcareous concretions. Strengths commonly reach 144 kPa (3,000 lb/ft^2).

Deltaic Plain Deposits

Classification. The Mississippi deltaic plain consists of the present delta and at least six additional deltas, the oldest of which was abandoned sometime after sea level reached its present stand about 5,000 years ago (Fig. 3). Environments of deposition are divided into (1) fluvial sediments deposited along the major active or abandoned streams that traverse the deltaic plain; (2) fluvial-marine deposits laid down off the mouths of the deltas as they advanced; (3) organic and minor inorganic deposits formed *in situ* or carried short distances and redeposited in a paludal environment; and (4) marine deposits resulting from reworking of the deltaic deposits by tides, wind, and waves.

The prevalence and the types of landforms within each of these groups vary depending from delta to delta. *Fluvial* environments and *fluvial-marine* environments predominate in deltas such as the Mississippi with high sediment input, high discharge, minor reworking of sediments by waves, and low tidal ranges. *Paludal* environments become increasingly important in deltaic plains in humid tropical areas such as the Mekong, the Chao

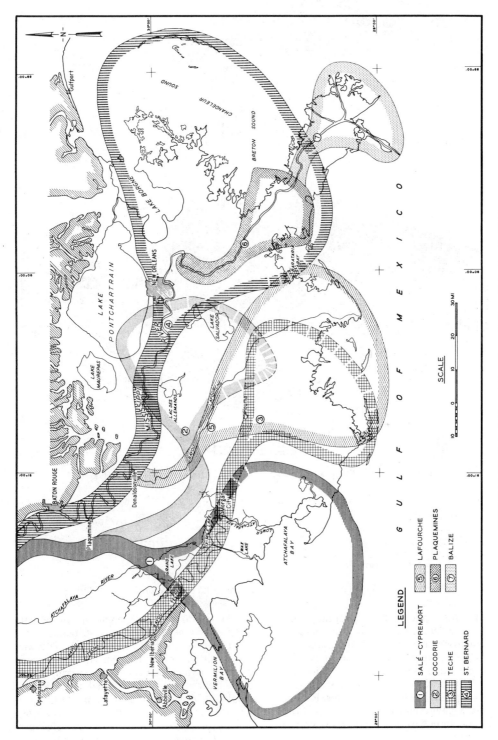

FIGURE 3. Mississippi River deltas (from Kolb and Van Lopik, 1958).

129

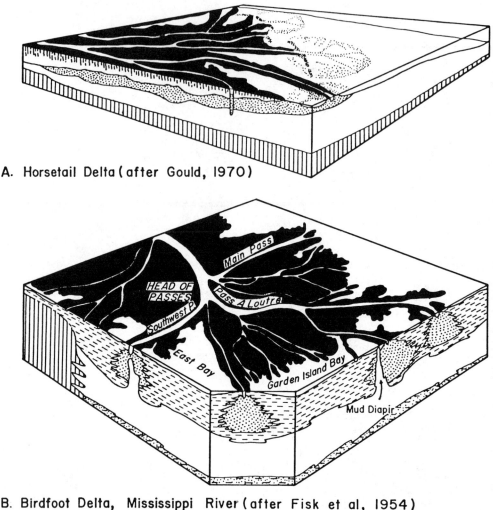

A. Horsetail Delta (after Gould, 1970)

B. Birdfoot Delta, Mississippi River (after Fisk et al, 1954)

Marsh	Prodelta Marine Clay	Strand Plain Sand
Delta Front Sheet Sand and Bar Finger Sand	Interdistributary Clay	Older Deposit

FIGURE 4. Ancient and modern Mississippi River delta types (after Gould, 1970).

Phraya, and the Orinoco. They form far smaller portions of the deltaic mass in the drier climates. *Marine* environments are more important volumetrically and more varied where wave and tidal energy is great and fluvial dominance is less pronounced. Evaporites form major constituents of such deltaic plains in the drier climates. Reworking of marine environments by the wind is also an important factor in the drier climates (see Vol. III: *Delta Dynamics*).

Fluvial Deposits. Fluvial environments in the Mississippi deltaic plain are conveniently divided into (1) natural levees, (2) point and lateral bars, and (3) abandoned courses and distributaries. *Natural levees* form slightly elevated areas flanking streams within the deltaic plain. They are the most conspicuous highs on otherwise strikingly level plains. Nearly all major inhabited and cultivated areas in the Mississippi deltaic plain are located on them. The levees are formed by the deposition of the coarsest sediments carried in suspension by flood waters that top the river banks. Figure 5 is a schematic representation of typical crest elevations, widths, thicknesses, and soil types forming

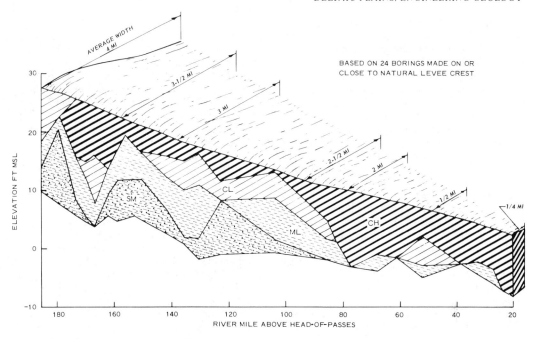

FIGURE 5. Longitudinal slope, width, thickness, and general composition (soil type) of natural levee deposits from Donaldsville, Louisiana, to Head of Passes (from Kolb, 1962).

the natural levees from Donaldsonville, Lousiana, to Head of Passes (see Fig. 3). Soil strengths of natural levee deposits are typically high, with cohesive strengths ranging between 183 and 575 kPa (800 and 1,200 lb/ft²). Desiccation and oxidation undoubtedly account for their high strengths. Natural levees are common to most streams transiting deltaic plains. Their crest heights and widths are generally proportional to the size—that is, the discharge and suspended load—of the streams they flank. In streams with high tidal ranges—such as the Ord, the Colorado, the Ganges-Brahmaputra, and the Klang—they are only poorly developed. Arctic rivers such as the Colville and the Ob are essentially void of landforms recognizable as natural levees in their deltaic plains.

Point bars or *lateral bars* are formed by meandering and lateral migration of the river or its distributaries. These environments normally contain the coarsest material carried by the river or moved along the bottom as bed load. Whereas point bar deposits make up more than 75 percent of the deposits forming the Mississippi alluvial valley upstream from the deltaic plain, they become progressively less prevalent along the river downstream from Baton Rouge; downstream from New Orleans, almost all accretions along the sides of the channel consist of occasional lateral bars. The term *point bar* is normally applied to accretions within a bend; *lateral bar* is used for the more or less straight accretionary segments that form on either side of the master stream or its larger

distributaries (see Vol. III: *Bars*). Samples from hundreds of borings in point bar deposits in the portion of the Mississippi deltaic plain between Donaldsonville and the Gulf consist principally of poorly graded fine sand. The upper portions of the deposits vary from clay to silty sand with a characteristic increase in grain size with depth. Depths to sand are highly variable, and soil types change rapidly both horizontally and vertically. Only the basal one-third to one-half of the point bar deposits can be expected to be fairly clean, fine-grained sand. Point and lateral bars are common along present and abandoned channels of many of the world's deltaic plains. Many rivers, however—such as the Mekong, Sao Francisco, Klang, Burdekin, and Ganges-Brahmaputra—typically develop middle bars and islands rather than the point or lateral bars common along the Mississippi.

Abandoned courses and *distributaries,* although mapped in some detail in the Mississippi deltaic plain, are only poorly known and delineated in most of the world's deltaic plains. Exceptions include such well-documented river deltas as the Po, the Rhone, the Niger, and the Colorado. In the Mississippi deltaic plain, sediments filling abandoned courses and distributaries form distinctive ribbons of relatively coarse sediments within the deltaic mass, sandy bodies that are comparatively small volumetrically but are of considerable significance to the engineering geologist. Abandoned course deposits fill the main channel left by the river when it is diverted at some point upstream

131

into a new, shorter course to the sea (Fig. 6). As shown in Figure 3, six such diversions of the Mississippi occurred in the recent geological past. Each course is now filled with distinctive sediments to depths and widths as great as the former channel—45 to 50 m deep, for example, and from 0.18 to 1.2 km wide. Fisk et al. (1952) showed that abandonment of a course is a slow process, but that a critical phase of the diversion process occurs when about 50 percent of the river's flow is diverted through the new channel. The former course is then plugged with sand just downstream from the point of diversion, and the new channel rapidly enlarges to take the entire flow. After abandonment, only high water is capable of breaching the sand wedge at the head of the abandoned course and sandy materials are distributed for some distance downstream from the point of diversion. The resulting body of sediment typically consists of a wedge of sand, gradually thinning downstream, overlain by a complementary clay wedge thinning upstream.

Abandoned distributaries are integral parts of deltaic advance, and hundreds, perhaps thousands, of narrow bands of distinctive sediments filling abandoned distributaries attest to the importance of this environment of deposition in the Mississippi deltaic plain. They range from a few meters of organic fill material—more logically considered part of the ubiquitous marsh deposits that flank or overlie them—to wedges of sandy organic soils 20 or more meters thick. In deltas characterized by high tides and high sand loads—such as the Colorado, the Ord, and the Burdekin—abandoned distributaries are typically filled with coarse clastics rather than with organic soils. Both ebb and flow tides sweep sand into the abandoned courses and distributaries, and thick, coarse-grained fill is not uncommon.

Fluvial-Marine Deposits. Three environments of deposition, each forming distinct lithological entities, characterize the advance of a delta: the prodelta, the intradelta, and the interdistributary. The *prodelta* consists of the first terrigenous sediments introduced into the depositional area. The basal portions of the prodelta sediments are deposited in an essentially marine environment, but as the subaerial delta advances, more and more fluvial deposits are introduced and very fine clays with many marine shells give way to leaner clays and silts, and shells progressively begin to make up less of the unit. Mississippi prodelta deposits are about 80 percent clay. Kolb and Kaufman (1967) have mapped thick wedges of prodelta clay that underlie the deltaic plain of the Mississippi and the immediate offshore areas. Strengths vary between 96 and 335 kPa (200 and 700 lb/ft^2). Thicknesses of the deposits range from a few meters to as much as 120 m. Oomkens (1970) records a maximum thickness of 50 m of prodelta clays in

the Rhone delta. Naidu and Mowatt (1975) describe a prodelta sequence off the Colville in Alaska that differs basically both in process and in lithology from the prodelta of temperate and tropical areas. The Colville and other arctic rivers are icebound for most of the year. At the spring breakup, water begins to issue from "strudels" (cracks and holes in the ice), carrying with it silts, sands, and sometimes gravel from the river bottom and spreading them out onto the more or less intact ice shelf that covers the delta. Essentially two prodelta facies result from rafting and melting of the ice. Inside the 20-m isobath, the deposits consist of a heterogeneous mixture ranging in grain size from clay to gravel; outside the isobath, the prodelta consists of very poorly sorted gravelly mud with molluskan and echinoderm remains.

Associated with and essentially a part of the prodelta clay environment are mud flats and mud lumps. *Mud flats* are principally prodelta clays, which, instead of being deposited in a broad fan seaward of the delta, are swept by littoral and wind currents onto sheltered areas along the coast. They sometimes become established as huge masses of organic clays overgrown with marsh vegetation. *Mud flats* are common in many world deltas, for example, the Colorado, the Irrawaddy, and the Klang. *Mud lumps* are prodeltaic clays that have worked their way upward through overlying coarser deltaic deposits—sometimes more than 100 m—as fingers of clay that reach the surface as mud islands within or immediately seaward of the Mississippi subaerial delta. Extensive studies—such as those by Morgan (1961) and Morgan et al. (1963)—detail these curious features that characterize the advance of the modern Mississippi delta. They probably occur in other world deltas (perhaps the Orinoco), but they have not been well documented.

Intradelta deposits are the coarser sediments associated with deltaic advance. As stream velocity decreases at the mouth of a distributary, the greater part of its load is deposited in distributary mouth bars. Sediments accumulate on the bar crest or on the seaward side of the bar. They also top underwater and subaerial natural levees on either side of a distributary as the delta builds seaward. Thus, an irregular wedge of clastic sediments builds on either side and below the distributary as it advances. The mass sinks into the underlying prodeltaic clay and, in the Mississippi, distinctive sandy units, sometimes 50 or more meters thick, are formed beneath and parallel to the major distributaries. Often called *bar fingers* (Fisk et al., 1954), they are evident in Figure 4B. Where many distributaries form channels in close proximity to one another, the coarse deposits tend to coalesce to form a single unit called a *delta front sand sheet,* also shown in Figure 4B. Delta front sand sheets have been reported from many

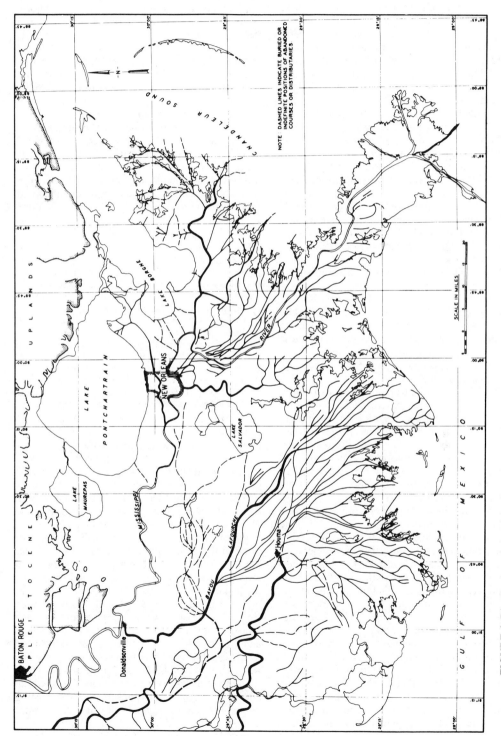

FIGURE 6. Distribution of abandoned courses and distributaries in the Mississippi deltaic plain (from Kolb and Van Lopik, 1958).

world deltas. The Mekong, Colorado, Klang, Niger, Nile, and Rhone are but a few of the deltas where delta front sand sheets have formed. The sand units that form as the deltas advance in some of these deltas (the Klang, the Colorado, and the Ord, for example) are reworked by strong tidal currents into elongate bars that extend seaward, principally underwater, in the offshore areas for several kilometers. These bars shift with the fluvial and tidal cycles and can range up to tens of meters in height. Coleman (1981) cites a height of 10 to 22 m for such *tidal bars* on the Ord. Lengths vary from 1 to 15 km. These bars result from convergence of flood- and ebb-dominated bed-load transport. On being buried during deltaic advance, these features form significant sand bodies in the deltaic mass. Variations of this process occur in wave-dominated deltas where the sandy bars arrange themselves at some angle to the direction of delta advance in the offshore area; some eventually parallel the shoreline. Where they reach the surface, the bars form alongshore beaches or barrier beaches.

Studies indicate that nearly 85 percent of the materials within the Mississippi bar fingers and the delta front sand sheets (that is, within the *intradelta* environment) consists of silt or fine sand. Clay and organic clay occur in moderate amounts. From a geotechnical standpoint, these units tend to flank the paths of abandoned courses and distributaries irregularly and to grade laterally sometimes into poorly consolidated clays, which form between the more widely spaced distributaries, the *interdistributary* environment.

Considerable thicknesses of interdistributary clays are deposited between the present Mississippi River distributaries. They form discrete bodies that grade downward into prodeltaic clays and upward into the richly organic clays of swamp or marsh deposits. The line of demarcation between the interdistributary and overlying marsh is indistinct. A true marsh or swamp forms when the watery area between distributaries or flanking the main channel has shallowed sufficiently to support vegetative growth. Interdistributary deposits are characteristically low-strength, high-water-content organic clays, some of the softest and least consolidated of the materials forming the deltaic mass. Interdistributary deposits in the more arid deltaic plains—such as the Ord, the Shatt al Arab, the Senegal, and the Indus—are commonly interfingered with evaporites. Tidal flats larger than 100 km^2 in area with gypsiferous and saline sediments 7 m thick are reported from the Ord.

Paludal Deposits. Paludal deposits are largely *in situ* organic sediments. The predominant paludal environments are swamps and marshes, areas half land, and half water, which seldom rise more than a half-meter above mean gulf level in the Mississippi delta. Complexly intermixed with these environments are clayey lacustrine and tidal channel deposits. The formation of lakes and tidal channels is an important part of marsh development in the Mississippi deltaic plain. The majority of these features, however, are shallow, insignificant water bodies in their original state that, when abandoned, leave behind sediments that for all practical geotechnical purposes can be classified as marsh. Nevertheless, a few are of such size and depth that they form deep channels or lakes during their active stages, and, on being abandoned, fill with fairly massive bodies of organic clay.

Deposits filling abandoned tidal channels or left by migrating tidal channels become volumetrically important in deltas characterized by moderate to high tides and significant amounts of fluvially introduced sediments. Coleman (1981) describes tidal channels in the Klang completely filled with sand. Allen (1965, 1970) describes these features in the Niger:

> There is a major reticulate pattern of interconnected meandering tidal creeks which surround flats whose platform-like upper surfaces lie between tide lines. Cutoff meander loops are not uncommon and many of the creeks give evidence of capture by others; evidently the creeks are unstable, shifting their position in a manner similar to river channels. . . . Each meander loop encloses a tidal point-bar. The bar deposits become finer grained upwards but at most levels muddy as well as sandy deposits can be found. The creeks reach a maximum depth of about 20 m. . . . The lower deposits consist of thick layers of fine- to coarse-grained sand in alternation with gray and black silty clays and clayey silts rich in finely divided organic matter and drifted debris.

The marsh forms flat, pervasive expanses of grasses and sedges which, together with the swamp environment, covers about 90 percent of the Mississippi deltaic plain. The vegetation of the marsh grows in close formation, providing a relatively firm surface underfoot, or the grasses may grow in tufts with mud and water between them. In some areas *flotant,* a mat of floating marsh, overlies black organic muck that has the consistency of thick gravy. Foundation problems in such areas are manifold. Interstate highways crossing them are often built on concrete pilings and are among the highest in per-mile cost for highways anywhere in the United States. Subsidence, the distance from the zone of active wave attack, and complex biological factors affect the distribution of land and water. Peats, organic oozes, and humus are formed as the marsh plants die and are covered by water. Normal subaerial oxidation processes are limited, decay is largely due to anaerobic bacteria, and in stagnant water, thick deposits consisting almost entirely of organic materials 20 or more meters deep are formed. Variable amounts of inorganic materials are found in the marsh deposits, the greatest amounts in what is termed *fresh-*

water marsh in areas subject to repeated inundation by flood flows or other fresh-water sources. In the New Orleans area, swamp and marsh deposits underlie large portions of the expanding city, and *subsidence* of the marsh due to ground-water withdrawal, compaction, and other reasons causes a host of foundation problems. Areas once above sea level now lie at elevations some 8 m below sea level and 15 m below the top of massive artificial levees that protect the city from recurrent floods on the Mississippi. Marsh deposits are common to most world deltas but are fairly unimportant in the drier climatic zones. In the arctic, they take on characteristics and landforms unique to the arctic environment. Arnborg et al. (1962) describe *patterned ground* in the Colville delta, wind-oriented lakes, and ice-wedge polygons that often contain clastic subsurface materials intercalated with the polygon ice. *Pingos,* thick masses of perennial ice, rise above the surface of the deltaic plain at irregular intervals in mounds a few hundreds of meters in diameter at their bases and many meters high.

Swamps are distinguished from marshes principally because of the dense growth of trees on the former and their absence on the latter. This is reflected in swamp deposits by the typical occurrence of partially decayed stumps and trunks of trees. Organic content of swamp deposits is high but is generally lower than that of marsh deposits. *Mangrove swamps* are essentially salt-water swamps. Only found in a few isolated instances in the Mississippi deltaic plain and other temperate-zone deltas, mangrove swamps are of significant areal and geotechnical importance in the humid tropical deltas of the world. Mangrove swamps (e.g., of the Irrawaddy, Mekong, Orinoco, Godavari, and Niger) are described as impenetrable jungles periodically inundated by the tides. The closely packed profuse vegetation, with its exposed intricate root systems, tends to trap clastic sediments brought in by flooding from fluvial sources as well as marine tidal sources. These sediments, mingled with finely divided organic matter and layers of drifted debris, form black organic muds and peats. These, in turn, are intercalated with the organic refuse left by decay of the mangrove jungle so that the mangrove swamps are sometimes built on organic platforms to elevations more than a meter higher than the banks of the tidal and distributary channels that traverse them. Information about the thickness and variability of the strata typical of mangrove swamps is meager; it is known, however, that the deltaic plain of the Klang contains 18 m of peat.

Marine Deposits. The three principal environments of marine deposition in the Mississippi deltaic plain are (1) bay-sound and the closely related nearshore gulf, (2) reef, and (3) beach. Sedimentation in these environments takes place exclusively under marine conditions; as such, it might be argued that they are not truly deltaic plain deposits. However, large portions of deltaic plains consist of marine deposits that formed beneath, at the margins, or in the paths of deltaic advance or retreat. In the case of the Mississippi, significant volumes of marine sedimentation are intercalated with the deltaic plain. Of the Holocene deposits in Louisiana, only the chenier plain of western Louisiana is considered entirely marine and is not considered part of the deltaic plain.

Nearshore gulf and *bay-sound* deposits are relatively coarse, shelly sediments that are difficult to distinguish lithologically when encountered in borings. Environmentally they are distinctive in that nearshore gulf deposits are laid down at the borders of the open ocean, while bay-sound deposits are laid down in relatively quiet waters protected by barrier beaches and carpet the floors of bays and sounds. The nearshore gulf deposits are generally sandier than their bay-sound counterparts. They form a continuous blanket seaward of the bays and sounds except where interrupted by active deposition at the mouth of the Mississippi. Nearshore gulf deposits in the Mississippi deltaic plain are associated with a transgressive sea. Thus each abandoned delta is characterized by a variable thickness of such deposits as these deltas subsided and succumbed to wave attack. One of the complicating factors in delineating these buried basal sandy deposits in borings is their frequent merging in the Mississippi deltaic plain with "strand plain" sands and shell at the base of the deltaic deposits immediately overlying the Pleistocene. Here a fairly persistent shell-and-sand horizon formed as the sea level rose in late Pleistocene and early Holocene times. Where identifiable, this horizon is more aptly referred to as the strand plain, not an integral part of the deltaic plain deposits. Not much is mentioned in the literature concerning nearshore or bay-sound deposits in other deltas. That comparable deposits exist is certain, however, particularly where deltaic plains are subsiding and undergoing attrition through wave attack.

Reefs are shell units that are undoubtedly associated with many deltaic plains, particularly those in tropical and subtropical areas; they have not, however, been widely reported in the literature. The only reef-forming mollusk of importance in the Mississippi deltaic plain is the oyster, *Crassostrea virginica.* Oysters typically build reefs in nearshore areas where regular influxes of fresh water bring about a mean salinity range of 10 to 30 percent, and where the mean temperature ranges between 10° and 25°C. Reefs of varying areas occur fairly extensively throughout the immediate-offshore areas of southern Louisiana, particularly in the more sheltered portions of the bay-sound environment. Thickness of the reef deposits averages between 1 and 5 m; some are many kilometers in

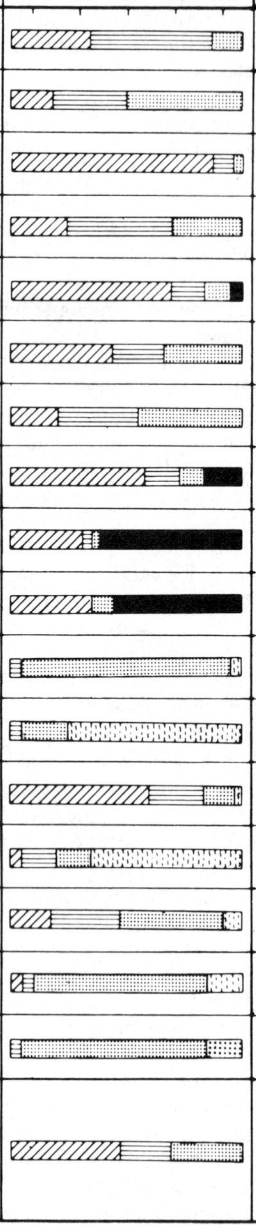

DEPOSITIONAL TYPES	LITHOLOGY PERCENT (0 25 50 75 100)	REMARKS
NATURAL LEVEES		Disposed in narrow bands flanking the Mississippi River and its abandoned courses and distributaries. Elevation varies from 25 feet near Baton Rouge to sea level.
POINT BAR		Usually found flanking the more prominent bends of present and abandoned courses. Thickness in excess of 100 feet.
PRODELTA CLAYS		Fat clay in offshore areas and at depth beneath deltaic plain. Thickness ranges between 50 and 400 feet.
INTRA DELTA		Coarse portion of subaqueous delta. Intricately interfingered deposits. Disposed in broad wedges about abandoned courses and major distributaries.
INTERDIS-TRIBUTARY		Forms clay wedges between major distributaries. Minor amounts of silts and fine sands typically occur in very thin but distinct layers between clay strata.
ABANDONED DIS-TRIBUTARY		Form belts of clayey sediments from a few feet to more than 1000 feet in width and from less than 10 to more than 50 feet in depth.
ABANDONED COURSE		Form belts of fairly coarse sediment in abandoned Mississippi River courses. Lower portion filled with sands, upper portion with silts and clays. Coarsest fill near point of diversion.
SWAMP		Tree-covered organic deposits flanking the inner borders of the marsh and subject to fresh-water inundation. Deposits 3 to 10 feet thick.
MARSH		Forms 90 percent of land surface in the deltaic plain. Ranges from watery organic oozes to fairly firm organic silts and clays. Average thickness 15 feet.
ABANDONED TIDAL CHANNELS		Found principally in peripheral marsh areas. Average depths on the order of 25 feet. Widths average 200 feet. Filling varies from peat to organic clay.
SAND BEACH		Border the open gulf except in areas of active deltaic advance. May be a mile or more wide and more than 10 miles long. Sand may pile as high as 30 feet and subside to depths 30 feet below gulf level.
SHELL BEACH		Border landward shores of protected bays and sounds and marshland lakes. Vary from 25 to 200 feet in width and from 2 to 6 feet in height. Lengths usually less than a mile.
LACUSTRINE		Deposits vary in thickness from 2 to 25 feet. Stratification in clayey lacustrine deposits is poorly developed or lacking.
REEF		Active reefs found principally in bay-sound areas. Buried reefs 5 to 10 feet thick a common occurrence within deltaic plain. Reach dimensions of 1/2-mile wide and 10-miles long.
BAY-SOUND		Relatively coarse sediments on bottoms of bays and sounds. Thickness between 3 and 20 feet.
NEARSHORE GULF		Found at the borders of the open ocean seaward of the barrier beaches. Thickness normally increases with distance from shore.
SUB-STRATUM		Massive sand and gravel deposits filling entrenched valley and grading laterally into nearshore gulf deposits. Material becomes coarser with depth.
PLEIS-TOCENE		Ancient former deltaic plain of Mississippi River. Consists of environments of deposition and associated lithology similar to those found in Recent deltaic plain. Depth of this ancient, eroded surface increases in a southerly and westerly direction in southeastern Louisiana.

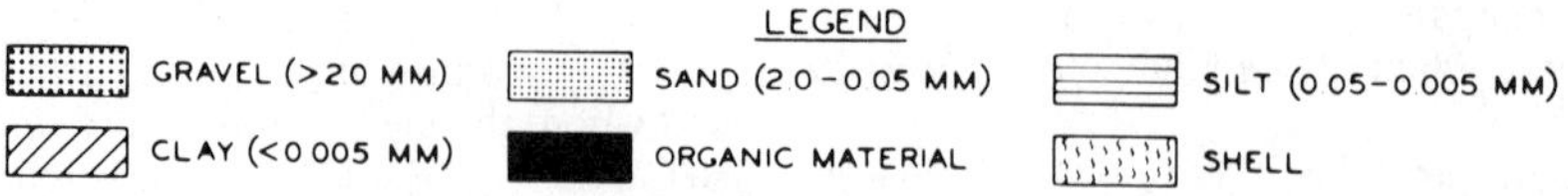

LEGEND

GRAVEL (>20 MM) SAND (2.0–0.05 MM) SILT (0.05–0.005 MM)

CLAY (<0.005 MM) ORGANIC MATERIAL SHELL

FIGURE 7. Summary of depositional types within the deltaic plain of the Mississippi River (left) and typical associated properties (right) (from Kolb and Van Lopik, 1958).

FIGURE 7. *(continued)*

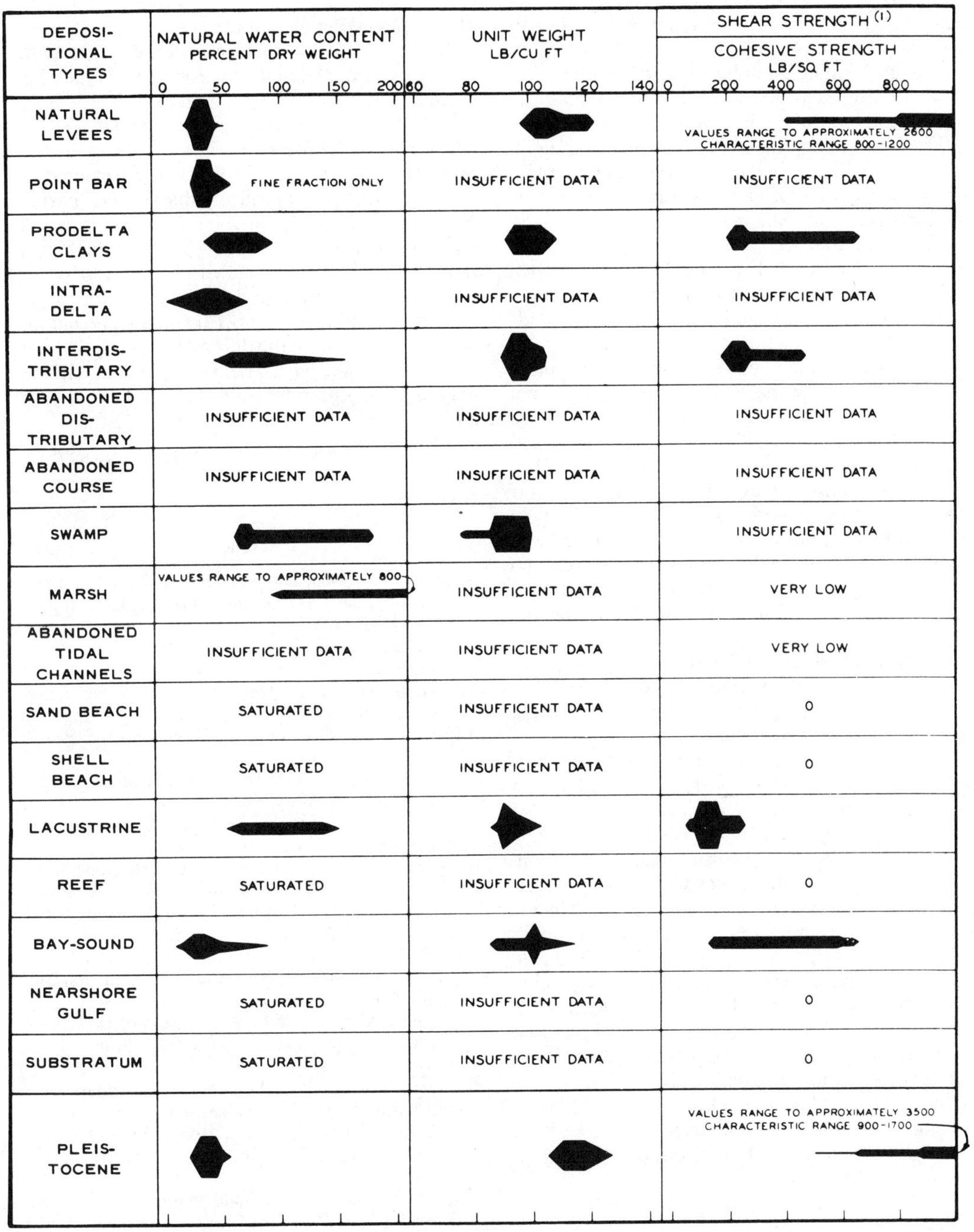

DEPOSI-TIONAL TYPES	NATURAL WATER CONTENT PERCENT DRY WEIGHT	UNIT WEIGHT LB/CU FT	SHEAR STRENGTH [1]
			COHESIVE STRENGTH LB/SQ FT
NATURAL LEVEES			VALUES RANGE TO APPROXIMATELY 2600 CHARACTERISTIC RANGE 800-1200
POINT BAR	FINE FRACTION ONLY	INSUFFICIENT DATA	INSUFFICIENT DATA
PRODELTA CLAYS			
INTRA-DELTA		INSUFFICIENT DATA	INSUFFICIENT DATA
INTERDIS-TRIBUTARY			
ABANDONED DIS-TRIBUTARY	INSUFFICIENT DATA	INSUFFICIENT DATA	INSUFFICIENT DATA
ABANDONED COURSE	INSUFFICIENT DATA	INSUFFICIENT DATA	INSUFFICIENT DATA
SWAMP			INSUFFICIENT DATA
MARSH	VALUES RANGE TO APPROXIMATELY 800	INSUFFICIENT DATA	VERY LOW
ABANDONED TIDAL CHANNELS	INSUFFICIENT DATA	INSUFFICIENT DATA	VERY LOW
SAND BEACH	SATURATED	INSUFFICIENT DATA	0
SHELL BEACH	SATURATED	INSUFFICIENT DATA	0
LACUSTRINE			
REEF	SATURATED	INSUFFICIENT DATA	0
BAY-SOUND			
NEARSHORE GULF	SATURATED	INSUFFICIENT DATA	0
SUBSTRATUM	SATURATED	INSUFFICIENT DATA	0
PLEIS-TOCENE			VALUES RANGE TO APPROXIMATELY 3500 CHARACTERISTIC RANGE 900-1700

Scale for NATURAL WATER CONTENT: 0, 50, 100, 150, 200. Scale for UNIT WEIGHT: 60, 80, 100, 120, 140. Scale for COHESIVE STRENGTH: 0, 200, 400, 600, 800.

(1) SHEARING STRENGTH OF CLAYS BASED ON UNCONFINED COMPRESSION TESTS.

TYPICAL RANGE OF VALUES INDICATED BY LENGTH OF BAR. BAR WIDTH INDICATES RELATIVE DISTRIBUTION OF VALUES.

length. They are widely used for building roads and as aggregate in an area where coarse material suitable for such purposes is at a premium. Locating and exploiting shell reefs buried within the Mississippi deltaic plain has become big business in southern Louisiana, where natural material coarser than fine sand is virtually nonexistent.

From the standpoint of their engineering significance, *beaches* associated with the deltaic plain of southeastern Louisiana fall into two classes: *sand beaches* and *shell beaches.* Sand beaches are by far the more common in fluvially and sediment-dominated deltas such as the Mississippi. Shell beaches in southern Louisiana form along the inner margins of bays and sounds and on islands within these water bodies. Waves typically fracture the shells, and such deposits can be readily distinguished in soil borings from the more intact remains typical of reefs. Sand beaches in southern Louisiana consist almost entirely of fine sand and range in height from 1 to 5 m above sea level. They compress the softer deltaic materials at their bases, and the sand bodies sometimes reach thicknesses of 13 m or more. As the various deltas wax and wane, new sets of beaches are formed and subsequently become buried within the deltaic mass. Although not nearly so valuable as the coarser shell deposits of the reef and shell beach environments, these sand units are often sought and exploited as permeable fill and for surcharging and compacting areas underlain by marsh peats and highly organic clays.

Compared with the majority of deltaic plains, sand beaches are only poorly developed within and flanking the Mississippi deltaic plain. *Stranded beach ridges* reach far inland within the Mekong delta, for example, and form conspicuous elongate highs. Low sand beaches with extensive spits and eolian dunes typify the Ebro, and the Niger is characterized by nearly continuous sand beaches along the shoreline of its deltaic plain. The Nile and the Senegal have conspicuous broad, high sand beaches and barrier beaches with well-developed wind-blown dunes. The Sao Francisco is characterized by extremely large eolian dunes attaining elevations in excess of 22 m. Beneath the dunes and inland on the delta plain, broad sandy beach ridges, plastered one against another, form the major landforms within the entire deltaic plain (Coleman, 1981).

Summary

Figure 7 summarizes and permits a comparison of soils characteristic of the environments of deposition in the Mississippi deltaic plain. Various ranges of selected physical or engineering properties are shown. The classification is predicated on its value in defining or delineating lithologically similar units wherein geotechnical properties of the soils can be estimated. At the same time, the classification has been kept within the framework of accepted geological terminology.

There are marked differences between geotechnically significant deltaic environments in the Mississippi and other world deltas; however, these differences are not so great that the framework developed for the Mississippi cannot be applied to other deltas. The grouping of deltaic environments into fluvial, fluvial-marine, paludal, and marine types can be applied to any deltaic plain. Principal differences in environmental types of geotechnical significance between the Mississippi and other world deltas are due chiefly to (1) discharge of the stream(s) forming the delta, (2) amount and nature of sediment load, (3) the depth of water into which the deltaic material is being deposited, (4) the magnitude of the tides and offshore wave action and currents, and (5) the climate.

CHARLES R. KOLB*

*Deceased

References

Allen, J. R. L., 1965. Late Quaternary Niger Delta and adjacent areas, *Am. Assoc. Petroleum Geologists Bull.,* **49,** 547-600.

Allen, J. R. L., 1970, Sediments of the modern Niger Delta, in J. P. Morgan (ed.), *Deltaic Sedimentation: Modern and Ancient, Soc. Econ. Paleontologists and Mineralogists Spec. Pub. 15,* 138-151.

Andel, T. H., 1967. The Orinoco Delta, *Jour. Sed. Petrology,* **37**(2), 297-310.

Arnborg, L. E.; Walker, H. J.; and Peippo, J., 1962. *Suspended Load in the Colville River, Alaska.* Baton Rouge: Coastal Studies Institute, Louisiana State University, Rept. 54, 131-144.

Bernard, H. A., and LeBlanc, R. J., 1965. Resume of the Quaternary geology of the northwestern Gulf of Mexico province, in W. E. Wright and D. G. Frey (eds.), *The Quaternary of the United States: Review Volume for the VII Congress of the International Association for Quaternary Research,* Princeton, N. J.: Princeton University Press, 137-186.

Coleman, J. M., 1981. *Deltas—Processes of Deposition and Models for Exploration,* 2nd ed. Minneapolis: Burgess, 124p.

Coleman, J. M., and Wright, L. D., 1975. Modern river deltas: variability of processes and sand bodies, in M. L. Broussard (ed.), *Deltas, Models for Exploration,* 2nd ed. Houston, Tex.: Houston Geological Society, 99-150.

Fisk, H. N.; Kolb, C. R.; and Wilbert, L. J., 1952. *Geological Investigation of the Atchafalaya Basin and Problems of Mississippi River Diversion.* Vicksburg, Miss.: U.S. Army Corps of Engineers, Mississippi River Commission, 145p.

Fisk, H. N.; McFarlan, E., Jr.; Kolb, C. R., and Wilbert, L. J., Jr., 1954. Sedimentary framework of the modern Mississippi Delta, *Jour. Sed. Petrology,* **24,** 76-99.

Frazier, D. E., 1967. Recent deltaic deposits of the Mississippi River: their development and chronology, *Gulf Coast Assoc. Geol. Socs. Trans.,* **17,** 287-315.

Gagliano, S. M., and McIntire, W. G., 1968. *Reports on the Mekong River Delta,* Baton Rouge: Coastal Studies Institute, Louisiana State University, Tech. Rept. 57, 143p.

Galloway, W. E., 1975. Process framework for describing the morphological and stratigraphic evolution of deltaic depositional systems, in M. L. Broussard (ed.), *Deltas, Models for Exploration,* 2nd ed. Houston, Tex.: Houston Geological Society, 87-98.

Gould, H. R., 1970. The Mississippi Delta complex, in J. P. Morgan (ed.), *Deltaic Sedimentation: Modern and Ancient, Soc. Econ. Paleontologists and Mineralogists Spec. Pub. 15,* 3-30.

Inman, D. L., and Nordstrom, C. E., 1971. On the tectonic and morphologic classification of coasts, *Jour. Geology,* **79,** 1-21.

Kolb, C. R., 1962. Engineering soils bordering the Mississippi from Donaldsonville to the Gulf, *U.S. Army Corps Engineers, Waterways Expt. Sta. Misc. Paper 3-481.*

Kolb, C. R., and Dornbusch, W. K., 1975. The Mississippi and Mekong deltas—a comparison, in M. L. Broussard (ed.), *Deltas, Models for Exploration,* 2nd ed. Houston, Tex.: Houston Geological Society, 193-207.

Kolb, C. R., and Kaufman, R. I., 1967, Prodelta clays of southeast Louisiana, *Marine Géotechnique,* 3-21.

Kolb, C. R., and Van Lopik, J. R., 1958. Geology of Mississippi River deltaic plain, southeastern Louisiana, *U.S. Army Corps Engineers, Waterways Expt. Sta. Tech. Rept. 3-483 and 3-484,* 2 vols.

LeBlanc, R. J., 1975. Significant studies of modern and ancient deltaic sediments, in M. L. Broussard (ed.), *Deltas, Models for Exploration,* 2nd ed. Houston, Tex.: Houston Geological Society, 13-85.

Morgan, J. P., 1961. Mudlumps at the mouths of the Mississippi River, *Genesis and Paleontology of the Mississippi River Mudlumps,* Louisiana Dept. of Conservation Geol. Bull. 35, 1-116.

Morgan, J. P., 1970. Depositional processes and products in the deltaic environment, in J. P. Morgan (ed.), Deltaic Sedimentation: Modern and Ancient, Soc. Econ. Paleontologists and Mineralogists, Spec. Pub. 15.

Morgan, J. P., Coleman, J. M., and Gagliano, S. M., 1963. *Mudlumps at the Mouth of South Pass, Mississippi River: Sedimentology, Paleontology, Structure, Origin, and Related Deltaic Processes.* Baton Rouge: Coastal Studies Institute, Louisiana State University, Coastal Studies Series No. 10, 190p.

Naidu, A. S., and Mowatt, T. C., 1975. Depositional environments and sediment characteristics of the Colville and adjacent deltas, northern arctic Alaska, in M. L. Broussard (ed.), *Deltas, Models for Exploration,* 2nd ed. Houston, Tex.: Houston Geological Society, 268-283.

Nelson, B. W., 1970. Hydrology, sediments dispersal, and recent historical development of the Po River Delta, Italy, in J. P. Morgan (ed.), *Deltaic Sedimentation: Modern and Ancient,* Soc. Econ. Paleontologists and Mineralogists, Spec. Pub. 15, 152-184.

Oomkens, E., 1970. Depositional sequences and sand distribution in the post-glacial Rhone Delta complex, in J. P. Morgan (ed.), *Deltaic Sedimentation: Modern and Ancient,* Soc. Econ. Paleontologists and Mineralogists, Spec. Pub., 15, 198-212.

Rodolfo, K. S., 1975. The Irrawaddy Delta: tertiary setting and modern offshore sedimentation, in M. L.

Broussard (ed.), *Deltas, Models for Exploration,* 2nd ed. Houston, Tex.: Houston Geological Society, 339-356.

Cross-references: *Alluvial Plains, Engineering Geology; Clay, Engineering Geology; Clays, Strength of; Coastal Engineering; Geomorphology, Applied; River Engineering.* Vol. XIV: *Alluvial Systems Modeling; Canals and Waterways; Cat Clays.*

DILATANT, DILATANCY—See RHEOLOGY, SOIL AND ROCK.

DISCONTINUUM MECHANICS—See FOUNDATION ENGINEERING; RHEOLOGY, SOIL AND ROCK; SOIL MECHANICS.

DIVING GEOLOGY—See OCEANOGRAPHY, APPLIED; SUBMERSIBLES; UNDERSEA TRANSMISSION LINES, ENGINEERING GEOLOGY.

DRAINAGE SURVEYS—See Vol. XII, Pt. 1: SOIL DRAINAGE. Vol. XIV: CANALS AND WATERWAYS; HYDROGEOCHEMICAL PROSPECTING; LAND DRAINAGE.

DREDGES, DREDGING—See CHANNELIZATION AND BANK STABILIZATION.

DRILL HOLE INFORMATION—See WELL DATA SYSTEMS. Vol. XIV: WELL LOGGING.

DRILLING TECHNOLOGY—See RAPID EXCAVATION AND TUNNELING; TUNNELS, TUNNELING. Vol. XIV: AUGERS, AUGERING: BOREHOLE DRILLING.

DURICRUST, ENGINEERING GEOLOGY

Definition

Duricrust (q.v. in Vols. VI, XII) is a peculiarly Australian term used to describe the case-hardened superficial mantle of rocklike material that outcrops over much of the arid and semi-arid parts of Australia (Finkl, 1979). Outliers of duricrust also occur in the better-watered parts of the continent. In some places the duricrust has been affected by tectonic movements and has been found covered by more recent sediments in boreholes (Wopfner and Twidale 1967; Finkl and Churchward, 1973).

The material of the duricrust has been formed by physicochemical processes involving reactions between the atmosphere, ground water, and soil and rock. It consists of soil or rock cemented or replaced either by oxides of silicon, iron, or aluminum or by such salts as calcium carbonate or sulfate. The solubilities in water of some of the substances involved in these reactions are shown in Figure 1. Calcium sulfate as gypsum has a constant solubility of 0.2 percent at all pH levels. Where cementation or replacement is complete, or almost complete, such materials as *silcrete* (silica), *ferricrete* and *laterite* (iron oxides—ferricrete rocklike and laterite soil-like—usually with some aluminium oxide), *bauxite* (aluminium oxide, usually with some iron oxides), *calcrete* (calcium carbonate), or *gypcrete* (calcium sulfate) have been formed. Where cementation or replacement is partial, materials have been produced that may possess unusual properties. Some of these

materials—such as laterite, calcrete, and partly silicified soil or rock—may harden on exposure to atmospheric conditions. This intrinsic property has made them useful for construction and engineering purposes (see *Caliche, Engineering Geology; Laterite, Engineering Geology*). Buchanan (1807) described examples of laterite being used for building construction in India. The laterite was easily dug by a spade but hardened to a bricklike consistency on exposure to the atmosphere.

Natural Occurrence and Uses

Materials used for engineering construction include not only the duricrust itself but also wasting products such as colluvial and alluvial gravels derived by disintegration of the duricrust, *gibber* (rounded or angular siliceous stone 10 to 200 mm in diameter, sometimes larger) derived by deflation of the duricrust and the *ironstone* mantle derived by leaching of silica and iron from topographic highs with deposition on the lower parts of the surface. (Ironstone is mostly siliceous but contains sufficient iron oxide to be colored red or brown.) Gibber and ironstone mantles cover large areas in arid and semi-arid Australia (Goudie, 1973; McFarlane, 1976).

The value of the materials of the duricrust and its disintegration and deflation products lies in their location in relatively unsettled areas where transport until recently has not been organized. The realization of the value in providing adequate means of transport for stock from the arid rangelands to the markets in eastern and southern Australia and for expansion of the mineral industry in the arid areas has combined to produce an embryonic network of roads and, to some extent, railways over a considerable proportion of the continent. Over much of the area, the terrain is underlain by soft sediments, which do not provide materials suitable for the construction of these communication paths; the only available suitable materials are either the silcretes, ferricretes, laterites, bauxites, and calcretes of the weathering mantle or the products derived from them. To a lesser extent, the ferricretes and laterites of the eastern coastal belt (Stephens, 1971) have been used in the past for similar purposes (Persons, 1970). Gypcrete is not usually used as a constructional material due to its inherent softness (although it has been used as a pavement on lightly trafficked unsealed roads), but it is worked as a source of gypsum for the manufacture of plaster. Similarly, bauxite is worked as a source of alumina for aluminum smelting (Valeton, 1972). Grant and Aitchison (1970) have discussed the formation and engineering significance of some of these materials.

The use of silcrete, ferricrete, laterite, and calcrete is enhanced by the easy manner in which these materials are usually won. Massive quartzitic silcrete

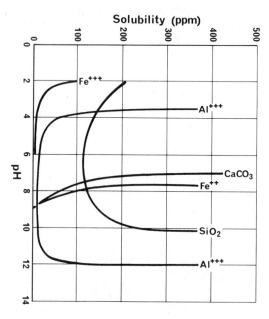

Solubility (ppm)

FIGURE 1. Solubilities in water of some substances and ions involved in the formation of duricrust.

is available for crushing for railway ballast, for concrete and bitumen aggregates, and for screenings for road sealing. It can be won from an *in situ* silcrete face, in which case, beneficiation is required to sort suitable silcrete from associated unsuitable less siliceous material and from associated ferricrete. Alternately, in areas with a gibber pavement, it may be simply scraped from the ground; then, after appropriate tests prove that the material is suitable for the desired purpose, crushing and sizing are all that is required. Indeed, use has already been made of crushed gibber for road-sealing screenings (Fig. 2) and for aggregate for minor concrete works in Queensland. Also, in some areas, the rounded siliceous ironstone mantle has been similarly used for road screenings. Silcrete rubble, mixed with a fine earthy fraction in the form of a colluvial gravel, is available as a road pavement material and has been used as such. Similar use has been made of partly silicified claystones, shales, and sandstones. Laterites of various forms (some are more correctly bauxites) are won from shallow depth in a soft state and are already in extensive use as road pavement materials; some self-hardening silcretes, similarly won, have also been used for this purpose. Calcrete is a common construction material used where it occurs.

Engineering Properties

The efficient use of these materials depends on a knowledge of their properties; unfortunately, engineers know little about these properties. The lack of definition of the materials and the absence of an adequate system for their description, coupled with the remoteness of their environment,

FIGURE 2. Crushed gibber used as screenings in road pavement sealing with bitumen.

have exacerbated the problem. Difficulties encountered during and after a road-building program in northern Australia exemplify this lack of knowledge.

In general, the building of the cheaply constructed, lightly trafficked "beef" roads has been successful, but some problems have been encountered. One major difficulty has been the selection of materials suitable for pavement construction. In one area, material was carried a considerable distance because local material, although known to be lateritic, would not pass conventional plasticity tests. Yet soon after the road was constructed, table drains—which had been easily cut by a grader blade during construction—had hardened to a rocklike consistency. This hardened local material appeared likely to perform in a pavement the same as, if not better than, the imported material. The self-bonding properties of laterite had been overlooked. In the reverse situation, partly silicified material expected to harden upon exposure to drying conditions did not harden; as a result, the pavement failed.

Other problems have concerned the fine fraction of some colluvial silcrete gravels used as pavement material. Although standard testing showed that the material would be suitable, once it was used in the pavement it was weaker than expected. In this case, the clay in the fine fraction was in the form of silt-sized aggregates. While working during testing had sufficed to break these aggregates down to a normal clay consistency, working during construction had failed to do so, and the material tended to behave as a silt. In addition, in some areas this type of clay produced a very smoothly polished surface on forming and rolling. This surface resisted the penetration of the bitumen priming coat used; consequently, the bitumen seal could be rolled off the pavement like a carpet.

In other areas silcrete and ferricrete screenings have given unsatisfactory service when used as the running surface after bitumen sealing (see *Laterite, Engineering Geology*). Bitumen did not adhere well to some crushed quartzitic silcrete, with a consequent loss of the running surface when the road was constructed by usual methods. In some instances precoating the screenings with bitumen overcame the problem. In other places, rounded siliceous ironstone, won directly by scraping from the ground surface, has worn and broken more rapidly than expected, again with a consequent deterioration of the running surface.

Engineering Tests and Performances

In assessing the potential use of these materials, a knowledge of their mineralogical composition and physicochemical properties is a valuable guide to determining the form of testing that should be applied to them. The greatest use of silcretes, ferricretes, laterites, and calcretes is likely to be in the construction of road pavements (Persons, 1970), and it is important that a correct testing procedure be adopted. Standard tests are appropriate in some cases, but in others—for example, with self-bonding silcretes, laterites, and calcretes—new testing procedures may have to be adopted.

The type of silcrete used is of critical importance. Little trouble may be expected from completely silicified material. Standard tests are appropriate, because while silica possesses a certain solubility within a limited environment, rates of solution and deposition are so slow that no discernible changes in silicification will occur during the life of any particular project. Partly silicified material may be subject to normal weathering processes; if argillaceous parent material is allowed to come into contact with ambient atmospheric conditions, it may break down. Attention should be paid to silicified materials that have the appearance of claystone, porcelanite, or sandstone. Tests should be designed to determine their weathering properties. With silcrete colluvial gravels, attention should be paid not only to the type of silcrete incorporated in the gravel, but also to the properties of the fine fraction. If the fine fraction shows a tendency to aggregate, tests should be run to determine the best working procedures to break down the aggregation.

With ferricretes and laterites, each case must also be considered on its merits. Only if segregation of iron and clay is complete will the ironstone gravel fraction be sufficiently hard for use under normal circumstances without stabilization.

Certain silcretes, laterites, and calcretes possess potential self-bonding properties. Therefore, tests should be designed to show the degree and rate of self-bonding that may be expected if they are used in a pavement, and the properties—such as hardness and abrasiveness—that they will exhibit during and after the formation of the bond. The conditions under which the bond might disintegrate must also be determined. Standard tests are not appropriate with self-bonding materials and may give misleading information about how the materials would perform when in use. For instance, results of plasticity tests may be considerably in excess of specification limits on the material as won but may be well within specification limits after self-bonding occurs. Self-bonding of silcrete, which is usually irreversible, occurs when partly silicified material in contact with a saturated solution of silica is allowed to dry completely. The silica precipitated from solution acts as the bonding agent.

Certain generalizations can be made about the self-bonding properties of laterite (see *Laterite, Engineering Geology*). If all the iron is in the ferric state and no material is present to promote a reducing environment, the material is essentially stable and no self-bonding can be expected. Standard tests are then appropriate. If all iron is not in the ferric state, or if conditions might promote a reducing environment that would allow for further mobilization of iron, self-bonding by reoxidation may occur under appropriate conditions of wetting and drying. Standard tests are then not appropriate.

Similar arguments can be applied to calcrete. If waters containing significant quantities of carbon dioxide in solution are available, the material may partly dissolve as calcium bicarbonate; when the carbon dioxide content of the water drops, the material may recement as calcium carbonate. This action of solution and cementation may be cyclic.

In using silcrete for concrete aggregate, first it must be tested to ensure that no unsilicified clay remains in the material. Second, as silcrete may contain amorphous and opaline silica, testing should be designed to show whether any cement-aggregate reaction is likely to occur, and if so, what the result of the reaction might be. Similarly, in using silcrete or siliceous ironstone for screenings for road sealing, appropriate testing should be designed to determine the wearing properties in use and how bitumen would adhere to the material.

KEITH GRANT

References

Buchanan, F., 1807. *A Journey from Madras Through the Countries of Mysore, Canava, and Malabar, etc.,* Vol. 2. London: East India Company, 436-461.

Finkl, C. W., Jr., 1979. Stripped (etched) landsurfaces in southern Western Australia, *Australian Geog. Studies,* 17(1), 33-52.

Finkl, C. W., Jr., and Churchward, H. M., 1973. The etched landsurfaces of southwestern Australia, *Jour. Geol. Soc. Australia,* 20(3), 295-307.

Goudie, A., 1973. *Duricrusts in Tropical and Subtropical Landscapes.* Oxford: Clarendon Press, 174p.

Grant, K., and Aitchison, G. D., 1970. The engineering significance of silcretes and ferricretes in Australia, *Eng. Geology,* 4, 93-120.

McFarlane, M. J., 1976. *Laterite and Landscape.* New York: Academic Press, 151p.

Persons, B. J., 1970. *Laterite: Genesis, Location, Use.* New York: Plenum, 103p.

Stephens, C. G., 1971. Laterite and silcrete in Australia: a study of the genetic relationships of laterite and silcrete and their companion materials, and their significance in the formation of the weathered mantle, soils, relief, and drainage of the Australian continent, *Geoderma,* 5, 3-52.

Valeton, I., 1972. *Bauxites.* Amsterdam: Elsevier, 226p.

Wopfner, H., and Twidale, C. R., 1967. Geomorphological history of the Lake Eyre Basin, in J. N. Jennings and J. A. Mabbutt (eds.), *Land Form Studies from Australia and New Guinea.* Canberra: Australian National University Press, 119-143.

Cross-references: *Arid Lands, Engineering Geology; Caliche, Engineering Geology; Laterite, Engineering Geology; Rocks, Engineering Properties; Soil Classification System, Unified.* Vol. XIV: *Engineering Geochemistry; Rock Weathering Classification.*

E

EARTHQUAKE ENGINEERING

The term *earthquake engineering* denotes a field that has become significant only in recent years. It subsumes a broad range of applied engineering sciences, including seismology (q.v.), geophysics, soil dynamics (q.v.), structural design (dynamics), and mathematical modeling. The importance of earthquake engineering is evidenced by the number of people—approximately 14 million in the past several centuries—who have lost their lives in earthquakes or in the aftereffects such as tsunamis, fires, and landslides, (Duke, 1958). The photographs in Figure 1 illustrate how destructive even a small (magnitude 5.7) earthquake can be if structures are not engineered to resist ground shaking. These photographs raise several questions: Can earthquake occurrence be reliably predicted? And can engineers design for more involved than just buildings to withstand this force of nature?

Terminology

Following is a list of standard terminology used in *engineering seismology* (Richter, 1958).

Hypocenter. Used synonymously with *focus,* the *hypocenter* is the initial location or starting place of an earthquake within the earth. Earthquakes with foci at depths less than 70 km are called *shallow,* or *normal.* If the hypocenter is deeper, but not over 300 km, the depth is called *intermediate.* Hypocenters deeper than 300 km are classified as *deep,* although sometimes this term is applied to all quakes of greater-than-normal depth of focus. Over 70 percent of all earthquakes are shallow.

Epicenter. The *Epicenter* is the point on the surface of the earth directly above the hypocenter. The instrument that writes a permanent, continuous record of earth motion is a *seismograph.* Its record is called a *seismogram.* For an example, see the accelerogram (seismogram recording acceleration) for the 1940 El Centro, California, earthquake (Fig. 2).

Magnitude. *Magnitude,* a rating of energy release for a given earthquake, is calculated from measurements on seismograms. It is defined in terms of the logarithm of the maximum amplitude of the seismogram. Of the many scales in use, the *Richter magnitude scale* is the best known. Because the scale is logarithmic, every upward step of one unit of magnitude means the recorded amplitude is about 10 times that of the lower magnitude (or about a thirtyfold increase in the energy release). A magnitude of 0 was assigned arbitrarily to the smallest recorded earthquakes at the time. With current instruments, shocks smaller than 0 can be recorded. The largest known earthquake magnitudes, the results of direct observation, are near 8½ on the Richter scale. Other scales may yield different magnitudes for the same shock.

Intensity. The term *intensity,* used to describe the degree of shaking at a specified place, is a rating assigned by an experienced observer using a descriptive scale, with grades generally indicated by Roman numerals from I to XII (Table 1).

P, S, and R Waves. Although other wave types are known in seismology, only P-, S-, and R-waves are of major significance to this discussion. *P- (Primary) waves* (also called *compressional* or *longitudinal (dilational) waves)* are the fastest traveling body waves. They are created by the back-and-forth motion of particles (tension and compression in an elastic solid) in the direction of propagation, characterized by change in volume but free from rotation. The velocity of propagation through deeper rock and the earth's core ranges from about 20,000 ft/sec to 45,000 ft/sec. *S- (secondary) waves* (also known as *shear* or *distortional waves)* are the slower-moving elastic body waves. A particle in the path of an S-wave may oscillate in any direction in the plane normal to the direction of the advance of the wave, characterized by no change in volume, but possessing rotational quality. Although slower than longitudinal waves, earthquake-S-induced waves generally transmit more energy than P-waves. When either P- or S- waves strike a surface, they are reflected and create surface waves that propagate along the boundary of the medium. Some of the most important waves in engineering seismology are the *R- (Rayleigh) waves,* which are sine waves of uniform amplitude with a retrograde elliptical motion. Displacements are both vertical and horizontal, and in the direction of propagation.

Spectra. In terms of earthquake engineering, a *spectrum* is defined as an envelope of maximum amplitudes of motion for the corresponding period of that motion. There are two basic types of spectra: (1) ground motion (Fourier analysis) spectra and 2) response spectra. A *ground motion*

FIGURE 1. Modern Saada Hotel in Agadir, Morocco, before and after 1960 earthquake (photo: American Iron and Steel Institute).

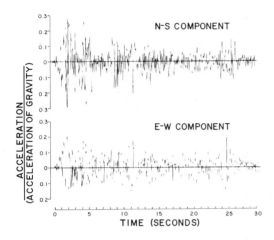

FIGURE 2. Accelerogram for El Centro earthquake (1940), showing acceleration peaks along north-south and east-west components.

ground motion. A ground motion spectrum would be the same as an undamped response spectrum if there were no successive cycles of ground movement to reinforce the oscillator motion. These envelopes of motion may be plots of displacement, velocity, or acceleration versus period. Generally, tripartite plots are the most common. The fundamental *period* of a structure refers to the time required for the uppermost discrete mass to complete a single cycle of free vibration. The time it takes the whole structure to make a full response is the fundamental mode of the structure. There are other (higher) modes, aside from the fundamental, which can also influence structural design. The length of the free period depends on the elastic properties and the physical dimensions of the structure (or in the higher modes, its components). *Seismicity* is a measure of frequency of occurrence of earthquakes in an area.

spectrum is a plot of the maximum amplitudes of the simple harmonic components of recorded ground movement against the period of the ground movement. A *response spectrum* is a plot of the maximum amplitudes of simple oscillators (of varying periods) produced at a recorded or assumed

Earthquake Mechanism

Howell (1959) defines an earthquake as a sudden transient motion or series of motions of the ground that originate in a limited region and spread from there in all directions. An earthquake usually has some definite beginning, continues for a time, and then gradually dies out. Earthquakes can be caused

TABLE 1. Modified Mercalli Intensity Scale of 1931 (abridged).

I. Not felt except by a very few under especially favorable circumstances. (I Rossi-Forel scale.)

II. Felt only by a few persons at rest, especially on upper floors of buildings. Delicately suspended objects may swing. (I to II Rossi-Forel scale.)

III. Felt quite noticeably indoors, especially on upper floors of buildings, but many people do not recognize it as an earthquake. Standing motorcars may rock slightly. Vibration like passing of truck. Duration estimated. (III Rossi-Forel scale.)

IV. During the day felt indoors by many, outdoors by few. At Night some awakened. Dishes, windows, doors disturbed; walls make creaking sound. Sensation like heavy truck striking building. Standing motorcars rocked noticeably. (IV to V Rossi-Forel Scale.)

V. Felt by nearly everyone, many awakened. Some dishes, windows, etc. broken; a few instances of cracked plaster; unstable objects overturned. Disturbance of trees, poles, and other tall objects sometimes noticed. Pendulum clocks may stop (V to VI Rossi-Forel scale.)

VI. Felt by all, many frightened and run outdoors. Some heavy furniture moved; a few instances of fallen plaster or damaged chimneys. Damage slight (VI to VII Rossi-Forel scale.)

VII. Everybody runs outdoors. Damage **negligible** in buildings of good design and construction; **slight** to moderate in well-built ordinary structures; **considerable** in poorly or badly designed structures; some chimneys broken. Noticed by persons driving motorcars. (VIII Rossi-Forel scale.)

VIII. Damage **slight** in specially designed structures; **considerable** in ordinary substantial buildings with partial collapse; **great** in poorly built structures. Panel walls thrown out of frame structures. Fall of chimneys, factory stacks, columns, monuments, walls. Heavy furniture overturned. Sand and mud ejected in small amounts. Changes in well water. Persons driving motorcars disturbed. (VIII+ to IX Rossi-Forel scale.)

IX. Damage **considerable** in specially designed structures; well-designed frame structures thrown out of plumb; **great** in substantial buildings, with partial collapse. Buildings shifted off foundations. Ground cracked conspicuously. Underground pipes broken. (IX+ Rossi-Forel scale.)

X. Some well-built wooden structures destroyed; most masonry and frame structures destroyed with foundations; ground badly cracked. Rails bent. Landslides considerable from river banks and steep slopes. Shifted sand and mud. Water splashed (slopped) over banks. (X Rossi-Forel scale.)

XI. Few, if any (masonry), structures remain standing. Bridges destroyed. Broad fissures in ground. Underground pipe lines completely out of service. Earth slumps and land slips in soft ground. Rails bent greatly.

XII. Damage total. Waves seen on ground surfaces. Lines of sight and level distorted. Objects thrown upward into air.

Source: Wood and Neumann, 1931.

by numerous factors, such as explosions, meteor impact, rock slides, or volcanoes. This discussion, however, is confined to natural earthquakes caused by tectonic forces.

The *elastic rebound theory* offers one explanation of the mechanism that produces an earthquake. Briefly, this theory states that earthquake faulting is the result of the gradual accumulation of stress, which builds up until the breaking strength of the rock is exceeded. This faulting may not exhibit a "trace" on the surface, for it can be entirely below ground. Rock stresses can accumulate very slowly; the strain on the San Andreas Fault in California, for example, is approximately 5 cm per year. The strain is stored as elastic energy in the deformed rocks until a threshold is reached and the rocks break. Figure 3 is an idealized diagram of the deformation that occurred along the San Andreas Fault near San Francisco (Anderson, 1971) prior to the 1906 earthquake.

The elastic rebound theory of earthquake movement is analogous to the bending of a leaf spring, fixed at one end and free at the other. The amount of energy released is proportional to the square of the deflection of the free end.

It is generally believed that faulting starts at one specific point and then spreads along the fault plane with a speed no greater than the velocity of propagation of elastic waves through the rock. This elastic slip along the fault plane may involve only a relatively small area, followed after a short interval by a shift over a much larger surface. If several different breaks can be distinguished before the main shock, they are called *foreshocks*. The foreshocks of a major earthquake usually cannot be distinguished from a minor shock that bears no relation to a forthcoming major earthquake, although there are suggestions that precursors may provide a means of short term prediction of earthquake occurrence.

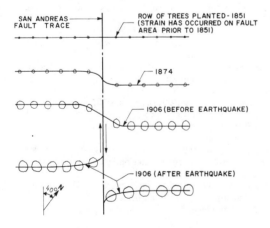

FIGURE 3. Lateral displacement along San Andreas Fault trace after the 1906 earthquake.

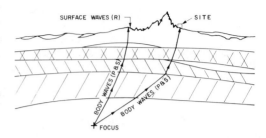

FIGURE 4. Curvature of earthquake wave paths as they pass through materials of different density.

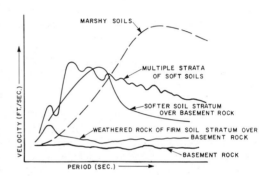

FIGURE 5. Ground motion spectra for different kinds of surface materials, ranging from bedrock to marsh soils.

Even if there were no other mechanism complicating the ground motion during a shock, shaking would continue until the two blocks of earth completed their shift relative to each other. Because the released energy travels in wave trains through the ground at several different velocities and can travel along various paths from one point to another (Fig. 4), the duration of an earthquake at any location will be much longer than that of the movement at the break. In general, the greater the distance from the fault, the longer the motion will last. Near the epicenter of an earthquake the duration can vary from a few seconds to several minutes.

Critical amplification of this incident seismic wave train within the surface materials of a site can occur from (1) a series of waves with a period similar to the predominant period (or periods) of a site, and (2) from a soil system with poor strength characteristics (elastic properties). A *resonance condition* will generally occur only with a relatively simple subsurface condition, such as a single soil strata overlying basement rock. Figure 5 presents an example of the spectra of ground motion that can be expected on various soil and rock strata.

Seismicity

Discussion of earthquake mechanisms does not, however, give any clues to the frequency or geographical distribution of earthquakes. Most of the world's destructive earthquakes occur in well-defined *seismic zones* (Iacopi, 1973). One of the most active is the Circum-Pacific Belt, which forms an almost continuous circle surrounding the Pacific Ocean. It extends from New Zealand, the Philippines, Japan, the Aleutians, the coasts of North, Central, and South America, and almost to the tip of Cape Horn. Another is the Alps—Caucasus—Himalayan seismic zone, or the Alpine Belt, which extends from the West Indies to the Azores across the Alps of Mediterranean Europe, and across Asia to Burma along the front of the Himalayas. One branch shoots northward toward the upper course of the Yellow River in China, and the other branch turns southward passing through Sumatra and Java and merging with the Circum-Pacific Belt. Figure 6 shows the distribution of major earthquakes throughout the world. The number of earthquakes occurring in a particular area is roughly indicated by the thickness of the line; for example, the Japanese Islands constitute one of the most seismically active areas in the world. There are other minor active zones in the Arctic, Atlantic, and Indian oceans. Earthquakes causing considerable damage also occasionally occur in areas considered to be stable (cratonic intraplate

regions). Examples include the 1904 quake near Oslo, Norway and the 1811-1812 series of shocks near New Madrid, Missouri.

In the United States, the Pacific Coast, including Alaska and the Aleutian Islands, is the most seismically active area; however, one of the largest North American earthquakes occurred in Southern Missouri and the Mississippi Valley in 1811-1812 (Fuller, 1912).

It is axiomatic that no area in the world may be considered completely free of the possibility of ground motion. The risk of a structure's being damaged by a major earthquake can vary greatly. In the areas of the greatest seismicity in the United States, it must be assumed that any man-made structure will experience the effects of destructive earthquake motion, unless the economic life of the structure is extremely short. In areas of lesser seismicity, structures of major importance, and those in which damage could result in great loss of life, still must be designed to withstand earthquake movement.

The location and magnitudes of earthquakes generally may be predicted only on the basis of an area's geological and seismic history (Howell, 1973; Press, 1975). Locations of earthquakes and earthquake damage have been noted for at least 1700

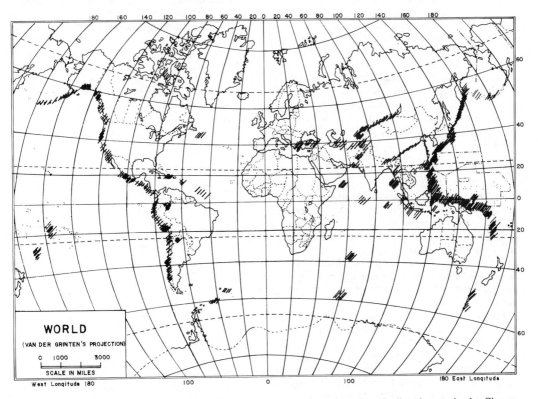

FIGURE 6. Worldwide earthquake distribution, showing concentration of earth disturbances in the Circum-Pacific Belt and along other plate boundaries.

years. Records have been well kept since the advent of sensitive seismographs early in this century. A knowledge of local and regional geology allows one to extrapolate the known seismic history over a greater period: the known geological history of a region (Scholz et al., 1973). A statistical evaluation of the seismic history of an area can yield fairly realistic predictions of the seismic hazard of a particular site in an area of great seismic activity; however, the use of statistical analysis in an area of low seismicity is in itself hazardous because of the statistical "sample size."

The selection of "design" earthquakes to help in the planning and analyses of major construction is extremely involved. Concepts and degree of conservatism vary with different structures, and with the amount of available information.

Earthquake Damage

Type of Structure-Subsurface Conditions. Experience from past earthquakes throughout the world indicates that damage to buildings and other types of structures is closely related to the design, quality of construction, types of construction, and soil or geological conditions. For the moment, let us neglect the effect of construction quality on earthquake damage. Poor construction naturally increases the susceptibility of any structure to damage and, in many instances, it may be the major cause of damage (Wallace, 1974). If construction quality can be neglected, however, certain generalizations can be drawn.

Reinforced concrete buildings of modern design and construction can stand up satisfactorily to earthquake motion. Naturally, the buildings must be reinforced to withstand lateral as well as vertical loads. Steel-frame buildings encased in reinforced concrete for fire protection, durability, and strength have also withstood earthquakes well. Steel-frame buildings with brick or tile walls, though, typically suffer the most from shocks. Because structures provided with seismic walls of reinforced concrete deform less, they suffer less damage. Steel frames can withstand some elastic and plastic displacement, but less flexible materials within or attached to the structure are susceptible to damage and even failure. Poorly constructed, unreinforced masonry, adobe, and similar materials, for example, are always severely damaged in larger earthquakes.

Pattern of Motion — Subsurface Conditions. The significance of subsurface conditions and the effect of earthquake motion structures can also be generalized. The amplitude of motion will generally be much larger on soft soils. Generally, the largest amplitudes (although not necessarily the highest accelerations) are experienced on marshy deposits, whereas unweathered basement rocks perform well as foundation material. Thus, firm founda-

tions make for safer structures because the magnitude of shaking is reduced (see *Foundation Engineering*). Certain secondary effects must also be considered, however. At the hypocenter, an earthquake may be a source of random impulses with no predominant period of motion. Figure 7 depicts the significance of periods, amplitude, and frequency in relation to simple harmonic motion. The random impulses emitted by the quake are assumed to comprise a number of superimposed components of simple harmonic motion, as an aid in analyses and calculations. The motion of the hypocenter can be altered by the effects of the geological conditions between the hypocenter and the site in question. The most obvious change, at longer epicentral distances, is that the short-period (high-frequency) waves are attenuated and only the longer period motion is recorded.

In addition, the soil conditions at a particular location may contribute to a natural (or predominant) period for the site. If the period of wave motion is the same as the period of a site, a resonance condition may occur. Complete resonance will not occur because of the damping effect of naturally occurring anisotropic soils. Partial resonance can occur, however, if the period of several successive wave trains is about the same as the natural period of the site. On firm, unweathered rock, the natural period is generally extremely short. Because many short-period waves are filtered out at distances greater than 40 to 80 km from the expicenter, the resonance condition on rock is usually not a major problem. Marshy soils or deep, layered, soft alluvium will probably not create a "predominant" period condition at a site because of the multiple layering and the resultant different resonance period for each layer. What

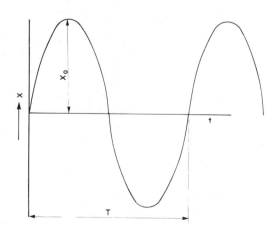

FIGURE 7. Simple harmonic motion sine wave, where the maximum displacement amplitude of vibration (X) is expressed as $X = X \sin t$, and the period of motion (T) is expressed as $T = 2\pi/w$, where w = circular frequency and f = frequency (in cycles/sec) = $1/T$.

does occur, however, is an increased *amplification* of the incident wave train over a very large range of periods.

An incident earthquake wave train can be amplified as a result of (1) a series of waves with a period similar to the predominant period (or periods) of a site, creating a resonance condition and thereby increasing the amplitude, and (2) a soil system with poor elastic properties, which tends to slow the passage of earthquake waves, thereby increasing the amplitudes of the wave motion. Resonance will generally occur only at sites with a relatively simple subsurface condition. For example, a single soil stratum overlying basement rock, as shown in Figure 5, can develop large ground-motion amplitudes at the single predominant period if several successive earthquake waves arrive at the same period.

The resonance problem becomes even more critical when one of the free periods of a structure (building, dam, transmission tower, wharf, etc.) is the same as the predominant period of the site. As shown in Figure 8, the movement of a structure at a site is extremely complicated and depends on interrelationships among physical properties of the structure, site, and materials between the earthquake origin and the site, as well as the energy release at the focus. A critical situation could evolve if a series of wave trains were to coincide with the natural period of the structure and the site. The resulting resonance or quasi-resonance condition would cause great structural damage.

Once the interaction of geological conditions and structures is appreciated, patterns of structural damage during earthquakes become clear. Wooden structures, which are quite flexible and have long free periods, experience damage on softer soils because longer-period wave motions cause large deflections and sometimes a partial resonance condition. This resonance effect is also responsible for damage to rigid masonry buildings on firmer soils. More rigid structures have shorter free periods, which may be the same as or near the natural period of a site underlain with firm soils or rock. Although the amplitude of ground motion may not be as great on a site with firmer soils, the resonant condition between the site period and the building period can cause great damage.

Type of Motion. The damaging effects of earthquakes on structures are generally produced by horizontal motion (S- and R-waves) rather than by vertical (P-wave) ground motions. This is because horizontal motion is usually greater everywhere but in the immediate vicinity of the epicenter, and because vertical motion operates against or with gravity. Structures are normally designed for relatively heavy material (vertical) loads, well above the design for lateral forces. The result is that structural damage primarily comes from S-waves, with a near-vertical direction of propagation (Fig. 4) creating essential horizontal distortion. Long-period, later-arriving R-waves (Fig. 4) may result in critical horizontal amplitudes because they tend to prolong the shaking of already-weakened buildings.

For all practical purposes, therefore, earthquake-resistant design must consider horizontal oscillations, although high vertical motion (caused by P waves) has been noted near the epicentral area of some earthquakes.

Effects on Structures. During the passage of a single wave train, the ground moves in one direction; at the end of travel, the motion stops and reverses direction. This reversal is characteristic of all recorded earthquake motion. With respect to man-made structures, therefore, the power of an earthquake is practically infinite. When the ground moves, everything attached to it must conform to that movement. This compulsion for movement is applied at the base of the structure. The effect of the compulsion and any mass above the base is to change the velocity of that mass. This change in velocity cannot occur until the compulsion becomes effective and a certain time elapses between the motions of adjacent masses along the height of the structure. The velocities of these successive masses vary; hence, differential displacement or distortion occurs. The total distortion of the entire structure is related to the amount of displacement of the base and the time required for a complete response (Eckel, 1970).

The response of a structure to imposed base displacement is a function of the free vibration period of the structure. The elapsed time between the application of a compulsion at the base of a structure and its effect at any portion of the structure above the base is almost one-fourth of the free fundamental period of the portion of the structure affected. Hence, the distortion will rise more rapidly in more rigid than in less rigid portions.

The period of a structure is not constant. Building periods are sometimes measured under the effect of small motions (usually wind). As a structure subjected to strong earthquake motion becomes damaged, the structural period increases as a result of the loss of rigidity. In a large earthquake, building periods can sometimes change more than 100 percent. When the *elastic resilience* of a structure is partially destroyed, the internal damping

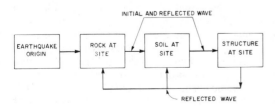

FIGURE 8. Earthquake-foundation-structure interaction.

will increase, possibly offsetting the weakened condition of the structure and allowing many structures to stand through an earthquake although severely damaged.

Because the period of a structure is a function of stiffness, any yielding of the foundation can also increase the natural period. For fairly rigid foundation conditions, however, building periods do not increase appreciably. Whatever the structural period, earthquake damage is evidence of excessive strains. In most structures, the framework that supports vertical loads and resists wind pressure is composed of materials of greater strength and ductility than those used for less important purposes. Much subsidiary construction in a building is relatively brittle and incapable of withstanding large deformations. It is not possile to control earthquake damage merely by limiting the stresses in the frame to safe levels. Buildings cannot be considered unrelated assemblies of parts in which the frame acts by itself independent of walls, partitions, and the material around it. The building must be considered as a unit, and the influence of walls, partitions, floors, and frame must be considered. The idea that the design of tall buildings is successful if the steel frame will remain standing even though the walls, partitions, and floors are shattered and wrecked is not a useful concept. Unless the steel frame has those elements around it that make a habitable building, it is a perfectly useless thing. Buildings are structures, and in their design they must be considered as integrated units.

Lateral forces are first resisted by the most rigid elements in the building, usually the walls and partitions. They accommodate shear or stress in proportion to their relative rigidities. In practical design, it is not always possible to apportion shears to resisting elements in a rigorous, theoretical way. This fundamental fact cannot be ignored; and earthquake-resistant designs thus should consider the relative rigidities of the various resisting elements.

Although not completely destroyed, many structures are badly damaged in earthquakes. Railway lines are damaged by the change of level in softer soils, usually due to slumping or subsidence. Rails are sometimes bent. Bridges are often seriously damaged by lateral shaking. Generally, tunnels receive little damage unless they are in the immediate vicinity of the fault zone. Here, thrusting of the fault blocks themselves can break and offset the tunnel structures. Damage to irrigation systems may be caused by slumping, the emergence of water and sand, or the direct breaking of canals and ditches by faulting. Water-supply systems may be damaged indirectly when transformers are thrown down from poles or electrical power is cut off from pumps used in irrigation.

Some of the worst damage occurs at waterfront facilities. Many times these structures have been built on "made" land, overlying soft organic or recent marine and alluvial sediments. Such soil conditions can result in high amplifications of incident earthquake motion. Seismograph records indicate that amplifications of 12 or more over basement rock motion can exist at sites with the aforementioned subsurface conditions. The high water tables found near shore areas also tend to increase the intensity of shaking. In addition, waterfront structures generally have long free periods which, when coupled with the long-period amplification of soft soils, will result in major structural damage. Added to this is the hazard of tsunamis resulting either from nearby earthquakes or from strong shocks that may occur thousands of kilometers away. Tsunamis from Chilean earthquakes, for example, have caused damage as far away as the West Coast of the United States, the Hawaiian Islands, and Japan (Bolt, 1978).

Large earthquakes also occasionally resulted in the failure of dams. In the comparatively moderate shaking of the 1925 earthquake at Santa Barbara, California, an earth-filled dam retaining the Sheffield Reservoir gave way. In addition to shaking the dam itself, the hydrodynamic effect of the water on the dam embankment or other structures that retain liquids can cause damage (Khattri and Wyss, 1978).

Even in small earthquakes, underground pipes are put out of service by slumping or subsidence (liquefaction) of soft or loose soils. The initial failure need only be small cracks, which can create even greater damage wherever the pipes handle the flow of liquids. Old and weakened pipes are commonly reported out of service after mild earthquakes, when local shaking was hardly perceptible.

Aseismic Design

Local building codes often specify requirements for earthquake-resistant design. Most codes have similar aseismic design specifications, which are based on a *lateral force hypothesis*. This lateral force, taken as some percentage of the weight of the structure, comes from Newton's second law of motion in which the force is equal to the mass times the acceleration. This is only an approximation, as the mass of the structure (unless it is perfectly rigid) does not follow the movement of the ground and the building acceleration is not necessarily that of the ground. In addition, this lateral force hypothesis neglects the effect of resonance between the ground and structure.

The lateral force hypothesis is admittedly inadequate, and building codes contain modifications based on results of post-earthquake investigations. In some instances, building codes may not be adequate for structures that must withstand earthquakes of greater magnitudes than those previously experienced in the area.

Formal dynamic approaches have also been taken to aseismic design. For background consider the effect of an earthquake on a rigid structure. Such a structure is shown in Figure 9. It is assumed that both the building and its foundations are rigid so that earthquake motions of the ground, U_g, are transmitted directly to the building. In this case, an equivalent force, F, will be developed in the structure equal to the product of the ground acceleration and the mass of the structure, $F = (\ddot{u}_g)W/g$, in which $\ddot{u}_g$ is the ground acceleration gravity. For convenience, this equation is usually rearranged so that the force is given as the product of the weight of the structure and the aseismic coefficient, C, which represents the ratio of the ground acceleration to the acceleration of gravity: $F = (\ddot{u}_g/g) W = CW_o$. For design purposes, it is common to express the earthquake force at the base of the structure. In this case, simple statics show that the base shear, V, is equal to the force F, and is given by $V = CW$.

The last equation $V = CW$ indicates that the dynamic analysis of a rigid structure is very simple. All that is required is an estimate of the maximum ground acceleration that will occur during an earthquake. This acceleration, expressed as a ratio to the acceleration of gravity, is the aseismic coefficient C.

The rigid-structure concept provides the basis for the lateral force provisions in some of the earliest earthquake codes, which specify that the structures should be designed for a certain percentage of gravity (say, 10 percent or 12 percent), regardless of the characteristics of the structure. Unfortunately, the dynamic response characteristics of actual structures are not simple. Their flexibility and mass impart vibration properties that directly affect the magnitude of the seismic forces to which they will be subjected during an earthquake.

The effect of a structure's flexibility on this response is exemplified by the simple one-story structure shown in Figure 10. The weight of the structure, W, is assumed to be concentrated at the roof level. Such a structure is said to have a single degree of freedom (considering plane motion only) because only one type of deformation is possible, represented here by the displacement, u. The significant dynamic properties of this structure, in addition to its weight, are the stiffness of the columns, k, which represents the force needed for a unit displacement, and the damping, c, which represents the force per unit of velocity. Although damping will be omitted for simplicity in the explanation that follows, the effect of damping must be included in the final analysis.

In the absence of damping, the base shear of this structure is the product of the displacement and the column stiffness, $V = ku$. Dynamic equilibrium conditions (using d'Alembert's principle) show that the base shear must balance the inertial force of the mass, that is, $(W/g)\ddot{u}_t + ku = 0$ (using the sign convention assumed in Figure 10). It will be noted the inertia force here depends on the total motion of the mass, u_t, rather than the ground motion only, as was the case in Figure 9. These two motions differ; the difference depends on the free period of the structure.

Difficulties arise in extending the preceding discussion to more complex structures. The force k is not directly resisted by shears in the columns. First, the various parts of a structure must be accelerated. Since the actual acceleration of portions of the structure takes time, and the shock wave travels through the building in finite time, the inertia of the structure and the time history of the imposed motion are important.

As previously stated, as a compulsion to movement is applied to a structure and becomes effective throughout the height of a structure, time elapses from the initiation of motion one portion of the structure to adjacent portions; hence, dif-

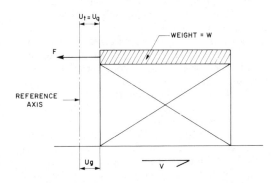

FIGURE 9. Studies of dynamic response problems are based on considerations of the effect of an earthquake on a rigid structure. It is normally assumed that both the building and its foundations are rigid, enabling ground movements to be transmitted directly to the building.

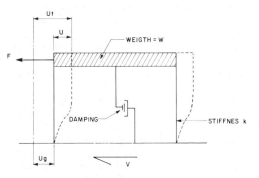

FIGURE 10. Flexible structures such as the one shown here have a single degree of freedom, considering planar motion only, because only one type of deformation is possible, represented here by the displacement (u).

ferential displacement occurs. Stresses occur in the structure as the result of these strains. However, the stress level reached at any time depends on the position of the element in the structure and its relative rigidity. Therefore, even though the design procedures treat each mode of building vibration separately and superimpose the results in multistory structures, the analysis becomes complex (in some procedures, the analysis can be done only with the aid of computers), although a liberal application of engineering judgment is vital.

In general, three basic design concepts arc in use in the United States today: (1) the straightforward lateral force concept (with the aid of empirical adjustment for the nonrigidity of structures); (2) dynamic analysis based on the use of response spectra information; and (3) the response (using computer analogs) of a structure when subjected to a recorded (computerized) earthquake acceleration record.

An account can be taken of the effects of earthquake motion in structural design. Similar concepts may be applied to dam design. The problem then becomes one of determining what type of motion will be experienced by the base of a structure at a particular location. This is based on many variables, including location, site conditions, and the magnitude and frequency of earthquakes.

A measure or knowledge of the size of the earthquakes that a structure must endure is obviously, very important in the design of a building. Tremendous sums of money would be involved in designing a structure to resist all earthquake motion in an elastic manner (i.e., causing little or no damage). Yet structures must be designed to withstand, without irreparable damage, a major shock. Therefore, it is wise to design for a large nearby earthquake in seismic areas but to allow for some damage to the structure. In other words, the building should be allowed to deform elastoplastically during the large nearby earthquake. Excessive strain (damage) will occur during plastic deformation of the structure, but the acceptance of plastic deformation will make the structure more flexible and tend to reduce the possibility of complete destruction.

In many parts of the world, structures must withstand local smaller earthquakes, or larger distant earthquakes, fairly frequently. If these structures are allowed to deform elasto-plastically for every shock, no matter how small or how far away, the costs of repair after each earthquake would be prohibitive. Therefore, an accepted procedure is to predicate an elastic design on smaller, more frequent earthquakes and to design a structure to respond elasto-plastically to the maximum earthquake that it may have to withstand during its lifetime. For moderate-size shocks, the structure will respond elastically and damage will be minimal. For larger shocks, although the damage will be significant, the structure should be reparable and the chance of inhabitants' deaths will be mitigated. Thus, the first step in a full *seismic design evaluation* should be to study the statistics of the occurrence of earthquakes in an area and their relation to geological structures, and to develop design criteria for both the elastic and the elastoplastic responses of the structure.

The next step is to determine the dynamic characteristics of the site and to estimate the amplitude of ground motion to be expected at that location. Although it is impossible to consider all variations of subsurface conditions at a general site and the focus of an earthquake, it is possible to satisfactorily define appropriate dynamic characteristics of a particular location from the surface down to basement rock. Because both mathematical and empirical relationships have been developed for the amplitude of movement on basement rock, it is possible to evaluate the effects of subsurface conditions at a site along the earthquake wave trains as they travel from basement rock to ground surface. Knowledge of dynamic site characteristics and a statistical study of earthquake occurrence in an area makes it possible to predict ground motion and/or the response spectrum to be used in the design of structures.

To develop the needed information, the site is investigated and its characteristics are compared with the characteristics of sites with known strong-motion records. The data from previous earthquakes—in terms of both for ground motion and response spectra—can then be used to estimate realistic spectra for the site in question. Using these procedures, it is possible to design for the impacts of a major earthquake, unless the structure lies directly across the fault trace.

Summary

Earthquake motions are exceedingly complicated phenomena, but it is possible to predict their occurrence approximately. Although the damage resulting from earthquakes can be costly, it is possible to design against the effects of ground shock. Earthquake-resistant design must take into account the type of structure, the rigidity of the structure, the free period of the structure, the subsurface conditions at the site, and the location, frequency, and size of the design earthquake(s). Many advances in both the science and art of earthquake engineering have occurred in recent years; however, it is still far from being a subject open to vast and accurate theoretical predictions. Much of the work that continues to be done will be based on empirical results, experience, and judgment.

JOSEPH A. FISCHER

References

Anderson, D. L., 1971. The San Andreas fault, *Sci. American,* **225**(5), 52-67.

Bolt, B. A., 1978. *Earthquakes: A Primer.* San Francisco: Freeman, 241p.

Cook. N. G. W., 1976. Seismicity associated with mining, *Eng. Geologist,* **10,** 99-122.

Duke, M. C., 1958. Effects of ground and destructiveness of large earthquakes, *Am. Soc. Civil Engineers Proc., Jour. Soil Mechanics and Found. Div.,* **84,** SM3.

Eckle, E. B., 1970. The Alaskan earthquake, March 27, 1964: lessons and conclusions, *U.S. Geol. Survey Prof. Paper 546,* 57p.

Fuller, M. L., 1912. The New Madrid earthquake, *U. S. Geol. Survey Bull. 494,* 119p.

Hadley, J. B., 1964. Landslides and related phenomena accompanying the Hebgen Lake earthquake of August 17, 1959, *U. S. Geol. Survey Prof. Paper 435-K,* 107-138.

Healy, J. H.; Hamilton, R. M.; and Rayleigh, C. B., 1970. Earthquakes induced by fluid injection and explosion, *Tectonophysics,* **9,** 205-214.

Howell, B. F., 1959. *Introduction to Geophysics.* New York: McGraw-Hill.

Howell, B. F., 1973. Earthquake hazard in the eastern United States, *Earth and Mineral Sci.,* **42,** 41-45.

Iacopi, R., 1973. *Earthquake Country.* Menlo Park, Calif: Lane Books, 160p.

Khattri, K., and Wyss, M., 1978. Precursory variation of seismicity rate in the Assam area, India, *Geology,* **6**(11), 685-688.

Press, F., 1975. Earthquake prediction, *Sci. American,* **232,** 14-23.

Richter, C. F., 1958. *Elementary Seismology.* San Francisco, Calif. Freeman.

Scholz, C. H.; Sykes, L. R.; and Aggarwal, Y. P., 1973. Earthquake prediction: a physical basis, *Science,* **181,** 803-810.

Wallace, R. E., 1974. Goals, strategy, and tasks of the earthquake hazard reduction program, *U. S. Geol. Survey Circ. 701,* 27p.

Wood, H. O., and Neumann, F., 1931. Modified Mercalli intensity scale of 1931, *Seismological Soc. America Bull.* **21**(4).

Cross-references: *Foundation Engineering; Residual Stress, Rocks; Rheology, Soil and Rock; Rocks, Engineering Properties; Seismological Methods; Soil Mechanics.* Vol. XIV: *Cities, Geological Effects; Environmental Engineering.*

EARTH SCIENCE, INFORMATION AND SOURCES

Within the past two decades, a growing awareness of the importance of the practical application of geology (see *Geology, Applied*) has generated an enormous amount of research literature. To undertake a thorough search of this literature within a reasonable length of time, the researcher should first become acquainted with the bibliographic sources of earth science information, that is, abstracts, bibliographic serials and compilations, catalogs, and data base services. These index current and/or retrospective primary source literature, such as books, monographs, journal articles, reports, maps, symposia proceedings, government documents, and theses. Data files also provide nonbibliographic numerical or factual data.

The most comprehensive English-language index of earth science literature is the *Bibliography and Index of Geology* (Falls Church, Va.: 1934-), American Geological Institute, which is worldwide in scope. Each monthly issue is divided into 21 intradisciplinary categories and contains subject and author indexes to its citations. The *Bibliography* is compiled from GeoRef, the AGI's computerized, indexed reference file.

Among other current bibliographic serials of a general nature are (1) the *Bibliographie des Sciences de la Terre* (Orleans, France: Bureau de Recherches Géologiques et Minières, 1968-), a monthly that offers worldwide coverage; (2) *Geotitles Weekly,* a "current awareness" service for geoscience (London: Geosystems, 1969-); (3) the monthly *Referativnyi Zhurnal* (Moscow, Akademiia Nauk SSSR, 1956-), whose worldwide, comprehensive citations are in the language of the original publications, with abstracts in Russian; and (4) Zentralblatt für Geologie und Paläontologie (Stuttgart: E. Schweizerbart'sche (1960-), which is published irregularly but offers international listings.

Retrospective bibliographic serials include the *Annotated Bibliography of Economic Geology* (Lancaster, Pa.: Economic Geology Publishing Company, 1929-1971); the U.S. Geological Survey's *Abstracts of North American Geology* (Washington, D.C., 1966-1971); *Bibliography of North American Geology* (Washington, D.C.: U.S. Geological Survey, 1923-1970); *Geophysical Abstracts* (Washington, D.C.: U.S. Geological Survey, 1929-1971), revived in 1977 (Norwich, England: Geo Abstracts Ltd) as a bimonthly publication; and *Neues Jahrbuch für Mineralogie, Geologie und Paläontologie* (Stuttgart: E. Schweizerbart'sche, 1830-1949).

Among the abstracts and indexes concerned with special subject fields in applied geology are the American Petroleum Institute's *API Abstracts of Refining Literature* (New York, 1961-); *Geotechnical Abstracts* (Essen: German National Society of Soil Mechanics and Foundation Engineering, 1970-); *IMM Abstracts* (London: Institution of Mining and Metallurgy, 1949-); *International Journal of Rock Mechanics and Mining Sciences* and *Geomechanics Abstracts* (Oxford, England and New York: Pergamon Press, 1974-); *Mineralogical Abstracts* (London: Mineralogical Society of Great Britain and the Mineralogical Society of America, 1920-); *Oceanic Abstracts*

(Louisville, Ky.: Data Courier, 1968-); *Petroleum Abstracts,* (Tulsa: University of Tulsa, Division of Information Services, 1961-); *Selected Water Resources Abstracts* (Washington, D.C.: U.S. Department of the Interior, Water Resources Scientific Information Center, 1968-); and *Zentralblatt für Mineralogie* (Stuttgart: E. Schweizerbart'sche, 1950-).

Recently published reports are cited in the U.S. Department of Commerce, National Technical Information Service's biweekly *Government Reports Announcement and Index* (section entitled "Earth Sciences and Oceanography"). The *Monthly Catalog of U.S. Government Publications* (Washington, D.C.: Superintendent of Documents) lists federal government publications. The U.S. Geological Survey's *Open-File Reports* are cited in the *Publications of the Geological Survey,* issued as monthly, annual, and cumulative volumes.

Theses in geology are listed in John Chronic's *Bibliography of Theses Written for Advanced Degrees in Geology and Related Sciences at Universities and Colleges in the United States and Canada through 1957* (Boulder, Colo.: Pruett Press, 1958) and *Bibliography of Theses in Geology, 1958-1963* (Washington, D.C.: American Geological Institute, 1965); Dederick C. Ward's "Bibliography of Theses in Geology, 1964," *Geoscience Abstracts,* 7 (12), pt. 1 (1965), 103-129; D. C. Ward and T. C. O'Callaghan's *Bibliography of Theses in Geology, 1965-1966* (Washington, D.C.: American Geological Institute, 1969); and D. C. Ward's "Bibliography of Theses in Geology, 1967-1970," *Geological Society of America Special Paper 143* (Boulder, Colo., 1973). *The Bibliography and Index of Geology* also cites theses indexed in the GeoRef data base since 1967.

The *Glossary of Geology,* 2nd ed., a dictionary edited by Robert L. Bates and Julia A. Jackson, second edition (Falls Church, Va.: American Geological Institute, 1980, 749p.) is an excellent reference tool. Dictionaries concerned with specific fields are mentioned in the lists of references accompanying this entry.

Useful guides to the services and products of the U.S. Geological Survey can be found in its Circular 777, "A Guide to Obtaining Information from the USGS," by Paul F. Clarke, Helen E. Hodgson, and Gary W. Noath (1981, 42p.), and *Circular 817,* "Scientific and Technical, Spatial, and Bibliographic Data Bases of the U.S. Geological Survey, 1979," prepared by the Survey's Office of the Data Base Administrator (1980, unpaged).

Space does not permit a comprehensive listing of reference publications relevant to any one subject field of applied geology. A selected number of titles representative of the kinds of information available are mentioned in the following reference lists. Additional suggestions can be found in an excellent guide to geologic literature: Dederick C.

Ward, Marjorie W. Wheeler, and Robert A. Bier, Jr., *Geologic Reference Sources: A Subject and Regional Bibliography of Publications and Maps in the Geological Sciences,* 2nd ed. (Metuchen, N.J.: The Scarecrow Press, 1981, 560p.).

Economic Geology

American Association of Petroleum Geologists, 1917- . *Bulletin.* Tulsa, published monthly.

American Association of Petroleum Geologists, 1962- . *Memoirs.* Tulsa.

American Institute of Mining, Metallurgical and Petroleum Engineers, 1975. *Industrial Minerals and Rocks (nonmetallics other than fuels),* 4th ed., ed. by Stanley J. Lefond. New York: AIME, 1360p. Available from Society of Mining Engineers, Salt Lake City.

Averitt, Paul, and Lopez, Lorreda, 1972. *Bibliography and Index of U.S. Geological Survey Publications Relating to Coal, 1882-1970, U.S. Geol. Survey USGS Bull. 1377,* 173p.

Battelle Memorial Institute, Columbus Laboratories, 1975. *Energy Information Resources: An Inventory of Energy Research and Development Information Resources in the Continental United States, Hawaii and Alaska,* comp. by Patricia L. Brown, Clarence C. Chaffee, Robert S. Kohn, and Joseph P. Miller. Washington, D.C.: American Society for Information Science, 207p.

Cagnacci Schwicker, Angelo, 1970. *International Dictionary of Metallurgy-Mineralogy-Geology: Mining and Oil Industries, English-French-German-Italian.* Milan: Technoprint International; New York: McGraw-Hill, 1530p.

Canada Department of Energy, Mines, and Resources, Mines Branch, 1959- . *Technical Bulletin* (TB). Ottawa.

Canadian Mining and Metallurgical Bulletin, 1898- . Montreal: Canadian Institute of Mining and Metallurgy, monthly.

Coal Research in CSIRO, 1957- . Melbourne: Australia Commonwealth Scientific and Industrial Research Organization, Division of Coal Research.

Colorado School of Mines, Golden, 1905- . *Quarterly.*

Dixon, Colin J., 1979. *Atlas of Economic Mineral Deposits.* London: Chapman and Hall, 143p.

Economic Geology, 1905- . Mount Pleasant, Mich.: Economic Geology Publishing Company, published monthly.

Fassett, James E. (ed.), 1978. *Oil and Gas Fields of the Four Corners Area.* Durango, Colo. Four Corners Geological Society, 2 vols., 762.

Forum on Geology of Industrial Minerals, 1965- . *Proceedings.* (Published by sponsoring organization).

Geoexploration, 1963- . Amsterdam: Elsevier, published quarterly.

Geoexploration Monographs, 1966- . Berlin: Gebruder Bornträger.

Geological Survey of Canada, 1926- . *Economic Geology Report.* Ottawa.

Geophysical Prospecting, 1953- . The Hague: European Association of Exploration Geophysicists, published quarterly.

Geophysics, 1936- . Tulsa: Society of Exploration Geophysicists, bimonthly.

Geothermics: International Journal of Geothermal Research, 1972- . Pisa, Italy: International Institute for Geothermal Research, published quarterly.

Govett, G. J. S. (ed.), 1981-1983. *Handbook of Exploration Geochemistry,* 3 vols. Amsterdam: Elsevier, 1156p.

Hall, Vivian S., comp., 1980. *An Eastern Gas Shales Bibliography. Selected Annotations: Gas, Oil, Uranium, Etc. Citations in Bituminous Shales Worldwide.* Washington, D.C.: U.S. Department of Energy, Technical Information Center, 532p. For sale by the National Technical Information Service. (Prepared for the U.S. Dept. of Energy, Morgantown Energy Technology Center; DOE/METC/United States Dept. of Energy; 12594-12.)

Hutchinson, V. V., 1964. *Selected List of Bureau of Mines Publications on Petroleum and Natural Gas, 1910-1962, U.S. Bur. Mines Inf. Circ. 8240,* 98p. (This is updated by *Circ. 8534* (1961-1970) and a second supplement to *Circ. 8240,* covering 1971-1975, published by the U.S. Energy Research Development Administrative Center.)

Illinois State Geological Survey, 1972- . *Illinois Minerals Notes.* Urbana.

Illinois State Geological Survey, 1919- . *Illinois Petroleum Series.* Urbana.

Illinois State Geological Survey, 1936- . *Oil and Gas Drilling Reports.* Urbana.

Industrial Minerals, 1967- . London: Metal Bulletin, published monthly.

Institution of Mining and Metallurgy, 1966- . *Transactions. Section B: Applied Earth Science.* London, published quarterly.

International Journal of Coal Geology, 1980- . Amsterdam: Elsevier, published quarterly.

Journal of Geochemical Exploration, 1972- . Amsterdam: Elsevier, published quarterly.

Journal of Geophysical Research, 1896- . Washington, D.C.: American Geophysical Union, published monthly.

Journal of Petroleum Geology, 1978- . Beaconsfield, Bucks. England: Scientific Press, published quarterly.

Meurs, Anton Pedro Hendrik van, 1971. *Petroleum Economics and Offshore Mining Legislation: A Geological Evaluation. Amsterdam:* Elsevier, 208p.

Nelson, Archibald and Nelson, K. D., 1967. *Dictionary of Applied Geology, Mining and Civil Engineering.* London: Newnes, 421p.

Ridge, John D. (ed.), 1968. *Ore Deposits of the United States, 1933-1967: The Graton Sales Volume.* New York: American Institute of Mining, Metallurgical and Petroleum Engineers, 2 vols., 1880p.

Rogers, Marianne P., 1969. *List of Bureau of Mines Publications on Oil Shale and Shale Oil, 1917-1968, U.S. Bur. Mines Inf. Circ. 8429,* 61p.

Roskill Information Services Ltd., 1980. *Roskill's Dictionary of Sources for Metals and Minerals Data,* 4th ed. London, 139p.

Sheriff, Robert E., 1973. *Encyclopedic Dictionary of Exploration Geophysics.* Tulsa: Society of Exploration Geophysicists, 266p.

Thrush, Paul W., 1968. *A Dictionary of Mining, Mineral, and Related Terms.* Washington, D.C.: U.S. Department of the Interior, 1269p.

U.S. Bureau of Mines, 1960. *List of Publications Issued by the Bureau of Mines from July 1, 1910 to January 1, 1960 with Subject and Author Index.* Washington, D.C., 826p. (Kept up-to-date with annual supplements.)

U.S. Bureau of Mines, 1932/1933- . *Minerals Yearbook.* Vol 1: *Metals and Minerals.;* Vol. 2: *Area Reports:*

Domestic; Vol. 3: *Area Reports: International.* Washington, D.C., published annually.

U.S. Geological Survey, 1935- . "Coal Investigations Map." Washington, D.C. Beginning with 1950, maps are designated as "C" series.

U.S. Geological Survey, 1950- . "Mineral Investigations Field Studies Map" (MF). Washington, D.C.

U.S. Geological Survey, 1952- . "Mineral Investigations Resource Map" (MR). Washington, D.C.

U.S. Geological Survey, 1944- . "Oil and Gas Chart." Washington, D.C. (Numbers 1-39 are designated "preliminary"; those from 40 onward are designated as "OC" series.)

U.S. Geological Survey, 1943- . "Oil and Gas Investigations Map." Washington, D.C. (Numbers 1-109 are designated "preliminary"; those from 110 onward are designated as "OM" series.)

Utah Geological and Mineralogical Survey, 1972- . *Oil and Gas Field Studies.* Salt Lake City.

Walker, Flora K., 1980. *Bibliography and Index of U.S. Geological Survey Publications Relating to Coal, January 1971 Through June 1978.* Falls Church, Va.: American Geological Institute, 64p.

White, Donald E., and Williams, D. L., eds., 1975. Assessment of Geothermal Resources of the United States— 1975, *U.S. Geol. Survey Circ. 726,* 155p.

Wolf, K. H. (ed.), 1981. *Handbook of Strata-bound and Stratiform Ore Deposits.* New York: Elsevier, 10 vols.

Engineering Geology/Soil Mechanics

American Society for Testing and Materials, Committee D-18 on Soil and Rock for Engineering Purposes, 1970. *Special Procedures for Testing Soil and Rock for Engineering Purposes: Suggested Methods, Standard and Tentative Methods, Definitions, and Nomenclature (By Reference Only),* 5th ed. Am. Soc. Testing and Materials Spec. Tech. Pub. 479, 630p.

Association of Engineering Geologists, 1964- . *Bulletin.* Berkeley, published semimonthly.

Clays and Clay Minerals, 1968- . New York: Pergamon Press, published bimonthly. (Supersedes Clay Minerals Society's *Clays and Clay Minerals: Proceedings of the Conference.)*

DePuy, G., 1969. *Tunnel Geology Bibliography.* Denver: Bureau of Reclamation, Office of Chief Engineer, 50p.

Developments in Geotechnical Engineering, 1972- . Amsterdam: Elsevier.

Engineering Geology—An International Journal, 1965- . Amsterdam: Elsevier, published bimonthly.

Engineering Geology and Soils Engineering Symposium, 1963- . *Proceedings.* Pocatello, Idaho.

Engineering Geology Case Histories, 1957- . New York: Geological Society of America, Division on Engineering Geology.

Fairbridge, Rhodes, and Finkl, Charles W., Jr. (eds.), 1979. *The Encyclopedia of Soil Science, Part I: Physics, Chemistry, Biology, Fertility, and Technology.* Encyclopedia of Earth Science Series, vol. XII. Stroudsburg, Pa.: Dowden, Hutchinson & Ross, 646p.

Géotechnique, International Journal of Soil Mechanics, 1950- . London: Institution of Civil Engineers, published quarterly.

Institution of Engineers, Australia, 1971- . *The Australian Geomechanics Journal of the Institution . . . and the Australasian Institute of Mining and Metallurgy.* Sydney, published annually.

International Journal of Soil Dynamics and Earthquake Engineering, 1982- . New York: Springer-Verlag, published quarterly.

Lancaster-Jones, P. F. F., 1966. *Bibliography of Rock Mechanics.* Croyden, Surrey, England: Cementation Company, 89p.

Quarterly Journal of Engineering Geology, 1967- . London: Geological Society of London.

Schuster, Robert L., and Krizek, Raymond J. (eds.), 1978. *Landslides: Analysis and Control, Natl. Acad. Sci. Natl. Res. Council, Transportation Bd. Spec. Rept. 176,* 234p.

Selby, M. J., 1982. *Hillslope Materials and Processes.* Fair Lawn, N.J.: Oxford University Press, 320p.

Soil Science Society of America, 1975. *Glossary of Soil Science Terms.* Madison, Wis., 34p.

Symposium on Field Instrumentation, London, 1973, 1974. *Field Instrumentation in Geotechnical Engineering: A Symposium Organized by the British Geotechnical Society.* New York: Wiley, 720p.

Vollmer, Ernst, 1967. *Encyclopedia of Hydraulics, Soil and Foundation Engineering.* Amsterdam: Elsevier, 398p.

Wahlstrom, Ernest E., 1973. *Tunneling in Rock.* Amsterdam: Elsevier, 250p.

Other useful sources of information are the individual county soil surveys published by the U.S. Department of Agriculture and state soil agencies.

Environmental Geology

Coates, Donald R., 1981. *Environmental Geology.* New York: Wiley, 736p.

Conference on Trace Substances in Environmental Health, 1967- . *Proceedings.* Sponsored by the University of Missouri Environmental Health Center and Extension Division. Columbia, Mo., published annually.

Eister, Margaret F. (comp.), 1978. Selected Bibliography and Index of Earth Science Reports and Maps Relating to Land-resource Planning and Management Published by the U.S. Geological Survey Through October 1976, *U.S. Geol. Survey Bull. 1442,* 76p.

Environmental Geology, 1970. Washington, D.C.: American Geological Institute, AGI Short Course Lecture Notes, 1 vol. (various pagings).

Environmental Science and Technology, 1967– . Washington, D.C.: American Chemical Society, published monthly.

Fairbridge, Rhodes Whitmore (ed.), 1972. *The Encyclopedia of Geochemistry and Environmental Sciences,* Encyclopedia of Earth Science Series, vol. IVA. New York: Van Nostrand Reinhold, 1321p.

Fried, Sherman (ed.), 1979. *Radioactive Waste in Geologic Storage.* ACS Symposium Series 100. Washington, D.C.: American Chemical Society, 344p. (Based on a symposium sponsored by the ACS Division of Nuclear Chemistry and Technology held in Miami Beach, Fla., in 1978.)

Galley, John E. (ed.), 1968. *Subsurface Disposal in Geologic Basins: A Study of Reservoir Strata, Am. Assoc. Petroleum Geologists Mem. 10,* 253p.

Geochemistry and the Environment, 1974-1979. Washington, D.C.: National Academy of Sciences, 3 vols., 476p.

Hall, Vivian S., 1975. Environmental Geology: A Selected Bibliography, *Geol. Soc. America Microform Pub. 1,* 4 cards, 4 sheets.

Illinois State Geological Survey, 1965- . *Environmental Geology Notes.* Urbana.

Northeastern Environmental Science, 1982- . Troy, N.Y.: Northeastern Science Foundation, published quarterly.

Robinson, G. D., and Spieker, Andrew M. (eds.), 1978. Earth-science Maps Applied to Land and Water Management, *U.S. Geol. Survey Prof. Paper 950,* 97p.

Schildhauer, Carole, 1972. *Environmental Information Sources: Engineering and Industrial Applications; a Selected Annotated Bibliography.* New York: Special Libraries Association, 45p.

Spangle, William, and Associates; Leighton, F. B., and Associates; and Baxter, McDonald & Company, 1976. Earth-science Information in Land-use Planning: Guidelines for Earth Scientists and Planners. *U.S. Geol. Survey Circ. 721,* 28p.

Tuwiner, Sidney Bertram, 1973. *Environmental Science Technology Information Resources.* Ed. in conjunction with Chemical International Information Center. Park Ridge, N. J.: Noyes Data Corp., 218p.

U.S. Council on Environmental Quality, 1970- . *Environmental Quality: Annual Report.* Washington, D.C.: U.S. Government Printing Office.

Wolff, Garwood R. (ed.), 1974. *Environmental Information Sources Handbook.* New York: Simon and Schuster, 568p.

Hydrogeology/Hydrology

American Geophysical Union, 1971- . *Water Resources Monograph.* Washington, D.C.

Biswas, Asit K., 1983. *History of Hydrology,* 2nd ed. Oxford: Pergamon Press.

Canada Department of Energy, Mines, and Resources, Inland Waters Branch, 1965- . *Surface Water Data: [province].* Ottawa, published annually.

Chapman, Richard E., 1981. *Geology and Water: An Introduction to Fluid Mechanics for Geologists.* Developments in Applied Earth Sciences, vol. 1, Boston: Martinus Nijhoff, 228p.

Freeze, R. Allan, and Cherry, John A., 1979. *Groundwater.* Englewood Cliffs, N.J.: Prentice Hall, 604p.

Geraghty, James J.; Miller, David W.; Van der Leeden, Frits; and Troise, Fred L., 1973. *Water Atlas of the United States.* Port Washington, N.Y.: Water Information Center, unpaged.

Giefer, Gerald J., 1976. *Sources of Information in Water Resources.* Port Washington, N.Y.: Water Information Center, 290p.

Giefer, Gerald J., and Todd, David K. (eds.), 1972. *Water Publications of State Agencies: A Bibliography of Publications on Water Resources and Their Management Published by the States of the United States.* Port Washington, N.Y.: Water Information Center, 319p. (Supplement, 1976, 189p.)

Gray, Donald M. (ed.), 1973. *Handbook on the Principles of Hydrology.* Port Washington, N.Y.: Water Information Center, 720p.

Ground Water, 1963– . Worthington, Ohio: Water Well Publishing Company for the National Water Well Association, published bimonthly.

International Glossary of Hydrology, 1974. Geneva, Switzerland: UNESCO Panel on Terminology, World Meteorological Organization Publication 385, 393p.

Journal of Hydrology, 1963- . Amsterdam: North-Holland, published quarterly.

Matthess, G., 1982. *The Properties of Groundwater.* Chichester, England: Wiley, 420p.

Nelson, Archibald, and Nelson, K. D., 1973. *Dictionary of Water and Water Engineering.* London: Butterworths, 271p.

Titelbaum, Olga Adler, 1970. *Glossary of Water Resource Terms.* Chicago: U.S. Department of the Interior, Federal Water Pollution Control Administration, 39p.

Todd, David Keith, 1980. *Groundwater Hydrology.* 2nd ed. New York: Wiley, 535p.

Todd, David Keith, 1970. *The Water Encyclopedia: A Compendium of Useful Information on Water Resources.* Port Washington, N.Y.: Water Information Center, 559p.

U.S. Geological Survey, 1953- . *Hydrologic Investigations Atlas* (HA) Washington D.C.

U.S. Geological Survey, 1967- . *Techniques of Water-Resources Investigations of the United States Geological Survey.* Washington, D.C.

U.S. Geological Survey, 1896- . *Water-supply Paper.* Washington, D.C.

U.S. Geological Survey. Water Resources Division, 1975- . *Water Data Report, Water Year.* Washington, D.C., annual. (Continues *Water Resources Data* for [state], published 1965-1975.)

U.S. Office of Water Data Coordination, 1979. *Catalog of Information on Water Data: Index to Water-data Acquisition.* Reston, Va.: U.S. Department of the Interior, Geological Survey, Office of Water Data Coordination, 21 vols. (by region). (This publication does not contain the actual data, but provides information on where and by whom data are being collected, the types of data acquired, and how these data can be obtained.)

Van der Leeden, Frits, 1975. *Water Resources of the World.* Port Washington, N.Y.: Water Information Center, 568p.

Van der Leeden, Frits, 1974. *Ground Water: A Selected Bibliography,* 2nd ed. Port Washington, N.Y.: Water Information Center, 146p.

Water Resources Research, 1965- . Washington, D.C.: American Geophysical Union, published quarterly.

Other sources of hydrologic data are the publications prepared by individual state geological surveys, state water agencies, and the federal government. Various universities also publish water research series.

REGINA BROWN

Cross-references: *Geological Communication. Geological Information, Marketing.* Vol. XIV: *Abbreviations, Ciphers, and Mnemonicons; Associations, Institutes, and Publications; Geohistory, Founding Fathers USA; Geological Cataloging; Geological Surveys, State and Federal; Map and Chart Depositories; Professional Geologists' Associations; Punch Cards, Geological Referencing; Remote Sensing and Photogrammetry, Societies and Periodicals.*

EARTH SCIENCE JARGON — See GEOLOGICAL COMMUNICATION.

ECONOMIC GEOLOGY

In the broadest and literal sense, economic geology is the application of earth sciences to the fulfillment of man's needs and desires (see *Geology, Applied.)* Most authorites, however, use a more restricted definition. An adequate definition of geology (when modified by the adjective *economic)* is that it is the science of the earth—its history, structure, and composition.

Geology comprises all the earth sciences, including many subdivisions such as geophysics, geochemistry, hydrology, and physical geography (see Vol. XIV: *Geology, Scope and Classification).* It is common to separate these branches of earth sciences from geology, not only in academia but also in government and industry. Furthermore, a distinction is made in industry and government between geologists, geophysicists, geochemists, and so on. A seismic geophysicist working for an oil company might not be pleased to be called an economic geologist; he would insist that he was a geophysicist.

The adjective *economic,* of course, refers to the noun *economics,* which is the social science concerned with the production, distribution, and consumption of goods for fulfilling human needs and desires. A professor teaching geology and conducting geological research with no obvious practical application would be satisfying a need or desire for knowledge and training, but he would not be considered to be practicing economic geology. Rather, *economic geology* refers to the search for and exploitation of mineral resources.

The terms *economic geology* and *applied geology* (q.v.) are sometimes used synonymously, but the term *applied geology,* is often used in a broader sense. As an example, engineering geology certainly is applied geology; nevertheless, per se it should not be considered economic geology even though it is economically useful applied geology. Locating a sand and gravel deposit or a rock quarry for use in an engineering project certainly is an example of economic geology, but not all authorities would consider the application of geological principles to the design and construction of a highway or a dam to be economic geology.

Mineral Deposits (Lindgren, 1933, p. 9) states that the term *economic geology* applies to the occurrence, composition, structure, and origin of geological bodies that can be technically utilized, and that economic geology shows where geological bodies can be searched for and how they can be evaluated. Thus, economic geology concerns a deposit that can be extracted and used. Lindgren (1933, p. 1) states that the text is confined to the principal deposits of metallic and nonmetallic minerals of economic importance and does not include coal, mineral oils, and structural materials of geological origin. In *Economic Mineral Depos-*

its, Bateman (1954) says, "Economic geology deals with the materials of the mineral kingdom that man wrests from the earth for his necessities of life and comfort" (p. l). He says that economic geology includes mining geology, petroleum geology, and metallurgy (to some extent).

The *Dictionary of Applied Geology* (Nelson and Nelson, 1967, p. 120) in defining economic geology, speaks of groundwater, ores, minerals, coal seams, and building and construction materials. The dictionary states that "petroleum geology" is another important branch.

Economic geology is the application of geological principles to the location, identification, description, evaluation, and commercial extraction of deposits of rocks, minerals, and fossil fuels (see Vol. XIV: *Exploration Geology* and prospecting topics). The location, identification, description, and evaluation of the deposits are primarily the domain of economic geologists, but they may be assisted by geophysicists, mining engineers, paleontologists, metallurgists, mineral economists, and other specialists. Although economic geologists may be concerned with the extraction of a valuable mineral commodity, the extraction is primarily the responsibility of engineers. No sharp line can be drawn between economic geology as this applied science merges with engineering, applied economics, and other disciplines.

Mining geology and *petroleum geology* are universally recognized as branches of economic geology. Since water is a mineral, *ground-water geology* is properly considered as a third branch of economic geology. By the same reasoning, the measurement and evaluation of surface waters for irrigation or other uses might be considered economic geology; but by common usage, such work is not considered to be economic geology; rather it is *hydrology* (q.v.).

JULES A. MACKALLOR

References

Bateman, Alan M., 1954. *Economic Mineral Deposits.* New York, Wiley, 916p.

Lindgren, Waldemar, 1933. *Mineral Deposits,* 4th ed. New York: McGraw-Hill, 930p.

Nelson, A., and Nelson, K. D., 1967. *Dictionary of Applied Geology:* New York: Philosophical Library Inc., 421p.

U.S. Bureau of Mines, 1968. *A Dictionary of Mining, Mineral, and Related Terms.* Washington, D.C.: U.S. Government Printing Office, 1269p.

Cross-references: *Geology, Applied; Hydrogeology and Geohydrology; Hydrology. Vol. XIV: Geology, Philosophy; Geology, Scope and Classification.*

ELECTRICAL, ELECTROMAGNETIC SURVEYS — See ELECTROKINETICS. Vol. XIV: MARINE MAGNETIC SURVEYS; SEA SURVEYS; VLF ELECTROMAGNETIC SURVEYING; WELL LOGGING.

ELECTROKINETICS

Electrokinetics is a term applied to physicochemical transport of charges, action of charged particles, and effects of applied electric potentials on formations and fluid transport in various porous media (see *Hydrodynamics, Porous Media.)*

More than a hundred years ago, R. W. Fox observed *earth potentials* in a mine shaft in Cornwall, England. Inasmuch as the earth's crust is composed of many electrical conductors, which are subjected to many external influences, electrical phenomena are observed almost everywhere in the ground. After all, matter is largely electrical in nature.

Electrical potentials in formations and interstitial waters mainly result from: (1) electrochemical action, (2) electromagnetic induction, (3) contact of dissimilar substances, (4) heat, and (5) pressure.

Electrokinetic Phenomena

As pointed out by Ambah (1963), electrokinetic phenomena are related to a tangential movement of two phases along each other. They may result from either (1) a movement of the phases along each other resulting in a transport of electricity (e.g., in the case of streaming potential and migration potential) or (2) an external electric field directed along the phase boundary and resulting in a movement of phases (e.g., electroosmosis and electrophoresis). (See Vol. XII, Pt 1: *Electrochemistry*). Electrokinetic phenomena include (1) migration potential, (2) electroosmosis (q.v. in Vol. XII, Pt. 1), (3) streaming potential, and (4) electrophoresis.

Migration Potential. *Migration potential* originates when particles suspended in a liquid are forced to move, and an electrical field is generated in the direction of the movement. If the two electrodes are placed in the liquid in such a manner that the path of the moving particles is enclosed between them, a potential difference — the migration potential — can be measured. This is known as the *Dorn effect* (Kruyt, 1952, p. 194).

Electroosmosis. *Electroosmosis* is defined as the movement of a liquid with respect to a solid wall as a result of an applied electrical potential gradient. If a porous plug or a capillary tube is placed between two electrodes in a fluid, the liquid will move from one side to the other when an *electromotive force* is applied. The direction and the velocity of the fluid depends on the properties of the plug and the flowing liquid, and also on the magnitude of applied electrical potential.

Streaming Potential. *Streaming potential* is the reverse of electroosmosis (q.v.). If a measurable amount of a liquid is forced through the porous plug by an external pressure, an electrical potential difference can be measured between the two electrodes. This resulting potential difference (streaming potential) is usually small when compared with the electrical potential used in electroosmosis.

Electrophoresis. Solids suspended in a liquid, including colloidal particles in the form of *sols* or *gels,* move when placed in an electrical field. This movement of suspended particles and sometimes liquid drops in an electrical field is known as *electrophoresis, or cataphoresis.* It is comparable to the mobility of electrolytic ions, and the velocity of the movement can be determined directly with the aid of an ultra-microscope. Tchillingarian (1952) proposed the use of electrophoretic phenomenon for separating very fine sediments (clays) into grades.

Classical Laws and Theories

The first electrokinetic experiment apparently was performed by the Russian physicist Ruess (1807, cited in Kruyt, 1952). His apparatus consisted of two pieces of glass tubing, which he drove into a block of wet clay and then filled with water. When an electrical potential was applied across the clay block, the water in the tube connected with the *anode* assumed a milky appearance due to the electrophoretic migration of colloidal clay particles. An increase in the volume of water (at the *cathode*), which remained clear, demonstrated the phenomenon of electroosmosis.

Wiedemann (1852, cited in Kruyt, 1952) made the first accurate quantitative measurement of liquid movement through porous clay diaphragms by means of a current of known strength. He showed that in the case of application of an electrical field to a liquid in capillaries, the flow volume is proportional to the electrical current. If the flow is prevented, the resulting electroosmotic pressure is also proportional to the current. Wiedemann also found an inverse relationship between the electroosmotic flow and the concentration of the electrolyte. At first he thought this phenomenon might be the direct action of mechanical forces produced by the current and acting on the liquid, and giving rise to the electroosmotic pressure. Later, however, he found that the material of the wall and the cross-sectional area of the plug have a definite influence.

Quincke (1859, 1861, cited in Kruyt, 1952) recognized that electroosmosis and streaming potential are inverse phenomena. He explained them on

the basis of an *electrical double layer* at the boundary between the liquid and the solid wall (Rieke and Chilingar, 1974, p. 229). It was assumed that the positive charge of the double layer is present in the liquid phase and the negatively charged part of the double layer is fixed on the wall surface. Upon application of an electrical potential, the positively charged mobile part of the double layer (Fig. 1) will move toward the cathode, dragging with it the liquid by viscous drag. On the other hand, if a liquid is forced through capillaries, a streaming potential or a current is produced which is proportional to the applied pressure differential. In 1879 Helmholtz (cited in Ambah et al., 1964, 1965) expressed these laws and theories into a mathematical form for electroosmosis (q.v. in Vol. XII, Pt. 1) in a single capillary tube. The migration potential or the potential of falling particles was observed by Dorn (1878, cited in Kruyt, 1952). He obtained reproducible sedimentation potentials with sand in distilled water placed in a long cylinder. Electrodes were placed at the top and bottom of the cylinder. Kruyt (1952) and Ambah (1963) present comprehensive theoretical reviews on the subject.

Electrical Potentials Observed in Rocks

The electrical potentials existing in rock formations include (1) telluric potentials, (2) electrochemical potentials, and (3) contact potentials.

Telluric Potentials. *Telluric Potentials* are induced in the ground as a result of variable magnetic fields. The earth is surrounded and permeated by a magnetic field, the intensity of which varies constantly. The potential gradient resulting from the diurnal and secular variations of the earth's magnetism is negligible, whereas the usual irregular surges of magnetism induce appreciable voltages. The horizontal gradient is much larger than the vertical component, averaging several millivolts per kilometer.

Electrochemical Potentials. Electromotive forces resulting from chemical reactions exist in the ground. For example, in the vicinity of some ore deposits (pyrite in particular) one can measure

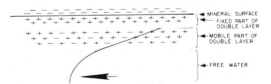

FIGURE 1. Schematic diagram of an electric double layer in a capillary (only half is shown). Flow occurs from positive to negative electrode. The solid curved line indicates the velocity distribution during the flow (after Chilingar et al., 1970, fig. 1, p. 831. Courtesy of the Society of Petroleum Engineers of AIME.)

electrical potentials. These *electrochemical potentials* probably result from oxidation of the mineral and are used for the location of the mineral deposits (see Vol. XIV: *Exploration Geophysics*).

Contact Potentials. A difference in potential can be observed when dissimilar substances are brought in contact. Thus, *contact potentials* exist in the ground because sedimentary formations consist of rocks and fluids of various natures. For example, when a sand and a shale are in contact, a potential difference will usually exist between them. When a liquid flows through pores (moves against a solid), a frictional electricity (streaming potential) is observed.

Besides these three main electrical potentials, other miscellaneous potentials also exist in the ground. Pressure and temperature modify to some extent the potentials of chemical origin that existed during the diagenesis of sediments. Compaction pressures probably also generated electrical potentials. Some minerals (quartz crystals, for example) develop electrical charges when subjected to pressure *(piezoelectricity)*. Other minerals (such as tourmaline crystals) become charged with positive electricity at one end and negative at the other when heated *(pyroelectricity)*. The polarity is reversed on cooling. Inasmuch as random crystal orientation exists in sedimentary rocks, pyroelectricity and piezoelectricity are also present in these formations.

Ion Filtration by Charged-Net Clay Membranes

Wyllie (1955), Davis (1955), McKelvey et al. (1957), von Engelhardt (1961), McKelvey and Milne (1962), and Bredehoeft et al. (1963) showed that buried waters may be subjected to ionfiltration by charged-net clay membranes. (See Rieke and Chilingar, 1974, for greater details and a list of references.)

Many shale beds are considered to be ideal membrane electrodes. The most suggestive argument for this assumption is the observed constancy of a "shale baseline" on "spontaneous potential" logs found in drill holes in every part of the world (Wyllie, 1955). A quantitative theoretical treatment of the electrochemical properties of clays is given by the *theory of membrane behavior* of Myer-Sievers-Teorell (Davis, 1955), which, according to calculations based on the spontaneous potential (SP) curve of electric logs approximates the behavior of shales *in situ* in the earth (also see Degens and Chilingar, 1967).

During the compaction of clay-containing sediments, the salt held back accumulates in the formation water retained in the strata. The process of salt removal or concentration depends on the large excess charge permanently attached to the clay membrane, which prevents the passage of like-charge ions. Thus, the separation is achieved

because of the electrical properties rather than the size of the electrolytes. This process yields a lower salt level in the filtrate as compared to the original solution. Thus, the salt is filtered by virtue of its electrolytic dissociation and the electrical properties of the clay membrane. The filtration of salt solutions through charged-net membrane has been suggested as the mechanism for producing fresh water from saline water. Rieke and Chilingar (1974) discussed this subject in detail.

Effect of Electrical Potentials During Diagenesis (Electrodiagenesis)

Experimental work by Serruya et al. (1967) suggests that electrical currents and potentials affect some diagenetic processes; this may be termed *electrodiagenesis*. Currents produced by ionic exchange processes, for example, flowing through sediments may stimulate cementation and the formation of authigenic minerals, such as gibbsite, limonite, calcite, hydrohematite, hydrogoethite (lepidocrocite), hisingerite, allophane, allophanoid, gypsum, hematite, magnetite, nontronite, trona, and natron ($Na_2CO_3 \cdot 10\ H_2O$). Currents may also cause *selective ion drive* and explain zonation of some trace elements and various minerals. Movement of fluids, due to compaction of sediments, for example, generate potential gradients, which in turn may affect diagenetic processes. Electrical currents cause interstitial solutions to flow through practically impervious muds and localize mineral neoformations in shales.

Utilization of Electroosmosis in Engineering Geology

Electroosmosis has long been applied in soil engineering (see Vol. XIV: *Engineering Soil Science*). Several patents on the removal of water from clayey and silty soils by electroosmosis were issued in Germany before World War II. Later the method was widely and successfully used in Germany, England, the USSR, and Canada in drying water-saturated soils before heavy construction. The development of these practical applications has been largely due to the work of Casagrande (1962), who conducted continuous research on their feasibility in relation to various soil characteristics.

The literature of civil engineering, soil mechanics, and highway research presents results of investigations made by Winterkorn (1947-1958), Casagrande (1937-1962), and others on the electrical and electrochemical treatment of soils.

Examples of Electroosmotic Treatment in Civil Engineering. Several attempts were made to excavate for a U-boat pen in Trondheim, Norway, about 14 m deep in a very thick stratum of clayey silt interspersed by seams of sand near the sea. These attempts failed, however, because of the very active uplift phenomenon. In order to cause the water to flow away from the excavation side, direct electrical current was utilized next. Presence of salts increased the electrical conductivity of the soil; consequently, the consumption of current was high.

Before the application of electrical current, the flow rate varied from 1 to 50 l/hr per well. The application of a 20-amp current and a 40-V potential increased the flow rate from 11 to 479 l/hr per well. The average power consumed was around 0.4 kWh/m³ of soil excavated (Ambah, 1963, p. 208).

In Salzgitter, Germany, difficulties arose during the construction of a double-track railway cutting in a loose-loam deposit, due to the flow of soft soil (Casagrande, 1947). The problem was solved by a large-scale drying operation using electroosmosis. Well electrodes 7.5 m deep and 10 m apart were used. Before the application of electrical potential, the average rate of flow of water was 0.4 m³/day per 20 wells. An electrical potential of 180 V, with an average current of 19 amps per well, was applied. During an eight-week period, the flow continued at an almost constant rate of 60 m³/day per 20 wells. This was 150 times greater than the flow rate before direct electrical current was applied. (Ambah, 1963).

Other tests were made by Casagrande (cited in Amba, 1963) in dewatering lime sludge deposits at Wulprath, Germany, having a uniform water content of 120 percent on a dry-weight basis. A 25 percent decrease in moisture content was obtained on application of 70-V electrical potential and a 50-amp current for 14 days. The electroosmotic dewatering process took place uniformly along the lines of equal potential strength.

Electrochemical Treatment of Soil and Weak Rocks

The first work on electrochemical treatment of soils was done by Casagrande (1930, cited in Casagrande, 1937) who noticed that a permanent stabilization of soil could be achieved by using aluminum electrodes (see Vol. XII, Pt. 1: *Soil Stabilization*). In the course of experiment he found that the aluminum electrodes were greatly corroded and that aluminum compound had been deposited around them. In 1937, Casagrande undertook a full-scale experiment and concluded that electrochemical treatment could be used for increasing the bearing capacity of piles.

In 1960, Casagrande and his coworkers attributed the increase in the bearing capacity to the following: (1) Metal derivatives were deposited around the anode and CaCO, around the cathode, acting as cementing agents between soil particles. (2) Water was transported from one electrode to another, causing a change in the water saturation. (3) The low-valence ions loosely attached to the clay-plate surfaces were displaced with higher-valence ions by base exchange, such as the replacement of Na^+ ions by Al^{3+} ions.

Probably the best example of the use of electroosmosis to increase the bearing capacity of piles was in the bridge foundation over the Big Pic River, near Marathon, Ontario, Canada (Casagrande et al., 1950, cited in Ambah, 1963, p. 210). The ultimate bearing capacity of piles driven 34 m into the ground was constant over a one-year period at 30 tons. The piles were about 7 m apart. A 100-V potential was used to give an average current of 15 amps per pile. The bearing capacity of the piles increased from 30 to 100 tons after maintaining this treatment for four weeks.

Soviet scientists have made significant contributions to the investigation of the electrochemical stabilization of soil and the induration of weak rocks, through the continued addition of different electrolytes at the anodes and by the use of electrodes of various materials. Zhinkin (1952, 1959), Rebinder (1957), Titkov et al. (1959), and others seem to agree that soil and weak rocks undergo major physicochemical changes upon electrochemical treatment. All investigations indicated that these physicochemical changes lead to the formation of a new soil structure (coagulational-crystallization structure). Zhinkin (1958, citied in Titkov et al., 1959) further indicated that these changes were irreversible. Greater strength was attained progressively by the newlyformed clay structure even after the discontinuation of electrochemical treatment.

Electrochemical Induration of Drill-Hole Walls. Titkov et al. (1959) introduced the idea of indurating weak rocks of the borehole walls by electrochemical treatment. This is carried out by continuous introduction of some electrolyte solution at the anode during the electrical treatment. These authors have thoroughly investigated the electrochemical treatment of different types of rocks, using various combinations of electrolyte solutions. The investigation was performed on a variety of rock samples in the laboratory; as well as at several field sites. In addition, they tested the use of different electrode material, such as aluminum, iron, and carbon.

Electrochemical induration of weak rocks basically depends on the formation of a new and stronger authigenic cementing material. Gibbsite, allophane, aluminite, limonite, hisingerite, calcite, and gypsum formed during electrochemical treatment (Titkov et al., 1959). Cylindrical movable electrodes with variable diameters were designed by Titkov and coworkers to test this treatment in the walls of boreholes. With stationary or circulating fluid, they obtained favorable results in indurating the walls of boreholes for different lengths of time at depths ranging from 10 to 116 m. The circulating fluid was varied from pure electrolyte solution to a chemically treated clay-cement mixture. After carrying out tests on the electrochemically indurated boreholes 13 months later, Titkov et al. (1959) concluded: (1) that reversing the electrode's polarity during electrochemical treatment speeds up the cementing process; (2) that the suitability of electrochemical treatment of weak rocks is quite promising, not only for the time of drilling, but also during well exploitation. (3) that the electrochemical induration of weak rocks is certainly an irreversible process; and (4) that the walls of a borehole may be indurated electrochemically either by (a) the creation of a crust of hardened mixture consisting of clay or loam and binding materials such as cement, or by (b) increasing the stability of shales by changing them through the electrical action.

Possible Application of Direct Electrical Current in Oil Production and Its Economic Feasibility

According to theoretical analyses of electrokinetic phenomena, technical applications of these phenomena in related engineering fields, and experimental investigations constructed by several investigators (Ambah, 1963; Adamson et al., 1963*a,b)*, it is possible that direct electrical current can be applied in promoting the flow of oil and water during oil production. There are various means and methods in which the application of direct electrical current can be used for different purposes.

Comprehensive research work on this subject by Ambah (1963), at the Petroleum Engineering Laboratories of the University of Southern California, led him to the following conclusions: (1) The flow rate of oil and water may be increased by the application of current during primary or secondary recovery (Fig. 2). (2) Chemical additives may be used in conjunction with electrical treatment to augment the flow rate of oil and water. (3) Electrochemical treatment may be used for well stimulation. (4) Electroosmosis may be used as a selective ion-drive process. Chilingar (1970) also concluded that direct electrical current probably can be used for well stimulation inasmuch as only short periods of time (several days) are necessary to increase the permeability. The x-ray analysis of clays after electrical treatment shows that their structures are drastically reduced as a result of imposing direct electrical current. Silicates seem to be destroyed upon treatment (Chilingar et al., 1970, p. 835). In addition, through base exchange, it is possible to convert highly swelling clays to low-swelling clays, such as Na-montmorillonite to Ca-montmorillonite.

GEORGE V. CHILINGARIAN
LUCAS G. ADAMSON
HERMAN H. RIEKE III

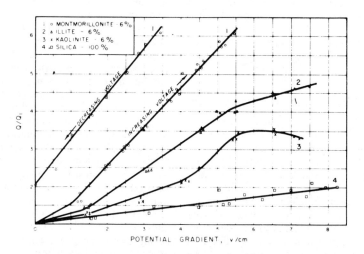

FIGURE 2. k_a = permeability to air; k_w = permeability to water; $\emptyset$ = porosity; L = length of the core; A = cross-sectional area of the core. Relationship between electrical potential gradient and normalized flow rate, q/qi, where q_i is the initial flow rate before application of direct electrical current.
Core 1: L = 17 cm; A = 5 cm^2, $\emptyset$ = 26%; k_a = 200 md; k_w = 3.1 md.
Core 2: L = 17.1 cm; A = 5 cm^2, $\emptyset$ = 25%; k_a = 150 md; k_w = 3.6 md.
Core 3: L = 17 cm; A = 5 cm^2; $\emptyset$ = 26%; k_a = 230 md; k_w = 7.1 md.
Flowing solutions: 0.5% by weight of NaCl and 0.5% CaCl$_2$.
(After Chilingar et al., 1970, fig. 3, p. 834. Courtesy of the Society of Petroleum Engineers of AIME.)

References

Adamson, L. G.; Ambah, S. A.; Chilingar, G. V.; and Beeson, C. M., 1963a. Possible use of electrical current for increasing volumetric rate of oil and water during primary or secondary recovery, *Chimika Chronika,* **28** (1), 1-4.

Adamson, L. G.; Chilingar, G. V.; and Beeson, C. M., 1963b. Some data on electrokinetic phenomena and their possible application in petroleum production, *Chimika Chronika,* **28** (10), 121-127.

Ambah, S. A., 1963. *Use of Direct Electrical Current for Increasing the Flow Rate of Reservoir Fluids During Petroleum Recovery.* Ph.D. Dissertation, University of Southern California, 255p.

Ambah, S. A.; Chilingar, G. V.; and Beeson, C. M., 1964. Use of direct electrical current for increasing the flow rate of reservoir fluids during petroleum recovery, *Jour. Canadian Petroleum Technology* **3** (1), 8-14.

Ambah, S. A.; Chilingar, G. V.; and Beeson, C. M., 1965. Application of electrical current for increasing the flow rate of oil and water in a porous medium, *Jour. Canadian Petroleum Technology* **4** (2), 1-8.

Bredehoeft, J. D.; Blyth, C. R.; Waite, W. A.; and Maxey, G. B., 1963. Possible mechanism for concentration of brines in subsurface formations, *Am. Assoc. Petroleum Geologists Bull.* **47,** 257-269.

Casagrande, I. L., 1937. Full scale experiment to increase bearing capacity of piles by electrochemical treatment, *Bautechnique* **15** (1), 14-16.

Casagrande, I. L., 1947. The application of electroosmosis to practical problems in foundations and earthworks, *Building Research Tech.* Paper No. 30. London, England: Department of Science and Industry Research.

Casagrande, I. L., 1962. La electroosmosis y fenomenos conexos, *Ingenieria,* **32** (2), 66p.

Casagrande, I. L.; Soderman, L. G.; and Loughney, R. W., 1960. *Increase of Bearing Capacity of Friction Piles by Electroosmosis.* Paper presented at ASCE Convention, Boston, Mass.

Chilingar, G. V.; El-Nassir, A.; and Stevens, R. G., 1970. Effect of direct electrical current on permeability of sandstone cores, *Jour. Petroleum Technology,* **7,** 830-836.

Davis, L. E., 1955. Electrochemical properties of clays, in J. A. Park and M. D. Turner, eds., *Clays and Clay Technology.* National Conference on Clays and Clay Technology (1st) Proceedings, State of California Division on Mines Bulletin 169, 47-53.

Degens, E. T., and Chilingar, G. V., 1967. Diagenesis of subsurface waters, in G. Larsen and G. V. Chilingar, eds., *Diagenesis in Sediments.* Amsterdam: Elsevier, 477-502.

Kruyt, H. R., 1952. *Colloid Science.* Amsterdam: Elsevier, 389p.

McKelvey, J. G., and Milne, I. H., 1962. Flow of salt solutions through compacted clay, *Clays Clay Mineralogy Proc., Natl. Conf. Clays Clay Mineralogy* **11,** 248-259.

McKelvey, J. G.; Spiegler, K. S.; and Wyllie, M. R. J., 1957. Salt filtering by ion-exchange grains and membranes, *Jour. Phys. Chemistry* **61,** 174-178.

Rebinder, P. A., 1957. Physicochemical mechanics as a new field of knowledge, *Akad. Nauk SSSR Vestnik* **27,** (10), 32-42.

Rieke, H. H. III, and Chilingar, G. V., 1974. *Compaction of Argillaceous Sediments.* Amsterdam: Elsevier, 424p.

Serruya, C.; Picard, L.; and Chilingar, G. V., 1967. Possible role of electrical currents and potentials during diagenesis ("electrodiagenesis"), *Jour. Sed. Petrology,* **37** (2), 695-698.

Tchillingarian, G. V., 1952. Possible utilization of electrophoretic phenomenon for separation of fine sediments into grades, *Jour. Sed. Petrology,* **22,** 29-32.

Titkov, N. I.; Korzhuev, A. S.; Smolyaninov, V. G., Nikishin, V. A.; and Neretina, A. Y., 1959. *Electrochemical Induration of Weak Rocks,* Moscow: Gostoptekhizdat, 1961, 52p. (Trans. by Consultants Bureau, New York.)

von Engelhardt, W., 1961. Zum Chemismus der Porenlösung der Sedimente, *Geol. Inst. Univ. Upsala Bull.* **40,** 189-204.

Winterkorn, H. F., 1947. Fundamental similarities between electroosmotic and thermoosmotic phenomena, *Highway Research Board Proc.* **27,** 443-455.

Winterkorn, H. F., 1958. Mass transport phenomena in moist porous systems as viewed from the thermodynamics of irreversible processes, *Highway Research Board Spec. Rept. 40,* 324-338.

Wyllie, M. R. J., 1955. Role of clay in well-log interpretation, in J. A. Park and M. D. Turner, eds., *Clays and Clay Technology.* National Conference on Clays and Clay Technology (1st) Proceedings, State of California Division of Mines Bulletin 169, pp. 282-305.

Zhinkin, G. N., 1952. Experiment of utilizing electrochemical induration of grounds for stabilization of railway beds, *Leningrad Inst. Inzhenerov Zheleznodorozh. Transporta Sb.* **144,** 64-78.

Zhinkin, G. N., 1959. Experiment on the improvement of properties of clayey soils using electrochemical stabilization methods, *Leningrad Inst. Inzhenerov Zheleznodorozh. Transporta Sb.* **151,** 164-180.

Cross-references: *Foundation Engineering; Geochemistry, Applied; Geotechnical Engineering; Hydrodynamics, Porous Media; Hydromechanics; Permafrost, Engineering Geology; Rocks, Engineering Properties; Soil Mechanics.* Vol XII, Part I: *Pore-Size Distribution; Soil Pores.* Vol. XIV: *Engineering Soil Science; Exploration Geophysics; Well Logging.*

ENGINEERING GEOLOGY—See ALLUVIAL PLAINS, ENGINEERING GEOLOGY; ARID LANDS, ENGINEERING GEOLOGY; CALICHE, ENGINEERING GEOLOGY; CLAY, ENGINEERING GEOLOGY; COASTAL ENGINEERING; COASTAL INLETS, ENGINEERING GEOLOGY; DAMS, ENGINEERING GEOLOGY; DELTAIC PLAINS, ENGINEERING GEOLOGY; DURICRUST, ENGINEERING GEOLOGY; EARTHQUAKE ENGINEERING; GEOTECHNICAL ENGINEERING; LATERITE, ENGINEERING GEOLOGY; PERMAFROST, ENGINEERING

GEOLOGY; PUMPING STATIONS AND PIPELINES, ENGINEERING GEOLOGY; REMOTE SENSING, ENGINEERING GEOLOGY; ROCK SLOPE ENGINEERING. Vol. XIV: LANDSLIDE CONTROL.

ENGINEERING GEOLOGY REPORTS

Engineering geology reports are the end-products of field investigations. They differ from conventional scientific publications and reports in the geological sciences in that they establish a communication link between geologists and engineers, that is, between people of different backgrounds who use different technical terminologies. Geologists' reports on participation in engineering projects are basically supporting, although very important, contributions to the overall engineering scheme.

The writing of engineering geology reports can be particularly difficult, perhaps even more so than the writing of scientific papers for peer-group review (see *Geological Communication*). These reports should attempt to use more prosaic or common well-known, rather than overly scientific, terms (see *Geological Communication*). Such popularization should not, however, be carried so far that the original intent is lost or confused. Phenomena that may be of interest and importance to geologists— such as fossils, age of formation, and mineralogical composition (to a certain degree)—but that are not directly related to the engineering problem at hand should be deemphasized in the text.

Each engineering geology report should be considered as a unique, special case. Current attempts to formalize and stylize reports for computerlike preparation appear hardly justified. There are, however, some general items that should be featured in every engineering geology report.

Most reports are judged by their content, not by their bulk. Because voluminous reports require much time to study, and because it is always a risk that some essential data might be lost in a lengthy and involved text, the best approach is to prepare a short report that is a critical condensation. Supporting data—such as logs, photographs, maps and line drawings, and other laboratory data—are important legal documents that should be separated from the general text and included in well-organized appendixes.

Engineering geology reports normally contain the following sections (1) Abstract, Summary, and/or Conclusions; (2) and/or Introduction; (3) Scope of Study; (4) Regional Geology; (5) Engineering Geology; (6) Recommendations; and (7) Appendixes.

Abstract, Summary, and/or Conclusions

A brief abstract or summary at the beginning of a report provides a general overview of the project and sets the scene for more detailed considerations. An abstract should be a specific statement rather than a general description; for example, "The Dry Creek site appears to be suitable for a 100-meter-high earthfill dam" is preferable to "The Dry Creek dam site is discussed." A more extensive introductory "Summary and Conclusions" section is sometimes preferable to a short abstract, especially in a report that summarizes large, multifaceted engineering projects.

If possible, the report should be written by the geologist who conducted the actual field work. If he cannot prepare the report himself, this fact should be clearly stated in the report, along with the reasons for such substitution. Such a statement may have a direct bearing on the quality and credibility of the report, especially if the original geologist was dismissed from the project because of, say, malfeasance.

Engineering geology reports, like other engineering and geological reports, should be reviewed by other qualified investigators who have a general knowledge of the subject. "In-house" reviews are valuable because the reviewers may detect ambiguous statements. (The name of the reviewer is usually included in the report.)

Introduction

An introductory section is essential to any geological report that is longer than a note. This section introduces the reader to the work under discussion and establishes the style and tone of the author.

There are six components of an introduction that prepare the reader for the material. The *status statement* states the problem or topic clearly. Because the engineering geology report assumes an intelligent reader, there is no history or definition of terms, only as much background as necessary to understand the problem. The *purpose statement* orients the reader to the project goal or objective. The *method statement* assumes that the reader is familiar with the usual techniques of engineering geology. This secion outlines the variables or conditions under which the work proceeded and indicates any unusual methodology. The *scope statement* describes the actual field work and organization of the engineering geology report itself. Limitations or omissions and the exact phases of the problem discussed in the report are identified here. The *recognition statement* equates the report with other similar works in the field to show the significance of the present engineering geology report. Lastly, the *significance statement* warns of limitations in sampling or other engineering techniques (see discussion in Mitchell, 1968).

Scope of Study

The "Scope of Study" section indirectly informs the reader of the report's value and reliability. A reconnaissance investigation that was based on one field trip and not supported by drilling or trenching is, for example, definitely less reliable than a more protracted study that included test drilling, trenching, and laboratory testing. Equally important is a statement regarding the time of the field study. For example, if an area was mapped during the winter, it is possible that the researchers may have overlooked rock outcrops or seeps covered by snow.

Regional Geology

The chapters describing regional geology are usually, although not always, based on published data. Even though these chapters may be brief, they should contain references to pertinent and comprehensive publications and other works related to the subject.

Engineering Geology

Descriptions of engineering geology and recommendations are the most important sections of a report. Because each report is not only a subjective evaluation of factual data but also a legal document, it is important that actual facts are clearly presented and differentiated from interpretations. Any generalization or theory should be supported by facts. For example, "Depth of decomposed rocks on the divide is over 10 meters" is less informative than the factually supported statement "Estimated depths of decomposed rocks on the divide, based on observations in test holes No. X and No. Y, is over 10 meters."

The importance of clear, interesting, and informative illustrations of good quality cannot be overemphasized. Clear, well-oriented photographs are worth many words in the text. A proper scaling object (hammer, person, compass) should be included in each picture. Numbers or letters can be superimposed on a print (and explained in the caption) to accent points of importance. Simplified line drawings on photographs showing, for example, the location of a proposed dam axis or road alignment may be instructive. Captions ideally should state the exact location, purpose, and date of exposure for each photograph. Photographs without proper dating can be misleading and can even result in misrepresentations; for example, a photograph taken during the dry season may create erroneous impressions that could offset recommendations for flood control.

Because borehole drilling (q.v.) is expensive and time-consuming, every effort should be made to collect all available data and to summarize them in properly complied geological logs. Because of their legal character, such logs should clearly separate actual observations from interpretations. A well-designed format is essential to proper logging. Examples of well-organized formats are the standard log forms used by the U.S. Bureau of Reclamation (1974, pp. 148-152). It is also generally recommended to illustrate logs of test holes with photographs of cores.

Engineering geology maps attached to reports are generally more detailed than regional maps (see *Maps, Engineering Purposes*). Such engineering maps usually deemphasize the geological age of earth materials, formation names, color, and, to a certain degree, the mineralogy of rocks because other factors—such as physical engineering properties, degree of weathering, strength, and slope stability—are relevant to the construction engineer (see Vol. XIV: *Map Symbols*). It is, for example, generally more important to state that the rock is "moderately weathered, loosely jointed, and rust stained" than to mention that it contains amphiboles and biotite. Several notes on a map, such as "2 meters high, nearly vertical stable cuts in a slopewash," can be highly instructive. On maps, in logs, and in the text common adjectives such as *hard* or *soft* usually require clarification. "Hard rock, cannot be scratched with a knife blade" is preferable to just "hard rock." Descriptions of rock formations should be oriented to emphasize engineering properties like strength, weathering, and characteristics of individual units. It is also advisable to illustrate geological maps and profiles with summary logs of test holes, trenches, or pits.

The writer of an engineering geology report should always bear in mind that his report may become a legal document (see Vol. XIV: *Legal Affairs*) and should be sure that in the text, on maps and profiles, and elsewhere he clearly separates his interpretations from observed data.

NIKOLA PROKOPOVICH

References

Mitchell, J. H., 1968. *Writing for Professional and Technical Journals.* New York: Wiley, 405p.
U.S. Bureau of Reclamation Staff, 1974. *Earth Manual: A Water Resource Technical Publication.* Washington, D.C.: U.S. Government Printing Office.

Cross-references: *Geological Communication; Geological Information, Marketing; Maps, Engineering Purposes.* Vol. XIV: *Legal Affairs; Map Symbols.*

ENGINEERING GEOLOGY SURVEY—See
ENGINEERING GEOLOGY REPORTS;
URBAN ENGINEERING GEOLOGY.

ENGINEERING MAPS—See MAPS,
ENGINEERING PURPOSES.

ENGINEERING SOIL CLASSIFICATION—
See SOIL CLASSIFICATION SYSTEM,
UNIFIED.

ENGINEERING STRUCTURES—See
COASTAL ENGINEERING; DAMS,
ENGINEERING GEOLOGY;
EARTHQUAKE ENGINEERING;
FOUNDATION ENGINEERING; PIPELINE
CORRIDOR EVALUATION; PUMPING
STATIONS AND PIPELINES,
ENGINEERING GEOLOGY;
REINFORCED EARTH; RIVER
ENGINEERING; ROCK SLOPE
ENGINEERING; SHAFT SINKING;
SOIL MECHANICS; TUNNELS,
TUNNELING; UNDERSEA
TRANSMISSION LINES,
ENGINEERING GEOLOGY; URBAN
ENGINEERING GEOLOGY; URBAN
HYDROLOGY; URBAN TUNNELS AND
SUBWAYS; WELLS, AERIAL.

EXCAVATIONS—See CAVITY
UTILIZATION; PIPELINE CORRIDOR
EVALUATION; RAPID EXCAVATION
AND TUNNELING; SHAFT SINKING;
TUNNELS, TUNNELING; URBAN
TUNNELS AND SUBWAYS. Vol. XIV:
BLASTING AND RELATED
TECHNOLOGY; CRATERING,
MAN-MADE.

EXPERIMENTAL GEOLOGY—See Vol.
XIV: ALLUVIAL SYSTEMS MODELING;
GEOLOGICAL METHODOLOGY;
RHABDOMANCY.

F

FIELD GEOLOGY

Definition and Scope

As the term implies, *field geology* means field work, geology as practiced by direct observation of outcrops, exposures, landscapes, and drill cores. Those engaged in field geology investigate rocks and rock materials in their natural environment. Field geologists thus attempt to describe and explain surface features, underground structures, and their interrelationships. Lahee (1961), however, emphasizes that although field geology is based on observation, many conclusions are predicated on inferences. He states that "the ability to infer and infer correctly is the goal of training in field geology" (p. 4). Proficiency as a geologist is largely measured by one's ability to draw reasonable conclusions from observed phenomena and to predict the occurrence of features, conditions or processes using field experience.

Field work, supplemented by laboratory studies, is critical to advances in knowledge of the geology of the earth. Whether in the acquisition of original data, primary reconnaissance, detailed surveys, geoexploration, or academic pursuits, field geologists are hired by private consulting firms, colleges and universities, mining companies, or national geological surveys. Hypotheses developed in the workroom (see Vol XIV: *Geology, Philosophy*) also must eventually be tested in the field, as geology is essentially an outdoor science that demands field training.

Much of the field work conducted in the public sector is formalized by geological surveys. Almost every country in the world has an official geological survey; the earliest was instituted in Great Britain in 1835. These geological surveys emphasize applied geology (q.v.) and use a multidisciplinary approach to solve problems. Typical activities of many surveys focus on geological mapping (see Vol. XIV: *Maps, Logic of*); geophysical surveys, such as seismological investigations or studies of geomagnetism (see *Magnetic Susceptibility, Earth Materials*); geochemical surveys, such as stream sediment, rock chip, and soil surveys (q.v. in Vol. XIV); engineering geology; and environmental protection (see Vol. XIV: *Environmental Geology; Environmental Management*). With increasing industrial demands for raw materials and safe sites for urban development, many surveys now con-centrate on the assessment of energy resources, ground-water supply, and the lessening of risk from earthquakes, slope failures (rockfall, landslip, and landslide), volcanic activity, mining operations, and the disposal of toxic waste. All these activities involve field work that is essential to understanding and developing the ground upon which we live and work. Although the tasks of field geologists are varied, they depend first on background preparation in the classroom and proper field training, and then on adequate field equipment, application of codes for geological field work (q.v. in Vol. XIV), and finally on the examination of exposures for the purpose of preparing geological maps.

Field Equipment

The field geologist's basic equipment (see Vol. XIV: *Field Work, Equipment for*) consists of a hammer with a pick or chisel at one end, a pocketknife, a hand lens, a compass and clinometer, a notebook, and writing utensils. Additional tools include a ruler or tape measure, a scale for photographs, sample bags (paper, cloth or plastic), and sample identification tags. These materials will fit into a modest-sized knapsack, but additional packs should be carried if numerous rock samples are to be collected and carried. A photographic record of sample sites or general shots of exposures are often useful in later reports or slide presentations. Many geologists carry a small camera for black-and-white photographs, as these are generally more suitable for reproduction in published reports. Conversion of color slides to black-and-white prints is often less than satisfactory, so an additional small lightweight camera for taking color photographs is recommended as part of the standard equipment. Some geologists use, in addition, an instant camera so that sample sites can be precisely annotated directly on the picture.

Topographic maps and appropriate aerial photographs are useful for locating sample sites and for preparing field maps. If a pocket stereoscope is available, stereo-paired aerial photos will make a convenient base for mapping, as mapping units can be delineated directly on the photos, or more preferably on clear plastic film overlays.

For a more comprehensive discussion of instruments and other equipment used in field mapping, see sections in Lahee (1961), King (1966), and

Dakombe and Gardiner (1983) that describe the compass-and-clinometer method, the hand-level method, the barometer method, and the plane table method; techniques of topographic survey; geomorphological mapping; and geophysical methods for subsurface investigation. See also the following entries in Volume XIV: *Compass Traverse; Plane Table Mapping; Surveying, General; Thickness Measurement.*

Examination of Exposures

Exposures provide clues to many significant sedimentary, igneous, and metamorphic structural features. Unconformities, joints, veins, faults, folds, bedding planes, graded bedding, chilled margins, and the like all seem obvious enough when described in the classroom or illustrated by clear photos or diagrams in textbooks. Such features, whether large or small, are often not obvious in the field, especially when quarry walls, road and railway cuts, or natural surfaces are partially masked by vegetative overgrowth, dust or slump material, or the effects of weathering. With experience, the observant field geologist learns to recognize *critical exposures* that provide unmistakable evidence of geological processes, relationships between rock units, or other properties that assist in the recognition and classification of rocks. The ability to "read exposures," as Bates and Kirkaldy (1977) characterize critical field observation, provides an opportunity for developing working hypotheses to be tested by closer field examination and enables the geologist to plan subsequent field and laboratory work effectively. Those hypotheses may then be developed into computer-based theoretical models, but to carry conviction they must be testable, and compatible with field evidence or "ground truth," as the remote-sensing specialists call it.

The *scale of observation* is an important consideration. First impressions are afforded by standing back and "reading" the main features as indicated by inclined strata, unconformities, faulting, folding, intercalation of paleosols, color, and other distinctive characteristics. Closer inspection should focus on rock colors, relationships of colors to surfaces of weakness, mineralogical composition, and meso- and microscale structures. Many surface features other than color are of importance for their interpretive value. Lahee (1961), for example, carefully describes interpretations of smoothed and polished rock surfaces and provides keys for the identification of scratches and grooves and for pits and hollows on rock surfaces. These and other smaller structures are fully described and illustrated by Shrock (1948) and Coneybeare and Crook (1968). The effects of fretwork or honeycomb weathering, subsurface corrosion, and phases of soil development should also be meticulously recorded in the field notebook, as they may be important considerations in subsequent interpretations.

The field notebook should contain full details of each day's work (see Vol. XIV: *Field Notes, Notebooks*). Although the type of data recorded in notebooks varies with the project, the following kinds of information should be recorded: date, names of other members in survey party, location, general characteristics of the area (topography, soil, vegetation, climate-microclimate, nature of outcrops), name of unit or brief rock name, thickness and overall structure of unit, fossils, description of rocks with most abundant type described first (color, induration, grain sizes, grain shapes, fabric, cement, porosity, mineralogy), and nature of contacts (Compton, 1962).

Rock samples are preferably broken directly from the outcrop (talus or scree samples should be avoided if possible), and the exact location should be noted and numbered (see Vol. XIV: *Samples, Sampling*). Specifications for sample sizes depend on grain size and homogeneity of the rock (size of homogeneous structural units), but samples $8 \times 10 \times 3$ cm are normally adequate for most purposes. When oriented specimens are important, samples should be marked before removal to indicate the top and structural attitude. Fossils require special care in trimming and should be well packed for transport in the knapsack. Each rock or fossil specimen must be marked with a number matching that used in the notes; the number should also correspond to a location on a map or aerial photograph. It can be inscribed on a piece of adhesive tape and squeezed onto the sample. Ballpoint pen ink is indelible, but felt pen ink is easily smeared or washed away by rain. When several geologists are sampling outcrops in the same area, it is often useful to record the initials of the collector with other numbers. Such efforts often assist in the retrieval of supplementary information.

Geological Mapping

Most primary geological maps (Thomas, 1979) are initially prepared in the field using a topographic map as a base. Some maps are prepared from remotely sensed data using radar, or from color-enhanced satellite imagery, but these maps are most useful when checked in the field by a process referred to as "ground truthing" (see *Photogrammetry; Remote Sensing, Engineering Geology;* Vol. XIV: *Photogeology; Photo Interpretation*). Still other kinds of geological maps may contain map units initially interpreted from remotely sensed images that are later correlated with actual conditions after field inspection. Field geologists often combine satellite imagery or aerial photography with ground operations in reconnaissance traverses to produce geological maps of both a general and a specific nature. Peters (1978) discusses techniques of geological mapping as they apply to surface mineral exploration and also as they pertain to mapping in underground mines.

Geological maps show the distribution of surficial deposits and solid rocks, sometimes together—depending on the depth and extent of overburden—and sometimes separately. The Institute of Geological Sciences in Great Britain, for example, sometimes publishes two editions for the same area: a "Solid and Drift Edition" delineating both surficial materials and exposed bedrock using color and a "Solid Edition" showing all bedrock by means of color but identifying overlying surficial deposits only by means of symbols. The Geologic Quadrangle Maps of the U.S. Geological Survey mostly correspond to the solid and Drift Edition of British maps (Roberts, 1982). Maps on scales of 1:250,000, 1:1 million and smaller scales are almost always "solid." The solid versions are very useful for solving structural problems, for mineral exploration, and for mining operations. The drift map is of course essential for surface engineering work and helpful for Quaternary stratigraphers and geomorphologists.

Although field maps generally show the nature of materials as they occur at the ground surface, geologists often distinguish between two slightly different approaches by using the colloquial terms "soft-rock" and "hard-rock" geology (see Vol. XIV: *Rock Geology, Hard vs Soft)*. Soft-rock geologists concentrate on sedimentary materials that either are loose and unconsolidated or may have undergone some lithification and diagenesis but are mostly unmetamorphosed. Hard-rock geologists, on the other hand, deal with igneous rocks as well as beds that have been folded, contorted, and metamorphosed, forming resistant hard-rock units (Bates and Kirkaldy, 1977).

Mapping Sedimentary Rocks. The rocks most commonly shown on geological maps are sedimentary in origin. Field work with sedimentary rocks implies a knowledge of stratigraphic-lithological units that are defined by physical characteristics and time-stratigraphic units as defined according to the fossils they contain (see Vol. XIV: *Sedimentary Rocks, Field Relations)*. A range of rock properties important to mapping is normally recorded in relation to textural and compositional characteristics, such as grain size, shape and sorting, fabric, porosity, mineralogical composition, and cements (see Vol. VI: *Clastic Sediments and Rocks)*. Rock units may be classified into general categories in the field, but if the field geologist is unsure of the category, or needs closer refinement, he or she should clearly number the specimen for later identification in the lab (see Vol. XIV: *Samples, Sampling)*. Maps may be prepared in this manner, provided that the various units are consistently identified and their relationships to one another noted. Many field maps may,

for example, simply show rock units A, B, C, and D and so on in the first attempt. Legends are later compiled and classification units correlated with map units. Last but not least come structural interpretations. Inclined and folded strata, faults, and unconformities often pose interesting challenges for the field geologist who must employ a range of techniques to determine sedimentary structure accurately (Freeman, 1971). Primary structures are the details of bedding, such as illustrated by Shrock (1948) and Coneybeare and Crook (1968). Secondary, or tectonic, strutures are those caused by diastrophic processes. Platt and Challinor's (1980) booklet dealing with simple geological structures adequately summarizes field methods applicable to completing outcrop maps, determining a bedding plane, finding the vertical thickness and dip of a bed or the displacement of a section due to faulting, and relating topography to geological structure, among other important field techniques.

Mapping Igneous Rocks. Rocks produced by the solidification of magma may take the form of intrusive bodies (Boyer, 1971), that solidify within the earth's crust or extrusive rocks that are erupted from a volcano or fissure before solidifying at the earth's surface. Both intrusive and extrusive rocks tend to occur in characteristic suites, as conditioned by their mode of emplacement or extrusion and chemical composition as acid, basic, or intermediate magmas (see Vol. XIV: *Igneous Rocks, Field Relations)*. Igneous rocks are usually depicted on geological maps using special colors and ornamentation. Maps of the U.S. Geological Survey, for example, use a capital letter to give the stratigraphic age of the intrusion followed by an abbreviated form of the rock name. Volcanic rocks are mapped in the same way as sedimentary formations, but there are several important distinctions. These rocks are often so heterogeneous that it is not possible to differentiate sequential formations, lava flows and ash falls tend to form lenticular bodies that are erupted from more than one vent, and lava flows and pyroclastic deposits are quickly produced in an instant of geological time (Roberts, 1982). Structures produced by volcanic intrusions—such as sills and dikes, laccoliths and domes, lopoliths, and necks and plugs—generally have to be interpreted indirectly. Field work with intrusive igneous rocks can be especially difficult, as Compton (1962) points out, because their interpretation often requires working back in time from the youngest or most obvious features at widely distributed outcrops to relics of former events. Intrusive events and sequences may be understood only after mapping is well advanced. Forms of erupted magma—for example, ropy (pahoehoe) lava and blocky (aa) flows, ignimbrites, ash-fall deposits, fissure lavas, and pillow lavas—are often more easily interpreted as genetic map units because of their continuous exposure in outcrop. Erupted rocks tend to follow the laws of succession appropriate to sedimentary rocks, whereas plutonic and dike rocks have cross-cutting relationships (the last cross-cut is the youngest).

Mapping Metamorphic Rocks. Metamorphic rocks have been modified in a solid state by recrystallization of their minerals by heat (thermal or contact metamorphism), pressure (dynamic metamorphism), combined heat and pressure (dynamothermal metamorphism), or percolation of hot fluids or gases through fractured rocks (hydrothermal metamorphism). Because a given metamorphism is typically the composite of several processes, Compton (1962) advises field geologists to map and classify the premetamorphic lithology, stratigraphic sequence, and structure; determine the amounts and kinds of deformation; and map zones of metamorphic minerals and textures (see Vol. XIV: *Metamorphic Rocks, Field Relations*). These procedures are applicable to the two basic kinds of metamorphic units, (1) metamorphosed sedimentary or igneous rocks and (2) metamorphic zones. Field work with metamorphic rocks (Romey, 1971) requires familiarity with metamorphic foliations and lineations, styles of folding, and regional deformation structures. Metamorphic conditions are thus generally mapped as a series of zones based on key minerals, textures, and structures, as described by Compton (1962) and Roberts (1982) for zones of contact metamorphism, mineral zones in regional terrains, structural zones, and regional zones of textural reconstruction.

The understanding and mapping of surficial features and formations is often no less important than getting to the bedrock. Both scientific and economic needs often call for Quaternary stratigraphic mapping and geomorphic cartography (King, 1966).

Geological Illustration

Maps that show the distribution of rocks and structures are, in the broadest sense, geological maps. They take many different forms, and the symbolization used depends on factors such as policies of national geological surveys, nature of the material depicted, and purpose of the map (see Vol. XIV: *Map Symbols*). Field sketches typically use line symbols for contacts (boundary lines between geological formations) folds, and faults. Solid lines denote accurate locations, whereas dashed and dotted lines respectively indicate approximate and concealed locations. Profile sections (see Vol. XIV: *Profile Construction*), block diagrams (q.v. in Vol. XIV), joint diagrams, and columnar sections provide useful insight into relations between rock units. Geological cross-sections (see Vol. XIV: *Cross Section*), panel or fence diagrams, skeleton diagrams, and correlations of lithological or formation logs (see *Well Data Systems*. Vol. XIV: *Well Logging*) are best prepared in the office, as they require extensive computations or computerized display (see discussions in Lahee, 1961; Langstaff and Morrill, 1981; and Roberts, 1982).

Topographic forms and relationships between geological structure and topography have, on occasion, been expressed in field sketches that border on art form. Notable early European efforts at field sketching include examples from the works of James Hutton (1726-1797), Horace Benedict de Saussure (1740-1799), John Playfair (1747-1819), William Buckland (1784-1856), and Sir Charles Lyell (1797-1875). Albert Heim (1849-1937) and Marcel Bertrand (1847-1907) were the first detailers to illustrate nappes and thrusts respectively. Field sketches by John Wesley Powell (1834-1902) made during advance explorations of the American West, especially his notebook sketches of the Grand Canyon, record remarkable details of structure, topographic form, and stratigraphy. The persuasive field sketches of William Morris Davis (1850-1934) still serve to illustrate the impressions of field relationships between rock units and topography in the interpretation of geomorphic history. Such artistic efforts of pioneer geologists to record details of field observations today represent what might be termed *geostenography* (q.v. in Vol. XIV), a form of data recording through field sketching that is literally almost a lost art. The camera, the substitute, can also be used to create art as well as to convey a scientific message, as illustrated by Shelton's (1966) collections.

Conclusion

Field geology is an important part of geological endeavor. It is so basic to research and routine surveys that field training is almost taken for granted. Most colleges and universities require at least one field course for those majoring in geological sciences, while many national or state geological surveys and private consulting firms offer on-the-job training in field techniques. Lahee's *Field Geology* (1961) served as the basic handbook for nearly two decades, but now there are other helpful guides as well (e.g., Berkman and Ryall, 1976; Bates and Kirkaldy, 1977; Moseley, 1981). The increasing attention given to proper training of field geologists shows concern for an area that was somewhat neglected in the past. Such interest will ensure a continuous supply of competent field geologists that are highly qualified to conduct field surveys using sophisticated equipment as well as master the use of simple tools that geologists have used for so long.

CHARLES W. FINKL, JNR.

References

Bates, D. E. B., and Kirkaldy, J. F., 1977. *Field Geology.* New York: Arco, 215p.

Berkman, D. A., and Ryall, W. R., 1976. *Field Geologists' Manual.* Parkville, Victoria: Australasian Institute of Mining and Metallurgy, 295p.

Boyer, R. E., 1971. *Field Guide to Plutonic and Metamorphic Rocks* (Earth Science Curriculum Project Pamphlet Series). Boston: Houghton Mifflin, 53p.

Compton, R. R., 1962. *Manual of Field Geology.* New York: Wiley, 378p.

Coneybeare, C. E. B., and Crook, K. A. W., 1968. Manual of Sedimentary Structures, *Australian Bur. Mineral Resources Geology and Geophysics Bull.* **102,** 327p.

DaKombe, R. V., and Gardiner, V., 1983. *Geomorphological Field Manual.* London: Allen and Unwin, 254p.

Freeman, T., 1971. *Field Guide to Layered Rocks* (Earth Science Curriculum Project Pamphlet Series). Boston: Houghton Mifflin, 44p.

King, C. A. M., 1966. *Techniques in Geomorphology.* New York: St. Martin's, 342p.

Lahee, F. H., 1961. *Field Geology.* New York: McGraw-Hill, 926p.

Langstaff, C. S., and Morrill, D., 1981. *Geologic Cross Sections.* Boston: International Human Resources Development Corporation, 108p.

Moseley, F., 1981. *Methods in Field Geology.* San Francisco: W. H. Freeman, 211p.

Peters, W. C., 1978. *Exploration and Mining Geology.* New York: Wiley, 696p.

Platt, J. I., and Challinor, J., 1980. *Simple Geological Structures.* London: Murby, 56p.

Roberts, J. L., 1982. *Introduction to Geological Maps and Structures.* Oxford, England: Pergamon Press, 332p.

Shelton, H., 1966. *Geology Illustrated.* San Francisco: W. H. Freeman, 434p.

Shrock, R. R., 1948. *Sequence in Layered Rocks: A Study of Features and Structures Useful for Determining Top and Bottom or Order of Succession in Bedded and Tabular Rock Bodies.* New York: McGraw-Hill, 507p.

Thomas, J. A. G., 1979. *An Introduction to Geological Maps.* London: Allen and Unwin, 67p.

Cross-references: *Geology, Applied; Maps, Engineering Purposes; Field Geology; Photogrammetry; Remote Sensing, Engineering Geology.* Vol. XIV: *Block Diagrams; Compass Traverse; Control Data for Surveys; Detailed Mapping; Exposures, Examination; Field Maps, Mapping; Field Work, Equipment for; Fractures, Fracture Structures; Geological Field Work, Codes for; Geological Structures; Geological Reports; Geostenography; Map Symbols; Plane Table Mapping; Profile Construction; Reconnaissance Mapping; Samples, Sampling; Thickness Measurement; Topography, Topographic Forms.*

FIELD METHODS — See ENGINEERING GEOLOGY REPORTS; FIELD GEOLOGY. Vol. XIV: PROSPECTING; RHABDOMANCY; SOIL SAMPLING.

FLAME TESTS — See XIV: BLOWPIPE ANALYSIS; MINERAL IDENTIFICATION, CLASSICAL FIELD METHODS.

FORENSIC GEOLOGY

The use of geological evidence in crime detection originated, as did the use of many other kinds of physical evidence, with Sherlock Holmes. The fictional detective — in stories written by Sir Arthur Conan Doyle between 1887 and 1893 — suggested many crime-detection methods that professional scientists later developed and applied in real cases. According to Holmes's "able assistant," Dr. Watson, the great detective could "tell at a glance from which part of London the various splashes of soil on his trousers had been picked up." In 1893 Hans Gross, an Austrian professor of criminology, published the *Handbook for Examining Magistrates,* which was to have a profound effect on the development and use of science in criminal investigation (see Thornwald, 1967). Although there were no actual cases involving forensic geology at the time of publication, Gross made the prophetic statement that "dirt on shoes can often tell us more about where the wearer of those shoes had last been than toilsome inquiries." Once the idea of geological evidence appeared in print, both in fiction and in nonfiction, it was not long before minerals, rocks, and fossils in the hands of a geologist would become clues and evidence in an actual criminal case.

In October 1904, Georg Popp, a chemist, microscopist, and earth scientist in Frankfurt, Germany, was asked to examine the evidence in a murder case in which a seamstress, Eva Disch, had been strangled in a bean field with her own scarf. A filthy handkerchief had been left at the scene of the crime, and the nasal mucus on it contained bits of coal, particles of snuff, and, most interesting of all, grains of minerals, particularly hornblende. A prime suspect was known to work both in a coal-burning gasworks and at a local gravel pit. Popp found coal and mineral grains, including hornblende, under the suspect's fingernails. It was also determined that the suspect used snuff. Examination of soil removed from the suspect's trousers revealed that minerals in a lower layer in contact with the cloth matched those of a soil sample taken from the place where the victim's body had been found. Encrusted on this lower layer, a second soil type was found. Examination of the minerals in the upper layer revealed a mineralogy and size of particle, particularly crushed mica grains, that Popp determined were comparable with soil samples collected along the path that led from the murder scene to the suspect's home. From these data it was concluded that the suspect picked up the lower soil layer at the scene of the crime, and that this lower layer—and thus earlier material—was covered by splashes of mica-rich mud from the path on his return home. When confronted with soil evidence the suspect admitted the crime, and the Frankfurt newspapers of the day carried headlines proclaiming "The Microscope as Detective."

It is impossible to determine from the distance of three-quarters of a century how a contemporary forensic geologist or a jury would evaluate the geological evidence amassed by Popp. Nevertheless, one fact is evident: minerals had been used in an actual case, fulfilling Hans Gross's prophecy and providing a real-life example worthy of Sherlock Holmes. Popp worked on many other criminal cases and made substantial contributions to forensic science. In fact, he probably should be considered the founder of forensic geology.

In 1906 Conan Doyle himself became involved in an actual criminal case, during which he applied some of the methods of his fictional sleuth. An English solicitor was accused and convicted of killing and mutilating horses and cows. After serving three years in prison, he was released but not given a pardon despite some evidence that he was actually innocent. Conan Doyle observed that the soil on the shoes worn by the convicted man on the day of the crime was black mud and not the yellow, sandy clay found in the field where the animals had been killed. This observation, combined with other evidence, ultimately led to a full pardon and contributed to the creation of a court of appeals in England.

Today, rocks, minerals, fossils, and other natural and synthetic materials are studied in connection with the thousands of criminal cases tried each year (Murray and Tedrow, 1975).The Federal Bureau of Investigation laboratory in Washington, D.C., one of the first forensic laboratories in the United States to have geologists study soils and related material as physical evidence, is a worldwide leader in forensic geology (Saferstein, 1982).

Physical evidence is divided into two general types: individual items and class items. Fingerprints, some tool marks, and spent ammunition are said to be *individual* items, meaning they have only one possible source. But most kinds of physical evidence—for example, blood, paint, glass, and hair—are grouped under the heading of *class* items because they could come from a variety of sources. In general, the value of a class item depends on how common that item is. Forty-three percent of the population has type O blood, whereas only 3 percent of the population has type AB. Type AB blood would thus be a more valuable bit of evidence than type O. Similarly, the paint from a 1932 Rolls Royce would be more valuable as evidence than that from a 1970 Ford.

Although geological materials can seldom be considered truly individual items, there are excep-

tions. In one vandalism case, for example, a concrete block was broken into fragments that were thrown through a number of store windows from a moving car. In that instance it was possible to piece together the fragments found in the stores and those remaining in the car to reconstruct the original block. Not only did the pieces fit together, but individual mineral grains lined up across the pieces and all the fragments were shown to be of the same kind of concrete.

Geological material is almost always class evidence, but its value lies in the fact that different kinds and combinations of rocks, minerals, fossils, and related materials are almost limitless. The evidential potential of geological materials is therefore greater than that of almost all the other kinds of physical evidence of the class type.

There are more than 2,200 different minerals q.v. (in Vol. IVB), many of which are not common. Almost all these minerals exist in a wide range of compositions; as a result, the number of recognizable kinds of minerals is almost unlimited. In addition, more than a million different kinds of fossils have been identified. Most fossiliferous rocks have populations of fossils that commonly reflect the environment or deposition in which the rock was formed. These groups of fossils provide a very large number of possible combinations.

Almost all rocks—igneous, sedimentary, and metamorphic—are composed of minerals. In any given igneous or metamorphic rock, the kinds and amounts of minerals, their size, and their texture represent a wide range of variations. The possible combinations of minerals, the sizes and shapes of minerals, and the kinds and amounts of cement between the grains in sedimentary rocks offer an almost unending diversity. The weathering processes that break up rocks and produce soil add yet new dimensions. Also, in most urban areas the soils contain particles contributed by man, which further increase the complexity and diversity.

Many commercially manufactured mineral products—such as face powder, cleaning powder, abrasives, masonry, and wallboard—become the study material for the forensic geologist. Hundreds of criminals have been brought to justice because of the minerals found on their burglary tools or clothing.

Most interesting is the insulation material used in safes and strong boxes. When fire-resistant safes are drilled, blown, cut, or pried open, the fire insulating material that fills the space between the outer and inner metal walls is disrupted. It commonly clings to the tools and clothing of the safe breaker. In a classic case, a man was arrested and brought to the police station on a routine minor charge. An observant detective, noticing that the suspect appeared to have a severe case of dandruff, examined his hair. Instead of finding dandruff, though, he found diatoms. On further examination it was learned that the diatoms were of the same species as those present in the insulation of a safe that had been blown the previous day. Accordingly, the suspect was charged with the burglary.

Geological maps can often be used in crime investigation to outline the areas where rocks and minerals associated with crimes or suspects could have originated. The owner of some valuable gems found chips of common rock instead of precious stones when she opened a cargo box that had been sent by air. Study of the chips indicated that they came from a foreign country that was a stopover point on the air route. Examination of the geological map for that area indicated the probable source of the rock chips. This evidence cleared the air-freight handlers at the final destination and led to the apprehension of those responsible for the substitution.

Even topographic maps (see Vol. XIV: *Topographic Mapping and Surveying*) have made their contribution. An informer reported that an illegal still was located somewhere between two towns in southern New Jersey in an area of swamps and higher gravel ridges, and that the water well at the site of the still reputedly had a water level 20 feet below the ground. Since the ground-water table and the swamps were on approximately the same level, the still had to have been on a ridge 20 feet above the local swamp level. A study of the topographic maps of the area showed there only one place on one ridge where the elevation met that requirement. A warrant was obtained, and the still was found in a church cellar.

RAYMOND C. MURRAY

References

Murray, R. C., and Tedrow, J. C. F., 1975. *Forensic Geology,* New Brunswick, N.J.: Rutgers University Press, 217p.
Saferstein, R., 1982. *Forensic Science Handbook.* Englewood Cliffs, N.J.: Prentice Hall, 725p.
Thorwald, J., 1967. *Crime and Science.* New York: Harcourt Brace and World, 494p.

Cross-references: *Geology, Applied.* Vol. XIV: *Geology, Scope and Classification; Popular Geology; Serendipity.*

FORMATION PRESSURES, ABNORMAL*

Detection and quantitative evaluation of *overpressured subsurface formations* are critical to exploration, drilling, and production operations involving hydrocarbon and geothermal resources.

---

*Referred to as *"geopressures"* by some investigators; this usage however, seems incorrect because formation pressures can be also normal or subnormal.

Hence, an interdisciplinary technical team approach is required to optimize the safety, engineering, and financial aspects of operating in such hostile subsurface environments (Fertl, 1976).

Worldwide experience indicates a significant correlation between the presence and magnitude of formation pressures and the shale/sand ratio of sedimentary sections. The distribution of oil and gas is related to regional and local subsurface pressure and temperature environments. Knowledge of the expected pore pressure and fracture gradients is the basis for (1) efficiently drilling wells with correct mud weights, (2) properly designing casing programs, and (3) completing wells effectively and safely, allowing for killing the well without excessive formation damage. In *reservoir engineering,* formation pressures influence compressibility and the failure of reservoir rocks, and can be responsible for water influx from adjacent overpressured shale sections as an additional driving mechanism in hydrocarbon production. Furthermore, statistical correlations have been developed between a shale-resistivity profile, as derived from geophysical well logs (see Vol. XIV: *Well Logging*) in clastic formations, and the distribution of commercially attractive oil and gas fields.

Abnormally high pore-fluid pressures are encountered in formations ranging from the Cenozoic era (Pleistocene age) to as old as the Paleozoic era (Cambrian age). Such pressures may occur from a hundred meters or so below the surface to depths exceeding 7,620m (25,000 ft) and can be present in shale-sand sequences and/or massive carbonate-evaporite sections. In the worldwide search for hydcarbon resources, both on- and offshore, abnormal formation pressures have been encountered on all continents.

Subsurface Pressure Concepts

Hydrostatic pressure, caused by the weight of interstitial fluids, is equal to the vertical height of a fluid column times the specific weight of fluid. The size and shape of this fluid column have no effect on the magnitude of this pressure. The hydrostatic pressure gradient is affected by the concentration of dissolved solids (salts) and gases in the fluid column and the magnitude of (varying) temperature gradients. An increase in dissolved solids (higher salt concentration) tends to increase the pressure gradient, whereas increasing amounts of gases in solution and higher temperatures would decrease the hydrostatic pressure gradient. For example, a pressure gradient of 0.1073 kg cm^{-2} m^{-1} (0.465 psi/ft)* assumes a water salinity of 80,000 ppm NaCl at a temperature of 25°C (77°F).

Typical average hydrostatic gradients that may

---

* 1 psi = 7.031 $\times$ 10^{-2} kg cm^{-2}

be encountered during drilling for oil and gas range from 0.10 kg cm^{-2} m^{-1} (0.433 psi/ft) in rocks containing fresh and brackish water to 0.1074 kg cm^{-2} m^{-1} (0.465 psi/ft) in formations containing salt water.

Overburden pressure originates from the combined weight of the formation matrix (rock) and the fluids (water, oil, and gas) in the pore space overlying the formation of interest. Generally, it is assumed that overburden pressure increases uniformly with depth. For example, average Tertiary formations on the U.S. Gulf Coast and elsewhere exert an overburden pressure gradient of 0.231 kg cm^{-2} m^{-1} (1.0 psi/ft) of depth. This corresponds to a force exerted by a formation with an average bulk density of 2.31 g/cm^3 (sp. gr. = 2.31). Worldwide experience also indicates that the probable maximum overburden gradient in clastic rocks may be as high as 0.312 kg cm^{-2} m^{-1} (1.35 psi/ft). Furthermore, field observations over the last few years have led to the development of the concept of a varying (not constant) overburden gradient for fracture gradient predictions used in drilling and completion operations.

Formation pressure is the pressure acting on the fluids (i.e., formation water, oil, and gas) in the pore space of the formation. Normal formation pressures in any geological setting are equal to the hydrostatic head (hydrostatic pressure) of water extending from the surface to the subsurface formation. Abnormal formation pressures, by definition, are then characterized by any departure from the normal trend line. Formation pressures exceeding hydrostatic are defined as abnormally high formation pressures *(surpressures)*, whereas

lower-than-hydrostatic formation pressures are called subnormal *(subpressures)* (Fig. 1.).

In normal pressure environments, the matrix stress supports the overburden load because of grain-to-grain contacts. Any reduction in the grain-to-grain stress will cause the pore fluids to support part of the overburden, resulting in abnormally high formation pressures.

Causes of Abnormal Formation Pressures

Abnormal formation pressures, both subpressures and surpressures, are frequently caused by a combination of superimposed factors. Although some of the effects may be minor, they are nevertheless present, either short- or long-lived on a geological time scale. To place the possible causes of abnormal formation pressures in proper perspective, it is necessary to understand the importance of petrophysical and geochemical parameters and their relationship to the stratigraphy and structural and tectonic history of a given area or basin.

Because conditions can vary widely, special care should be taken not to assume that the cause(s) of overpressures established from experience in a well-known geological area is necessarily the cause of identical or similar conditions in a nearby basin, which may not yet have been adequately studied and tested by drilling. Table 1 summarizes some of the major possible causes for abnormal subsurface formation pressures (Fertl and Chilingarian, 1981).

Evaluation of Subsurface Pressure Regimes

Extensive field experience indicates that costly misinterpretations are best avoided by studying a combination of several measured formation-pressure indicators, including seismic, drilling, and well-logging data (Fig. 2). Although several such indicators should be monitored at the well site, not all will always turn out to be usable or necessarily needed in any one drilling operation. Drilling and well logging overpressure detection and evaluation techniques are summarized in Table 2. The following are parameters affecting such practical formation-pressure detection techniques (after Fertl and Chilingarian, 1976).

Geological Factors. Geological age changes. Compaction effects on regional (basin edge versus basin center) and local scale and differential compaction across structures. Sand/shale ratio in clastic sediments. Lithology effects: pure shales (soft, hard); limey and silty shales; bentonitic markers; gas-bearing ("shale gas"), organic-rich, bituminous shales; types and amounts of clay minerals in shales (depending on depositional environment and/or diagenesis). Heavy minerals (siderite, pyrite, mica, etc.) Drastic variations in the formation water salinity variations in subsurface. High geothermal gradients. Stratigraphic and tectonic features (acting as overpressure continuities or barriers),

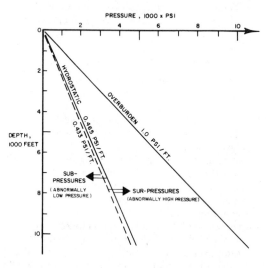

FIGURE 1. Schematic diagram of subsurface formation-pressure environment concepts. 0.433 psi/ft = 0.1 kg cm^{-2}/m; 0.465 psi/ft = 0.11 kg cm^{-2}/m; 1.0 psi/ft = 0.231 kg cm^{-2}/m (from Fertl and Chilingarian, 1976, p. 348; copyright © 1976 by SPE-AIME).

TABLE 1. Possible causes for abnormal formation pressures

Origin	Comments
Piezometric fluid level	Effect of regional potentiometric surface. Artesian water system, such as Artesian Basin, Florida, U.S.A., Great Artesian Basin, Australia, and North Dakota Basin, U.S.A.
Structure of permeable reservoir	Pressure transmission to shallower part of reservoir. Large anticlines, steeply-dippping beds, etc.
Rate of sedimentation and depositional environment	High depositional rates in clastic sequence and high shale/sand ratios.
Tectonic activities	Local and regional faulting, folding, lateral sliding and slipping; squeezing caused by downdropping of fault blocks; diapiric salt, sand, or shale movements; earthquakes; etc.
Osmotic and salt-filtering phenomena	On regional basis, e.g., San Juan Basin, New Mexico; Western Sedimentary Basin, Canada; San Joachi Valley, California; U.S.A. Gulf Coast; and Paradox Permian Basin, Illinois, U.S.A.
Digenetic phenomena	Postdepositional alterations: (1) montmorillonite → mixed layer clays → illite, with resulting water release; (2) gypsum → anhydrite + water; volcanic ash → clay minerals + carbon dioxide.
Secondary precipitation of cementing materials	Calcium sulfates, sodium chloride, dolomite, siderite, calcite, silica, etc. may act as (1) sealing barriers ("pressure caps"), and (2) directly cause overpressure by decreasing pore space due to crystal growth in closed reservoirs (e.g., NaCl in Markovo oil pool in the Osinskiy Series, USSR).
Repressuring of reservoir rocks	Caused by hydrocarbon production, drilling operations, and massive water injection progrtams (secondary recovery). May occur across faults, behind casing, or may be due directly to planned injection.
Massive areal rock salt deposition	Presence of salt (NaCl) beds which are impermeable. Examples present in U.S.A., USSR, North Africa, Middle East, North Germany, etc.
Paleopressures	Old, completely sealed-off reservoir rocks. Depth change due to uplifting or erosion.
Thermodynamic effects	Formation temperature changes cause fluid pressure variations.
Biochemical effects	About 2- to 3-fold volume increase caused by breakdown of hydrocarbon molecules.
Permafrost environment	Formation of giant frost heaves (pingos). Freeze-back pressure around shut-in arctic oil wells. Gas hydrate reservoirs (e.g., in Mackenzie Delta, Canada).

Source: From Fertl and Chilingarian (1981).

including unconformities, pinchouts, and faults; proximity to large salt masses, mud volcanoes, geothermal "hot" spots, etc. Steep, thin, overturned beds. Pore pressure gradients within single thick shale interval.

Borehole Environment. Borehole size, shape, and deviation. Shale alteration and hydration (exposure time of open hole to drilling mud). Type of drilling mud (fresh-water, salt-water, oil-base). Type and amount of weighting material (barite, etc.) and/or lost circulation material (mica, etc.). Degree of "gas cutting" in mud.

Drilling Conditions. Hole size, shape, and deviation. Mud programs and mud hydraulics (circulation rate). Rotary speed. Bit type (button, diamond, insert, etc). Ratio of bit weight to bit diameter. Bit wear (sharp, new bits vresus dull, old bits). Degree of overbalance. Floater ("heave" action) versus fixed on- or offshore rig.

Sample Selection. Type and size of sample (avoid sand, cavings, recirculated shales). Sampling technique. Sampling frequency. Analysis methods, for example, cutting density-variable density column, multiple-density solution technique (float-and-sink method), mercury pump technique, mud balance technique. Proper calibration of utmost importance.

Geophysical Well Logging. Well Logging: Difference in basic measuring principles (and shales are anisotropic). Sonde spacing. Depth of tool investigations. Temperature ratings. Proper tool calibration. Tool malfunction (overlapping repeats or reruns).

Parameter Plotting Techniques. Interval (or sample) selection. Sampling frequency. Linear, logarithmic plots. Plots of comparable data (not compatible are bulk density from logs versus cuttings, short normal versus induction log resistivities). Proper selection of "normal" compaction trend lines (discrepancies become enhanced with increasing depth of wells). Important to use all information available and to hire experienced, properly trained personnel.

The best plan for drilling operations in over-

TABLE 2. Drilling and well-logging overpressure detection and evaluation techniques

Drilling operations	Drilling parameters (instantaneous)	Drilling rate, torque, drag, hole fill (→ rearming), modified *d*-exponent, analytical drilling model concepts.
	Logging-while-drilling (instantaneous or semi-instantaneous)	Electrical and acoustic-type transmission concepts (using cables, drill string, mud system); analysis of drill string vibrations; downhole gas detection tool.
	Drilling mud parameters (while drilling, but delayed by circulation lag time)	Mud weight, gas content, temperature, flow rate, hole fill-up, pit level and total pit volume, salinity (resistivity, conductivity), well kicks.
	Drill cuttings analyses (while drilling, but delayed by time required for sample return)	Density, shape, size, color, volume over shale shaker, and moisture content, "lithofunction" plots, shale factor of cuttings (cation exchange capacity). Cuttings slurry and/or filtrate: resistivity, color, pH, redox potential, bicarbonate content, specific anion and cation concentrations, and filtration rate.
Well logging	Formation parameters from geophysical well logs	Electrical surveys: resistivity, conductivity, salinity, shale formation factor; acoustic (sonic) surveys: interval transit time and wave train presentations (VDL, signature log, etc.); bulk density surveys: density log, downhole gravity meter; hydrogen index (neutron-type logs); thermal neutron capture cross section (pulsed neutron logging); nuclear magnetic resonance; gamma ray spectral analysis logs.

Source: From Fertl and Chilingarian (1976). Copyright © 1976 by SPE-AIME.

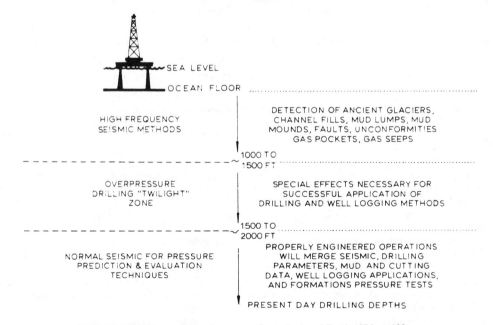

FIGURE 2. Offshore exploration scene (from Aud and Fertl, 1976, p. 122).

pressured environments incorporates (1) gathering and interpreting all known information (i.e., planning the well), (2) gathering and interpreting all known information (i.e., planning a preliminary drilling program), and (3) the flexibility to deviate from the plan whenever necessary (i.e., revising the drilling program).

Numerous developments and refinements in the acquisition, processing, and interpretation of geophysical data, such as seismic data, have made it possible for the oil industry to study not only the structural configuration of formations in sedimentary basins but also the depth and magnitude of abnormal formation-pressure environments prior to drilling a well. These concepts have a profound impact, especially on wildcat well planning (properly engineered mud and casing programs, predicting penetration rates, etc.) for high-cost, deepwater drilling and/or exploration of remote, untested regions.

The proper interpretation of seismic data can predict the top of overpressures to within an average of ±152 m (±500 ft.) and pore-pressure magnitudes can be estimated to within ±0.12 kg/dm³ (±1.0 lb/gal) mud-weight equivalent in more than 50 percent of field cases studied (Reynolds, 1973).

Drilling parameters and logging-while-drilling methods provide instantaneous or at least semi-instantaneous information. Drilling fluid parameters and shale-cutting analyses can be obtained while drilling, but they are delayed by the time required for circulation and sample return. Interpretation of recorded data is not always straightforward, and accuracy depends on geological factors, the borehole environment, sample selection at certain depth increments, and plotting techniques.

Some of the most reliable quantitative overpressure detection and evaluation techniques are based on geophysical wireline-logging methods, even though these methods are "after-the-fact" techniques, (i.e., the wellbore has to be drilled prior to logging). Hence, several short logging runs are sometimes necessary. Logs that can detect abnormally high formation pressures are summa-

rized in Table 2; they have been discussed in detail by Fertl (1976, pp. 177-230). Figure 3 illustrates the schematic response of several well-logging parameters to hydrostatic and overpressured environments.

Except for the SP curve data, all parameters that are recorded in shales, rather than in sands, are plotted versus depth. Trend lines are then established for normal compaction. Interpretation of the logs depends on the magnitude of departure from the normal trend, that is, the divergence of the formation pressure from the normal hydrostatic pressure at a specific depth. The acoustic and short normal resistivity logs are the tools most valuable for quantitative *in-situ* pressure evaluation.

Newly developed wireline formation testers, which allow the recording of a large number of pressure tests over the entire length of the uncased borehole, have been rapidly accepted by the drilling industry. Such measurements can then be used to "calibrate" other drilling and/or log-derived pressure indicators.

Drilling Concepts in Overpressured Environments

Maximum well control and minimum cost are key factors in present-day drilling operations. To

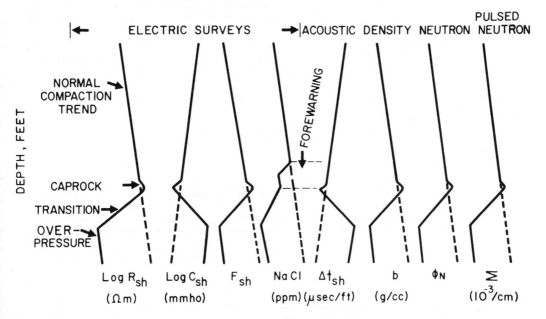

FIGURE 3. Schematic response of well-logging parameters to normal (hydrostatic) and overpressured environments. C_{sh} = shale conductivity (mmho); R_{sh} = resistivity (Ω), resistivity log; F_{sh} = formation resistivity factor, dimensionless; Δ_{sh} = acoustic transit time (μsec/ft), acoustic log; ρ = density (g/cm³), density log; ϕ_n = porosity index (%), neutron log; Σ = neutron capture cross-section of shale (10^{-3}cm⁻¹), pulsed neutron log. (from Fertl and Timko, 1972-1973).

achieve this, a basic understanding of two key formation pressures,—formation pore pressure and fracture pressure—is a prerequisite. Formation pressure can be determined from drilling and logging parameters. Formation fracture pressures (or gradients) can be calculated using one of the several models available (Table 3) or can be determined by actual tests in the well. Pressure testing below the casing shoe can (1) test the cement job, and (2) determine the integrity (fracture gradient) in the first weak (permeable) zone below the casing seat. Further drilling objectives define the possible limits for such a pressure leak-off test.

Such tests are carried out by pumping drilling fluid into the borehole against a closed annular choke at a slow constant rate and plotting fluid volume pumped versus surface pumping pressure. The surface pressure at the point of leak-off plus the static mud-column pressure shows the maximum pressure the borehole may tolerate (contain). As Moore (1974) candidly pointed out, however, in drilling deeper it becomes obvious that if the "operator exceeds the leak-off pressure from his last test, he is inviting problems and in all probability he will not be disappointed."

Casing in the borehole protects penetrated for-

TABLE 3. Prediction methods for fracture pressure gradients

Authors and formulas*	Comments
Hubbert and Willis (1957) $$(FP/D)_{min} = 0.5\,[(P_o/D) +)P_f/D)]$$ $$(FP/D)_{max} = P_o/3D + 2P_f/3D$$	Difference between minimum and maximum fracture pressure gradients decreases with increasing pore pressure. Hence, importance of surge pressures increases with depth.
Matthews and Kelly (1967) $$FP/D = P_f/D + K_i\delta/D$$	Introduction of variable matrix stress coefficient (K_i), which can be empirically determined from fracture initiation (breakdown) pressure. This method is used widely.
Eaton (1970) $$FP/D = P_f/D + \left(\frac{\mu}{1-\mu}\right)\delta/D$$	Extension of concept proposed by Matthews and Kelly above. Eaton introduced Poisson's ratio into the mathematical model. This method is used widely.
Taylor and Smith (1970)	Empirical correlation of injection test data in normally pressured formations only. Location: offshore of Louisiana near Mississippi delta. Minimum propagation test data correspond with fracture initiation pressure gradient given by Matthews and Kelly (1967).
MacPherson and Berry (1972)	Developed quantitative correlation between ratio of sand elastic modulus to overburden pressure and fracture pressure gradient. Based on data from 10 oil fields offshore of west Louisiana.
Christman (1973) $$FP/D = K_i[(P_o/D) - (P_f/D)] + P_f/D$$ where	Expanded technique of Matthews and Kelly (1967).
$$P_o/D = \frac{0.4335}{D}(\rho_w D + \bar{\rho}_b D)$$	Effect of water depth can be taken into account.
Anderson et al. (1973) $$FP/D = \left(\frac{2\mu}{1-\mu}\right)(P_o/D) + \alpha\left(\frac{1-3\mu}{1-\mu}\right)(P_f/D)$$	Based on Terzaghi's (1923) concept of effective stress. $\alpha = 1 - C_r/C_b$. Standard deviation = 0.083 psi/ft, between calculated and measured fracture pressure gradients based on 29 field tests in U.S.A. Gulf Coast area.
Althaus (1975) $$FP/D = K_i\,(19.2 - P_f/D) + P_f/D$$ where $$K_i = 0.23375 + 4 \bullet 10^{-5}D$$	Based on method by Matthews and Kelly (1967). K_i is determined by regression-type analysis of field data.

*Where: D = depth of zone of interest (ft); FP = fracture pressure at point of interest (psi); P_o = overburden pressure (psi), P_f = formation pressure, psi; $\delta = P_o - P_f$; K_i = matrix stress coefficient, dimensionless; μ = Poisson's ratio; ρ_w = density of water (g/cm^3); $\bar{\rho}_b$ = average bulk density (g/cm^3); C_r = intrinsic compressibility of solid rock material; C_b = bulk compressibility or rock.

Source: From Fertl and Chilingarian (1981).

mations and the wellbore from contaminations and drilling hazards (see Vol. XIV: *Borehole Drilling*). In both hydrostatic and overpressured zones, the recommended mud weight does not exceed formation pressure by a margin of more than 0.024 to 0.479 g/cm³ (0.2 to 0.4 lb/gal). In addition, control of drilling mud must also focus on other mud properties, including gel strength, viscosity, and water loss.

Mud-weight requirements in drilling through overpressured formations can be met by the use of high-density solids in the fluids. Barite ($BaSO_4$), which is a nontoxic, nonreactive solid, is frequently used. American Petroleum Institute specifications require a minimum specific gravity of 4.2, because the use of low-grade barite (specific gravity < 4) results in mud viscosity problems, requires a higher concentration of chemical thinner, and causes excessive circulating pressures in deep, high-temperature wells.

Under special circumstances, such as drilling in superpressured environments (e.g., in the Middle East) and/or fighting wells already out of control, other heavy minerals (some with specific gravities of about 7) have been used successfully. These include various types of iron ores and lead sulfide. However, they are expensive, must be ground to extremely fine particle size, giving rise to a high viscosity mud system.

Casing point selection under perfectly balanced conditions that is, the formation (pore) pressure is equal to the hydrostatic pressure of the mud column) is shown in Figure 4. Indispensable, basic requirements are a good cement job, and the absence of weak zones somewhere along the open hole, including lignite and other coal beds, lime streaks (in cap rocks), carbonate shell beds, and small faults or joint systems intersecting the wellbore (Fertl, 1976, pp. 233-235).

Drillpipe sticking occurs due to underbalanced conditions (Fig. 5). In the overpressured section, the equivalent mud weight exceeds the pore pressure down to a depth, *D*, where the drillpipe will get stuck. As the drilling bit approaches this "stuckpoint", the borehole starts caving in and the torque increases simultaneously.

Generally speaking, pipe-sticking forces depend on (1) the difference between hydrostatic drilling mud and formation pressures (differential pressure), (2) the permeability of the formation, (3) the thickness of the zone, (4) the thickness and slickness of the filter cake, (5) the length of time the pipe remains motionless against the formation, (6) hole and pipe size, and (7) pipe shape. Inasmuch as the latter is very critical, special drillcollar configurations have been developed, including spiral collars with circulation grooves, square drillcollars, and shouldered drillcollars. In addition, *key-seating* is recognized as one of the primary reasons for pipe sticking.

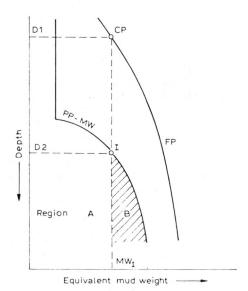

FIGURE 4. Casing point selection under perfectly balanced conditions. Region A = "safe" drilling and well-killing conditions. Region B = well kicks cannot be handled and circulation is lost. CP = casing point; FP = fracture pressure; PP = pore pressure; MW = specific weight of drilling fluid (from Fertl, 1976, p. 234).

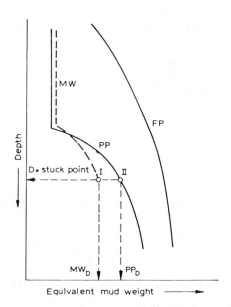

FIGURE 5. Schematic diagram (depth versus equivalent mud weight) showing reasons for drillpipe sticking due to an underbalanced mud system. Symbols defined in Figure 4. D = well depth at which pipe sticking occurs (from Fertl, 1976, p. 236).

Lost circulation may occur for several reasons, including excessively weighted drilling fluid (i.e., an overbalanced mud condition) and the presence of very permeable fractured and vugular formations.

Surge and swab pressures are also important parameters in well control. For example, running the drillstring into a fluid-filled borehole causes a pressure surge (increase in pressure), whereas pulling the string out of a liquid-filled borehole causes a swab pressure (reduction in pressure). According to Moore (1974), over 25 percent of all *blowouts* result from swab pressure, whereas excessive surge pressures are frequently responsible for initiating lost circulation and/or physical formation damage (i.e., blocking of pores by solids).

Properly engineered well-design plans *(casing programs)* must be effective, safe, and soundly applied for optimum results. Whereas drilling programs are designed from the surface to the bottom of the hole, well planning is a reverse process and starts at the bottom of the hole. It is important to note, however, that there is no single, unique casing program. Each one has to be tentative, flexible, and continuously reappraised for each well. Hence, optimum casing programs are the result of continuous evaluation and incorporation of additional data and new experiences, and are not the outgrowth of a one-time, single decision. Generalized typical casing programs in several abnormally pressured U.S. basins are listed in Table 4.

Geological, technological, safety, and economic considerations influence size, setting depth, and design factors (burst, collapse, and tension) in the production string in overpressured environments. Numerous additional related problems frequently are encountered, particularly in high-pressure sour-gas environments, including hydrate formation, salt precipitation, packer fluid selection problems, linear cementing problems, tubing leaks, and so on.

Reservoir Engineering Concepts in Abnormal-Pressure Environments

Reservoir engineering has emerged as a highly technical and specialized field of petroleum engineering, with the objective of describing reservoir and production behavior based geological and petrophysical data, reservoir fluid properties, and so on (see Vol. XIV: *Petroleum Geology*). Since the early 1960s, increased attention has been focused on the analysis of reservoir reserves, the behavior of reservoirs, and possible mechanisms important to production from abnormally high-pressured reservoir rocks. Frequently, such reservoirs, containing mainly gas, are of limited areal extent and/or have limited aquifers, and often do not behave as volumetric gas reservoirs. This complicates gas-in-

place estimates and usually results in reserve figures that are much too optimistic (by as much as 100 percent).

Mechanisms proposed for overpressured reservoirs include (1) shale-water influx from shale zones into adjacent potential pay water sands (Wallace, 1969; Bourgoyne et al., 1972), (2) rock compressibility and rock failure (Harville and Hawkins, 1969), and (3) water influx into the reservoir from limited aquifers. A typical graph of p/Z versus cummulative production for an overpressured gas reservoir in clastic sequence is shown in Figure 6.

Exploration Concept in Clastic Overpressure Environments

Numerous studies have related the occurrence of oil and gas deposits to single parameters or to a combination of parameters, such as source rock potential, geological time and age, depth, pressure, temperature, sand/shale ratio, shale-resistivity ratio, and specific geochemical and hydrodynamic criteria. In nature, a combination of several superimposed factors generally prevails, with individual parameters varying in importance. For example, important findings are also shown in cross-plots of formation pressure gradient and temperature (Fig. 7). A temperature range of 102°C to 143°C (215°F to 290°F) coincides with the range of highest-pressure gradients in hydrocarbon zones. This is particularly noteworthy because this temperature range lies in the bulk part of second-stage clay dehydration as proposed by Burst (1969); that is, it lies in the zone of maximum fluid distribution. At the same time, extremely high-pressure gradients are encountered in high-temperature aquifers and zones with noncommercial oil and gas shows (Timko and Fertl, 1971).

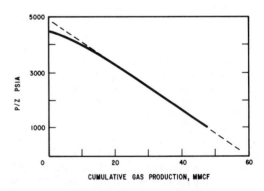

FIGURE 6. Typical graph of p/Z versus cumulative production for an overpressured gas reservoir in sand-shale sequence; p = reservoir pressure in psi, Z = compressibility factor, and MMcf = million of cubic feet (from Fertl and Chilingarian, 1976, p. 351; copyright © 1976 by SPE-AIME).

TABLE 4. Generalized typical casing programs in several abnormally pressured U.S. basins

Region*	Typical casing sizes (inches)	Approximate average casing point depth (ft)			
Delaware Basin (West Texas)		*Ward Co.*	*Reeves Co.*	*Pecos Co.*	*Winkler Co.*
	(1) 20″ conductor	5,000	5,000	1,700	5,000
	(2) 13⅜″	11,000	11,000	10,500	13,000
	(3) 9⅝″ or 10¾″	14,500	16,000	15,500	17,000
	(4) 7⅝″ liner	20,000	21,000	22,000	22,500
	(5) liner tie-back** to surface, 7″ (7⅝″) × 5″ (5½″) or 7⅝″ only				
U.S. Gulf Coast (Texas, Louisiana)		*South Texas*	*Upper Texas*	*Onshore Louisiana*	*Offshore Louisiana*
	(1) 20″ (locally plus 30″)	9,000	3,000	3,000-4,000	4,000
	(2) 13⅜″ or 10¾″	12,500	12,000	13,000-15,000	13,000
	(3) 9⅝″ or 7⅝″			15,000-18,000	20,000
	(4) 7″ liner and/or 5½″ (or 5″)	15,000	15,500	(plus)	(plus)
Anadarko Basin (Texas Panhandle, Western Oklahoma)	For 20,000-ft. wells:	*Oklahoma*	*Texas*		
	(1) 24″ conductor	no	yes		
	(2) 20″ (or 16″) conductor	2,000	4,000		
	(3) 13⅜″ (or 10¾″)	12,000	13,500		
	(4) 7⅝″ (or 9⅝″) liner	19,000	16,700		
	(5) 5″ (or 7″) liner	to TD	to TD		
Rocky Mountain area (Utah, Colorado)		*Uinta Basin (Utah)*			
	(1) 20″ conductor				
	(2) 13⅜″	1,500			
	(3) 9⅝″	9,500-12,500			
	(4) 7″ liner	to TD			
		Piceance Basin (Colorado)			
	(1) 13⅜″ conductor				
	(2) 9⅝″	5,000			
	(3) 7″	14,500			
	(4) 5″ liner	to TD			

*Schematic is shown in Fig. 16-17, Fertl and Chilingarian (1981), p. 609.
**Typical liner tie-back strings in deep Ellenburger wells.
Source: From Fertl (1976), p. 260.

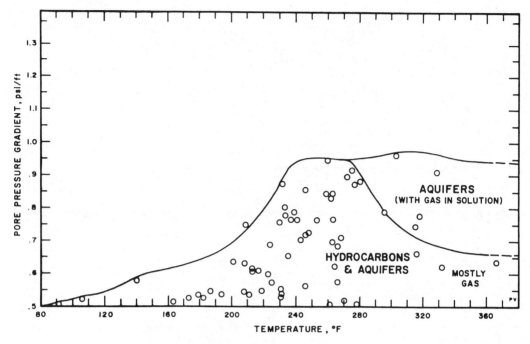

FIGURE 7. Formation-pressure gradient temperature in 60 overpressured Gulf Coast wells (from Timko and Fertl, 1971; copyright © 1971 by SPE-AIME).

A generalized correlation has been developed between typical Gulf Coast shale-resistivity profile and the distribution of oil and gas fields in the subject area (Fig. 8). Application of such findings indicates whether it is possible for commercial production to exist below the depth to which the well has already been drilled and logged, and whether it is economically attractive to continue drilling a borehole below a given depth in shale-sand sequences (Timko and Fertl, 1971).

A shale-resistivity ratio (ratio of normal R_{sh} to observed R_{sh}) parameter, which is a function of thermodynamic and geochemical effects, is often used. Initially developed for the gulf Coast area, additional experience has shown the model to hold true in California and several other Tertiary basins with shale-sand sequences throughout the world, including Canada, South America, Africa, and the Far East. The concept is used by the petroleum industry in many areas as a supplementary completion guide and is a decisive factor in "dry-hole" money negotiations between companies. Correlations similar to those in the Gulf Coast area, however, have to be established.

Based on the shale-resistivity-ratio method and regardless of measured formation-pressure gradients, the following guidelines can be drawn for shale and sand sequences (but not for massive carbonate sections): (1) Most commercial oil sands exhibit shale-resistivity ratios lower than 1.6 in adjacent shales and generally can be reached without a

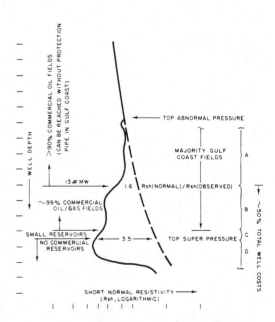

FIGURE 8. Typical Gulf Coast shale-resistivity profile, based on the short-normal curve, correlated to distribution of gas-oil reservoirs. Profile is based on hundreds of commercially productive wells (from Timko and Fertl, 1971; copyright © 1971 by SPE-AIME).

string of protection pipe. (2) Most commercial gas-sand reservoirs exhibit ratios in adjacent shales of about 3.0 and lower. These wells can have extremely high (measured) pressure gradients. (3) Wells with ratios of 3.0 to 3.5 can be commercially gas-productive and generally will produce as one- or two-well reservoirs. (4) No commercial production is found when the shale-resistivity ratio reaches or exceeds 3.5, no matter what the actual pressure gradient is. These wells are often highly productive initially and are characterized by an extremely fast pressure depletion. It is important to mention here that shale-resistivity ratios can be used for classifying source rocks and in predicting subsidence-proof areas in the world.

Global Occurrence and Economics in Overpressure Environments

In the worldwide search for hydrocarbon resources, both on- and offshore, abnormal formation pressures have been encountered as shallow as a hundred meters or so below the surface or at depths far exceeding 7,620 m (25,000 ft). Exploration costs in normal-pressure environments shows generally predictable trends for time, cost, and

risk. However, the presence of abnormal pressures, especially surpressures, is a very critical factor. Time, cost, and risks can increase drastically, greatly affecting profit (Figure 9).

GEORGE V. CHILINGARIAN
WALTER H. FERTL

References

Althaus, V. E., 1975. A new model for fracture gradient, *5th CWLS Form. Eval. Symp.,* Paper M, Calgary, Alberta, 12p.

Anderson, R. A.; Ingram, D. S.; and Zanier, A. M., 1973. Determining fracture pressure gradients from well logs, *Jour. Petroleum Technology* **25**, 1259-1268.

Aud, B. W., and Fertl, W. H., 1976. Overpressure in the Twilight Zone, *Petroleum Engineer,* **48**(5), 122-218.

Bourgoyne, A. T.; Hawkins, M. F.; Lavquial, F. P.; and Wickenhauser, T. L., 1972. *Shale Water as a Pressure Support Mechanism in Superpressured Reservoirs.* Baton Rouge: Louisiana State University, SPE 3851, 3rd Symposium on Abnormal Subsurface Pore Pressure, May, 11p.

Burst, J. F., 1969. Diagenesis of Gulf Coast clayey sediments and its possible relation to petroleum migration, *Am. Assoc. Petroleum Geologists Bull.* **53**(1), 73-93.

Christman, S.A., 1973. Offshore fracture gradients,*Jour. Petroleum Technology* **25**, 910-914.

Eaton, B. A., 1970. How to drill offshore with maximum control, *World Oil* **171**(5), 73-77.

Fertl, W. H., 1976. *Abnormal Formation Pressures: Implications to Exploration, Drilling and Production of Oil and Gas Resources.* Amsterdam: Elsevier, 382p.

Fertl, W. H., and Chilingarian, G. V., 1976. *Importance of Abnormal Formation Pressures to the Oil Industry.* Paper presented to Spring Meeting of the European Society of Petroleum Engineers of AIME, Amsterdam, April, SPE 5946, 10p.

Fertl, W. H., and Chilingarian, G. V., 1981. Drilling through overpressured formation, in G.V. Chilingarian and P. Vorabutr (eds.), *Drilling and Drilling Fluids.* Amsterdam: Elsevier, 581-620.

Fertl, W. H., and Timko, D. J., June 1972-Mar. 1973. How downhole temperatures, pressures affect drilling, *World Oil* (10-part series of articles).

Harville, D. W., and Hawkins, M. F., Jr., 1969. Rock compressibility and failure as reservoir mechanisms in geopressured gas reservoirs,*Jour. Petroleum Technology* **21**, 1528-1530.

Hubbert, M. K., and Willis, D. G., 1957. Mechanics of hydraulic fracturing, *Am. Inst. Mining Eng. Trans.* **210**, 153-168.

MacPherson, L. A., and Berry, L. N., 1972. Prediction of fracture gradients from log-derived elastic moduli, *Log Analyst* **13**(5), 12-19.

Matthews, W. R., and Kelly, J., 1967. How to predict formation pressure and fracture gradient, *Oil Gas Jour.* **65**(8), 92-106.

Moore, P. L., 1974. *Drilling Practices Manual.* Tulsa, Okla.: The Petroleum Publishing Co., 448p.

Reynolds, E. B., 1973. *The Application of Seismic Techniques to Drilling Techniques.* Society of Petroleum Engineers of AIME, 48th Annual Fall Meeting, Las Vegas, Nev., Sept. 30-Oct. 3, SPE 4643.

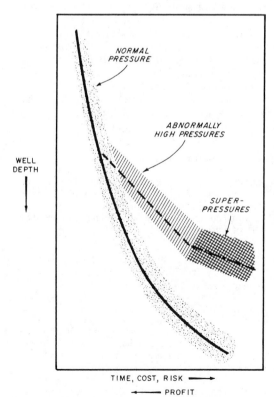

FIGURE 9. Generalized trends of deep-drilling costs and profits that show drastic deviations where abnormal pressures are encountered.

Taylor, D. B., and Smith, R. K., 1970. Improving fracture gradient estimates in offshore drilling, *Oil Gas Jour.* **68**(15), 67-72.

Terzaghi, K. V., 1923. Die Berechnung der Durchlässigkeitsziffer des Tones aus dem Verlauf der Hydrodynamischen Spannungserscheinungen, *K. Akad. Wiss. Wien, Math. Naturwiss. Kl. Abts.* **2A**, 105-132.

Timko, D. J., and Fertl, W. H., 1971. Relationship between hydrocarbon accumulation and geopressure and its economic significance, *Jour. Petroleum Technology,* August, pp. 923-932.

Wallace, W. E., 1969. Water production from abnormally-pressured gas reservoirs in south Louisiana, *Jour. Petroleum Technology* **21**, 969-983.

Cross-references: *Electrokinetics; Hydrodynamics, Porous Media; Hydromechanics; Well Data Systems.* Vol. XIV: *Exploration Geophysics; Petroleum Geology; Well Logging.*

FOUNDATION ENGINEERING

The term *foundation* originates from the Latin *fundatio,* from the verb *fundare,* to found, to set, to place. Thus *foundation* refers to the *artificially* laid base on which a structure—namely, superstructure—stands, or upon which any erection is built. Generally the term refers to all those parts of the weighty structure laid below the ground surface (or water surface, as in the case of a bridge pier), and upon which a structure of any kind is to be built and supported. Hence engineers speak of a foundation as an artificial structural element and of the *laying of foundations.* The term *laying of a foundation* agrees with the spirit of most of the languages of the Western culture.

The purpose of a foundation is twofold: (1) to receive and transmit structural loads and externally applied loads (wind, vibrations, seismic shocks, and tremors) to the superstructure, directly to the underlying soil (rock) at a given depth below the ground or water surface, whichever is the case, safely without causing any distress to the superstructure, and (2) to distribute pressure at the base of the footing of the foundation to an intensity allowable on the soil (rock).

Foundation engineering treats soil-foundation-load systems. It is concerned with the dual problem of soil (rock) bearing capacity and the sound design and construction of the foundation, that is, the mutual interaction of the soil-foundation-load system as an integral unit. Foundation engineering has now become a universally accepted collective engineering concept that comprehends the science and technology of the various methods of analysis for the safe and economical design and laying of foundations.

Some Historical Notes about Foundations

Work in foundation engineering dates far back historically; because the need for a good, safe foundation of a structure has been recognized throughout the ages, foundation building is one of the oldest arts. Timber piles were used from prehistoric times onward for construct foundations in and above water. It is known that prehistoric lake dwellers built their huts on piles driven into lake bottoms. Timber piles under structures built during the Roman Empire and medieval era have been discovered, still in good condition, below the groundwater table. Even in the Bible, one reads that King Solomon built his temple on cedar piles from Lebanon.

The long span of life of the oldest structures of the Eastern and Western worlds, which have survived until the present in a fair state of preservation, may be attributed to the inherent soundness of their foundations (and also probably partly to favorable climatic conditions).

The practice of well (caisson) sinking, it is interesting to notice, is also centuries old. For example, in Egypt in about 2000 B.C. the cutting edges of caissons were used to sink wells or shafts through sand and conglomerate (*Engineering News Record,* 1933). To make the cutting edge, a hole was cut in a circular limestone slab. The open caisson then sank by its own weight. To reduce frictional resistance during penetration, the outside surface of the caisson was made smooth.

The Cheops pyramid, which is 228.6 m × 228.6 m (750 ft × 750 ft) at its base and 146.30 m (480 ft) high, was built in about 3000 B.C. (Rawlinson, 1945) on a mat of limestone blocks, the latter being supported by limestone bedrock.

According to Steinman and Watson (1941), bridge piers in China (from 200 B.C. to approximately A.D. 220) were founded as follows:

Part of the river was closed by a double-row cofferdam made of bamboo piles fastened together with ropes. Then bamboo mats were put on each row of piles, the intervening space being filled with clay. The whole cofferdam was curved against water pressure. The water was pumped out by means of crude wooden tread pumps worked by pairs of men.

About foundations, the first-century Roman engineer Vitruvius wrote in his *Ten Books on Architecture :*

Let the foundations of those works be dug from a solid site and to a solid base if it can be found. But if a solid foundation is not found, and the site is loose earth right down, or marshy, then it is to be excavated and cleared and remade with piles of alder or of olive or charred oak, and the piles to be driven close together by machinery, and the intervals between are to be filled with charcoal. (cited in Morgan, 1914).

Vitruvins added that even "the heaviest foundations may be laid on such a base." (cited in Gwilt, 1874).

In the medieval period (about A.D. 400-1400), the only structures of significance associated with laying of foundations were castles, city fortification walls, cathedrals, and campaniles (bell towers) (Jumikis, 1962). As magnificent as medieval cathedrals are, their foundations were so poorly designed that many of them were laid on poor soil, and they usually were laid at insufficient depth to transmit the heavy loads of the superstructure of the cathedrals to a firm soil at a proper depth.

Some of the noteworthy examples of foundations laid in the fifteenth to seventeenth centuries are those of the Rialto single-arch bridge in Venice, Italy (completed in 1591), and some of the bridges in Paris built in the sixteenth and seventeenth centuries. Because of the marshy site and adjacent large buildings, pile driving for the Rialto Bridge foundations presented a problem of great acuteness.

The foundations of the Pont Royal (a seventeenth-century bridge in Paris) were built for the very first time by means of the open caisson, with watertight timber sides. After the excavation, the caisson was sunk to the bed, but the top was kept above water level. The masonry work of the pier was then built up inside the chamber.

The French engineer Triger is credited with sinking in 1839 the pneumatic caissons for laying of deep foundations in waterlogged soils. According to Jacoby and Davis (1941), the first pneumatic caisson in America was built in 1852 for sinking cast iron cylinders ("pneumatic piers") of bridges over the Pee Dee and Santee rivers in South Carolina. One of the first large wooden pneumatic caissons in the United States was used in 1870 for building the piers of the Eads' arch bridge in St. Louis, Missouri. There the compressed air pressure attained a level as high as 3.42 phys atm = 3.44 kgf/cm^2 (49 psi) (White, 1962).

The Brooklyn Bridge in New York City, built in 1871, represents the first large-scale use of pneumatic caissons for bridge foundations in the United States. The caissons were built of timber. Thirty-one m × 52.43 m (102 × 172 ft), it was sunk to a depth of 23.77 m (78 ft) below the high-water table. The method of laying foundations by means of the pneumatic caisson is limited to a depth of about 36.58 m (120 ft). (The endurance limit for man is about 3 to 4 atm under pressure). For greater depths, open caissons are used.

After World War I, the engineering discipline, now known as foundation engineering—a branch of geotechnical engineering (q.v.)—developed and progressed rapidly, especially in conjunction with the scientific development of the engineering discipline soil mechanics (q.v.) since 1922 (Statens Järnvägars Geotekniska Kommission [Swedish State Railways Geotechnical Commission], (1922); Terzaghi, 1925; Fellenius, 1929; Fröhlich, 1934; Terzaghi and Fröhlich, 1936; Krey and Ehrenberg 1936; Casagrande, 1942; Borowicka, 1943).

The piers of the Transbay Bridge (completed in 1936) between San Francisco and Oakland, California, were sunk with the aid of dome-capped caissons 73.15 m (240 ft) below the water surface. A caisson for the San Francisco-Oakland Bridge (completed in 1937) was sunk to a depth of 73.8 m (242 ft) (Purcell, 1934). One of the open caissons for the Mackinac Straits Bridge, Michigan, is founded 61.0 m (200 ft) below water (*Engineering News Record*, 1955).

The tower foundations of the Verrazano-Narrows Bridge in New York City (opened to traffic in 1964) are laid deep beneath the waters of the Narrows—32.0 m (105 ft) below the mean high-water level at the Staten Island side and 51.82 m (170 ft) at the Brooklyn side (Triborough Bridge and Tunnel Authority, 1964). At these locations, the rock surface was found to be from about 48.77 m (160 ft) below mean sea level at the Staten Island anchorage to more than 91.44 m (300 ft) at the Brooklyn tower. Overlying the rock are various soft to firm layers of silt, clay, sand, and gravel and a layer of decomposed rock immediately above the solid bed of gneiss and Manhattan schist (Ammann, 1963; Just and Obrician, 1966).

The sinking depth of one of the several dome-capped caissons for the construction of foundations and piers of the Tagus River suspensions bridge at Lisbon, Portugal (completed in 1966), was 99.3 m (260 ft) below the water surface to bedrock (Riggs, 1966).

Nowadays, the geotechnical discipline of foundation engineering has attained a high degree of perfection. The recent output of books, conference proceedings, research reports, periodicals, and various articles on the various theoretical and technological aspects of foundation engineering, soil mechanics, and rock mechanics may be regarded as evidence of realization of the importance of foundation engineering in construction technology and industry. Also, many national and international conferences, congresses, symposia, and short courses (see *Soil Mechanics, History of*) have contributed to, and broadened the scope of, foundation engineering.

General Notes about Foundations

A foundation must be designed and laid to support the weight of the superstructure, live loads, the weight of the snow, wind load, earth pressure, hydrostatic and hydrodynamic loads, seismic forces, and surcharge loads. Besides, the foundation design should provide for the static moments and torsional vibration brought about by the loads.

As stated previously, the purpose of a foundation is to transmit structural loads to the soil or rock safely without causing any distress to its superstructure. The superstructure exerts a resultant load, *R,* on the foundation. Thus the foundation must be designed so that the resultant load (1) does not bring about any lateral sliding of the foundation base on soil (rock); (2) does not bring about rotation, namely, tilting of the structure; (3) does not cause any groundbreak or ground collapse underneath the structure; and (4) does not produce any intolerable settlement of the structure. Also, the stresses within the foundation materials and those of the loaded soil (rock) formations induced by the loading from the superstructure should not exceed the allowable limits usually stipulated by building codes.

In regions of seismic activity (see *Earthquake Engineering*), the design of any structure and its foundations should meet all the requirements prescribed by the corresponding seismic building codes for a seismic design. Foundation analysis and design assist the engineer in the safe and economical construction of substructures.

Particular methods of design and construction involve geology; soil mechanics and soil dynamics; the science, technology, and mechanics of materials (concrete, reinforced concrete, and steel); construction statics; building dynamics; and a host of other engineering disciplines. Hence a discussion of these is beyond the scope of this entry. They are dealt with in other, special, technical literature.

Depending on local geological, hydrological, soil, or rock conditions at the construction site, as well as on the techological and architectural importance and significance of the structure, foundations may be grouped into three broad categories: shallow, deep, and special foundations. There is, however, no sharp demarcation between these three groups of foundations, and one group blends into the other.

Shallow Foundations

Shallow or *ordinary* foundations are sometimes called *spread foundations*—the most common type of building foundations. They are laid at a relatively shallow depth on soil or rock of competent bearing capacity below the ground surface, above as well as below the groundwater table. Shallow foundations may be classed into (1) footings; (2) mats or rafts; and (3) combined footings.

Footings. A *footing* is a unit of foundation laid at a shallow depth. It is designed to provide a bearing area on a supporting firm soil or rock formation, and to transmit and distribute structural loads safely to soil and/or rock, whichever is the case. Specifically, a footing is the enlargement of the base of a column or wall to distribute the load on soil (rock) to the limit of, or below, its allowable bearing capacity. The bearing capacity of a soil (rock) is the maximum load per unit area that the soil (rock) can support without rupture.

Soil and rock bearing capacities can be obtained from (1) soil and rock loading tests *in situ;* (2) laboratory testing; (3) analytical methods; and (4) building codes, official regulations, and civil engineers' handbooks.

A footing, or a foundation, is said to be *direct* when it rests in direct contact with the soil (no piles). Thus we arrive at the concept of the integral *soil-foundation-load* system. A footing supporting a single column is called an *isolated* footing or a *spread* footing (Fig. 1). An elongated spread footing that supports a wall or a row of three or more columns is called a *wall* footing or a *continuous,* or *strip,* footing (Fig. 2). Continuous footings can "span over" local soft spots in the soil or smaller depressions in the rock surfaces, thus providing a fairly uniform support for a row of columns or a long wall.

Mats or Rafts. A *mat* or *raft* is a shallow, single-slab foundation laid directly on the ground, covering a large area. In a way, the mat is a footing laid over the entire area beneath a structure. It supports many columnar and/or all the wall loads. Thus the purpose of a mat is to distribute soil (rock) contact pressures uniformly from structural loads on the ground over a large enough area so that the soil (rock) can support the loads safely and without any excessive, intolerable settlement (or without being crushed, in the case of a weak rock support). Obviously, the mat material must be strong enough to resist the load received from the superstructure and the soil (rock) reaction.

Mats are used (1) when the soil bearing capacity is too low to support the loads by other kinds of foundations; (2) to distribute loads from the periphery of the structure over the entire area on the

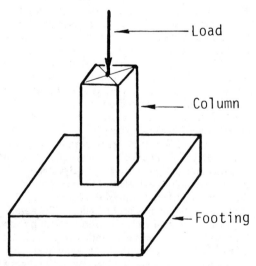

FIGURE 1. Isolated or individual column footing.

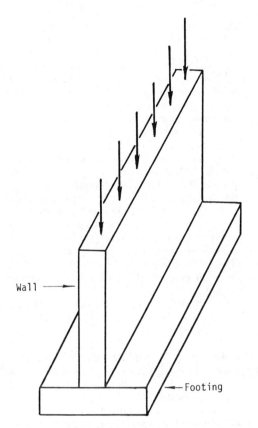

FIGURE 2. Wall or strip footing.

ground beneath the structure; (3) to reduce concentrations of high contact pressures on soil; (4) to resist hydrostatic pressure head uplift; (5) to "bridge over" weak spots in the soil; and (6) at property lines of adjacent sites and/or buildings.

Large mats are reinforced by ribs above the mat for stiffness (Fig. 3) and thus for a reasonably good pressure distribution on soil. Stiffness may also be attained by a cellular construction.

Combined Footings. *Combined footings* are intermediate between single spread footings and mats. In essence, several individual footings are merged to form a little mat supporting several columns (Fig. 4). When several footings are joined in a straight line by means of a slab or a beam (Fig. 5), they are called *connected,* or *strapped,* foot-

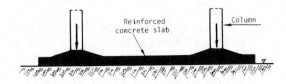

FIGURE 3. Flat-slab type of mat foundation.

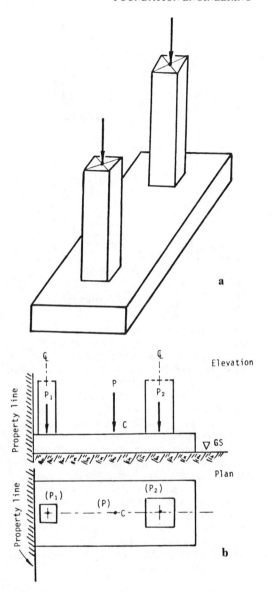

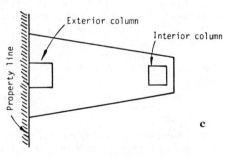

FIGURE 4. (a) combined footing; (b) combined asymmetrical rectangular footing; (c) combined trapezoidal footing for unequal loads (plan).

187

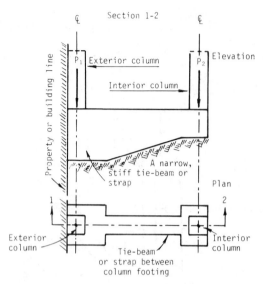

FIGURE 5. Connected or strap footing.

ings. A combined footing usually supports an exterior column and one or more interior ones.

In congested areas such as in cities the design and laying of foundations may become complex with respect not only to property line restrictions and limitations but also to adjacent buildings. If property rights prevent the use of footings projecting beyond the exterior walls of the designed building, combined footings, connected footings, and mats are built so that they do not project beyond the wall columns. Combined footings are also used when the soil has a low bearing capacity, and also where single footings under heavily loaded columns would overlap or merge or nearly merge.

General Requirements of Foundations. Like any element of a structure, the foundation must satisfy certain stability requirements. Among the many requirements for a solid foundation, the following are important:

1. The foundation should be laid at a proper depth below the ground surface to avoid lateral expulsion of soil from underneath the base of the foundation, to avoid damage due to freezing and thawing (where it applies), and to protect it from scour and washout by erosion of soil by water.
2. The foundation should resist to groundwater and other aggressive water relative to foundation material.
3. The foundation should be strong as a whole, as well as in its component parts. That is, the deformations of the foundation should be no larger than those allowable under the condition of its exploitation.
4. The foundation should be stable against any lateral (horizontal) sliding, against any rotary

movement, and against intolerable differential settlement to avoid any distress to the structure.
5. The soil-foundation system must be safe against rupture of soil (groundbreak); this requirement pertains to the exhaustion of the shear strength (namely, the bearing capacity) of the soil, and may be regarded as a natural consequence of the requirement mentioned in item 1.
6. The foundation must be designed and laid with the view of future excavations of the soil around the foundation for eventual repairs of the foundation, for installation of service ducts and pipes (for example), and for foundations of new additions to the initial structure. Hence the stability of the soil-foundation system must be analyzed with no foundation backfill.
7. The foundation should be durable and should function properly during its assigned service period.
8. The foundation should be economical; the laying of it should lend itself to mechanization of work.

Obviously, all the preceding requirements pertain equally to the foundation-supporting soil as well as to the structural foundation as an integral soil-foundation system.

To summarize, in designing structures such as foundations, the engineer must meet simultaneously two independent requirements for foundation stability: (1) there should be adequate safety against a shear failure within the soil mass, viz., groundbreak (not to exceed the bearing capacity of the soil being built upon); and (2) the probable maximum and differential settlements of the soil—viz., various parts of the foundations—must be limited to a safe, tolerable magnitude.

Depth of Shallow Foundations. In foundation design, the following factors are to be considered: (1) the depth of the footing; (2) the size and form of the footing; and (3) the factor of safety against groundbreak.

The depth to which shallow foundations are to be laid is of paramount importance, and depends in turn on a series of factors such as the following:

1. technological significance of the structure (for example, subsurface structures such as tunnels, or monumental, or profane, or temporary structures and their safety requirements);
2. functional requirements of the structure (underground structures; need for cellars or basements; service utilities; adjoining and neighboring structures);
3. the kind of foundation material (e.g., timber, stone, bricks, plain concrete; reinforced concrete; steel);

4. the kind and magnitude of loads to be transmitted to the foundation-supporting soil (centric, eccentric, vertical, inclined, static, dynamic loads);
5. geological conditions at the construction site (faults; nature and types of soil and/or rock and their stratification and inclination; degree of weathering; cavities in karst regions; frost-proneness of soil; swelling and/or shrinkage of soil; "quick" condition of soil) (Jumikis, 1962, 1966, 1977, 1979);
6. hydrological conditions at the construction site (precipitation regimen; possibility of inundation of site and foundations; surface runoff; soil erosion by water; scouring of foundations; drainage conditions; presence, position, and fluctuation of groundwater table);
7. seasonal variations in climatic conditions (frost penetration depth in soil and its associated heaving, where applicable; depth of thawing in permafrost regions; swelling and shrinkage of soils brought about by seasonal changes in moisture content in soil) (Jumikis, 1966, 1977);
8. the allowable bearing capacity of the soil, viz., shear strength of the soil (Jumikis, 1965, 1983);
9. consolidation properties and tolerable settlement of the soil (Terzaghi and Fröhlich, 1936; Jumikis, 1983);
10. the tolerable settlement of the structure;
11. the distance and depth of the foundations of existing adjacent structures, where applicable (pressure overlap, settlements), (Jumikis, 1971); the method of laying the foundation; and
12. the cost of the foundation.

Other factors may also be considered (Vesić, 1973).

What all these considerations indicate is that the footing of the foundation should be laid down to firm soil (where available or attainable). Also, if possible, the footing should be kept above the groundwater table. If not, drainage of the soil and/or lowering of the groundwater table should be pursued (Jumikis, 1964, 1966, 1971, 1983).

The safety of a structure is the most important and imperative requirement. It is to be strived for with the least expense possible. As stated previously, the soil-foundation system must be safe against groundbreak (Jumikis, 1965, 1983). For reasons of durable stability, monumental structures are founded deeper than small buildings with lighter loads. To avoid the decaying of a timber grillage of a foundation, the grillage should be laid below the lowest elevation of the groundwater table.

The greater the load transmitted by the foundation, the deeper the foundation must be laid to provide for a lateral counterweight (overburden) of the adjacent soil against groundbreak, namely, the lateral expulsion of soil from underneath

the base of the footing (Jumikis, 1964, 1965, 1969, 1983).

Usually, building codes specify how deep foundations may be. Shallow foundations must be laid on clean, natural mineral soil. All kinds of artificial fill, garbage and trash, rubble, organic soil, remnants of organic matter, and topsoil covering the good soil must be removed before foundations are laid.

Calculations to determine the ultimate bearing capacity of soil for shallow foundations can be found in *Soil Mechanics* (Terzaghi, 1943; Meyerhof, 1951, 1963, 1970; Jumikis, 1965, 1971). The design of various parts of various reinforced concrete foundations are described in standard texts on fundamentals of design of concrete structures. Where firm soil is unattainable, deep foundations are used (Kézdi, 1965).

Deep Foundations

A foundation is considered, arbitrarily, to be "deep" if its depth, *d,* below the ground surface exceeds approximately twice the width, *B,* of the base of the footing. Again, the classification of deep foundations is only relative; there are no sharp demarcations between the "shallow" and "deep" categories. Deep foundations are laid at considerable depth below the ground surface and/or below the lowest part of the superstructure to reach, to rest on, and to transmit structural loads safely to a firm geological stratum of adequate bearing capacity. This is usually practiced when the soil beneath the level at which a shallow foundation would usually be laid has too weak a bearing capacity. Therefore the structural loads must be transferred to a good load-supporting strata at a greater depth by means of deep foundations such as piers, caissons, or piles. Deep foundations may be laid in open foundation pits (excavations), in dry, or below the water table.

Deep foundations are usually used in any one or a combination of the following instances:

1. where soil mechanical properties are inferior for shallow foundations;
2. where structural loads must be transmitted through weak, nonuniform, compressible soil and/or deep water to a firm soil or rock;
3. when a groundbreak or rupture of soil underneath a shallow foundation can be anticipated;
4. when the soil conditions are such that a washout and erosion of soil from underneath the base of a shallow foundation may take place;
5. when a competent or firm load-bearing soil lies at great but attainable depth consistent with safety, economy, and present-day technology; or
6. when the design of a structure calls for unequal and nonuniform loads on soil (rock).

The type of foundation to use depends on the bearing capacity of the soil beneath the foundation, not only of that at the base of the footing but also of the soils forming the underlying strata.

The commonest types of deep foundations are (1) deep piers; (2) various types of caissons; and (3) various kinds of piles.

Piers. Generally, a *pier* may be defined as a vertical, column-like engineering structure on land as well as in open water, whose purpose is to transmit safely structural load onto a deep-lying, firm, and load-bearing stratum of soil or rock covered by soft soil materials. (In bridge engineering, the term *pier* denotes an intermediate, vertical support for the adjacent ends of two bridge spans, that is, a superstructure.)

Piers are prismatic or cylindrical columns. In principle, the purpose of a pier is the same as that of a footing and/or pile, namely, to transmit structural loads to deeper strata. Also, the function of piers is basically the same as that of piles—to transfer loads to soil. Therefore, no sharp distinction between piers and piles can really be made. The main difference between piers and piles is the method of their installation into the ground (Peck, 1965; Schousboe, 1972).

Piers may be built in open excavations or open caissons. Alternatively, they can be *drilled piers,* installed in vertical holes or *shafts* from approximately 1.5 to 3 m in diameter, drilled to a depth of a firm soil-bearing strata (see *Shaft Sinking*). The holes are filled with concrete—cased or uncased. Small-diameter piers may be installed in sand by means of vibration. Piles are installed by driving, or by vibrating the stuctural member and displacing the soil. Concrete piles are prefabricated; they are also cast-in-place. A drilled pier in homogeneous soil derives its bearing capacity from a combination of the frictional resistance on the shaft (side surface) and the end-bearing resistance (Woodward et al., 1972). Pier foundations, laid on or socketed into a rock formation, are normally designed to carry and transmit heavy loads (White, 1943).

Use of Piers. Piers and caissons are utilized for laying foundations at greater depths than shallow foundations. Drilled piers are frequently used as an alternative to piles.

Piers and caissons are used (1) when the upper layers of soil are too soft or compressible; (2) when the overlying soil materials may be eroded away; or (3) if a soft soil material contains obstacles—such as sunken logs, trunks of trees, and boulders—that piles cannot penetrate.

Piers are founded on the top surface of a stratum of dense sand, gravel, or rock encountered at a considerable depth below the ground surface overlaid by a soft, compressible soil material.

Bearing Capacity of a Pier Foundation. The ultimate bearing capacity, P_{ult} of a deep foundation such as a pier (Fig. 6) may be considered to be composed of two parts, namely:

$$P_{ult} = P_B + F_S = q_{ult} \bullet A + U \bullet s \bullet D \quad (1)$$

where $P_B = q_{ult} \bullet A$ = the total ultimate soil bearing capacity underneath the entire base area A of the pier

q_{ult} = the ultimate unit-bearing capacity of the soil

F_S = the skin friction force along the embedded part *(D)* of the vertical side surface *(U • D)* of the pier

U = the perimeter of the pier

$s = \sigma_h \tan \phi_1 + c_{adh} \quad (2)$
= the average value of skin friction between the soil and the vertical mantle (side) surface of the pier at failure

D = the embedment depth of the deep foundation, namely, the pier

σ = the average horizontal (normal) pressure on the vertical side (mantle) surface of the embedded part of the pier at failure

ϕ_1 = the angle of friction between the soil and the pier material

c_{adh} = the adhesion (if any) of the soil to the foundation (pier) wall material, all in consistent units

The safe, allowable soil (rock) bearing capacity values σ_{all} are based on two considerations, namely: (1) that the factor of safety against shear failure of the soil must be adequate, and (2) that settlement under allowable soil bearing pressure should not exceed tolerable manitudes.

The allowable safe bearing capacity value, σ_{all}, to use is obtained by applying on ultimate soil bearing capacity σ_{ult} an appropriate factor of safety of η (between 2 and 3 for dead load plus normal live load) (Jumikis, 1967; Sowers, 1969):

$$\sigma_{all} = \frac{\sigma_{ult}}{\eta} \quad (3)$$

Caissons. By definition, a *caisson* (French *caisson,* box) is a watertight box or chamber used for laying foundations under water (as in harbors, rivers, and lakes), and even on land.

Caissons are utilized for laying foundations at greater depths than shallow foundations. The heavy loads of bridges and multistory structures, concentrated at certain points, can be transmitted to deep-lying, firm soil and/or rock by means of

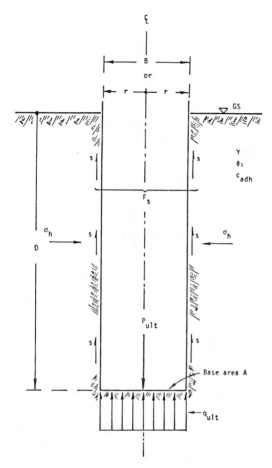

FIGURE 6. Vertical section through a pier. B = side of square pier or diameter of pier; r = radius of pier; s = skin friction; σ_h = horizontal average pressure; CL = center line; ∇GS = ground surface; γ = unit weight of soil; ϕ_1 = angle of wall friction; and c_{adh} = adhesion of soil to pier.

caissons. In open water and at shallow depth, caissons can be sunk from an artificial sand island (Ammann, 1963; Just and Obrician, 1966).

There are two principal types of caissons, open and pneumatic. Open caissons can be grouped into two subgroups, *well caissons* (top and bottom open to air), and *floating caissons* (open top and closed bottom). During sinking operations, the chamber of an open caisson is exposed to atmospheric pressure, whereas in the chamber of the pneumatic caisson there is maintained a pressure (compressed air) greater than atmospheric pressure in order to keep water and soil from entering the caisson.

Caissons may be circular, elliptical, or rectangular in plan. The fundamental differences among the types of caissons is illustrated in Figure 7.

Open Caissons. Open caissons are hollow cyl-

inders made of timber, brick, masonry, concrete, reinforced concrete, or steel. The basic elements of an open caisson are (1) its *shell* (well), also called the *shaft;* (2) a knife-edge or *cutting edge* at the lower end of the caisson; and (3) its *back-fill* (ballast).

The purpose of the cutting edge is to facilitate the sinking (penetration) of the caisson through the weak strata down to a firm, load-bearing soil or rock. Also, it strengthens the lower part of the well (shaft).

An open caisson is sunk into soil by its own weight plus that of its next build-up (this assists sinking of the caisson as the excavation proceeds). Alternatively, a dead weight can be placed on top of the caisson and then soil excavated through the open shaft of the caisson through water by means of a clamshell or grab bucket. On removing the soil from underneath the cutting edge, the caisson sinks deeper into the soil. The concrete, as the ballast, is filled into the open caisson under water by the *tremie method* (see Fig. 8), or by the use of special bags.

A tremie, or an elephant trunk, is used for placing large quantities of plastic concrete under-water. The tremie consists of a large metal pipe or tube about 30 cm or more in diameter at the top with a hopper at the top end, and flares out slightly at the bottom. It must be long enough to reach the bottom of the excavation. The tube can be closed or opened at its lower end (the submerged end) by a remote control. The tremie is filled with concrete by means of belt conveyors or pumps from concrete mixing plants.

When concrete is placed in a caisson under water, the tremie is always kept full with plastic concrete with the lower end of the tremie immersed in the concrete just deposited. The concrete slides down in the tube by gravity, and is kept below the surface of the already deposited concrete. Thus the plastic mass of concrete is pressed laterally and upward from within the already deposited mass of concrete (Fig. 8).

The tremie is raised as the concrete rises in the caisson. To reduce side friction between the soil and the shaft upon sinking, the outside walls of caissons are sometimes tapered longitudinally with a slope of 1/15 to 1/7.5 of their height, flaring out at the bottom.

Some advantages of open caissons as compared with pneumatic caissons are: (1) they can be sunk more quickly; (2) small obstructions encountered during sinking can be removed by divers; (3) they can penetrate to a great, practically unlimited depth; and (4) they are more economical. Therefore, as a kind of deep foundation, the open caisson—as compared with all other types of deep foundations—is superior not merely technically but also economically.

Open caissons do, however, have disadvantag-

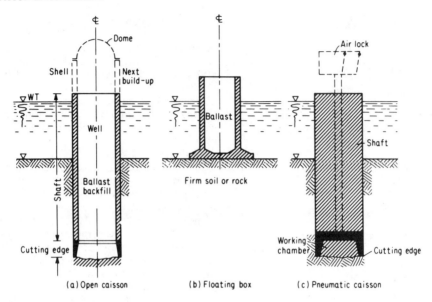

FIGURE 7. Types of caissons (from Jumikis, 1971, p. 482; copyright © 1971 by Harper & Row, Publishers, Inc., by permission of the publisher).

es: (1) ineffectiveness where soil enters into the excavation more rapidly than it can be dredged out (in coping with such conditions, the pneumatic caisson—for depths corresponding to less than 4 atm pressure, which is the endurance limit of humans working under the pressure of compressed air—is more effective than the open caisson); (2) the difficulty of cleaning at the bottom of caissons through water; (3) problems in placing a concrete seal in the caisson through water; (4) delay in sinking when obstacles to penetration are encountered in soil; and (5) lessened control and less accurate positioning of caissons as compared to pneumatic caissons, nor as accurately positioned. This last disadvantage pertains especially to the difficulty of pitching the caisson in position on a submerged site. Generally, the larger the caisson, the easier it is to keep plumb and the more accurately it can be landed. To keep small, open caissons in plumb, they must be shored.

Depending on the size and weight of open caissons, and the load they have to sustain, cutting edges are made of timber or angles and plates of structural steel, reinforced concrete, or a combination of timber and steel, or reinforced concrete and steel (Fig. 9).

Other types of open caissons are (1) belled-out caissons (Fig. 10), a caisson with an enlarged base; (2) step caissons (Fig. 11), which have telescoping steel cylinders as the shaft lining; (3) Chicago caissons, a large-diameter shaft with vertical timber sheeting sunk in increments as bracing, and additional sheeting installed; (4) inclined caissons; and (5) special caissons. All caissons, however, have a common element, namely, a permanent shaft, which is an integral part of the foundations of the bridge, building, or hydraulic structure.

By the terms caisson, drilled-in-caisson, and bored pile some practicing foundation engineers frequently denote drilled pier foundations.

Large, rectangular open caissons are provided with circular dredging wells. Figure 12 shows the 6.10-m (20-ft) high cutting edge and 4 of 28 dredging wells 5.48 m (18 ft) in diameter of an all-walled caisson that is 46.02 m × 26.81 m (151 ft × 88 ft) and weighs 907.18 metric tons or 1,000 short tons. The caisson—fabricated at Dravo Corporation's Neville Island plant, near Pittsburgh—formed the base of Pier 2 of the Mississippi River Bridge at New Orleans, built by the company in the mid-1950s. The caisson rests on a stratum of hard clay 54.86 m (180 ft) below water level, and was placed on a fascine mattress to prevent scour around the bridge pier (Jumikis, 1971). The open-caisson suspension cable support pier No. 18 of the Mackinac Straits Bridge has 21 dredging wells, each 2.74 m (9 ft) in diameter.

A large, circular open caisson for the North Main Tower pier of the Mackinac straits Bridge is shown in Figure 13. This caisson, 35.36 m (16 ft) in outer diameter, is formed by two concentric shafts. The inner space, 26.21 m (86 ft) in diameter, is thus one huge dredging well. This pier was sunk 60.96 m (200 ft) below the water surface to bedrock.

Open rectangular caissons with circular cells over square-cell dredging wells were used for the construction of the tower piers of the Verrazano-Narrows Bridge in New York City. The caissons, 69.80 m × 38.41 m (229 ft × 129 ft) in plan (Ammann, 1963; Just and Obrician, 1966), were

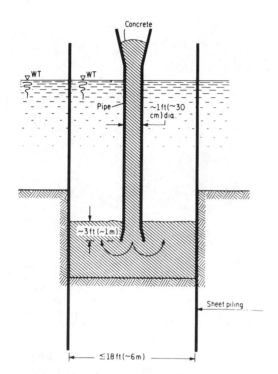

FIGURE 8. Underwater concreting (tremie method) (from Jumikis, 1971, p. 193; copyright © 1971 by Harper & Row, Publishers, Inc., by permission of the publisher).

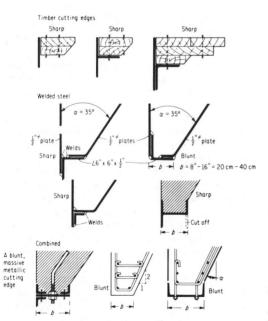

FIGURE 9. Cutting edges. α = angle of the inside slope of the cutting edge; b = width of the blunt cutting edge (from Jumikis, 1971, p. 490; copyright © 1971 by Harper & Row, Publishers, Inc., by permission of the publisher).

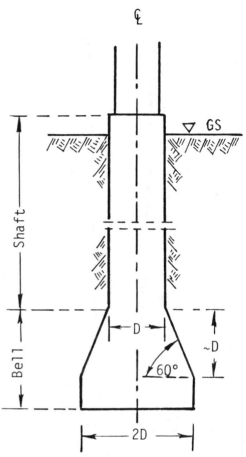

FIGURE 10. Belled-out caisson. D = diameter of caisson.

sunk to the required elevation by dredging the overburden with clamshell buckets through the dredging wells of the caisson, which were open top and bottom. Later the bottom of the caisson was sealed with tremie concrete, 6.10 m and 6.71 m (20 ft and 22 ft) thick in the Staten Island and Brooklyn caissons, respectively. The wells were then closed at the top with precast concrete covers. The wells were left filled with water. The caissons were sunk 32.81 m (107.65 ft) below the water surface at the Staten Island side, and 52.62 m (172.65 ft) deep below the water surface at the Brooklyn side. The walls of the caissons are 1.52 m (5 ft) thick. The sinking of the caissons was started out from artificial "sand islands." The sand-island fill was enclosed and kept by a series of sand-filled cellular sheet-pile cofferdams, and was kept dry by a wellpoint dewatering system.

Sometimes the ordinary dredging wells encountered in all open caissons are capped with removable domes to allow compressed air pressure to be

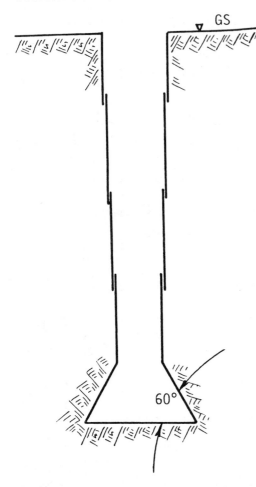

FIGURE 13. Dredging of bottom sediment and sinking of the circular caisson for the North Main Tower of the Mackinac Straits Bridge to bedrock 60.96 m (200 ft) below the water surface (from Jumikis, 1971 p. 496; courtesy and permission of Steinman, Boynton, Gronquist & Birdsall, Consulting Engineers, New York).

FIGURE 11. Step caisson constructed of telescoping steel cylinders. (Gow method. Note belled-out excavation at base.)

put in the wells. This aids the floating, sinking, and stabilization operation of the caisson by increasing or decreasing the pressure in selected groups of well cylinders. Also, this method allows easy conversion into pneumatic-type caissons. Such dome-capped cylinder caissons were used for the foundations of the San Francisco-Oakland Bay Bridge piers (Purcell et al., 1934). This type of caisson is credited to D. E. Moran (Purcell et al., 1934); see Figure 14.

Another conspicuous use of dome-capped caissons was in the construction of the foundations and piers of the Tagus River Bridge in Lisbon, Portugal (Riggs, 1966; see Fig. 15). There the sinking depth was 79.24 m (260 ft) below the water surface to bedrock, or 5.49 m (18 ft) deeper than one of the main tower piers of the San Francisco-Oakland Bay Bridge.

Floating caissons, sometimes called *floating box caissons,* have closed bottoms and open tops. Because the floating caisson does not penetrate the soil but rests on it, this kind of caisson usually has a wide base and is therefore suitable for floating.

A floating caisson can be quickly and conveniently constructed on land (on slips, in docks, on floating barges, or on sand islands). It is relatively inexpensive to transport such caissons by floating them to the site. Because their parts are prefabricated, floating caissons enjoy the indisputable advantage of quality construction.

Floating caissons may be constructed of timber, concrete, reinforced concrete, steel, or a combination of these materials. Floating caissons are used where quick progress in construction is desired, or where, with great depth of water, great resis-

FIGURE 12. All-welded steel cutting edge (from Jumikis, 1971, p. 494; courtesy of Dravo Corporation).

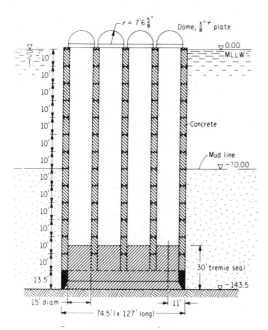

FIGURE 14. Moran's dome-capped caisson (from Jumikis, 1971, p. 497; copyright © 1971 by Harper & Row, Publishers, Inc., by permission of the publisher).

FIGURE 15. Dome-capped caisson for the Tagus River Bridge at Lisbon, Portugal (from Jumikis, 1971, p. 498; courtesy and permission of Steinman, Boynton, Gronquist & Birdsall, Consulting Engineers, New York).

tance to horizontal forces (wave action, hydrostatic pressure, earth pressure) is necessary, such as in the construction of harbor piers, moles (breakwaters), wharves, quays, or bridge piers. They are also used for closures in dike construction, and are suitable for laying foundations in deep-lying strata, on gravel strata, and where no danger of scour exists.

The main advantages of a floating caisson are (1) its quick and convenient construction; (2) prefabrication possibilities; (3) inexpensive transport on water; and (4) its relatively small construc-

tion cost. The main disadvantages are (1) that it is too costly for deep excavations, and (2) that it is necessary to prepare a level soil or rock surface for supporting the caisson.

Pneumatic Caissons. A *pneumatic,* or compressed-air-pressure, caisson is popularly described as a rigid, inverted box with the bottom omitted (Fig. 16). The presence of compressed air keeps the box free of water and prevents the influx of soil and mud during the sinking. In this pneumatic box, soil excavation work and the pouring of concrete is carried out in dry, under compressed air. The caisson is sunk as the excavation proceeds, and becomes a permanent part of the monolith of a deep foundation.

The space where workers enter and work under compressed air is called the *working chamber.* The least pressure p_a necesssary to keep the water out of the working chamber is calculated as $p_a = \gamma_w H,$ where γ_w = the unit weight of water and H = the pressure head of water from its surface to

Pneumatic caissons are used for building deep foundations for bridge piers and abutments, lighthouses, wharves, quays and other waterfront facilities, building foundations, hydraulic structures, and underwater tunnels. Pneumatic caissons are also used in open water, when there is a great influx of water, when difficult obstructions in soil (tree trunks, boulders), can be anticipated, or when other types of caissons and foundations are infea-

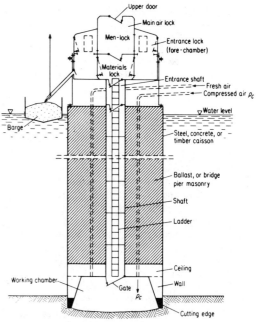

FIGURE 16. Sketch of a pneumatic caisson (from Jumikis, 1971, p. 548; copyright © 1971 by Harper & Row, Publishers, Inc., by permission of the publisher).

sible because of danger of scour or erosion. In these repects, the pneumatic caisson is the *ultima ratio* for the execution of a deep foundation.

The pneumatic caisson is used for average depths between about 12.19 m and 33. 53 m (40 ft and 110 ft). The maximum depth of a pneumatic caisson is dictated by the endurance limit of man in working under compressed air pressure, which is 3 to 4 atm, or about 30.48 m to 36. 58 m (100 ft to 120 ft) under water. (Work in a pneumatic caisson under compressed air pressure is regulated by special labor laws, and requires the supervision of health of the workers by a special physician [Seelye, 1956].)

Among the advantages of pneumatic caissons are the following:

1. The bottom of the caisson is easily accessible.
2. Obstructions under the cutting edge encountered in sinking a pneumatic caisson are relatively easily and quickly removed.
3. Excavation and pouring of concrete can be readily done in the dry.
4. At the design elevation, the soil in the working chamber can be conveniently inspected.
5. Soil samples can be taken.
6. The soil bearing capacity in the working chamber can be determined by the method of loading plates and jacks.
7. The working chamber can be concreted in the dry.
8. Sinking the caisson brings about no vibrations.
9. The position of the groundwater table remains unaltered; hence no settlement of the caisson or adjacent structures occurs because of the groundwater.
10. The construction site requires less steel than a sheet-pile foundation pit, and less equipment (cranes, excavators) than in the case of open caissons.

The following are some of the disadvantages of pneumatic caissons:

1. It is inconvenient to work in the caisson under compressed air pressure.
2. The penetration depth of the pneumatic caisson below the water table is limited to man's endurance limit of working under pressure of 3 to 4 atm.
3. Construction costs for wages for work under pressure are higher than costs incurred for installing open caissons.
4. A great deal of manual work is involved.
5. Caisson disease.
6. Special labor laws must be observed.
7. Work in a caisson can be dangerous during seismic activities.

Caisson Disease. Caisson disease is a condition resulting from a too rapid decompression in atmospheric pressure after a stay in a compressed atmosphere, for example, as in a pneumatic caisson. Too rapid change in atmospheric pressure from high to normal cause nitrogen bubbles to form in the blood and body tissues. Caisson disease is marked by neuralgic pains and paralysis, and sometimes by fatal disorders in the human body. The *bends* and *aeroembolism* are some of the designations of caisson disease.

Upon leaving the caisson too fast, or with rapid decompression, air is liberated in the human body. While oxygen is readily absorbed by the blood, the nitrogen is not. Nitrogen must be expelled or it forms bubbles, thus impeding circulation of blood, muscular movements, and the senses. The nitrogen bubbles may lodge in the heart, brain, or spinal column; or enlarged bubbles expand and may rupture blood vessels, resulting in paralysis or even death. If the human body is decompressed slowly, the nitrogen gas does not form bubbles, but escapes gradually and harmlessly from the surface of the lungs. Length of time for work in caissons under air pressure is reduced considerably to avoid health problems.

Pile Foundations

A *pile* is a relatively long, slender columnar construction whose cross-sectional size is small compared to its length. It is embedded into soil to receive, support, and transmit vertical, horizontal, and inclined loads into a firm soil or to rock below the ground surface at an economically feasible depth in such a way that these soil strata (or rock) can sustain the loads without causing intolerable settlements to the soil (rock) and structure. Piles are made of wood, precast and cast-in-place concrete, reinforced concrete, metal (H-piles or hollow piles), or a combination thereof.

Pile foundations are among the oldest kinds of foundations. Timber piles were used to support the (albeit primitive) structures of the lake dwellers, and to transmit these loads to firm soil, and ancient bridge builders succeeded in building bridge piers on piles of considerable length and depth, thus avoiding difficult cofferdam work, pumping problems, erosion, scour, and other problems associated with laying of foundations in river beds. Julius Caesar's *Commentaries* mention bridges founded on piles (across the Rhine), which were built by Roman legions prior to the Christian era.

It is interesting to note that in Italy (specifically, Venice), the Netherlands, and in many other countries, wooden piles were and still are used for foundations.

Depending on the mode in which piles transfer loads to the soil, piles are usually classed into end- or point-bearing piles, and friction or floating piles (Fig. 17). *End-bearing piles* derive their support mainly from the underlying firm layer of soil

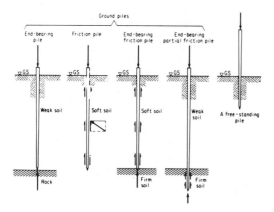

FIGURE 17. Types of piles (from Jumikis, 1971; copyright © 1971 by Harper & Row, Publishers, Inc., by permission of the publisher.)

(rock). These piles transmit their loads through their bottom tips. *Friction* piles derive most or all of their load support mainly from side (skin) friction and cohesion developed between the side surface of the pile (shaft) and the soil through which the pile is driven.

By the geometric form of their longitudinal profile, piles may be cylindrical, tapered (conical), and prismatic. The form of transverse cross-sections of piles may be square, circular, octagonal, hexagonal, or triangular. (See Fig. 18).

Some of the main functions of piles are (1) to support superimposed structural loads; (2) to transfer structural loads to layers of soil or rock of good bearing capacity, and below the extent of scour and erosion where this applies; (3) to assist and

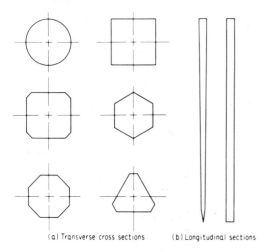

(a) Transverse cross sections   (b) Longitudinal sections

FIGURE 18. Cross-sections of piles (from Jumikis, 1971; copyright © 1971 by Harper & Row, Publishers, Inc., by permission of the publisher.)

transfer lateral loads such as those from structures, soil, water, wind, ice, and lateral impact, as well as against uplift by water; (4) to densify the soil to increase its bearing capacity; (5) to form a sheet-pile wall against the influx of water, dry running sand, and/or plastic soil; (6) to reduce excessive settlement of the structure.

Many types of piles are used today. Almost every pile manufacturer has its own patented or trademarked pile, and many patented methods of pile-making techniques, and they are too numerous for all to be treated in this entry. Detailed description of the various types of piles and pile-driving equipment may be read from the pile manufacturers' leaflets, flyers, brochures, and catalogs (see also Grand, 1970).

Pile foundations are generally used (1) where a shallow foundation is out of the question because it would be too large or because intolerable settlement is expected; (2) where soil erosion around and underneath an ordinary foundation can be anticipated; (3) where inclined forces of considerable magnitude act on the foundation; and (4) where soil density—that is, bearing capacity— must be increased.

Piles are used especially for laying foundations of civil and industrial engineering structures, bridge piers and abutments, earth retaining walls, machine foundations, locks, docks, sluices, piers, landings, and lighthouses; and for bracing and strutting excavation walls and other hydraulic and waterfront structures. All in all, it is sound to say that piles should be used as foundation elements only when it is uneconomical and infeasible to construct ordinary kinds of foundations directly supported on the soil (Chellis, 1961).

Some of the factors affecting pile foundations are the following: (1) unimpregnated timber piles must remain below the lowest level of the fluctuating water level to prevent decay; (2) in some harbors, marine borers attack wood piles; (3) consolidation of compressible soil layers below the pile-bearing stratum will cause the pile foundation to settle; (4) the pile-bearing stratum must be checked to see whether it will support the entire pile system; and (5) consolidation of soil layers along the pile due to surcharge on the ground surface (fill on clay, for example, and partly due to self-weight of the clay) causes negative side (skin) friction or down-drag on the pile. The magnitude of the load on piles from negative skin friction may approach or even exceed the load transferred from the superstructure.

Bearing Capacity of a Simple Pile. One of the most important factors in the design and construction of a pile foundation is its bearing capacity—the maximum load a pile can sustain by soil resistance. The bearing capacity of pile foundations is estimated based either on the bearing capacity of a single pile, or on that of a group of piles.

The *Manual of Engineering Practice No. 27* (American Society of Civil Engineers, 1946) defines the term *bearing capacity* as "that load which can be sustained by a pile foundation without producing objectionable settlement or material movement—initial or progressive—resulting in damage to the structure or interfering with its use."

The *ultimate bearing capacity* of a pile is the load at which the pile begins to penetrate into the soil without increase in load. According to some specifications, the ultimate load on a pile is that which brings about the settlement of the pile to an amount greater than 10 percent of the diameter of the pile.

A *safe load* is a load less than that which would bring about objectionable settlement. This load includes a load factor, or factor of safety, accounting for variation of some irregularities in pile material, workmanship, and loading.

The *allowable* or *working load* on any vertical or batter (inclined) pile, applied concentrically along its axis, is that which is safe with regard to ultimate bearing capacity, negative skin friction, the allowable load based on penetration, pile spacing, overall bearing capacity of the soil below the tips of the piles, and the allowable settlement demonstrated by load test divided by a factor of safety, or the basic maximum load prescribed by the applicable pertinent design standard or building code.

The bearing capacity of a pile depends on (1) the type and properties of the soil; (2) the surface and/or groundwater regimen; (3) the geometry of the pile (solid, hollow, rectangular, straight, or tapered); (4) the pile material (timber, concrete, steel); (5) the size of the pile (cross-section, length); (6) the roughness or smoothness of the side surface of the pile; (7) the driving depth of the pile; (8) the method of embedding the pile into the soil (driving, jacking, jetting, vibrating, casting in place); (9) the position of the pile (vertical or inclined); and (10) the spacing of piles in a pile group.

There are three main methods of determining the bearing capacity of a single pile: (1) by a static loading test in the field (one of the most reliable methods, but very expensive; (2) analytically, by means of static pile bearing-capacity formulas; and (3) by a dynamic pile-driving test *in situ*.

Static Pile Bearing-Capacity Formulas. These formulas are based on soil mechanics principles. If safe, allowable soil frictional (shear) stress values τ_{ave} on a slender pile are used, then the allowable load P_{all} on the slender pile is calculated approximately as

$$P_{all} \leq R_t + R_f = \underbrace{(\sigma_{all} + \gamma_{ave}L_c)A}_{\substack{\text{tip resistance} \\ R_t}} + \underbrace{UL_c\tau_{ave}}_{\substack{\text{side friction} \\ R_f}} \quad (4)$$

where σ_{all} = the allowable soil bearing capacity at the depth of the tip of the pile below the ground surface

γ_{ave} = the weighted average unit weight of the soil

L_c = the computation length of the pile

A = the cross-sectional area of the pile

$U = \pi d_{ave}$ = the perimeter of the cross-section of the pile

d_{ave} = the weighted average of the diameter of the pile

τ_{ave} = the average weighted specific frictional (shear) stress on the pile

The bearing capacity P of a bulbed-end pile is the soil reaction R on the bulb (Fig. 19):

$$P = R = \sigma_{all} A = \frac{\pi}{4} \sigma_{all} D^2 \quad (5)$$

where σ_{all} = the allowable soil bearing capacity and D = the diameter of the horizontal projection of the bulb (perpendicular to R; see Fig. 19).

Among the commonly known static pile bearing-capacity formulas (Bénabenq, 1921; Dörr, 1922) are those by Vierendeel (1927) and Krey and

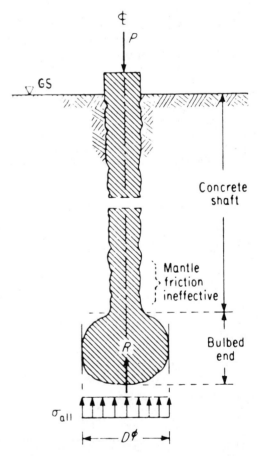

FIGURE 19. Forces on a bulbed-end pile (from Jumikis, 1971; copyright © 1971 by Harper & Row, Publishers, Inc., by permission of the publisher.)

Ehrenberg (1936). These formulas, based on earth pressure theory, consider pile mantle (side) friction.

With *Vierendeel's formula* (1927-1928) the pile bearing capacity P is determined by test data of the soil and soil/pile properties:

$$P = (\tfrac{1}{2})\mu_1\gamma\pi L_c^2 \tan^2(45° + \phi/2) \qquad (6)$$

where μ_1 = the coefficient of friction between the soil and the pile.
γ = the unit weight of the soil
d = the diameter of the pile
L_c = the computation length of the pile
ϕ = the angle of internal friction of the soil.

For cast-in-place piles and driven timber and concrete piles with rough side surfaces, Vierendeel recommends using $\mu_1 = 0.33$. For all other types of piles in wet and plastic soils, $\mu_1 = 0.25$. The formula implies static friction piles.

With *Krey's Formula*, the static bearing capacity P of a pile is the sum of the tip resistance R_t and that of the side friction R_f between the soil and the pile:

$$P = R_t + R_f = \gamma_a L_c \tan^2(45° + \phi_1/2) \qquad (7)$$

$$K_p + (\tfrac{1}{2})\,\gamma\mu_1 U L_c^2 K_p$$

where γ = the unit weight of the soil
a = the cross-sectional area of the pile
L_c = the computation length of the pile
ϕ_1 = the angle of side friction between the soil and the pile material
K_p = $\tan^2(45° + \phi/2)$
= the coefficient of earth resistance (= passive earth pressure)
ϕ = the angle of internal friction of the soil
μ_1 = $\tan\phi_1$
= the coefficient of friction between the soil and the pile material
U = the perimeter of the pile

Dynamic Pile-Driving Formulas. Using the data obtained from the pile-driving tests, such as the driving resistance to penetration of the pile into the soil, and other pertinent data, the pile bearing capacity is then calculated by means of the dynamic pile-driving formulas (Chellis, 1961). These formulas are derived either by means of the work-energy relationship or by the impact-momentum theory. The bearing capacity of the pile is thus calculated by equating the energy necessary to drive the pile to the mechanical work done by the pile in penetrating the soil. Also, this in its turn means that the bearing capacity of the pile depends not only on the strength of the pile material itself but also on the strength of the pile-supporting soil.

The dynamic pile-driving formulas—there are several of them—consider the weight and fall of the pile-driving hammer and the number of blows on the pile to drive it 1 in.

After a succession of blows from a falling weight of the pile-driving hammer, the pile penetrates into the soil. Theoretically, the kinetic energy E of the falling hammer for performing work may be written as

$$E = Wh - \Sigma \text{ (energy losses)} \qquad (8)$$

where W = the weight of the falling hammer
h = the height of fall of the hammer
Wh = mechanical work

Figure 20 illustrates the concept of pile driving. When the pile is struck, there occur considerable losses of energy (friction, heat, temporary elastic deformation of the pile-driving cap, the soil, elastic compression of the pile, hammer rebound, vibration, and other possible losses) that reduce the available energy for performing useful work.

It is assumed that the resistance to penetration of the pile into soil may be considered a single force of an average resistance R, and that with each stroke (blow) of the hammer the pile penetrates a distance s against R. Thus the mechanical

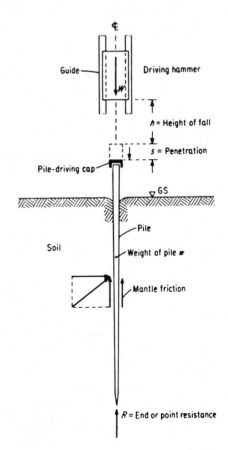

FIGURE 20. Concept of pile driving (from Jumikis, 1971; copyright © 1971 by Harper & Row, Publishers, Inc., by permission of the publisher.)

work performed is Rs. Here s is called the *set of penetration* of the pile per blow. Sometimes s is recorded in number of blows per 1 in. of penetration.

The safe bearing capacity P of the pile is calculated as

$$P = \frac{R}{\eta} \qquad (9)$$

where η is a factor of safety, usually taken from $\eta = 6$ to $\eta = 8$.

For a description of the various dynamic pile-driving formulas, see American Society of Civil Engineers (1946), Chellis (1951), Cummings (1940) Gates (1957), Jumikis (1962, 1971) and Michigan State Highway Commission (1965). Cummings (1940) reviewed the various pile-driving formulas and found that some of them contained questionable assumptions, and that dynamic theories are mixed with static ones. All these formulas assume that the dynamic resistance of the soil to pile driving is equal to the static bearing capacity of the pile. This is an assumption with no theoretical justification. Therefore, in general, dynamic pile-driving formulas are unreliable.

If, however, their estimates can be examined in light of past experience on similar sites or in conjunction with the results of carefully performed loading tests, dynamic pile-driving formulas may serve as a guide in estimating reasonably safe and uniform results over the entire pile-driving operation on a construction site with reasonably uniform soil conditions (cohesionless soils).

One of the simpler, and most commonly used, dynamic pile-driving formulas for estimating the static load-bearing capacity of a driven pile is the so-called *Engineering News formula* published in 1888 by A. M. Wellington of *Engineering News*. This formula allegedly gives the safe bearing capacity P of a pile for a presumed factor of safety of $\eta = 6$ and for a drop hammer (to be a free-falling body) as

$$P = \frac{Wh}{\eta(s + 1.0)} \qquad (10)$$

where $s =$ the average penetration of the pile, per stroke, for the last five strokes of a drop hammer. This formula is obtained for a weightless pile, and is extensively used in the United States.

The modified *Engineering News formula* (Michigan State Highway Commission, 1965) is a variation of the original *Engineering News* formula, and contains a nominal factor of safety $\eta = 6.0$:

$$R = \frac{2Wh}{s + 0.1} \cdot \frac{W + e^2 w}{W + w} \qquad (11)$$

where $w =$ the weight of the pile,

$e =$ the coefficient of restitution,* and all other symbols are the same as before. For timber, $e \approx 0.2$; for metal piles, $e > 0.5$; for a perfect elastic blow, $e = 1.0$; and for a perfect inelastic blows, $e = 0$.

During the course of time, engineers learned that dynamic pile-driving formulas are of limited value in piled foundation work mainly because the dynamic resistance of soil does not represent its static resistance, and because often the results obtained from the use of dynamic equations are of questionable dependability.

Gates's formula (Gates, 1957) is a strictly empirical relationship between hammer energy, final set, and measured design test load with a safety factor of $= 3.0$:

$$R = (2000) \, (\tfrac{1}{7}) \sqrt{E_n} \, (\log \frac{s}{10}) \qquad (12)$$

where $R =$ the dynamic resistance of the soil (i.e., the ultimate bearing capacity of the pile in soil) and $E_n =$ the manufacturer's maximum rated energy, in ft-lb.

The *Gates relationship,* equation (12), does not have rational limits and does not apply to "refusal" conditions when $s \to 0$.

Michigan Studies on Piles. One of the purposes of the so-called Michigan research on piles in the 1960s was "to determine the correlation between bearing capacity of the load-tested piles and estimated pile bearing capacity as obtained by [eleven] selected pile-driving formulas."

In its final report (1965), the Michigan State Highway Commission studied the following eleven dynamic pile-driving formulas:

1. *Engineering News* (EN)
2. Hiley (extensively used in Great Britain)
3. Pacific Coast Uniform Building Code (PCUBC)
4. Redtenbacher
5. Eytelwein
6. Navy-McKay
7. Rankine
8. Canadian National Building Code (CNBC)
9. Modified Engineering News (Modified EN)
10. Gates
11. Rabe

These formulas may be found enumerated in Chellis (1961, 525-538).

The report cautions that it is the consensus of informed engineers that no one dynamic formula, relating dynamic to static resistance, affords a reliable means of estimating the long-time bearing capacity of piles in general.

---

*The constant e, which is the ratio of the relative velocity of two elastic spheres after direct impact to that before impact can vary from 0 to 1, with 1 equivalent to an elastic collision and 0 equivalent to a perfectly elastic collision.

For the specific conditions considered, some of the more important and pertinent findings were the following:

1. In several instances the design capacities of the *Engineering News,* Navy-McKay, and Rankine formulas had true safety factors of less than unity with respect to test load capacity. Design capacity or design load is the load that the pile is intended to carry without excessive movement and with an acceptable factor of safety against plunging failure (Deep Foundation Institute, 1981).
2. In several instances (Hiley, PCUBC, Redtenbacher, CNBC), formula design capacities had a true safety factor of 9 or more with respect to test load capacity.
3. In general, the modified *Engineering News* and Gates's formulas gave design capacity values with true safety factors substantially all falling in the range of 1.5 to 6 with respect to associated test load capacities.

On the basis of the aforementioned comprehensive studies of field tests and their results in regard to determining the load-bearing criteria for driven piles, the Michigan State Highway Department adopted the modified *Engineering News* formula for specifications and job control.

Embedment of Piles by Vibration. Piles, sheet piles, and caissons may also be embedded into soil by vibration. Vibration is less noisy than impact driving, and it creates and transmits no shock in the soil. The latter fact may be critical where pile work must be performed in a congested area.

Vibration of piles is an efficient, expedient method of embedment in foundation engineering. The ultimate load R_u that a vibrated pile will support may be calculated by the following equation as derived by Davisson (1970) for the Bodine Resonant Driver (BRD) as

$$R_u = \frac{(550)\,N + Wr_p}{r_p + fs_L} \tag{13}$$

where N = the number of horsepowers delivered to pile, in HP

W = 22,000 lb = weight of BRD-1,000 type of vibrator

r = final rate of pile penetration, in ft/sec

f = frequency, cycles/sec

s_L = loss factor, ft/cycle, to be determined empirically (for closed-end pipe and H-piles in medium-dense sand or sand and gravel, $s_L = 0.025$ ft/cycle)

Fig. 21 shows a two-mass vibrator attached to a steel pile; Fig. 22 shows determination of forces in piles by Culmann's graphical method.

Basically in Culmann's graphical method the externally applied resultant force R (drawn to a

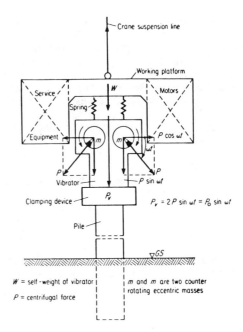

FIGURE 21. Two-mass vibrator attached to a steel pile (from Jumikis, 1971; copyright © 1971 by Harper & Row, Publishers, Inc., by permission of the publisher.)

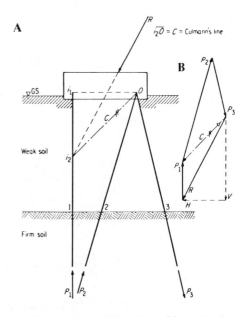

FIGURE 22. Determination of forces in piles by Culmann's graphical method (from Jumikis, 1971; copyright © 1971 by Harper & Row, Publishers, Inc., by permission of the publisher.)

201

force scale; see Fig. 22*b*) is resolved graphically into its components P_1, P_2, P_3 parallel to the directions of piles 1, 2, and 3. The forces in the piles, P_1, P_2, and P_3, are then scaled off from the force polygon in Fig. 22*b*.

The following is the procedure to determine pile loads by Culmann's graphical method (Fig. 22):

1. Intersect the externally applied resultant load R with pile 1 to obtain point of intersection i_2.
2. Connect point i_2 with point O by means of the so-called Culmann's line C.
3. contruct the force polygon as shown in Fig. 22*b* and scale of the pile load P_1.
4. Resolve (force) C into components P_2 and P_3, and scale off the pile loads P_2 and P_3 from the force polygon.

Now that the three forces, C, P_2, and P_3 pass through a common point of intersection O (system's center), these three forces are in static equilibrium.

Settlement of a Pile Group. The settlement analysis of a pile foundation, such as shown in Figure 23, may be performed along the same lines as that performed for soils under loaded bearing areas (unpiled foundations) resting directly on soil, and as dealt with in soil mechanics. The settlement of a pile group depends mainly on the compressibility of the soil below the tips of the piles. It is assumed that the soil between the piles in the pile group does not deform. The stress distribution in soil below the tips of the piles is assumed to take place either according to Boussinesq (1885), or according to the usual straight line (Fig. 19), whichever method is used. The depth of the stressed zone ($H \approx (1.5) B$) below the plane of the tips (Fig. 23) is assumed to be the same as that used in settlement analysis of foundations without piles.

The settlement of a pile group is larger than that

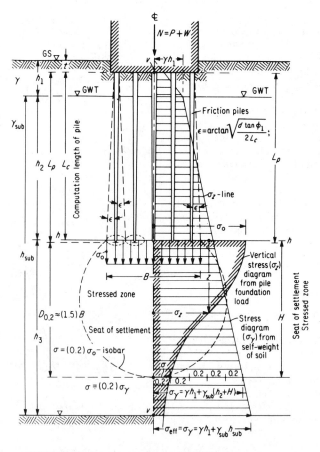

FIGURE 23. Load-settlement system of a pile group of a floating pile foundation (from Jumikis, 1971; copyright © 1971 by Harper & Row, Publishers, Inc., by permission of the publisher.)

of a single pile because of pressure overlap from adjacent piles. Furthermore, with equal loads on piles, the settlement of a pile group increases with an increase in the number of piles in the group. With an increase of spacing between piles in a pile group, its settlement decreases, approaching the settlement of a single pile. With a spacing of about 6 diameters, the settlement of a pile group is approximately equal to the single pile. The soil-foundation system must also be checked for the strength of the soil bearing capacity. If the shear strength of the soil should become exhausted, a groundbreak in the soil would occur and might cause the piled structure to collapse. The stability analyses of soil-pile foundation systems are performed by the same methods as used in soil mechanics for unpiled foundations.

Special Foundations

Generally *special foundations* are foundation systems, the analyses, design, construction, and function of which differ more or less from the conventional foundations described previously. Special foundations pertain to their design and construction under difficult environmental conditions (Zeevaert, 1972).

Design of special foundation systems differ from conventional design in accordance with the nature and geology at the site, the degree of complexity of the structure to be built, and the kind of construction materials and machinery available. Some foundations may require a costly amelioration of deficient foundation soils and/or rocks. Also, some designs of special foundations may require the use of advanced and even complex theories and more extensive physical-mathematical-engineering analyses than designs of conventional foundation (beams on elastic support; plastic bending of beams and plates, for example).

To the group of special foundations the following may belong:

1. foundations for atomic power plants (sensitive to settlement);
2. foundations for parabolic dishlike telecommunications antennas (sensitive to differential settlement);
3. foundations for large mirror-type solar energy collector stations;
4. foundations for tall structures (smokestacks, lighthouses, electric transmission line towers, radio and television transmission towers);
5. foundations for subsurface and overland pipe lines;
6. foundations for port, marine, internal waterway, and offshore structures (uplift, wave action, erosion);
7. foundations for various hydraulic structures such as locks, docks, massive dams, and hydraulic power plants;
8. foundations that are subjected to dynamic loads, such as machine foundations (forge hammers and vibrations) (Sheta and Novak, 1982);
9. foundations in seismic regions;
10. foundations on elastic support;
11. foundations for vehicular and aqueous tunnels;
12. foundations for structures in permafrost and on frozen sites;
13. foundations for other kinds of structures of specific nature (for refrigerated structures; structures founded on expansive soils, for example);
14. foundations for funicular towers and suspension bridges;
15. foundations for tall, high-voltage transmission towers;
16. foundations for fortifications and air-raid shelters;
17. foundations for high-rise buildings; and
18. foundations for underpinning of structures, and other possible applications.

Choosing the Kind of Foundation

The choice of a particular kind of foundation depends on the character of the soil and the presence of water at the site. The compilations in Table 1 may serve as a guide for tentatively choosing, based on preliminary soil exploration, a kind of foundation suitable for various soil and water conditions at the site. The kind of foundation chosen dictates, then, the nature of a more detailed site exploration, the kind of soil-sampling program to follow, and the kinds of soil tests to be performed.

The contents of Table 1 are not to be construed as "cookbook" recipes. The table merely may help in narrowing down the choice of a type of foundation, eliminating those that will not be suitable in solving a particular foundation problem. Usually there will be several acceptable types of foundations, and all may render a correct solution depending on the designer's approach consistent with safety and economy. It is also important to remember that no foundation engineering problem is exactly like another, and that therefore each foundation engineering problem must be treated and solved according to its own requirements and merits. In other words, there are no set, rigid rules available for choosing the "right" kind of foundation.

ALFREDS RICHARDS JUMIKIS

TABLE 1. Suitable Kinds of Foundations for Various Site Conditions

Site Conditions		Suitable Kinds of Foundations	Nature of Excavation and Coping with Water
Soil	Water Conditions Relative to Excavation		
1. Firm soil near the ground surface or at moderate depth	a. Open, dry excavation	Direct, shallow foundations	Open excavation with vertical or sloped walls
	b. Excavation through open water:	Individual footings, combined footings, strip footings, mats	
	1. Shallow depth of water	Direct foundations, high-piled grillage	Open excavation between sheet-piling enclosure or cofferdam
	2. Great depth of water	Caissons Long tubular piles; shafts	
2. Firm soil at attainable depth	a. Open, dry excavation and excavation with moderate influx of water	Direct foundations	Dewatering of excavation by pumping
	Seams and springs	Reinforced concrete piles; pier foundations	Sealing off springs
	b. Excavation in ground water and through open water	High-piled grillage with sheet-pile enclosure; reinforced concrete slabs	
		Caissons	Expulsion of water by compressed air
		Direct foundations between steel sheet piling or cellular cofferdams	Open excavation between sheet piling or cofferdam Lowering of ground-water table
3. Firm soil, overlying soft material	a. Open, dry excavation	Direct foundations	
	b. Open excavation with water present	Mat Friction piled	Dewatering by pumping Lowering of ground-water table
4. Weak soil overlying firm soil	a. Open, dry excavation	Bearing piles or piers	
	b. Open excavation with water present	Bearing piles or piers	
5. Thick stratum of weak soil present; firm soil unattainable	a. Open, dry excavation	Mat: reinforced concrete slab Slab grillage Soil stabilization Friction piles Concrete and bulb-end reinforced concrete piles	
	b. Construction in and through open water and ground water	Mat: piled slab grillage Caissons	Cofferdams Expulsion of water by compressed air
		Foundations in sheet-pile enclosures	Sheet-pile enclosures or cofferdams

References

American Society of Civil Engineers (ASCE), 1946. *Pile Foundations and Pile Structures.* New York: ASCE, Manuals of Engineering Practice, No. 27, 72p.

Ammann, O. H., 1963. Planning and design of the Verrazano-Narrows Bridge, *New York Acad. Sci. Trans.*, ser. II, **25**(26), 598-620.

Bénabenq, F., 1921. Résistance des pieux, *Annales Ponts et Chaussées* **91**(4), 5-67.

Borowicka, H., 1943. Über ausmittig belastete, starre Platten auf elastisch-iostropem Untergrund, *Ingenieur-Archiv* **14**(1), 1-8.

Boussinesq, J. V., 1885. *Application des potentiels à l'étude de l'equilibre et du movement des solides élastiques.* Paris: Gauthier-Villars, 721p.

Casagrande, A., 1942. Application of soil mechanics in designing building foundations, *Am. Soc. Civil Engineers Proc.* **68**, 1487-1520.

Chellis, R. D., 1961. *Pile Foundations,* 2nd ed. New York: McGraw-Hill, 704p.

Cummings, A. E., 1940. Dynamic pile driving formulas, in *Contributions to Soil Mechanics 1925-1940.* Boston, Mass.: Boston Society of Civil Engineers, 392-413.

Davisson, M. T., 1970. *BRD Vibratory Driving Formula, Foundation Facts.* New York, N.Y.: Raymond Concrete Pile Division of Raymond International, Inc., vol. 6, pp. 9-11.

Deep Foundation Institute, 1981. *Glossary of Foundation Terms.* Springfield, N..J., 62p.

Dörr, H., 1922. *Die Tragfähigkeit der Pfähle.* Berlin: Wilhelm Ernst und Sohn, 68p.

Engineering News Record, 1933. December 7, 675.

Engineering News Record, 1955. Mackinac Bridge—incredible but true. January 27, 35-44.

Fellenius, W., 1929. Jordastatiska beräkningar för vertical belastning på horisontal mark under antagande av cirkulär-cylindriska glidytor, *Teknisk Tidskrift* **5** and **6,** 57-63 and 75-80.

Fröhlich, O. K., 1934. *Druckverteilung im Baugrunde.* Vienna: Springer, 185.

Gates, R. D., 1957. Empirical formula for pile bearing capacity, *Civil Engineering* (ASCE), **27**(3), 183-184.

Grand, B. A., 1970. Types of piles: their characteristics and general use, in *Pile Foundations, Highway Research Rec. 333.* Washington, D.C.: Highway Research Board, Division of Engineering, National Research Council, National Academy of Sciences-National Academy of Engineering, 3-32.

Gwilt, J. (transl.), 1874. *The Ten Books of Vitruvius.* London: Lockwood, Book III, Chapter III, p.72.

Jacoby, H. S., and Davis, R. P., 1941. *Foundations of Bridges and Buildings.* New York: McGraw-Hill, 535p.

Jumikis, A. R., 1962. *Soil Mechanics.* Princeton, N.J.: D. Van Nostrand, 791p.

Jumikis, A. R., 1964. *Mechancis of Soils—Fundamentals for Advanced Study.* Princeton, N.J.: D. Van Nostrand, 483p.

Jumikis, A. R., 1965. *Stability Analyses of Soil-Foundation Systems.* New Brunswick, N.J.: Rutgers University, College of Engineering, Bureau of Engineering Research, Engineering Research Publication No. 44, 55p.

Jumikis, A. R., 1966. *Thermal Soil Mechanics.* New Brunswick, N.J.: Rutgers University Press, 267p.

Jumikis, A. R., 1967. The factor of safety in foundation engineering, in *Classification, Safety Factor and Bearing, Highway Research Rec. 156.* Washington, D. C.: Highway Research Board, National Academy of Sciences-National Academy of Engineering, 23-32.

Jumikis, A. R., 1969. *Theoretical Soil Mechanics.* New York: Van Nostrand Reinhold, 432p.

Jumikis, A. R., 1971. *Foundation Engineering.* Scranton, Pa.: Intext Educational Publishers, 828p.

Jumikis, A. R., 1977. *Thermal Geotechnics.* New Brunswick, N.J.: Rutgers University Press, 375p.

Jumikis, A. R., 1979. *Rock Mechanics.* Clausthal-Zellenfeld: Trans. Tech. Publications, 356p.

Jumikis, A. R., 1983. *Soil Mechanics.* Melbourne, Fla.: Robert E. Krieger, 583p.

Just, L. H., and Obrician, V., 1966. Verrazano-Narrows Bridge: design of tower foundations and anchorages, *Am. Soc. Civil Engineers Proc., Jour. Construction Div.* **92**(CO2), 71-93.

Kézdi, A., 1965. Deep foundations, *Proceedings of the 6th International Conference on Soil Mechanics and Foundation Engineering,* vol. 3. Ottawa, Canada: National Research Council of Canada, 256-264.

Krey, H. D., and Ehrenberg, J., 1936. *Erddruck, Erdwiderstand und Tragfähigkeit des Baugrundes,* 5th ed. Berlin: Wilhelm Ernst und Sohn, 347p.

Meyerhof, G. G., 1951. The ultimate bearing capacity of foundations, *Géotechnique* **2**(4), 301-332.

Meyerhof, G. G., 1963. Some recent research on the bearing capacity of foundations, *Canadian Geotech. Jour.* **1**(1), 16-26.

Meyerhof, G. G., 1970. Safety factors in soil mechanics, *Canadian Geotech. Jour.* **7**(4), 349-355.

Michigan State Highway Commission, 1965. *A Performance Investigation of Pile Driving Hammers and Piles: Final Report of a Study in Cooperation with the Bureau of Public Roads, U.S. Department of Commerce; Michigan Road Builders Association; Wayne State University; and Representative Hammer Manufacturers.* Lansing: Michigan State Highway Commission, Office of Testing and Research, Research Project 61 F-60, 38p.

Morgan, M. H. (transl.), 1914. *The Ten Books on Architecture.* Cambridge, Mass.: Harvard University Press, 331p.

Peck, R. B., 1965. Pile and pier foundations, *Am. Soc. Civil Engineers Proc., Jour. Soil Mechanics and Found. Div.* **91**(SM2), 33-38.

Purcell, C. H.: Andrew, C. E.; and Woodruff, G. B., 1934. Bay bridge foundations built with unique domed caissons, *Engineering News Record* **112**(14), 431-436.

Rawlinson, G. (transl.), 1910. *The History of Herodotus,* vol. 1. New York: E.P. Dutton, 366p.

Riggs, L. W., 1966. Tagus River Bridge-Tower Piers, *Civil Engineering* (ASCE), **66**(2), 41-45.

Schousboe, I., 1972. Suggested design and construction procedures for pier foundations, *Am. Concrete Inst. Jour. Proc.* **69**(8), 461-480.

Seelye, E. E., 1956. *Foundations.* New York: Wiley, 11-16.

Sheta, M., and Novak, M., 1982. Vertical vibration of pile groups, *Geotech, Eng. Div. Jour.* (ASCE), **108** (GT-4), 570-590.

Sowers, G. F., 1969. The safety factor in excavations and foundations, *Highway Research Rec. 269,* 23-34.

Statens Järnvägars Geotekniska Kommission (1922). *Slutbetänkande 1914-1922.* Stockholm: Statens Järvägars, 180p. and 42 plates.

Terzaghi, K., 1925. *Erdbaumechanik auf bodenphysikalischer Grundlage.* Leipzig and Vienna: Franz Deuticke, 399p.

Terzaghi, K., 1943. *Theoretical Soil Mechanics.* New York: Wiley, 124-136.

Terzaghi, K., and Fröhlich, O. K., 1936. *Theorie der Setzung von Tonschichten.* Leipzig and Vienna: Franz Deuticke, 168p.

Triborough Bridge and Tunnel Authority, 1964. *Spanning the Narrows.* New York, 48p.

Vierendeel, A., 1927, 1928. Étude du pouvoir des terrains meubles, *La Technique des Travaux* **3,** 1927, 623-636 and **4,** 1928, 28-36.

Wellington, A. M., 1888. Formulae for safe loads of bearing piles, *Eng. News* **20,** 509-512.

White, R. E., 1943. Heavy foundations drilled into rock, *Civil Engineering,* (ASCE), **13**(1), 19-22.

White, R. E., 1962. Caissons and cofferdams, in G. A. Leonards (ed.), *Foundation Engineering.* New York: McGraw-Hill, 894-964.

Woodward, R. J.; Gardner, W. S.; and Greer, D. M., 1972. *Drilled Pier Foundations.* New York: McGraw-Hill, 287p.

Zeevaert, L., 1972. *Foundation Engineering for Difficult Subsoil Conditions.* New York: Van Nostrand Reinhold, 320p.

Cross-references: *Atterberg Limits and Indices; Clays, Strength of; Geotechnical Engineering; Grout, Grouting; Residual Stress, Rocks; Rheology, Soil and Rock; Rock Mechanics; Rocks, Engineering Properties; Rock Structure Monitoring; Soil Mechanics; Urban Engineering Geology; Wells, Water.* Vol. XII, Pt. 1: *Soil Mechanics.* Vol. XIV: *Artificial Fill; Borehole Drilling; Rock Geology, Hard vs Soft.*

G

GAS SAMPLING—See Vol. XIV:
ATMOGEOCHEMICAL PREOSPECTING.

GEOANTHROPOLOGY,
GEOARCHAEOLOGY—See Vol. XIV.

GEOBIOLOGY—See Vol. XIV:
GEOMICROBIOLOGY.

GEOBOTANY—See Vol. XIV:
GEOBOTANICAL PROSPECTING.

GEOCHEMICAL EXPLORATION—See
GEOCHEMISTRY, APPLIED.Vol. XIV:
ATMOGEOCHEMICAL PROSPECTiNG;
BIOGEOCHEMISTRY; EXPLORATION
GEOCHEMISTRY; HYDROCHEMICAL
PROSPECTING; HYDROGEOCHEMICAL
PROSPECTING; INDICATOR ELEMENTS;
LAKE SEDIMENT GEOCHEMISTRY;
LITHOGEOCHEMICAL PROSPECTING;
MARINE EXPLORATION
GEOCHEMISTRY; PEDOGEOCHEMICAL
PROSPECTING.

GEOCHEMISTRY, APPLIED

Geochemistry in practical service to man and
the environment by the identification and resolution of problems can be termed *applied geochemistry.* Water, food, fibers, construction materials,
energy sources, natural resources for industry,
and capacity for the correct and safe disposal of
wastes are all necessary for faunal and floral survival on earth. Applied geochemistry is concerned
with these needs, and others.

Background Literature

Much of the modern literature in geochemistry
has developed since the early 1950s in both the
Western and Eastern hemispheres. During 1956,
Geochimica et Cosmochimica Acta began publi-

cation, as did *Geokhimia* (in Russian); in 1966,
Chemical Geology commenced publication. In
addition, papers in which geochemistry was a
fundamental or principal part were (and still are)
published in most geological journals. Geochemical papers also became important in publications
of the International Geological Congresses, and
in those of state and federal geological surveys.

As research in geochemistry flourished, it was
realized that science would best be served by
providing focal journals for specific problems in
geochemistry especially in applied studies. The
most obvious application of geochemistry is to
supply indications of, or discover, natural resources
(see Vol. XIV: *Exploration Geochemistry*): metals,
nonmetals, and hydrocarbons. Two early volumes
on mineral exploration via geochemistry—both
titled *Principles of Geochemical Prospecting*—
were by Ginzburg (1957) in Russian (translated to
English in 1960) and by Hawkes (1957). The latter,
a bulletin of the U.S. Geological Survey, served
as a basis for early geochemical exploration programs in many developed and developing countries. Subsequently, the classic book by Hawkes
and Webb (1962), *Geochemistry in Mineral Exploration,* served exploration geochemists for more
than a decade and was then complemented by
other texts, including *Introduction to Geochemical Exploration* by Levinson (1974) and *Applied
Geochemistry* by Siegel (1974). Siegel's book considers aspects of geochemistry in health and pollution problems and geochemistry in marine natural
resource exploration, subjects not previously treated
in such texts (see Vol. XIV: *Marine Exploration
Geochemistry*). Second editions of the former
two texts have recently been published (Levinson,
1980; Rose, et al., 1979; Siegel, 1982). A translation to English of the Russian book *Geochemical
Exploration Methods for Mineral Deposits* (Beus
and Grigorian, 1975) was published during 1977.
Specialized texts on the use of flora in prospecting programs (see Vol. XIV: *Biogeochemistry;
Geobotanical Prospecting*) have also been published (Malyuga, 1964; NASA, 1968; Brooks, 1972,
and Kovalevskii, 1979, translated from the 1974
Russian edition), and on geochemical prospecting
for hydrocarbons (Kartsev et al., 1959). In addition, Hood (1979) contributed a volume that
updated many topics related to geochemistry and
geophysics in the search for metal ores.

In 1966, the first International Congress on

Exploration Geochemistry, in Ottawa, Canada, brought together specialists in the field of applied geochemistry to exchange ideas and experiences. This extremely successful meeting was followed by congresses held at two-year intervals. The 1980 congress was held in Hanover, Germany, in 1980, and the 1982 and 1983 congresses were held in Finland and Brazil, respectively. The published proceedings of each congress represent international contributions to geochemical prospecting.

The interest in these congresses clearly indicated a need for an organization dedicated to the advancement of geochemical exploration. Thus in 1970 the Association of Exploration Geochemists was chartered. In 1972, this association sponsored publication of the *Journal of Geochemical Exploration* as a medium for bringing together international studies on geochemical prospecting. In addition, it has provided the scientific community with bibliographies of exploration geochemistry and other aspects of applied geochemistry (Hawkes, 1972, 1976, 1977, 1979, 1982), which are updated annually in the *Journal.*

Several volumes on applied geochemistry in health and pollution studies have been published. In recent years, Cannon and Hopps (1971, 1972), Hopps and Cannon (1972), Cannon (1974), and Freedman (1975) have edited volumes on geochemistry in relation to health and disease, and the U.S. National Academy of Sciences has published several general and specific reports on geochemistry and the environment (Cannon, 1974; Mertz, 1977; Hopps, 1978; Angino, 1979; Wixson, 1980). Since 1967, there have been annual symposia on trace substances in environmental health at the University of Missouri, the proceedings of which have been published yearly since 1968. A direct result of these symposia was the formation in 1971 of the Society for Environmental Geochemistry and Health, with the object of the furthering interest in and knowledge of the effects of the geochemical environment on the health and disease of plants and animals, including humans, and to advance knowledge in this important scientific area, emphasizing a multidisciplinary approach. This society sponsors the publication *Interface.* Finally, it should be noted that Webb (1975) has reviewed environmental problems and the exploration geochemist, and Siegel (1979) edited a UNESCO volume, *Review of Research on Modern Problems in Geochemistry,* in which applied and basic problems are considered.

Geochemical Prospecting

Exploration for, and evaluation of, natural resource deposits is the major effort in applied geochemistry. The premise for geochemical prospecting is that natural materials that serve as prospecting samples (Table 1) will yield chemical signals highlighting localized or regional areas that contain significantly greater than "background" (normal) amounts of an element or assemblage of elements. Such signals may be subjectively or quantitatively designated as "anomalies", and may provide information about the positions of hidden or concealed mineralized zones. Such geochemical information, however, cannot properly indicate whether the prospect being studied is economically exploitable. This can be done only by drilling out the deposit and determining the amount of ore present plus probable and possible ore volumes, and the ore tenor, after which an evaluation must be made in terms of existing and predicted economic parameters.

Both exposed and hidden (concealed) mineralization is sought by geochemical prospecting. In some situations, the elements being analyzed are major elements in ore deposits; in others, an associated minor or trace element, often termed a *pathfinder,* is used. Whether a principal element or associated elements are used in prospecting depends on several factors, among which is an evaluation of element mobility in the geological environment being studied, the probable size and shape of the target (ore deposit), and the analytical capabilities available to the project geochemist.

The environmental parameters are a function of the geology of the study area, its geomorphological stage of development, the genesis of the processes through which that stage had been achieved, and the prevailing climatic regime. Finally, the chemical response of the elements in the regional or local environments must be assessed in terms of the elemental geochemistry in its environmental niche. For example, for sulfide minerals in a siliceous weathering environment, zinc is commonly very mobile and copper has intermediate mobility, but in a calcareous weathering environment, zinc has an intermediate mobility and copper is essentially immobile. In both weathering environments, lead is rather immobile.

In exploration geochemistry (q.v.), the applied geochemist commonly uses chemical signals from one or more sample types to detect hidden mineralization under overburdens of varying thicknesses and genetic developments. The elements used in the prospecting program are selected in accordance with the knowledge of the geology and an evaluation of how secondary-surface or near-surface environments may be expected to influence element mobility and dispersion from the original sites of mineralization. The dispersion factor is extremely important since a knowledge of element association and distribution as a result of primary and/or secondary and later functions will, together with an assumption of probable mineralization zone size and orientation, influence the density of the sampling scheme.

This is true for both regional and localized

mineral prospecting. The effect of certain orientations on sampling patterns that would give a maximum "target-hit" potential for cost effectiveness-time restrictions has been considered in a paper by Sinclair (1975). For the localization of mineral deposits after general areas of maximum mineralization potential or probability have been targeted by regional studies, a premise is made that the localizer sample type (e.g., soil, vegetation) will contain the guide element or elements in proportion to their contents in the concealed or hidden rock systems (see Vol. XIV: *Geobotanical Prospecting; Lithogeochemical Prospecting; Pedogeochemical Prospecting*). For example, considering residual (*in situ*) soils developed from underlying rock material of which part is mineralized, the soils overlying mineralization with high or higher contents of an element than the enclosing rock may also contain high or higher quantities of the element being sought. Deviations from such rather direct relations are common in residual soil-rock systems and can be related and evaluated via geological observations that would account for displacement of an anomalous accumulation of an element. Mass movement (creep or slide processes) may cause a physical displacement of an anomaly; a secondary dispersion via chemical processes and responses may result in

TABLE 1. Samples used in applied geochemistry

Rocks and Minerals

> Fresh host rock
> Weathered host rock
> Minerals separated from the host rock
> Ore
> Minerals separated from the ore

Sediments

> Stream sediments
> Alluvium
> Minerals separated from stream sediments and alluvium
>   (magnetic and nonmagnetic heavy minerals.)
> Ultrasonic concentrates of material films (fines) adhered
>   to detrital fragments
> Lake sediments
> Marine bay environment bottom sediments
> Marine (inner neritic zone) suspended sediments

Soils

> Humus related to soils
> Bacteria related to soils

Glacial materials

> Peat

Snow and Water Samples

> Stream waters
> Lake waters
> Subsurface waters
> Spring waters

Vapors and Aerosols

> Air
> Soil gases
> Rock gases
> Drilling mud gases

Vegetation (Flora) Including Seaweed

Animals (Fauna) Including Lacustrine Fish and Marine Shellfish

Humans

Source: After Siegel (1974).

an element concentration away from the immediate mineralization but still in the residual system; structural characteristics of the mineralization-containing rock system—especially the dip of fissures, fractures, joints, or faults—may result in a considerable displacement of a geochemical anomaly laterally from the location of the mineralization in the subsurface, with the degree of displacement being a function of the depth of the deposit and the angle of dip of the channelway systems cited above.

Direct relationships between mineralization in the subsurface environment and surface or near-surface localizer samples may be disturbed by other processes as well. If a direct (residual) anomaly or near-surface anomaly were in the path of a glacier, the anomaly would be physically dispersed away from the mineralization in a down-ice direction. In a similar way, anomalies in vegetation may be the result of subsurface waters carrying elements derived from mineralized areas down-dip under the influence of gravity to a position where the chemically anomalous waters are available to the root system, allowing the vegetation to pick up the element(s) and develop the floral anomalies displaced from the point of origin of the chemical elements providing the anomaly signal (see Vol. XIV: *Biogeochemistry*). The subject of conceptual models and their relationships to actual observations in geochemical prospecting has been treated in some detail by Siegel (1974) and for specific regions, such as the Canadian Cordillera and Canadian Shield (Bradshaw, 1975), on Norden and East Greenland (Kauranne, 1976), the basin and range area of the western United States and northern Mexico (Lovering and McCarthy, 1978), and Australia (Butt and Smith, 1980).

Thus, it is obvious that the applied geochemist must be aware of and able to evaluate the effect of various geological events and geochemical processes that have influenced the distribution and dispersion of elements being analyzed in a given study zone; in some cases multiple events must be evaluated.

Element Analysis in Prospecting

Analytical techniques being most used by exploration geochemists are atomic absorption spectrometry, emission spectography, and direct-reading emission spectrometry, with the first being the most favored. The use of ICP (inductively coupled plasma) with the equipment cited is resulting in more accurate and precise data. Each technique has its own detection limits for the elements being sought and its own unique accuracy and precision capabilities. In general, such equipment is set up in permanent laboratories, although it may be mobilized for work in remote but accessible areas. In many areas where geo-

chemical prospecting is now going on or will be concentrated in the future (e.g., in jungle, mountainous, and desert areas), such accessibility would be the exception rather than the normal situation. It is desirable, however, to have *in situ* analytical capability in such areas so that the field geochemist can make decisions concerning program directions and emphasis. This capability can be achieved via field analyses for selected metals using field prospecting kits in which the metals may be extracted from a sample with an ammonium citrate solution (with a given pH) and reacted with a dithizone solution that serves as an indicator for the presence of, and relative amounts of, the metals in the sampling medium. Specific ion electrodes (see Vol. XII, Pt. I: *Electrochemistry*) are also being used in some research studies to determine their field capabilities, especially in areas where water can serve as the sample.

Much recent effort has gone into studies of selective and sequential phase extractions to identify the contribution of each phase to the metal content of a total sample or a sample fraction (Bradshaw et al., 1974; Rose, 1975; Bolviken and Paus, 1976; Gatehouse et al., 1977; Hoffman and Fletcher, 1979; Filipek and Theobald, 1981). A very unique idea has been developed in Scandinavia: to use dogs as field indicators. The canine olefate (nose) is more sensitive to gas emissions (10^{-18} g) than any analytical equipment available to the geochemist. Alsatians, for example, have been conditioned to react to specific gas emissions that may derive from minerals reacting with the surface or near-surface environment (e.g., SO_2, CS_2, or COS from sulfide chemical weathering; Taylor et al., 1982). This technique has been successful in some cases (Ekdahl, 1976) and research is continuing on the subject (Brock, 1972).

Geochemical Data—An Adjunct to Geological Mapping

Since a principal aim of geochemistry is to determine the geographical distribution of the chemical elements comprising the earth (and to interpret why such variations exist), it is clear that geochemistry is applicable to geological mapping problems involving subsurface rock units in areas where residual soils have developed (see Vol. XIV: *Lithogeochemical Prospecting*). Kilpatrick (1969) worked in an area of Guyana underlain by a complex assemblage of pelitic and mafic metamorphic rocks intruded by serpentinites in the subsurface by anomalously high contents of nickel, chromium, cobalt, and manganese in residual soils of the area. Similarly, Lecomte et al. (1975) were able to use geochemical soil surveys (see Vol. XIV: *Pedogeochemical Prospecting*) over Cambrian and Lower Devonian formations in the Belgian Ardennes where copper and lead distributions permit the precise location of the boundary between Cambrian and Lower Gedinnian formations, and where nickel distributions seem to be a good stratigraphic indicator for distinguishing Upper from Lower Gedinnian.

At the Imperial College, London, Webb and his colleagues in the Applied Geochemistry Research Group have prepared the Wolfson geochemical atlas of England and Wales (Webb et al., 1978). The information presented in this magnificent atlas has been used in regional geochemical mapping and interpretation in Britain (Webb and Howarth, 1979; Plant and Moore, 1979) and in geochemistry and health research in the United Kingdom (Thornton and Webb, 1979; Thornton and Plant, 1980). Regional geochemical mapping for prospecting and for geomedicine studies is also being carried out systematically in Norway (Lag, 1980).

Geochemical Parameters for Earthquake Prediction

Applied geochemistry has multiple functions in environmental studies of physical, chemical, and biological nature, including health problems. In recent years, several countries have suffered through destructive earthquakes and attendant consequences (Peru, Nicaragua, Yugoslavia, China, the United States, Iran, Turkey, Guatemala, and Italy). Geologists and geophysicists are making basic and sophisticated studies in an effort to attain at least a limited, if not extended, capability to predict earthquakes.

Geochemical monitoring of the concentration levels of natural gas emanation along known fault zones indicate, for example, that increasing levels of selected gas concentrations can be correlated with increased heat flow and seismic activity previous to major earthquakes (see Vol. XIV: *Atmogeochemical Prospecting*). This technique might be a sensitive indicator of potential earthquake motion, especially when coupled with other parameters that are potential predictors of earthquake motion (e.g., type and intensity of seismic motion, strain gauge reading, earth dilation measurements).

Russian scientists have identified geochemical parameters that may be used to predict earthquakes (Eremeev et al., 1973; Fridman, 1970; Fursov et al., 1968; Gorbushina et al., 1971; Ovchinnikov et al., 1973). Geochemical prospecting programs based on gas samples have indicated that there is a continuous degassing of the earth and that both the amount and composition of the emissions are irregularly distributed over the earth's surface and the dominant gas varies from region to region. Areas of strong emanations are apparently related to zones of deep tectonic fracturing. King (1978, 1980) found that

in California, subsurface radon emanation monitored in shallow (0.6 = m to 0.8 = m) dry holes showed spatially coherent temporal variations that seemed to correlate with local seismicity and that episodic radon charges may be caused by a changing outgassing rate in a fault zone in response to some episodic strain changes, which incidentally caused the earthquakes. Birchard and Libby (1980) corroborated this latter observation in their study of two faults that had quadrants of compression and dilation that correlated with radon concentration increases and decreases, respectively. Gas outflow may occur in regions of compression, and inflow in areas of dilation, thus giving the observed radon changes.

During the 1966 Anapa earthquake in the northern Caucasus, monitoring of a CO_2 anomaly over a mercury deposit showed both a threefold increase in the CO_2 flow and an isotopic change in Carbon-13 content from the "normal" 2.92 to 2.37 on the day of the earthquake. Irwin and Barnes (1980) point out that the presence of CO_2-rich springs may indicate a potentially hazardous seismic region and propose that monitoring of CO_2 discharges could be useful in earthquake prediction. Similarly, during the 1966 Tashkent earthquake, the helium contents of subsurface waters (in an observational drillhole) increased twelvefold, and that of randon increased threefold. However, Teng (1980) wrote that in contrast to Russian observations showing a long-term buildup of radon emission before an earthquake, several recent reports from China and Japan showed that ground-water radon anomalies can be short in duration and occur only a few days before the main shocks. Therefore, there is a need for a continuous monitoring system as described by Shapiro et al. (1980).

In other areas it has been demonstrated that the flow of free helium reached maximum concentrations in areas of tectonic crushing, fissuring, and high permeability. Diurnal changes in soil-gas helium concentration do occur because of surface and near-surface soil moisture. However, Reimer (1980) finds that this would not impose any severe limitations on the use of helium-concentration data to help predict earthquakes. Most recently, Sugisaki (1981) found that variations of the helium/argon ratio of gas bubbles in a mineral spring along a fault zone coincided with fluctuations of areal dilation induced by the earth tide. This suggests that deep-seated gases characterized by higher helium/argon ratios are squeezed out by stress preceding an earthquake. The prediction of volcanic activity is also being complemented by geochemical studies of gaseous emissions from volcanoes. Rate of flow, type, and concentration of gases are being related to stages of volcanic activity, including eruption.

Geochemistry in Epidemiological, Health and Pollution Research

In the area of environmental applications, geochemists are establishing a data bank that contains baseline levels of elemental concentrations in different media (e.g., waters, suspended and deposited sediments, soils, rocks, crops, animals, the atmosphere, and even humans). Such information is derived from the investigation of variations in the areal distribution of the chemical elements and their isotopes (e.g., the work of the Imperial College Applied Geochemistry Research Group previously cited). After the so-called base-level concentrations have been established for natural uncontaminated environments, deviations may be identified. Geochemists can then speculate about the causes of such variations and the processes through which they developed; they may even attempt a time-fix on the advent of anomaly concentrations in a given system. The significance of such geochemical information serves as an alert to the possible existence of insidious marginal imbalances that result in such subclinical conditions as infertility and loss of production (Webb, 1975). Lakin (1979), and Fortescue (1979) summarize the role of major and minor elements excesses and deficiencies in rocks and soils in the nutrition of plants, animals, and humans. People may be more seriously debilitated due to nutrient imbalances of trace metals than is presently realized (see *Medical Geology*).

Deviations in the bioavailability of trace metal nutrients may result from natural or man-influenced causes, or both, and these can be recognized and potential detrimental effects on the environment projected by the geochemist working, for example, with medical researchers, nutritionists, and epidemiologists. The ability to recognize such factors increases, of course, with the increasing experience of interdisciplinary environmental evaluation teams. In the same manner, areas with a high incidence of a disease may be subjected to a total evaluation, of which geochemistry is a part. The disease incidence may then be related to a geochemical parameter for which a therapy may be suggested.

In the case of nutrition, optimum ranges of element concentration are required for the proper functioning of life systems. Toxicities may result from the ingestion of large amounts of a particular element or assemblage of elements, and deficiencies from the lack of ingestion; both fauna and flora are affected by such toxicities and deficiencies of nutrient elements (see Vol. XII, Pt. I: *Macronutrients; Micronutrients*). Other elements, especially those not essential for growth (e.g., lead, mercury, arsenic) are also toxic to life forms in concentrations above certain levels. Deficiency-related problems in plant and animal nutrition can often be alleviated by supplementing fertilizers or foodstuffs with the required element. Humans also suffer from nutritional deficiencies, for example, goiter (iodine deficiency); dental caries (fluorine deficiency); anemia (iron deficiency); and endocrine abnormalities such as dwarfism, hypogonadism, and marked hepatosplenomegaly (zinc deficiency).

Toxicity problems also occur, but they are generally more difficult to isolate and cope with. The incidence of multiple sclerosis, for example, appears to be associated with high lead contents in soils (Warren, 1961). Foodstuffs grown in soils with abnormally high lead contents absorbed the metal, which was later ingested as food. After studying data from Ireland, Norway, Scotland, and Sweden, Warren et al. (1967) recognized that zones with a low prevalence of the disease generally corresponded with areas of Precambrian gneisses and schists. In Northern Ireland, areas with a high incidence of multiple sclerosis correspond with the distribution of Paleozoic rocks that contained lead and zinc mineralization. After reevaluating the data from Norway, Sweden, and northern Scotland and studying some small areas in detail, Warren et al. (1967) noted that where schistose and gneissic rocks and overlying soils contained anomalously high concentrations of base metals, there was an elevated incidence of multiple sclerosis.

Examples of man-caused pollution investigated by the geochemist are many and varied and increasingly recognizable as more multidisciplinary programs in pollution research are carried out. Many of the pollution-health-geological-geochemical relations come to light via epidemiological studies that reveal disease-incidence levels as a function of geographical distribution. As mentioned previously, attempts are made to explain the incidence levels in terms of climate, topography, diet, occupation, and race as primary factors, and then by other factors, such as pollutants.

Not only can applied geochemists indicate via geochemical maps (single and multi-element representations) natural and man-induced elevated chemical element concentrations in the life environment, but they can also work on methods to alleviate or eliminate their influence. Agriculturists may change crops or effect an environmental treatment to be able to use a natural system that is otherwise detrimental to health. Proper treatment and/or storage (containment) of mining wastes and other anthropogenic wastes such as sludge and fly ash can also eliminate health hazards. The same problems that exist on land exist in the marine environment. The bioavailability of an element and the capacity of a member of the food chain to concentrate it can ultimately lead to a concentration in the food chain and thus cause human health problems.

Other pressing environmental problems exist;

one that involves geochemistry is related to energy requirements and nuclear power facilities. Geochemists are doing research on discharge from nuclear power plants both from "normal" operation and from the potential "accident" event and major discharge of radionuclides into the environment. The applied geochemical research is on the geochemical response of the elements in different environments (e.g., major rivers, estuarine systems, and the marine system) with respect to the sedimentary regime and life forms. In addition, applied geochemists are working with other scientists on the investigation of uptake of radionuclides by natural and man-prepared materials in an effort to come up with the best material that could be reacted with radionuclides accidentally discharged into the environment to prevent them from moving significant distances away from the point of discharge.

Finally, the storage, containment, and environmental monitoring of radioactive wastes is one of the major disposal problems being researched by the applied geochemist. A new technique for inground radioactivity monitoring that obviates the problems associated with short-term gamma-ray spectrometer and scintillometer measurements has been field-tested by Siegel et al. (1981) and is based on the use of buried lithium fluoride thermoluminescence badges as integrating detectors.

FREDERICK R. SIEGEL

References

Angino, E. E. (committee chairman), 1979. *Geochemistry of Water in Relation to Cardiovascular Disease.* Washington, D. C.: U.S. National Academy of Sciences, 98p.

Beus, A. A., and Grigorian, S. V., 1977. *Geochemical Exploration Methods for Mineral Deposits* (transl. from 1975 Russian publication). Wilmete, Ill.: Applied Publishers Ltd.

Birchard, G. F., and Libby, W., 1980. Soil radon concentration changes preceding and following four magnitude 4.2-4.7 earthquakes on the San Jacinto fault in Southern California, *Jour. Geophys. Research,* 85, 3100-3106.

Bolviken, B., and Paus, P. E., 1976. Snertingdal II: extraction of lead from various size fractions of stream sediments, *Jour. Geochem. Exploration* 5, 331-335.

Bradshaw, P. M. D. (ed.), 1975. Conceptual models in exploration geochemistry—the Canadian cordillera and the Canadian shield, *Jour. Geochem. Exploration* 4, 213p.

Bradshaw, P. M. D.; Thomson, I.; Smee, B. W.; and Larsson, J. O., 1974. The application of differential analytical extractions and soil profile sampling in exploration geochemistry, *Jour. Geochem. Exploration* 3, 209-225.

Brock, J. S., 1972. The use of dogs as an aid to exploration for sulphides, *Western Miner* (December), 28-32.

Brooks, R. R., 1972. *Geobotany and Biogeochemistry in Mineral Exploration.* New York: Harper & Row, 90p.

Butt, C. R. M., and Smith, R. E. (comps. and eds.), 1980. Conceptual models in exploration chemistry: Australia, *Jour. Geochem. Exploration* 12, 365p.

Cannon, H. L. (workshop chairman), 1974. *Geochemistry and Environment. Vol. 1: The Relation of Selected Trace Elements to Health and Disease.* Washington, D. C.: U.S. National Academy of Sciences, 113p.

Cannon, H. L., and Hopps, H. C. (eds.), 1971. Environmental geochemistry in health and disease, *Geol. Soc. America Mem. 123,* 230p.

Cannon, H. L., and Hopps, H. C. (eds.), 1972. Geochemical environment in relation to health and disease, *Geol. Soc. America Spec. Paper 140,* 77p.

Ekdahl, E., 1976. Pielavesi: the use of dogs in prospecting, *Jour. Geochem. Exploration* 5, 296-298.

Eremeev, A. N.; Sokolov, V. A.; Solovov, A. P; and Yanitskii, I. N., 1973. Application of helium surveying to structural mapping and ore deposit forecasting, in M. J. Jones (ed.), *Geochemical Exploration 1972.* London: Institution of Mining and Metallurgy, 183-192.

Filipek, L. H., and Theobald, P. K., Jr., 1981. Sequential extraction techniques applied to a porphyry copper deposit in the Basin and Range Province, *Jour. Geochem. Exploration* 14, 155-174.

Fortescue, J. A. C., 1979. Role of major and minor elements in the nutrition of plants, animals and man, in F. R. Siegel (ed.), *Review of Research on Modern Problems in Geochemistry* Paris: UNESCO, 57-87.

Freedman, J. (ed.), 1975. Trace element chemistry in health and disease, *Geol. Soc. America Spec. Paper 155,* 118p.

Fridman, A. I., 1970. *Natural Gases of Ore Deposits.* Moscow: Nedra (in Russian).

Fursov, V. Z.; Vol'fson, N. B; and Khvalovskiy, A. G., 1968. The results of the study of mercury vapours in the zone of the Tashkent earthquake, *Akad. Nauk SSSR Doklady* 179, 208-210.

Gatehouse, S.; Russell, D. W.; and VanMoort, J. C., 1977. Sequential soil analysis in exploration geochemistry, *Jour. Geochem. Exploration* 8, 483-494.

Ginzburg, I. I., 1960. *Principles of Geochemical Prospecting.* (Transl. from 1957 Russian publication). New York: Pergamon Press, 311p.

Gorbushina, L. V., et al., 1971. On the effect of geological-tectonic factors on the content of gases in ground water of Tashkent artesian basin, in *Tashkent Earthquake.* Tashkent: FAN.

Hawkes, H. G., 1957. Principles of geochemical prospecting, *U.S. Geol. Survey Bull.* 1000-F, 225-355.

Hawkes, H. E. (comp.), 1972. *Exploration Geochemistry Bibliography, Period January 1965 to December 1971,* Spec. Vol. No. 1. Toronto: The Association of Exploration Geochemists, 118p.

Hawkes, H. E. (comp.), 1976. *Exploration Geochemistry Bibliography, Period January 1972 to December 1975,* Special Vol. No. 5. Rexdale, Ont.: Association of Exploration Geochemists, 195p.

Hawkes, H. E. (comp.), 1977. *Exploration Geochemistry Bibliography, Period January 1976 to June 1977.* Rexdale, Ont.: Association of Exploration Geochemists, 63p.

Hawkes, H. E. (comp.), 1979. *Exploration Geochemistry Bibliography, Period July 1977 to December 1978.* Rexdale, Ont.: Association of Exploration Geochemists, 85p.

Hawkes, H. E., 1982. *Exploration Geochemistry Bibliography,* Spec. Vol. No. 11, Rexdale, Ontario: Association of Exploration Geochemists, 388p.

Hawkes, H. E., and Webb, J. S., 1962. *Geochemistry in Mineral Exploration.* New York: Harper & Row, 415p.

Hoffman, S. J., and Fletcher, W. K., 1979. Selective sequential extraction of Cu, Zn, Fe, Mn, and Mo from soils and sediments, in J. R. Watterson and P. K. Theobald, Jr. (eds.), *Geochemical Exploration 1978.* Rexdale, Ont.: Association of Exploration Geochemists, 289-299.

Hood, P. J., 1979. Geophysics and geochemistry in the search for metallic ores, *Canada Geol. Survey Econ. Geology Rep. 31,* 811p.

Hopps, H. C., 1978. *Geochemistry and the Environment. Vol. III: Distribution of Trace Elements Related to the Occurrence of Certain Cancers, Cardiovascular Diseases, and Urolithasis.* Washington, D.C.: U.S. National Academy of Sciences, 200p.

Hopps, H. C., and Cannon, H. L. (eds.), 1972, Geochemical environment in relation to health and disease, *New York Acad. Sci. Annals* **199,** 352p.

Irwin, W. P., and Barnes, I., 1980. Tectonic relations of carbon dioxide discharges and earthquakes, *Jour. Geophys. Research* **85,** 3115-3121.

Kartsev, A. A., et al., 1959. *Geochemical Methods of Prospecting and Exploration for Petroleum and Natural Gas* (transl. from Russian publication). Berkeley: University of California Press, 349p.

Kauranne, L. K. (ed.), 1976. Conceptual models in exploration geochemistry—Norden 1975, *Jour. Geochem. Exploration* **5,** 420p.

Kilpatrick, B. E., 1969. Nickel, chromium and cobalt in tropical soils over serpentinites, Northwest District, Guyana, *Colorado School Mines Quart.,* **64**(1), 323-332.

King, C-Y., 1978. Radon emanation on San Andreas fault, *Nature* **271,** 516-519.

King, C-Y., 1980. Episodic radon changes in subsurface soil gas along active faults and possible relation to earthquakes, *Jour. Geophys. Research* **85,** 3065-3078.

Kovalevskii, A. L., 1979. *Biogeochemical Exploration for Mineral Deposits:* (transl. from 1974 Russian publication) New Delhi: Amerind Publishing Co., 135p.

Laq, J., 1980. Geomedicine in Norway, *Geol. Soc. London Jour.* **137,** 559-563

Lakin, H. W., 1979. Excesses and deficiencies in rocks and soils as related to plant and animal nutrition, in F. R. Siegel (ed.), *Review of Research on Modern Problems in Geochemistry.* Paris: UNESCO, 89-110.

Lecomte, P.; Sondag, F.; and Martin, H., 1975. Geochemical soil surveys over Cambrian and Lower Devonian formations in the Belgian Ardennes as a tool for geological mapping, *Jour. Geochem. Exploration,* **4,** 215-229.

Levinson, A. A., 1974. *Introduction to Exploration Geochemistry.* Alberta: Applied Publishers Ltd., 612p.

Levinson, A. A., 1980. *Introduction to Exploration Geochemistry: Supplement.* 309p.

Lovering, T. G., and McCarthy, J. H., Jr. (eds.), 1978. Conceptual models in exploration geochemistry— The Basin and Range Province of the Western United States and northern Mexico, *Jour. Geochem. Exploration* **9,** 276p.

Malyuga, D. P., 1964. *Biogeochemical Methods of Prospecting.* New York: Consultants Bureau, 205p.

Mertz, W. (ed.), 1977. *Geochemistry and Environment. Vol. II: The Relation of Other Selected Trace Elements*

to Health and Disease. Washington, D. C.: U.S. National Academy of Sciences, 163p.

NASA, 1968. Application of Biogeochemistry to Mineral Prospecting, *NASA Spec. Publ. 5056,* 134p.

Ovchinnikov, L. H.; Sokolov, V. A., Fridman, A. I., and Yanitskii, I. N., 1973. Gaseous geochemical methods in structural mapping and propecting for ore deposits, in M. J. Jones (ed.), *Geochemical Exploration 1972.* London: Institution of Mining and Metallurgy, 177-182.

Plant, J., and Moore, P. J., 1979. Regional geochemical mapping and interpretation in Britain, *R. Soc. London Philos. Trans.* **B 288,** 95-112.

Reimer, G. M., 1980. Use of soil gas helium concentrations for earthquake prediction: limits imposed by diurnal variation, *Jour. Geophys. Research* **85,** 3107-3114.

Rose, A. W., 1975. The mode of occurrence of trace elements in soils and stream sediments applied to geochemical exploration, in I. L. Elliot and W. K. Fletcher (eds.), *Geochemical Exploration 1974* New York: Elsevier, 691-705

Rose, A. W.; Hawkes, H. E.; and Webb, J. S., 1979. *Geochemistry in Mineral Exploration,* 2nd ed. New York: Academic Press, 657p.

Shapiro, M. H.; Melvin, J. D.; Tombrello, T. A.; and Whitcomb, J. H., 1980. Automated radon monitoring at a hard-rock site in the southern California Transverse Ranges, *Jour. Geophys. Research* **85,** 3058-3064.

Siegel, F. R., 1974. *Applied Geochemistry.* New York: Wiley, 353p.

Siegel, F. R. (ed.), 1979. *Review of Research on Modern Problems in Geochemistry.* Paris: UNESCO, 290p.

Siegel, F. R., 1982. *Applied Geochemistry.* 2nd ed. New York: Wiley.

Siegel, F. R.; Lindholm, R. C.; and Vaz, J. E., 1981. In-ground environmental radioactivity monitoring in the Culpeper Basin, Virginia using LiF thermolumine/cence dosimeters, *Environ. Geol.* **4,** 67-74.

Sinclair, A. J., 1975. Some considerations regarding grid orientation and sample spacing, in I. L. Elliot and W. K. Fletcher (eds.), *Geochemical Exploration 1974.* New York: Elsevier, 133-140.

Sugisaki, R., 1981. Deep-seated gas emission induced by the earth-tide: a basic observation for geochemical earthquake prediction, *Science* **212,** 1264-1266.

Taylor, C. H.; Kesler, S. E.; Cloke, P. L., 1982. Sulfur gases produced by the decomposition of sulfide minerals: application to geochemical exploration, *Jour. Geochem. Exploration* **17,** 165-185.

Teng, T., 1980. Some recent studies on groundwater radon content as an earthquake precursor, *Jour. Geophys. Research* **85,** 3089-3099.

Thornton, I., and Plant J., 1980. Regional geochemical mapping and health in the United Kingdom, *Geol. Soc. Lond. Jour.* **137,** 575-586.

Thornton, I.; and Webb, J. S., 1979. Geochemistry and health in the United Kingdom, *R. Soc. London Philos. Trans.* **B 288,** 151-168.

Warren, H. V., 1961. Some aspects of the relationship between health and geology, *Canadian Jour. Pub. Health* **52,** 157-164.

Warren, H. V.; Delavault, R. E., and Cross, C. H., 1967. Possible correlations between geology and some disease patterns, *New York Acad. Sci. Annals* **136,** 657-710.

Webb, J. S., 1975. Environmental problems and the exploration geochemist; in I. L. Elliot and W. K.

Fletcher (eds.), *Geochemical Exploration 1974.* New York: Elsevier, 5-17.

Webb, J. S., and Howarth, R. J., 1979. Regional geochemical mapping and intepretation in Britain, *R. Soc. London Philos. Trans.* **B 288,** 95-112.

Webb, J. S.; Thornton, I.; Thompson, M.; Howarth, R. J.; and Lowenstein, P. L., 1978. *The Wolfson Geochemical Atlas of England and Wales.* Oxford, England: Oxford University Press, 70p.

Wixson, B. G. (ed.), 1980. *Trace-Element Geochemistry of Coal Resource Development Related to Environmental Quality and Health.* Washington, D.C.: U.S. National Academy of Sciences, 153p.

Cross-references: Vol. XIV: *Atmogeochemical Prospecting; Biogeochemistry; Exploration Geochemistry; Hydrochemical Prospecting; Hydrogeochemical Prospecting; Indicator Elements; Lake Sediment Geochemistry; Lithogeochemical Prospecting; Marine Exploration Geochemistry; Pedogeochemical Prospecting.*

GEOCHRONOLOGY

Geochronology is a branch of isotope geology concerned with the analytical measurement of time. Fossils are widely used for the correlation of Cambrian and younger rocks, but in the Precambrian rocks the fossil record is meager or may be completely lacking. For these rocks, which represent 85 percent of geological time, radiometric ages affected a major breakthrough. A second spectacular breakthrough came through radiocarbon (^{14}C) dates, which permit dating from historical through archeological and into Pleistocene time (see Vol. XIV: *Geoanthropology*). A casual examination of the literature reveals the tremendous impact radiometric age measurements have had on geological research since about 1955.

Radiometric age determinations have many applications, not only in geology but also in planetology and cosmology. They aid the field geologist in preparing maps, the stratigrapher in correlating rock formations, the structural geologist in analyzing tectonic development, the geochemist in formulating models to explain the evolution of the crust and mantle, the economic geologist in studying the origin of ore deposits, and so forth. The time parameter is fundamental in studying rates of geological processes, for example, the rates of uplift and erosion. Without the time parameter, geological history, like human history, tends to become somewhat muddled.

Geochronology is a relatively young discipline. One can, of course, relate the origins of radiometric ages to the discovery of radioactivity around the turn of the century; investigations in considerable depth were published shortly after, for example, by Boltwood (1907). Boltwood determined lead and uranium contents of samples of the mineral uraninite and showed that the lead/uranium ratio

in samples is time-dependent. Professor Alfred Nier of the Department of Physics, University of Minnesota, can perhaps be considered the "Founder of Geochronology." Nier (1939) published a paper in *The Physical Review,* "The Isotopic Constitution of Lead and the Measurement of Geological Time." Since then his research has contributed to the development of mass spectrometry and the determination of the masses and atomic abundances of the isotopes of the elements. Results published by Nier and his students provided the numerical data for Arthur Holmes's (1947) geological time scale.

Although the work of Nier and of Holmes is widely appreciated, less well known is the work of Nier's student, L. T. Aldrich, in the development of geochronology. At the Department of Terrestrial Magnetism of the Carnegie Institution, Aldrich and his colleagues devised simple, straightforward methods of age measurements.

The profession of geology has gradually accepted radiometric age determinations, and it is now common to find a laboratory in the geology rather than in the physics department. Earlier, isotope-dilution and mass spectrometric techniques were identified with "black-box" art. Measured ages were thought to be absolutely reliable, and the only qualification necessary was a statement of the probable analytical error of the measurement. Undoubtedly, the careful assessment of probable errors was partially responsible for the discovery that the measured ages do not always give the real age of the rock or mineral. The isotopic systems do not always remain closed but may be interrupted or disturbed by geological processes. Some geologists viewed these findings as evidence of the unreliability of age determinations, but others quickly grasped the geological significance and saw a new technique for probing and investigating the geological processes. Aldrich was a pioneer in this area; He repeatedly urged the use of more than a single radiometric technique for the investigation of geological processes. To illustrate, Aldrich and coworkers (1965) dated a variety of minerals from the metamorphic and igneous rocks in northern Michigan and showed that different minerals have varying degrees of resistance to alteration of the different decay schemes. For example, potassium-argon age determinations on feldspar were lower than on biotite from the same rock, and generally much lower than rubidium-strontium ages on K-feldspar or muscovite. Since this work was published, considerable progress has been made.

The decay shemes used in geochronology are listed in Table 1, which also includes recent determinations of the decay constants and references to the literature. For details of laboratory procedures and interpretation of results, a number of reference books are available, for example: for

TABLE 1. Decay Systems and Constants Used in Geochronology

Parent	Daughter	Decay Constant	Reference
^{238}U	^{206}Pb	1.55125×10^{-10} yr^{-1}	See Steiger and Jäger (1977) for documentataion of con-
^{235}U	^{207}Pb	9.8485×10^{-10} yr^{-1}	stants for U, Th, Rb, and Sr.
^{232}Th	^{208}Pb	4.9475×10^{-11} yr^{-1}	
^{87}Rb	^{87}Sr	1.42×10^{-11} yr^{-1}	
^{40}K$_{\beta^-}$	^{40}Ca	4.962×10^{-10} yr^{-1}	
^{40}K$_e$	^{40}Ar	0.581×10^{-10} yr^{-1}	
^{147}Sm	^{143}Nd	6.54×10^{-12} yr^{-1}	Lugmair and Marti (1978)
^{176}Lu	^{186}Hf	1.96×10^{-11} yr^{-1}	Patchett and Tatsumoto (1980)
^{187}Re	^{187}Os	1.62×10^{-11} yr^{-1}	Luck, Birck, and Allègre (1980)
^{14}C	^{14}N	1.21×10^{-4} yr^{-1}	$T\frac{1}{2} = 5,730$ yr; by convention
			$T\frac{1}{2} = 5,570$ yr is used by all laboratories (Godwin, 1962).

K-Ar, Dalrymple and Lanphere (1969); for Rb-Sr, Faure and Powell (1972); and for U-Th-Pb, Doe (1970). Broader treatments have been published by York and Farquhar (1972) and Faure (1977). Many of the problems that have been encountered in geochronology, as well as recent developments in methodology, are well summarized in a volume edited by Jäger and Hunziker (1979).

K-Ar System

The K-Ar system (q.v. in Vol. IVA) involves two types of radioactive decay of ^{40}K. Capture of an electron by the nucleus *(K-capture)* with the emission of a gamma ray produces ^{40}Ar; loss of an electron *(beta decay)* forms ^{40}Ca. Statistically, out of 100 atoms of ^{40}K, 11 will decay to ^{40}Ar and 89 to ^{40}Ca. The relationships are shown in Table 2, in which Z is the number of protons, N the number of neutrons, and A the mass or the sum of the protons and neutrons. The proton number *(Z)* determines the chemical species; the different number of neutrons give rise to the isotopes. The atomic weight of an element is calculated from the atomic abundances of the isotopes with suitable mass corrections. The recent determination for potassium (Garner et al., 1975) gives an atomic weight of 39.0983 based on the atomic abundances of the isotopes of potassium given in Table 2.

As can be seen in Table 2, most of natural calcium is composed of the ^{40}Ca isotope, and, therefore, it is analytically difficult to distinguish the radiogenic ^{40}Ca formed by decay of ^{40}K from the ^{40}Ca present in silicate minerals and rocks at the time of their formation. This basic problem has limited the use of ^{40}Ca/^{40}K in geochronolgy whereas ^{40}Ar/^{40}K has been widely used. ^{40}Ar makes up most of the argon, approximately 1 percent, in the atmosphere, where it is always accompanied by ^{36}Ar. Atmosphereic argon that contaminates the radiogenic ^{40}Ar extracted from minerals can be monitored and corrected on the basis of ^{36}Ar. The analytical determinations of ^{40}Ar and potassium are the basis of the K-Ar method. The age is calculated from the equation

$$t = \frac{1}{\gamma_\beta + \gamma_\epsilon} \ln \left[\frac{^{40}Ar}{^{40}K} \left(\frac{\gamma_\beta + \gamma_\epsilon}{\gamma_\epsilon} \right) + 1 \right]$$

The widespread distribution of a variety of potassium-bearing minerals made the K-Ar method versatile and popular, so that 50 or more laboratories at one time were turning out hundreds of ages, principally on minerals such as biotite, muscovite, lepidolite, K-feldspar, glauconite, and hornblende, but also on igneous, sedimentary, and metamorphic rock samples. As the number of age determinations increased, conflicting data became apparent, and geological factors that might affect the isotopic system were considered. For example, the K-Ar ages determined on K-feldspar and on biotite from the same rock were found to differ considerably, with the feldspar age usually, but not always, lower than the biotite age. Efforts to apply corrections for loss of argon were not successful, and the term *apparent age* became commonplace.

The application of K-Ar dating to potassium-poor minerals such as pyroxenes (Hart, 1961) brought home another lesson earlier learned but not properly evaluated. ^{40}Ar may be incorporated in a mineral at the time of its formation, whether

TABLE 2. Isotopic Abundances and Relationships Between Potassium, Argon, and Calcium

Element	Z	N	A	Atom%
Argon	18	18	36	0.337
		20	38	0.063
		22	40	99.600
Potassium	19	20	39	93.2581
		21	40	0.01167
		22	41	6.7302
Calcium	20	20	40	96.97
		22	42	0.64
		23	43	0.145
		24	44	2.06
		26	46	0.0033
		28	48	0.185

by crystallization from a magma or by recrystallization in the solid state. The amount of the incorporated ^{40}Ar may be so small that in relatively old potassium-rich minerals it is not significant, but in minerals like the pyroxenes with low potassium contents, the inherited ^{40}Ar may be a large part of the total ^{40}Ar, and the calculated age can be too old by a considerable amount.

Attempts to use the isochron technique in which the ^{40}Ar/^{36}Ar ratio is plotted against ^{40}K/^{36}Ar have not been particularly productive. The ^{40}Ar-^{39}Ar technique introduced in the early 1960s produced interesting data on lunar samples and now is being used extensively on terrestrial materials. The method involves long-period neutron irradiation of the samples to make ^{39}Ar from ^{39}K. The process requires standard reference samples that are irradiated with the unknown samples. Corrections for radiation products and for the radioactivity of the ^{39}Ar must be made.

Rb-Sr System

The decay of ^{87}RB → ^{87}Sr + β^{-} provides a flexible geochronometer with many applications. The early promise, however, has been somewhat dissipated through analytical and geological problems that became apparent as the number of laboratories and studies increased (q.v. in Vol. IVA). Normal or common strontium contains approximately 7 percent ^{87}Sr so that a correction must be made to arrive at the ^{87}Sr produced by radioactive decay of ^{87}Rb in the rock or mineral. As the amount of ^{87}Sr has increased with geological time, it follows that the correction for common strontium will vary with the age of the sample. In rocks with high strontium and low rubidiumb contents, the radiogenic ^{87}Sr and the normal ^{87}Sr are difficult to differentiate and thus pose a serious problem in younger rocks. For this reason the method was first used for the analysis of minerals such as biotite, lepidolite, and muscovite, which usually contain large amounts of rubidium but only small amounts of strontium. In this case the assumption of how much ^{87}Sr was present initially is not critical in the age calculation. Errors in estimating the initial strontium (^{87}Sr$_i$) may result in large errors in the age of young rocks and minerals. The basic equation follows:

$$t = \frac{1}{\gamma} \ln \left(\frac{^{87}Sr_t - ^{87}Sr_i}{^{87}Rb} + 1 \right)$$

Because atomic ratios can be measured with great precision with the mass spectrometer, a convenient form is

$$t = \frac{1}{\gamma} \ln \left(\frac{(^{87}Sr/^{86}Sr) - (^{87}Sr/^{86}Sr)_i}{^{87}Rb/^{86}Sr} + 1 \right)$$

The initial ^{87}Sr/^{86}Sr ratio may be determined in a low-rubidium mineral phase such as apatite, but if the rock has undergone metamorphism, the ratio may not be valid for the original rock. During metamorphism the elements can migrate short distances, and apatite, a high-calcium phosphate, is a favored host for ^{87}Sr migrating from a rubidium-rich mineral.

Investigations of the problem of variations in the initial strontium in the Bernard Price Institute of the University of the Witwatersrand and in laboratories of the Australian National University and the Massachusetts Institute of Technology led to the whole-rock and rock-mineral isochron methods, which are now widely used. A number of samples with varying Rb/Sr ratios, if the samples are genetically related, can be plotted in a diagram (Fig. 1) from which the initial ^{87}Sr/^{86}Sr ratio can be determined. Minerals separated from the rock samples can be analyzed, and if the system has not been disturbed, the data points for the minerals will lie, within analytical error, along the whole-rock isochron.

During metamorphism some strontium and rubidium may migrate, in which case the rock-mineral isochron will be rotated, and a secondary isochron with a higher initial will result. This method has been used to determine the time of the metamorphism (Fig.1). Computer programs include esti-

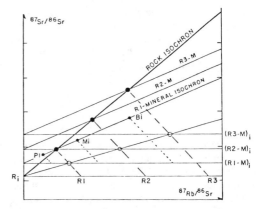

FIGURE 1. Schematic ^{87}Sr evolution diagram for whole-rock samples R1, R2, R3, and minerals plagioclase (P1), microcline (mi), and biotite (Bi) from R1. The rock isochron gives the age or the time since crystallization, and R$_i$ is ^{87}Sr/^{86}Sr at the time of crystallization. The rock-mineral (R1-M) isochron gives the time since the metamorphism. During the metamorphism strontium in the minerals was homogenized to give an initial ratio of (R1-M)$_i$. The rock-mineral isochron are parallel lines with successively higher initial rations. The basic assumption is that rock system remained closed and the metamorphic event was negligibly short compared to the total time.

mates of the accuracy of the $^{87}Sr/^{86}Sr$ and $^{87}Rb/^{86}Sr$ determinations and give the isochron age and the initial ratio with a statement of the probable error in each. Scatter of points or departures from the isochron greater than the analytical uncertainty indicate geological problems that might result from samples that are not genetically related or have had a different geological history. Commonly the nonconforming samples are omitted from the age calculation. A somewhat different problem is that of the apparently sound isochron in which the data points fit the regression line but fail to give the proper age. This failure has been demonstrated in a number of areas where stratigraphic control or U-Pb ages showed that the Rb-Sr whole-rock isochron ages were too low. Loss of radiogenic ^{87}Sr by ground-water leaching appears to be the mechanism involved. Similarly, low-grade metamorphism may affect the Rb-Sr system, for example, in Archean granitic rocks for which the Rb-Sr whole-rock isochron age may be as much as 150 million years younger than the U-Pb zircon age (Peterman et al., 1972).

U-Th-Pb Systems

Uranium and thorium are closely associated, and the conbined U and Th isotopic systems are valuable chronometers. The three isotopic systems $^{206}Pb/^{238}U$, $^{207}Pb/^{235}U$, and $^{208}Pb/^{232}Th$ provide flexibility and application to a variety of geological problems. The decay of uranium and thorium involves a number of steps and intermediate products formed by alpha decay (emission of ^4_2He) and beta decay (emission of e^-). To illustrate, the first few steps of the decay of ^{238}U follow:

$$^{238}U \rightarrow {}^{234}Th + \alpha$$

$$^{234}Th \rightarrow {}^{234}Pa + \beta^-$$

$$^{234}Pa \rightarrow {}^{234}U + \beta^-$$

The complete series from ^{238}U to ^{206}Pb involves a mass depletion of 32 amu; for ^{235}U and ^{232}Th, the mass depletions are 28 and 24 amu, respectively. When the ages determined by the three systems fail to agree within the analytical uncertainty, the variations are commonly referred to as *age discordance.*

The early work by Nier (1939) was done on pitchblende, uraninite, thorite, and related minerals that occur largely in pegmatites and veins. The usefulness of the U-Th-Pb decay systems was greatly expanded by applications to uranium- and thorium-bearing accessory minerals such as allanite, apatite, monazite, sphene, and zircon. Zircon is commonly used because of its widespread occurrence as an accessory mineral in rocks. A common misconception is that zircon, a durable mineral, is resistant to alternation, and for this reason more suitable for geochronology than minerals such as

biotite, hornblende, and so forth. Although some concordant ages have been determined for zircon, most of the ages are discordant. For zircon the pattern of age discordance usually is $^{207}Pb/^{206}Pb > {}^{207}Pb/^{235}U > {}^{206}PB/^{238}U > {}^{208}Pb/^{232}Th$. The $^{207}Pb/^{206}Pb$ age is not an independent determination. It is derived from the two uranium series that have bery different half lives (Table 1); hence, it is time dependent.

As in the Rb-Sr system where a correction must be made for common strontium, The U-Th-Pb systems also must be corrected for common lead incorporated in the minerals or rocks at the time of formation. The correction is made on the basis of ^{204}Pb for which no present-day radioactive parent is known. Laboratory contamination is a serious problem, and laboratories and procedures for purification of the water and reagents are constantly checked by determination of blanks.

Mineral separation predecures, decomposition techniques, and mass spectrometric measurements gradually have been improved. A significant innovation was the introduction by Silver and Deutsch (1963) of the technique of separating zircon concentrates into fractions on the basis of size and magnetic susceptibility. Commonly the small size fractions contain more uranuim than the larger sizes, and the more magnetic frations are relatively enriched in uranium. Radiation adamage and the resulting metamict structure developed in zircon are dependent on the uranium and thorium contents and on the time since formation. Loss of lead from metamict zircon is closely related to radiation damage and is a major cause of discordant ages.

Most laboratories use a dissolution technique devised by Krogh (1973) in which small samples of zircon, as well as other minerals, are decomposed with nitric and hydrofluoric acids in teflon capsules enclosed in stainless steel pressure vessels at a temperature of approximately 200°C for a period of several days or longer. The size of the sample depends on the uranium content and the age. Samples of about 10 mg are often used, but for older Precambrian zircon the sample may be reduced to less than 1 mg. Analyses have been made of single zircon grains (Lancelot et al., 1976), but the precision of much of this data is low and greatly complicates the interpretation of the results.

The development of a sensitive high-mass resolution ion microprobe at the Australian National University, Canberra, by Compston and associates (1982) has carried the reduction in sample size even further, permitting the determination of Pb isotopic composition and apparent ages in different parts of a single crystal. Ion microprobe analyses have provided dramatic confirmation of complex geological history of some old Precambrian rocks, of the existence of areas (domains) of high-uranium content, and of cores of older zircon.

The approach to the interpretation of discordant U-Pb ages differs greatly with individuals and to a large extent with their trainsing and experience (e.g., physicist, chemist, geologist). Discordant ages are a clear signal that something has happened so that the U-Pb systems were open at some time. The event may have been of relatively short duration or it may have involved millions of years. As a result, lead was lost from the zircon, partically or completely; uranium may have been added in the form of newly formed crystals or as overgrowths on older crystals, or a combination of these processes may have been operative. The systematics of the U-Th-Pb systems permit some reconstruction of the history.

Most useful for the interpretation of discordant U-Pb ages is a graphical scheme formalized by Wetherill (1956) now known as the concordia diagram (Fig. 2). In Figure 2 the atomic ratio $^{206}Pb/^{238}U$ is plotted against $^{207}Pb/^{235}U$. Points on the concordia curve give concordant ages. If because of some geological process, the systems are open and lead is lost at some time, the data points will plot along a line (discordia). The upper intercept (Fig. 2) gives the original age of the zircon, and the lower intercept is the time of the event causing the lead loss. If a high-temperature event (e.g., intrusion of a high-temperature magma) is involved, some of the original zircon (A) may be completely recrystallized, eliminating the radiogenic lead, so that it dates the time of the intrusion (B). The discordia line, then, can be considered a mixing line with two zircon populations (A and B). Some intermediate points (C, D, and E) are shown in order of decreasing lead contents.

Linear arrays of zircon data on a concordia diagram may be apparent rather than real. Figure 3 is an idealized case in which zircon of an original age (A) was altered by an event at time (B). Regardless of the nature of the event and whether we are dealing with lead loss or with a mixed

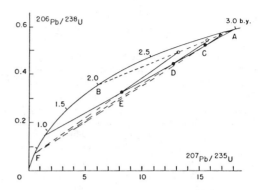

FIGURE 3. A second relatively young event (F), 0.4 billion years ago, is shown superimposed on the metamorphic event (B) in Figure 2, with lead losses for points C, D, and E similar to Figure 2. A line through these points intercepts concordia at (A), but the lower intercept is well above (F) and has no geological significance.

zircon population, the altered zircon will plot on line A-B. None may plot at either A or B. In the diagram (Fig. 3) the points C, D, and E have been plotted to corresponds to lead losses of 10, 20, and 40 percent, respectively. If the rock was much later subjected to a second, relatively recent geological process during which the lead losses were essentially of the magnitude of the first event, the points would be shifted toward the origin to give a linear array. With the assumptions made, the upper intercept is the original age (A), but the lower intercept has no geological significance. The conditions assumed in Figure 3 may not be approached in nature, but may be closely simulated. The apparent linear array may suggest an age that is older or younger than the original age (A).

Two or more geological events can result in a complicated zircon U-Pb age pattern. In the older Precambrian rocks of Wyoming and Montana, for example, apparent ages generally show the effects of a young geological event that can be related to the Laramide Orogeny or to a younger magmatic activity (~20 m.y. ago). In the Lake Superior region and adjacent parts of the Canadian Shield the young event is not readily identifiable. Goldich and Mudrey (1972) have interpreted the discordant U-Pb ages as the result of the chemical alteration of metamict zircon. Water is strongly adsorbed on the colloidal particles of secondary minerals formed during the metamict process. Quartz-rich rocks expand and shatter when they are uncovered and brought close to the surface by erosion. The dilatancy effect results in loss of part of the adsorbed water and dissolved radiogenic lead. An appreciable length of time elapses between original crystallization and development of the metamict structure, and time is required for uplift and erosion of area to bring the

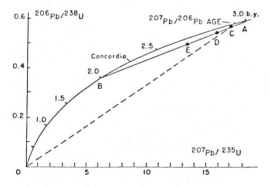

FIGURE 2. Schematic U-Pb concordia diagram for zircon fractions from a rock that formed 3.0 billion years ago (A) showing the effects of lead loss during a metamorphic event 2.0 billion years ago (B). The points C, D, and E are plotted to correspond to lead losses of 10, 20, and 30 percent, respectively.

rocks close to the surface. This explains why the $^{207}Pb/^{206}Pb$ ages commonly approach the time of origin.

Krogh (1982) introduced an air-abrasion technique, which removes the outer part of metamict zircon that may have lost lead as well as the badly altered and cracked grains. The U-Pb ages on the abraded zircon are more nearly concordant and the data points are shifted upward toward concordia. With this technique and with careful selection of the least magnetic grains, Krogh has been able to eliminate most of the age discordance and to reduce the probable age error to ±1 to 3 m.y. for 2700-m.y.-old zircons.

In old Precambrian rocks with a complex history that involved a number of younger geological events, the U-Pb systems may have been disturbed to an extent that permits only an approximation of the original age. It is important to differentiate between an interpretative age and a measured one. It is equally important to recognize that a precisely measured age, for example, on sphene or monazite, may date a younger metamorpic event and not the time of origin.

Sm-Nd System

The decay of $^{147}Sm \rightarrow {}^{143}Nd + \alpha$ provides a radiometric system that Lugmair (1974) and Lugmair and Marti (1978) worked out by the analysis of meteorites. G. J. Wasserburg, students, and associates at the California Institute of Technology developed analytical procedures for the application of the Sm-Nd system to terrestrial rocks and a method for calculating model ages on individual samples; see Jacobsen and Wasserburg (1980) for references.

The rare earth elements (REE) are relatively immobile and have been less enriched in continental crustal rocks than potassium, rubidium, uranium, and thorium. In the Sm-Nd system the initial ratio of $^{143}Nd/^{144}Nd$ in granitic rocks has increased at a lesser rate than in the ocean crust. The low concentrations of ^{147}Sm and the restricted range in the $^{147}Sm/^{144}Nd$ ratio presents analytical problems, but the lesser mobility of the REE compared to potassium, argon, rubidium, strontium, uranium, and lead results in greater stability of the Sm-Nd isotopic system under conditions of metamorphism and weathering, and the system has great promise for use in conjunction with other methods.

Lu-Hf System

The decay of $^{176}Lu \rightarrow {}^{176}Hf + \beta^-$ has promise as a radiometric system on the basis of the work done by M. Tatsumoto and associates in the laboratories of the U.S. Geological Survey in Denver (Patchett and Tatsumoto, 1980). The decay system has the advantages, but also some of the analytical problems, of the Sm-Nd system. Model ages on zircon, in which hafnium is concentrated, appear to be particularly useful.

Re-Os System

The decay of $^{187}Re \rightarrow {}^{187}Os + \beta^-$ has been investigated by Luck and coworkers (1980) in meteorites. Applications to terrestrial rocks can be expected, for example, in the analysis of molybdenite (MoS_2) in which rhenium commonly is concentrated. Molybdenite occurs in rocks ranging in age from the Tertiary to the Archean.

Radiocarbon System

The ^{14}C or radiocarbon dating system (q.v. in Vol. IVA) is different from the preceding systems in a number of respects. Unlike the radioactive uranium, rubidium, and potassium nuclides, which have a finite abundance in the earth, ^{14}C is generated by an $\eta\rho$ reaction in the upper atmosphere.

$$^{14}_{7}N + {}^{1}_{0}\eta \rightarrow {}^{14}_{6}C + {}^{1}_{1}\rho$$

The ^{14}C, however, is unstable and decays:

$$^{14}_{6}C \rightarrow {}^{14}_{7}N + \beta^-$$

The half-life is approximately 5,730 years, but to avoid confusion an earlier value of 5,570 years, which has been used for some time, has been retained by agreement among radiocarbon laboratories. The radiocarbon method depends on the assumption that the production of ^{14}C has been constant over the past 50,000 years, the effective range controlled by the short half-life.

The ^{14}C generated in the atmosphere is oxidized to $^{14}CO_2$ and mixes rapidly with the CO_2 in the atmosphere, from which it is taken up by plants and finds its way into animal life. So thorough is the process that all present-day organisms have a constant proportion of ^{14}C. When the plant or animal dies, however, the ^{14}C decays and after a period of approximately eight to nine half-lives the amount of remaining ^{14}C is so small that it no longer can be determined with the required precision.

Radiocarbon dating, like all methods, has its problems. Many of the limitations have been found and assessed (Olsson, 1970), and ^{14}C determinations are produced in thousands, exceeding all other radiometric dating. Among the variety of materials that have been dated are wood, charcoal, human hair, linen, rope, grain, shells, bones, and water. The method was developed by Libby and his associates at the University of Chicago. Libby's (1955) book and a wealth of literature reflect the widespread use and interest in the method. The *American Journal of Science* publishes a semiannual supplement of radiocarbon dates.

Other Methods

Other methods of dating have been used, at times with very good success. The isotopic composition of common lead varies with time, and a number of models have been devised to explain the evolution of lead. In these studies the isotopic composition of lead in galena and in minerals that contain little or no uranium and thorium is determined. Patterson (1956) obtained 4.55 billion (10^9) years for the age of the earth by comparing the abundance ratios of lead isotopes in meteorites with terrestrial lead. Early studies attempted to use helium produced by the decay of uranium and thorium, but failed to obtain reliable ages because of helium leakage. More recently, disequilibrium series in the uranium and thorium systems have been used to date geologically young material ranging back to approximately 250,000 years. Methods have been developed in a number of laboratories; John Rosholt of the U.S. Geological Survey is a pioneer in this work. Some of the Survey's work has been described by Szabo and Rosholt (1969).

Spontaneous fission provides a number of interesting possibilities, of which the most productive has been *fission-track dating* (q.v. in Vol. IVA). ^{238}U undergoes spontaneous fission, a rare process compared to alpha decay. The energy involved, however, is large, and considerable damage is done in minerals, forming fission tracks that become visible on etching a cleaved or polished surface. Heating anneals the crystals asnd destroys the tracks, limiting the applicability of the method. The temperature of erasure of the tracks, however, differs in minerals, and discordant ages may be obtained on apatite, zircon, and sphene. These comparative ages permit interpretation of the thermal history. Applications by Naeser (1979) and Zeitler et al. (1982) show great promise for the study of rates of uplift and erosion.

Oldest Rocks

The Precambrian shield areas of all the continents contain Archean rocks remnants that were formed between 3,00 and 3,800 million years ago. Probably the most studied of these rocks (3,600-3,800 m.y. old) are in the Godthaab District, West Greenland. Old rocks have been found in the Pilbara Block, West Australia; Newfoundland, Canada; Antarctica; South Africa; India; USSR; and the United States (Michigan, Minnesota, Wyoming, and Montana).

Time Scale

The Geological time scale undergoes constant study and revision (see *International Geochronological Time Scale*). The time scale in Table 3 is a tentative one currently used by the U.S. Geological Survey. Recent studies of the Phanerozoic time scale are summarized in two volumes edited by G. S Odin (1982). This work confirms earlier experience; approximate ages can be determined for rock boundaries, but the precise ages are extremely difficult to establish.

SAMUEL S. GOLDICH

TABLE 3. Phanerozoic Time Scale

Era	Period	Epoch	Millions of Years Ago
Cenozoic	Quaternary	Holocene	0.01
		Pleistocene	
			2
	Tertiary	Pliocene	
			5
		Miocene	
			24
		Oligocene	
			38
		Eocene	
			55
		Paleocene	
			63
Mesozoic	Cretaceous		
			138
	Jurassic		
			205
	Triassic		
			240
Paleozoic	Permian		
			290
	Pennsylvanian		
			330
	Mississippian		
			360
	Devonian		
			410
	Silurian		
			435
	Ordovician		
			500
	Cambrian		
			570
Precambrian			

References

Aldrich, L.T.; Davis, G. L.; and James, H. L., 1965. Ages of minerals from metamorphic and igneous rocks near Iron Mountain, Michigan, *Jour. Petrology* **6**, 445-472.

Boltwood, B. B., 1907. On the ultimate disintegration products of radio-active elements, Part II. The disintegration products of uranium, *Am. Jour. Sci.*, ser. 4, **33**, 77-88.

Compston, W., and Williams, I. S., 1982. Protolith ages from inherited zircon cores measured by a high mass-resolution ion microprobe, *5th International Conference on Geochronlogy, Cosmochronology, Isotope Geology.* Nikko, Japan, 63-64.

Dalrymple, G. B., and Lanphere, M. A., 1969. *Potassium-Argon Dating.* San Francisco: W. H. Freeman, 258p.

Doe, B. R., 1970. *Lead Isotopes.* New York: Springer-Verlag, 137p.

Faure, G., 1977. *Principles of Isotope Geology.* New York: Wiley, 464p.

Faure, G., and Powell, J. L., 1972. *Strontium Isotope Geology.* New York: Springer-Verlag, 188p.

Garner, E. L.; Murphy, T. J.; Gramlich, J. W.; Paulsen, P. J.; and Barnes, I. L., 1975. Absolute isotopic abundance ratios and the atomic weight of a reference sample of potassium, *U.S. Natl. Bur. Standards Jour. Research* **79A,** 713-725.

Godwin, H., 1962. Half-life of radiocarbon, *Nature* **195,** 984.

Goldich, S. S., and Mudrey, M. G., Jr., 1972. Dilatancy model for discordant U-Pb zircon ages, in A. I. Tugarinov (ed.), *Contributions to Recent Geochemistry and Analytical Chemistry* (Vinogradov Volume). Moscow: Nauka, 415-419.

Hart, S. R., 1961. The use of hornblendes and pyroxenes for K-Ar dating, *Jour. Geophys. Research* **66,** 2995-3001.

Holmes, A., 1947. The construction of a geologic time-scale, *Geol. Soc. Glasgow Trans.* **21,** 117-152.

Jacobsen, S. B., and Wasserburg, G. J., 1980. Sm-Nd isotopic evolution of chondrites, *Earth and Planetary Sci. Letters* **50,** 139-155.

Jäger, E., and Hunziker, J. C. (eds.), 1979. *Lectures in Isotope Geology.* New York: Springer-Verlag, 329p.

Krogh, T. E., 1973. A low-contamination method for hydrothermal decomposition of zircon and extraction of U and Pb for isotopic age determinations, *Geochim, et Cosmochim. Acta* **37,** 485-494.

Krogh, T. E., 1982. Improved accuracy of U-Pb zircon ages by creation of more concordant systems using an air abrasion technique, *Geochim. et Cosmochim. Acta* **46,** 637-649.

Lancelot, J.; Vitrac, A.; and Allègre, C. J., 1976. Uranium and lead isotopic dating with grain by grain zircon analysis: a study of complex geological history within a single rock. *Earth and Planetary Sci. Letters* **29,** 357-366.

Libby, W. F., 1955. *Radiocarbon Dating,* 2d ed. Chicago: University of Chicago Press, 175p.

Luck, J. M.; Birck, J. L.; and Allègre, C. J., 1980. [187]Re-[187]Os systematics in meteorites: early chronology of the solar system and age of the galaxy, *Nature* **283,** 256-259.

Lugmair, G. W., 1974. Sm-Nd ages: a new dating method, *Meteoritics* **9,** 369.

Lugmair, G. W., and Marti, K., 1978. Lunar initial [143]Nd/[144]Nd: differential evolution of the lunar crust and mantle, *Earth and Planetary Sci. Letters* **39,** 349-357.

Naeser, C. W., 1979. Fission-track dating and geologic annealing of fission tracks, in E. Jäger and J. C. Hunziker (eds.), *Lectures in Isotope Geology.* New York: Springer-Verlag, 154-169.

Nier, A. O., 1939. The isotopic constitution of radiogenic leads, and the measurement of geological time, II, *Phys. Rev.* **55,** 153-163.

Odin, G. S. (ed.), 1982. *Numerical Dating in Stratigraphy, Parts I and II.* New York: Wiley, 1040p.

Olsson, I. U. (ed.), 1970. *Radiocarbon Variations and Absolute Chronology: Proceedings of the 12th Nobel Symposium at Uppsala University.* New York: Wiley, 652p.

Patchett, P. J., and Tatsumoto, M., 1980. Lu-Hf total-rock isochron for the eucrite meteorites, *Nature,* **288,** 571-574.

Patterson, C. C., 1956. Age of meteorites and the earth, *Geochim, et Cosmochim. Acta* **10,** 230-237.

Peterman, Z. E.; Goldich, S. S.; Hedge, C. F.; and Yardley, D. H., 1972. Geochronology of the Rainy Lake region, Minnesota-Ontario, in B. R. Doe and D. K. Smith (eds.), *Geol. Soc. America Mem.* **135,** 193-215.

Silver, L. T., and Deutsch, S., 1963. Uranium-lead isotopic variations in zircons: a case study, *Jour. Geology* **71,** 721-758.

Steiger, R. H., and Jäger, E., 1977. Subcommission on geochronology: convention on the use of decay constants in geo- and cosmochronology, *Earth and Planetary Sci. Letters* **36,** 359-362.

Szabo, B. J., and Rosholt, J. N., 1969. Uranium-series dating of Pleistocene molluscan shells from southern California—an open system model, *Jour. Geophys. Research* **74,** 3253-3260.

Wetherill, G. W., 1956. Discordant uranium-lead ages, I, *Am. Geophys. Union, Trans.* **37,** 320-326.

York, D., and Farquhar, R. M., 1972. *The Earth's Age and Geochronology.* New York: Pergamon Press, 178p.

Zeitler, P. K.; Johnson, N. M.; Naeser, C. W.; and Tahirkheli, R. A. K., 1982. Fission-track evidence for Quaternary uplift of the Nanga Parbat region, Pakistan, *Nature* **298,** 255-257.

Cross-references: *International Geochronological Time Scale.* Vol. IVA: *Carbon-14 Dating; Fission Track Dating; Geochronometry; Geologic Time Scale; Ionium-Thorium Dating; Potassium-Argon Age Determination; Radionuclides; Rubidium-Strontium Dating Method; Uranium-Helium Isotopic Age Method; Uranium-Thorium-Lead Age Determination.* Vol. XIV: *Lichenometry.*

GEOCHRONOMETRY—See GEOCHRONOLOGY; INTERNATIONAL GEOCHRONOLOGICAL TIME SCALE.

GEOCRYOLOGY

Geocryology (from the Greek *geos,* earth; *kryos,* cold; and *logos,* discourse) literally means a discourse on the cold portion of the earth. It is the study of ice and snow on the earth, especially the study of *permafrost.* By convention, the term has not generally included the study of ice in the atmosphere, yet snow and ice cover a major part of the earth. Terminology is complicated (see e.g., Everdingen, 1976). For emphasis and ease of handling, various subdivisions of geocryology are discussed elsewhere in this volume (q.v. *Permafrost, Engineering Geology*). Overlap also will be found with other topics (Washburn, 1980).

The "world of ice" is a fascinating place for skiers and others, but a scourge to many. Avalanches from a local overabundance of snow are to be avoided or reduced in their destructiveness by periodic artificial triggering, yet elsewhere techniques are sought to augment snow accumulation for an ultimate increase in water supply. Perennial and seasonal snow and ice modify the work of water, wind, gravity processes, and organisms (including humans) and superimpose on the landscape their own distinctive marks. To combat their effects, most man-made works in frost environments must be designed accordingly, and the timetable for construction and transportation must allow for their role. All are familiar with the delays induced by snow storms and icy streets. Snow removal is estimated to cost more than $100 million annually in the United States alone. Road icings from ground water, ice jams in streams, cutting or breaking of seasonal ice for ship and barge traffic, air bubbling in water to reduce ice damage to structures, and frost heave and breakup in roads are common in the cold regions.

Freezing of water produces snow and hail in the atmosphere, which may fall on the earth, and frost or various aggregates of ice directly on and in streams, lakes, oceans, and the organic and lithic portions of the earth proper (Kingery, 1963). Perhaps only 2 percent of the total water on the earth occurs as ice, but it is widespread in the polar and high alpine regions throughout the year and in the temperate regions in winter.

The cold regions may be defined as those portions of the earth within the O°C isotherm of mean temperture for the coldest month of the year (Fig. 1). The cold regions thus constitute about half the land mass of the Northern Hemisphere. Snow and frost also occur outside the cold regions, and sea ice covers up to 12 percent of the oceans.

Ice is a compound of hydrogen and oxygen, whose proportions make up about 11 percent and 89 percent, respectively, by weight. The normal phase of ice in nature is hexagonal. Clathrate hydrate of ice with "quest" substances in the lattice may occur at depth in antarctic ice. The hydrogen bonding plays an important role in many properties of ice. However, varying amounts of impurities—such as air, dirt, organic matter, micro-meteorites, and trace elements and compounds of many kinds—are also present and are equally or more important. Normally more than one each of the several isotopes of hydrogen and oxygen is present. The isotopes of oxygen are used to reconstruct past temperature changes in the atmosphere where snow formed that fell on old glaciers, as in Greenland and Antarctica. The ratios of oxygen-16 and oxygen-18 differ signifi-

cantly over long periods and from summer to winter.

Snow accumulates on the surface of the earth at densities of less than 0.1 to more than 0.4. The time required in the transition from snow to ice may be only hours or days to many centuries. The natural transition of snow to ice is promoted by warm temperature, incident radiation, pressure, and circulating air. The distinction between snow and ice is based commonly on the transition from permeable to impermeable conditions to air. In an aggregate, the transition occurs at a density of about 0.83 (pure ice is 0.917 at 0°C and 1 atm).

Old snow increases in strength with time even in the absence of pressure. Bonds between grains grow by means of vapor transfer, and density increases by means of volume diffusion. The process can be speeded up in cold regions by artificial milling to reduce particle size. Processed snow (Peter snow where the Peter machine has been used) ages more rapidly than normal snow and is used in buildings, runways, and roads. Where greater strength is needed, brush, timbers, wood pulp, and artificial fibers are added. These increase resistance to bending and fracture, and wood pulp and other fibers can eliminate or drastically reduce brittleness. In the aging process, *sintering* that is, regulation at ice contacts, takes place. Ice also adheres to other substances with *adfreeze strengths,* which vary considerably because of the nature of the material and surface conditions. The heaving of piles is a common result of seasonal freezing.

Many mechanical properties are related directly to the crystallographic orientation of ice. A single ice crystal, for example, glides readily in any direction along its basal plane, but with much more difficulty in other directions. The lattice responds like a deck of cards that slide over one another in only one plane. The crystal lattices of an aggregate of ice crystals deform less easily than single crystals but similarly in visco-plastic flow, being more "viscous" at low stresses and more "plastic" at high stresses. Movements of dislocations are a major mechanism of flow in cold ice, and grain boundary movements increase in importance in warm ice. *Pressure melting* may occur at some junctions of grains. Clearly, rate of stress, time, and temperature play important roles in the flow of ice. The amount and kinds of fractures also are especially dependent on temperature and impurities. Ice is very brittle at cold temperatures.

A liquid-like layer a few tens of angstroms thick and with properties intermediate between ice and water may be found on the surface of ice and between grains. When ice recrystallizes, impurities tend to be rejected and concentrated at grain boundaries, lowering the melting point at grain boundaries. When abundant, as in sea ice, the impurities can drain off in time, increasing the hardness and potability of the ice. The response of ice to stress is markedly nonlinear. It is a type of

FIGURE 1. Climatic zones of the cold regions of the Northern Hemisphere as defined by the isotherms for 0°C (32°F), −17.8°C (0°F), −31.7°C (−25°F) for the mean temperature for the coldest month of the year (from Army Cold Regions Research and Engineering Laboratory, 1969).

creep and differs in behavior in compression or tension from shear. Granulation and recrystallization commonly are present in displacements that are very temperature-dependent. At low temperatures, the hardness of ice increases, and strain rates are reduced with equivalent stresses.

The *coefficient of thermal expansion* of ice is 5 to 10 times that of most common rock-forming minerals. Ice is weak under tension, and temperature drops readily induce cracking. Because water expands on freezing, by approximately 9 percent, its liquid and solid phases are opposite to most other natural substances. In confinement, ice can develop enough expansive force to fracture any rock. Freezing of water in joints, cracks, parting planes, cleavage planes, and the like promote granular disaggregation, spalling, exfoliation, and a variety of angular fragments. In the ground, the growth of ice lenses and the cryostatic pressure of confined water may cause the surface to heave. The *latent heat of fusion* of ice at 0°C and standard pressure is 333.5 kj · kg⁻¹ and the *latent heat of sublimation* at the triple point of ice, water, and vapor is 2,838 kJ · kg⁻¹. Both are important facets in the thermal environment of the freeze/thaw process. Furthermore, the heat capacity of ice is only half that of water, but ice conducts heat about four times as quickly as water.

The unusual electrical properties of ice have been put to use in attempts to devise mechanisms for finding crevasses under snow cover, for determining the thickness of glaciers, for measuring the density of ice, and so on. Preferential absorption of snow and ice in the infrared region and strongly varying *albedo* of pure to bubbly to dirty ice and of fresh to old snow alters the thermal regime accordingly. Coal dust and other dark substances have been used to decrease the albedo of snow and ice and speed up the rate of thaw in spring to lengthen navigation, to clear roads and runways, and to permit earlier spring planting. Mechanical and chemical means of snow and ice removal are commonplace.

Radio echo sounding of cold glaciers, especially the thick ice of Antarctica and Greenland, utilize frequencies in the range of 30 to 50 MHz for airborne equipment. Greater distances can be covered in a few hours than in an entire field season by surface transport. Warm glaciers, where water is present, are less suitable. Water beneath glaciers, however, provides the best reflection. Radio echo sounders on the surface of flowing glaciers have been used to measure velocity by recording the shape of the returning pulses from the base. Side-looking airborne radar and infrared are among various techniques that have been used to determine sea ice conditions in seaways.

A host of special applied problems are introduced by ice in its various forms in water and on and in the ground. An understanding of the physics of ice is fundamental to the solutions of the problems (Hobbs, 1974). Snow cover, glaciers, and ice in rivers, lakes, and oceans create problems more obvious to most people than does ice in the ground. However, ice in the ground is especially important in geomorphic processes, such as slope movements and frost action, and also makes construction much more difficult and expensive than in areas outside the cold regions (National Research Council of Canada, 1978). For economic reasons alone, we must learn to live with the geocryological problems rather than fight them. The "winters of the world" have been with us throughout earth history and will continue to be (John, 1979).

ROBERT F. BLACK

References

Army Cold Regions Research and Engineering Laboratory, 1969, *Cold Regions Science and Engineering Monograph 1-A.*

Everdingen, R. O. van, 1976. Geocryological terminology, *Canadian Jour. Earth Sci.* **3,** 862-867.

Hobbs, P. V., 1974. *Ice Physics.* Oxford, England: Clarendon Press, 837p.

John, B. S. (ed.), 1979. *The Winters of the World—Earth under the Icebergs.* New York: Wiley, 256p.

Kingery, W. D. (ed.), 1963. *Ice and Snow—Properties, Processes, and Applications.* Cambridge, Mass.: M.I.T. Press, 684p.

National Research Council of Canada, 1978. *Proceedings of the Third International Conference on Permafrost,* Vols. I and II. Ottawa, 1202p.

Washburn, A. L., 1980. *Geocryology.* New York: Wiley, 406p.

Cross-references: *Geotechnical Engineering; Remote Sensing, Engineering Geology; Rheology, Soil and Rock.* Vol. II: *Climate, Geomorphology.* Vol. III: *Frost Action; Patterned Ground; Permafrost; Solifluction.*

GEODATA MANAGEMENT—See COMPUTERIZED RESOURCES INFORMATION BANK. Vol. XIV: GEOLOGICAL CATALOGING; GEOSTATISTICS; PUNCH CARDS, GEOLOGICAL REFERENCING.

GEODESY—See Vol. XIV: SATELLITE GEODESY AND GEODYNAMICS.

GEOELECTRICAL SURVEYS—See Vol. XIV: EXPLORATION GEOPHYSICS; VLF ELECTROMAGNETIC SURVEYING.

GEOENGINEERING TESTS—See ATTERBERG LIMITS AND INDICES; CLAY, ENGINEERING GEOLOGY; CLAYS, STRENGTH OF; CONSOLIDATION, SOIL; FOUNDATION ENGINEERING; LATERITE, ENGINEERING GEOLOGY; MARINE SEDIMENTS, GEOTECHNICAL PROPERTIES; PIPELINE CORRIDOR EVALUATION; ROCK STRUCTURE MONITORING; SOIL CLASSIFICATION SYSTEM, UNIFIED; SOIL MECHANICS.

GEOEPIDEMIOLOGY—See MEDICAL GEOLOGY. Vol. XII, Pt. I: DENTAL CARIES AND SOILS.

GEOEXPLORATION— See Vol. XIV: EXPLORATION GEOCHEMISTRY; EXPLORATION GEOLOGY; EXPLORATION GEOPHYSICS; PROSPECTING.

GEOLOGICAL COMMUNICATION

Meaning of Communication

Communication is defined as an act of transmitting information, the information transmitted, and the form in which it is transmitted. Transmission alone requires only the sending, but for communication to be accomplished, reception and comprehension are essential. O. N. Skulberg (1972), commenting on scientific communication, states, "It is sometimes forgotten . . . that communication is effective only when information is conveyed to and understood by another human mind. Unless this basic goal of understanding is achieved, [scientists] will fail in the primary purpose" (p. 372). It is imperative for the communicator of scientific information to recognize the intended audience's capacity to understand the contents of the message.

Audiences for Geological Communication

Geological information (q.v.) is disseminated to a variety of audiences. The primary audience consists of geologists, but, in this day of specialization, that audience is divided into many groups. Thus, many communications are now addressed to fellow specialists, who belong to an informal network for exchange of information on research in progress and first announcement of findings and conclusions. The published (formal) communication of the announcement for placement in the permanent record, principally the scientific literature, and for general availability, follows. Geologists working outside the same special area can be regarded as a different audience, as can scientists in other major fields. Further audiences consist of engineers, technologists, students being educated and trained in a science, other students, laymen with a need for special scientific information, and laymen with a casual interest in science.

Requirements

The growth of applied geology (q.v.) in this century has increased the importance of communication between geologists and audiences consisting mainly of laymen, who are not geologists, seeking assistance in dealing with practical problems (see Whitmore, 1959). The requirements of the laymen are varied, not only because of the

different purposes in seeking geologists' assistance but because of differences in the type of information needed, the time and cost allotted to the investigation, and the type of communication desired (see Vol. XIV: *Popular Geology*). The substance of communications may consist of data relevant to the problem, which the requester intends to evaluate; of interpretations or recommendations by the geologist, without background information; or of interpretations or recommendations with appended explanations, which the requester may wish to have for reference or other interpretation.

Groups of laymen that make regular use of geological assistance often are well-informed about geological factors that influence their activities and can specify the type of information needed and the form in which it is to be communicated. Geologists trained in such fields as petroleum geology (q.v.) and engineering geology are similarly informed about the requirements of requesters and are equipped to communicate effectively with them. It is in the areas of occasional use that difficulties can arise. Some groups of laymen may have a rudimentary knowledge of geology, but geologists are generally uninformed about the nature of their activities. For example, most geologists drawn into the service of armies in time of war have had no exposure to the concepts and terminology of military science. While no permanent resolution of the difficulty is achieved through such temporary connections, effective relationships have been developed in World War I and World War II (see *Military Geoscience*).

Groups that have essentially no knowledge of geology do not usually seek help from geologists; when they do, they have difficulty specifying the type of information they require. The widespread adoption of environmental protection and land-use policies, and the enforcement of regulations requiring the evaluation of impacts of proposed actions on the environment, have brought new groups of laymen into increasingly frequent contact with geologists. Real estate developers and builders are an example. Mathewson and Ruckman (1974) found that the members of this group "are not proficient in engineering or geology, nor do they wish to be." As a group they are not able to make use of geological, soils, and hydrological reports and maps to evaluate sites for land use. Instead of data, they request "interpretive information specifying the suitability of land for various uses," presented in a form intelligible to them. In this case, engineering geologists are better informed about the natural suitability of sites than are the requesters. However, environmental protection regulations demand consideration of other factors, especially costs, risks, and benefits. It is becoming necessary for geologists to acquire sufficient knowledge of the other factors, which are often more important than the geological factors, and relate their interpretations more directly to those factors (see Vol. XIV: *Environmental Geology*).

Sources of Information

The answers to practical problems involving geological factors are based mainly on information already recorded or on new information obtained through investigations undertaken specifically for the purpose. The recorded knowledge is contained largely in the geological literature, on maps, and in collections of data (see *Geological Information, Marketing;* Vol. XIV: *Geological Cataloging; Map and Chart Depositories*). Despite highly developed systems for retrieving material from libraries and computerized storage, probably or possibly useful material is often difficult to identify in any of the existing types of indices (see *Computerized Resources Information Bank*). The quantity of available information may be insufficient to provide answers to the problems, and the reliability of some information may be questionable. Those obstacles should be cited in the communication to the layman.

Many continually active types of applied geology depend chiefly on observations on the ground and in the laboratory for data. Geophysical, geochemical, and other special methods, as well as interpretation of aerial photography (see Vol. XIV: *Photogeology*) and other imagery (see *Remote Sensing, Engineering Geology*), have made the procurement of original observations a source of information for both applied and basic scientific purposes.

Texts

Geological writing, whether for geological or other audiences, has been criticized by geologists and editors for wordiness, excessive use of jargon, bad style, and poor grammar since the early days of the science (see Betz, 1963, 1979). Mather and Mason (1950) observed, "The literature of geological science is extraordinarily voluminous, partly because of the tendency displayed by geologists . . . to set forth their ideas *in extenso* and to give lengthy, detailed descriptions of the phenomena with which they [deal]."

Armstrong (1962) remarked on the increasingly specialized usages of existing words, and the development of new words peculiar to the science. "To the extent that the understanding of these words is restricted to smaller and smaller groups of people they become jargon in the ears of all but the experts. However justifiable the use of such jargon by the initiate, it becomes increasingly difficult for the uninitiate to understand." Particularly pertinent to applied geology is the further comment: "Jargon, surely; specialized speech, yes, indeed, but only where they belong. Communication with the public involves cerebration at the desk—a job that must be done."

Speaking of scientists generally, an editor, J. D. Elder (1954) stated:

No scientific writer will admit that he uses confused, unintelligible language, or that he writes gibberish, or even that his vocabulary is secret (no matter what others may say of it). But if the technical vocabulary of a science is jargon, then writers of or on science use jargon. . . . Whether or not to use jargon in a particular piece of writing depends on the audience for whom the writing is intended. . . . Jargon is good when the reader can reasonably be assumed to know what it means, and bad when he cannot. Even when he does know the meaning of jargon, however, simple words can often be substituted for technical ones.

G. O. Smith (1922), one-time director of the U.S. Geological Survey, admonished geologists as producers of geological communications addressed to the public to remember the difference in vocabularies, adding, "I have told practical men of business that they should give little credence to the geologist who cannot tell his story in common language."

Half a century later, the involvement of geologists in preparing environmental impact statements (see Vol. XIV: *Environmental Geology*) for laymen has brought further comment. Lessing and Smosna (1975) reviewed geological contributions to such statements and found that "many contain erroneous, irrelevant, misleading, and grammatically poor geological statements." They found also examples of "geological trivia" cluttering the statements. The problem, as in the careless use of jargon that does not convey information, represents inattentiveness to the requirements of useful of communication. The basic difficulty, whatever the mode of communication, is expressed by Lüttig (1978): "(Even now many geoscientists . . . cannot translate scientific knowledge into layman's data which can be directly applied" to solving practical problems.

Maps

Laymen often prefer maps to texts for various reasons. Maps can be a more direct means of access to information, which might be scattered through many pages of text. They can be referred to rapidly whenever needed. Maps emphasize the relationship between the area or place and the information. For some uses, the portability of maps is a distinct advantage (see Vol. XIV: *Maps, Logic of*).

Basic geological, soil, hydrologic, and other geoscientific maps are not designed primarily for the untrained user and do not focus on practical applications. Annotations can make some of these maps usable by the layman. However, geoscientists have responded to the need for maps that are designed specifically for nonscientists. Broadly characterized, they are either data maps or interpretive maps. Some of both types convey the information wholly or largely by the graphic presentation. The map legend may be accompanied by short, simple explanations (see Vol. XIV: *Map Series and Scales*). Other maps serve chiefly as an index to information supplied in accompanying tables or matrices, or in longer texts. There are innumerable variations, which are developed to meet such conditions as the requirements of the intended users, the characteristics of the problem, the availability and reliability of data, and the conditions under which the map will be used.

Maps are especially useful for the presentation of data relating to engineering problems, (see *Maps, Engineering Purposes*), and their development for civil and military applications can be traced through many examples produced since the beginning of this century (see, for example, Stremme and Moldenhauer, 1921; Reuter and Thomas, 1961; Lutzen and Williams, 1968). While trained engineers may be able to use data maps with appended explanations, builders and developers of real estate, who must comply with environmental-protection regulations and land-use policies, appear to be unable to derive information from them. For the latter group, Mathewson and Piper (1975) proposed a type of geo-economic map on which each geological environmental unit is defined in terms of value, based on an appraisal of costs and risk, which represents the meaning of physical environment to that audience.

During World War II, terrain evaluation (q.v.) for the purpose of determining trafficability for tracked vehicles was a forerunner of comprehensive evaluation of environment for different, nonmilitary purposes (see *Military Geoscience;* Vol. XIV: *Environmental Geology*). Various types of maps were created by military geologists and geographers, some of them entirely interpretive, others highlighting evaluations, but with important supporting information on the map sheet. Some solutions were ingenious combinations of easily read interpretations and syntheses of environmental data.

Evaluation of the suitability of areas for specific human uses and the impact of the uses on the environment and humans is a problem for multidisciplinary investigations in which geoscientists are participants. Since the problem became a major feature of applied geoscience in the 1970s, experimentation with different types of communication than the prescribed formats of environmental-impact statements has been in progress. Different solutions have been offered, which make use of maps as an essential element of communication (see, for example, Cendrero, 1975; Environmental Analysis Group, 1980; Fisher et al.,

1972; Lüttig, 1978; and Montgomery, 1969). The number of factors to be considered has been an obstacle to designing effective graphic presentations. The communication of information depends on the accompanying detailed tables and matrices. Lately, computers have been used to produce maps, graphs, and analog models for regional-planning and land-use projects (see Vol. XIV: *Cartography, Automated*). The capacity of computers to store and treat data provides a means for preparing and revising the maps rapidly. They differ markedly from conventional types of maps, and it is not certain that they are more effective. (For a discussion of the computer as a tool in a land use project, see St-Onge et al., 1975).

Because of the wide range of human uses considered in planning projects (see Montgomery, 1969), a heterogeneous audience can be expected for the communications. As the content becomes more complex and the style of presentation more complicated, it is likely that geoscientists, bioscientists, and engineers, as well as economists and other specialists, will have to be called on to interpret the communications for many laymen. The objective of communicating environmental information for practical purposes in a simple manner has not yet been attained.

Scientific Value

Since World War II, the term *research and development (R&D)* has come into use in the military-industrial complex to represent the linking of scientific research with the application of scientific knowledge to an array of practical problems. The demands placed on the sciences for new information has caused the boundaries between applied research and basic research to become indistinct in many areas of science (see *Geology, Applied*). The communications—which may consist of texts, tables, and maps—prepared by scientists for R&D projects are referred to as *technical reports*. In the structure of scientific communication, their status has not been clearly defined. For a longer time, technical reports were regarded as *informal* communications, which could not be classed as part of the scientific literature.

Technical reports are often concealed for reasons of national security or industrial confidentiality, and their distribution is carefully controlled. It was claimed, when technical reports first became known as a type of communication, that the contents would not generally warrant publication in the open literature because of being largely derivative information or tentative information, which would require further development and critical evaluation to determine publishability. When unclassified technical reports were released for general distribution, it was soon recognized that many contained publishable information, including the first communication of some new findings of fundamental research and of advances in applied research. A sampling of more than 800 technical reports in physics (Gray and Rosenborg, 1957) showed that all the publishable material in half of the sample had been published in the open literature within two or three years. Easy access to a great many technical reports in different sciences and citation in standard bibliographic and abstract journals have reduced the amount of republication that authors formerly considered necessary. Technical reports for the purposes of applied science have become an important source of basic scientific information (see *Engineering Geology Reports*).

The recent, growing demand for environmental information has required the development of new concepts and methods of multidisciplinary research and of criteria for evaluation of data, which have been reported largely in communications dealing with practical applications, but are important fundamental contributions to the geosciences for use in basic research on the environment. While initial communication is usually directed to nonscientific audiences, information of scientific value, including case histories, is being published in the scientific literature. Similar transfers of information from applied to basic science have occurred in the past, but are becoming increasingly important as applied science moves into the forefront of scientific activity.

FREDERICK BETZ, JR.

References

Armstrong, H. S., 1962. Jargon and geology, *Royal Soc. Canada Trans.* **56** (ser. 3), 101-110.
Betz, F., Jr., 1963. "Geologic communication," in C. C. Albritton (ed.), *The Fabric of Geology.* Reading, Mass.: Addison-Wesley, 193-217.
Betz, F., Jr., 1979. Geology presented in collections of papers from the published literature, Part 1, in A. P. Harvey, and J. A. Diment (eds.), *Geoscience Information — A State-of-the-Art Review.* Heathfield, Sussex: Broad Oak Press, 247-254.
Cendrero, A., 1975. Environmental geology of the Santander Bay area, northern Spain, *Environmental Geology* **1**, 97-114.
Elder, J. D., 1954. Jargon — good and bad, *Science* **119**, 536-538.
Environmental Analysis Group, University of Santander, 1980. Environmental survey along the Santander-Unquera coastal strip, northern Spain, and assessment of its capacity for development, *Landscape Planning* **7**, 23-56.
Fisher, W. L., et al., 1972. *Environmental geologic Atlas of the Texas Coastal Zone: Galveston-Houston Area.* Austin: Bureau of Economic Geology, University of Texas, 91p.
Gray, D. E., and Rosenborg, S., 1957. "Do technical reports become published papers?" *Physics Today* **10**(6), 18-21.

Lessing, P., and Smosna, R. A., 1975. Environmental impact statements — worthwhile or worthless? *Geology,* **3**, 241-242.
Lüttig, G., 1978. Geoscience and the potential of the natural environment, *Nat. Resources and Development* **8**, 93-107.
Lutzen, E. E., and Williams, J. H., 1968. Missouri's approach to engineering geology in urban areas, *Assoc. Eng. Geologists Bull.* **5**, 109-121.
Mather, K. F., and Mason, S. L., 1950. *A Source Book in Geology.* New York: McGraw-Hill.
Mathewson, C. C., and Piper, D. P., 1975. Mapping the physical environment in economic terms, *Geology* **3**, 627-629.
Mathewson, C. C., and Ruckman, D. W., 1974. Geologic needs and knowledge of real estate brokers and builders, *Geology* **2**, 539-542.
Montgomery, H. B., 1969. Environmental analysis in local development planning, *Appalachia* **3**(3), 1-11.
Reuter, F., and Thomas, A., 1961. Die ingenieurgeologische Kartierung in Deutschland, *Zeitschr. Angew. Geologie* **7**, 116-122.
Skulberg, O. N., 1972. Discussion, in N. Polunin (ed.), *The Environmental Future.* London: Macmillan, 372.
Smith, G. O., 1922. Plain geology, *Econ. Geology* **17**, 34-39.
St-Onge, D. A., et al., 1975, Geoscience and planning in the future, in F. Betz, Jr. (ed.) *Environmental Geology.* Stroudsburg, Pa.: Dowden, Hutchinson & Ross., Benchmark Papers in Geology, 25, 354-365.
Stremme, H., and Moldenhauer, E., 1921. Ingenieurgeologische Baugrundkarte der Stadt Danzig, *Zeitschr. Prakt. Geologie* **29**, 97-100.
Whitmore, F. C., Jr., 1959. Geologic writing for the nongeologist, *Jour. Geol. Education* **7**(1), 25-28.

Note: The papers by Lutzen and Williams (1968), Montgomery (1969), and Stremme and Moldenhauer (1921), listed above, are reprinted in F. Betz, Jr. (ed.), 1975, *Environmental Geology.* Stroudsburg, Pa.: Dowden, Hutchinson & Ross, Benchmark Papers in Geology, 25.

Cross-references: *Conferences, Congresses, and Symposia; Earth Science, Information and Sources; Engineering Geology Reports; Geological Information, Marketing; Geology, Applied; Maps, Engineering Purposes; Military Geoscience; Terrain Evaluation, Military Purposes.* Vol. XIV: *Cartography, Automated; Cartography, General; Environmental Geology; Geological Cataloging; Geological Highway Maps; Geomythology; Geostenography; Map and Chart Depositories; Maps, Environmental Geology; Popular Geology; Terrain Evaluation Systems.*

GEOLOGICAL ENGINEERING METHODS — See ARID LANDS, ENGINEERING GEOLOGY; CHANNELIZATION AND BANK STABILIZATION; COASTAL ENGINEERING; EARTHQUAKE ENGINEERING; FOUNDATION

ENGINEERING; GROUT, GROUTING; LIME STABILIZATION; RAPID EXCAVATION AND TUNNELING; RIVER ENGINEERING; ROCK SLOPE ENGINEERING.

GEOLOGICAL FIELD SURVEYS, MAPPING—See FIELD GEOLOGY; MAPS, ENGINEERING PURPOSES. Vol. XIV: ENVIRONMENTAL GEOLOGY; EXPLORATION GEOLOGY; MAPS, PHYSICAL PROPERTIES.

GEOLOGICAL INFORMATION, MARKETING

The recent national realization that natural processes such as landslides, expanding soils, floods, hurricanes, earthquakes, and volcanoes are disasters only when people are victims has given impetus to include geology in land use decision making. The national flood insurance program, the USGS earthquake hazard reduction and landslide hazard reduction programs, and numerous other national, state, and local government programs now require input of geological information. Many major land developers who plan communities now consider the local geology (see Vol. XIV: *Land Capability Analysis*). Unfortunately, many programs and plans do not fully utilize geological information (q.v.) because a basic communication gap exists between geologists and land-use decision makers.

Many professional efforts to encourage wider application of environmental geology (see Vol. XIV), land use geology, or urban geology (q.v.) appear to be written by geologists for geologists. Consequently, planners, developers, or decision makers tend to ignore the data because of the unfamiliar vocabulary, or because they lack the expertise required to make a meaningful analysis. In addition, users not familiar with geological concepts are unable to ask the "right" questions (see Vol. XIV: *Popular Geology*).

As shown by a survey by Mathewson and Font (1974) of 150 federally funded urban plans (in the HUD-701 Planning Program), there is a general disregard of geological factors in land use decision making (Fig. 1). Considering Hurricane Agnes

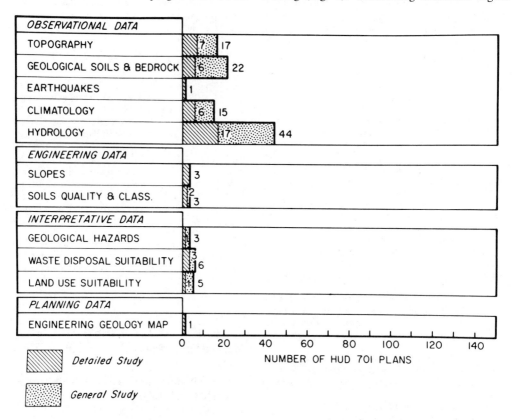

FIGURE 1. Bar diagram of geological information reported in 150 Urban Plans (from Mathewson and Font, 1974, p. 24).

(1972), Hurricane Eloise (1975), the Mississippi River and Rapid City floods, and other recent flood disasters, it is discouraging to note that more than 70 percent of 150 city plans do not even mention local hydrologic conditions. In an effort to determine why this lack of geological information exists, Mathewson and Ruckman (1974) surveyed the geological knowledge of Texas real estate brokers and builders. It is believed that if users such as they were more aware of geology, they would insist that local governments have geological data available.

The survey by Mathewson and Ruckman (1974)

was divided into three parts. Part A asked the respondent to rate the significance of engineering geological factors as they apply to his evaluation of a homesite (Fig. 2). Part B asked the respondent to evaluate the suitability of a land tract based on the given geological description, and Part C asked the respondent to identify the specific geological information that would be desired for land suitability analysis. The results of Part A are shown in Table 1 and the results of Part C in Figure 3. Results from these two parts suggest that most correlations appear compatible because low or insignificant ratings correspond to low demands.

Part A: Factor Rating

Kindly rate the significance of each of the engineering geology factors listed below as they apply to your evaluation of a homesite. Remember, it is as important to our study that you identify those factors that you do not know as it is that you rate the significance of the other factors. Please assume that all of the factors are available to you for your site evaluation.

	Don't Know	Insignificant	Importance Low	Medium	High	Extreme
fault zones	_	_	_	_	_	_
surface drainage	_	_	_	_	_	_
depth to the groundwater table	_	_	_	_	_	_
subsurface drainage	_	_	_	_	_	_
surface topography	_	_	_	_	_	_
local hydrology	_	_	_	_	_	_
soil permeability	_	_	_	_	_	_
modern-holocene geology	_	_	_	_	_	_
bedrock composition	_	_	_	_	_	_
biologic assemblies	_	_	_	_	_	_
regional geologic structure	_	_	_	_	_	_
soil mineralogy	_	_	_	_	_	_
thickness of the A horizon	_	_	_	_	_	_
soil pH	_	_	_	_	_	_
bedrock topography	_	_	_	_	_	_
soil water holding capacity	_	_	_	_	_	_
100 yr flood elevation	_	_	_	_	_	_
soil plasticity	_	_	_	_	_	_
Atterberg Limits	_	_	_	_	_	_
soil compressibility	_	_	_	_	_	_
soil corrosivity	_	_	_	_	_	_
AASHO classification	_	_	_	_	_	_
excavation characteristics	_	_	_	_	_	_
subsoil characteristics	_	_	_	_	_	_
soil profile	_	_	_	_	_	_
soil swell potential	_	_	_	_	_	_
soil drainage	_	_	_	_	_	_
depth to the C horizon	_	_	_	_	_	_
slope stability	_	_	_	_	_	_
field density of soil	_	_	_	_	_	_
soil consolidation characteristics	_	_	_	_	_	_
Unified soil classification	_	_	_	_	_	_
soil unconfined compressive strength	_	_	_	_	_	_
soil shear strength	_	_	_	_	_	_
PVR (potential vertical rise)	_	_	_	_	_	_
free swell index	_	_	_	_	_	_
soil infiltration capacity	_	_	_	_	_	_
agricultural soil classification	_	_	_	_	_	_
soil bearing capacity	_	_	_	_	_	_
earthquake zone	_	_	_	_	_	_

FIGURE 2. Part A of the Real Estate Broker and Homebuilder Survey questionnaire (from Mathewson and Ruckman, 1974, p. 539.)

TABLE 1. Engineering and Geological Knowledge of Real Estate Brokers and Builders

(Percent of Responses)

Engineering Geology Factor	Real Estate Brokers						Builders					
	Don't Know	Insig-nificant	Low	Medium	High	Extreme	Don't Know	Insig-nificant	Low	Medium	High	Extreme
Hydrology												
Surface drainage	0	0	0	2	44	54	0	0	0	17	39	43
Local hydrology	43	5	11	13	17	8	43	4	13	13	26	0
100-year flood elevation	6	6	5	24	32	27	4	13	13	9	22	39
Subsurface drainage	0	6	16	32	38	8	9	9	17	26	17	21
Soil drainage	9	2	9	21	37	24	13	0	22	17	17	30
Depth to ground-water table	2	16	14	32	25	9	9	13	4	34	21	17
Geology												
Fault zones	24	5	16	11	9	30	9	13	22	9	17	26
Earthquake zones	14	14	17	3	14	37	9	26	22	4	0	39
Bedrock composition	19	14	21	29	16	2	4	13	30	9	30	13
Bedrock topography	25	22	24	17	8	2	13	30	30	9	13	4
Regional geologic-structure	25	17	29	19	8	0	9	13	48	17	13	0
Modern holocene geology	78	6	11	6	0	0	65	4	9	13	4	0
Other												
Biological assemblies	57	13	19	6	3	0	52	13	26	9	0	4
Surface topography	2	2	5	32	46	15	4	0	9	17	39	30
Thickness of A horizon	71	6	11	3	6	0	65	9	9	13	0	0
Depth to C horizon	79	5	5	6	6	0	69	4	9	4	9	4
AASHO classification	86	3	6	6	0	0	83	4	9	4	0	0
Unified soil classification	57	14	11	14	2	0	61	4	13	13	4	0
Agricultural soil classification	14	22	22	30	6	3	17	35	35	13	0	0

(continued)

TABLE 1. *(continued)*

(Percent of Responses)

Engineering Geology Factor	Real Estate Brokers						Builders					
	Don't Know	Insig-nificant	Low	Medium	High	Extreme	Don't Know	Insig-nificant	Low	Medium	High	Extreme
Soil Engineering Properties												
Soil permeability	27	5	17	29	14	3	13	0	36	22	22	17
Soil water holding capacity	5	9	29	25	22	11	9	9	22	26	9	26
Soil drainage	9	2	9	21	37	24	13	0	22	17	17	30
Soil infiltration capacity	41	6	17	21	8	2	39	13	13	9	13	13
Soil mineralogy	17	16	33	21	5	2	17	30	30	13	4	4
Soil pH	38	17	21	21	3	2	26	22	22	9	13	9
Soil corrosivity	22	14	16	30	14	0	26	13	26	4	13	13
Soil field density	41	6	21	22	6	0	30	13	22	13	17	4
Soil profile	22	6	21	22	17	2	17	17	9	22	22	9
Subsoil characteristics	9	5	14	44	21	3	13	4	22	22	22	17
Atterberg limits	87	2	2	6	2	2	69	9	13	9	4	4
Soil plasticity	29	5	21	27	11	8	13	9	17	26	22	17
Potential vertical rise	54	6	9	11	16	3	39	9	17	17	13	4
Free swell index	56	6	13	11	11	2	52	13	4	4	9	4
Soil swell potential	25	3	11	38	14	11	13	13	9	9	43	17
Soil compressibility	17	16	14	32	13	2	4	0	4	22	30	13
Soil unconfined compressive strength	63	11	14	14	3	2	43	4	30	22	13	4
Soil consolidation characteristics	48	5	17	24	2	0	52	13	13	13	9	4
Soil shear strength	54	6	21	13	6	0	43	4	22	13	13	4
Soil bearing capacity	25	3	14	24	19	11	9	9	17	22	22	22
Slope stability	21	3	9	27	27	11	17	0	17	30	26	9
Excavation characteristics	19	5	11	41	21	0	4	9	13	26	30	17

Source: Mathewson and Ruckman, 1974, p. 541.

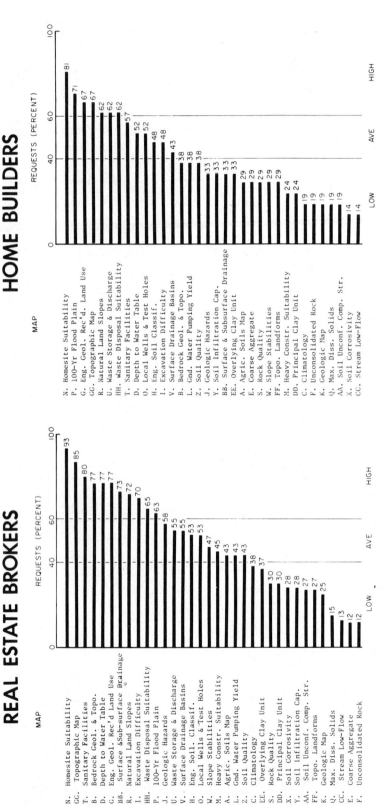

FIGURE 3. Bar graph showing requested information from the survey (after Mathewson and Ruckman, 1974, p. 542).

However, 77 percent of the real estate brokers requested a geological bedrock map but rated bedrock composition and topography low, and only 25 percent of the real estate brokers and 19 percent of the builders requested a geological map but rated information about fault zones and earthquake zones highly significant. The inconsistencies are paradoxical, possibly because the respondents did not understand what information is available on what map.

The survey showed that private-industry decision makers, real estate brokers, and home builders are not proficient in engineering geology and that they do not feel it necessary to be so. They did, however, express a strong wish to have geological information available to them, provided it be represented intelligibly. This survey of a large potential market for geological information demonstrated that the market exists but that few products are presently available.

Marketing Geological Information

The basic objective of marketing is to move a product—in this case, geological information—from the producer (or geologist) to the consumer or general public. This objective requires a marketable product. Unfortunately, much geological information is presented in technical terms that most consumers do not understand well. As a result, very little geological information exists in a marketable form. The problem might be overcome by launching a nationwide geological education program or by developing geological communication techniques. The complexity of the first alternative suggests that it would be more efficient for geologists to alter their product to fit the market. A publication by Varnes (1974), however, makes a significant step toward providing a basic educational tool for engineers (see Vol. XIV: *Maps, Logic of*).

For geological information to be marketable, it also must be presented in an easy-to-use format. The report or discussion must be written in such a manner that understanding requires little more than a normal command of English.

Maps must be presented at useful scales (see Vol. XIV: *Map Series and Scales*). Three basic scales are employed for general use, each meeting a specific need. *Regional scales,* 1:250,000, provide a broad overview for large-scale planning. *Local scales,* 1:24,000, provide basic information about land planning and development tract selection. *Site scales,* 1:1,200, provide detailed information for the design of particular tracts of land. Real estate decisions are made at local and site scales, whereas urban and regional plans are made at regional or local scales. Maps also must be published in convenient page sizes and convenient formats.

Both maps and reports must be designed to answer specific, practical questions about factors such as construction suitability and hazardous areas, and must not provide baseline data for geological units, soil types, and engineering properties.

Finally, practical information must be available at reasonable cost. If the previous objectives are met and a large market can be developed, then the unit cost of the map or report can be reduced.

The geologist must remember that land use and investment decisions are based on complex planning processes and that information concerning the physical environment is only a part of this entire process. It is necessary to understand the basic components of the planning process in order to appreciate the role of geological information in land investment decision making.

The planning process (Fig. 4) applies to major urban city planners and land developers alike, for goals, objectives, constraints, and opportunities always must be analyzed prior to policy development. Goals and objectives establish the intent and direction of land use decisions while constraints and opportunities inhibit or promote the goals and objectives (see Vol. XIV: *Environmental Geology; Legal Affairs*). Geological information can be classed under constraints and opportunities, because geological conditions do not control land use decision making; rather, economics does. The decision is based on a cost-benefit analysis of an engineering design that is safe under given geological conditions. The marketability of geological information, therefore, depends on its being easily and economically incorporated into the constraints and opportunities phase of the decision-making process.

Geological Information for Planning

The majority of the geological information presently available for land use planning has been produced by the U.S. Geological Survey or state geological surveys. Considering the size of the market and the limited personnel of federal or state surveys, private land use geological consultants appear to have a necessary role. The difference between government agencies and industry is that the product must show a profit in the

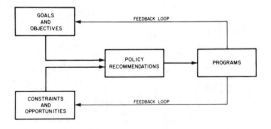

FIGURE 4. Schematic flow diagram of the planning process (from Mathewson and Font, 1974, p. 25).

private sector. To be profitable, the product must sell; to sell, the product must be economical.

Price constraints require that new communication techniques be developed that meet marketability requirements and cost limitations. Mathewson and Font (1974) proposed a series of ordered maps (Fig. 5) to meet the long-term needs of urban areas. Ruckman (1978) applied this series to the Bryan-College Station areas of Texas and found it economically feasible to present geological information for urban areas. The ordered maps are, however, beyond the price capability of private land developers.

To meet the needs of private industry, Mathewson and Piper (1975) applied economic principles to a 12.95-km² (3,200-acre) tract along the Texas coast (Fig. 6). The analysis, requiring about 25 aerial photographs and about 40 man-hours to complete, evaluates costs and risks inherent in each geological environment and identifies those areas of high costs and risks as land with low geo-economic value (Fig. 7). Thus, land with high geo-economic value is best suited to development ranging from single-family dwellings to industrial parks because

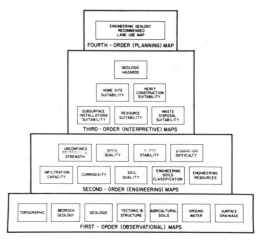

FIGURE 5. Schematic diagram of an ordered map series to provide a complete system of geological information for an urban area. First- and second-order maps are technical maps for the professional geologist and engineer, whereas third- and fourth-order maps are interpretive maps for the general user (from Mathewson and Font, 1974, p. 27).

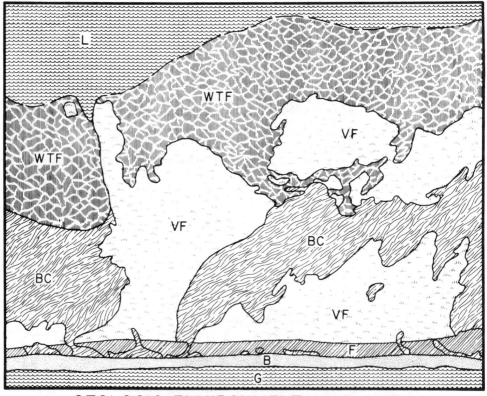

GEOLOGIC ENVIRONMENT MAP - 1973

FIGURE 6. Geological map of 12.95-km² area along the Texas coast. L = lagoon, WTF = wind tidal flats, VF = vegetated flats, BC = banner complex (dune field), F = foredunes, B = beach, G = Gulf of Mexico. As presented, this map is difficult for a nongeologist to interpret (after Mathewson and Piper, 1975, p. 629).

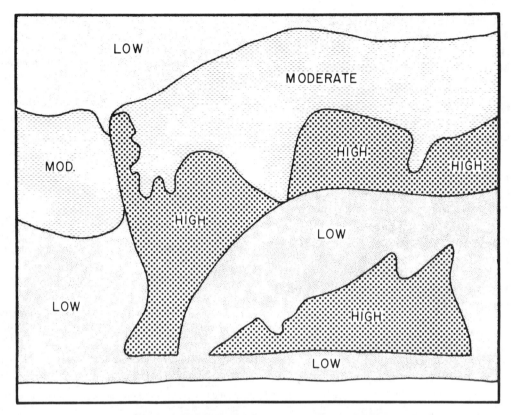

GEO-ECONOMIC MAP - 1973

FIGURE 7. Geo-economic map of the area shown in Figure 6, in which high-value land has the lowest relative cost and risk and low-value land has the highest. The developer now has a simple map that interprets the geological environment (from Mathewson and Piper, 1975, p. 629).

these sites have the lowest geological risks and costs. Because lands with low geo-economic value have high use limitations due to great risks and costs, only high-cost structures may be economical.

The unique feature of these geological communication techniques is the lack of technical geological jargon on the interpretative maps (see *Geological Communication*). Professional-quality geological studies must still be made and technical maps produced, but these items form a data base for the ultimate product: marketable geological information.

CHRISTOPHER C. MATHEWSON

References

Mathewson, C. C., and Font, R. G., 1974. Geologic environment: forgotten aspect in the land use planning process, in H. F. Ferguson (ed.), *Geologic Mapping for Environmental Purposes*. Geol. Soc. America Eng. Geology Case Histories No. 10, 23-28.

Mathewson, C. C., and Piper, D. P., 1975. Mapping the physical environment in ecomomic terms, *Geology* 3, 627-629.

Mathewson, C. C., and Ruckman, D. W., 1974. Geologic needs and knowledge of real estate brokers and builders, *Geology* 2, 539-542.

Ruckman, D. W., 1978. *Geologic Land Use Mapping of a Part of Brazos County, Texas,* unpublished M.S. thesis, Department of Geology, Texas A&M University, Tex., 129p.

Varnes, D. J., 1974. The logic of geological maps, with reference to their interpretation and use for engineering purposes, *U.S. Geol. Survey Prof. Paper 837,* 48p.

Cross-references: *Earth Science, Information and Sources; Engineering Geology Reports; Geological Communication.* Vol. XIV: *Environmental Geology; Legal Affairs; Maps, Environmental Geology; Maps, Logic of; Map Series and Scales; Popular Geology.*

GEOLOGICAL LOGGING—See WELL DATA SYSTEMS. Vol. XIV: WELL LOGGING.

GEOLOGICAL OCEANOGRAPHY— See MARINE SEDIMENTS, GEOTECHNICAL

PROPERTIES; SUBMERSIBLES. Vol. XIV: ACOUSTIC SURVEYS, MARINE; HARBOR SURVEYS; MARINE MAGNETIC SURVEYS; SEA SURVEYS.

GEOLOGICAL SURVEYS, U.S.—See Vol. VIII, Pt. 1. Vol. XIV: GEOLOGICAL SURVEYS, STATE AND FEDERAL.

GEOLOGIC INVESTIGATIONS— See DAMS, ENGINEERING GEOLOGY; ENGINEERING GEOLOGY REPORTS; NUCLEAR PLANT SITING, OFFSHORE; PIPELINE CORRIDOR EVALUATION; PUMPING STATIONS AND PIPELINES, ENGINEERING GEOLOGY; URBAN ENGINEERING GEOLOGY. Vol. XIV: SOIL SAMPLING.

GEOLOGIC TIME SCALE— See INTERNATIONAL GEOCHRONOLOGICAL TIME SCALE.

GEOLOGY, APPLIED

The term *applied geology* refers to the use of geology to solve, or aid in solving, human problems in which geological factors play a part. The problems may be economic, social, political, or cultural in origin. As the technological age progresses and the demands of growing populations increase, the variety and complexity of problems multiply. Virtually all aspects of geological science (see Vol. XIV: *Geology, Scope and Classification*) have become involved in varying degrees.

If limits cannot be placed on the scope of applied geology, it is due in part to the flexible definition of the science of geology. A committee (Albritton et al., 1972) was created by the Executive Committee of the International Union of Geological Sciences in 1970 to "establish a clear definition of contemporary geology, including both academic and practical aspects." In the report it presented to the 24th International Geological Congress in 1972, the committee promptly shifted its focus from geology per se to *geoscience*, which it identified as "a comprehensive term embracing all the sciences that concern the earth, including its atmosphere, hydrosphere, biosphere and spheres of the solid earth, insofar as these sciences treat of present and past configurations within the continuity of time."

Albritton et al. went on to comment that geo-science, or any scientific field can be divided into subfields, according to criteria such as "(a) the object of study, (b) the progression of stages in study, (c) the mode of study, (d) pure and applied study, and (e) administrative convenience." These criteria, however, "are useful for different purposes, but all are, in the last analysis, artificial. There are no intellectual discontinuities dividing geoscience into independent compartments."

In reference to criterion (d), the report goes on, "The distinction so commonly drawn between the *pure* studies on the one hand and the *applied* arts on the other is an artifact which has long obscured the true working relationships between these modes of thought."

Auger (1961) finds that, in the modern world, it has become more and more difficult to distinguish between pure and applied research in the natural sciences by reference to the problem they set out to solve or even by reference to the methods used. In practice,

alternation between application and pure science is almost a general rule. . . . It becomes obvious that pure research can hardly be undertaken without continuous resort to applications of science—utilitarian applications for the purpose of achieving a particular aim, no matter how "pure" it may be. It is clear, too, that progress in the practical application of science cannot long be sustained without continual advances in the field of disinterested knowledge.

The classification in two categories has become inadequate. Auger sees pure (fundamental) science divided into *free research* and *oriented research*, and application into *applied research* and *development work*.

Free research, according to Auger, is "directed toward fuller understanding of nature and discovery of new fields of investigation, with no practical purpose in mind." The prospect for practical application in unpredictable.

Oriented research focuses on extending human knowledge and comprehension of clearly defined problems, "without entering the field of concrete utilitarian application." It provides "the basis not only for great theoretical discoveries, but also for major practical applications." There are two subdivisions of oriented research. One, *field-centered research*, concentrates on a well-defined objective. The other, *background research*, is aimed at increasing the knowledge of a field by the collection of data, observations, and measurements. *Applied research* is often an outgrowth of oriented research. It is "deliberately aimed at a concrete result that will help to meet a specific human need." *Development work* seeks to obtain benefits from applied research, and "does not, in principle, involve the utilization of any new scientific results, but it often calls for much empirical, technical, and scientific knowledge in a wide variety of fields."

Another important distinction between pure and applied research, discussed by Barber (1952), lies in the ideals that are expected to be upheld by different kinds of scientific activity. In general, pure science takes greater cognizance than does applied science of rationality, universality, individualism, and disinterest. The distinction has come under scrutiny by both the scientific community, which is most concerned with ideals of pure science, and the public, which is concerned with the effects of applied science. Referring to one of the critical developments in recent time, Auger comments,

The growth of applied research has been so great that, in some cases, there has been a danger of its squeezing out disinterested research. . . . The achievement of an optimum balance between these types of research is at present a social problem of great importance.

The problem is complicated by another factor, which Barber observed: disinterest and the other ideals are not without limits in pure science.

Emergence and Acceptance of Applied Geology

As a formal science, geology is slightly more than two centuries old. "As in other branches of learning," Woodward (1911) noted,

the knowledge which led to the foundation of geology was acquired by degrees, sometimes at long intervals; but many of the observations on which are based the principles of the science were of necessity or even forcibly brought to the notice of mankind in the earliest days of human existence.

These observations centered on sources of useful rocks and minerals, fuels, and water, and on the behavior of the land and water bodies under the influence of natural processes and human action. Miners, particularly, had acquired an understanding of geological features and conditions important to their activity. Thus, there was an extensive store of knowledge when in 1556 Agricola summarized the known characteristics of ore minerals and their deposits in his great work *De re metallica.* Zittel (1901) remarked that when mining schools were founded at Freiberg in 1763, Schemnitz in 1770, St. Petersburg in 1783, and Paris in 1790, they were the chief seats of mineralogical and geognostic teaching. (Werner, at Freiberg, reserved the term *geognosy* for the systematic study of earth materials, their occurrence, structures, and mutual relationships, and *geology* for consideration of the origin and structure of the earth. The distinction disappeared in the early nineteenth century by abandonment of the term *geognosy,* except in continental European countries, where its use continued for several more decades.)

Practical application was, therefore, one of the original purposes of gaining the knowledge that became an integral part of the science of geology, but Zittel observed that speculation occupied "The leaders of thought, whose activities towards the close of the eighteenth, and in the first twenty years of the nineteenth century, won for geology an acknowledged place as a scientific study." However, "only a limited number of the founders of geology and palaeontology belonged to teaching bodies. The universities were unwilling to countenance young and indefinite sciences, and only tardily incorporated them in their academical curricula." The details of geology were gathered not so much by trained geologists as by chemists, mineralogists, and other scientists, as well as those in non-scientific professions, including clergymen, physicians, and soldiers. Application of geology to the solution of practical problems was not their primary concern, although some recognized the scientific value of information to be derived from practical pursuits. Woodward (1911) pointed out that

The progress of knowledge, as of old, was due largely to the development of economic products, and to the information thereby acquired. . . . In England, records of strata passed through in mineshafts and well-sinkings were from time to time published; important information on soils, strata, and minerals being gathered by writers who contributed to the old Board of Agriculture.

The use of such sources of information may have caused John Paris, one of the physician-geologists, to tell the Royal Geological Society of Cornwall in 1818 that "the fabulous and romantic age of geology may be said to have passed away; its disciples, no longer engaged in support of whimsical theories, direct all their attention to the discovery of facts, and to their application to purposes of extensive utility." Paris's conclusion does not appear to have been valid. Twenty-five years later, Nicholas Whitley, stated in the introduction to his book on the application of geology to agriculture:

The interesting science of Geology has gnerally been studied with a view either to discover the changes which the crust of the earth has undergone, and the causes by which these changes have been effected; — or to bring to light its hidden records of those various races of plants and animals which have successively lived and perished on the surface of our globe. But, as a science applicable to the ordinary and practically useful avocations of life, it has not received the attention which its importance merits. It is true that in one calling, — that of the miner, — the value of geological knowledge has so obviously appeared, that, to a certain extent it has been appreciated and acted on; especially with reference to the coal formation. But in a sphere of much greater importance, and more universal application, — the cultivation of the soil, — man has not hitherto availed himself of the aid afforded by this science; although its value appears scarely less obvious in this art, than in that of Mining. (Whitley, 1843)

The development of applied geology did not depend solely on geologists' interest in dealing with practical problems, but it also demanded an awareness of the value of geology among those who then might seek the assistance of geologists. The stimulation of awareness continued throughout the nineteenth century. For example, military officers and students in military schools were introduced to the rudiments of geology and presented with illustrations of the usefulness of that science for their purposes (see *Military Geoscience*). Occasional use of geologists by armies was reported during the 1800s and early 1900s. Civil engineers were encouraged to learn enough about geological materials, features, and conditions to know when to call on trained geologists for aid. A sound working relationship between engineers and geologists was not formed easily, owing to differences in approach to investigation and interpretation of geological factors in engineering problems (Wagner, 1884). Nevertheless, interest in application of geology to construction problems grew steadily in Europe, with the result that by 1900 *engineering geology* was accepted by engineers. In the United States, however, very few cases of its use had been recorded until highway construction and road improvement programs were under way in the early twentieth century (Kiersch, 1955).

It was not as difficult to induce more use of geology in the search for deposits of minerals and fossil fuels required increasingly for industrialization and urbanization, since the value of geology to mining had already been well established. In the English-speaking world, applied geology concerned with resources became known as *economic geology* (q.v.). The term focuses not on the subject of application, but on value judgments made by recipients of the geological and other information relevant to exploitation of resources. In some interpretations of its scope, the term *economic geology* was allowed to refer to all types of applied geology in which economic considerations were the critical factors. That made it possible, for example, to classify all applications of geology to engineering practice as a subdivision of economic geology, but in most usage, a distinction between engineering geology and economic geology is recognized.

Woodward (1911) observed that application of geology to useful purposes justified the establishment of geological surveys by governments (see Vol. XIV: *Geological Surveys, State and Federal*), and Walcott (1895) stated that one of the basic purposes of founding the U.S. Geological Survey in 1879 had been "examination of the geological structure and mineral resources and products of the national domain." Although the mission of that organization was redefined in 1882 to place systematic mapping and description of the country in the foreground, the identification of resources contributing to the economic development of the

nation and welfare of the people was a visible objective. The primary program of the Geological Survey could be regarded, therefore, as both basic research and background research for application of geology to practical problems. Applied geology was confined to economic geology in the strict sense. Thus Walcott, speaking authoritatively as director of the Geological Survey, could say that, in the case of highway construction in the United States, the organization was concerned with supplies of construction materials, but *not* with the engineering problems of construction.

In a retrospective view of the progress of applied geology, A. H. Brooks (1913) examined the subject matter of literature on North American geology and concluded that in 1890, 12 percent of all the publications were devoted in part, or wholly, to applied geology and that in 1909 the figure had risen to 47 percent. Considering only those of the U.S. Geological Survey, less than 1 percent dealt with applied geology in 1890, while 98 percent could be classified in that category in 1910.

The growing attention paid to applied geology was reflected in the acceptance of applied geology in the curriculum of some universities and technical schools—first, of economic geology before the end of the nineteenth century, and then, in the early twentieth century, of engineering geology. In Germany, the textbook on methods of practical geology by Konrad Keilhack, which appeared first in 1896, proved to be a durable publication, appearing in revised, enlarged editions until 1922 and in a Spanish translation in 1927.

Papers on applied geology had generally not been accepted for publication in established geological journals until the early twentieth century. Now, journals were founded for the purpose of providing a forum for discussion of applied geology and placing a record in the geological literature. They included the *Zeitschrift für praktische Geologie* in Germany (1893), the *Giornale di geologia pratica* in Italy (1903), and *Economic Geology* in the United States (1906).

It was also a time for taking stock of the standing of applied geology. In the first issue of *Economic Geology*, F. L. Ransome (1906) discussed the subject and found that, despite having established the identity of applied geology, geologists on the whole "displayed a rather languid interest" in practical problems. In the same years, D. W. Johnson (1906) noted that the interest of academic geologists in applied geology lay in research not immediately concerned with the commercial value of the findings. Economic geologists in universities were particularly identified with developing theories of origin of ore deposits. Johnson referred to such research as an intermediate stage between pure and applied geology. In later terminology (cited previously), it could be called *oriented fundamental research*. In 1907, practical geology was the subject of the presidential address by Federico Sacco to the Italian Geological Society. Describing the possible scope of applications, Sacco anticipated the inclusion of other fields with geology that would come to characterize modern applied geoscience.

Influences on the Status of Applied Geology in the Twentieth Century

The twentieth century has seen the spread of applied geoscience and the improvement of its standing in the various activities of geoscientists. First, the greatly increased demand for the services of geologists and other geoscientists to aid in the discovery of resources and in the conduct of major engineering projects has changed the composition of the geoscientific community. Today, more geologists are engaged in the search for energy resources, especially petroleum (see Vol. XIV: *Petroleum Geology*), than in any other geological activity. To meet the demand, the training of geologists in applied specialties has become a significant feature in university curricula.

Second, the interest in utilitarian activity has stimulated many scientists to seek participation in work on practical problems. For example, Kiersch (1955) commented, "The attitude of 50 years ago when geologists showed little interest in the problems of engineering and heavy construction has reversed itself. Of late, the geologist has begun to claim his rightful share of the borderland between geology and engineering." The trend represents "a gradual breakdown of the barriers that separated pure geological research from practical engineering application." It cannot be claimed, however, that there has not been some hesitancy among geologists to follow the trend toward applied geology. Although interest in engineering and other types of applied geology appeared to have been present in France in the late nineteenth and early twentieth centuries, Brooks (1920) found during World War I that "in comparison with other countries France has not to any great extent applied geology to industrial problems. The ideals of her leading geologists have been to advance the science than to show its practical utility." As late as 1978, Gerd Lüttig in Germany remarked that geologists, in many cases, "have been educated to think that geoscientific research is mainly activity within the academic sphere. They see it as pure basic research, useful only for determining natural laws."

Third, the experience of geologists who were drawn into service with armies in World War I to deal with application of geology to military problems was conveyed to the larger community of geologists after the war. They showed that the experience could be translated into applications for nonmilitary purposes and, thus, that it was an important factor in the development of engineering geology. The wartime experience also demonstrated the need for communicating more effectively with users who were not at all, or only slightly, trained in geology by addressing them in terms that they could understand easily and by keeping the problem in the foreground and omitting irrelevant matter. The merit of that requirement has been kept before the geoscientific community. As the number of geoscientists engaged in applied activities increases, effective communication is treated increasingly as ultimately the most important requirement to be observed.

Fourth, in the 1960s, the widespread public concern over the degradation of the environment in many parts of the world, caused by human action, led to the enunciation of environmental protection policies by governments (see Vol. XIV: *Environmental Geology*). Earlier application of geological knowledge to evaluation of the environment had concentrated on determining suitability of geological materials and conditions for human uses without regard for the effect of alterations of the environment, such as those resulting from strip mining or the extensive paving of the land surface in developed regions. With the initiation of programs for evaluation of impacts of human activities on the environment, a new factor was introduced in most types of applied geology. At the same time, the ramifications of the human-environment relationships were discovered to be so extensive that virtually every aspect of the science of geology has been enlisted in what is now termed *environmental geology* (see Vol. XIV). Furthermore, recognition of the environment as an entity, which must be analyzed as such in predicting the full impact of actions taken by man, has required the participation of all concerned sciences. The transition from a group of compartmentalized disciplines to a unified environmental science is taking place as concepts of environment and environmental quality are clarified.

Fifth, even more than in the early nineteenth century, the basic scientific reward derived from information obtained for applied purposes cannot be overlooked. As pointed out by Albritton et al. (1972), less than a century after Lyell was setting forth the principles of stratigraphy

as an exercise of the intellect, . . . stratigraphy was more widely practiced by geologists concerned with discovery of oil and gas than with opening the doors to the distant past. Moreover, the drilling of more than 2,000,000 boreholes in search of fossil fuel has yielded new and unexpected stratigraphic data in such abundance as to require a rewriting of earth history for the past 600,000,000 years.

FREDERICK BETZ, JR.

References

Albritton, C. C., et al., 1972. Geoscience and man, in *Earth Sciences and the Quality of Life,* J. E. Gill (ed.), 24th Internat. Geol. Cong., Montreal, Symp. 1, 25-29.

Auger, P., 1961. *Current Trends in Scientific Research.* New York: United Nations, 245p.

Barber, B., 1952. *Science and the Social Order.* Glencoe, Ill.: Free Press, 228p.

Brooks, A. H., 1913. Applied geology, *Smithsonian Inst. Ann. Rept. (1912),* 329-352.

Brooks, A. H., 1920. The use of geology on the Western Front, *U.S. Geol. Survey Prof. Paper 128-D,* 85-124.

Johnson, D. W., 1906. The scope of applied geology, and its place in the technical school. *Econ. Geology* 1, 243-256.

Keilhack, K., 1896. *Lehrbuch der praktischen Geologie.* Stuttgart: F. Enke, 638p.

Kiersch, G. A., 1955. Engineering geology—historical development, scope and utilization, *Colorado School Mines Quart.* **50**(3), 123p.

Lüttig, G., 1978. Geoscience and the potential of the natural environment, *Nat. Resources and Development (Tübingen, West Germany),* **8,** 93-107.

Paris, J. A., 1818. Observations on the geological structure of Cornwall, with a view to trace its connection with, and influence upon its agricultural economy, and to establish a rational system of improvement by scientific application of mineral manure, *Royal Geol. Soc. Cornwall Trans* **1,** 168-200.

Ransome, F. L., 1906. The present standing of applied geology, *Econ. Geology* **1,** 1-10.

Sacco, F., 1907. La funzione practica della geologia, *Boll. Soc. Geol. Ital.* **26,** lxxi-cii.

Wagner, C. J., 1884. *Die Beziehungen der Geologie zu den Ingenieur-Wissenschaften.* Vienna: Spielhagen & Schurich, 88p.

Walcott, C. D., 1895. The United States Geological Survey, *Pop. Science Monthly* **46,** 479-498.

Whitley, N., 1843. *The Application of Geology to Agriculture and to the Improvement and Valuation of Land.* London: Longmans.

Williams, S. G., 1886. *Applied Geology.* New York: Appleton, 386p.

Woodward, H. D., 1911. *History of Geology.* New York: G. P. Putnam, 204p.

Zittel, K. A. von, 1901. *History of Geology and Palaeontology to the End of the Nineteenth Century.* London: W. Scott 562p.

Cross-references: *Economic Geology; Geological Communication; Geotechnical Engineering; Military Geoscience.* Vol. XIV: *Environmental Geology; Geohistory, Founding Fathers, USA; Geological Methodology; Geological Surveys, State and Federal; Legal Affairs; Prospecting.*

GEOLOGY, PUBLIC INFORMATION AND EDUCATION—See EARTH SCIENCE, INFORMATION AND SOURCES; GEOLOGICAL COMMUNICATION; GEOLOGICAL INFORMATION, MARKETING. Vol. XIV: POPULAR GEOLOGY.

GEOMAGNETICS, GEOMAGNETIC SURVEYS—See MAGNETIC SUSCEPTIBILITY, EARTH MATERIALS. Vol. XIV: EXPLORATION GEOPHYSICS; MARINE MAGNETIC SURVEYS; VLF ELECTROMAGNETIC SURVEYING; WELL LOGGING.

GEOMECHANICS—See CLAY, ENGINEERING GEOLOGY; CONSOLIDATION, SOIL; EARTHQUAKE ENGINEERING; REINFORCED EARTH; RESIDUAL STRESS, ROCKS; RHEOLOGY, SOIL AND ROCK; ROCK MECHANICS; ROCK STRUCTURE MONITORING; SOIL MECHANICS. Vol. XIV: LANDSLIDE CONTROL; MINE SUBSIDENCE CONTROL.

GEOMEDICINE—See MEDICAL GEOLOGY. Vol. XII, Pt. 1: DENTAL CARIES AND SOILS.

GEOMETRONICS—See REMOTE SENSING, ENGINEERING GEOLOGY; SEISMOLOGICAL METHODS. Vol. XIV: ACOUSTIC SURVEYS, MARINE; AERIAL SURVEYS, GENERAL; EXPLORATION GEOPHYSICS; HARBOR SURVEYS; MARINE EXPLORATION GEOCHEMISTRY; MARINE MAGNETIC SURVEYS; PHOTOGEOLOGY; PHOTO INTERPRETATION: SEA SURVEYS; SURVEYING, ELECTRONIC; VLF ELECTROMAGNETIC SURVEYING; WELL LOGGING.

GEOMORPHIC ENGINEERING—See ALLUVIAL PLAINS, ENGINEERING GEOLOGY; ARID LANDS, ENGINEERING GEOLOGY; BEACH REPLENISHMENT, ARTIFICIAL; COASTAL ENGINEERING; DELTAIC PLAINS, ENGINEERING GEOLOGY; GEOMORPHOLOGY, APPLIED;

**GLACIAL LANDSCAPES,
ENGINEERING GEOLOGY;
URBAN GEOMORPHOLOGY. Vol. XIV:
LANDSLIDE CONTROL.**

**GEOMORPHIC MAPS, MAPPING — See
GEOMORPHOLOGY, APPLIED. Vol. III.**

GEOMORPHOLOGY, APPLIED

Applied geomorphology may be defined as geomorphology in the service of man. There is a long history of such service, because every use made of land is influenced by the form of the ground, its associated materials, and its processes of development. Hence man has adapted to that form and exploited his knowledge of its natural process relationships. For example, the original locations of towns and cities were commonly determined by geomorphological features, including fords for crossing rivers, coastal and estuarine inlets providing shelter for harbors, and entrances to upland passes that had military or commercial advantages (Schmid, 1974). From ancient times defense needs resulted in elevated townsites, or sites protected by rivers and marshes. The intersections of major valleys have traditionally been growth points for transport networks and larger towns. To the present day, smaller settlements are normally positioned in geomorphologically advantageous sites such as better-drained "rises" in low-lying plains, slopes with sunnier aspects in mountain valleys, and spring-line benches in scarplands. In these as in all other cases, water supply too has always been a vital consideration.

Similarly in agriculture, the most essential of human activities, land use patterns and practices have been closely adapted to their topographic setting. Geomorphological process systems have been manipulated to increase productivity since time immemorial. Slopes have been corrugated or trenched to improve the drainage of wet lands; streams have been controlled, diverted, and spread over dry lands; surfaces have been regraded, terraced, and ditched to reduce erosion, conserve soil, increase infiltration, or disperse potentially damaging runoff. Applied geomorphology is thus a long-established feature of engineering practices, because in such manipulation man has had to modify landforms and drainage networks, or construct new ones. The possibilities for these man-made landforms have been greatly extended with the development of modern technology. For instance, steep valley heads and hillslopes in semi-arid areas were rarely terraced in earlier times, but are now speedily transformed by present-day excavating machinery (Fig. 1). Likewise, most tradi-

FIGURE 1. Level terraces cut into bedrock (marl with conglomerate capping) on the site of a hillside valley head in Murcia, southeastern Spain. The terrace bank is 4 m high (note crestal ridge.)

FIGURE 2. Traditional and modern applications of geomorphology near Ullapool, Scotland. The old road (center left) followed the contour on well-drained, relatively stable hill-footslopes around the margins of a tributary valley floor. It thus avoided steeper, eroding hillslopes and poorly drained valley floor sites. The new road (right center) utilizes man-made landforms: an embankment across the valley floor, and an excavated pass through the hillside.

tional routeways were closely adapted to surface configuration and composition, whereas new landforms and materials introduced by man now carry highways across the grain of the land (Fig. 2).

In modern times geomorphology (q.v. in Vol. III) has come to be an essential tool in many fields besides engineering. Land is common ground for diverse scientific disciplines, including geology, geochemistry, hydrology, pedology, physical geography, ecology, and agriculture — and there are close relationships between landforms and the objects of study in these disciplines. The analysis of such relationships receives emphasis by different specialists not only in fieldwork but also in the interpretation of remote sensing imagery, which is of great value in terrain studies, because "topography in one form or another is often the main information derived from remote sensing data" (Lee, 1975, p. 826). Moreover, the need to understand geomorphological processes is of cardinal importance in environmental management generally. It is not surprising, therefore, that geomorphology has found numerous applications and practitioners in other fields.

Having established the diversity of the field of applied geomorphology, this entry will cover only the following applications, which are fundamentally of the greatest importance: the control of natural hazards, soil and water management, and the planned exploitation of natural resources. It must be remembered, however, that the basis of all applied geomorphology lies in the close connections that exist between landforms and other elements of the natural environment. Therefore, it is pertinent first to outline the nature of those connections.

Environmental Interrelationships

The shape of the ground has far-reaching implications in relation to landscape character as a whole. Surface morphology is the product of degradational and aggradational processes working in given parent material through time. Hence local morphological contrasts testify to variations in parent material or physiographic history, or both. As well as being an expression of past events, however, the shape of the ground is also a major factor influencing the achievements of current land-forming processes, so surface morphology is a central consideration in any attempt to understand those processes and in predicting their consequences. For these reasons geomorphology has valuable applications in exploitation of natural environments. The interactions between land-forming processes and earth materials that engendered landforms also governed the formation of the associated soils, surficial water systems, and hence habitats for plant growth. Geomorphology thus provides a framework for the integrated study of the primary natural resources that constitute land. In particular, geomorphological process systems strongly influence the possibilities and effects of our use of land, so geomorphology has diverse applications in land use research and environmental management (see Vol. XIV: *Terrain Evaluation Systems*).

Geomorphological features, in detail and regionally, are expressive of underlying parent materials and the nature and duration of land-forming processes. Detailed slope characteristics reflect the rate of rock weathering relative to its removal, which varies according to the intrinsic properties of the rock and to the combination of weathering and transporting agencies at work on the slope. Within a particular climatic regime, therefore, each lithology tends to have a characteristic range of slope forms and gradients. There are similarly close associations between surface configuration and bedrock on a regional scale. Changes in the dip and stratigraphic arrangement of rocks are etched into major relief forms such as tablelands, cuestas, and ridge-and-valley topography. These regional differences are emphasized by distinctive orientations and densities of slope, valley, and channel networks within the major forms, reflecting spatial variations in rock jointing and strike, for example, as well as in dip.

Surface configuration and land-forming processes are closely connected, as already implied. Hence different models of slope evolution have been related to contrasted process systems: parallel retreat of steeper slopes under sheetwash, for instance, and progressive decline of slope angles where creep or solifluction was dominant. Likewise there are close links between lithology, the nature of mass movements, and surface forms, and also between slope development and the activity of adjacent streams. The relationship between landforms and the agents of earth sculpture is a reciprocal one, however. Ground shape is not just the product of materials, land-forming processes, and time, but is itself an important factor influencing the achievements of the processes. Thus the vigor of most of these processes depends on the downslope component of the force of gravity, which is proportional to the sine of the slope angle. Moreover, the angle, aspect, microrelief, and elevation of a slope influence the amount of incident solar radiation, humidity, ground temperature, and moisture conditions. Form and composition are important here because steeper, eroding surfaces mostly have rock at or near the surface, and hence have higher conductivities, heat capacities, and albedos than gentler slopes with deeper weathering mantles. Steeper surfaces are commonly also less pervious, leading to more rapid dispersal of rain and snow and thus to distinctive moisture and heat budgets. On a broader scale, the disposition and dimensions of interfluves and valleys govern regional variations in precipitation, temperature, atmospheric pressure, winds, thermal currents, water-table relationships, surficial water movement, and land drainage. In these ways, landform characteristics influence local and regional changes in runoff and infiltration relationships, ground climate, and the nature and intensity of weathering, eroding, and transporting agencies (Flohn, 1965; Geiger, 1965; Carson and Kirkby, 1972; Young, 1972; Yoshino, 1974; Oke, 1978; Atkinson, 1981; Barry, 1981).

The relationships just outlined underlie the close links between landforms and soils. Physically inseparable, slopes and soils share common parent material and processes of development. In particular, slope variations influence the effects of atmospheric agencies at the ground surface, and so local weathering contrasts are associated with geomorphological differences. Furthermore, just as slope evolution is a function of the rate of production of weathering products relative to their removal, so this relationship also governs the degree of contemporaneous soil formation. Hence the need to study soils in relation to their geomorphological setting has long been recognized. Toward the end of the nineteenth century, for instance, Dokuchaev affirmed the importance of "recurrent associations of soils in relation to regularities of topography" (Volobuev, 1864, p. 87). Similarly, Hilgard (1911, pp.10-15) advocated that soils should be classified according to not only their "physical constituents" but also their "genetic relationships" ("the origin and the adaptation of lands"), the latter being essentially geomorphological. These ideas were elaborated in the United States by C. F. Marbut (the founder of the American School of pedology), who related major soil variations and finer subdivisions of them to relief differences, mainly slope changes. It was accepted that soil and slope could not be regarded as being distinct from one another, and that soil-slope units were environmental types each with a characteristic climate, drainage, tendency to erosion, and native plant association (Ableiter, 1940). Such environmental interconnections between landforms and soils have since been verified by numerous workers.

Geomorphological features exert an indirect yet powerful influence over vegetation because of their effects on soil conditions, surface stability, and ground climate. Therefore, ecologists such as Tansley (1949, p. 156) regarded slope gradient, altitude, and aspect as major "physiographic factors" of plant habitat "affecting the temperature, rainfall, air moisture and insolation . . . to which a given piece of vegetation is exposed, and therefore to a large extent the particular species of plants that form it." Examples given by Polunin (1960, p. 30) verified that micro-climates were commonly "engendered by physiographic change," and that microhabitats were their environmental result. He questioned the notion of climatic climax communities because microhabitat differences resulted from variations in relief, and hence no region would have a uniform vegetation cover. This implies that geomorphology should facilitate understanding of vegetation patterns, as Willis et al. (1959) demonstrated in a study of coastal dunes in southern England. The dune habitats were differentiated according to topographic position relative to the watertable and to form-process relationships determining relative surface instability. The results confirmed that such geomorphological differentiation elucidated spatial variations in the relative distribution of species. Similarly, Loucks (1962)

recognized that geomorphology influenced habitat conditions, and hence forest community changes, in New Brunswick. Therefore, by referece to soil and landform characteristics, including slope gradient and position in toposequence, he compounded "scalars" to segregate sites in terms of their moisture relationships as a basis for testing a numerical analysis of variations in vegetation composition. Bormann and Likens (1969) drew attention to another potentially useful approach. As nutrient cycles (q.v. in Vol. XII, Pt. 1) were closely related to the hydrologic cycle, Bormann and Likens regarded drainage basins as appropriate units for ecological study. They demonstrated this by instrumentation of catchments that enable input-output relationships within the corresponding ecosystems to be quantified.

These diverse interrelationships involving climate, earth materials, and living things are nature's dominant feature. Landforms and their associated process systems occupy a central position within them. As will be illustrated, therefore, applied geomorphology is of primary importance in our adaptations to, and management of, natural environments.

Natural Hazards

Natural hazards are natural events that create damage or loss of life. Most of the more dramatic and devastating of these are geological phenomena, including many that are specifically geomorphological. The latter differ notably from other geological hazards, such as volcanic eruptions and earthquakes, in that they can be triggered or intensified by improper use of land and disregard of natural processes. They include catastrophic events such as floods, and continuing degradation by shifting sand, for instance. Applied geomorphology is an essential element in their prediction and managerial perception, and in planning how best to minimize their consequences. Flooding and landsliding are selected for more detailed coverage here, for several reasons. Running water and associated mass movements are the chief agents of landscape sculpture in most parts of the world, and so floods and landslides are extremely common. In terms of loss of life and damage, they figure high on any historical list of natural disasters. Furthermore, they are catastrophic events and, while they may recur in a particular area, their actual timing is unforeseen. On all counts, therefore, their prediction and control are of major importance.

Floods. Floods are among the most damaging of natural hazards, and indeed Newson (1975) cites work that lists them first in a world table of natural disasters during the 20 years after 1947. They are most common on river flood plains and over alluvial fans. Thus the frequency of overbank flow in flood plains is often in the range of 6 months to 2 years, with 1.5 years a common value (Cooke and Doornkamp, 1974). Likewise, flooding is a natural feature on alluvial fans that occur chiefly on the flanks of uplands in semi-arid areas, forming where the ephemeral flows of mountain streams debouch onto the adjoining footslopes and plains (Figs. 3, 4, and 5). Although flooding is a serious hazard in humid regions, it can be devastating also in semi-arid regions, where high rates of runoff following storms over sparsely vegetated slopes produce widespread, costly flood damage down-valley (Bigger, 1974; Rantz, 1975). Recurring floods are also typical in coastal and estuarine zones (Ward, 1978).

Damage from flooding is increasing in many areas despite expensive efforts to reduce it (Cooke and Doornkamp, 1974; U.S. Water Resources Council, 1974). This seems related to the increasing use of drainage basins and flood plains, commonly made possible by partial protection works, which created major changes in the hydrologic system (see *Alluvial Plains, Engineering Geology*). Urbanization leads to especially rapid change, replacing permeable by impermeable surfaces and a natural system of channels by storm sewers and other drains (see *Channelization and Bank Stabilization*). The effects on flood hydrograph parameters are well documented, producing increases in peak discharge values and decreases in time between peaks, accompanied by rapid changes in channel geometry and sediment yield, for example (Leopold, 1974; Newson, 1975; Fox, 1976; Lazaro, 1979).

The nature and occurrence of floods are governed by diverse factors, including rainfall characteristics, intrinsic properties of the drainage catchment, and land use in the catchment. Because of the complex interactions involved, much work has been directed toward predicting* floods by identifying the most important variables that are relatively simply derived (see Vol. XIV: *Alluvial Systems Modeling*). Geomorphological variables figure prominently in this work. Hence, using the drainage basin as the basic study unit, numerous investigations have related flood characteristics to such variables as catchment area, drainage density, and main channel slope. Analysis of the characteristics of floods themselves is important in planning preventive measures. Thus the statistical probability of flood events of a given magnitude can be quantified through magnitude and/or frequency analyses, which provide a yardstick by which the hazard and the cost of preventing it can be assessed. Similarly, flood hydrographs showing peak flow, total runoff, rate of change of discharge, and flood-to-peak interval, for example,

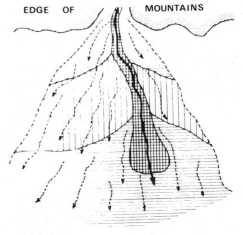

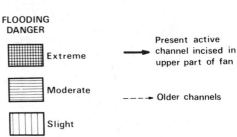

FIGURE 3. Relative flooding danger on a typical alluvial fan (after Cooke and Doornkamp, 1974).

have direct application in planning flood prevention. Rainfall analyses are also basic to flood analyses, because heavy rainfall is the major cause of flooding, and rain gauging commonly has both a longer duration and a denser network than river gauging.

In many drainage basins, however, both rain gauge networks and river gauging are inadequate, so planning for flood prediction and management must be based on the study of catchment characteristics. Common approaches to estimating river flows in this case include precise surveys of channel cross-sections, of flood water and surface slope as evidenced by debris lines, and of hydraulic roughness. Instantaneous peak discharge has been assessed by the "slope-area" method, and simple "flood formula" have been developed to relate the maximum peak flow at a site to the catchment area, with a coefficient and exponents being fixed for the area in question (Newson, 1975; Ward, 1978). This has led to the development of many versions of the "rational formula," which uses a figure for rainfall intensity as well as catchment area together with a dimensionless constant, the "run-off coefficient." Equations used for calculating time of concentration have been based on catchment area, average slope, and straight-line distance from outflow to remotest divide.

---

*The term is used here to cover both *flood prediction* (for engineering design purposes) and *flood forecasting* (for warning purposes), although there are major distinctions between them (see Ward, 1978).

FIGURE 4. Apron of coalescing alluvial fans between mountain range (top) and cultivated valley plain (bottom), near Quetta, Baluchistan. The village is positioned at the distal margin of a fan, avoiding flood-prone areas in higher sectors and in the zone of concentrated flow that passes immediately to the left of the village. The arrows indicate linear series of shafts connecting man-made tunnels, termed *karez,* which channel water underground from natural aquifers in upper parts of the fans to settlements and irrigated fields downslope (photo: Surveyor General of Pakistan).

The application of the rational formula depends on the selection of an appropriate value for the runoff coefficient. Theoretically this coefficient incorporates all factors controlling the proportion of rain that becomes runoff, including surface configuration, vegetation, infiltration rate, soil storage capacity, and drainage patterns. But all these data are rarely available, so many coefficient values

place emphasis on the two most important and most readily identified features, landforms and land use. Slope gradient is a major consideration, usually in association with the nature of the surface, surface storage, and degree of saturation (Gray, 1970; Hudson, 1971).

As Hudson explains, the advantage of the rational formula is that it can always be used for

FIGURE 5. Farmers taking water from one of the karez shafts shown in Figure 4. Note the line of shaft extending upslope over the fan surface.

estimating maximum runoff rates no matter how little recorded information is available. Hence it is of value in areas lacking rainfall or hydrologic records. However, several methods have been evolved for estimating rates of runoff where more detailed information is available. The approach used by the U.S. Soil Conservation Service considers the effect of four principal factors in the catchment, two of which are geomorphological— "relief" and surface storage—and the others soil infiltration and vegetal cover. Each factor has four categories ranging from "low" to "extreme" as runoff-producing characteristics, and with corresponding numerical "weightings" of 10 to 40 for relief and 5 to 20 for the other variables. The sum of the four weightings is the "watershed characteristic," which, by reference to prepared tables or charts, is used to predict peak flow values.

Hudson describes a similar method developed for use in some African conditions and further refinements where the effect of management practices is added. The U.S. Geological Survey and others have taken this catchment analysis further to obtain multiple-regression expressions for mean annual flood and a variety of catchment characteristics. Values obtained for specific flood recurrence intervals can then be used in engineering design. Orsborn (1976) discusses such regression models and relatively sophisticated "hydrogeological" methods whereby drainage basin properties—

especially geomorphological characteristics—are analyzed to identify parameters having the best correlations with gauged streamflows so that flows for ungauged streams can be predicted using their basin parameters and the gauged correlations.

Geomorphological considerations are also important in more elaborate approaches as applied, for example, by the U.S. Department of Agriculture's Hydrograph Laboratory (Holtan and Lopez, 1971). Here a model has been evolved to provide a mathematical continuum from drainage divide to catchment outlet. The framework for the model consists of dividing a catchment into "hydrologic response zones" corresponding to the three major geomorphological divisions: uplands (i.e., interfluve crests), hillsides, and bottomlands. Soils within each zone are grouped according to land capability classes and infiltration, with evaporation and overland flow computed for each zone using published data and conventional formulas.

Input to the model consists of a continuous record of rainfall weighted to represent the catchment. Vegetation, land use, and soil properties are considered in computing evapotranspiration and infiltration. Detailed geomorphological characteristics also receive emphasis. For instance, the model incorporates "depression storage" of the surface water that is held until dissipated by infiltration, which is extremely important where techniques such as terracing and contour furrowing are practiced. Rainfall in excess of infiltration is routed in the model across each hydrologic response zone and cascaded, subject to further infiltration, to the channel across subsequent zones at lower elevations. The equation for computing overland flow includes a coefficient dependent on roughness, length, and degree of slope; channel flows and subsequent return flows are routed by simultaneous solutions of this equation and a storage function. In this way, flow from different parts of the catchment may be routed separately through watershed storage and then summed to predict watershed outflow. Ward (1978) outlines a number of other computer simulation models incorporating such drainage basin, or hydrogeological, characteristics and meteorological inputs.

Four main approaches have been developed to help control flooding: (1) watershed management, (2) protective structures, (3) landuse control in flood plains, and (4) calculated risk and absorption of loss (Cooke and Doornkamp, 1974; U.S. Water Resources Council, 1974; U.S. Army Corps of Engineers, 1975; Noble, 1976; Ward, 1978).

Human activities in river catchments, including vegetation clearance and improper agricultural methods, have greatly increased the dangers of flooding in urban or other areas down-valley. Consequently, land-use management practices, termed *watershed management,* are necessary to reduce and delay run-off to rivers in these catchments.

They include the construction of terraces, contour furrowing and contour strip cropping, crop rotation, and reforestation. Whereas these measures reduce peak flows from many smaller storms, however, they may have little effect on large floods. Consequently, protective structures have been designed with the object of reducing overflow in the flood plain and ensuring the relatively harmless passage of water through the threatened areas. They mostly involve the construction of storm channels and reservoirs to divert or retard excessive run-off, levees or floodwalls to confine flood flow, various bank protection structures, and channel modifications to increase the river's water-transmitting capability. These latter include straightening, steepening, widening, or deepening channels, locally lining them with concrete, and using dikes to direct flow into desired alignments (see *Alluvial Plains, Engineering Geology; Channelization and Bank Stabilization*).

The third approach mentioned, landuse control in flood plains, often referred to as *flood-plain regulation,* does not attempt to reduce or eliminate flooding but rather to minimize its damaging effects by controlling the use made of flood-prone areas, especially the type and location of buildings. It includes designation of floodways, zoning prohibitions, subdivision regulations, building codes, and other ordinances.

The final approach is that of calculated risk and absorption of loss. Here property owners anticipate that they will have to bear little loss, or temporarily evacuate to prevent greater loss, or rely on some form of public relief subsidy or insurance to meet their losses. The extent of such losses is strongly influenced by the availability of flood forecasting and warning services, and by flood proofing, or adjusting buildings and their contents to withstand flooding.

Applied geomorphology figures prominently in these control measures. Thus the two most common approaches, watershed management and the building of protective structures along the river, are largely applied geomorphology—albeit mostly practiced by engineers and agriculturalists—in that they involve building new landforms, ranging from levees and diversion channels to dams and terraces, or modifying existing ones by deepening and straightening channels, contour furrowing, and so forth. Likewise, flood-plain regulation is a geomorphological exercise in terms of defining floodways, identifying the extent and likely frequency of previous floods, areas of potential inundation, and tracts of varying degree of risk that are thus suited for different subdivision and building purposes. Furthermore, flood prediction and associated flood frequency reports and inundation maps have major geomorphological components, as noted earlier.

The factors influencing the selection of reme-

dies to deal with flooding vary markedly from place to place. As Newson (1975) explains by reference to U.K. examples, there may be little choice as to the methods employed because local circumstances of geomorphology and urban conditions may preclude the application of many measures. Alternatively, costly construction can be avoided where the risks are less great. In this latter case, a system of public adjustment to flooding could be based on flood danger warnings, which requires some idea of the typical flood hydrograph for the particular catchment, to time the flood's arrival downstream and its duration at sites of risk. Flooding problems are extremely complex in many areas, however, and it is difficult to find permanent solutions, as in southern California (Rantz, 1975). Because the population in this region is growing rapidly, there is great pressure on the land for building and other development purposes. But floods here are of tremendous violence, and so their control necessitates a variety of interrelated remedies, including construction of flood-water reservoirs and debris basins; diversion of flood waters onto areas where sediment can be deposited and excess water can percolate underground; realigning, enlarging, and paving permanent channels to accommodate excess runoff; and retarding erosion and surface runoff in catchments by means of various slope and channel modifications.

Despite the complexities of resolving problems with flood control, certain general principles are evident. In the first place, satisfactory results can be achieved only if the geomorphological process systems within the flood-prone basin are properly understood and allowed for in the design of control measures. Second, watershed management and protective structures in the danger areas down-valley should be regarded as complementary rather than alternative approaches. Third, management of flood-plain use must be a continuing priority together with the devising of systems to regulate streamflows for all beneficial purposes, not only flood control but also water use and conservation generally. The complementary nature of the main flood control methods is again apparent here: reservoirs, which are among the most expensive measures, are advantageous because flood prevention becomes part of general water resource planning by checking excess runoff at or near its source in the uplands (see *Dams, Engineering Geology*). Thus "regulating reservoirs" and "trans-basin aqueducts" have been chosen as a key element of water resource development in England and Wales (Newson, 1975). These general principles, and especially the need to understand the dynamics of geomorphological process sytstems, are well illustrated in examples given by Rantz (1975) and by Noble (1976), Kolb (1976), Keller (1976), and Palmer (1976).

Landslides. The term *landslide* is used here to refer to all types of rapid mass movements. Landslides, like flooding, regularly cause major disasters with much loss of life, and are frequently recurring hazards that are expensive in terms of the physical damage they produce. They may be the actual agent of destruction, as when masses of debris sweep over settlements, or the initial cause, as in Norwegian fjords when the resultant waves destroy shoreline villages, and as in the worst dam disaster in history, which killed 2,600 people at the Vaiont Dam in Italy in 1963 (Cooke and Doornkamp, 1974; Morton and Streitz, 1975; and Kiersch, 1975). In many areas landslides occur regularly as a continuing sequence of natural events. Elsewhere slopes may be comparatively stable until landslides are triggered by disturbance factors that are either relatively infrequent, such as earthquakes, or new developments, such as land use changes. In particular, landslides are an increasing hazard in areas of urban development, chiefly because slopes that appear naturally stable may become unstable if moisture conditions, loadings, or gradients are changed through such developments.

Landslides are related to gravity-produced shearing stresses (see *Rheology, Soil and Rock*), which increase with slope gradient, and slope height, and unit weight of the underlying materials. Surficial processes of differential volume change— such as freezing and thawing, and shrinking and swelling— produce further shearing stresses. Landslides occur, therefore, when the shear stress forces creating downslope movement exceed the shear strength of the materials. Hence for each kind of material— varying in terms of bulk density, angle of internal friction, and cohesion—there is a limiting slope gradient above which rapid mass movement will occur from time to time, and below which the slope is relatively stable. The steepness of a slope influences not only the susceptibility to sliding but also the volume of the slide. In engineering work it has also been found that the time lag between slope excavation and subsequent failure in clays, for instance, is related to slope steepness. "External" geomorphological relationships are also most important. Thus shear stress may be increased through the removal, by natural erosion or man-made excavations, of "supporting" material along the "toe" of a slope, or by the accumulation of talus in the higher parts. Similarly, shear strength may be decreased by increased pore-water pressure due to man-induced changes in surficial water systems through reservoir construction, producing higher ground-water tables, or land use practices resulting in increased run-on. Therefore the location of landsliding is strongly influenced not only by the form and composition of slopes, but also by their geomorphological setting.

Geomorphology has valuable applications in both predicting areas of potential landsliding and controlling the causes of landslides (see Vol. XIV: *Landslide Control*). Thus Cooke and Doornkamp (1974) compiled a checklist whereby, with reference to readily observable features of slope form and composition, any slope unit in an area of potential landsliding could be given a stability rating on a scale, ranging from stability through increasing degrees of potential instability to failure. Inherited features can also be important because potentially unstable relict landforms, which developed under conditions no longer operative, are quite common and can be reactivated during excavations for road work, for instance. However, they can often be identified by geomorphological ground survey and airphoto interpretation because even these ancient landslides commonly have distinctive slope forms within and around them. Leggett's (1973) case studies illustrate such diverse geomorphological considerations in areas prone to landsliding. They demonstrate that the toe of a landslide is a critical location, because it requires little interference with this distal margin to reduce the length of the arc on which are acting two major forces: the weight of the mass of material that is tending to move downslope, and the shear strength that provides the resistance to movement. Predictably, therefore, many valley-side slopes above eroding streams or road cuts are especially prone to landsliding.

Bailey (1973) illustrated the importance of environmental interrelationships in landsliding and the value of applied geomorphology in elucidating them. He identified the connections between fluvial processes and slope stability in a study of landslide hazards in Wyoming, and found that the soil and rock mantles of slopes were in precarious balance with the physical environment, and road construction and logging activities disturbed that balance. He identified the location of unstable areas, the associated environmental factors and processes, and the limitations to management practices. Active landslides, inactive landslides, and active taluses were recorded on a landslide hazard map, and guidelines for logging and road construction were described in each case. Oversteepened slopes characterized by active or inactive debris avalanches and mudflows, and potential landslide areas where no apparent movement had yet taken place, were also differentiated and similar guidelines laid down. Slope gradient was recognized as a characterizing feature for mass movements of all types, and was positively correlated with road-related slides (see Vol. XIV: *Slope Stability Analysis*).

The extent and importance of applied geomorphology in the engineer's dealing with landslide problems are seen in the manual produced by the U.S. Highway Research Board's Committee on Landslide Investigations (Eckel, 1958). Procedures used in the identification and interpretation of both actual and potential slides are largely based on geomorphological considerations. Thus a checklist for use by engineers to recognize different kinds of active or recently active landslides depends on detailed examination of the form, microrelief, and surface composition of predefined sets of geomorphological units for both "stable parts surrounding the slide" and "parts that have moved" (Ritchie, 1958, pp. 56-57). In identifying areas of potential slides, "special attention should be given to the slopes, changes in slope, and their relationship to the different materials involved" (Ritchie, 1958, p. 51). Geomorphological interpretation of airphotos plays a major role (see Vol. XIV: *Photo Interpretation*). This is because landform differences can be easily identified on aerial photographs and because of their close bedrock relationships, as well as slope-process associations, which are important in delimiting areas of actual and potential landslides. Each type of landform, distinguished by a specific kind of geological material, overburden, and topographic expression, "poses relatively distinct problems for the engineer, particularly from the standpoint of landslide susceptibility" (Ta Liang and Belcher, 1958, p. 71). In more detailed airphoto interpretation, individual landslides are indicated by "the sharp line of break at the scarp; the hummocky topography of the sliding mass below it; the elongated, undrained depressions in the mass; and the abrupt differences in vegetative and tonal characteristics between the landslide and the adjoining stable slopes" (Ta Liang and Belcher, 1958, p. 72). Airphoto clues to sites of potential landsliding include banks undercut by streams, steep slopes, drainage lines on higher ground contributing seepage waters downslope, and seepage depressions.

For such reasons, detailed *geomorphological mapping* (q.v. in Vol. III) provides an ideal framework for planning geotechnical or soil engineering studies prior to highway design, for example. This approach was applied by Brunsden et al. (1975) (see also Cooke and Doornkamp, 1974), who demonstrated that mapping of slopes that are potentially unstable, or have already failed, is useful in the planning states of engineering projects. Routes can be aligned to avoid these slopes, or appropriate remedial measures can be designed and costed to enable them to be crossed. Such mapping also facilitates planning the deployment of costly research effort by identifying potentially unstable slopes in which field sampling and laboratory analysis should be concentrated, with less intensive coverage of relatively stable surfaces. The geomorphological basis of terrain evaluation for landslide investigations is demonstrated also by Schuster and Krizek (1978).

Most methods for controlling landslides have been devised by engineers in constructing roads, buildings, and other structures. They have been developed because of the widespread occurrence of slopes that become unstable once man disrupts their natural state. Moreover, the cuttings and embankments of many "artificial" landforms are inherently unstable without such control methods. Four main groups of methods are employed, other than avoiding the danger area: excavation, drainage, restraining structures, and miscellaneous methods (Eckel, 1958). To be effective, they depend on the application of geomorphological knowledge in identifying the critical causative factors of landsliding in a particular locality, and how best to modify these to induce relative stability. They involve the design and construction of new stable slopes (examined in detail by Schuster and Krizek, 1978) or the modification of the original ones and their surficial water systems (see Vol. XIV: *Landslide Control*), as to be explained.

Preventive measures in excavation consist of designing slopes, varying according to the material, whose gradients and drainage will minimize sliding. Such slopes may be produced by removing part of the original head or upper part of the slide, by grading the slope to a gentler inclination, or by cutting benches. Particular attention has to be given to surficial water conditions because increases in pore-water pressure are a common trigger mechanism in landsliding. Appropriate measures include *draining* the slide and creating *diversion channels* upslope. This may involve construction of paved ditches, installation of flumes or conduits, and paving or bituminous treatment of slopes, in order to minimize the possibility of water from run-on or runoff percolating the unstable area. Major landform modifications may be necessary, such as regrading overloaded slopes at the head of a slide, building structures where support is needed, or diverting channels undercutting the toe of the slope. Many restraining structures are concentrated at the critical toe of a slide, including buttresses to provide added weight and thus to increase the resistance to movement. Among miscellaneous methods, artificial cementation has been used to increase the shear strength of the soil, and partial removal of the toe of a landslide can be a temporary expedient to protect a structure until more permanent safeguards can be provided.

These methods for controlling slope instability are among the more widely used of traditional engineering applications of geomorphology because, in Leighton's words "Slope, including its underlying material, is the most important geologic and engineering element in hillside (urban) development" (Leighton, 1974, p. 207). He illustrates this by citing examples of both predicting the occurrence of landslides and of treating areas of actual or potential movement—unstable slopes being made stable by resculpturing and "grading" (Fig. 6). Other case studies described by Leighton (1976) underline the crucial role of geomorphology in

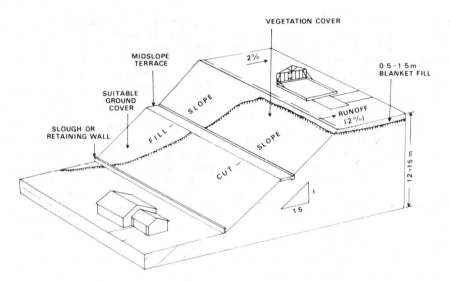

FIGURE 6. Graded benched slopes for urban building purposes. Additional devices to protect against mass movements and runoff erosion include: impermeable blanket fill with reverse gradient to carry runoff away from major slopes; vegetation cover on steeper graded slopes; paved mid-slope terrace to carry slope runoff; basal retaining wall (after Leighton, 1974).

the identification, interpretation, and engineering control of landslide problems.

Soil and Water Management

Soil Erosion by Water. The widespread and sometimes devastating effects of soil erosion by water, commonly induced or accelerated by unwise agricultural practices, are well documented. Such erosion was a factor in the downfall of some empires (Stallings, 1957; Hudson, 1971), continues to be a serious problem even in technologically advanced parts of the world, and is one of the chief obstacles to agricultural efficiency, and hence to economic progress, in most developing countries. It tends to be a regional rather than a localized problem because accelerated erosion not only denudes the affected areas but produces much detritus deposited by streams at lower levels, where it impairs soil drainage and fertility, infills stream channels and reservoirs, and damages roads, buildings, and other structures. This erosion and related damage is enormously expensive (Stallings, 1957; Stall, 1973).

The main agents of soil erosion (q.v. in Vol. XII, Pt. 1) by water are raindrop splash and run-off moving over the surface as either sheet flow or channelized flow in rills and gullies. Numerous studies have demonstrated that the rate and amount of soil lost due to these agents vary according to slope gradient and length, and are also influenced by slope curvature, contour curvature, and slope aspect. The steeper the slope, the greater the erosion for various reasons: there is more splash downhill, there will be more runoff, and it will flow faster. The amount of this erosion is not simply proportional to the slope gradient, but rises rapidly as the gradient increases. The length of slope has a similar effect on soil loss. Thus the amount, velocity, and depth of runoff are greatly increased on a longer slope. This produces scour erosion, which would not occur on a shorter slope, or where the effective slope for runoff is reduced, as between terraces. Values derived for these exponential relationships are summarized by Smith and Wischmeier (1962) and Hudson (1971). (See also Stallings, 1957; Carson and Kirkby, 1972; Young, 1972; Strahler, 1973; and Kirkby and Morgan, 1980.)

Soil erosion is a function of erosivity and erodibility. *Erosivity* is governed by rainfall and so cannot be controlled. *Erodibility* is the susceptibility of a soil to erosion. It depends on many factors, primarily including the nature of the soil itself, ground configuration, and vegetation cover. Vegetation is especially important because its disturbance or removal reduces the proportion of rainwater that infiltrates the soil, permits splash erosion, increases runoff, and hence leads to accelerated erosion by surface flow. Such disturbance is inevitable in land use, however, and so the amount of subsequent erosion depends largely on the quality of management.

Control of Run-off and Erosion. The term *management* is used here to include land management and crop management. These correspond respectively with the two main kinds of erosion control measures: mechanical measures dealing with the ground itself, and nonmechanical measures concerned with crops and animals. Mechanical measures are essentially applied geomorphology in that they consist of reshaping surface form to manipulate the associated geomorphological processes. In particular, they aim to modify the gradient, length, and microrelief of slopes in order to regulate the concentration and velocity of surface water flow. Terraces are one of the most widely used of these mechanical measures. They consist of benchlike earthworks with banks along their downslope margins, constructed at right angles to the direction of maximum gradient, that is, along the contour. In this way the original slope is divided into several small catchments, corresponding to a number of short slope segments, each with a gentler gradient than originally. On each terrace, or catchment, this gentler gradient and shorter slope length not only decrease the amount of runoff by increasing the proportion of rainfall that infiltrates the soil, but also, by lowering the velocity of runoff, reduce soil loss and cause more flowing water to be absorbed as it moves slowly over the ground.

Terraces may be divided into two categories according to their primary function: *level,* or *absorption, terraces* to conserve moisture, and *graded* or *dispersion, terraces* for the orderly disposal of water during periods of excess rainfall. Level terraces are common in drier areas where rainfall is inadequate for maximum crop growth, and graded terraces are suited to more humid conditions. A level terrace is level along the contour in order to hold rainwater so that it infiltrates the soil. In contrast, a graded terrace is constructed as a very shallow channel with a longitudinal gradient slightly oblique to the contour so that surplus run-off flows away at nonerosive velocities to a place where it can be safely discharged (Fig. 7a, b). Level terraces may be built either by excavation (Fig. 7b) or, in drainage floors, by impounding natural sedimentation (Fig. 7c). Although approximately level along the contour, the shelf of a level terrace may have a slight downslope gradient—that is, in the direction of the original slope—or a reverse gradient in the opposite direction to that of the original slope. The bank along the downslope edge of such a level terrace is relatively steep and is commonly, although not always, capped by a raised lip of earth to prevent water on the shelf from overflowing the bank. The downslope margin of a graded terrace, however, normally consists of a gentler-sided, levee-like embankment constructed to impound water flowing in the shallow channel immediately upslope.

A simpler form of earthwork designed to intercept surface runoff on gentle slopes is the *contour bund* (Fig. 7d). This is a low ridge of soil thrownup along the contour, commonly planted with grass or shrubs to stabilize it and also to assist in trapping silt washed downslope by runoff. In many areas the chief function of contour bunds is to hold runoff until it infiltrates the soil, and hence they

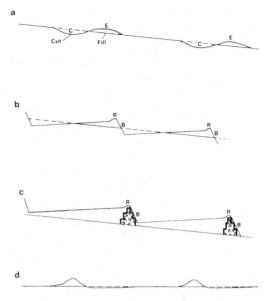

FIGURE 7. (a) Graded terraces; (b) excavated level terraces; (c) level terraces formed by natural infilling of alluvium-colluvium behind dams in a drainage trench; (d) contour bunds.

are especially common in semi-arid areas. Other forms of such man-made microrelief include tied ridging, and ridges and furrows. Tied ridging consists of ridges closely spaced in two directions at right angles so that the ground is covered with small rectangular depressions intended to prevent runoff and wash erosion. However, the well-known *ridge-and-furrow method* is used primarily to improve surface drainage, but it also affords some measure of erosion control.

As graded terraces are intended to protect arable land by leading away surplus runoff, they are commonly designed in conjunction with two other components of such a man-made geomorphological system, stormwater drains and grass waterways. A *stormwater drain* is a ditch constructed along the contour upslope from the terraces to intercept runoff, which would otherwise flow down from higher ground on to the arable land. The runoff in both graded terraces and stormwater drains must be led away from the area, and so the *grass waterway* is designed to fulfill that function. It is excavated down the slope with discharge inlets from the terraces and their stormwater drain.

Other control measures that are adapted to surface configuration, but intended to modify its character, include contour plowing and contour strip cropping. *Contour plowing* is one of the most effective control measures for cultivated cropland, because it increases the proportion of rain that infiltrates the soil, disrupts surface flow, and reduces its velocity by increasing surface roughness. Deeper plowing is also a feature of semi-arid areas, in order to impede as much runoff as possible by creating pronounced microrelief, and to facilitate infiltration by breaking surface crusts and hardpans, thus enabling the soil to absorb and retain more water. *Contour strip cropping* involves alternating strips of cultivated row crops and close-growing or sod crops. The row crops induce erosion, of course. However, the sod crops protect the underlying soil from erosion by both raindrop splash and surface flow. Moreover, during all but extremely heavy or prolonged rains, a sod strip can absorb the water flowing from the row crop immediately upslope, and thus prevent run-on to the adjoining row crop downslope. This increases the infiltration rate (q.v. in Vol. XII, Pt. 1) and reduces the overall amount and erosive capacity of runoff.

The design of these various management measures is largely determined by geomorphological considerations. Thus terraces are planned as small drainage catchments that can be handled through one outlet or system of outlets and can accommodate the anticipated rainfall and runoff intensities without overtopping or breaching of their banks. The uppermost terrace of a sequence is especially important here because its capacity must not be overtopped by runoff, or the whole system will be at risk. Therefore it must be built with a suitably small drainage catchment near the crest of the slope or be protected by diversion ditches at its upslope margin. These considerations—including the construction of channels to collect, divert, or lead away water—are described by Stallings (1957), Hudson (1971), and Bennett and Chapline (1973), who give standard design formulas in which the governing factors for terraces are the slope and configuration of the ground, and for channels, the size, shape, gradient, and bed roughness of the proposed structure in relation to estimated discharge. Similarly, the planning and effectiveness of contour cultivation and strip cropping are strongly influenced by the gradient and length of slope as well as by soil type (Food and Agriculture Organization, 1965).

Traditional Applications. Such methods of soil and water management, and geomorphological controls in their location and design, have received much attention since about the beginning of this century (Glenn, 1973). Geomorphology has been applied in this way since time immemorial, however, and many twentieth-century methods are refinements of long-established techniques. These traditional applications of geomorphology characterize agricultural land use in the semi-arid province of Murcia in southeast Spain, for example. The district around the town of Mula is typical. Except on steeper slopes the landscape has been completely transformed by terraces, bunds, and related structures for soil and water management (Fig. 8). These man-made landforms are closely attuned to their geomorphological setting

FIGURE 8. Typical semi-arid landscape near Mula, Murcia Province, southeastern Spain. Note level terraces on less steep hillslopes and tributary drainage zones in background, and in shallow drainage trenches in foreground.

and reflect the farmers' traditional understanding of the associated natural processes. The main aim is to exploit all gently or moderately sloping surfaces (gradients up to about 12 percent) receiving run-on and influent seepage from upslope, mostly by building level terraces and bunds. On interfluves these surfaces occur chiefly on footslopes adjoining low hills, cuestas, and tablelands, and on the flanks of low rises in undulating plains. In their original form they comprised concave slopes up to about 9 percent. However, gradients were commonly 5 to 6 percent on the shorter, more curved footslopes of hills and flanks of rises, whereas longer, near-planar surfaces of about 4 to 5 percent characterized the footslopes of cuestas and tablelands.

These morphological differences are reflected in terrace design (Fig. 9a-9h). Thus *hillfoot terraces* are more closely spaced (10 to 15 m) and have relatively steep reverse gradients to impound as much water as possible (Figs. 9b, 10). In contrast, terraces on the longer scarpfoot slopes are more widely spaced (25 to 30 m), with slight downslope gradients to ensure that water will flow gently over the whole surface (Figs. 9c, 11). Terraces are relatively uncommon on hillslopes with gradients greater than about 18 percent because of severe erosion problems. On less steep hillslopes these problems are countered by using narrow terraces, with shelves less than 10 m across, and reducing runoff by keeping a vegetated strip of the original surface immediately upslope from each terrace shelf (cf. Fig. 9a).

Here as elsewhere, drainage floors have the greatest concentration of run-on and seepage waters. Therefore, all floors throughout the area have been converted for cultivation by using terraces and bunds. The floors comprise drainage trenches and drainage zones. The former include trunk

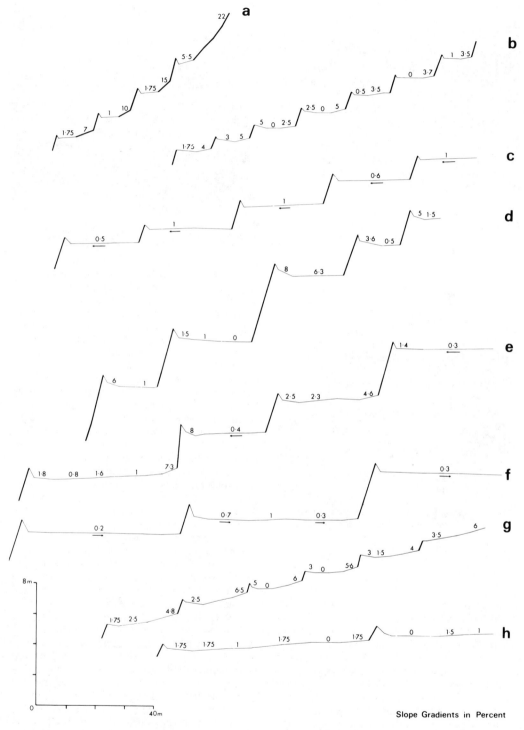

FIGURE 9. Longitudinal profiles of level terraces near Mula, Spain: (a) hillslope; (b) hill-footslope; (c) scarp-footslope; (d) drainage trench in upper sectors; (e) trench in mid-sectors; (f) trench in lower sectors. Longitudinal profiles of contour bunds: (g) tributary drainage zone; (h) main drainage zone. Thin lines indicate plowed surfaces, and thick lines are slopes covered with native grasses and shrubs. (Vertical exaggeration × 5.)

FIGURE 10. Level terraces on hill-footslope planted with olives and almonds near Mula, Spain. Note contour plowing.

FIGURE 12. Level terraces in tributary drainage trench along edge of badlands shown in Figure 11.

FIGURE 11. Scarp-footslopes dissected into badlands (background), and with level terraces cultivated for cereals and olives (foreground), near Mula, Spain.

drainage floors along the main rivers where irrigated terraces have been constructed. They also include the floors of tributary drainage trenches occurring especially in "badlands," which periodically have large amounts of inflowing waters from densely branching networks of tributary gullies and their steep side slopes (Figs. 11, 12). The trenches receive much fine-grained sediment brought in by these waters and by mass movements. Hence terrace construction here simply involves harnessing these geomorphological processes by building dams to impede sediment transport and thus infill the floor behind each dam (Fig. 7c). In this way the farmers not only construct water-retaining structures but make deep soil where

none existed before. Figure 13 illustrates the magnitude of these transformations: a badland trench almost 1,500 m long has been completely infilled and terraced except for a narrow gorge section in upper sectors. The character of such a badland floor is reflected in terrace design. Hence in steeper upper sectors (original longitudinal gradients about 12 percent), terraces are more closely spaced, with shelves having pronounced reverse gradients and the highest banks; in mid-sectors (original gradients 6 to 7 percent) they are more widely spaced, with slight down-valley gradients; and in the lowest sectors (original gradients 2 to 3 percent) they are most widely spaced, nearly flat or with gentle reverse gradients (Figs. 9d, e, f, 14).

The area also includes extensive branching networks of very shallow, unchanneled drainage "zones" occurring on undissected interfluve plains and footslopes. These were probably sites of natural infilling by colluvium and alluvium, in contrast to the eroding nature of tributary trenches, and phased downcutting and alluviation in the main trenches. Original gradients ranged from about 2 percent in the trunkline drainage zones to about 5 percent in tributaries. Their surfaces are now characterized by bunds as well as terraces, which are more closely spaced in tributary zones, where greater runoff and colluvation lead to soil accumulation immediately upslope from each bund (Figs. 9g, h, 15). In many parts the very shallow depressions corresponding to these drainage zones may not be perceived by visual inspection. However, detailed ground measurement and airphoto analysis reveal that throughout the whole area their occurrence and boundaries have been identified with great precision by the local farmers in constructing these bunds. Both terraces and bunds are built in association with various forms of microrelief designed to maximize infiltration and minimize erosion. Surfaces around tree crops are kept rough

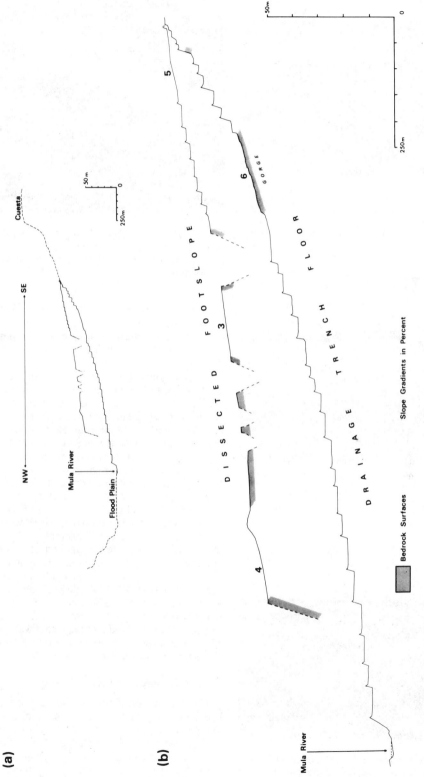

FIGURE 13. (a) Longitudinal profiles of badland drainage trench incised in footslope between cuesta escarpment and Mula River flood plain, southeastern Spain. (b) Level terraces in drainage trench and on less closely dissected upper part of footslope. (Vertical exaggeration × 5.)

FIGURE 14. Old level terraces constructed as in Figure 7c, in formerly deep drainage trenches near Mula, Spain. Note mid-sectors of main trench in foreground, and steeper tributary trench in middle distance.

FIGURE 17. Young almond tree planted in hillside pit as in Figure 16. The pit is open to receive runoff from upslope but enclosed by a bank to retain water on its downslope edge.

FIGURE 16. Man-made microrelief of pits and banks to decrease runoff and increase infiltration on hillside near Mula, Spain.

and broken, young trees are grown in circular depressions until they are properly established (Figs. 16 and 17), and contour plowing is practiced even on the most gently inclined terrace surfaces (Figs. 10, 14).

The foregoing discussions of natural hazards, soil erosion, and water management underline the central role of water in human activities. Water is not only a most essential need, together with food, but also commonly a limiting factor in food production, a major hazard when out of control, and a key element in such contrasted problems as environmental pollution and landslide prevention. It is the single most specific determinant of many physical processes in the natural environment and is indispensable to all biological processes. Lack of space prevents consideration of these diverse questions here. However, further examples to illustrate the primary importance of applied geomorphology in manipulating surficial water systems include gully control (Jepson, 1973), irrigation (Stallings, 1957; Schwab et al., 1966; Askochensky, 1973; Withers and Vipond, 1974; Currey, 1977), groundwater extraction (Figs. 4, 5), ground-water recharge (Freeman, 1974), land drainage (Schwab et al., 1966), and disposal of waste and refuse (Schneider, 1975; Foose and Hess, 1976).

Natural Resource Surveys and Land Evaluation

The exploitation of natural resources has always been, and remains, of the greatest importance. In many countries it provides the basis of national economies; elswhere it has laid the foundation for commercial and industrial development. The primary resources that constitute "land"—rocks, soils, surficial water features, and native vegetation—are the chief concern here because land provides the immediate setting for most human activities, especially food production. Applied geomorphology has made notable contributions in the investigation and planned use of these resources.

Geology. Geomorphology is a traditional tool in geological surveys. In most erosional terrain, lithological and structural variations are expressed in detailed slope changes. Therefore a standard survey approach is "feature mapping," the identification of such bedrock and slope relationships in order to trace the topographic expression of bedrock variations between exposures. This is evident, for instance, in Lahee's (1961) treatise on survey methods. Landform and geology relationships are especially clear on aerial photographs, and so geomorphological interpretation is the basis of airphotogeology, which is an indispensable tool in modern surveys (Fig. 18) (see *Remote Sensing, Engineering Geology;* Vol. XIV: *Photo Interpretation*). In Allum's (1966, p. 31) words, "photogeological interpretation is ultimately dependent on relative tone . . . and morphological expression . . . which is a measure of the relative resistance to

FIGURE 18. Airphoto showing subhorizontal quartzites forming tablelands, with shale beds etched into benches (top), and granitic basement rocks forming hill lands (bottom), locally overlain by Tertiary deepweathering deposits forming mesas (e.g., right center). Drainage networks reflect jointing patterns in both tablelands and hill lands. Orientations of slopes on higher hills (lighter tones, bottom of photo) also reflect jointing patterns. (Photo by the Royal Australian Air Force in the West Kimberley area, northwestern Australia.)

erosion of the different rocks." As he points out, many morphological lineaments reflecting geological features appear more clearly defined in airphotos than on the ground; some features may even be visible in photos but not readily apparent on the ground. By reference to the form, pattern, relief amplitude, relative occurrence, and boundary relationships of slopes and drainage networks, Allum demonstrates how geomorphological interpretation of airphotos enables lithological and structural variations to be identified within sedimentary, igneous, and metamorphic categories of rocks (see also Abdel Rahman and Wright, 1979).

For such reasons, "photogeologic exploration is conducted chiefly by application of geomorphic principles, and . . . [geomorphology is] . . . an essential technique of lithologic and structural mapping and mineral exploration" (Tator, 1960, p. 172). Hence as Miller (1968, p. 56) puts it, "the photogeologist is only as good as his grasp of geomorphology" (see also Schumm, 1968). Equally, with respect to the varied techniques of remote sensing generally, Reeves et al. (1975, p. 1108) assert that "the geological interpretation of remote-sensor pictorial data is in large part applied geomorphology." This is well illustrated in diverse examples provided by these authors. They show that detailed interpretation of structure and lithology is feasible even in areas of dense forest, because there is a high degree of correlation between geology and drainage pattern, and in these areas the latter can be mapped accurately using radar imagery, for instance.

Soils. Traditionally geomorphology has also been applied in soil surveys. Thus soil-slope units were adopted as taxonomic units ("soil types") in the American system developed under Marbut, because they provided a useful basis for assessing land according to its productivity "for different plants according to alternative types of management" (Ableiter, 1940). *Catenary analysis,* conceived by Milne (1935), also had a seminal influence on the growth of soil survey methods (Wright, 1972a, Young, 1976). Milne recognized that spatial groups of soils, *catenas,* occurred in downslope sequences from interfluve crests to valley floors. In his view, therefore, soils were linked in their occurrence by conditions of topography, a corollary being that a soil-landform pattern would recur within a particular landform region. Consideration of catenary variations is now standard survey procedure.

Completely random or rigorously systematic sampling is relatively uncommon in soil surveys. A more pragmatic approach is usually necessary, in which sampling is guided by variations in landscape features known to be related to soil changes, and progressively refined until the mapping attains the desired level of detail. Thus major landform differences, and their associated bedrock contrasts, generally provide the framework for soil mapping on a broad scale, followed by more intensive work in small areas representative of the main landform units.

For detailed mapping, the initial density of sampling varies according to the anticipated soil complexity as perceived from known geological differences, surface configuration, and land use, for example. This may involve "grid" sampling, linear traverses, or "free surveys." Areas are then selected for the most intensive sampling, which aims to confirm, and position the boundaries of, soil mapping units. Such selection and boundary location is largely based on geomorphological considerations, and especially on the relationships between soil changes and "breaks" of slope along *toposequences,* or catenas. Hence according to Young (1976, p. 345), "landform regions are frequently the basis for mapping units" in reconnaissance surveys. "At the opposite extreme, that of intensive surveys, soil distribution on erosional relief is dominated by the catenary pattern, and on depositional relief by the pattern of alluvial landforms." Consequently, the geomorphological build of an area is reflected in most soil maps in terms of the character, extent, and spatial distribution of the constituent soil mapping units. Such mapping has been carried out traditionally for agricultural and other land research purposes, but Olson et al. (1976) and Schmidt and Pierce (1976) illustrate how the practical value, for land use planning, of a combined geomorphological and soil survey can be enhanced when supplemented by appropriate engineering data.

Geomorphology is also emphasized in airphoto interpretation (q.v.). Soil properties are seldom directly perceived in an aerial view, although spectral reflectance contrasts have complex relationships with soil mineral content, texture, moisture, color and organic matter, for example. These contrasts can be analyzed with the aid of multispectral scanner imagery, but their soil relationships vary greatly other than when the soil is nonvegetated and uniformly cultivated (Myers, 1975). Because of the close genetic links beween soils and landforms, however, some of these properties can be predicted—and ground information for many more can be extrapolated—by means of airphoto interpretation of the associated landforms. Hence geomorphological "elements" form the backbone of all "interpretations" of soils on airphotos (Vink, 1968, p.120). This is substantiated by Frost (1960, p. 347) who gives many examples demonstrating that airphoto interpretation of soils "requires thorough knowledge of geomorphology." In addition, ground-truth data can be extrapolated on the basis of such interpretation because "any two soil materials derived from the same parent rock, deposited in the same way, and occupying similar topographic positions . . . have similar properties and appear in aerial photographs in similar patterns" (Frost, 1960, p. 343). Therefore, "at all stages" of a soil survey, "the recognition and delineation of mapping units is largely on the basis of landform differences" (Young, 1976, p.359). Correlations between local variations in landforms and soils are examined by Wright (1972b) and Wright and Wilson (1979). Wright and Wilson also use a numerical method to assess the validity of geomorphologically defined, or any other, soil mapping units.

Vegetation. Practical interest in the ecological associations of landforms has received increasing attention in recent years, chiefly in integrated resource studies but also in more specialized surveys. For example, Poore and Robertson (1964) mapped Jordanian pasture lands within a geomorphological framework. In their view this was necessary because soils and landforms, reflecting bedrock geology also, were the chief determinants of habitat differences and hence pasture changes. Similarly, Lacate (1961) described widespread applications of physiographic approaches to forest survey in which "landtypes" were identified primarily on geomorphological grounds. This was "a practical and valuable starting point in the assessment of the capabilities of a specific land area to produce forest crops" (Lacate, 1961, p. 278). Rey (1963) adopted an equivalent standpoint for the contrasted land use setting of Gascony. To him, surface morphology was the fundamental factor that created order in biogeography ("l'ordonnateur biogeographique"): "It is the form of the surface relief which, at one and the same time, expresses the diversity of landscapes, explains ecological differences, and conditions the practical possibilities of land productivity" (Rey, 1963, p. 71). Rey's ecological "situations" were defined in terms of slope gradient, aspect, and susceptibility to erosion, and they provided a readily identifiable framework within which complex biogeographical and land use problems could be examined. Likewise, Troll's (1966) primary units for landscape differentiation in ecological surveys, "ecotopes," were distinguished largely on geomorphological grounds.

Airphotos and other forms of remote sensing are indispensable in such surveys, for many vegetation features can be seen on them. Nevertheless, landform analysis plays an important role because vegetation varies in response to many different factors, and so community boundaries are commonly problematic. Moreover, an aerial view provides only limited information about community composition, and so without a dense network of ground observations—which is possible only in localized studies—communities can rarely be compared on airphotos in sufficient detail for classification and mapping purposes (see Vol. XIV: *Vegetation Mapping*). In contrast, the character and spatial relationships of the corresponding landform types can be readily identified in all but the most densely vegetated terrain. Therefore, the geomorphological setting of plant communities provides a valuable means of extrapolating field observations and mapping vegetation on airphotos, as has been demonstrated in a range of environments (Wright, 1971, 1972b). Geomorphological relationships have also warranted close attention in the analysis of remote-sensor imagery for forest surveys, because of recent trends toward "environmental awareness" and the need for forest management to include management of land in addition to timber. Thus a "modern classification" of forest lands for use with image-interpretation techniques "might describe the entire land system, including vegetation cover, slope position, slope grade, slope length, slope configuration, soil cover, temperature zonation, surface rock and stone content, drainage configuration and stream order" (Thorley, 1975, p. 1353).

Land Capability Classification. For decades geomorphology has received emphasis in land classification and evaluation (see Vol. XIV: *Land Capability Analysis*), due to the ecological and land use associations mentioned previously. The most notable early work in such classification was carried out in the United States between the two world wars. Geomorphology figured prominently in this work because close links had been recognized between landform and soil, the principal raw material for land use (Wright, 1972a). In *land classification,* above all, soil had to be considered as a unit of landscape that supported plants, and not simply as a soil profile. Hence the definition of *soil units* for such classification included the range of surface gradients and degree of erosion, because land use was the ultimate concern. Some soil types would be mapped in two or more *phases* to indicate differences important in use and management, for example, an undulating phase with 2- to 7-percent gradients and a 7- to 14-percent rolling phase (Stallings, 1957).

A land capability classification, as pioneered by the U.S. Soil Conservation Service, is still one of the best aids to land management, and so the U.S. scheme has been widely adapted for use in other countries (Hudson, 1971; Young, 1976). In this approach soils were grouped on the basis of erosion hazards and other limitations so that suitable cropping systems could be planned to protect the soil and ensure sustained yields. Geomorphology was a major consideration. Thus the scheme required a map to be made of the study area showing not only soil types but slope and other important land characteristics (Stallings, 1957). The soils were then combined into capability units whose specified limits included soil type, slope, and degree of erosion. Where necessary, these units were grouped into subclasses, each with a particular kind and degree of permanent limitation, and

into broader classes according to the degree of such limitation. Geomorphological factors also determined these groupings. For instance, among the four *capability classes* suited for cultivation, *Class I* included soils that were nearly level and with no risk, or only slight risks, of damage; *Class II* had soils with gentle slopes that were subject to moderate erosion; *Class III* had moderately steep slopes subject to more severe erosion; and *Class IV* had steep slopes subject to severe erosion. In an African version of the scheme, for example, these four classes had maximum permissible slopes of 2 percent, 5 percent, 8 percent, and 12 percent, respectively (Hudson, 1971).

Engineering. Geomorphology has been applied extensively in land classification and evaluation for engineering purposes (see *Remote Sensing, Engineering Geology*). As Flawn (1970, p. 12) explains, the typical landforms and drainage features of an area may be apparent from maps and airphotos, and so with little or no ground information military geologists have been able to infer from these sources "data on construction sites for airfields and other buildings, sources of construction materials, water supply and transportation routes." Likewise, the engineering geology maps of the U.S. Geological Survey describe "terrain, natural slope, and slope stability" as well as various features of rocks and surface conditions (Flawn, 1970, p. 15). Leggett (1973) presents examples from several countries of "engineering-geological" surveys in which mapping units were similarly defined in terms of geomorphological features such as surface form, relief amplitude, drainage, erosion, and relative slope instability (see Vol. XIV: *Map Symbols*). This not only facilitated the interpretation of site conditions but also provided a basis for extrapolating known conditions into unknown areas and for planning appropriate construction methods (see also Black, 1973). The genetic relationships of landforms, which in most areas can be inferred fairly readily, have also received much emphasis because, according to Kreig and Reger (1976, p. 58), "similar geologic processes usually result in landforms with similar characteristics and engineering problems." Consequently, the terrain analysis advocated by these latter authors depended on a genetic classification of landforms because this gave a reliable means of arranging and correlating borehole and soil test properties with each landform having a distinctive range of such properties.

Largely because of the geological, soil, and natural-process relationships, landform interpretation of airphotos is standard practice in these engineering surveys. As in other contexts, "landform is the dominant element of the aerial-photographic pattern, and a knowledge of petrology and geomorphology forms the basis for its proper evaluation" (Lueder, 1959, p. 123). Leuder illustrated the value of such evaluation in numerous airphoto examples and related engineering questions. Belcher (1960) also reviewed many case studies showing that airphoto interpretation in engineering was based largely on geomorphological inferences about subsurface as well as surface conditions, as did Rib (1975), whose examples covered a wide variety of engineering projects and diverse techniques of remote sensing. Similar case studies are included in UNESCO (1968) and Coates (1976).

Integrated Surveys. The planned exploitation of land resources for agricultural development is of cardinal importance in the poorer countries of Africa, Asia, and Latin America. All depends on making the best use of land in these countries. This is the only sure way of maintaining even present living standards and is essential if productivity is to increase to the point where agricultural and commercial sectors can maintain growth in the economy as a whole. For huge areas, however, appropriate scientific information is grossly inadequate, and only very limited funds and scientific personnel are available. Hence reconnaissance studies are needed to plan the most efficient deployment of the intensive research effort needed for land evaluation and development planning. Ideally these studies should comprise "integrated surveys" based on interdisciplinary teamwork and aiming for a comprehensive appraisal of natural resources, because understanding environmental interrelations is necessary for sound land use planning, rather than traditionally separate, resource surveys. Recognition of this is vitally important in developing countries because of the need for ecological efficiency in land use (Wright, 1972b). Information from integrated surveys should also find more ready application to regional planning problems than would the results of several more or less uncoordinated studies. And these surveys, involving interdisciplinary work that avoids duplication of effort and technical facilities, enable more ground to be covered more economically as well as more effectively.

Such integrated surveys were organized after the Second World War in Australia, Canada, and the USSR, where there was a need to determine land use potential for vast underdeveloped territories, and they have subsequently been applied in many developing countries (Wright, 1971). They comprised projects in which land complexes rather than individual resource attributes were mapped and described by multidisciplinary teams of scientists. The basic mapping units were "land systems," "land types," and "landscapes" in the Australian, Canadian, and Russian schemes, respectively. These were areas or groups of areas that were characterized by repetition of particular landforms with associated patterns of soils, vegetation, and land use. Detailed internal variations

termed "land units" (Australia), "physiographic sites" (Canada), and "facies" (Russia) were described but usually not mapped. Airphotos were an essential tool, enabling large areas to be covered quickly by extrapolating field data on the basis of airphoto interpretation.

Geomorphology has been of central importance in the development of these surveys—almost inevitably, in view of its long-established applications in more conventional resource studies. Thus the environmental significance of landform variations received emphasis from those who pioneered the approach. According to Christian (1958, pp.75-76), the Australian land system concept visualized that land was the product of its physiographic evolution, during which the surface had been shaped into landforms, "each developing in the process its own hydrological features, soil mantle, vegetation communities, animal populations, and range of micro-environments." Similarly, Hills and Portelance (1960, p. 113) in Canada recognized that the "physical" controls of crop production varied with landform patterns: these provided "the basis for establishing patterns of soil moisture, nutrient availability within soil profile types (and) also patterns of local climates." Moreover, geomorphological analysis was of primary concern in airphoto interpretation at all stages of the surveys; in the initial planning, throughout fieldwork, and in the final extrapolations and mapping. For such reasons, landforms were the foundation for identifying terrain units. In the land system approach, as described by Christian (1959, pp. 591-592), a land unit was a "particular land form" which at each of its various occurrences had associated with it "the same group of soils and vegetation communities." The land system was "a naturally occurring pattern of land units, geomorphologically associated and morpho-genetically related." Closely similar applications of geomorphology were employed in integrated surveys of diverse areas ranging from the United Kingdom to India, from Argentina to Botswana, as well as from Australia to Canada (Wright, 1971; Cooke and Doornkamp, 1974).

The role of geomorphology in these surveys is thus a particularly vital one, for it can provide the necessary framework whereby environmental variations may be viewed synthetically to ensure truly integrated interdisciplinary effort, rather than simply a multidisciplinary collation of different specialist studies. This was illustrated by Wright (1972b), who questioned some survey methods because they produced general descriptions of broad-scale terrain units with little indication of local variations, and constituted an inventory rather than a classification of land. These methods involved progressive subdivision of large areas according to imprecise criteria and governed by more evident airphoto contrasts. Therefore the final subdivisions comprised terrain units of different kinds and different orders of magnitude, and were described in terms of a general summation of the findings of different specialists instead of a synthesis of resource characteristics. Wright advocated an alternative, synthetic method based on geomorphological differentiation of small environmental units, termed "sites," at a level that was also appropriate for planning intensive agricultural research. Such primary units were equivalent to the taxonomic individuals needed in systematic classification and in interdisciplinary teamwork. They facilitated a methodical approach to the differentiation of broader land complexes, but these were built up from within by aggregation of classified individuals rather than delimited from without in terms of preconceived boundaries.

Surveys in a wide range of environments have been carried out using this alternative approach. Geomorphological sites were defined as patches of ground of relatively uniform shape internally, regularly curved or nearly planar, and delimited by relative discontinuities in rate of change of gradient. Site analysis involved detailed *slope-profiling* along transects chosen to include all the variability within the area being investigated, as deduced from cross-country traverses and intensive airphoto interpretation. Coordinated records of soil, vegetation, and land use characteristics were compiled with each measured site, and the principal characteristics confirmed laterally away from each slope profile for up to about 100 m (Fig. 19). Boundaries of sites or small clusters of sites, depending on local complexity, were recorded on airphotos. *Site characteristics* were correlated with airphoto tonal variations to permit site data to be extrapolated between the measured slope profiles. Although in plainlands many site differences on the ground were apparent only on measurement, they were reflected in detailed tonal and textural variations on airphotos (Fig. 20). From this groundwork and airphoto interpretation "site-types" were identified, each comprising several individuals with closely similar landform, soil, and vegetation characteristics. Spatial assemblages of sites could thus

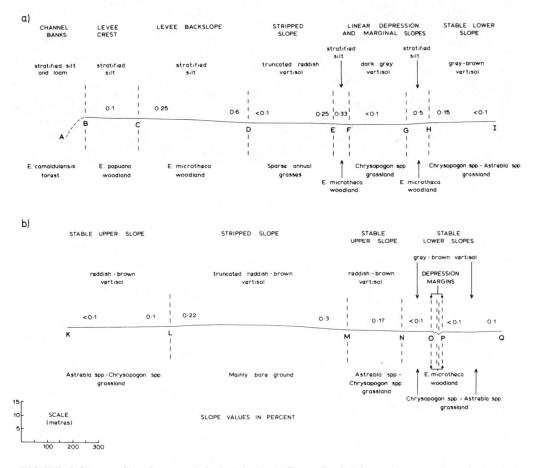

FIGURE 19. Slope profiles of geomorphological sites in the Fitzroy flood plains, northwestern Australia (vertical exaggeration × 10.) (From Wright, 1972b.)

FIGURE 20. Airphoto of part of the Fitzroy flood plains. The area largely consists of grasslands with relief amplitude of less than 2 m with tonal contrasts reflecting slope, soil, drainage, and vegetation differences (photo: Royal Australian Air Force; from Wright, 1972*b*).

be delimited, each dominated by recurring members of a few site-types. Assemblages distinguished by the same dominant site-types were then grouped into *land systems* whose natural resources and agricultural potential were described in detail in the survey reports to provide a basis for planning the deployment of more intensive land evaluation projects (Wright, 1972*b*).

These site assemblages are the topographic expression of spatial contrasts in land-forming processes. For instance, one flood-plain assemblage in Figure 21 comprises areas of active deposition traversed by drainage zones, and characterized by levee sites as in part in profile *a* in Figure 19, whereas the other flood-plain assemblage largely consists of stable weathering surfaces interspersed with stripped sites, as in profile *b* in Figure 19. In detail, however, both assemblages include a range of distributary channels and flood-water depres-

sions; aggrading sites within drainage zones, on levees and at the termination of distributary channels; degrading sites associated with channel incision, overbank flooding, gullying, or stripping; and relatively stable weathering sites with contrasted water-table relationships and drainage conditions. Such spatial variations are of primary importance ecologically, and hence in land use planning, because they represent radical changes in natural-process systems and their associated surficial water conditions and microclimates, which are expressed in the distinctive soils and native vegetation of the resultant sites.

Conclusions

Geomorphology has numerous and varied practical applications, many more than could be examined in this entry. Mention was made earlier of

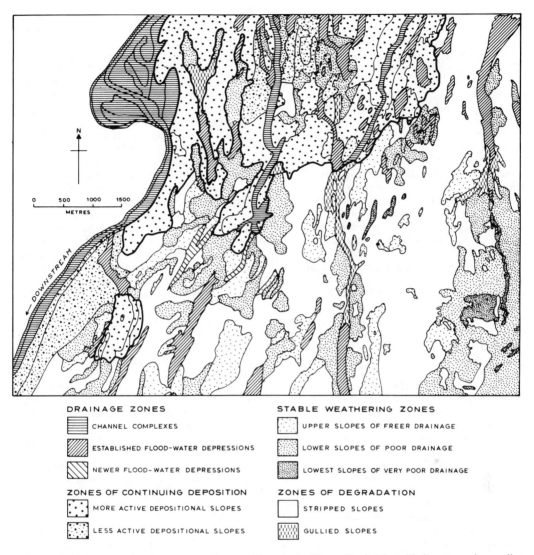

DRAINAGE ZONES

CHANNEL COMPLEXES

ESTABLISHED FLOOD-WATER DEPRESSIONS

NEWER FLOOD-WATER DEPRESSIONS

ZONES OF CONTINUING DEPOSITION

MORE ACTIVE DEPOSITIONAL SLOPES

LESS ACTIVE DEPOSITIONAL SLOPES

STABLE WEATHERING ZONES

UPPER SLOPES OF FREER DRAINAGE

LOWER SLOPES OF POOR DRAINAGE

LOWEST SLOPES OF VERY POOR DRAINAGE

ZONES OF DEGRADATION

STRIPPED SLOPES

GULLIED SLOPES

FIGURE 21. The boundary between two site-assemblages in the Fitzroy flood plains with sites grouped according to dominant land-forming processes as inferred from slope microforms and composition of surficial materials. Most of the mapped area is shown in Figure 20 (from Wright, 1972*b*).

applications relating to water use that had to be excluded. Other notable omissions include management of coastal zones (q.v. in Vol. XIV); conservation of wind-eroded soils, surveys of sand, gravel, and mineral deposits, terrain analysis in permafrost areas (see *Permafrost, Engineering Geology*); reclamation of mined lands (see Vol. XIV: *Mine Subsidence Control; Mining Preplanning*); and landscape evaluation for recreational purposes (see Vol. XIV: *Open Space*) (Stallings, 1957; Hudson, 1971; Coates, 1973; Cooke and Doornkamp, 1974; McKenzie and Utgard, 1975; Clark, 1977; Coates, 1980). In all these examples, as in those considered here, geomorphology has been applied by many different specialists: agri-

culturalists, ecologists, engineers, hydrologists, soil scientists, and others. In addition, it has provided a framework for the coordination of multidisciplinary effort, which is essential because of the complexity of environmental problems and because many disciplines must be involved in problem solving and decision making.

Such diverse applications of geomorphology stem from the recognition that landforms have close interconnections with other landscape features and, through their natural-process relationships, are a key element in the functioning of natural systems. Consequently, applied geomorphology is an indispensable tool in environmental management, including the control of natural hazards. This is readily apparent in dealing with flood problems, for example. Here all depends on possessing a comprehensive knowledge of geomorphological processes within the integrated structure of the drainage basin, their dynamics, causative factors, and spatial variations. This knowledge is an essential prerequisite in identifying and positioning effective flood control measures in catchments and along rivers. It is also needed in defining natural units for the regulation of floodplain use. In particular, it provides a rational basis for assessing the relative merits of control schemes, in environmental as well as economic terms. Without that basis, the selected measures could have undesirable environmental repercussions, ranging from habitat deterioration for wildlife to widespread erosional loss of farmland and damage to roads and buildings (Cooke and Doornkamp, 1974; Emerson, 1975; Keller, 1976).

This underlines the importance of spatial relationships in environmental management. Measures to reduce flood damage in one locality have often produced damage down-valley, as noted earlier. Similarly, the effects of sea-defense structures have demonstrated the need to consider the whole coastal system. Thus seawalls, while providing protection from direct erosion, also tend to increase backwash and can produce a destructive wave effect. Likewise, by trapping transported material, groins and breakwaters can be detrimental to areas downdrift, where erosion may occur. Hence environmental systems must be examined in their *spatial dimensions,* because conditions at any locality reflect not only internal interreactions but also external influences, both local and regional. *Landform analysis* facilitates the study of such spatial relationships because the geomorphological structure of these natural systems is readily perceived. For example, the drainage basin—with its interfluve catchments, stream network, flood plains, and their constituent site-types—is the logical, natural entity for studying floods or other problems linked to fluvial processes. This has been demonstrated in numerous applied studies of basin morphometry and dynamics, referred to previously. Similarly, coastal tracts, landslides, wind-blown dunes, and other zones of natural hazards are composed of geomorphological units, each varying in character according to the processes and materials that form it. Therefore, differentiation of these units provides an appropriate framework for investigating the associated process systems as a basis for planning control measures.

For corresponding reasons, geomorphology has a special role in the planned exploitation of land resources. In the first place, landforms and landforming processes are major elements of the natural systems governing plant growth and crop production. Second, man can modify these forms and to some extent manipulate these processes, as

seen in examples discussed earlier. The geomorphological structure of an area is thus of major concern in evaluating land use possibilities, as in dealing with natural hazards. Geomorphological "sites," for instance, because of their process relationships, are appropriate functional units for the study of biological and physical phenomena involved in primary productivity. Furthermore, site assemblages provide a broader structure within which the spatial organization of these biophysical phenomena may be analyzed. Therefore, differentiation of geomorphological sites and their spatial patterns of variation provides an appropriate framework for organizing intensive research for land use development and management planning. As ecological units they constitute a common base for diverse studies. They also serve to indicate the limits of application of information subsequently derived from these studies, by the extrapolation within assemblages of knowledge gained about particular site-types. In this way the potential of large areas may be assessed more rapidly than would be possible otherwise.

Spatial considerations have been dwelt on to underline the fact that local heterogeneity in enviromental conditions is a characteristic feature even in plainlands, as described in an earlier section. Hence much groundwork is essential in order to plan the deployment of an intensive land-use research effort. This groundwork warrants far greater attention than is usual before arduous and costly development projects are embarked on, because scientific expertise and funds for research are scarce in many countries and so maximizing the efficiency of their usage is a vital objective. Furthermore, spatial variations in both social and biophysical conditions are a major consideration in subsequent agricultural development planning. As emphasized by Wharton (1970), most development plans have been severely restricted because they were framed without reference to these variations.

Such environmental principles apply equally to the management of urban areas. Thus Coates (1974, p. 362) identifies the need for a comprehensive and integrated systems approach to urban planning (see *Urban Geomorphology*). In particular, methodologies are needed to adapt this planning to natural land-water ecosystems. Geomorphology is most important because geomorphologists are especially schooled in understanding such ecosystems, and their expertise is essential for planning management procedures to minimize human deformation of nature's equilibrium systems (Coates, 1974, p. 360; see also Flawn, 1970; Leggett, 1973; McKenzie and Utgard, 1975). Implicit here is the need for public control of urban land use, and so Flawn (1970) considers how authorities may take steps to protect lives and property against geomorphological hazards, for example. In his view, city and county zoning, as a traditional method of controlling land use, should be guided by "knowledge of processes, resources, and behavior of rocks and soils." Swain (Flawn, 1970, p. 221; Coates, 1974, p. 408) and other landscape architects stress the need for such guidance. As he explains, appropriate building codes should be devised and enforced to avoid problems "arising out of the topography."

Geomorphology also has many hypothetical applications. Much has been written in the past about these potential applications (Tricart, 1964, 1965; Steers, 1971; Cooke and Doornkamp, 1974; all articles in Hails, 1977) and a large body of more recent research is of direct relevance to environmental problems though all too rarely applied to them. Such research includes quantitative studies, through mathematical modeling, of the environmental interrelationships of landforms, and of the processes and mechanisms that are operating. To maximize their practical value, however, these studies must be directed specifically toward land use problems because they are ultimately the chief concern in an applied context.

ROBERT L. WRIGHT

References

Abdel Rahman, M. A., and Wright, R. L., 1979. Airphoto lineations, joints and bedding in part of southeast Spain, *Second International Conference on Basement Tectonics, Proceedings.* Denver: Basement Tectonics Committee Inc., 225-235.

Ableiter, J. K., 1940. Productivity rating of soil types, *Missouri Agric. Exp. Station Bull.* **421**, 13-24.

Allum, J. A. E., 1966. *Photogeology and Regional Mapping.* Oxford: Pergamon Press, 107p.

Askochensky, A. N., 1973. Basic trends and methods of water control in the arid zones of the Soviet Union, in D. R. Coates (ed.), *Environmental Geomorphology and Landscape Conservation, III: Non-Urban Regions.* Stroudsburg, Pa.: Dowden, Hutchinson & Ross, 207-216.

Atkinson, B. W., 1981. *Meso-scale Atmospheric Circulations.* London: Academic Press, 495p.

Bailey, R. G., 1973. Forest land use implications, in D.R. Coates (ed.), *Environmental Geomorphology and Landscape Conservation, III: Non-Urban Regions.* Stroudsburg, Pa.: Dowden, Hutchinson & Ross, 388-413.

Barry, R. G., 1981. *Mountain Weather and Climate.* London: Methuen, 313p.

Belcher, D. J., 1960. Photo interpretation in engineering, in R. N. Colwell (ed.), *Manual of Photographic Interpretation.* Washington, D. C.: American Society of Photogrammetry, 169-342.

Bennett, H. H., and Chapline, W. R., 1973. Soil erosion—a national menace. Part I: Some aspects of the wastage caused by soil erosion, in D. R. Coates (ed.), *Environmental Geomorphology and Landscape Conservation, III: Non-Urban Regions.* Stroudsburg, Pa.: Dowden, Hutchinson & Ross, 57-83.

Bigger, R., 1974. The flood problem, in D. R. Coates (ed.), *Environmental Geomorphology and Landscape Conservation, II: Urban Areas.* Stroudsburg, Pa.: Dowden, Hutchinson & Ross, 187-196.

Black, R. F., 1973. Permafrost, in D. R. Coates (ed.), *Environmental Geomorphology and Landscape Conservation, III: Non-Urban Regions.* Stroudsburg, Pa.: Dowden, Hutchinson & Ross, 160-184.

Bormann, F. H., and Likens, G. E., 1969. The watershed-ecosystem concept and studies of nutrient cycles, in G. M. Van Dyne (ed.), *The Ecosystem Concept in Natural Resource Management.* New York: Academic Press, 49-76.

Brunsden, D.; Doornkamp, J. C.; Fookes, P. G.; Jones, D. K. C.; and Kelly, J. M. H., 1975. Large scale geomorphological mapping and highway engineering design, *Quart. Jour. Eng. Geology* **8**, 227-253.

Carson, M. A., and Kirkby, M. J., 1972. *Hillslope Form and Process.* Cambridge, England: Cambridge University Press, 475p.

Christian, C. S., 1958. The concept of land units and land systems, *Ninth Pacific Sci. Congr. Proc.* **20**, 74-81.

Christian, C. S., 1959. The eco-complex in its importance for agricultural assessment, in *Biogeography and Ecology in Australia.* The Hague: Junk, 587-605.

Clark, J. R., 1977. *Coastal Ecosystem Management.* New York: Wiley, 928p.

Coates, D. R. (ed.), 1974. *Environmental Geomorphology and Landscape Conservation, II: Urban Areas.* Stroudsburg, Pa.: Dowden, Hutchinson & Ross, 454p.

Coates, D. R.(ed.), 1976. *Geomorphology and Engineering.* Stroudsburg, Pa.: Dowden, Hutchinson & Ross, 360p.

Coates, D. R. (ed.), 1980. *Coastal Geomorphology.* London: Allen and Unwin, 404p.

Cooke, R. U., and Doornkamp, J. C., 1974. *Geomorphology in Environmental Management.* Oxford: Clarendon Press, 413p.

Currey, D. T., 1977. The role of applied geomorphology in irrigation and groundwater studies, in J. R. Hails (ed.), *Applied Geomorphology.* Amsterdam: Elsevier, 51-83.

Eckel, E. B. (ed.), 1958. Landslides and engineering practice, *Highway Research Board Spec. Rep. 29,* Washington D.C., NAS-NRC Publication 544, 232p.

Emerson, J. W., 1975. Channelization: a case study, in G. D. McKenzie and R. O. Utgard (eds.), *Man and His Physical Environment: Readings in Environmental Geology,* 2nd ed. Minneapolis: Burgess, 58-60.

Flawn, P. T., 1970., *Environmental Geology.* New York: Harper & Row, 313p.

Flohn, H., 1969. Local wind systems, in H. Flohn (ed.), *World Survey of Climatology, Volume 2: General Climatology.* Amsterdam: Elsevier, 139-171.

Food and Agriculture Organization, 1965. Soil eorison by water—some measures for its control on cultivated lands, *Agricultural Development Paper 87,* 284p.

Foose, R. M., and Hess, P. W., 1976. Scientific and engineering parameters in planning and development of a landfill site in Pennsylvania, in D. R. Coates (ed.), *Geomorphology and Engineering.* Stroudsburg, Pa.: Dowden, Hutchinson & Ross, 289-312.

Fox, H. L., 1976. The urbanizing river: a case study in the Maryland piedmont, in D. R. Coates (ed.), *Geomorphology and Engineering.* Stroudsburg, Pa.: Dowden, Hutchinson & Ross, 245-271.

Freeman, V. M., 1974. Water spreading as practiced by the Santa Clara water-conservation district, Ventura County, California, in D. R. Coates (ed.), *Environmental Geomorphology and Landscape Conservation, II: Urban Areas.* Stroudsburg, Pa.: Dowden, Hutchinson & Ross, 111-117.

Frost, R. E., 1960. Photo interpretation of soils, in R. N. Colwell (ed.), *Manual of Photographic Interpretation.* Washington, D. C.: American Society of Photogrammetry, 343-402.

Geiger, R., 1969. Topoclimates, in H. Flohn (ed.), *World Survey of Climatology, Volume 2: General Climatology.* Amsterdam: Elsevier, 105-138.

Glenn, L. C., 1973. Denudation and erosion in the southern Appalachian region and the Monongahela basin, in D. R. Coates (ed.), *Environmental Geomorphology and Landscape Conservation, III: Non-Urban Regions.* Stroudsburg, Pa.: Dowden, Hutchinson & Ross, 36-56.

Gray, D. M., 1970. *Handbook on the Principles of Hydrology.* New York: Water Information Center, 591p.

Hails, J. R. (ed.), 1977. *Applied Geomorphology.* Amsterdam: Elsevier, 418p.

Hilgard, E. W., 1911. *Soils.* New York: Macmillan, 596p.

Hills, G. A., and Portelance, R., 1960. *The Glackmeyer Report on Multiple Land Use Planning.* Ontario: Department of Lands and Forests, 193p.

Holtan, H. N., and Lopez, N. C., 1971. USDAHL-70 Model of Watershed Hydrology, *U.S. Dept. Agriculture Research Service Tech. Bull. 1435,* 84p.

Hudson, N., 1971. *Soil Conservation,* London: Batsford, 320p.

Jepson, H. G., 1973. Prevention and control of gullies, in D. R. Coates (ed.), *Environmental Geomorphology and Landscape Conservation, III: Non-Urban Regions.* Stroudsburg, Pa.: Dowden, Hutchinson & Ross, 283-298.

Keller, E. A., 1976. Channelization: environmental, geomorphic, and engineering aspects, in D. R. Coates (ed.), *Geomorphology and Engineering.* Stroudsburg, Pa.: Dowden, Hutchinson & Ross, 115-140.

Kiersch, G. A., 1975. The Vaiont reservoir disaster, G. D. McKenzie and R. O. Utgard (eds.), *Man and His Physical Environment: Readings in Environmental Geology,* 2nd ed. Minneapolis: Burgess, 71-75.

Kirkby, M. J., and Morgan, R. P. C. (eds.), 1980. *Soil Erosion.* Chichester, England: Wiley, 312p.

Kolb, C. R., 1976. Geologic control of sand boils along Mississippi river levees, in D. R. Coates (ed.), *Geomorphology and Engineering.* Stroudsburg, Pa.: Dowden, Hutchinson & Ross, 99-113.

Kreig, R. A., and Reger, R. D., 1976. Preconstruction terrain evaluation for the trans-Alaska pipeline project, in D. R. Coates (ed.), *Geomorphology and Engineering.* Stroudsburg, Pa.: Dowden, Hutchinson & Ross, 55-76.

Lacate, D. S., 1961. A review of landtype classification and mapping, *Land Economics 37,* 271-278.

Lahee, F. H., 1961. *Field Geology,* 6th ed. New York: McGraw-Hill, 926p.

Lazaro, T. R., 1979. *Urban Hydrology.* Ann Arbor, Mich.: Ann Arbor Science, 249p.

Lee, K., 1975. Ground investigations in support of remote sensing, in F. J. Janza (ed.), *Manual of Remote Sensing, I: Theory, Instruments and Techniques.* Falls Church, Va.: American Society of Photogrammetry, 805-856.

Leggett, R. F., 1973. *Cities and Geology.* New York: McGraw-Hill, 624p.

Leighton, F. B., 1974. Landslides and hillside development, in D. R. Coates (ed.), *Environmental Geomorphology and Landscape Conservation, II: Urban Areas.* Stroudsburg, Pa.: Dowden, Hutchinson & Ross, 206-223.

Leighton, F. B., 1976. Geomorphology and engineering control of landslides, in D. R. Coates (ed.), *Geomorphology and Engineering,* Stroudsburg, Pa.: Dowden, Hutchinson & Ross, 273-287.

Leopold, L. B., 1974. Hydrology for urban land planning, in D. R. Coates (ed.), *Environmental Geomorphology and Landscape Conservation, II: Urban Areas.* Stroudsburg, Pa.: Dowden, Hutchinson & Ross, 69-86.

Loucks, O. I., 1962. Ordinating forest communities by means of environmental scalars and phytosociological indices, *Ecol. Monographs 32,* 137-166.

Lueder, D. R., 1959. Aerial Photographic Interpretation. New York: McGraw-Hill, 462p.

McKenzie, G. D., and Utgard, R. O. (eds.), 1975. *Man and His Physical Environment: Readings in Environmental Geology,* 2nd ed. Minneapolis: Burgess, 388p.

Miller, V. C., 1968. Aerial photographs and land forms (photogeomorphology), in *Aerial Surveys and Integrated Studies.* Paris: Unesco, 41-69.

Milne, G., 1935. Some suggested units of classification and mapping particularly for East African soils, *Soil Research 4,* 183-198.

Morton, D. M., and Streitz, R., 1975. Landslides, in G. D. McKenzie and R. O. Utgard (eds.), *Man and His Physical Environment: Readings in Environmental Geology,* 2nd ed. Minneapolis: Burgess, 58-60.

Myers, V. I., 1975. Crops and soils, in L. W. Bowden (ed.), *Manual of Remote Sensing, II: Interpretations and Applications.* Falls Church, Va.: American Society of Photogrammetry, 1715-1813.

Newson, M. D., 1975. *Flooding and Flood Hazard in the United Kingdom.* Oxford: Oxford University Press, 60p.

Noble, C. C., 1976. The Mississippi River flood of 1973, in D. R. Coates (ed.), *Geomorphology and Engineering.* Stroudsburg, Pa.: Dowden, Hutchinson & Ross, 79-98.

Oke, T. R., 1978. *Boundary Layer Climates.* London: Methuen, 372p.

Olson, G. W., 1976. Land use contributions of soil survey with geomorphology and engineering, in D. R. Coates (ed.), *Geomorphology and Engineering.* Stroudsburg, Pa.: Dowden, Hutchinson & Ross, 23-41.

Orsborn, J. F., 1976. Drainage basin characteristics applied to hydraulic design and water resources management, in D. R. Coates (ed.), *Geomorphology and Engineering.* Stroudsburg, Pa.: Dowden, Hutchinson & Ross, 141-171.

Palmer, L., 1976. River management criteria for Oregon and Washington, in D. R. Coates (ed.), *Geomorphology and Engineering.* Stroudsburg, Pa.: Dowden, Hutchinson & Ross, 329-346.

Polunin, N., 1960. *Introduction to Plant Geography.* London: Longmans, 640p.

Poore, M. E. D., and Robertson, V. C., 1964. *An Approach to the Rapid Description and Mapping of Biological Habitats.* London: Sub-Commission on Conservation of Terrestrial Biological Communities of the International Biological Program, 68p.

Rantz, S. E., 1975. Urban sprawl and flooding in southern California, in G. D. McKenzie and R. O. Utgard (eds.), *Man and His Physical Environment: Readings in Environmental Geology.* 2nd ed. Minneapolis: Burgess, 45-52.

Reeves, R. G.; Kover, A. N.; Lyon, R. J. P.; and Smith, H. T. U., 1975. Terrain and minerals: assessment and evaluation, in L. W. Bowden (ed.), *Manual of Remote Sensing, II: Interpretations and Applications.* Falls Church, Va.: American Society of Photogrammetry, 1107-1351.

Rey, P. (ed.), 1963. *Recherches Experimentales et Essai de Synthese Biogeographique dans la Region des Coteaux de Gascogne.* Paris: Centre National de la Recherche Scientifique, 110p.

Rib, N. T., 1975. Engineering: regional inventories, corridor surveys and site investigations, in L. W. Bowden (ed.), *Manual of Remote Sensing, II: Interpretations and Applications.* Falls Church, Va.: American Society of Photogrammetry, 1881-1945.

Ritchie, A. M., 1958. Recognition and identification of landslides, in E. B. Eckel (ed.), *Landslides and Engineering Practice.* Highway Research Board Spec. Rept. 29. Washington D.C.: NAS-NRC Publication 544, 48-68.

Schick, A. P., 1971. A desert flood: physical characteristics; effects of man, geomorphic significances, human adaptation—a case study of the southern Arava watershed, *Jerusalem Studies in Geography* **2,** 91-155.

Schmid, J. A., 1974. The environmental impact of urbanization, in I. Manners and M. W. Mikesell (eds.), *Perspectives on Environment.* Washington, D.C.: Association of American Geographers, 213-251.

Schmidt, P. W., and Pierce, K. L., 1976. Mapping of mountain soils west of Denver, Colorado, for land use planning, in K. R. Coates (ed.), *Geomorphology and Engineering.* Stroudsburg, Pa.: Dowden, Hutchinson & Ross, 43-54.

Schneider, W. J., 1975. Hydrologic implications of solid-waste disposal, in G. D. McKenzie and R. O. Utgard (eds.), *Man and His Physical Environment: Readings in Environmental Geology.* 2nd ed. Minneapolis: Burgess, 125-134.

Schumm, S. A., 1968. Aerial photographs and water resources, in *Aerial Surveys and Integrated Studies.* Paris: Unesco, 70-79.

Schuster, R. L., and Krizek, R. J., (eds.), 1978. *Landslides Analysis and Control.* Washington, D. C.: National Academy of Sciences, 234p.

Schwab, G. O.; Frevert, R. K.; Edminster, T. W.; and Barnes, K. K., 1966. *Soil and Water Conservation Engineering,* 2nd ed. New York: Wiley, 683p.

Smith, D. D., and W. H., Wischmeier, 1962. Rainfall erosion, *Advances in Agronomy* **14,** 109-148.

Stall, J. B., 1973. Man's role in affecting the sedimentation of streams and reservoirs, in D. R. Coates (ed.), *Environmental Geomorphology and Landscape Conservation, III: Non-Urban Regions.* Stroudsburg, Pa.: Dowden, Hutchinson & Ross, 103-119.

Stallings, J. H., 1957. *Soil Conservation.* Englewood Cliffs, N.J.: Prentice-Hall, 575p.

Strahler, A. N., 1973. The nature of induced erosion and aggradation, in D. R. Coates (ed.), *Environmental Geomorphology and Landscape Conservation, III: Non-Urban Regions.* Stroudsburg, Pa.: Dowden Hutchinson & Ross, 18-35.

Swain, W. G., 1974. Man vs. gravity: making up for 200-year loss in Pittsburgh, in D. R. Coates (ed.), *Environmental Geomorphology and Landscape Conservation, II: Urban Areas.* Stroudsburg, Pa.: Dowden, Hutchinson & Ross, 408-412.

Steers, J. A. (ed.), 1971. *Applied Coastal Geomorphology.* London: Macmillan, 227p.

Ta Liang, and Belcher, D. J., 1958. Airphoto interpretation, in E. B. Eckel (ed.), *Landslides and Engineering Practice.* Highway Research Board Spec. Rept. 29. Washington D.C.: NAS-NRC Publication 544, 69-92.

Tansley, A. G., 1949. *Introduction to Plant Ecology.* London: Allen and Unwin, 260p.

Tator, B. A., 1960. Photo interpretation in geology, in R. N. Colwell (ed.), *Manual of Photographic Interpretation.* Washington, D.C.: American Society of Photogrammetry, 169-342.

Thorley, G. A., 1975. Forest lands: inventory and assessment, in L. W. Bowden (ed.), *Manual of Remote Sensing, II: Interpretations and Applications.* Falls Church, Va.: American Society of Photogrammetry, 1353-1426.

Tricart, J., 1964. Panorama de la géomorphologie appliqúe dans le monde, *Rev. Gén. Sciences* **71,** 345-361.

Tricart, J., 1965. *Principes et Methodes de la Geomorphologie.* Paris: Masson et Cie, 496p.

Troll, C., 1966. Landscape ecology, *Publications of the ITC-Unesco Centre for Integrated Surveys, No. S4,* 23p.

UNESCO, 1968. *Aerial Surveys and Integrated Studies.* Paris: Unesco, 575p.

U.S. Army Corps of Engineers, 1975. Guidelines for reducing flood damages, in G. D. McKenzie and R. O. Utgard (eds.), *Man and His Physical Environment: Readings in Environmental Geology,* 2nd ed. Minneapolis: Burgess, 53-57.

U.S.Water Resources Council, 1974. Floods and flood damages, in D. R. Coates (ed.), *Environmental Geomorphology and Landscsape Conservation, II: Urban Areas.* Stroudsburg, Pa.: Dowden, Hutchinson & Ross, 158-167.

Vink, A. P. A., 1968. Aerial photographs and the soil sciences, in *Aerial Surveys and Integrated Studies.* Paris: Unesco, 117-125.

Volobuev, V. R., 1964. *Ecology of Soils.* Jerusalem: Israel Program for Scientific Translations, 260p.

Ward, R., 1978. *Floods: A Geographical Perspective.* London: Macmillan, 244p.

Wharton, C. R., 1970. The execution of agricultural development: case studies of planned change, in C. R. Wharton (ed.), *Subsistence Agriculture and Economic Development.* London: Cass, 387-392.

Willis, A. J.; Folkes, B. F.; Hope-Simpson, J. F.; and Yemm, E. W., 1959. Braunton Burrows: the dune system and its vegetation, Part I, *Jour. Ecology* **47,** 1-24; Part II, *Jour. Ecology* **47,** 249-288.

Withers, B., and Vipond, S., 1974. *Irrigation Design and Practice.* London: Batsford, 306p.

Wright, R. L., 1971. The role of integrated surveys in developing countries: review and reappraisal, in R. L. Wright (ed.), *Seminar on Integrated Surveys, Range Ecology and Management, Proceedings.* New Delhi: UNESCO, 47-105.

Wright, R. L., 1972a. Principles in a geomorphological approach to land classification, *Zeitschr. Geomorphologie* **16,** 351-373.

Wright, R. L., 1972b. Some perspectives in environmental research for agricultural land-use planning in developing countries, *Geoforum* **10,** 15-33.

Wright, R. L., and Wilson, S. R., 1979. On the analysis of soil variability, *Geoderma* **22,** 297-313.

Yoshino, M., 1974. *Climate in a Small Area,* Tokyo: University of Tokyo Press, 549p.

Young, A., 1972. *Slopes.* London: Longmans, 288p.

Young, A., 1976. *Tropical Soils and Soil Survey.* Cambridge, England: Cambridge University Press, 468p.

Cross-references: *Alluvial Plains, Engineering Geology; Channelization and Bank Stabilization; Coastal Engineering; Consolidation, Soil; Deltaic Plains, Engineering Geology; Permafrost, Engineering Geology; Remote Sensing, Engineering Geology; River Engineering; Rock Slope Engineering; Soil Mineralogy, Engineering Applications; Urban Engineering Geology; Urban Geology; Urban Geomorphology.* Vol. XIV: *Alluvial Systems Modeling; Coastal Zone Management; Land Capability Analysis; Maps, Environmental Geology; Open Space; Photogeology; Photo Interpretation; Slope Stability Analysis, Terrain Evaluation Systems; Topographic Mapping and Surveying; Vegetation Mapping.*

GEONOMY—See Vol. XIV.

GEOPHYSICAL PROSPECTING—See Vol. XIV: ACOUSTIC SURVEYS, MARINE; EXPLORATION GEOPHYSICS; MARINE MAGNETIC SURVEYS; PETROLEUM GEOLOGY; VLF ELECTROMAGNETIC SURVEYING.

GEOPRESSURES—See FORMATION PRESSURES, ABNORMAL.

GEOSCIENCE INFORMATION—See CONFERENCES, CONGRESSES, AND SYMPOSIA; EARTH SCIENCE, INFORMATION AND SOURCES; GEOLOGICAL COMMUNICATION; GEOLOGICAL INFORMATION, MARKETING. Vol. XIV: MAP AND CHART DEPOSITORIES.

GEOSCIENCE MAPS—See MAPS, ENGINEERING PURPOSES. Vol. XIV: MAPS, ENVIRONMENTAL GEOLOGY; MAPS, PHYSICAL PROPERTIES.

GEOTECHNICAL ENGINEERING

Geotechnical engineering is a branch of civil engineering that is concerned with the uses of soil and rock as materials of construction (Reese, 1977; Hunt, 1983). The American Society of Civil Engineers (1978) defines *geotechnical engineering* as "that part of civil engineering involving the interrelationship between the geological environ-

ment and the works of man." The term *geotechnics,* or *geotechnique,* has been in common use in Europe and South America for many years. Only recently has it become accepted in the United States. It was adopted by the American Society of Civil Engineers in 1974 as a replacement for the "Soil Mechanics and Foundation Engineering Division" to give a broader connotation to the discipline and particularly to encompass the subject of rock mechanics. The term *soil and foundation engineering* (q.v.) is still used more commonly as the title of many technical texts, but it is only one element of geotechnical engineering.

The general term *geotechnical engineering* signifies studies and designs for civil engineering works such as foundations for structures (see *Foundation Engineering*), retaining structures (see *Reinforced Earth*), stabilization of slopes (see Vol. XIV: *Landslide Control; Slope Stability Analysis*), tunnels (see *Tunnels, Tunneling*), erosion control (see Vol. XIV: *Land Drainage*), airfield and roadways pavement support, earth and rockfill dams (see *Dams, Engineering Geology*), and embankments (see *Channelization and Bank Stabilization; Coastal Engineering*) whether constructed in soil or rock.

The development of designs requires application of the principles of soil mechanics (q.v.), rock mechanics (q.v.), and foundation engineering (q.v.) to soil and rock materials. Many geological sciences are important to an understanding of the physical properties and stratigraphy of an area under study, the more significant of these being structural geology, petrology, geomorphology (see *Geomorphology, Applied*), historical geology, physical geology, geohydrology (q.v.), and geophysics (see Vol. XIV: *Exploration Geophysics*).

Engineering geology (q.v.) is the branch of geology that provides geological information to civil engineers for project studies and the preparation of designs. Environmental geology (see Vol. XIV) on the other hand, is a relatively new science that involves the study of geological impacts on the human environment as well as the effects of construction or development on the natural environment (see Vol. XIV: *Cities, Geologic Effects*).

ROY E. HUNT

References

American Society of Civil Engineers, 1978. (Personal communication from executive director, ASCE, New York.

Hunt, R. E., 1983. *Geotechnical Engineering Investigation Manual.* New York: McGraw-Hill, 982p.

Reese, L. C., 1977. *Geotechnical Engineering and the Engineering Profession.* 2nd annual lecture, Joe J. King Professional Engineering Achievement Award, University of Texas at Austin, February 25, 16p.

GLACIAL LANDSCAPES, ENGINEERING GEOLOGY

Cross-references: *Alluvial Plains, Engineering Geology; Arid Lands, Engineering Geology; Caliche, Engineering Geology; Channelization and Bank Stabilization; Clay, Engineering Geology; Coastal Engineering; Coastal Inlets, Engineering Geology; Dams, Engineering Geology; Deltaic Plains, Engineering Geology; Duricrust, Engineering Geology; Earthquake Engineering; Foundation Engineering; Glacial Landscapes, Engineering Geology; Hydrodynamics, Porous Media; Laterite, Engineering Geology; Marine Sediments, Geotechnical Properties; Permafrost, Engineering Geology; Pipeline Corridor Evaluation; Pumping Stations and Pipelines, Engineering Geology; Rapid Excavation and Tunneling; River Engineering; Rock Mechanics; Rocks, Engineering Properties; Rock Slope Engineering; Shale Materials, Engineering Classification; Urban Engineering Geology; Urban Tunnels and Subways. Vol. XIV: Blasting and Related Technology; Cities, Geologic Effects; Engineering Geochemistry; Engineering Soil Science; Slop Stability Analysis.*

GEOTECHNICS—See GEOTECHNICAL ENGINEERING.

GEOTECHNIQUE—See GEOTECHNICAL ENGINEERING.

GEOTECHNOLOGY—See RAPID EXCAVATION AND TUNNELING; TUNNELS, TUNNELING. Vol. XIV: BLASTING AND RELATED TECHNOLOGY; BOREHOLE DRILLING.

GLACIAL LANDSCAPES, ENGINEERING GEOLOGY

The practical applications derived from glacial studies differ in at least one significant manner from those of other surface processes, such as rivers, mass movements, and coastal waters. Whereas man alters and manipulates these other processes, he has not chosen to modify active glaciers. Thus of all four components of the system—the ice mass, the process, the sediment, and the landform—only the last three are relevant in a discussion of *applied glacial geology*. Of course, certain schemes—such as towing icebergs to water-deficient sites and placing lampblack on ice to accelerate its melting rate—have been proposed, but none have been prosecuted. Furthermore, it is important to understand how the glacial process worked because such knowledge is crucial in predicting the occurrence, distribution, and types of the features that have resulted. Countless engineering failures have occurred and costly mistakes have

been made when the results of the glacial process—the sediments and the landforms—have been incorrectly interpreted. During Quaternary times, glaciers covered more than 30 percent of the earth's land surface, and they still cloak more than 10 percent with ice that ranges to as much as 5 km thick. However, considerably more land was effected by glaciation because of *outwash* from the ice and *loess* blown several hundreds of kilometers from the margin.

The cardinal rule in dealing with *glacial deposits* is to realize that no other types or groups of sediments change so much in such short distances, both horizontally and vertically. This lack of site homogeneity must always be considered in programs that involve glacial deposits. Such differences can occur in all sediment properties, including texture, fabric, and composition. The mode of deposition, from ice to meltwater sediment, can also produce abrupt changes and contacts. Similarly, no other sediment-based landforms can form by such a wide range of differing genetic processes, and be underlain by such contrasting materials. Glacial landforms comprised of sediments can be extremely complex. Indeed, no other landforms contain such diverse sets of materials that have formed from so many different genetic processes. For example, Kaye (1976) shows the engineering difficulties that occurred because the Beacon Hill (Boston, Massachusetts) *end moraine* was originally interpreted as a series of *drumlins.* This type of misinterpretation proves the *principle of equifinality:* landforms that look alike are not all formed by the same process.

A practical knowledge of glacial geology can be applied in many fields, including engineering, economic resources, land use planning, and law.

Engineering Applications

Although engineering covers a broad spectrum of human endeavors, this entry will concentrate on the properties of glacial sediment as an influence on engineering structures. The correct assessment of the type and properties of the glacial materials is of primary concern. Because materials can run the entire gamut of physical character, accurate mapping and borehole information is crucial to the success of any operation. Mollard (1973) presents a regional approach for analysis of important features, and Kreig and Reger (1976) explain the technique that was used for aid in aligning the Alaska Pipeline. Of course, not only must such work be done prior to construction, but the data must also be correctly interpreted. For example, Legget (1974) discusses the building of the St. Lawrence waterway project and the millions of dollars that were lost by contractors who failed to interpret the drilling data of glacial materials accurately. The unifying theme of glacial deposits is that they are similar in their lack of homogeneity. There are three different large groups of glacial deposits: till, ice contact, and outwash.

Tills have often been considered by engineering generalists to be uniform in composition, physical properties, and even thickness. They show exceptional ranges in all these characteristics, however. Their geotechnical quality depends primarily on four factors: (1) grain-size distribution and mineralogy; (2) nature of the sequences of materials; (3) stress history during and after deposition; and (4) number, type, and orientation of joint planes. Of course, many other factors are also involved in determining whether tills are suitable for use as fill and embankment materials, as substrate for structures, and for cuts to establish stable repose angles. These include water content and water-table conditions, thickness and fabric of different units and other glacial deposits, and till type. For example, lodgment till, and most glacial clays, are highly overconsolidated (see *Consolidation, Soil*), flow till is only lightly overconsolidated, and ablation (melt-out) till is normal consolidated.

The engineering quality of tills ranges from compact and nonplastic to weak, plastic, clayey tills. Their *modulus of elasticity* may vary from 145 kg/cm^2 to 670 kg/cm^2, and their *undrained compressive strength* from 1.5 to 30 t/ft^2 (1.4 to 29 kg/cm^2). Silty or sandy tills can be compacted to a more dense state provided the soil has the proper water content at the time of compaction. A water content 2 to 3 percent above the optimum of the soil is generally too soft for compaction. Degrees of compaction above several percent more than the original volume are attained only with a substantial increase in compactive forces. Furthermore, the compaction of till to higher densities does not appreciably change compressibility and shear strength properties. Thus the compressibility of till is relatively independent of its initial density except for tills with unusually large amounts of plastic fines. Till texture also determines overall strength. A change in water content of only 2 percent may change the remolded, undrained strength by 50 percent in clayey tills, whereas relatively little change occurs in sandy tills. Great caution must be exercised when translating laboratory and field testing of till strength to operational situations. The *in situ* till mass is anisotropic with respect to deformational modulus, so that many testing procedures overestimate the operational strength of the sediment. The *in situ* coefficient of consolidation can run as high as 1,000 times greater than the fill value from the same till. Table 1 compares four different types of glacial sediment in terms of their suitability and behavior for various engineering uses.

Slope stability of tills is governed by many factors, including the bulk undrained and drained shear strength, consolidation or swelling properties, permeability and ground-water conditions, and degree of homogeneity and its interruptions (see Vol. XIV: *Slope Stability Analysis*). The repose angle for some tills in England is determined by the presence of polygonal systems of ice-wedge clast relics, which create vertical planes of weak-

TABLE 1. Engineering and economic uses of glacial materials.

Type of Glacial Deposit	Bearing Capacity	Settlement	Slope Stability
Lodgement till	Best. Can contain irregular soft units. Silts subject to frost heave.	Smallest, but long-term changes and differential displacement can occur.	Best with highest repose angles. However, when wetted and dried subject to desiccation and movement along planes of discontinuity.
Ablation till	Good but can be highly variable.	Slight to moderate.	Moderate but lower cohesion than lodgement. Variable permeability can produce differential slope movement.
Ice-contact cohesionless materials	Generally good but interbedded till and openwork gravel produce variability.	Moderate. Most occurs during construction but long-term consolidation possible and continuing changes due to till and clay in interbeds.	Generally at angle of repose but instabilities can occur when gross textural sizes occur and with presence of till and clay.
Outwash cohesionless materials	Generally good to very good.	Slight, but dependent on sedimentology of deposit.	Invariably at angle of repose.

Source: After Fookes et al. (1975)

ness. In British Columbia some dense tills have stability slopes shallower than 20° due to thin zones of highly plastic, slickensided clay materials. Driving of well points to increase stability proved impractical because of the erratic character of interbeds.

Eden (1976) has shown the importance of water content of till in determining its construction suitability. Many problems were encountered in the cut-and-fill operation during construction of the alignment for the Quebec North Shore and Labrador Railway. The route traversed a ground moraine with drumlinized ridges for about 240 km. In the cut sections the equipment became mired in sediment that liquefied (see Vol. XII, Pt. 1: *Thixotropy, Thixotropism*), and nearly equal difficulties were encountered in fill areas. The unstable till had a 14.3-percent water content, so that about 4 percent of water content needed to be extruded to reach the optimum dry density of 10.2 percent. Due to the impermeabel nature of the till, however, attempts at densification were initially unsuccessful because high pore-water pressures were induced, with attendant unstable behavior of the sediment. Dikes were necessary in the construction of the Churchill Falls Power Project (Canada) in 1969-1970. Till was the major available borrow material but proved too unstable. It could not be sufficiently compacted to design density and permeability because the water content exceeded test specifications. Drier materials had to be transported from much greater distances, and thus project costs increased.

Glaciolacustrine and glaciomarine sediments are notorious as producers of landslides (see Vol. XIV: *Landslide Control*) and unstable terrain. Coates (1977), Dunn and Banino (1977), and Mollard (1977) discuss landslides in the extrasensitive fine-grained sediments that may undergo liquefaction when disturbed, in such areas as New York, Norway, Alaska, and Canada. Such sediment was responsible, for example, for most of the hundreds of millions of dollars of damage done in Anchorage, Alaska, during the 1964 Good Friday Earthquake.

Glacial aspects of dam siting (see *Dams, Engineering Geology*) and construction are discussed by Fluhr (1964), Legget (1974), and Philbrick (1976). Fluhr shows the importance of knowing the environmental setting as well as character of the materials for successful dam construction of four dams for the New York City water supply in the Catskill Mountains. At the Downesville site, till was used for the impervious core with 2 to 23 percent clay, whereas kame deposits with glaciofluvial sands and gravel were used for the semipervious shell material with 4 to 7 percent clay. The optimum density proved to be 2 g/cm^3 at a 9.3-percent moisture content. The Portage Mountain Dam on the Peace River in British Columbia (Legget, 1974) is 180 m high and 1,140 m long. It is an immense earth-fill dam that contains 44 million m^3 of glacial materials thereby using glacial deposits as an economic resource. Philbrick shows the importance of making a complete evaluation of all factors associated with glaciation in construction of the Kinzua Dam in Pennsylvania. Here site-specific

Excavation	Use for Fill	Use for Aggregate
Very difficult. When wet, hindered by plasticity. Can be nearly brick-hard when dry. Boulders a deterrent.	Especially suitable when impermeable materails are needed. May contain water content above optimum. Sensitive to moisture changes.	Unsuitable due to excessive fines.
Moderately difficult. Boulders are special problem and composition and texture can provide wide variability in ease of handling.	Good to very good. Silt particles can be sensitive to moisture changes. When selectively mined, suitable for impermeable foundation.	Generally unsuitable due to excessive fines and variability. Processing required exceeds economic merits.
Generally easy except in till and clay zones. Occasional boulders may present problems.	Generally good granular and free draining. Some caution necessary to prevent piping if materials are not blended.	Can provide important sand and gravel, but needs washing and screening. Kame deltas and Kame terraces provide better sources than kames and eskers.
Easiest materials to remove and process.	Good granular materials that are free-draining.	Best source of building and construction material. Sediments are best sorted, with less deleterious rock.

studies were made that indicated the type of dam that should be built, specifications for permeability of the foundation and embankments, the source of aggregate, and foundation and slope problems associated with highway and railroad relocations. Unfortunately, several of these factors were not analyzed in construction of the Lake Lee Dam near Lee, Massachusetts. The dam broke on March 24, 1968, killed two people, and caused damages of $10 million. The failure can be attributed to piping, which produced collapse after seepage was channelized under the structure. A similar catastrophe was averted in the Garrison Dam, North Dakota (Arnold, 1964), by construction of relief wells to prevent piping and the development of sand boils.

Economic Resources

Minerals. Glaciation has produced and distributed a variety of materials and features of economic value. Knowledge of the processes that have produced such benefits is often essential for their successful use and exploitation. Glacial deposits range from exotic occurrences of diamonds and gold to more prosaic sand and gravel resources. Since 1863 diamonds have been discovered at 82 sites in glacial drift in the Great Lakes region. From the early reports by Hobbs (1894) to the later work by Gunn (1968), many have attempted to trace such erratics to a Canadian "mother lode," but with no success. Gold in minute amounts occurs in many sand and gravel stream beds throughout the Great Lakes, transported there from Canadian source deposits, and is especially prevalent in Indiana (Blatchley, 1902), although it does not occurr in sufficient abundance to be classed as an ore deposit. There are many theories that have been advanced to explain the existence of the lead-zinc district of Wisconsin, Illinois, and Iowa; one proposed mechanism for its origin involves meteoric waters generated by a glacial-artesian system (McGinnis, 1968). Such a theory holds that the weight of the ice sheet distorts and displaces normal ground-water, causing it to flow to discharge sites, where heat loss, pressure loss, and solubility changes cause the ore-bearing solution to precipitate.

Analysis of till is being increasingly used as a method for ore exploration (Shilts, 1976; Alley and Slatt, 1976; Kujansuu, 1976), although certain till distribution principles were in use 240 years ago. The systematic geochemical mapping program for ore prospecting in Finland started in 1971. These surveys include fabric analysis, petrographic analysis, trace-metal content, granulometric analysis, and microfossil analysis (see *Lithogeochemical Prospecting*). Results to date have differentiated five different tills, which will lead to the development of a new glacial chronology and dispersal directions of ice flow. In Newfoundland the probable source area of chalcopyrite-pyrite mineralization was determined on the basis of a float dispersion fan; concentrations of copper, iron, and sulfur; clast lithologies; and fabric of till. Alley and Slatt (1976) point out the importance of studying the substrate by means of 2- to 2.5-m-deep pits, instead of placing complete reliance on surficial materials. Shilts (1976) provides guidelines for how a mineral exploration of till should be conducted.

Glaciofluvial sand and gravel deposits constitute significant economic resources throughout the world. The large majority of sand and gravel in Scotland is of glacial origin, as are 25 percent of similar resources in England and Wales. In New York State, 95 percent of marketed sand and gravel comes from glacial deposits. Glacial sand and gravel occurs in ice-contact deposits such as kames, kame terraces, and eskers, and as outwash. The outwash deposits generally consist of high-quality materials because they are more uniform, have fewer deleterious sediments, and contain more resistant rocks. Due to the longer transport distances, outwash materials are better sorted and stratified, contain harder rocks that have withstood comminution, possess less local rock and soil, and thus need less processing before use as a resource. Kame deltas can also serve as important aggregate sources. Proglacial lake beach environments may contribute valuable sand deposits if the lake existed long enough for winnowing action of sediments to occur along the strandline. For example, Glacial Lake Iroquois beaches comprise significant sand deposits throughout the eastern Great Lakes. The search for appropriate sand and gravel resources can be shortened when there is knowledge of the glacial history of the site area. For example, John Arborio, Inc., had the contract to build 50 km of the four-lane highway Interstate 88 near Sidney, New York, and needed a sand and gravel source for the highway. This area is in the Susquehanna River valley, and it was known that the local Catskill-type rocks had become incorporated into the glacial deposits and formed such a high percentage of the deposits that they would not pass state specifications for use due to large losses in the $MgSO_4$ soundness test. This test is made to determine the loss percentage of deleterious materials. For example, if loss exceeds 45 percent in the five-cycle test, the aggregate cannot be used for bituminous concrete highways, and an 18-percent loss for a ten-cycle test causes rejection for use on concrete highways. Thus both time and money were saved by restricting the search for a suitable resource to other areas. In this region only sand and gravel in through valleys (those with communication through the Akllegheny Escarpment) was of sufficiently high quality for use on interstate roads; it was in this glacial environment setting that an appropriate deposit was discovered. Superior materials exist in the through valleys because the sediments contain fewer local rocks, contain harder rocks transported from the northern part of the state, and have higher percentages of outwash-type material.

Sand and gravel resources influence road construction in other ways. In the hilly, glaciated Appalachian Plateau, most government roads follow stream valleys, where the majority of glacial stratified sediments are located. In construction of new highways, private landowners displaced by the development may beleive payment for their land is inadequate because of mineral resources. They initiate lawsuits against the government in the search for a higher monetary award (see Coates, 1971, 1976). In such cases the geological testimony is crucial regarding the claim and assessment of the quality and quantity of the materials. Thus, it is important to establish whether the deposit has a unique quality, whether it will pass specifications for use, and whether a market is available. In some cases (Coates, 1971) it was important to map all sand and gravel sources in a 16-km radius to demonstrate that there was a great abundance of material of a similar grade in the Catskill region (237 million m³), thus showing that the material on the condemned property could not be claimed to be unusual or unique.

Water. Glaciation has been instrumental in producing significant water resources in all glaciated regions. In North America glacial lakes contain 25 percent of all the fresh water on the earth. The Great Lakes supply water for more than 25 million people and countless industries. Waterfalls created by glacial processes, such as Niagara Falls, are significant sites for the development of hydroelectric power plants. Ground water in glacially originated aquifers of sand and gravel constitutes a vital resource to numerous communities and industries. It is common for glaciofluvial sands and gravels to produce some of the highest-yield wells; pumping rates often exceed 3.8 m³ per minute. The ANSCO plant was located in the Binghamton, New York, region because of the occurrence of abundant cold, clean, clear ground water in the glacial aquifers. Such cities as Endicott, Johnson City, Vestal, Elmira, and Corning in New York derive municipal water supplies from glacial aquifers.

The siting of productive wells can depend on information about the manner in which glacial processes operated in the area. For example, it is important to know that in the valleys of southern New York and northern Pennsylvania the strongest aquifers occur in valleys with south-flowing streams and the weakest aquifers are in valleys with north-flowing streams. The cause for the great discrepancy is that the glacial fill in south-flowing stream valleys contains higher percent-ages of sand and gravel. Thus when Frito Lay Industries decided to place their plant in a north-draining valley they were cautioned that at such a site they would be unable to develop a strong well system. Such advice was unheeded, and thousands of dollars were wasted in an unsuccessful water-drilling program. However, C. J. Martin & Sons were able to drill successful wells at Whitney Point, New York, when shown the proper site for exploration in a glacial moraine, and at Apalachin, New York, when shown the best possibility for water in a kame and kettle complex. The water from Whitney Point was used for the cement in Interstate 81, and at Apalachin the water was used for the Southern Tier Expressway.

Soil. The soil can be a valuable economic resource. Soils derived from glacial materials show an unusually wide range for crop productivity types, engineering capabilities, and home-site developments. On Long Island, New York, farmers who recognized loess soil areas and established their fields at such locations obtained much higher crop yields than those who farmed the less fertile sand and gravel plains.

Land Use

Glacial geologists can play a significant role in land use planning (see Vol. XIV: *Land Capability Analysis*) of glaciated terrain (LaFleur, 1974; Legget, 1974). Sand and gravel resources should be mapped and placed into conservation zoning before urban sprawl development makes them unobtainable, as in much of the Chicago area. Soils and the character of the substrate should be classified into different use sequences. For example, prior to purchase of an expensive lot, a property owner should learn whether the soil will pass the appropriate percolation tests and permit installation of a sewage system. In similar manner, sites for sanitary landfills (see Vol. XIV: *Artificial Fill*) must be chosen with utmost care. Prior to the 1960s the location of landfills at abandoned gravel pits was standard practice in many communities. The substrate at such localities is often highly permeable, however, so that the resulting landfill leachate moves readily into the surrounding water-supply systems. It is now recognized that landfills should be sited where the subsurface materials are impermeable and where there is an available source of cover material that is also impermeable. Till is often the preferred type of deposit, so the character of the drift sheets, their thickness, and orientation is an important prerequisite in siting a successful sanitary landfill operation.

For larger developments and engineering projects, the land use mapping should contain information about the slope stability of materials, especially in permafrost terrain and in areas underlain by glaciolacustrine and glaciomarine strata.

Even when information is available, some land developers ignore it, in some cases with disastrous results. The U.S. Geological Survey, for example, had correctly interpreted the nature of the Bootlegger Cove Clay but part of Anchorage, Alaska, was built on it anyway. In somewhat similar fashion, the Leda Clay, a notorious landslide producer throughout the St. Lawrence region, was known to exist at St. Jean Vianney, Canada. In spite of this, a housing development was sited there and on May 4, 1971, 31 lives were lost when the material liquefied and produced landslide chasms to depths of 30 m. Thus the cooperation of earth scientists and community planners is important in land use management strategies, such as occurred at Saskatoon, Canada (Christiansen, 1970). Here the thorough analysis of the glacial geology as reported by the Canadian government scientists was instrumental in avoiding unnecessary losses of valuable lands.

Glaciation has affected land use in many other ways, some minor and some major. Due to the till shadow influence on hills of southern New York, the thickness of glacial drift is often 8 to 10 times greater on southern than on northern slopes. In these uplands ground water for wells is found in bedrock, so that homeowners who select sites on the southern sides of hills must drill much deeper, and at greater expense, to obtain water than those on the northern hillslopes. The grandeur of glacial scenery is unsurpassed in places such as the fiords of Norway, the Swiss Alps, Yosemite Valley, Niagara Falls, and the Finger Lakes. Because of their beauty, such locales have become highly commercialized, have led to land use developments, and unfortunately, exploitation. However, parks and wilderness sites have also been set aside for the enjoyment of the glaciated terrain and to enhance the human spirit.

DONALD R. COATES

References

Alley, D. W., and Slatt, R. M., 1976. Drift prospecting and glacial geology in the Sheffield Lake-Indian Pond area, Northcentral Newfoundland, in R. F. Legget (ed.), *Glacial Till,* Royal Soc. Canada Spec. Pub. 12, p. 249-266.

Arnold, A. B., 1964. Relief well on the Garrison Dam and Snake River Embankment, North Dakota, *Geol. Soc. America Eng. Geology Case Histories* 5, 45-52.

Blatchley, W. S., 1902. Gold and diamonds in Indiana, *27th Ann. Rept. of Indiana Dept. Geology and Nat. Resources,* 11-47.

Christiansen, E. A. (ed.), 1970. Physical environment of Saskatoon, Canada, *Saskatchewan Research Council, NRC Pub. No. 11378,* 68p.

Coates, D. R., 1971. Legal and environmental case studies in applied geomorphology, in D. R. Coates (ed.), *Environmental Geomorphology.* Binghamton: State University of New York, 223-242.

Coates, D. R., 1976. Geomorphology in legal affairs of the Binghamton, New York, metropolitan area, in D. R. Coates (ed.), *Urban Geomorphology.* Geol. Soc. America Spec. Paper 174, 111-148.

Coates, D. R., 1977. Landslide perspectives, in D. R. Coates (ed.), *Landslides.* Geol. Soc. America Reviews in *Eng. Geology* 3, 3-28.

Dunn, J. R., and Banino, G. M., 1977. Problems with Lake Albany "clays," in D. R. Coates (ed.), *Landslides.* Geol. Soc. America Reviews in *Eng. Geology* 3, 133-136.

Eden, W. J., 1976. Construction difficulties with loose glacial tills on Labrador Plateau, in R. F. Legget (ed.), *Glacial Till.* Royal Soc. Canada Spec. Pub. 12, 391-400.

Fluhr, T. W., 1964. Earth dams in glacial terrain, Catskill Mountain Region, New York, in G. A. Kiersch (ed.), *Geol. Soc. America Eng. Geology Case Histories* 5, 15-29.

Fookes, P. G.; Gordon, D. L.; and Higginbottom, I. E., 1975. Glacial landforms, their deposits and engineering characteristics, in *The Engineering Behaviour of Glacial Materials.* Birmingham, England: Midland Soil Mechanics and Foundation Engineering Society, University of Birmingham, 18-51.

Gunn, C. B., 1968. Relevance of the Great Lakes discoveries to Canadian diamond prospecting, *Canadian Mining Jour.* 89, 39-42.

Hobbs, W. H., 1894. On a recent diamond find in Wisconsin and on the probable source of this and other Wisconsin diamonds, *Am. Geologist* 14, 31.

Kaye, C. A., 1976. Beacon Hill end moraine, Boston: new explanation of an important urban feature, in D. R. Coates (ed.), *Urban Geomorphology.* Geol. Soc. America Spec. Paper 174, 7-20.

Kreig, R. A., and Reger, R. D., 1976. Preconstruction terrain evaluation for the Trans-Alaska Pipeline Project, in D. R. Coates (ed.), *Geomorphology and Engineering.* Stroudsburg, Pa.: Dowden, Hutchinson & Ross, 55-76.

Kujansuu, R., 1976. Glaciogeological surveys for ore-prospecting in northern Finland, in R. F. Legget (ed.), *Glacial Till,* Royal Soc. Canada Spec. Pub. 12, 225-239.

LaFleur, R. G., 1974. Glacial geology in rural land use planning and zoning, in D. R. Coates (ed.), *Glacial Geomorphology.* Binghamton: State University of New York, 375-388.

Legget, R. F., 1974. Glacial landforms and civil engineering, in D. R. Coates (ed.), *Glacial Geomorphology.* Binghamton: State University of New York, 350-374.

McGinnis, L. D., 1968. Glaciation as a possible cause of mineral deposition, *Econ. Geology* 63, 390-400.

Mollard, J. D., 1973. *Landforms and Surface Materials of Canada: A Stereoscopic Atlas and Glossary.* Regina, Sask.: J. D. Mollard, 336p.

Mollard, J. D., 1977. Regional landslide types in Canada, in D. R. Coates (ed.), *Landslides.* Geol. Soc. America Reviews In Eng. Geology 3, 29-56.

Philbrick, S. S., 1976. Kinzua Dam and the glacial foreland, in D. R. Coates (ed.), *Geomorphology and Engineering.* Stroudsburg, Pa.: Dowden, Hutchinson & Ross, 175-197.

Shilts, W. W., 1976. Glacial till and mineral exploration, in R. F. Legget (ed.), *Glacial Till.* Royal Soc. Canada Spec. Pub. 12, 205-224.

Cross-references: *Dams, Engineering Geology; Geocryology; Geomorphology, Applied; Permafrost, En-*

gineering Geology; Soil Mechanics. Vol. XII, Pt I: *Thixotropy, Thixotropism.* Vol. XIV: *Groundwater Exploration; Slope Stability Analysis; Soil Fabric; Terrain Evaluation Systems.*

GOSSAN, BOXWORK—See MINERAGRAPHY.

GRAVITY SURVEYS—See Vol. XIV: EXPLORATION GEOPHYSICS.

GROUND CONDITIONS—See ARID LANDS, ENGINEERING GEOLOGY; TERRAIN EVALUATION, MILITARY PURPOSES. Vol. XII, Pt. 1 : THIXOTROPY, THIXOTROPISM. Vol. XIV: CAT CLAYS; DISPERSIVE CLAYS; EXPANSIVE SOILS; TERRAIN EVALUATION SYSTEMS.

GROUT, GROUTING

Grouting involves the injection of appropriate materials under pressure into certain parts of the earth's crust through specially constructed holes in order to fill, and therefore seal, voids, cracks, seams, fissures, or other cavities in soils or rock strata. The result is to ensure watertightness by establishing very low or negligible permeability, thus reducing or preventing uplift and other movements deleterious to engineering structures. Compaction grouting is a recent invention, which relies not on infilling, but on densification of suitable soils by displacement in order to overcome settlement and other problems. The term *grouting* may also be applied to the sealing of cracks in manmade structures such as dams, tunnels, and mines.

Almost certainly, the most widely used grout is made up of cement and water, probably because of the relatively low cost. Additionally, clay, clay and cement, asphalt and various chemical solu-

tions, and a number of other materials in differing proportions may be used in sealing operations.

Grouting is particularly valuable in foundation work before, during, or after construction. Before construction, it may be used to control water problems during boring or sampling, to infill voids to eliminate possible future settlement, and to allow an increase in permissible soil pressures relative to the untreated soil for new structures and extensions to already existing ones. During construction, proper grouting can control groundwater flow, prevent loose-sand densification below adjacent structures due to pile driving, and increase the stability of granular soils below existing structures, thus reducing the need for lateral support. After construction, the possible applications include underpinning, reduction of machine foundation vibrations, and the elimination of seepage through openings.

Classification and Characteristics

The classification of grouts is difficult because of their diversity. At the beginning of the century, engineers had far fewer to use and could therefore make a broad distinction between cement-based grouts and those based on sodium silicate, which were called "chemical" grouts. Now, however, it is necessary to separate nondiluted silicates having a viscosity around 100 cP from extremely fluid aqueous resins of the acrylamide type with a viscosity close to 1 cP, both of these being *chemical grouts.* In Europe, it has become the practice to refer to grouts having a certain viscosity as *gels* and to use the term *resins* for those with a viscosity close to that of water. This is not entirely satisfactory because chemical composition is ignored. Jones (1964) asserted that the precise meaning of the term *chemical grout* is difficult to understand. His proposed division between organic and inorganic grouts would, however, appear to be almost equally inadmissible.

As shown in Table 1, grouting materials can be grouped into three basic categories: suspensions, solution, and emulsions.

1. *Suspensions* are multiphase systems capable of forming subsystems after being subjected to

TABLE 1. Interrelationships of some grouts

Single Injections			
Suspensions	Emulsions	Solutions	Two Fluid Injections
Normal e.g., cement-clay mixes	e.g., bitumen	One-shot sodium silicate and coagulant	Two-shot successive injections of sodium silicate and an electrolyte
Thixotropic e.g., bentonite			

natural sieving processes. Their chemical properties must be carefully scrutinized to ensure that they do not conflict with controlled properties of setting and strength. Examples are grouts made from water and Portland cement, or from clays.

2. *Solutions* are intimate one-phase systems retaining an originally designed chemical balance until completion of the relevant reactions. Solutions in which the solute is present in the colloidal state are known as *colloidal solutions.* Chemical grouts fall into this category.

3. *Emulsions* are two-phase systems in which the disperse phase comprises minute droplets of liquid, such as water and bitumen.

The desirable properties of all grouts include suitable rheological characteristics with appropriate viscosity (normally as low as possible), correct setting time, maximal volume with minimal weight, strength, stability, and durability. A clear understanding of the individual problem is mandatory for correct application of any type of grout; otherwise costly mistakes may be made. For instance, while cement can seal water-bearing fissures in rock strata quite admirably, it cannot be used to consolidate fine sands. Generally speaking, coarser and moderately permeable soils require high-viscosity grouts whereas low-viscosity grouts are used with finer soils having lower permeability. A special subject of study in grouting is *groutability,* that is, the specification of the conditions under which a particular grout may be expected satisfactorily to penetrate into permeable materials and seal all voids.

Cement and Clay Grouting Materials

High-strength Portland cement is made from limestone, clay, and a small amount of iron oxide, all mixed and heated to produce a clinker, which is subsequently ground. It is particulate, as are clays, forming suspensions comprising non-Newtonian fluids. Clays are often locally obtainable and so tend to be cheaper. Clay-cement combinations are sometimes used, and bentonite as a cement additive increases lubrication. Cements are primarily useful in fractured ground and soils with openings wider than 0.1 mm, while clays are better suited to finely fissured materials.

Chemical Grouting

Chemical grouts, which are Newtonian fluids, were developed for use in finely fissured ground; they enable materials such as fine sands to be treated successfully.

Silicates. Silicates, first used as grouts in 1887, are applied by means of the Joosten two-shot technique, ("Joosten I"), that is, successive injections of a concentrated solution of sodium silicate and a strong electrolytic salt solution. This causes gelation with accompanying permanent solidifi-

cation of granular materials, giving crushing strengths of up to $6,900 \, kN/m^2$. A newer approach, termed "Joosten II," uses silicate, alkali dilution, and calcium chloride. There is now also a "Joosten III" method (silicate, heavy metal salt, and ammoniacal colloid), a one-shot adaptation. The difference between one- and two-shot methods is shown in Figure 1.

Commercial sodium silicate-based grouts include Siroc grout, which may be injected by a single-shot batch system (see Fig. 2). Another type is Earthfirm GVS, but it does not last as long as Siroc.

Lignochrome. A Newtonian fluid whose viscosity increases with time, lignochrome is made from lignosulfite, a residual product resulting from the production of cellulose from wood pulp by the bisulfite process. The common lignochrome grout is TDM, which has a low viscosity (2.5 to 4 cP) and a gel time ranging from a few minutes to several hours. It can be used to seal and also to consolidate.

Newtonian Low-Viscosity Grouts. These grouts are ideal for fine sands and possess a viscosity close to that of water. AM-9, polymerized, is an elastic resin without application to soil or rock stabilization, usually being controlled by water flow. It can be pumped anywhere that water will flow. Other commercial products include Cyanaloc grouts (applicable where strength as well as water shutoff is necessary), Geoseal grouts (one of which, MQ-4, is suitable for combating saline groundwater conditions) and Terranier grouts are low-molecular-weight polyphenolic polymers which, dissolved in water in the presence of appropriate catalysts, form insoluble, high-strength gels.

Emulsions. The most common emulsion grout is bitumen, a term covering many mixtures of hydrocarbons, particularly solid or tarry ones, all of which are soluble in carbon disulfide. They occur naturally or can be manufactured. Other emulsion grouts include latex and salt water, pinewood resin in a suitable alkali, and some materials that are used when molten, such as sulfur, naphthalene vapor, and some metals.

Grouting Methodologies

Basically, grouting activities are essential for strengthening and water sealing. Strengthening can be achieved by means of *compaction grouting.* The basic technique is to inject an expanding bulb of highly viscous grout with high internal friction into a compactible soil so that, acting as a radial hydraulic jack, it can physically displace the soil particles and move them closer together, thus achieving controlled densification (Graf, 1969; Brown and Warner, 1973). Blanket, contact, and curtain grouting can cut off water effectively, for example, in dam construction. There are two main grouting methods, stage and packer. In *stage grouting* (Fig. 3), holes are drilled down to the seam closest to the surface and grouting inserted.

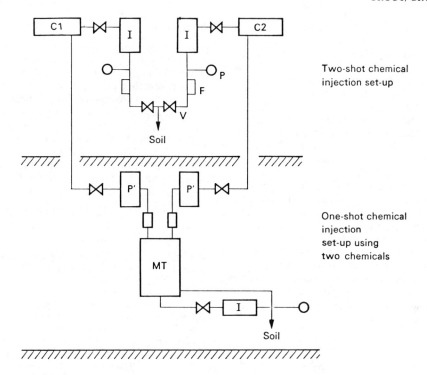

FIGURE 1. Two-shot and one-shot methods of chemical grouting. C1, C2 = chemicals 1 and 2, I = injection pumps, P' = pump, P = pressure gauge, F = flow meter, V = inlet valve, MT = mix tank.

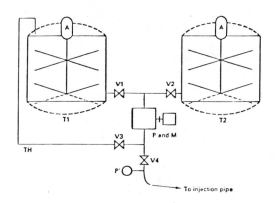

FIGURE 2 (left). Batch system. A = agitator, T1, T2 = tanks number 1 and 2, V1-V4 = valve numbers 1-4, P & M = pump and motor, TH = transfer hose, P' = pressure gauge.

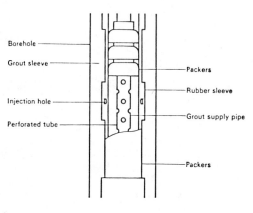

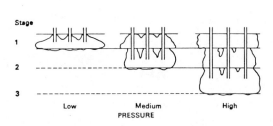

FIGURE 3. Stage grouting by depth.

FIGURE 4. Cup leather type of packer.

The holes are then cleaned, and drilling continues to the next seam down. In *packer grouting,* holes are immediately drilled down to the maximal planned depth. A zone of specified thickness is then grouted, and a packer is inserted into the hole to a level corresponding to the top of the grouted zone. The overlying zone is then grouted for a specified thickness and the process repeated until the uppermost seam is grouted. Figure 4 shows a cup leather type of packer, which is best used in hard rock where smooth-walled boreholes of moderate size can be drilled.

Grouts to be used vary. With chemical ones, the compositions are predeterminable and, with cement, the quantity of water used is critical. Chemicals act like glues and may be harmful; two workers were admitted to a hospital after exposure to epoxy resin, a thermosetting type of grout, in a pipe-manufacturing plant near Pittsburgh, Pennsylvania. Such glues do not have high crushing strengths in the raw condition, but develop them optimally in sands. Caron (1973) has given an empirical relationship:

$$R_s \quad \frac{R_p}{(1 + k(S)^{1/2})}$$

in which R_s = the crushing strength of the grouted soil material, R_p = the specific strength of the raw product, S = the specific surface of the soil, and k = a factor that depends on the chemical nature of the grout.

Conclusion

In sum, grouting today is an invaluable tool in engineering practice, aimed at sealing off water flow and consolidating soils. Radiosotopy may be used in preliminary investigations, such as in locating leakage from reservoirs. The practice of grouting is, to a degree, an art. It certainly cannot be considered an exact science because too many imponderables are involved. A knowledge of soil and rock mechanics, fluid mechanics, and engineering geology is indispensable. A detailed exposition of the subject is given in *Grouting in Engineering Practice* (Bowen, 1975).

ROBERT BOWEN

References

Bowen, Robert, 1975. *Grouting in Engineering Practice.* London: Applied Science Publishers, 187p.
Brown, Douglas R., and Warner, James, 1973. Compaction grouting, *Am. Soc. Civil Engineers Proc., Jour. Soil Mechanics and Found Div.* **99** (SM8), 589-601.
Caron, Claude, 1973. Strength properties of chemically solidified soils, *Am. Soc. Civil Engineers Proc., Jour. Soil Mechanics and Found Div.* **99** (SM10), 891-892.
Graf, Edward D., 1969. Compaction grouting technique and observations, *Am. Soc. Civil Engineers Proc., Jour. Soil Mechanics and Found. Div.* **95** (SM5), 1151-1158.
Jones, G. K., 1964. *Grouts and Drilling Muds in Engineering Practice.* London: Butterworths, 164.

Cross-references: *Engineering Geology Reports; Foundation Engineering; Geotechnical Engineering; Hydrodynamics, Porous Media; Hydromechanics.*

H

HYDRODYNAMICS, POROUS MEDIA

The study of flow through porous media is of importance in connection with many geological applications. Such diversified fields as soil mechanics (q.v.), ground-water hydrology, and petroleum engineering (see Vol. XIV: *Petroleum Geology)* rely heavily on it as basic to their individual problems.

Porous Media

Porous media are solid bodies that contain "pores," small void spaces (see Vol. XII, Pt. 1: *Pore Space),* which are distributed more or less frequently throughout the material. The problem of complete geometric characterization of a porous medium has not yet been solved. One is able only to define some geometric parameters of a porous medium that are based on averages. The first of these is the *porosity, P,* which is equal to the average ratio of the *void volume* to the *bulk volume* of the porous medium. The second is the *specific surface area, S,* which is the average ratio of internal surface to the bulk volume of the porous medium.

Darcy's Law

About a century ago, Henri Darcy (1856) made some experiments to investigate the flow of water through the sand filters of the water purification plant at Dijon, France. The following law can be deduced from Darcy's experiments:

$$q = - \left(\frac{k}{\mu}\right) (\text{grad } p - \rho)$$

Where q is the *filtration velocity* (the quotient of the total volume of flow per unit of time through an infinitesimal cross-section of porous medium divided by that cross-section), μ is the *viscosity* of the fluid, ρ is the *density* of the fluid, and **g** is the *gravity vector* (of magnitude **g** and direction downward). Finally, k has been termed the "permeability" of the porous medium (see Vol. XII, Pt. 1: *Permeability).*

To apply the basic equations of flow through porous media to practical cases, one has to add a continuity condition

$$-P\left(\frac{\partial \rho}{\partial t}\right) = \text{div } (\rho q)$$

where, in addition to the symbols already defined, t is time, and an equation of state for the fluid

$$\rho = \rho(p)$$

As is evident, this system of equations is nonlinear, which makes it difficult to obtain solutions. Fortunately, however, it is often possible to linearize the system of basic equations to a heat conductivity equation for which many solutions are known (Carslaw and Jaeger, 1959).

A peculiar problem occurs if one studies gravity flow with a free surface in a porous medium. The free surface is an *equipressure surface;* any streamline having one point in common with it must lie entirely within it. If the free surface intersects an "open" surface of the porous medium, then a surface of seepage will be formed below the line of intersection. The problem of finding the free surface, even under steady-state conditions, is thus a problem with a floating boundary condition. Various analytical methods have been adapted to this end; the most comprehensive survey is probably that in a monograph by Polubarinova-Kochina (1962).

Limitations of Darcy's Law

Darcy's law is valid only in a certain *seepage velocity domain,* outside which more general flow equations must be used to describe the flow correctly.

To characterize this seepage velocity domain, it is customary to introduce a *Reynolds number,* as follows:

$$Re = \frac{q\rho\delta}{\mu}$$

where, in addition to the symbols defined earlier. δ is a microscopic diameter associated with the porous medium ("pore diameter"). The universal critical Reynolds number, beyond which Darcy's law is no longer valid, ranges between 0.1 and 75. Forchheimer suggested in 1901 (Scheidegger, 1974) that Darcy's law be modified for high-flow velocities by including a second-order term in the velocity, which leads to (in a linear system):

$$\frac{\partial p}{\partial x} = aq + bq^2$$

Solutions of the preceeding high-velocity flow equations for particular cases are difficult to obtain. Notwithstanding the difficulties, Engelund discussed some possibilities in 1953 (see Scheidegger, 1974).

A breakdown of Darcy's law also occurs in gases at low pressures, due to the various molecular effects that may come into prominence in rarefied gas dynamics. Thus, Knudson (slip) flow or even molecular flow may occur.

Theoretical Models of Porous Media

The flow laws just discussed are essentially empirical laws, deduced from a series of experiments. However, since the fluids flowing through a porous medium can in most instances be regarded as ordinary Newtonian, that is, viscous fluids, it should be expected that their motion can be described by obtaining a suitable solution of the *Navier-Stokes equation.* This could be done if it were possible to formulate the boundary conditions of the problem correctly, stating that the fluid must stick to the walls of the pores. Unfortunately, porous media are of an extremely complex nature. Correspondingly, it is entirely impossible to treat the flow of a fluid confined within the pore space in any manner that could claim to be microscopically exact. This seems to preclude forever the understanding of such things as Darcy's law from a microscopic standpoint.

A way out of this difficulty is to represent an actual porous medium by something that the human brain can comprehend. One therefore makes gross simplifications in a porous medium to be able to formulate the boundary conditions and to integrate the basic Navier-Stokes equations. Such simplified versions of porous media are called *models.*

The simplest model consists of a bundle of parallel capillaries of circular cross-section, all with the same diameter. A serious drawback of such a model is that all the pores are supposed to go from one face of a porous medium right through to the other. This is evidently a picture far removed from actuality.

The opposite extreme picture would be obtained by assuming that the pore space is lined up serially, so that each particle of fluid would have to enter at one pinhole at one side of a porous medium and travel through very tortuous channels through all the pores, and then emerge at only one pinhole at the other face of the medium. Obviously, this picture is just as unreal as a parallel-type model; a realistic model lies somewhere between the extremes.

A different type of model has been suggested by Emersleben in 1925 (see Scheidegger, 1974). It may be noted that, in the models just discussed, the porous medium has been visualized essentially as a piece of solid material with holes in it. Emersleben took the opposite view, in that he visualized the porous medium as a fluid-filled space with a few obstacles in it. The drag exercised by all the obstacles on the fluid, then, represents the resistance of the medium to flow. It turns out that this type of model satisfactorily describes the flow through highly porous media, such as an agglomeration of fibers.

All these models are unsatisfactory on general grounds, because they attempt to describe disordered porous media by well-ordered models. The proper modification of the above models, therefore, seems to be a recourse to statistical mechanics (Scheidegger, 1954, 1974). This leads to a statistical treatment of the hydrodynamics in porous media, in which either the flow paths of the individual fluid particles or the flow channels in a porous medium are assumed to be randomly distributed. The main prediction is that of occurrence of mechanical dispersion in porous media by which individual fluid particles become intermixed not by the molecular diffusive motion, but by the splitting up and rejoining of flow channels.

Displacement Processes

Of particular interest in connection with geological applications is the *theory of displacement processes* in porous media. Such displacement processes are important in the production of oil from underground strata and in the exploitation of ground-water resources.

Displacement occurs in an entirely different fashion, depending on whether the two fluids involved are miscible or not. It is possible, however, to set up an elementary theory that is the same for miscible and immiscible fluids. In this elementary theory, no mixing between the two flowing phases is assumed to take place at all. The two phases are confined to domains whose extent changes with time. Each fluid moves within its own domain according to Darcy's law.

This "elementary" displacement theory also shows that instability may result if a fluid is displaced from a porous medium by a less viscous one. In this case, fingering is liable to occur, which means that the displacement front becomes unstable: fingers form that shoot at relatively great speed through the porous medium. The calculations are fairly involved; see Scheidegger (1974) for details.

Some theories do truly account for the simultaneous flow of different fluids at any one point in the porous medium. Turning first to *immiscible displacement,* we note that a theory can be obtained by writing down Darcy's law for each phase, introducing a "relative permeability" k_1 (as a fraction of total permeability k) to the phase *(i)* that is assumed to be a function of saturation only. This, in conjunction with usual continuity equations and equations of state for both flowing phases, leads to a complete system of differential equations that describes the displacement processes. Unfortunately, it is nonlinear, and solutions have been obtained only for the one-dimensional case.

Finally, turning to *miscible displacement,* we note that there are no fronts; the predominant effect (as long as no fingering occurs) is a progressive blurring of the juncture of the two fluids, which has been called *dispersion.* It is described by a diffusivity equation referring to a moving coordinate system connected with the overall motion of the mixture of the two fluids. Solutions have been obtained for the linear case. The mechanical dispersion of the two fluids is caused by the interconnections of the flow channels. This shows that the statistical models discussed earlier have a direct application to miscible displacement theory (Scheidegger, 1974).

ADRIAN E. SCHEIDEGGER

References

Carslaw, H. S., and Jaeger, J. C., 1959. *Conduction of Heat in Solids.* Oxford: Clarendon, 510p.

Darcy, H., 1856. *Les fontaines publiques de la ville de Dijon.* Paris: V. Dalmont, 647p.

Polubarinova-Kochina, P. Y., 1962. *Teoriya dvizheniya gruntovykh vod.* Moscow: Gosodarstvennor Izdateistvo Tekhnicheskoi-Teoreticheskoi Litevcitury, 676p. (English translation by R. De Wiest, Princeton, University Press.)

Scheidegger, A. E., 1954. Statistical hydrodynamics in porous media, *Jour. Appl. Physics* **25,** 994-1001.

Scheidegger, A. E., 1974. *The Physics of Flow through Porous Media.* Toronto: University of Toronto Press, 373p.

Cross-references: *Hydrology; Hydromechanics; Soil Mechanics.* Vol. IVA: *Hydrology.* Vol. XII, Pt. I: *Conductivity, Hydraulic; Flow Theory; Imbibition; Percolation; Pore Space, Drainable; Water Fluxes; Water Movement; Wetting Front.*

HYDROGEOLOGY AND GEOHYDROLOGY

Definitions, Terminology, and Use

Known as the "science of water," *hydrology* (q.v.) broadly refers to the study of water as it occurs above, on, or below the earth's surface as water vapor, precipitation, streamflow, soil moisture, and ground water. Some authorities regard hydrology as a branch of civil engineering, but Ward (1967) suggests that this application may be too restrictive. Indeed, specialists in many different fields have contributed to the development of hydrology as a subject. Hydrologic techniques have been developed in response to needs in such essential fields as agriculture, forestry, geology (see *Urban Hydrology),* geomorphology (see *Geomorphology, Applied),* and soil science (q.v. in

Vol. XII). Examples of other applications, as discussed by Prickett (1975), Krenkel and Novotny (1980), and Chow (1981), consider significant advances in stochastic hydrology, reservoir storage theory, computer-assisted recognition of simulated hydrologic time series, and infiltration theory (q.v. in Vol. XII). Urban stormwater hydrology, a rapidly expanding subdiscipline, deals with stormwater management insofar as it relates to urban planning, pollution control, water quality, and better understanding of urban runoff processes (Kibler, 1981). Hydrology, then, is an applied earth science concerned with the occurrence of water in the earth, its role in biophysical and geochemical reactions, and its relations to life on earth (see Vol. IVA: *Hydrology).*

Applications of hydrologic principles, data, and techniques to geological problems are encompassed by *hydrogeology,* the science that deals with subsurface waters and with related geological aspects of surface waters. Important advances as well as critical turning points in the development of physical and chemical hydrogeology have recently been summarized in a series of Benchmark papers by Freeze and Back (1983) and Back and Freeze (1983). Although originally defined by Mead in 1919 as the study of the laws of the occurrence and movement of subterranean waters, the term *hydrogeology* has more recently been used interchangeably with *geohydrology* (Bates and Jackson, 1980). The latter term was proposed by Meinzer (1942) as a branch of hydrology dealing with subterranean waters, but the term has also been used in reference to all hydrology on the earth without restriction to geological aspects (Stringfield, 1966). *Ground-water hydrology,* the science of subsurface water, emphasizes geological aspects and is therefore frequently used to advantage with less confusion and inference.

At a recent symposium considering recent trends in hydrology (Narasimhan, 1982), conferees generally seemed to agree that the subject of hydrogeology deals with all geological processes that involve water near the land surface, including surface-water/ground-water interaction, and at depth (e.g., geopressured systems; see *Formation Pressures, Abnormal).* Perhaps the most basic of all hydrogeological concepts is that of regional ground-water motion. Subterranean flow dynamics has, for example, been related to hydrocarbon accumulations and wellfield completion (see Vol. XIV: *Petroleum Geology),* geothermal resource exploration, possible impacts of geologically disposed nuclear wastes to ground-water systems, the use of shallow ground-water systems for energy storage, as well as the hydrodynamics of ground-water supply for agricultural, urban, and industrial applications. The scope of hydrogeology is extremely wide, as is the scale of observation, which ranges from capillary flow to regional or basin analysis. The "International Hydrogeological Map of Europe, 1:500,000" is an excellent example of a coordinated effort to gather regional information pertaining to aquifer exploitation and water management. Struckmeier (1978) reports that these small-scale maps provide a basis for multipurpose scientific investigations, that they indicate gaps in hydrogeological knowledge and point to new directions for future research, and finally that they may be used as models for other national small-scale hydrogeological maps.

Reports dealing with water resources, particularly quantity and quality of surface and ground waters, stream measurements, underflow, artesian pressure, water analyses, well completions, the geology of reservoir and dam sites, and water-power resources—as well as numerous other special publications such as hydrographic atlases—are produced by the U.S. Geological Survey. The first water-supply paper, "Pumping Water for Irrigation" by H. M. Wilson, was published as early as 1896; such reports continue to be issued as an important part of survey activities in hydrogeology and geohydrology. Most state geological surveys also produce useful reports dealing with aspects of water resources development and supply. In addition to information provided by other national geological surveys, the following organizations compile water statistics: The U.S. Agency for International Development (AID), Canadian International Development Agency (CIDA), U.S. Food and Agriculture Organization (FAO), International Association of Hydrogeologist (IAH), International Commisson on Irrigation and Drainage (ICID), International Bank for Reconstruction and Development (IBRD), Inter-American Development Bank (IDB), International Hydrological Decade (IHD), International Hydrological Program (IHP), United Nations Development Program (UNDP), United Nations Educational, Scientific and Cultural Organization (UNESCO), World Health Organization (WHO), and World Meteorological Organization (WMO). Computerized data bases available for online information searches include, for example, Lockheed Corporations's Dialog Information Retrieval Service. The following files contain references to published works that deal with aspects of water resources management: AGRICOLA, AQUALINE, GEOARCHIVE, GEOREF, POLLUTION ABSTRACTS, and WATER RESOURCES ABSTRACTS.

The Hydrologic Cycle

The hydrologic cycle (q.v. in Vol IVA), the never-ending circulation of water that penetrates the total earth system through the atmosphere, hydrosphere, and lithosphere (Chow, 1964), was regarded by Meinzer (1942) as the central concept in hydrology. The large-scale model of the hydrologic cycle (Fig. 1) incorporates smaller cycles that are interlocking subsystems. If any one of them varies, others change in response (Holzman, 1941); for example, when water evaporates from inland lakes, it falls again as precipitation in the catchment area and is returned by surface streamflow or as ground water into the lake from which it evaporated. The hydrologic cycle thus provides a useful summary of interrelationships among the processes of evaporation, precipitation, evapotranspiration, infiltration, seepage, storage, and runoff, but it is often difficult to apply to specific areas (Ward, 1967).

A more direct approach that relates the amount of water going through the hydrologic cycle during a given period for an area may be found in the

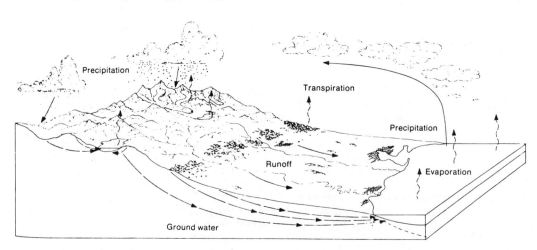

FIGURE 1. Diagrammatic illustration of the hydrologic cycle, showing the never-ending circulation of water in the total earth system. Numerous smaller subcycles are not shown in this simplified model of a complex reality (from Hamblin, 1975, p. 53).

hydrologic, or water balance, equation (also called the continuity equation):

$$I = O \pm \Delta S$$

where I is the total inflow of surface runoff, ground water, and total precipitation; O is the total outflow (including water loss from evapotranspiration, surface, and subsurface runoff from the area); and ΔS is the change in storage (Ward, 1967; Hordon, 1972). Factors in the equation vary according to the size of the area being considered and also with the time period applied. Thus, a run of data for a few minutes may suffice for small plots, but data for a year or long-term averages covering scores of years may be required for large regions.

Table 1 summarizes the world water balance by continent in terms of precipitation, total river runoff, ground-water runoff, surface-water runoff, total soil moisture, and evaporation per unit area. It is interesting to note that for each continent ground-water runoff accounts for approximately one-third of the total runoff. Other models based on subsystems in the "great cycle"—such as precipitation simulation, evaporation, infiltration, runoff, and subterranean flow—have been devel-

oped for hydrologic prediction of flood control, irrigation schemes, water supply, and resource planning in general. More's (1967) review of paradigms in physical hydrology, ground-water models, and overall catchment models features interpretations of subsystem interrelationships in attempts to solve practical problems. Models of projected water use in the United States from 1980 to 2020 (Table 2), for example, predict alarming increases in withdrawals for industrial and energy-related activities by the end of the second decade of the twenty-first century.

Management of Ground-Water Supply

Water in the land phase of the hydrologic cycle is derived from precipitation. After reaching the land surface, a small portion of the water of precipitation is returned to the atmosphere by evaporation, but most flows overland to the ocean or infiltrates the soil. Part of the water entering the soil is stored temporarily as soil moisture; that in excess of field capacity (see Vol. XII, Pt. 1: *Soil Moisture Management*) percolates (see Vol. XII, Pt. 1: *Imbibition; Infiltration; Permeability*) down-

TABLE 1. Estimates of the world water balance, by continent

Water Blance Elements	Europe[a]	Asia	Africa	North America[b]	South America	Australia[c]	Total Land Area[d]
Area, millions of km²	9.8	45.0	30.3	20.7	17.8	8.7	132.3
			in mm				
Precipitation	734	726	686	670	1,648	736	834
Total river runoff	319	293	139	287	583	226	294
Ground-water runoff	109	76	48	84	210	54	90
Surface-water runoff	210	217	91	203	373	172	204
Total soil moisture	524	509	595	467	1,275	564	630
Evaporation	415	433	547	383	1,065	510	540
			in km³				
Precipitation	7,165	32,690	20,780	13,910	29,355	6,405	110,303
Total river runoff	3,110	13,190	4,225	5,960	10,380	1,965	38,830
Ground-water runoff	1,065	3,410	1,465	1,740	3,740	465	11,885
Surface-water runoff	2,045	9,780	2,760	4,220	6,640	1,500	26,945
Total soil moisture	5,120	22,910	18,020	9,690	22,715	4,905	83,360
Evaporation	4,055	19,500	16,555	7,950	18,975	4,440	71,475
			Relative values				
Ground-water runoff as percent of total runoff	34	26	35	32	36	24	31
Coefficient of ground-water discharge into rivers	0.21	0.15	0.08	0.18	0.16	0.10	0.14
Coefficient of runoff	0.43	0.40	0.23	0.31	0.35	0.31	0.36

[a] Including Iceland.
[b] Excluding the Canadian archipelago and including Central America.
[c] Including Tasmania, New Guinea, and New Zealand.
[d] Excluding Greenland, Canadian archipelago, and Antarctica.
Source: From Lvovitch (1973), p. 33; *Copyright © by the American Geophysical Union.*

TABLE 2. Projected water use for the United States, 1980–2020 in billion gallons per day

Type of Use	Projected Withdrawals			Projected Consumptive Use		
	1980	2000	2020	1980	2000	2020
Rural domestic	2.5	2.9	3.3	1.8	2.1	2.5
Municipal (public supplied)	33.6	50.7	74.3	10.6	16.5	24.6
Industrial (self-supplied)	75	127.4	210.8	6.1	10	15.6
Steam-electric power						
Fresh	134	259.2	410.6	1.7	4.6	8
Saline	59.3	211.2	503.5	.5	2	5.2
Agriculture						
Irrigation	135.9	149.8	161	81.6	90	96.9
Livestock	2.4	3.4	4.7	2.2	3.1	4.2
U.S. total	442.6	804.6	1,368.1	104.4	128.2	157.1

Source: From American Water Works Association (1970), in van der Leeden (1975).

ward in porous rocks to the water table, where it enters the zone of saturation (Miller, 1977). Subsurface water in this saturated zone, defined as ground water by Meinzer (1923), is under *hydrostatic pressure* (atmospheric pressure or greater) and is free to flow laterally under the influence of gravity (see Vol. IVA: *Groundwater Motion in Drainage Basins. Vol. XII, Pt. 1: Flow Theory*). If water in an aquifer (q.v. in Vol. IVA) is under sufficient pressure to rise above the base of a confining layer, *artesian conditions* exist.

Aquifers serve as natural storage reservoirs and as distribution conduits. Pumping (see *Wells, Water*) in accordance with good water management practices reduces the natural discharge without excessively lowering water tables or degrading water quality. Overpumping, or excessive drawdown (see Vol. IVA: *Drawdown, Cone of Depression*), in seaboard regions may lead to salt-water encroachment in coastal aquifers as experienced in southern Florida (Parker et al., 1955) or on Long Island, New York (Cohen et al., 1968) or to land subsidence where parts of large metropolitan areas such as London, Mexico City, Tokyo, and Venice slowly sink as ground water is withdrawn (Legget, 1963) to meet increasing demands for this liquid resource. Overdevelopment of deep aquifers may also lower the piezometric surface, which in turn leads to decreased flows in springs and artesian wells, as experienced in the Paris basin and in important hydrogeological basins in southern Algeria and Tunisia (Castany, 1981). In many municipalities the total water supply is drawn from relatively shallow subterranean reservoirs. In the United States, for example, major metropolitan areas rely on wellfields for 100 percent of their fresh (potable) water in areas of California, Florida, Nevada, New Mexico, Texas, Washington, and Wisconsin (Table 3).

The southeastern Florida coastal corridor is a particularly good example of a highly urbanized region with total dependence on ground water.

The Biscayne aquifer (Schroeder et al., 1958), the primary drinking-water supply for the region, is recharged from Lake Okeechobee by percolation from overland flow across conservation areas lying south of this lacustrine storage facility (which is also the largest fresh-water lake in the conterminous United States, exclusive of the Great Lakes). This reservoir system, composed of lakes, everglades, and shallow aquifers (Fig. 2), is managed by the South Florida Water Management District to control flooding, reduce salt-water intrusion in coastal aquifers, provide irrigation water for agriculture, and supply coastal urban areas with potable water. The availability of adequate supplies of good-quality ground water in southern Florida will probably limit urban growth and industrial development more than any other single factor.

Ground-Water Quality

The quality of ground water is partly related to the mineral composition of the water-bearing strata from which it is derived. Dissolved natural constituents include the bicarbonates, sulfates, and chlorides of sodium, calcium, and magnesium as well as lesser amounts of silica, potassium, iron, aluminum, manganese, fluoride, boron, nitrate, and phosphate, and frequently the gases of hydrogen sulfide and carbon dioxide. Highly mineralized or salty ground water is unsuitable for most uses, but changes in water quality brought about by human activities are beginning to far outweigh natural limitations. Many hydrogeologists are becoming increasingly concerned with the deterioration of ground-water quality due to contamination from industrial effluents, agricultural byproducts (e.g., fertilizers, biocides, unconsumed water from irrigation), and urban waste disposal in sanitary landfills (Thomas, 1956; Legget, 1963). Disposal of septic tank sludge, grease-trap waste, and solid waste materials is particularly troublesome in regions with shallow water tables because

TABLE 3. Municipal water supply systems for the United States (based on operating data of selected water utilities as of 1970)

City or Utility	Total Population Served	Quantity Available, 10^6 gal/yr					Average Production million gal/day	Delivered to Distribution System, 10^6 gal/yr	Consumption, gal/capital/day
		Production							
		Surface Water	Ground Water	Purchased	Total				
Alabama									
Birmingham	525,000	20,368	522	8,258	29,148		79.86	27,983	152
Huntsville	160,000	2,383	4,413	719	7,515		20.59	7,515	129
Mobile	200,000	–	–	–	7,289		19.97	7,289	100
Montgomery	160,000	3,550	3,803	–	7,353		20.15	7,253	126
Alaska									
Anchorage	60,000	3,120	1,323	–	4,443		12.17	–	203
Arizona									
Phoenix	633,000	35,153	17,575	–	52,728		144.46	52,728	228
Arkansas									
Little Rock	159,728	12,065	–	–	12,065		33.05	11,792	207
California									
Bakersfield[a]	130,600	–	17,004	–	17,004		46.59	17,004	357
Fresno	180,000	–	21,147	–	21,147		57.94	21,147	322
Long Beach	361,000	–	8,854	12,047	21,261		58.25	21,254	161
Los Angeles	2,862,000	147,988	24,062	17,005	189,055		517.96	190,185	181
Los Angeles[b]	663,700	2,130	22,500	18,450	43,080		118.03	39,650	178
Oakland[c]	1,080,000	77,053	128	–	77,181		211.45	77,181	196
Pasadena	137,187	990	4,157	6,270	11,417		31.28	11,417	228
Riverside	122,000	–	12,141	7	12,148		33.28	12,148	273
Sacramento	270,000	21,000	3,000	–	24,000		65.75	24,000	244
San Diego	723,000	2,788	–	37,519	40,307		110.43	40,307	153
San Francisco	2,000,000	274,754	–	–	274,754		752.75	88,927	376
San Jose Waterworks	525,000	5,052	21,377	8,525	34,954		95.76	34,954	182
Colorado									
Colorado Springs	150,000	13,900	–	1,600	15,500		42.47	13,930	283
Denver	772,790	70,976	–	–	70,976		194.45	59,571	252

[a]California Water Service.
[b]Southern California Water Co.
[c]East Bay Municipal Utilities.

(continued)

287

TABLE 3. *(continued)*

City or Utility	Total Population Served	Quantity Available, 10⁶ gal/yr Production Surface Water	Ground Water	Purchased	Total	Average Production million gal/day	Delivered to Distribution System, 10⁶ gal/yr	Consumption, gal/capital/day
Connecticut								
Bridgeport[d]	340,000	24,160	891	—	25,051	68.63	25,051	202
Hartford[e]	393,000	21,304	—	—	21,304	58.37	20,740	149
New Haven	400,000	19,996	744	—	20,740	56.82	20,740	142
District of Columbia		73,008	—	—	73,008	200.02	73,007	185
Florida								
Fort Lauderdale	139,000	—	13,728	—	13,728	37.61	13,728	271
Miami	500,000	—	58,199	—	58,199	159.45	51,176	319
Orlando	177,000	—	11,820	—	11,820	32.38	11,820	183
St. Petersburg	235,000	—	10,067	—	10,067	27.58	9,901	117
Georgia								
Atlanta	700,000	38,000	—	—	38,000	104.11	32,000	149
Columbus	145,000	9,020	—	—	9,020	24.71	8,805	170
Hawaii								
Honolulu	535,000	—	40,197	54	40,251	110.28	40,251	206
Illinois								
Chicago	4,506,000	377,636	—	—	377,636	1,034.62	377,636	230
Peoria	163,040	456	7,896	—	8,352	22.88	8,352	140
Indiana								
Fort Wayne	194,670	10,741	—	—	10,741	29.43	10,741	151
Gary[f]	250,000	11,260	—	—	11,260	30.85	10,968	123
Indianapolis	680,000	30,996	2,250	—	33,246	91.08	33,805	134
Iowa								
Cedar Rapids	110,640	—	6,372	—	6,372	17.46	6,297	158
Des Moines	243,770	12,534	—	—	12,534	34.34	12,124	141
Kansas								
Kansas City	180,000	11,561	—	—	11,561	31.67	—	176
Topeka	133,000	6,010	—	—	6,010	16.47	6,010	124
Wichita	284,720	4,160	9,732	—	13,892	38.06	13,760	134
Kentucky								
Louisville	673,000	42,377	—	—	42,377	116.10	40,214	173

[d]Bridgeport Hydraulic Co.
[e]Hartford Metropolitan District.
[f]Gary Hobart Water Co.

(continued)

TABLE 3. *(continued)*

City or Utility	Total Population Served	Quantity Available, 10⁶ gal/yr Production				Average Production million gal/day	Delivered to Distribution System, 10⁶ gal/yr	Consumption, gal/capital/day
		Surface Water	Ground Water	Purchased	Total			
Louisiana								
Shreveport	200,000	10,022	—	—	10,022	27.46	10,021	137
Maine								
Augusta	22,000	1,311	125	—	1,436	3.93	1,436	179
Portland	146,480	7,984	200	—	8,184	22.42	8,184	153
Maryland								
Baltimore	1,500,000	93,803	—	—	93,803	256.99	89,556	171
Hyattsville	1,200,000	44,678	524	—	45,202	123.84	44,414	103
Massachusetts								
Boston	608,000	—	—	51,714	51,714	141.68	51,714	233
Springfield	200,000	14,612	—	—	14,612	40.03	14,612	200
Michigan								
Detroit	3,790,500	244,562	—	—	244,562	670.03	244,562	177
Lansing	146,360	—	8,002	—	8,002	21.92	7,946	150
Saginaw	92,570	10,226	—	—	10,226	28.02	10,015	303
Minnesota								
Duluth	100,000	6,565	—	—	6,565	17.99	6,504	180
St. Paul	404,380	20,122	—	—	20,122	55.13	19,737	136
Mississippi								
Jackson	170,000	7,068	—	—	7,068	19.36	5,931	114
Missouri								
Kansas City	632,724	35,739	—	—	35,739	97.92	34,469	155
St Louis	622,500	68,660	—	—	68,660	188.11	68,078	302
St. Louis County	764,512	38,566	—	—	38,566	105.66	37,960	138
Montana								
Billings	73,800	5,365	—	—	5,365	14.70	5,365	199
Nebraska								
Lincoln	149,520	—	11,604	—	11,604	31.79	11,604	213
Omaha	425,000	19,576	10,867	—	30,443	83.41	25,484	196
Nevada								
Las Vegas[g]	200,000	—	15,638	4,387	20,025	54.86	20,025	274
Reno[h]	109,000	10,522	1,839	—	12,361	33.87	12,361	311

[g] Las Vegas Valley Water District.
[h] Sierra Pacific Power Co.

(continued)

TABLE 3. *(continued)*

City or Utility	Total Population Served	Quantity Available, 10^6 gal/yr			Total	Average Production million gal/day	Delivered to Distribution System, 10^6 gal/yr	Consumption, gal/capital/day
		Production						
		Surface Water	Ground Water	Purchased				
New Jersey								
Elizabeth[i]	1,251,670	30,000	8,000	—	38,000	104.11	34,000	83
Short Hills[j]	232,000	—	—	—	10,110	27.70	10,110	119
Weehawken[k]	1,000,000	32,370	5,287	2,099	39,756	108.92	39,756	109
New Mexico								
Albuquerque	250,000	—	18,000	—	18,000	49.32	18,000	197
New York								
Lynbrook[l]	262,400	—	10,637	—	10,637	29.14	10,511	111
New Rochelle	160,900	1,084	—	7,106	8,190	22.44	8,190	139
New York City	7,898,000	795,708	—	—	795,708	2,180.02	486,371	276
Oakdale[m]	650,000	—	24,277	—	24,277	66.51	24,277	102
Rochester	291,270	18,599	—	—	18,599	50.96	18,599	175
Syracuse	197,210	18,826	—	53	18,879	51.72	18,005	262
North Carolina								
Asheville	125,000	7,360	—	—	7,360	20.16	7,306	161
Greensboro	150,000	7,100	—	—	7,100	19.45	7,100	130
North Dakota								
Fargo	53,460	2,663	—	—	2,663	7.30	2,482	136
Ohio								
Akron	380,000	18,053	—	—	18,053	49.46	17,684	130
Dayton	350,000	—	25,698	—	25,698	70.41	25,372	201
Oklahoma								
Tulsa	395,000	22,626	—	—	22,626	61.99	22,290	157
Oregon								
Eugene	73,350	8,191	—	—	8,191	22.44	8,191	306
Portland	650,000	70,805	—	—	70,805	193.99	33,444	298

(continued)

[i]Elizabethtown Water Co.
[j]Commonwealth Water Co.
[k]Hackensack Water Co.
[l]Long Island Water Co.
[m]Suffolk Water Authority.

TABLE 3. (continued)

City or Utility	Total Population Served	Quantity Available, 10^6 gal/yr				Average Production million gal/day	Delivered to Distribution System, 10^6 gal/yr	Consumption, gal/capital/day
		Production						
		Surface Water	Ground Water	Purchased	Total			
Pennsylvania								
Bryn Mawr[n]	808,000	19,649	7,157	27,701	54,507	149.33	27,292	185
Chester	115,000	10,509	—	—	10,509	28.79	10,357	250
Erie	200,000	—	—	—	16,764	45.93	16,764	230
Pittsburgh[o]	513,240	23,902	—	—	23,902	65.48	23,902	128
South Carolina								
Charleston	175,900	11,040	—	—	11,040	30.25	10,700	172
Greenville	200,000	12,500	—	—	12,500	34.25	12,500	171
South Dakota								
Sioux Falls	72,440	56	4,050	—	4,106	11.25	4,106	155
Tennessee								
Knoxville	185,000	10,728	—	—	10,728	29.39	10,210	159
Memphis	623,530	—	32,959	—	32,959	90.30	32,959	145
Nashville	360,000	21,011	—	—	21,011	57.56	20,607	160
Texas								
Amarillo	127,010	6,513	2,429	—	8,942	24.50	8,942	193
Austin	250,000	18,397	—	—	18,397	50.40	17,614	202
Corpus Christi	274,720	26,959	—	—	26,959	73.86	26,959	269
Dallas	1,130,000	63,575	—	2,739	66,314	181.68	60,744	161
El Paso	330,000	3,109	19,811	—	22,920	62.79	22,920	190
Houston	1,076,000	46,436	51,708	—	98,144	268.89	98,114	250
Lubbock	170,000	8,519	1,660	—	10,179	27.89	9,936	164
San Antonio	578,860	—	34,633	—	34,633	94.88	34,633	164
Utah								
Salt Lake City	400,000	20,105	—	—	20,015	55.08	4,949	138
Virginia								
Annandale	477,000	16,470	406	1,174	18,050	49.45	18,050	104
Norfolk	500,000	21,885	—	—	21,885	59.96	21,379	120
Richmond	250,000	14,961	—	506	15,467	42.38	14,967	170

[n]Philadelphia Suburban Water Co.
[o]South Pittsburgh Water Co.

(continued)

TABLE 3. *(continued)*

City or Utility	Total Population Served	Quantity Available, 10⁶ gal/yr Production — Surface Water	Ground Water	Purchased	Total	Average Production million gal/day	Delivered to Distribution System, 10⁶ gal/yr	Consumption, gal/capital/ day
Washington								
Seattle	573,150	60,247	—	—	60,247	165.06	49,706	288
Spokane	171,800	—	20,716	—	20,716	56.76	20,716	330
Tacoma	175,000	23,020	4,840	—	27,860	76.33	27,860	436
West Virginia								
Morgantown	43,500	2,300	—	—	2,300	6.30	2,218	145
Wisconsin								
Madison	184,000	—	10,615	—	10,615	29.08	10,615	158
Milwaukee	933,000	59,243	—	—	59,243	162.31	59,243	174
Wyoming								
Cheyenne	50,000	3,036	1,011	—	4,047	11.09	4,047	222

Source: From American Water Works Association (1970) in van der Leeden (1975).

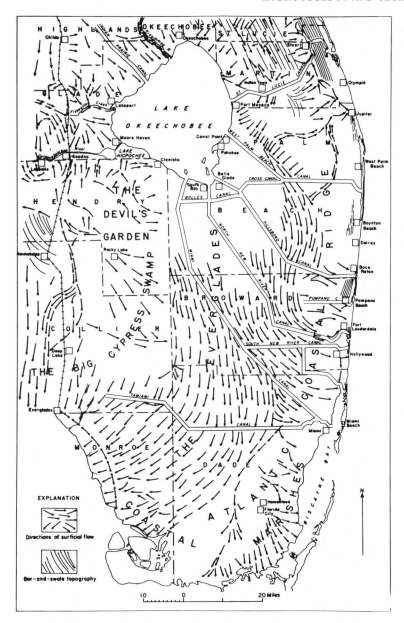

FIGURE 2. Map of southern Florida showing directions of overland flow across the everglades, rechange area of shallow coastal aquifers, and the main drainage canals used for flood control (from Parket et al., 1955).

of the potential for ground-water pollution. Soakage pits, canal discharge, and construction of deep borrow pits in cones of depression areas of public wellfields further complicate protection of aquifers used for municipal water supply. Such factors impair the ability of engineers to ensure adequate reserves of good-quality ground water, affect the structural integrity of the water-supply system (see *Pumping Stations and Pipelines, Engineering Geology),* and seriously compromise natural cleansing

of ground water as it percolates through porous formations (see *Hydrodynamics, Porous Media*).

Exploration and Development

Methods of ground-water exploration (q.v. in Vol. XIV) commonly focus on the electrical resistivity of earth materials, small variations in the earth's gravity and magnetic fields, response of rocks to vibrations from explosive charges, and

TABLE 4. Conversion factors for well and pumping test units

Discharge Rate

	litres/sec	m³/day	m³/sec	Imp. gal/day	U.S.gal/day	ft³/day
1 litre/sec	1.000	86.40	1.000×10^{-3}	1.901×10^4	2.282×10^4	3.051×10^3
1 m³/h	0.2777	24.00	2.777×10^{-4}	5.279×10^3	6.340×10^3	8.476×10^2
1 m³/day	1.157×10^{-2}	1.000	1.157×10^{-5}	2.200×10^2	2.642×10^2	35.32
1 m³/sec	1.000×10^3	8.640×10^4	1.000	1.901×10^7	2.282×10^7	3.051×10^3
1 Imp. gal/day	5.262×10^{-5}	4.546×10^{-3}	5.262×10^{-8}	1.000	1.201	0.1605
1 U.S. gal/day	4.381×10^{-5}	3.785×10^{-3}	4.381×10^{-8}	0.8327	1.000	0.1337
1 ft³/day	0.3277	2.832×10^{-2}	3.277×10^{-7}	6.229	7.481	1.000

Hydraulic Conductivity

	m/day	m/sec	cm/h	Imp.gal/day-ft²	U.S. gal/day-ft²	Imp. gal/min-ft²	U.S. gal/min-ft²
1 m/day	1.000	1.157×10^{-5}	4.167	20.44	24.54	1.419×10^{-2}	1.704×10^{-2}
1 m/sec	8.640×10^4	1.000	3.600×10^5	1.766×10^6	2.121×10^6	1.226×10^3	1.472×10^3
1 cm/h	0.2400	2.777×10^{-6}	1.000	4.905	5.890	3.406×10^{-3}	4.089×10^{-3}
1 Imp. gal/day-ft²	4.893×10^2	5.663×10^{-7}	0.2039	1.000	1.201	6.944×10^{-4}	8.339×10^{-4}
1 U.S. gal/day-ft²	4.075×10^2	4.716×10^{-7}	0.1698	0.8327	1.000	5.783×10^{-4}	6.944×10^{-4}
1 Imp. gal/min-ft²	70.46	8.155×10^{-2}	2.936×10^2	1.440×10^3	1.729×10^3	1.000	1.201
1 U.S. gal/min-ft²	58.67	6.791×10^{-2}	2.445×10^2	1.195×10^3	1.440×10^3	0.8326	1.000

Transmissivity

	m²/day	m²/sec	Imp. gal/day-ft	U.S. gal/day-ft	Imp. gal/min-ft	U.S. gal/min-ft
1 m²/day	1.000	1.157×10^{-5}	67.05	80.52	4.656×10^{-2}	5.592×10^{-2}
1 m²/sec	8.64×10^4	1.000	5.793×10^6	6.957×10^6	4.023×10^3	4.831×10^3
1 Imp. gal/day-ft	1.491×10^{-2}	1.726×10^{-7}	1.000	1.201	6.944×10^{-4}	8.339×10^{-4}
1 U.S. gal/day-ft	1.242×10^{-2}	1.437×10^{-7}	0.8326	1.000	5.783×10^{-4}	6.944×10^{-4}
1 Imp. gal/min-ft	21.48	2.486×10^{-4}	1.440×10^3	1.729×10^3	1.000	1.201
1 U.S. gal/min-ft	17.88	2.070×10^{-4}	1.199×10^3	1.440×10^3	0.8326	1.000

Source: From Berkman and Ryall (1976), p. 232.

differences in heat radiation from the earth's surface. Test drilling (see Vol. XIV: *Borehole Drilling*) is the most diagnostic of all exploration methods but is also the most expensive and slowest. Well tests designed to determine the properties of ground-water aquifers provide data for quantitative estimates of well conditions, formation properties, and hydrologic characteristics of the rock formation in which the well is completed. Conversion factors for pumping tests (discharge rate, hydraulic conductivity, transmissivity) are given in Table 4. Most ground-water well tests measure pressure, flow rates, and concentration, properties that are adequate for even sophisticated experiments (Schroeder, 1982). Borehole probes used to obtain physical parameters of the formation (e.g. lithology, structure, permeability) may, for convenience, be grouped into optical methods, mechanical methods, acoustic imaging methods, and electrical methods (see Vol. XIV: *Well Logging*). Descriptions of the use and application of mechanical calipers, video recorders, borehole televiewers, and dipmeters are summarized by Nelson (1982), who also considers applications of computer processing of borehole data. Other less scientific methods are espoused by water diviners, who claim the ability to locate underground water using curious techniques (see Vol. XIV: *Water Divining*).

CHARLES W. FINKL, JNR.

References

Back, W., and Freeze, R. A., (eds.), 1983. *Chemical Hydrogeology,* Benchmark Papers in Geology, Vol. 73. Stroudsburg, Pa.: Hutchinson Ross, 432p.

Bates, R. L., and Jackson, J. A. (eds.), 1980. *Glossary of Geology.* Falls Church, Va.: American Geological Institute, 751p.

Berkman, D. A., and Ryall, W. R., 1976. *Field Geologists' Manual.* Parkville, Victoria: Australasian Institute of Mining and Metallurgy, 295p.

Castany, G., 1981. Hydrogeology of deep aquifers: the hydrogeological basin as the basis of groundwater management, *Episodes* **3,** 18-22.

Chow, V. T., 1964. Hydrology and its development, in V. T. Chow (ed.), *Handbook of Applied Hydrology.* New York: McGraw-Hill, 1-1-1-22.

Chow, V. T. (ed.), 1981. *Advances in Hydroscience.* New York: Academic Press, 440p.

Cohen, P.; Franke, O. L.; and Foxworthy, B. L., 1968. An atlas of Long Island's water resources, *New York Water Resources Comm. Bull.* **62.**

Freeze, R. A., and Back, W. (eds.), 1983. *Physical Hydrogeology,* Benchmark Papers in Geology, Vol. 72. Stroudsburg, Pa.: Hutchinson Ross, 448p.

Hamblin, K. W., 1975. *The Earth's Dynamic Systems.* Minneapolis: Burgess, 578p.

Holzman, B., 1941. The hydrologic cycle, in *Climate and Man* (U.S. Department of Agriculture 1941 Yearbook of Agriculture). Washington, D.C.: U.S. Government Printing Office, 532-536.

Hordon, R. M., 1972. Hydrologic cycle, in R. W. Fairbridge (ed.), *The Encyclopedia of Geochemistry and Environmental Sciences.* New York: Van Nostrand Reinhold, 515-519.

Kaufman, M. I.; Goolsby, D. A.; and Faulkner, G. L., 1973. Injection of acidic industrial waste into a saline carbonate aquifer: geochemical aspects, *2nd Internat. Symposium Underground Waste Management and Artificial Recharge* (New Orleans, Sept. 26-30, 1973), **1,** 526-551.

Kibler, D. F., 1981. Urban Stormwater Hydrogeology, *Am. Geophys. Union, Water Resources Mon. 7,* 280p.

Krenkel, P. A., and Novotny, V., 1980. *Water Quality Management.* New York: Academic Press, 684p.

Leeden, F. van der, 1975. *Water Resources of the World.* Port Washington, N.Y.: Water Information Center, 568p.

Legget, R. F., 1963. *Cities and Geology.* New York: McGraw-Hill, 624p.

Lvovitch, M. I., 1973. The Global water balance, *Eos* **54** (1), 33.

Mead, D. W., 1919. *Hydrology: The Fundamental Basis of Hydraulic Engineering.* New York: McGraw-Hill, 647p.

Meinzer, O. E., 1923. The Occurrence of Ground Water in the United States: With a Discussion of Principles, *U.S. Geol. Survey Water Supply Paper 489,* 321p.

Meinzer, O. E., 1942. *Hydrology.* New York: Dover, 712p.

Miller, D. H., 1977. *Water at the Surface of the Earth,* International Geophysics Series, Vol. 21. New York: Academic Press, 557p.

More, R. J., 1967. Hydrological models in geography, in R. J. Chorley and P. Haggett, (eds.), *Models in Geography.* London: Methuen, 145-185.

Narasimhan, T. N. (ed.), 1982. Recent trends in hydrogeology, *Geol. Soc. America Spec. Paper 189,* 448p.

Nelson, P. H., 1982. Advances in borehole geophysics for hydrology, *Geol. Soc. America Spec. Paper No. 189,* 207-219.

Parker, G. G.; Ferguson, G. E.; and Love, S. K.; 1955. Water resources of southern Florida, *U.S. Geol. Survey Water Supply Paper 1255,* 965p.

Prickett, T. A., 1975. Modeling techniques for groundwater evaluation, *Advances in Hydroscience* **10,** 1-143.

Schroeder, M. C., 1982. Instrumentation for well tests, *Geol. Soc. America Spec. Paper 189,* 199-206.

Schroeder, M. C.; Klein, H.; and Hoy, N. D., 1958. Biscayne aquifer of Dade and Broward counties, Florida, *Florida Geol. Survey, Rept. Investigations No. 17,* 56p.

Stringfield, V. T., 1966. Hydrogeology—definition and application, *Ground Water* **4,** 2-4.

Stuckmeier, H. K., 1978. The hydrological map of Europe, *Episodes* **4,** 16-18.

Thomas, H. E., 1956. Changes in quantities and qualities of ground and surface waters, in W. L. Thomas, Jr. (ed.), *Man's Role in changing the Face of the Earth.* Chicago: University of Chicago Press, 542-563.

Ward, R. C., 1967. *Principles of Hydrology.* London: McGraw-Hill, 403p.

Cross-References: *Arid Lands, Engineering Geology; Channelization and Bank Stabilization; Coastal Inlets, Engineering Geology; Dams, Engineering*

Geology; Formation Pressures, Abnormal; Geomorphology, Applied; Hydrodynamics, Porous Media; Hydrology; Pumping Stations and Pipelines, Engineering Geology; River Engineering; Urban Hydrology; Wells, Aerial; Wells, Water. Vol. IVA: Aquifer; Artesian Water; Drawdown, Cone of Depression; Geophysical Methods for Hydrologic Search; Groundwater Motion in Drainage Basins; Hydrologic Cycle; Hydrology; Subsurface Water; Porosity and Permeability; Water Table. Vol. XII, Pt 1: Flow Theory; Imbibition; Infiltration; Moisture Management; Permeability; Surface Soil Water Content. Vol. XIV: Borehole Drilling; Canals and Waterways; Groundwater Exploration; Water Divining; Well Logging.

HYDROLOGIC MAPS—See Vol. IVA.

HYDROLOGY

There is no universally accepted definition of hydrology, but the U.S. Federal Council for Science and Technology (1962) says it is "the science that treats of all the waters of the earth, their occurrence, circulation, and distribution, their chemical and physical properties, and their reaction with their environment including their relation to living things. The domain of hydrology embraces the full life history of water on the earth." Hydrology is, therefore, one of the most comprehensive earth sciences (q.v.), involving a multidisciplinary synthesis of a wide range of environmental studies. To reduce the scope of the subject to manageable proportions, some scientific studies of water are conventionally, if somewhat arbitrarily, excluded. For example, although hydrology is concerned with the occurrence and distribution of all forms of precipitation on the earth, certain aspects of atmospheric moisture fall mainly within the province of meteorology. Similarly, oceanic water comes within the largely separate science of oceanography (q.v. in Vol. 1) and frozen water in the permanent ice sheets is primarily part of glaciology (q.v. in Vol. XVII). Even so, a complete knowledge of hydrology requires a formidable breadth of expertise; many hydrologists have specialized interests that often draw directly on a formal training gained previously in an allied discipline (see Hydrogeology and Geohydrology). Ground-water hydrology necessarily depends heavily on a knowledge of geology, soil moisture studies on pedology, and surface-water hydrology on hydraulics and fluid mechanics. Other supporting sciences include physics, chemistry, biology, mathematics, and statistics. Other hydrologists may specialize in the study of water in either particular types of physical environments, such as arid zones or karst areas, or particular land use areas, such as forests or towns.

The scope of hydrology is further widened because it is not just a pure science but also has many important practical applications, especially in relation to the assessment, use, and overall management of the world's water resources. These applications have been stressed in the definition of hydrology by Wisler and Brater (1959, 1-2) as "the science that deals with the processes governing the depletion and replenishment of the water resources of the land areas of the earth. It is concerned with the transportation of water through the air, over the ground surface, and through the strata of the earth." The practical aspects of hydrology embrace subjects such as agriculture, forestry, and all branches of engineering hydrology dealing with structural design, water supply, irrigation, flood control, water pollution control, and waste water treatment and disposal. In many of these areas, hydrology begins to overlap with certain social and behavioral sciences such as economics and sociology. Despite detailed demarcation problems, a basic distinction can often be maintained between systematic hydrology, which is concerned with moisture processes in the natural environment, and regional hydrology, which deals with drainage basins and applied operational studies relating to water (Fig. 1).

As with so many other scientific disciplines, the most rapid strides in hydrologic understanding have taken place in the present century. Progress was particularly stimulated by the organized activity conducted during the International Hydrological Decade, which began on January 1, 1965. On the other hand, Biswas (1970), in his definitive history of the subject, traces the first recorded evidence of water resources work back to 3200 B.C.

The Hydrologic Cycle

The conceptual framework of hydrology is organized around the hydrological cycle (Fig. 2). It illustrates the dynamic nature of water movement but also shows how water may occur in each of its three natural states as a liquid, a solid, or a gas, and how water is widely distributed through the earth-atmosphere system in various storage phases. The hydrologic cycle is powered by solar energy and controls the continuous circulation of moisture, which extends from a height of over 15 km in the atmosphere to an average depth of some 0.8 km in the lithosphere. Although there is no real beginning or end to the cycle, the atmospheric storage phase is a convenient starting point: all water enters by the evaporative processes, which alone utilize 22 percent of the total solar radiation received at the top of the earth's atmosphere, and all water exits through precipitation. When precipitation reaches the land areas of the earth, it immediately begins its return journey to the atmosphere. Some precipitated water is actually evaporated as it falls,

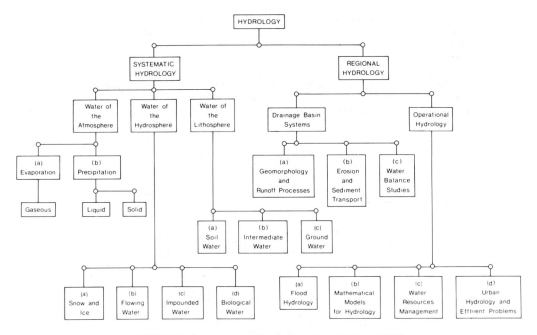

FIGURE 1. Components of hydrology (from Meinzer, 1923).

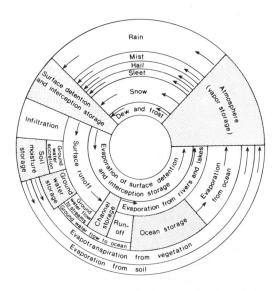

FIGURE 2. A diagrammatic representation of the hydrologic cycle. The diagram should be read counterclockwise (after Wisler and Brater, 1959).

whereas intercepted precipitation is quickly re-evaporated from vegetation and building surfaces. Surface runoff carries water in the rivers directly back to the oceans, which in turn produce over 80 percent of the world's evaporation. Infiltrated water may be returned to the atmosphere by evapotranspiration from plants and the upper soil

layers, or it may percolate to the water table. This ground water will then move slowly underground to reappear at the surface in springs, or by means of seepage it will eventually reappear in river channels or ocean basins.

In hydrology the main interest is in the transfer of water between the storage phases; the cycle concept implies a simplicity and smoothness that does not really exist. First, as indicated in Figure 2, water can follow numerous, alternative pathways, especially in the land-based part of the cycle, and there are therefore many hydrological cycles, not just one. Second, the cycle is not always regular, and although, for example, water is continuously being evaporated from the ocean surfaces, an arid area may not experience precipitation for several years. Similarly, the time water remains in the various storage units varies enormously. Thus intercepted water is likely to be retained for only a few minutes before it is re-evaporated; the average residence time for a water molecule in the atmosphere is 10 days, while other water molecules may be held for hundreds of years under virtually static conditions in icecaps and in deep aquifers.

The basic water transfer process is between the atmosphere and the earth, and the hydrologic cycle may also be seen as a downstroke, or *input,* which depends on precipitation, and an upstroke, or *output,* which results from evaporation. Within recent years this approach has been rationalized into a systems view of the hydrologic cycle. As a system, the cycle may be defined as a dynamic, physical, and cascading system in which subsystems are linked together by the sequential transfer of water from one phase to another. This sequential movement between the various storage subsystems of the vegetation, ground surface, soil moisture, channel, and ground water is schematically illustrated in Figure 3. The system normally has a single input, precipitation, and two major outputs, evapotranspiration and runoff, but in detail the output from one subsystem becomes the input for the next.

The hydrologic cycle is not only a highly complex natural system but is also progressively complicated by human interference, both deliberately and inadvertently. As shown in Figure 4, virtually every subcomponent of the system has been modi-

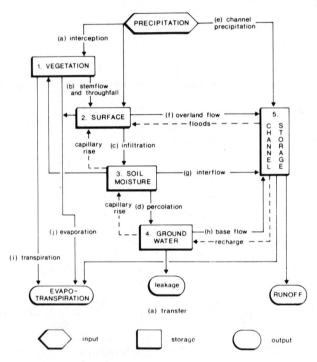

FIGURE 3. A systems representation of the natural hydrologic cycle within a drainage basin (after Ward, 1975).

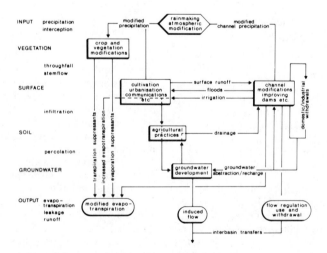

FIGURE 4. A systems representation of the basin hydrological cycle as modified by man (after Ward, 1975).

fied in an attempt to secure a better temporal and spatial distribution of water resources, either by speeding up the circulation process by rainmaking or land drainage or by retarding the removal of available water in the land-based phase of the cycle by the construction of storage dams or the artificial recharge of ground-water reserves. Most of these changes are eventually reflected in streamflow; there can be few, if any, large rivers remaining in the world that still possess an entirely natural regime. Some of the modifications may be both unexpected and adverse, such as the water-logging and salinization of soils that may result from certain inadequate irrigation schemes.

The Water Balance

If the hydrologic cycle is viewed in the systems concept of storage units influenced by inputs and outputs, and if these factors can be quantified, it follows that a series of water balances can be drawn up for different parts of the cycle. These balances are expressed mathematically by the *hydrologic* or *continuity, equation* which, in its simplest form, is written

$$I = O \pm \Delta S$$

where I is the input over any area during any given period, O is the output over the same area and period, and ΔS represents the resulting change in storage. The basic equation can, at least in theory, be applied to any of the storage subcompartments of the hydrologic cycle, such as interception or ground water, and can also be applied on virtually any scale of space or time. Thus, according to the nature of the problem to be solved, it could refer to a small experimental plot over a few hours, to a large drainage basin on a monthly basis, or even to

a continent in terms of seasonal or annual data. Generally speaking, the time scale increases with the areal dimensions in order to maintain a given level of resolution or accuracy.

Despite the fundamental simplicity of the water balance equation, it is frequently difficult to apply successfully. Essentially, the hydrologist is interested in the water balance of the land areas and, in practice, the equation is normally solved on a drainage basin scale to obtain the relationships between the incoming precipitation and the outgoing evaporation and runoff. In this situation the equation would have to be rewritten as

$$P = E + R \pm \Delta S$$

where P is precipitation in all its forms, E is the total evapotranspiration process, R is runoff, and ΔS represents all the storage changes on the land surface such as in interception, surface detention, soil moisture, and ground water. In practice it is impossible to obtain uniformly reliable and accurate areal measurements for all these factors on a basin scale. Often the equation has been simplified to include data for widely available data, such as precipitation and runoff, whereas other values have been assumed rather than measured. One of the first reasonably complete water balance calculations on a basin scale in Britain was made by Penman (1950) when he employed monthly values of precipitation, runoff, and also evapotranspiration calculated from meteorological variables to estimate changes in ground-water storage over the chalk basin of the River Stour. He was able to compare his estimate of storage with monthly measurements of rest-water levels in an observation well; the correspondence over a 16-year period is shown in Figure 5. The seasonal fluctuations

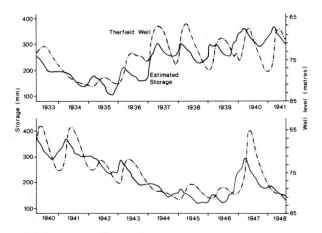

FIGURE 5. Estimated changes in monthly storage over a chalk catchment in eastern England compared with observed water-level changes in a well. The graph has been rezeroed at the beginning of 1940 to reduce the significance of cumulative errors (after Penman 1950).

and longer trends of estimated storage are reflected quite satisfactorily in the well-level fluctuations. There is a slight difference in phasing owing to the delayed response of the watertable to surface events. Such lag effects are a common complicating factor in the water balance.

The concept of water balance can be applied on a global scale to quantify the average volumes of water held in the various storage units. According to Nace (1969), the total capital stock of water in the earth-atmosphere system is 1,384,000 $km^3 \times 10^{-3}$, of which 97.6 percent is located in the ocean basins. A further 1.9 percent, comprising almost 80 percent of the world's fresh-water reserves, is locked up in glaciers and icecaps. The next largest storage unit is ground water, which totals some 0.5 percent of all water and represents by far the largest directly utilizable source available to man.

Sellers (1965) has shown that there are large geographical variations in the water balance of the earth's surface; the annual water balance for the various oceans and continents is given in Table 1. As a whole, the oceans lose more water by evaporation than they gain by precipitation, the deficit being made good by runoff from the continents, where precipitation is generally higher than evaporation. The earth's wettest continent is South America, which receives 1,350 mm of precipitation annually, of which nearly two-thirds is regained by the atmosphere by means of evaporation from the land surface. In the dry continents of Australia

TABLE 1. Annual Water Balance of the Oceans and Continents (mm year^{-1})

Ocean	E	r	ΔfL	Δfo	Δf
Atlantic Ocean	1,040	780	−200	−60	−260
Indian Ocean	1,380	1,010	−70	−300	−370
Pacific Ocean	1,140	1,210	−60	130	70
Arctic Ocean	120	240	−230	350	120
All Oceans	1,250	1,120	−130	0	−130

Continent	E	r	Δf	$\Delta f/r$
Europe	360	600	240	0.40
Asia	390	610	220	0.36
N. America	400	670	270	0.40
U.S.A.	560	760	200	0.26
S. America	860	1,350	490	0.36
Africa	510	670	160	0.24
Australia	410	470	60	0.13
Antarctica	0	30	30	1.00
All land	410	720	310	0.43

E = evaporation
r = precipitation
ΔfL = inflow from surrounding lands
Δfo = inflow from Arctic and Pacific Oceans
Δf = runoff for land areas and horizontal redistribution of water for oceans

Source: From Sellers (1965), p. 85, by permission of The University of Chicago Press.

and Africa, this evaporative loss exceeds 75 percent of the annual rainfall.

Drainage Basin Hydrology

In practice the most convenient areal unit for hydrologic studies is the drainage basin, or watershed. Its primary attribute is that of a drainage divide, which, except in special geological circumstances, is assumed to coincide with the topographic perimeter. Thus all the precipitated water that is surplus to evaporation requirements, both surface and subsurface, eventually flows gravitationally to leave the basin through runoff at the lowest downstream point on the main river. Some water will inevitably be gained or lost to the basin by deep subterranean leakage, but in most cases the amount will be small in relation to the other variables.

The first stage of most hydrologic investigations is to collect field observations within the basin from an optimally designed network (Dawdy, 1979). In many studies streamflow measurements are the most valuable data. This is partly because rivers constitute the part of the hydrologic cycle that is most easily modified by man, but also because streamflow, since it is a residual element, reflects the previous hydrologic behavior of the basin. Quite apart from this, measurements of river discharge provide a unique areal integration of *basin hydrology,* since a point measurement of streamflow obtained at the basin outlet can readily be converted into a mean areal value of depth equivalent for comparison with other variables such as precipitation, evaporation, and so on. Fortunately, the rate of flow in open channels can be measured relatively accurately using one or more of a variety of techniques (Herschy, 1978). The most widespread method is that adopted at *velocity-area stations,* where the stream cross-section is divided into a number of parts for each of which the area, the velocity and the discharge are separately determined. The mean section velocity is obtained by a *current meter* and, by adding together the partial discharges, a total is achieved for the stream. Other methods include the use of *gauging structures* such as weirs or flumes. These are rigid structures of known cross-sectional area that enable direct relationships to be established between the height of water passing throught the control and the discharge of the river. Use is also made of chemical techniques, whereby the downstream dilution of a tracer injected into the river at a known concentration provides a measure of streamflow. Newer techniques are based on electromagnetic and ultrasonic principles.

All other hydrologic measurements depend on point samples. The sample size is often small, such as in the case of an individual rain gauge, and the spatial distribution of measurements within the basin is rarely satisfactory. For example, much higher sampling requirements are necessary for local, short-term hydrologic studies than for deriving long-term averages over large basins. Thus, Huff and Shipp (1969) have shown that, for a minimum acceptance of 75 percent explained variance between sampling points, a gauge spacing of only 0.5 km is needed for 1-min rainfall rates compared with a 12-km distance for total precipitation in summer storms. Rain gauges are also subject to certain systematic errors due to defects in design and performance. It has been known for over 200 years that standard rain gauges exposed above ground level catch less precipitation than actually reaches the ground.

Evaporation measurements may be obtained from selected sites more or less directly by means of evaporation pans, such as the U.S. Class A pan illustrated in Figure 6, or they may be estimated through the use of an evaporation formula, such as that originally devised by Penman (1948), which depends, in turn, on meteorological measurements of solar radiation, temperature, humidity, and wind speed. Such meteorological measurements are increasingly made within the drainage basin by automatic weather stations; recent technical advances in remote sensing have greatly increased the potential availability of all hydrologic data. These advances include the long-distance interrogation of more or less standard instruments by telemetry and even satellites, together with entirely different alternatives such as precipitation measurement by radar (Wiesnet, 1976).

Once the data have been collected, they must be analyzed to provide required answers. Empirical hydrologic studies depend on monitoring the various water processes within selected drainage basins, which may often be designated either representative or experimental. The former are intended to represent typical assemblages of basin geometry, geology, climate, soils, vegetation, or land use that remain largely constant over the observation period. Experimental basins, on the other hand, are established to measure the hydrologic consequences of deliberately induced modifications, such as deforestation or urbanization. Sopper and Lull (1967) have documented many examples of the effects of changes in forest hydrology.

The advent of electronic computers has stimulated a great deal of hydrologic analysis based on the theoretical modeling of drainage basin behavior (see Vol. XIV: *Alluvial Systems Modeling*). According to Linsley (1967), runoff forecasting is the principal objective in hydrology, and many models are mathematical expressions of the basin response to rainfall input. The models are based on the physical laws of water movement, insofar as they are known, and aim at simulating future runoff behavior, especially during flood and drought conditions (Weeks and Hebbert, 1980).

Applied Hydrology

From a practical viewpoint, the main uncertainties in hydrologic behavior center on the extremes of water availability. Much of applied hydrology is therefore concerned either with water resource studies, which emphasize the nature and consequences of a deficiency in supply, or with drainage and flood studies, where excess water is the problem. In both cases, one of the most important applications is in providing engineering design criteria for water control structures such as dams, storage reservoirs, sewers, bridges, and irrigation systems. Since the early 1970s, it has also become apparent that hydrologists increasingly recognize the importance of the quality dimension of water, as detailed by Gower (1980).

In water resource studies a recurrent problem is how to determine the safe yield of various sources, such as wells or reservoirs. The size of the dam and the subsequent capacity of any reservoir depends on the degree to which the natural fluctuations in streamflow can be modified by storage to correspond with the required demands. The minimum annual or seasonal runoff that can be expected is always an important hydrologic factor; in the usual absence of long-term gauging records at the proposed site, estimates have to be prepared based on simulation or synthesis techniques. Low runoff may be associated with high evaporative demands, so in all such studies allowance must be made for evaporation and seepage losses from the reservoir.

In arid regions or during dry spells, agriculture relies heavily on irrigation, and successful irrigation depends on a clear understanding of water balance principles. Ideal conditions occur when soil moisture is maintained within the range available to the crop rooting system, that is, between *field capacity* and the *wilting point*. Precise water

FIGURE 6. The American Class A open-water evaporation pan.

management requires a knowledge of the existing soil moisture deficit that irrigation must improve. Lvovich (1980) has reviewed the general importance of soil conditions for hydrology.

Floods are the most obvious of all hydrologic hazards. Since they can be controlled by engineering works in a way that extreme droughts, for example, cannot, flood studies represent probably the most important aspect of applied hydrology. The ultimate aim of most *flood hydrology* is to predict the peak volume and maximum height of water in a river channel during a particular event, either real or assumed. The statistical concept of frequency occurrence is basic to all flood studies because, without some idea of the probability of a stated risk, engineering structures can neither be designed nor operated safely and economically. Some of the special features related to the statistical analysis of both floods and droughts have been discussed by Gumbel (1958).

Since exceptional rainfall is the immediate cause of most floods, much emphasis is placed on an understanding of the hydrometeorological processes associated with storm rainfall. The characteristics of storm rainfall are often expressed in terms of frequency, intensity, and areal extent since the intensity of precipitation increases as both the duration of time and the spatial scale contract. These attributes reflect the importance of concentrated, short-lived convectional storms rather than the more widespread precipitation associated with frontal activity. Figure 7 shows a rainfall frequency graph for various storm durations at Cleveland, Ohio, from 1902 to 1947. The graph indicates the average time period within which a rainfall of specified amount can be expected to

occur once; the straight-line relationships allow some extrapolation.

In certain circumstances, such as the design of a major dam, it may be desirable to estimate the magnitude of the largest possible flood that can physically occur at that point. Such an extreme event, known as the *probable maximum flood* (PMF), represents the upper limit of flooding that the climate regime in a stated drainage basin can produce in a basin that size. It will result from the maximum possible combination of precipitation and snowmelt together with minimum evaporative losses. Wiesner (1970) has detailed the various hydrometeorological approaches to PMF estimation, which in turn depend largely on estimates of the *probable maximum precipitation* (PMP). A distinction can be drawn between the PMP and the probable maximum storm, in that the PMP for basin areas of smaller than 260 km^2 and storm durations of less than 6 hours is likely to come from a thunderstorm, whereas frontal storms or tropical cyclones will usually cover larger areas and have longer durations.

KEITH SMITH

References

Biswas, A. K., 1970. *History of Hydrology.* Amsterdam: North-Holland, 336p.

Dawdy, D. R., 1979. The worth of hydrologic data, *Water Resources Research* **15**, 1726-1732.

Gower, A. M. (ed.), 1980. *Water Quality in Catchment Ecosystems.* New York: Wiley, 335p.

Gumbel, E. J., 1958. Statistical theory of floods and droughts, *Inst. Water Engineers Jour.* **12**, 157-184.

Herschy, R. W., (ed.), 1978. *Hydrometry.* New York: Wiley, 511p.

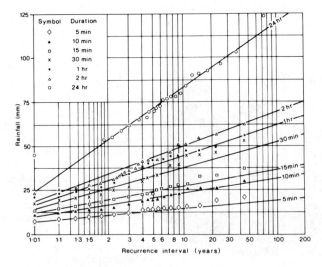

FIGURE 7. Rainfall-frequency graph for different storm durations at Cleveland, Ohio, 1902-1947 (after Linsley and Franzini, 1955).

Huff, F. A., and Shipp, W. L., 1969. Spatial correlations of storm, monthly and seasonal precipitation, *Jour. Appl. Meteorology* **8**, 542-550.

Linsley, R. K., 1967. The relation between rainfall and runoff, *Jour. Hydrology* **5**, 297-311.

Linsley, R. K., and Franzini, J. B., 1955. *Elements of Hydraulic Engineering.* New York: McGraw-Hill, 238p.

Lvovich, M. I., 1980. Soil trend in hydrology. *Hydrol. Science Bull.* **25**, 33-45.

Meinzer, O E., 1923. Outline of groundwater hydrology, *U.S. Geol. Survey Water Supply Paper 494.*

Nace, R. L., 1969. World water inventory and control, in R. J. Chorley (ed.), *Water, Earth and Man.* London: Methuen, 31-42.

Penman, H. L., 1948. Natural evaporation from open water, bare soil and grass, *Royal Soc. (London) Proc.,* Ser. A., **193**, 120-145.

Penman, H. L., 1950. The water balance of the Stour catchment area, *Jour. Inst. Water Engineers* **4**, 457-469.

Sellers, W. D., 1965. *Physical Climatology.* Chicago: University of Chicago Press, 272p.

Sopper, W. E., and Lull, H. W. (eds.), 1967. *Forest Hydrology.* Oxford: Pergamon Press, 539p.

U.S. Federal Council for Science and Technology, Ad Hoc Panel on Hydrology, 1962. *Scientific Hydrology.* Washington, D.C.

Ward, R. C., 1975. *Principles of Hydrology,* 2nd ed., London: McGraw-Hill, 367p.

Weeks, W. D., and Hebbert, R. H. B., 1980. A comparison of rainfall-runoff models, *Nordic Hydrology* **11**, 7-24.

Wiesner, C. J., 1970. *Hydrometeorology.* London: Chapman & Hall, 232p.

Wiesnet, D. R., 1976. Remote sensing and its application to hydrology, in J. C. Rodda (ed.), *Facets of Hydrology.* London: Wiley, pp. 37-59.

Wisler, C. O., and Brater, E. F., 1959. *Hydrology,* 2nd ed. New York: Wiley, 408p.

Cross-references: *Alluvial Plains, Engineering Geology; Channelization and Bank Stabilization; Deltaic Plains, Engineering Geology; Geomorphology, Applied; Hydrogeology and Geohydrology; River Engineering; Urban Hydrology.* Vol. IVA: *Hydrology, Coastal Terrain; Limestone Terrains; Semiarid Regions; Subsurface Water; Volcanic Terrain.* Vol. XIV: *Alluvial Systems Modeling; Canals and Waterways.*

HYDROMECHANICS

The term *hydromechanics* is generally applied to the fluid mechanics of incompressible flows. In the geological field, these flows include those of the oceans, rivers, and ground waters. Two phenomena are encountered in certain hydromechanical situations that are unique to flowing liquids: the existence of a *free surface* and the occurrence of *cavitation,* or low-pressure boiling.

Analyses of hydromechanical flows are based on the laws of conservation of mass, energy, and momentum. These analyses can be found in the works of Lamb (1945), Milne-Thomson (1955), McCormick (1973), Schlichting (1960), and Val-

entine (1959). For reference, some of the equations needed in hydromechanical analyses are presented here. The reader should consult the listed references for more complete coverage.

Hydrostatics

As the term implies, *hydrostatics* refers to fluids at rest. Included in this area are problems in ship stability, pressure-hull design, and dam design (see *Dams, Engineering Geology*).

The basis equation of hydrostatics is the following (McCormick, 1973):

$$p = - \int_0^{-h} \gamma dz = \begin{cases} \gamma h, \ \gamma \ \text{invariant} \\ F(h), \ \gamma \ \text{variable} \end{cases} \quad (1)$$

where p is the *static pressure,* γ is the specific weight of the fluid, z is measured vertically upward from the free surface, and h is the depth of the fluid as in Figure 1. In deep-ocean problems, γ of salt water does vary due to the enormous pressures experienced at great water depths.

Hydrodynamics

As previously mentioned, the hydrodynamic analysis involves the conservation of mass, momentum, and energy. In two-dimensional flows, laws of conservation are expressed as follows.

The *conservation of mass* for an incompressible flow of velocity (Fig. 1) is

$$v = u\mathbf{1} + w\mathbf{k} \quad (2)$$

The conservation of mass is expressed as

$$\nabla \cdot v = 0 \quad (3)$$

If the flow is irrotational, the velocity can be represented by a velocity potential ϕ as

$$v = \nabla(\phi) \quad (4)$$

and equation (2) can be expressed as

$$\nabla^2 (\phi) = 0 \quad (5)$$

which is Laplace's equation.

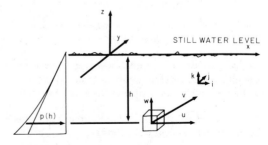

FIGURE 1. Notation for hydrostatics and hydrodynamics.

The general expression for the *conservation of momentum* of a flowing fluid, called the *Navier-Stokes* equation, is expressed as

$$\rho \left(\frac{\partial v}{\partial t} + v \cdot \nabla v \right) = -f - \nabla(p) + \mu \nabla^2(v) \quad (6)$$

where ρ is the mass-density of the fluid, f is the resultant body where force on a fluid element, such as the gravitational force, and μ is the coefficient of viscosity. Equation (6) has no general solution since it is *nonlinear* due to the term $v \cdot \nabla v$. There are, however, many situations in which simplifications of equation (6) can be made and solutions obtained (Schlichting, 1960).

If the viscosity can be neglected, then equation (6) reduces to

$$\rho \left(\frac{\partial v}{\partial t} + v \cdot \nabla v \right) = f - \nabla(p) \quad (7)$$

which is called *Euler's equation.*

If the nonlinear term in equation (7) is replaced by using the identity

$$v \cdot \nabla v = \nabla \left(\frac{v^2}{2} \right) - v \times (\nabla \times v) \quad (8)$$

and the flow is assumed to be irrotational so that $\nabla \cdot v = 0$ and equation (4) can be applied, then the resulting equation can be integrated to obtain *Bernoulli's equation,* which is the expression of the conservation of energy for an ideal flow:

$$\rho \frac{\partial \phi}{\partial \phi} zrx + \frac{1}{2} \rho v^2 + \rho gz + p = f(t) \quad (9)$$

where the body force f of equation 8 is assumed to be the gravitational force in the z- direction only.

Cavitation

When Bernoulli's equation is applied to flow situations in the ocean, the time function $f(t)$ in equation (9) can be assumed to be zero simply by assuming that the still-water level—that is, $z = 0$—is the energy datum. If the flow is steady, then $\partial \phi / \partial t = 0$ and equation (9) can be rewritten as

$$p = -\rho gz - \rho v^2 \quad (10)$$

For a given value of z, say, $z = -d$, one sees that as the velocity, v, increases the pressure, p, decreases. It is possible, therefore, for the pressure to reduce to the vapor pressure, p_v, by increasing velocity. This situation is encountered on the tips of high-speed propellers such as that shown in Figure 2.

A measure of cavitation susceptibility is given by the *cavitation index* or *number:*

$$\sigma = \frac{p - p_v}{\frac{1}{2} \rho v^2} \quad (11)$$

When $\sigma = 0$, cavitation or low-pressure boiling will occur.

Waves

Most waves on the surfaces of the oceans are caused by the wind and are thus called *wind waves.* The turbulent pressure fluctuation of the wind on the free surface of the water causes small deformations such as those sketched in Figure 3a. Note the narrow trough and broad crest of this wave. This wave profile is caused by the strong influence of *surface tension.* The wave, called a *capillary wave,* is the first created by the wind. These small waves then grow due to the combined action of turbulence and viscous shear on the surface. Soon the profile resembles that of Figure 3b, that is, *a sinusoid.* The wave increases in height, H, and length, λ, and eventually has a profile like that sketched in Figure 3c. Note the difference between the profiles of Figures 3a and 3c.

The wave moves at a velocity, c, which is called *phase velocity,* or *celerity.* For a irrotational and sinusoidal wave, the celerity in deep water—that is, where $\lambda/2 < h$, is approximately

$$c = \left(\frac{g\lambda}{2\pi} \right)^{1/2} \quad (12)$$

and the *wavelength* is

$$\lambda = \frac{gT^2}{2\pi} \quad (13)$$

where T is the *wave period.*

The water particles in this wave travel in nearly circular orbits with the velocity components of

$$u = Hg/2c \cos(kx - wt)$$
$$\text{and } w = Hg/2c \sin(kx - wt) \quad (14)$$

where $k = 2\pi/\lambda$ is the *wavenumber,* and $w = 2\pi/T$ is the *circular frequency.* When the horizontal particle velocity at a crest equals the celerity, then the wave is said to *break,* that is,

$$\frac{u}{\frac{H}{2}} = c \quad (15)$$

After the wave breaks, turbulence occurs, dissipating most of the wave energy.

Refer to Lamb (1945), Milne-Thomson (1955), and McCormick (1973), and the *Shore Protection Manual* (U. S. Army Staff, 1974) for more complete discussions of the mathematical and physical descriptions of waves. Bascom (1980) gives an excellent nonmathematical discussion of waves.

MICHAEL E. MCCORMICK

FIGURE 2. A 61-cm (24-in.) propeller mounted in a 91-cm (36-in.) variable-pressure water tunnel shows heavy tip cavitation. This design was tested in various sizes in several water tunnels belonging to the International Towing Tank Conferences. The design, once used on the old destroyer Hamilton, is now a classic research propeller (photo: Naval Ship Research and Development Center).

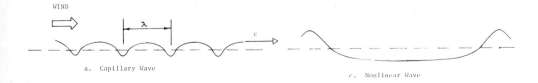

FIGURE 3. Three wind-generated wave forms: (a) capillary wave; (b) sinusoidal wave; (c) nonlinear wave.

References

Bascom, W., 1980. *Waves and Beaches,* Garden City, N.Y.: Doubleday, 366p.

Lamb, H., 1945. *Hydrodynamics.* New York: Dover, 738p.

McCormick, M. E., 1973. *Ocean Engineering Wave Mechanics.* New York: Wiley-Interscience, 179p.

Milne-Thomson, L., 1955. *Theoretical Hydrodynamics.* New York: Macmillan, 632p.

Schlichting, H., 1960. *Boundary Layer Theory.* New York: McGraw-Hill, 647p.

U.S. Army Staff, 1974. *Shore Protection Manual.* Washington, D.C.: U.S. Government Printing Office, 3 vols.

Valentine, H., 1959. *Applied Hydrodynamics.* London: Butterworths, 272p.

Cross-references: *Dams, Engineering Geology; Hydrodynamics, Porous Media.* Vol. XII, Pt. I: *Flow Theory.*

I

INFORMATION CENTERS

During the last two decades, the combined technologies of the telephone, television, and computer have merged to form an integrated information and communication system that collapses the information float from weeks or days to seconds. Communication in the earth sciences (see *Geological Communication*) depends on the rapid and accurate acquisition of data from a variety of sources (see *Earth Science, Information and Sources*). Many traditional information sources are increasingly taking advantage of sophisticated information technology. Even though many societies, professional associations, and publishers are offering more incentives for geoscientists to avail themselves of "high-tech" information strategies, most detailed geotechnical data is available on a parochial basis through local, state, or national organizations. Some of the more important information geoscience centers are identified on a worldwide basis in this entry (see Tables 1, 2, and 3).

Data, the lifeblood of science, are essential to many fields of interest. The acquisition of data normally involves expenditure of time and effort, often without immediate renumeration for the researcher. Researchers, teachers, and consulting geoscientists involved with practical applications of geotechnical data (such as engineering geologists, exploration geologists, geophysicists, and geochemists) must, in addition to generating their own data, peruse the literature to make use of other data. This task is complicated by several factors that, to varying degrees, inhibit the dissemination of geoscience information. Dowden et al. (1979), for example, point out that 45 percent of all geological literature is published, on a worldwide basis, in languages other than English. Thus, most important geological literature is in languages most English-speaking geoscientists do not read; furthermore, a given "foreign" geological publication is less likely to be available in the average library than a publication of no greater significance. Geological literature also differs from other sciences in that it is historical, is often descriptive and voluminous, is visual (relies on maps, photographs, and drawings), and has a wide subject distribution (including sources in the physical, biological, and social sciences) (Dvorzak, 1980).

Serials are the most important form of geoscience literature, although monographs, treatises, textbooks, and conference proceedings are also essential reference sources (Mason, 1953). Most geoscience publishing societies and agencies do not advertise widely, whereas nontrade publications belong to the ephemeral literature and are often difficult to obtain (Pruett, 1979). Some of these difficulties are partly nullified by computerized databanks, but these in turn present their own special problems, and Ward (1979) reports that they are typically underused, at least by North American geologists. Information sources of practical concern are listed in the Preface of this volume. The proceedings of the First International Conference on Geological Information (London, 1978) (Harvey and Diment, 1979) contains a worldwide review of geological information sources that should be of interest to information managers. Other standard guides to documentation centers and geoscience information include, for example, the following works: *Bibliography of Congressional Geology* (Pestana, 1972), *An Encyclopedia of Geology* (Hallam, 1977), *Geological Map Acquisitions* (Diment, 1979), *Geologic Reference Sources* (Ward, 1967), *Geologic Reference Sources* (Ward and Wheeler, 1972), *Geologists' Year Book* (1977), *Guide to Geologic Literature* (Pearl, 1951), *A Guide to Information Sources in Mining, Minerals and Geosciences* (Kaplan, 1965), *Introduction to United States Public Documents* (Morehead, 1978), *List of Serial Publications Held in the Library* (Geological Society of London, 1978), *The Literature of Geology* (Mason, 1953), *Library Guide to Geology* (Haylock, 1973), *The Use of Earth Sciences Literature* (Wood, 1973), *U.S. Government Scientific and Technical Periodicals* (Yannarella and Aluri, 1976), and the *World Geo Guide* (BYGGINFO Staff, 1981).

World Data Centers

The World Data Centers (WDC) system, developed during the International Geophysical Year (1957-1958), has served the international geoscience community in a variety of disciplines since 1957. Both the United States and the Soviet Union agreed to operate WDCs in all disciplines, forming World Data Centers A (WDC-As) and B (WDC-Bs), respectively (Lander, 1979). Other countries cover individual disciplines for WDC-Cs in earth, ocean atmosphere, solar-terrestrial, and space sciences. The WDC system is coordinated by a panel from the nongovernmental International Council of Scientific Unions (ICSU), which issues a "Guide to International Data Exchange Through WDC's." Most data are freely exchanged with members on a barter system, but non contributors may acquire data on a cost-of-copying basis. All WDCs share a common core collection and have access to more specialized data (Shapley and Hart, 1982). These modern data centers perform a wide range of functions by providing facilities for duplicating, microfilming, digitizing, computer graphics, and publishing. Locations of a coordination office and seven subcenters for World Data Center A (United States) are given in Table 1.

National Geological Surveys

Institutions that operate under the aegis of a "geological survey" attempt, in various ways, to meet the needs of governments for information about their landmasses and contained resources. Earth surveys have long supported mankind's occupation and use of Planet Earth. As early as 2000 B.C., for example, the Egyptians were mining for gemstones and metals to use for ornamentation and implements. Even earlier, in the Third Dynasty, the Egyptians used stone to build funeral monuments to immortalize their rulers, as in the

TABLE 1. Offices and Subcenters of World Data Center A (WDC-A), United States

World Data Center A
Coordination Office
National Academy of Sciences
2101 Constitution Avenue, N.W.
Washington, D.C. 20418

World Data Center A: Meteorology
National Climatic Center
Federal Building
Asheville, North Carolina 28801

World Data Center A: Oceanography
National Oceanic and Atmospheric Administration
Washington, D.C. 20235

World Data Center A: Rockets and Satellites
Goddard Space Flight Center, Code 601
Greenbelt, Maryland 20771

World Data Center A: Rotation of the Earth
U.S. Naval Observatory
Washington, D.C. 20390

World Data Center A: Solar-Terrestrial Physics
Environmental Data and Information Service
National Oceanic and Atmospheric Administration
Boulder, Colorado 80303

World Data Center A: Solid-Earth Geophysics
Environmental Data and Information Service
National Oceanic and Atmospheric Administration
Boulder, Colorado 80303

Source: After Lander (1979).

construction of the first great stone structure, which was erected for King Zoser in about 2900 B.C. (Poss, 1975). During the Industrial Revolution, the search for additional earth resources quickened, and geology turned more and more into an exact science, especially in the area of geological survey and the use of mineral resources (see Vol. XIV: *Geology, Philosophy; Geology, Scope and Classification*).

The activities of modern geological surveys now encompass a wide spectrum of scientific and technical endeavor to provide data, information, and advice on the use, protection, and government of the national landmass (Smith, 1974). Most surveys attempt to meet their mandates by publishing results of their investigation in the form of reports, circulars, bulletins, monographs, books, leaflets, and maps. The U.S. Geological Survey, in addition, maintains "open-file reports" that are available for public inspection at selected depositories. These reports included manuscript copy, basic data, field notes, and other data not reproduced through formal publication series. Other geological surveys provide similar services, but because it is impossible to detail the scope of coverage, interested readers should contact the appropriate geological survey directly.

The addresses of state geological surveys in the United States are given in *The Encyclopedia of World Regional Geology, Part 1: Western Hemisphere* (Volume XIII, The Encyclopedia of Earth Science Series) under the entry "United States—General" (Wilson, 1975). Table 2 provides the addresses of government geoscience agencies that have functions similar to one or more operating divisions of the U.S. Geological Survey.

Major International Organizations

Professional geoscience societies (see Vol. XIV: *Professional Geologists' Associations*) provide a wide range of services including the collection, recording, organization, and dissemination of information. Many reports issued by local societies provide practical information of interest to geoscientists working in the vicinity. Although local associations and societies compile detailed geotechnical data for specific geographic areas, interest here is focused on major international organizations that serve broader needs. Addresses of local "mini-information centers" have been given, for example, in *Geotimes* as a "Directory of Societies in Earth Science" (1982). Table 3 lists major international organizations that, to varying degrees, represent information centers in applied geoscience. This list incorporates organizations that are broadly concerned with aspects of applied geology such as mineral exploration, geodesy, and hydraulics.

TABLE 2. World-wide Directory of National Geological Surveys

Afghanistan
Department of Mines and Geology
Ministry of Mines and Industries
Darulaman
Kabul

Albania
Ministry of Industry and Mining
Tirane

Algeria
Sous-Direction de la Géologie
Ministère de L'Industrie Lourde
Rue Ahmed Bey de Constantine
Algiers

Angola
Direcção de Servicos de Geologia e Minas
Caixa Postal 1260-C
Luanda

Argentina
Servicio Geologico Nacional
Secretaria de Estado de Mineria
Avenida Santa Fe 1548
Buenos Aires

Australia
Bureau of Mineral Resources, Geology, and
Geophysics
Post Office Box 378
Canberra City, A.C.T.

Austria
Geologische Bundesanstalt
Rasumofskygasse 23
A-1031 Vienna

Bangladesh
Geological Survey
Pioneer Road
Segun Bagicha
Dacca

Belgium
Service Géologique de Belgique
13 rue Jenner
1040 Brussels

Benin (formerly Dahomey)
Direction des Mines, de la Geologie, et des
Hydrocarbures
Ministere de l' Equipment
B.P. 249
Cotonou

Boliva
Servicio Geológico de Bolivia (GEOBOL)
Casilla de Correo 2729
La Paz

Botswana
Geological Survey Department
Ministry of Mineral Resources and Water Affairs
Private Bag 14
Lobatsi

Brazil
Instituto de Pesquisas Especiais (INPE)
Caixa Postal 515
São José dos Campos
São Paulo, SP

Bulgaria
Geologischeski Institut
Bulgarska Akademiya na Naukite
Akademik Bonchev St., Blok 2
Sofia

Burma
Department of Geological Survey and Exploration
Ministry of Mines
Kanbe Road, Yankin P.O.
Rangoon

Burundi
Ministère de la Géologie des Mines
Boite Postale 1160
Bujumbura

Cameron
Direction des Mines et de le Géologie
Ministry of Mines and Energy
B.P. 70
Yaoundé

Office National de la Recherche Scientifique et
Technique
(ONAREST)
B.P. 1457
Yaoundé

Canada Geological Survey of Canada
Department of Energy, Mines, and Resources
601 Booth Street
Ottawa, Ontario K1A 0GI

Chad
Direction des Mines et de le Géologie
B.P. 816
Fort-Lamy

Chile
Dirección de Geologia y Minas
Corporacion Nacional del Cobre de Chile
Huerfanos 1189, 8 Piso

China
Academia Sinica
5th Bureau
Earth and Environmental Science
Institute of Geology
Lanzhou

Columbia
Instituto Nacional de Investigaciones Geológico-
Mineras
(INGEOMINAS)
Carrerra 30, No. 51–59
Apartado Aéreo 4865
Bogota

Congo
Service des Mines et de la Géologie
B.P. 12
Brazzaville

continued

TABLE 2. *(continued)*

Costa Rica
Departmento de Geologia
Instituto Costarricense de Electricidad (ICE)
Apartado 10032
San José

Cuba
Instituto de Geologia
Academia de Ciencias de Cuba
Ave. Van-Troi, No. 17203
La Habana

Cyprus
Geological Survey Department
Ministry of Agriculture and Natural Resources
Nicosia

Czechoslovakia
Ústave Geologický
Ceskoslovenská Akadémie Ved
Lysolaje 6
Praha 6

Denmark
Geological Survey of Denmark
Thoravej 31
DK 2400 Kobenhavn N.V.

Ecuador
Dirección General de Geologiá y Minas
Ministerio de Recursos, Naturales y Energéticos
Carrion #1016 y Paez
Quito

Egypt
Geological Survey and Mining Authority
Ministry of Industry
3, Salah Salem Street
Abbasia, Post Office Building
Cairo

El Salvador
Centro de Investigaciones Geotécnicas
Ministerio de Obras Públicas
Avenida Peralta, final, costado Oriente Talleres
El Coro
San Salvador

Ethiopa
Ethiopian Institute of Geological Survey (EIGS)
Ministry of Mines, Energy, and Water Resources
P.O. Box 486
Addis Ababa

Fiji
Ministry of Lands and Mineral Resources
Mineral Resources Department
Private Mail Bag, G.P.O.
Suva

Finland
Geologinen Tutkimuslaitos
Kivimiehentie 1 02150 Espoo 15

France
Bureau des Recherches Géologiques et Minières
(BRGM)
Avenue de Concyr, B.P. 6009
45018 Orleans Cedex

French Guiana
Bureau des Recherches Géologiques et Minières
B.P. 42
Cayenne

German Democratic Republic
Zentrales Geologisches Institut
Invalidenstrasse 44
104 Berlin

Germany, Federal Republic of
Bundesanstalt für Geowissenschaften und Rohstoffe
(BGR)
(Geobund)
Alfred-Bentz-Haus
Postfach 510153
Stillweg 2
3000 Hannover 51

Ghana
Geological Survey of Ghana
P.O. Box M 32
Accra

Greece
Institute of Geological and Mining Research
70 Messoghion Street
Athens 608

Greenland
Grønlands Geologiske Undersøgelser
Ostervoldgade 5-7, Tr. KL
DK- 1350 Copenhagen, K
Denmark

Guadeloupe
Office de la Recherche Scientifique et Technique
Outre-Mer (OSTOM)
B.P. 504
97165 Pointe-A-Pitre
Cedex F 1960

Guatemala
División de Geologia
Instituto Goegráfico Nacional
Avenida las Américas 5:76, Zona 13
Guatemala City

Guinea
Direction Générale des Mines et Géologie
Conarky
Guinea-Bissau (formerly Portuguese Guinea)
Direccao Geral, Geologia e Minas
Comissariado Dos Recursos Naturais
Bissau

Guyana
Geological Surveys and Mines Department
P.O. Box 1028, Brickdam
Georgetown

Haiti
Department of Mines and Energy Resources
Village Willy Lamothe
Delmas 9
P.O. Box 2174
Port-au-Prince

TABLE 2. *(continued)*

Honduras
    Exploracion Geotérmica
    Empressa Nacional de Energia Electrica (ENEE)
    Apartado 99
    Tegucigalpa

Hong Kong
    Mines Department, Branch Office
    Canton Road Government Offices
    Canton Road
    Kowloon

Hungary
    Központi Földtani Hivatal
    H-1251 Budapest 1
    Iskola u. 13

Iceland
    Division of Geology and Geography
    Museum of Natural History
    Laugavegi 105 and Hverfisgata 116
    P.O Box 5320
    105 Reykjavik

India
    Geological Survey of India
    27 Jawaharal Nehru Road
    Calcutta 700016

Indonesia
    Direktorat Sumber Daya Mineral
    Jalan Diponegoro 57
    Bandung

Iran
    Geological and Mineral Survey of Iran
    Ministry of Industry and Mines
    P.O. Box 1964
    Tehran

Iraq
    Directorate General of Geological Survey and
      Mineral
    Investigation
    State Organization for Minerals
    P.O. Box 2330
    Baghdad

Ireland
    Geological Survey of Ireland
    14 Hume Street
    Dublin 2

Israel
    Geological Survey of Israel
    30 Malchei Israel Street
    Jerusalem

Italy
    Servizio Geologico d'Italia
    Salita S. Nicoló da Tolentino 1B
    00187 Roma

Ivory Coast
    Direction des Mines et de la Géologie
    Ministère des Mines
    BP. V 28
    Adidjan

Jamaica
    Geological Survey Division
    Ministry of Mining and Natural Resources
    Hope Gardens
    Kingston 6

Japan
    Geological Survey of Japan
    Ministry of International Trade and Industry (MITI)
    3, Higashi, Yatabe-cho 1-chome
    Tsukuba-gun, Ibaraki-Ken 300-21

Jordan
    Department of Geological Resources and Mining (NRA)
    P.O. Box 39
    Amman

Kenya
    Geological Survey of Kenya
    Mines and Geological Department
    Ministry of Natural Resources
    P.O. Box 30009
    Nairobi

Korea (North)
    Geology and Geography Research Institute
    Academy of Sciences
    Mammoon-dong
    Central District
    P'yongyang

Korea (South)
    Korea Research Institute of Geoscience and Mineral
      Resources
      (KIGAM)
    219-5 Garibong-dong
    Youngdeungpo-gu
    Seoul 150-06

Kuwait
    Kuwait Institute for Scientific Research
    P.O. Box 12009 Shamiah
    Kuwait

Laos
    Departement de Géologie et des Mines
    Ministère de l'Industrie et du Commerce
    Vientiane

Lebanon
    Directorate General of Public Works
    Ministry of Public Works
    Beirut

Lesotho
    Department of Mines and Geology
    P.O. Box 750
    Maseru

Liberia
    Liberian Geological Survey
    Ministry of Lands and Mines
    P.O. Box 9024
    Monrovia

Libya
    Geological Research and Mining Department
    Industrial Research Center
    P.O Box 3633
    Tripoli

continued

TABLE 2. *(continued)*

Liechtenstein
Landesbauamt des Fürstentums Liechtenstein
Städtle 49
9490 Vaduz

Luxembourg
Service Géologique
Ponts et Chaussées
4 bd Roosevelt
Luxembourg

Madagascar
Service Geologique
Ministère des Mines et de l'Economie et du
Commerce
B.P. 322, Ampandrianomby
Antananarivo

Malawi
Geological Survey Department
Ministry of Natural Resources
P.O. Box 27, Liwonde Road
Zomba

Malaysia
Geological Survey Department
Bangunan Ukor
Jalan Gurney
Kuala Lumpur

Mali
Direction Nationale des Mines et de la Géologie
B.P. 223
Loulouba, Bamako

Malta
Public Works Department
Beltissebh

Martinique
Arrondissement Mineralogique de la Guyane
B.P. 458
Fort-au-France

Mauritania
Direction des Mines et de la Géologie
Ministère de l'Industrialisation et des Mines
B.P. 199
Nouakchott

Mauritius
Ministry of Agriculture and Natural Resources and
the Environment
Port Louis

Mexico
Instituto de Geologia
Universidad Nacional Autónoma de México (UNAM)
Cuidad Universitaria
México 20, D.F.

Mongolia
Institute of Geology
Academy of Sciences
UI, Leniadom 2
Ulaanbaatar

Morocco
Division de la Géologie
Ministère de l'Energie et des Mines
Quartier Administratif
Rabat

Mozambique
Direcção Nacional de Geologia é Minas e Defensa
Praca 25 de Junho/Maputo
P.O. Box 217
Maputo

Namibia (Southwest Africa)
Geological Survey
P.O. Box 2168
Windhoek

Nepal
Department of Mines and Geology
Ministry of Industry and Commerce
Lainchaur
Kathmandu

Netherlands
Rijks Geologische Dienst
Spaarne 17, P.O. Box 157
2000 AD Haarlem

New Caledonia
Service des Mines et de la Geologie
Rte. No. 1
B.P. 465
Noumea

New Herbides
Geological Survey
British Residency
Port Villa

New Zealand
New Zealand Geological Survey
Department of Scientific And Industrial Research
(DSIR)
P.O. Box 30-368
Lower Hutt

Nicaragua
Corporación Nicaraguense de Mines e Hidrocarburos
(CONDEMIA)
Apartado Postal No. 8
Manague, D. N.

Niger
Direction des Mines et de la Géologie
Ministère des Mines et de l'Hydraulique
B.P. 257
Niamey

Nigeria
Geological Survey Department of Nigeria
Ministry of Mines and Power
P.M.B. 2007
Kaduna South, Kaduna State

Norway
Norges Geologiske Undersøkelse
P.B. 3006 Ostmarkneset
Leiv Erikssons Vei 39
7001 Trondheim

Oman
Ministry of Petroleum and Minerals
P.O. Box 551
Muscat

TABLE 2. *(continued)*

Pakistan
  Geological Survey of Pakistan
  P.O. Box 15
  Quetta

Panama
  Corporación de Desarrollo Minero-Cerro Colorado
    (CODEMIN)
  Apartado 5312
  Panama 5

Papua New Guinea
  Geological Survey
  Department of Minerals and Energy
  P.O. Box 2352
  Konedobu

Paraguay
  Direccion de Recursos Minerales
  Ministerio de Obras Publicas y Communicaciones
  Calle Alberdi y Oliva
  Asuncion

Peru
  Instituto Geologico Minero y Metalúrgico
    (INGEMMET)
  Pablo Bermudez 211
  Apartado 211
  Lima

Philippines
  Bureau of Mines and Geosciences
  P.O. Box 1595
  Pedro Gill Street
  Manilla 2801

Poland
  Instytut Nauk Geologicznych PAN
  Al. Zwirki i Wigury 93
  02-089 Warszawa

Portugal
  Direcção General de Mines e Serviços Geologicos
  Ministèrio de Industria
  Rua Antonio Enes, 5
  Lisboa 1000

Qatar
  Industrial Development Technical Center (IDTC)
  P.O. Box 2599
  Doha

Reunion
  Services des Travaux Publiques
  St. Denis

Romania
  Institutul de Cerecetari Geologice şi Geofizice
  Str. Caransebeş No. 1
  Sector 7, Bucharest

Rwanda
  Ministère des Ressources Naturelles, des Mines et des
    Carrieres
  B.P. 413
  Kigali

Saudi Arabia
  Ministry of Petroleum and Mineral Resources
  Directorate General of Mineral Resources
  P.O. Box 345
  Jiddah

Senegal
  Direction des Mines et de la Geologie
  Ministere du Developpement Industrial
  Route de Ouakam
  B.P. 1238
  Dakar

Sierre Leone
  Geological Survey Division
  Ministry of Lands, Mines, and Labor
  New England, Freetown

Singapore
  Public Works Department
  Structural Design and Investigation Branch
  Ministry of National Development
  National Development Building
  Maxwell Road
  Singapore 0106

Solomon Islands
  Ministry of Natural Resources
  Geology Department
  P.O. Box G24
  Honiara

Somalia
  Geological Survey Department
  Ministry of Minerals and Water Resources
  P.O. Box 744
  Mogadishu

South Africa
  Geological Survey of South Africa
  Department of Mines
  233 Visagie Street (Private Bag X112)
  Pretoria 0001

Spain
  Servicio de Geologia
  Ministerio de Obras Públicas y Urbanismo
  Avenida de Portugal, 81
  Madrid-11

Sri Lanka
  Geological Survey Department
  48 Sri Jinaratanana Road
  Colombo 2

Sudan
  Geological and Mineral Resources
  Ministry of Energy and Mining
  P.O. Box 410
  Khartoum

Suriname
  Geologisch Mijnbouwkundige Dienst
  Klein Wasserstraat 1 (2-6)
  Paramaribo

Swaziland
  Geological Survey and Mines Department
  P.O. Box 9
  Mbabane

Sweden
  Sveriges Geologiska Undersökning (SGU)
  Box 670
  S-751 28 Upsala

continued

TABLE 2. *(continued)*

Switzerland
  Geologische Kommission der Schweizerischen
  Naturforschende Gesellschaft
  Bernoullianum
  4056 Basel

Syria
  Directorate of Geological Research and Mineral
    Resources
  Ministry of Petroleum
  Fardos Street
  Damascus

Taiwan
  Institute of Geology
  National Taiwan University
  1 Roosevelt Road, Section 4
  Taipei

Tanzania
  Geology
  Ministry of Water, Energy, and Minerals
  P.O Box 903
  Dodoma

Thailand
  Department of Mineral Resources
  Ministry of Industry
  Rama VI Road
  Bangkok 4

Togo
  Direction des Mines et de al Géologie
  Ministère des Mines, et Ressources Hydrauliques
  B.P. 356
  Lome

Tonga
  Ministry of Lands, Survey and Natural Resources
  P.O. Box 5
  Nukúalofa

Trinidad and Tobago
  Ministry of Energy and Energy-Based Industries
  P.O. Box 96
  Port-of-Spain

Tunesia
  Service Géologique de Tunisie
  95 Avenue Mohamed V

Turkey
  Mineral Research and Exploration Institute of Turkey
  Eskisehir Yolu
  Ankara

Uganda
  Geological Survey and Mines Department
  P.O. Box 9
  Entebbe

United Kingdom
  Institute of Geological Sciences
  Exhibition Road
  London SW7 2DE

United States
  U.S. Geological Survey
  National Center
  12201 Sunrise Valley Drive
  Reston, Virginia 22092

Upper Volta
  Direction de la Geologie et des Mines
  B.P. 601
  Ouagadougou

Uruguay
  Instituto Geologico del Uruguay
  Calle J. Herrera y Obes 1239
  Montevideo

U.S.S.R.
  All-Union Scientific Research Geological Institute
    (VSEGEI)
  Sredniy Prospekt 72B
  199026 Leningrad

  Institute of Geology
  Akademiya Nauk USSR
  109017 Moscow ZH-17

Venezuela
  Dirección de Geologia
  Dirección General Sectorial de Mines y Geologia
  Ministerio de Energia y Minas
  Torre Norte, Piso 19
  Centro Simon Bolivar
  Caracas

Vietnam
  Geologic Section
  State Committee of Sciences
  Hanoi

Western Sahara (formerly Spanish Sahara)
  Dirección General de Plazas y Provincias Africanas
  Servicio Minero y Geológico
  Castellana No. 5
  Madrid 1, Spain

Yemen (Aden)
  Overseas Geological Surveys of London, England
  c/o Aden Public Works Department
  Aden

Yemen (San'A)
  Geological Authority
  Yemen Oil and Mineral Resources Corporation
    (YOMICO)
  P.O. Box 81
  Saña

Yugoslavia
  Zavod za Geoloskli i Geofizicka Istrazivanja
  Karadjordjeva 48
  Belgrade

Zaire
  Service Geologique du Zaire
  Minisatry of Mines
  B.P. 898
  44 Avenue des Huileries
  Kinshasa

Zambia
  Geological Survey Department
  Ministry of Mines
  P.O. RW 135
  Ridgeway, Lusaka

Zimbabwe
  Geological Survey of Rhodesia
  Ministry of Mines and Lands
  P.O. Box 809, Causeway
  Salisbury

Source: After Bergquist et. al (1981).

TABLE 3. Directory of Major International Organizations
that Emphasize Aspects of Applied Geology

American Association of Petroleum Geologists (AAPG)
P.O Box 979
Tulsa, Oklahoma 74101

American Congress on Surveying and Mapping
210 Little Falls Street
Falls Church, Virginia 22046

American Geological Institute (AGI)
5205 Leesburg Pike
Falls Church, Virginia 22041

American Geophysical Union (AGU)
2000 Florida Avenue
Washington, D.C. 20009

American Institute of Mining, Metallurgy and
    Petroleum Engineers
345 East 47th Street
New York, New York 10017

American Society of Photogrammetry
210 Little Falls Street
Falls Church, Virginia 22046

Association of Geoscientists for International
    Development
    (AGID)
Central American Research Council for Industry
Avenida La Reforma 4-47, Zona 10
Guatemala City, Guatemala

Association Internationale pour l'Etude des Argiles
Institute für Bodenkunde
Technische Hochschule München
D-8050 Freising-Weihenstephan
West Germany

Association of Women Geoscientists
P.O. Box 1005
Menlo Park, California 94025

Australasian Institute of Mining and Metallurgy (AIMM)
P.O. Box 310
Carlton South, Victoria 3053
Australia

Cartographic Information Society
143 Sciences Hall
University of Wisconsin
Madison, Wisconsin 53706

Centro Regional de Seismologia para America del Sur
    (CERESIS)
Av. Arenales 431 Oficina 702
Apartado 3747
Lima, Peru

Circum-Pacific Council
Halbouty Center
5100 Westheimer Road
Houston, Texas 77207

Commission for the Geological Map of the World
51, Boulevard de Montmorency
75026 Paris, France

Earthquake Engineering Research Institute
2620 Telegraph Avenue
Berkeley, California 94704

European Association of Exploration Geophysicists
30 Carel van Bylandtlaan
The Hague, The Netherlands

European Geophysical Society
The Royal Society
6 Carlton House Terrace
London SW1Y 5AG England

Geological Society of America (GSA)
P.O. Box 9140
3300 Penrose Place
Boulder, Colorado 80301

International Association for Advancement of Earth
    and Environmental Sciences
Department of Earth Sciences
Northeastern University
Northeastern Illinois State University
Bryn Mawr at St. Louis Avenue
Chicago, Illinois 60625

International Association of Engineering Geology
Geologisches Landesamt NW
De Greiff Strasse 195
415 Krefeld, West Germany

International Association of Geodesy (IAG)
39ter rue Gay Lussac
7500S Paris, France

International Association of Hydrogeological Sciences
    (IAHS)
Research Unit of Water Resources Development
Rakoczi Ut 41
Budapest 8, Hungary

International Association of Hydrological Sciences
19, rue Eugene Carrier
F-75018 Paris, France

International Association for Mathematical Geology
    (IAMG)
Department of Geology
Syracuse University
Syracuse, New York 13210

International Association of Sedimentologists (IAS)
Department of Geology
Parks Road
Oxford OX1 3PR England

International Cartographic Association (ICA)
Flottbrovagen 16
S-112 64 Stockholm, Sweden

International Commission on Large Dams
22 et 30 Avenue de Wagram
75008 Paris, France

International Council of Scientific Unions (ICSU)
51 Boulevard de Montmorency
75016 Paris, France

International Development Association (IDA)
1818 H Street, N.W.
Washington, D. C. 20433

International Federation of Societies of Economic
    Geologists
Comandante Fortea 7-3
Madrid 8, Spain   *continued*

315

TABLE 3. *(continued)*

International Geographical Union (IGU) Department of Natural Resources UN University Toho Seime Building 15-1, Shibuya 2-chome Shibuya-ku Tokyo 150, Japan	International Seismological Center 5 rue Rene Descartes 67 Strasbourg, France
International Geological Congress (IGC) Maison de la Geológie 77-79 rue Claude-bernard F 75005 Paris, France	International Society for Photogrammetry (ISP) U.S. Geological Survey 917 National Center Reston, Virginia 22092
International Geologica Correlation Program (IGCP) UNESCO 7, Place de Fontenoy 75700 Paris, France	International Society for Rock Mechanics (ISRM) Laboratorio Nacional de Engenharia Civil 101 Av. Do Brasil 1799 Lisboa CEDEX, Brazil
International Glaciological Society Lensfield Road Cambridge CB2 1ER England	International Union of Geodesy and Geophysics (IUGG) Observatoire Royal de Belgique 1180 Bruxelles, Belgium
International Hydrographic Bureau (IHB) 7 Avenue President J.F. Kennedy B.P. 345 Monte Carlo, Monaco	International Union of Geological Sciences (IUGS) Geological Survey of Canada 601 Booth Street Ottawa, Ontario K1A 0E8 Canada
International Hydrological Programme UNESCO, Division of Water Sciences 7 Place de Fontenoy 75700 Paris, France	Seismological Society of the South-West Pacific (SSSWP) Seismological Observatory P.O. Box 8005 Wellington, New Zealand
International Institute for Aerial Survey and Earth Sciences (ITC) 144 Boulevard 1945 P.O Box 6 Enschede, The Netherlands	Smithsonian Science Information Exchange, Inc. 1730 M Street, N.W., Room 300 Washington, D.C. 20036
International Institute for Applied Systems Analysis Schloss Laxenburg 2361 Laxenburg, Austria	Society for Geology Applied to Mineral Deposits Bureau de Recherche Geologiques et Minieres B.P. 6009 45018 Orleans CEDEX, France

Source: After Bergquist et al. (1981).

Conclusion

National geological surveys and professional geologists' associations (q.v. in Vol. XIV) provide a wide range of services to the geoscience fraternity and to the public sector. Local geological societies compile geological data for geographic areas within their spheres of interest. Site-specific information, essential to many geotechnicians, is, however, commonly retained in proprietary records of consulting engineering firms and becomes available only with difficulty in the public domain. The development of World Data Centers and major international geoscience organizations has encouraged data-sharing schemes, but the greatest advances in the dissemination of geoscience information have come hand-in-hand with computerized databanks. Bibliographic data bases feature online computer searching and are also used in the production of abstracting and indexing journals (Leigh, 1981). Major online data bases of interest to geologists are GeoRef (American Geological Institute), GeoArchive (Geosystems, London), and the earth sciences subfile of PASCAL-GEODE (Centre de Documentation Scientifique et Technique, CNRS, Paris). The exchange of bibliographic information between GeoRef and PASCAL-GEODE data bases, initiated in 1981, produced a joint bilingual global data base that greatly expanded thesauri and indexing capabilities (Rassam and Gravesteijn, 1982). Leigh (1981) reports that over 50 bibliographic data bases in science and technology are accessible through the Science Reference Library, part of the British Library Reference Division. North American geological information systems are, according to Bie and Gabert (1981), only in an early stage of development but show promise as scientific tools and as profitable economic investments. Although in the past large amounts of data have been transferred from original paper records, future trends point to marked increases in the use of computerized data bases to storm and retrieve geological data. Traditional information centers probably will continue to provide practical information in the form of paper copy, but many professional organizations will commit themselves to computerized data bases, to the advantage of applied geologists.

CHARLES W. FINKL, JNR.

References

Bergquist, W. E.; Tinsley, E. J.; Yordy, L.; and Miller, R. L., 1981. Worldwide directory of national earth-science agencies and related international organizations, *U.S. Geol. Survey Circ. 834,* 87p.

Bie, S. W., and Gabert, G., 1981. Review of North American geological information systems, *Geol. Soc. London Jour.* **138,** 629-630.

BYGGINFO Staff, 1981. *World Geo Guide.* Stockholm, 280p.

Diment, J. A., 1979. Geological map acquisitions: a guide to the literature, *Geoscience Information Soc. Proc.* **9,** 111-144.

Dowden, A. M.; Goodman, R. L.; and Howell, G. D.; 1979. Book publishing in the geosciences: problems and prospects, *Geoscience Information Soc. Proc.* **9,** 21-31.

Dvorzak, M., 1980. Collection development in geoscience libraries, *Geoscience Information Soc. Proc.* **10,** 3-16.

Geological Society of London, 1978. *List of Serial Publications Held in the Library.* London, 172p.

Geologist's Year Book, 1977. Dorset, England: Dolphin Press, 299p.

Geotimes, 1982. Directory of Societies in Earth Science, August, 19-28.

Hallam, A. (ed.), 1977. *An Encyclopedia of Geology.* Oxford: Elsevier-Phaidon, 320p.

Harvey, A. P., and Diment, J. A. (eds.), 1979. *Geoscience Information.* Heathfield, England: Broad Oak Press, 287p.

Haylock, J. R. (comp.), 1973. *Library Guide to Geology—Geology: A Guide to Publications and Sources of Information.* Sunderland, England: Sunderland Polytechnic Library, 18p.

Kaplan, S. R., 1965. *A Guide to Information Sources in Mining, Minerals and Geosciences.* New York: Interscience, 599p.

Lander, J. F., 1979. The World Data Center system, *Episodes* **2,** 26-28.

Leigh, B., 1981. Online searching of bibliographic geological databases and their use at the Science Reference Library, *Geol. Soc. London Jour.* **138,** 589-597.

Mason, B., 1953. *The Literature of Geology.* Ann Arbor, Mich.: Edwards Brothers, 155p.

Morehead, J., 1978. *Introduction to United States Public Documents.* Littleton, Colo. Libraries Unlimited, 276p.

Pearl, R. M., 1951. *Guide to Geological Literature.* New York: McGraw-Hill, 239p.

Pestana, H. R., 1972. *Bibliography of Congressional Geology.* New York: Hafner, 285p.

Poss, J. R., 1975. *Stones of Destiny.* Houghton, Mich.: Michigan Technological University, 253p.

Pruett, N. J., 1979. Collection development in a geology-geophysics research collection, *Geoscience Information Soc. Proc.* **9,** 145-152.

Rassam, G. N., and Gravesteijn, J., 1982. Cross-database, cross-national geologic indexing: problems and solutions, *Geology* **10** (11), 600-603.

Shapley, A. H., and Hart, P. J., 1982. World Data Centers *Am. Geophys. Union Trans.* **63** (30), 585.

Smith, C. H., 1974. Geological surveys in the public service, *U.S. Geol. Survey Prof. Paper 921,* 2-6.

Ward, D. C., 1967. Geologic reference sources, *Colorado Univ. Studies Ser. Earth Sci.* **5** (12), 1-144.

Ward, D. C., 1979. State-of-the-art in geoscience information—USA in A. P. Harvey and J. A. Diment (eds.), *Geoscience Information.* Heathfield, England: Broad Oak Press, 1-8.

Ward, D. C., and Wheeler, M. W., 1972. *Geological Reference Sources.* Metuchen N.J.: Scarecrow Press, 453p.

Wilson, L., 1975. United States—General, in R. W. Fairbridge (ed.), *Encyclopedia of World Regional Geology, Part 1: Western Hemisphere (Including Antarctica and Australia.)* Stroudsburg, Pa.: Hutchinson Ross, 502-513.

Wood, D. N. (ed.), 1973. *The Use of Earth Sciences Literature.* London: Butterworths, 459p.

Yannarella, P. A., and Aluri, R., 1976. *U.S. Government Scientific and Technical Periodicals.* Metuchen, N.J.: Scarecrow Press, 271p.

Cross-references: *Computerized Resources Information Bank; Earth Science, Information and Sources; Geological Communication; Geological Information, Marketing. Vol. XIV: Association, Institutes, and Publications; Geological Cataloging; Geological Methodology; Geology, Philosophy; Geology, Scope and Classification.*

INFORMATION AND DOCUMENTATION—See COMPUTERIZED RESOURCES INFORMATION BANK; EARTH SCIENCE, INFORMATION AND SOURCES; GEOLOGICAL INFORMATION, MARKETING; WELL DATA SYSTEMS. Vol. XIV: GEOLOGICAL SURVEYS, STATE AND FEDERAL; MAP AND CHART DEPOSITORIES; REMOTE SENSING AND PHOTOGRAMMETRY, SOCIETIES AND PERIODICALS.

INFORMATION RETRIEVAL—See COMPUTERIZED RESOURCES INFORMATION BANK; WELL DATA SYSTEMS.

INSTRUMENTATION, GROUND CONTROL—See ROCK STRUCTURE MONITORING.

INTERNATIONAL GEOCHRONOLOGICAL TIME SCALE

In 1972 the International Union of Geological Sciences approved the preparation of an international geochronological time scale at the meeting of the International Geological Congress in Montreal. The scale was to include all of geological time and provide the isotopic ages for the time-stratigraphic units in the geological column. In 1976; a symposium was organized in conjunction with the 25th International Geological Congress, held in Sydney, Australia, to develop the international time scale.

The symposium was established by an organizing committee consisting of specialists throughout the world in the field of stratigraphy and isotopic age dating. The first of four sessions was an introductory one, covering the methods of isotopic dating, geochronological scales in general, magnetic polarity time scales, biochronology, and stratotypes and problems of correlation between stratotypes and dated sections. The second session, organized by the Subcommission on Geochronology of the Commission on Stratigraphy, International Union of Geological Sciences, dealt with the physical time scale and related geochronological problems. The subject of decay-constant usage and computational conventions was discussed because of the importance of establishing precise and accurate values of decay constants for naturally occurring radioactive isotopes used in determining the ages of rocks. The third and fourth sessions considered age-dating problems and isotopic dates for the various geological systems and their boundaries.

Many geochronological scales based on *radiometric dating methods* have been prepared by scientific organizations and individuals since 1907, when Professor B. B. Boltwood, a radiochemist at Yale University, published a list of geological ages based on radioactivity (Newman, 1975).

Although Boltwood's ages have since been revised, they did show that geological time would be measured in terms of thousands of millions of years. Some of these time scales are discussed in Harland et al. (1964), Berggren (1972), Holmes (1960), Kulp (1961), Lambert (1971), Harland and Francis (1971), and Van Eysinga (1972).

Rapid development took place in isotopic dating in the 1950s and early 1960s; since then, only minor changes have been made (see *Geochronology*). Because of the reliability of the analytical methods and the data obtained, it is doubtful that any future significant age changes will be made in the major divisions of the geochronological time scale. Some minor changes will undoubtedly arise in smaller divisions of geological time, especially within the Tertiary and Quaternary periods.

Geological Time Divisions

Geological time is divided into two *Eons*. The *Precambrian Eon* extends from the beginning of the Earth to about 570 million years ago. Estimates place the birth of the Earth at about 4.5 billion years ago. The *Phanerozoic Eon* includes all of the geological time since the end of the *Precambrian Eon.*

Geological time is further divided into Eras, which are broad divisions based on the general character of life that existed during those times, and Periods, which are shorter spans of time based in part on evidence of major disturbances in the Earth's crust. Periods are the fundamental units of the standard geological time scale.

On the time scale of the Phanerozoic Eon, published by the Geological Society of London in 1964, the geological Eras, from youngest to oldest, are: the Cenozoic (duration, 65 million years), the Mesozoic (duration, 160 million years), and the Paleozoic (duration, 345 million years). Table 1 lists the major divisions of the Phanerozoic Eon.

GEORGE V. COHEE

References

Berggren, W. A., 1972. A Cenozoic time-scale—some implications for regional geology and paleobiogeography, *Lethaia* 5 (2), 195-215.

Harland, W. B.; Smith, A. G.; and Wilcock, B. (eds.), 1964. The Phanerozoic time-scale; a symposium, *Geol. Soc. London Quart. Jour., Supp.,* **120s,** 458p.

TABLE 1. Major Divisions of the Phanerozoic Eon

Era	Period	Age Estimates of Boundaries (in millions of years)
Cenozoic	Quaternary	Present to 2
	Tertiary	2-65
Mesozoic	Cretaceous	65-136
	Jurassic	136-195
	Triassic	195-225
Paleozoic	Permian	225-280
	Pennsylvanian	280-320
	Mississippian	320-345
	Devonian	345-395
	Silurian	395-440
	Ordovician	440-500
	Cambrian	500-570

Harland, W. B., and Francis, E. H. (eds.), 1971. The Phanerozoic time scale: A supplement, *Geol. Soc. London Spec. Pub. 5,* 356p.

Holmes, Arthur, 1960. A revised geological time-scale, *Edinburgh Geol. Soc. Trans.* **17,** Pt. 3, 183-216.

Kulp, J. L., 1961. Geologic time scale, *Science* **133** (3459), 1105-1114.

Lambert, R. St. J., 1971. The pre-Pleistocene Phanerozoic time-scale—a review, in W. B. Harland and E. H. Francis (eds.), *The Phanerozoic Time Scale: A Supplement.* Geol. Soc. London Spec. Pub. 5, 9-31.

Newman, W. L., 1975. *Geologic Time.* Washington, D. C.: U. S. Geological Survey, 20p.

Van Eysinga, F. W. B., 1972. *Geological Time Scale,* 2nd ed. New York: American Elsevier, 1p.

Cross References: *Geochronology.* Vol. IVA: *Geochronometry; Radioactive Isotopes.* Vol. VI: *Radioactivity in Sediments.* Vol. XII, Pt. I: *Radioisotopes* Vol. XIV: *Lichenometry; Tephrochronology.*

ISOTOPE GEOLOGY—See GEOCHRONOLOGY

L

LAND CLASSIFICATION—See TERRAIN EVALUATION, MILITARY PURPOSES. Vol. XIV: TERRAIN EVALUATION SYSTEMS.

LAND SUBSIDENCE—See Vol. XIV: MINE SUBSIDENCE CONTROL; MINING PREPLANNING.

LATERITE, ENGINEERING GEOLOGY

Laterite, an irreversibly indurated material, should not be confused with *lateritic soils,* which harden slightly on drying but soften when wetted (Finkl and Fairbridge, 1978; Netterberg, 1975). Other misunderstandings concern laterite quarries that display a cap or face of indurated rock. Most often, the materials used from the pit were not indurated, but easily excavated soils. Testing and comparison of used material and remaining quarry rock is sometimes necessary to uncover the discrepancy. A more comprehensive discussion of laterites that focuses on the engineering properties of these materials may be found in Persons (1970) and De Graft-Johnson (1975).

Uses of Laterite

Building Material. When used as a building material, laterite is most desirable in the form of well-joined, small, globular cuirasses and is most readily quarried in convenient predetermined shapes. Laterite as a building stone is limited to facings, small gravity structures such as retaining walls (see *Reinforced Earth*) or simple dams (see *Dams, Engineering Geology*), and single course block construction such as headwalls, culverts, canals, or stone-surfaced courtyards.

Road and Airfield Subgrade. Roads and airfields, because they are extensive, rather than intensive, usually require both cut and fill. Road and airfield engineers strive to effect similiarity of performance within cut and fill sections of construction. Thus they determine the ultimate behavior of cut subgrade and design fill subgrades to behave similarly. In rain forests, cuts may range from mature laterite to partially lateritized mate-

rials to materials that would never be subjected to lateritization. The variety of eventual performances of these subgrades influences, in large measure, the determination of the required performance of filled areas (De Graft-Johnson, 1975). The materials that underlie potential cuts directly affect the ultimate performance of a cut surface. One must understand the consolidation characteristics of the embankment in order to select carefully the grade line of embankment, which, subsequent to settlement, must coincide with adjacent cut surfaces (Persons, 1970).

Ground-moisture characteristics of natural subgrade materials are highly significant in determining soil usage in rain forests. The most significant cause of subgrade failure in the rain forest is inordinate and undesirable moisture in the subgrade materials. The forest road or airfield should always be designed to assure a totally and continuously well-drained subgrade. In assuring proper drainage, not only is the sizing of drainage structures that underlie fill areas important, but it is also necessary that forest vegetation be kept clear of the immediate environs of the construction. If possible, this area should be sloped away from the construction to ensure a continuously drained subgrade. Sometimes these provisions and attendant costs are inadvertently omitted in preliminary cost studies or in final design plans. Forest clearing and subgrade drainage are more essential than proper surfacing.

For subgrade fill, the engineer should select lateritized materials that will perform adequately under different soil moisture conditions during the life of the project. Some lateritic soils are moisture-sensitive and lose strength with increasing moisture content. Thus the sensitivity (loss of strength through moisture increase) should be the measure for determining the acceptability of the soil for the construction (see *Soil Mechanics*). Sensitivity of embankment materials can be estimated by preparing compacted samples at the lowest density appropriate to the construction and subjecting them to strength tests at varying degrees of moisture up to saturation. The minimum strength of the soil is appropriate for design use. The engineer should beware of a soil that loses more than 25 percent of its strength upon soaking (Persons, 1970).

Because laterite per se is normally too scarce

for use in subgrade fills, most fills are constructed from lateritic soils (Netterberg, 1975). The soil engineer should note moisture contents of embankment materials during dry and wet seasons and relate them to moisture contents necessary for minimum compaction of the design subgrade (see Vol. XIV: *Engineering Soil Science*). If an embankment material has a moisture content in excess of that required for a desired degree of compaction, it must be sufficiently aerated so that compaction can be achieved (Peck et al., 1974).

In summary, most soil engineers will normally deal with highly sensitive lateritic soils rather than insensitive soil or laterite. These materials naturally possess at moisture contents that are close to, if not in excess of, the moisture contents at which desirable degrees of compaction can be achieved. Because of high ambient moisture contents and frequent forest rains, drying is often impractical. Subgrades must be designed within these parameters for construction to be completed in a reasonable time.

Just as soil materials for subgrade must be realistically used to obtain optimum strength and behavior, so must base and surface materials be used to negate subgrade deficiencies. Road or runway failures in high-moisture areas are usually the result of subgrade failure, which principally results from moisture infiltration into the subgrade soils from below or from above (Persons, 1970).

The installation of drainage blankets below subgrade fills is necessary if a positive head is expected to force water into the fill from below or if subsurface drainage within the fill is impeded without such a blanket. The wearing surface must receive wheel loads without immediate failure. Basal and wearing surfaces must, however, remain impermeable, or else moisture from above will infiltrate the subgrade, causing it to fail. This infiltration will in turn weaken surface support, causing repetitive failures. A loss of subgrade strength and an increase in compressibility due to an increase in soil moisture in lateritic soils signals the engineer to use a slightly thicker impermeable base and wearing surface than normally employed with an insensitive soil. This increased factor of safety translates to an increased thickness of the impermeable base and wearing surface.

In determining material requirements and evaluating shrinkage through crushing and compaction, the engineer should study the materials in test sections from laterite at potential borrow pits. Such tests should be preceded by laboratory tests to obtain data on *crushing characteristics* and optimum moisture and strength values (Ingles and Metcalf, 1973). As a general guide, thickness requirements (see Fig. 1) for base and wearing surfaces should not be reduced for base course CBR (California bearing ratio) values of less than 80 (see "Testing Placed Materials" in this entry).

For values in excess of 80, the total required thickness may be reduced by 27 percent.

Prepared subgrade surfaces should be sloped and crowned for proper surface drainage during preparation of base material for compaction. Because laterite dehydrates when exposed, aeration of the borrow will reduce the moisture content for proper compaction. Watering in arid climates is more difficult due to rapid evaporation of surface moisture. Careful scheduling of watering, with immediate compaction, assures densification and sealing before evaporation.

For an optimum thickness of 15 cm, tandem sheepsfoot rollers weighing 8 tons usually will achieve a density of 95 percent of the Modified AASHO Test with four to six passes. An 8-to 10-ton steel-wheeled vibratory roller is effective with about the same effort. *Pellet laterite* is most easily compacted with smooth-wheeled or pneumatic tire rollers in the 5-to 10-ton range. For base courses requiring multiple lifts of compaction, the lower compacted surface is roughened and moistened before further uncompacted material is applied. This procedure forces the base to curve as a single unit and not as a stack of separate plates. Furthermore, a uniformly compacted and properly joined base will properly inhibit moisture infiltration.

A laterite-surfaced construction, when properly maintained, will perform adequately in desert, grassland, or rain forest. Compacted laterite is sufficiently impermeable to resist surface water, and thus no surface treatment is necessary. Under conditions of wear that smooth a surface, wet laterite surfaces may be less skid-resistant than desired. Because laterite is more subject to surface wear than asphalt-sealed roads, laterite roads require regular maintenance. This need is offset in primitive locations by casual maintenance practices attendant to paved road surfaces. There is a fallacious belief that once a surface has been blacktopped it will forever be maintenance-free. Surface failures in blacktop are less expeditiously repaired than those in compacted-rock surfaces, because abraded blacktop's potholes can be avoided by traffic and are not appropriate reminders of need for maintenance, as are the abraded rock surfaces.

In remote locations in undeveloped regions and on lightly trafficked roads, minimum-thickness laterite wearing surfaces are common. A thickness of several centimeters is usually sufficient for roads. A corresponding minimum thickness is adequate for blacktopped surfaces of light-duty runways or hardstands. Surfaces of heavy-duty runways, on the other hand, should be thicker than in nonlaterite areas. This requirement assures an extra margin of surface strength and impermeability, attendant with unsophisticated maintenance practices and the critical nature of landings on heavy-duty run-

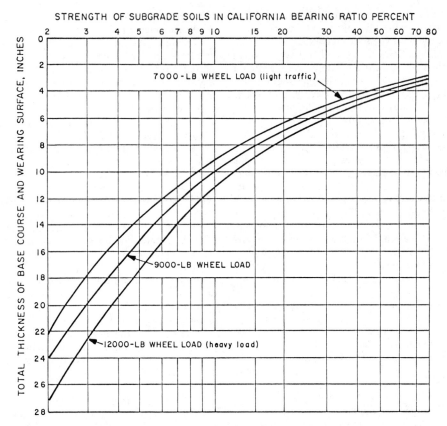

STRENGTH OF SUBGRADE SOILS IN CALIFORNIA BEARING RATIO PERCENT

FIGURE 1. Strength of subgrade (for total depth of wheel load influence) versus basecourse and wearing-surface thickness (after Persons, 1970, p. 44).

ways. The choice of wearing surfaces for heavy-duty use depends not only on the availability or the cost of importing materials but also on the ability of a surface to display symptoms that show need for maintenance and the ease with which maintenance can be achieved without disrupting operations. Generally, asphalt surfaces are more amenable to construction over subgrades of minimum soil strength in which failures are more likely. It is best to design adequately and protect subgrades against failure.

Dam Embankments. Dams present special problems when laterite is used as a construction material. Rockfill and rock-material dams are constructed at locations where suitable earth materials are scarce. Rocks suitable for use in dam embankments include sandstones, shales, decomposed granite, gneisses, schists, ore tailings, and some slags. When compacted, these materials are more permeable than silty clay or clay soil of low plasticity and cannot be used as impermeable blankets (Lee et al., 1968). Laterite, however, has recementing characteristics similar to those of coquina and is therefore useful as an impermeable surface (see *Dams, Engineering Geology*).

Laterite deposits are scarce, and the cost of moving and placing laterite material is great compared with other lateritic or nonlateritized soils locally available. Consideration of laterite as a select engineering material should focus on permeability characteristics in relation to other materials available for the construction. Before laterite is selected for embankment fill, the engineer must ascertain whether its strength and permeability parameters are reasonably uniform and whether the material can be placed, crushed, moisture-conditioned, and compacted economically. Because permeability tests cannot be conducted in the field, samples must be shipped to a laboratory.

Laterite materials have great compacted strength, so they are effective in improving the total strength of a dam embankment. Because of the different strength and permeability characteristics of soil materials that must be mixed with laterite fill, it is difficult to summarize what ratios of mixes or proper balances should be used. Laterite is most effective when it is crushed and compacted and used as core material, but it is also employed as *dam facing, riprap,* and *toe drain materials.*

In tropical climates dam sections are frequently

protected by grassed embankments, but laterite is used as a surface covering in arid areas, where grass is insufficient. It is important that the materials selected for facing do not harden in a short period because indurated surfaces inhibit the growth of surface covers (McFarlane, 1976). Grass can be effectively grown on earth materials where the ratio of silica to sesquioxides exceeds 2:5. These materials remain soft as long as the surface remains moist. Blanketing materials sometimes supply sesquioxides and induce hardening in underlying soils. Tests should, therefore, be performed on underlying soils to determine susceptibility to hardening.

Construction Materials

Borrow Development. Laterite borrow pits are developed much the same way as borrow for other materials used in construction. The surface is first cleared of vegetation and organic soils. Mineral soils overlying laterite rock are removed to expose a fresh working surface. Sidehill borrow operations, beginning at lower extremities of the deposit and working uphill, provide a well-drained work area and afford an increasing thickness of rock exposed on the cut face. Generally, laterite deposits are thicker at upper extremities than at the edges (Goudie, 1973; McFarlane, 1976).

Although it is possible to dislodge and secure laterite with quarry excavators, the cost of transporting specialized equipment to remote locations and maintaining it there is usually prohibitive. In the nonporous laterites, skillful blasting can produce material for direct loading. Once a vertical cut face about 2 m high has been developed in the borrow area, experimental blasting can establish effective means for securing dislodged materials.

Often, a front-end loader or a power shovel can be used to dislodge and load laterite. When such equipment is available, its use is preferred to blasting. The method selected for dislodging laterite rock should be expeditious and effective in reducing the size of dislodged pieces for transport and placement on the construction surface without additional breaking.

Induration in Air. Some lateritic soils become indurated when subjected to periods of desiccation. The exposed soil surface should be cleared and properly sculptured immediately to assure good drainage throughout the wet season.

Sometimes it is more effective to dislodge partially indurated materials. When placed in a loose pile, aerated, and well drained, such materials become indurated in one or two seasons. Materials that are not irreversibly indurated can be stockpiled or transported to other locations and allowed to harden during several seasons of desiccation.

Rarely will the cost of laterite be so expensive to render artificial induration economical. Many lateritic soils that have not been indurated will, under the proper conditions, become indurated during several seasons of drought (McFarlane, 1976). Several methods of artificial induration have been suggested for soils having the potential to become laterite. The major objective of artificial induration is to reduce the relative concentrations of silica, which inhibit lateritization, and to increase the concentrations of iron and aluminum sesquioxides. This may be accomplished by inducing an alkaline environment so that the silica may readily be removed by solution. The remaining lateritic soils are then more receptive to induration through drying.

Crushing. To produce a strong and impermeable covering blanket, it is necessary to crush the laterite prior to compaction. In the normal course of engineered construction, it is presumed that a surfacing material will be transported from the borrow area to the work site in a workable form, that is, when amenable to compaction. Hard-rock base course material, for example, cannot be compacted on a subgrade with normal crushing equipment. Even one attempt to crush such materials can damage the subgrade by forcing angular cobbles and boulders into the base. Crushed laterite chunks are sometimes overridden by heavy crushing equipment to reduce particle sizes for compaction. Sufficient fines are also needed to render an appropriate grain-size distribution. Specifications currently in vogue in parts of Brazil, the Ivory Coast, Gambia, Liberia, Senegal, and Nigeria stipulate that, prior to compaction, the base course laterite material for the road or airfield must be reduced to distributions within the limits shown on the grain-size distribution chart in Figure 2. This range shows a bulge between 1 mm and 0.25 mm, indicating an absence of this fraction. The deficiency of sand sizes indicates a need for adding moderate amounts of sand-sized material to the mix. Materials having a grain-size distribution within the limits cited are most amenable to compaction. CBR strength characteristics required for a particular subgrade can then be readily achieved. Laboratory studies as well as studies performed for the particular construction will indicate grain-size distribution requirements of compacted base course materials. Furthermore, grain-size distribution requirements and specifications can be determined by studying successful completed construction projects for which similar grain-size distributions were used.

Compaction. Compaction should be achieved with minimum effort, in the minimum time, and without damage to the soils that support the materials being compacted. These requirements are directed to the selection of compaction equipment with particular reference to its weight, use, and location. In West Africa, for example, the smooth-wheeled roller is the most popular machine

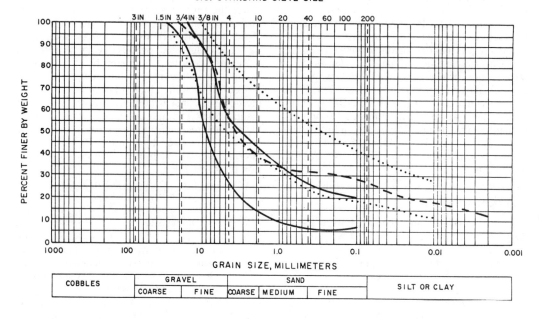

FIGURE 2. Grain-size distribution chart for Ivory Coast and Senegalese laterite (after Persons, 1970, p. 60).

for compacting laterite and volcanic material and for engineering construction. If a 15-cm thickness of laterite is required, this machine has no superior. Because it produces a smooth upper surface in the compacted medium, it is not a desirable piece of equipment if multiple compacted lifts are to be placed. The sheepsfoot or grill-type roller leaves a highly irregular surface on a compacted lift that is well suited to receive the next lift of material. Five to eight passes with a sheepsfoot roller is normally sufficient to compact a 20-to 25-cm thickness of uncompacted laterite or similar material to 90 percent of the maximum density achieved by the Modified AASHO method of compaction. The motor patrol is a useful machine for spreading and windrowing placed laterite. The scarifying attachments are particularly suited to roughening the surface at a compacted laterite lift that has previously been smoothed with a smooth-wheeled roller. Other scarifying devices, while effective, cannot be used with the care and precision gained with the motor patrol.

Testing Placed Materials. The compaction characteristics of laterite materials (both soil and rock) may be determined by a simple compaction test. The Modified AASHO Test (ASTME-1557-66T and AASHO T180-61-C) is appropriate for determining maximum density under the compactive effort that most closely resembles strenuous and diligent construction endeavors. Compaction testing indicates the optimum moisture content. For a uniform compaction effort, the resulting density increases as the moisture content is increased to

optimum, and thereafter decreases as the moisture content increases (see discussion in Peck et al., 1974, chap. 1).

The *strength* of a compacted material may be determined by the California bearing ratio (CBR) test, in which the crushed material is compacted at optimum moisture content (Lee et al., 1968; Ingles and Metcalf, 1973). Subsequent to the compaction, a surcharge weight is placed on the soil surface, and the mold in which the soil was compacted is soaked. Once the soaking process has been completed, the mold is placed in a testing machine, and a 19.35-cm^2 circular plunger is forced 2.5 mm into the soil. The resistance encountered for the penetration is recorded. The ratio of that resistance to the resistance of crushed rock to the same penetration is taken to represent the CBR value (Goodwin, 1965). A CBR value of 100 indicates excellent material with supporting properties as good as those of crushed rock, while lower CBR values indicate materials of lesser strength and more deflective characteristics. It is from data such as these that the design relationship for strength of base course materials are derived.

The *modulus of subgrade reaction* is the slope of the stress-strain curve (within the elastic range) for a soil or rocky material. It is normally expressed in pounds per square inch and is useful in assaying the reaction of subgrade to wheel load, providing the load produces a reaction within the elastic range of the subgrade material.

The *field moisture content* of construction materials must be obtained in a timely manner. The time-honored method of determining the moisture content is by oven-drying overnight. It is recommended that some other method be found for immediately determining moisture content at the time natural or borrow soils are densified.

In-place density of soil and rock materials compacted in engineering construction must be ascertained by field density tests. At remote locations and in adverse circumstances, one of the most suitable tests for ascertaining density is ASTM Test D 1556, the standard test for determining the density of soil in place by the sand-cone method. Density of soil in place may also be determined by ASTM Method D-2167, a test of soil in place using the rubber balloon method. A template is placed over a prepared surface and a hole is dug in the material to a depth of approximately 15 to 20 cm. Over this is placed a rubber balloon, together with a calibrated vessel containing water (American Society for Testing and Materials, 1970).

The use of a nuclear moisture-density meter can be justified for extensive fill operations. Because of the uncertainty of keeping such a piece of equipment operating, however, it cannot be recommended for use in remote locations where repair and replacement would be difficult.

Maintaining Laterite

Surfacing. As laterite constructions are normally confined to remote areas lacking sophisticated surveillance and maintenance, the use of all-weather surfacing gives a sense of longevity and permanence that can be not only misleading but also dangerous. In the case of extensive laterite construction such as a rainforest road, those who use it and those who maintain it too often believe it has been constructed to last forever—maintenance-free. Surface failures, when they occur, are often ignored and are costly to repair; sometimes irreparable damage is done to the base and subgrade by surface water and/or applied loads.

Road Surveillance. A designer should decide to blacktop a road in a rain forest only after careful study, with knowledge of the sometimes dangerous sense of security that a blacktop lends. The first requirement of a properly designed and constructed rain forest road is that it function under traffic load without impermeable surface failure or moisture penetrating to its subgrade. When the blacktop fails and ulcers form through its base into the subgrade, traffic is directed around the ulcer, often striking its edges, toward the shoulders. This traffic diversion causes other failures because loads are introduced at locations that were not intended to receive them. The pattern is progressive as the entire surface becomes pockmarked with ulcers and the road edges abrade toward the center. If the blacktop has been omitted from the construction, persons using the road and those exercising surveillance do not expect quite as much from the surface and regard the construction more carefully. Potholes that develop in a laterite wearing surface are usually attended to more readily than potholes that occur in blacktop.

By the time failures become obvious, the subgrade has been seriously damaged and requires extensive repair. Bleeding of the blacktop surface; appearance of longitudinal, transverse, or diagonal series of spider-web cracks in the road surface or base course; localized depressions in the wearing surface; unusual tire noise or corduroy surface characteristics; poor surface drainage or the appearance of water through surface cracks from a pumping subgrade—all these are early symptoms of failure and require immediate correction.

Surveillance inspection on a rain forest road should be performed every week or 10 days, if possible. A narrative record of the reconnaissance should be submitted as part of the surveillance report as well as detailed accounts of locations and observations of symptoms of failures, along with recommendations of the maximum time allowable for repairs before serious damage to the base and subgrade occurs. Deficiencies noted in the surveillance should serve as the impetus for preventive measures to be taken to avoid reoccurrence.

Traffic Control. Design criteria prove meaningless if the authority to implement load controls on vehicular traffic cannot be delegated and exercised. Load control can usually be effective where owners are willing to include traffic barriers. Careful attention by the designer can result in selection of curved radii, which limit the length of vehicles using the road. Bridge superstructures can also limit the external dimensions of vehicles. While these measures may seem archaic, their effectiveness should not be discounted. If the designer recognizes the value of built-in load-carrying restrictions, appropriate measures may be executed at some future date should the road subgrade, base course, and wearing surface prove capable of sustaining heavier loads. This foolproof method of protection appears reasonable to assure the design life of the road. The design engineer is strongly advised to build safeguards into any rain forest road.

In upgrading a road, the designer should become acquainted with all types and patterns of vehicular traffic using the road. Special attention should be given to other roads of higher order in the region so that the designer may visualize the type of traffic that the current road may be expected to receive. Other considerations should include projections of load and vehicular characteristics during the design life of the road.

Treatment. The well-designed, constructed, and disciplined road or airfield may be expected to resist the abrasion of traffic and the effect of weather during its design life. It must be anticipated that the surface and edges will be scarred by physical and mechanical weathering. Drainage may possibly become impaired, causing other problems. Treatment must be anticipated in the maintenance program and carried out before damage occurs. Drainage structures and courses must be inspected continuously during the rainy season and, where stoppages are noted, the offending materials removed. Bulldozers can remove large tree trunks and boulders provided that maintenance crews use extreme caution not to inflict further damage on the road. Where construction traverses terrain containing volcanic ash and lateritic soils, slopes are frequently designed to be vertical. Vegetation frequently grows above the slopes and hangs down over the cut face. The vegetation overloads the unprotected face and provides root paths for surface water to percolate into the soil, causing face sloughing. Vegetation must be removed from the top of the slope continuously.

The treatment of wearing surfaces should be planned and executed on schedule, regardless of the appearance of the surface to be treated. For compacted surfaces of laterite, the surface may be rejuvenated by shallow scarifying (2 to 5 cm) topped by a well-crushed blanket of laterite. The scarified material and the newly placed blanket should be thoroughly compacted with a smooth-wheeled roller, with care taken to assure that the design crown and curve superelevation are maintained.

Treated laterite surfaces should be retreated at least every two years. If the surface has been abraded, it should be returned to original grade by the introduction of additional material. Single shots of cutback asphalt may be employed repeatedly if the road must be kept open during treatments. Such a mixture cures readily and quickly during dry weather but is of little value if employed during rainy seasons.

Surface drainage is important and must be maintained through crowning, ditching, and culverting. Impermeable barriers such as the wearing surface must prevent downward percolation of water. Where failure occurs, a waterlogged soil must be suspect. Subsurface water percolating upward through a subgrade frequently can be traced to a geological or topographic feature that is directing the water into the earth mass through hydrostatic pressure. Subsurface drainage through interception ditches or French-type underdrains is the simplest method of diverting subsurface water. If the offending aquifer is found to be fractured supporting rock, drainage can sometimes be effected and pressure released by locating and opening a drainage path at an elevation below (downslope from) the saturated soil mass. If the source of subsurface flow is located upslope of the road, suitable drainage can

be installed upslope and the offending soil mass stabilized.

BENJAMIN S. PERSONS

References

American Society for Testing and Materials, 1970. Special procedures for testing soil and rock for engineering purposes, *Am. Soc. Testing and Materials Spec. Tech. Pub. 479.*

De Graft-Johnson, J. W. S., 1975. Laterite soils in road construction, *Soil Mechanics and Found. Engineering, Reg. Conference for Africa, Proc.* 1(6), 89-98.

Finkl, C. W., and Fairbridge, R. W., 1978. Duricrust, in R. W. Fairbridge and J. Bourgeois, (eds.), *Encyclopedia of Sedimentology.* Stroudsburg, Pa.: Dowden, Hutchinson & Ross, 274.

Goodwin, W. A., 1965. Bearing capacity, In C. A. Black (ed.), *Methods of Soil Analysis.* Madison, Wis.: American Society of Agronomy, 485-498.

Goudie, A., 1973. *Duricrusts in Tropical and Subtropical Landscapes.* Oxford: Clarendon Press, 174p.

Ingles, O. G., and Metcalf, J. B., 1973. *Soil Stabilization.* New York: Wiley, 374p.

Lee, I. K.; Lawson, J. D.; and Donald, I. B., 1968. Flow of water in saturated soil and rockfill, In I. K. Lee (ed.), *Soil Mechanics.* Sydney: Butterworths, 82-194.

McFarlane, M. J., 1976. *Laterite and Landscape.* New York: Academic Press, 151p.

Netterberg, F. (ed.), 1975. Pedogenic materials, *Soil Mechanics and Found. Engineering, Reg. Conference for Africa, Proc.* 1(6), 291-295.

Peck, R. B.; Hanson, W. E.; and Thornburn, T. H., 1974. *Foundation Engineering.* New York: Wiley, 514p.

Persons, B. S., 1970. *Laterite: Genesis, Location, Use.* New York: Plenum Press, 103p.

Cross-references: *Arid Lands, Engineering Geology; Caliche, Engineering Geology; Consolidation, Soil; Geotechnical Engineering; Rocks, Engineering Properties; Soil Mechanics. Vol. VI: Duricrust. Vol XII, Pt. 2: Duricrust, Ferricrete.* XIV: *Engineering Geochemistry; Rock Weathering, Engineering Classification.*

LIME STABILIZATION

Soil ("earth") is one of the oldest and most available construction materials (see Vol. XII, Pt. 1: *Soil*). Engineering properties of different soil types vary greatly, and some soils may be unsuitable or of marginal quality as borrow or foundation materials. Highly plastic clays (fat clays, or "CH," according to the Unified Soil Classification, q.v.) are one of the most troublesome soil types. Fortunately, engineering properties of many soils, including some fat clays, can be improved by mechanical or chemical treatments.

Numerous forms of such treatments have been proposed and successfully used. Some of the treatments are relatively simple, for example, wetting or drying of borrow materials prior to or after excavation or during their placement in order to obtain *optimum moisture content* for compaction, screening of undesirable *oversize* fractions, blending (mixing) of different soil types, and so on. Among other sophisticated treatments are lime stabilization, electrostabilization (see Vol. XII, Pt. 1: *Soil Stabilization*), and grouting (q.v.) or mixing of soils with different chemicals and other materials such as fly ash and cement.

Background

Lime stabilization is one of the most widespread and relatively inexpensive forms of chemical treatments. This process is not a new approach. It was well known and used in road construction in Roman and, to some extent, in pre-Roman civilizations. Presently, lime stabilization is widely used in the United States in the construction of roads and flight strips. It also has been adopted for hydraulic structures, foundation improvements, and so forth. Several pre-World War II lime-treatment efforts were rather ineffective, primarily because of ignorance about geological-geochemical factors. These failures somewhat delayed the use of this potentially valuable technique in the United States. In postwar years, intense usage of lime treatment has, however, become increasingly popular.

Literature dealing with basic principles and case histories of lime stabilization is rather extensive but widely scattered. Numerous papers were published in different highway publications and by the U.S. Lime Institute. Probably one of the best summaries of data and essential literature was made by Herrin and Mitchell (1961).

Lime treatment is used for improving engineering properties of clays, including fat clay, as well as clayey-silty gravels or sands and other soil types. It involves mixing the original materials with a certain amount of lime. The exact optimum amount of lime required for the treatment varies in individual jobs and should be determined experimentally prior to the treatment. It varies from 1 to 10 percent, but usually is about 3 to 4 percent.

Both quicklime (CaO) and hydrated lime $(Ca(OH)_2)$ are used. The quicklime appears to be, at least in some cases, more desirable because of its high affinity for water (its *drying power*) and because it weighs relatively less and thus is cheaper to haul than hydrated lime. Both high-grade calcium and high-magnesium (dolomitic) lime are used for the treatment. The effect of magnesium-rich lime versus calcium-rich lime on end products is not yet clear.

Soil mixed with the lime and properly wetted, after a certain period of *curing* and *ripening,* is placed (as roadbed material, canal lining, foundation pad, etc.) and compacted. The time for curing

and the degree of compaction appear to be critical in determining the quality of the end products. The hardness of such material usually increases with time, which is an indication of post-placement cementation. In typical cases, lime treatment notably decreases soil plasticity, soil swelling by wetting, shrinkage by drying, and ability to slump, but increases soil workability, porosity-permeability, shear strength, penetration resistance, slope stability, and so on.

Theoretical Considerations

The basic theory of lime treatment is not fully understood. Many aspects of the efficiency of the method, mineralogy, geochemistry, and geological control are disputed or uncertain. Theoretically, the effects of lime treatment are attributed to four processes: (1) *Flocculation* (q.v. in Vol. XII, Pt. 1) of clay soil particles by excessive amounts of electrolytes, particularly Ca^{2+}. The process should decrease plasticity, swelling, and shrinkage of soils, but increase their permeability and grain size. (2) *Substitution* in clay particles of exchangeable sodium ions by more desirable calcium ions (see Vol. XII, Pt. 1: *Exchange Phenomena*). (3) *Cementation* of some soils with calcium carbonate developed by the following reactions:

$$Ca(OH)_2 + CO_2 = CaCO_3 + 2H_2O$$
$$\text{or} \quad Ca(OH)_2 + CO_2 = Ca(HCO_3)_2 + H_2O$$
$$\text{and} \quad Ca(HCO_3)_2 = CaCO_3 + H_2O + CO_2$$

The important component in these reactions is carbon dioxide, which can be available either from air or from water solution, including soil moisture. (4) *Cementation* of mineral grains by *pozzolanic reactions,* that is, by calcium, sodium, alumina, and other silicates developed through weathering of some clays in a highly alkaline media, created by calcium hydroxide. The existence of pozzolanic reactions is of particular importance for lime treatment of earth-lined canals. Deterioration of lime-treated soils due to leaching of carbonate cement by flowing water in such canals is more probable than leaching of carbonates in dry roadbeds. Pozzolanic cementation should not be affected by leaching and will provide a durable effect. The chemistry and final results of pozzolanic reactions are not fully understood. The development of some new aluminosilicates during lime treatment is, however, proved by direct x-ray and chemical analyses and some indirect observations.

Limitations

Each of the preceding processes will be effective only in a certain geological environment. Additional flocculation of clay particles will take place, for example, only in sediments relatively low in soluble salts (i.e., electrolytes). No calcium ion exchange will take place in highly calcareous sediments (e.g., in the "B" horizon in calcareous soils). Unweathered clays with relatively large amounts of unstable mineral grains are more susceptible to weathering than already weathered clays. Such weathering provides silica, alumina, and iron for the pozzolanic reaction. On the other hand, deposits rich in humus and/or soluble sulfate are well recognized as poor material for lime treatments. Relatively little studied is lime treatment of material that will be in constant contact with moving water (for example, canal lining).

NIKOLA P. PROKOPOVICH

Reference

Herrin, M., and Mitchell, H., 1961. Lime-Soil mixture, in *Lime Stabilization: Properties, Mix Design, Construc-tion, Practice and Performance.* Highway Research Board Bull. 304, 99-138.

Cross-references: *Alluvial Plains, Engineering Geology; Caliche, Engineering Geology; Clay, Engineering Geology; Clays, Strength of; Grout, Grouting; Soil Mechanics.* Vol. XII, Pt. 1: *Lime, Liming; Soil, Soil Mineralogy; Soil Conditioners; Soils, Nonagricultural Uses; Soil Stabilization; Soil Structure.* Vol. XIV: *Dispersive Clays; Engineering Soil Science.*

LITHOBIOLOGY—See Vol. XIV: GEOMICROBIOLOGY; LICHENOMETRY.

LOGS, LOGGING—See WELL DATA SYSTEMS. Vol. XIV: WELL LOGGING.

M

MAGNETIC SURVEYS— See Vol. XIV: AERIAL SURVEYS, GENERAL; VLF ELECTROMAGNETIC SURVEYING.

MAGNETIC SUSCEPTIBILITY, EARTH MATERIALS

Magnetic susceptibility is the ratio of induced magnetization to the strength of the magnetic field causing the magnetization. *Magnetic (susceptibility) anisotropy,* in minerals with low crystal symmetry or in rocks with planar or linear fabric, refers to magnetic susceptibility that is not perfectly parallel with the inducing magnetic field because it depends on direction and induced magnetization. *Ferromagnetism* is a type of magnetic order in which all magnetic atoms in a domain have their moments aligned in the same direction. *Ferrimagnetism,* another type of magnetic order, macroscopically resembles ferromagnetism. Magnetic ions at different crystal sites are opposed, that is, antiferromagnetically coupled. There is nevertheless a net magnetization because of inequality in the number or magnitude of atomic magnetic moments at the two sites. This type of magnetic order occurs in magnetite. *Paramagnetic* minerals— those having a small positive magnetic susceptibility, for example, olivine, pyroxene, or biotite— contain magnetic ions that tend to align along an applied magnetic field but do not have a spontaneous magnetic order. *Diamagnetic* materials have a small negative magnetic susceptibility and do not show paramagnetism or magnetic order. Typical diamagnetic minerals are quartz and feldspar. Coal is a diamagnetic rock.

In other words, the magnetic susceptibility, K, of a substance is the ratio of intensity of magnetization, I, to the magnetizing field, H. $K = I/H$, defined with respect to unit volume. The specific susceptibility, X, is defined with respect to unit mass.

Most minerals are paramagnetic or diamagnetic with positive or negative susceptibilities on the order of 10^{-6} cgs. (Lindsley et al., 1966). The susceptibility of such minerals is essentially independent of the applied field. On the other hand, for ferromagnetic materials such as iron or nickel, the intensity of magnetization reaches a maximum value within a finite strength of applied field and the susceptibility approaches or exceeds unity. Each substance has a characteristic temperature known as the *Curie point,* at and above which the magnetic susceptibility disappears.

Measurements of Magnetic Susceptibilities

Over the years many different experimental arrangements have been developed for the measurement of the susceptibilities of liquids and solids. Three main methods for the measurement of these susceptibilities need be considered. Provided that an adequate quantity, say, 10 to 15 cm^3, of the material under investigation is available, the *Gouy method* is preferred. This method permits absolute measurements of a high order of accuracy to be made with the ordinary equipment of laboratory quality, and permits comparison measurements to be made with ease. When only small quantities of a substance are available or measurements have to be made at elevated temperatures, the *Curie method* is recommended. It requires the use of a strong field that varies rapidly over a short distance. This method is also referred to as the *Faraday method.* Absolute measurements used to be impossible by this method. However, the incorporation of micro and electrodynamometer balances have made it possible to achieve great accuracies. *The Vibrating Sample Magnetometer* (VSM) permits a new approach to the measurement of susceptibilities. It incorporates the advantages of the previous methods in addition to versatility. It provides means to measure anisotropy as well as high and low temperature.

The Gouy Method. The Gouy method (Gouy, 1889) was first used by Pascal (1910). The material to be investigated must be in the form of a rod of uniform cross-section, or as a solution or very finely ground powder that can be placed inside a glass tube of uniform cross-section. The specimen or its container is suspended with its axis vertical from one arm of a sensitive balance, so that its lower end is near the mid-point of the field between two flat pole tips, while its upper end is in a region well outside the gap, as shown in Figures 1, 2, and 3. The lower end is thus in a strong uniform field, H, while the upper end is in a very much weaker field, H_0 (fringe field). According to Bates (1951), the magnetic force acting on an element of solid sample of length dx and volume dv is given by:

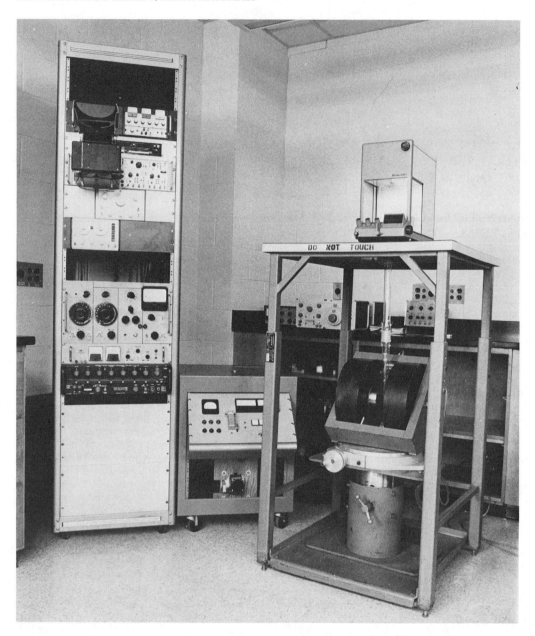

FIGURE 1. Test assembly using the Gouy or the Faraday method.

$$dF = \tfrac{1}{2}(k_2 - k_1) \frac{\partial}{\partial x} [H_x^2 + H_y^2 + H_z^2] \, dv \qquad (1)$$

where k_2 and k_1 are the respective susceptibilities of the sample and the surrounding media, usually air, and the H's are the component of the field in the region lying between x and $x + dx$. The k's are related to the mass susceptibility χ by

$$\chi = k/\rho \qquad (2)$$

where ρ is the density.

From the symmetry of the system, it can be concluded that the dominant field component is the one in the $y =$ direction. Hence, it follows by integration that the total downward magnetic force acting on the whole sample is

$$F = \frac{A}{2} (k_2 - k_1) (H^2 - H_0^2) \qquad (3)$$

where A is the sample cross-section, and that $H >> H_0$ in practice.

To execute the experiment, the sample is weighed

FIGURE 2. Close-up of pole face, sample, and protective tubing for the Gouy method.

before and after the application of H. The difference should be the magnetic pull or push on the sample.

Accordingly:

$$k_2 - k_1 = 2(M_f - M_i) \frac{g}{AH_y^2} \qquad (4)$$

or

$$X_2 - X_1 = 2(M_f - M_i) \frac{gl}{MH_y^2} \qquad (5)$$

where M_i is the initial mass, M_f is the mass after

applying the field, M is the sample mass, l is its length, and g is the gravitational acceleration.

Pyrex glass may be used in the measurement, but quartz is preferred because it introduces fewer impurities to the system and can stand much higher temperatures. All system accessories should be made of quartz, including the suspending thread. It is also advisable to check or calibrate the field measurements using a standard sample such as a solution of nickel chloride of known concentration or powdered crystals of manganese sulfate ($MnSO_4 \cdot 4H_2O$). The usual accuracy is 1 part in

or

$$\chi_2 - \chi_1 = \eth(M_f - M_i)\frac{gl}{M}\frac{\partial H_y^2}{\partial x} \qquad (8)$$

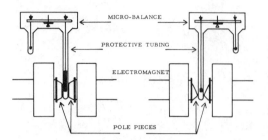

FIGURE 3. Layout procedure for the Gouy and the Faraday methods (after El-Ashry and Nejib, 1972).

1,000 — 1 part in 100 in the case of powders of uniform packing — is attainable.

The Curie or Faraday Method. Consider a situation in which the two pole pieces of an electromagnet are set in exaggerated positions, producing a curved line of forces. Then a small spherical body placed at 0 will tend to move along the χ-direction due to the force.

$$F_x = \frac{v}{2}(k_2 - k_1)\frac{\partial}{\partial x}|H_x^2 + H_y^2 + H_z^2|$$

$$= \frac{v}{2}(k_2 - k_1)\frac{\partial}{\partial x}(H_y^2) \qquad (6)$$

$$= v(k_2 - k_1)H_y\frac{Hy}{\partial x}$$

Consequently, the method requires that the curve of $\eth(Hy^2)/\eth x$ against χ must exhibit a reasonably flat or extended maximum, so that the value is fairly constant over a short distance. Errors and stability problems may arise through the failure to maintain the body in a standard position. This defect may be avoided by using pole tips of special design so that the force is constant over distances of about 1 to 2 cm (Fig. 3). Faraday (1931) designed such poles, which have constant values of $H\eth H/\eth x$ on the order of 10^6 over a distance of about 1.5 cm. Other shapes of pole faces achieving similar objectives were introduced by Davy (1942).

The advantage of this method is its high accuracy when used with micro balances, which is due to the fact that only a small amount of the sample is needed. This will reduce sample contamination errors as well as making it more practical in cases where large quantities of the specimen are difficult to obtain or are expensive. As with the Gouy method, the system is preferred to be made of quartz. The measuring procedure is also similar but uses

$$k_2 - k_1 = \eth(M_f - M_i)\frac{g}{A}\frac{\partial H_y^2}{\partial x} \qquad (7)$$

The Magnetometer Method. Various types of static magnetometers are in use for a variety of purposes such as magnetic anomalies of geological structures or sedimentary fabric estimation. However, the introduction of the Vibrating Sample Magnetometer (VSM) has allowed a great degree of flexibility into the measurement of material susceptibilities.

When a sample material is placed in a uniform magnetic field, a dipole moment proportional to the product of the sample susceptibility times the applied field is induced in the sample. If, at the same time, the sample is made to undergo sinusoidal motion, an electrical signal can be induced in a pickup coil situated in the area. This signal is proportional to the magnetic moment and the vibration frequency. A typical VSM is shown in Figures 4 and 5.

A few millimeters of the material being tested are usually contained in a sample holder made of good quartz. The sample holder is centered between the poles of an electromagnet. A thin, vertical rod then connects the sample holder with the assembly that will vibrate the sample: a transducer. The rod is long enough to allow the transducer assembly and other hardware to be outside the magnet. The sample is sometimes attached directly to the rod by glue or cement.

The sample follows a vertical sinusoidal motion within the magnetic field. The pickup coils, mounted on the pole faces, will identify the signal induced in the sample. This signal, which is at the vibration frequency and proportional to the magnitude of the moment induced in the sample, is transferred to the processing unit.

The electronic sophistication of this procedure allows direct display of the susceptibility values. Chart recorder output can be provided very easily. The VSM system lends itself to a great degree of flexibility due to its structure as well as the electronic sophistication that is built into it. High as well as very low sample temperatures can be used. The sample mounting techniques allow anisotropy to be measured. Consequently, continuous values of susceptibility as a function of field or temperature can be displayed and plotted.

Magnetic Susceptibility of Rocks and Minerals

The magnetic susceptibility of rocks has been investigated since the end of the nineteenth century (Slichter, 1942). Nagata (1953, 1966) has extensively studied and reported on the magnetic susceptibility of rocks and rock-forming minerals. Other good reviews and accounts of magnetic susceptibility of rocks and minerals include those by Slichter (1942),

FIGURE 4. Test assembly using the Vibrating Sample Magnetometer.

Mooney and Bleifuss (1953), Runcorn (1956), and Lindsley et al. (1966).

A marked correlation between susceptibility and rock type is well attested. The extent of this correlation is illustrated in Table 1. The table shows what Slichter (1942) has demonstrated, namely, that the major rock types may be graded as follows, in order of decreasing magnetic susceptibility: (1) basic effusives; (2) basic plutonics; (3) granites; (4) gneisses, schists, and slates; and (5) sedimentary rocks. Mooney and Bleifuss (1953), from examination of rocks from Minnesota, found that the mean susceptibilities of acid intrusives, basic intrusives, and basic extrusives are in the ratio 1:2:7.

Rocks have magnetic properties because of the presence of ferrimagnetic and paramagnetic minerals scattered among a nonmagnetic (basically diamagnetic) matrix. Minerals having paramagnetic properties include olivine, pyroxene, amphibole, biotite, and garnet. Ferrimagnetic materials, on the other hand, can be classified into two groups: metallic oxides and metallic sulfides (Nagata, 1966). The oxides—such as magnetites, titanomagnetites, maghemites, titanomaghemites, hematites, hemo-ilmenites, and their natural solid solutions—are commonly contained in most rocks. The ferrimagnetic iron sulfides, known as pyrrhotites, are usually localized in nature because of the special conditions under which the minerals are produced. It follows then that the metallic oxides are the most important cause of ferrimagnetism in rocks.

TABLE 1. Range of Magnetic Susceptibility in Major Rock Types susceptibility in cgs emu/cm^3

		Percentage of Samples with Susceptibility			
Rock Type	Number of Samples	Less than 10^{-4}	Between 10^{-4} and 10^{-3}	Between 10^{-3} and 4×10^{-3}	Greater than 4×10^{-3}
Mafic effusive rocks	97	5	29	47	19
Mafic plutonic rocks	53	24	27	28	21
Granites and allied rocks	74	60	23	16	1
Gneisses, schists, slates	45	71	22	7	0
Sedimentary rocks	48	73	19	4	4

Source: After Lindsley et al. (1966).

FIGURE 5. Close-up of VSM system showing pickup coils, pole faces, and sample protective tubing.

Of the oxides, however, magnetite appears to play the major role in determining susceptibility (Runcorn, 1956). Slichter (1942) reports the susceptibility of magnetite as between 0.3 and 0.8, and that of the other iron oxides between 1/10 and 1/100 of these (Table 2). He states that the susceptibility of a rock may be well estimated by multiplying a susceptibility value of 0.3 by the proportion of magnetite present in the rock. Mooney and Bleifuss (1953) found a definite dependence of susceptibility on magnetite content, giving a

value of the susceptibility of the magnetite near to that given by Slichter.

The correlation between the susceptibility of magnetic minerals and that of the rocks containing them has been well investigated as shown before. It has also been established by several investigators that many rocks have an anisotropic susceptibility, which is influenced by forces acting during the rock formation (e.g. Stacey, 1960; King and Rees, 1962). This property has been used by Rees (1965) in estimating sedimentary fabrics. Other correlations have been suggested also, but very little has been reported on the relationship between magnetic susceptibility and elemental composition of sedimentary rocks.

El-Ashry and Nejib (1972) and Nejib and Anderson (1972) reported on the magnetic susceptibility of sedimentary shales and claystones from northeastern Pennsylvania and the relationship of the suscepti-

bility values to the concentration of the elements nickel, cobalt, chromium, iron, and manganese in the individual rock samples. The magnetic susceptibility of the rocks was measured using the Gouy method with spot checking by the Faraday method. Table 3 shows the susceptibility values and elemental concentrations of black and gray shales from the "Coal Measures" of the Pennsylvanian system.

TABLE 2. Susceptibility of Magnetic Minerals

Mineral	Susceptibility (cgs emu/cm^3)	Field Strength (Oe)
Magnetite	0.3-0.8	0.6
Pyrrhotite (Sudbury, Ont.)	0.028	0.6
Pyrrhotite	0.007	—
Ilmenite	0.031	—
Ilmenite	0.044	0.6
Specularite	0.004	0.6
Specularite	0.003	—
Franklinite	0.036	—

Source: After Slichter (1942).

TABLE 3. Magnetic Susceptibility and Concentrations of Ni, Co, Cr, Mn, and Fe in Shales from Northeastern Pennsylvania

Susceptibility ($\times 10^{-6}$)	Ni (ppm)	Co (ppm)	Cr (ppm)	Mn (ppm)	Fe (%)
14.9	129	72	263	526	4.77
2.3	175	30	136	74	1.23
11.1	105	31	115	494	4.88
18.2	140	54	198	270	3.92
2.1	113	28	140	1291	7.09
5.4	51	65	118	184	1.89
4.2	40	55	168	188	1.11
12.2	104	46	114	824	5.23
4.5	150	36	97	276	2.14
4.6	93	46	107	262	3.69
2.2	40	41	138	167	3.52
8.6	93	42	208	144	1.75
4.7	46	50	125	262	2.98
9.3	96	53	170	992	6.38
6.1	111	53	120	189	1.23
3.4	61	33	103	385	3.34
3.2	58	43	122	252	3.76
5.1	100	43	100	266	2.06
6.7	104	64	124	531	4.78
3.3	74	37	144	893	4.70
8.5	123	60	212	253	2.93
8.9	91	23	200	380	3.07
4.7	70	37	170	510	3.79
13.2	115	13	199	133	1.05
6.7	110	66	259	371	4.32
7.2	77	59	115	880	6.14
9.9	98	25	194	325	3.20
3.4	26	22	61	152	1.44
4.0	55	25	149	335	3.43
7.0	73	43	121	285	3.57
3.0	76	27	222	122	1.04
5.1	96	42	134	225	2.89
9.7	106	70	157	932	5.97
6.7	107	58	140	256	3.56
13.7	140	57	126	232	3.82

The studies show that the susceptibility values correlate well with the concentration of nickel in the rock samples and, to a certain extent, with that of chromium and cobalt. Iron concentrations did not correlate well with the measured susceptibility values, while manganese showed a minor degree of correlation. In examining the nickel-susceptibility relationship, one finds a linear correlation between the susceptibility value and the nickel concentration (Fig. 6).

From their study, El-Ashry and Nejib (1972) suggested that magnetic susceptibility measurements can provide a first approximation of the concentrations of certain trace elements (particularly nickel) in shales and similar rocks and that susceptibility measurements can be established as a simple and fast analytical method.

The first measurements of the diamagnetic susceptibility of coal were carried out by Wooster and Wooster (1944). Wooster and Wooster used monolithic samples in the shape of a cube with a 4-mm edge. This had the advantage that the magnetic anisotropy could also be examined. They found, however, that only anthracites show anisotropic effects. All samples proved to be diamagnetic with the exception of two, one of which had paramagnetic mineral impurities attached to one face. After these had been scraped off, the

sample appeared to be diamagnetic. El-Ashry and Nejib (1972) also reported on the diamagnetic susceptibility of anthracite coal from Pennsylvania. Van Krevelen and Schuyer (1957) suggested that measurements of the diamagnetic susceptibility may offer possibilities for the analysis of coal constitution.

Magnetic Susceptibility of Fish Fossils

One of the great interests in fish fossils is due to the capacity of bone phosphate for selectively concentrating a number of rare elements, principally rare earths and yttrium. These elements may be accumulated in considerable amounts in mineralized fish remains under favorable conditions. Fish fossils generally have a high magnetic susceptibility (Polushkina and Sharkov, 1974). The high susceptibility is attributed largely to the presence of a mechanical admixture of a ferromagnetic mineral called melnikovite.

Polushkina and Sharkov (1974) investigated the nature of the magnetic susceptibility of fish remains from the Upper Oligocene deposits of the eastern Caspian region. Relatively clean bone fragments from the surfaces of fractions with different magnetizations were selected with the aid of a binocular magnifier, and segregated by means of a

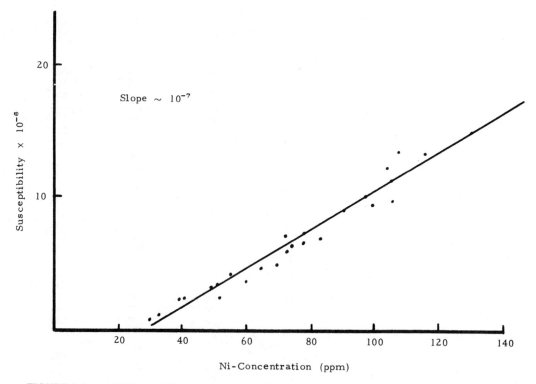

FIGURE 6. Susceptibility vs. nickel concentration in shales from northeastern Pennsylvania (after El-Ashry and Nejib, 1972).

five-pole magnet. Four magnetic fractions were obtained: a "true magnetic" fraction and three electromagnetic ones (the first and second electromagnetic fractions are included together under the index II). In Table 4, the symbols for the II, III, and IV electromagnetic fractions are given in order of decreasing degree of magnetization. Magnetic susceptibility values for the different fractions were obtained using the Faraday method; the results are shown in Table 4. As shown in the table, the magnetic susceptibility of the bone fragments from the magnetic and electromagnetic fractions varies with the field intensity, thus indicating the presence of ferromagnetic minerals. Polushkina and Sharkov concluded that the high magnetic susceptibility of bone remains, usually detected during their magnetic separation, is due largely to the presence of melnikovite; a ferromagnetic mineral, previously identified as iron sulfide, with the formula Fe_3S_4.

Factors Affecting Magnetic Susceptibility of Rocks

The magnetic susceptibility of a rock varies according to several factors, including the strength of the inducing field; the type, composition, and amount of ferrimagnetic minerals; grain size; fabric; temperature; and pressure (Lindsley et al., 1966).

Strength of the Inducing Field. Figure 7 shows a magnetization curve typical of a lava and illustrates the magnetic hysteresis that is characteristic of ferrimagnetic and ferromagnetic minerals.

Type, Composition, and Amount of Ferrimagnetic Minerals. The principal ferrimagnetic minerals in rocks are the iron-titanium oxides. Magnetite (Fe_3O_4) is the most important, as mentioned before, because of its high susceptibility and common occurrence. The susceptibility value for most rocks is also proportional to magnetite content.

Grain Size of Ferrimagnetic Minerals. Several investigators—including Akimoto, Gottschalk and Wartman, Herroun, Koenigsberger, and Petrova (see references in Lindsley et al., 1966)—have shown that the magnetic susceptibility of ferri-

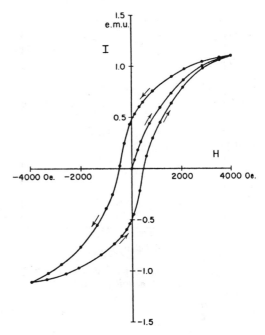

FIGURE 7. Magnetization curve typical of a lava (after Nagata, 1953).

magnetic minerals is related to grain size. The magnetic susceptibility decreases in general with decreasing grain size (Fig. 8).

Fabric. The maximum and intermediate elongation axes of ferrimagnetic minerals tend to align along the bedding planes of detrital rocks and along the foliation planes of metamorphic rocks. Such alignment results in a plane of maximum susceptibility parallel to these planar structures.

Temperature. Nagata (1953) determined the relationship of magnetic susceptibility to temperature for many rocks and minerals. For many igneous and metamorphic rocks, and most sedimentary rocks, the magnetic susceptibility first changes slowly with increasing temperature, then

TABLE 4. Magnetic Susceptibility of Fossil Fish Bones

Spec. No.	Fractions	$X \cdot 10^6$ at different H (Oe)				$x_{meas} \cdot 10^6$	$x_{calc} \cdot 10^6$
		550	8,800	13,000	14,800		
143.2	Magnetic	600	Specimen highly contaminated by ferromagnetic admixtures			—	0.364
	Electromagnetic						
143	II	166.0	102.0	71.2	62.5	0.4	0.457
143	III	50.0	30.7	21.4	18.6	0.45	0.356
143	IV—1	23.1	14.5	10.2	9.0	0.5	0.545
160	IV—2	7.2	4.55	3.55	3.30	0.75	0.704
4	Nonelectromagnetic	1.20	1.24	1.19	1.17	1.20	1.05
6	Nonelectromagnetic	1.06	1.08	1.08	1.03	1.06	0.865

Source: After Polushkina and Sharkov (1974).

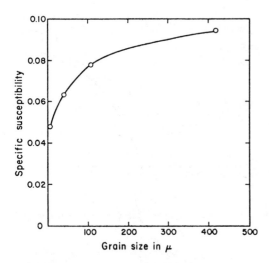

FIGURE 8. Specific susceptibility of rock-forming tita-niferous magnetites as a function of particle size. External field = 1.35 Oe (after Nagata, 1953).

decreases rapidly to zero as the Curie point is approached. In the absence of physical or chemical changes resulting from heating the sample, the curve is reversible upon cooling. Rocks containing more than one ferrimagnetic mineral typically display composite susceptibility-temperature curves, showing a break in slope at each Curie point.

Pressure. Magnetic susceptibility of rocks varies in response to mechanical stress. In general, compression of rocks produces a decrease in susceptibility measured in a field parallel to the direction of stress. Susceptibility measured perpendicular to the compression axis, however, may decrease or increase with compression. This information suggests that the susceptibility measured from a core sample may differ from that of the sample in place because of differences in mechanical stress.

M. T. EL-ASHRY
U. R. NEJIB

References

Bates, L. F., 1951. *Modern Magnetism,* 3rd ed. Cambridge, England: Cambridge University Press.

Davy, N., 1942. The design of pole pieces of electromagnets and the forces acting on small bodies placed in their magnetic field, *London, Edinburgh and Dublin Phil. Mag. and Jour. Sci., Series 7,* **33,** 575.

El-Ashry, M. T., and Nejib, U. R., 1972. Magnetic susceptibility studies of shales from Northern Anthracite Field of Pennsylvania, *Am. Assoc. Petroleum Geologists Bull.* **56**(3), 616-617.

Faraday, R. A., 1931. An improved method for the comparison of small magnetic susceptibilities. *Phys. Soc.* [London] Proc. **43,** 383.

Gouy, L. G., 1889. *Compte Rendus Hebdomadaires des Séances de l'Academie des Sciences.* Paris, 109, 935p.

King, R. F., and Rees, A. I., 1962. The measurement of the anisotropy of magnetic susceptibility of rocks by the Torque Method, *Jour. Geophys. Research* **67** (4), 1565.

Lindsley, D. H.; Anderson, G. E.; and Balsley, J. R., 1966. Magnetic properties of rocks and minerals, in S. P. Clark, Jr. (ed.), *Handbook of Physical Constants.* Geol. Soc. America Mem. 97, 543-552.

Mooney, H. M., and Bleifuss, R., 1953. Magnetic susceptibility measurements in Minnesota—Pt. II: analysis of field results, *Geophysics* **18,** 383-393.

Nagata, T., 1953. *Rock Magnetism.* Tokyo: Maruzen & Co., 230p.

Nagata, T., 1966. Magnetic properties of rocks and minerals, in *Handbuch der Physik* [Encyclopedia of Physics], Vol. III. Berlin: Springer-Verlag, 248.

Nejib, U. R., and Anderson, J., 1972. Magnetic characteristics of multi-colored shales from northeastern Pennsylvania, *Pennsylvania Acad. Sci. Proc.* **46,** 123-126.

Pascal, P., 1910. *Comptes Rendus Hebdomadaires des Séances de l'Academie des Sciences.* Paris, 150, 1054p.

Polushkina, A. P., and Sharkov, A. A., 1974. Nature of magnetic susceptibility of fish fossils, *Internat. Geol. Review* **16**(11), 1214-1219.

Rees, A. I., 1965. The use of anisotropy of magnetic susceptibility in the estimation of sedimentary fabric, *Sedimentology* **4,** 257.

Runcorn, S. K., 1956. Magnetization of rocks, in *Handbuch der Physik* [Encyclopedia of Physics], Vol. I. Berlin: Springer-Verlag, Berlin, 470-497.

Slichter, L. B., 1942. Magnetic properties of rocks, in *Handbook of Physical Constants.* Geol. Soc. America Spec. Paper 36, 293-298.

Stacey, F. D., 1960. Magnetic anisotropy of igneous rocks, *Jour. Geophys. Research* **65**(2), 2429.

Van Krevelen, D. W., and Schuyer, J., 1957. *Coal Science: Aspects of Coal Constitution.* Amsterdam: Elsevier, 257.

Wooster, W. A., and Wooster, N., 1944. *Extra-Fine Structure of Coals and Cokes Conf. Proc.,* 322p.

Cross-references: Vol. VI: *Anisotrophy in Sediments.* Vol. XIV: *Exploration Geophysics; Harbor Surveys; Marine Magnetic Surveys; VLF Electromagnetic Surveying; Well Logging.*

MAN-MADE CAVES—See CAVITY UTILIZATION.

MAPS, ENGINEERING PURPOSES

Special geological maps have been developed in an attempt to present information about geological materials and processes in a format most usable for engineering analysis. This has resulted in a modification of conventional geological mapping procedures and concepts so that the most significant properties of the geological environment, applicable to engineering, are the basis for mapping (Montgomery, 1968).

One of the limitations of conventional geological maps, from the viewpoint of engineering application, is that the map units are defined on the basis of stratigraphic or time-sequence parameters (see Vol. XIV: *Map Symbols*). Since engineering properties are not used in the identification of geological map units, rocks or unconsolidated materials with different engineering properties may be mapped together because they are of the same age or origin (Galster, 1977). As a result, it is difficult to adapt conventional geological maps to engineering use. Another shortcoming is that conventional geological maps do not provide quantitative information about the physical properties of rocks, the nature of discontinuties, or dynamic process in a form applicable to engineering analysis (see Vol. XIV: *Maps, Physical Properties*). In addition, conventional geological maps do not include soil or shallow deposits of unconsolidated materials as map units, although these materials are often very significant in engineering studies.

The objective of special geological maps prepared for engineering purposes is to avoid these limitations by emphasizing factors significant to the engineer, and by presenting them in a format suitable for engineering investigation (Varnes, 1977). The mapping units represent specific geological conditions that may be delineated and evaluated for engineering analysis; in addition, these map units may be evaluated with respect to their suitability for various activities (Legget, 1968, 1973).

Concepts

The approaches to special geological mapping range widely in direction, scope, and content depending on the ultimate purpose of the investigation (see Vol. XIV: *Maps, Logic of*). One end of the spectrum is illustrated by the broad, regional-scale programs designed to present, from an engineering viewpoint, geological conditions over a large area. An example of this approach is the report *Engineering Geology of the Northeast Corridor, Washington, D.C. to Boston, Massachusetts* (U.S. Geological Survey, 1967). This study represents one of the first examples of large-scale regional engineering geological mapping for land use planning. It was undertaken at the request of the Department of Transportation with the intent of summarizing all available information about geological factors as they would affect construction of a proposed high-speed ground transportation facility. This study was presented as a series of maps at a scale of 1:250,000, cross-sections, and tables compiled from existing published and unpublished data. No original mapping or field investigations were done specifically for this report. This approach satisfied the criteria of its intended use, and provided a regional summary of geology as required by planners, geologists, and engineers for early feasibility studies and subsequent planning of detailed investigations.

At the other end of the spectrum have been the detailed studies of specific sites, in which thorough field and laboratory investigations are included. The extent of these investigations is of necessity much more limited in scope than those that essentially involve only summarization and interpretation of existing data. Such maps are closely coordinated with engineering studies and describe specific geological conditions that will influence the development of subdivisions, industrial parks, or new towns. The application of this approach is demonstrated by the engineering geological mapping program carried out during the planning stages of Round Rock, Texas (Hunt, 1973). Round Rock is to be a new town, planned for 1,375 hectares with a projected population of 30,000 by the year 2000. The site investigation for this project, carried out by a private consulting firm, required detailed engineering geological mapping at a scale of 1:12,000 and an extensive field and laboratory testing program. The objective was to evaluate the engineering characteristics of underlying soils and bedrock, and ground-water conditions at the proposed site, and to interpret their impact on the development. The results of this study were incorporated into the planning and design of Round Rock. Geological conditions were evaluated for preliminary foundation design criteria and to prepare cost estimates for site grading, excavation for utilities, and the construction of structures and pavements.

The Northeast Corridor and Round Rock studies illustrate the wide range of application of special geological maps prepared for engineering purposes. As the importance of geological factors has become recognized in engineering analysis, many different approaches to special-purpose mapping have been developed. Because of the diversity of federal, state, and private agencies involved in special geological mapping, however, and because of the variety of potential applications, a uniform procedure or standardization of this type of mapping has not been accepted. Nevertheless, certain fundamental concepts are applicable to all approaches, which generally involve three components: (1) a definition of the special-purpose engineering geological map units, based on the engineering properties of the geological environment; (2) a qualitative and quantative evaluation of these units with respect to engineering parameters; and (3) an interpretation of the suitability of these units for various engineering activities (although this component is not necessarily included as a basic part of the map).

Special-Purpose Engineering Geological Map Units

Geological conditions that exhibit similar engineering characteristics can be grouped to represent a unique or specific unit, and such units constitute special-purpose geological maps. They generally include several levels of classification or subdivision. Under most situations, the engineering characteristics of a specific geological environment (soil-bedrock association) are governed by the physical properties of the bedrock. The strength and stability parameters required for engineering study depend on these properties, as do the physical properties of the soils and weathered materials formed from it. Thus, the first level of classification of engineering geological map units generally is made from an evaluation of bedrock properties. However, when extensive deposits of unconsolidated materials such as alluvium or loess cover the bedrock to the extent that bedrock influence is insignificant, their engineering properties become the controlling factors for engineering design, and these deposits are then also designated as first-level map units to show their importance. Table 1 illustrates an example of the relationship between conventional geological units and special geological units defined for engineering purposes.

The first-level classification units are then subdivided with respect to the impact of topography, drainage, weathering, and soil distribution in modifying the basic engineering properties of the bedrock (or extensive deposit of unconsolidated materials). These factors generally will vary over the area of a particular first-level unit. Therefore, subunits are defined on the basis of the *interaction* among these factors, and the characteristic engineering geological condition this interaction produces. These second-level units represent a specific engineering geological environment based on parameters directly applicable to engineering analysis. Figure 1 is an example of special geo-

logical map units and subunits defined for engineering purposes.

Figure 2 compares a conventional geological map with a special geological map prepared for engineering purposes. It is apparent from this illustration that, although the boundaries between highly different geological materials are the same, the significance given to the engineering properties in defining the mapping units will result in a different map pattern.

Quantitative Analysis

Another basic component of special geological maps for engineering purposes is the presentation of quantitative data on the engineering properties of the geological materials and dynamic processes. These data may be included with the description of the map units or presented in summary tables. The type and extent of these data depend on the scope of the study and testing facilities available because they are evaluated from intact samples tested under laboratory conditions. Typically, for unconsolidated materials, the most useful information would include plastic indices, grain-size distribution, density, permeability, swelling potential, and some form of strength parameter. For consolidated bedrock, the most frequently required data are degree of induration, density, permeability, and strength.

In addition to quantitative laboratory data on the engineering properties, descriptive data of the field or *in situ* engineering characteristics of the bedrock/soil complex making up the mapping units are usually presented. These data may be

TABLE 1. Example of Relationship Between Conventional Geological Units and Special Engineering Geological Units

Formation Name/Age	Conventional Geological Description	Engineering Geological Description	Special Engineering Geological Classification
St. Genevieve Fm. Meramecian Series (Mississippian)	Limestone, light gray, clastic, medium-thick bedded, generally coarsely crystalline and oölitic with beds of finely crystalline limestone.	Limestone, light gray, thin- to massive-bedded, highly jointed, solution enlargement of joints and bedding planes common near surface, may develop karst topography, weathers to a red blocky, highly cherty clay soil over an irregular bedrock surface.	Ca
St. Louis Fm. Meramecian Series (Mississippian)	Limestone, light gray, thin- to medium-bedded, fine-grained to sublithographic, cherty.		
Salem Fm. Meramecian Series (Mississippian)	Limestone, light gray, thin- to medium-bedded, fine- to medium-grained, cherty, argillaceous, grades into siltstone and shale at base.	Interbedded shale, siltstone and limestone, thin- to medium-bedded, jointed, limited solution activity in limestone. Weathers into highly plastic clay soil	Sh/Ca
Warsaw Fm. Meramecian Series (Mississippian)	Interbedded argillaceous limestone and calcareous shale and siltstone.		

Source: Adapted from Rockaway and Lutzen (1970).

MAP UNITS
FIRST LEVEL   SECOND LEVEL
DESIGNATION   DESIGNATION   DESCRIPTION

AL	2	Alluvium, moderately deep (10-40 ft.) located in major river channels, stratified and sorted, coarse to fine grained materials, organics possible.
	T	Alluvium, river terrace deposits, stratified, fine sands, silts and clays. Located above flood plain level.
Ca	2	Carbonate bedrock, residual soil developed on irregular bedrock surface. Soil depth greater than 3 ft. Little solution activity or joints and bedding planes.
	3	Carbonate bedrock, residual soil developed on highly irregular bedrock surface. Solution features extensive, sinkholes, springs common.
Sh	1	Shale bedrock, clay-rich residual soil greater than 3 ft. thick.

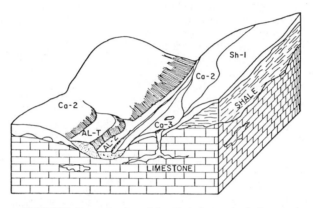

FIGURE 1. Example of special engineering geological map units (adapted from Rockaway and Lutzen, 1970).

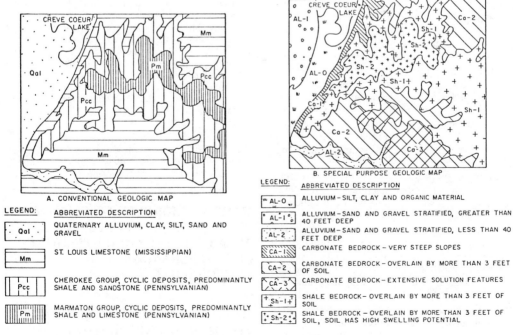

A. CONVENTIONAL GEOLOGIC MAP

LEGEND: ABBREVIATED DESCRIPTION

Qal — QUATERNARY ALLUVIUM, CLAY, SILT, SAND AND GRAVEL

Mm — ST. LOUIS LIMESTONE (MISSISSIPPIAN)

Pcc — CHEROKEE GROUP, CYCLIC DEPOSITS, PREDOMINANTLY SHALE AND SANDSTONE (PENNSYLVANIAN)

Pm — MARMATON GROUP, CYCLIC DEPOSITS, PREDOMINANTLY SHALE AND LIMESTONE (PENNSYLVANIAN)

B. SPECIAL PURPOSE GEOLOGIC MAP

LEGEND: ABBREVIATED DESCRIPTION

AL-O — ALLUVIUM–SILT, CLAY AND ORGANIC MATERIAL

AL-1 — ALLUVIUM–SAND AND GRAVEL STRATIFIED, GREATER THAN 40 FEET DEEP

AL-2 — ALLUVIUM–SAND AND GRAVEL STRATIFIED, LESS THAN 40 FEET DEEP

CA-1 — CARBONATE BEDROCK – VERY STEEP SLOPES

CA-2 — CARBONATE BEDROCK–OVERLAIN BY MORE THAN 3 FEET OF SOIL

CA-3 — CARBONATE BEDROCK–EXTENSIVE SOLUTION FEATURES

Sh-1 — SHALE BEDROCK– OVERLAIN BY MORE THAN 3 FEET OF SOIL

Sh-2 — SHALE BEDROCK – OVERLAIN BY MORE THAN 3 FEET OF SOIL, SOIL HAS HIGH SWELLING POTENTIAL

FIGURE 2. Comparison of conventional (A) and special-purpose (B) geological maps (adapted from Rockaway and Lutzen, 1970).

qualitative or quantitative in nature, and the extent of presentation should reflect the geological environment of the study region. Factors to be considered would include description of the landforms or geomorphic conditions under which the units have developed, their stratification and/or homogeneity, the nature and extent of weathering, structural properties, density, orientation and type of discontinuities, and the association with the ground-water regimen. The descriptive data should be of sufficient thoroughness to define the geological environment of the study region in terms suitable for engineering analysis.

Also included in the quantitative analysis will be a discussion and/or evaluation of geodynamic processes, such as slope stability, erosion, sedimentation, solution, and seismicity. These processes are quite variable, not present in all areas, and in many cases need not be investigated, as they may have no impact on engineering analysis. In other cases, they may be the dominant geological factor and thus warrant a detailed study. The method for qualitative analysis is unique for each process; the presentation of the data should follow the format of the accepted practice in each field.

Interpretative Analysis

The third component of most special geological maps is engineering interpretation. In this stage, the special-purpose engineering geological units are evaluated with respect to their suitability for various engineering activities. Typical activities would include foundation support, ease of excavation, slope stability, mineral resource potential, waste disposal, or any other consideration that is of particular engineering significance in the study area.

The suitability rating system is based on a ranking scale that describes the anticipated level of influence that geological parameters will have on the specified activity—for example, *none, slight, moderate,* or *severe*. Other terminology or categories may be used, but the objective of all systems is to establish limits or ranges to describe the impact of geological factors in anticipated engineering problems.

An extension of the engineering interpretation phase is to prepare derivative maps in which engineering geological map units with the same suitability ratings are presented on a map in the same color code or pattern. In this manner, *derivative maps* may be prepared for each separate engineering activity considered important to the investigation.

The inclusion of interpreted data provides the additional advantage of presenting information in a form compatible to computer analysis (Turner and Coffman, 1973). Those factors (geological or nongeological) critical to determining the suitability of an area for a given use one can be compiled. First individual factor matrices are constructed on a grid, and then, for each factor, matrix cells are assigned numerical values proportional to the suitability of the factor for the intended use. By varying input criteria and weighing parameters, one can rapidly and inexpensively evaluate an alternative engineering analysis.

Special geological maps for engineering purposes differ from conventional geological maps only in that different geological parameters are considered significant to the investigation. Stratigraphic or time-sequence relationships among geological materials have little importance in engineering analysis. The physical properties of the materials and active dynamic processes are the factors that are most critical and must be presented to the engineer via these maps.

JOHN D. ROCKAWAY

References

Galster, 1977. A system of engineering geology mapping symbols, *Assoc. Eng. Geologists Bull.* **14** (1), 39-47.
Hunt, R. E., 1973. Round Rock, Texas, new town: geologic problems and engineering solutions, *Assoc. Eng. Geologists Bull.* **10** (3), 231-243.
Legget, R. F., 1968. Engineering-geological maps for urban development, in G. A. Kiersch (ed.), *Engineering Geology Case Histories.* Boulder, Colo.: Geological Society of America No. 6, 19-21.
Legget, R. F., 1973. *Cities and Geology.* New York: McGraw-Hill, 624p.
Montgomery, H. B., 1968. What kinds of geologic maps for what purposes? in G. A. Kiersch (ed.), *Engineering Geology Case Histories.* Boulder, Colo.: Geological Society of America No. 1-8.
Rockaway, J. D., and Lutzen, E. E., 1970. Engineering geology of the Creve Coeur Quadrangle, St. Louis County, Missouri, *Missouri Div. Geol. Survey and Water Resources Eng. Geology Ser. No. 2* 18p.
Turner, A. K., and Coffman, D. M., 1973. Geology for planning: a review of environmental geology, *Colorado School Mines Quart.* **68** (3), 1-27.
U.S. Geological Survey, 1967. Engineering geology of the Northeast Corridor, Washington, D.C. to Boston, Massachusetts, *U.S. Geol. Survey Misc. Geol. Inv. Map I-514 A, B, C.*

Cross-references: *Environmental Engineering; Map, Environmental Geology; Maps, Physical Properties; Map Series and Scales; Map Symbols.*

MARINE ENGINEERING—See Vol. XIV: FLOATING STRUCTURES; NUCLEAR PLANT PROTECTION, OFFSHORE; OCEAN, OCEANOGRAPHIC ENGINEEERING.

MARINE GEOCHEMISTRY—See Vol. XIV: MARINE EXPLORATION GEOCHEMISTRY.

MARINE GEOLOGY—See MARINE SEDIMENTS, GEOTECHNICAL PROPERTIES; SUBMERSIBLES. Vol. I: MARINE GEOLOGY, TECHNIQUES AND TOOLS; MARINE SEDIMENTS, MINERAL POTENTIAL OF THE OCEAN. Vol. III: SUBMARINE GEOMORPHOLOGY. Vol. VI: MARINE SEDIMENTS. Vol. XIV: MARINE EXPLORATION GEOCHEMISTRY; MARINE MAGNETIC SURVEYS; MARINE MINING.

MARINE SEDIMENTS, GEOTECHNICAL PROPERTIES

The engineering properties of marine sediments depend on the interactions and interrelationships among the component parts of the sediment acting as a whole. Components include the solid minerals or biogenous sediment framework and the fluid or gaseous pore filling. Important engineering properties of marine sediments for which significant amounts of data are available include bulk density, water content, porosity, permeability, plasticity, state of consolidation, and shear strength. Other mass physical properties such as elastic wave velocities, attenuation, and thermal conductivity are discussed elsewhere, as they are generally regarded to be in the realm of geophysics (see Vol. XIV: *Explorations Geophysics*).

Shear strength and consolidation are perhaps the most basic of the properties, as they allow prediction of the bearing capacity of the sea floor under an applied load, its resistance to penetration, as well as later settlement under loading over a prolonged period. Knowledge of these parameters is vital to the successful placement of structures or the operation of vehicles on the sea floor. Aside from the more obvious engineering applications, a knowledge of the strength and consolidation characteristics of marine sediments and variation of these properties with burial depth is important in fundamental considerations of diagenesis, of erosion, and of the redistribution of sediments by gravity-induced mass movement.

Density and Porosity

The *bulk density* of a sediment is a function of the density of the solid constituents and of the fluid and/or gas entrapped in the pore spaces. In most deep-sea sediments of relatively low organic content, gas or air entrapment can be safely disregarded. Most of the solid constituents are normally made up of a relatively few varieties of the common rock-forming minerals. To these are usu-

ally added lesser amounts of authigenic minerals formed by *in situ* precipitation or alteration of other components and biological detritus. In certain environments, biogenous components dominate, which results in the occurrence of highly calcareous or siliceous deposits.

In detrital mineral studies it is customary to separate the minerals in a dense liquid such as bromoform; this will allow *light minerals* with grain densities smaller than about 2.89 g/cm^3 to float and allow the *heavy minerals* to sink. Light minerals, which make up the bulk of most sediment, are mainly quartz (2.66 g/cm^3) and feldspar (2.57 to 2.77 g/cm^3). The dominance of light minerals results in average grain densities of about 2.65 g/cm^3 for most sands. Densities greater than 2.66 g/cm^3 indicate an unusually large admixture of heavy minerals. Similarly, average grain densities less than 2.60 g/cm^3 indicate sediments with unusually large amounts of biogenous materials. The grain density of a deep-sea sediment containing appreciable amounts of montmorillonite, opaline silica, or volcanic glass would generally also fall below the level of 2.60 g/cm^3.

Computation of *in situ* bulk densities also requires a knowledge of bottom-water densities. Bottom-water densities in the shallow, open marine environment off southern California, for example, usually range between 1.02 and 1.03 g/cm^3, according to Hamilton et al. (1956). The density of normally saline bottom waters will increase to about 1.05 g/cm^3 at depths of 5,000 to 6,000.

Natural bulk density is usually determined by weighing a known volume of sediment at its natural water content and allowing for the density of the contained seawater, which is assumed to buoy up the sediments in their natural environment. Table 1 lists *in situ* saturated densities of surficial sediments representative of many determinations in a number of sea-floor environments. These are a function of porosity and degree of cementation as well as of grain and pore fluid densities.

Porosity is defined as the ratio of the volume of voids (pore space between grains) to the total sample volume; it is normally expressed as a percentage. In natural sediments porosity is inversely related to grain size. The complex relationship depends on several factors. In the sedimentation process, sand grains settle rapidly to the bottom, where they are oriented among other grains under the influence of gravity and bottom currents. Sand and coarse silt-sized particles thus assume a single-grained or mixed-grained structure (Figs.1*a* and 1*b*). Naturally deposited sands of uniform size have porosities ranging from 23 percent to about 50 percent. With materials of fine silt and clay size occupying interstices between grains, porosity is diminished.

Porosity determinations of fine silt, and particularly clay-sized materials, are much more complex

TABLE 1. Geotechnical Properties of Submarine Sediments—Mean Values (0-4 m)

Property Texture	Continental Shelf			Marginal Basin Clayey Silt	Continental Slope Clayey Silt	Clay	Abyssal Plain	
	Sand	Sandy Silt; Silty Sand	Silt				Calc. Ooze	Silic. Ooze
Bulk density (g/cm^3)	2.03	1.79	1.80	1.46	1.52	1.50	1.41	1.17
Porosity (%)	45	55	73	73	71	70	66	88
Water content (% dry wt.)	35	40	60	125	88	175	104	340
Atterberg limits								
Liquid	—	—	55	111	79	100	60	—
Plastic	—	—	30	47	36	41	—	—
Shear strength (g/cm^2)	—	12	11	50	85	45	87	—
Sensitivity	—	5	5	2	3	4	7	—
Bearing capacity for square footing at surface (g/cm^2)	—	89	82	370	630	333	644	—

344

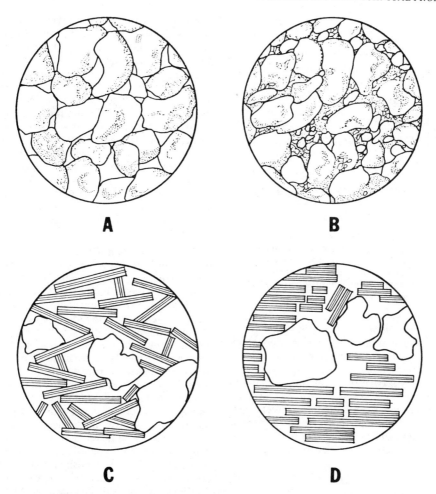

FIGURE 1. (A) Single-grained structure of high porosity formed as result of single sediment source and/or good sorting. (B) Mixed-grain structure of porosity lower than (a) forms as a result of two dominant sources and/or poor sorting. (C) Undisturbed marine clay with edge-to-face orientation (random arrangement of domains) resulting in strong attractive forces. (D) Remolded, or completely disturbed, condition of face-to-face and edge-to-edge, which results in loss of attractive forces and greatly lowered strength.

than those of sands. Clayey sediments have very great surface areas on a particle-to-particle basis, and for this reason the fabric (randomly arranged domains, Fig. 1c), and hence porosity, is controlled mainly by molecular forces. These forces in turn have an important influence on the associated water. Fine-grained sediments possess three types of water rather than the one, pore water, found in sands. Clays contain pore water, absorbed water around the particles, and bound water within the crystal lattice of the minerals themselves. Owing to the manner of determining porosity (oven-drying a fully saturated sample until no more pore water is lost), drying temperatures are critical to ensure that only pore water is driven off and not the other waters, which require relatively higher temperatures. Bulk densities and porosities of clays

are erroneous, therefore, unless a proper drying temperature is used, commonly 105°C.

Density and porosity are interdependent in a well-established straight-line relationship, in which wet density has a negative correlation with porosity between the extremes of no water and 100-percent water for any given grain density. Porosity versus wet density for a large number of samples of marine sediments is shown in Figure 2, and Table 1 lists average wet densities, porosities, and water contents for a number of depositional environments. The amount of water contained in sediment pore spaces is expressed in various ways by workers in the different fields; for example, engineers report water content as the ratio of the weight of water to the weight of the oven-dried sediment, whereas geologists commonly report

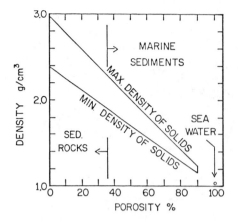

FIGURE 2. Relationships of density and porosity for marine sediments and sedimentary rocks. Most marine sediments and sedimentary rocks will plot in the zone between maximum and minimum density lines of solid sedimentary components.

the ratio of the weight of water to the weight of the total (wet) sediment. In addition, porosity—which we have referred to as the ratio of the volume of voids to total sample volume expressed as a percentage—is also expressed as *void ratio* in soil mechanics and as *water content* in certain sedimentological studies. Assuming 100-percent saturation, interrelationships among these expressions in terms of porosity *(n)*, water content *(w)*, void ratio *(e)*, and specific gravity of solids *(G)* are:

$$n = \frac{V_v}{V}, \quad w = \frac{W_w}{W_s}, \quad e = \frac{V_v}{V_s}$$

$$n = \frac{Gw}{(1 + Gw)} = \frac{e}{(1 + e)}, \quad e = \frac{n}{(1 - n)} \quad (1\text{-}5)$$

where V = the total bulk volume, V_v = the volume of voids, V_s = the volume of solids, W_w = the weight of the water, and W_s = the weight of the solids.

Permeability

Permeability expresses the ease with which water will flow through a sediment. It is a function of several sediment properties, such as the size, shape, and arrangement of grains; the shape and size of voids; and the void ratio, or porosity, of the sediment. Properties of pore fluid are also important; in the case of marine sediments, the chief property is viscosity. The rate at which water flows through a given cross-sectional area of sediment was shown experimentally by Darcy over a hundred years ago to be a direct function of the imposed pressure

gradient. The coefficient of permeability (usually expressed in units of cm/sec) is thus taken as equal to the rate of flow divided by the product of the pressure gradient and the cross-sectional area. Highly permeable sands and gravels generally have coefficients well above 10^{-1} cm/sec, whereas plastic clays may fall below 10^{-3} cm/sec. Sand-silt-clay mixtures are normally in the range of 10^{-2} to 10^{-4} cm/sec. Compared to sand and coarse silt-sized particles, the clays are therefore relatively impermeable. Clay permeabilities are of the greatest significance in foundation or other settlement studies where the rate at which a sediment will consolidate is directly dependent on permeability. This will be discussed in *Consolidation, Soil.*

Plasticity

The *plasticity* of a sediment is a function of its textural, mineralogical, and chemical characteristics. Plasticity is usually expressed as the *plasticity index,* which is equal to the *liquid limit* minus the *plastic limit* (the Atterberg limits). Atterberg limits (see *Atterberg Limits and Indices*) are used for soil classification and comparison purposes and can, in some cases, be correlated with other physical properties. In general, these limits define the range of water contents (expressed as percentages of oven-dry sample weights) through which a soil is in a plastic state, that is, the state in which it can undergo considerable shearing deformations without rupture. Above the liquid limit the soil is supposed to be in a liquid state, and below the plastic limit the sediment is in a semisolid state. Actually a naturally deposited marine sediment commonly has a water content well above the "liquid limit" yet still has an appreciable shear strength due to the cohesive properties of the material. Values of the *plasticity index* for marine sediments of various environments range from 10 to 30 percent for inorganic clays of low to medium plasticity and sandy or silty clays, and from 20 to 80 percent for inorganic clays of high plasticity (Richards, 1962). Variations of Atterberg limits (liquid and plastic) for different depositional environments are given in Table 1.

Consolidation

To the geologist, the term *consolidation* is synonomous with *lithification.* In soil mechanics (q.v.), however, consolidation refers to the reduction of the volume of a sediment under an applied load, either by natural overburden pressure or by the implacement of a structure. The amount and rate of consolidation depend on the amount and rate of loading, the permeability of the sediment, and its shear strength. The *consolidation test* is one of the basic tests required in foundation studies because it yields information about the rate and ultimate amount of settlement expected under a planned structure. To the earth scientist, this test provides some indication of the *loading history* for a particular deposit. True depth-porosity relationships are also obtained from the test.

The consolidation test is performed by confining a small cylindrical sediment sample in a ring with porous stones at top and bottom, subjecting it to increasing load increments, and noting the rate and amounts of compression, or volume reduction. A load placed on the sample is initially borne by the semi-confined pore water alone; the excess pore-water pressure equals the imposed load. As the water seeps through and out of the pore, the load is gradually transferred to the solid particle structure of the sediment. Primary consolidation is completed when the excess pore pressure is dissipated. A secondary consolidation may continue for a considerable period thereafter; it is believed to be the result of plastic deformation of the mineral fabric. Plots of primary consolidation are conventionally shown as void ratio *(e)* versus the log of imposed pressure (log *P*), as shown in Figure 3. Usually there is a relatively horizontal reloading portion of the curve (A to B), an inflection *(c)*, and, after the inflection, a steeper straight section. A perfect test on an undisturbed sample would result in the reloading curve passing through the *in situ* void ratio and pressure of the field consolidation line *(ef, pf)*; in practice it never does, but

instead passes below and to the left. The more disturbed the sample, the farther away from the ideal curve and the less pronounced is the deflection. Figure 3 shows an example of tests on both highly disturbed and relatively undisturbed samples. Regardless of disturbance, field and laboratory curves converge at about 42 percent of the initial laboratory void ratio (e_0).

If a sediment is fully consolidated under its present effective overburden pressure, in soil mechanics terminology it is said to be *normally consolidated.* If the sediment has been normally consolidated under greater-than-present *overburden pressures,* it is said to be *overconsolidated* and can be detected because the inflection point (C in Fig. 3) on the consolidation curve is displaced to the left. *Underconsolidated* sediments result when there has been insufficient time for dissipation of excess pore pressures resulting from rapid loading. This causes the inflection of the consolidation curve to be displaced well to the right, or the high-pressure side, of the normally expected inflection point (C).

The rate at which a sediment will consolidate is a direct function of its permeability, or the rate at which pore water will escape under pressure. In very fine-grained sediments the coeficient of permeability is commonly derived by computing it from the time-compression data of consolidation tests. Hamilton (1964) and Bryant et al. (1975) report coefficients of permeability for sediments of the experiment Mohole off Guadalupe Island, Mexico, and the Gulf of Mexico, respectively, as varying between 1.0×10^{-6} and 1.0×10^{-8}, which they considered normal for the silty clays tested. Relatively few data are yet available concerning the *in situ,* or field, consolidation characteristics of contemporaneous saturated marine sediments, particularly from depths of more than a few meters below the sea floor. Studies involving the experimental Mohole (Hamilton, 1964), and various sites of the Deep Sea Drilling Project (Keller and Bennett, 1973; Trabant et al., 1975, Davie et al., 1978) are some of the relatively few studies that have provided data on the consolidation characteristics of samples from more than 100 m below the deep-sea floor. Figure 4 shows the reconstruction of the field consolidation curve for hemipelagic clays collected 136 m below the sea floor in comparison with data for marine shales from an onshore boring.

Shear Strength

Sediment shear strength is one of the most critical of the engineering properties and is also one of the most important of the properties reflecting diagenetic changes that occur with time and burial. The *shear strength* of a sediment mass may be defined as the summation of the forces of

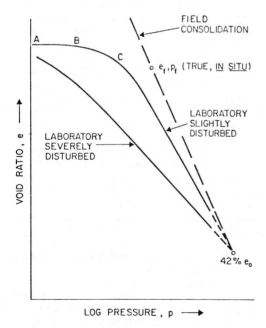

FIGURE 3. The shapes of typical laboratory consolidation curves of highly disturbed and slightly disturbed samples. The field consolidation curve was reconstructed by computations from laboratory data.

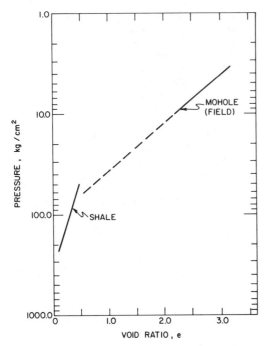

FIGURE 4. Field consolidation curve of hemipelagic silty clay at preliminary Mohole site near Guadalupe Island, Mexico. Sample from a depth of 136 below the sea floor in a water depth of 3,475 m (data from Hamilton, 1964).

friction, cohesion, and bonding, which combine to resist failure by rupture along a slip surface or by excessive plastic deformation under applied stress. A fully saturated cohesive marine sediment consists of a compressible framework of solid particles enclosing interconnected voids, or pores filled with seawater. When this mass of sediment is subjected to external forces, the stresses can be resolved into shear stress acting along the plane of potential shear, and normal stress acting perpendicular to this plane. As a fluid, the pore water cannot support any shear stress, and this component is borne entirely by the particles in the solid framework. In some cases, however, the normal stress may be partially or entirely supported by the confined pore water. Most commonly, this occurs in fine-grained, relatively impermeable sediments where stress to the sediment mass is applied more rapidly than of the semi-confined pore water is drained from the stressed mass; excess pore water pressures are thus generated.

Realization of the profound effects of pore-water pressures in determining the strength of sediment, or soil masses, has led to the concept of *effective* stresses, wherein the shear strength of a sediment mass is a function of the cohesion of the particles and the effective stress normal to the potential plane of failure. These relationships are conventionally expressed as:

$$s = \tau_f = c + (\sigma - u)\tan \phi \qquad (6)$$

where s is the shear strength in g/cm^2, Tf is shear stress on the failure plane at failure (in g/cm^2), c is apparent cohesion (in g/cm^2), σ is the total load normal to the failure plane (in g/cm^2), u is the pore-water pressure (in g/cm^2), and ϕ is the angle of internal friction.

The expression $\sigma - u$ is the effective stress, commonly designated by the symbol $\bar{\sigma}$. The shear strength of a sediment is therefore dependent first on pore pressure, which in turn is controlled by the rate of application of stresses and the permeability of the sediment. Secondly it depends on the magnitude of its intrinsic resistive forces. This concept was introduced in 1932 by the late Carl Terzaghi, a pioneer and important contributor to the new science of soil mechanics.

Intrinsic forces in clays and other sediments having high clay contents are complex and related to molecular bonds (see *Clays, Strength of*). Investigations into the constitution of clays has shown that the flaky particles of the clay minerals form part of double layers of bound and partially bound water. Ions in the water and the bipolar character of the water molecules serve to transmit or modify electrochemical forces between the clay particles or, more commonly, the groups of particles *(domains)*. Because these forces are an exponential function of particle separation, the geometry of particle arrangement *(fabric)* rather than porosity alone is effective in controlling intrinsic strength forces (Bennet et al., 1977; Bohlke and Bennett, 1980). Attraction between edges and faces of particles or domains and repulsion of edges and faces are believed to be responsible for the randomly arranged domains commonly found to be characteristic of submarine clays (Figures 1C and 1D).

Because shear strength is dependent on the pore pressures that control effective stresses within sediments, it is also dependent on the degree or state of consolidation of the sediments. This in turn is largely related to the kinds of sediments and the rates at which they accumulate. Underconsolidated sediments are a characteristic of environments of very rapid deposition of fine-grained, relatively impermeable clays and silty clays such as those found in large delta complexes. These underconsolidated sediments commonly display an excess pore pressure because they accumulate overburden pressure more rapidly than they relieve pore pressure by draining of the pore water. The strength is thus equal to cohesion alone and is independent of load normal to the plan of shear; see equation (6). As a result, little or no increase in strength occurs with burial depth. Very *slow* deposition, on the other hand, may lead to overconsolidation as a result of molecular or ionic bonding and/or incipient cementation of materials near the sediment-water interface. The latter condition leads to high values of shear strength with relatively rapid increases in strength with burial depth. Porosity is similarly controlled by depositional rate, in that in certain cases there are two extreme rates of accumulation that lead to little porosity change with depth of burial: very rapid accumulation results in underconsolidation, and very slow accumulation causes considerable strength to be imparted near the surface by incipient cementation and by molecular bonding.

Earlier shear strength measurements of marine sediments were made on surficial deposits (Moore, 1962; Richards, 1962). More recently, the Deep Sea Drilling Project has afforded the opportunity to investigate engineering properties down to depths of 1,000 m below the sea floor (von Rad et al., 1979). Two tests are commonly used for these measurements: the unconfined compression test, in which shear strength equals half the unconfined compressive strength, and the vane shear test, wherein a vane is rotated in the sample until sediment failure occurs on a cylindrical surface. Both tests, when performed on saturated cohesive sediments, are equivalent to a test in which drainage is not allowed and shear strength is equal to cohesion, as normal stress is borne by the semiconfined pore pressures of the relatively impermeable sediments. Table 1 lists average values of shear strength of surficial sediments (0 to 4 m) from major marine provinces.

Bearing Capacity. The ultimate value of contact pressure between a bearing surface and seafloor sediment that will produce shear failure within the sediment mass is termed the *bearing capacity.* Engineering experience on land has shown that complete shear failures under overloaded foundations occur only in plastic clays. Characteristic properties of these clays are a shearing strength equal to half the unconfined compressive strength and an angle of internal friction (ϕ) of zero. Approximate values of bearing capacity (P_{max}) in terms of cohesion, or undrained shear strength in plastic clays (see equation [6]), have been developed for various footing shapes. For footings having a length much greater than breadth, $P_{max} = 5.14c$; for square footings, $P_{max} = 7.4c$; and for circular footings, $P_{max} = 6.18c$. For loads seated beneath the sediment surface a depth factor must be considered which will increase the bearing capacity value. Representative values of surface sediment bearing capacity in major marine environments are given in Table 1.

Variation of Shear Strength with Burial Depth. The rate of increase in shear strength with burial depth is a function of the degree of consolidation and lithification of the sediment. In general, slowly deposited sediments will show a higher rate of increase than those rapidly deposited, as a result of adjustment of pore-water pressure to overburden pressure. Compositional characteristics also affect strength where incipient cementation is effective. The latter should be most important in environments of very slow accumulation such as those of the deep-sea pelagic deposits. Until recently, relatively few data have been available on the strength of marine sediments at depths of more than a few meters below the sea floor. Indications—yet to be substantiated owing to the small data base—are that there may be a correlation between original depositional rate and the rate of increase in strength with burial depth (Fig. 5). As more data become available, especially through the Deep Sea Drilling Project and continental margin drilling programs, additional light should be shed on this concept.

Sensitivity. *Sensitivity,* the ratio of natural to remolded shear strength, provides a measure of strength loss due to disturbance of the sediments (see *Foundation Engineering*). Average sensitivities of submarine sediments are found to range from 4 to 10 (Keller and Bennett, 1970), indicating a loss of strength on the order of 75 to 89 percent upon remolding. This property is of importance in the use of various vehicles and dredging devices on the sea floor.

Stability of Submarine Slopes. Failures of submarine sedimentary slopes on either a very large scale (cubic kilometers of material) or local slumps (a few cubic meters of sediment) are basically controlled by the shear strength of the materials and regionally by tectonic activity. Local slumping is most common on delta front deposits off large river mouths where rapid deposition of cohesive sediment leads to the accumulation of underconsolidated materials whose shear strengths increase only a little with burial depth. Even on gentle slopes the downslope shear stress component of

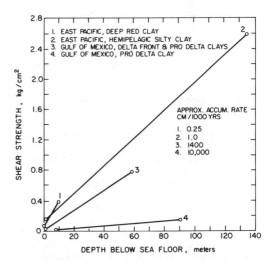

FIGURE 5. Rates of increase in shear strength with burial depth at localities of contrasting depositional rates from Moore (1964). Curve 2 is based on shear tests on the same samples as used for the consolidation tests of Figure 4. Note that the rate of increase in strength is apparently a function of the original depositional rate.

overburden pressure commonly exceeds sheer strength and results in slumping caused solely by sedimentation. Extensive slumping both on large and small scales is a rather pronounced process commonly found associated with continental slopes (Embley and Jacobi, 1977; Richards, 1977; Knebel and Carson, 1979; Summerhayes et al., 1979; Bunn and McGregor, 1980; Embley, 1980). Large-scale slumping may be expected in regions of intermittent tectonic activity. Combined with a large sediment supply, earthquakes could cause rapid accumulation of metastable deposits as on continental slopes.

DAVID G. MOORE
GEORGE H. KELLER

References

Bennet, R. H.; Bryant, W. R.; and Keller, G. H., 1977. *Clay fabric and Geotechnical Properties of Selected Submarine Sediment Cores from the Mississippi Delta.* U.S. Dept. Commerce, NOAA Prof. Paper 9, 86p.

Bohlke, B. M., and Bennett, R. H., 1980. Mississippi prodelta crusts: a clay fabric and geotechnical analysis, *Marine Geotechnology* **4,** 55-82.

Bryant, W. R.; Hottman, W.; and Trabant, P., 1975. Permeability of unconsolidated and consolidated marine sediments, Gulf of Mexico, *Marine Geotechnology* **1,** 1-14.

Bunn, A. R., and McGregor, B. A., 1980. Morphology of the North Carolina continental slope, western North Atlantic, shaped by deltaic sedimentation and slumping, *Marine Geology* **37,** 253-266.

Davie, J. R.; Fenske, C. W.; and Serocki, S. T., 1978. Geotechnical properties of deep continental margin soils, *Marine Geotechnology* **3,** 85-119.

Embley, R. W., 1980. The role of mass transport in the distribution and character of deep-ocean sediments with special reference to the North Atlantic, *Marine Geology* **38,** 23-50.

Embley, R. W., and Jacobi, R., 1977. Distribution and morphology of large submarine sediment slumps on Atlantic continental margins, *Marine Geotechnology* **2,** 205-228.

Hamilton, E. L., 1964. Consolidation characteristics and related properties of sediments from experimental Mohole (Guadalupe site), *Jour. Geophys. Research* **69,** 4257-4269.

Hamilton, E. L.; Shumway, G.; Menard, H. W.; and Shipek, C. J., 1956. Acoustic and other physical properties of shallow water sediments off Southern California, *Acoust. Soc. America Jour.* **28,** 1-15.

Keller, G. H., and Bennett, R. H., 1970. Variations in the mass physical properties of selected submarine sediments, *Marine Geology* **9,** 215-223

Keller, G. H., and Bennett, R. H., 1973. Sediment mass physical properties—Panama Basin and Northeastern Equatorial Pacific, in T. H. van Andel, G. R. Heath et al. (eds.), *Initial Reports of the Deep Sea Drilling Project,* Vol. 16. Washington, D.C.: U.S. Government Printing Office, 499-512.

Knebel, H., and Carson, B., 1979. Small-scale slump deposits, middle Atlantic continental slope off eastern United States, *Marine Geology* **29,** 221-236.

Moore, D. G., 1962. Bearing strength and other physical properties of some shallow and deep-sea sediments from the North Pacific, *Geol. Soc. America Bull.* **73,** 1163-1166.

Moore, D. G., 1964. Shear strength and related properties of sediments from experimental Mohole (Guadalupe site), *Jour. Geophys. Research* **69,** 4271-4291.

Rad, U. von; Ryan, W.; Arthur, M.; Lopatin, B.; Weser, O.; Sarnthein, M.; McCoy, F.; Cita, M.; Lutze, G.; Hamilton, N.; Cepek, P.; Wind, F.; Mountain, G.; Whelan, J.; and Cornford, C., 1979. *Initial Reports of the Deep Sea Drilling Project,* Vol. 47, P. 1. Washington, D.C.: U.S. Government Printing Office, 835p.

Richards, A. F., 1962. Investigations of deep-sea sediment cores, II: Mass physical properties, *U.S. Naval Hydrograph. Office Tech. Rep. 106,* 145p.

Richards, A. F., 1977. Marine slope stability: an introduction, *Marine Geotechnology* **2,** 1-8.

Summerhayes, C.; Bornhold, B.; and Embley, R. W., 1979. Surficial slides and slumps on the continental slope and rise off Southwest Africa, *Marine Geology* **31,** 265-277.

Trabant, P. K.; Bryant, W. R.; and Bouma, A. H., 1975. Consolidation characteristics of sediments from leg 31 of the Deep Sea Drilling Project, in D. E. Karig, J. C. Ingle et al. (eds.), *Initial Reports of the Deep Sea Drilling Project,* Vol. 31. Washington, D.C.: U.S. Government Printing Office, 569-572.

Cross-references: *Atterberg Limits and Indices; Clay, Engineering Geology; Clays, Strength of; Consolidation, Soil; Oceanography, Applied; Soil Mechanics; Undersea Transmission Lines, Engineering Geology.* Vol. VI: *Marine Sediments.* Vol. XII, Pt. 1: *Bulk Density; Particle Density; Permeability; Pore Space; Soil Mechanics.* Vol. XIV: *Harbor Surveys; Soil Fabric.*

MATERIALS HANDLING—See CALICHE, ENGINEERING GEOLOGY; DURICRUST, ENGINEERING GEOLOGY; LATERITE, ENGINEERING GEOLOGY. Vol. XIV: COAL MINING; PLACER MINING.

MATHEMATICAL GEOLOGY—See Vol. XIV: GEOSTATISTICS.

MEDICAL GEOGRAPHY—See Vol. IVA.

MEDICAL GEOLOGY

Although *geomedicine* has not yet been recognized as a subfield specialty within the geological sciences, it constitutes a subject area that is becoming increasingly important throughout the world. The first medical geology textbook is still on the drawing boards, but many articles have been written on the topic and it does form a chapter in some recent textbooks (Cargo and Mallory, 1977; Keller, 1982). Although there is no unanimity as to what constitutes the field, geomedicine is usually considered to be the study that relates mineral, rock, and water properties to human health.

A wide range of disciplines are involved when factors of human health become influenced by the total geological environment. Medical pathologists must determine the character of the health problem. Analytical chemists must provide the measurements that prove the abnormalities. Geologists must discover the source of contamination intrusion as traced from rocks, soils, and waters. Thus the geologist acts as a detective in determining the location of the deleterious materials and how they become mobilized and transported through the natural environment. Although it is the doctor's responsibility to cure the patient, it becomes the geologist's responsibility to undertake measures of prevention that will eliminate, or reduce continued recurrence of, the problem. This is similar to geoengineering, whose applications prevent loss of life by landslides; geologists discover not only the root of the problem but how best to mitigate it.

Nearly 60 different chemical elements occur in biological systems; of these, more than 30 are essential for human life and health. Oxygen, carbon, hydrogen, nitrogen, calcium, and phosphorus constitute 98.5 percent of the human body. Even though trace elements make up less than 0.05 percent of the body, at least 14 have been determined to be important to health: chromium, cobalt, copper, fluorine, iodine, iron, manganese, molybdenum, nickel, selenium, silicon, tin, vanadium, and zinc (Cannon and Davidson, 1967; Cannon and Hopps, 1971). There is a narrow range of tolerance in the human system for these elements; too much or too little can produce dangerous health problems. For example, although the body requires selenium, too much selenium can cause toxicity and produce such ailments as motor ataxia and damage to blood-forming organs.

Geologists can provide four services in the field of geomedicine: (1) determine properties of the geological environment that are capable of producing a danger to human health; (2) evaluate the pathways of movement—in surface water, ground water, and even air—of hazardous substances; (3) provide mineralogical and petrological analysis of hard parts of the human body; and (4) develop programs that will aid in the attenuation or elimination of the harmful substances. To accomplish these goals, the geologist must study not only the natural environment but also harmful products that society places within the geological habitat (Fig. 1).

Of special interest in geological studies is the evaluation of *trace elements* (see Vol. XII, Pt.1: *Micronutrients*). It is important to understand their stable and bonded form in rocks and minerals and how through weathering and erosion processes they become mobilized to forms that eventually become part of the food chain. The pathways of change are investigated along with the mechanisms for getting into the soil, waters, and living systems (both plant and animal). Predictions must be made concerning the concentrations and chemical compounds that result, and their geographical distribution. Such factors are vital in determining the siting of community facilities or remedial actions that may be necessary to alleviate potential problems.

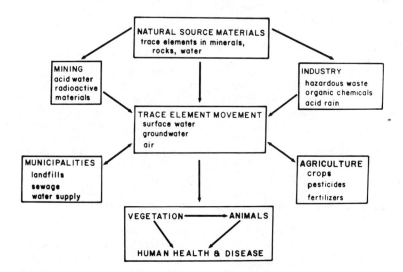

FIGURE 1. Pathways of movement of trace elements and materials that influence human health and disease.

Another aspect of geomedicine relates to hazardous and toxic wastes that society introduces into otherwise natural systems. Mankind is greatly accelerating health problems by introducing a bewildering array of harmful and lethal substances into the geological environment. These dangerous materials include not only deleterious trace elements, but also many organic compounds that have been shown to be carcinogenic and debilitating to the human body. Thus geologists need to be involved in the siting of sanitary landfills (see Vol. XIV: *Artificial Fill*) and other dumping areas, in the monitoring of water systems adjacent to waste and industrial plants, and in the location of water supplies that will be free from contamination.

A rather specialized area of geological involvement with medicine is the mineralogical and petrographic study of hard parts of the human body, such as the bones and teeth. Identification of the mineral composition and structural details can indicate abnormalities and provide important information about dietary deficiencies or harmful food and water intake (Stevenson and Stevenson, 1972).

Historical Background

Hippocrates (ca. 460-370 B.C.) was the earliest person of renown to try to relate health, and even physical characteristics and racial traits, to environmental factors. His 72 works discuss both practical and theoretical aspects of man, his illnesses, diseases, and character. These beginnings of comparative ethnography tell how important it is for those who wish to pursue a course in medicine to have also a knowledge of hot and cold winds, properties of the waters, and even the risings of the sun (Jones, 1957). Hippocrates reported that the physique of the Phasians was different from others, and that their complexion was yellowish "...as though they suffered from jaundice" (Jones, 1957, p. 113). He attributed these characteristics to their living in marshes, drinking stagnant water, and eating imperfect fruit. The physical geographer Ellsworth Huntington (1917), repeatedly asserted in his writings in the early 1900s that climatic factors were an important contributor to the health, strength, vigor, and intellectual achievement of the human races.

Various early Greek and Roman writings refer to diseases that were related to various professions, but it wasn't until 1472 that a German pamphlet showed that mercury and lead poisoning was especially prevalent in goldsmiths (Eckholm, 1977). In 1556, the mineralogist Agricola wrote about the health problems of miners and how they were related to mining activities.

The quality of water has always been important in the health of mankind. For example, Frank (1955) shows that not only were poor people struck down with water-borne diseases and plague, but many notables also fell victim to contaminated water, including Louis VIII of France, Charles X of Sweden, Prince Albert of England, his son Edward VII, and his grandson George V. Ironically, Louis Pasteur's two daughters reportedly died of typhoid fever. Unfortunately, the lack of water purity continues to be a serious problem throughout the world. There were 130 reported outbreaks of disease attributed to contaminated water in the United States during the 1961-1970 decade.

Geomedicine is still a science in search of an integrated identity. A coherence for the many complexities involved in geomedicine has only started to emerge within the past two decades. In spite of knowledge that such geological environmental factors as mining and water use contribute to ill health, preventive programs (other than construction of sewage disposal plants) largely date since World War II. Since then, numerous new toxins have been introduced to exacerabate health and environmental problems. Several factors have contributed to the delay in the development of the medical geology field: (1) Plastics, largely developed from organic compound, were not common until after the 1950s. (2) A great many chemicals are in use, and many more are invented yearly. More than 70,000 different chemicals, many toxic, are commercially produced in the United States, and their numbers grow by about 1,000 each year. The Environmental Protection Agency tests an average of 7,000 chemicals yearly; in 1977, 1,500 were found to be carcinogenic. (3) Analytical techniques for determining the chemistry of substances in parts per billion did not become widespread until the 1960s. (4) The Atomic Age did not start until World War II, so the possible introduction of radioactive contaminants is only a recent phenomenon. (5) An understanding of the importance of trace elements in humans has emerged only during the past few decades.

Trace Elements

Trace elements will be considered those chemical elements in living tissue that, although present in minute amounts, are necessary to regulate vital life processes. They may be bonded and incorporated into chemical compounds, and some may be metabolized, whereas others cannot. If trace elements are required for health maintenance but cannot be metabolized because of their mode, they become stored within the organism. If they accumulate, they can eventually lead to toxicity and disease. There can be a very narrow margin of safety, and a major problem arises when man-induced waste moves into the environmental system, increasing trace element concentrations. The following is a representative listing of some of the trace elements and the way they affect humans and animals: *Cadmium* can be a factor in hypertension and can cause osteromalacia (deterioration and softening of bones) and pulmonary emphysema. Wastes from mines, smelters, and landfills are causing alarming amounts of this metal to be released into the biosphere. *Chromium* is related to glucose tolerance and may be a factor in diabetes. Bronchitis and emphysema may also be related to a chromium imbalance. *Cobalt,* nickel, molybdenum, and vanadium are all important enzyme catalysts when present in trace amounts, but they have toxic effects when they are too abundant. *Copper* is required for cell respiration, but a deficiency may cause a loss of muscular control and anemia. *Fluorine* is especially important as an inhibitor of caries (tooth decay cavities) in young children. The optimum amount is about 1 ppm in water, but too little can produce excessive tooth decay and too much causes mottling of the enamel. Highly deficient amounts increase incidence of osteoporosis. A deficiency of *iodine* can cause goiter and can increase the incidence of breast cancer. Stunted physical and mental development (cretinism) may also result. A *lead* imbalance can impair protein synthesis. Large amounts of lead can accumulate in bone, damage the central nervous system, and cause anemia. A *lithium* deficiency can lead to mental disorder (mania). *Manganese* is an enzyme catalyst, but toxic in large amounts. *Mercury* is cumulative so that animals higher in the food chain have larger amounts. It attacks the central nervous system and has caused disastrous results in some Japanese fishing villages. *Molybdenum* is known especially to affect animals; molybdenosis in cattle that interferes with normal growth, nutrition equilibrium, reproduction, and kidney processes. Toxic amounts of *Selenium* cause motor ataxia, damage the alimentary canal, and produce disease in blood-forming organs. Selenium may also be related to caries, and to cancer. *Silicon* is especially needed in proper amounts in calcification processes and cartilage formation. It is thought to play a role in the aging process. *Zinc* promotes cell growth; its absence will stunt growth. Zinc also aids regeneration of tissue and promotes healing, perhaps even inhibiting some cancers. It is toxic when present in too high amounts.

Geology and Mobility

Trace elements are transported through the soil, water, and air. Since these elements are vital for plant growth, they become major paths for movement through natural systems into biological systems. The distribution and occurrence of trace elements are also functions of bedrock and the environmental setting. There can be large differences among different rock types, and even within certain similar-appearing rocks. For example, Warren (1964) reports that some slates of the same age within a small area show trace element differences as great as 20:1.

Silicate minerals do not contribute much in cycling of trace elements. Instead, non-silicates produce much more activity. Ultrabasic rocks generally yield high amounts of cobalt, nickel, and chromium. The release of trace elements and their mobilization into the biosphere is related to the weathering processes. Their suitability for transfer becomes a function of their solubility, exchangeable ionic relationship, and bonding types. Once incorporated within soil, their further release depends on such variables as pH, Eh (reduction-oxidation potential), moisture, drainage, clay particle presence, their mode, and temperature. In general, the combination of factors that favor chemical reactions are those that also produce soils with higher amounts of mobilized trace elements.

Health and Society

Society is its own worst enemy. The failure to adopt stringent environmental management measures for control of wastes that intrude into natural systems has led to monumental health problems and places much of humankind in jeopardy. Many endeavors in society such as mining and industry produce excessive release of trace elements and other materials that can become pathogenic.

Mining. The notorious black lung disease, endemic to many coal miners, costs the U.S. government more than $1 billion per year in compensation payments to its victims. In 1975, mining inspectors reported that 30 percent of mines still had conditions conducive to black lung disease. Silicosis is another disease that afflicts miners who must work with ore bodies or resources with silicate-rich minerals. (Perhaps the most harmful of all silicates are those of the fibrous variety, those that produce asbestos. Breathing these fibers and their dust residue has been demonstrated to lead to cancer. Many asbestos products in the United States have been taken off the market and removed from public places, such as ceiling tile.) Waste from mines can lead to acid drainage, and to increased mobility of trace elements from spoil banks and exposed rock faces.

Radioactive wastes *(radwastes)* pose problems not only from nuclear plants, but also from mining operations. For example, radwastes occur on 51 ha in Salt Lake City, and at Grand Junction, Colorado, more than 300,000 tons of radwaste tailings were used both as fill and as building materials. The U.S. government has identified 25 additional sites in the West where radwaste contamination poses a level of risk that warrants expensive cleanup.

Manufacturing. Manufacturing processes produce staggering amounts of wastes, many of which become entrained in water systems and in the air. These by-products may be either trace elements or a broad range of organic contaminants, many of which have been shown to be carcinogenic. The infamous case of the Love Canal area of Niagara Falls, New York, is symptomatic of what will become an increasing problem in the disposal of hazardous wastes. The number of examples of contaminated waters from industrial sites are now legion. Acid rain is another result of industrial contamination: sulfur, nitrogen, and trace elements are released into the air with the burning of coal. Disastrous effects of this were first felt in the Fennoscandinavian region, but are now reaching menacing proportions in the eastern United States and Canada.

Agriculture. Farming activities provide non-point (diffuse entry of pollutants) sources of possible environmental contamination and resultant human health problems. Rachel Carson (1962) and others have especially documented the injurious and lethal effects of the *organochlorides* (the base, along with the organophosphates, for many pesticides) on fish, birds, and other wildlife. These materials contain both organic and trace elements, which become entrained and move through surface and ground-water systems and ultimately through the food chain. Mineral fertilizers can also contain substantial amounts of trace elements. On-going problems as well as potential problems from such man-induced contaminants are a major focus of the PLUARG (Pollution from Land Use Activities project) (Parizek, 1980). This joint team of Canadian and American scientists has been evaluating the complex physical, geochemical, and biochemical interactions and time dimensions of the nature and sources of pollutants throughout the Great Lakes watershed. They have also addressed the problem involved with designing remedial measures in an attempt to control the diffuse sources of contamination.

Municipal and Domestic Environments. Municipalities also provide source materials that lead to human health disorders. Waste products in sanitary landfills and other disposal systems may introduce trace elements or other noxious to toxic materials into waters that humans ultimately ingest. In addition, it is possible that even public water supplies may contain ingredients capable of harming citizens.

Disease Types

Heart Disease. An increasing number of studies have evaluated possible relationships between mortality by health failure and the geological environment. Some studies have concentrated on the relative hardness of drinking water, and others on the character of soils, rocks, and trace elements. Some studies in Japan, England, Wales, Sweden, and the United States have concluded that there is a direct relationship between heart failure and soft water, that is, the softer the water the greater the likelihood for heart disease (Winton and McCabe, 1970). Many other variables are obviously involved with heart problems, however. A study of water

quality and heart failure in the Binghamton, New York, metropolitan region concluded that a relationship could not be proved. The incidence of death from heart attacks in the Binghamton, which uses "soft" Susquehanna River water, was not statistically different from the adjacent communities of Endicott and Johnson City, which use "hard" ground water (McDuffie et al., 1974).

Shacklette et al. (1970, 1972) conducted extensive studies of the geochemical environments of Georgia and cardiovascular mortality rates. They investigated various parts of Georgia bedrock types and soil groups and identified 30 elements in the geochemical systems. Their principle conclusion was that areas with low death rates could be correlated more closely with beneficial trace elements. Thus heart disease seemed to result more from a deficiency of these elements than a super-abundance of them.

Cancer. Mankind's habitat is becoming increasingly occupied with potential carcinogenic materials, as both trace elements and organic compounds increase as a result of human activities. In some studies, however, even natural environments have been blamed as the principal culprit for increasing the incidence of cancer. Spencer (1970) has shown that an iodine-deficient zone in the United States correlates with both higher rates of goiter and breast cancer. At West Devon, England, Allen-Price (1960) attributed higher stomach cancer incidence to differences in mineralized water. The older Devonian rocks, being more highly mineralized than the younger Carboniferous rocks, seemed to correlate with higher cancer rates. Other studies in Wales and England (Stocks and Davies, 1960) concluded that abnormal rates of stomach cancer were related to larger amounts of organic material in the soil. They found that zinc, cobalt, and chromium were also associated with greater stomach cancer rates. In northern Iran, esophageal cancer rates varied greatly within short distances, but most associations of high rates indicated relationships with soil differences, vegetation, and agricultural practices. The highest rates occurred where soils were saline. Since climate influences moisture availability, this too could be viewed as a controlling parameter (Kmet and Mahboubi, 1972).

Epilogue

Before conclusions can be drawn, and the first blame placed on geological environments, the case history of health in the Adirondack Mountains should be examined. In 1959 a New York medical report stated there was a relationship between the high rate of malformed children in the Adirondacks and the granitic rocks, which were assumed to have high levels of radioactivity (Coates, 1981). Further studies, however, showed the incidence of malformities had many contributory causes that instead were in the realm of socioeconomic factors.

DONALD R. COATES

References

Allen-Price, E. D., 1960. Uneven distribution of cancer in West Devon, *Lancet* **1**, 1235-1238.

Cannon, H. L., and Davidson, D. F. (eds.), 1967. Relation of geology and trace elements to nutrition. *Geol. Soc. America Spec. Paper 90,* 64p.

Cannon, H. L., and Hopps, H. C. (eds.), 1971. Environmental geochemistry in health and disease, *Geol. Soc. America Mem. 123,* 230p.

Cargo, D. N., and Mallory, B. F., 1977. *Man and His Geologic Environment,* 2nd ed. Reading, Mass.: Addison-Wesley, 581p.

Carson, R., 1962. *Silent Spring.* Boston: Houghton Mifflin, 368p.

Coates, D. R., 1981. *Environmental Geology.* New York: Wiley, 701p.

Eckholm, E. P., 1977. *The Picture of Health:* New York: W. W. Norton, 256p.

Frank, B., 1955. The story of water as the story of man, in *Water: The Yearbook of Agriculture.* Washington, D.C.: U.S. Government Printing Office, 1-8.

Huntington, E., 1917. Climatic change and agricultural exhaustion as elements in the fall of Rome, *Quart. Jour. Economics* **31**, 173-208.

Jones, W. H. S. (ed.), 1957. *Hippocrates,* Vol. I. Cambridge, Mass.: Harvard University Press, 361p.

Keller, E. A., 1983. *Environmental Geology,* 3rd. ed. Columbus, Ohio: Charles E. Merrill, 526p.

Kmet, J., and Mahboubi, E., 1972. Esophageal cancer in the Capian littoral of Iran, *Science* **175**, 846-853.

McDuffie, B.; Kempner, S. M.; and Goltz, R. D., 1974. Cardiovascular death rates in three neighboring upstate New York communities with differing water hardness, in D. D. Hemphill, (ed.), *Trace Substances in Environmental Health,* Vol. VIII. Columbia: University of Missouri Press, 75-80.

Parizek, R. R., 1980. Non-point-source pollutants within the Great Lakes—a significant international effort, in D. R. Coates and J. D. Vitek, (eds.), *Thresholds in Geomorphology.* London: Allen & Unwin, 435-472.

Shacklette, H. T.; Sauer, H. I.; and Miesch, A. T., 1970. Geochemical environments and cardiovascular mortality rates in Georgia. *U.S. Geol. Survey Prof. Paper 574C,* 39p.

Shacklette, H. T.; Sauer, H. I.; and Miesch, A. T., 1972. Distribution of trace elements in the occurrence of heart disease in Georgia. *Geol. Soc. America Bull.* **83**, 1077-1082.

Spencer, J. M., 1970. Geologic influence on regional health problems, *Texas Jour. Sci.* **21**, 459-469.

Stevenson, J. S., and Stevenson, L. S., 1972. Medical geology, in R. W. Fairbridge (ed.), *The Encyclopedia of Geochemistry and Environmental Sciences.* New York: Reinhold, 696-699.

Stocks, P., and Davies, R. I., 1960. Epidemiological evidence from chemical and spectrographic analysis that soil is concerned in the causation of cancer, *British Jour. Cancer* **14**, 8-22.

Warren, H. V., 1964. Geology, trace elements and epidemiology, *Geog. Jour.* **130**, 525-528.

Winton, E. F., and McCabe, L. J., 1980. Studies relating to water mineralization and health, *Am. Water Works Assoc. Jour.* **62**, 26-30.

Cross-references: *Forensic Geology; Geology, Applied.* Vol. XIV: *Artificial Fill; Environmental Geology;*

MILITARY GEOGRAPHY—See Vol. IVA.

MILITARY GEOLOGY—See MILITARY GEOSCIENCE.

MILITARY GEOSCIENCE

Terminology

The idea of making use of geology for military purposes is almost as old as the science of geology itself (von Bülow et al., 1938). Applied to problems of military operations affected by geological factors, the function became known as *military geology.* The scope of military geology was necessary to the operational capability of military forces. When technological advances have improved or increased available resources, the requirements placed on military geology have also changed. At the same time, the capability of military geology to deal with both old and new problems has advanced with the growth of geoscientific knowledge and the development of methodology and tools of investigation (cf. Whitmore, 1960).

It became evident in World War II that mechanized military operations increasingly required coordinated consideration of all the elements of the physical environment, not only the geological. Thus soil science, hydrology, botany, forestry, and meteorology have become subdisciplines of what might better be called *military geoscience* (see *Geology, Applied*).

While the specific problems have changed, the basic functions of military geoscience remain much the same. *Terrain evaluation* is, and has been, the focal point of military geology or geoscience (see *Terrain Evaluation, Military Purposes*). In the military definition, *terrain* is an area of ground of unspecified size, and also its physical features, especially topography, as they relate to the area's suitability for a specified military purpose, such as maneuvering or the construction of fortifications (cf. Stamp, 1961). Erdmann (1943) observed that "Terrain is the common denominator of geology and war." In military geoscience, *terrain* signifies the sum of all physical features and conditions at or near the earth's surface which must be taken into account when determining the usefulness of an area for a specified use (see Betz and Elias, 1957; Van Lopik, 1962). As such, terrain evaluation for military or other purposes is an aspect of environmental science (see Vol. XIV: *Environmental Geology*).

It should be noted that in wartime, geologists and other geoscientists also become involved with problems that are not usually encountered in the framework of military geology. The most important of them is procurement of mineral supplies for use in industries essential to prosecution of the war (see Committee on War Effort, 1942). Responsibility for dealing with that problem is generally regarded as a function of *economic geology* (q.v.).

Before World War I

Military men were aware of the new science of geology in the late eighteenth century, which may account for Napoleon's appointment of Déodat de Dolomieu, a mineralogist-geologist, to the commission of 167 experts in many fields that accompanied his expedition to Egypt in 1798. However, A. H. Brooks (1920) points out that this was part of scientific investigations of a little-known region and was not a recognition of the military value of geology. Scientific investigations made in connection with military and semimilitary expeditions, as originated by Napoleon, have since been widely adopted. Much of the first knowledge of the western part of the United States was gained in this manner.

Probably the first time a geologist helped plan a military operation was in 1813, when the Prussian general von Blücher consulted Karl August von Raumer, a professor of mineralogy and one-time student of Abraham Werner, for information on the terrain of Silesia in advance of the successful battle at Katzbach, in which Napolean's army was defeated.

Another of Werner's students, Johann Samuel von Grouner, chief inspector of mines in Switzerland, cartographer, head of the government printing office of Switzerland, and engineer officer in the Bavarian army, was an ardent student of the Alps. One purpose of his many field trips there was to consider the influence of a rugged moutainous terrain on the planning and conduct of military operations. In his recorded observations, which were published posthumously in 1826, von Grouner stated that, as far as he knew, *geognosy* (geology) had never before been studied in its relation to military science. Another early study was conducted in the different terrain of Luxemburg by Rudolf von Bennigsen-Förder (1843), partly for the purpose of discovering geological information of possible military significance.

In Great Britain, several geologists attempted to encourage the use of geology by armies by presenting lectures to officers, and offering courses for military students, and publishing papers in military journals to show how geological knowledge could be valuable. Leading advocates were Joseph Portlock, a Royal Engineer officer and geologist, and T. Rupert Jones, a physician, palaeontologist, and professor of geology at Sandhurst Military Academy. Their aim was to impress soldiers sufficiently to cause them to pursue one of two courses: to call on trained geologists for assistance when necessary, or to become trained to solve problems without outside help. The same alternatives were presented by geologists to civil engineers.

Geology was also taught to military officers as part of courses on *military geography* a subject that had long occupied military historians and strategists, and had been one of their continuing pursuits during peacetime. Voluminous studies of countries and regions, replete with descriptions and statistics that might be useful in strategic planning, were produced by General Staff officers. According to Carlo Porro (1898), an Italian general and military geographer, geology was eagerly adopted by military geographers and assigned a supporting role in explaining terrain and surface conditions. That was particularly true in France, where courses in military geography stressed the effects of geology on terrain, as shown, for example, by textbooks for the Artillery and Engineer School written by Marga (1880-1882) and Barré (1897-1901).

The practice of military geology remained an occasional activity throughout the nineteenth and early twentieth centuries. Geological assistance was reported to have been provided in the construction of permanent fortifications in peacetime, and in building field fortifications, dealing with water supply problems, and countering enemy mining operations in wartime. Also, during wars, some armies appear to have made use of information on the influence of terrain on communications, fields of fire, cover, and concealment, and on terrain obstacles in zones of combat, which was supplied by military officers, geologists, and geographers.

World War I

In the years shortly before World War I, individual geologists had addressed military establishments on the need for organized military geological services in armies. A paper by Walter Kranz, a German geologist and engineer reserve officer, published in a military journal in 1913, is considered to have presented the most persuasive arguments, which had the desired effect on the German military authorities. The British army, the U.S. Army, and other armies also formed geological services, but the British and German geologists established the credentials of military geology. The prolonged periods of stabilization of the war front in Western Europe made it possible for the geologists to demonstrate the value of applying their knowledge and skills to engineering problems of trench and mine warfare. In the area of strategic terrain evaluation, the Allied Headquarters had a military geographical adviser, who was an American geomorphologist.

Between The World Wars

The use of geological information in World War I was considered by the military to have been a success, but it did not assure a role for military geology in peacetime armies. In the next two decades, the field existed mainly in publications by former military geologists, who discussed its history, the problems and methods of investigation in the past war (illustrated with excerpts of reports and samples of maps), and the geology of the war zones. These authors were interested in displaying accomplishments, which might be used as guides for some future activity. Among the valuable contributions were those of W. B. R. King (1919) and various authors of a series prepared for the Institution of Royal Engineers (1921-22) in Great Britain; by Hans Philipp (1923) and Walter Kranz (1921, 1934) in Germany; and by A. H. Brooks (1920) and D. W. Johnson (1921) in the United States.

In the 1930s, geologists were once again employed by the German army to assist in the building of border defenses, and, as the war neared, several summaries of the scope, purpose, and procedures of military geology were published, no doubt in anticipation of renewed activity. The volume written by Kurd von Bülow and several other geologists (1938) for German users reached foreign countries; in at least one of them it served as a first guide for geologists brought into service in the unfamiliar field of military geology.

World War II

The second opportunity for military geology to become a viable form of applied geology came in World War II, this time with more fully organized services in some of the armies than in World War I. In all, many more geologists were involved, especially in the U.S. and German activities. Information on the number of people engaged in military geology in the USSR indicates that many were actually working on problems assigned to branches of economic geology.

Mobile warfare created the necessity for providing answers to problems of both planning and operations much more rapidly than was generally required by the static warfare of World War I. On some war fronts, determination of trafficability of ground for cross-country movement of tanks and other military vehicles was the chief problem. Accompanying engineering problems were presented by the need for prompt action to facilitate mobility. Advance planning allowed more deliberate evaluations, as in the study of possible landing sites on the islands of the Pacific and on the

beaches of Normandy (concerning the latter, see King, 1951).

Two organizations were prominent in the development of terrain evaluations during the war: one, the Military Geology Unit, led by geologists, and the other, the Forschungsstaffel by geographers. Since early in the war, the U.S. Geological Survey had dealt with a wide range of problems presented by the U.S. Army. As continuing use of the Geological Survey's services indicated the need for an organized "Military Geology Unit," such a group was formally established in mid-1942 through an agreement with the Corps of Engineers. Previously experience showed that terrain evaluation, the basis for providing answers to many of the questions asked, would require the joint effort of geologists, soil scientists, botanists, foresters, and engineering specialists. Thus, the designation Military Geology Unit obscured the multidisciplinary composition of the group and scope of its investigations.

The Military Geology Unit created a type of report (see Vol. XIV: *Geologic Reports*), a terrain intelligence folio, which contained maps and tables that furnished a wide range of information applicable to both broad planning and solution of tactical field problems. Among special problems, trafficability assumed great importance, and a format for cross-country movement maps was devised, which stressed cartographic simplicity so that the user could recognize the evaluations easily. Further information in brief, simple statements was available in the legend.

Terrain evaluation (q.v.), also termed *terrain appreciation,* was expedited by availability of aerial photographs for many of the areas being studied. Used in conjunction with reliable geological, soils, vegetation, and hydrological maps, a high degree of accuracy and detail could be achieved in determining terrain conditions and identifying sources of construction materials and water supplies without direct ground reconnaissance (see Hunt, 1950; Military Geology Unit, 1945).

The second organization was the Forschungsstaffel (in English, Research Detachment), a special unit of the German Wehrmacht, created in mid-1943 after a period of operation under another name in North Africa in 1942. It concentrated on preparation of terrain-evaluation maps with chief attention paid to trafficability for tanks. The Forschungsstaffel based its operations on a concept termed the *combination method,* developed prior to the war by a small group of German geographers. The method called for investigations in the field by teams of physical geographers, plant ecologists, foresters, geologists, soil scientists, meteorologists, and other specialists, supported with all available logistical and technical resources. Photo interpretation and aerial reconnaissance were combined with ground reconnaissance to produce the terrain evaluations. The maps produced by the

Forschungsstaffel were ingenious presentations of data and interpretations. The combination method and the maps arouse much interest among military terrain specialists in other countries when they became known after the war (see Smith and Black, 1946; Wilson, 1948).

After World War II

The hiring of geoscientists by military establishments in the postwar years proceeded in various directions. Some programs similar to those of the war period were carried on. Other programs concentrated on the development of criteria for, and methodology of, terrain evaluation. Still others were concerned with scientific investigation of regions that had not been studied thoroughly in the past, which might be regarded as background research in the event of future application to military use (see *Geology, Applied*). Where military engineering agencies have a responsibility in national environmental management, geologists are also engaged.

Since current applications of geoscience for military purposes are concealed in wartime—and often in peacetime as well—information for the period since the end of World War II is necessarily fragmentary. Some of the applied research is discussed in the open literature, however, and can be found indexed under such headings as *engineering geology, hydrology, environmental geology,* and *geomorphology*.

Contributions to Geoscience

The intensive application of geology to military engineering problems in World War I was an important factor in projecting engineering geology into prominence among geological activities in the postwar period. The many publications by former military geologists on the wartime experience provided engineering geology with convincing evidence of its value, which was translatable into nonmilitary applications (see, for example, a textbook of engineering geology by Kranz, 1927).

After World War II there was no immediate place for terrain evaluation in geologists' activities. Thus there was no urgent need for dissemination of the concepts and methodology of terrain evaluation in the geological community. Geographers, however, continued to build on the experience of military terrain specialists, developing applications to regional planning and landuse, as well as formats for presentation of information to different audiences. Since the 1960s, geologists have become involved in studying the suitability of areas for different human uses and in predicting impacts of human actions on the environment (see Vol. XIV: *Environmental Geology*). The multidisciplinary approach to conducting the investigations has been rediscovered. Although the full benefit of the wartime experience in terrain evaluation has not been realized, its place as a forerunner of the new environmental geology has not escaped notice.

Before World War II, aerial photographs had not come into wide use as a geological research tool. Limited coverage of areas being studied and lack of training in photointerpretation were obstacles. The valuable experience gained during the war stimulated interest in the use of aerial photography and, later, other types of imagery, which have become invaluable sources of information for basic and applied geoscientific research.

Scientific by-products of wartime activity have been countless publications on the geology and geomorphology of war areas, based on material gathered in the process of obtaining information for military purposes. Following World War II, a vast amount of geoscientific data was obtained in fulfillment of programs sponsored by some military establishments, which might be regarded as background research for possible military application in the future, but which also was an immediate contribution to the store of scientific knowledge.

FREDERICK BETZ, JR.

References

Barre, O., 1897-1901. *Cours de géographie: Croquis géographiques.* 2 vols. Fontainebleau: Ecole d'Applic. Artill. et Génie.

Betz, F., Jr., and Elias, M. M., 1957. Relationship of geology to terrain (abstract), *Geol. Soc. America Bull.* **68,** 1700-1701.

Brooks, A. H., 1920. The use of geology on the Western Front, *U.S. Geol. Survey Prof. Paper 128 (D),* 95-124.

Committee on War Effort, 1942. *Utilization of Geology and Geologists in War Time.* Boulder, Colo.: Geological Society of America, 8p.

Erdmann, C. E., 1943. Application of geology to the principles of war, *Geol. Soc. America Bull.* **54,** 1169-1194.

Hunt, C. B., 1950. Military geology, in S. Paige (chmn.), *Application of Geology to Engineering Practice.* New York: Geological Society of America, 295-327.

Institution of Royal Engineers, 1921-1922. *The Work of the Royal Engineers in the European War 1914-1919,* 7 vols. Chatham, England: Mackay. See especially vol. 7, T. W. E. David, and W. B. R. King (1922), *Geological Work on the Western Front.* 71p.

Johnson, D. W., 1921. *Battlefields of the World War— Western and Southern Fronts.* Am. Geog. Soc. Res. Ser. no. 3. New York: Oxford University Press, 648p.

King, W. B. R., 1919. Geological work on the Western Front, *Geog. Jour.* **54,** 201-215 (disc., 215-221).

King, W. B. R., 1951. The influence of geology on military operations in north-west Europe, *Advances in Science* **8,** 131-137.

Kranz, W., 1913. Militärgeologie, *Kriegstech. Zeitschr.* **16,** 464-471.

Kranz, W., 1921, 1934. Beiträge zur Entwicklung der Kriegsgeologie, *Geol. Rundschau* **11,** 329-349; **25,** 194-201.

Kranz, W., 1927. *Die Geologie im Ingenieur-Baufach.* Stuttgart: F. Enke, 425p.

Marga, A., 1880-1882. *Géographie Militaire.* Fountainebleau: Ecole d'Applic. Artill. et Génie.

Military Geology Unit, 1945. *The Military Geology Unit.* Washington, D.C.: U. S. Geological Survey and Army Corps of Engineers, 22p.

Philipp, H., 1923. Die Methoden der geologischen Aufnahme, in E. Abderhalden (ed.), *Fortschritte der biologischen Arbeitsmethoden, Abt. 10—Methoden der Geologie, Mineralogie, Paläobiologie, Geographie.* Berlin: Urban & Schwarzenberg, 395-484.

Porro, C., 1898. *Guido allo studio della geografia militare.* Turin, Italy: Unione Topogr.-Edit., 391p.

Smith, T. R., and Black, L. D., 1946. German geography: war work and present, *Geog. Rev.* **36,** 398-408.

Stamp, L. D. (ed.), 1961. *A Glossary of Geographical Terms.* London: Longmans, 537p.

Van Lopik, J. R., 1962. Optimum utilization of airborne sensors in military geography, *Photogramm. Engineering* **28,** 773-778.

von Bennigsen-Förder, R., 1843. Geognostische Beobachtungen im Luxemburgischen, *Archiv Minralogie, Geognosie* **17,** 3-51.

von Bülow, K., et al., 1938. *Wehrgeologie.* Leipzig: Quelle & Meyer, 170p.

von Grouner, J. S., 1826. Verhältnis der Geognosie zur Kriegs-Wissenschaft, *Neues Jahrb.* **6,** 187-233.

Whitmore, F. C., Jr., 1960. Terrain intelligence and current military concepts, *Am. Jour. Sci.* **258-A,** 375-387.

Wilson, L. S., 1948. Geographic training for the postwar world: a proposal, *Geog. Rev.* **38,** 575-589.

Note: The papers by Hunt (1950) and King (1951) are reprinted in F. Betz, Jr. (ed.), *Environmental Geology.* Benchmark Papers in Geology, Vol. 25, Stroudsburg, Pa.: Dowden, Hutchinson & Ross, 1975, 101-119.

Cross-references: *Geology, Applied; Geomorphology, Applied; Terrain Evaluation, Military Purposes.* Vol. XIV: *Environmental Geology; Land Capability Analysis; Terrain Evaluation Systems.*

MINE DEVELOPMENT, PLANNING—See Vol. XIV: MINING PREPLANNING

MINERAGRAPHY

Mineragraphy (ore microscopy), the study of opaque and translucent minerals by means of reflected incident light in a polarizing microscope, is complementary to petrology (q.v. in Vol. V), which is concerned with the study of transparent minerals through the use of transmitted light in a polarizing microscope.

Mineragraphic methods require materials to be studied as highly polished sections set in plastic mounts. Materials examined include sulfides, oxides, metallic and other varieties of ores, as well as oxidized equivalents. The oxidized equivalents are important components of *gossans* and of some oxidized outcrops. Although mineragraphy is also extensively involved in the quantitative examination of the products of mineral beneficiation, and is therefore an important part of both operational and research ore-dressing procedure, the following discussion will focus on the interpretation of gossans.

Gossan

The term *gossan* is applied to oxidized, leached, and commonly cellular exposures of sulfides either at the surface or beneath superficial cover, but within a zone of oxidation. *Cap rock,* although an alternate term, usually refers to outcrop expressions of sulfide ore bodies.

An appreciation of the significance of gossans in particular, and of leached rock outcrops in general, emerged from investigations conducted by Locke (1926) and Boswell and Blanchard (1927) soon after the turn of the century. In 1968 the Nevada Bureau of Mines produced a comprehensive, illustrated publication, *Bulletin 66* (Blanchard, 1968), concerning the "interpretation of leached outcrops." This major reference work has been complemented by recent studies, which provide further data on specific ore deposits (Loghry, 1972; Groves and Whittle, 1976; Blain and Andrew, 1977).

The application and further improvement of the techniques of gossan interpretation were interrupted by the introduction and extensive use of geochemistry (see Vol. XIV: *Exploration Geochemistry*) and geophysics (see Vol. XIV: *Exploration Geophysics*) in the immediate postwar period. There followed a period during which it was recognized that geochemical—and, in some cases, geophysical—data could be more positively interpreted when oxidized outcrops in anomalous areas were concurrently studied by means of microscopic methods. In particular, the mineragraphic examination of both gossans and gossanous rocks was found to provide the information to account for spurious geochemical or geophysical results.

Many companies involved in mineral exploration (see Vol. XIV: *Prospecting*) have now adopted gossan interpretation as one of the methods of examining new terrain and for reassessing previously explored areas. The interpretations are particularly useful in the extensive arid regions of Australia, where weathering, oxidation, and leaching may extend to depths of hundreds of meters and thus where depleted metal values may render geochemistry unreliable.

On the other hand, significant increases in the magnitudes of certain metal contents in surface or near-surface leached outcrops may result from the progressive deposition of exotic limonite, with or without manganese oxides, during long periods of gossan exposure. This dual accretionary and metal scavenging phenomenon, superposed on the original leaching, produces a complex geochemistry in ferruginous rock, ironstone, and gossan outcrops. This complexity does not in most cases express the inherent chemical characteristics of the rock outcrop, or of the oxidized ore at the site. A mineragraphic examination of the outcrops may resolve the complexity and may provide a reasonably reliable interpretation of the inherent characteristics of the rock.

The interpretations of leached outcrops include consideration of sulfide gossans and of other rocks, which may be gossanous to greater or lesser degrees. This latter case therefore involves the distinction between the leached expressions of silicates, carbonates, and so on, and those of the ore-forming minerals. The interpretations thus become more complicated, as for example in mineralized ultramafic rock outcrops where the nickel content may be in part attributable to sulfides, and in part to a silicate phase of nickel involved originally in the ferromagnesian silicate components of the host rock.

Common Characteristics of Gossans

Gossans commonly exhibit positive relief in the terrain. If the gossan has not been secondarily ferruginized, it will retain a cellular structure, and its color will usually be some shade of reddish brown. Dependent on the amounts of earthy limonite, jarosite, manganese oxides, or various secondary encrusting metal oxysalts such as malachite, the gossan may appear yellowish black or bluish green, but in the general case there is a hard skeletal framework of reddish brown goethite. This framework is known as a *boxwork.*

Gossans may be concealed by deposits of sand, soil, and other detritus. Cover of this type is common in arid regions, where it may be of considerable thickness, as at the South Windarra nickel deposit in Western Australia. There the gossan occured under 50 m of soil and sand. The presence of concealed gossans is usually noted from subsurface geochemical and geophysical anomalies and from samples obtained from percussion drilling, mineragraphic examination, and interpretation.

Examination of different types of boxwork that contribute to the complex cellular structure of gossans provides a basis for interpretation. The skeletal boxwork framework consists mainly of goethite, but hematite and lepidocrocite may also be present. In special cases maghemite may occur when magnetite was a component of the mineralizing assemblage. The inherent color of the boxwork depends mainly on the relative proportions of these several iron oxides, but it is further modified by encrustations of yellow or brown earthy limonite, jarosite, metal oxysalts, and the several different species of manganese oxides.

Indigenous and Exotic Oxides

The several hydrated iron oxides that constitute the boxwork are indigenous oxides, which formed by hydrolysis of ferric sulfate and by other reactions during oxidation of ore-forming minerals (Park and MacDiarmid, 1970; Garrels and Christ, 1965). The distinction between indigenous iron oxides and other iron and manganese oxides of exotic and subsequent origins is most important in gossan interpretations. The latter are commonly colloform banded or botryoidal in texture, and are therefore distinct as fillings and encrustations on the quasi-geometric boxwork structures adopted by indigenous iron oxides. However, there are also certain lamellar, radiating spherulitic, and polygonal goethite cell structures, which although very similar to certain sulfide boxwork structures, are in fact forms adopted by colloidal gel goethite after its precipitation from ground waters. These cellular expressions of exotic goethite lead to difficulties in gossan interpretation. If exotic manganese oxides are associated with these structures, it is often the case that zinc, lead, silver, and other metals scavenged from ground waters progressively build up.

In addition to indigenous goethite, and more or less exotic goethite, gossans usually contain remnants of the relatively insoluble gangue mineral components such as quartz, barite, and various silicates, as well as portions of the mineralized host rock. Careful examination of these is a further requirement for complete gossan interpretation, particularly in the case of vein assemblages that contained, in addition to coarse phases of sulfides and the like, very fine phases of some of the ore mineral suite. Some minerals of the fine phase will have persisted through prolonged oxidation and leaching when protected as inclusions in quartz, baryte, rhodonite, tourmaline, and other refractory gangue components. The minor but economically significant ore minerals—such as gold, silver, silver-bearing sulfosalts, and molybdenite—commonly exist as a fine phase in association with the coarser, more common iron and base metal sulfides, which are usually oxidized and leached to boxwork. Depending on the mode of analysis used in geochemical determinations, the presence of such fine-phase ore components may therefore escape detection, except through microscopic examination.

The complex structure of a gossan will be the combined expression of two or more varieties of boxwork, except when only one ore mineral was present. Several different boxwork structures may exist separately among relict gangue or the gangue mineral boxwork and exotic limonite, or they may become complexed together in cases where the original ore minerals occurred in binary, ternary, or more complex intergrowths. Provided that the ore minerals were larger than 0.1 mm, a boxwork will usually have developed; at finer sizes, however, goethite replicas rather than cellular structures are usually manifestations. The replicas of very fine pyrite are generally isometric in shape, and may therefore be distinguished from the common ovoidal or lobulate forms of anhedral minerals such as pyrrhotite, sphalerite, and chalcopyrite. Further distinction between replicas of sulfides of the latter type is possible microscopically only in cases where a proportion persists unoxidized as a protected inclusion in quartz, or other refractory gangue minerals. Identification of the replicas is usually possible if the polished section is examined with an electron probe analyzer, but this involves additional cost and is done only in critical cases.

Boxwork and Replica Structures

Various quasi-geometrical configurations of common sulfide boxworks may be distinguished with the aid of a hand lens or low-power stereobinocular microscopes if the original sulfides had grain sizes in excess of several millimeters. Examination of sulfide boxworks of finer size is best conducted on specially prepared polished sections using incident light. Boxwork identification may be further facilitated by the use of oil immersion objectives, a half-shadow diaphragm in the collimator, or an interference contrast objective to provide adequate contrast in either relief or color.

Configurations of common sulfide boxworks at megascopic scale have been illustrated by Blanchard (1968). His text also includes numerous sketches of the more detailed structure of boxwork cells in the magnified sample.

In the common case, the boxwork or replicas of the sulfides, as well as those of the oxides, silicates, and carbonates, consist principally of goethite, with only small proportions of iron and manganese oxides. When indigenous goethite becomes complexed with microcrystalline silica, it is referred to as *siliceous goethite*. In extreme cases of replacement by supergene silica, the configuration of the boxwork may be perpetuated in silica alone. This is common in arid regions; and because of the near-complete removal of iron and the occluded base metal oxides, completely silicified gossans may exhibit extremely low and entirely misleading geochemical characteristics.

The range of boxwork and replica structures defined to date is small, but because common mineral species are included, it is possible to interpret gossans from many ore deposits, as well as the approximate former compositions of both the associated gangue assemblages and the host rocks.

The illustrations in this entry provide a few examples of the boxwork of sulfides, oxides, silicates, and carbonates as seen in polished sections used for mineragraphic examination. Although the detail protrayed may seldom be seen in the megascopic view, some of it may be revealed by stereobinocular examination.

Boxwork of Isometric Minerals

The distinction between the boxwork structures adopted by the various isometric minerals relies on paragenesis, and on inherent characteristics such as morphology, cleavage, twinning, and the usual degree of idiomorphism adopted by a given species. The identifications may not be positive unless these factors and the geochemistry are taken into account.

The boxwork of *pyrite* (Figs. 1, 2, 3) expresses its common idiomorphic habit and lack of distinct cleavage. Internal partitions have irregular configuration, continuity, and thickness. Highly irregular structures are presented by the boxwork of compact aggregates of subidiomorphic granular pyrite when the cells have been filled by microcrystalline goethite.

Although generally xenomorphic in habit, *pentlandite* boxwork (Figs. 4 and 5) are occasionally

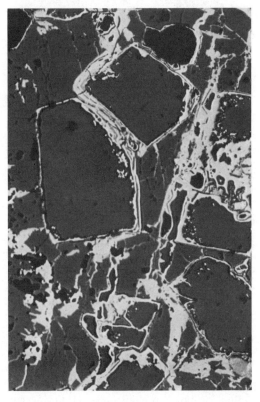

FIGURE 1. Pyrite boxwork: open cells and idiomorphic outlines.

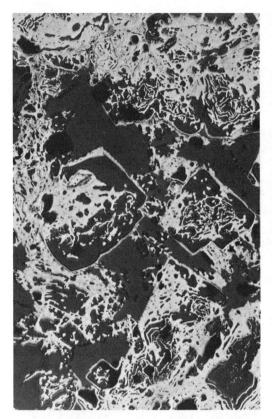

FIGURE 2. Pyrite boxwork: irregular or variably concentric internal partitions, and idiomorphic outlines.

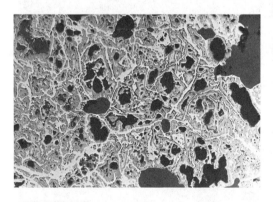

FIGURE 3. Composite pyrite boxwork: irregular partitions, irregular grain boundary contacts of the original pyrite anhedra, and advanced cell filling by exotic goethite.

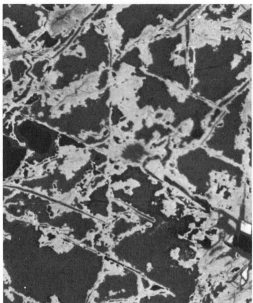

FIGURE 4. Pentlandite boxwork: widely spaced cell partitions expressing the (111) cleavage of pentlandite.

FIGURE 5. Pentlandite boxwork: almost completely silicified in the supergene zone, but thin remnants of goethite among silica outline the original cell partitions.

isometric. The distinctive feature of pentlandite boxwork is the regular and widely spaced intersecting partitions, which express (111) cleavage. The close association with chromite, and with the boxwork of both martite (Figs. 6, 7) and ferromag-

nesian silicates (see Fig. 20), as well as frequent inclusion within the boxwork of pyrrhotite (see Figs. 16, 21, 22), provide additional evidence toward confirming its identification.

Figure 5 illustrates an example of the highly silicified boxwork of pentlandite, contained in a silicified pyrrhotite-pentlandite gossan with less than 500 ppm nickel. This is a good example of the depression of inherent geochemical characteristics through supergene silicification.

The external outlines of the boxwork of *magnetite* are commonly isometric because the mineral is commonly euhedral (Figs. 6 and 7). Although its cleavage is indistinct, there are octahedral partings as well as lamellar twinning parallel to (111).

FIGURE 6. Magnetite boxwork: characteristic closely spaced partitions representing (111) planes. The boxwork is in juxtaposition with a goethite replica of pyrrhotite, and both are contained in quartz.

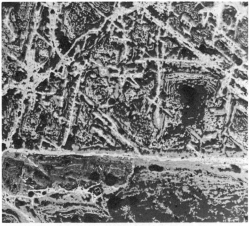

FIGURE 8. Galena boxwork: boxwork from coarsely crystallized galena; the partitions reflect most of the original crystallographic characteristics.

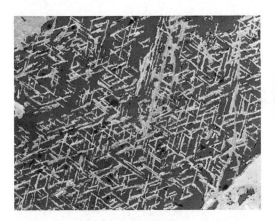

FIGURE 7. Magnetite boxwork: detail of the (111) partitions.

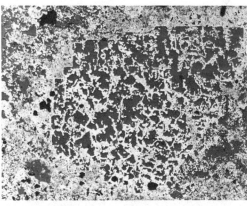

FIGURE 9. Galena boxwork: boxwork from medium-grained galena showing stepped crystal outline, and partitions reflecting original (100) cleavages and octahedral partings.

The development of the boxwork of magnetite follows on its progressive martitization along the (111) crystallographic planes, producing closely spaced partitions. This close spacing of partitions and the isometric outlines of the boxwork, if present, provide distinctions from the boxwork of pentlandite.

The boxwork of *galena* (Figs. 8, 9, 10) is not easily distinguished because the mineral commonly exists in its finer-grained phases as lobulate individuals, and as xenomorphic granular aggregates. More coarsely crystallized galena usually displays isometric outlines, some of which are characteristically stepped (Fig. 9). Except in cases in which galena was very fine-grained, the boxwork is characterized by partitions reflecting the highly perfect (100) cleavage, the octahedral parting, and the tabular twinning on (111).

The acutely angular intersections between the major partitions in coarse *sphalerite* boxwork probably reflect the perfect dodecahedral cleavage

(Fig. 11). The mineral is usually xenomorphic; hence, the external form of the boxwork is irregularly lobulate (Fig. 12). The less common variety known as hieroglyphic boxwork is present in the highly leached boxwork from ore assemblages that were strongly pyritic (Fig. 13).

The boxwork of *garnet* is generally contained within polygonal or dodecahedral outlines because of the mineral's common idiomorphism (Fig. 14). The internal partitions are usually randomly oriented because of the absence of cleavage, but there may be some sets of parallel-disposed intersecting partitions related to the (110) partings. Hence garnet boxwork may be confused with the boxwork from binary chalcopyrite-sphalerite grains.

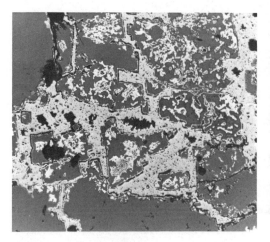

FIGURE 10. Galena-tetrahedrite boxwork: highly leached galena boxwork with thin ferruginated remnant partitions. Contour boxwork from small tetrahedrite inclusions remains within the galena cells.

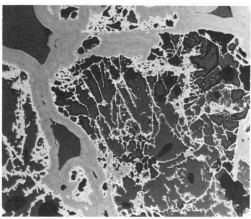

FIGURE 12. Sphalerite sponge boxwork: outlines of the boxwork structures are irregularly lobulate and the delicate thin partitions intersect at acute angles. The sphalerite was associated with copious pyrite: there was extensive leaching followed by the deposition of colloform-banded limonite around the sphalerite.

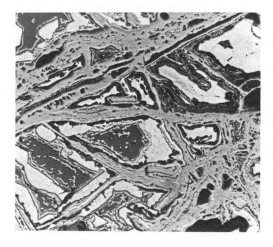

FIGURE 11. Sphalerite boxwork: common style of boxwork with thick walls and partitions intersecting at sharp angles.

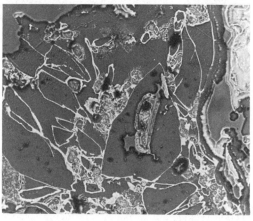

FIGURE 13. Hieroglyphic sphalerite boxwork: conditions for formation were similar to those noted in Figure 12.

Boxwork of Minerals of Lower Orders of Crystal Symmetry

Minerals belonging to systems other than the isometric produce boxwork with contrasting structures. The approximately quadrangular structure of *chalcopyrite* boxwork, for example, probably relates to the structure of the unit cell rather than to poorly developed (201) cleavage. It is a clearly defined structure with long, wavy, thick subparallel walls, which are linked at irregular intervals by shorter thick partitions inclined at steep angles (Fig. 15). The thinner internal partitions follow similar patterns, but less consistently.

The boxwork of single-phase *hexagonal* pyrrhotite is typically six-sided (Fig. 16) and often contains radiating internal partitions. It is similar to the boxwork of gel goethite (see Fig. 24). Hexagonal pyrrhotite boxwork is less common than that of two-phase monoclinic-hexagonal pyrrhotite (see Fig. 22).

The rhombic cleavage boxwork of the *carbonates* is distinctive; however, its origin in calcite, dolomite, siderite, and cerussite cannot be distinguished from the boxwork (Fig. 17). Its association with other minerals and its geochemistry usually provide the clue to the nature of the original carbonate.

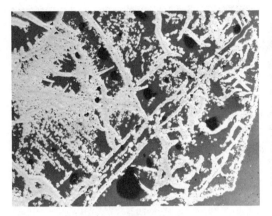

FIGURE 14. Garnet boxwork.

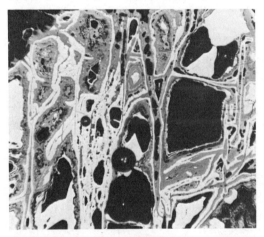

FIGURE 15. Chalcopyrite boxwork: typical quadrangular structure.

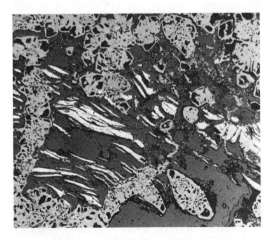

FIGURE 16. Hexagonal pyrrhotite boxwork: groups of sixsided cells developed from veinlets of fine grained pyrrhotite contained in sheared magnetite-bearing ultrabasic greenschist. The long white grains are sheared slivers of martite.

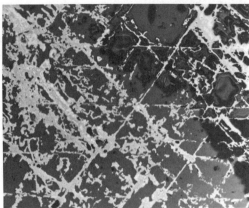

FIGURE 17. Carbonate boxwork: at this orientation, complete sets of partitions represent the rhombohedral cleavages.

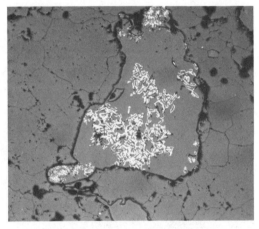

FIGURE 18. Jarosite boxwork: remnants of the boxwork inside a void in quartz at the site of leached pyrite.

The boxwork of aggregates of fine-grained *jarosite* is often present in pyritic gossans. The structures express the prominent (0001) cleavage of the mineral (Figs. 18 and 19).

The structure of the boxwork of the *amphiboles* is a reflection of the strong (110) cleavages (Fig. 20). At certain orientations the pattern of intersections is similar to that of the boxwork of the carbonates (see Fig. 17).

The boxwork of *monoclinic pyrrhotite* is characterized by sets of thin, parallel partitions that are expressions of the (001) cleavages (Fig. 21). Changes in the orientation of these partitions define the original grain boundaries of pyrrhotite individuals in pyrrhotite intergrowths. The boxwork of *two-phase monoclinic-hexagonal pyrrhotite* is a complex of the sets of parallel partitions of the monoclinic phase, separated by the six-sided cells of the hexagonal phase (Fig. 22). In the original

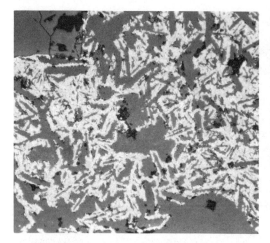

FIGURE 19. Jarosite boxwork: detail of the jarosite boxwork as determined by the (0001) cleavages of differently oriented individuals.

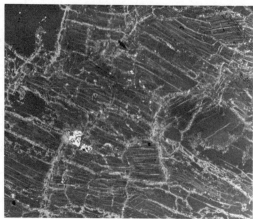

FIGURE 21. Monoclinic pyrrhotite boxwork: enclosing small original inclusions of martite (white).

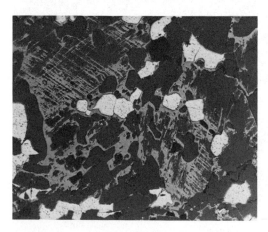

FIGURE 20. Amphibole cleavage boxwork: in this case rather like that of carbonate boxwork, but lacking the third set of parallel partitions.

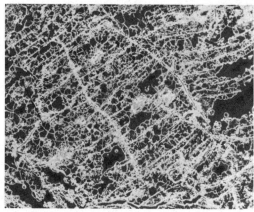

FIGURE 22. Monoclinic-hexagonal pyrrhotite boxwork: a combination of the cell structures shown in Figures 21 and 16.

mineral the two phases usually coexist as exsolution intergrowths.

False Gossans

Some cellular ironstone outcrops have the appearance of typical sulfide gossan, but in general they are distinct in their dark color and evenly fine cellular texture. These false gossans are commonly characterized by metal contents of significant magnitudes, especially zinc, lead, silver, copper, cobalt, and manganese.

When drilled, false gossans are usually found to have little depth extension. They tend to merge into country rock rather than underlying ore mineral assemblages. When examined microscopically, noncellular false gossans are found to consist either of massive structureless aggregates of microcrystalline goethite or of aggregates of colloform banded and botryoidal goethite with or without colloform manganese oxides (Fig. 23). Cellular varieties consist of extensive aggregates of minute polygonal goethite cells, which are referred to as gel goethite boxwork (Fig. 24).

Difficulties in Gossan Interpretations

Highly silicified gossans require petrographic and mineragraphic examination. Completely silicified sulfide gossans characterized by only a few hundred ppm of base metals are occasionally encountered in arid areas. In such cases the relicts of sulfide boxwork may be distinguished in thin section by contrasting patterns of supergene silica, as revealed by crossed polars, and by thin trails of remnant unreplaced goethite.

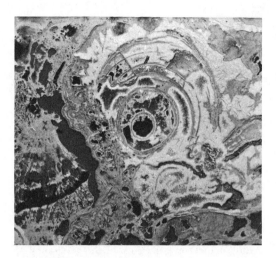

FIGURE 23. Supergene exotic goethite and psilomelane: the common association of colloform-banded goethite (gray) and psilomelane (white), as found in false gossans.

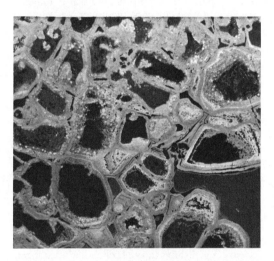

FIGURE 24. Supergene exotic gel goethite: a network of polygonal cells developed during the deposition of colloidal iron hydrates from ground waters.

Supergene silicification will reduce base metal geochemical characteristics of sulfide gossans. It is therefore important to determine from petrographic methods the extent of silicification by either microcrystalline silica or by colloform hyalite. The silicification may not, however, remove all trace of the original goethite boxwork structure. In such cases the magnitude of geochemical anomalies relates to the amount of residual goethite.

While it is generally possible to achieve a useful conclusion in the case of silicified gossans, greater difficulty exists in the interpretation of secondarily ferruginated or manganese oxide-impregnated

gossans. Most boxwork structures contain small amounts of exotic colloform goethite (or psilomelane) as linings on the indigenous goethite cell walls. In areas where ground waters deposit large amounts of supergene limonite or manganese oxides, the cell structures become progressively filled and finally obscured. At this stage the gossan is no longer cellular and appears megascopically as a massive, structureless ironstone. The mineragraphic study of this type of gossan involves the distinction between colloform exotic goethite and indigenous goethite. More difficulty obtains in cases where microcrystalline exotic goethite merges with inherent cell structures. In such cases, oil immersion objectives or interference contrast objectives provide definition of inherent goethite forms within the dominant enclosing mass of exotic goethite. The contrast depends on the relative hardness characteristics and the presence or absence of inclusions in successive generations of goethite. The indigenous goethite is usually harder because it is siliceous, whereas the exotic goethite usually contains clay mineral and organic inclusions.

The greatest difficulty exists in cases in which exotic goethite was deposited from colloidal suspension as minute globules, probably as ovoidal or spheroidal films or bubbles enclosing air or water. After deposition, dehydration, and shrinkage, the mass of globules adopts a polygonal texture resembling forms of pyrrhotite boxwork (cf. Figs. 16 and 24). This polygonal-structured gel goethite appears in small amounts in many gossans and in large proportions in long-weathered gossanous outcrops in base leveled terrains. Alone, it constitutes very fine cellular false gossans, which are frequently found in deeply weathered ultrabasic terrains. Drilling establishes that these accretionary cellular ironstone masses have only shallow depth extension. In addition to its common manifestation as extensive masses composed of minute polygonal cells, gel goethite also appears as small lamellar structures, as polygonal cells with concentric internal partitions, and as more or less ovoidal forms with radiating internal partitions.

Whether the exotic goethite be colloform, microcrystalline, or in pseudo-boxwork cell form, it usually embodies various metals that are of exotic origin. These metals will have been progressively scavenged from ground waters that deposited the goethite. Large proportions of scavenged metals will be found in cases where psilomelane, cryptomelane, and specific manganese compounds such as coronadite, hetaerolite, and chalcophanite are present in the secondary ferruginized gossan, or in the false gossan. Experience has shown that levels of zinc in particular—and, to lesser extents, levels of lead, cobalt, nickel, silver, and copper—may be greatly enhanced and do not reflect the existence of the relevant sulfides at the site of the ironstone.

A third factor of importance in relating present geochemical characteristics to the structures present in a gossan involves the proportion of pyrite, pyrrhotite, or marcasite in the original ore assemblage. Excessive proportions of the boxworks of the iron sulfides among those of the base metal sulfides can usually be considered to have led to significant reductions in the levels of metals that are mobile in an environment of low pH—for example, copper and zinc. The probable magnitudes of the proportions of chalcopyrite and sphalerite in the original ore can usually be provided in these cases from estimates of the abundances and sizes of the boxworks from these sulfides, as seen in polished sections of the gossans. In the general case, except when reactive carbonate gangue was present in the ore assemblage, it can readily be demonstrated that the proportions of base metal sulfide boxwork are greater than, and therefore incompatible with, the present copper or zinc contents of the gossan.

Summary

Mineragraphic techniques, especially aspects of gossan interpretation, are useful methods in mineral exploration. Complemented by the development of exploration geochemistry and geophysics (q.v. in Vol. XIV) in the postwar period, mineragraphy has found wide application in extensive arid regions where weathering, leaching, oxidation, reduction, and other epidiagenetic processes tend to mask indications of mineralization at the ground surface. Mineragraphy, now applied to the geological examination of new terrain and previously explored areas, facilitates the never-ending search for additional mineral resources as demanded by modern technological societies.

ALICK W. G. WHITTLE

References

Blain, C. F., and Andrew, R. L., 1977. Sulphide weathering and the evaluation of gossans in mineral exploration, *Minerals Sci. and Engineering* 9 (July), 119-150.
Blanchard, R., 1968. Interpretation of Leached Outcrops, *Nevada Bur. Mines Bull. 66.*
Boswell, P. F., and Blanchard, R., 1927. Oxidation Products derived from sphalerite and galena, *Econ. Geology* 22 (5), 419-453.
Garrels, R. M., and Christ, C. L., 1965. *Solutions, Minerals, and Equilibria.* New York: Harper & Row.
Groves, D. I., and Whittle, A. W. G., 1976. Mineragraphy applied to Mineral Exploration and Mining, *Australian Mineral Foundation Lecture Series.*
Locke, A., 1926. *Leached Outcrops as Guides to Copper Ore.* Baltimore: Williams and Wilkins, 175p.
Loghry, J. D., 1972. Characteristics of favourable cappings from several south-western porphyry copper deposits, unpublished M.Sc. thesis, University of Arizona, Tuscon.
Park, C. F., and MacDiarmid, R. A., 1970. *Ore Deposits.* San Francisco: W. H. Freeman, 522p.

Cross-references: *Geochemistry, Applied.* Vol. XIV: *Exploration Geochemistry; Exploration Geology; Indicator Elements; Lithogeochemical Prospecting; Pedogeochemical Prospecting; Prospecting.*

MINERAL ECONOMICS

The discipline that applies principles of economic theory to problems involving mineral resources, *mineral economics,* specifically relates concepts and ideas of general economics to the various aspects of the occurrence, exploitation, and final use of minerals. In the past, the major concern of mineral economists has been the study of supply and demand for energy, metals, and nonmetallic or industrial mineral resources. Such studies normally include investigations of the geography of mineral occurrences, valuation of mineral deposits, exploration and production cost analysis, transportation and marketing studies, as well as inquiries into historical use, future markets, and substitution potential for the various mineral commodities. In more recent years, the mineral economist has also been called on for public policy evaluation, specifically to measure or estimate the private and public costs and benefits of mineral resource development as compared with the costs and benefits of alternative uses for land areas such as public parks and recreation (Krutilla and Fisher, 1975; Fisher, 1981).

In many ways mineral economics can be viewed as a specialized area within the general economic study of natural resources (Herfindahl and Kneese, 1974). The unique aspects of mineral economics largely revolve around the nonrenewable nature of mineral deposits as opposed to the renewable nature or regeneration potential of agriculture, forestry, and fishing resources. In addition, mineral resources have become increasingly important in improving the productivity of agriculture, forestry and fishing through chemical fertilizers, pest and disease control, and processing technology. Unfortunately, the production and use of some mineral resources have also had detrimental impacts on these same industries through air and water pollution (Howe, 1979).

Many of the developing nations of the world depend on the production and sale of mineral commodities as their principal source of national income. An understanding of how changes in mineral commodity markets affect national economic growth is a vital part of government planning. The advent of substantially higher petroleum prices has focused new attention on the international oil industry and its economic impact on developed as well as developing nations. The continuing energy crises of the mid-1970s have created new demand for even more expertise in mineral economics.

The range of activities encompassed by the discipline of mineral economics has been described as "sufficiently broad and complex that, for effective work, an understanding of both technical and economic factors is essential" (Buck, 1972, p. 1). Therefore the typical mineral economist normally studies a technical discipline such as geology, mining engineering, mineral engineering (beneficiation of mineral ores), petroleum engineering, metallurgy, or some other earth science or engineering field prior to turning to economics and its application to mineral problems. While it is possible to enter the field directly after academic or practical training in science, engineering, or economics, more and more colleges and universities are developing specialized programs in mineral economics as an independent field. Schools such as Columbia University, the University of Minnesota, Stanford University, and other universities in major mineral-producing states have offered programs in mineral economics for many years, but they have normally organized the study of mineral economics as an adjunct to existing programs in economics or mining engineering. The Pennsylvania State University has perhaps the oldest and most consistent program for specifically training mineral economists. More recently developed programs in mineral economics at the Colorado School of Mines, West Virginia University, and the University of Arizona attest to the growing importance of this profession. The University of Queensland in Australia has also initiated a program in mineral economics, indicative of the growing international recognition of this field. Specific areas of investigation in mineral economics have been outlined as follows (Buck, 1972):

1. Resource availability: quantity, quality, and location of mineral deposits.
2. Regional, national, and international supply and demand factors for mineral commodities.
3. Exploration, development, production, and processing functions.
4. Mineral (commodity) markets and mineral uses.
5. Factors of substitution, secondary and recycled materials, and competition.
6. Transportation methods and costs.
7. Technological change and its economic impact on mineral utilization.
8. Environmental quality as it relates to mineral exploitation.
9. Mineral commodity analysis.
10. Methods of finance and corporate structure of the mineral industry and its components.
11. Mineral policy analysis—government policies, taxation, laws, and regulations.
12. Mineral policy formulation.

Detailed discussion of these activities may be found in Buck (1972) and Vogley (1976). The interested reader may also wish to review one or more of the publications in the following list of references for further information on mineral economics.

HENRY N. McCARL

References

Barnett, H. J., and Morse, C., 1963. *Scarcity and Growth: The Economics of Natural Resource Availability.* Baltimore: Johns Hopkins University Press for Resources for the Future, 288p.

Buck, W. K., 1972. *Mineral Economics: Its Definition and Application.* Canada Dept. Energy, Mines and Resources Mineral Resources Div., 10p.

Fisher, A. C., 1981. *Resource and Environmental Economics.* Cambridge: Cambridge University Press, 284p.

Herfindahl, O. C., 1969. *Natural Resource Information for Economic Development.* Baltimore: Johns Hopkins University Press for Resources for the Future, 212p.

Herfindahl, O. C., and Kneese, A. V., 1974. *Economic Theory of Natural Resource.* Columbus, Ohio: Charles E. Merrill, 405p.

Howe, C. W., 1979. *Natural Resources Economics: Issues, Analysis, and Policy.* New York: Wiley, 350p.

Krutilla, J. V., and Fisher, A. C., 1975. *The Economics of Natural Environments: Studies in the Valuation of Commodity and Amenity Resources.* Baltimore: Johns Hopkins University Press for Resources for the Future, 292p.

McDivitt, J. F., 1965. *Minerals and Men: An Exploration of the World of Minerals and Its Effect on the World We Live In.* Baltimore: Johns Hopkins University Press for Resources for the Future, 158p.

Proceedings of the Council of Economics of the American Institute of Mining, Metallurgical and Petroleum Engineers, various years (published annually since 1966). New York.

Robie, E. H. (ed.), 1964. *Economics of the Mineral Industries,* 2nd ed. New York: American Institute of Mining, Metallurgical and Petroleum Engineers, 787p.

Vogley, W. A. (ed.), 1976. *Economics of the Mineral Industries.* New York: American Institute of Mining, Metallurgical, and Petroleum Engineers, 863p.

World mineral economics, in W. B. Stephensen, K. L. Fetters, and T. C. Frick, (eds.), 1971. *Centennial Volume, American Institute of Mining, Metallurgical, and Petroleum Engineers.* 1871-1970. New York, 42-87.

Cross-references: *Economic Geology.* Vol. XIV: *Mineralogy, Applied.*

MINERAL EXPLORATION—See MINERAGRAPHY. Vol. XIV: ATMOGEOCHEMICAL PROSPECTING; BIOGEOCHEMISTRY; BLOWPIPE ANALYSIS; EXPLORATION GEOCHEMISTRY; EXPLORATION GEOLOGY; EXPLORATION GEOPHYSICS; GEOBOTANICAL PROSPECTING; GROUNDWATER

EXPLORATION; HYDROCHEMICAL PROSPECTING; HYDROGEOCHEMICAL PROSPECTING; INDICATOR ELEMENTS; LAKE SEDIMENT GEOCHEMISTRY; LITHO-GEOCHEMICAL PROSPECTING; MARINE EXPLORATION GEOCHEMISTRY; MARINE MAGNETIC SURVEYS; MINERAL IDENTIFICATION, CLASSICAL FIELD METHODS; PEDOGEOCHEMICAL PROSPECTING; PLATE TECTONICS, MINERAL EXPLORATION; WELL LOGGING.

MINING ENGINEERING GEOLOGY— See SHAFT SINKING. Vol. XIV: BOREHOLE MINING; COAL MINING; MARINE MINING; MINE SUBSIDENCE CONTROL; MINING PREPLANNING; PLACER MINING.

MINING GEOPHYSICS—See Vol. XIV: EXPLORATION GEOPHYSICS.

MINING METHODS—See CAVITY UTILIZATION; SHAFT SINKING; TUNNELS, TUNNELING; URBAN TUNNELS AND SUBWAYS. Vol. XIV: BOREHOLE MINING; COAL MINING; MARINE MINING; PLACER MINING.

MINING PRACTICE—See CAVITY UTILIZATION.

MUSEUMS AND DEPOSITORIES—See Vol. IVB: MUSEUMS, MINERALOGICAL. Vol. XIV: MAP AND CHART DEPOSITORIES.

N

NUCLEAR PLANT SITING, OFFSHORE

Any major site selection study involves both a wide range of considerations and a complex choice of alternatives. Thus the selection of possible, feasible, and finally optimum sites can be made only through systematic evaluation. The key, in turn, to successful evaluations lie in developing a valid set of weighted criteria by which sites must be judged.

Unfortunately, any subjective procedure is limited by the system used. These limitations may be minimized through team participation by developing criteria, screening of areas, and selecting locations for detailed investigations prior to selecting sites. Subjective differences are reconciled using the *Delphi method* (subsequently described).

When studying siting for floating nuclear power plants off the east and Gulf coasts of the United States, the authors followed the aforementioned procedure, which is a standard evaluation technique used in making this sort of system engineering decision (see, e.g., Asimow, 1962). Fischer and Fox (1973) and Fischer and Ahmed (1974) postulated similar concepts for evaluating sites for nuclear power plants. In outlining the siting approach that follows, emphasis was placed on general worldwide practical applications.

Methodology

The overall methodology used in what we will call the U.S. study was adapted largely from that described by Fischer and Ahmed (1974), and McHarg (1969). In designing the system, prime emphasis was placed on relative simplicity, since a common failing of many complicated, subjectively oriented systems has been their subsequent misuse, often owing to ambiguity. Any methodology should also include an efficient filtering mechanism that enables the user to consider a large number of criteria in different geographic environments. In

addition, it should be replicable; conducive to quantification, which may be debated and refined if opinions differ; and flexible enough to accept new input on changing considerations. Finally, a means to reach compromises should be built into the system.

Siting Team. The technical siting team selected should consist of scientists with specific knowledge in the general areas of socioeconomic/environmental appraisal (especially marine ecosystem sensitivity), geology and seismic evaluation (see *Seismological Methods*), engineering soils and foundations (see *Foundation Engineering*), ocean engineering (see *Oceanography, Applied*), coastal processes (see *Coastal Engineering*) and hydrology (see *Hydrogeology and Geohydrology*). In selecting individuals to coordinate and manage the various discipline activities, preference should be given to scientists with broad knowledge and experience in (1) investigations for nuclear power plants on shore and/or (2) investigations for large structures offshore. Overall coordination should be provided by a project manager experienced in environmental and earth science aspects of nuclear plant siting.

Besides coordinating team efforts, the project manager should be responsible for maintaining stringent liaison with an independent review board, which, in the U.S. study, included civil design engineers, representatives of the nuclear plant manufacturer, and prospective owners.

All key personnel on the team must be thoroughly familiar with general siting concepts, all Nuclear Regulatory Commission guidelines and appendixes (e.g., 10 CFR 100, and Appendix A, USAEC, 1974; USAEC, 1973 and 1972 and notably Appendix M to 10 CFR 50 relating to standardization of design and the manufacture of reactors away from the site of eventual operation) and the National Environmental Policy Act of 1969 and amendments (or the equivalent in countries other than the United States).

The appointed technical team should be responsible for establishing criteria and weighting them in order of importance, and making all major decisions in the site-screening process. The responsibility of the review board is to monitor the progress of the technical team through each developmental stage in the site-selection study and to provide perspective on and orientation toward both a mega- and a detailed mini-scale.

Basically, therefore, the technical team should consist of idealists who are largely free from the burden of economic considerations and concerned only with the optimum site(s) in terms of their science. The review board, on the other hand, should be concerned with the practicality and economics of propositions. It is their prerogative to question a site rating, for example, because of its distance from a load center or existing transmission facility, or, in the case of an inshore site (behind a barrier island or beach site, say), because of the dredging cost for installation. Their function revolving around cost-benefit and engineering design includes examining the economic feasibility of, for example, a possible three-unit (3,450-MW) power plant, as opposed to a two-unit (2,300-MW) plant at a favorable location, the conceptual design and costs of different breakwater systems, and so on. The review board, therefore, has the power of persuasion, but cannot eliminate or alter the rating designated to a particular site without the agreement of the technical team.

Resolving Conflict. The Delphi method, often used in scientific studies to reconcile subjective differences among a group of people, is suggested. Differences in opinion in developing criteria, assign relative importance factors to these criteria and disciplines, to rate sites, and so on.

The Delphi method is essentially an iterative question-and-answer procedure with the following basic characteristics: (1) anonymity, which eliminates domination of any one individual; (2) feedback, which allows an individual to judge the views of others; and (3) statistical response, which lessens the pressure for group conformity and allows each participant to be represented in the final result.

Criteria

Criteria for judging sites fall into two categories: (1) exclusion criteria, which eliminate from further consideration obviously unsuitable sites, and (2) rating criteria, which facilitate a common ground for comparison and relative evaluation of sites. Each criterion is assigned an importance factor according to the degree to which it is intended to influence the judgment of sites. The disciplines and criteria discussed here were developed to identify potential sites for floating nuclear power plants on a regional scale, in the first stages of site selection.

Environmental Assessment

Exclusion Criteria. The presence of one or more of the following features is deemed to conflict directly with siting (i.e., a go-no-go situation): (1) designated shipping lanes and intracoastal waterways and maintained areas in which established activities would preclude siting; (2) obstructions—

such as pipelines, cable routes, fish havens, and similar designated-use areas—in which existing structures may present a hazard to, or be impaired by, the siting of a plant; (3) military restricted and warning areas, where inherent dangers may be presented to a floating nuclear power plant (FNP): for example, air-to-air missles, intense flight activity, and gunnery practice; (4) aquatic preserves and designated preservation areas of highly redeeming social and/or ecological value (restricted in the sense that any sort of development is prohibited or at least strongly discouraged); (5) areas immediately adjacent to large population centers.

It is possible for specific variances to be granted to site an FNP near the periphery of a larger restricted area if warranted by advantageous conditions in the other disciplines.

Environmental Rating Criteria. The U.S. study established the environmental criteria and importance factors in Table 1.

Geology/Seismology Rating Criteria. See Table 2.

Soils/Foundation Rating Criteria. Many of the criteria itemized are applicable not so much to the plant itself—which is mounted on a floating barge—but rather to the protective breakwater and mooring system for the plant (see *Nuclear Plant Protection, Offshore*). This is particularly true in the case of seismic and foundation design. See Table 3.

Hydrology Rating Criteria. See Table 4.

Ocean Engineering/Coastal Processes Rating Criteria. See Table 5.

Definition of Terms

To avoid ambiguity in the use of criteria when implementing the rating system, all subdiscipline terms must be accurately defined. The following definition of liquefaction will serve as an example.

The term *liquefaction* describes the phenomenon in which certain soils, under dynamic loading (e.g., earthquakes) lose strength and acquire a degree of mobility sufficient to cause excessive settlement, slope instability, or foundation failure. During an earthquake, a soil mass is subjected to repetitive cyclic loads of varying stress levels. The ground motions induce both normal and shear stresses that may vary with time and location within the soil strata. In saturated soils, if there is

TABLE 1. Environmental Rating Criteria

	Rating Criteria	Importance Factor
1.0	Population density	15%
2.0	Airports	25%
3.0	Conservation area or proximity to preservation area	25%
4.0	Ecosystem sensitivity	35%

TABLE 2. Geology/Seismology Rating Criteria

Rating Criteria	Subdivision Importance Factor	Subdiscipline Importance Factor
1.0 STRATIGRAPHY		30%
1.1 Substrate materials and conditions	40%	
1.2 Continuity of Strata	30%	
1.3 Bedforms, coastal forms, and dynamics	30%	
2.0 STRUCTURE		30%
2.1 Regional	40%	
2.2 Local	60%	
3.0 SEISMOLOGY		20%
3.1 Local	40%	
3.2 Regional	60%	
4.0 NEED FOR ADDITIONAL INFORMATION		20%

TABLE 3. Soils/Foundation Rating Criteria

Rating Criteria	Importance Factor
1.0 Strength	40%
2.0 Liquefaction	40%
3.0 Compressibility	20%

Note: Foundation problems related to potential solution cavities in limestones were included under strength and compressibility.

TABLE 4. Hydrology Rating Criteria

Rating Criteria	Importance Factor
1.0 Areas of large ground water withdrawal	40%
2.0 Areas where the dredging operations should be limited (where confining layers prevent salt-water intrusion to aquifers)	40%
3.0 Offshore disposal of wastes and subsurface disposal through wells	20%

TABLE 5. Ocean Engineering/Coastal Processes Rating Criteria

Rating Criteria	Importance Factor
A. Ocean Engineering	
1.0 Areas of anomalous surge and wave height (probable maximum hurricane, tsunami)	100%
B. Coastal Processes	
1.0 Dynamics of the coastline in terms of routing with respect to: (a) circulating water structures from confined areas; and (b) transmission lines, (Watson, et al., 1974)	50%
2.0 Coastal and nearshore effects, erosion, and deposition	50%

no drainage, the pore-water pressure increases. If the pore-water pressure builds up to a point where the effective (intergranular) stress becomes zero, the saturated granular soil loses its strength completely and a liquefied state develops (see *Consolidation, Soil; Foundation Engineering*). The U. S. study supplemented this definition with a plot showing the gradation characteristics most commonly associated with liquefiable soils. This graph emphasized the susceptibility of silts and poorly graded fine sands to liquefaction. A further graph showed the effect of particle size on the liquefaction potential of soils. This plot again emphasized that: (1) the soils susceptible to liquefaction are the silty sands and very fine sands; and (2) gravels and insensitive clays are, for all practical purposes, not susceptible to liquefaction.

Primary Screening Process

Before potential sites are rated according to the criteria, a primary screening process is conducted. If location maps (e.g., nautical charts with scales of about 1:50,000 with soundings) are displayed in a work room, each disipline team can delineate on clear, acetate overlays on the maps, all major data that might influence siting. For

373

example, the geology/seismology team might point out a major fault system.

After all data are collected, the technical team should examine the overlays to isolate areas to be removed from further consideration and delineate which areas might be rated as potential sites. In the U.S. study, for instance, the only factors deemed serious enough to preclude FNP siting were in the environmental category: shipping lanes, physical obstructions, military restricted areas, preservation areas, and areas with high population density.

Rating of Areas

All areas delineated by the primary screening process are rated according to the criteria developed by the siting team. Naturally, before this procedure can begin, extensive technical data must be collected in each subdiscipline category. As information is added to the bank of knowledge for each site, sensitivity of subsequent cycles of the methodology should be improved.

All areas not excluded in the primary screening are rated on a point scale. Following is an example of such a scale, which could be used in all disciplines:

0—not to be considered; rule out regardless of qualification
1—poor; negative factors predominate
2—has negative considerations
3—average; no negative considerations
4—has some factors that are better than average
5—excellent, under terms of consideration

Where this scale cannot be directly applied in a meaningful way, supporting scales such as the one in Table 6 can be constructed. The scale shown

TABLE 6. An Example of a Supporting Scale for Runway Lengths

FNP Proximity (Mi)	Rating Based on Runway Length		
	Over 7,000 Ft.	3,500- 7,000 Ft.	Less than 3,500 Ft.
>8	5		
>7	4		
>6	3	5	
>5	2	4	5
>4	1	2	4
>3	0	1	2
>2	0	0	0

here was used to rate areas with airports—the major criteria being the proximity of a proposed FNP site to airports having different runway lengths and a measure of the size aircraft they could support.

The overall site rating is determined by arithmetic, for example, see Table 7.

General Discussion

Methodology. Although computers were not used in the U.S. study, grid-based information systems have been successfully used in many similar regional studies. The computer recording of multidisciplinary data about geographical locations is recommended in studies where extensive manipulation of data is anticipated, to assist in the primary screening of areas (e.g., a regional study of the coastal zone for general development). Other computer techniques such as factor analysis and discriminant function analysis (Harbaugh and

TABLE 7. Site Rating Determination

foundation evaluation	etc.	site rating factor
$IF_a \times \sum_{i=1}^{3} 1Q = U_a$ +	(similar procedure for → each discipline)	$\sum_{i=e}^{e} U = R$

$$\uparrow$$
$$\text{strength}$$
$$IF_1 \times Q_1 = 1Q_1$$
$$+$$
$$\text{liquefaction potential}$$
$$IF_2 \times Q_2 = 1Q_2$$
$$+$$
$$\text{compressibility}$$
$$IF_3 \times Q_3 = 1Q_3$$

where IF = importance factor, Q = quality factor, and R = rating factor.
$$\sum_{i=1}^{n} 1Q \leq 5$$

Merriam, 1968) have also been used successfully (e.g., Watson, 1971) to discern simplifying relationships in complicated data.

Rating of Areas—Step 1. Step 1 in any study is aimed at isolating areas for further investigation. In the U.S. study, for example, the siting team selected criteria from only three disciplines—environmental assessment, geology, and seismic evaluation and foundation engineering, since these were considered to influence most significantly the early stages of site selection. Each of these three disciplines was given equal weight.

Examples of Rated Areas. The area rated highest in step 1 was given a multidiscipline rating factor of 4.1 (on a 5-point scale). The area, situated in an open space between an aircraft restricted zone and an aquatic preserve, was assigned an environmental rating of 4.8 contingent on obtaining a variance to enter military (aircraft) restricted space. It was given a geological rating of 3.6, upgraded on the grounds of the lack of tectonic structure in the area, but downgraded on the basis of probable Pleistocene channeling and sparse data on stratigraphic conditions. It was assigned a rating of 4 in the soils category, as favorable data extrapolated from a nearby bridge boring suggested the presence of dense sands at a depth of about 12 m.

A poor multidiscipline rating (2.9) in another area reflected: (1) proximity to a national seashore, a shipping lane, and a buried pipeline (environmental rating, 2.2); (2) a location within 8 km of a known salt dome, where ring and radial faulting associated with dome emplacement was likely, and within 8 km of an earthquake epicenter (geology/seismology rating, 2.6); and (3) the aforementioned soils rating.

Boundaries of the Study Area. The offshore limit set by the team combined consideration of siting within State Territorial Sea Limits (e.g., New Jersey, 3 nautical mi.; Florida, 3 marine leagues or 9 nautical mi.; Mississippi, 3 nautical mi., which means 3 mi. seaward of barrier islands) and in less than 60 ft of water. A water depth in the range of 45 ft was considered to represent the optimum for siting, on one hand, in terms of satisfying, without dredging, the manufacturer's plant design envelope for siting in a minimum depth of water, and on the other in terms of not incurring excessive breakwater construction costs (Watson, et al., 1975).

The inshore limit for siting was conservatively established at approximately 1 nautical mi. inland of any body of seawater (shoreline, bay).

Environmental Considerations. In the U.S. study, the siting team tentatively concluded that, generally, the most attractive siting area is offshore or in open ocean—as opposed to (1) in a bay, in an estuarine, or behind a barrier island site, or (2) on a beach or in an inland location. In restricted waters, the sensitive ecosystem would be affected both by the dredging necessary for the emplacement of units and thermal pollution during operation. Marine life in nearshore waters, in turn, has a tremendous influence on the productivity and general well-being of the entire marine ecosystem. For example, Gunter (1967) estimates that more than 97 percent of the entire commercial fisheries catch in the Gulf is ultimately estuarine-dependent. The Gulf fishery catch in turn represents 35 percent of the total U.S. production (Riley, 1970). A growing national awareness of these facts will, with passing time, make it increasingly difficult to develop potentially harmful industry in those areas.

Geological and Seismological Considerations. The better part of the east and Gulf coasts of the United States lies in a region of low seismic risk; as such, the coasts are well suited to be nuclear plant locations.

Stratigraphically, potential siting areas were confined to the seaward extention of the semi-consolidated and unconsolidated Tertiary sediments of the Coastal Plain. The structural relationship, while relatively simple on a macro scale, is complicated on a small (engineering) scale; a well-directed program of subsurface investigation is essential to outline, in sufficient geotechnical detail, the precise nature and extent of shallow (30 m) interbedded cohesive and noncohesive sediments (see *Marine Sediments, Geotechnical Properties*).

Local siting considerations of a geological nature may include, for example, the active but slow geosynclinal sedimentation and associated salt diapirism, salt doming, and growth faults similar to those on the eastern Gulf coast and the potential subsidence associated with fluid withdrawals (widely publicized in the Houston-Baytown area of Texas, where elevation changes of several feet have been recorded). This form of subsidence, incidently, is predictable and considered not to pose an insurmountable problem to siting.

Foundation Materials. Even in the preliminary stages of site selection, some field data are essential to predict foundation conditions. (see *Foundation Engineering*). In more detailed follow-up studies considerable economy may be achieved through a rigorously coordinated program of (1) vibratory corings that provide a shallow ($\pm$12 m) but continuous stratigraphic profile; (2) continuous seismic reflection coverage, which facilitates correlation between borings; and (3) relatively undisturbed sampling. (Laboratory and *in situ* test data quantify and extend the geotechnical subsurface model constructed on the basis of data available from 1 and 2 just discussed.

Coastal Process Considerations. From findings of other U.S. studies, the study team concluded that although the hydrodynamic regime in many parts of the East and Gulf coasts is relatively active, the mechanics of coastal processes would not, in general, preclude the siting of FNPs. Findings in this area did, however, persuade them to recommend that in the best interests of long-term economy, coastal process-related considerations be taken into account in all site-selection facets beyond the initial rating of areas. It is significant in planning any offshore construction to note, for instance, that within historical times entire barrier island systems have migrated as far as 7 nautical mi. (e.g., the Mississippi Sound barrier island system), and that new islands have formed while previously inhabited islands have completely disappeared (e.g., Tuckers Island, New Jersey). Similar highly active erosion is documented in many parts of the world. Such movement, if not taken into account, could jeopardize the integrity of a buried transmission system of an FNP, (see *Undersea Transmission Lines, Engineering Geology*) and/or the safety of a protective breakwater (see Vol. XIV: *Nuclear Plant Protection, Offshore*) designed to incorporate the instability of the barrier island. Further consideration relates to areas where subaerial erosion has taken place, since these tend to have buried channel systems that complicate foundation analysis.

In summary, a good knowledge of longshore drift systems, sediment transport rates, wave energy levels, and erosion rates is important from the standpoint of both the design of the plant facility and an assessment of the potential environmental influence resulting from construction of a proposed power plant (see *Coastal Engineering;* Vol. XIV: *Coastal Zone Management*).

Ocean Engineering Considerations. The main purpose of this phase of study is to analyze the general ocean environment—including tides, currents, waves, and storms—and to generate preliminary design information such as whether storm surge will occur and what sizes and types of waves associated with the most severe storm that can conceivably occur in any particular region.

The design basis of hurricane protection for a nuclear-powered electrical generating station was established by the Atomic Energy Commission (the forerunner of the Nuclear Regulatory Commission) with the aim of preventing the loss of function of safety-related structures, systems, and components as the result of the combined action of surge and waves associated with a "Probable Maximum Hurricane" (PMH). As the parameters of the PMH are defined in terms of meterological characteristics, the evaluation of PMH flooding involves a choice of appropriate meterological parameters and the use of mathematical techniques for the prediction of storm surge and wave action (Noble, 1975).

Tentative calculations suggest that throughout the U.S. study region, the design parameters discussed above do not vary greatly.

Ground Water Hydrology Considerations. Although ground water hydrology criteria are listed in a separate category, this area of consideration is regarded to relate to more localized rather than regional siting. In the early stages of the study, therefore, hydrological data may be collected along with geological data.

Conclusion

An important conclusion to be drawn from the U.S. study is that the procedure improved judgment and formalized documentation resulting from the use of a systematic, quantitative approach to siting. This was evident in terms of overall economics, not only in selecting the best site(s), but also in defending the site during licensing and promoting cost-effective engineering during construction. It is important to note that in the study described, the value of a systematic methodology was greatly enhanced by both the standarized plant and envelope of site parameters of the FNP (Bonsack, 1975).

The methodology outlined in the present text is considered to be sufficiently flexible to be used on smaller studies, for validating, invalidating, or for adding credibility by quantification to sites selected by less rigorous techniques. Any methodology is in itself merely a tool, however, and not a substitute for sound engineering and management, for rigorous data collection and meticulous analysis, or in fact for comprehensive reporting on data collected. For example, sites initially eliminated through the pure application of the method can be made viable through creative engineering or by petitioning for the adjustment of institutional restrictions. Lastly, it is emphasized that no methodology can improve judgments merely by quantifying and manipulating them (least of all by computer application).

IAN WATSON
JOSEPH A. FISCHER
DAVID P. MANIAGO

References

Asimow, M., 1962. *Introduction to Design,* Englewood Cliffs, N.J.: Prentice-Hall, 135p.

Bonsack, Frederic E., 1975. Siting alternatives for floating nuclear plants: an update, in *Ocean 75.* New York: Institute of Electrical and Electronics Engineers, 643-651.

Code of Federal Regulations, 1974. 10 Part 100, *Reactor Site Criteria.* Washington, D.C.: Atomic Energy Commission, 409.

Code of Federal Regulations, 1974. 10 Part 100, *Seismic and Geologic Siting Criteria for Nuclear Power Plants,* Appendix A, 412-419.

Code of Federal Regulations, 1974. 10 Part 50, *Standardization of Design: Manufacture of Nuclear Power Reactors; Construction and Operation of Nuclear Power Reactors Manufactured Pursuant to Commission License,* Appendix M, 323-325.

Fischer, J. A., and Ahmed, R., 1974. *A Systematic Approach to Evaluate Sites for Nuclear Power Plants.* Cranford, N.J.: Dames & Moore, 23p.

Fischer, J. A., and Fox, F. L., 1973. Siting constraints for an offshore nuclear power plant, *Dames & Moore Eng. Bull.* **42,** 3-6.

Gunter, G., 1967. Some relationships of estuaries to the fisheries of the Gulf of Mexico, in G. H. Lauff (ed.), *Estuaries,* Am. Assoc. Adv. Sci. Pub. 83, 621-638.

Harbaugh, J. W., and Merriam, D. F., 1968. *Computer Applications in Stratigraphic Analysis.* New York: Wiley, 282 p.

McHarg, I. L., 1969. *Design with Nature,* Garden City, N.Y.: Natural History Press, 197p.

Noble, R. M., 1975. Design basis flooding at offshore reactor sites, *Ocean Engineering III Conference,* American Society of Civil Engineering, Newark, Delaware, June 12-15, 89-95.

Raiffa, H., 1968. *Decision Analysis: Introductory Lecture on Choices Under Uncertainty.* Reading, Mass.: Addison-Wesley.

Riley, F., 1971. *Fisheries of the United States.* Washington, D. C.: U.S. Department of Commerce, National Marine Fisheries Service, Current Fishery Statistics No. 5600, 101p.

Watson, I., 1971. A preliminary report of new photogeological studies to detect unstable natural slopes, *Quart. Jour. Eng. Geology* **4,** 133-137.

Watson, I., et al., 1974. Transmission lines probe a new frontier, the ocean, in *Electric Power and the Civil Engineer.* New York: American Society of Civil Engineering, 557-578.

Watson, I., et al., 1975. Geotechnical aspects of rock borrow for large breakwaters, in *Offshore Technology Conference, OTC-2392.* Houston: Offshore Technology Conference, 553-563.

Cross-references: *Coastal Engineering; Foundation Engineering; Marine Sediments, Geotechnical Properties; Undersea Transmission Lines, Engineering Geology.* Vol. XIV: *Coastal Zone Management; Nuclear Plant Protection, Offshore; Ocean, Oceanographic Engineering.*

NUCLEAR REACTOR SITING, ENGINEERING, GEOLOGY—See NUCLEAR PLANT SITING, OFFSHORE.

O

OCEANOGRAPHY, APPLIED

Oceanographic programs can be viewed as consisting of two kinds of activities: basic science-oriented activities, and applied, or mission-oriented activities.

Basic science activities are those whose primary goal, as viewed by both scientists and sponsoring agencies, is the advancement of *ocean science* as a field of knowledge. Their purpose is to increase understanding of oceanic phenomena in physical, chemical, geological, or biological terms. The great hypotheses or theories about which such programs cluster are concerned with such questions as how the continents and ocean basins developed, how the oceans were filled, why and how life began in the sea and why it has evolved there as it has, and how the oceans, the atmosphere, and the hydrologic cycle interact to create weather, climate, and ocean currents.

In contrast, *applied oceanography* is more closely identified with the application of principles and concepts to a specific and generally limited problem. Many applied programs are mission-oriented and have as their purpose the solution of practical problems faced by an agency in meeting its public responsibility.

Problem areas may be in engineering, defense, exploiting fish or mineral resources, protecting public health, developing recreational facilities, or providing seismic sea-wave or hurricane warnings, to name a few. If the contribution to the solution of specific problems is direct and immediate, the program may be considered problem-oriented; where the contribution is indirect, deferred, or only speculative, it may be classified as subject-oriented. Although the oceanographer is concerned with all aspects of scientific study of the oceans—including the physical, chemical and biological aspects—it is the physical parameters that have the most bearing on offshore engineering. Examples of specific applications to engineering considerations are shown in Table 1.

The parameters of main interest in oceanography include temperature, depth, salinity, sound velocity, current (value and direction), wave height and direction, tide, dissolved oxygen, and pH. There are several ways of measuring each of these and, in most cases, a wide range of equipment to use. Before assessing the overall capability of modern oceanography to support offshore engi-

TABLE 1. Physical Parameters — Use in Offshore Engineering

Parameter	Important Uses
Temperature	Diving operations
	Determination of velocity of sound
	Effect of marine growth on structures
	Effect on corrosion/fatigue, e.g., on risers
Depth	Diving operations, submersible operations
	Profiling of data through water column
Salinity	Determination of velocity of sound
Sound velocity	Direct measurement of velocity of sound for precision navigation and survey data
Current	Knowledge of loading on structures
	Determination of seabed movements that might lead to scouring
Wave	Knowledge of loading on structures
Tide	Correction of bathymetric data

neering, let us summarize the present tools of the oceanographer's trade.

Temperature

The most common instrument to measure temperature is a type of transducer known as a *thermistor*, which is simply a highly stable resistor (usually made of platinum), the resistance of which changes in a known manner with temperature. The thermistor forms one part of a *Wheatstone bridge* and therefore gives a voltage reading corresponding to the temperature.

Depth

The two main types of instruments used to measure depth are the *strain gauge* and the *piezoelectric crystal*. The principle behind the operation of the strain gauge transducer is that the resistance of a wire changes with elongation, which measures less accurately than the more expensive piezoelectric crystal systems. The crystal systems work on the principle that the resonant frequency of the piezoelectric crystal (quartz) varies with pressure-induced stress. Quartz crystals are used in almost every frequency standard application because of their excellent stability, superior elastic properties, and insensitivity to effects such as temperature. One of the best known systems employing this principle is the Digiquartz unit.

Salinity

The measurement here is, in fact, that of the conductivity of seawater. One method is simply to measure the resistance between two electrodes, but its disadvantage is that it is essential to keep the electrodes clean and free from grease, corrosion, and, for long-term immersion, marine growth.

More satisfactory is the use of *conductivity sensors*. In one British-made brand of sensors, for example, the transducers are two toroidal transformers protected from the environment by a metal, encapsulated pressure housing. During immersion, the toroidal transformers are inductively coupled by the conducting loop formed by the surrounding seawater. A voltage applied to the input winding of one transformer induces a current to flow in the seawater loop, impressing a magnetomotive force on the second transformer.

Sound Velocity

While the velocity of sound in water can be determined by measuring temperature, salinity, and pressure (depth) and by applying these values to established empirical formulas, a more satisfactory method is to measure the velocity directly by simply measuring the time for a pulse to travel a known distance in seawater (Fig. 1.). Accuracies of ± 0.15 m/sec are achievable with this method (see Vol. XIV: *Acoustic Surveys, Marine*).

Current

Two main types measuring devices for current are those designed for measuring particle velocity and those for measuring ocean currents. For par-

ticle velocity, again, two main types of unit are available: the electromagnetic type, which works on the principle of electromagnetic induction caused by the movement of a conductor (seawater) over electrodes; and the acoustic type, which works on the principle that movement of the water will increase or decrease the time taken for a signal to pass through a known distance.

For the measurement of ocean currents, quick response is not necessary, so rotor-and-propellor units are more commonly used. Direction of current is almost invariably measured by a magnetic compass.

Waves

Some devices and principles used for wave measurement are summarized in Table 2. The choice of which type to use depends on the application. Wave direction is more difficult to measure; the options are measurement of surface slope (IOS pitch/roll buoy), photography, and radar.

TABLE 2. Wave Measurement

Type of Device	Principle Used
Floating body	Mechanical movement Vertical accelerometer
Electrical	Resistance between conductors Capacitance
Pressure	Pressure transducer (strain gauge) Pressure transducer (quartz crystal)
Optical	Laser—continuous wave, amplitude modulation Laser—pulsed
Radar	Pulsed FM
Acoustic	Upward-looking echo sounder

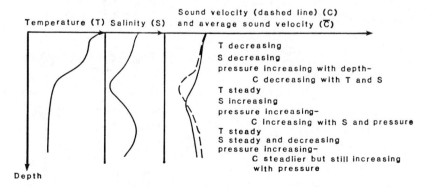

FIGURE 1. Typical curves for temperature, salinity, sound velocity, and average sound velocity against depth.

Tides

There are several proven methods of tide measurement, but the primary requirement is some form of self-recording capability, often in remote locations. Considering the large range of oceanographic instruments available today, the obvious shortcoming is the lack of an integrated approach to the problems of data collection, recording and processing. Individual manufacturers of ranges of equipment have gone some way to solving the problem for their particular products. One British firm, for instance, has designed all its products on a modular "black box" basis, so its telemetering and ocean profiling system, for example, can be extended to meet any requirement for shipboard use. There is still, however, considerable scope for development of general-purpose integrated systems. Another firm is developing its acoustic telemetry systems for calling up a subsea data device (e.g., current meter, tide gauge) and asking for the recorded data to be transmitted acoustically to the surface in much the same way as telephone-answering machines can be interrogated remotely to play back any messages. The main advantage of this system is that it is not necessary to recover the current meter or tide gauge to gather the data.

There is an increasing demand for the use of releasable transponders as releases for current meters. Being able to range in on the current meter string before release, and to track the string on its ascent and when it reaches the surface, virtually guarantees the recovery of data.

Other Developments

Among other developments is the use of radio transmitters for sending data to ships. Both the Van Essen Tide Gauge and the Datawell Waverider Buoy use this method, but the development of a general purpose radio transmitter/receiver, which is under way, could enable the survey ship to log and process, in real time, oceanographic and survey data. A block diagram of the complete integrated circuit is shown in Figure 2, illustrating a method by which this additional data could be taken into an existing integrated navigation system.

Data Collection and Processing

Possibly no other technological innovation promises greater benefit to oceanography than the application of digital concepts to collecting, organizing, interpolating, processing and classifying oceanographic data. These benefits would include relief from the routing and drudgery associated with many present operations.

A second real benefit is time saving. Present progress is due to faster data collection using sensors with shorter response times. For many types of measurement this means data can be obtained while the ship is in motion, possibly at cruise speed. Probably more significant is the ability to process data aboard ship in order to present a "quick-look" at the data while on station. It is within our present capability to devise sensors, systems, and concepts for collecting an

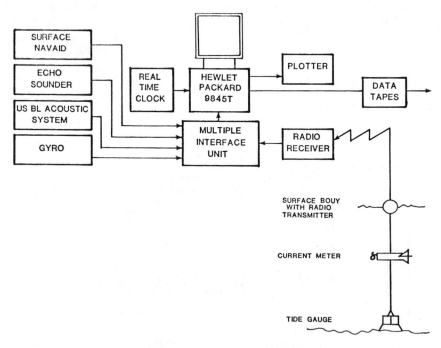

FIGURE 2. Real Time Data Acquisition.

increased volume of data that may be processed into valuable charts, statistics, and predictions on an around-the-clock basis.

Standardization presents a third benefit. Advances in applied data processing systems ensure that a data format and processing technique can evolve that will accommodate a great majority of oceanographic parameters. For further discussion see Ketchum and Stevens (1961).

Examples of Applied Oceanography

The forms of applied oceanography are so varied that no neat summary is possible. A few examples, though, may illustrate the role of applied oceanography today. At times nature becomes dangerous or damaging to people or to our environment. Natural phenomena may take the form of electrical storms, hurricanes, cyclones, floods, tsunamis, or earthquakes. Reporting, warning, and protective services apply to several areas: (1) the design, development, and production of instrumentation; (2) investigations and research; and (3) routine data production and analysis. Since the above-mentioned natural disasters recur at frequent intervals, the reporting, forecasting, and warning services are maintained throughout the world 24 hours per day. Applied research and development has led to improvement in forecasting, increased utilization of satellite cloud photographs, and improved and more economical communication facilities.

Communications. The application of the art of communications to the task of implementing oceanographic research takes place in two media—under the sea (acoustics) and in the atmosphere (electromagnetics). In each media there have been developed rather sophisticated technologies that largely resolve the problems peculiar to oceanographic requirements (Frasch, 1962).

One of the basic problems that has been given considerable attention even on an international scale is that of obtaining suitable frequency allocations that will permit full-time communications. The preponderance of oceanographic traffic appears to require frequencies in the high-frequency (HF) band because of the long-distance propagation characteristics combined with good signal-to-noise performance. Unfortunately, this band is also in demand by many other users for the same reason.

The problem of communications through the air/sea interface appears to offer little hope of substantial success because of the diverse transmission characteristics of the two media for any selected method. Limited success with shallow penetration of radio-frequency energy at very low frequencies has been obtained. The more usual method employs acoustic frequencies in the denser water medium, and radio frequencies above the water, with some form of transducer acting as a conversion device. Some researchers hope the laser may provide some improvement in this area. The role of acoustics is well established for all underwater operations (see Vol. XIV: *Acoustic Surveys, Marine*). Despite the inherent limitations of acoustics, there appears to be nothing capable of providing better performance. Basic and applied research in underwater communications should continue, however, with the hope of achieving breakthroughs (Horton, 1957).

Some of the more unique communications techniques, such as ionosphere scatter and meteor trail modes, appear to offer some potential for ocean communication of stored data. The meteor trail principle involves storage and time compression of data at the data platform, while the communications system monitors propagation conditions. These unique modes, although attractive and apparently feasible, have serious operational and practical problems yet to be solved.

Navigation. Under the broad heading of navigation are two less general operations—course keeping and positioning. In general, the older methods of navigation—which might be called unaided celestial and elementary dead reckoning—in which only a clock, a compass, and a log are used, have little value today for ocean research because of insufficient accuracy. There is no single system based on high-accuracy modern technology that is by itself sufficient. Moreover, requirements exist today that barely can be met with optimal combination of the most accurate systems, supplementing and cross-checking each other.

The advent of modern electronics, particularly its contribution to hyperbolic, ranging, and satellite navigation systems, has been responsible for major improvements in navigation accuracy (see Vol. XIV: *Surveying, Electronic*). Other methods that approach or surpass the accuracies attainable with the systems mentioned above are variations of the inertial technique and, under some conditions, Doppler methods (Mourad and Fubara, 1974).

Energy Source. The technical nature of this section logically leads to an outline breakdown into energy sources—nuclear, chemical, and solar—and types of conversion methods for each source.

Considerable research is being conducted in materials and techniques for all the energy sources. In the nuclear field, new processes are being considered (fission fragment and plasma-magnetohydrodynamics), and new methods of using radionuclides are appearing. In the field of chemical sources, the fuel cell promises the highest efficiencies and probably is the most active research area in materials and methods. Closed-cycle systems also can fill an important gap in power sources for longer mission times than can be supplied by batteries.

Power generation from the sea also promises energy for remote, unattended stations as well as major power possibilities from harnessing temperature differentials. This field is not very active at present (Anon., 1970).

Waste Disposal. There are many competitive uses for both fresh and marine water, and often an area of water may be used for many purposes at the same time. Water pollution by sewage, industrial and agricultural wastes, and silt remains the largest problem affecting the use of water for municipal, industrial, or commercial purposes. In addition, sewage in particular and other forms of pollution in general reduce the desirability and availability of water use for other recreational purposes such as swimming and boating. It has been noted that the most common causes of fish kills in both ocean and lacustrine environments are (1) domestic wastes, (2) industrial wastes, (3) mining or dredging operations, (4) agricultural poisons, and (5) sediment. All these, of course, affect both recreational and commercial aspects of this resource. To alleviate these problems, applied coastal and sanitary engineers have joined efforts to design facilities for combined treatment of both municipal and industrial wastes.

Investigations have shown that certain industrial toxic wastes, previously thought untreatable biologically, can now be destroyed when mixed with sanitary sewage. Reuse of municipal sewage water is also being encouraged. In some cities involved in coastal sanitation efforts, the use of reclaimed sewage water is cheaper, and requires less treatment, than other available local water supplies. In many cases, the cost of waste water treatment can be offset by reuse and sale of by-products, for example, the sale of sludge as fertilizer (National Academy of Sciences, 1970).

Undersea Construction, Habitation, and Vehicles. Man's ability to live, work, and play in an undersea environment can be enhanced with the aid of applied oceanography. One aspect of applied oceanography particularly interesting to geologists is soil mechanics and the ocean floor. (For a discussion of development of underwater facilities, construction methods, and construction equipment, see Vol. XIV: *Ocean, Oceanographic Engineering.*

Extremely important factors involved with underwater construction and operations are related to submarine topography and the relief of the sea floor, the strength of bottom sediments, and water clarity (Brahtz, 1968). It follows that a tremendous amount of effort must be devoted to the study of the ocean floor to increase our knowlege of its physical properties.

From the standpoint of underwater construction, the main property of importance is the bearing capacity of the sediment. This is calculated from the shear strength by means of formulas and constants that have been determined empirically for terrestrial soils, but which have no proven relationship to deep-sea sediment conditions (Austin,

1966). In fact, very few strength measurements have been made on deep-sea sediments, and these probably do not represent true *in situ* values since these values are subject to change as a result of disturbance during sampling and handling. So in addition to the obvious need for complete ocean surveys, better corers must be developed, handling and processing techniques must be improved, and every effort must be made to correlate the geological setting and index properties with the *in situ* shear strength. Moreover, *in situ* dynamic load tests should be made in both shallow and deep waters in conjunction with in situ shear strength measurements to derive a relationship between the measured shear strength and dynamic bearing capacity of deep-sea sediments (Hamilton, 1970).

Coastal Engineering. Shorelines are constantly changing as a result of many factors. The processes that cause these shifts require constant study, and the land requires constant replenishment or removal. Storms and currents cause beach erosion and a continual shift of sand along the beach or on- and offshore. The Army Corps of Engineers is continually studying this problem to develop methods of sand replenishment to halt or reduce beach erosion (Bretschneider, 1969, Finkl, 1981). An excellent study of these problems with recommended research areas has been compiled by the Wave Research Foundation of the University of California at Berkeley. Further discussion of the problems and their solutions can be found in Wiegel (1964).

Mineral Exploitation. Available data indicate that fossil fuel resources of the lithosphere are scarce relative to those of other minerals. Oil and gas production capabilities may begin to decline before the end of this century. Coal is more abundant, but if coal is eventually used to synthesize oil and gas, or otherwise displace the fluid fuels, depletion of solid fuel resources will be accelerated significantly. Total resources of other minerals in the lithosphere are exceedingly large in terms of present and potential rates of use. As currently estimated, however, commercial reserves, while adequate for the immediate future, are for some commodities grossly inadequate to meet longer-range requirements. For example, copper reserves are equivalent to a 54-year supply at current rates of use. If all the world were to match the United States in per capita consumption, the reserve would dwindle to only a 12-year supply; assuming increasing population, it would be even smaller. Commercial reserves of most of the common metals are of the same magnitude as those of copper. Total resources of the lithosphere, on the whole, are many thousands to millions of times greater and thus present ample opportunity for discovery and technical progress to make available adequate metal supplies for the indefinite future. Reserves of the nonmetallic elements, particularly those essential to maintaining agriculture, guarantee that world needs can also be met.

Evidence shows, however, that the process of converting theoretical resources into exploitable resources needs to be accelerated for some commodities. Past experience has demonstrated that science and technology are essential to the achievement of this objective. Nevertheless, it is imperative that present research designed to improve techniques for finding hidden ore deposits and for extracting and treating lower-grade materials be speeded up (Cruickshank et al., 1968) (see Vol. XIV: *Marine Mining*).

The store of dissolved elements in the ocean also is large compared with annual world requirements. With a few exceptions, however, this store is considerably smaller than that existing in the lithosphere. Rocks to a depth of only 3 km contain 59 times more potassium and 218 times more gold than the entire ocean. Similar factors for other commodities are: uranium, 1,170; copper, 20,000; manganese, 440,000; iron, 4,400; and aluminum, 7,100,000. The ocean contains more chlorine, iodine, and bromine than the rock crust. It also has high concentrations of sodium, magnesium, and calcium, but the concentrations and total quantities of these three elements are greater in the lithosphere. Seawater contains no fossil fuels. However, vast energy is available in its currents, waves, tides, marine life, and thermal gradients, and through fission of elements found in seawater. Exploitation of these resources in the near future is problematic. Fusion energy derived from hydrogen and deuterium, elements that are more abundant in the sea than on the land, may make a significant contribution in future world energy supplies.

The extremely low concentration of elements in seawater is a major deterrent to commercial production of minerals now obtained from rocks. For example, much gold is produced profitably from rocks containing 10 parts per million (ppm), which is 2.5 million times richer than the average gold content of seawater. The average yield of copper ores mined today is 15,000 ppm—a disparity of 5 million in favor of the lithosphere (Mero, 1965).

Extraction of many additional elements (except about seven) from the sea is improbable until new technologies are developed. The limiting factor is one of economics; elements and compounds can be obtained less expensively from other sources than from the oceans or from brines. Because elements are so dilute in seawater, tens to hundreds of thousands of cubic miles of seawater would have to pass through extraction plants to supply just the United States' needs. Using present technology, the cost of building the necessary extraction plants would not be financially feasible.

Fundamental research on the basic chemistry of seawater is necessary. The prospect of recovering materials from seawater will be improved as our general knowledge of the sea improves. Understanding marine biology is also essential to planning for filtering equipment, control of fouling organisms or sulfate-reducing bacteria in seawater piping, and the protection of exposed structures against boring organisms. The choice of construction materials for marine exposure represents a problem of considerable magnitude in the design, construction, and operation of plans on, in, or near the oceans. The highly corrosive nature of seawater is compounded by the presence of destructive marine life, sometimes scouring currents, and often severe mechanical strains. Thus, each new development that reduces the cost of handling seawater improves the outlook for economic recovery of dissolved minerals and organics (Howard and Pagan, 1966).

The prospect for mineral recovery from seawater will be further enhanced by combination with other operations. For example, many power plants along the coastal areas use seawater for cooling. The effluent from these plants should be available at incremental costs, providing significant savings to any recovery process. Perhaps even more important in the future will be the availability of concentrated effluents from desalinization plants, which represent more attractive targets for recovery operations than seawater itself. The study of the chemistry and biochemistry of these concentrated effluents should be planned to keep pace with the growing need for fresh-water facilities.

The technology is well developed for mining, refining, and marketing near-shore and beach deposits such as ilmenite, rutile, zircon, oil, gas, and sulfur (see Vol. XIV: *Placer Mining*). Further work will undoubtedly be undertaken as the need arises, largely by firms already established in this business. Technology that has been developed and proved for obtaining these offshore minerals should provide a valuable background for the exploitation of certain other marine deposits in shallow water.

Some economic incentive to exploit the beach and near-shore deposits of magnetite may exist if processes can be developed for either removing unwanted minerals from the ore or tolerating the contaminant found during metallurgical processes. Japan has had commercial success in the use of iron sands with considerable quantities of titanium. Certain other heavy minerals may be profitable to mine.

A need for process development exists in the cases of glauconite and phosphorite. If processes can be developed for economically recovering potash from glauconite, and if offshore phosphorite can be economically beneficiated for use in manufacturing fertilizers, it might be feasible to exploit offshore deposits of the minerals. An enormous reserve of these ores may exist offshore in many parts of the world. It must be admitted, however, that we simply do not know how much phosphorite (or any other potential ore) exists anywhere in the offshore area. Although present indications are that deposits of phosphorite ores are extensive off the west coast of the United States, further exploration will be necessary to determine their locations and sizes more precisely.

Preliminary studies by mining companies interested in offshore exploitation indicate that only the higher-valued minerals or metals would yield profitable returns on the investment. Obviously, all such operations will weigh carefully the comparative costs of onshore and offshore recovery. Probably a few companies will locate sufficiently rich deposits and make the necessary capital investment, but it is unlikely that miners or mining companies will turn the continental shelf or deep sea into another California gold rush type business boom.

Major questions that now inhibit exploitation of deep-ocean manganese nodules concern techniques and economics for recovering and refining valuable metals from these ores (see Vol. XIV: *Marine Mining*). Another question concerns the environment in which the potential deep-ocean miner must work, as well as further defining locations and sizes of nodule deposits. A third important problem area concerns the feasibility of marketing the large-volume production that may be available from such ventures.

It would appear that studies on substituting abundant marine elements (such as magnesium) for other metals in short supply are warranted. If magnesium metal, which is virtually inexhaustible in seawater, could be substituted for iron and similar metals, we would be assured of a supply for all foreseeable time on earth.

Concurrent with technological and economic studies relating to ocean resource exploitation, certain legal problems require attention (Commission on Marine Science, 1969). It is not unreasonable, perhaps, to expect that recovery processes, to be acceptable, would have to avoid gross pollution of the ocean or any of its significant parts. What organization would have jurisdiction and what recourse is there in case gross contamination occurs outside continental limits? A still greater legal problem is associated with a dream of recovering minerals or producing special organics *in situ*, for example, by cultivating and harvesting marine organisms in certain areas of the high seas where temperatures, nutrients, and the like, are favorable. Would areas of this type be designated? If so, where would one file a claim, and what protection would one have against intruders?

Serious exploitation of offshore mineral deposits depends on further work in development, mining, and processing—in short, all phases of mining. Exploration activities on the continental shelves

should provide much more precise information concerning the size, location, and environment of various deposits. Such information will provide a much sounder base upon which decisions can be made on whether capital should be invested in ocean mining ventures. Furthermore, the development of processing techniques for recovering valuable minerals from seafloor ores and from sea water may also lead to the exploitation of these substantial reserves.

CHARLES P. GIAMMONA

References

Anonymous, 1970. New concept for harnessing ocean waves, *Ocean Industries* **5**, 62-63.

Austin, Carl F., 1966. Manned undersea structures—the rock site concept, *Naval Ordinance Test Station Tech. Paper No. 4162.*

Brahtz, J. F. (ed.), 1968. *Ocean Engineering.* New York: Wiley, 720p.

Bretschneider, C. I. (ed.), 1969. *Topics in Ocean Engineering.* Houston: Gulf Publishing, 420p.

Commission on Marine Science, Engineering and Resources, 1969. *Marine Resources and Legal-Political Arrangements for Their Development: Panel Reports, III.* Washington, D.C.: U.S. Government Printing Office.

Cruickshank, M. J.; Romanowitz, C. M.; and Overall, M. P., 1968. Offshore mining—present and future, *Eng. and Mining Jour.* **169**, 84-91.

Finkl, C. W., 1981. Beach nourishment, a practical method of erosion control, *Geo-Marine Letters* **1** (2), 155-161.

Frosch, R. A., 1962. Underwater sound, *Internat. Sci. and Technology* **9**, 40.

Hamilton, E. L., 1970. Sound velocity and related properties of marine sediments, North Pacific, *Jour. Geophys. Research* **75**(23), 4423-4446.

Howard, T. E. and J. W. Pagen, 1966. Problems in evaluating marine and mineral resources, *Mining Eng.* **18**, 57-61.

Ketchum, David D., and Stevens, Raymond G., 1961. A data acquisition and reduction system for oceanographic measurements, *Marine Sci. Instrumentation* **1**, 55.

Mero, J. L., 1965. *The Mineral Resources of the Sea.* Amsterdam: Elsevier, 312p.

Mourad, A. G., and Fubara, M. J., 1974. Requirements and applications of marine geodesy and satellite technology to operations in the ocean, in *Proceedings of the International Symposium on Applications of Marine Geodesy.* Battelle Memorial Institute, Columbus, Ohio, June 3-5.

National Academy of Sciences, National Acedemy of Engineering, 1970. *Waste Management Concepts of the Coastal Zone.* Washington, D.C., 126p.

Wiegel, R. L., 1964. *Oceanographical Engineering.* Englewood Cliffs, N. J.: Prentice-Hall, 523p.

Cross-references: *Beach Replenishment, Artificial; Coastal Engineering; Coastal Inlets, Engineering Geology; Marine Sediments, Geotechnical Properties; Nuclear Plant Siting, Offshore; Undersea Trans-*

mission Lines, Engineering Geology. Vol. XIV: *Acoustic Surveys; Coastal Zone Management; Floating Structures; Harbor Surveys; Marine Exploration Geochemistry; Marine Magnetic Surveys; Marine Mining; Nuclear Plant Protection, Offshore; Ocean, Oceanographic Engineering; Sea Surveys; Surveying, Electronic.*

OFFSHORE WORKS — See NUCLEAR PLANT SITING, OFFSHORE;

UNDERSEA TRANSMISSION LINES, ENGINEERING GEOLOGY. ORBITAL, AERIAL PHOTOGRAPHY — See PHOTOGRAMMETRY; REMOTE SENSING, ENGINEERING GEOLOGY. Vol. XIV: AERIAL SURVEYS, GENERAL; PHOTOGEOLOGY; PHOTO INTERPRETATION; REMOTE SENSING, GENERAL; SATELLITE GEODESY AND GEODYNAMICS.

P

PERMAFROST, ENGINEERING GEOLOGY

Permafrost, or permanently (perennially) frozen ground, is a subsurface thermal condition that is independent of texture, water or ice content, and lithology (Gold and Lachenbruch, 1973; Brown and Kupsch, 1974). The essential characteristic of permafrost is that the ground remains continuously below 0° C for several consecutive years. Terms such as *dry, saturated,* and *supersaturated* occasionally are used to describe the water (ice) content of permafrost. Other related terms include: *active layer,* the layer subjected to annual freezing and thawing; *ground ice* or *underground ice,* ice occurring as particle coatings, grains in soil and rock pores, lenses, or massive bodies; *unfrozen water content,* that portion of the total water present existing at interfaces and phase boundaries or in capillaries and brine pockets in a liquid-like state; *taliks,* unfrozen layers within or above the permafrost; and *subsea permafrost* (offshore or submarine), permafrost occurring beneath the

sea bottom (Muller, 1947; Black, 1954; Washburn, 1956; Pewe, 1966, 1969; Ferrians et al., 1969; Brown, 1970a, 1970b). Essential aspects of the permafrost condition are shown in Figure 1.

Occurrence and Extent

Permafrost is widespread in North America, Eurasia, and Antarctica. In the Northern Hemisphere, permafrost usually is divided into two zones, a northern zone of continuous permafrost where temperatures at the depth of zero annual change are generally below −5° C and a southern zone of discontinuous permafrost where subsurface temperatures range between −5° C and 0° C or above (Fig. 2a). It is estimated that about 20 percent of the earth's land surface is underlain by permafrost, and that 0.2 to 0.5 million km³ of underground ice is involved (Bird, 1957; Brown, 1970a; Washburn, 1970). Table 1 indicates the distribution of permafrost in the Northern Hemisphere.

In the continuous zone, permafrost generally underlies the ground everywhere and extends uninterrupted vertically to its lower boundary and laterally in all directions except under and adjacent to large rivers and deep lakes.

In the discontinuous zone, the occurrence of unfrozen islands or strips of earth increase in areal extent southward. At the southern fringe of the discontinuous zone, permafrost is limited to small masses of earth underlying peat, or at high elevations. In high mountainous areas, permafrost extends south of the general latitudinal boundary. In general, permafrost is thicker and extends to lower latitudes in the USSR than in North America; Figure 2b shows the zones of maximum depth of permafrost in the USSR. The presence of layers of unfrozen ground within or above the permafrost (taliks) are common in the warmer permafrost of

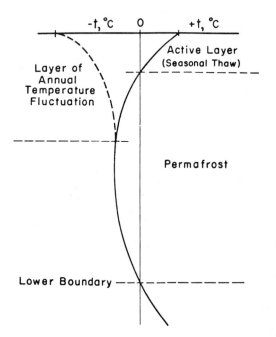

FIGURE 1. Essential aspects of permafrost condition.

TABLE 1. Distribution of Permafrost in the Northern Hemisphere (in millions of km²).

	Continuous Zone	Discontinuous Zone
Eurasia	3.6	7.4
North America	3.9	7.1
Total	7.5	14.5
Grand total = 22.0		

Source: From Black (1954).

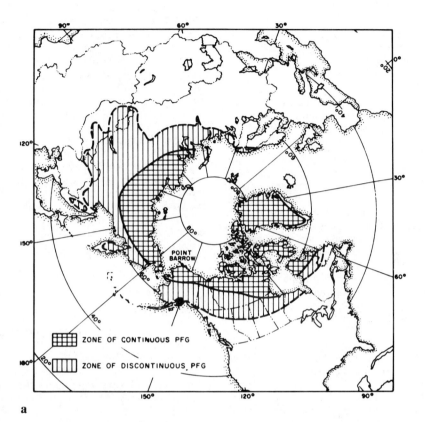

a

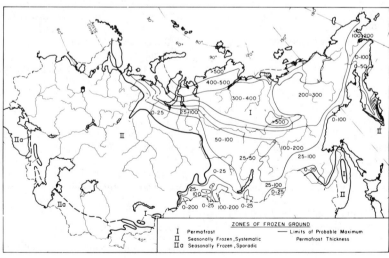

b

FIGURE 2. (a) Main zones of continuous and discontinuous permafrost in the Northern Hemisphere. (b) Zones of seasonally and perenially frozen ground in the USSR, with limits of maximum permafrost thicknesses, in meters (from U.S. Geological Survey Map I-445; Geological Survey of Canada Map 1246A; and Gold and Lachenbruch, 1973).

TABLE 2. Permafrost Thickness at Some
Representative Locations in North America

Location	Thickness (meters)
Barrow, Alaska	200-400
Fairbanks, Alaska	30-100
Prudhoe Bay, Alaska	650
Resolute Bay, Canada	390
Norman Wells, Canada	30-65
Mackenzie River delta, Canada	100

Source: Data from U.S. Geological Survey Map I-445;
Geological Survey of Canada Map 1246A; and Gold and
Lachenbruch (1973).

the discontinuous zone and reflect thermal conditions currently unfavorable for further growth of the permafrost. A knowledge of the occurrence and location of taliks is of importance in the selection of sites for cities, dams, and structures of all kinds due to the special types of engineering designs and practices that must be employed when unfrozen and frozen layers are proximate. Measured permafrost thicknesses at some representative locations are shown in Table 2.

Of the many environmental parameters governing the existence of permafrost, the mean annual air temperature usually is taken as the single most important and convenient one. Although the point is controversial and there are important deviations, the southern limits of permafrost coincide roughly with the $-1°$ C air temperature (mean annual) isotherm. The boundary between the continuous and the discontinuous zones corresponds roughly to the $-8°$ C air temperature isotherm ($-5°$ C mean annual ground temperature). In the discontinuous zone, permafrost commonly underlies north-facing slopes and aspect is also evident in the continuous zone where the permafrost is thicker and depth of summer thaw is shallower in north-facing than in south-facing slopes. Vegetation, particularly mosses, provide insulation and inhibit freezing and thawing; snow cover has a similar effect (Brown, 1970a, 1970b; Gold and Lachenbruch, 1973). Differences in soil properties have noticeable effects on the permafrost regime; well-drained sands and gravels freeze more quickly and deeply and also thaw more rapidly than silty or clayey soils. The existence of *subsea permafrost* has only recently been reliably established. Recent investigations support the belief that its occurrence under the ice-covered portions of the Artic Ocean may be extensive.

Formation and Duration

The formation of permafrost is a consequence of cold climate. Permafrost forms when the net heat balance of the earth's surface over a period of time produces a ground temperature continuously below $0°$ C. If the period of climatic cold is sufficiently long, this negative temperature regime results in the formation of permafrost many hundreds of meters thick. Although permafrost is generally considered a product of past ground temperatures brought into equilibrium with present climatic conditions and the geothermal gradient, numerous examples of permafrost formation in recent sediments are known. Frozen plant and animal remains that have been radiocarbon-dated as Wisconsin ages or older testify to the formation of permafrost in the ice-free areas during the Pleistocene glaciation. Aggradation and degradation of permafrost undoubtedly occurred throughout the Pleistocene. These processes continue today (Pewe, 1969). Conditions for formation of permafrost to considerable depths exist presently throughout the zone of continuous permafrost. At the southern boundary of permafrost, there is evidence for both the advance and retreat of permafrost. In offshore areas, subsea permafrost exists either as a relict condition from the drowning or erosion of coastal areas, or as a result of past or present negative seawater temperatures in shallow waters.

Depth of Thaw

The annual depth of thaw over permafrost varies both spatially and temporally in any given region. Major factors influencing thaw depth are vegetation, soil type, water content, surface topography and aspect and, most importantly, the annual and summer climates. Whenever the vegetation is disturbed or when heat-producing structures are emplaced without proper preventative measures, deterioration of the permafrost is likely (Lachenbruch, 1957; Gold and Lachenbruch, 1973; Jumikis, 1973). Because of its high latent heat of freezing, soil water retards both freezing and thawing. However, running or seeping subsurface waters aid in thawing by acting as efficient heat exchangers. Total yearly thaw penetration can be estimated for a given set of soil properties from the summation of the seasonal degree-days above $32°$ F ($0°$ C). Representative thawing indexes are, in degree-days, northern Alaska, ($280-550°$ C); interior Alaska, $1,100-1650°$ C; and the southern boundary of permafrost in Canada, approximately $2,200°$ C. Observed depths of thaw for several locations in North America are presented in Table 3.

Characteristic Surface Features

A number of surface features are uniquely characteristic of permafrost terrain. Most may be classified into "sorted" or "nonsorted" types according to the degree of particle sorting brought about by repetitive freezing and thawing. The most conspicuous and striking of the sorted types are *sorted stone circles* (Fig. 3), in which a circular border of stones surrounds a plug or mass of fine-grained

TABLE 3. Thaw Depths in Permafrost Regions.

Location	Latitude	Soil Type	Thaw Depth (meters)
Churchill, Canada	58°	Sand	2.5-3.8
		Clay	1.0-2.5
Southampton Island, Canada	64°	Gravel	1.0
Barrow, Alaska	71°	Silt	0.4
Resolute Island, Canada	75°	Sand/silt	0.5
Axel Heiberg Island, Canada	80°	Sand/gravel	0.6

FIGURE 3. Sorted stone circles found in the vicinity of Thule, Greenland.

FIGURE 4. Aerial oblique photograph of ice-wedge polygons, northern Alaska (courtesy of R. I. Lewellen, Littleton, Colo.).

soil. A widely occurring nonsorted form is the ice-wedge polygon (Fig. 4). The troughs are underlain by ice wedges commonly ranging in dimension up to a meter or more in width to several meters in depth. Ice wedges form over extended time intervals, and might be considered somewhat analogous to tree growth in that lateral growth occurs in recognizable increments accompanying each thermal contraction-expansion cycle. As the near-surface permafrost cools during the winter, it contracts and finally cracks when its tensile strength is exceeded. Cracking normally occurs in a polyg-onal pattern similar in appearance to desiccation cracking. Most of the cracks fill with ice prior to the expansion of the ground during seasonal warming the following spring. The zone of weakness established by the initial cracking repeatedly cracks and fills so that in time, a V-shaped ice mass is formed by the addition of successive increments of ice, averaging about 2 mm per cycle (Black, 1954; Washburn, 1956, 1970).

Pingos, another feature unique to permafrost, are large mounds or hills, each containing a massive ice core. Two types may be distinguished: an open system, normally formed on sloping ground, and a closed system, commonly found in old lake beds. Heights range up to 50 m and diameters up to 700 m. Pingos originate from the arching of a permafrost layer caused either by the intrusion or expulsion of water under pressure. In open-system pingos the water pressure results from a hydraulic head developed in a soil layer that acts as an aquifer confined by frozen layers above and below, whereas in closed-system pingos it develops when pore water is confined on all sides by its frozen surroundings and then is progressively frozen (Washburn, 1956, 1970; Brown, 1970*a).* The well-known pingos of the Mackenzie delta are of the closed-system type and were formed when isolated masses of unfrozen ground became surrounded by permafrost and were exposed for many years to mean annual air temperatures low enough that eventually freezing was complete. Pingo-like mounds recently were reported offshore also.

Since permafrost commonly contains more ice than is required, when melted, to saturate the soil pore space completely, thawing usually is accompanied by subsidence and loss of soil strength. Sink holes result and create a topography termed *thermokarst* (Shumskii, 1964; Washburn, 1970). Most of the sinks contain water so that a thermokarst topography typically is one of many small, roughly circular lakes.

Ice Segregation

Ice segregation and soil water movement is of primary importance to engineering works. In general, consequences arising from ice segregation and associated earth heaving are secondary in importance only to the subsidence that accompanies thawing of permafrost. Adsorption forces and the osmotic effects of ions and solutes concentrated adjacent to the surfaces of soil minerals prevent the water immediately proximate to mineral surfaces from freezing. The unfrozen interface separating the mineral surfaces from ice are quite thick (more than 100 Å) near 0° C but rapidly diminish in thickness with decreasing temperature. Soils containing a large proportion of very fine particles have high specific surface areas; consequently, near 0° C the amount of unfrozen interfacial water in such soils is large, ranging up to 0.4 g per gram of soil. In coarser soils the amount of unfrozen water present under comparable conditions of temperature is lower (Anderson and Morganstern, 1973). In soils having moderate to high specific surface areas, the unfrozen water films are sufficiently thick and extensive at temperatures near 0° C to permit the thermal transport of considerable quantities of water. Transport occurs along the temperature gradient in the direction of falling temperature. Because the soil particles are surrounded by liquid-like water films, they tend to be excluded from growing ice crystals. This process leads to migration and sorting of particles whenever they are free to move individually (see Fig. 3) and to frost heaving (as distinguished from mere expansion of confined water on freezing) when they are restrained by overburden. Frost heaving is a potentially destructive process any time water-bearing soils or earth materials are subjected to alternate freezing and thawing. Heaving of the overburden 0.5 m or more and the development of heaving pressures of several hundred pounds per square inch are extremes recorded in laboratory experiments. The principal requirements for frost heaving include (1) a temperature gradient such that temperatures are below freezing at the soil surface but above freezing at some depth below, (2) soil water available at depth for transfer to the freezing zone, and (3) soils containing more than 3 percent of less than 20-μ-diameter particles. The rate and extent of frost heaving depend critically on the existence of a favorable balance between the movement of soil water to the growing ice lens and the removal of the latent heat of freezing. This balance is governed by the kind of soil, the initial soil water content, depth to a water table, the hydraulic and thermal conductivites of the soil, the thermal regime during freezing, and overburden pressure. If all other factors remain constant, heaving rates diminish rapidly with an increase in overburden pressure or in water-table depth.

Engineering Aspects

For engineering purposes, permafrost can be divided into (1) *hard frozen*, in which ice acts as a cement between grain boundaries, (2) *plastic frozen*, in which some water remains unfrozen, and (3) *granular frozen,* ground in which the grains are in mutual contact and excess ice is not present. Where hard frozen or plastic frozen ground is encountered, the usual approach is either to maintain thermal stability of the permafrost throughout and following construction, or preconstruction thawing and removal of excess ice. In addition to conventional methods, electrical thawing and removal of excess water by electroosmosis (q.v. in Vol. XII, Pt. 1) is employed experimentally or whenever it is economically expedient. The preservation of permafrost is accomplished by (1) placing an insulating pad of gravel over the building or road site, (2) placing buildings on piles with ventilated air space between the ground and building; (3) artificially refrigerating foundations, and (4) installing pipes and utilities in insulated below-ground conduits or above-ground insulated utilidors (Muller, 1967; Linell, 1957; Saltykov, 1959; Sanger, 1969, Jumikis, 1973). Technology developed in devising engineering methods of achieving these objectives also are being employed in the artificial freezing of soils in temperate and tropical regions to facilitate such activities as excavation (q.v.), tunneling (q.v.), soil stabilization (q.v. in Vol. XII, Pt. 1), and underground storage of cryogenic liquids. Granular frozen ground usually does not present serious engineering problems. This kind of permafrost therefore is preferable to the other two for most types of construction; in this respect frozen bedrock of low water content may be included in this category also.

The importance of adequate exploration for the purpose of selecting construction sites in permafrost areas cannot be overstressed. Once the site is selected, a thorough knowledge of the local distribution and properties of the permafrost is required prior to final design and the beginning of construction. Aerial photos are particularly valuable for assessing permafrost conditions when the presence of massive ground ice is usually accompanied by such indicators as ice-wedge polygons (see Fig. 4), pingos, and thermokarst pits. Unstable ground and frost-susceptible soils are indicated by the presence of frost scars, solifluction lobes, mudflows, and landslides. "Drunken" spruce forests are indicative of permafrost at shallow depth and poor drainage. Interpretations obtained from photos, geological maps, and analysis of available climatic data should be followed by detailed site investigations, including coring, to determine (1) the local composition and properties of the permafrost, (2) seasonal variation in thaw and freezing depths, and (3) temperature profiles.

Following site selection, attention turns to structural design and construction methods. Under present climatic conditions, disturbance of the natural vegetative cover invariably leads to accelerated thawing of buried ice. Canal-like, water-filled channels frequently develop after a single pass of a tracked vehicle. Hastily laid roadbeds quickly subside, becoming rutted and eventually impassable. To ensure satisfactory performance, upwards of 2 m of non-frost-susceptible, sub-base material is laid down to insulate the permafrost against thaw and disturbance when suitable materials are available. Otherwise, closed-cell expanded polystrene panels sufficient to provide the necessary thermal barrier are incorporated into sub-base material.

Large buildings constructed over permafrost generally have an air space between floors and the ground surface. Forced or naturally circulating cold air prevents the transfer of heat from the building into the ground, maintaining or, in some cases, even building up the cold reservoir beneath. Pile and pier foundations are used in this form of construction. Augering, drilling, and occasionally steam points are employed during their emplacement. Afterwards in marginal situations the piles are refrigerated, by active or passive means, to accelerate freezeback and to ensure against thawing during the life of the building. The design of the trans-Alaska pipeline system, a "hot" pipeline that traverses both the continuous and discontinuous zones of permafrost, has stimulated many new techniques to maintain the integrity of the permafrost and to minimize the chances of adverse environmental impacts.

The impermeable nature of ice-rich permafrost creates serious problems in the disposal of raw or treated sewage through conventional drainage fields; a satisfactory solution is yet to be found. Until recently, sewage was frequently disposed of in rivers or lagoons. In coastal areas, disposal was accomplished by hauling 55-gallon drums of collected waste out onto the sea ice. Low temperatures slow down biological breakdown and inhibit buildups of dangerous populations of infectious organisms. Currently, portable or semi-portable sewage treatment systems are being widely used in areas where construction projects are active. New techniques of using sewage lagoons are receiving attention, particularly in the Canadian Artic.

Water, an abundant commodity during spring and summer, becomes difficult to obtain during winter. Surface water supplies such as shallow lakes, streams, and rivers freeze to the bottom during winter, and the formation of thick ice covers on deeper bodies of water causes the concentration of dissolved solutes in the water to remain unfrozen during late winter, which often makes them unpalatable. Unfrozen river gravels can be developed into year-round water supplies. Many remote installations store spring and summer water in large, insulated tanks for winter consumption. Melting of ice and snow during winter by efficient heat exchangers is another method used to obtain pure water. Subpermafrost waters obtained from deep wells provide reliable flows

of water, but frequently require treatment to remove objectionable quantities of minerals and organic matter.

Permafrost exists on extraterrestrial planetary bodies, for example, Mars. The landing of space vehicles and construction of space stations will necessarily involve many of the same engineering and construction principles and techniques developed in dealing with these problems in the arctic and subarctic regions.

DUWAYNE M. ANDERSON
JERRY BROWN

References

Anderson, D. M., and Morganstern, N. R., 1973. Physics, chemistry and mechanics of frozen ground: a review, in *Permafrost: The North American Contribution to the Second International Conference.* Washington, D.C.: National Academy of Sciences, 257-288.

Bird, J. Brian, 1957. *The Physiography of Arctic Canada.* Baltimore: John Hopkins University Press, 336p.

Black, R. F., 1954. Permafrost—a review, *Geol. Soc. America Bull.,* **65,** 839-856.

Brown, R. J. E., 1970a. *Permafrost in Canada.* Toronto: University of Toronto Press, 261p.

Brown, R. J. E., 1970b. *Permafrost in Canada: Its Influence on Northern Development.* Toronto: University of Toronto Press, 234p.

Brown, R. J. E., and Kupsch, W. O., 1974. Permafrost terminology, *Nat. Resource Council Canada Tech. Memo No. 111.* 62p.

Ferrians, O. J., Jr.; Kachadoorian, R., and Green, G. W., 1969. Permafrost and Related Engineering Problems in Alaska, *U.S. Geol. Survey Prof. Paper 678,* 37p.

Gold, L. W., and Lachenbruch, A. H., 1973. Thermal conditions in permafrost—A Review of the North American literature, in *Permafrost: The North American Contribution to the Second International Conference.* Washington, D.C.: National Academy of Sciences, 3-25.

Jumikis, A. R., 1973. *Influence value and charts for temperature distribution from heated rectangular structures on permafrost.* Paper 73-WA/HT-10 presented before the American Society of Mechanical Engineers (Professional Division) at the Winter Annual Meeting, Detroit, Michigan, November 11-15, 16p.

Lachenbruch, A. H., 1957. Three-dimensional heat conduction in permafrost beneath heated buildings, *U.S. Geol. Survey Bull.* **1052-B,** 51-69.

Lachenbruch, A. H., 1970. Some Estimates of the Thermal Effects of a Heated Pipeline in Permafrost, *U.S. Geol. Survey Circ. 632.*

Linell, K. A., 1957. Airfields on permafrost, *Am. Soc. Civil Engineers Proc.* **83,** 1326-1-1326-15.

Muller, S. W., 1967. *Permafrost or Permanently Frozen Ground and Related Engineering Problems.* Ann Arbor, Mich.: Edwards Brothers, 230p.

Pewe, T. L., 1966. Permafrost and its effect on life in the north, in H. P. Hanse, (ed.), *Arctic Biology.* Corvallis: Oregon State University Press, 3-40.

Pewe, T. L. (ed), 1969. *The Periglacial Environment, Past and Present.* Montreal: McGill-Queens University Press, 487p.

Saltykov, N. I. (ed.), 1959. *Principles of Geocryology,* Parts I and II. Moscow: Akademiya Nauk SSSR, 459p. (Part I), 365p. (Part II). (Partial Translation Series, National Research Council of Canada, Ottawa.)

Shumskii, P. A., 1964. *Principles of Structural Glaciology.* New York: Dover, 497p.

Washburn, A. L., 1956. Classification of patterned ground and review of suggested origins, *Geol. Soc. America Bull.* **67,** 823-865.

Washburn, A. L., 1970. *Periglacial Processes and Environments.* London: Edward Arnold Ltd.

Cross-references: *Foundation Engineering; Geomorphology, Applied.* Vol. XII, Pt. 1: *Electroosmosis; Soil Stabilization.* Vol. XIV: *Engineering Soil Science; Terrain Evaluation Systems.*

PETROLEUM ENGINEERING—See Vol. XIV.

PHOTOGEOLOGY—See Vol. XIV.

PHOTOGRAMMETRY

Photogrammetry may be considered a system of measuring data recorded on "photograms." As such, it is applicable to all sciences that depend on reliable geometric measurements (Moffitt, 1967). The term *photograph* is often used as a synonym for *photogram.* A photogram is a photograph taken with a photogrammetric camera, a precision camera with "fiducial" or "collimating marks" and with a fixed distance between the negative plane and the lens. Closely related to photogrammetry is *photographic interpretation,* or photo interpretation (q.v.), the process of examining recorded photographic data for purposes of identification, evaluation, and classification (Spurr, 1960).

Branches of Photogrammetry

Although Aimé Laussedat made the first attempt to use photographs for measuring purposes in 1850, photogrammetry is a comparatively modern science or art. Photography from ground stations, *terrestrial photogrammetry,* was predominant initially, but today *aerial photogrammetry,* in which the photogrammetric camera is mounted on an aircraft, is far more common. *Space photogrammetry,* a relatively new branch, refers to all aspects of extraterrestrial photography, whether the camera is fixed on earth, placed on the moon or a planet, or contained in an artificial satellite.

Classification of Aerial Photographs

Aerial photographs taken with the optical axis of the camera in a vertical or nearly vertical position are classified as *vertical photographs (*Fig. 1). Because it is difficult to keep the camera axis

FIGURE 1. Vertical aerial photograph of the Fort Lauderdale area, Florida. The original runs were flown in 1945 and produced at a scale of approximately 1:24,000 (photo: National Oceanographic and Atmospheric Administration).

exactly vertical, such photographs are often termed *near-vertical* to account for small amounts (usually less than 3°) of tilt. An *oblique photograph* is taken with the optical axis of the camera intentionally inclined (Fig. 2). A *high oblique* shows the apparent horizon, whereas the horizon is not imaged in a *low oblique.* (The words *high* and *low* are not related to the altitude of the exposure station.) *Horizontal photographs,* used in terrestrial photogrammetry, may be regarded as a particular type of high oblique photograph in which the camera axis is directed horizontally.

Considering the angular coverage of the camera as a function of its focal length and the diagonal measure of the film used, photographs may be classified as *normal-angle* (about 60°), *wide angle* (about 90°), and *super-wide-angle* (about 120°).

Other types of photographs—for example, *convergent* photographs, *trimetrogon* combinations of obliques and vertical photographs—are also occasionally used. Most common in photogrammetric practice are wide-angle vertical photographs in black-and-white panchromatic film. Black-and-white infrared and color photography (true color, false color, color infrared), are mainly used for photo interpretation purposes (see Vol. XIV: *Photo Interpretation*).

Applications of Photogrammetry

The primary aim of photogrammetry is to aid in terrain mapping. Measurements on photographs replace field surveys to a large extent, although

FIGURE 2. Oblique aerial photograph of Hillsboro Inlet, Florida. This high-angle oblique shows the beach configuration prior to inlet stabilization in 1952. (photo: Birdseye Aerial Photograph).

photogrammetry must rely on geodetic methods for fundamental control surveying (see Vol. XIV: *Surveying, General*). Consequently, the acquisition and use of photographs and photogrammetric techniques in mapping and interpretation is referred to as *aerial survey*.

The original, and still most important, use of photogrammetry is in the production of planimetric maps, topographic maps, and large scale engineering maps and town plans. These may be presented in classical form as *line maps* using conventional symbols to represent ground details, or as *orthophotomaps*, which are orthographic representations of the ground in continuous-tone pictures. For relatively flat terrain, *photomosaics* can be prepared by assembling individual photographs systematically into one continuous-tone picture (Thompson, 1966).

Photogrammetric measurements can also meet demands for extreme high accuracy, which is characteristic of *cadastral surveys* (see Vol. XIV: *Surveying, Electronic*). In addition, photogrammetry has a wide range of other applications ranging from astronomy to microscopy (nontopographic photogrammetry). *Nontopographic photogrammetry* is useful for such tasks as the measurement of perishable or changing objects; of objects in motion (water waves) as well as objects difficult to reach;

of sensitive objects, which ordinary measurements might deform (such as swellings in the human body); of very rapid or very slow events (such as explosion of an atomic bomb or plant growth in an experimental field); and so on (Hallert, 1960).

X-ray photogrammetry is used in the diagnosis and treatment of certain medical conditions, such as the location and spread of a tumor or growth, the locations of fractures and foreign matters in the human body, and the like. Other, less striking nontopographic applications include the evaluation of timber stands and volumes of coal piles, the reconstruction of cultural-historical objects, crime and accident investigation, and so on.

Basic Principles

Geometry of the Aerial Photograph. An aerial photograph that has been exposed in a calibrated photogrammetric camera has certain metrical characteristics, and can be treated analytically. The *focal length* of the camera lens is one of the most important calibration factors.

At the moment of exposure, a potential perspective projection exists between points on the ground and the negative plane. This projective relation is, however, upset by atmospheric refraction and lens aberrations as the light rays pass through the atmosphere, and the lens system, and then are imaged in the focal plane on the negative emulsion. A further distortion takes place during film processing. Disregarding these influences, it is assumed that a photograph geometrically records a "central," or "perspective," projection of the object. This assumption is the foundation for the analytical treatment of the whole theory of photogrammetry, in which projective geometry is of fundamental importance. Image formation is thus considered to take place via "straight lines" or "bundles of rays" from the "points" of the object through a common "point," the projection or perspective center that schematically represents the camera lens. The principles are analogous to those of human sight (Moffitt, 1967).

The image plane can be at "negative" and "positive" positions. In central projection, the negative produces a reversed image of all object points, whereas the positive gives a direct image of the object.

Definitions. The *exposure station* is the position of the camera lens at the moment of exposure. Schematically it is represented by the projection center. *Principal distance,* the distance from the projection center to the negative plane, it is not always equal to the focal length of the lens (Fig. 3). *Flying height* refers to the elevation of the exposure station above a selected datum, usually mean sea level. *Nadir point,* or *plumb point,* is the point at which a plumb line passing through the perspective center intersects the negative plane. The *principal point,* the point at which a perpendicular dropped from the perspective center strikes the negative plane, can be obtained by reference to the fiducial marks imaged on the photograph at the moment of photography. The term *tilt,* or *angle of tilt,* refers to the angle formed between the optical axis of the camera and a vertical line through the perspective center. The *isocenter* is the point at which the bisector of the angle of tilt intersects the negative plane. The *principal line,* or *line of maximum tilt,* indicates the line in the plane of the photograph defined by the nadir point, principal point, and isocenter. The *isoline,* or *axis of tilt,* marks the horizontal line in the plane of the photograph perpendicular to the principal line through the isocenter. On a truly vertical photograph, the nadir point and the isocenter coincide with the principal point.

Difference Between a Vertical Photograph and a Map. A photograph is a *central projection,* whereas a map is an *orthogonal* projection of the terrain. Both are similar when the terrain is perfectly flat and horizontal and the photograph is exactly vertical. Although a vertical photograph may resemble a map, it differs essentially from a map because the negative, at the moment of the

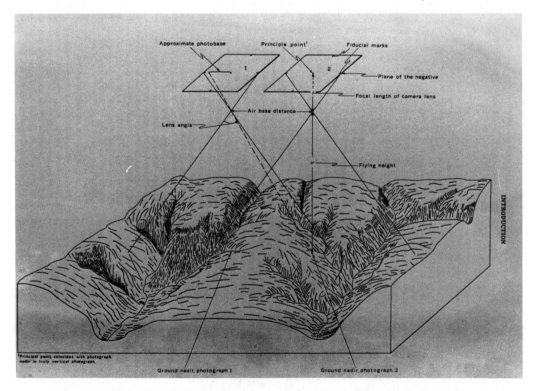

FIGURE 3. Terminology and geometry of the vertical aerial photograph (from Ray, 1960).

exposure, is seldom parallel to the ground, and because of the elevation differences on the ground.

Images of object points that are above or below a chosen reference plane on the ground will be displaced on the photograph from their *orthographic*, or *map*, position with respect to the reference plane. This effect, known as *elevation* or *relief displacement*, takes place along directions radial to the nadir point. Only object points located on the reference plane will appear on the photograph in their "true" position. Furthermore, if the camera axis is tilted at the instant of exposure, *tilt displacements* occur in a direction radial to the isocenter.

Since tilt and relief displacements are radial from different points, their combined effect in a single photograph result in a lateral displacement of images. This displacement is significant in photographs of rough terrain. The positions of points, measurement of distances, angles, and areas from the photograph are influenced by these displacements.

Oblique photographs are more economical than verticals because of their increased ground coverage. Even though they present a familiar landscape picture, they register a distorted view of the terrain, which differs radically from a map.

Although relief displacement constitutes an important disadvantage when using *individual* photographs as maps, it is nevertheless of greatest value to photogrammetry when viewed in stereo.

Scale of the Vertical Photograph. A map, an orthographic or orthogonal projection of the ground surface, has a uniform scale regardless of relative elevations. This property is not true for a photograph because it is a perspective projection, where areas closer to the camera at the instant of exposure appear larger than areas lying farther away.

The scale of the photographic image varies irregularly according to ground elevations. The situation is further complicated by tilt of the camera's axis, which causes regular variations in scale. This variation is negligible for near-vertical photographs. Consequently, the photo scale will differ from point to point. To determine the scale at a particular point, it is necessary to know the elevation of the point, its position in the photograph, and the angle of tilt.

In planning the flight mission, it is common practice to use an approximate value known as the *average* or *mean photo scale*. For vertical photographs, the *representative fraction (RF)* is expressed as the ratio between the principal distance (a camera constant) and the flying altitude above a horizontal plane representing the average elevation of the terrain.

Stereoscopy and Parallax

Stereoscopic Vision. Estimation of distances, awareness of space, and perception of depth are mainly the result of visual observation. *Monocular vision,* seeing with one eye, permits fairly good depth perception. *Binocular vision,* however, noticeably improves the quality of human vision, especially over short distances. *Stereoscopic vision* is related to the ability to conceive a plastic, three-dimensional impression of the objects imaged on a pair of photographs or drawings. Stereoscopic vision, made possible through binocular vision, is critical for photogrammetric work (Zorn, 1981).

In natural binocular vision, each eye perceives a slightly different image of the observed object, because the eyes view the object simultaneously but from two different positions. Both images are coordinated in the brain and blend together perfectly. The image dissimilarities are transformed into depth perception, and the object is perceived in a plastic sense. This perception is a very sensitive and complicated mechanism (Hallert, 1960).

In "artificial" binocular vision or stereoscopic vision, as used in photogrammetry, instead of looking at the original object each eye observes a central projection of the object from two photographs with projection centers in two different exposure stations, which correspond schematically to the eye positions. The perceived three-dimensional image of the object is called a *stereoscopic model,* or *stereo model* for short. This pair of photographs must fulfill specific requirements, however; that is, they must form a *stereoscopic pair.*

Stereoscopic Viewing. In natural binocular vision, one observes in *epipolar planes,* where a plane passes through the eye-base and the point looked at. Each eye sees its "own" image of the object, but both images are viewed simultaneously. When observing a stereo pair of photographs, one should attempt to imitate natural vision as completely as possible. The photographs have to be positioned in such a way that their *epipolar lines* are extensions of each other and parallel to the eye-base.

It is possible to observe a stereo pair with convergent or parallel eye axes. Observation with convergent eye axes is the most natural because *accommodation* (focusing of the eye lenses) and *convergence* (directing the lines of sight at a certain point) take place at the same distance, just as in natural vision. The two photographs, however, have to be printed or projected on top of each other; therefore, image separation is needed because each eye should see only one photograph (Zorn, 1981).

The most common methods for image separation are color separation by color filters (*anaglyphs),* Polaroid 3D vision with polarizing filters, and separation by intermittent light with rotating shutters. Although initially popular, observation with convergent eye axes is now used in only a few photogrammetric instruments.

Observation with parallel eye axes is currently

applied in most photogrammetric instruments. The eye axes have to be directed almost parallel, but at the same time the accommodation must be adjusted for vision at close distance. Because the "automatic link" between accommodation and convergence is disconnected, most viewers seek the convenience of an auxiliary aid such as a *stereoscope.*

Stereoscopes. The first stereoscope was invented in 1838, and today similar instruments are still extensively used in photogrammetry. There are two basic types, the *lens,* or *pocket, stereoscope* and the *mirror stereoscope;* each has both advantages and disadvantages.

The simplest, the pocket stereoscope, consists of a pair of simple magnifying lenses that are mounted in a frame with a separation equal to the average eye distance (65 mm) (Fig. 4). In some pocket stereoscopes this separation can be adjusted for individual users. A great disadvantage of the pocket stereoscope is that the separation between conjugate images on the photographs can be no larger than the user's interpupillary distance. This means that photographs will cover each other to a certain extent, requiring the photographs to be folded over in order to examine the entire common area. On the other hand, the pocket stereoscope has a large field of view, its image quality is very good, it is inexpensive and portable, and it is well suited for field work.

The *mirror stereoscope* has, in addition to the magnifying lenses, a combination of two small mirrors or prisms and two large wing mirrors. This construction enlarges the visual base, allowing one to examine the entire stereoscopic overlap without the need to fold the photographs (Fig. 5). Mirror stereoscopes are usually equipped with removable binoculars, which enlarge particular areas for detailed study but reduce the field of vision.

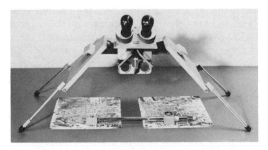

FIGURE 5. Mirror stereoscope with removable binoculars and parallax bar. This multipurpose instrument is especially suitable for detailed interpretation of aerial photographs (courtesy of Sokkisha Co., Ltd., Tokyo).

There is enough clearance between the instrument and the photographs so that it is possible to draw on the photographs. The image quality is, in many cases, not as good as that produced by a pocket stereoscope. The binoculars also require the observer to tilt his head in a rather tiring position. Finally, the mirror stereoscope is not portable, and its use is mainly confined to the office.

The principle behind the mirror stereoscope applies to a large number of photogrammetric instruments, although its practical realization can prove to be more complicated than in the stereoscope. A more complex type of stereoscope is the *scanning stereoscope,* which combines some of the advantages of both the lens and the mirror stereoscopes. The so-called *zoom stereoscope* has provision for continuous change in magnification of both eyepieces and variation of magnification between eyepieces. This differential magnification allows two photographs having different scales to be viewed stereoscopically.

Before a three-dimensional impression of the object can be obtained, the stereo pair has to be properly oriented under the stereoscope. Particular attention should be given to avoid reversing the positions of the photographs, otherwise ground relief appears reversed, causing a "pseudoscopic" effect (hills appearing as valleys, valleys appearing as ridges, buildings showing as cavities in the ground). Shadows have to be taken into account for the same reason.

Stereoscopic Parallax. In a broader context, *parallax* refers to the apparent displacement of an object relative to another or to its background, caused by a change in the position of observation. In photogrammetry, the displacement of the image of an object point on two successive exposures, due to the change in position of the camera, is called the *absolute,* or *stereoscopic, parallax* of the point. The difference between the displacements of the images of two object points on successive exposures is named the *parallax difference,* or *differential parallax,* between the two points.

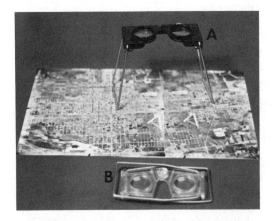

FIGURE 4. Examples of pocket stereoscopes: (A) a sturdy metal-framed scope with adjustable interpupillary distance between 60 and 70 mm; (B) a student pocket stereoscope with plastic lenses and metal frames.

These displacements take place along or parallel to the line of flight. The *flight line* is approximately defined on the photograph by the *photo base,* the line joining its principal point and the transferred principal point from the adjacent photograph. The parallax difference is directly related to the relief displacement on the photographs, and it is the principal cause of the perception of depth obtained by stereoscopic observation. Parallax differences are used to determine elevations of objects and to draw contour lines with aerial photographs by means of stereoscopic instruments.

Measurement of Stereoscopic Parallaxes. The stereoscopic parallax of an object point is a measure for its height, and it can be determined in various ways. Assuming a pair of near-vertical photographs properly oriented for stereoscopic observation, parallax can be determined by measuring the distance between the principal points of the photographs and subtracting the distance between conjugate images. This, however, is seldom done; instead, parallax differences are measured. A satisfactory method for making simple parallax measurements involves use of a *parallax bar* in combination with a stereoscope. Other devices are the *parallax wedge* and the *parallax ladder.* A parallax bar consists of two pieces of glass or plastic engraved with a reference mark, connected by a bar. The separation between the reference marks can be changed using a micrometer screw attached to the bar. The micrometers are mostly numbered increasingly as the distance between the corresponding images of an object point decreases. This means that a point with a larger parallax gives a higher reading, which corresponds with a point of greater elevation. Parallax differences can be measured and then elevations of points and height differences may be determined using the *parallax formula.*

The parallax formula assumes ideal situations where a pair of exactly vertical photographs have been taken with a horizontal air base. If this formula is applied to other situations, the results will be inaccurate. Accuracy of these computations is therefore hampered by tilt and the effect of tilt on measured parallaxes and parallax differences. When appropriate photogrammetric instruments are used, tilt is automatically taken into account and the results, in terms of elevation differences, are directly generated without computation and with high accuracy (Thompson, 1966).

Parallax measurements may be refined using a *comparator.* The comparator is the basic measuring instrument for the analytical solution of photogrammetric problems. The measurements are automatically read out and recorded using automatic registration devices. All other phases are handled computationally by sophisticated programs and electronic computers.

The Principle of the Floating Mark. For accurate measurements in the three-dimensional model, the "principle of the floating mark" is used. Two identical marks, called *floating marks* or *measuring marks,* are fused in the stereoscopic image into one mark, which occupies a definite position in the three dimensional space. The floating mark is the actual measuring tool in stereo photogrammetry, and when combined with stereoscopic vision, allows accurate settings and precise measurements of model points in all three dimensions (Hallert, 1960). Floating marks are used in nearly all types of stereoscopic instruments, although the application of the principle may be somewhat different.

The Stereo Model. The *stereo model* is the spatial model observed stereoscopically, as opposed to the *geometrical model,* which is defined as the locus of the intersections of corresponding rays after the two bundles of rays have been properly oriented in the photogrammetric instrument. In this model, very accurate measurements can be performed. The stereo model, the subjective visual impression observed under a stereoscope, may look very similar to the original scene, but in practice it is rarely so. A certain amount of relief exaggeration is sensed in the stereoscopic image. The parallax differences influence the impression conveyed by the model. Thus, the bigger the parallax is for a certain relief, the stronger the relief impression.

The *stereoscopic,* or *vertical, exaggeration* is caused by real or apparent changes in parallax differences between points on the photograph. The *vertical exaggeration factor* is the apparent proportion between vertical and horizontal scales in a stereo model.

In general, it should be noted that the stereoscopic model will be deformed during stereoscopic viewing if the photographs have a position different from the corresponding conditions during the photographic process. Such deformations occur if tilted photographs are observed stereoscopically as if they were vertical photographs. The height conditions in such a model can be considerably distorted.

Estimation of Slopes. Because stereoscopic exaggeration is difficult to determine, it may be difficult to estimate slope or dip angles directly from the stereo model. In the stereoscopic model, the tops of mountains seem to point always toward the observer. Thus, slope estimation is highly dependent on the position of the observer's eyes. It is therefore advisable to avoid determination of slope angles based on estimated values. Several methods have been developed to determine slope angles in the stereoscopic model (Mekel et al., 1977). The determination of slopes, as well as other precise measurements, poses no problem when using instruments where the geometrical model is accurately formed, as in most photogrammetric instruments currently used.

Categories of Photogrammetry

The main task of photogrammetry is the production of an orthogonal projection (map) of the terrain on a certain plane and scale by means of one or more central projections (photographs) of the terrain. Generally, two photographs are required to produce the orthogonal projection, the process being called *double-image photogrammetry* or *stereo photogrammetry*. In special cases when the terrain is nearly a plane and orthogonal projections are obtained from individual photographs, one speaks of *single-image photogrammetry*.

Single-Image Photogrammetry. Single-image photogrammetry is applicable only for comparatively flat terrain. *Rectification* is the process by which a tilted photograph taken from a given exposure station is transformed into an equivalent vertical photograph taken from the same exposure station. This "rectified" photograph then becomes similar to an orthogonal projection and may be enlarged or reduced. This operation is called *scale ratioing,* and its purpose is to bring a series of photographs with different scales, due to varying flying heights, to the same scale at a particular elevation, usually the average elevation of the terrain. Rectification may be performed with numerical, graphical, or optical-mechanical methods. The last procedure is most important for practical purposes. In optical-mechanical rectification, the original negative is reprojected onto a plane through a lens in special instruments called *rectifiers.*

In principle, a rectifier is a photographic enlarger with a number of degrees of freedom to change the shape and scale of the image. Thus, the projector can be given, relative to the projection plane, the same tilt it had during exposure. Four reference points, "control points," located near the corners of the photograph, are used to find the correct orientation of the projector. The negative is exposed again, and the result is a recified photograph in which the tilt displacements have been eliminated but not the displacements caused by the relief of the terrain. From here that rectification is suitable only for plane terrain where the relief displacement is negligible. Rectified photographs are used for the preparation of *photo mosaics* and *photo maps* (Spurr, 1960; Thompson, 1966).

Photo Mosaics. A photo mosaic is an assembly of a series of overlapping vertical photographs that have been cut and matched systematically to form a composite view of the area. The assembly is usually rephotographed and may be reprinted at desired scales. A mosaic assembled from fully rectified and ratioed photographs to fit a control network is called a *controlled mosaic.* It fulfills certain map accuracy requirements but still contains local relief displacements. When a coordinate grid system is drawn and contour lines added to the mosaic with names and boundaries, it becomes a *photo map.* If certain details are delineated for emphasis and clarity, an *annotated photo map* is produced.

Mosaics assembled using contact prints without regard to reference or control points are referred to as *uncontrolled mosaics* (Fig. 6). The scale is not uniform, and mismatching of images due to tilt and relief displacements and to scale differences between successive photographs cannot be avoided. If the available reference points are limited or rudimentary, the assembled photographs form a *semicontrolled mosaic.*

Although a photo mosaic differs from a map in many respects, it can occasionally be substituted for a planimetric map. A vast amount of detail is shown and terrain features can be recognized more readily than on a map. The relative position of geological phenomena and forest boundaries will show only local discrepancies of limited character. The mosaics may be compiled in a shorter time and more economically than a conventional line-map.

A basic disadvantage of the mosaic is its limited accuracy in the horizontal positions of features due to relief, tilt, and unequal scales. (The controlled mosaic is an exception: it can be considered a satisfactory map.) Other disadvantages are that mosaics do not provide elevations or configurations of the terrain in the vertical direction. Their compilation is restricted to reasonably flat terrain. Finally, the construction of a satisfactory mosaic demands a good deal of skill and effort.

For hilly and mountainous terrain, the only acceptable method for preparing precise mosaics meeting normal map specifications is to use special instruments such as *orthophotoscopes* to achieve "differential rectification." The result is an *ortho-photograph,* in which the influence of both tilt and relief displacements have been eliminated. A mosaic assembled with orthophotographs, contour lines, coordinate grid system, and names and boundaries makes an *orthophoto map,* which may be regarded as an exact map. The differential rectification phase is, however, part of stereo photogrammetry.

Double-Image Photogrammetry (Stereo Photogrammetry)

Acquisition of Aerial Photography. During aerial photography, the survey area is covered by several runs (strips) of photographs taken in a planned sequence along parallel flight lines and from a predetermined height above a *datum,* usually mean sea level. Along a given flight line, photographs are taken at frequencies that allow an overlap of about 55 to 90 percent between successive photographs (Fig. 7). Because this *forward*

FIGURE 6. Uncontrolled mosaic of the Port Everglades area, Florida. Note mismatching of images due to tilt and relief displacements along the coast (photo: National Oceanographic and Atmospheric Administration).

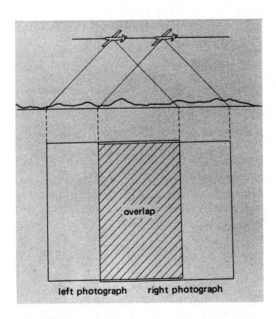

FIGURE 7. Two aerial photographs taken in succession usually overlap approximately 60 percent (courtesy of Sokkisha Co., Ltd., Tokyo).

overlap—or, simply, *overlap*—is greater than 50 percent, there will always be a common area between any three consecutive photographs, thus providing for stereoscopic viewing and extension of horizontal and vertical control by photogrammetric methods (Fig. 8). Between adjacent flight lines the *lateral overlap,* or *sidelap,* amounts to about 15 to 60 percent, providing connection between runs. The actual amount of overlap and sidelap depends on the nature of the photogrammetric project and has to be determined at the planning stage. For most cases, overlaps of about 60 percent and sidelaps of about 20 percent are used (Thompson, 1966).

Fundamental Principles. After acquisition of the photographs, the bundles of rays that have produced these photographs have to be reconstructed in the laboratory. The simplest procedure is to place the developed negative in the camera in its original position with respect to the lens. If the photograph is illuminated from above, the image points are projected through the camera lens with a bundle of rays that become congruent with the original bundle from the photography process. In this reconstruction of the bundle of rays, called *inner orientation,* the ray path is reversed. Usually,

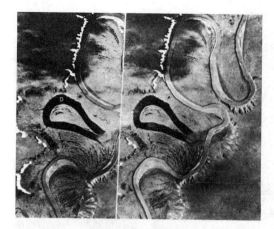

FIGURE 8. Overlap between consecutive aerial photographs provides for stereoscopic viewing of this polygonally patterned ground in a permafrost area of northern Alaska (from Ray, 1960).

projectors that are congruent with the taking camera are used rather than the camera itself.

If a stereo pair is placed in the projectors, the reconstructed bundles will not intersect at a chosen horizontal reference plane, the *projection plane*. A terrain point such as a house will be projected in two different points on the projection plane, one from each projector; that is, parallaxes will occur. The separation between the two projected images is the parallax p, which may be resolved into two components, parallel with a given horizontal reference system xy. The x-axis is approximately parallel to the flight line, with the y-axis perpendicular to it. The two parallax components are then respectively named the *x-parallax (px)* and the *y-parallax (py)*.

The x-parallax, perceived as height, can be eliminated by lowering or raising the projection plane. To reconstruct the model of the terrain, all conjugate rays must intersect each other, which is equivalent to the condition that the y-parallax must simultaneously be corrected in all points common to the two photographs. For two "perspective" bundles of rays, projective geometry states that if five pairs of corresponding rays are made to intersect, all corresponding rays will intersect. Our photogrammetric bundles of rays will be perspective bundles if the inner orientation is correct. Accordingly, the y-parallax must be simultaneously corrected in at least five points. These points are chosen in specific locations, regularly distributed on the overlap between the photographs.

By rotating the projectors around three orthogonal axes and by displacing them along these axes, following a certain systematic procedure known as *relative orientation*, one can place the two projectors in the same position relative to each other as the camera had in the two consecutive stations during exposure. The corresponding rays—that is, rays originating from the same terrain points—will now intersect, forming a *model* of the terrain. This model is geometrically similar to the terrain, but it has a hitherto unknown scale and its position relative to the vertical is also unknown.

If a number of control points is available, an *absolute orientation* may be accomplished in two distinct phases, called *scaling* and *leveling*. The model can be given a desired scale by changing its "base," that is, the separation between the two projectors. The scale is determined by comparing the distances between the control points plotted on a map-sheet and their corresponding distance in the model. By common rotations of the projectors, the model can be leveled with respect to the projection plane to correspond to the level of the terrain with respect to its datum plane. This is done with the aid of the known elevations of the control points. This *geometrical model* can then be regarded as the original object (terrain) and accurate measurements can be performed in it instead of in the object, taking advantage of stereo viewing.

With the help of the measuring marks (floating marks) appearing in the optical system of the instrument, the operator can take vertical and horizontal measurements. The position of the measuring mark in the model space can be read from linear scales or graduations. The movement of the measuring mark throughout the model is transmitted to a plotting pencil, which traces on a map-sheet, placed on the plotting table attached to the instrument, the correct planimetric positions of the features appearing in the model.

To trace a contour line, the required elevation is set on the height-scale; maintaining the same level, the operator drives the measuring mark in a horizontal plane, connecting points of equal elevation by a continuous trace along the surface of the model. The contour lines and the planimetric details are thus plotted in orthographic projection on the map-sheet at the required scale (Moffitt, 1967). In terrain mapping, whole areas can be "moved" into the laboratory as geometrical models and mapped at ease (Hallert, 1960).

Photogrammetric Triangulation. Photogrammetry is extensively used in terrain mapping and is closely related to geodesy. For each photogrammetric project a relatively dense network of points with known ground coordinates, *control points,* is needed. These control points are necessary to establish the position and orientation of each photograph in space relative to the ground. This in turn allows use of the photographs in compilation of photo mosaics, planimetric and topographic maps, and other special-purpose maps. For the basic ground control, photogrammetry must depend on geodetic methods. On the other hand, a considerable amount of the expensive and time-consuming field survey is replaced by photogrammetric methods.

The photogrammetric technique of establishing a network of control points using a minimum of ground surveying is termed *aerial triangulation,* or *photogrammetric triangulation.* The control points determined by photogrammetric methods are usually called *minor control points.* At present, the accuracy of photogrammetric triangulation is so high that it can be used for cadastral surveys. Although aerial triangulation is just one of the many phases in the total mapping process, the accuracy and economy of the whole mapping project lean heavily on triangulation.

With respect to the type of measurements and the instruments used, the methods of photogrammetric triangulation may be divided into two main groups: *radial triangulation,* or *planimetric triangulation,* providing only horizontal control (X and Y coordinates), and *spatial triangulation,* which provides both horizontal and vertical control (X, Y, and H coordinates). The most popular methods of radial triangulation are slotted template triangulation (a mechanical method) and analytical radial triangulation. The spatial triangulation methods are subdivided into three groups: (1) aeropolygon; (2) independent model triangulation, the most popular; and (3) analytical triangulation.

Photogrammetric Instruments for Natural Resources Surveys. Available equipment suitable for natural resources surveys may be classified into two groups: stereoplotting equipment and equipment for the projection of single photographs (Jerie, 1970). A brief survey of some of these instruments follows.

The stereo plotting equipment includes, among other items, a stereosketch, radial line plotters, Deville-type plotters (e.g., SOFELEM, Stereoflex, KEK, Multiscope), stereometer-type instruments (Abrams contour finder, Stereocomparagraph, Stereopret, etc.), a Santoni Stereomicrometer, a Zeiss Stereotope, optical projection-type instruments (e.g., Kelsh, Balplex, Multiplex) and topographic plotters of various makes (e.g., Wild B-8, Kern PG-2, Galileo-Santoni Stereosimplex, Zeiss Planicart). Equipment available for the projection of single photographs includes the Sketchmaster, optical pantographs (e.g., Caesar-Saltzman, Map-O-Graph, Grant Projector, Klimsch Antescope) the Epi diaprojector, and mechanical pantographs (e.g., Perspektomat). In a separate group is the Zeiss-Jena Interpretoskop.

Application of Photogrammetry to Geology

Geologists were among the first earth scientists to realize the importance of photogrammetry to their discipline. Although geology is an earth science that relies on field investigation, some of its specific problems may be partially treated by photogrammetric and photo interpretation methods.

Photogrammetry may serve geology in three basic ways: (1) in the preparation of geological base maps; (2) in the transfer of geological data interpreted on aerial photographs to existing base maps; and (3) in the measurement and interpretation of geological features (Fischer, 1955).

The ideal base maps for the compilation of geological data are topographic maps. Where these are not available, preliminary base maps or map substitutes must be made with sufficient relative accuracy to permit later map-to-map transfer. For example, mechanical radial triangulation methods can be used for expanding the horizontal control to produce low-cost maps of constant scale and sufficient accuracy for preliminary geological compilation. In some works, photo mosaics may be used as substitutes for planimetric maps.

For the transfer of geological data from aerial photographs to a base map, stereo plotting equipment of the direct optical projection type have been used. Or, in its absence, less accurate instruments, such as Stereomicrometer, Stereotope, or radial planimetric plotters—and even low accuracy instruments such as the Stereometers and Stereosketch— may be used. When orthophotographs are available, they can be used as base map substitutes and for accurate transfer of data from photo to map.

In the interpretation and measurement of geological features the basic instrument is the stereoscope. In the office, various types of mirror stereoscopes are used and in some cases a scanning stereoscope may be useful. The pocket stereoscope is handy for field work. Three basic measurements often made by geologists include: (1) the composite thickness of a series of rock units, (2) the relative vertical movement along a fault and the inclination of the fault plane, and (3) the strike and dip of beds (Fischer, 1955). Measurements of this type are normally made with simple stereometer—type instruments (e.g., Abrams contour finder, Stereocomparagraph, and Stereopret). Some geological problems demand greater precision, for example, the study and mapping of variations of thickness of a formation or rock unit over large areas. Higher-precision instruments, such as the Kelsh Plotter and the Wild B-8, are then required. A common geological mapping procedure involving contouring of a selected surface within the bedrock sequence can be done by photogrammetric means.

Angles of slope, strikes and dips, and inclination of fault planes are important measurements in geological studies. These can be calculated from horizontal and vertical measurements but can also be read directly in the stereoscopic model. In the latter case, the stereoscopic exaggeration of relief may influence results, causing considerable error.

Oblique photographs have found valuable application in geological mapping.

New Developments

In the past, geologists have made use of remotely sensed data for geological applications. More recently, the development of Interactive Graphical Design Systems can provide the earth scientists with new possibilities for data management and data retrieval.

I. ENGELSTEIN

References

Fischer, W. A., 1955. Photogeologic instruments used by the U.S. Geological Survey, *Photogramm. Eng.* **21** (1), 32-39.

Hallert, B., 1960. *Photogrammetry.* New York: McGraw-Hill, 340p.

Jerie, H. G., 1970. *Photogrammetry for Natural Resources Surveys.* Enschede, The Netherlands: International Institute for Aerial Survey and Earth Sciences (ITC), 19p.

Mekel, J. F. M.; Savage, J. F.; and Zorn, H. C., 1977. *Slope Measurements and Estimates from Aerial Photographs.* Enschede, The Netherlands: International Institute for Aerial Survey and Earth Sciences (ITC), 32p.

Moffitt, F. H., 1967. *Photogrammetry.* Scranton, Pa.: International Textbook Company.

Ray, R. G., 1960. Aerial photographs in geologic interpretation and mapping, *U.S. Geol. Survey Prof. Paper 373,* 230p.

Spurr, S. H., 1960. *Photogrammetry and Photo-Interpretation.* New York: Ronald Press, 472p.

Thompson, M. M., 1966. Manual of Photogrammetry. Falls Church, Va.: American Society of Photogrammetry, 1,199p.

Zorn, H. C., 1981. *Binocular Vision for Photogrammetrists and Photointerpreters.* Enschede, The Netherlands: International Institute for Aerial Survey and Earth Sciences (ITC), 82p.

Cross-references: *Remote Sensing, Engineering Geology.* Vol. XIV: *Photogeology; Photo Interpretation; Remote Sensing, General; Remote Sensing and Photogrammetry, Societies and Periodicals; Surveying, Electronic; Surveying, General; Topographic, Mapping and Surveying.*

PHOTOGRAPHIC INTERPRETATION—See PHOTOGRAMMETRY. Vol. XIV: PHOTOGEOLOGY; PHOTO INTERPRETATION.

PHOTO INTERPRETATION—See Vol. XIV.

PIPELINE CONSTRUCTION—See PIPELINE CORRIDOR EVALUATION.

PIPELINE CORRIDOR EVALUATION

The design of any excavation and its associated temporary support system, whether for pipelines, trenches, or any subsurface structure, is an engineering problem more complicated than the structural design of permanent facilities. Many reasons may be cited for this. Soils come in an endless variety of natural formations. Their properties are nonhomogeneous, nonisotropic, and nonlinear. There is no unique stress-strain relationship, and a soil's behavior depends on its stress history, applied stress, time, and the environment. All these factors and conditions change continuously from location to location.

In nearly all cases, the soil mass to be evaluated is underground. It cannot be seen in its entirety, but must be evaluated on the basis of limited number of small samples obtained from isolated locations. In addition, most soils are sensitive to disturbance from sampling; thus their behavior and properties measured by laboratory tests may be unlike that of the in-place soil. As an excavation is opened, more information becomes available; the geotechnical problem must be resolved and the design modified (Lambe and Whitman, 1969).

The solution to trenching or excavation problems involves a thorough understanding of geotechnical engineering (q.v.); geology; the mechanics of soils and rocks (see Rock Mechanics; Soil Mechanics); and above all, experience and judgment. Cave-ins may result if engineers and geologists are not permitted to apply their art, skill, and judgment to trenching and excavation problems, or when industry, government, and the public ignore the fact that these professional skills exist (Thompson and Tanenbaum, 1975).

A complete engineering analysis and design for shallow excavations less than 6.1 m (20 ft) deep, which is the limit for most pipelines and trenches, contains the following major components: site evaluation, design analysis, and observation during and after construction. The purpose herein is to present a summary of these components emphasizing those methods, procedures, equipment, theory, and special considerations directly applicable to shallow trenches with vertical or near-vertical walls.

Evaluation of prospective sites for the construction of linear structures, such as sewer systems and pipelines, is basically the same as that required for all other engineering structures. However, two points directly related to linear structures must be remembered during the investigation of a site. A linear structure, by definition, is extremely narrow relative to its length. A normal tendency for engineers is to restrict evaluation to within actual construction boundaries of the project. Because of the narrowness of the site, such an approach may be costly and dangerous. Important site properties may be overlooked such as strike and dip of underlying strata, ground-water flow patterns, presence of weak materials nearby which may influence overall stability of the system, and so on.

The second point involves the method of construction, usually a temporary, open trench. The nature of a temporary trench requires that it be designed in such a manner to remain open just long enough for construction and backfilling to be completed without adversely affecting adjacent structures. To do more may not be economical and to do less might be fatal. Thus the project is a slope stability-earth pressure-settlement problem whose success strongly depends on such factors as vegetation, methodology, and so forth. The influence of such factors must be evaluated during the site investigation phase.

Before slopes can be analyzed or support systems designed, an exploration program is required to determine the lateral and vertical limits of soil and/or rock deposits and layers, and the position of the water table. Ordinarily, this is done by drilling boreholes (see Vol. XIV: *Well Logging*). Soil samples taken from borings are tested in the laboratory to determine the kind of soil and its condition. If auger borings (see Vol. XIV: *Augers, Augering*) are used, the soil can only be classified. Undisturbed soil samples are required if the soil conditions, as exemplified by shear strength, density, and permeability, are to be determined. These soil parameters are required input for a slope stability analysis.

Site Evaluation

Reconnaissance Investigations. Reconnaissance investigations are conducted to determine project feasibility, to plan exploration programs, and, in some cases, to select the best site among possible alternatives. Thorough background research, the first step in evaluation, consists of obtaining and examining old and recent topographic maps; geological maps and reports; mining maps; soil surveys; aerial photographs; subsurface exploration reports; records of government agencies and private firms; university publications and theses; articles from journals and professional publications; information from public utility companies, planning agencies, and engineering departments; well logs; and federal, state, and local legal and code requirements. Such data facilitate planning and will often decrease the required extent and expense of the actual subsurface exploration.

Topographic maps provide information on site accessibility, terrain, and preliminary evaluation of site geology, geomorphology, soils, rock structure, drainage patterns, ground-water conditions, and existing land use (location of structures, utilities, and transportation facilities). Sources in the

United States include the U.S. Geological Survey, Army Map Service, National Ocean and Atmospheric Administration, U.S. Army Corps of Engineers, U.S. Forest Service, Hydrographic Office of the Department of the Navy, planning and zoning boards, private surveyors, and city and county engineers (see Vol. XIV: *Map and Chart Depositories; Map Series and Scales*).

Geological maps and reports provide pertinent geological information including lithology, stratigraphy, areal extent, structural aspects, history, and depositional environments of formations found in and adjacent to the job site. Sources of geological information include maps, reports, bulletins, circulars, monographs, and papers published by universities, the U.S. Geological Survey, National Ocean and Atmospheric Administration, U.S. Bureau of Mines, U.S. Bureau of Reclamation, Geological Society of America, American Geological Institute, professional societies and journals, local government agencies, and local societies.

Mining maps and reports will provide information relative to the location, depth, and extent of surface (strip or pit) and subsurface (room and pillar; longwall; primary, secondary, and tertiary recovery) mining, which is important for predicting the potential for problems related to surface subsidence and ground disturbances. Information of this type if available from the U. S. Bureau of Mines, U. S. Geological Survey (see Vol. XIV: *Geological Surveys, State and Federal*), and individual mining companies.

Soil surveys, particularly those prepared by the U.S. Department of Agriculture (Soil Conservation Service) and the Federal Highway Administration, present information concerning the areal extent and type of surface soil deposits, physiography, relief, drainage patterns, climate, and vegetation. More recent reports incorporate pertinent geotechnical engineering data and descriptions. Two additional sources of soil surveys are state soil surveys and university extension services.

Aerial photographs are particularly useful for large projects or in areas where little or no preexisting data are available. Such photos, which require interpretation, are used for topographic, geological, and soil mapping. A combination of landform, drainage, erosion, vegetation and photographic-tone analyses is employed to evaluate the type of bedrock, rock structure, type and thickness of overburden, surface and subsurface drainage, depth to ground water, and relative percentages of sand and gravel. Major sources of aerial photographs include the U.S. Department of Agriculture, U.S. Geological Survey Map Information Office, National Ocean and Atmospheric Administration, National Atmospheric and Space Administration, and private firms (See Vol. XIV: *Photo Interpretation; Remote Sensing and Photogrammetry, Societies and Periodicals*).

Gathering and evaluating reconnaissance information consists, primarily, of developing a "feel" for what is there, that is, a mental picture of the site. Most of this "picture" will be confirmed or negated through subsequent sources of information and activity. With good mental picture to begin with, the first site visit will be more efficient and have more meaning. It will also help plan a more efficient and economical site investigation. The more "up front" work done, the better equipped is the engineer to handle unforeseen conditions and to avoid costly construction delays.

A visit by an observant engineer or geologist— with a sound knowledge of geology, geomorphology, hydrology, and soils—is mandatory prior to any subsurface exploration. Numerous interviews should be systematically examined and recorded for later evaluation. A general list of these items is presented in Table 1.

In addition to this information, supplemental data, related to the site, should be gathered. This should include an evaluation of local experience from highway officials, railroad engineers, river and hydrologic engineers, local consulting engineers, mining engineers, city and county records, newspapers, farmers, ranchers, and "old-timers." Talk is cheap, yet valuable. Also, knowledge of local practices, costs, and special problems may aid in developing the most economical and safe design.

The precise locations of utility lines are of particular importance in trench design. Numerous cave-ins and deaths have been directly related to broken water lines, electrocutions, and leaking gas lines. Finally, local building codes and zoning ordinances, and federal and state safety codes, such as the Occupational Safety and Health Administration regulations, should be examined to determine their effect on job design and safety, and the legal responsibilities of all parties involved.

Exploration for Preliminary Design. A preliminary exploration program is conducted to obtain adequate subsurface data to permit selection of types, locations, and principal dimensions of all major structures; estimate costs; to establish the depth, thickness, and areal extent of all major soil

---

TABLE 1. Information Gathered During Preliminary Site Visit

1. Site Conditions
   a. Topography—flat, rolling, hilly, mountainous, etc.
   b. Vegetation—marsh, lightly wooded, heavily wooded, tree diameters, grass, etc.
   c. Slopes—percent slope and evaluate stability, evidence of slides and creep (scarps, hummocky ground, bowed trees, or tilted fence posts, misaligned fences, tilted or warped strata, etc.).
   d. Surface drainage—water bodies, springs, brooks; high flood or tide marks; seasonal fluctuation; overall drainage network.
   f. Existing structures—on site and in vicinity
      i. Appearance—signs of stress from settlement, swell, shear failure, etc.
      ii. Foundations—type, depth, allowable bearing value.
      iii. History of problems.
      iv. Sources of vibrations—machinery, traffic, blasting, seismic.
   g. Pavements—locations, purpose, type, probable cross-section, and condition.
   h. Climate—precipitation (in./yr. and in./mo.), evaporation (in./yr. and in./mo.), and degree of frost penetration
   i. Utilities on site—if present, location and height of overhead powerlines.
   j. Accessibility for exploration equipment—effects of weather, availability of drilling water.
   k. Other signs of previous excavations or trenches.
   l. Evidence of surface or subsurface mining (spoil dumps, highwalls, adits, abandoned equipment, acid-mine drainage, subsidence, etc.).

2. Geological Conditions
   a. Geomorphic province—coastal plain, alluvial, colluvial, aeolian, glacial, and modifiers (lake bed, till plain, etc.).
   b. Soil classification—Unified Soil Classification System, Burmister Classification System.
   c. Observations—describe, sketch, photograph and note location:
      i. Surface—soil type on surface, including topsoil, rubbish, fill, etc.
      ii. Cuts—material exposed in road cuts, stream embankments, etc.
      iii. Excavation—excavations for sewers, buildings, etc.
      iv. Bedrock—outcrop locations.
   d. Evaluation
      i. Depth of Topsoil, etc.—auger or probe to determine.
      ii. Underlying profile—as a summary, describe the anticipated strata.
      iii. Depth of ground water—estimate from U.S.G.S. topographic map and elevation of nearby water bodies.
      iv. Depth of hard soil or rock.
      v. Depth to mine, and overburden thickness and condition.
      vi. Availability of construction materials.

Source: After Hunt (1972).

and rock strata; the location of the ground-water table, and the location of existing utilities; to identify the sources, quantity, and quality of construction materials; and to plan the final exploration program. If not previously available, field mapping is conducted at this stage.

In addition, an important component of the preliminary exploration is the detailed evaluation of existing, adjacent structures, utilities, and properties that might experience distress when the excavation is made. A thorough study of these facilities should be made, which would include present elevation reference marks, physical condition, and sketches and photographs for future reference.

Caution against the blind, rigid use of the rules and guidelines to be presented must be emphasized. These guidelines are suggested as planning guides. As each site is unique, modifications are required to tailor the program to fit the unique job requirements and site conditions encountered. Design should not remain static.

Spacing of Borings. Even though a boring layout cannot be precisely dictated for all sites, some general guidelines are recommended for linear structures. Based on experience, Dunlap (1975) recommends a 76-m (250-ft) to 152-m (500-ft) spacing for highways and taxiways, and 23-m (75-ft) to 38-m (125-ft) spacing for runways. Other recommendations are available (Hvorslev, 1949; Navy Facilities Engineering Command, 1971).

For shallow, linear excavations, such as sewer lines and pipelines, spacings for preliminary borings have not been previously recommended. If not governed by building code regulations, the following minimum layout is suggested, which is similar to layouts for highways.

Preliminary borings should be spaced a minimum of 91 m (300 ft); staggered in a zig-zag pattern along the limits of the expected wall instability equal to 1½ times the trench depth *(H)* from the edge of the trench; the first boring located at the centerline, the next 1½*H* to one side of the centerline, then back to the centerline, then 1½*H* to one side of the centerline, and so on. This would result in borings spaced 366 m (1,200 ft) along the limits of instability. If a trench is less than 183 m (600 ft) long, borings should be placed along the centerline at each end of the trench and one in the middle, preferably offset a distance of 1½*H* to one side. Figure 1 is a plan sketch depicting the recommended spacing. The reason for staggering the borings is to determine the strike and dip of any underlying strata. Of course, limitations presented by property owners and rights-of-way may prohibit the staggering of borings. If, through the reconnaissance investigation, visual changes of soil or rock types have been noted, borings should be made to locate these contacts.

Depth of Borings. As with spacing, no guidelines exist for boring depth in shallow excavations. In relation to the previous spacing guidelines, the following depths are recommended. Borings along the trench centerline should be carried to a depth equal to one-half time the bottom width *(B)* of the excavation below the bottom of the excavation. Three-fourths to one time the bottom width has been recommended for deep excavations (Hvorslev, 1949); however, one-half should be adequate for shallow excavations. Remaining borings should be carried to a minimum of 0.3 m (1 ft) below the gradeline. If the trench is located on fill material, borings should be drilled through the fill until virgin soil is encountered to delineate the fill's vertical extent and characteristics.

Nascimento (1970) developed an equation for analyzing the change in vertical stress acting on a horizontal plane at some depth below the bottom of excavations. This change in stress was found to be small compared to the original stress at that depth and became even smaller as the depth increased, assuming no external forces such as

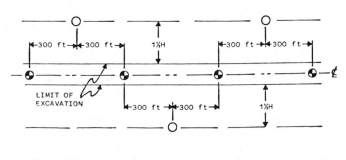

⊕ STANDARD PENETRATION TESTS IN SANDS, AND 3-INCH DIAMETER
   THIN-WALL SHELBY TUBES IN CLAYS AND SILTS

○ DISTURBED AUGER BORINGS

H = DEPTH OF EXCAVATION

FIGURE 1. Recommended type and spacing of borings for linear excavations.

artesian pressures. Thus, the recommended minimum boring depth of 0.5B below the bottom of the excavation seems reasonable. These are *minimum* recommendations only, however. Situations will arise in which borings must be drilled deeper, for example, when locating good bearing material below a layer of organic silt, or delineating the thickness of a clay layer overlying an artesian sand body to determine the potential for a "blowout" when the excavation is made.

Boring Methods. For shallow excavations, the following guidelines are recommended as being economical and providing adequate data for preliminary design (see Fig. 1). Centerline borings should be advanced by *rotary drilling* with capabilities for conducting standard penetration tests and obtaining undisturbed samples for laboratory testing. *Power-flight auger borings* will be adequate for remaining borings. In accessible areas, rigs should be truck-mounted to reduce mobilization time. In inaccessible areas, skid-rigs, with portable auger and rotary equipment, should be used (see Vol. XIV: *Augers, Augering; Borehole Drilling*).

Geophysical Methods. Unfortunately, in most situations, the high cost of geophysical methods precludes their economical use to date in trench investigations. The development of economic exploration techniques that can delineate lateral stratigraphic changes and the location of utility lines would be a welcome addition to present subsurface exploration methods. One such method, acoustics, is presently under investigation (Ash et al., 1974; Rubin et al., 1974).

Sampling Equipment. For undisturbed samples, a thin-wall Shelby tube sampler, with a low area ratio, should be adequate in a trench exploration. A continuous-flight auger and split-spoon sampler are recommended for disturbed samples and the standard penetration test (SPT).

Laboratory Tests. Laboratory tests for trench design can be divided into two parts: classification tests and structural properties tests. Classification tests (i.e., sieve-hydrometer analysis, moisture content, Atterberg limits, and specific gravity) should be conducted on representative samples from all auger and split-spoon sampled borings. Structural properties and classification tests should be conducted on all undisturbed samples to determine their shear strength characteristics, consolidation characteristics if settlement of adjacent structures poses a problem, and permeability characteristics if seepage-pressure problems are anticipated. Consolidation tests on samples, at their natural moisture content and saturated, should be run to determine the potential for large volume changes in soils suspected of having collapsing characteristics. Laboratory vane shear or unconfined compression tests are a rapid and economical method

for the evaluation of unconsolidated, undrained shear strength characteristics for short-term stability analysis, the major concern in trench design.

In Situ Tests. A number of *in situ* methods exist for obtaining data in the field. Although these tests would be used primarily during the final exploration and observations-during-construction stages, they can be employed successfully during the preliminary exploration stage. The standard penetration test can provide data on the compactness (relative density), consistency, and unconfined compressive strength of the soil. Hand-held penetrometers and field vane shear tests can also provide data on the *in situ* shear strength. In addition, the moisture content and density can be evaluated in the field. The *Cohron Sheargraph,* although not yet proven reliable, can evaluate a soil's unconsolidated, undrained total cohesion and angle of internal friction. Where seepage pressure may be a problem, or when dewatering or filter placement is required because an excavation is made below the water table, field permeability tests, although expensive, may be justified. Finally, portable metal detectors can be used to help locate existing, buried utility lines.

Other Tests and Evaluations. Utility lines are subject to corrosion and attack by chemicals, such as sulfates, resulting possibly in sudden breaks or leaks from old lines into the new trench due to their weakened condition. Thus, soil and water samples should be sent to the laboratory to determine the presence of deleterious chemicals (D'Appolonia et al., 1972). The potential seismic activity of the region should be evaluated to determine the influence of seismic vibrations on trench stability. The physical condition of surrounding foundations and structures should be evaluated to determine future effects due to excavation. Reference marks should be placed on the structures for settlement analysis, and all existing cracks should be mapped and recorded prior to opening the excavation. *Test pits* are not always an economic investigation method for trenches unless they are used in conjunction with the evaluation and monitoring of a proposed shoring system in a test section of the trench.

Exploration for Final Design. Unlike the first two stages of site evaluation, no rigid or specific guidelines can be given for the final exploration program. The type, location, and depth of borings, the type and quantity of samples recovered, and the type of laboratory or field tests conducted are functions of the quality and quantity of data retrieved from the first two stages coupled with the economic status of the total program. It is purely a judgment decision that must be made by the engineer. The end product of exploration should be a thorough report including all maps, boring logs, generated cross-sections, findings, and evaluations, which should be made available to all concerned parties prior to bid letting.

Design Analysis and Trench Support

It is not the intent herein to provide a detailed presentation of the soil mechanics' theory relative to earth-pressure analysis and the stability of slopes. The reader is assumed to possess adequate knowledge in this area or can obtain the desired background from any standard geotechnical engineering text. The objective is to present only those theories, equations, and the like that can be applied directly to the design of shallow excavations without getting involved in their derivation.

The action of gravity in an infinite soil mass at rest with a level surface causes forces in the vertical direction. The soil is squeezed vertically. When a material is shortened vertically it expands horizontally if unrestrained (see *Soil Mechanics*). In an infinite mass, the horizontal motion is restrained and horizontal forces develop on vertical planes.

Initially, the forces on horizontal planes are equal to the unit weight of the material times the depth. The forces on the vertical planes are equal to the coefficient of earth pressure at rest times the horizontal force. The coefficient of earth pressure at rest is a function of the material stiffness. The stiffer it is, the smaller the coefficient of earth pressure at rest.

There are no shearing stresses on horizontal and vertical planes for this at-rest, or *geostatic*, condition because there is no relative motion between vertical columns of soil. Thus the horizontal and vertical planes are the planes of principal stress. Shearing forces exist on planes not horizontal or vertical.

By trenching or excavating, horizontal forces on vertical planes are removed and the unrestrained soil expands into the void. Shearing forces are developed on the horizontal and vertical planes. If the soil strength is insufficient, failure occurs by the opening of vertical extension cracks and shear failure along inclined surfaces passing through the trench bottom. The ultimate result is a slide of soil into the excavation.

The solution of a mechanics problem predicts motions. Required information for a solution include loads and body forces; boundary conditions; field or conservation equations (equation of motion, energy, etc.); and a stress-strain law for material behavior. A stress-strain law for soils in various conditions has never been developed; therefore, neither motion of soil masses, soil strains, nor soil stresses can be predicted.

A failure stress criterion developed by Otto Mohr in 1882 (Lambe and Whitman, 1969) from the work of Coulomb in 1776 has been used for soil since the 1930s. Supposedly when the shearing stress τ on a plane reaches a constant plus the normal stress σ on that plane times another constant, large strains, plastic flow, or fracture can be expected. Mathematically,

$$\tau \le c + \sigma \tan \phi \qquad (1)$$

where the first constant is the cohesion *(c)* and the second is the angle of shearing resistance (ϕ). The Mohr-Coulomb failure criterion also allows directions of fracture, slip, or flow planes through a point to be calculated if the directions of the stresses are known at the point. Since soil stresses generally cannot be measured, the failure-stress criterion must be used as a force-failure criterion.

Slope Stability. Because motions cannot be predicted, safety factors against sliding are calculated for soil masses above surfaces adjacent to the excavation. The calculation of the safety factor of an earth slope requires a knowledge of (1) the soil profile (the lithology of the various soil layers or deposits throughout the excavation site or route, and the water-table position); (2) the kind and condition of each soil layer as given by the unit weight of the soil (γ), its cohesive strength *(c)*, its angle of shearing resistance (ϕ), and its coefficient of permeability *(k)*; (3) the slope and depth of the proposed cut; (4) the loads, both static and dynamic, that will be applied to the surface; and (5) a theoretical procedure for slope stability analysis.

The general concepts of slope stability analysis (q.v.) are as follows: four forces tend to cause motion of soil adjacent to an excavation (weight of soil, loads placed on the surface, dynamic forces, and buoyancy and seepage forces of water in the soil). By selecting a possible individual slip surface, the motion-causing forces become known (Fig. 2a). Three forces contribute to resistance to soil motion above the possible slip surface (cohesion force, normal force, and friction force) (Fig. 2b). Each of these forces can be described by its magnitude, direction, and lever arm.

The ratio of the available cohesive force to the required cohesive force to keep the soil body in equilibrium is termed the *cohesive safety factor.* The ratio of the available friction force to the required friction force is termed the *friction safety factor.* The resultant factor of safety referred to in the analysis of the stability of slopes is obtained by resolving the problem until these two individual safety factors are equal to each other.

Three forces with three descriptions for each force produce a total of nine unknowns. There are only three equations of static equilibrium, and the failure criterion equation; therefore, five independent assumptions are required for the solution

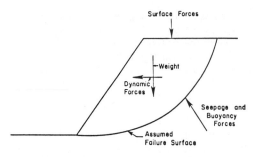

(a) FORCES TENDING TO CAUSE MOTION

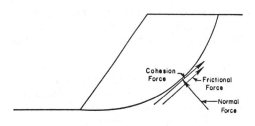

(b) FORCES TENDING TO RESIST MOTION

FIGURE 2. Forces acting on a slope (from Tanenbaum, 1975, p. 63).

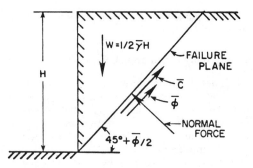

FIGURE 3. Vertical slope with planar, active, failure surface with no surcharge, seepage, hydrostatic, or vibrational forces acting (from Tanenbaum, 1975, p. 65).

set equal to 1.0, the limiting equilibrium case, it is found that the critical depth or height, H_c, is:

$$H_c = \frac{4\bar{c}}{\bar{\gamma}} \tan \left(45° + \frac{\bar{\phi}}{2} \right) \qquad (3)$$

The use of a $45° + \bar{\phi}/2$ inclination of the slip plane originates from the Rankine theory of active failure, and is borne out by laboratory triaxial tests.

These are the simplest, general forms of a limiting equilibrium analysis of slope stability. This approach can be extended for use in other analysis techniques, such as the method of slices, using failure surfaces of various shapes (planar, logarithmic spiral, cycloidal arc, or an irregularly shaped failure plane).

The safety factors calculated using either homogeneous or nonhomogeneous properties seem to be about the same for steep slopes (Fang, 1975). Critical depth calculations are usually confined to very steep slopes; therefore, it seems reasonable to assume homogeneous and isotropic conditions for trench analyses.

For dry, clean, cohesionless soils, the critical slope angle, β_c, is about 34° for loose sand and about 38° for dense sand. Since sand will not stand unsupported at angles greater than β_c, steeper slopes always require support, including close sheeting to prevent the soil from running into the excavation under the influence of either gravity or seepage forces. For example, if the seepage of water is parallel to the slope, β_c should be reduced to 15°.

Because trenches remain open for only short periods of time, the stability analysis of its walls in clays is assumed to be a short-term (undrained) rather than a long-term (drained) problem. For the undrained case,

$$H_c = \frac{4c_u}{\gamma} \qquad (4)$$

of a two-dimensional, static equilibrium, slope-stability problem.

There are at least 36 different slope stability analysis methods depending on the assumptions proposed (Tanenbaum, 1975). It is unnecessary to evaluate the theory behind, and advantages and disadvantages of, each method, as other studies (Hsieh, 1971; Wright, 1969) provide such evaluations.

The limiting equilibrium method is useful because it is simple yet flexible, even though it ignores problems of soil strain compatibility and kinematics (Bailey, 1966); it can be applied to any surface shape, in homogeneous or nonhomogeneous soils using either total or effective stress analysis methods. Although the simplifying assumptions sometimes do not satisfy statics, they are justified by demonstrating that the resultant safety factor is not significantly different from the safety factors derived from more rigorous methods. If a vertical slope with a planar, active, failure surface inclined at an angle of $45° + \bar{\phi}/2$ above the horizontal is assumed, and no surcharge, seepage, hydrostatic or vibrational forces are acting (Fig. 3), then:

$$F = \frac{4c}{\gamma H} \tan \left(45° + \frac{\bar{\phi}}{2} \right) \qquad (2)$$

where F is the factor of safety, $\bar{c}$ is the effective cohesion, $\bar{\gamma}$ is the effective unit weight, H is the height of the vertical slope, and $\bar{\phi}$ is the effective angle of shearing resistance. If the safety factor is where H_c is the critical depth and c_u is the undrained shear strength. This is the basic equation for critical depth.

Based on the need to be conservative in determining the critical depth, and realizing that tension cracks frequently form, the following general equations are recommended for design purposes: (1) for vertical walls with or without surcharge loads:

$$H_c = \frac{2(\bar{c} - q)}{\bar{\gamma}} \tan\left(45° + \frac{\bar{\phi}}{2}\right) \quad (5)$$

(2) for nonvertical walls with or without surcharge loads:

$$H_c = \frac{2(\bar{c} - q) \sin \beta \cos \bar{\phi}}{\gamma \left[1 - \cos(\beta - \phi)\right]} \quad (6)$$

where q is the unit surcharge load and β is the wall angle. Both these equations assume that the factor of safety is 1.00. The calculated critical height should be divided by a factor of safety deemed appropriate by the engineer. According to Thorson (1973), the minimum set-back distance (X or critical X_c) from the edge of the trench, beyond which it is safe to stockpile excavated soils, is determined by the following expressions:

If $H < H_c$ $$X = H\left[\frac{\theta - \sin \theta}{1 - \cos \theta}\right] \quad (7)$$

$$\theta = \arccos\left[1 - \left(\frac{\bar{c}}{\bar{\gamma}H}\right)^{-1} \frac{1 - \sin \bar{\phi}}{2 \tan(45° + \bar{\phi}/2)}\right] \quad (7a)$$

If $H \geq H_c$ $$X_c = H_c\left[\frac{\theta_c - \sin \theta_c}{1 - \cos \theta_c}\right] \quad (8)$$

$$\theta_c = 90° - \bar{\phi} \quad (8a)$$

The relationship between X/H (or X_c/H_c) versus θ (or θ_c) is presented in Figure 4. A minimum set-back distance of 0.5 is recommended for design.

Handy (1974) also determined the set-back factor, using a cycloidal arc as a function of ϕ, and compared this with those values obtained by Terzaghi (1943), who assumed a planar failure surface intersecting the toe of the slope, at an angle of $(45° + \frac{\bar{\phi}}{2})$ measured from the horizontal, and the bottom of a vertical tension crack with depth $z = H_c/2$. The results, compared graphically in Figure 5, indicated that the cycloidal arc failure surface provides more conservative values for X_c than Terzaghi's method. A relatively new method, at variance

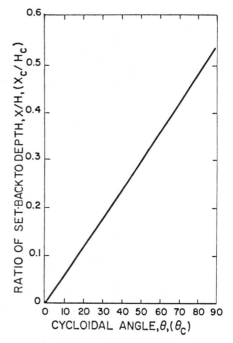

FIGURE 4. Graph of set-back factors, X/H and X_c/H_c, as a function of θ and θ_c (after Thorson, 1973, p. 1028).

with the limiting equilibrium method, has recently been developed. Called the *limit analysis technique*, it is based on the concept of a yield criterion and its associated flow rule while taking into account the soil's stress-strain relationship (Fang, 1975). Chen, et al. (1969) used this method, along with a logarithmic spiral failure surface, to analyze the stability of a slope. They developed the following expression for the critical height for a homogeneous soil:

$$H_c \leq \frac{c_u}{\gamma} N_s \quad (9)$$

or

$$N_s = \frac{\gamma H_c}{c_u} \quad (10)$$

where N_s is defined as a stability number.

Chen et al. also compared the limit analysis method with the stability numbers derived from Taylor (1937) and Fellenius (1936) where the backfill is horizontal. For vertical cuts, the results are identical. As the slope angle decreases, a slight variance is found. At $\beta = 45°$ and $\phi = 25°$, a maximum difference of 2 percent is found between the limiting equilibrium method and the limit analysis method. Thus, for shallow excavations with vertical or near-vertical walls, the limit analysis

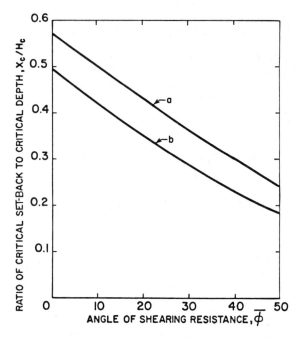

curve a = cycloidal arc failure surface

Where $X_c = H_c \left[\dfrac{\frac{\pi}{2} - (\frac{\pi}{180}) \, \phi - \cos \phi}{1 - \sin \phi} \right]$ (after Handy, 1974)

curve b = planar failure surface with tension crack

Where Z = depth of tension crack = $H_c/2$

$X_c = H_c/2 \left[\tan(45 - \phi/2) \right]$ (after Terzaghi, 1943)

FIGURE 5. Critical distance, X_c, from edge of unsupported cut (from Tanenbaum, 1975, p. 74).

method is just as valid as limiting equilibrium methods. Chen et al. (1969) believe that limit analysis will be equally valid for nonhorizontal backfills even though no limit equilibrium solutions for such a case are available for comparison.

There is another reason, besides the comparability of results, for recommending the limit analysis approach. It combines, in a convenient, easy-to-use fashion, the relationship among the angle of shearing resistance, slope angle, and backfill angle for all homogeneous şoils, except sands ($c = O$), which do not have a stability number and will not stand unsupported in vertical cuts.

A numerical solution for obtaining N_s from equation (10) can be accomplished using a computer. The results are presented in tabular form in Table 2, and in graphical form in Figure 6.

When it is determined from the stability analysis that the slope will not stand unsupported and that it cannot be laid back to a safe angle, then a supported system must be designed and used. This involves the theory of earth pressures exerted against the supports.

Earth Pressure. Three basic types of earth pressure are considered: at rest, active, and passive.

The *coefficient of earth pressure* at rest is defined as the ratio of the effective horizontal to vertical stress exerted on a soil element where no deformation occurs. In equation form for *in situ, at-rest* conditions:

$$K_0 = \frac{\overline{\sigma}_{H0}}{\overline{\sigma}_{V0}} \tag{11}$$

where K_0 is the coefficient of earth pressure at rest and $\overline{\sigma}_{H0}$ and $\overline{\sigma}_{V0}$ are the effective horizontal and vertical normal stresses, respectively, acting on a soil element. If, through lateral movement, the soil were allowed to expand sufficiently, the earth pressure would drop to a minimum value ($\overline{P}_A$). This is the active condition and the coefficient of active earth pressure (K_A) is defined as

$$K_A = \frac{\overline{\sigma}_{HA}}{\overline{\sigma}_{VA}} = \tan^2(45° - \frac{\overline{\phi}}{2}) = \frac{1 - \sin \overline{\phi}}{1 + \sin \overline{\phi}} \tag{12}$$

where $\overline{\sigma}_{HA}$ and $\overline{\sigma}_{VA}$ are the effective horizontal and vertical active normal stress, respectively, acting on a soil element. If forces were applied so that the soil became compressed horizontally, the earth

TABLE 2. Stability Factor, $N_s = H_c\bar\gamma/\bar c$, by Limit Analysis

Friction Angle ϕ, Degrees	Slope Angle α, Degrees	Slope Angle β, Degrees															
		90	85	80	75	70	65	60	55	50	45	40	35	30	25	20	15
0	0	3.83	4.081	4.325	4.57	4.789	5.026	5.25	5.462	5.760	5.86	6.063	6.249	6.51	6.602	6.787	7.35
5	0	4.19	4.502	4.818	5.14	5.469	5.807	6.17	6.526	6.920	7.33	7.839	8.414	9.17	10.130	11.668	14.80
	5	4.14	4.436	4.740	5.05	5.366	5.691	6.03	6.384	6.764	7.18	7.645	8.194	8.93	9.821	11.271	14.62
10	0	4.59	4.971	5.375	5.80	6.249	6.732	7.26	7.844	8.515	9.32	10.298	11.606	13.53	16.636	23.137	45.53
	5	4.53	4.907	5.300	5.72	6.153	6.625	7.14	7.717	8.375	9.14	10.129	11.416	13.26	16.638	22.785	45.15
	10	4.47	4.829	5.207	5.61	6.031	6.487	6.98	7.543	8.180	8.93	9.872	11.109	12.97	15.839	21.957	44.56
15	0	5.02	5.498	6.012	6.57	7.176	7.854	8.64	9.537	10.642	12.05	13.972	16.829	21.71	32.108	69.404	
	5	4.97	5.437	5.940	6.49	7.084	7.754	8.52	9.418	10.513	11.91	13.816	16.652	21.50	31.850	69.047	
	10	4.90	5.363	5.853	6.39	6.971	7.628	8.38	9.262	10.339	11.73	13.591	16.383	21.14	31.378	68.256	
	15	4.83	5.270	5.743	6.28	6.825	7.460	8.18	9.045	10.088	11.42	13.228	15.916	20.59	30.254	65.173	
20	0	5.51	6.099	6.751	7.48	8.299	9.253	10.39	11.799	13.628	16.18	19.996	26.655	41.27	94.632		
	5	5.46	6.040	6.681	7.40	8.212	9.157	10.30	11.687	13.506	16.04	19.850	26.485	41.06	94.377		
	10	5.40	5.969	6.596	7.31	8.105	9.038	10.15	11.542	13.346	15.87	19.641	26.232	40.73	93.776		
	15	5.33	5.882	6.496	7.20	7.970	8.886	9.98	11.347	13.122	15.59	19.322	25.818	40.16	92.896		
	20	5.25	5.773	6.366	7.04	7.793	8.681	9.78	11.066	12.785	15.17	18.770	25.011	39.19	88.632		
25	0	6.06	6.793	7.624	8.59	9.696	11.048	12.75	14.972	18.098	22.92	31.333	50.059	120.0			
	5	6.01	6.735	7.556	8.52	9.611	10.955	12.65	14.864	17.961	22.78	31.188	49.887	119.8			
	10	5.96	6.666	7.475	8.41	9.508	10.842	12.54	14.727	17.829	22.60	30.966	49.635	119.5			
	15	5.89	6.584	7.378	8.30	9.382	10.700	12.40	14.547	17.623	22.37	30.687	49.234	118.7			
	20	5.81	6.483	7.258	8.16	9.220	10.514	12.17	14.297	17.325	21.98	30.198	48.503	117.4			
	25	5.71	6.354	7.104	7.97	9.003	10.257	11.80	13.922	16.851	21.35	29.245	46.759	115.5			

continued

TABLE 2. Stability Factor, $N_s = H_c\bar{\gamma}/\bar{c}$, by Limit Analysis—continued

Friction Angle ϕ, Degrees	Slope Angle α, Degrees	Slope Angle β, Degrees															
		90	85	80	75	70	65	60	55	50	45	40	35	30	25	20	15
30	0	6.69	7.607	8.675	9.96	11.485	13.439	16.11	19.712	25.413	36.63	58.274	144.199				
	5	6.63	7.550	8.607	9.87	11.400	13.348	16.00	19.607	25.298	35.44	58.127	144.011				
	10	6.58	7.483	8.529	9.79	11.301	13.239	15.87	19.475	25.151	35.25	57.924	143.738				
	15	6.53	7.404	8.436	9.67	11.180	13.104	15.69	19.305	24.956	34.99	57.629	143.307				
	20	6.44	7.309	8.323	9.54	11.029	12.931	15.48	19.076	24.682	34.64	57.159	142.538				
	25	6.34	7.190	8.181	9.37	10.833	12.700	15.21	18.744	24.265	34.12	56.302	140.842				
	30	6.22	7.038	7.995	9.15	10.561	12.369	14.81	18.216	23.544	33.08	54.252	134.524				
35	0	7.43	8.581	9.969	11.68	13.857	16.774	20.94	27.448	39.109	65.53	166.378					
	5	7.38	8.524	9.902	11.60	13.774	16.685	20.84	27.344	38.995	65.39	166.220					
	10	7.32	8.458	9.825	11.51	13.676	16.578	20.71	27.216	38.851	65.22	166.003					
	15	7.26	8.382	9.735	11.41	13.560	16.448	20.55	27.053	38.662	65.03	165.720					
	20	7.18	8.291	9.627	11.28	13.417	16.285	20.36	26.836	38.401	64.74	165.188					
	25	7.11	8.180	9.494	11.12	13.234	16.072	20.07	26.533	38.015	64.18	164.296					
	30	6.99	8.041	9.325	10.93	12.990	15.778	19.73	26.071	37.384	63.00	162.333					
	35	6.84	7.858	9.098	10.66	12.641	15.337	19.21	25.271	36.150	60.80	166.378					
40	0	8.30	9.771	11.608	14.00	17.152	21.724	28.99	41.857	71.485	185.6						
	5	8.26	9.713	11.541	13.94	17.069	21.635	28.84	41.784	71.370	185.5						
	10	8.21	9.649	11.465	13.85	16.974	21.530	28.69	41.657	71.226	185.3						
	15	8.15	9.574	11.377	13.72	16.860	21.405	28.54	41.498	71.038	185.0						
	20	8.06	9.487	11.273	13.57	16.723	21.249	28.39	41.290	70.780	184.6						
	25	7.98	9.382	11.147	13.42	16.551	21.049	28.16	41.002	70.406	184.0						
	30	7.87	9.252	10.989	13.21	16.326	20.779	27.88	40.578	69.812	183.2						
	35	7.76	9.086	10.784	12.95	16.016	20.391	27.49	39.885	68.728	182.3						
	40	7.61	8.863	10.501	12.63	15.551	19.773	26.91	38.525	66.119	181.1						

Source: From Fang (1975), p. 368.

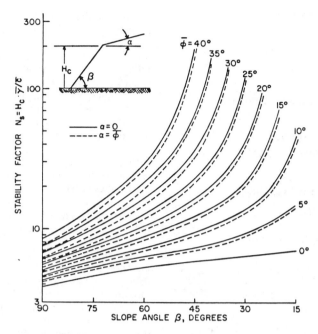

FIGURE 6. Effect of slope angle and angle of shearing resistance on stability factor (from Fang, 1975, p. 369).

pressure would rise to a maximum value called the passive (P_P) earth pressure. The coefficient of passive earth pressure $(\overline{K}_P)$ is defined as

$$K_P = \frac{\overline{\sigma}_{HP}}{\overline{\sigma}_{VP}} = \tan^2(45° + \frac{\overline{\phi}}{2}) = \frac{1 + \sin \overline{\phi}}{1 - \sin \overline{\phi}} \qquad (13)$$

where $\overline{\sigma}_{HP}$ and $\overline{\sigma}_{VP}$ are the effective horizontal and vertical passive normal stress, respectively, acting on a soil element.

These three conditions are depicted diagrammatically in Figure 7. Part (a) shows the condition of earth pressure at rest. Parts (b) and (c) show the conditions of active and passive earth pressures with a sheet pile wall employed to apply the expansive and compressive reactions. From the above equations, it is obvious that $K_P > K_0 > K_a$.

Historically, earth pressure theory was first developed by Coulomb in 1776 and Rankine in 1857, after which it was extended and modified. The general equations for the passive and active pressures developed from Coulomb, which considers cohesion and wall friction (Rankine's theory does not) and a linear failure surface, based on the diagram in Figure 8, are as follows: Active case (Fig. 8a):

$$\overline{P}_A = \frac{\overline{\gamma} H^2}{2} K_A \qquad (14)$$

$$K_A = \qquad (15)$$

$$\frac{\sin^2(\alpha + \overline{\phi})}{\sin^2\alpha \sin(\alpha - \delta) \left[1 + \left\{ \frac{\sin(\overline{\phi} + \delta) \sin (\overline{\phi} - \beta) \frac{1}{2}}{\sin (\alpha + \delta) \sin (\delta + \delta)} \right\} \right]^2}$$

Passive case (Fig. 8b):

$$P_P = \frac{\overline{\gamma} H^2}{2} K_P \qquad (16)$$

$$K_P = \qquad (17)$$

$$\frac{\sin^2(\alpha + \overline{\phi})}{\sin^2\alpha \sin(\alpha - \delta) \left[1 + \left\{ \frac{\sin(\overline{\phi} + \delta) \sin (\overline{\phi} - \beta) \frac{1}{2}}{\sin (\alpha + \delta) \sin (\delta + \beta)} \right\} \right]^2}$$

where α is the wall angle (previously referred to as β in equation 6), δ is the effective angle of wall resistance, and β is the surface slope angle as shown in Figure 8.

In dealing with braced excavations (see *Reinforced Earth*), a major theoretical problem arises. The classical earth pressure theories apply to the rotation of a rigid wall about its toe, which results in a linear distribution of earth pressure with depth. Because of the method of construction and the application of the first top strut, which restrains wall movements, a braced wall is not rigid and does not rotate about its toe. The distribution of earth pressure with depth in this case deviates markedly from the linear distribution of classical theories.

Lacking a rational theory to describe this distribution for braced excavations, empirical earth-pressure diagrams have been developed based on recorded field strut-load data. Several very important points must be made concerning the use of these diagrams. They do not represent actual earth pressure distributions. Based on a compilation of all reported data, they represent

413

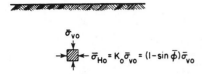

$$\overline{\sigma}_{Ho} = K_o \overline{\sigma}_{vo} = (1 - \sin \overline{\phi}) \overline{\sigma}_{vo}$$

(a) AT-REST EARTH PRESSURE CONDITION

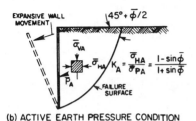

(b) ACTIVE EARTH PRESSURE CONDITION

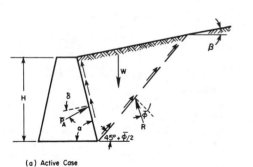

(c) PASSIVE EARTH PRESSURE CONDITION

FIGURE 7. Basic earth-pressure conditions (from Tanenbaum, 1975, p. 80).

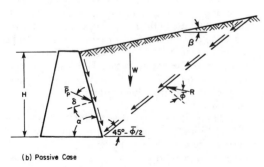

(a) Active Case

(b) Passive Case

FIGURE 8. General Coulomb earth pressures (after Bowles, 1968, pp. 326, 328).

the maximum anticipated loads that a strut could possibly experience, under *normal* soil conditions, at any time during the life of the structure. All the data come from excavations having sheet piling, which may or may not be required in shallow excavations.

Another problem arises in the use of these diagrams. They were determined from data obtained from deep excavations only, that is, for excavations in excess of 6.1 m (20 ft) and under unique site conditions. Virtually no data, theory, or apparent earth-pressure diagrams exist for strutted excavations less than 6.1 m (20 ft) deep for all site conditions. This is disconcerting, since a majority of excavation work falls in the 2.4-to-4.6-m (8-to-15-ft) depth range, namely, sewer lines. It is believed, however, that their use, until full-scale field tests in shallow excavations can be run to verify these diagrams, will be highly conservative; hence a design safety factor only slightly more than 1.00 is recommended in conjunction with these diagrams, backed up by monitoring of the system and planning of alternate designs should problems arise.

Various lateral earth stress distribution diagrams were recommended at different times for cohesionless (Fig. 9) and cohesive (Fig. 10) soils. The uniform lateral earth stress against a vertical wall (Fig. 9g) is the maximum stress that could exist at any depth in a cohesionless soil, not the instantaneous distribution of lateral stress. The horizontal stress of $1.3 \overline{P}_A H = 0.65 \overline{\gamma} HK_A$, where K_A is the coefficient of active earth pressure (Fig. 11). Water pressures, if present, must be added into the analysis.

Parameter K (Fig. 10) is not a coefficient of lateral earth pressure, but an empirical constant, in total stress terms, determined from comparisons of measured strut loads with calculated values. The maximum lateral earth stress for normally consolidated soils is given as $K\gamma H$ (Fig. 10f), where $K = 1 - m (4c_u/\gamma H)$, and $n\gamma H$ for overconsolidated fissured clays (Fig. 10g), where n is taken as 0.15 for shallow excavations (Chapman, et al., 1972). The values of K can be related to the stability number as depicted in Figure 12. The diagrams in Figures 9g, 10f, and 10g are recommended for use in design (Flaate, 1966; Flaate and Peck, 1973) until better data can be obtained from load measurements on shallow strutted excavations.

Prior to determining *lateral earth stresses,* the need for *shoring* must first be evaluated for cohesive soils; cohesionless soils will not stand unsupported much beyond 30°. The decision can be made graphically (Fig. 13) by plotting the data in Table 2 for various safety factors. In sewer lines, H is normally fixed. By plotting the calculated value of N_s against $\overline{\phi}$ for a desired safety factor (1.3 recommended), the need for shoring can be evaluated.

A shallow excavation is normally a short-term stability problem; it is assumed that there is

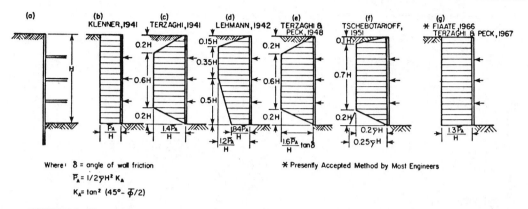

FIGURE 9. Historical development of apparent lateral earth-pressure diagrams for cohesionless soils (from Tanenbaum, 1975, p. 85).

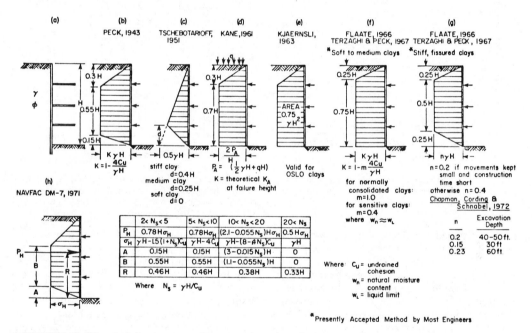

FIGURE 10. Historical development of apparent lateral earth—pressure diagrams for cohesive soils (from Tanenbaum, 1975, p. 86).

insufficient time for water to drain from cohesive soils during the excavation. Unconsolidated, undrained tests are used for determining the short-term shear strength of undisturbed samples. In effective stress terms, only one Mohr circle is generated because of the subtraction of the pore-water pressure from the total confining pressure. The effective cohesion approaches zero, and $\bar{\phi}$ has some finite value.

Since K cannot be converted to effective stress terms, however, a total stress analysis is required. In the unconsolidated, undrained test, the total confining pressure changes with each increase in applied pressure such that several Mohr circles,

all of equal diameter, are developed. Thus the total unconsolidated, undrained angle of shearing resistance, ϕu, is zero and the total shear strength, τ, equals the unconsolidated undrained cohesion, c_u. This analysis method has also been termed the $\phi = 0$ method, but it is valid only in total stress terms. The technique is employed in Figure 12, where K is related to N_s.

Once the need for *strutting* is determined, and K_A or K is determined from the apparent lateral earth pressure diagrams in Figures 9g, 10f, and 10g in conjunction with Figures 11 and 12, the calculation of the loads on each strut can be described (Lambe and Whitman, 1969). The load carried by each

415

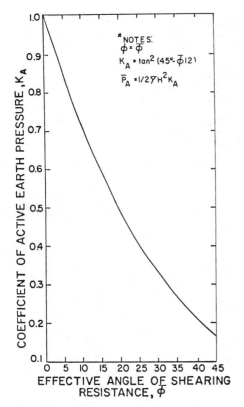

FIGURE 11. Design curve of K_A as related to $\bar{\phi}$ for cohesionless soils ($\bar{c} = 0$): see Figure 9g (from Tanenbaum, 1975, p. 88).

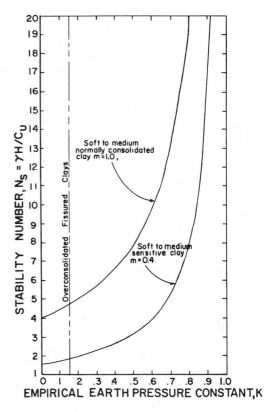

FIGURE 12. Design curves for evaluating empirical lateral earth-pressure constants, K, for cohesive soils ($\phi = 0$): see Figures 10f and 10g (from Tanenbaum, 1975, p. 89).

strut is assumed to contain that portion of the diagram midway between each strut (Fig. 14).

Knowing the earth pressures, strut sizes and allowable loads, this procedure can be reversed to find the maximum strut spacing. In the absence of a thorough engineering design, the worst possible conditions must be anticipated. Here, the minimum design should require the support system to resist a uniform horizontal loading of 146 kg/m² (30 lb/ft²) times the depth of the ditch in feet or meters for the entire vertical face, plus the loads caused by surface surcharges. This uniform load is determined for sand at 95-percent saturation, where the unit weight is 1.9 kg/m³ (120 lb/ft³) $\bar{\phi} = 30°$; the horizontal stress is determined from Figure 14. The uniform load times the horizontal and vertical spacing will determine the loads the struts must carry. The individual supporting members should be able to resist this load with a safety factor of 1.5 against failure of buckling. Structural design should also include all requirements for underpinning of structures when necessary.

A more accurate solution is to assume that the wall is an elastic beam undergoing flexure due to the applied loads. This is a statically indeterminate problem where the equation for the elastic curve

must be determined, redundants are selected, and the forces (strut loads) are resolved.

Another solution, which is more conservative than the statistically indeterminate solution, but which is more accurate than the area method, is to assume that the wall is a rigid beam with hinges at the strut locations. The beam can then be divided into separate free-bodies, which permits a statistically determinate solution to the forces (strut loads).

In addition to the diagrams shown in Figures 9 and 10 for cohesionless and cohesive soils, Armento (1972) has suggested some apparent lateral earth pressure diagrams for cohesive-granular (c-ϕ) soils, in particular dense clayey sands and stiff, sandy days. These diagrams are presented in Figure 15 and are recommended for use where appropriate.

Settlement and Bottom Heave. Besides the failure of an excavation wall, two other types of failures can occur: failure at the excavation bottom, and settlement of the adjacent ground surface and associated structures.

The removal of soil by excavation results in the redistribution of stresses outside the trench, which may result in soil movement and heave of the excavation bottom. This point must be considered when determining the critical depth of an excavation.

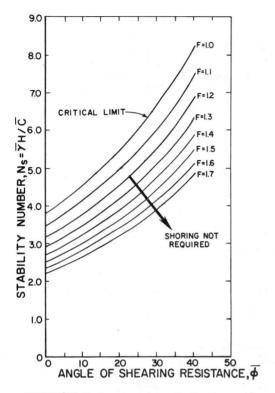

FIGURE 13. Design curves for evaluating the need for shoring for cohesive ($\phi = 0$) and cohesive-granular ($c - \phi$) soils (from Tanenbaum, 1975, p. 87).

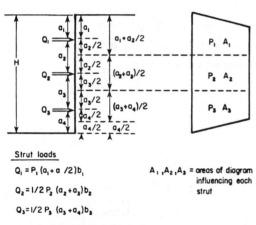

Strut loads

$Q_1 = P_1 (a_1 + a/2) b_1$

$Q_2 = 1/2 P_2 (a_2 + a_3) b_2$

$Q_3 = 1/2 P_3 (a_3 + a_4) b_3$

A_1, A_2, A_3 = areas of diagram influencing each strut

Where: b_1, b_2, b_3=horizontal distance between struts at the respective strut level.

FIGURE 14. Calculation of strut loads using apparent lateral pressure diagrams and the area method (after Lambe and Whitman, 1969, p. 188).

Cohesionless soils are generally incompressible and do not exhibit plastic flow due to stress redistribution, which is observed in cohesive soils. The problem of bottom instability of cohesionless soils is primarily related to the seepage pressures

exerted on the soil. If these pressures are sufficiently high, the weight of the soil grains can be overcome, resulting in their flotation (a quick condition). This produces what is known as *sand boils*. The end result can be softening of the excavation bottom, flooding, undermining the walls because of piping, loss of ground and adjacent settlement, and total excavation instability. The solution to this problem involves the investigation of groundwater and soil permeability conditions, evaluation through conventional methods of the seepage pressure conditions, and development of a design that will alleviate or avoid seepage pressure conditions. Some solutions include dewatering around the excavation, placement of soil filters to prevent the development of quick conditions, ground freezing, or flooding the excavation and working "under water," as was done in Oslo. Any of these solutions must be examined in terms of their effects on other stability aspects of the excavation, particularly the settlement of adjacent ground.

According to Teng (1962), analysis of the stability of the excavation bottom can be determined, based on the diagram shown in Figure 16, in the following manner. If the soil is cohesive, the shear resistance, S, is

$$S = \tfrac{1}{2} q_u (H - q_u/\gamma) \tag{18}$$

If the soil is cohesionless, then

$$S = \tfrac{1}{2} K_A \bar{\gamma} H^2 \tan \bar{\phi} \tag{19}$$

where q_u is the unconfined compressive strength of the soil. Ignoring the weight of the soil below line bc and on both sides of line be, and taking moments about point b,

$$\tfrac{1}{2}(\gamma H + q)B_1{}^2 - SB_1 = \frac{\pi}{2} B_1{}^2 q_u \leq \tfrac{1}{2}q_u B_1{}^2 \tag{20}$$

where q is the uniform surcharge load if present, and $q_u B_1$ is the total passive earth pressure. If $q_u B_1$ is less than the left-hand side of the equation, then the sheet piling or soldier beams must be driven to an additional depth (at least $\tfrac{2}{3}D_1$) to provide additional resistance. For a square, rectangular, or circular plan excavation, or where $H > B$, use Skempton's stability chart (Fig. 17). The factor of safety against bottom heave is

$$F = N_c \frac{c_u}{\gamma H + q} \tag{21}$$

According to Tschebotarioff (1973), bottom failure in clay will follow a cylindrical surface and is caused by the weight of the overlying soil, that is, the pressure, P', exerted on plane AB in Figure 18. The relationship for P' is as follows:

$$P'_{max} = 2.57 q_u (1 + 0.44D/L) \tag{22}$$

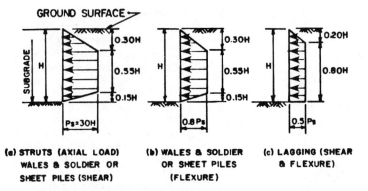

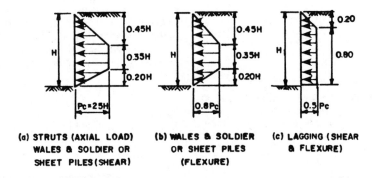

FIGURE 15. Apparent lateral earth-pressure diagrams for two cohesive-granular ($c - \phi$) soils (after Armento, 1972, p. 1,301).

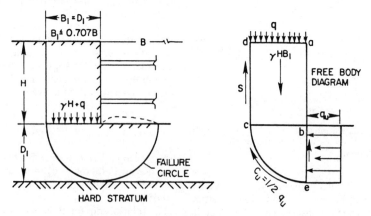

FIGURE 16. Bottom-heave stability according to Teng (after Teng, 1962, p. 399).

SCHEMATIC DRAWING

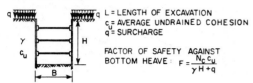

L = LENGTH OF EXCAVATION
c_u = AVERAGE UNDRAINED COHESION
q = SURCHARGE

FACTOR OF SAFETY AGAINST
BOTTOM HEAVE : $F = \dfrac{N_c\, c_u}{\gamma H + q}$

STABILITY NUMBER, N_c

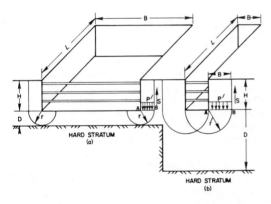

FIGURE 17. Factor of safety against bottom heave (after Flaate and Peck, 1973, p. 11).

FIGURE 18. Bottom-heave stability according to Tschebotarioff (after Tschebotarioff, 1973, p. 433).

The pressure P', acting on plane AB, is set equal to the unit weight of the soil above the plane reduced by the shearing stress along the vertical back and side faces of the clay block overlying the plane:

$$S = q_u/2 \qquad (23)$$

Thus, at limiting equilibrium,

$$P'_{\max} = H(\gamma - q_u)(\frac{1}{2D} + \frac{1}{L}) \qquad (24)$$

and the maximum (or critical) depth of the cut compatible with equilibrium against bottom heave is defined as

$$H_{\max} = \frac{2.57q_u(1 + 0.44\,\dfrac{D}{L}}{\gamma - q_u\,\dfrac{1}{2D} + \dfrac{1}{L}} \qquad (25)$$

If $L \geq D$, local squeezing may occur in soft clays at the center of L. Thus, if the end effects are ignored, equation (25) becomes

$$H_{\max} = \frac{2.57q_u(1 + 0.44\,\dfrac{2D}{L}}{\gamma - q_u\,(\frac{1}{2}D + \dfrac{2D - L}{DL}} \qquad (26)$$

which is valid only if $D < L < 2D$ and $B > D$. If $L = D$, the two equations above are equal to each other; if $L \geq 2D$ and $B > D$, and ignoring the end effects:

$$H_{\max} = (2.57q_u)/\gamma - (q_u/2D) \qquad (27)$$

If $D > B$ (Fig. 17b), use equation (27), but substitute B for D. The value of q_u in these equations is actually the measured value of q_u divided by the safety factor, which varies from 1.6 for nonsensitive clays to 2.5 for highly sensitive clays. Additional discussions concerning these equations can be found in Bjerrum and Eide (1956), Finn (1963), and King (1967).

Assuming the soil beneath the excavation behaves as a frictionless material under undrained conditions, and the surrounding soil acts as a surcharge leading to base failure, stability numbers can be used to evaluate the potential for bottom heave (Peck, 1969a). Given:

$$N_b = \gamma H/S_{ub} \qquad (28)$$

where S_{ub} is the undrained shear strength of the soil below the bottom of the cut, then the following conclusions are recommended: (1) If $N_b < 3.14$, bottom heave is largely elastic and minimal. (2) If $3.14 \leq N_b < 5.14$, bottom heave increases significantly. (3) If $N_{cb} = N_b \geq 5.14$, base failure or failure by heave occurs. Bjerrum and Eide (1956) suggest replacing $N_{cb} = 5.14$ with $N_{cb} = 6.5$ to 7.5 for cuts of ordinary shape.

As with other aspects of trench stability analysis, the potential for failure due to bottom heave in overconsolidated clays is a function of their formation and condition. Here more heave than originally predicted by elastic rebound can occur, even if $N_b < N_{cb}$. This is possibly due to passive failure resulting from large prestored (residual) lateral pressures (Moretto, 1969).

In preconsolidated clays, no danger of real bottom failure exists. If $K_0 > 1.0$, the stress release through excavation may shift the state of stress nearer to the Mohr-Coulomb failure condition; however, should this condition be reached, only a small deformation is required to release the horizontal stress and thus ease the soil. The only possible consequences are the opening of a few fissures and local bulging.

In desiccated clays, where $K_0 < 1.0$, the excavation process shifts the state of stress closer to a hydrostatic condition, creating a more stable condition in the soil at the bottom. The only deformation might be in the form of a slight elastic rebound. Thus bottom heave in overconsolidated soils is not a serious problem (see *Consolidation, Soil*).

Settlement of adjacent ground can become an extremely detrimental and costly problem, particularly if adjacent structures and utilities are involved. Settlement can be broken down into two parts: (1) settlement resulting from displacement of the wall toward the excavation, and (2) settlement due to consolidation of soil.

Consolidation settlement, usually brought about by the addition of overburden loads or the removal of water through drainage or dewatering processes, can be determined through the use of textbook consolidation theory and laboratory tests. Its influence is usually small during the construction process, but it becomes progressively more important during the extended time period after construction. Also, during dewatering, significant rapid settlements can occur in sands, particularly loose sands. Settlement due to movement of the wall is considered to be equal to the area of wall displacement. Since wall displacement is a function of the bracing system and care of installation, this type of settlement cannot be computed, but only forecast.

In cohesionless soils, if properly braced, settlement in loose sands will not exceed 0.5 percent of the depth of the cut, which may, in some cases, be too much. If the surface carries a load, settlement will not extend horizontally more than the depth of the cut. If there is no load, this lateral distance is reduced to $\frac{1}{2}H$ (Terzaghi and Peck, 1967). If the water table is below the sand layer, or lowered in a controlled manner, settlement in dense sands will be minimal. If the ground water is not controlled, erratic settlements can occur because of flow of water plus soil into the cut, which cannot be predicted because of the dependence on freak soil conditions (Peck, 1969a).

In cohesive granular soils, movement is generally small. The cohesive component reduces the sensitivity of the soil to seepage pressures. In saturated plastic clays, large settlements related to wall movement plus consolidation can occur and may extend as much as $3H$ to $4H$ from the edge of the cut. The relationship between settlement and distance from the edge of the cut for plastic clays is shown in Figure 19. The settlement behavior of plastic clays is strongly influenced by minor changes in construction detail. This type of soil is most susceptible to bottom heave.

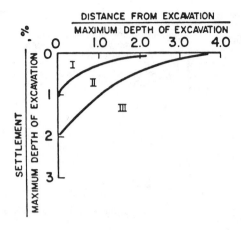

FIGURE 19. Adjacent settlement as related to distance from edge of excavation. Zone I: sand and soft to hard clay; average workmanship. Zone II: very soft clay—(1) limited depth of clay below bottom of excavation and (2) significant depth of clay below bottom of excavation but $N_b < N_{cb}$; settlements affected by construction activities. Zone III: Very soft to soft clay to a significant depth below bottom of excavation and with $N_b < N_{cb}$. Note that all data shown are for excavations using standard soldier piles or sheet piles braced with cross-bracing or tie-backs (after Peck 1969a, 266).

The reduction in vertical pressure due to excavation, at a given depth, is less important to settlement as the strength and stiffness of the soil increase. This is the case for stiff, overconsolidated, plastic clays. The ground surface may actually experience an elastic rise instead of settlement. A reduction of the ground-water level reduces the potential of ground surface rise (Peck, 1969a). However, prolonged lowering of the water table can cause undesired settlement due to consolidation, or drying of a cohesionless soil, which destroys its critical capillary forces and thus its apparent cohesion (Sowers, 1957). Another theory for the relationship between settlement and the distance from the edge of the cut can be found in Caspe (1966), which assumes a log-spiral failure surface and ignores consolidation.

Support Systems

Excavations will encounter any of three basic ground conditions: hard, compact soil; soil likely to crack or crumble; and/or loose or running material (see Vol. XIV: *Rock Geology, Hard vs. Soft*). The most dangerous of these conditions is often hard, compact soil. Although it would be correct to assume hard, compact soil to be the safest of the three categories, it is the most ignored in terms of minimum shoring. Trench walls in loose or noncohesive soils are obviously dangerous and thus require appropriate precautions. The trenches that "look good" are often the most dangerous.

There are various types of support systems, which are often termed "standard" support systems. There is, however, no such thing as a standard support, considering the uniqueness of each site and the variability of the soils present. The following discussion is based on the work of Petersen (1963-1964). Not all methods of support are discussed.

The simple, upright brace (Fig. 20, Table 2) is installed as soon as the trench section has been exposed, usually in shallow trenches with good cohesive walls. The system is designed to alleviate small, localized weaknesses in the wall, and it is of no value in soils exhibiting caving potential. Under caving conditions, such bracing is actually dangerous, providing a false sense of security for the workers.

In trenches deeper than 1.8 m (6 ft), the uprights must be moved closer together; and, at depths exceeding 4.6 m (15 ft), the uprights would be in direct contact with each other. If the excavation is not wider than about 1.5 m (5 ft), extensible metal pipes, called *trench braces,* are commonly used instead of timber. If the excavation is too wide for the use of struts extended across the entire width, the wales may be supported by inclined struts known as *rakes* or *rakers.* Their use requires the soil in the base of the excavation to be firm enough to provide adequate bearing support for the inclined members.

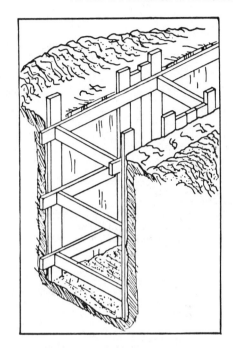

FIGURE 21. Skeleton shoring (after Petersen, Nov. 1963, p. 29).

Skeleton shoring is selected for more unstable ground (Fig. 21, Table 3). Where necessary, vertical sheeting can be driven behind the stringers to improve the overall safety. In addition, it is recommended that the upper few feet be protected with an extra waler since this material is usually less stable than the ground underneath it, due to weathering and possible disturbance.

Skeleton shoring may be used for excavations in clays, because clay is highly dependent on its moisture content for stability. The moisture content of exposed clay may decrease through evaporation, and the clay may shrink, crack, and spall off in chunks. Rainfall may later fill the cracks with water, creating hydrostatic pressures behind the wall and increasing the lateral force on the support system.

It may be possible to eliminate the uprights and replace them with a series of H-piles spaced 1.2 to 2.4 m (4 to 8 ft) apart. The H-piles, known as *soldier piles,* are driven with their flanges parallel to the sides of the excavation. As the soil next to the piles is removed, horizontal boards, known as *lagging,* are introduced and are wedged against the soil outside the cut. As the depth of excavation advances from one level to another, wales and struts are installed.

Close sheeting (Fig. 22, Table 2) is the strongest and safest support method for soils exhibiting a high potential for cave-ins. Continuous support is provided to the face with close-fitting planks. It is

FIGURE 20. Simple upright braces (after Petersen, 1963, p. 28).

TABLE 3. Some Specifications for Shoring

A: Minimum Shoring Under the Best Conditions

| Trench | | Uprights | | Stringers | | Struts | |
Width	Depth	Size	Horizontal Spacing	Size	Vertical Spacing	Size	Horizontal Spacing
Up to 42″	4′ to 10′	2″ × 6″	6′ c-c	None		2″ × 6″[a]	6′ c-c
Over 42″	4′ to 10′	2″ × 6″	6′ c-c	4″ × 6″	4′ c-c	4″ × 6″[a]	6′ c-c
Up to 42″	10′ to 15′	2″ × 6″	4′ c-c	None		2″ × 6″[b]	4′ c-c
Over 42″	10′ to 20′	2″ × 6″	Close	6″ × 6″	4′ c-c	4″ × 6″[b]	6′ c-c
Up to 42″	Over 15′	2″ × 6″	Close	4″ × 12″	4′ c-c	4″ × 12″	6′ c-c
Over 42″	Over 20′	2″ × 6″	Close	6″ × 8″	4′ c-c	6″ × 8″	6′ c-c

Close: Close uprights up tight.
c-c: Center to center.
[a]Minimum: Two struts to 7′ depth and three to 10′.
[b]Minimum: Three struts to 13′ depth and four to 15′.
Source: From Petersen (1963), p. 28.

B: Specifications for Skeleton Shoring

| Trench | | Uprights | | Stringers | | Struts | |
Width	Depth	Size	Horizontal Spacing	Size	Vertical Spacing	Size	Horizontal Spacing
Up to 42″	4′ to 10′	2″ × 6″	3′ c-c	2″ × 6″	[a]	2″ × 6″[b]	6′ c-c
Over 42″	4′ to 10′	2″ × 6″	3′ c-c	4″ × 6″	4′ c-c	4″ × 6″[b]	6′ c-c
Up to 42″	10′ to 15′	2″ × 6″	3′ c-c	2″ × 6″	[c]	2″ × 6″[d]	6′ c-c
Up to 42″	Over 15′	2″ × 6″	Close	4″ × 12″	4′ c-c	4″ × 12″	6′ c-c

Close: Close uprights up tight.
c-c: Center to center.
[a]Minimum: two stringers, one on top and one on bottom.
[b]Minimum: two struts to 7′ depth and three to 10′.
[c]Minimum: three stringers, placed top, bottom and center.
[d]Minimum: three struts to 13′ depth and four to 15′.
Source: From Petersen (1963), p. 29.

C: Specifications for Close Sheeting

| Trench | | Uprights | | Stringers | | Struts | |
Width	Depth	Size	Horizontal Spacing	Size	Vertical Spacing	Size	Horizontal Spacing
Up to 42″	4′ to 10′	2″ × 6″	Close	4″ × 6″	[a]	4″ × 6″	6′ c-c
Over 42″	4′ to 10′	2″ × 6″	Close	4″ × 6″	[a]	4″ × 6″	6′ c-c
Up to 42″	10′ to 15′	2″ × 6″	Close	4″ × 6″	[b]	4″ × 6″	6′ c-c
Up to 42″	Over 15′	2″ × 6″	Close	4″ × 12″	4′ c-c	4″ × 12″	6′ c-c

Close: Close uprights up tight.
c-c: Center to center.
[a]Minimum: two stringers to 7′ depth and three to 10′.
[b]Minimum: three stringers to 13′ depth and four to 15′.
Source: From Petersen (1963), p. 33.

necessary to place the sheeting in position as quickly as possible to prevent slumping of the sides. The walls can be made watertight through the use of tongue-and-groove boards or overlapping sheeting with a packing of dry straw or hay behind the sheeting.

Telescopic shoring (Fig. 23) is used where added protection is desired at lower depths, or where the trench must extend below available uprights and sheeting. The initial cut, which is several inches wider than the planned width, is excavated to the maximum practical depth and shored. This process

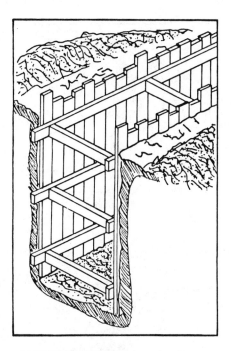

FIGURE 22. Close sheeting (after Petersen, Nov. 1963, p. 33).

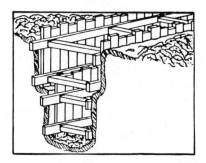

FIGURE 23. Telescopic shoring (after Petersen, Dec. 1963, p. 46).

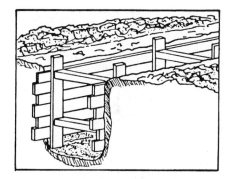

FIGURE 24. Box shoring (after Petersen, Dec. 1963, p. 46).

FIGURE 25. Prefabricated shoring (after Petersen, Dec. 1963, p. 47).

is repeated for the next lower cut, as shown in the figure. This shoring method is very effective in loose or free-running material.

Shoring contemporaneously with increasing depth can be accomplished with the box shoring method (Fig. 24). The trench is excavated to a maximum depth of 3.5 m (12 ft). Horizontal planks are positioned and braced. Standard-size plywood panels can replace the planks in box shoring, as a time and cost saver. This type of shoring should provide protection in almost any soil type.

Prefabricated timber frames (Fig. 25) are devised for use in peat bogs and similar bad ground, making it possible to shore from the ground level before any workers enter the trench. The pre-assembled frames are lowered into the trench from outside the trench. The method is effective for trenches extending to a depth of 4.9 m (16 ft) or less. For deeper trenches in poor material, *telescopic shoring* is recommended.

Hydraulic shoring (Fig. 26) provides a way of installing standard shoring speedily and safely. The method illustrated is comprised of two aluminum uprights connected with hydraulic jacks powered by a ground-based hydraulic pump, which forces the uprights against the walls when the assembly is placed in the trench. Standard shoring can be installed safely behind the temporary hydraulic shoring, which can be advanced without fear of collapse. Another option, conditions

423

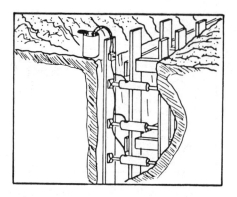

FIGURE 26. Hydraulic shoring (after Petersen, Dec. 1963, p. 48).

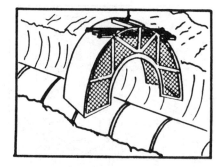

FIGURE 28. Bell hole protection (after Petersen, Dec. 1963, p. 48).

FIGURE 27. Foundation shoring (after Petersen, Dec. 1963, p. 48).

FIGURE 29. Sliding trench shield (after *World Construction,* 1969, p. 29).

A *trench shield* (Fig. 29) is a prefabricated, movable box composed of steel plates welded to a heavy steel frame. OSHA standards permit the use of a trench shield as long as the protection it provides is equal to or greater than the protection that would be provided by the appropriate shoring system.

Where working space must be maximized, and where cross-bearing will interfere with the work, a system of *tie-backs* (Fig. 30), either rock or soil anchors, may be installed in place of standard shoring. If the tie-backs form a component of the permanent structure, they should be suitably protected against corrosion. For temporary use, corrosion protection is not required. Tie-backs are normally tensioned to resist at least the active lateral earth pressure, which would include hydrostatic or seepage pressure if present. Care must be taken not to overtension the tie-backs, which could push the walls into the passive earth-pressure failure zone.

Occasionally, exterior walls are constructed in a trench filled with a slurry or heavy fluid consisting of a clay suspension similar to drilling mud (Fig. 31). The slurry stabilizes the walls of the trench and permits excavation without the need for sheeting or bracing. Cages of reinforcement steel are lowered into the slurry; the slurry is then displaced by tremie concrete. Special equipment is needed for

permitting, is to use hydraulic shoring for the entire trench and eliminate the standard shoring.

Foundation shoring with *raker struts* (Fig. 27) may be required for large excavations, and can safely restrain ground that is likely to cave in. Raker struts should be set in concrete if they are short and steep, or if they must remain in place for a long period.

Bell hole protection (Fig. 28) is a special cage designed to protect bell hole welders on large-diameter pipe. The heavy steel-plate frame follows the curvature of the pipe, protecting welders working in the cage. The wire mesh sides permit air ventilation and dissipation of welding fumes.

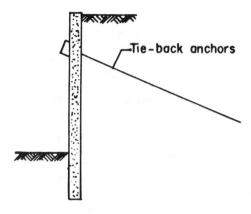

FIGURE 30. Tie-back anchors (after Xanthakos, 1974, p. 175).

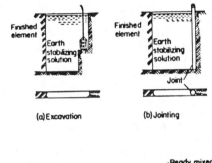

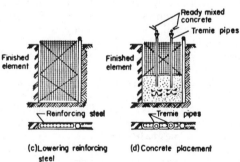

FIGURE 31. Process of fluid trench construction (from Xanthakos, 1974, p. 2).

the various operations and occasional imperfections must be anticipated and repaired. The use of fluids increases the critical depth of the excavation as well as the factor of safety against failure. A complete discussion of this method is available from Xanthakos (1974).

Whatever support system is used, workers should always apply shoring by starting from the top of the trench or excavation and working down. In installing the shoring, care must be taken to place the cross-beams or trench jacks in true horizontal position and to space them vertically to appropriate intervals. The braces also must be secured to

prevent sliding, falling, or kickouts. All materials used for shoring must be in good condition, free of defects, and of the right size. Timbers with large or loose knots should not be used.

As soon as the work is completed, the trench should be *backfilled* as the shoring is dismantled. After the trench has been cleared, workers should remove the shoring from the bottom up, taking care to release jacks or braces slowly. In unstable soil, ropes should be used to pull out the jacks or braces.

A greater awareness of the safety problems to be overcome in excavations, on the part of the person who designs the protection and the person who installs it, will help end cave-in hazards in construction. The previous discussion of support methods is intended to introduce the novice to some, but not all, of the available methods. Thorough engineering design and quality construction are imperative. Details of the design and installation of support systems can be obtained from consulting and construction firms specializing in this work.

External Factors Influencing Trench Stability

Besides the four major theoretical considerations for trench stability (slope stability, earth pressures, bottom heave, and settlement), numerous external factors can influence the stability of a slope. These include surface surcharge, vibrational loads, seepage forces, surface-water flow, climate, vegetation, and man-induced factors.

Surface Surcharge. The application of any source of pressure above and/or behind the edge of an excavation can result in a prolonged stressing of the soil and associated plastic deformation or failure. Surcharge is usually considered as a point load (such as equipment), a line load (such as a railroad), or a uniformly distributed load (spoil, building, pond, snow). If these loads encroach on the edge of the excavation between the edge and the intersection of the ground surface with the possible failure surface, they must be added to the weight of the soil behind the excavation wall and considered in the total stability analysis of the excavation.

Superimposed loads in the vicinity of an excavation also increase the pressure on walls. Heavy equipment and materials such as pipes or timbers should be kept as far back from the excavation as possible. When heavy loads must be located near an excavation, they must be braced, sheetpiled, or shored to support safely the additional lateral pressure generated by the surcharge loads.

Surcharge can also influence settlement of the ground surface. For example, the stockpiling of spoil material adds an additional vertical stress to the soil, which can squeeze out additional water, which in turn result in increased consolidation and an extension of settlement farther back from the edge of the cut. Existing structures can apply additional vertical stresses to the soil during the dewatering process, causing excessive consolidation and associated detrimental settlement beneath the structure. In collapsing soils, the addition of a surcharge load coupled with saturation is particularly hazardous, as this type of soil can experience an instantaneous reduction in volume by as much as 40 percent. In some soils that have reached a critical state of stress due to excavation, the mere presence (weight) of a person near the edge of the cut can produce failure of the excavation.

The analysis of surcharge loads in the design of any structure is one of the weak points in soil mechanics requiring the use of elastic theory and simplifying assumptions. Soil mechanics methods may be used to analyze surcharge loads (Bowles, 1968; Kezdi, 1975; Lambe and Whitman, 1969; Spangler and Handy, 1973).

Vibration Loads. When vibrating machinery, vehicles, pile drivers, blasting or other dynamic loads are in the vicinity of an excavation, stress waves are propagated through the soil. When pressure waves are reflected off the face of an excavation, the soil is placed in tension and cracking or spallation may occur. This can be particularly dangerous in the brittle materials, such as clayey sand or gravel, that can have a high static strength. The basic reason for the danger is that soil has inertia and very little tensile strength. Also the effects of vibrations are cumulative over periods of time. Motions can become large if the vibrating equipment has the proper frequency.

Vibrations can cause individual soil grains to move in some soil types. If confined, saturated cohesionless soils are subjected to vibrations, a quick condition can result, destroying what little cohesion remains in the mass (see Vol. XII, Pt. 1: *Thixotropy, Thixotropism*). In general terms, vibrations alter the state of stress in a soil in a very complex, and as yet poorly understood manner. The services of a specialist are mandatory when vibrations appear to represent a problem in design or in construction.

Seepage. When trenches or excavations are carried below the water table, a *dewatering procedure* must be developed. This may be as simple as pumping the water out of the excavation or as elaborate as a well point or ground-freezing system. But in all cases the amount of water flowing into an excavation and the effect on the soil must be calculated.

When water flows upward into an excavation through noncohesive soil, with sufficient velocity to overcome the gravitational forces, the soil loses all strength and a quick condition results. The seepage forces can remove the fines in the soil, undermining and weakening the trench, a process called *piping*. To prevent such situations, the water

table may be lowered below the excavation bottom by interceptors or well points, or may be solidified by freezing to eliminate flow.

Pumping to lower the water table can also cause consolidation with associated damage to adjacent structures. When the coefficient of permeability of the soil is known, most seepage and consolidation problems can be solved by the use of Darcy's law and the principle of the conservation of mass (see *Hydrodynamics, Porous Media*). These are analogous to heat-flow problems. If the boundary conditions can be defined, the rate of flow and the water pressure can be calculated throughout the field. (For a discussion of analysis methods, see Casagrande, 1937; Cedegren, 1968; Harr, 1962; and Hvorslev, 1949.) The usual solution technique is termed *flow-net sketching.* Conformal mapping and direct integration can be used. Models, electric analogs, and computer methods are also standard procedures used to analyze the quantity of flow, gradient, and seepage pressures. Water is the number-one villain in trench stability; thus the analysis of ground-water conditions is of primary importance.

Surface-Water Flow. As with ground water, surface water can also be a hazardous aspect of trenching. The source (streams, ponds, rainfall, snowmelt, etc.) and location of surface water, its present and anticipated volume and flow velocity, the direction of flow and drainage patterns in the vicinity of the excavation, and the available methods for safely diverting surface water away from the excavation and adjacent structures must be considered in the design evaluation (see Vol. XIV: *Land Drainage*). Surface water can have several detrimental effects. First and foremost, it can flood the excavation. It can add weight to the walls of the excavation by saturating the soil, and possibly decrease the effective stress. The same situation can occur because of rapid drawdown, that is, the rapid dewatering of a flooded excavation. Softening of the soil and a corresponding reduction in shear strength may also occur. In certain soils, swelling may increase the loads on support systems and adjacent structures. Cementitious materials can be dissolved and removed from between soil grains, thus reducing the overall strength of the mass. Finally, flowing water is an erosive force that can destroy the walls of a cut carrying material to the bottom, undermine the walls of an excavation or stockpiled material, and remove stockpiled material from the site of the excavation, resulting in economic losses and environmental degradation downstream. Thus the force of surface-water flow requires advanced evaluation and use of storage or diversion systems to avoid the detrimental effects.

Climate. The influence of climate can be considerable and unpredictable. Sudden, excessive quantities of water resulting from rainfall (also broken sewers, storm drains, waterlines, and high tides) can, for example, lead to flooding of the excavation along with soil saturation, increased unit weight, and decreased effective stress. Seepage forces and rates can increase, along with softening, swelling, spallation, erosion, and undercutting of the soil. The melting of snow has the same effects as excessive rainfall. It also results in a surcharge load, potential freezing effects, and local softening of exposed soil. Excessive drying can produce desiccation cracks and zones with little or no cohesion, initiate slaking and spallation, create avenues for water movement with its associated problems, induce planes of weakness, and reduce strut loads and strut effectiveness. Air movement can influence the temperature of the trench system, increase the rate of drying of the soil, move soil particles in the form of blowing dust, and create vibrational forces on adjacent structures and the soil system.

Temperature changes, which strongly influence the stability and effectiveness of the soil and support system, can take several forms, including seasonal, diurnal, and sudden changes due to direct sunlight and shade. Temperature may, for example, directly influence the soil particles by causing cyclic changes of volume and strength. The major hazard is the freeze-thaw effect, especially in silty soils. Sands usually have a sufficiently high permeability to prevent the accumulation of freezable water, whereas a clay's permeability is very low and freeze-related expansion is resisted by internal cohesive forces. A silt's permeability is low enough to permit water accumulation but not high enough to allow rapid, free drainage. In addition, the cohesive forces present in clays are nonexistent in silts. Even though a hard frost may appear to solidify a trench, the freezing at depth will create cracks and lead to progressive failure (Petersen, 1963-1964). The formation of ice lenses will cause the soil-water system to expand drastically, resulting in bulging of the soil and excessive loads on the support system. With sudden thawing due to direct exposure to sunlight or other heat sources, the ice turns to water with a shear strength of zero, which decreases the effective soil strength and leads to failure.

Frost penetration causes a tremendous increase in strut loads and may result in bending and buckling of struts and wales (Bjerrum et al., 1965). DiBiagio and Bjerrum (1957) found that freezing of the clay surface caused loads in the top two struts to drop by 50 percent, probably the result of vertical frost heaving of the soil. The pressure on the bottom four struts increased drastically from 3 to 10 times the normal load, or as much as 13,620 kg (30,000 lb), causing some of the struts to buckle. During the subsequent thaw, the struts had to be tightened to maintain their load. Thus, freeze-thaw conditions can exert a commanding influence on the system.

There is also a direct relationship between strut load and air temperature. Steel struts, which expand with an increase in temperature, experience a 10- to 20-percent increase in load for an average daily temperature variation of only 8°C (Chapman et al., 1972; Flaate, 1966; Morgenstern and Eisenstein, 1970).

As previously mentioned, ground freezing may be used to control ground-water flow into an excavation. Because of the potential detrimental effects of frozen ground, the ground-freezing procedure should be completed prior to any excavation, and the excavation should be backfilled prior to removing the ground-freezing system and allowing the soil to thaw.

Vegetation. The presence or removal of vegetation affects the microclimate near the soil surface as well as the erosive stability of the soil. A more important effect is the influence that vegetation exerts on the moisture regime of the soil. Through the respiration process, vegetation continuously removes vast quantities of water from the soil. After removal of vegetation, the soil will absorb water and a new state of equilibrium will be established. This often results in softening, swelling, and strength-reduction of the soil.

Man-Induced Factors. Excavation cave-ins are frequently reported to be the result of human activity. This is an erroneous assumption because there is only one cause of cave-ins, weakening the soil by applied stresses. In other words, whether an excavation will fail is dependent primarily on the soil properties. Some man-induced factors may be influential while others, resulting from ignorance, can actually lead to fatalities and injuries. These factors, which are self-explanatory, are listed in Table 4.

Specific Soil Characteristics

Due to the nature and properties of various kinds of soil materials, and altering processes occuring during and after deposition, specific trench stability problems associated with these deposits can be anticipated. An awareness of these unique characteristics, on the part of the geologist or engineer, can mean the difference between failure and success. The following is a brief presentation of salient soil characteristics of importance to engineering works and trenches in particular.

Cohesionless granular soils (mainly sands and gravels) rarely stand unsupported beyond an angle of 30° measured from the horizontal. Thus they normally require close sheeting throughout the entire depth of the cut, particularly if the water table is above the trench bottom. As the water seeps into the trench, slides, quick conditions, and boils may develop.

Cohesive granular soils (silty sands, clayey sands, clayey silts, etc.) may stand unsupported at angles

TABLE 4. Man-Induced Factors Leading to Cave-ins and Fatalities

Engineering Activity	Percent[a]
No shoring	50%
Improper or defective shoring	35%
Backfill placed too close to edge of excavation	10%
Accidentally broken utility lines	3%
Lack of proper safety warnings and barricades	2%
Excavation and construction technique and workmanship	
Men working ahead of shoring and bracing	
Men working too close to excavating equipment	
No ladders—use of struts as ladders	
No sheet piling to supplement shoring in unstable soil	
Lengthy delays in shoring installation to permit ease of construction	
Inadequate initial geotechnical exploration	
Inadequate location of underground utilities	
Improper installation or removal of shoring	
Improper maintenance of shoring system	
Creation of voids when trying to salvage sheeting	
Undermining	
Inadequate control of drainage and seepage	
Operation of vehicles too close to edge of cut	

[a]Percentage of 125 fatalities over a 24-month period ending in June 1967.
Source: After Land (1968).

greater than their cohesionless counterparts. Their apparently safe slopes are deceptive, however, because these soils easily lose their cohesive strength under the influence of external factors. They are also highly susceptible to deformations resulting from freeze-thaw conditions. Trench designs for these soils must be approached with extreme caution.

Collapsing soils (loess, basin alluvial fill, mudslides, loose sands) contain large void spaces inherent from their environment of deposition. Such soils may, upon saturation and loading, experience an instantaneous volume loss of as much as 40 percent; yet under normal, undisturbed conditions they can remain stable in vertical cuts to great depths. It is sometimes desirable to induce collapse prior to trench construction through surcharge and saturation in order to work with a denser, more stable soil.

Quick, sensitive clays (see Vol. XI, Pt. 1: *Thixotropy, Thixotropism*) are equally deceptive. Because of their high moisture content and open structure, sudden vibrations or loads can cause instantaneous flowage. Sensitive clays, and the detrimental properties associated with them, can be identified only through careful engineering evaluation. Both undisturbed and remolded sample tests should be conducted to determine sensitivity if quick clays are suspected.

Normally consolidated clays come closest to fitting the assumption of homogeneity. The maximum safe depth to which a vertical cut can be made is a function of shear strength. Some clays may stand at depths of 9.2 m (30 ft), whereas others will fail at depths shallower than 1.5 m (5 ft).

The stability analysis and trench design of such clays is fairly straightforward. They are most susceptible to failure immediately after construction and gain strength with time because of drainage and consolidation.

Overconsolidated, nonfissured clays, often referred to as stiff clays, differ from normally consolidated clays in that they have experienced greater loads in the past than at present. Such loads could have been induced by previous overburden material now removed by erosion, glacial ice, or desiccation due to drying. They can be highly stable initially after construction but lose strength in time with the release of residual strains and the absorption of moisture. These soils have, unfortunately, been referred to by contractors as appearing "solid as a brick wall" shortly before it collapsed.

In many theoretical considerations, the soil is regarded as a homogeneous, isotropic material. Thus, for analytical purposes it is assumed that the properties of a soil, under a given state of stress, are uniform, continuous, and unchanging throughout the soil mass. Unfortunately, soils are nearly always nonhomogeneous and nonisotropic, are layered, and contain planes of weakness. Layering is a function of a change in grain size resulting from a change in the depositional or erosional environment. The angle or dip at which these layers rest can be altered by normal geological processes. Bedding planes, fissility surfaces, and contraction cracks are developed during sedimentation. Tectonic fractures and faults, shrinkage cracks, and desiccation cracks are developed after sedimentation.

An *overconsolidated fissured clay* is an example of a soil type whose behavior is controlled by its structure. If a fissure is open, no resistance to shear exists along its surface. According to Duncan and Dunlop (1969), only the disturbed strength can be mobilized along closed fissures, and thus fissures may adversely influence stresses within slopes, increasing the likelihood of progressive failure. Open fissures permit the entrance of water, and the soil is free to expand along the fissure, resulting in soil swelling and softening. Swelling pressures may induce sufficient force to open deeper fissures, leading to progressive failure of the cut. Hydrostatic pressures within the soil may also increase and thus lead to failure (Esu, 1966). If the water freezes, expansive forces can dominate all other considerations in trench stability.

The *shear strength* (τ) of a clay is defined by the Mohr-Coulomb failure criterion (equation 1) in total stress terms. The normal stress (σ) at any depth below the surface for geostatic conditions can be defined as the soil's total unit weight (γ) times the depth *(z)*, and c and ϕ can be determined in the lab. Thus, τ can be evaluated.

Plotting, from a laboratory test, the deformation, δ, against the shear strength, for a normally consolidated soil, the results would appear as shown in Figure 32a. The shear strength would increase as the deformation increased up to a peak point after which, with further deformation, there would be a slight drop in strength. The maximum strength is the *peak strength*; the final strength is the *residual strength*. For a normally consolidated

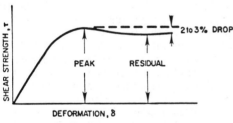

(a) NORMALLY CONSOLIDATED CLAY

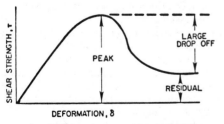

(b) OVERCONSOLIDATED CLAY

FIGURE 32. Peak and residual strength of clay (from Tanenbaum, 1975, p. 128).

clay, the difference between the two amounts to only 2 to 3 percent. For an overconsolidated clay, this drop in strength is much greater (Fig. 32b).

In a *normally consolidated clay,* the difference between the two strengths is so small as to become unimportant for design purposes. In an *overconsolidated fissured clay,* the strength along the fractures has dropped to perhaps its lowest value, owing to previous deformation, and will control the stability of the cut. In *overconsolidated nonfissured (intact) clays,* no major deformation has occurred. This clay could maintain stability up to its peak strength value; however, time begins to play an important role.

The *pore-water pressure (μ)* in a soil changes with time—the difference between the concepts of effective and total stress (see Vol. XII, Pt. 1: *Soil Mechanics*). The effective normal stress can be stated simply as the difference between the total normal stress and pore-water pressure. Thus, equation (1) can be rewritten as:

$$\tau = \bar{c} + (\sigma - \mu) \tan \bar{\phi} \qquad (29)$$

where $\tau, \bar{c},$ and $\bar{\phi}$ are now effective values of shear strength, cohesion, and angle of shearing resistance, respectively. Obviously, as pore pressures increase, the shear strength of the soil and the safety of the slope decrease.

For normally consolidated soils the pore pressure is at its highest point just after the cut is completed and drops as a function of time. Hence, the strength of a normally consolidated soil increases with time according to equation (29).

In overconsolidated clays, horizontal stresses predominate and the clay will expand toward the excavation. As a result of this expansion, the pore pressure becomes negative. To relieve this "vacuum," the clay will absorb water from the surrounding area over a period of time, causing the pore pressure to increase once again. Thus, immediately after excavation, overconsolidated clays are at their highest strength in accordance with equation (29). If the change in pore pressure is plotted in relation to time (Fig. 33a), this concept is well illustrated for the two types of clay. A similar plot of safety factor as related to time (Fig. 33b) delineates that, at the end of the excavation, a normally consolidated clay is at its most dangerous stage, whereas an overconsolidated nonfissured clay is at its safest stage.

Terzaghi (1936) first pointed out that the release of large horizontal pressures through excavation in overconsolidated fissured clays can result in opening of these fissures. Once these fissures exist, analysis of slope stability becomes a matter of determining the correct cohesion for the soil. Skempton (1964) showed that, as the strength of a soil drops from its peak to disturbed value, the

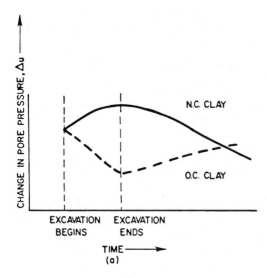

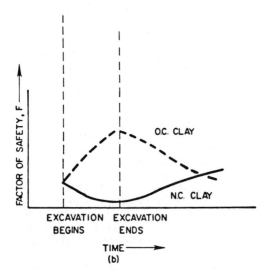

FIGURE 33. Influence of time on the stability of normally consolidated (N.C.) and overconsolidated (O.C.) clays (from Tanenbaum, 1975, p. 130).

cohesion, as well as the angle of shearing resistance, drops. Measurements of $\bar{c}$ and $\bar{\phi}$ on intact undisturbed samples could result in overestimating the stability of the slope by as much as 80 percent.

The evaluation of cohesion is of critical importance yet extremely difficult in overconsolidated fissured clays. The strength along the fissures is considerably lower than that of the mass. The use of undisturbed samples 3.9 cm (1.5 in.) in diameter may result in higher-than-actual *in situ* values of cohesion, as the samples may not contain any fractures because of their size. It has, for example, been shown that as the sample size

increases, more fissures are included in the sample and the estimate of cohesion decreases (Skempton and LaRochelle, 1965). Also, even over a short period of time (a few days) after opening the excavation, water migrates toward the fissures, causing an additional reduction in strength.

Because measurement of actual shear strength properties along a fissure is a difficult, expensive, and time-consuming process, and because little is known about the influence of fissures on strength reduction in particular soils, it is suggested that the calculated shear strength determined from normal lab tests should be multiplied by a reduction factor of 0.5. This procedure should take into account the influence of fissures in overconsolidated clays.

Colluvial soils (soils deposited by gravity-mass movement) can also be considered in light of the previous discussion of residual strength. Colluvial soils that are primarily fine-grained will probably possess a plane of weakness along which movement has occurred. It appears that the percentage of clay along this plane, which frequently exhibits slickensides, is higher than the remaining soil mass. The movement along this plane under the influence of gravity has reduced the shear strength to the residual value. In addition, this plane acts as a passageway for the movement of water, which further softens the soil. If encountered in an excavation wall, large masses of colluvial soil may move swiftly into the excavation.

Organic silts and *clays (peat)* are highly compressible soils. Their presence in the vicinity of an excavation normally can lead to excessive ground-surface settlement and deformations. Organic soils are unsatisfactory for trenching and must be removed, bypassed, or supported.

Trenching in *man-made fills* (see Vol. XIV: *Artificial Fill*) is very treacherous. Because older fills normally consist of a heterogeneous mixture of soils and construction debris, their physical properties are highly anisotropic and virtually impossible to predict. A great many trench cave-ins can be traced to the presence of fill. Fill material rarely achieves strength properties exhibited by natural soils (Petersen, 1963-1964), and there is no way to define the strength of such materials, which must be assumed to be very low or zero for design purposes. Recent fills are usually placed in layers and compacted to achieve high strengths, but this is not always the case (Van Horn, 1969). Recently backfilled or puddled areas should be avoided where possible, but this may not be possible because many trench excavations are made to fill areas along older utility lines.

Besides the specific characteristics associated with various types of materials, some additional properties of several geological deposits and rocks, faults, cements, and gaseous and chemical impurities should also be considered in the analysis and design of excavations. *Buried stream deposits* can be a source of excessive water flow into an excavation requiring dewatering or diversion. A knowledge of the local geological history and geomorphology will aid in locating these deposits. Aerial photographs can be used to locate aligned vegetation (phreatophytes), linear soil (silt) deposits, old terraces, and other geomorphic forms indicative of streams. Meander loop and cutoff deposits are abandoned channels filled with very soft, compressible organic silts and clays. The problems are the same as those associated with organic soils. Aerial photography can be used to determine their location (see Vol. XIV: *Photogeology; Photointerpretation*).

Clay deposits immediately above limestone are frequently soft and compressible (D'Appolonia et al., 1972). Limestone may be weakened by solutioning, which can create seams of weak clay, caverns, and sinkholes. These regions require detailed surface and photographic mapping. Although not immediately apparent, weathered rock behaves as poorly as noncemented soil and should be treated as such.

Geologists are continuously aware that faults bring underlying materials to the surface or laterally displace materials into the path of a trench, besides being planes of weakness in their own right. For example, a fault in Salt Lake City, Utah, brought a layer of loose, weakly cemented gravel to the surface (Van Horn, 1965). When the trench traversed this fault, gravel raveled into the trench, excessively widening the upper part. Vibrations of equipment resulted in extreme danger of cave-ins, so workers were not allowed in the trench during the excavation process. In this case, the influence of the fault had been recognized; however, engineers and contractors must always be aware of geological processes that can alter normal site conditions.

The degree and type of *cementation* can influence soil properties. Common cements include calcium carbonate, gypsum, clay, and iron oxide. When exposed to the elements through excavations, cements can be partially or completely removed, reducing soil strength. The designer must plan for this process of weathering and strength reduction.

Noxious and explosive gases (e.g., methane) can seep into an excavation, creating toxic or flammable conditions. Sources of these gases should be located and if such a problem is suspected, soil and water samples should be chemically tested. Proper ventilation in the trench should always be maintained. In addition, utility lines are subject to corrosion and attack by chemicals, such as sulfates, that exist in the soil-water system; sudden breaks or leaks into a new excavation from old, weakened lines can result. Thus soil and water samples should be tested for the presence of deleterious chemicals.

Observations During Construction

Observations during construction are required to verify the anticipated soil profile; obtain new data for unusual or unanticipated conditions and prepare a new design; control ground motions or ground-water problems; observe and correct support system conditions, motions, and distress; observe and correct distress induced to adjacent structures; and prevent injury and loss of life due to trench instability (Peck, 1969b).

Warning Signs of Failure. An observant, trained individual can normally prevent fatalities and injuries due to cave-ins by being aware of the many warning signs of imminent trench instability. Adjacent ground subsidence (q.v.), tension cracks, spalling and sloughing of soil, softening, changes in wall slope or wall bulge, or a sudden increase or decrease in strut loads are indicative of failure of the wall of the cut. Softening, adjacent subsidence, or the removal of more soil than predicted may indicate bottom heave conditions. Increased water seepage, erosion, small boils, dirty water, or slight bottom heave followed by large boils are indicative of quick and piping conditions. Such conditions and problems frequently follow rainstorms, requiring additional close monitoring beyond normally scheduled daily inspections.

Monitoring Methods. The best person to monitor the stability of an excavation is a trained, experienced engineering inspector. Mechanical devices are available to aid the engineer (see *Rock Structure Monitoring*). General surveying techniques can be used to monitor wall movement, bottom heave, and adjacent settlement. In conjunction with surveying, settlement plates, heave points, inclinometers, and lasers can be installed to increase the accuracy of the measurements. The temperature of the air, soil, and struts should be recorded to determine the distress of the system. Strut loads can be monitored using portable hydraulic jacks, micrometer strain gauges, electrical-resistance strain gauges, vibrating-wire gauges, and load cells incorporating vibrating-wire gauges, which can also be used to monitor earth pressures exerted against the support system. Changes in the water table and pore pressures can be monitored using well points and piezometers. Bjerrum et al. (1965) present a discussion of some of these methods for monitoring the stability of braced excavations.

Observations after Construction

Most effects of trenching cease once backfilling is completed. Settlement of adjacent ground and structures may persist long after backfilling, however, and should be evaluated periodically. The trench must be backfilled with care. Clean fill should be placed in a maximum loose-lift thickness of 23 cm (9 in.) and vibrated to a minimum 70-percent relative density with a hand-operated tamper, or heavier equipment if it is determined that it will not result in any distress to the trench or structures in the trench. One reason for taking such care in backfilling is not only to protect the pipe from settling and cracking, but to ensure the stability of future trenches that may traverse the fill.

Summary of Geotechnical Recommendations

Exploration procedures, analysis methods, and design criteria available to the engineer have been evaluated for shallow excavations. These recommendations, by themselves, do not represent a complete detailed procedure for an engineer's use in the total design of shallow excavations. Such a detailed procedure would fulfill the requirements for a manual of practice for engineers. For example, no designs of the structural support components

of the trench have been presented however, the design procedure has been brought to the point of enabling specification of the loads a support system must carry. The design of this system is assumed to be the structural engineer's responsibility. The recommended design procedures presented are believed to represent the most up-to-date methodology readily available to the geotechnical engineering profession. For convenience, these recommendations are abstracted in the following series to 12 flow charts (Figs. 34-45). The first chart (Fig. 34) is an index diagram depicting the order in which each proceeding flow chart is encountered during the geotechnical analysis of a trench. Appropriate design equations, figures, and tables referred to in these charts can be found in the main body of this section.

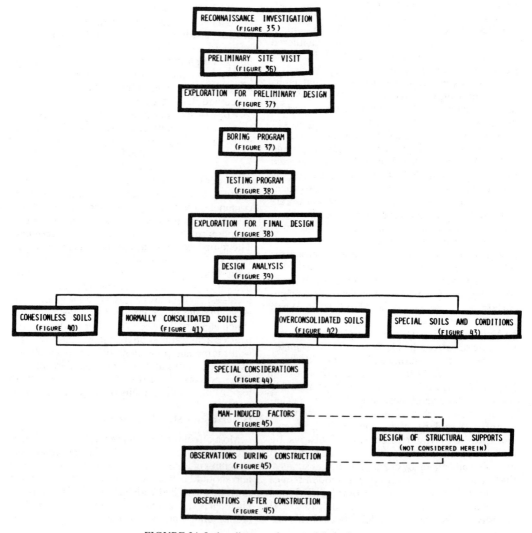

FIGURE 34. Index diagrams for use of flow charts.

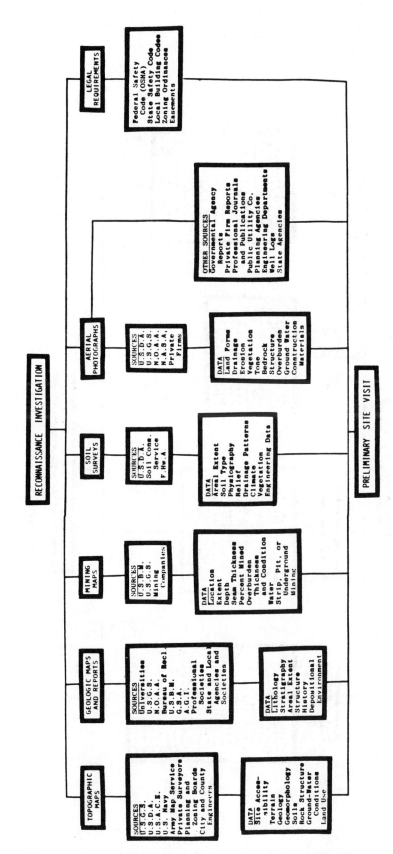

FIGURE 35. Flow chart: reconnaissance investigation.

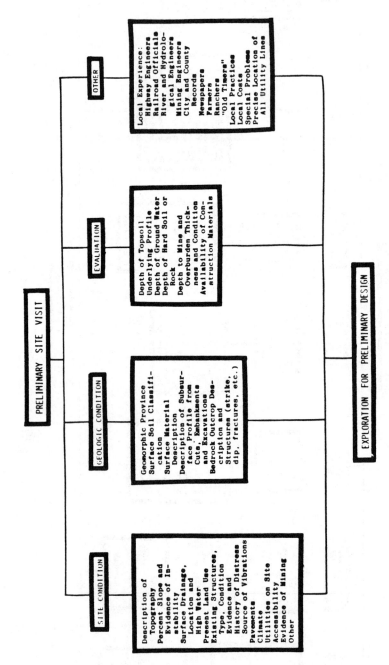

FIGURE 36. Flow chart: preliminary site visit.

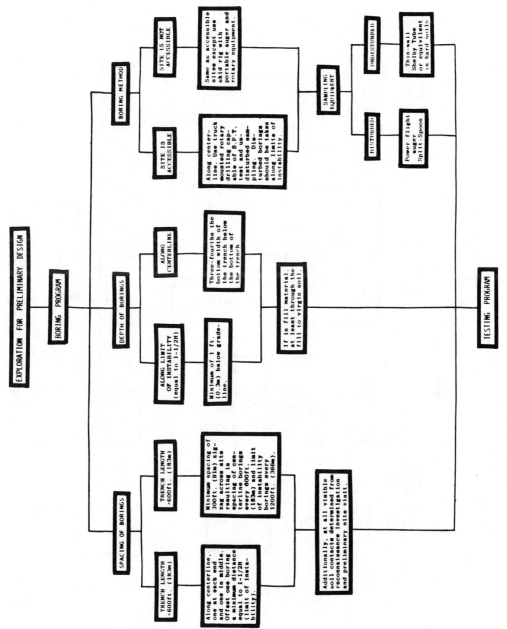

FIGURE 37. Flow chart: exploration for preliminary design — boring program.

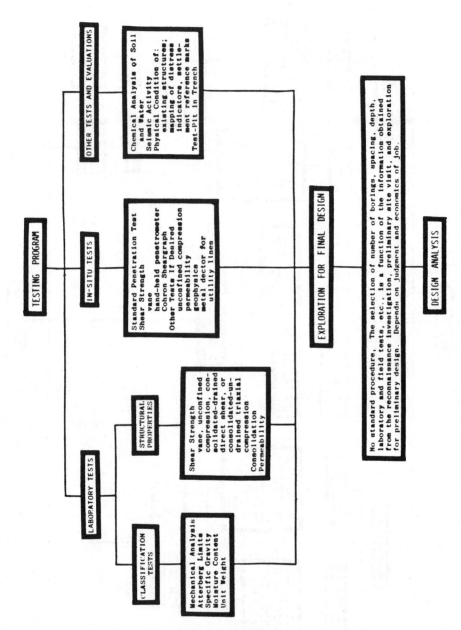

FIGURE 38. Flow chart: testing program—exploration for final design.

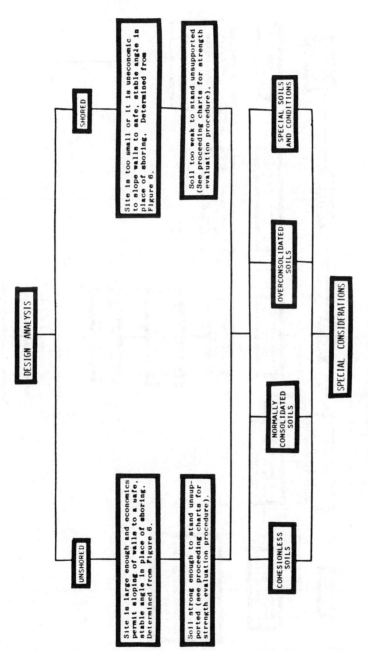

FIGURE 39. Flow chart: design analysis—shoring systems.

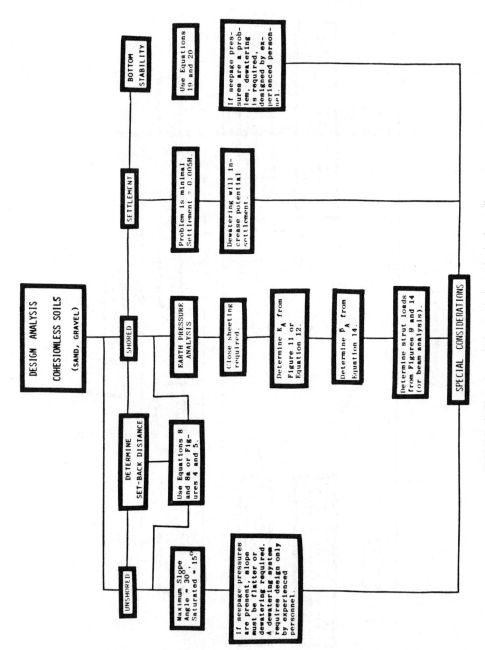

FIGURE 40. Flow chart: design analysis—cohesionless soils.

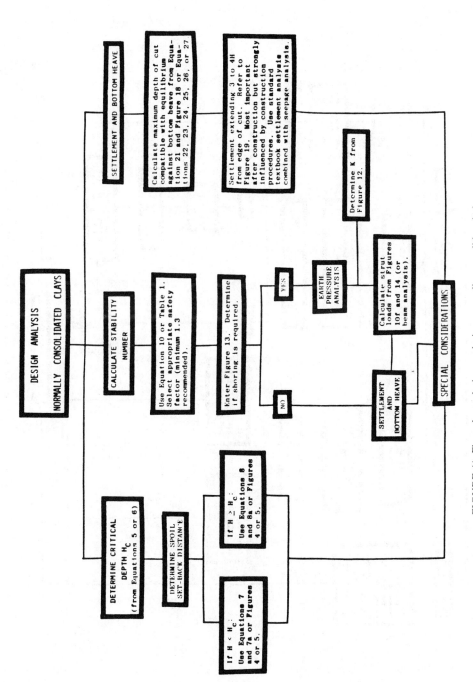

FIGURE 41. Flow chart: design analysis—normally consolidated clays.

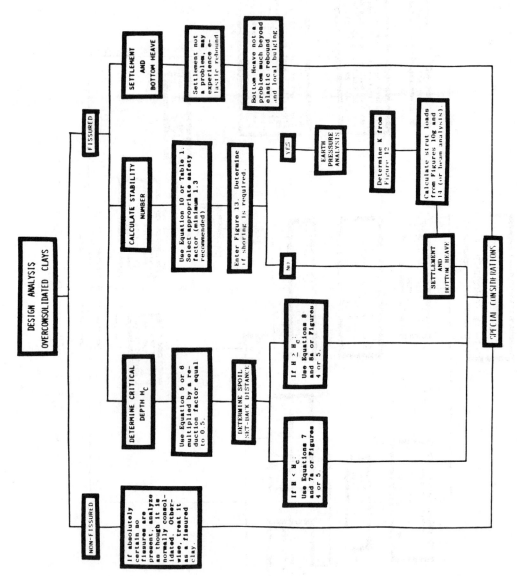

FIGURE 42. Flow chart: design analysis—overconsolidated clays.

440

Conclusion

The successful analysis and design of a trench encompasses a total environmental problem involving the interrelationships of engineering geology, climate, and man. To ignore any single part can lead to failure of the whole. Thus the geotechnical engineer or engineering geologist cannot isolate himself to particular site boundaries or specialties. He must be completely informed of all influential factors and be constantly aware of changing conditions. Involvement is total, from beginning to end, from the initial site evaluation, through construction inspection and monitoring of trench stability, to the final monitoring of adjacent structure settlement following backfilling of the excavation.

RONALD J. TANENBAUM

References

Armento, W. J., 1972. Criteria for lateral pressures for braced cuts, *Am. Soc. Civil Engineers Proc., Spec. Conference on Performance of Earth and Earth-Supported Structures* **1** (Part 2), 1283-1302.

Ash, J. L.; Russel, B. E.; and Rommel, R. R., 1974. *Improved Subsurface Investigation for Highway Tunnel Design and Construction. Volume 1: Subsurface Investigation System Planning.* Washington D.C.: National Technical Information Service.

Bailey, W. A., 1966. *Stability Analysis for Limiting Equilibrium.* Ph.D. thesis. Massachusetts Institute of Technology: Cambridge, Mass.

Bjerrum, L., and Eide, O., 1956. Stability of strutted excavations in clay, *Géotechnique* **6** (1), 32-47.

Bjerrum, L.; Kenny, T. C.; and Kjaernsli, B., 1965. Measuring instruments for strutted excavations, *Am. Soc. Civil Engineers Proc., Jour. Soil Mechanics and Found Div.* **91** (SM1), 111-141.

Bowles, J. E., 1968. *Foundation Analysis and Design.* New York: McGraw-Hill, 326, 328, 355-361.

Casagrande, A., 1937. Seepage through earth dams, *New England Water Works Assoc. Jour.* **51** (2), 295-336.

Caspe, M. S., 1966. Surface settlement adjacent to braced open cuts, *Am. Soc. Civil Engineers Proc., Jour. Soil Mechanics and Found. Div.* **92** (SM4), 51-93; **92** (SM6), 255-256; **93** (SM5), 320-322.

Cedergren, H. R., 1968. *Seepage, Drainage, and Flow Nets.* New York: Wiley, 489p.

Chapman, K. R.; Cording, E. J.; and Schnable, H., Jr., 1972. Performance of a braced excavation in granular and cohesive soils, *Am. Soc. Civil Engineers Proc., Spec. Conference on Performance of Earth and Earth-Support Structures,* **3**, 271-293.

Chen, W. F.; Giger, M. W.; and Fang, H. Y., 1969. *On the Limit Analysis of Stability of Slopes.* Bethlehem, Pa.: Lehigh University Institute of Research, Fritz Engineering Laboratory Report No. 355.4, 14p.

D'Appolonia, E., et. al, 1972. Subsurface investigation for design and construction of buildings, *Am. Soc. Civil Engineers Proc., Jour. Soil Mechanics and Found. Div.,* **98** (SM5), 481-490; **98** (SM6), 557-578; **98** (SM7), 749-764; **99** (SM8), 771-785.

DiBiagio, E., and Bjerrum, L., 1957. Earth pressure measurements in a trench excavated in stiff marine clay, in *Proceedings of the 4th International Conference on Soil Mechanics and Foundation Engineering, Vol. 2.* Paris, 395-401.

Duncan, J. M., and Dunlop, P., 1969. Slopes in stiff-fissured clays and shales, *Am. Soc. Civil Engineers Proc., Jour. Soil Mechanics and Found. Div.* **95** (SM2), 467-492.

Dunlap, W. A., 1975. Personal communication, Civil Engineering Department, Soils Division, Texas A&M University, College Station, Tex.

Esu, F., 1966. Short-term stability of slopes in unweathered jointed clays, *Géotechnique* **16** (4), 321-328.

Fang, H. Y., 1975. Stability of earth slopes, in H. F. Winterkorn and H. Y. Fang (eds.), *Foundation Engineering Handbook.* New York: Van Nostrand Reinhold, 354-372.

Fellenius, W., 1936. Calculation of the stability of large dams, in Transactions of the 2nd Congress on Large Dams, Vol. 4, 445.

Finn, W. D. L., 1963. Stability of deep cuts in clay, *Civil Engineering* **33** (6), 67.

Flaate, K. S., 1966. *Stresses and Movements in Connection with Braced Cuts in Sand and Clay.* Ph.D. thesis, University of Illinois, Urbana, Ill., 264p.

Flaate, K. S., and Peck, R. B., 1973. Braced cuts in sand and clay, *Norwegian Geotech. Inst.* **96**, 7-29.

Handy, R. L., 1974. Use of cycloidal arcs for estimating ditch safety—discussion, *Am. Soc. Civil Engineers Proc., Jour. Geotech Eng. Div.* **100** (GT1), 81-83.

Harr, M. E., 1962. *Groundwater and Seepage.* New York: McGraw-Hill, 315p.

Hsieh, K. S., 1971. *Changes in Failure Modes in Soil Slopes Due to Changes in Slope Geometry, Soil Strength Parameters, and Water Effects.* M.S. thesis, University of Lousiville, Louisville, Ky.

Hunt, R. E., 1972. *Site Reconnaissance Report Form.* Caldwell, N.J.: Converse, Ward, Davis, Dixon and Associates, Consulting Engineers.

Hvorslev, M. J., 1949. Subsurface exploration and sampling of soils for civil engineering purposes, *U.S. Army Corps of Engineers Waterways Expt. Sta. 15-21,* 72-81.

Kane, H., 1961. *Earth Pressures on Braced Excavations in Soft Clay.* Ph.D. thesis, University of Illinois, Urbana, 139p.

Kezdi, A., 1975. Lateral earth pressure, in H. F. Winterkorn and H. Y. Fang (eds.), *Foundation Engineering Handbook.* New York: Van Nostrand Reinhold, 197-220.

King, G. J. W., 1967. Base heave in braced excavation in saturated clays, *Civil Engineering and Pub. Works Rev.* (April), 433-444.

Kjaernsli, B., 1963. *Review of Measurements on Srutted Ecavations in Clay in Oslo, 1957-1962.* Norwegian Society of Professional Engineers, Short Course on Foundation Engineering.

Klenner, C., 1941. Versuche Überverteilung des erddruckes über die Wände ausgesteifter baugruben (Experiments with the distribution of earth pressure on walls of braced cuts), *Die Bautechnik* **19**, 316-319.

Lambe, T. W., and Whitman, R. V., 1969. *Soil Mechanics.* New York: Wiley, 553p.

Land, W. C., 1968. Protection against trench failures, *Natl. Safety Congres Constr. Div. Trans.* **8**, 6-9.

Lehmann, H., 1942. Dive Verteilung erdangriffs an einer Obendrehabar gelagerten Wand (The distribution of the earth pressures on a wall which can rotate about the top strut), *Die Bautechnik* **20**, 273-283.

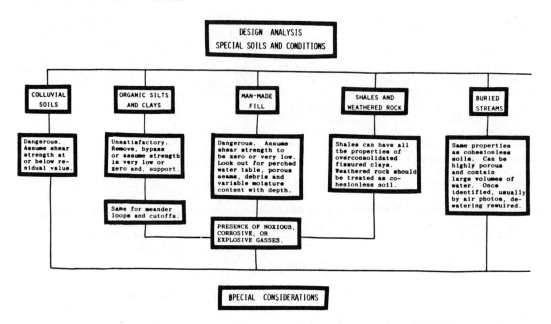

FIGURE 43. Flow chart: design analysis—special soils and conditions.

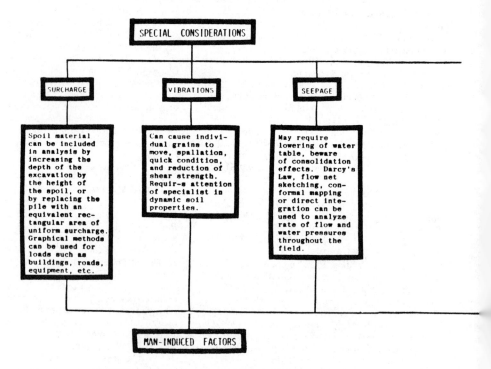

FIGURE 44. Flow chart: special considerations.

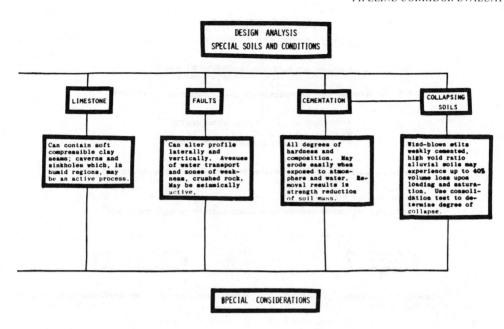

DESIGN ANALYSIS
SPECIAL SOILS AND CONDITIONS

LIMESTONE

Can contain soft compressible clay seams; caverns and sinkholes which, in humid regions, may be an active process.

FAULTS

Can alter profile laterally and vertically. Avenues of water transport and zones of weakness, crushed rock. May be seismically active.

CEMENTATION

All degrees of hardness and composition. May erode easily when exposed to atmosphere and water. Removal results in strength reduction of soil mass.

COLLAPSING SOILS

Wind-blown silts weakly cemented, high void ratio alluvial soils may experience up to 40% volume loss upon loading and saturation. Use consolidation test to determine degree of collapse.

SPECIAL CONSIDERATIONS

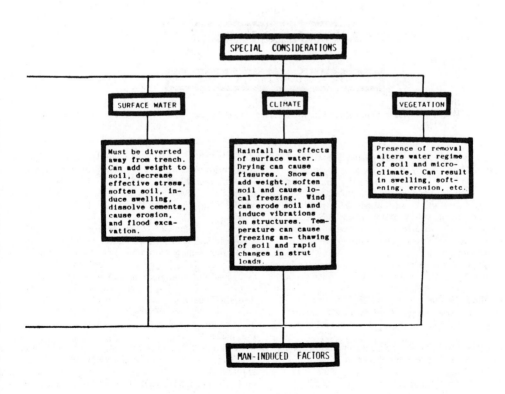

SPECIAL CONSIDERATIONS

SURFACE WATER

Must be diverted away from trench. Can add weight to soil, decrease effective stress, soften soil, induce swelling, dissolve cements, cause erosion, and flood excavation.

CLIMATE

Rainfall has effects of surface water. Drying can cause fissures. Snow can add weight, soften soil and cause local freezing. Wind can erode soil and induce vibrations on structures. Temperature can cause freezing an- thawing of soil and rapid changes in strut loads.

VEGETATION

Presence of removal alters water regime of soil and micro-climate. Can result in swelling, soft-ening, erosion, etc.

MAN-INDUCED FACTORS

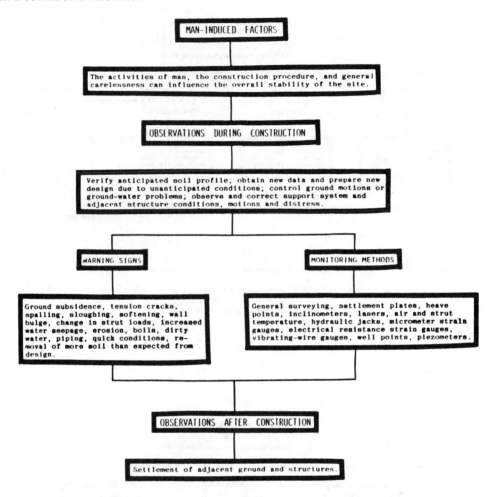

FIGURE 45. Flow chart: man-induced factors—observations during construction—observations after construction.

Moretto, O., 1969. Deep excavations and tunnelling in soft ground (discussion), in *Proceedings of the 7th International Conference on Soil Mechanics and Foundation Engineering,* Vol. 3, 357-359.

Morgenstern, N. R., and Eisenstein, Z., 1970. Methods of estimating lateral loads and deformations, in Specialty *Conference on Lateral Stresses in Ground and Design of Earth Retaining Structures.* American Society of Civil Engineers, 51-102.

Nascimento, U., 1970. A method of stress analysis in excavation slopes, *Laboratorio Nacional de Engenharia Civil, Memoria No. 358., Lisboa.*

Navy Facilities Engineering Command, 1971. *Design Manual—Soil Mechanics, Foundations and Earth Structures,* NAVFAC DM-7, 7-2-11-7-2-13, 7-10-1-7-70-28.

Peck, R. B., 1943. Earth pressure measurements in open cuts, Chicago Subway, *Am. Soc. Civil Engineers Trans.* **108,** 1008-1036.

Peck, R. B., 1969a. Deep excavations and tunnelling in soft ground, in *Proceedings of the 7th International Conference on Soil Mechanics and Foundation Engineering: State-of-the-Art Volume,* 225-290.

Peck, R. B., 1969b. Advantages and limitations of the observational method in applied soil mechanics, *Géotechnique* **19** (2), 171-187.

Petersen, E. V., 1963-1964. Cave-ins! *Roads and Engineering Constr.* (Nov.), 25-33; (Dec.), 45-49; (Jan.), 40-46.

Rubin, L. A.; Hipkins, D. L.; and Whitney, L. A., 1974. *Improved Subsurface Investigation for Highway Tunnel Design and Construction. Vol. 2: New Acoustic Techniques Suitable for Use in Soil.* Washington, D.C.: National Technical Information Service.

Skempton, A. W., 1964. Long-term stability of clay slopes, *Géotechnique* **14** (2), 75-102.

Skempton, A. W., and LaRochelle, P., 1965. The Bradwell slip: a short-term failure in London clay, *Géotechnique* **15** (3), 221-242.

Sowers, G. F., 1957. Trench excavation and backfilling, *Roads and Engineering Constr.* **95** (5), 49, 51, 53-54, 151-154.

Spangler, M. G., and Handy, R. L., 1973. Soil Engineering, 3rd ed. New York: Intext Educational.

Tanenbaum, R. J., 1975. *Recommendations for the Analysis and Design of Shallow Excavations with Vertical* or *Near Vertical Walls,* Ph.D. Thesis, Texas A & M University College Station, Tex., 175p.

Taylor, D. W., 1937. Stability of earth slopes, *Boston Soc. Civil Engineers Jour.* **24** (3), 197.

Teng, W. C., 1962. *Foundation Design.* Englewood Cliffs, N.J.: Prentice-Hall.

Terzaghi, K., 1936. Distribution of the lateral pressure of sand on the timbering of cuts, in *Proceedings of the 1st International Conference on Soil Mechanics and Foundation Engineering,* Vol. 1. Cambridge, 211-215.

Terzaghi, K., 1941. General wedge theory of pressures, *Am. Soc. Civil Engineers Trans.* **106,** 68-97.

Terzaghi, K., 1943. *Theoretical Soil Mechanics.* New York: Wiley, 152-181.

Terzaghi, K., and Peck, R. B., 1967. *Soil Mechanics in Engineering Practice,* 2nd ed. (1st ed., 1951). New York: McGraw-Hill, 433-435.

Thompson, L. J., and Tanenbaum, R. J., 1975. Excavations, Trenching and Shoring: The Responsibility for Design and Safety. College Station, Tx.: *Texas A&M University Research Foundation Rept. No. RF-3177,* 3-10.

Thorson, B. M., 1973. Use of cycloidal arcs for estimating ditch safety: Discussion, *Am. Soc. Civil Engineers Proc., Jour. Soil Mechanics and Found. Div.* **99** (SM11), 1028-1029.

Tschebotarioff, G. P., 1973. *Foundations, Retaining and Earth Structures,* 2nd ed. (1st ed., 1951). New York: McGraw-Hill, 433-435.

Van Horn, R., 1965. Geologic factors effect excavation projects along SLC's Wasatch fault. Reprint from *Intermountain Contractor.*

Van Horn, R., 1969. Anatomy of cave-in. Reprint from *Intermountain Contractor.*

World Construction, 1968. Main causes of accidents in construction. Feb., 26-29.

Wright, S. C., 1969. *A Study of Slope Stability and the Undrained Shear Strength of Clay Shales,* Ph.D. thesis, University of California, Berkeley.

Xanthakos, P. P., 1974. *Underground Construction in Fluid Trenches.* Chicago: National Education Seminar, University of Illinois at Chicago Circle, 291p.

Cross-references: *Clay, Engineering Geology; Clays, Strength of; Consolidation, Soil; Foundation Engineering; Hydrodynamics, Porous Media; Hydromechanics; Permafrost, Engineering Geology; Rapid Excavation and Tunneling; Reinforced Earth; Rheology, Soil and Rock; Rock Structure Monitoring; Soil Mechanics; Tunnels, Tunneling; Urban Tunnels and Subways.* Vol. VII, Pt. 1: *Soil Mechanics.* Vol. XIV: *Artificial Fill; Engineering Soil Science; Expansive Soils; Slope Stability Analysis; Soil Sampling.*

POPULAR GEOLOGY—See GEOLOGICAL INFORMATION, MARKETING.

POWER PLANT SITING—See NUCLEAR PLANT SITING, OFFSHORE.

PUMPING STATIONS AND PIPELINES, ENGINEERING GEOLOGY

Pumping Stations

Engineering geological investigations for pumping stations should be aimed at determining a "geotechnical model" of the site, for example, a geological picture plus the measured or judged properties of the various rock or soil units (Stapledon, 1973). From this model the following criteria can be supplied to the designing engineers.

Foundation Conditions. An accurate assessment of engineering properties of the foundation mass is essential. Permissible settlement, which must be a negligible factor, is tied to correct pump alignment. The possibility of breakage at pipeline connections is also an important consideration, as is intense vibration during station operation. Methods of investigation include several lines of attack. Geological mapping should be conducted in detail (scale of 1:1,000 or 1:2,000) close to the proposed site or alternative sites. Mapping should focus on regional scales or a special study should be made of existing regional geological maps. Aerial photos should also be used where available (see *Remote Sensing, Engineering Geology*). This information will provide the basis for the "model," and will determine what other investigations might be required. Where deep-seated foundations are required, seismic refraction surveys may indicate the natural subsurface conditions. For stations founded close to the ground surface, a trenching program is usually sufficient to indicate the nature of near-surface foundation materials or the presence of significant discontinuities. For structures with deeper-seated foundations, drilling is often useful to indicate likely variations in conditions with depth and across the site (see *Pipeline Corridor Evaluation*). The location, angle, and depth of drill holes depends on results of previous studies and should be aimed at providing as much information as possible for geotechnical models of the site. Water-pressure testing of the rock should give an indication of openness of discontinuities and whether grouting (q.v.) is required to prevent possible settlement.

Ground Water. The ground water regime in the area of the pumping station must be established, particularly for below-surface structures close to a river or water storage. Ground water can, for example, affect (1) the design of the structure due to possible uplift pressures on the floor, and horizontal pressures on the side walls, (2) the type of cement used in concrete works if chemical reaction with ground water salts is possible, (3) the design of open cuts below the water table, especially as they relate to the possibility of dewatering requirements (Fig. 1), and (4) the type of steel used for construction if corrosion seems likely.

FIGURE 2. Back-hoe trench to investigate foundation conditions at an intermediate pumping station. Pipeline alignment in center of photograph.

FIGURE 1. Pumping station excavation in bryozol limestone. Solution effects, as exposed in walls, required special dewatering arrangements.

Excavation Conditions. The machinery necessary for excavation—for example, scraper or heavy-duty tractor with ripper—should be determined on the basis of seismic refraction and drilling results. Excavation experience elsewhere with similar rock units can give some indication of conditions that might be encountered. Local site investigations, however, provide the most reliable results.

Natural Slopes. Existing hill slopes should be studied to determine their overall stability and possible impacts of earthworks for pumping stations or pipelines. A trenching program may be necessary to locate possible ancient slip surfaces. The feasibility of seismic activity should also be considered. The history of similar hill slopes can be used to anticipate possible instability.

Man-Made Slopes. Recommendations for earthworks batter angles should be provided to the design engineers. Detailed geological mapping of surface and trench exposures and drill cores plus examination of existing natural slopes provide a basis for batter angle recommendations for rock slopes (McMahon, 1974). Laboratory testing of undisturbed soil samples is necessary for batters in soil (Terzaghi and Peck, 1967).

Pipelines. Pipelines may be above ground, fully buried, or a combination of both. For the above-ground pipeline, the main requirements are determination of foundation conditions and likely stability problems. At anchor blocks, both loading and uplift forces may occur during pipeline operation and should be considered during foundation investigations. Normally, trenching and surface mapping are sufficient to provide the required information (Fig. 2), but in certain cases where

bearing strengths of the foundation are uncertain, it may be necessary to conduct a plate bearing test (see *Foundation Engineering*).

For below-ground pipelines, the main requirement is ease of excavation. This aspect can be assessed from trenching and seismic refraction surveys. The stability of trench walls (see *Reinforced Earth*), particularly in relation to potential water problems, should also be investigated at this stage.

For surface and buried pipelines, as well as pumping stations, ground-water regimes and chemistry should be established. Details differ for particular situations: for example, a surface pipeline over rocky, hilly country may require only surface inspections at valley crossings, whereas a buried pipeline passing through poorly drained, swampy country may require detailed hydrologic and hydrochemical studies. Possible active faults crossing a pipeline must be located prior to pipeline design. For pipelines traversing ground subject to freezing, consideration must be given to possible effects of freeze-thaw conditions (see *Geocryology*) as they may affect pipeline operation.

The geological investigations and conclusions drawn from them should be presented to design engineers in the form of a full geotechnical report (see *Engineering Geology Reports*) (Stapledon, 1973). Environmental impacts of the proposed works should be considered during investigations and discussed in this report.

W. R. P. BOUCAUT

References

McMahon, B. K., 1974. Design of rock slopes against sliding on pre-existing fractures, in *Proceedings of the 3rd Congress of the International Society of Rock Mechanics, Denver*, Vol 2, Pt. B. Washington, D.C.: National Academy of Sciences, 803-808.

Stapledon, D. H., 1973. *Workshop Course: Engineering Site Investigations.* Adelaide, S.A.: Australian Mineral Foundation, 73p.

Terzaghi, K., and Peck, R. B., 1967. *Soil Mechanics in Engineering Practice.* New York: Wiley, 729p.

Cross-references: *Dams, Engineering Geology; Engineering Geology Reports; Foundation Engineering; Geotechnical Engineering; Grout, Grouting; Pipeline Corridor Evaluation; Reinforced Earth; Rock Slope Engineering; Soil Mechanics; Undersea Transmission Lines, Engineering Geology.*

R

RADIOACTIVITY SURVEYS—See Vol.
XIV: AERIAL SURVEYS, GENERAL;
WELL LOGGING.

RADIOGEOLOGY—See GEOCHRON-
OLOGY. Vol. XIV: EXPLORATION
GEOPHYSICS.

RAPID EXCAVATION AND TUNNELING

Technological advances in tunnel boring machines (TBMs, or "moles") have progressed rapidly
in the past two decades (see discussion in Lane
and Garfield, 1972), so much so that most people
outside the field of tunnel construction are not
aware of such progress. For example, more than
400 tunnels in the world have been excavated by
TBMs, which have been built by at least 26 TBM
manufacturers. The largest machine—12 m in
diameter—excavated a highway tunnel near Kobe,
Japan, and the smallest excavates a 1.3-m swath.

Boring Rates

The best progress for an unsupported TBM
tunnel (in Navajo Sandstone) was 127 m in one day
for the 3.3-m-diameter Oso Tunnel in Colorado.
The best progress for a TBM tunnel, which was
completely supported during excavation by precast
concrete segments, was 83 m in one day (in old
alluvium) for the 7-m-diameter San Fernando
Tunnel in California (Proctor, 1969).

Some extremely hard rock types have been
successfully bored by TBMs Deere, 1970). The
hardest was dolerite 100,000-psi compressive
strength in Australia. Other hard rocks bored by
especially designed TBMs are 57,000-psi granite in
Austria, 51,000-psi quartzite in Idaho, and rhyo-
dacite in Nevada with a hardness of 7 on the Mohs
scale. The 31,000-psi Manhattan Schist under New
York City was bored at the rate of 35 m per day.
The extensive Chicago sewer system was almost
entirely excavated by TBMs in dolomitic limestone
as strong as 31,000 psi. The daily excavation
progress, from a sample of 54 TBM projects for
which data are available, reveals average progress
of 20.5 m per 24-hour work day.

Most TBMs are designed for rapid excavation
in "soft ground," and in fact most tunnels in the

world have been bored through soft sandstone,
shale, and limestone with compressive strengths
less than 8,000 psi.

Speed of excavation, and the cost savings resulting therefrom, are the main advantages of boring
machines. Machines are not yet a panacea for
rapid excavation in all kinds of rocks, however.
Some hard, granular igneous and metamorphic
rocks are presently more economical to excavate
by means of the conventional excavation cycle
(drilling, loading explosives, blasting, ventilating,
mucking, erecting supports). This is true mainly
because of high bit-replacement and maintenance
costs in abrasive rocks. Conversely, in very soft
sediments, or flowing ground as in subaqueous
tunneling, a cutting edge on a full-circle shield
pushed by hydraulic jacks may be all that is required for economical tunneling (Pattison and
D'Appolonia, 1974).

Soil or rock containing certain clays or serpen-
tinite may cause squeezing or swelling around a
TBM and cause it to become wedged. In such
cases, steel tunnel supports may be designed to
yield, or the tunnel bore may be initially over-
excavated by extending the gage cutting bits on
the perimeter of the TBM. Swelling clay problems
in a tunnel are mainly caused by the availability of
water. Such clay usually appears first as firm, stiff
clay, and only after hours or days does swell occur;
if unchecked it may continue until the tunnel is
closed (Terzaghi, 1950).

Tunneling Equipment

The first mole of record was built and briefly
worked in a railroad tunnel in Massachusetts, in
1856. It was 2.13 m in diameter and powered by
compressed air. Colonel Beaumont's English
Channel Tunneling Machine, also 2.13 m in diameter and powered by compressed air, fared better
and was able to tunnel 2,499 m under the Channel
in 1884, at a best rate of 15 m per day. Politics
forced abandonment of the project because Great
Britain feared it could be invaded through the
tunnel by armies from the continent.

There exist several types of moles other than
the common rotating wheel cutterhead: One type
has a shield with heavy forepoling plates that can
be extended a meter or so forward (Fig. 1). Inside
the shield is a large, swiveling backhoe-type bucket
equipped with a ripper tooth on the bottom and
large auger teeth on top. Another machine has a

FIGURE 1. Modern 6-m diameter tunnel boring machine (courtesy James S. Robbins Company, Seattle, Washington).

small-diameter rotating cutterhead mounted on a large rotating arm. The cutterhead can slide along the arm, and thus a rectangular-shaped tunnel is possible. In addition, several manufacturers build a modification of a single wheel; for example, the TBM used for a Moscow subway extension had six rotating cutterheads mounted under a shield, and another model has four rotating bucket augers under a shield (Rapid Excavating and Tunneling Conference Staff, 1972, 1974, 1976, 1979, 1981, 1983).

The heaviest TBM weighs 500 tons and recently excavated part of the Paris subway. The most powerful TBM to date has a cutterhead power of 1,500 horsepower, used in sandstone and chalcocite-shale of the White Pine Copper Mine, Michigan.

RICHARD J. PROCTOR

References

Deere, D. U., 1970. Indexing rock for machine tunneling, in D. H. Yardley (ed), *Rapid Excavation—Problems*

and Progress. New York: Society of Mining Engineers, 32-38.

Lane, K. S., and Garfield, L. A. (eds), 1972. *Rapid Excavation and Tunneling Conference (1st) Proceedings,* 2 vols. New York: American Institute of Mining, Metallurgical and Petroleum Engineers, 1664 p.

Pattison, H. C., and D'Appolonia, E. (eds), 1974. *Rapid Excavation and Tunneling Conference Proceedings,* 2 vols. New York: American Institute of Mining, Metallurgical and Petroleum Engineers, 1843 p.

Proctor, R. J., 1969. Performances of tunnel boring machines. *Assoc. Eng Geologists Bull.* **2,** 105-117.

Rapid Excavation and Tunneling Conference Staff, 1972, 1974, 1976, 1979, 1981, 1983. *Proceedings,* II vols. New York: American Institute of Mining Engineers, American Society of Civil Engineers.

Terzaghi, K., 1950. Geologic aspects of soft-ground tunneling, in P. D. Trask (ed.), *Applied Sedimentation.* New York: Wiley, 193-209.

Cross-references: *Rock Mechanics; Rocks, Engineering Properties; Tunnels, Tunneling; Urban Tunnels and Subways.*

**RECONNAISSANCE GEOLOGY—
See FIELD GEOLOGY. Vol. XIV:
EXPLORATION GEOLOGY;
PROSPECTING.**

**REFERENCE RETRIEVAL SYSTEMS—
See COMPUTERIZED RESOURCES
INFORMATION BANK; EARTH
SCIENCE, INFORMATION AND
SOURCES. Vol. XIV: MAP AND CHART
DEPOSITORIES.**

**REFLECTION PROFILING—See Vol. XIV:
ACOUSTIC SURVEYS, MARINE.**

REINFORCED EARTH

The technique of reinforcing earth in the direction of maximum tensile stress with a material of high tensile strength is not new. Nature's example of plant roots strengthening the soil has been imitated by engineers throughout the ages; twigs and sticks have long been used to reinforce dams and embankments (Schlosser and Vidal, 1969; Schlosser, 1973). Although a patent in 1930 was granted to Andreas Munster in the United States for a reinforced earth retaining wall, present methods are based on the work of the French engineer Henri Vidal in the early 1960s. Because of their economy, this patented method and others have become widely used in many countries for earth retaining structures.

Generally, reinforced earth construction involves the building up of alternate layers of compacted, free-draining soil and reinforcement strips, the strips holding in place a light facing material to prevent the loss of soil (Fig. 1) (Lee, et al., 1973). Lateral displacement of the soil due to its own weight and any imposed loads is resisted by friction between the reinforcement strips and the soil. Soil not in direct contact with the reinforcement is assumed to be held in place by the development of arches in the soil (Harrison and Gerrard, 1972).

The reinforced earth mass can be viewed as a composite material with orthotropic material properties, the unification being provided to the soil mass by frictional effects, both between the soil particles and reinforcement material. The essential requirements are: (1) flexible reinforcement strips that are strong in tension and corrosion resistant (e.g., galvanized or stainless steel, aluminium alloy, glass fiber, plastic); (2) a free-draining, granular fill material that exhibits a good friction value and causes no corrosion of the reinforcement; and (3) a flexible facing material

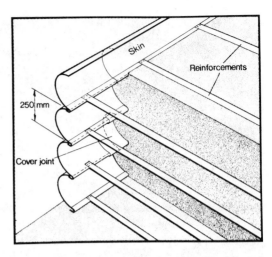

FIGURE 1. Cross-section of steel-faced reinforced earth wall.

(e.g., semi-elliptical galvanized steel sheets, precast concrete units).

Internal failure (either insufficient friction between reinforcement and soil, or tension failure of the reinforcement) is conservatively analyzed by assuming an active failure condition and using classical earth-pressure theory. Although this may provide conservative answers, observed stresses in full-scale structures have approximated those calculated by assuming an active failure wedge. Other unknowns and assumptions do not appear to invalidate classical theory as an upper bound, although it has been suggested that at-rest earth pressures be used for design in seismic areas (Richardson and Lee, 1975).

Sliding or rotation can cause external failure of the reinforced earth mass, which can be analyzed in a similar manner to conventional retaining structures. Although simple design formulas have been developed, the number of simplifying assumptions does mean that further field studies are required and a more rational design technique is evolved. Laboratory model studies provide little information because the reduction in scale modifies the soil-reinforcement interaction behavior. Evidence from field studies indicates that the tension in the reinforcement rises from a small value at the face of the wall to a maximum near the face, and then drops to zero at the free end of the strip. These points of maximum tension define the transition between an active and passive zone and point to fundamental differences between the classical earth-pressure analyses and actual behavior (i.e., the material behaves like an orthotropic composite material, maximum tension is not at the wall face, and active pressure does not act on the wall).

Further research into these questions, and others

such as corrosion and the use of backfill with some cohesion will undoubtedly lead to the more widespread use of reinforced earth, which already is showing its economic advantage over high conventional retaining walls.

R. A. FRASER

References

Harrison, W. Jill, and Gerrard, C. M., 1972. Elastic theory applied to reinforced earth, *Am. Soc. Civil Engineers Proc., Jour. Soil Mechanics and Found. Div.* **98** (SM12), 1325-1345.
Lee, K. L.; Adams, B. D.; and Vagneron, J. J., 1973. Reinforced earth retaining walls, *Am. Soc. Engineers Proc., Jour. Soil Mechanics and Found. Div.* **99** (SM10), 745-765.
Richardson, G. N., and Lee, K. L., 1975. Seismic design of reinforced earth walls, *Am. Soc. Civil Engineers Proc., Jour. Geotech. Div.* **101** (GT2), 167-188.
Schlosser, F., and Vidal, H., 1969. La terre amée, *Liaison Lab. P. et Ch. Bull. 41.*
Schlosser, F., 1973. La terre Armeé dans l'échangeur de Sete, *Liaison Lab. P. et Ch. Bull. 63.*

Cross-references: *Foundation Engineering; Rheology, Soil and Rock; Urban Engineering Geology.* Vol. XIV: *Engineering Soil Science*

REMOTE SENSING, ENGINEERING GEOLOGY

The concept of remote (noncontact) collection of terrain data is not new to the engineering profession. *Photogrammetry* and *airphoto interpretation* were established as practical engineering analysis procedures in the late 1920s and early 1930s. Traditional engineering applications had consisted of such examples as route reconnaissance, site evaluation, flood control surveys, and topographic mapping (q.v.). A more comprehensive utilization of remotely collected data evolved in the 1940s. Principles of engineering terrain and soils mapping were developed, as were methods for interpreting geological and geomorphic conditions.

With the development of new hardware capable of sensing emitted and reflected energy throughout the electromagnetic spectrum and with new and improved film types for use in better cameras, many fresh applications of remote sensing to engineering endeavors are practical.

Remote-Sensing Instrumentation

Hardware has been developed that can detect, measure, and record energy having a variety of wavelengths. The equipment can be classified simply as consisting of optical cameras, optical-mechanical scanners, and antenna devices. Cameras directly record spectral data on film emulsions, whereas scanners and antenna sensors contain electronic sensing configurations that produce data initially in the form of a voltage signal. There is more to remote sensing than hardware, however. The term *remote sensing* has evolved into a general descriptor of all technologies related to the detection, measurement, recording, and use of emitted and reflected electromagnetic energy. Sensor platforms are an integral part of remote sensing, as are data analysis techniques.

The optimum remote-sensing system for a given application should be determined with due regard for energy wavelength to be sensed, sensor design, type of platform, and potential analysis techniques. Figure 1 illustrates the range of energy wavelengths comprising the electromagnetic spectrum, the relative atmospheric transmissibility of energy at a particular wavelength, and types of instrumentation used for sensing within various energy bands.

General Data Analysis Techniques

The data obtainable from a remote-sensing system ideally consist of a magnitude of reflected or emitted electromagnetic energy and the location of the phenomena that affected or produced that energy. Although it would be desirable to have precise accurate magnitudes and spatial positions of such energies, most operational remote-sensing systems do not provide both. Furthermore, the reflected and/or emitted energies may be modified by phenomena extraneous to those of interest. Considering the complete electromagnetic spectrum, the energy available for sensing is the sum of reflected and emitted energy. Radiation that is directed at an object or feature may come from the sun, sky, laser, or microwave transmitter. A portion of this energy will usually be reflected in specific directions. The remaining energy will be absorbed. Since objects with a temperature greater than absolute zero emit electromagnetic radiation, there will also be a component of emitted energy directed toward the sensor system. That portion of reflected and emitted energy remaining after atmospheric attenuation will be available for sensing.

The most basic information obtained by a remote-sensing system is a discrete measure of magnitude or wavelength of reflected and/or emitted electromagnetic energy, the source of which can be located with some degree of accuracy. The magnitude or wavelength of energy may be expressed initially as a tone or color on film emulsions or as voltage outputs from electronic sensors. Coordinate positioning information may be derived from visual associations on images or angular measurements associated with a nonimaging sensor.

A geometrically rectified aerial photography can

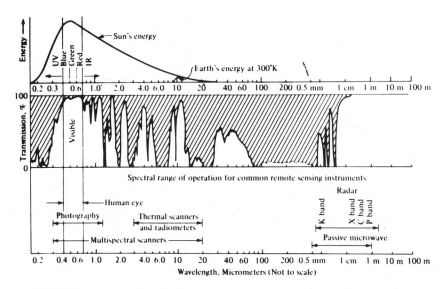

FIGURE 1. Composite chart showing source energy, atmospheric transmission, and range of operation of various sensors (from Wolf, 1974).

provide for very accurate three-dimensional terrain-surface measurements (American Society of Photogrammetry, 1966). The relative position of features and objects can be determined precisely. Because of exposure and developing parameters, however, it is often difficult to record unique, reproducible terrain spectral values. On the other hand, electronic systems such as optical/mechanical scanners or radiometers have the capability to detect and record electromagnetic energy with high levels of reproducibility but generally exhibit some degradation of geometric fidelity.

Image analysis may consist of qualitative interpretation, film density or color mapping, or spatial mensuration. The interpretation of imagery is usually accomplished with the identification of discrete objects or features or with a detailed description and mapping of more regional natural and/or man-made phenomena. Such interpretations are usually based on identification keys, visual recognition of characteristic patterns, or a deductive reasoning process representing the interpreter's total experience. In cases where approximate spectral signatures evidenced as film densities or colors are of prime significance, quantification through densitometry may be of value. Black-and-white and color display image-density scan devices make it possible to enhance film density variations and to map areas of similar film density. Macro- and microdensitometers can be used to quantify film density. The purpose of film density quantification is usually to correlate density in some unique manner with a terrain phenomenon of interest.

One can interpret imagery that varies drastically in quality. Moderate geometric distortion as well as non-optimum exposure and processing conditions can be tolerated and accounted for in the extraction of information from remotely sensed data. Thus a precise mathematical model establishing the physics (and chemistry) of origin, transmission, detection, and recording of remotely sensed electromagnetic energy is not a prerequisite for its use in solving practical problems.

However, owing to the sheer bulk of imagery necessary for sequential national or worldwide remote-sensor coverage and to the problems associated with extracting quantitative spectral data from photographic records, some applications are facilitated with automatic pattern recognition and spectral signature analysis techniques that interpret data recorded on magnetic tape. All those portions of the electromagnetic spectrum applicable to remote sensing (including the visible) can be monitored with electronic sensors using magnetic tape for data storage. There is potential for applying automatic spectral signature analyses in solving some engineering problems. Spectral data analyses by computer can accommodate many times the amount of data than a human operator can evaluate visually. A spectral signature of an object or feature can be developed which consists of a unique combination of reflected and/or emitted energies lying within discrete portions of the electromagnetic spectrum. It is possible for objects or features to be identified from such signatures as well as being automatically delineated and mapped from a regional matrix of spectral data.

Engineering Applications

The many applications of remote sensing to engineering involve the detection, interpretation, and/or mensuration of a number of terrain and environmental parameters. A partial listing of terrain parameters that pertain to engineering endeavors is presented in Table 1. Their practical extraction from remote sensor data is documented by the corresponding reference citations. From such terrain surface expressions (see Vol. XIV: *Terrain Evaluation Systems*), it is possible to infer the character of a much wider range of engineering parameters. Since many of the basic engineering parameters apply to a variety of applications, there is a greater return from remote-sensor data than that of solving a single, specific problem. Furthermore, when remote-sensing technology is used in a timely fashion as a part of a comprehensive data-gathering endeavor, economy and efficiency can be achieved.

One of the most significant general applications of remote-sensing technology to engineering is that of providing for economical field sampling and testing programs. In any engineering application, remotely sensed data can be used to efficiently determine the optimum number, location, and type of field sampling stations.

Environmental Quality

Water. It cannot be debated that the best measures of water quality come from the analysis of field samples. However, in trying to characterize the quality of a body of water or tracing the movement of effluents, most field sampling techniques are very inefficient. Typically, a constant-interval gridded sampling pattern is used. This often means that many samples are obtained that

add little to the body of data being collected. Indeed, water characteristics of vital significance but limited distribution may well be completely overlooked if the sampling interval is greater than the extent of the feature.

The advantage of using remote-sensing techniques as a part of an overall water-analysis program is that the remotely sensed data can provide a synoptic map of the apparent spectral characteristics of a body of water. Even though water-quality characteristics may not be directly measurable or even identifiable with high levels of accuracy, the very important aspect of locating zones or boundaries of changing water conditions is readily obtainable from remote-sensor data (Kiefer and Scherz, 1970).

Water quality can often be inferred from aerial photography. For detailed studies of local areas, image-scale ratios of 1:10,000 or larger should be used. Smaller-scale photography including satellite photography can be useful for regional studies of large bodies of water. *Turbidity* caused by sediment or other large particulates will be indicated as lighter tones on black-and-white photography or as colors other than dark blue or black on color and color infrared photography. Other effluents that change the reflectivity of the receiving body of water can also be interpreted as a change in water character (James and Burgess, 1970; Klemas et al. 1973; Wobber, 1969, 1970).

Qualitative and quantitative water surface-temperature data can be obtained from infrared scanners or radiometers operating in the 3.5- to 14-micron μ range of the electromagnetic spectrum. The best

TABLE 1. Reference from Appendix on Extraction of Terrain Parameters from Remote Sensor Data

Terrain Parameter (1)	Remote Sensing System			
	Aerial Photography (2)	Aircraft-Mounted Multispectral or Infrared Scanner (3)	SLAR (4)	Earth Orbital Satellite (5)
Topography	44[a]			
Site geometry	3, 5	6	50	65
Existing land use	15, 24, 40, 44, 71	12	10, 25	34
Existing water bodies	21, 64		5	
Distribution of vegetative cover	49	36		4, 37
Classification of vegetation	9, 13	56		1, 33
Vegetative health and vigor	47	19, 56, 69		
Regional geomorphology and geology	2, 8		5	42
Local geology—parent material soil	20, 22, 28, 38, 57, 60	35, 46, 62	7, 26	14
Mass-wasting phenomena	18, 39	11		
Local erosion and deposition	30, 45, 61			
Local soil moisture conditions	70	51		
Regional hydrology	41, 52			55
Local hydrology	27, 48, 59	63		
Pollutional discharge characteristics	29, 31, 53, 54	58, 66		32, 72, 73
Flooding potential	17			16
Air pollution factors	43, 67, 68			
Geological hazards	23, 74			

[a] See numbers at ends of References.

data concerning thermal emissions will come from 8- to 14-μ systems. The surface extent of thermal plumes and their temperature gradients can be measured from calibrated thermal imagery. With uncalibrated systems, only relative temperature data can be obtained. Dispersal patterns, however, can be analyzed with acceptable accuracy (Robinsove, 1965; Van Lopik et al., 1968).

Aerial photography and scanner imagery provide the ideal means for planning field sampling programs. Rather than establishing a repetitive, fixed-interval sampling grid, zones of similar condition and boundaries can be delineated from the remote-sensor imagery to provide for optimal sampling. Complex areas can be intensely sampled and analyzed, whereas areas of similar condition and little change can be characterized with a minimum of sampling. The net result is that of efficient sampling with the number of samples in direct proportion to the complexity of the situation (Piech and Walker, 1971).

Point sources of effluents different in character from those of the receiving body of water can be identified and located. In many cases, patterns of effluent dispersal can be evaluated. In addition to providing a mapping base, the remote-sensor data can be obtained repetitively over a period of time for temporal studies (Piech and Walker, 1972).

The ability to obtain and store repetitive remote-sensor coverage is of great importance for water-quality surveillance. Once a base water condition has been established, it is possible to monitor the relative ability of industry or government units to comply with clean-water standards. Even though the composition of effluents may not be determined remotely, an inference of quality based on color or temperature of effluent can be made and compared with the base data. The remote-sensor data can then provide a permanent record of changes in the data base.

Urban Runoff. Large-scale color photography is one of the best sensor records for the extraction of information about urban land use and surface characteristics. Although panchromatic photography is entirely sufficient for planimetric mensuration, topographic mapping, and interpretation of general land use, color photography is superior for the interpretation of relative permeabilities of natural and man-made features. Color infrared photography can increase the level of knowledge relative to urban runoff, in that the vigor of vegetation and surface moisture conditions can be infrared (Lohman and Robinsove, 1964; Witenstein, 1972).

Small-scale photography may be of value for preliminary regional runoff studies. Although scale and resolution will impose a limit on the precision of interpretation, general land use and landforms within a region can be delineated. Color infrared photography is the optimum high-altitude film type for runoff studies (Bryan, 1975; Henderson, 1975).

Regardless of image scale, the final product from a remote-sensor study will be a map of inferred land surface permeabilities and pollution sources correlated with natural and man-made drainage networks. The precise geometric location of the various components of the urban surface as well as terrain slopes and gradients can be determined photogrammetrically (Colwell, 1970; Parry and Turner, 1971).

Field investigation is obviously a necessity. However, with remote sensing used as a part of the overall analysis procedure, field sampling and measuring can be minimized. The field work can be used to verify the identity of mapped surface cover types and to establish quantitative measurements of permeability at specific points for extrapolation throughout a mapped feature.

Sanitary Landfills. Remote-sensing techniques are ideal for the preliminary selection of potential sanitary landfill sites. Rather than conducting extensive field investigation over an entire region in search for potential sites, one can interpret terrain characteristics and infer land use features from remote-sensor imagery to eliminate unsuitable areas. The remaining potential sites can then be field-checked, sampled, and tested. The expense of time-consuming field work is held to a minimum level.

The remote-sensing phase of a sanitary landfill site selection should consist of a regional analysis of terrain conditions followed by a more detailed study of each potential site (see Vol. XIV: *Artificial Fill*). With procedures of pattern analysis and logical deduction, one can infer landform and parent material type, drainage patterns, vegetation, and associated cultural features from aerial photography. The geographic relationships of potential sites to surrounding features can also be assessed. Although adequate geological and geographic data can be inferred from black-and-white and/or color photography, color infrared is the single best form of data. Additional information about soil moisture and vegetation conditions can be inferred from color infrared. Infrared imagery in the 8- to 14-μ range can provide additional soil moisture information when the significance of apparent soil temperature changes is evaluated. Passive microwave radiometer sensors may potentially provide data relative to soil moisture conditions from which ground-water conditions and perhaps subsurface geological information can be inferred.

The regional evaluation should be based on the interpretation of small-scale photography. Overlays of inferred geological engineering parameters and cultural features can be constructed. A composite of the pertinent overlays will graphically indicate the relative suitability of portions of the region for sanitary landfill construction. Areas indicating high potential can then be studied in more detail (Hunter and Bird, 1970).

The detailed studies should be conducted from large-scale photography and/or imagery. Soils of inferred similar engineering properties and surficial drainage patterns can be delineated. Detailed topographic maps can be constructed photogrammetrically. The detailed study of each potential site will provide the information necessary for planning an efficient field sampling program and specifying needed laboratory tests (American Society of Photogrammetry, 1966; Wobber et al., 1975).

The ultimate selection of a landfill site can be based on the results of field exploration and laboratory tests, results supplemented by the detailed remote-sensor study. Some of the design parameters of the landfill can come from earth cut-and-fill volumes and mass diagrams as computed from the photogrammetrically derived topographic maps (Barr and Hensey, 1974).

Atmospheric Quality. Air pollutants that reflect light in the visible range of the spectrum can be mapped with acceptable accuracy using a combination of vertical and oblique aerial photographs. Three-dimensional monitoring of such polluted air masses can be accomplished by means of conventional photogrammetric techniques. Color photography is the best form of data, although black-and-white and color infrared photography also record the dispersal patterns of visible air pollutants. The most desirable scale of photography for air pollution mapping will depend on the level of accuracy required and type of pollutant to be monitored. Heavy concentrations of reflective pollutants will be detectable and mappable on satellite photography, whereas entirely gaseous pollutants will not be detectable at any scale or resolution (Ludwig et al., 1969). Indeed, the application of photogrammetry to air pollution detection and monitoring is limited basically to certain types of particulate emissions. With the exception of steam, an inference as to relative concentration of pollutant can be made by evaluating the density of gray scale or intensity of color of a smoke plume (Veress, 1970a, 1970b).

Water Resources

Precipitation, Runoff, and Infiltration. Remote sensors provide the means for significantly upgrading runoff determinations. An aerial photograph or image is actually a map of the state of the ground. With measurements and inference, many parameters pertinent to precipitation runoff can be determined. The ideal approach is to use the remote-sensor data as a basis for extrapolation from location of measured field parameters.

In large watersheds, regional landforms and inferred parent material, soil, or rock type can be delineated on satellite images and SLAR (side-looking airborne radar) imagery. Using a knowledge of geological processes and limited field testing, approximate permeabilities of the various delineated earth materials can be determined. For more localized analyses, a similar procedure of delineation can be followed using panchromatic, color, and/or color infrared photography or scanner imagery at scales as small as 1:120,000. The relative influence of vegetative cover on surface permeabilities can be evaluated on color and color infrared photography. Gross vegetation types can be mapped on SLAR imagery, but such inferences would be of value only for regional estimates. Although it has not been demonstrated as a practical technique, it is potentially feasible to use microwave radiometer data to assess the degree of moisture saturation of surface soils. Stream patterns and channel characteristics that determine the rate at which runoff will become concentrated into trunk streams can be measured. Accurate measurements of terrain slopes and watershed areas can be made photogrammetrically (Howe, 1958; Meyers et al., 1963; Lohman and Robinsove, 1964; Schneider, 1968; Taylor and Stingelin, 1969; Powell et al., 1970; Parry and Turner, 1971).

Remote-sensing techniques provide for the efficient collection of physical watershed parameters. When these parameters are considered in calculations of precipitation runoff, the design of hydraulic structures in ungaged drainage basins can be based on rational criteria rather than best guesses. The designs can be more efficient, functional, and safe.

Ground-Water Hydrology. The inference of landform and parent material from remote-sensor images can, in a preliminary sense, indicate whether or not geological processes and landforms favor shallow aquifer development (Winkler, 1966). A more detailed study of drainage pattern development as indicated on color infrared photography can be used to delineate areas of terrain exhibiting internal drainage and thus permeable soil or rock conditions. Color infrared photography, thermal infrared imagery, and potentially microwave radiometer data are useful for inferring soil moisture conditions.

Thermal infrared imagery has proved useful for detecting surface springs seeping at the ground surface and into existing bodies of water. The presence of near-surface aquifers has been discovered in some locations where spring water is several degrees cooler than the adjacent water or ground surface. In coastal areas of Florida and Mexico, studies have been undertaken to locate springs discharging fresh water into salt or brackish water. The ability to detect such springs so that the fresh water could be used before it mixed with salt water would be a significant contribution to the preservation of water resources. Some estimates of spring flow rates have been attempted by measuring the size of cool-water discharges from calibrated thermal infrared imagery (Ory, 1968).

Relatively large-scale imagery or photography is required for the study and inference of ground-water conditions. A scale of 1:10,000 would be ideal for general studies.

Stream Flow and Flood Mapping. Accurate measurements of the plan geometry of streams can be made from black-and-white photography at scales as small as 1:40,000. Meander frequency, stream width, flood-plain width, and stream pattern relationships (see *Hydrology*) all can be used to calculate other parameters of interest empirically. An estimate of stream gradient can be made photogrammetrically.

Stream velocity can be estimated from large-scale aerial photography in situations where obstructions in the stream create wave refraction patterns. Panchromatic photography at scales of 1:20,000 to 1:10,000 is entirely adequate for this application. Since the technique can provide at best only gross estimates, its use should be limited to the preliminary study of relatively inaccessible bodies of water.

Sediment load and sources of sediment can be inferred from aerial photography. Turbidity can be inferred from color and color infrared photography. Turbidity will be indicated as brown to yellow colors or as gray to white tones on color infrared photography as those areas exhibiting erosional rills and scars with minimal vegetative cover.

The extent of flooding within a drainage basin can be adequately measured from any type of aerial photography provided that cloud cover is not too great and the area is flown in the day (Dill, 1955). SLAR imagery, however, would seem to be the best sensor for real-time flood mapping. With its nearly all-weather capacity and day or night operation, sequential SLAR flights over a flooded watershed can provide data invaluable for downstream flood prediction. The delineation of all water-covered surfaces at any time from the onset of a storm is feasible with SLAR imagery (McBeth, 1965; Rib and Miles, 1969).

Transportation Engineering

Route Location. Optimum transportation system-route location can be facilitated with the use of remote-sensing techniques. A logical application of remote sensing is to analyze synoptic coverage of a region for potential transportation corridors followed by the use of more detailed remote-sensor data from each potential corridor. Finally, field data for the selection of the best route can be obtained. Remote-sensor coverage of the best corridor can be used to aid in the selection of a specific centerline and for final design (Belcher, 1948).

Small-scale color infrared photography from high-altitude aircraft or satellites, and SLAR imagery (1:120,000 to 1:1,000,000) can provide a good synoptic map base from which to analyze potential corridors. The SLAR imagery is, perhaps, one of the best forms of synoptic coverage, in that the critical parameters of topography, local terrain roughness, concentrations of cultural features, and gross vegetation types are exhibited on the imagery without the complex local land use tone and color variations that would be evident on small-scale photography. Since corridor evaluation is a preliminary function (see *Pipeline Corridor Evaluation*), the lower resolution of available SLAR sensors is an advantage. In cloud-covered areas of the world, SLAR imagery is the only available synoptic form of data.

Small-scale photography, Landsat imagery, and SLAR imagery can be interpreted to yield information about landforms and parent material. This geological information, along with topographic measurements and delineations of surface drainage, vegetation type, and cultural features, will provide an adequate means for establishing the most feasible corridors within the region of concern (see Vol. XIV: *Photogeology*).

A detailed analysis of each potential route can be accomplished most efficiently with a multisensor approach correlated with field measurements. One good sensor combination is an aerial mapping camera system with film such as Kodak's Aero-Neg- and a multispectral scanner operating in the near- and far-infrared portions of the spectrum. Panchromatic film transparencies from the Aero-Neg type film will provide the coverage from which photogrammetric measurements of topography can be made. Also, panchromatic photography will be adequate for most landform, cultural, and drainage interpretations. In areas where interpretations of black-and-white photography are ambiguous and in areas where vegetation greatly influences terrain patterns, color prints can be made from the Aero-Neg film. The added information exhibited on the color photography and the near-infrared multispectral imagery should allow unambiguous interpretations of vegetative influences to be made (Lauer, 1969; Northrop and Johnson, 1970; Tanguay and Miles, 1970). The thermal infrared imagery can be interpreted to yield information about probable soil moisture conditions. In certain geological environments, such as karst topography, these inferences may be extrapolated to include subsurface soil and rock conditions. Subsurface caverns or filled voids can sometimes be inferred from color infrared photography. Vegetation often will show distress due to dry conditions caused by subsurface voids. After all the geological and physiographic domains likely to pose different design or construction problems have been identified and mapped along the corridor, a specific program of exploratory drilling, sampling, and testing can be efficiently planned. The field data should then be used to verify, modify, or correct the engineering information inferred from the remote-sensor data (Rib and Miles, 1969).

A significant amount of geometric design data can be obtained photogrammetrically from the aerial photography. Dimensions of the terrain surface geometry and locations of critical features can be measured before the system is designed so that the facility can be constructed to fit with the terrain surface in the most economical and efficient manner. In the case of highway design, volumes of cut-and-fill material can be calculated from the stereo model of the terrain. In addition, construction planning and bidding parameters such as sources of construction materials, water, and points of access or egress can be obtained from the remote-sensor data (Anschutz and Stallard, 1967; Barr, 1969).

Transportation System Surveillance and Maintenance. High-resolution data are needed for all surveillance applications. Imaging sensors have the widest range of applicability, although non-imaging microwave radiometer systems do have potential for detecting subsurface phenomena such as buried pipeline leaks. All the imaging sensors record data in a form that is easily adapted for keeping records of the status of the system.

There are several specific indicators of highway pavement and appurtenance quality. The condition of a highway sub-base can be inferred from the relative ability of the right-of-way to drain freely. Mud holes along the edge of a pavement and/or mud squirted from beneath a rigid pavement (pavement pumping) are sure signs that the base has ceased to be free draining and that pavement cracking and deterioration will follow. This information can best be inferred from large-scale (1:5,000) color photography. The major indicator will be the contrast in color between pavement and pumped soil. Small cracks in a pavement can be readily identified on large-scale color infrared photography obtained a short time after a rain. The cracks will retain moisture after the unimpaired pavement has dried, thus establishing a color contrast. Large pavement cracks can be seen on any large-scale aerial photography. The condition of painted pavement markings and the general quality of the pavement wearing surface can be inferred from large-scale color photography. Also, color photography can be interpreted with respect to the condition of painted bridge surfaces and other structures. Large-scale black-and-white photography is adequate for the monitoring of major damage to pavements, structures, or signs.

General conditions of highway and rail beds can be assessed from remote-sensor data. Problems such as blocked drainage or the initial phases of soil creep and landsliding will often be evident as terrain pattern changes detectable from the air but obscured from a single vantage point on the ground. Large-scale panchromatic or color photography is useful in monitoring the physical character of a right-of-way.

With respect to waterway routes, color aerial photography can be used to infer sedimentation patterns and trends. Also, bank erosion can be detected. Although of a qualitative nature, such data are useful for planning channel maintenance programs.

Generally, buried and surface pipeline rights-of-way are periodically inspected visually from low-flying aircraft. The only advantage of remote sensing would be the detection of leaks not visually evident. Color infrared photography, thermal infrared imagery, and potentially microwave radiometer systems are applicable to the problem. Vegetation distressed by pipeline leaks likely would be detectable as having decreased near-infrared reflectance before there would be visible signs of trouble. A change in apparent temperature of the land surface due to leakage of a pipeline fluid or gas may be detectable on thermal infrared imagery. There are research indications that microwave radiometer systems will detect subsurface changes in moisture or fluid content that may be correlated with pipeline leaks. In all cases of pipeline surveillance by remote sensing, however, the detection of potential leaks is only a means for indicating locations to be field-checked. Many natural conditions could create anomalies as well.

Large-scale panchromatic photography can be used to monitor transmission towers for structural damage. It would be especially applicable after a natural disaster (Hackman, 1969). Also, it is feasible that airborne thermal radiometers could be used to detect anomalously hot transformers in local power-distribution systems.

Geological Engineering and Soil Mechanics

Soil Exploration and Classification. Regional and/or detailed local analyses of remote-sensor data can be undertaken for engineering soil-exploration purposes. With a technique of pattern analysis and inferences based on a knowledge of geological and pedologic processes, areas of similar parent material can be delineated and described. All features of the terrain surface add to the inference technique. In particular, however, drainage patterns, erosion surfaces and channel shapes, distribution of vegetation, terrain surface color, and soil moisture content all add to a body of information from which engineering soil characteristics can be logically deduced (Frost and Woods, 1948; Leuder, 1959; Gerberman et al., 1971).

Landsat imagery, SLAR imagery, and high-altitude color infrared photography at scales of 1:120,000 to 1:1,000,000 are optimum forms of regional remote-sensor data to be used for soil classification (see *Soil Classification System, Unified*). Landforms can be identified and delineated. An inference of engineering soil type associated with each landform can be made with an acceptably high level of accuracy (Rib and Miles, 1969).

Detailed studies of specific local areas can go

farther than the inference of parent material associated with a landform. The degree of soil variability within each landform unit is possible from large-scale photography and scanner imagery.

The variability of engineering soil characteristics is influenced most significantly by soil particle size, density, organic material content, and soil moisture content. Inference of these conditions from remote-sensor data is based on the evaluation of surface indicators of these conditions. Panchromatic photography is adequate for many applications. Erosion and drainage patterns as well as surficial organic and moisture conditions are expressed on the pan photograph as tonal and textural patterns. Additional information can, however, be obtained from color and color infrared photography. Soil color variability can be evaluated from color photography, whereas vegetation conditions and variations in soil moisture can best be inferred from color infrared photography. Thermal infrared imagery will provide apparent soil temperature data from which surface and near-surface soil moisture conditions can be inferred (Colwell, 1946; Boon, 1960; Matalucci and Abdel-Hady, 1969; National Academy of Sciences, 1970; Kristof and Zachary, 1975).

An understanding of geological and soil-forming processes will allow an interpreter to make acceptable inferences about the engineering character of the surface soil (see Vol. XIV: *Engineering Soil Science*). At best, the mapped soil units can be classified according to the Unified Soil Classification System (Smith, 1941; Ulaby and McNaughton, 1975).

Mass Wasting. A prime advantage of using remotely sensed data for evaluation of mass-wasting phenomena is realized from the synoptic coverage of terrain. Surface mass-wasting patterns not detectable from spot ground observations may be evidenced on photography or imagery. Also, specific local indicators of erosion or potential mass wasting such as deranged surface-water flow, springs, or distressed vegetation may be inferred from remotely sensed data (Jones and Keech, 1966; Coker et al., 1969).

Color infrared photography is perhaps the single best film type for mass-wasting studies. Thermal infrared imagery is a valuable adjunct for determining apparent ground temperatures from which ground-water conditions can be inferred. Color infrared photography can provide stereoscopic terrain models from which topographic measurements can be made. In addition, drainage, soil moisture, and vegetation patterns can be analyzed. From these data, terrain surfaces that exhibit indications of mass wasting can be mapped, described, and specified as warranting field investigation (Liang and Belcher, 1957; Dishaw, 1967).

Thermal infrared imagery is valuable as a source of inferred soil moisture information. Apparent soil surface temperatures vary with moisture content; the variability is extreme in the case of some phenomena such as subsidence in karst opportunity.

Site Selection. Remote-sensor data now exist for most regions of the world. Panchromatic photography has been obtained by various government agencies. Remote sensing is also the most efficient means of collecting additional data for preliminary site evaluations. Practically all the available sensor systems are applicable to this problem. However, depending on the nature of the site requirements and the scope of the region to be investigated, there will be an optimum sensor configuration.

A logical approach to the problem is to study a region of interest using sensors with synoptic ground coverage and then analyze potentially suitable local areas with more detailed sensor data. A regional study of geological structure, landform, associated parent material, and drainage can be performed adequately from satellite imagery, high-altitude aircraft photography, and/or SLAR imagery. In addition, regional geography, land use, and transportation networks can be mapped from small-scale imagery (Meyer and Calpouzos, 1968; Lowman, 1974). Factor overlay maps provide a means for eliminating portions of the region as unsuitable (Anschutz and Stallard, 1967; Nunnally, 1969; Aldrich, 1971).

The study of individual potential sites can be as detailed as time and economics allow. Color and color infrared photography are the best sensors for detailed site analysis although only a little information vital to engineering site selection will be lost if panchromatic photography is used. Contour maps can be constructed photogrammetrically with maximum scale and minimum contour interval, depending on the photo scale and resolution. Slopes and landscaping cut-and-fill data can be calculated from such maps measured directly from a stereoplotter. Distances and areas of concern can also be measured photogrammetrically.

Cultural features and objects associated with a potential site can be identified by recognition of shape, size, and associated patterns (Hadfield, 1963; Davis, 1966; Lindgren, 1971). Dichotomous keys are often used for such an analysis. Terrain parameters dealing with such engineering factors as foundation stability, slope stability, aesthetic environment, water supply, and waste disposal must be inferred by deducing the probable geological history of the area as indicated by the remote-sensor data. Detailed geological structure, landform, and parent material delineations can be made on large-scale black-and-white, color, or color infrared photography (Conway and Holz, 1973). The extent and vigor of vegetation as well as surface soil moisture conditions can be best inferred from color infrared photography (Colwell, 1946; Boon, 1960; Meyer and Calpouzos, 1968; Edwards et al.,

1975). Additional soil moisture and geological structure information can be inferred from a study of thermal contrasts on thermal infrared imagery. There are research data to indicate that subsurface soil structure and moisture conditions can be measured with microwave radiometer sensors (American Society of Photogrammetry, 1966; Weber and Polcyn, 1972).

The information derived from a detailed study of remote-sensor data should be sufficient to identify the poorest of several alternate sites. Final selections, however, must be based on a planned program of field observation sampling and laboratory testing. The planning of the field program can be made using the remote-sensor data as a base. Thus, field data can be collected that will be complimented but not duplicated by the remote-sensor data (Barr and Hensey, 1974).

Construction Material Location and Inventory. Synoptic remote-sensor data such as high-altitude photography SLAR imagery or Landsat imagery can be interpreted to produce regional landform maps. Areas likely to contain economical construction material deposits can be identified from photography or imagery with little difficulty. Thus the fact that many areas of the world are topographically and geologically unmapped does not preclude regional surveys of construction material (Belcher, 1948; Homes, 1967; Henderson, 1975).

Once those areas exhibiting high potential for containing construction materials have been identified, either from previously existing maps or from a regional remote-sensing survey, more detailed study of local remote-sensor data is warranted. In arid environments, black-and-white or color photography at a scale of 1:20,000 is sufficient to provide detailed delineations of parent material. In all other environments color infrared photography is the optimum sensor. The evidence of surface or near-surface construction material deposits as indicated on photography includes landform shape, drainage pattern, soil moisture variation, type of vegetation cover, and relationships with associated landforms (Smith, 1941). Additional information of value for delineating buried deposits can be obtained from thermal infrared imagery. Subsurface soil moisture conditions and material type can often be inferred from variations in apparent surface temperature. There is a potential for the quantitative determination of subsurface soil profile characteristics from microwave radiometer and long-wavelength radar data (Hunter and Bird, 1970).

As in any other engineering application of remote sensing, detailed delineations of construction material deposits must be verified and perhaps modified from field data. The initial delineations, however, provide an efficient means of planning a field drilling and sampling program.

Coastal Engineering. Remote sensing is of value for inferring existing coastal conditions as well as for establishing the historical development of coastal features. Small-scale color photography and imagery (1:120,000 to 1:1,000,000) is of value in evaluating regional sediment patterns produced by turbid water discharging from estuaries. Also, the general zones of beach development and/or erosion will be evidenced by landform patterns (see Vol. XIV: *Coastal Zone Management*). SLAR imagery can also be used for the mapping of coastal landforms (Geary, 1968; Klemas et al., 1973, 1974, 1975).

Large-scale photography (panchromatic and color) is valuable for more detailed analysis of erosion and deposition zones. Also, wave patterns produced by or refracted from existing features can be analyzed with respect to surface-water velocities (Terwinkel, 1963; Stafford, 1971).

One of the most significant applications of remote sensing to coastal engineering (q.v.) is that of determining the historical development of coastal features. Most coastal areas have been photographed many times in the last several decades. The analysis of such existing photography with respect to the detection of change is a practical way in which to predict future coastal events. These predictions can be based on inferences or photogrammetrically measured parameters (Klemas et al., 1975).

DAVID J. BARR

References

Aldrich, R. C., 1971. Space photos for land use and forestry, *Photogramm. Eng.* **37** (4), 389-401. (1)

Allum, J. A., 1962. *Photogeology and Regional Mapping.* New York: Pergamon Press, 107p. (2)

American Society of Photogrammetry, 1966. *Manual of Photogrammetry.* Menosha, Wis.: George Banta, 1,199p. (44)

Anschutz, G., and Stallard, A. H., 1967. An overview of site evaluation, *Photogramm. Eng.* **33** (12), 1,381-1,396. (3)

Ashley, M. D., and Rea, J., 1975. Seasonal vegetation differences from ERTS imagery, *Photogramm. Eng.* **41** (6), 713-719. (4)

Barr, D. J., 1969. *Use of Side-Looking Airborne Radar Imagery (SLAR) for Engineering Soils Studies.* Fort Belvoir, Va.: U.S. Army Engineer Topographic Laboratories, Technical Report TR-46, 156p. (5)

Barr, D. J., and Hensey, M. D., 1974. Industrial site study with remote mensing, *Photogramm. Eng.* **40** (2), 79-85. (6)

Barr, D. J., and Miles, R. D., 1970. SLAR Imagery and Site Selection, *Photogramm. Eng.* **36** (11), 1,115-1,170. (7)

Belcher, D. J., 1948. The engineering significance of landforms, *Highway Research Board Bull.* **13**, 9-29. (8)

Boon, D. A., 1960. Interpretation of vegetation: report of Working Group 4, Commission VII, *Internat. Soc. Photogrammetry* **26** (4), 283-302. (9)

Bryan, M. L., 1975. Interpretation of an urban scene using multi-channel radar imagery, *Remote Sensing Environment* **4** (1), 49-66. (10)

Coker, A. E.; Marshall, R.; and Thomson, N. S., 1969. Application of computer processed multispectral data to the discrimination of land collapse (sinkhole) prone areas in Florida, in *Proceedings of the 6th Symposium on Remote Sensing of the Environment.* Ann Arbor: University of Michigan, 65. (11)

Colwell, J. E., 1970. *Multispectral Remote Sensing of Urban Features: Final Report,* 1 June-31, October 1969. Ann Arbor: Infrared and Optics Laboratory, Willow Run Laboratories, University of Michigan. (12)

Colwell, R. N., 1946. The estimation of ground conditions from aerial photographic interpretation of vegetation types, *Photogramm. Eng.* 12 (1), 151-161. (13)

Conway, Dennis, and Holz, R. K., 1973. Use of the near-infrared photography in the analysis of surface morphology of an Argentine alluvial floodplain, *Remote Sensing Environment* 2 (4), 235-242. (14)

Davis, J. M., 1966. Uses of aerial photos for rural and urban planning, *U.S. Dept. Agriculture, Agriculture Handbk. 315,* 409p. (15)

Deutsch, M., and Ruggles, F., 1974. Optical data processing and projected applications of the ERTS-1 imagery covering the 1973 Mississippi River Valley floods, *Water Resources Bull.* 10 (5), 1023-1039. (16)

Dill, N. W., 1955. Photo interpretation in flood control appraisal, *Photogramm. Eng.* 21 (1), 112-114. (17)

Dishaw, H. E., 1967. Massive landslides, 1976. *Photogramm. Eng.* 33 (6), 603. (18)

Edwards, G. J.; Schehl, T.; and DuCharme, E. P., 1975. Multispectral sensing of young citrus tree decline, *Photogramm. Eng.* 41 (5), 653-657. (19)

Frost, R. E., and Woods, K. B., 1948. Airphoto patterns of soils of the western United States, *U.S. Dept. Commerce Civil Aeronaut. Admin., Tech. Div. Rept. No. 85,* 100p. (20)

Geary, E. L., 1968. Coastal hydrography, *Photogramm. Eng.* 34 (1), 44-50. (21)

Gerbermann, A. H.; Gausman, H. W.; and Weigand, C. L., 1971. Color and color-IR films for soil identification, *Photogramm. Eng.* 37 (4), 359-364. (22)

Hackman, R. J., 1969. Interpretation of Alaskan post-earthquake photographs, *Photogramm. Eng.* 31 (7), 604-610. (23)

Hadfield, S. M., 1963. *An Evaluation of Land Use and Dwelling Unit Data Derived from Aerial Photography.* Chicago: Chicago Area Transportation Study, Report UR-1, 29p. (24)

Henderson, F. M., 1975. Radar for small-scale land-use mapping, *Photogramm. Eng.* 41 (3), 307-319. (25)

Homes, R. F., 1967, Engineering materials and side-looking radar, *Photogramm. Eng.* 33 (7), 767. (26)

Howe, R. H., 1958. The application of aerial photographic interpretation to the investigation of hydrologic problems, *Photogramm. Eng.* 26 (1), 85-95. (27)

Hunter, G. T., and Bird, G., 1970. Critical terrain analysis, *Photogramm. Eng.* 36 (9), 939-952. (28)

James, J., and Burgess, F. J., 1970. Ocean outfall dispersion, *Photogramm. Eng.* 36 (12), 1,241-1,250. (29)

Jones, R. G., and Keech, M. D., 1966. Identifying and assessing problem areas in soil erosion surveys using aerial photographs, *Photogramm. Rec.* 5 (27), 189-197. (30)

Kiefer, R., and Scherz, J., 1970. Applications of airborne remote sensing technology, *Am. Soc. Civil Engineers Proc., Jour. Surveying and Mapping Div.* 96 (SUI), 57-80. (31)

Klemas, J.; Borchardt, J. F.; and Treasure, W. M., 1973. Suspended sediment observations from ERTS-1, *Remote Sensing Environment* 2 (4), 205-221. (32)

Klemas, V., et al., 1974. Coastal and esturine studies with ERTS-1 and Skylab, *Remote Sensing Environment* 3, 153-174. (33)

Klemas, V.; Bartlett, D.; and Rogers, R., 1975. Coastal zone classification from satellite imagery, *Photogramm. Eng.* 41 (4), 499-513. (34)

Kristof, S. J., and Zachary, A. L., 1975. Mapping soil features from multispectral scanner data, *Photogramm. Eng.* 40 (12), 1,427-1,434. (35)

Lauer, D. T., 1969. Multispectral sensing of forest vrgetation, *Photogramm. Eng.* 35 (4), 346-354. (36)

Leamer, R. W.; Weber, D. A.; and Wilgand, C. L., 1975. Pattern recognition of soils and crops from space, *Photogramm. Eng.* 41 (4), 471-478. (37)

Leuder, D. R., 1959. *Aerial Photographic Interpretation: Principles and Applications.* New York: McGraw-Hill, 462p. (38)

Liang, T., and Belcher, D. J., 1957. Landslides and engineering practice, in *Airphoto Interpretation,* Natl. Research Council Spec. Rept. 2, Pub. 554, 232p. (39)

Lindgren, D. T., 1971. Dwelling unit estimation with color-IR photos, *Photogramm. Eng.* 37 (4), 373-377. (40)

Lohman, S. W., and Robinsove, C. J., 1964. Photographic description and appraisal of water resources, *Photogrammetria* 19 (3), 83-103. (41)

Lowman, P. D., 1974. Geologic Structure in California: Three Studies with ERTS-1 Imagery, *U.S. Natl. Aeronautics and Space Adm. Tech. Rept. NASA-TM-X-70799,* 19p. (42)

Ludwig, C.; Bartle, R.; and Washington, M., 1969. *Study of Air Pollution Detection by Remote Sensors.* San Diego, Calif.: General Dynamics Corporation, Report GDC-DBE, 11p. (43)

McBeth, F. H., 1965. A method of shoreline delineation, *Photogramm. Eng.* 22 (2), 400-405. (45)

Matalucci, R. V., and Abdel-Hady, M., 1969. Surface and subsurface exploration by infrared surveys, *Highway Research Board Spec. Rept.* 102, 1-12. (46)

Meyer, M. P., and Calpouzos, L., 1968. Detection of crop disease, *Photogramm. Eng.* 34 (6), 554-557. (47)

Meyers, V. I.; Ussery, L. R.; and Rippert, W. J., 1963. Photogrammetry for detailed detection of drainage and salinity problems, *Am. Soc. Agric. Engineers Trans.* 11 (4), 332-334. (48)

National Academy of Sciences, 1970. *Remote Sensing.* Washington, D.C., 424p. (56)

Northrop, K. G., and Johnson, E. W., 1970. Forest cover type identification, *Photogramm. Eng.* 36 (5), 483-490. (49)

Nunnally, N. R., 1969. Integrated landscape analysis with radar imagery, *Remote Sensing Environment,* 1 (1), 1-6. (50)

Ory, R. T., 1968. Seepage detection by remote sensing, in *Proceedings of the 2nd Seepage Symposium, United States Department of Agriculture.* Phoenix, Ariz., 75p. (51)

Parry, J. T., and Turner, H., 1971. Infrared photos for drainage analysis, *Photogramm. Eng.* 37 (10), 1,031-1,038. (52)

Piech, K. R., and Walker, J. E., 1971. Aerial color analyses of water quality, *Am. Soc. Civil Engineers Proc., Jour. Surveying and Mapping Div.* 97 (SU2), 185-197. (53)

Piech, K. R., and Walker, J. E., 1972. Outfall inventory

using airphoto interpretation, *Photogramm. Eng.* **38** (9), 907-914. (54)

Powell, W. J.; Copeland, C. W.; and Drahovzal, J. A., 1970. Delineation of Linear Features and Applications to Reservoir Engineering Using Apollo 9 Multispectral Photography, *Alabama Geol. Survey Inf. Ser. 41,* 37p. (55)

Rib, H. T., and Miles, R. D., 1969. Multisensor analysis for soils mapping, *Highway Research Board Spec. Rept. 102,* 22-37. (57)

Robinsove, C. J., 1965. Infrared photography and imagery in water resources research, *Am. Water Works Assoc. Jour.* **57** (7), 834-840. (58)

Schneider, W. J., 1968. Color photographs for water resources studies, *Photogramm. Eng.* **34** (3), 257. (59)

Smith, H. T. U., 1941. Aerial photographs in geomorphic studies, *Jour. Geomorphology* **4**, 171-205. (60)

Stafford, D., 1971. An aerial photographic survey of coastal erosion, *Photogramm. Eng.* **37** (6), 565-575. (61)

Tanguay, M. G., and Miles, R. D., 1970. Multispectral data interpretation for engineering soils mapping, *Highway Research Rec.* **319,** 58-77. (62)

Taylor, J. I., and Stingelin, R. W., 1969. Infrared imaging for water resources studies, *Am. Soc. Civil Engineers Proc., Jour. Hydraulics Div.* **95** (HY1), 175-189. (63)

Terwinkel, G. C., 1963. Water depths from aerial photographs, *Photogramm. Eng.* **30** (11), 1,037-1,042. (64)

Ulaby, F. T., and McNaughton, J., 1975. Classification of physiography from ERTS imagery, *Photogramm. Eng.* **41** (8), 1,019-1,027. (65)

Van Lopik, J. R.; Pressman, A. E.; and Ludlom, R. L., 1968. Mapping pollution with infrared, *Photogramm. Eng.* **34** (6), 561-564. (66)

Veress, S. A., 1970*a. Study of the Three-Dimensional Extension of Polluted Air.* Seattle: University of Washington, Department of Civil Engineering, 50p. (67)

Veress, S. A., 1970*b.* Air pollution research, *Photogramm. Eng.* **36** (8), 840-848. (68)

Weber, F. P., and Polcyn, F. C., 1972. Remote sensing to detect stress in forests, *Photogramm. Eng.* **38** (2), 163-175. (69)

Winkler, E. M., 1966. Moisture measurements in glacial soils from airphotos, *Ecology* **27** (1), 156-158. (70)

Witenstein, N. M., 1972. The application of photo interpretation to urban area analysis, *Photogramm. Eng.* **18** (3), 490-492. (71)

Wobber, F. J., 1969. Environmental studies using earth orbital photography, *Photogramm.* **24** (3/4), 107-165. (72)

Wobber, F. J., 1970. Orbital photos applied to the environment, *Photogramm. Eng.* **36** (8), 852-864. (73)

Wobber, F. J., et al., 1975. Coal refuse site inventories, *Photogramm. Eng.* **41** (9), 1,163-1,171. (74)

Wolf, P. R., 1974. *Elements of Photogrammetry.* New York: McGraw-Hill, 562p.

Cross-references: *Alluvial Plains, Engineering Geology; Coastal Engineering; Hydrogeology and Geohydrology; Pipeline Corridor Evaluation; River Engineering; Rock Slope Engineering; Soil Classification System, Unified; Urban Geology; Urban Geomorphology; Urban Hydrology. Vol. XIV: Aerial Surveys, General; Coastal Zone Management; Engineering Soil Science; Photogeology; Photo Interpretation; Remote Sensing, General; Remote Sensing and Photogrammetry, Societies and Periodicals; Satellite Geodesy and Geodynamics; Surveying, General.*

RESERVOIR ENGINEERING — See ELECTROKINETICS; FORMATION PRESSURES, ABNORMAL. Vol. XIV: PETROLEUM GEOLOGY.

RESIDUAL STRESS, ROCKS

Early Observations

The behavior of massive rocks in quarries, in mines, and in many outcrops commonly indicates that they are highly strained and are supporting stresses of large magnitude. The actions and effects of these stresses have no doubt been observed during many millenia of stoneworking and mining, but records of directed study and measurement appear to start about the middle of the last century with the notes of Johnston (1854) on the movements of sandstone strata in a quarry at Portland, Connecticut. When channels excavated by pick in an east-west direction were nearly completed through a sandstone stratum, the remaining 23 to 30 cm of rock ruptured violently and the walls of the channels became closer by about 2 cm. Johnston observed that channels cut in a north-south direction were little or not at all affected, and that when the spontaneous movements in an east-west trench occurred, only the north wall moved, toward the south. He supposed the reason for this was that the channel was cut near the southern wall of the deep excavation. Observations in quarries were continued and were the subject of an article by Niles (1871) concerning expansion of quarried rock at Monson, Massachusetts, and elsewhere.

The reports of spontaneous movements of rocks in quarries and mines were reviewed comprehensively in papers by Hankar-Urban (1907) and by Rzehak (1906). One of the most interesting observations was that of Pilkington (in Strahan, 1887, p. 401), who described in 1789 the explosive property of rocks of Derbyshire, England, and added that the rocks were said to lose this property "very soon after they are taken out of the mine." Hankar-Urban recognized that if this report were true, such delayed behavior would help explain the explosive character of rocks. Indeed, the spontaneous expansion or disintegration of isolated blocks of rock still remains at the center of any discussion of residual stress.

Strahan (1887, p. 408) concluded his account of the explosive slickensides and vein material in the lead mines of Derbyshire by saying, "The explanation, which perhaps best satisfies the requirements of the problem, appears to be that the spars are in a state of molecular strain, resembling that of the Rupert's Drop, or of toughened glass, and that this condition of strain is the result of the earth-movements, which produced the slicken-

461

sides." Clearly, Strahan perceived that rock stresses, although having their origin in what we would now call tectonic movements, were subsequently locked up in the rock in a state of metastable equilibrium.

About the time of Hankar-Urban's work, Dale (1923) began his study on the granites and other building stones of the New England region and reported many observations on the state of compressive strain in the quarries. He also summarized current ideas regarding granites and their structure, including those on the origin of the more or less horizontal jointlike fractures (or sheeting) that so facilitate the quarrying of otherwise massive granite (see Fig. 1). Dale noted the work of Gilbert (1904) on the origin of apparently similar sheetlike structures in the granitic terrain of the Sierra Nevada in and near Yosemite Valley, California, particularly those that enclose the imposing granitic domes.

Gilbert examined the hypothesis that sheeting in granite was due either to thermal effects of the sun's radiation or to the physical-chemical effects of weathering. He came to favor a third process— that of dilation of the rocks upon unloading as the superjacent material was removed by erosion.

Gilbert noted that dome structure appears to develop only in massive rock and that systems of closely spaced joints, by furnishing other avenues for relief of strain, prevent formation of exfoliation shells. Furthermore, he found that the general

FIGURE 1. Sheeting joints at Dell Hitchcock Quarry, Quincy, Massachusetts (from Dale, 1923).

parallelism of exfoliation joints with the ground surface was not restricted to domes but extended also to concave forms, and that curved sheets resembling synclines of sedimentary strata underlay the bottoms of some valleys carved in granite.

The state of stress in the rock that is necessary to produce the observed sheeting and other features in massive rocks has remained largely speculative until the recent development of methods to actually measure the stresses present in in-place rock. These are taken up below, but the point to be made here is that such measurements have been subject to two divergent interpretations: that the measured stresses (1) are produced by presently active tectonic loads impressed on whole regions of rock from exterior sources or by gravity, or (2) may in part or whole have been produced by forces no longer acting and "locked" within the rock until released by the cutting operations necessary to perform the measurement.

The measured stress, in-place, has been referred to variously as *in-place stress* and *residual stress;* it is commonly ascribed to now-active tectonic processes, but usually without consideration of the second possibility. Confusion of the terminology of stress and the supposed origins of the stress has resulted.

We here follow the use of the term *residual stress* as it has long been used in metallurgy and as explicitly applied to rocks by Voight (1966, p. 45), who defined residual stresses as "systems of stresses on the inside of a body which are in equilibrium, or approach equilibrium, when neither normal nor shear stresses are transmitted through its exterior surfaces."

Storage of Residual Stresses in Rocks

Residual stresses and the resulting stored elastic strains are introduced into rock masses by various natural loads—such as gravity, tectonism, glaciers, or thermal gradients—and then locked in by some type of permanent nonelastic deformation, such as plastic or viscous flow, volume changes, and chemical changes. McClintock and Argon (1966) recognized that residual stresses in metals occur on a microscopic scale as a result of anisotropic plastic behavior, dislocations, and inclusions; and on a macroscopic scale can arise from forced alignment of parts, from loads or thermal gradients causing non-uniform plastic flow or creep, and from volume changes caused by metallurgical or chemical processes. Whether on a microscopic or a macroscopic scale, residual stresses are part of a complicated, balanced internal stress system that consists of both intergranular and intragranular stresses. In the absence of thermal and chemical changes, the stress systems must be in a state of equilibrium and satisfy the conditions of compatibility; otherwise, rocks containing residual stresses would be unstable.

Residual Stresses as a Result of Deformations on a Microscopic Scale

In general, rocks are polymineralic rather than monomineralic and, in many cases, have a heterogeneous grain geometry and size distribution. The various minerals found in rocks have a wide variety of physical properties, including crystalline anisotropies that greatly complicate the intragranular plastic, viscous, and elastic deformations caused by external loads, thermal gradients, and chemical changes. Nur and Simmons (1970) demonstrated the complex nature of stresses within an intrusive rock induced by *thermal cooling* and removal of overburden. On the basis of the differences of individual mineral bulk moduli and thermal coefficients of expansion, they calculated grain shear stresses within an augite-quartz assemblage of 2 and 12 kilobars (kbars), caused, respectively, by the removal of 10 kbars of overburden pressure and by contraction associated with a temperature drop of $6 \times 10^{2}°C$. These stored stresses are caused merely by unloading and cooling of an intrusive rock mass, but further complications are introduced by viscous and plastic deformations such as dislocations, microfractures, recrystallization, and grain-boundary slip. These deformations produce keying effects that lock in elastic or *visco-elastic strain energy*. Friedman (1972) thought of residual stresses as being composed of two parts: (1) crystal distortions reflecting previous external loads that are locked in the aggregate, called the locked-in stresses, and (2) the constraining or locking stresses. Carter (1969) calculated the stress required for specific edge-dislocation spacings genetically associated with quartz deformation lamellae. A spacing of 315 Å, or 3×10^{5} edge dislocations per centimeter, can account for the observed change of birefringence across the lamellae of 0.002. The *differential stress* necessary for a stress-optical effect of this magnitude is about 5.4×10^{3} bars. Thus, owing to permanent deformation, large stress differences may be stored within the grains of the rock. The stored stress associated with the deformation lamellae can be relieved if the dislocations migrate so as to reduce their density along the lamellae. It is not known whether the dislocations move at room temperature upon relaxation of constraints around a given grain. Lamellae do, however, exist in freed, isolated quartz grains.

Consolidation (q.v.) and *cementation* in sedimentary deposits are other processes that contribute to the storage of residual stresses in rock masses. Gallagher (1971) demonstrated, using a two-dimensional photoelastic model simulating sandstone, the manner in which residual stress can be introduced into sedimentary rocks. He loaded photoelastic disks resembling sand grains in shape, and then cemented the loaded disks with an epoxy glue. Upon removal of the external loads, both the disks and the intergranular glue retained part of the strain energy introduced by the external loads. Gallagher's models also demonstrated the complex nature in which stresses are locally distributed in granular masses as a function of grain geometries. Voight (1974b) gave an example in which he postulated the physical conditions of sedimentation and burial to 1 km, diagenesis, and denudation of a sedimentary rock. Based on these conditions he calculated horizontal surface stresses of 115 bars in tension that would result after denudation, provided the tensile strength of the rock is not exceeded.

These examples demonstrate some of the ways in which residual stresses may be stored in polycrystalline rock masses and also indicate the complexity of distribution of these stresses.

Residual Stresses as a Result of Deformations on a Macroscopic Scale

On a larger, macroscopic scale, the *stress fields* in rocks are probably analogous to those in metals; that is, when the scale of observation is sufficiently large so that granular anisotropies can be disregarded because they can be treated as being statistically homogeneous, the stress fields can then be analyzed as being in a continuous medium within the defined boundaries of that medium. Internally stored stresses on this scale are controlled by the boundaries or the geometry of the body as well as the gross material properties of the body. The boundaries of discrete blocks are defined by joints, faults, and chemical alteration zones; and the mass material properties that control deformation are defined by the rock fabric, mineralogy, and density distribution.

In large rock masses residual stresses other than those attributed to microscopic mechanisms are also stored through elastic, viscous, and plastic deformations (see *Rheology, Soil and Rock*). Permanent deformations are generally caused by (1) displacements along large discontinuities such as faults and joints; (2) tilting displacements; (3) volume changes caused by dilation, microfracturing, chemical changes, and thermal gradients; and (4) flow, either plastic or viscous.

Voight (1974a) has given possible explanations for storing residual stresses in the development of Appalachian décollement zones. One of the mechanisms he cites in his argument is the viscous deformation of large thrust sheets, and he postulates that measured horizontal stresses in excess of those attributed to overburden may have been locked in the thrust sheets by such deformation. Also Savage (1978) has given an explanation for storing residual stresses in plutons and their adjacent intruded rocks. These are *thermal stresses* caused by cooling of the plutons.

General State of Residual Stresses

In any rock mass, then, residual stresses may consist of an internal distribution of stresses within a particular geologic domain, which is further complicated by added stresses originating from externally applied loads. Holzhausen and Johnson (1979) have alluded to the complexity of such stress systems. Examples of stress superposition may be seen in some of the rock masses of the Paleozoic mountain chains of the eastern United States. Surface stress measurements demonstrate horizontal stresses to be in excess of those that can be explained by *lithostatic loading* (Sbar and Sykes, 1973). Measurements in freed blocks demonstrate the existence of residual stresses in the Devonian Barre Granite of Vermont that may account for part of the *horizontal stresses* determined in that area (Nichols, 1975). In the Cambrian Potsdam Sandstone of northeastern New York, Engelder et al. (1975) have also determined that residual stresses are present at some locations. In addition to residual stresses, externally applied active tectonic loads probably contribute significantly to the excessive horizontal stresses. In an experiment conducted near Cedar City, Utah, Swolfs et al. (1974) obtained results indicating that the *in situ* stresses within a quartz diorite rock mass are composed of major components of both residual and *tectonic* stresses.

Relief of Residual Stresses

The relief of residual stresses in rock and the resulting deformations are poorly understood phenomena, even though much information is available on *in situ* stress determinations and the methods by which such determinations have been made. Commonly used methods rely on the actual measurement of deformations caused by cutting rock surfaces or the measurement of fluid pressures necessary to cause rock fracture (see *Rock Structure Monitoring*). In both methods, measurements are made at or near a rock surface, and assumptions must be made as to how the stresses are generated and transmitted through the mass. It has been assumed in the past that residual stresses are balanced within very small volumes and that most rock deformations produced by cutting new surfaces are caused by external loads acting at some distance from the location of measured deformation. Emery (1964), Swolfs et al. (1974), Nichols (1975), and Bock (1979), however, demonstrated that recoverable residual strains of a high order can exist in freed rock samples and that they vary in magnitude from point to point. Varnes and Lee (1972) demonstrated, using a two-dimensional mechanical model containing elastic, frictional, and viscous restraints, that residual strain energy of sufficient quantities, released by relaxation of internal stresses, may do work on the surroundings.

They also suggested that the rate of relaxation decreases with the distance from a newly created surface. Nichols (1975) also demonstrated this on a specimen of Barre Granite. Nichols and Abel (1975) and Nichols and Savage (1976) have shown that the strain energy released from laboratory rock specimens can be as large as 2.11×10^5 ergs/cm^3 and that the release of energy may be in large part time-dependent.

These examples are concerned with the release of strain energy on a small scale as compared to the much greater scale of release that probably exists in the large volumes of rock of the continental crust. We lack quantitative knowledge of large-scale deformations occurring within geological terranes that are now considered tectonically inactive. Hence, we can only infer how deformations resulting from residual stresses take place, using geological field evidence, small-scale laboratory investigations, and model studies of residual stress. The work done in the creation of new rock surfaces and the removal of external constraints appears to be a major factor in triggering the release of internally stored energy (Fig. 2). As previously mentioned, this is accomplished in natural rock masses by uplift and denudation, where uplift is a result of continuing *isostatic adjustment* and *orogenic forces* and denudation is produced by erosion. As surface rocks are removed, so also is the confinement that they provided to the underlying rock masses. In addition, new surfaces have been formed. The removal of confinement allows internal readjustment of stored residual stresses, and the creation of a new surface mobilizes near-surface strain energy that also causes

FIGURE 2. A brittle failure in granite at Mount Airy, North Carolina, produced by quarrying the adjacent constraining rock (from Nichols and Abel, 1975).

an internal adjustment. Both processes contribute to deformations near the newly created surface.

These internal adjustments take place by means of inelastic and elastic deformations on intergranular and intragranular scales, with the rock mass attaining a lower energy level as the process continues. The relieving strains do work on constrained portions of the rock mass and may create additional stress concentrations sufficient to cause new failures, thereby relieving more energy that deforms free surfaces. In such a manner, a rock mass eventually relieves itself of all internal stresses and, in so doing, goes through many deformations and possibly failures.

THOMAS C. NICHOLS, JR.
DAVID J. VARNES

References

Bock, H., 1979. Experimental determination of residual stress field in a basaltic column, in *Proceedings of the 4th Congress of the International Society of Rock Mechanics,* Vol. 1. Rotterdam, Netherlands: Balkema, 45-49.

Carter, N. L., 1969. Flow of Rock-Forming Crystals and Aggregates, in R. E. Riecker (ed.), *Rock Mechanics Seminar,* Vol. 2. Bedford, Mass.: Air Force Cambridge Research Laboratory, 509-594.

Dale, T. N., 1923. Commercial granites of New England, *U.S. Geol. Survey Bull.* **738,** 488p.

Emery, C. L., 1964. Strain energy in rocks, in W. R. Judd (ed.), *State of Stress in the Earth's Crust: Proceedings of the International Conference.* New York: American Elsevier, 234-279.

Engelder, T.; Sbar, M. L.; and Lelyveld, P., 1975. Preliminary analysis of in-situ strain in New York and Vermont, *Am. Geophys. Union Trans.* **56** (12), 1,057.

Friedman, M., 1972. Residual elastic strain in rocks, *Tectonophysics* **15** (4), 297-330.

Gallagher, J. J., Jr., 1971. Photomechanical model studies relating to fracture and residual elastic strain in granular aggregates, in *Studies in Rock Fracture.* U.S. Army Corps Engineers Quart. Tech. Rept. 3, 127p.

Gilbert, G. K., 1907. Domes and dome structure of the High Sierra, *Geol. Soc. America Bull.* **15,** 29-36.

Hankar-Urban, A., 1905. Troisième les autoclases ou ruptures note sur des mouvements spontanées de roches dans les mines, les carrières, etc., *Soc. Belge Gèologie, Paléontologie et Hydrologie Bull.* **23,** 260-270.

Holzhausen, G. R., and Johnson, A. M., 1979. The concept of residual stress in rock, *Tectonophysics* **58,** 237-267.

Johnston, John, 1854. Notice of some spontaneous movements occasionally observed in the sandstone strata in one of the quarries at Portland, Conn., *Am. Assoc. Adv. Sci. Proc.* **8,** 283-286.

McClintock, F. A., and Argon, A. W., 1966. *Mechanical Behavior of Materials.* Reading, Mass.: Addison-Wesley, 770p.

Nichols, T. C., Jr., 1975. Deformations associated with relaxation of residual stresses in a sample of barre granite from Vermont, *U.S. Geol. Survey Prof. Paper* **875,** 32p.

Nichols, T. C., Jr., and Abel, J. F., Jr., 1975. Mobilized residual energy—a factor in rock deformation, *Assoc. Eng. Geologists Bull.* **12** (3), 213-225.

Nichols, T. C., Jr., and Savage, W. Z., 1976. Rock strain recovery—factor in foundation design, in *Rock Engineering for Foundations and Slopes: American Society of Civil Engineers Specialty Conference Proceedings,* Vol. 1. New York, 34-54.

Niles, W. H., 1871. Some interesting phenomena observed in quarrying, *Boston Soc. Nat. History Proc.* **14,** 80-87.

Nur, Amos, and Simmons, Gene, 1970. The origin of small cracks in igneous rocks, *Internat. Jour. Rock Mechanics and Mining Sci.* **7** (3), 307-314.

Rzehak, A., 1906. Bergschläge und Verwandte Erscheinungen, *Zeitschr. Prakt. Geologie* **14,** 345-351. Translation into French in *Soc. Belge Géologie, Paléontologie et Hydrologie Bull.* **21** (1907), 25-36.

Savage, W. Z., 1978. The development of residual stress in cooling rock bodies, *Geophys. Research Lett.* (Washington) **5,** 633-636.

Sbar, M. L., and Sykes, L. R., 1973. Contemporary compressive stress and seismicity in eastern North America: an example of intra-plate tectonics, *Geol. Soc. America* **84** (6), 1,861-1,881.

Strahan, A., 1887. On explosive slickensides, *Geol. Mag.* **4,** 400-408.

Swolfs, H. S.; Handin, J.; and Pratt, H. R., 1974. Field measurements of residual strain in granitic rock masses, in *Proceedings of the 3rd Congress of the International Society of Rock Mechanics,* Vol. 2-A. Washington, D.C.: National Academy of Sciences, 563-568.

Varnes, D. J., and Lee, F. T., 1972. Hypothesis of mobilization of residual stress in rock, *Geol. Soc. America Bull.* **83** (9), 2,863-2,865.

Voight, Barry, 1966. Restspannungen im Gestein, in *Proceedings of the 1st Congress of the International Society of Rock Mechanics,* Vol. 1. Lisboa, Portugal: Bertrand, 45-50.

Voight, Barry, 1974a. A mechanism for "locking in" orogenic stress, *Am. Jour. Sci.* **274** (6), 662-665.

Voight, Barry, 1974b. Stress history and rock stress, in *Proceedings of the 3rd Congress of the International Society of Rock Mechanics,* Vol. 2-A. Washington, D.C.: National Academy of Sciences, 580-582.

Cross-references: *Consolidation, Soil; Foundation Engineering; Marine Sediments, Geotechnical Properties; Rapid Excavation and Tunneling; Rheology, Soil and Rock; Rock Mechanics, Rocks, Engineering Properties; Rock Slope Engineering; Rock Structure Monitoring.*

REVERSE CIRCULATION DRILLING—
See Vol. XIV: BOREHOLE DRILLING.

RHEOLOGY, SOIL AND ROCK

According to what is commonly accepted as the first axiom of rheology (study of flow), everything flows; hence, soil and rock must and indeed do exhibit time-dependent deformations due to their self-weight and the application of external loads. As with virtually all materials, the type and rate of deformation is highly dependent on the nature of the material and the conditions to which it is subjected. To render a problem amenable to analysis, the principles of rheology are frequently used to describe the mechanical behavior of the materials included in the system. In general, soils are extremely complex multiphase (solid, liquid, and gas) materials whose mechanical behavior is influenced by a large variety of poorly understood and often overlooked factors, and rock masses are usually composed of blocks of intact rock separated by oriented families of joints that exhibit some degree of roughness. In many cases both soil and rock masses are *dilatant* (that is, shear deformations are accompanied by volume changes), and this is perhaps the most complicated aspect of their behavior.

The particulate nature of soil and rock masses covers a very wide scale, ranging from submicron clay particles to rock blocks with dimensions on the order of meters; materials of intermediate size include silt, sand, gravel, boulders, and disintegrated rock. Extremely large masses of intact rock without joints or fractures usually do not cause problems of engineering significance (except possibly drilling or cutting), and the deformations associated with such conditions are generally much smaller than those exhibited by jointed rock masses. Despite the manifested particulate nature of soil and rock masses, continuum theories are often used with reasonable success to describe their behavior; however, the results achieved from continuum theories become increasingly inadequate as the dimensions of the particulate constituents approach the overall dimensions of the problem. In the latter case, *discontinuum mechanics* or the mechanics of clastic (from the Greek *klastos,* meaning fragmentary) media provide an alternative approach; in principle, once the basic behavior of an assembly of particles has been established, this latter system can be expressed in terms of continuum mechanics and analogous methods of solution can be used.

The conduct and interpretation of laboratory and field tests offer a prime opportunity to apply the principles of rheology in soil and rock mechanics. Unfortunately, they also present one of the most challenging problems that confront the profession: such tests are usually conducted to determine the stress-strain-time and perhaps strength response of the material in question, but, due to the generally complicated boundary conditions that prevail during these tests and the resulting nonhomogeneous distributions of stress and strain that exist in the material being tested, a knowledge of the stress-strain-time behavior of the material is a *prerequisite* to interpret the tests properly. Essentially the same situation is faced in attempts to

determine the strength of a material; nonhomogeneous stress and strain distributions frequently cause progressive failure in the material and may lead to erroneous interpretations. As a consequence, virtually all conventional tests that are performed to characterize the mechanical properties of a soil or rock mass do not produce the type of fundamental relationships that are sought; rather, they provide indices, which, when used in conjunction with collateral analyses and tempered with experience and judgment, lead to acceptable designs.

The field applications of rheological principles and techniques in soil mechanics (q.v.) are manifold. The *theory of elasticity* has long been used to estimate stress distributions (especially vertical) in masses due to applied external loads, but it has not been very successful in predicting the resulting deformations. Although *visco-elastic theory* would likely provide an improved approximation (provided the material parameters are adequately known), the degree of improvement in many cases may not justify the effort involved. The classical *theory of consolidation* advanced in the early 1920s provides an excellent example of the application of a rheological (Kelvin-Voigt) model in soils engineering; this theory predicts the volumetric deformations that take place in a clay soil as a consequence of the pore water being ejected with time due to an imposed pressure gradient (see *Consolidation, Soil*). For certain soils under particular conditions, *creep* (time-dependent increase in shear deformation under constant load) plays a significant role in the deformation of slopes, settlement under a surface load, and the pressure exerted on retaining walls (Suklje, 1969; Krizek, 1971a, 1971b). On the other hand, *relaxation* effects (time-dependent decreases in load under constant deformation) may serve to decrease the pressure acting on a retaining wall, and they can explain (at least qualitatively) the relationship between the duration of an applied load and the degree of compaction obtained for a soil. In most problems of soil dynamics, the use of rheology is virtually indispensable, because energy is dissipated and damping coefficients must be determined for the materials in question.

Intact rock and jointed rock masses generally give the impression of rigidity, but in reality they do deform with time under an imposed load (which may be only their self-weight). Although these time-dependent deformations may be caused by the collapse of pore space or sliding on joint planes rather than any specific viscous property of the rock, the theory of visco-elasticity can nevertheless frequently be used to characterize the time-dependent flow in rocks and to conduct stress analyses of rock masses in terms of fundamental rheological principles. Based on the assumption that the flow or creep of rock under normal engineering stress levels (as opposed to tectonic stress levels) is not affected by the confining stress, but only the difference between horizontal and vertical stresses, uniaxial stress tests are generally performed. The creep rate usually depends on the magnitude of the applied stress, and in many practical cases the amount of creep that precedes failure is so large that the actual point of failure does not have major significance as a design factor. Under normal stress conditions, the primary creep mechanism is probably governed at first by slip along pre-existing planes of weakness in the rock and by brittle fracture and cracking inside the rock. Subsequent deformations, however, are likely to be caused by flow mechanisms such as twinning, translation gliding, recrystallization, and dislocation.

Despite its importance in many problems, the creep of rocks is still a poorly understood phenomenon. Some rocks, such as gabbros and granites, show little time-dependent strain, while in others, such as rock salt and other evaporites, the creep strains may greatly exceed the instantaneous elastic deformations; for example, steady-state creep has been measured over long periods of time in the roofs and pillars of underground mines. A knowledge of creep in rock is necessary to determine the closure rate of underground cavities in rock formations and to estimate the rate of movement of a rock slope. On a larger scale, nonlinear creep may help explain many geodynamic phenomena, such as the magnitude and frequency patterns in aftershock sequences of large earthquakes, the damping of seismic waves, the phase shifts in the tides of the solid earth, and the characteristics of certain postglacial uplift patterns.

RAYMOND J. KRIZEK

References

Krizek, R. J., 1971a, Rheologic behavior of clay soils subjected to dynamic loads, *Soc. Rheology Trans.* **15** (3), 433–489.
Krizek, R. J., 1971b, Rheologic behavior of cohesionless soils subjected to dynamic loads, *Soc. Rheology Trans.* **15** (3), 491–540.
Suklje, L., 1969, *Rheological Aspects of Soil Mechanics.* New York: Wiley-Interscience, 571p.

Cross-references: *Clays, Strength of; Foundation Engineering; Residual Stress, Rocks; Rock Mechanics; Soil Mechanics* Vol. XIV: *Engineering Soil Science.*

RIVER ENGINEERING

Rivers as Dynamic Systems

Frequently, environmentalists, river engineers, and others involved in river development, navigation, and flood control consider a river to be static, that is, unchanging in shape, dimension, and pattern. An alluvial river, however, generally changes its position and shape continually as a consequence of hydraulic forces acting on its bed and banks and related biological forces interacting with these physical forces. These changes may be slow or rapid and may result from natural environmental changes or from changes caused by human activities. When a river channel is modified locally, the change frequently causes modification of channel characteristics both up and down the stream. A river often responds to man-induced changes in spite of attempts to keep the anticipated response under control.

It must be understood that a river is dynamic, that a man-induced change frequently sets in motion a response that can be propagated for long distances, and that, in spite of their complexity, all alluvial rivers are governed by the same basic forces (see Vol. XIV: *Alluvial Systems Modeling*). Successful river engineering requires an understanding of these natural forces. It is absolutely necessary that river system design be based on a competent knowledge of: (1) geological factors, including soil conditions; (2) hydrologic factors, including possible changes in flows, runoff, and the hydrologic effects of changes in land use; (3) geometric characteristics of the stream, including the probable geometric alterations that will be activated by the changes that development will impose on the channel; (4) hydraulic characteristics such as depth, slope, and velocity of streams and the changes that may be expected in these characteristics in space and time; and (5) ecological and biological changes that will result from physical change and in turn will induce or modify physical changes.

Evidence of the Natural Instability of Rivers

Historical evidence clearly indicates the inherent dynamic qualities of river channels and demonstrates that rivers, glaciers, sand dunes, and seacoasts are highly susceptible to change with time. Over a relatively short period, perhaps in some cases as long as a human lifetime, components of the landscape may seem to be relatively stable. Nevertheless, stability cannot be automatically assumed. Rivers are, in fact, the most actively changing of all geomorphic forms.

Figure 1 compares a section of the Mississippi River near Commerce, Missouri, as it appeared in 1884 with the same section as observed in 1968. In the lower part of the reach, the surface area has been reduced approximately 50 percent during this 84-year period. Some of this change has been natural and some has been the consequence of river development work (Simons et al., 1975).

In alluvial river systems, it is the rule rather than the exception that banks will erode, sediments will be deposited and flood plains, islands, and side channels will undergo modification with time. Changes may be very slow or dramatically rapid.

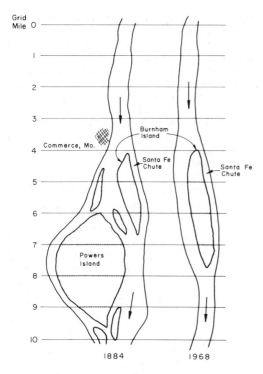

FIGURE 1. Comparison of the 1884 and 1968 Mississippi River channel near Commerce, Missouri (from Simons et al., 1975).

Fisk's (1944) report on the Mississippi River and his maps showing river position through time (Fig. 2) are sufficient to convince anyone of the innate instability of the lower Mississippi River.

The Mississippi is the largest and most impressive river in the United States, and because of its dimensions it has sometimes been considered unique. This is, of course, not so. Hydraulic and geomorphic laws apply at all scales of comparable landform evolution. The Mississippi may be thought of as a much larger than prototype model of many alluvial sandbed rivers.

A Survey of River Morphology and River Response

As previously stated, rivers are dynamic and respond to natural and man-induced change. The direction and extent of the change depends on the forces acting on the system. The major complicating factors in river mechanics are: (1) the large number of interrelated variables that can simultaneously respond to natural or imposed changes in a river system and (2) the continual evolution of river channel patterns, channel geometry, bars, and forms of bed roughness with changing water and sediment discharge. To provide a glimpse of the factors controlling the response of a river to the

actions of man and nature, a few simple hydraulic and geomorphic concepts are presented here.

Rivers can be classified broadly in terms of channel pattern, that is, the configuration of the river as viewed on a map or from the air. Patterns include straight, meandering, braided, or some combination of these (Fig. 3).

A *straight channel* can be defined as one that does not follow a sinuous course. Leopold and Wolman (1957) have pointed out that truly straight channels are rare in nature. Although a stream may have relatively straight banks, the *thalweg,* or path of greatest depths along the channel, is usually sinuous (Fig. 3b). As a result, there is no simple distinction between straight and meandering channels.

The *sinuosity* of a river—the ratio between thalweg length and down-valley distance—is most often used to distinguish between straight and meandering channels. Sinuosity varies from a value of 1 to a value of 3 or more. Leopold et al. (1964) took a sinuosity of 1.5 as the division between meandering and straight channels. It should be noted that in a straight reach with a sinuous thalweg developed between alternative bars (Fig. 3b), a sequence of shallow crossings and deep pools is established along the channel.

A *braided river* is generally wide with poorly defined and unstable banks, and is characterized by a steep, shallow course with multiple channel divisions around alluvial islands (Fig. 3a). Leopold and Wolman (1957) studied braiding in a laboratory flume. They concluded that braiding is one of many patterns that can maintain quasi-equilibrium among the variables of discharge, sediment load, and transporting ability. Lane (1957) concluded that, generally, the two primary factors that may be responsible for the braided condition are (1) overloading, that is, the stream may be supplied with more sediment than it can carry, resulting in deposition of part of the load; and (2) steep slopes, which produce a wide, shallow channel where bars and islands form readily.

Either of these factors alone, or both in concert, could be responsible for a braided pattern. If the channel is overloaded with sediment, deposition occurs, the bed aggrades, and the slope of the channel increases in an effort to maintain a graded condition. As the channel steepens, the velocity increases, and multiple channels develop and cause the overall channel system to widen. The multiple channels, which form when bars of sediment accumulate within the main channel, are generally unstable and change position with both time and stage.

Easily eroded banks can also cause braiding: the stream widens at high flow, and at low flow, bars form that become stabilized and form islands. In general, then, a braided channel has a steep

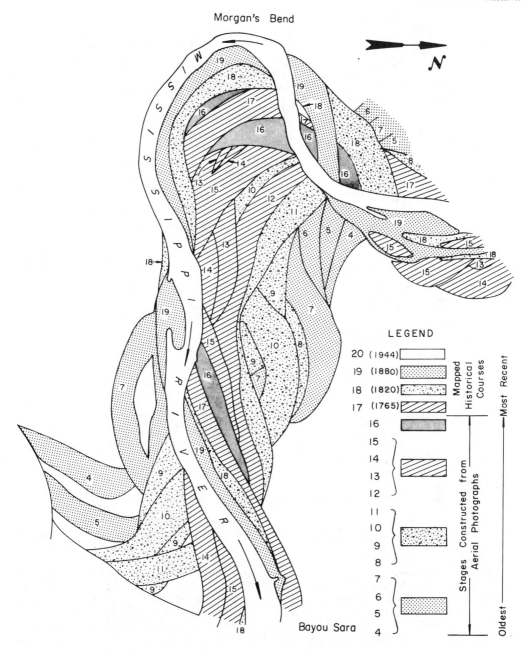

FIGURE 2. Channel changes, lower Mississippi River (after Fisk, 1944).

slope, a large bed-material load in comparison with its suspended load, and relatively small amounts of silt and clay in the bed and banks. The braided stream is difficult to manage in that it is unstable, changes its alignment rapidly, carries large quantities of sediment, is very wide and shallow even at flood flow, and is in general unpredictable.

A *meandering channel* consists of alternating bends, giving an S-shape appearance to the plan view of the river (Fig. 3c). More precisely, Lane (1957) concluded that a meandering stream is one whose channel alignment consists principally of pronounced bends, the shapes of which have not been determined predominantly by the varying nature of the terrain through which the channel passes. The meandering river consists of a series

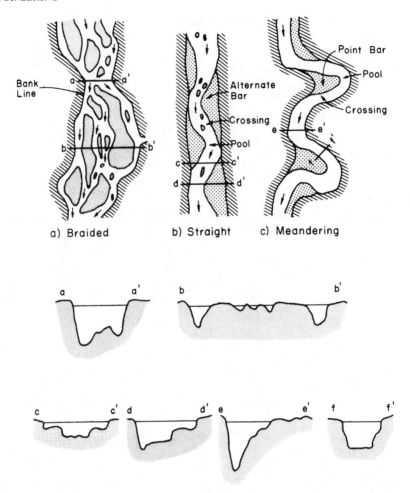

FIGURE 3. River channel patterns.

of deep pools in the bends and shallow crossings in the short, straight reach connecting the bends. The thalweg flows from a pool through a crossing to the next pool, forming the typical S-curve of a single meander loop. Figure 4 shows the classic Greenville meanders on the lower Mississippi River.

As shown schematically in Figure 3, the pools tend to be somewhat triangular in section with point bars located on the inside of the bend. In the crossing the channel tends to be more rectangular, widths are greater, and depths are relatively shallow. At low flows the local slope is steeper and velocities are larger on the crossing than in the pool. At low stages the thalweg is located very close to the outside of the bend. At higher stages, the thalweg tends to straighten. More specifically, the thalweg moves away from the outside of the bend and encroaches on the point bar to some degree. In the extreme case, the shifting of the current can cause chute channels to develop across the point bar at high stages.

Rivers, then, are broadly classified as straight, meandering, or braided or some combination of these classifications, but any changes that are imposed on a river may change its form (Khan, 1971). The dependence of river form on the slope of the channel bend, which may be imposed independent of the other river characteristics, is illustrated in Figure 5. By changing the slope, it is possible to change the river from a meandering character that is relatively tranquil and easy to control to a braided river that varies rapidly with time, has high velocities, is subdivided by sand bars and carries relatively large quantities of sediment. Such a change could be caused by a natural or artificial cutoff. Conversely, it is possible that a slight decrease in slope could change an unstable braided river into a more stable meandering one.

Important changes in river morphology are caused by modification of *discharge* and *sediment load*. One of the first to consider this important problem

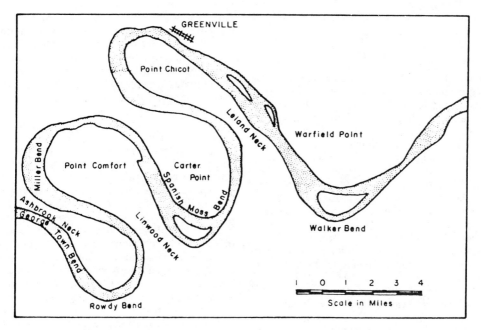

FIGURE 4. Meanders in Mississippi River near Greenville, Mississippi.

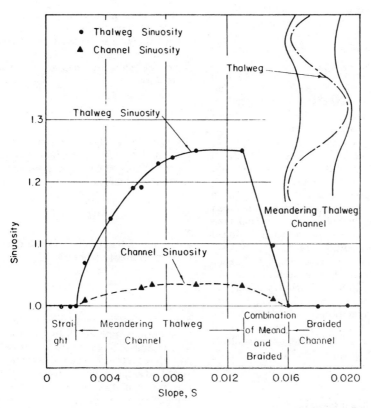

FIGURE 5. Sinuosity versus slope for a constant discharge of 0.15 ft³/s (from Khan, 1971).

was Lane (1955), who presented a relationship of the form

$$QSaQ_sD_{50} \qquad (1)$$

in which Q is the water discharge, S is the slope of the channel bed, Q_s is the bed-material discharge, and D_{50} is a measure of the size of the channel bed material. This relation is both simple and useful. For example, if a dam is constructed across a river and traps bed-material sediments moving through the system, clear water is discharged immediately downstream, indicating a decrease in Q_s. Looking at *Lane's relation* for the river downstream of the dam, the S must reduce on the left side of the equation to balance the decrease in Q_s on the right side, assuming constant Q and D_{50}. This implies degradation downstream of the structure. Lane's relation is very useful for qualitative analyses of stream response to both climatic and man-induced changes.

The significantly different channel dimensions, shapes, and patterns associated with different quantities of discharge and amounts of sediment load indicate that as these independent variables change, major adjustments of channel morphology can be anticipated. Further, if changes in sinuosity and meander wavelength or in width and depth are required to compensate for a hydrologic change, then a long period of channel instability can be envisioned, with considerable bank erosion and lateral shifting of the channel before stability is restored.

Changes in sediment and water discharge at a particular point or reach in a stream may have an effect ranging from some distance upstream to a point downstream where the hydraulic and geomorphic conditions can absorb the change. Thus it is necessary to consider a channel reach as part of a complete drainage system. Artificial controls that could benefit the reach may, in fact, cause problems in the system as a whole. For example, flood control structures can cause downstream flood damage to be greater at reduced flows if the average hydrologic regime is changed so that the channel dimensions are actually reduced (see *Alluvial Plains, Engineering Geology*). Also, where major tributaries exert a significant influence on the main channel by introducing large quantities of sediment, upstream control on the main channel may allow the tributary to intermittently dominate the system, with deleterious results. If discharges in the main channel are reduced, sediments from the tributary that previously were eroded will no longer be carried away and serious aggradation with accompanying flood and navigation problems may arise.

Continuum of Channel Patterns

Because of the physical characteristics of straight, braided, and meandering streams, all natural channel

patterns intergrade. Although braiding and meandering patterns are strikingly different, they actually represent extremes in a continuum of channel patterns. On the assumption that the pattern of a stream is determined by the interaction of numerous variables whose range in nature is continuous, one should not be surprised by the existence of a complete range of channel patterns. A given reach of a river, then, may exhibit both braiding and meandering, and alteration of the controlling parameters in a reach can change the character of a given stream from meandering to braided or vice versa.

A number of studies have quantified this concept of a continuum of channel patterns; Khan (1971), for example, related sinuosity, slope, and channel pattern (Fig. 5). Any natural or artificial change that alters channel slope, such as the cutoff of a meander loop, can result in modifications to the existing river pattern. A cutoff in a meandering channel shortens channel length, increases slope, and tends to move the plotting position of the river to the right on Figure 5. This indicates a tendency to evolve from a relatively tranquil, easy-to-control meandering pattern to a braided pattern that varies rapidly with time, has high velocities, is subdivided by sand bars, and carries relatively large quantities of sediment. Conversely, a slight decrease in slope could change an unstable braided river into a more stable meandering pattern.

Lane (1957) investigated the relationship among slope, discharge, and channel pattern in meandering and braided streams, and observed that an equation of the form

$$SQ^{1/4} = K \qquad (2)$$

fits a large amount of data from meandering sand streams. Here, S is the channel slope, Q is the water discharge, and K is a constant. Figure 6 summarizes Lane's plots and shows that when

$$SQ^{1/4} \leq 0.0017 \qquad (3)$$

a sandbed channel will tend toward a meandering pattern. Similarly, when

$$SQ^{1/4} \geq 0.01 \qquad (4)$$

a river tends toward a braided pattern. Slopes for these two extremes differ by a factor of almost 6. The region between these values of SQ can be considered a transitional range where streams are classified as intermediate. (Many U.S. rivers fall in this intermediate category.) If the discharge and slope of a meandering river border on transitional, a relatively small increase in channel slope could initiate a tendency toward a transitional or braided character.

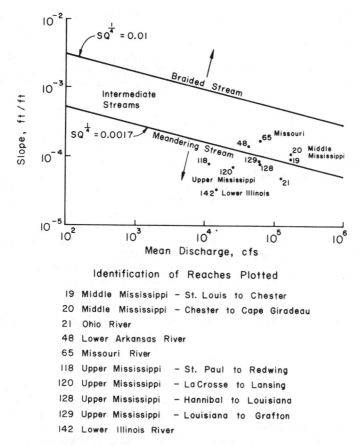

FIGURE 6. Slope-discharge relation for braiding or meandering in sandbed streams (from Lane, 1957).

Identification of Reaches Plotted

19 Middle Mississippi – St. Louis to Chester
20 Middle Mississippi – Chester to Cape Giradeau
21 Ohio River
48 Lower Arkansas River
65 Missouri River
118 Upper Mississippi – St. Paul to Redwing
120 Upper Mississippi – La Crosse to Lansing
128 Upper Mississippi – Hannibal to Louisiana
129 Upper Mississippi – Louisiana to Grafton
142 Lower Illinois River

Flow in Alluvial Channels

Most streams flow on sandbeds for the greater part of their length and nearly all large rivers have sandbeds. In sandbed rivers, the bed material is easily eroded and is continually being moved and shaped by the flow. The interaction between the flow of the water-sediment mixture and the sandbed creates different bed configurations which change the resistance to flow and rate of sediment transport. The gross measures of channel flow such as the flow depth, river stage, bed elevation, and flow velocity change with different bed configurations. In the extreme case, the change in bed configuration can cause a threefold change in resistance to flow and a 10- to 15-fold change in intensity of bed-material transport. For a given discharge and channel width, a 3-fold increase in a roughness coefficient such as Manning's n results in a doubling of the flow depth.

The interaction between the flow and bed material makes the analysis of flow in alluvial sandbed streams extremely complex. However, with an understanding of the different types of bedforms that may occur and a knowledge of the resistance

to flow and sediment transport associated with each bedform, alluvial channel flow can be analyzed.

Bedforms and Resistance. The bed of an alluvial river seldom forms a smooth, regular boundary; rather, it is characterized by shifting forms that vary in size, shape, and location under the influence of changes in flow, temperature, sediment load, and other variables. These bedforms constitute a major part of the resistance to flow exhibited by an alluvial channel, and exert a significant influence on flow parameters such as depth, velocity, and sediment transport. While the detailed mechanics of the interrelationships involved are essentially unknown, it is recognized that variation in bedforms permits an internal adjustment of a channel to accommodate relatively large changes in discharge, sediment load, and other variables without requiring a corresponding change in other channel boundary conditions (Simons and Richardson, 1966).

The bedforms that may occur in an alluvial channel are plane bed without sediment movement, ripples, ripples on dunes, dunes, plane bed with sediment movement, antidunes, and chutes and pools. These bend configurations are listed as they occur sequentially with increasing values of stream power (τV or $\gamma y S V$) for bed material with a D_{50} less than 0.6 mm. For coarser bed material, dunes form instead of ripples after the beginning of motion. Here τ is the bed shear stress, V is the flow velocity, γ represents the specific weight of the fluid, and S is the slope of the channel bed.

The different forms of *bed roughness* are not mutually exclusive in time and space in a stream. Bed roughness elements may form side by side in a cross section or reach of a natural stream, giving a multiple roughness; or they may form in time sequence, producing variable roughness.

Multiple roughness is related to variations in shear stress ($\gamma y S$) and stream power ($V\gamma y S$) in a channel cross-section. The greater the width-depth ratio of a stream, the greater the probability of a spatial variation in shear stress, stream power, or bed material. Thus, the occurrence of multiple roughness is closely related to the width-depth ratio of the stream.

Bed Configuration Without Sediment Movement.
If the bed material of a stream moves at one discharge but not at a smaller discharge, the bed configuration at the smaller discharge will be a remnant of the bed configuration formed when sediment was moving. Prior to the beginning of motion, the problem of resistance to flow is one of rigid-boundary hydraulics. After motion begins, the problem relates to defining bed configuration and resistance to flow.

Plane bed without movement has been studied to determine the flow conditions for the beginning of motion and the bed configuration that would form after motion begins (Gessler, 1971). In general, *Shields's relation,* Figure 7, for the beginning of motion is adequate. After the beginning of motion, for flat slopes and low velocity, sand material smaller than 0.6 mm in the plane bed will change to ripples, and coarser material will form dunes. Resistance to flow, which is due solely to sand grain roughness, is small for a plane bed

without sediment movement. Values of Manning's n range from 0.012 to 0.014, depending on the size of the bed material.

Bed Configuration with Sediment Movement.
Typical bedforms and their relationship to the water surface (in phase or out of phase) are shown in Figures 8 and 9. With these bedforms as basic criteria, flow in alluvial channels is divided into two regimes of flow separated by a transition zone. These two flow regimes are characterized by similarities in the shape of the bedform, the mode of sediment transport, the process of energy dissipation, and the phase relation between the bed and water surface (Simons and Richardson, 1963). These two regimes and their associated bedforms are the following:

1. lower flow regime—small stream power
   a. ripples
   b. ripples superposed on dunes
   c. dunes
2. transition zone
3. upper flow regime—large stream power
   a. plane bed (with sediment movement)
   b. antidunes
   c. chutes and pools

In the *lower flow regime,* resistance to flow is high and sediment transport is small. The water surface undulations are out of phase with the ripples or dunes that constitute the bed, and there is a relatively large separation zone downstream from the crest of each ripple or dune. The resistance to flow is primarily form roughness. The most common way that bed material is transported is for individual grains to move up the back of a ripple or dune and then avalanche down its face. After coming to rest on the downstream face of the ripple or dune, the particles are buried and remain at rest until the downstream movement of the dune exposes them again.

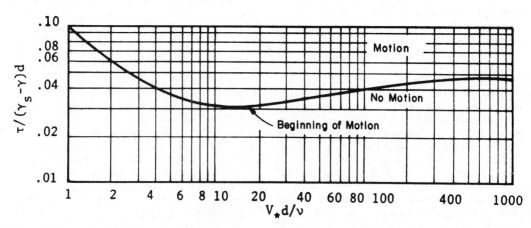

FIGURE 7. Shield's relation for beginning of motion (after Gessler, 1971).

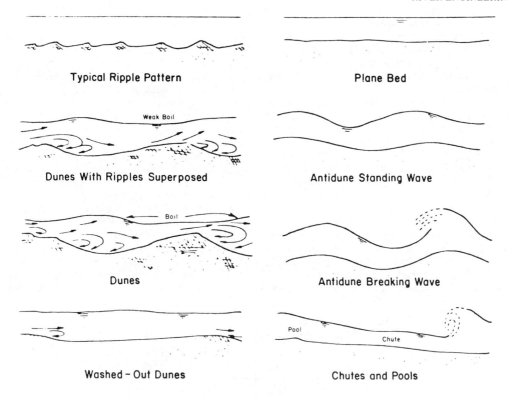

FIGURE 8. Forms of bed roughness in sand channels (from Simons and Richardson, 1966).

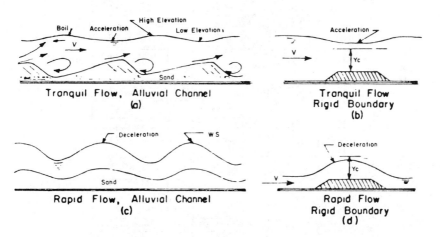

FIGURE 9. Relationship between water surface and bed configuration (from Simons and Richardson, 1966).

In the *upper flow regime,* resistance to flow is low and sediment transport is large. The most common bedforms are plane beds or antidunes, and the water surface is in phase with the bed surface except when an antidune breaks. Normally, there is little separation of the fluid from the bed surface. Resistance to flow is primarily the result of grain roughness; however, wave formation and

subsidence and energy dissipation when antidunes break also contribute to resistance. The dominant mode of sediment transport is continuous rolling of individual grains downstream in sheets several grain-diameters thick. *Antidunes* are so named because under certain conditions they can move upstream against the flow. They form as trains of waves that gradually build up from a plane bed and plane water surface, and may break like surf as they become unstable. As antidunes break, large quantities of bed material can be briefly suspended, stopping momentarily the continuous motion of sediment particles associated with upper-regime flow.

In the *transition zone* the bed configuration is erratic, ranging between conditions of lower- and upper-regime flow, as dictated primarily by antecedent conditions. Resistance to flow and sediment transport also exhibit the same variability as bedforms in the transition zone. In many instances of transition flow, the bed configuration oscillates between dunes and plane bed.

Simons and Richardson (1966) developed a graphical relation among stream power (τV), median fall diameter, and bedform using both flume and stream data. This relation (Fig. 10) gives an indication of the form of bed roughness to be anticipated if the depth, slope, velocity, and fall diameter of the bed material are known. Another useful graphical

relation (Fig. 11) shows schematically the effect of bedform on a roughness coefficient such as Manning's n. As the bed configuration makes its transition from the lower regime to the upper regime, Manning's n changes from a typical value of 0.012 to 0.014 for a plane bed without sediment motion to values as high as 0.04 for a dune bed. Increasing stream power and transition to upper-regime plane-bed conditions can produce a decrease in roughness to values as low as 0.010 to 0.015. The consequent effect on flow velocity can be seen in Figure 12. These data pertain to a single sand size ($D_{50} = 0.19$ mm) and were determined in the 8-ft flume at Colorado State University.

In a natural stream it is possible to experience a large increase in discharge with little or no change in stage as a result of a shift in bed configuration from dunes to plane bed. Figure 13 shows a typical break in the depth-discharge relation resulting from this phenomenon. Conversely, it has been shown that an increase in depth, with constant

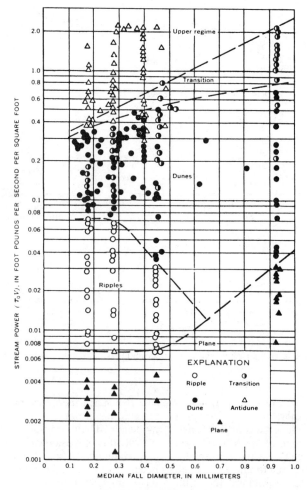

FIGURE 10. Relation of stream power and median fall diameter to form of bed roughness (Simons and Richardson, 1966).

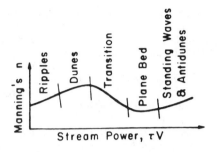

FIGURE 11. Resistance to flow versus bed-form (from Karaki et al., 1974).

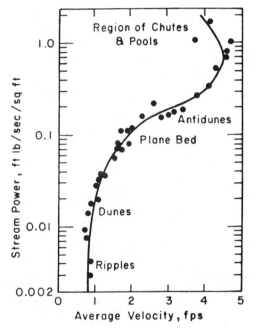

FIGURE 12. Relation of bedform to stream power and velocity (from Simons and Richardson, 1971).

slope and bed material, can change a dune bed to plane bed or antidunes, and that a decrease in depth can reverse the process.

Bars in Alluvial Channels. In natural or field-size channels, other bed configurations are also found. These bed configurations are generally called bars and are related to the plan-form geometry and the width of the channel. Bars are bedforms having lengths of the same order as the channel width or greater, and heights comparable to the mean depth of the generating flow.

Point bars occur adjacent to the convex bank of channel bends. Their shape may vary with changing flow conditions, but point bars do not move relative to the bends. *Alternate bars* occur in straighter reaches of channels and tend to be distributed periodically along the reach, with consecutive bars on opposite sides of the channel. Their lateral extent is significantly smaller than the channel width. Alternate bars move slowly downstream. *Transverse bars* (middle bars) also occur in straight channels and occupy nearly the full channel width. They occur both as isolated and as periodic forms along a channel, and move slowly downstream. *Tributary bars* occur immediately downstream from points of lateral inflow into a channel.

In longitudinal section, bars are approximately triangular, with very long, gentle upstream slopes and short downstream slopes that are approximately the same as the angle of repose. Bars appear as small barren islands during low flows. Portions of the upstream slopes of bars are often covered with ripples or dunes.

Manning's *n* Values for Natural Sandbed Streams. Observations on natural sandbed streams with bed material having a median diameter ranging from 0.1 to 0.4 mm indicate that the bed planes out and that resistance to flow decreases whenever high flow occurs. Manning's *n* changes from values as large as 0.040 at low flow to as small as 0.012 at

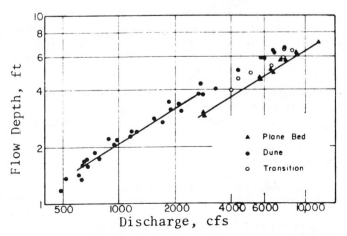

FIGURE 13. Relation of depth to discharge for Elkhorn River near Waterloo, Nebraska (from Simons and Richardson, 1971).

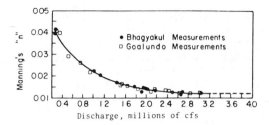

FIGURE 14. Change in Manning's *n* with discharge for Padma River in Bangladesh (from Karaki et al., 1974).

TABLE 1. Ranges in Manning's *n* for Various Sandbed Configurations under Different Flow Regimes

Lower Flow Regime	Upper Flow Regime
Ripples $(0.018 \le n \le 0.028)$	Plane bed $(0.010 \le n \le 0.013)$
Dunes	Antidunes
$(0.020 \le n \le 0.040)$	Standing waves $(0.010 \le n \le 0.015)$
	Breaking waves $(0.012 \le n \le 0.020)$
	Chutes and pools $(0.018 \le n \le 0.035)$

high flow. An example is given in Figure 14.

The range in Manning's *n* for the various bed configurations is given in Table 1.

Qualitative Response of River Systems

Many rivers have achieved a state of approximate equilibrium throughout long reaches. For practical engineering purposes, these reaches can be considered stable and are known as *graded streams* by geologists and as *poised streams* by engineers. This does not, however, preclude significant changes over a short period of time or over a period of years. Conversely, many streams contain long reaches that are actively aggrading or degrading.

Regardless of the degree of channel stability, man's local activities may produce major changes in river characteristics both locally and throughout an entire reach. All too frequently the net result of a river improvement is a greater departure from equilibrium than that which originally prevailed (see *Channelization and Bank Stabilization*). Good engineering design must invariably seek to enhance the natural tendency of the stream toward poised conditions. To do so, an understanding of the direction and magnitude of change in channel characteristics caused by the actions of man and nature is required. This understanding can be obtained by: studying the river in a natural condition, having knowledge of the sediment and water discharge, being able to predict the effects and magnitude of man's future activities, and applying to these a knowledge of the geology, soils, hydrology, and hydraulics of alluvial rivers.

Predicting the response to channel development is a very complex task. A great many variables are involved in the analysis, which are interrelated and can respond to changes in a river system and in the continual evolution of river form. The channel geometry, bars, and forms of bed roughness all change with changing water and sediment discharges. Because such a prediction is necessary, useful methods have been developed to predict both qualitative and quantitative responses of channel systems to change.

Prediction of General River Response to Change. Response can be quantitatively predicted if all the required data are known with sufficient accuracy. Usually, however, the data are not sufficient for quantitative estimates, and only qualitative estimates are possible.

The response of channel pattern and longitudinal gradient to variation in selected parameters has been discussed in previous sections. In more general terms, Lane (1955) studied the changes in river morphology in response to varying water and sediment discharge. Similarly, Leopold and Maddock (1953) and Schumm (1971) have investigated channel response to natural and imposed changes. These studies support the following general relationships:

(1) Depth of flow (y) is directly proportional to water discharge (Q) and inversely proportional to sediment discharge (Q_s).
(2) Channel width (W) is directly proportional to both water discharge (Q) and sediment discharge (Q_s).
(3) Channel shape, expressed as a width-to-depth (W/y) ratio, is directly related to sediment discharge (Q_s).
(4) Channel slope (S) is inversely proportional to water discharge (Q) and directly proportional to both sediment discharge (Q_s) and grain size (D_{50}).
(5) Sinuosity (s) is directly proportional to valley slope and inversely proportional to sediment discharge (Q_s).
(6) Transport of bed material (Q_s) is directly related to stream power (τV) and concentration of fine material (C_F), and inversely related to the fall diameter of the bed material (D_{50}).

A very useful relation for predicting system response can be developed by establishing a proportionality between bed-material transport and several related parameters:

$$QS \alpha Q_s D_{50}$$

which was introduced previously as equation (1). Equation (1) is essentially the relation proposed by Lane (1955), except that fall diameter, which includes the effect of temperature on transport, has been substituted for the physical median diameter used by Lane.

Applications of Qualitative Analysis

Equation (1) is most useful for qualitative prediction of channel response to natural or imposed changes in a river system. To use the classic example alluded to previously, consider the downstream response of a river to the construction of a dam (Fig. 15). Aggradation in the reservoir upstream of the dam will result in relatively clear water being released downstream of the dam, that is, Q_s will be reduced to Q_s^- downstream. Assuming that fall diameter and water discharge remain constant, slope must decrease downstream of the dam to balance the proportionality of equation (1).

$$Q_s D_{50}^o \, a \, Q^o S$$

In Figure 15 the original channel gradient between the dam and a downstream geological control (line CA) will be reduced to a new gradient (line C'A) through gradual degradation below the dam. With time, of course, the pool behind the dam will fill and thus sediment would again be available to the downstream reach. Then, except for local scour, the gradient C'A would increase to the original gradient CA to transport the increase in sediment load. Upstream, the gradient would eventually parallel the original gradient, offset by the height of the dam. Thus dams with small storage capacity may induce scour and then deposition over a relatively short period.

As another example, consider a tributary that is relatively small but carries a large sediment load entering the main river at point C (Fig. 16). This increases the sediment discharge in the main stream from Q_s to Q_s^+. It is seen from equation (1) that for a significant increase in sediment discharge Q_s^+ the channel gradient S below C must increase if Q and D_{50} remain constant. The line CA (indicating the original channel gradient), therefore, changes with time to position C'A. Upstream of the confluence the slope will adjust over a long period of time to the original channel slope. The river bed will aggrade from C to C'.

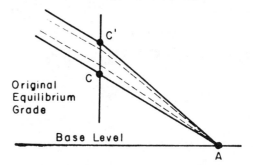

Final Equilibrium Grade

FIGURE 16. Changes in channel slope in response to an increase in sediment load at point C.

The river engineer is interested in quantitative results in addition to qualitative indications of trends. The geomorphic relation $QS \, a \, Q_s D_{50}$ is only an initial step in analyzing long-term channel response problems. This initial step is useful, however, because it warns of possible future difficulties in designing channel improvement and flood protection works and provides a good first-order estimate of response to development.

To summarize the qualitative approach to the analysis of river response, consider the problem of designing a river system to meet the conflicting demands of flood control and navigation. To minimize both flood damage and the height of levees required to provide flood protection, it is desirable to reduce the stage associated with a given flood discharge. Conversely, to minimize navigation channel dredging requirements it may be desirable to increase the stage associated with a given low flow. A qualitative analysis can point the way toward satisfying these apparently conflicting requirements.

Figure 17a illustrates the low-stage and high-stage thalwegs in a typical meandering river system. The low-stage thalweg (indicated by ⓛ) generally follows the outside or concave bank of the meander

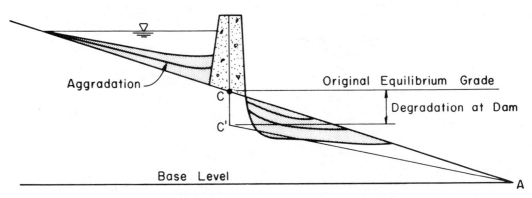

FIGURE 15. Channel adjustment above and below a dam.

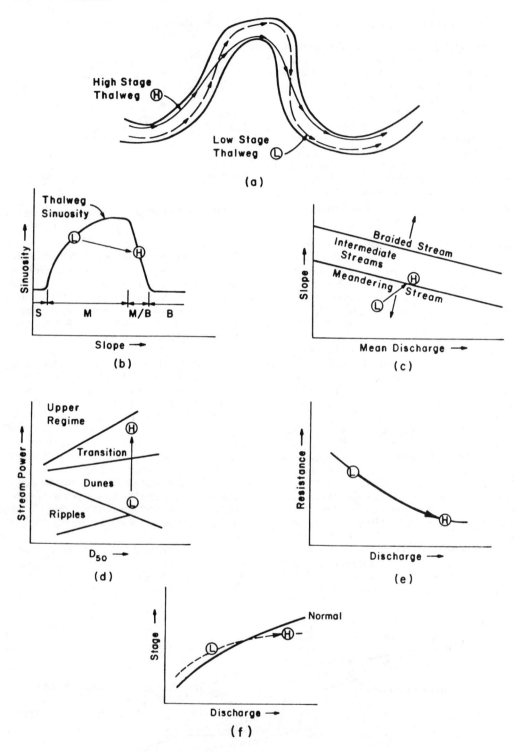

FIGURE 17. River system design.

bends, thus traversing a relatively long flow path through the reach. Because of increased velocity and momentum, the high-stage flow line (indicated by Ⓗ) generally short-circuits the meander pattern, cutting across the tips of the point bars on the inside or convex bank of the meander bends, and following a shorter flow path through the reach. This reduced length of flow path for high-stage flows implies an increase in slope through the reach with effects similar to those associated with artificial cutoff of meander loops. The low-stage and high-stage characteristics of a river system designed to satisfy both flood protection and navigation requirements must be held within certain geomorphic and hydraulic limits. A qualitative analysis can assist in establishing these limits.

For example, the sinuosity of the low-flow river (plotted as Ⓛ on Figure 17b) must be selected to ensure that the increase in slope associated with high-stage flows does not plot in the braided region (Ⓗ in Figure 17b). Note that Figure 17b is a schematic presentation of Figure 5. The limiting slope can be established on Figure 17c (compare with Fig. 6), where Ⓛ marks the desired plotting position well in the meander region under low-flow conditions and Ⓗ shows the upper limit under high-flow conditions. A plotting position located near the intermediate zone would be considered acceptable; however, the high-flow river should plot well below the braided region.

As indicated in Figure 10, the flow regime of an alluvial channel is closely related to stream power (τV) and, in turn, resistance to flow is a function of the type of bedform in the channel (Figs. 11 and 14). Figure 10 is drawn schematically as Figure 17d. To provide a high resistance to flow and thus increase the stage associated with a given low-flow discharge, the low-flow river should plot within the dunes region of Figure 17d. The high-flow river should plot in the transition region but below upper-regime conditions, thus assuring a decreased resistance to flow and lower stages. The direct effect of controlling flow regime on the resistance coefficient of an alluvial river can be seen in Figure 17e (compare with Fig. 14).

The end result of river system design to meet the conflicting demands of flood control and navigation is shown in Figure 17f. Here the normal stage-discharge relation (with constant resistance) is compared with a stage-discharge plot for controlled low-flow and high-flow conditions. The low-flow stage-discharge relation Ⓛ has been achieved by controlling the flow regime to ensure high resistance to flow and thus increased stages for a given low-flow discharge. The reduced stage for a given flood flow Ⓗ was obtained by increasing slope, velocity, and stream power to ensure transition flow, decreased resistance, and decreased depth of flow. A careful control of slope and sinuosity was required to prevent a change of river form

from the relatively easy-to-control meandering pattern to the less stable braided configuration. It must be emphasized that the plots used in a qualitative analysis such as this are applicable only to a specific river or river system. These plots must be developed from data derived from the particular river system of concern; qualitative relations used to supplement such an analysis must be checked with field data from the river in question.

PETER L. LAGASSE

References

Fisk, H. N., 1944. *Geological Investigation of the Alluvial Valley of the Lower Mississippi River.* Vicksburg, Miss: Mississippi River Commission, 78p.

Gessler, J., 1971. Beginning and ceasing of sediment motion, in H. W. Shen (ed.), *River Mechanics,* Vol. I. Fort Collins, Colo.: Water Resources Publishing, 220p.

Karaki, S.; Mahmood, K.; Richardson, E. V.; Simons, D. B.; and Stevens, M. A., 1974. *Highways in the River Environment: Hydraulic and Environmental Design Considerations,* Report prepared for the Federal Highway Administration, Colorado State University, Fort Collins, Colorado, 384p.

Khan, H. R., 1971. Laboratory study of alluvial river morphology, unpublished Ph.D. dissertation, Colorado State University, Fort Collins.

Lane, E. W., 1955. The importance of fluvial morphology in hydraulic engineering, *Am. Soc. Civil Engineers Proc.* **81** (745), 17p.

Lane, E. W., 1957. *A Study of the Shape of Channels Formed by Natural Streams Flowing in Erodible Material,* Missouri River Division Sediment Series No. 9, Omaha, Neb.: U.S. Army Corps of Engineers, Engineering Division, Missouri River, 106p.

Leopold, L. B., and Maddock, T., Jr., 1953. The hydraulic geometry of stream channels and some physiographic implications, *U. S. Geol. Survey Prof. Paper 252,* 57p.

Leopold, L. B., and Wolman, M. G., 1957. River channel patterns: braided, meandering, and straight, *U. S. Geol. Survey Prof. Paper 282-B,* 85p.

Leopold, L. B.; Wolman, M. G.; and Miller, J. P., 1964. *Fluvial Processes in Geomorphology.* San Francisco: W. H. Freeman, 522p.

Shumm, S. A., 1971. Fluvial geomorphology—the historical perspective, in H. W. Shen (ed.), *River Mechanics,* Vol. I. Fort Collins, Colo.: Water Resources Publishing.

Simons, D. B., and Richardson, E. V., 1963. A study of variables affecting flow characteristics and sediment transport in alluvial channels, in *Proceedings of the Federal Inter-Agency Sedimentation Conference,* 193-206.

Simons, D. B., and Richardson, E. V., 1966. Resistance to Flow in Alluvial Channels, *U. S. Geol. Survey Prof. Paper 422-J,* 61p.

Simons, D. B., and Richardson, E. V., 1971. Flow in alluvial sand channels, in H. W. Shen (ed.), *River Mechanics,* Vol. I. Fort Collins, Colo.: Water Resources Publishing.

Simons, D. B.; Lagasse, P. F.; Chen, Y. H.; and Schumm, S. A., 1975. *The River Environment.* Reference document prepared for U.S. Department of the Interior, Fish and Wildlife Service, Colorado State University, Fort Collins, Colorado, 520p.

Cross-references: *Alluvial Plains, Engineering Geology; Channelization and Bank Stabilization; Coastal Inlets, Engineering Geology; Geomorphology, Applied; Hydrology.* Vol. XIV: *Alluvial Systems Modeling; Canals and Waterways; Environmental Management.*

ROCK, ROCK CHIP SAMPLING — See Vol. XIV: LITHOGEOCHEMICAL PROSPECTING.

ROCK CHIP GEOCHEMISTRY— See Vol. XIV: LITHOGEOCHEMICAL PROSPECTING.

ROCK CLASSIFICATION— See SHALE MATERIALS, ENGINEERING CLASSIFICATION. Vol. XIV: ROCK WEATHERING, ENGINEERING CLASSIFICATION.

ROCK-COLOR CHART— See Vol. XIV.

ROCK DRILLING— See RAPID EXCAVATION AND TUNNELING; SHAFT SINKING. Vol. XIV: BOREHOLE DRILLING.

ROCK ENGINEERING— See ROCK MECHANICS; ROCKS, ENGINEERING PROPERTIES; ROCK SLOPE ENGINEERING; ROCK STRUCTURE MONITORING; SHAFT SINKING; TUNNELS, TUNNELING. Vol. XIV: BLASTING AND RELATED TECHNOLOGY; BOREHOLE DRILLING.

ROCK MECHANICS

Rock mechanics is the theoretical and applied science of the mechanical behavior of rock; it is that branch of mechanics concerned with the response of rock to force fields imposed by the physical environment. This broad definition suggests the manifold applications within rock mechanics. Scientific consumers—including structural geologists, tectonophysicists, and seismologists—focus on interpreting the chaos of deformed rocks. Engineers work to find better ways of breaking rocks or keeping them from breaking. They deal with a host of engineering problems such as surface

foundations (see *Foundation Engineering*), excavations, slope stability (see Vol. XIV: *Slope Stability Analysis*), underground openings (see *Cavity Utilization*) and boreholes (see Vol. XIV: *Borehole Drilling*), construction materials, comminution by drilling, blasting (see Vol. XIV: *Blasting and Related Technology*), crushing, and subsidence (see Vol. XIV: *Mine Subsidence Control*).

Paleolithic man must have used rock mechanics concepts when he considered the stability of the rock roof over his head or ways to optimize the efficiency of his tools. But rock mechanics really is a modern frontier. One early conceptual formulation appears in Coulomb's discussion of earth pressure in 1773. As recently as 1936, the physics Nobel laureate P. W. Bridgman noted, that geology seems to be approaching a point where an ultimate problem can no longer be avoided namely to determine the physical and mechanical behavior of materials that constitute the earth's crust. Because rock mechanics spans so many important fields, we delimit this discussion to rock mechanics in earth science, and direct attention toward experimental work and attendant tectonophysical applications.

Experimental rock mechanics serves as an adjunct to and not a replacement for field studies. The earth itself is a full-scale rock-deformation laboratory, following the concepts of the new global tectonics, in which cold lithospheric plates descend into the high-pressure, high-temperature region of the mantle at continental margins. Field study therefore evinces kinematic testimony for modes of deformation, but the origin of forces ultimately responsible for surficial deformation (see *Rheology, Soil and Rock*) and states of stress in the crust remains a mystery (see *Residual Stress, Rocks*). Also, except for information from elastic waves generated by natural or artificial seismic sources, little is known about the mechanical behavior of the deeper earth. The outermost skin of the earth can be sampled and observed, but study of deep processes is like attempting to examine the properties of the upper atmosphere before the advent of balloons and satellites.

The Domain of Rock Mechanics

Modern laboratory devices simulate for short times almost all deep earth environments from crust to core. On the other hand, the deepest drill holes penetrate only 9 km of the earth's 6,400-km radius, a little more than 0.1 percent, and 99 percent may never see a drill bit. Deep mines do not exceed 5 km.

Pressure increases about 0.3 kb/km (1 bar = 10^6 dynes/cm^2 = 14.5 psi), from about 30 kb at a depth of 100 km to almost 3,570 kb at the center of the earth. For comparison, the pressures in the deepest ocean trenches approach 1 kb, the impact of a rifle bullet on armor plate produces 100 kb, and the pressure at the center of the sun nears 100 million kb.

The increase of temperature with depth is poorly known, with uncertainties of several hundred degrees at depths as shallow as 300 km. Crustal gradients average about 30° C/km, but this rate must decrease rapidly away from the surface. Gradients also differ according to position in lithospheric plates, being steeper near spreading centers.

Time is the most difficult variable to estimate in rock mechanics. The earth is about 10^{17} seconds old. Figure 1 shows that relevant strain rates span 25 orders of magnitude from rapid shock rates to exceedingly slow geological rates on the other end of the spectrum. Strain rate is defined as

$$d\epsilon/dt = \frac{d\left(\dfrac{L}{L_0}\right)}{dt} = \dot{\epsilon}$$

where L is length and t is time.

Most laboratory tests are performed at rates between 10^{-1}/s and 10^{-8}/s, although creep apparatus permits tests as slow as 10^{-10}/s. That's still 10,000 times faster than geological rates of 10^{-14}/s frequently derived from warping of Pleistocene Lake Bonneville or calculated from distortion of triangular grids crossing active plate-boundary faults. Rock mechanics experiments, therefore, most often deal with two strain-rate regions: the dynamic realm increasing from 10^2/s and the static range decreasing from 10^{-4}/s. Little is known about effects on rocks at intermediate rates, or about behavior at the slowest rates of ultimate geological interest. In summary, then, rock mechanics experiments cover pressures to thousands of kilobars. Temperatures to thousands of degrees Centigrade, and strain rates over a range of almost 10^{25}.

Apparatus

The types of high-pressure "vise" used to simulate deep earth environments vary widely. Equipment is available now to duplicate the pressure-temperature conditions existing throughout the moon, or in the outer few hundred kilometers of the earth. Shock techniques, employing

explosives or the firing of projectiles at specimens, generate earth-core pressures to 3.5 Mb for millionths of a second duration.

Figure 2 shows major apparatus designs that apply high pressures to rocks. The simplest is the *opposed anvil configuration* built by Bridgman for low-temperature tests (Fig. 2A). It can achieve pressures and temperatures to 400 kb and 1,000° C, but not simultaneously. A truncated conical taper provides massive support, which prevents anvil failure. Equally simple in design is the piston-cylinder device (Fig. 2B). Here the pressure exerted on the sample can be determined accurately from the force per unit area on the piston, after suitable corrections for friction. Piston-cylinder devices generally cannot attain pressures beyond 30 kb.

Tracy Hall invented the *belt* apparatus (Fig. 2C) two decades ago. An annular belt separates opposed pistons and girdles the specimen. It in turn is enclosed by a suitable, low-strength gasket such as pyrophyllite. The flow of the gasket and presence of the belt give maximum support to highly stressed work pieces. Its capacity exceeds 100 kb and 1,000°C.

Figure 2D illustrates the fourth apparatus, also designed by Hall, the *tetrahedral press*. As the name suggests, the press has four separate anvils, each of which is an equilateral triangle of tungsten carbide. When they are pushed together by four different rams, they compress samples into tetrahedrons. Test conditions equal those of the belt apparatus.

Some experimenters use additional rams with 6, 12, and 20 faces. High-pressure apparatus employs either external resistance and induction heaters, or internal furnaces. Most work in opposed anvil, belt, and multiram devices (Fig. 2C) is done on geochemical problems such as mineral synthesis or phase stability; piston-cylinder presses using solid, liquid, and gas-pressure transmitting media serve as work horses for rock mechanics experiments.

Elastic properties of rocks and minerals under high pressure and temperature are of special interest to geophysics. Most experiments use an ultrasonic technique to measure the delay time required for an elastic pulse to travel the known length of a specimen. Pressures usually do not exceed 10 kb combined with low temperatures. Elastic parameters such as Poisson's ratio, compressibility, and the shear modulus are derived from measured shear and compressional wave velocities.

Rock-deformation tests are of two types: one is the *constant strain-rate test,* which yields a stress-strain curve, and the other is the *constant stress-variable strain-rate experiment,* which gives a creep or strain-time curve. Figure 3 illustrates typical stress-strain curves for rocks showing behaviors ranging from brittle to ductile. In the most brittle case, the initial linear elastic portion is terminated

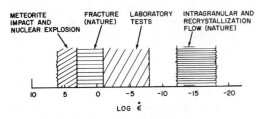

FIGURE 1. Range of strain rates encountered in rock mechanics, from dynamic on left to static "geological" conditions on right.

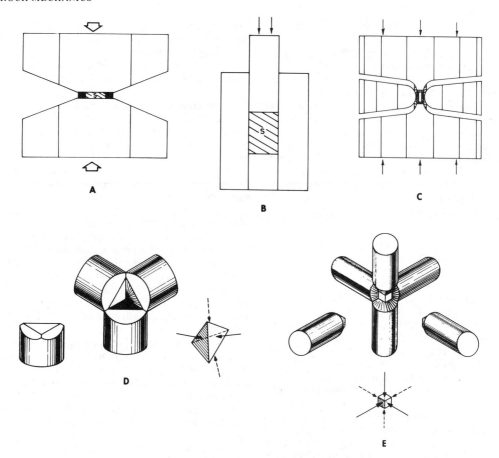

FIGURE 2. Major apparatus designs used in rock mechanics: A—opposed anvil; B—piston-cylinder; C—belt; D—tetrahedral; E—cubic (from Handin, 1966).

abruptly by fracture. With increasing ductility, the linear elastic portion passes through a yield point, followed by a region of strain hardening and then uniform flow.

Figure 4 illustrates a typical strain-time curve. Creep tests require very low stress levels applied for long periods of time generally at high temperatures, conditions that occur in the asthenosphere.

Stress

Stress sometimes is generalized as force per unit area on a plane. Stress, however, is a second-order tensor, and it involves the concept of force per unit area on planes of all possible orientations through a point. Figure 5 shows an enlargement of an infinitesimally small-volume element in an arbitrary force field. Nine stress components act parallel and normal to three orthogonal reference axes. Three stress elements act normal to planes and are called *normal stresses.* The other six lie in the planes with directions either parallel or normal to axes. These are the *shear stress components.* If rotational equilibrium prevails, then six of the shear components reduce to three, since

$$\tau_{xy} = \tau_{yx,} \quad \tau_{zy} = \tau_{yz}, \quad \tau_{xz} = \tau_{zx}$$

If we choose appropriate coordinate axes for the remaining three, shear components vanish and we find only three normal stresses—δ_{xx}, δ_{yy}, and δ_{zz} —parallel to the new axes. These are the principal stresses.

Sign convention varies, but we define $\delta_{xx} = 1$, $\delta_{yy} = 2$, $\delta_{zz} = 3$, with compressive stresses positive, and $\delta_1 > \delta_2 > \delta_3$. The following equations give the normal stress δ, and the shear stress τ, on any plane parallel to δ, inclined at an angle θ to δ_1.

$$\delta_N = \left(\frac{\delta_1 + \delta_3}{2}\right) - \left(\frac{\delta_1 - \delta_3}{2}\right) \cos 2\theta$$

$$\tau = \left(\frac{\delta_1 - \delta_3}{2}\right) sin\ 2\theta$$

Stress changes in the earth frequently lead to permanent deformation. Strain refers to changes in the position of particles relative to each other.

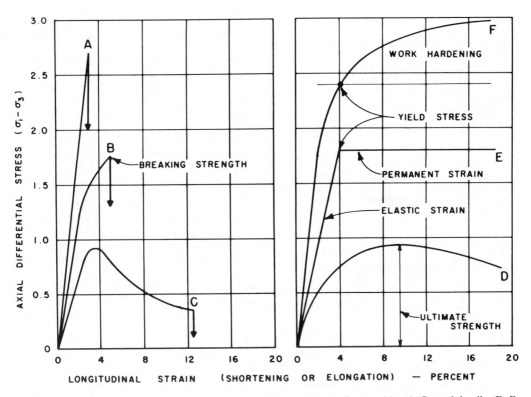

FIGURE 3. Typical stress-strain curves for rocks showing brittle (A, B), transitional (C), and ductile (D, E, F) behavior.

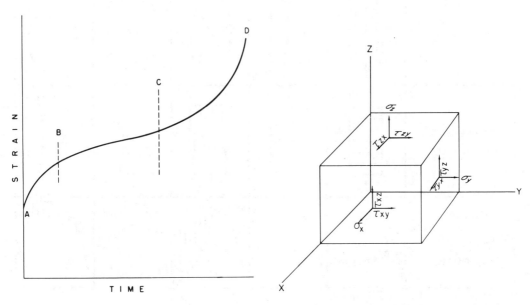

FIGURE 4. Representative strain-time or creep curve showing primary (decelerating), pseudoviscous (steady), and tertiary (accelerating) behavior.

FIGURE 5. Three-dimensional representation of stress components acting on an infinitesimal-volume element at an arbitrary point on a body.

Strength

Handin (1966) defines *strength* as the resistance to failure, continuing flow, or fracture. *Flow* is deformation, not instantly recoverable, without permanent loss of cohesion. *Fracture* refers to loss of cohesion, loss of resistance to stress difference, and separation into parts with release of stored elastic energy.

Curves A through F in Figure 3 illustrate variations in *ductility*, which is a measure of a rock's ability to undergo large permanent deformation without fracture. Figure 6 demonstrates typical changes in specimen geometry and the shape of stress-strain curves with increasing strain.

Although the theory of elasticity accounts for mechanical behavior of minerals up to the yield point on a *stress-strain diagram,* no theory of strength adequately interprets behavior beyond the yield point where fracture and flow manifest. Earth scientists do not yet fully understand the rules that govern *anelastic behavior* of minerals, and how the mechanical behavior of component minerals affects deformation of rock. Most rocks are composed of minerals, matrices, and cements exhibiting widely varying strengths and therefore behavior. Also grain boundaries, pores, cracks, and other flaws and defects must be as important to macroscopic deformation as they are for microscopic deformation.

Most common silicate rocks exhibit compressive strengths near 10 to 15 kb at 5 kb confining pressure and 27°C (Fig. 7). Compressive strengths for carbonates are much lower and the strength of pure quartz is very much higher. As shown in the figure, temperature weakens rocks, but the rate of weakening varies. The effect of the strain rate is not pronounced for most silicates until rates less than 10^{-4}/s are encountered. Rock strength also depends on the nature of the stresses; rocks exhibit maximum strength in compression, least strength in tension, and intermediate strengths with shear. Rock parameters such as grain size, anisotropy, homogeneity, chemical and phase composition, cohesion, and porosity also influence strength.

Mechanical properties of only a few minerals have been determined systematically. The few include quartz, olivine, calcite, and dolomite. Many other rock-forming minerals remain to be deformed experimentally. Investigations of the roles of pore pressure, phase transformation, shear melting, creep fracture, brittle fracture, and water in reducing strength require special attention.

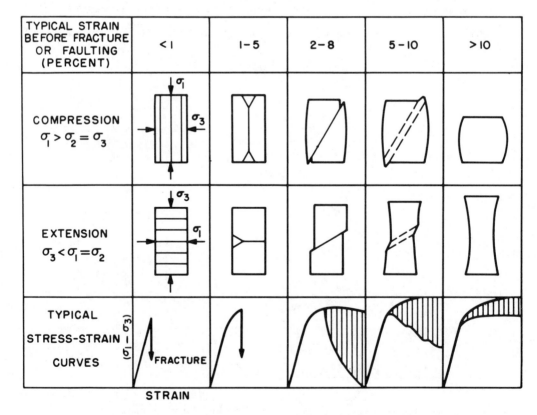

FIGURE 6. Representation of typical deformation behavior and stress-strain curves for rocks under compression and tension (from Griggs and Handin, 1960).

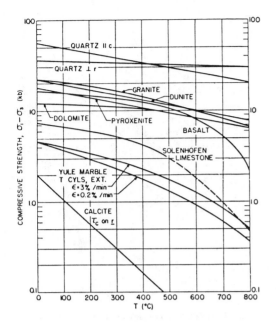

FIGURE 7. Typical compressive strengths of rocks under confining pressure of 5 kb (after Griggs et al., 1960).

Modes of Deformation

Three phenomena characterize rock deformation: extension fracture, faulting, and uniform flow. *Extension fracture* describes separation along a plane that lies normal to the direction of least principal stress. *Faulting* refers to relative displacement along a plane oriented at 45° or less to the greatest principal stress. Figure 8 illustrates these relations.

Uniform flow ideally covers homogeneous deformation in ductile materials. Flow includes several distinct deformation mechanisms. One is *cataclastic flow,* which involves the smashing and rupture of individual grains. It occurs in nature near the earth's surface and in the laboratory under conditions of low confining pressure, high pore pressure, low temperatures, and short times, usually with high stress fields and fast strain rates. *Gliding flow* predominates at deeper earth levels such as in the asthenosphere, where higher effective pressures, temperatures, and lower strain rates exist. Glide involves both mechanical twinning and translation within individual crystals, with displacements confined to definite crystallographic planes and directions within the plane. Distance of displacement along planes distinguishes twin from translation glide. In the former case, each layer must move a constant fraction of the interatomic distance; in the latter case, each layer moves an integral number of interatomic spaces.

Finally, *recrystallization flow* refers to molecular rearrangement by at least one of several processes:

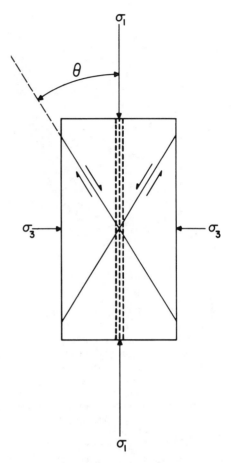

FIGURE 8. Orientation of fractures and faults relative to stress directions. Fracture is parallel to compression direction, faulting is at angle θ to direction of maximum principal compression.

solution, local melting, or solid diffusion. Recrystallization usually erases previous strain history and is the least understood and probably the most important process of deformation within the deeper earth. Its domain is at high temperatures and low strain rates, and it frequents the creep test in the laboratory.

Deformation depends on the presence of defects in mineral crystals, the most important of which is the *dislocation* (Fig. 9). Dislocations arise from disorder in the atomic lattice. Movement of edge and screw dislocations in shear planes produces uniform flow. Most natural crystals contain numerous dislocations, but significant macroscopic flow requires generation and multiplication of defects. Stresses required to move existing dislocations, to generate new ones, and to keep them moving differ. The first two govern nonductile behavior, while the last operates primarily during glide. The shape of the stress-strain curve depends

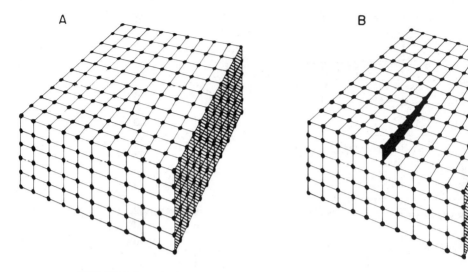

FIGURE 9. Imperfections in crystal lattice: A—edge dislocation; B—screw dislocation.

on the density, distribution, movement, blocking, climb, and annihilation of dislocations.

Two primary methods for interpreting fundamental behavior of deformed rocks are (1) etching and examination with the electron microscope, and (2) thin-sectioning and scrutiny with the polarizing microscope. Both depend on the visibility of edge components of slip bands, in the first case by location of etch pits, and in the second case by optical birefringence.

Structures induced by intragranular flow in calcite, dolomite, quartz, micas, ortho- and clino-pyroxenes, amphiboles, and olivines help determine orientations of principal stresses that lead to deformation. Methods depend on Turner's discovery that individual grains in an aggregate reflect total deformation of the polycrystal. Deformation produces internal and external rotations in affected grains, and these structural rotations become visible as extinction bands under the polarizing microscope. Measurement of rotations and determination of their relation to crystallographic elements then permits identification of stress directions inducing the deformation.

Applications

Although laboratory apparatus simulates only relatively shallow earth environments, nevertheless a significant number of important geological phenomena owe their origins to the mechanical behavior of rocks composing the outermost earth. For example, 85 percent of all earthquake foci occur at depths shallower than 200 km and the *Gutenberg low-velocity channel*, which may represent the weak zone separating the lithosphere from the deeper asthenosphere, lies in environments reproducible in the laboratory. Other sig-

nificant processes that may be investigated experimentally include the origin of magmas, the depth and nature of isostatic compensation, the nature of ocean floor spreading and convection, the mechanical basis for transform faulting, and underthrusting of the lithosphere.

Perhaps one of the most important immediate applications for rock mechanics research concerns earthquake prediction (see *Earthquake Engineering*). The classical view of the earthquake mechanism postulates a build-up of stresses from unspecified sources until some limit is reached and fracture occurs. This hypothesis accounts for observed slip, strain energy release, and the seismic radiation field. But overburden pressure precludes common faulting at all but shallow earth depths, because increasing frictional resistance exceeds rock strength. Earthquakes release as much as 10^{25} ergs of stored elastic energy, usually in a few seconds time, with stress drops of less than 100 b, and they may affect hundreds of thousands of square kilometers of the earth's surface.

In the laboratory, most silicate minerals exhibit great strength under shallow crustal environments, whereas the small stress drops associated with earthquakes suggest weakness. Study of the earthquake source mechanism in an attempt to learn what takes place mechanically during, before, and after an earthquake constitutes an important current frontier in rock mechanics.

ROBERT E. RIECKER

References

Griggs, D. T., and Handin, John, 1960. Observations on fracture and a hpothesis of erthquakes, in D. T. Griggs and John Handin (eds.), *Rock Deformation*. Geol. Soc. America Mem. 79, 349.

Griggs, D. T.; Turner, F. J.; and Heard, H. C., 1960. Deformation of rocks at 500° to 800°C, in D. T. Griggs and John Handin (eds.), *Rock Deformation.* Geol. Soc. America Mem. 79, p. 100.

Handin, John, 1966. Strength and ductility, in S. P. Clark (ed.), *Handbook of Physical Constants.* Geol. Soc. America Mem. 97, 223-289.

Cross-references: *Earthquake Engineering; Residual Stress, Rocks; Rheology, Soil and Rock; Rocks, Engineering Properties; Rock Slope Engineering; Rock Structure Monitoring.*

ROCK PROPERTIES—See ROCKS, ENGINEERING, PROPERTIES.

ROCKS, ENGINEERING PROPERTIES

The engineering properties of rocks are those that are important to the behavior or integrity of a structure or an excavation. The function required of the rock should determine the properties that must be investigated and the tests that would be appropriate. For reasons of economy and usefulness, a testing program should be designed to fit specific needs.

Rock properties may be determined in the laboratory from cores or blocks, or they may be determined in the rock mass, usually in boreholes and preferably at the construction site. Experience has shown that test values may depend on the size of the sample and on the equipment used. We know also that many rocks deteriorate when removed from the rock mass through loss or gain of moisture, through strain relief, or from the mechanical disturbance of sampling.

Many serious failures have involved displacements along pre-exisiting discontinuities, such as joints or faults, rather than along newly formed surfaces. A knowledge of the important structural attitudes and their effect on stability may be more useful than information regarding the behavior of a small, intact sample. The following list of properties is necessarily incomplete; it includes those properties most useful for construction in rock masses.

Density

A knowledge of the mass per unit volume, or *density*, is needed to calculate gravitational rock loads and to calculate elastic constants from seismic measurements. Density may vary as a result of composition, alteration, or fracturing.

Porosity

Porosity, the ratio of the volume of pore space to the volume of the specimen or rock mass, is expressed as a percentage of the rock volume. Porosity in crystalline rocks is largely a function of fracturing, whereas in sedimentary rocks it is more a function of the pores between grains. Porosity influences the *compressibility* of rock and determines the amount of fluid or gas in the rock.

Permeability, Water Content, and Water Pressure

The *coefficient of permeability* is a measure of the capacity of rock to transmit fluid under pressure. Permeability is best measured in the field to incorporate effects of fractures (Louis and Maini, 1970); it varies with the properties of the mass and of the fluid and with the deformational characteristics of the rock mass. Theory and methods for ground-water investigations were given by Ferris et al. (1962). Adequate theory does not exist, however, to treat the flow of water in a discontinuous system such as exists in a complexly fractured medium. Water content influences rock properties (Broch, 1974) and it is important to follow standard test procedures to minimize moisture effects on testing results.

The effects of ground-water pressures on the stability of dam foundations and landslides have been discussed by Wahlstrom (1974). Uplift due to the build-up of pore-water pressure lowers the *shear strength.* Seasonal changes in water level may provide the triggering mechanism for slope failure. Ground-water control is sometimes the only economical means of arresting hazardous slope movement. *Water pressure,* or head, in underground works may also be difficult to predict. Tunnels in fractured rock have been known to tap surface reservoirs at least 459 m (1,400 ft) above the tunnel grade, creating difficult flows and pressures (Hurr and Richards, 1966). Alpine tunnels have exhibited notoriously treacherous extremes in water pressure, flow, and temperature; these are no doubt caused by great topographic relief, complex geological structure, and a wide range of lithologies, including evaporites, carbonates, granites, and metamorphic rocks—all variously altered. Within an interval of only a few meters in the Simplon Tunnel in Switzerland, for example, 60°C water rose from the floor, while 15°C water descended from the arch (Sandström, 1963).

Rock Stresses and Strength Properties

Stress in rock, although not an intrinsic rock property, is important to engineering operations because it deforms rock (see *Rheology, Soil and Rock*). Furthermore, values of elastic "constants," such as Young's modulus and Poisson's ratio, depend on the magnitude of the applied stress. Three kinds of stresses are important in analyzing the behavior of rock: compressive stresses, which act to decrease the volume of the material; shear stresses, which tend to slide one surface along another surface; and tensile stresses, which pull material apart and thereby increase its volume. In a like manner, rock has compressive, shear, and tensile strengths.

Compressive Strength. *Compressive* strength represents the highest strength of rock under a given confining pressure. This upper bound, however, is seldom significant in foundation analysis. The compressive strength of intact material may be important if joints are widely spaced or discontinuous, or, in the case of tunneling, if a tunneling machine is used (see *Rapid Excavation and Tunneling*), as a guide for the design of the cutting head.

Shear Strength. The strength of a rock in *shear* is the internal resistance offered to shear stress. Most slope movements are failures in shear, although yield may occur along old discontinuities (see Vol. XIV: *Landslide Control; Slope Stability Analysis*). The properties of joint and bedding surfaces are significant in such cases. Shear strength is largely controlled by frictional properties, including friction between shear surfaces and bond strength between particles. It is a more complicated property than is indicated by the term *coefficient of internal friction*.

Cohesion is that component of the shear strength on the incipient failure surface when the normal stress on the surface is zero. It has the nature of an intergranular binding force; and values are low, or nearly zero, in highly fractured rocks.

Tensile Strength. *Tensile* strength is the resistance of rock to rupture by stresses pulling the rock apart. Commonly, the tensile strength of intact rock is many times less than the compressive strength and is negligible in fractured rock masses.

In Situ Stresses

To understand rock behavior, researchers have made stress measurements for many years around underground openings and in quarries. Recently, however, stress measurements and large-scale, expensive, *in situ* tests have been used to provide basic design data. *Gravitational stresses* are those that arise from the force of gravity on a rock mass, and at any point the vertical component is the product of density and depth. *Tectonic stresses* are stresses applied at the boundaries of rock masses and are generally thought to cause earthquakes, mountain uplift, and faulting. These stresses will be modified by topography and by variations in rock properties. *Residual stresses* are those that remain in the rock after it has been isolated from exterior loads. Such internally balanced force systems can be mobilized to do work on adjacent rock and add their effects to gravity or tectonic stresses. These stresses are remnants from a period of higher loading and account for otherwise anomalously high vertical stresses measured in some rocks (see *Residual Stress, Rocks*).

Deformational Properties

Critical strains or deformations of rock masses are often more important in defining stability relations of slopes and underground excavations than stress criteria. Testing of rock and rock masses indicates that failure strains are more consistent and therefore more easily predictable than failure stresses (Everell et al., 1974). Because of the uncertainty involved in predicting the stability of high, steep slopes and underground openings, rock-strain monitoring devices are used. Such monitoring requires knowledge of the magnitudes of strains or deformations appropriate for the rock mass and its function, and the interpretation of stress from strain measurements requires a knowledge of the Young's modulus and Poisson's ratio of the rock.

Young's Modulus. *Young's modulus* is the ratio of normal stress to normal strain for a rock sample under given loading conditions; it is numerically equal to the slope of the stress-strain curve. If the curve is linear, the material is said to be linearly elastic. *In situ* tests show that the deformation of rock is generally far from linear.

Modulus of Deformation. *Modulus of deformation* refers to deformation that is not perfectly elastic and implies inelastic deformation. This term would be appropriate for the stress-strain curve of a jointed rock mass. Experiments have shown that the modulus of deformation for a rock mass can be much less than the Young's modulus of an intact sample. Deformation during testing is controlled by sample size, previous relaxation, and stress magnitude (Wallace et al., 1972; Kruse, 1970). Static values of Young's modulus for core samples will result in unrealistically high safety factors for the rock mass.

Poisson's Ratio. *Poisson's ratio* is the ratio of the transverse normal strain to the longitudinal normal strain of an elastic body under a uniaxial load. The larger the Poisson's ratio of a rock, the larger is its lateral expansion. As is the case with Young's modulus, values of this ratio depend on the test and sample conditions. Knowledge of Poisson's ratio sometimes can be used to predict pillar or wall deformation in mines. The theoretical maximum value is 0.5; however, larger values have been calculated from tests on some rocks.

Creep. The foregoing short-term static rock behavior does not characterize the ability of rock materials to deform slowly with time, or creep. The creep rate under a specified load should be determined and compared to the life and purpose of the construction. Creep occurs in most earth materials and on very gentle slopes.

Dynamic Rock Properties

The information presented so far has dealt with the reaction of rock to static stresses. If a rock mass is subjected to earthquakes or explosions, however, transient dynamic stresses may exceed the existing static stresses by many orders of magnitude.

A *dynamic load* is defined by the speed of its application. An explosion applied over a period of microseconds will create pressures far in excess of the static compressive strength of the rock (Farmer, 1968, p. 81). Because rocks can resist a much higher magnitude of dynamic stress than static stress, failure predictions based on some contribution of dynamic stresses and applied to a structure are complex and must be applied with caution. Farmer (1968), Hayashi et al. (1974), and Godfrey (1974) have supplied detailed information that is beyond the scope of this article. A concise account of dynamic testing is given by Jaeger and Cook (1969, 171-176).

Fractures in Rocks

The importance of fractures to the engineering behavior of a rock mass can scarcely be overemphasized; the majority of rock failures associated with engineering works do not occur in the intact rock but follow pre-existing weaknesses.

Fractures, including joints and faults, are discontinuities resulting from the failure of rock under tensile and/or shear stress. Fault zones or shear zones refer to zones of intense deformation in which the engineering conditions are significantly different from those in the adjoining rocks. These differences may be expressed by changed groundwater conditions, swelling or nonswelling clay, and much weaker rock.

Three aspects of joints are particularly important to construction in rock: their surface roughness, attitude, and continuity and spacing. The *surface roughness* controls the sliding characteristics of planes of weakness. Undulations, not removed by shearing, cause an overriding of joint blocks above the failure surface; the stronger the sliding block, the higher the amount of shear resistance, owing to the resistance of the rock mass to dilatancy, that can be realized. Values from direct shear tests and observations of natural surfaces indicate that the effective angle of friction can be increased by 10° to 15° because of joint-surface features (Everell et al., 1974).

The *attitude* of fractures and their continuity, if critically oriented, can determine surfaces of movement, both on the surface and underground. Because prediction of displacements along geological discontinuities is inexact, high-risk situations call for careful field monitoring. The degree of joint *continuity* and *spacing* controls the mobility of a rock mass. As joint spacing is reduced, the rock behaves less as a "monolith"; the deformation modulus decreases, and rock-substance properties have less and less influence on rock-mass

behavior. Sufficiently intense jointing produces a mass resembling a coarse-grained soil. Bieniawski (1974) has related such factors as standup time, favorable tunneling directions, and support requirements to the spacing and orientation of fractures.

Lithology

The lithological classification of the immense variety of igneous, metamorphic, and sedimentary rocks may not reflect their engineering properties. A "granite," for example, may vary from quarry rock to thoroughly weathered material (saprolite) over a short distance (see Vol. XIV: *Rock Weathering, Engineering Classification*). A limestone may have strength or deformational properties identical to a sandstone or a gneiss. For this reason, geological-genetic rock classifications are of little use in engineering problems. The material should be described from the point of view of its proposed service, that is, building stone, foundation material, or concrete aggregate. See Varnes (1974) regarding the preparation of engineering geology maps.

Index Properties

Index properties are used to classify rocks rapidly and inexpensively, using many test repetitions. Such tests are unlike the more complex and expensive engineering tests that provide data for design calculations. Hoek and Bray (1974) argued that gaps in design test data can be filled by index tests. However, the numbers obtained should be treated as qualitative guides to rock behavior; for example, the fact that a feature important to the strength of the rock may have little effect on its modulus should be recognized.

Rock-Quality Designation (RQD). The RQD (Deere et al., 1967) is a "modified core recovery percentage in which all the pieces of sound core over 4-in. long (NX or larger diameter) are counted as recovery." Despite several drawbacks, this index is widely used as a guide to rock quality (not rock condition). It is not sensitive to amount or type of fracture fillings or fault gouge, or to the tightness of joints. Poor drilling techniques could also produce a low RQD. The descriptive technique given by Rankilor (1974) includes fracture inclinations and zones of gouge and crushed rock.

Point-Load Strength Index. The point-load strength index is the ratio of the force required to fracture a core to the squared distance between the loaded points. The object is not to predict tensile strength but to obtain a value that will generally represent the strength of the material (Franklin, 1970). The point-load index values are about 1/25 of the corresponding uniaxial compressive strength (Bieniawski, 1974).

Slake Durability. Slake durability is the

resistance of a rock to accelerated wetting and drying. The test subjects the sample to two cycles of drying and wetting, with loss of fines passing the standard mesh (Franklin, 1972). It is defined as the percentage ratio of final to initial dry-sample weights. Most susceptible to serious slaking are soft rocks that have a high clay content.

Swelling and Shrinkage. Rocks with a high clay content are commonly prone to swell or shrink, owing to a change of water content. Tests necessary to evaluate swelling are given by Franklin (1972). The heave produced by swelling of rocks has caused widespread foundation damage. Wet, clayey rocks can shrink on drying, with equally undesirable results. Volume changes of as much as 2,000 percent in sodium-montmorillonite clay have been recorded (Tourtelot, 1974).

FITZHUGH T. LEE

References

Bieniawski, Z. T., 1974. Geomechanics classification of rock masses and its application to tunneling, in *Advances in Rock Mechanics*. Proceedings of the 3rd International Society of Rock Mechanics Congress, Vol. 2, Pt. A, Washington, D.C.: National Academy of Sciences, 27-32.

Broch, E., 1974. The influence of water on some rock properties, in *Advances in Rock Mechanics*. Proceedings of the 3rd International Society of Rock Mechanics Congress, Vol. 2, Pt. A, Washington, D.C.: National Academy of Sciences, 33-38.

Deere, D. U.; Hendron, A. J., Jr.; Patton, F. D.; and Cording, E. J., 1967. Design of surface and near-surface construction in rock, in C. Fairhurst (ed.), *Failure and Breakage of Rock*. Proceedings of the 8th Symposium on Rock Mechanics, New York: American Institute of Mining, Metallurgical, and Petroleum Engineers, 237-302.

Everell, M. D.; Herget, G; Sage, R.; and Coates, D. F., 1974. Mechanical properties of rocks and rock masses, in *Advances in Rock Mechanics*. Proceedings of the 3rd International Society of Rock Mechanics Congress, Vol. 1, Pt. A, Washington, D.C.: National Academy of Sciences, 101-108.

Farmer, I. W., 1968. *Engineering Properties of Rocks*. London: E. and F. N. Spon, 180p.

Ferris, J. G.; Knowles, D. B.; Brown, R. H.; and Stallman, R. W., 1962. Theory of aquifer tests, *U.S. Geol. Survey Water-Supply Paper 1536-E,* 69-174.

Franklin, J. A., 1970. Observations and tests for engineering description and mapping of rocks, in *Intrinsic Properties of Rock Masses*. Proceedings of the 2nd International Society of Rock Mechanics Congress, Vol. 1, 11-16.

Franklin, J. A., 1972. Suggested methods for determining water content, porosity, density, absorption, and related properties, and swelling and slake-durability index properties, *Internat. Soc. Rock Mechanics, Comm. Lab. Tests, Doc. No. 2,* 36p.

Godfrey, Charles, 1974. Dynamic strength of in situ rock, in *Advances in Rock Mechanics*. Proceedings of the 3rd International Society of Rock Mechanics Congress, Vol. 2, Pt. A, Washington, D.C.: National Academy of Sciences, 398-403.

Hayashi, Masao; Kitahara, Yoshihiro; Fuziwara, Yoshikazu; and Komada, Hiroya, 1974. Dynamic deformability and viscosity of rock masses, in *Advances in Rock Mechanics*. Proceedings of the 3rd International Society of Rock Mechanics Congress, Vol. 2, Pt. B, Washington, D.C.: National Academy of Sciences, 713-718.

Hoek, Everet, and Bray, J. W., 1974. *Rock slope engineering*. London: Institution of Mining and Metallurgy, 309p.

Hurr, R. T., and Richards, D. B., 1966. Ground-water engineering of the Straight Creek tunnel (pilot bore), Colorado, *Eng. Geology* 3 (1-2) 80-90.

Jaeger, J. C., and Cook, N. G. W., 1969. *Fundamentals of Rock Mechanics*. London: Methuen, 513p.

Kruse, G. H., 1970. Deformability of rock structures, California state water project, in *Determination of the In Situ Modulus of Deformation of Rock*. Am. Soc. Testing and Materials Spec. Tech. Pub. 477, 58-88.

Louis, C., and Maini, Y. N., 1970. Determination of in situ hydraulic parameters in jointed rock, in *Intrinsic Properties of Rock Masses*. Porceedings of the 2nd International Society of Rock Mechanics Congress, Vol. 1, 235-245.

Rankilor, P. R., 1974. A suggested field system of logging rock cores for engineering purposes, *Assoc. Eng. Geologists Bull.* 11 (3), 247-258.

Sandström, G. E., 1963. *Tunnels.* New York: Holt, Rinehart and Winston, 427p.

Tourtelot, H. A., 1974. Geologic origin and distribution of swelling clays, *Assoc. Eng. Geologists Bull.* 11 (4), 259-275.

Varnes, D. J., 1974. The logic of geological maps, with reference to their interpretation and use for engineering purposes, *U.S. Geol. Survey Prof. Paper 837,* 48p.

Wahlstrom, E. E., 1974. *Dams, Dam Foundations, and Reservoir Sites*. New York: Elsevier, 278p.

Wallace, G. B.; Slebir, E. J.; and Anderson, F. A., 1972. Radial jacking test for arch dams, in K. E. Gray (ed.), *Basic and Applied Rock Mechanics*. Proceedings of the 10th Symposium on Rock Mechanics, New York: American Institute of Mining, Metallurgical, and Petroleum Engineers, 633-660.

Cross-references: *Geotechnical Engineering; Rock Mechanics; Rock Slope Engineering; Rock Structure Monitoring. Vol. XIV: Rock Weathering, Engineering Classification.*

ROCK SLOPE ENGINEERING

The purpose of rock slope engineering is to produce a slope that is stable for the duration of its working life (see Vol. XIV: *Slope Stability Analysis*). This will vary with its use: a cut for a highway would be designed to have a working life of perhaps many tens of years, whereas that for a temporary excavation as dug for foundation, or open-pit extraction, may have a life that is measured in terms of weeks or months. Time is an important element in the design of rock slopes. Displacement is another, for a slope can be considered to have failed if its displacements, however small, prevent it from performing its required task. Large displacements can be tolerated in certain circumstances, for example, in abandoned slopes of an open pit, whereas very small displacements may be critical to a slope upon which some installation is founded. Time and displacement are therefore the criteria on which the success of most rock slope engineering is assessed, whether the engineering involves the generation of a new slope or the improvement of an existing slope.

To engineer with success, it is essential to investigate the ground in which slope is to be cut, and for a rock slope this demands a carefully considered site investigation. This should be designed to identify the structure of the rock mass, the pattern of fractures within it, the presence of weak strata, and the presence of ground water. In many investigations some core drilling is required with particular attention being given to core recovery (see Vol. XIV: *Borehole Drilling*). Representative samples of the rocks are required, especially of those that are weak or weathered (see Vol. XIV: *Rock Weathering, Engineering Classification*). Clay and other soft material that may be filling discontinuities within the rock mass should be collected. This phase of the investigation may also include surface mapping and other coventional geotechnical surveys (see *Geotechnical Engineering*) to obtain proof of the quality of the rock within the volume of ground that will become, or is already, the slope. The specimens collected are used for calculating the weight of material within the slope and determining its shear strength. This data must be augmented by information of rock structure, for rock is characterized by systematic sets of discontinuities, which can take the form of bedding surfaces, join sets, and faults. These can exert a considerable influence on the manner in which the slope may deform and eventually fail (see Vol. XIV: *Slope Stability Analysis*). Information on rock structure can be obtained from carefully collected and oriented cores. It is also useful to inspect the walls of boreholes as they provide information on the dip and strike of discontinuities, together with their spacing and separation. In important slopes it is often advantageous to drill at least one hole that has a diameter large enough to permit a geologist to be lowered in a cage to inspect and record the rock mass and, where necessary, to collect samples of strata not recovered during other drilling. Great care should be taken with the ventilation of these holes—they should not be entered during periods of falling barometric pressure, as deoxygenated air can be discharged into them from the surrounding ground. All investigation holes, pits, and adits (where used) should be aligned so that the total investigation intersects all likely structural directions.

These data alone must not be used in ignorance of the geological environment of the area, and here much can be learned from a study of existing slopes in the region. The most important character is usually ground-water, two aspects of which must be considered: the *pressure head* of water within the ground, and the *permeability* of the ground (see *Hydrodynamics, Porous Media*). The former controls the effective normal load that can be expected to provide frictional resistance within the rock mass, and the latter controls the rate at which the pressure head within rock mass will vary with either drainage or recharge. Not only are absolute values for each of these parameters required, but also information on their variability within the rock mass and with time. Most ground-water regimes respond to the seasons. With this information it is possible to predict the pressure head within the slope and the seepage force that will accompany any flow. This force is a body force, like gravity, that acts in the direction of flow; it can be sufficient to destabilize an otherwise stable slope. Attention should also be given to the environmental factors that affect the movement of water to and from the rock mass, in particular the periods when ice formation may prevent the free drainage of water from the face of the slope (see *Permafrost, Engineering Geology*). Water quality is occasionally significant to the engineering design of a rock slope, especially when the solution rate of the rock permits an unacceptable reduction in rock mass strength to occur during the working life of the slope and where the chemical precipitation of minerals in solution blocks drains drilled into the slope to remove water.

In Situ stress is another aspect of the geological environment that can be important. It can be considered to take two forms: active and residual. *Active stresses* are those that are being generated at the present, for example, by gravitational or tectonic loading. *Residual stresses* (q.v.) are embodied in the rock, but the external stimulus that generated them has been removed. Because these stresses can be very high, they are capable of influencing the strength of rock under stress. The measurement of *in situ* stress is difficult and generally unsuccessful. Stress distributions are usually calculated (see *Rock Mechanics*) and an acceptable margin of safety added in the slope design to provide a factor of safety against failure by this means. From these general investigations it is possible to decide either the best slope angle to develop for any slope direction within the ground, or the stability of an existing slope. See Hoek and Bray (1977) for a comprehensive text and relevant references.

There are three major parts to an assessment of slope stability: (1) defining the boundaries of a moving mass, or of a mass that is likely to move; (2) resolving the forces for the shape chosen; and (3) selecting the correct parameters for use in the analyses. Site investigation and structural interpretation of the ground usually defines the shape of a moved or moving mass. To define the shape of a mass that has not yet moved is complicated by the fact that it requires a prediction of the movement process that is likely to operate, for example, sliding, toppling, and rockfalls (see Vol. XIV: *Slope Stability Analysis*). In many cases no estimate is made of the lateral boundaries of a potential failure. Instead, efforts are concentrated on ascertaining the likely mode of failure (e.g., circular slip or straight translation), and the most likely position of the major failure surface or surfaces.

Slopes therefore tend to be considered in two dimensions rather than in three. The common exception to this is the *wedge failure,* in which the sliding surfaces define fairly well the lateral limits of movement. Analyses exist for sliding on planar, noncircular, and circular surfaces, and stability charts can be used to enable a variety of conditions to be assessed quickly on site. These charts are normally for idealized situations and should be used only as a guide.

Selecting the correct parameters to use in the calculations requires experience. This can best be gained by back-analyzing previous movements, and should be considered as a vital aspect of rock slope engineering.

Excavation techniques used in the production of slopes in rock generally use some system of drilling and blasting (see Vol. XIV: *Blasting and Related Technology; Borehole Drilling*). Numerous theories exist to assist the prediction of blast effects, and can be applied with some success, but experience is required, especially when producing a blast face by pre-splitting or similar means. The rock types that blast well are those that are reasonably strong: there is little point in blasting weak rock that can normally be excavated with less trouble by other means. Some indication of the problems in stone can be obtained from using one of the blasting indices. These are based on the ratio of the maximum and minimum sonic velocities of rocks, but they also take into account the tensile strength of the rock and its structure in the field.

In the engineering of remedial works, it is helpful to note that such work invariably falls under one of three headings: restraint, redistribution of forces, and ground improvement. *Restraint* is commonly achieved with the aid of a retaining wall at the toe of the slope. Ground anchors and piles can also be used to pin the moving mass to firmer ground. Translation, circular, and noncircular movements can be held by these means; flow and creep cannot. Rockfall areas are frequently restrained by the use of either individual rock bolts to hold large blocks, or a net, bolted at regular intervals over the face.

The *redistribution of forces* commonly involves reducing the angle of slope, and often this is prohibitively expensive.

Ground improvement is usually the most eco-nomical way to stabilize moving slopes. If achieved, it will prevent all forms of mass movement. The two parameters of interest are cohesion and friction. Cohesion is usually zero in a failed slope and is rarely worth improving. Grouts and similar bonding substances can be injected, but the treated areas are often restricted to a short distance from the injection holes, making such a process uneconomical compared with the improvements than can be gained by increasing frictional resistance. Here the key to success lies in reducing ground water pressures. Drainage wells, gravity drainage holes, drains, and sometimes tunnels are used to lower the pressure head of water in slopes.

M. H. DE FREITAS

Reference

Hoeck, E., and Bray, J., 1977. *Rock Slope Engineering.* London: Institution of Mining and Metallurgy, 402p.

Cross-references: *Geotechnical Engineering; Hydrodynamics, Porous Media; Pipeline Corridor Evaluation; Residual Stress, Rocks; Rheology, Soil and Rock; Rocks, Engineering Properties; Rock Structure Monitoring.* Vol. XIV: *Blasting and Related Technology; Borehole Drilling; Landslide Control.*

ROCK STRUCTURE MONITORING

Monitoring, in a geotechnical context, is the surveillance of structures in rock or soil, either visually or with the help of instruments. The visual element should not be overlooked. The human eye is an economical and usually reliable instrument, and it should not be supplemented by more contrived forms of instrumentation unless there is some positive advantage in so doing. As with all other monitoring methods, it is essential to plan the observations ahead of time, to set up a program and checklist for the nature and sequence of observations to be made, and to determine the actions necessary should certain types of rock or soil behavior be observed.

Monitoring has several applications in connection with engineering projects. The objectives of the monitoring program should be carefully defined and the instrumentation system meticulously designed to meet them, no more and no less. Four of the more common functions of monitoring are the following:

1. To record natural variations in the environment before the start of an engineering project. The data are obtained for use in design and also as a base of comparison to evaluate the effect of the engineering works.
2. To ensure safety during construction; to warn of ground-water pressure fluctuations that may

endanger stability; and to check the build-up of loads on cable anchors and rockbolts.

3. To check the data and assumptions used in design, for example, in the recording of changes in ground stress and of tunnel lining movements.

4. To control the implementation of ground treatment and remedial works; to control movements in the vicinity of ground anchors; and to measure ground-water pressures and inflow in works where the ground is being grouted or drained.

Sometimes a monitoring program may be designed to fulfill two or more of these objectives. When instruments are installed well ahead of construction, they can be used to provide design input data, to check the safety of works during construction, to check the validity of the design itself, and subsequently to determine if there is a need for supplementary ground support.

Principles of Measurement

An instrument usually consists of three components: (1) a sensor or detector to measure the property of interest; (2) a transmitting system—for example, rods, electrical cables, or telemetry devices—to transmit the readings to the readout location; and (3) a readout unit such as a dial gage with a digital or graphic display of the measured quantity. There are many variations on this theme.

Mechanical Systems. The simplest and usually the most reliable way to measure movement is to use a mechanical system for detection and reading. Mechanical movement detectors, for example, consist of a steel tape or rod, which is fixed to the rock at one end and is in contact with a dial gauge micrometer at the other.

Mechanical systems are usually the most accurate, being simple and free from the drift and distortion that often impair the performance of electrical devices. The main disadvantage is that they do not lend themselves to remote reading or to continuous recording, which are essential in some circumstances.

Pneumatic/Hydraulic Systems. The *pneumatic/ hydraulic diaphragm transducer* has been used in a number of monitoring systems including those designed for measurement of water pressure, rock and soil pressure, settlement of rock structures, and loads on cable anchors. It is perhaps best known as a component of the Gloetzl cell for measurement of rock and soil pressures. In each of its various applications the hydraulic diaphragm transducer is identical, and operates as follows:

The actual measured quantity is fluid pressure. The pressure to be measured acts on one side of a flexible diaphragm made of steel, rubber, or plastic. Twin tubes connect the readout instrument to the other side of the diaphragm. Air, nitrogen, or hydraulic oil pressure is supplied from the readout unit through one of the tubes to the diaphragm. When the supply pressure is sufficient to balance the pressure to be measured, the diaphragm acts as a valve and allows flow along the return line to a detector in the readout unit. The balance pressure is recorded, usually on a standard Bourdon tube pressure gauge or a digital display.

The hydraulic diaphragm transducer may be used directly as a piezometer, simply by protecting the measuring side of the diaphragm with a porous plastic or ceramic element. It may be used in a pressure cell by connecting the measuring side of the diaphragm to a flatjack, forming a Gloetzl pressure cell as described later. It may also be used to measure settlements if the measuring side of the diaphragm is connected to a mercury U-tube and reservoir. The reservoir is embedded in the structure to be monitored so that settlements are proportional to mercury pressure at the transducer location.

Photoelastic Instruments. Certain kinds of transparent material have a property known as stress-birefringence, that is, they exhibit a pattern of colored fringes when stressed and viewed in polarized light. The number of fringes and their color are proportional to the magnitude of the applied load.

Perhaps the most common application of this photoelastic phenomenon in measuring instruments is the *photoelastic rockbolt load cell.* This cell incorporates glass disks mounted between steel loading platens, such that the load in the rockbolt is carried across the diameter of the two disks. To take readings, the disks are illuminated with polarized light. The number of fringes are then counted, and this reading is converted to an estimate of the applied load.

Electrical Resistance Strain Gauge Systems. Of the electrical instruments, perhaps the majority incorporate resistance strain gauges. One type of electrical resistance strain gauge consists of a length of wire mounted in a zig-zag pattern on a plastic backing material, allowing an appreciable length of wire in a small area. As an alternative, the same zig-zag pattern of conductor may be manufactured by etching a piece of thin metal foil with acid.

The strain gauge is fixed to the rock, steel, or concrete surface using special adhesive. Tensile or compressive strain of the surface is accompanied by an equal strain in the wire. The wire extension is accompanied by a proportionate increase in its electrical resistance, which may be measured by a Wheatstone bridge readout unit. The readout may be calibrated in terms of resistance, or directly in terms of strain, load, or pressure. Many load cell designs use four strain gages cemented to the surface of a steel column or cylinder. The load in the cylinder, applied by a rockbolt or by a steel rib, for example, places in a proportionate strain on the steel cell, and the strain is measured as a change in resistance of the gauges. In pressure transducer applications, the resistance strain gauges are cemented to the face of a flexible diaphragm. Ground-water or hydraulic pressure acting on the opposite face of the diaphragm causes deflection and strain. Some types of inclinometer use strain gauges, cemented to the face of a flexible metallic cantilever pendulum. Changes in tilt of the inclinometer probe cause the cantilever to bend, and this bending is measured as proportional to the sum of strain on the two opposing faces of the cantilever; the instrument may be calibrated to read directly in terms of tilt.

Vibrating-Wire Systems. A further measuring principle that finds application in various types of monitoring device is the *vibrating-wire strain gauge*. The tension in a vibrating wire is proportional to the square of its natural vibration frequency, so that if this frequency is measured, the tension in the wire may be determined.

The vibrating-wire load cell is typical application. In this case, one or more stainless steel wires (usually three wires mounted at 120° around the circumference of the cell) are tensioned, and held in place by nonslip end clamps. The wires run parallel to the axis of the cell and to the direction of the applied load. When load is applied to the cell, the cell body shortens and causes a reduction of tension in the wire, proportional to the applied load. An electromagnet with permanently magnetized poles is located close to the center of each wire. To set the wire in motion, a direct-current pulse is supplied from the readout unit. The wire vibrates in a magnetic field, inducing in the coils an alternating-current signal of frequency equal to that of the vibration. The signal is detected in the readout, filtered, amplified, and displayed. The tension in the wire is initially adjusted to ensure that at the full working capacity of the load cell, the wire remains under a tension equal to approximately half the tension that exists when the cell is unloaded.

The vibrating-wire method has an important advantage over other electrical methods in that readings are obtained in terms of frequency rather than voltage or current. Frequency is easier to transmit without distortion over long distances, requiring a digital count of pulses per second rather than the interpretation of an analog signal. Frequency counts can be transmitted and detected reliably even in the presence of heavy background noise.

Types of Instruments

The instruments available for monitoring may be conveniently subdivided into four main classes: those instruments used for monitoring ground movements; those for measuring ground-water pressures and fluctuations in the ground-water table; those for measuring soil and rock pressures and loads on anchor supports; and finally, those for monitoring ground vibrations.

Movement Monitoring. The monitoring of ground movements provides a direct check on the stablity of engineering works. Instability and failure of engineering structures involves movement, and the larger movements are nearly always preceded by smaller displacements that may be detected by sufficiently sensitive instrumentation.

The time intervals between the onset and the detection of movement and the eventual collapse of a rock or soil structure vary, depending on the characteristics of the ground and on the sensitivity and reliability of the instruments. In most circumstances, however, the warning period may range from several days to weeks, or even months. Structural failure is commonly associated with a gradual acceleration of displacements, whereas structures that move at a steady or decelerating rate will generally become stable in the long term. This, of course, is an oversimplification, and when interpreting movement records one must take into account the peculiarities of the individual project.

Conventional Survey. Conventional surveying techniques are economical and of proven reliability. Precise leveling is both rapid and accurate. Large areas of ground can be covered in a short space of time. This method is often used to establish the settlement of structural foundations, and of earth and rockfill embankments, and to record the pattern of subsidence at the ground surface above an advancing tunnel face (Kobold, 1961; Ashkenazi, 1973; Cheney, 1973).

Electro-Optic Distance-Measuring Instruments. Electro-optic instruments employ a modulated light or laser beam projected onto reflecting targets fixed in the ground. The time taken for the light beam to travel from the instrument to the target is measured so that the precise distance between the instruments and targets may be calculated. Target coordinates may then be determined by trilateration calculations (Froome and Bradsell, 1966; Romaniello, 1970; St. John and Thomas, 1970; Tomlinson, 1970).

The accuracy of electro-optic distance measuring instruments is generally between 1 and 10 mm, depending on the make and model of the instrument selected. These instruments are primarily designed for surveying and not for movement monitoring, and their limited accuracy may be a problem in the monitoring application (see Vol. XIV: *Surveying, Electronic*).

Photogrammetry. Photogrammetric methods of surface contouring are generally less accurate than either conventional survey or electro-optic distance measurements, but they have the great advantage of covering a complete field of view rather than a set of prelocated targets. Hence it is not necessary to make accurate forecasts of the nature, location, and direction of potential movements. A sequence of photographs taken at suitable time intervals can be compared to indicate and measure movements wherever these might develop. Inaccessible faces can be surveyed, and there is no interference with construction activities. The photographs obtained can also be extremely useful for evaluating the pre-existing ground topography (see Vol. XIV: *Photo Interpretation*), as a basis for engineering geological mapping, and for the back-analysis of previous landslides (Borchers, 1968; Planicka and Nosek, 1970).

The accuracy of contouring is inversely proportional to the distance between the camera and the ground surface. In general, therefore, air photographs, although suitable for surveying applications, are inadequate for movement measurements. Terrestrial photogrammetry is generally used. Photographs taken from ground stations at an object distance of 100 m, for example, when measured in a stereocomparator, can give an accuracy better than 20 mm in the plane of the photograph, and better than 30 mm in the perpendicular direction. A much greater precision can be obtained using close-up photographs; however, the field of view is correspondingly restricted and a greater number of photographs will be required.

Measurements at Cracks, Joints, and Faults. The presence and pattern of cracking, such as in a concrete structure, in a shotcrete tunnel lining, or around the crest of an open pit mining excavation, can provide a substantial amount of useful information concerning the mechanisms and directions of movements. Simple instruments are available to measure the expansion in aperture of these cracks; however, before such instruments are installed there is much to be learned from the pattern of cracking alone. It is helpful to mark the entire length of each crack with spray paint and to delineate the pattern of cracking on plans and cross-sections of the structure. At each subsequent date of observation, it will then be possible to note the appearance of new cracks and the elongation of ones previously marked.

Embedded Strain Gauges. Strains in soil embankments and in concrete linings to tunnels, for example, are frequently measured by strain gauges embedded within these materials. Perhaps the most common application of this type of instrument is the use of embedded vibrating-wire strain gauges in concrete, such as at construction joints in concrete dams or in a continuous ring around the circumference of a concrete tunnel lining. Strains in earth embankments are frequently monitored using a strain gage of much greater travel. One version consists of a wire-wound linear variable potentiometer in a stainless steel housing, filled with oil to prevent the ingress of moisture from the soil. The housing of the potentiometer and its sliding plunger are secured in the earth embankment by steel anchor plates, so that any relative movement of the anchors may be recorded as a change in potentiometer resistance. Several potentiometers may be installed in line in a trench to give a complete record of movements along the trench axis.

Convergence-Monitoring and Surface-Mounted Extensometers. Convergence instruments are most commonly used in underground applications to measure the inward movements or convergence of the walls of the excavation. The same instruments may be used to measure extension as opposed to inward movement, and also to measure movements along the surface of a rock slope or foundation (Fig. 1) (Milner, 1969).

Perhaps the most common convergence meter design is a stainless steel tape with a tensioning spring and a mechanical measuring device. The convergence meter is usually portable and measures between any number of pairs of fixed targets, although tapes can also be permanently installed, anchored at one end and with a freely suspended weight at the other. In the portable versions of the instrument, tape tensioning is usually achieved by a springloading device. The tape tension is critical to the measuring accuracy, and when selecting a convergence instrument particular attention should be given to tension reproducibility.

Convergence targets for both rod and tape convergence meters generally consist of a steel or aluminum anchor bolt secured in the ground by an expanding shell anchor or by grouting. Grouting with polyester resin gives a more reliable anchor

FIGURE 1. A convergence monitoring extensometer in use to measure movements in a shale tunnel, Toronto, Ontario.

than the mechanical variety, and at the same time the resin is fast-setting so that readings can be taken within minutes of installation. The targets are provided, depending on the design of the individual instrument, with a spherical or flat reference surface or with a hook to connect with the portable convergence meter.

Settlement Gauges. By definition, a settlement gauge measures vertical movements only, usually on the principle of the "U tube." The level of fluid in one end of a plastic tube is compared with the level of fluid in the other, one end being mounted in a stable instrument house and the other in the structure to be monitored. The simple U-tube method is convenient when it is possible to have an instrument house at very nearly the same elevation as that of the monitoring location. The permissible difference in elevation can be increased by placing two liquids—for example, water and mercury—in one tube, or by providing a measured and constant back-pressure to the end of the tube located in the instrument house. The tube in the instrument house is mounted on a panel alongside a measuring scale. If water is used as a fluid, it should be de-aired to prevent the formation of bubbles that may give erroneous readings; in cold climates the water should contain an antifreeze additive.

Borehole Extensometers. The displacement-measuring instruments described in the previous paragraphs have in the main been designed for measurements at the surface of the soil, rock, or concrete rather than at depth in the ground. Superficial measurements are simple but may not be sufficient. "In-depth" readings are required to locate the base of a landslide, for example or to assess the extent and depth of ground movements as they spread outward from an underground excavation. Superficial movements can sometimes be misleading. Convergence anchors installed in the walls of an underground excavation at shallow depth may give readings that reflect the local movement of loose blocks; the superficial movements of excavated or natural slopes may be the result of shallow soil creep. The extensometers and inclinometers described in this and the subsequent paragraph are designed for installation in boreholes, and are used to provide this missing information (Heilbron and Saylor, 1936; Whittaker and Hodgkinson, 1970; Burland et al., 1972).

Extensometers measure movements in the direction of the axis of the borehole (for example, settlement when the borehole is vertical), whereas inclinometers measure movements in a perpendicular direction. Borehole extensometer instruments usually record the differential movements of anchor points installed at various depths, from which rods or tensioned wires extend to a measuring instrument at the borehole collar. As many as 10 to 20 anchor points can be monitored in any one hole, or a larger number of simple two-position

instruments can be used. Usually the deepest anchor is located below the greatest predicted depth of movement to act as a stable reference. It is advisable to monitor movement of the borehole collar with respect to an external datum point.

The *multiple-rod* instruments (Fig. 2) are perhaps the most reliable and accurate. Usually each rod is protected by an individual plastic tube, and a number of such protected rods are installed side by side in the borehole, which is then backfilled with grout. The end of each rod protrudes from its protective sleeve and is secured to the surrounding rock by the grout. The remaining length of rod is free to move within its sleeve so that measurement of the movement of the rod at the borehole collar gives a reliable measurement of anchor movement. Measurements are usually made using a portable dial indicator gauge; however, when continuous-monitoring or automatic warning systems are required, an electric transducer may be installed permanently at the head of each rod. A disadvantage of the multiple-rod system is that the required diameter of the borehole increases according to the number of rods in the hole; multiple-wire extensometers are preferred in smaller drill holes. The wires are fixed to the rock or soil using mechanical expanding anchors. They must be tensioned, and the reproducibility of tensioning will affect the accuracy of measurement. Accuracy is also to some extent limited by the potential for creep and kinking of the wires. The use of multiple-wire instruments is generally restricted to applications where large movements are expected.

The *multiple-magnet extensometer* is a more recent development that uses for reference targets a series of ring magnets mounted on the outside of a plastic guide tube and fixed to the rock by grouting or mechanical anchors. To take readings, a probe instrument incorporating reed switches is inserted in the guide tube. When a reed switch enters the field of a magnet it closes, completing an electric circuit and activating a lamp or buzzer in the readout instrument. The probe is connected to the borehole collar by a device that accurately measures the distance between adjacent pairs of magnets. The probe instrument can therefore measure at any number of locations along the borehole and can be used to read in any number of holes. The diameter of magnets is small so that any number may be installed in a small-diameter drill hole. The sensitivity of the instrument is generally in the range of 0.1 to 1.0 mm; rod-mounted probes are used for greater accuracy and for horizontal or upward-inclined measurements, whereas tape-suspended probes are generally used for less precise measurements, such as of soil settlements.

Borehole Inclinometers. Inclinometers designed for permanent installation in a drill hole usually consist of a chain of pivoted rods. Rotations are measured at the pivot between each pair of

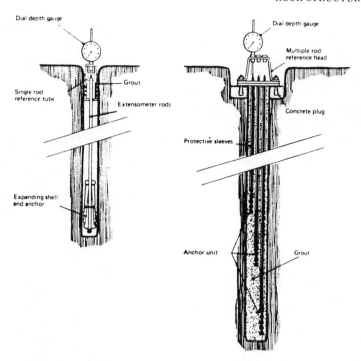

FIGURE 2. Installation details of the single- and multiple-rod extensometer system (courtesy of Rock Instruments Ltd, England).

rods, by means of resistance strain gauges mounted on cantilevers, or by inductance or capacitance transducers. Perhaps the simplest type of borehole inclinometer is the *shear strip,* which records movement perpendicular to the borehole axis. The shear strip consists of a printed circuit strip onto which electrical resistors are soldered in parallel. When the strip is sheared by ground movements the depth of shearing is recorded as a change in the total electrical resistance; the depth of movement can be determined but not its magnitude.

Probe inclinometers ("slope indicators," Fig. 3) are inserted into the borehole each time a set of readings is to be taken. They travel along a special borehole casing or lining tube with internal guide grooves to align the probe. The sensing device in the probe is usually a cantilever pendulum with resistance strain gages or vibrating-wire or inductive transducers to monitor cantilever deflection. Other types of probe incorporate electrolevels or servo-accelerometers. These probe inclinometers generally measure up to an angle of 10° or 20° either side of vertical, although instruments for use in horizontal holes (measuring vertical movement) are now also available. They can detect differential movements of from 0.5 to 1.0 mm per 10 m length of hole (Kellstenius and Bergam, 1961; Wilson, 1962; Henderson and Matich, 1963; Bromwell et al., 1971; Cornforth, 1973; Green, 1973; Phillips and James, 1973).

FIGURE 3. A Sinco portable inclinometer probe connected by cable to the readout unit. Also shown are the borehole collar and pulley assembly.

Horizontal movements can also be measured using pendulum or inverted pendulum devices (Fig. 4). In the pendulum type of instrument, a wire is clamped to the top of the structure with a heavy weight at the lower end. The weight is usually suspended in a water or oil tank to damp oscilla-

499

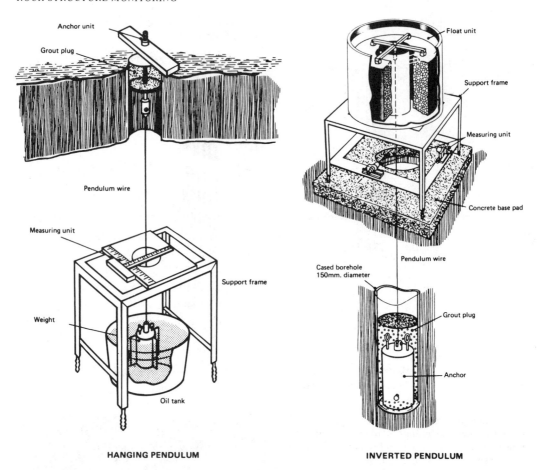

FIGURE 4. Hanging and inverted pendula used to measure horizontal displacements (courtesy of Rock Instruments, Ltd, England).

tions. Horizontal movements are measured with reference to the upper wire clamp at any number of elevations along the wire, either by using calibrated scales fixed to plates mounted on the structure or by using a vernier microscope sighted onto the wire. The inverted pendulum has the wire anchored at the lower end, usually in stable ground at the base of a hole. The upper end of the wire is secured to a float in a water tank. The float tensions the wire and keeps it vertical. Measurements of horizontal movement are made with respect to the wire, as is the case with the suspended pendulum. Inverted pendula are frequently used to provide a stable datum for geodetic surveying of first-order accuracy. Suspended and inverted pendula are often used for routine monitoring of the horizontal movements of large dams.

Ground-water Instrumentation. Ground-water has an important influence on the settlement and stability of rock and soil structures, and drainage can be a very effective method for stabilization. Monitoring of water pressures (piezometric moni-

toring) is often required to give information for design, and for control during construction (Cooling, 1962; Penman, 1956; Vaughan, 1969, 1973).

Instability problems can be associated with either excessive water discharge or excessive water pressure. It is important to recognize that, if the ground is impermeable, excessive pressures can exist without large discharges being evident, so that a tight rock mass may appear superficially dry, yet may benefit considerably from drainage. At the other extreme, if abundant water is available and the rock mass open, then substantial discharges can occur under quite small pressures. In such cases, the water pressures may be no great hazard, but the discharge may be inconvenient and can lead to progressive erosion or undermining. Measurements of ground-water pressures and water-table elevations are most often required for the design calculations. However, the ground-water regime may be greatly modified by engineering activities, particularly by excavation of soil or of rock. Because these changes are difficult to predict, the construction phase should be monitored periodically to check the predictions made in design. The results are used to control the rate of excavation or placing of fill and to determine the success of drainage, grouting, ground freezing, and other remedial works.

Standpipe Piezometers. Water pressures are monitored using piezometers, which in their simplest form consist of standpipes installed in a drillhole. The standpipe, or well point, consists of a tube perforated at its lower end or fitted with a porous plastic or ceramic element. A hole is drilled and the standpipe is inserted. The end, or "tip," of the standpipe is surrounded by filter material, usually coarse sand or gravel immediately around the tip and fine sand above. The piezometer tip is then hydraulically isolated, usually by placing a bentonite clay plug above the fine sand and by backfilling the remainder of the hole with clay or cement grout. The ground water rises in the standpipe until the water head is in equilibrium with water pressure in the ground around the piezometer tip.

The ground-water pressure at the tip can then be monitored using a simple water-level probe, or "dipmeter." The *dipmeter* consists of a twin conductor cable with the end of each conductor connected to a brass cylinder, the two cylinders being separated by insulating material. Entry of the dipmeter tip into the water completes an electric circuit and activates a buzzer or lamp indicator in the cable reel. The cable is graduated along its length so that the distance between the water level in the standpipe and the collar of the standpipe tube may be measured.

Hydraulic Twin-Tube Piezometers. The comparatively large internal diameter of the standpipe tube means that water may take a considerable time to rise or fall in response to changes of water pressure at the tip. The "response time" of a piezometer may be critical when relatively rapid fluctuations in ground-water pressure are expected and when the ground around the tip is impermeable. Hydraulic twin-tube piezometers overcome this problem to some extent, having a much smaller diameter of tube. They can also be monitored at a centrally located instrument house, and can be flushed periodically with de-aired water to prevent the formation of air bubbles that in other types of piezometers may lead to reading inaccuracies.

Pneumatic Piezometers. These consist of a porous plastic or ceramic element connected to a pneumatic transducer, which in turn is connected by flexible plastic tubes to a terminal panel or gauge house at the ground surface. Changes in ground-water pressure in the vicinity of the piezometer tip are measured by injecting air pressure down one of the twin tubes, sufficient to balance the water pressure on the reverse side of the transducer diaphragm. The pressure, which is sufficient to cause a return flow of air along the second plastic tube, is recorded; it is equal to the ground-water pressure. Only a very small diaphragm deflection is necessary to permit return flow, so that the volume change is minimal. This type of transducer has a rapid response to ground-water pressure fluctuations, even in impermeable ground. A further advantage is that the pneumatic piezometer may be installed at any elevation below or above the gauge house location; any magnitude of pressure can be measured.

Electric Piezometers. The electric types of piezometers consist of a porous plastic or ceramic element connected to an electric transducer, commonly of the vibrating-wire or resistance strain gauge type. Electric leads from the transducer connect the piezometer tip to the readout instrument at the ground surface. Electric piezometers have a rapid response to ground-water pressure fluctuations and may be used as an alternative to the pneumatic type in materials of low permeability, such as clays. Electric instruments are essential if automatic or continuous chart recording of pressure fluctuations is required.

Soil and Rock Pressures and Loads on Supports. Retaining walls, anchors, and rockbolts develop their prescribed working pressure or load only as the rock or soil starts to move against the retaining system. An attempt is usually made at the design stage to predict the pressures or loads that will act, but these predictions are often unreliable due to the accuracy limitations of input data and to the inherent limitations of calculation procedures. Therefore monitoring is often specified to check and confirm the actual pressures that are developed (Panek, 1961; Kruse, 1965; Brown, 1971).

Gloetzl Pressure Cells. The Gloetzl type of cell consists of a flatjack connected to a pneumatic or hydraulic transducer, which in turn is connected by twin flexible plastic tubes to a terminal panel or gauge house. The flatjack may be circular or rectangular. A variety of sizes are available, from small, rectangular ones for installation in a comparatively thin concrete tunnel lining, to circular flatjacks of 1 m or greater diameter for installation in earth embankments and foundations. The flatjack consists of two thin sheets of metal, welded around the circumference and filled with oil or with mercury to form a hydraulic pillow. The flatjack is connected by a short length of metal tube to the hydraulic transducer. The complete unit is installed at the location where pressure is to be measured, for example, by placing the cell in a carefully excavated pocket in a soil embankment, or by securing the cell to a rock surface or to concrete reinforcement prior to the pouring or spraying of concrete at that location.

Electrically Operated Pressure Cells. Perhaps the most common type of pressure cell operating on an electrical principle employs a vibrating-wire system. The cell consists of two circular plates of steel, each with the central portion removed by machining to form a thin and flexible diaphragm. Metal pillars are mounted on the internal face of each diaphragm at the circle of maximum angular deflection. The action of earth or rock pressure on the external faces of the cell causes a small angular rotation of the pillars, between which a vibrating wire is stretched under tension. Each of the two faces of the cell acts as a measuring device. The two halves are bolted together around the circumference and the complete unit is waterproofed. Electric leads connect the cell to a terminal panel or gauge house.

Load Cells. Load measurements are often required to monitor the tension in cable anchors and rockbolts, and to monitor the loads acting in steel ribs, columns, and arches. "Solid" load cells monitor compressive forces—for example, beneath piles—while "hollow" or center-hole, load cells are used in rockbolting and cable-anchoring applications. In the latter case, the bolt or cable passes through the center of the cell and the tensile force in the cable is transmitted as a compressive load through the load cell to the surface of the rock or concrete.

Monitoring of Ground Vibrations. Cyclic variations in movement, caused, for example, by daily temperature changes or seasonal fluctuations in ground-water level, are usually of sufficiently long period to be detected without difficulty using conventional forms of instrumentation. Cyclic variations with periods of seconds or fractions of a second may be experienced as, for example, structural oscillations due to wind loading; the recording of these dynamic movements generally requires specially designed instrumentation. Oscillations of even shorter wavelength, such as those caused by blasting or by earthquakes, require special dynamic detection and recording devices.

Most large dams are nowadays instrumented to record the magnitude and waveform characteristics of earthquakes that may affect their stability. Automatic three-axis seismographs and very sensitive tiltmeters (e.g., 10^{-9} rad) monitor ground motions; the recording is triggered by the onset of an earth tremor with amplitude greater than a predetermined minimum. Generally, a seismograph consists of a suspended inertia mass that remains stationary with respect to the housing of the instrument. A trace of the oscillation waveform is produced on a chart recorder.

A further application of dynamic movement monitoring is the detection of rock noise with the object of predicting rock bursts underground or landslides at the ground surface. At the instant of a rock burst or landslide, the noise is audible, but subaudible noises are generated at much earlier stages in the development of instability. Listening devices can be used to detect these subaudible

noises, and comprise in their simplest form a detector (geophone), an amplifier, and a counter or chart recorder.

Monitoring Requirements for Surface and Underground Works

Tunnels. The principal reason for monitoring in tunnels is to ensure safety of the construction workers in the tunnel and in the case of shallow tunnels, buildings, and services located above the tunnel along its line. Within the tunnel, convergence measurements are the most common, made in profiles that are typically spaced at 50- to 100-m intervals along the line of advance. Because of its simplicity and speed, convergence monitoring lends itself to frequent and closely spaced observations at minimal expense and without inconvenience. At each profile, horizontal and vertical convergences are usually recorded, with supplementary measurements across oblique diameters if considered necessary (Fig. 5) (Wallace and Ortel, 1971; Jones and Mahar, 1973; Schmidt, 1974; Schmidt and Dunnicliff, 1974).

A small percentage of convergence-monitoring profiles, selected to represent typical rock and tunnel-lining conditions, are chosen for more intensive monitoring as a check on primary and secondary support design. At these locations multiple-rod extensometers may be installed in holes drilled from within the tunnel. If possible, the deepest anchor should be at least two tunnel diameters from the wall of the tunnel to provide a stable reference for movement measurements. Where the depth of tunnel is shallow, multiple-position extensometers may be installed in vertically drilled holes from the ground surface, the base of the drill hole being just above the crown of the tunnel. Reference anchors are installed at various positions between the crown and the ground surface. The instrument is generally installed before tunneling begins so that the growing pattern of ground subsidence can be observed as the tunnel passes beneath it. The measured ground movements are compared with those predicted by design calculations, to warn of any major departures from the assumptions made by design engineers. In addition, pressure cells are generally installed at the time of placing the final lining, to give confirmation of lining stresses. Typically, these are located in pairs around the perimeter of the profile, one cell in each pair measuring the contact pressure between the rock and concrete and the second measuring the tangential or hoop stress in the concrete.

In addition to measurements relating to the behavior of the tunnel itself, a program of monitoring is generally required at the portals and at the ground surface along the tunnel line. The portals are often located in weathered rock or in

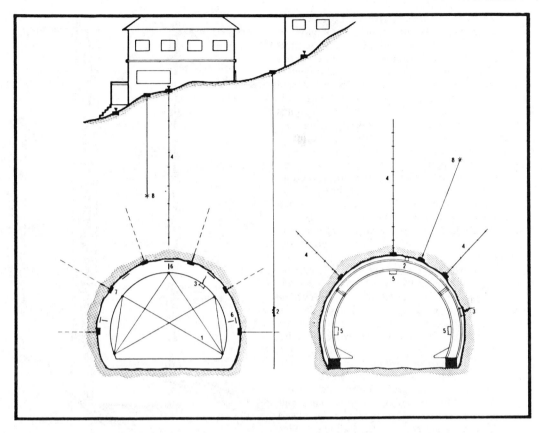

FIGURE 5. Tunnel instrumentation: (1) convergence measurements; (2) inclinometers; (3) crack monitoring; (4) multiple extensometers; (5) strain gages on and within the lining; (6) pressure cells; (7) rockbolt load cells; (8) piezometers (courtesy of Rock Instruments Ltd, England).

soil; this, combined with the relatively shallow cover that is usual at the portals, may give rise to stability problems greater than those encountered when the tunnel has penetrated a substantial distance. The presence of a slope will aggravate problems of portal stability; conversely, excavation into the toe of a slope will frequently endanger the stability of that slope.

At the ground surface, all major structures beneath which the tunnel is to pass should be visually inspected before the start of construction operations to record pre-existing cracking and structural damage that might otherwise be subsequently blamed on the tunneling works. A program of settlement monitoring may be needed if the depth of cover above the tunnel is limited, such that settlements may spread to the ground surface. Conventional leveling survey methods may be used, with a pattern of survey benchmarks established along the line of the tunnel and extending an appropriate distance to either side of the center line. Precise leveling will generally be required, with an accuracy on the order of 1 mm at each station. After the benchmarks have been located

and marked and an initial set of readings taken at all points, attention may be concentrated on an area immediately ahead of and behind the advancing tunnel face, where the greatest rate of movement is to be expected. Conventional survey methods may be supplemented by the monitoring of cracks in structures and in paved surfaces, and if necessary by the installation of permanent settlement gauges at critical locations.

Tunneling activities may have an appreciable effect on the local ground-water regime, and it is often necessary to install piezometers along the tunnel line to monitor drawdown of the water table. Where there is limited cover, these piezometers may be installed from the ground surface in vertical drill holes; the standpipe type of instrument will usually suffice. For deeper tunnels, piezometers must be installed in holes drilled from within the tunnel, in which case pneumatic or electric transducer instruments will be needed. Piezometers may also be required to measure water pressures that develop behind the final lining, since these pressures can contribute significantly to the lining stresses.

Shafts. The control of ground movements in and around vertical shafts in soil or rock may also rely on convergence-monitoring devices such as described for use in tunnels. It is often more convenient, however, to use probe inclinometers with the guide tubing located in drill holes external to the shaft lines. The inclinometer holes are best drilled before sinking the shaft so that readings may be taken continuously before, during, and after construction, without in any way interfering with construction work. Control of horizontal movements is particularly important in ground that is highly stressed in a horizontal direction, since stress relief by excavation of the shaft may be followed by appreciable inward movements. Movement monitoring may be augmented, to provide a check on shaft lining design, by the installation of pressure cells in and behind the lining (Fig. 6).

Frequently shafts pass through soil and into rock, meeting a wide variety of ground and ground-water conditions. Often the excavations require dewatering, freezing, or grouting to provide satisfactory working conditions to maintain shaft stability. Dewatering and compressed-air works require careful control of the ground-water table using a pattern of piezometers installed in vertical drill holes at various distances from the shaft axis. The response time of the selected piezometer system should match the permeability of the ground, and each piezometer tip should be hydraulically isolated in a selected stratum since the drainage effects caused by shaft excavation will invariably be different at different horizons.

Ground-freezing operations require ground temperature and careful monitoring, using sensors in drill holes over a range of distances from the shaft axis to monitor the thickness of the frozen membrane surrounding the shaft, and the temperature of material within this zone. In addition, both freezing and grouting operations can give rise to ground heave, so that conventional leveling survey benchmarks should be established at critical locations on the ground surface in the vicinity of these activities.

Underground Chambers. As the size of an underground excavation increases, the requirements for support also increase, and there may be a greater danger that instability will occur. Large underground excavations are generally used to house important equipment and facilities, for example, hydroelectric turbines, oil or gas storage reservoirs, or subway stations. The impact of instability can be far greater than in extended tunnels or excavations of smaller size. In general the requirements for monitoring increase rapidly as the excavation becomes larger (Cording, 1968). The methods are in some respects similar to those used for tunnels; however, they differ in a number of important respects. The excavation generally

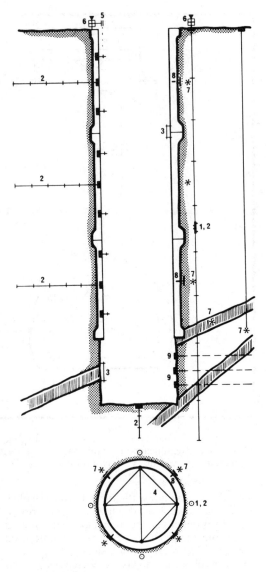

FIGURE 6. Shaft instrumentation: (1) inclinometers; (2) extensometers; (3) surface strain gauges; (4) convergence monitoring; (5) optical plumbing; (6) conventional surveying; (7) piezometers; (8) lining pressure cells; (9) anchor load cells (courtesy of Rock Instruments Ltd, England).

proceeds in stages, starting with the driving of a small adit or heading. This adit provides access to the immediate area of the excavation and may be used for rock testing, for confirmation of geological conditions, and for the installation of instruments in advance of the main works. Convergence measurements are difficult in an excavation that is steadily increasing in size; although they are generally made, greater reliance is placed on multiple-position extensometers installed from within the pilot heading or from the ground surface above. The ground conditions in the vicinity of a large underground excavation are generally explored in far greater detail than would be possible for an extended excavation such as a tunnel, so that instruments can be located to monitor the development of potential movements whose mechanisms have been identified and analyzed in advance. Through-going joints and faults may form the boundaries of potential wedge instability, and each potentially unstable zone is individually monitored during construction.

Soil and Rock Slopes. Slope failure can occur in a number of ways: circular or wedge sliding, planar sliding of rock slabs, and toppling or rotation of blocks. The cause will depend on such factors as the stratigraphy and bedding orientations of rocks and soils within the slopes, on the mechanical characteristics of the joints and bedding planes, and on the geometry of the slope, its inclination, and height. Usually these factors can be predicted and the monitoring system designed accordingly (Fig. 7) (Goodman and Blake, 1966; Wilson, 1970; Franklin and Denton, 1973).

Ground-water pressures are a critical factor. Piezometer installations will generally be needed,

whether to establish the pre-existing ground-water configuration in an existing slope whose stability is to be assessed, or to provide confirmation of changes in the ground-water regime where the slope is to be created by excavation. For an existing natural slope, the measurement of ground-water pressures may be a one-time or possibly seasonal operation, whereas for an excavated slope the readings will continue throughout the works. In either case, if drainage is to be used for improving slope stability, a continuing program of ground-water observations will be needed to monitor the efficiency of the drainage operations.

When the slope is extensive—such as an elongated road cut in a mountainous region, or a stretch of potentially unstable slope along the perimeter of a reservoir—movement monitoring must generally be confined to superficial observations, using conventional survey, electro-optic methods, or photogrammetry. The aim should be to identify critical slopes in the course of site exploration. Localized areas of critical slope, such as at tunnel portals, at the abutments of a dam, or at the perimeter of a basement excavation for a high-rise building, may require a more intensive program of monitoring using a variety of instruments both at

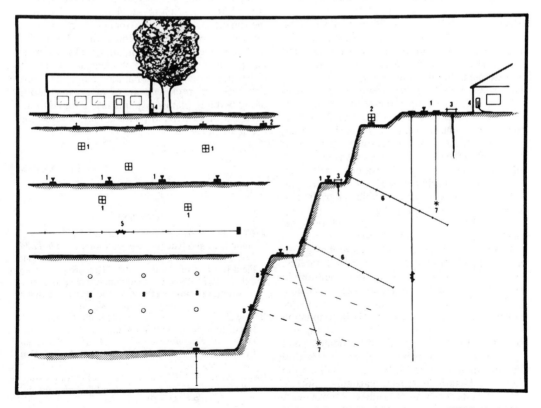

FIGURE 7. Slope instrumentation: (1) geodetic or electro-optic distance measurement (EDM) surveying; (2) line-of-sight survey offset measurements; (3) tension crack monitoring; (4) settlement measurements; (5) inclinometers; (6) extensometers; (7) piezometers; (8) anchor load cells (courtesy of Rock Instruments Ltd, England).

the ground surface and in depth. Conventional or electro-optic surveying should be used to provide a frame of reference. In addition, it is often helpful to install an inclinometer casing in holes drilled vertically through the potential planes of sliding, using a probe inclinometer or "slope indicator" to detect horizontal movements. Tension cracks that may have developed at the crest of the slope, or major rock joints and faults that may become involved in sliding, should be instrumented using crack-monitoring techniques. Borehole extensometers may also be useful, particularly in detecting the movements of isolated and potentially unstable blocks and wedges.

Embankments. Earth and rockfill embankments and embankment dams are generally designed using well-established soil mechanics techniques, to ensure stability and to check the anticipated settlements and lateral displacements. Monitoring may be useful as a check on the design calculations, to confirm the quality of earth materials used in their construction, and also to provide information for control of construction procedures (Wilson, 1967; Wilson and Marano, 1968; Irwin, 1973; Marsland, 1973).

In the case of a small or a critical embankment such as an embankment dam, the entire structure may be monitored, but more often monitoring is required for full-scale trials of test sections. Piezometers are often needed to measure the rise in pore-water pressure during construction, and its dissipation when construction is delayed. Embankment failures are frequently associated with excess pore-water pressures developed due to rapid or uncontrolled placing of fill. Settlements are generally measured along the embankment crest using standard leveling methods. In addition, it is generally advisable to measure settlements at depth within the structure using extensometers or settlement gages. In-depth measurements of settlement are particularly necessary when the foundations are on weak or compressible materials such as organic deposits or soft clay.

While settlement gages may be used to record the progress of embankment consolidation, measurements of horizontal displacement may give a better indication of the development of instability. Probe inclinometers are often used, operating in vertical guide tubes installed as close as possible to the crest of the embankment. The inclinometer casing should be extended deep into the foundation to distinguish reliably between embankment spread that is a natural consequence of settlement, and the development of incipient slide planes within the embankment or the foundation itself. In the case of embankment dams in particular, electro-optic methods may be used to measure movements on the downstream face of the dam as a result of both settlement and the action of water pressure on the upstream face.

Dams. Methods for monitoring the abutments and foundations of a dam are essentially similar to those for monitoring slopes and foundations. Much reliance is placed on the design calculations and on the assessment of rock, soil, and concrete properties, and it is usually considered imperative that the reliability of data and assumption should be checked at every stage (Hosking and Hilton, 1963; Planicka and Nosek, 1970).

Most dams employ a combination of grouting and drainage in the foundation and abutment rock and soils in order to control leakage from beneath and around the dam structure, and at the same time to limit the magnitude of pore-water pressures that are developed. These pressures are predicted by calculation and are subsequently checked by piezometers. In the case of embankment dams, piezometers are installed not only in the foundations and abutments, but also in the dam materials themselves, to guard against the possibility of embankment failures due to inadequate drainage.

The greatest attention is generally given to displacement components in the downstream direction, since it is in this direction that the greatest force acts and that failures are likely to develop. Measurements may be made on the exposed downstream face of the dam and on the rock or soil abutments using electro-optic distance measuring or theodolite triangulation. Inclinometers may be used in boreholes drilled from the dam crest deep into the foundation rocks, to detect any possible development of shear displacement either in the dam itself, or at the contact between concrete and rock or soil, or at greater depth along bedding planes or joints in the foundation. Inverted or suspended pendula may be used for the same purpose.

J. A. FRANKLIN

References

Ashkenazi, V., 1973. The measurement of spatial deformations by geodetic methods, in *Conference in Field Instrumentation in Geotechnical Engineering*. London: Field Institute of Geotechnical Engineering, 1-12.

Borchers, P. E., 1968. Photogrammetric measurements of structural movements, *Am. Soc. Civil Engineers Proc., Jour. Surveying and Mapping Div.* **94** (SU2).

Bromwell, L. G.; Ryan, C. R.; and Toth, W. E., 1971. Recording inclinometer for measuring soil movements, in *Proceedings of the 4th Pan American Conference on Soil Mechanics and Foundation Engineering*, Vol. 2.

Brown, S. F., 1971. The performance of earth pressure cells for use in road research, *Civil Engineering and Pub. Works Rev.* **66,** 160-165.

Burland, J. B.; Moore, J. F. A.; and Smith, P. D. K., 1972. A simple and precise borehole extensometer, *Géotechnique* **22,** 174-177.

Cheney, J. E., 1973. Techniques and equipment using the surveyors level for accurate measurement of building movement, in *Conference in Field Instrumentation in Geotechnical Engineering.* London: Field Institute of Geotechnical Engineering 85-99.

Cooling, L. F., 1962. Field measurements in soil mechanics, *Géotechnique* **12**, 75-104.

Cording, E. J., 1968. The stability during construction of three large underground openings in rock, *U.S. Army Corp Engineers Waterways Expt. Sta. Tech. Rept. 1-818.*

Cornforth, D. H., 1973. Performance characteristics of the slope indicator series 200B inclinometer, in *Conference in Field Instrumentation in Geotechnical Engineering.* London: Field Institute of Geotechnical Engineering, 126-135.

Franklin, J. A., and Denton, P. E., 1973. The monitoring of rock slopes, *Quart. Jour. Eng. Geology* **6** (3/4), 259-286.

Froome, K. D., and Bradsell, R. H., 1966. A new method for the measurement of distance up to 5000 ft. by means of a modulated light beam, *Jour. Sci. Instruments* **43**, 129-133.

Goodman, R. D., and Blake, W., 1966. Rock noise in landslides and slope failures, *Highway Research Rev.* **119**, 50-60.

Green, G. E, 1973. Principles and performance of two inclinometers for measuring horizontal ground movements, in *Conference in Field Instrumentation in Geotechnical Engineering.* London: Field Institute of Geotechnical Engineering, 166-179.

Heilbron, C. H., and Saylor, W. H., 1936. An invar tape extensometer, *Civil Engineering*, 99-101.

Henderson, R. P., and Matich, M. A. J., 1963. Use of slope indicator to measure movement in earth slopes and bulkheads, *Am. Soc. Testing and Materials Stand.* **322**, 166-186.

Hosking, A. D., and Hilton, J. I., 1963. Instrumentation of earth dams on the Snowy Mountain scheme, in *Proceedings of the 4th Australian-New Zealand Conference on Soil Mechanics and Foundation Engineering,* 251-262.

Irwin, M. J., 1973. Instruments developed by the TRRL for studying the behavior of earthworks, in *Conference in Field Instrumentation in Geotechnical Engineering.* London: Field Institute of Geotechnical Engineering, 194-206.

Jones, R. A., and Mahar, J. W., 1973. Instrumentation to monitor behavior of shotcrete support systems: use of shotcrete for underground structural support, in *Proceedings of the Engineering Foundation Conference, South Berwick, Maine.* ASCE-ACI, ACI Pub. SP-45, 297-319.

Kellstenius, T., and Bergam, W., 1961. In situ determination of horizontal ground movements, in *Proceedings of the 5th International Conference on Soil Mechanics and Foundation Engineering,* Vol. 1. 481-485.

Kobold, F., 1961. Measurement of displacement and deformation by geodetic methods, *Am. Soc. Civil Engineers Proc., Jour. Surveying and Mapping Div.* **87** (SU2), paper 2873.

Kruse, G. H., 1965. Measurement of embankment stresses on a hundred-foot-high retaining wall: instruments and apparatus for soil and rock mechanics, *Am. Soc. Testing and Materials Stand.* **392**, 131-142.

Marsland, A., 1973. Instrumentation of flood defense banks along the River Thames, in *Conference in Field Instrumentation in Geotechnical Engineering.* London: Field Institute of Geotechnical Engineering, 297-303.

Milner, R. M., 1969. Accuracy of measurements with steel tapes, *Building Res.,* CP51/69.

Panek, L. A., 1961. Measurement of rock pressure with an hydraulic cell, in *Proceedings of the American Institute of Mechanical Engineers Tech. Paper 61.*

Penman, A. D. M., 1956. A field piezometer apparatus, *Géotechnique* **6**, 57-65.

Phillips, S. H. E., and James, E. L., 1973. An inclinometer for measuring the deformation of buried structures with reference to multi-tied diaphragm walls, in *Conference in Field Instrumentation in Geotechnical Engineering.* London: Field Institute of Geotechnical Engineering 359-369.

Planicka, A., and Nosek, L., 1970. Terrestrial photogrammetry in measurement of deformations of rockfill dams, in *Transactions of the 10th International Congress on Large Dams,* Vol. 3. 207-215.

Romaniello, C. G., 1970. Advancing technology in electronic surveying, *Am. Soc. Civil Engineers Proc., Jour. Surveying and Mapping Div.,* **96** (SU2), paper 7566.

St. John, C. M., and Thomas, T. L., 1970. The NPL Mekometer and its application in mine surveying and rock mechanics, *Am. Inst. Mining and Metall. Engineers, Trans.* **79**, A31-A36.

Schmidt, B., 1974. Prediction of settlements due to tunneling in soil: three case histories, in *Proceedings of the Rapid Excavation and Tunneling Conference,* Vol. 2. Chicago: American Institute of Mechanical Engineers, 1,179-1,200.

Schmidt, B., and Dunnicliff, C. J., 1974. *Construction Monitoring of Soft Ground Rapid Transit Tunnels.* Washington, D.C.: U.S. Department of Transportation, UMTA-TSC 661.

Tomlinson, R. W., 1970. Distance measuring instruments, *Am. Soc. Civil Engineers Proc., Jour. Surveying and Mapping Div.* **96** (SU), paper 7521.

Vaughan, P. R., 1969. A note on sealing piezometers in boreholes, *Géotechnique* **19**, 405-413.

Vaughan, P. R., 1973. The measurement of pore pressure with piezometers, in *Conference in Field Instrumentation in Geotechnical Engineering. London:* Field Institute of Geotechnical Engineering, 411-422.

Wallace, G. B., and Ortel, W. H., 1971. Tests for tunnel support and lining requirements, in *Proceedings of the 12th Symposium on Rock Mechanics.* Rolla, Missouri: University of Missouri, 933-959.

Whittaker, B. N., and Hodgkinson, D. R., 1970. Strata displacement measurement by multi-wire borehole instrumentation, *Colliery Guard* **218**, 445-449.

Wilson, S. D., 1962. The use of slope measuring devices to determine movements in earth masses: field testing of soils, *Am. Soc. Testing and Materials Stand.* **322**, 187-197.

Wilson, S. D., 1967. Investigation of embankment performance, *Am. Soc. Civil Engineers Proc., Jour. Soil Mechanics and Found. Div.* **93** (SM4), paper 5311.

Wilson, S. D., 1970. Observational data on ground movements related to slope instability (6th Terzaghi lecture), *Am. Soc. Civil Engineers Proc.* (SM5), paper 7508.

Wilson, S. D., and Marano, D., 1968. Performance of muddy run embankment, *Am. Soc. Civil Engineers Proc.* **94** (SM4), 859-881.

Cross-references: *Foundation Engineering; Reinforced Earth; Remote Sensing, Engineering Geology; Residual Stress, Rocks; Rheology, Soil and Rock; Rock Mechanics; Rocks, Engineering Properties; Rock Slope Engineering.* Vol. XIV: *Landslide Control; Photogeology; Photo Interpretation; Surveying, Electronic.*

ROTARY DRILLING—See Vol. XIV: BOREHOLE DRILLING.

S

SCANNING SYSTEMS—See REMOTE SENSING, ENGINEERING GEOLOGY. Vol. XIV: AERIAL SURVEYS, GENERAL; PHOTOGEOLOGY; PHOTO INTERPRETATION; SEA SURVEYS.

SEISMIC DESIGN—See EARTHQUAKE ENGINEERING.

SEISMOLOGICAL METHODS

Seismological methods are the means by which the nature and origin of seismic events are determined. The main elements used in this determination are detection, recording interpretation, and analysis. Through the application of seismological methods, we can attempt to learn about the structure of the earth, about the causes of earthquakes, and about how we may in the future predict the location, size, and time of occurrence of earthquakes.

Detection and Recording of Seismic Events

Energy from the earthquake focus is transmitted by means of the propagation of elastic waves. It is the arrival of these waves at each location that causes surface disturbances such as shaking and lurching. These waves move out in all directions from the source and travel great distances, losing much of their energy in the process. At large distances, very sensitive detecting devices are required to receive these waves and obtain a picture of the wave pattern (Richter, 1958).

A *seismometer* is a detector used by seismologists to measure the amplitude, frequency, and duration of ground motion waves. Modern seismometers are based on the fundamental principle of the pendulum. The pendulum support is embedded in a concrete pier attached to the earth and thus moves in harmony with the ground, while the suspended mass remains in place due to its inertia. This relative differential movement of the mass in relation to the pier is the direct detection of the seismic energy wave. Figure 1 shows four basic types of seismometers used to measure the horizontal or vertical component of ground motion.

The next step, recording this differential

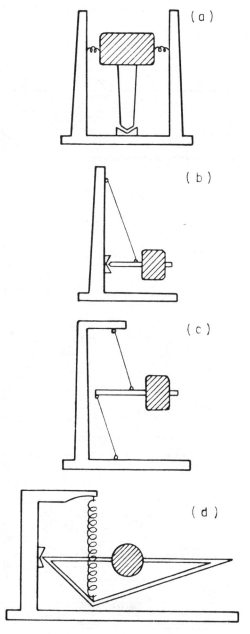

FIGURE 1. Four basic types of seismometers: (a) inverted pendulum, (b) conventional horizontal pendulum; (c) Zollner suspension; (d) vertical seismometer (after Hodgson, 1964).

movement, is accomplished through the use of a *seismograph,* which consists of an amplifier and a recording apparatus. The first seismographs consisted of pens attached to the suspended mass, which recorded on a rotating drum attached to the support. The large amount of friction introduced in this type of system prohibited the detection of nearby small events or large events at great distances. In later seismographs, a beam of light reflected off a mirror attached to the suspended mass, and a record was made on a moving sheet of photographic paper. Today's seismographs record electromagnetically through the use of a coil moving through a magnetic field, thus generating a current. This current operates a galvanometer, which records photographically or by pen on a moving drum. The continuous record produced is known as a *seismogram.* Figure 2 shows a series of seismograms recorded at the Berkeley, California, seismograph station of the September 1965 earthquake in Antioch, California. Trace 1 is from a foreshock of magnitude 2.1; traces 4, 9, 10, and 11 are from aftershocks of magnitudes 2.6, 2.8, 2.4, and 2.3, respectively.

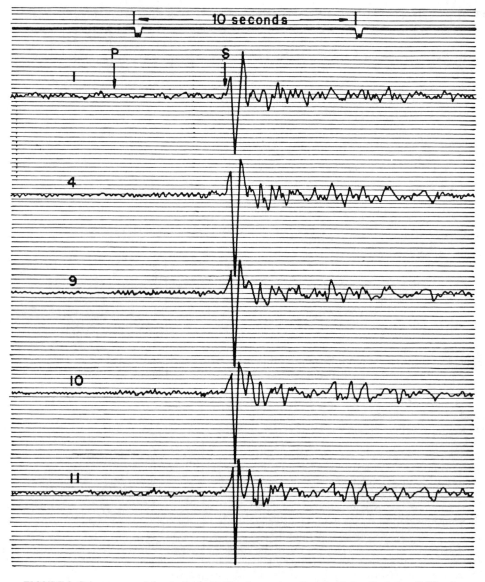

FIGURE 2. Seismograms of the Antioch, California, earthquake recorded at Berkeley, California on September 10, 1965. Trace 1 is from a foreshock of magnitude 2.1; traces 4, 9, 10, and 11 are from aftershocks of magnitudes 2.6, 2.8, 2.4, and 2.3, respectively (after Bolt, 1970).

Two other important factors must be considered in a discussion of the detection and recording of seismic events: period of vibration and orientation. Any moving system, such as a seismograph, has a natural period of vibration over which it will give generally the same response. However, since seismic waves have periods ranging from tenths of a second to several minutes, more than one seismograph is needed to detect and record the events accurately. Most well-equipped seismograph stations have one set of instruments designed to respond to short-period waves with periods from 0.5 to 2 seconds, and a second set for long-period waves of 15 to 100 seconds.

To describe ground motion completely, one must measure three components at right angles. Most seismograph stations therefore have one vertical (z-direction) seismograph, one north–south (x-direction) seismograph, and one east–west (y-direction) seismograph to permit the detected wave to be broken down into its various components.

The newest, most sophisticated recording equipment uses magnetic or digital tape to allow direct feed into a computer for analysis. These systems are generally limited to research centers, however, and the vast majority of seismograph stations still use paper recorders.

Locating the Epicenter

After the arriving seismic wave has been detected and appropriately recorded, the next step is to interpret the record to determine the location of the epicenter. A disturbance (such as a seismic event) in an elastic material (the earth) sets up two types of body waves: longitudinal and transverse. *Longitudinal waves* are compressional, in that individual particles move back and forth in the direction of wave propagation. *Transverse waves* are shear waves since the particles move at right angles to the direction of propagation.

When the seismic event occurs, the longitudinal and transverse waves are generated simultaneously, but the longitudinal wave travels at almost twice the velocity of the transverse; thus with distance the two waves become separated into two distinct pulses. On any seismic record located far enough from the seismic event to separate the two waves, the longitudinal wave is called the *P-wave* (primary, or first-arriving, wave) while the transverse wave is called the *S wave* (secondary wave). An example of the P- and S-wave separation is shown in Figure 3. By measuring the time lag between the P- and S-waves, the distance from the epicenter to the station can be calculated. This defines only a circle of computed radius, however, so two other stations are needed to "triangulate" and pinpoint the epicenter. In reality, hundreds of seismograph stations contribute to the determination of the epicenter location for any major event (Bolt, 1970).

Magnitude Versus Intensity

The terms *magnitude* and *intensity* are used to describe the size of an earthquake. *Magnitude* refers to the absolute measurement using scientific instruments of the strength of an earthquake, irrespective of the distance to the epicenter. To do this, a number of identical seismographs, accurately calibrated and distributed in a network, record the amplitude of the resulting ground motion. By plotting the log of the amplitude versus the distance to the epicenter, a point on a graph is obtained for each station. Connecting these points gives a smooth curve. As a reference datum, we also plot a curve of the log of the amplitude versus distance of the smallest event we are consistently able to detect without instruments and call this the "zero" earthquake.*

The average distance between our "zero" earthquake curve and the curve of the earthquake under study is called the magnitude of the earthquake. It is expressed as

$$M = \log A - \log A_0$$

*The accepted value is an amplitude A_0 of 0.001 mm at a distance of 100 km from the epicenter (Richter, 1958).

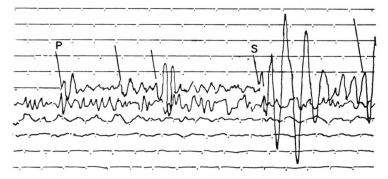

FIGURE 3. Seismogram showing the variation in arrival times of P- and S-waves (after Hodgson, 1964).

where M is the magnitude, A is the amplitude of the earthquake under study, and A_0 is the amplitude of the "zero" earthquake. This technique of magnitude determination, developed by Dr. C. F. Richter of the California Institute of Technology, is known as *Richter Magnitude;* the resulting scale is the *Richter Scale.*

The *intensity* of an earthquake refers to its relative strength through its surface effects on structures and people. As such, it is not an instrumentally measurable quantity and is closely related to distance from the epicenter, soil conditions, quality of construction practices, density of population, and above all the reliability of reports of people who experienced the event. While all these factors add to the inaccuracy of the determination of the intensity, the values assigned to various areas for a specific event allow the creation of an intensity map with isoseismals (contour lines of equal intensity), which often helps define the epicenter and allow an evaluation of the soil conditions and building practices for any given area (Guttenberg and Richter, 1956). See Table 1 in *Earthquake Engineering* for the *Modified Mercalli Scale* for earthquake intensities, which is currently in most common use.

The Mechanism of Earthquakes

One seismological method used in studying the mechanism of earthquakes involves plotting compressions and dilations in the vicinity of the epicenter. *Compressions* are defined as forces that appear to be moving toward the observer during the first instant of the earthquake, while in *dilation* areas the feeling is that the ground is being pulled away from the observer. A plot of these results almost always gives a quadrant distribution of compressions and dilations, as shown in Figure 4.

Of course, the plot of the zones of compressions and dilations is three-dimensional and the contacts are not straight lines but circles. These circles are known as *nodal planes* since the direction

of motion changes as one crosses these planes and so the motion must be zero on the plane. One of these represents the fault plane, while the other represents the plane perpendicular to the fault plane. A sufficiently concentrated seismograph network in the vicinity of an epicenter can therefore define the two nodal planes solely from the compression or dilation response of the various individual stations.

Applications for Understanding the Earth's Interior

Elastic waves travel out from the source in all directions at a velocity dependent on the propagation properties of the medium through which they travel. As a material of higher-velocity propagation is entered, the waves are bent, or refracted, along the higher-velocity layer and will arrive at a distant receiver before waves that travel only in the lower-velocity layer. This principle, which has been applied for many years in the field of exploration geophysics, is the basis of the *seismic refraction survey.* In such a survey, the energy is provided by explosives or some form of percussion and the waves are detected by receivers known as "geophones." The waves are then recorded on a special type of seismograph; from the resulting record the exploration geophysicist can interpret the shallow subsurface conditions. This survey has primarily been applied in shallow explorations for engineering projects and moderately deep explorations for petroleum.

Another type of survey, the *seismic reflection survey,* is used in exploration geophysics. In this type of work, the generated waves are reflected from layers of different velocities and are then detected using geophones and recorded on a seismograph. One of the primary applications of this type of survey has been in the field of petroleum exploration when detailed information about the subsurface structure at depth is required.

On a larger scale, both refraction and reflection interpretation methods of earthquake-generated

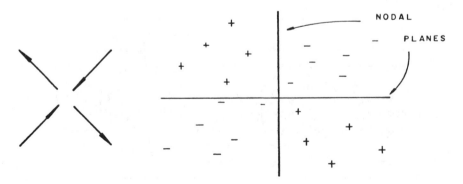

FIGURE 4. Quadrant distribution of earthquake compressions and dilations (after Hodgson, 1964).

waves are used by seismologists to learn more about the composition and structure of the earth's interior. From this research, scientists have been able to define more accurately the thickness and nature of the earth's crust, mantle, and core. This information has provided important substantiating evidence for the currently evolving theories of plate tectonics and sea-floor spreading.

Earthquake Prediction

The most rapidly developing and certainly the most potentially beneficial seismological methods are concerned with the prediction of location, time of occurrence, and magnitude of an earthquake (see *Earthquake Engineering*). While seismologists have been trying for many years to develop a means of predicting seismic events, only recently have they achieved limited success. The two methods that have been most successful in the past and that hold the greatest potential for the future are concerned with measurements of subtle *tilting* of the ground surface preceding the event and with gradual changes in the *propagation velocities* of vibrations in bedrock in the vicinity of the causative fault.

The first method is in reality more of a geological method applied to seismology studies rather than a strictly seismological method. It had been observed that minute changes in regional topography resulting in a general tilting of the land surface often precede a seismic event. If a sophisticated topographic control network with periodic resurveying is set up in a seismically active area, these subtle topographic variations can be monitored and warn of an impending earthquake.

The second method is based on what is usually called the *dilatancy theory*. The propagation velocity of P-waves decreases gradually but significantly preceding an earthquake. The velocity then quickly returns to normal and is followed by the earthquake. The time period from the return to normal of the P-wave velocity until the actual seismic event is roughly one-tenth the time of the period of decreasing velocities. In this way, if the time of initial velocity decrease and return to normal is known, a prediction can be made as to when the earthquake will occur. The magnitude of the resulting event is also proportional to the time period of decreasing velocities and return to normal. If the decrease and return to normal take place in a few days or weeks, the quake can be expected to be relatively small, whereas if the velocity decrease is stretched out over many months or years, the magnitude will probably be quite high.

The explanation of the dilatancy theory lies in the behavior of rock under high stress. As the crustal pressures preceding an earthquake approach the failure point of the rock mass, millions of tiny, often microscopic cracks open up. The opening of these cracks is what causes the decrease in P-wave velocities since the waves travel more slowly across the air-filled openings than they do in solid rock. As ground water seeps into the cracks, the wave velocity increases until all the openings are water-filled and the P-wave velocity returns to normal. The presence of the ground water serves as lubrication for the rock and allows the rock to fail, causing the quake. Figure 5 is a plot of velocity versus time and shows the general relationships of velocity changes to actual events.

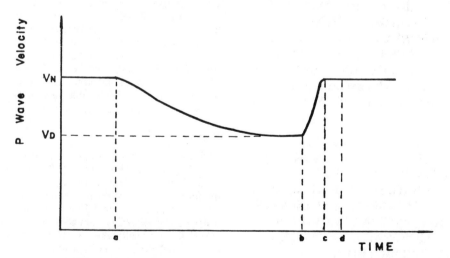

FIGURE 5. The dilatancy theory of P-wave velocity changes preceding an earthquake: V_N is the normal P-wave velocity; V_D is the decreased velocity due to dilatancy; a is the point at which P-wave velocity begins to decrease; b is the point at which the velocity begins to increase; at c the velocity returns to normal; and at d the earthquake occurs. Time period $cd = \frac{1}{10} \, ab$.

As more and more seismic monitoring stations are established in seismically active areas and more data on P-wave velocity changes become available, the day may come when seismologists can routinely predict the time, place, and severity of every earthquake using these seismological methods.

ANTHONY JOHN SABATINO

References

Bolt, B. A., 1970. Elastic waves in the vicinity of the earthquake source, in *Earthquake Engineering.* Englewood Cliffs, N.J.: Prentice-Hall, 1-20.
Guttenberg, B., and Richter, C. F., 1956. Earthquake magnitude, intensity, energy and acceleration (second paper), *Seismol. Soc. America Bull.* **46**, 105-145.
Hodgson, J. H., 1964. *Earthquakes and Earth Structure.* Englewood Cliffs, N.J.: Prentice-Hall, 166p.
Richter, C. F., 1958. *Elementary Seismology.* San Francisco: W. H. Freeman, 768p.

Cross-references: *Earthquake Engineering.* Vol. XIV: *Exploration Geophysics.*

SENSORS, SENSOR PLATFORMS — See ELECTROKINETICS; MAGNETIC SUSCEPTIBILITY, EARTH MATERIALS; OCEANOGRAPHY, APPLIED; REMOTE SENSING, ENGINEERING GEOLOGY; ROCK STRUCTURE MONITORING; SEISMOLOGICAL METHODS; TERRAIN EVALUATION, MILITARY PURPOSES. Vol. XIV: ACOUSTIC SURVEYS, MARINE; AERIAL SURVEYS, GENERAL; ATMOGEOCHEMICAL PROSPECTING; HARBOR SURVEYS; PHOTOGEOLOGY; PHOTO INTERPRETATION; SATELLITE GEODESY AND GEODYNAMICS; SEA SURVEYS; SURVEYING, ELECTRONIC; VLF ELECTROMAGNETIC SURVEYING.

SERVICE ENGINEERING — See CHANNELIZATION AND BANK STABILIZATION; COASTAL ENGINEERING; DAMS, ENGINEERING GEOLOGY; LATERITE, ENGINEERING GEOLOGY; PIPELINE CORRIDOR EVALUATION; RAPID EXCAVATION AND TUNNELING; RIVER ENGINEERING; UNDERSEA TRANSMISSION LINES, ENGINEERING GEOLOGY; URBAN TUNNELS AND SUBWAYS.

SHAFT SINKING

A *shaft* is a vertical or inclined excavation giving access to a mine or other underground workings; the inclination should not be less than about 40°; otherwise the correct term is *drift* (inclined) or *adit* (horizontal). Most shafts are sunk from the surface, but *interior shafts* are frequently used in large mines, for extending the workings without interfering with the main shaft operations or to reach depths or locations beyond easy reach of the main shaft. It is customary to restrict the term *interior shaft* to an opening provided with hoisting facilities; otherwise such an opening is called either a *winze* or a *raise,* according to whether the direction of advance is downward or upward (Cornish, 1967). *Compound shafts* start vertically from the surface and then change gradually to a steep inclination, usually to follow the direction of an ore body; hoisting is continuous through the transition. *Subvertical shafts* are vertical interior shafts usually in close association with a shaft sunk from the surface.

Prospect Shafts

Prospect shafts may be a prerequisite of a mining lease, or may be required to allow visual inspection of the ground at depth, and no further lateral development may take place. Although sinking methods for a well intended for water supply may approximate those used for small shafts, wells are not true shafts as defined. Boreholes may be used to transfer cables or pipework underground, or to provide ventilation, and if they are large enough (1 to 2 m in diameter) they may be termed *drilled shafts;* drilling is by shot (calyx) or by rotary coring drill (see *Borehole Drilling*). A drilled shaft may be used for prospecting deep foundations, as at dam sites, where *in situ* rock testing is needed in comparatively undisturbed ground. As an alternative to sinking, shafts may be *raised* from below where access is available, either manually or with a "raise-borer" guided by a cable passing through a borehole (see *Rapid Excavation and Tunneling*).

History

Shafts 9 m deep, with galleries radiating outwards for 10 m, have been found at Grimes Graves, England, where early man followed seams of flint in chalk. Similar Neolithic mines have been discovered in France and Belgium. Bronze Age shafts were sunk in Egyptian gold mines in about 1200 B.C. In the salt mines at Hallstatt in Austria, vertical and inclined shafts leading to workings more than 100 m below surface date from 1000 B.C. At Mount Laurion, Greece, where lead and silver occur in limestone and slate overlain by schist, more than 2,000 Iron Age shafts exist. One of the deepest, stopped by water at 118 m, dates from about 500 B.C. Greek shafts were used in the draining of Lake Copais, near Thebes, in 350 B.C. The Romans sank intermediate shafts up to 120 m deep to hoist spoil from the 5½-km tunnel driven in A.D. 50 to drain Lake Fucino. Agricola describes medieval Saxon shafts "2 × 2/3 fathom" (about 4.0 × 1.3 m), which were timbered and provided with a windlass and shaft-house. Of uncertain age are those in the Middle East where water has long been conducted underground through *qanats*, or underground tunnels, with a series of intermediate shafts. That at Aleppo in Syria is typical: it is 12 km long and has shafts up to 90 m deep every 50 m (Sandstrom, 1963).

Usage

In mining practice shafts were developed from 9 m at Grimes Graves to the latest 3,000-m ventilation shafts in South Africa. They are used for access for people, materials and services, ventilation, drainage, and hoisting, and are frequently sited in pairs to give a flow of air. In civil engineering, shafts are usually adjuncts to tunnels, to facilitate spoil removal, ventilation, and surveying, and to enable work to proceed on several faces simultaneously. Two famous English canal tunnels were Brindley's 2.5-km Harecastle Tunnel (1766-1777), in which 14 shafts were used, and the 5-km Standedge Tunnel (1794-1811), in which the deepest shaft was 180 m. The 7.6-km Hoosac Railway Tunnel in Massachusetts (1855-1876) required a 7-m-diameter shaft 315 m deep for additional faces and accurate surveying (Sandstrom, 1963). Recent underground hydroelectric stations include the 336-m shafts at Tumut I in the Snowy Mountains of Australia and the 460-m inclined pressure shaft at Cruachan, Scotland. Offshore shafts are used nowadays for power station intakes or sewer outfalls. Shafts have also been required for underground gas storage, nuclear test sites, missile silos, and underground reservoirs for pumped storage hydroelectric schemes. Shafts for water supply may be ancillary to tunnels such as the Catskill Aqueduct in New York or large wells, when their yield may be improved by the addition of radiating galleries.

Methods of Sinking in Rock

The Neolithic antler picks and shoulder-blade shovels could be used only in soft rock such as chalk, but wooden wedges, driven by stone or metal hammers and then soaked in water, would break up most types of rock. Heating the rock with fire and then dousing with water would assist penetration. The salt mines at Hallstatt were worked from inclined timber-lined shafts 1 to 2 diameter lined with moss and clay. These were sunk using bronze tools, while for the shafts at Mount Laurion iron hammers and wedges were used; here the long axis of the rectangular 1.9 × 1.3-m shafts

twists to accommodate the tree-trunk ladders better. The Saxon shafts were timbered and lagged in loose soil and the ladderway was screened off for safety reasons; in all essentials these were identical with small shafts used today, although rock-breaking was simplified when supplies of gunpowder became adequate in the late sixteenth century. The development of dynamite and pneumatic drills led to sinking methods that are still in use for the majority of shafts. These are usually rectangular with two or more hoisting compartments and separate manways, and are sunk either full-face or by benching. A round of holes is drilled and blasted, and then the spoil removed either by hand or using a mechanical grab such as the Hydromucker. Rectangular timber frames are then installed, each hanging from the one above. The shaft is usually oriented with its long axis normal to the bedding or schistosity of the country rock to resist ground pressure at depth. Many modern shafts, especially in Europe and South Africa, are circular in section and lined with concrete tremied from the surface. With the use of sliding shutters, and if the lining is kept well above the bottom, drilling and lining can proceed simultaneously. *Galloway* stages of up to 10 decks provide a working platform, and perhaps a support for a mechanical mucker such as the cactus grab.

By these means, shafts 3,000 m deep have been sunk at rates of up to 300 m per month. Such high sinking rates are possible only in sound rock where about 25 m can be left unsupported. In weaker ground, temporary support with concrete rings, or rockbolts and wire mesh, may be necessary. Concrete lining provides less resistance to air flow, while circular concrete-lined shafts provide the strongest resistance to rock pressures; however, the proportion of the sectional area available for hoisting may be less than that for rectangular shafts.

Sinking in Soft Ground

Where water is not present, this presents little difficulty, although lining is required. In 1825 I. K. Brunel devised a scheme for the Rotherhithe Shaft of the Thames Tunnel (see *Tunnels, Tunneling*). He laid an iron curb-ring, 15 m in diameter, on the ground and constructed on this a brick cylinder 12.5 m high and 1.0 m thick. When the ground beneath the ring was excavated by hand the cylinder sank into position, new courses of brick being added as necessary. Later variations to this method are the substitution of iron or concrete rings for brickwork, and more sophisticated methods of excavation (Cleasby et al., 1975). Excessive skin friction is overcome by jacking against kentledge, or may be reduced by injecting bentonite mud outside the lining above the cutting edge.

When water is present it may be practicable to control it by pumping. Failing this, compressed air can be used to balance the water pressure. An airlock is built at the shaft collar and an air tight diaphragm (airdeck) either immediately below the airlock or just above shaft bottom (leaving a working space). As sinking proceeds, lining rings may be added either at the top or the bottom. As higher pressures are used kentledge may be superimposed to prevent extrusion of the lining. Clay pocketing may reduce air losses in coarser-grained soils (McWilliams and Erikson, 1960).

Shafts may be drilled by the Honigmann method, which was introduced into Holland in 1930. Carboniferous strata there are overlain by up to 480 m of soft water-bearing overburden, and the shafts are drilled through this, by using mudflush. A 2-m-diameter pilot bit is followed by successively larger bits and reamers to give finished diameters of up to 7.65 m. A double-walled steel lining with a false bottom is then "floated" into position and backfilled with concrete. Normal sinking can continue in the bedrock.

Geotechnical Processes

Compressed-air working is costly and hazardous to health, while excessive pumping may induce settlement in nearby structures. Other means of excluding water, in both soft and hard rocks, are freezing and cementation (see *Grout, Grouting*) (Swaisgood and Versaw, 1974).

Freezing involves formation of an annulus of frozen ground through which the shaft is sunk, and has been used to depths of over 600 m. A ring of boreholes is drilled from surface or from an intermediate level, and cold brine is circulated until thermocouples indicate that an adequate "ice-wall" has formed. This may take two to four months, depending on the thermal conductivity of the strata. The brine is usually a solution of calcium chloride, but lithium chloride may be used in highly saline ground water. Liquid propane or nitrogen has been tried experimentally to shorten the period. Accurate boreholes are necessary because leakages may develop during thawing. A special lining is used in the frozen zone (Ruedy, 1955).

Cementation involves drilling a pattern of raking holes for up to 50 m ahead of the shaft bottom, and pumping in a grout based on cement or chemicals according to the size of fissures or pore spaces expected. When the grout has set, the shaft is advanced to within 5 to 6 m of the limits of the "cover" and the process repeated. To minimize delays, pretreatment from the surface for up to 1,250 m has been used in South Africa, using three to four boreholes. Both freezing and cementation may be ineffective (see *Grout, Grouting*) if strong ground-water flows persist.

Geological Considerations for Shaft Location

Mineshafts are frequently located on the foot-wall side to avoid damage from subsidence. Alternatively, a shaft should pass through a low-grade zone of the ore body if possible. A "shaft pillar" of unworked ore must be left for a distance around the shaft depending on the depth from surface, thickness of deposit, and nature of the rock. Typically this may be one-third of the depth and sterilize 250,000 tons of ore. Shafts should avoid passing through an active fault or one that may be reactivated by mining activities. They should not be sited where unfavorable ground conditions such as nearby old landslides, karst topography, possible subsidence areas, or soils unsuitable as foundations for the headframe and associated structures exist at the surface. Valleys apparently suitable for intermediate shafts along a tunnel route may conceal faults. Careful core drilling will facilitate both the location and the sinking of the shaft. Data from the core such as rock quality, fracture spacing, degree of weathering, and rock strength obtained from Schmidt hammer or point-load tests (Rankilor, 1974) will define the mechanical nature of the ground to be traversed. Down-the-hole geophysical surveys (see *Well Logging*), drill stem tests, and permeability tests of core samples will identify zones of potential high water inflows and indicate whether special geotechnical processes are necessary (Adamson and Scott, 1973). The nature of the ground water, and its rate of flow, must be determined if grouting or freezing is being considered.

Geological Hazards During Sinking

The main geological hazards are water and weak or running ground. The preliminary geological investigations will have given a general warning of conditions; a suitable lining and reduced unsupported lengths of shaft will be adopted in bad ground, and *ad hoc* measures will be used to deal with isolated caving.

In water-bearing ground, even if grouting or freezing has been undertaken, probe holes must be drilled from the shaft bottom before each round, safety precautions being taken against encountering water at high pressure. Where an inflow occurs, the shaft may have to be abandoned temporarily and allowed to flood until hydrostatic equilibrium is reached, or a plug may be constructed part way up the shaft. Once the inrush has been stopped, the shaft may be recovered by grouting. Running ground may have to be frozen before sinking is resumed.

P. F. F. LANCASTER-JONES

References

Adamson, J. A., and Scott, R. A., 1973. Borehole investigations and logging methods in shaft sinking, *Mining Engineer* **148** (132), 181-189.

Cleasby, J. V.; Pearse, G. E.; Grieves, M.; and Thorburn, G., 1975. Shaft-sinking at Boulby Mine, Cleveland Potash Ltd., *Inst. Mining and Metallurgy Trans.* **84**, A7-A28.

Cornish, E., 1967. Vertical shafts, *Colorado School Mines Bull.* **10** (5), 1-23.

McWilliams, J. R., and Erikson, E. G., 1960. Methods and costs of shaft-sinking in the Coeur d'Alene District, Idaho, *U.S. Bur. Mines Inf. Circ. 7961.*

Rankilor, P. A., 1974. A suggested field system of logging rock cores for engineering purposes, *Assoc. Eng. Geol. Bull.* **11**, 247-258.

Ruedy, R., 1955. Shaft sinking by freezing methods, *Nat. Resources Council (Canada) Tech. Inf. Svc. Rept. 43.*

Sandstrom, G. E., 1963. *History of Tunneling—Underground Workings Through the Ages.* London: Barrie & Rockliffe.

Swaisgood, J. R., and Versaw, R. E., 1974. Geotechnical investigations for mineshafts, *Mining Eng.* **26** (6), 37-40.

Cross-references: *Pipeline Corridor Evaluation; Rapid Excavation and Tunneling; Tunnels, Tunneling; Urban Tunnels and Subways.* Vol. XIV: *Borehole Drilling; Rock Geology, Hard vs. Soft.*

SHALE MATERIALS, ENGINEERING CLASSIFICATION

Shale is the most common rock in the world. It constitutes about 50 percent of the rock types exposed on the surface of the earth and comprises about 70 percent of all the sedimentary rocks (Ackenheil, 1969). Sedimentary rocks are formed from the accumulation of sediments that have been transported by water, air, or ice or precipitated chemically or biochemically, and are subsequently compacted into hard, firm, and stratified rocks.

Because shale is so abundant, it has produced major problems in the design and construction of foundations (Ackenheil, 1969), cut slopes (Chenevert, 1969; Fisher et al., 1968), and embankments (Sherard et al., 1963) in many parts of the world. Thus the engineer tends to view shale with suspicion and often recommends conservative design and construction procedures, for example, extra rolling to fragment the material, placing another material between the shale and the atmosphere (encasement), flattening slopes, and using berms. It is probable, though, that current practice is generally too conservative; some usable shales are being wasted, and the intrinsic strengths of relatively high-quality shales are not being used.

Shales exhibit a wide variety of characteristics that affect engineering. For example, some shales slake almost immediately in moist air (Underwood, 1965), while others can withstand numerous cycles of wetting and drying and are approximately as durable as sandstone or limestone. *Slaking* is the process through which a material disintegrates or crumbles into small particles or flakes when exposed to moisture, and especially when dried and immersed in water.

Since most shales exhibit characteristics intermediate between those of soil and those of rock, the tests that suitably classify soils and rocks are not adequate to classify shales. The researcher is thus faced with the need to modify existing tests or to evolve new tests. Such tests would be both sufficiently simple and discriminating to allow geotechnical engineers to guide the design and construction of shale embankments in a sound and economical manner.

Current Classification Systems for Shales

Shale is a rather loosely used term, applied to sedimentary rocks that have the fundamental property of laminated stratification (more or less developed), and that are generally clayey or argillaceous. The rock is sufficiently consolidated and lithified so that it has some ability to maintain its structure even when subjected to weathering (Pettijohn, 1957). The percentage of silt and clay can vary, but for the most part, shale should contain at least 50 percent of material finer than the sand size.

A complicating factor in the development of a classification system for shales is that there are two basic spheres of interest in this problem: the geologist's and the engineer's. Geological classifications use descriptive terms to group rocks of like lithology, texture, composition, and structure (see Vol. XIV: *Rock Weathering, Engineering Classification*). The engineer, on the other hand, is more interested in the numerical rock property descriptors that can be used in design or in predicting performance.

Various shale classification systems have been proposed by different investigators, but because each researcher had a differing objective, there is considerable variation in the test procedures used and also in the types and ages of geological materials that were evaluated. A number of these shale classification systems are summarized here.

Basically, shales can be divided in two broad groups: (1) compaction, or "soil-like," shales, which have been consolidated primarily by the weight of overlying sediments, and lack significant amounts of intergranular cement; and (2) cemented, "rocklike" shales, which are both consolidated and significantly lithified; the cementing material may be calcareous, siliceous, ferruginous, gypsiferous, or the like. If a cementing material is lacking,

TABLE 1. Classification of Shales

Soil-Like Shale (Compaction Shales)		Rocklike Shale (Cemented or Bonded)
Clayey shale (clay shale)	50% or more clay-size particles, which may or may not be true clay minerals.	Calcareous shale
Silty shale	25-45% silt-size particles. Silt may be in thin layers between shale bands.	Siliceous shale Ferruginous shale
Sandy shale	25-45% sand-size particles. Sand may be in thin layers between clayey shale bands.	Carbonaceous shale Clay-bonded shale
Black shale	Organic rich; splits into semi-flexible sheets.	

Source: After Underwood (1967).

the shale may be bonded by recrystallization of its clay minerals. The two general shale types can be further subdivided; see Table 1.

Shales are "bad actors" to various degrees. Some compaction shales slake almost immediately when exposed to air. Others can withstand numerous cycles of wetting and drying, without structural change. One investigator, Philbrick (1950), recommended a simple test of five successive cycles of wetting and drying with water or $100N$ ammonium oxalate to separate "durable" and "nondurable" shales.

Eigenbrod (1972) proposed a classification system for inorganic, noncalcareous sedimentary material based on the change in strength as a result of a soaking-in-water cycle. Such a system is depicted in Figure 1. To group inorganic sedimentary materials further, Eigenbrod combined magnitude and rate of slaking to obtain the 15 categories presented in Figure 2.

The Pennsylvania Department of Transportation has recently completed a study concerning the suitability of shales "for use as a granular material in highway construction" (Reidenouer, 1970). To determine whether a shale would rate as durable or nondurable, samples are subjected to a three-step examination:

(1) The lamination thickness of the shale is determined; if the lamellae are less than 0.5 mm thick, the shale should be rejected. (2) A 10 lb sample of the shale is soaked for 40 hours in ethylene glycol and visually inspected. If more than three pieces are cracked, the shale should be rejected. (3) The bulk specific gravity and absorption should be determined on samples that have not been rejected on the basis of the preceding

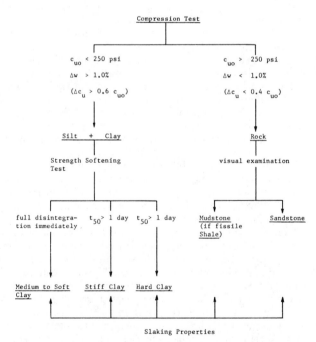

FIGURE 1. Classification system for inorganic, noncalcareous, sedimentary materials on the basis of the standard compression softening test results. The terms *rock* and *stone* may be used alternatively (after Eigenbrod, 1972).

	Amount of Slaking: $w_S = w_L$				
	very low VL $w_L < 20$	low L $20 < w_L < 50$	medium M $50 < w_L < 90$	high H $90 < w_L < 140$	very high VH $w_L > 140$
slow S $\Delta I_L > 1.4$	VL S	L S	M S	H S	VH S
fast F $0.8 < \Delta I_L < 1.4$	VL F	L F	M F	H F	VH F
very fast VF $\Delta I_L < 0.8$	VL VF	L VF	M VF	H VF	VH VF

Rate of Slaking: $\Delta I_L = \dfrac{I_{L1} - I_{L0}}{2h \text{ water immersion}}$

FIGURE 2. Slaking properties for inorganic, noncalcareous, sedimentary materials according to the quantitative slaking test results and the rate of slaking test (after Eigenbrod, 1972).

two tests. The shale is rated as durable if the bulk specific gravity is above 2.715, and nondurable if the bulk specific gravity is below 2.625. For shales whose bulk specific gravity lies between 2.715 and 2.625, the absorption test results are applied. If the percent absorption is above 2.35, the shale is rated as nondurable; otherwise it is rated as durable.

Underwood (1965) discussed many pertinent properties of shales and recommended an engineering evaluation of shales with respect to pore pressures, bearing capacity, tendency to rebound, behavior in cut slopes, slaking behavior, erosion potential, and tunnel-support problems. He recommended that basic engineering properties conventionally determined in the laboratory for major engineering projects be used to make this determination. Underwood's system is presented in Table 2.

Gamble (1971) tested 120 shales from many areas of the United States. He discussed the shortcomings of geological classification systems for shales and proposed the system illustrated in Figure 3. The primary factors are grain size and breaking characteristics. Gamble suggested that "slaking characteristics for mud rocks or shales be determined from a slaking durability test and modifiers to describe the relative durability be used with the suggested terms," for example, *low-durability claystone* or *high-durability silty shale*. Based on extensive laboratory testing. Gamble also proposed an engineering classification for shales and other argillaceous rocks using the criteria of slaking durability after two cycles of testing and the

Atterberg limits. The classification system developed from this approach is presented in Figure 4.

In a study conducted by Deo (1972), it was determined that shales could be suitably rated with only four tests: a slaking test of one cycle in water; a slake durability test on dry samples; a slake durability test on soaked samples; and a modified soundness test.

Slaking in One Cycle of Wetting. A broken piece of shale was immersed in water so that it was at least 12 mm below the water surface. After immersion, the shale piece was observed continuously during the first hour; after that, the condition of the piece was checked at 2, 4, 8, 12, and 24, hours. The condition of the piece was recorded as "complete breakdown," "partial breakdown," or "no change." If the piece seemed intact, the cloudiness of the water was also noted. The test was repeated on any shale that slaked completely or partially.

Slake Durability Test. The slaking test discussed in the preceding paragraph produces rather qualitative results. The slake durability test, on the other hand, measures a weight loss in water that can be expressed as a durability number. The apparatus was developed by Franklin and others at Imperial College, London in 1970 (Franklin et al., 1970).

A sample of 10 representative shale pieces, each weighing 50 to 60 g, were oven-dried and placed in the test drum. The drum was then half-immersed in the water bath and rotated. Material detached

TABLE 2. An Engineering Evaluation of Shales

Laboratory Tests and in-situ Observations	Physical Properties		Probable in-situ Behavior						
	Average Range of Values		High Pore Pressure	Low Bearing Capacity	Tendency to Rebound	Slope-Stability Problems	Rapid Slaking	Rapid Erosion	Tunnel-Support Problems
	Unfavorable	Favorable							
Compressive Strength	50-300 psi	300-5000 psi	✓	✓					
Modulus of Elasticity	20,000-200,000 psi	$200,000$-2×10^6 psi		✓					✓
Cohesive Strength	5-100 psi	100 psi to >1500 psi			✓	✓			
Angle of Internal Friction	10-20°	20-65°			✓	✓		✓?	
Dry Density	70-110 pcf	110-160 pcf	✓		✓	✓		✓	✓
Potential Swell	3-15%	1-3%				✓			
Natural Moisture Content	20-35%	5-15%	✓						

(continued)

TABLE 2. *(continued)*

Laboratory Tests and in-situ Observations	Physical Properties		Probable in-situ Behavior						
	Average Range of Values		High Pore Pressure	Low Bearing Capacity	Tendency to Rebound	Slope-Stability Problems	Rapid Slaking	Rapid Erosion	Tunnel-Support Problems
	Unfavorable	Favorable							
Coefficient of Permeability	10^{-5}–10^{+10} cm/sec	>10^{-5} cm/sec	✓				✓		
Predominant Clay Minerals	Montmorillonite, Illite	Kaolinite, Chlorite	✓			✓			
Activity Ratio = $\dfrac{\text{P.I.}}{\% \text{ Clay}}$	0.75 to >2.0	0.35 to 0.75				✓			
Wetting and Drying Cycles	Reduces to grain sizes	Reduces to flakes					✓	✓	
Spacing of Rock Defects	Closely Spaced	Widely Spaced		✓		✓		✓?	✓
Orientation of Rock Defects	Adversely Oriented	Favorably Oriented		✓		✓			✓
State of Stress	>Existing Overburden Load	≅Overburden Load			✓	✓			✓

Source: After Underwood (1965).

UNINDURATED	INDURATED GROUP		AFTER INCIPIENT METAMORPHISM	METAMORPHIC EQUIVALENTS
	MUDROCKS (SHALES OR MUD-STONES)			
	BREAKING CHARACTERISTICS			
	MASSIVE	FISSILE OR SHALY		
Silt ⟶	Siltstone	Silty Shale		
Mud (Mixture or un-determined amounts of silt and clay, with minor amount of sand)	Mudstone	Shale	⟵Argillite⟶	Slate, Phyllites, or Schist
Clay ⟶	Claystone	Clayey Shale		

FIGURE 3. Classification for argillaceous rock (after Gamble, 1971).

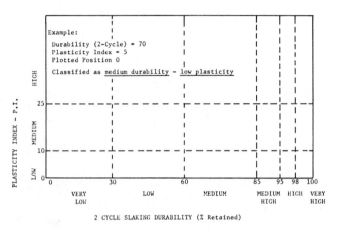

FIGURE 4. Durability-plasticity classification for shales and other argillaceous rocks (after Gamble, 1971).

from the pieces passed through the mesh, that is, became a sample weight loss. The durability number was calculated as the percentage ratio of final to initial dry-sample weights.

The durability number for 500 revolutions of the drum was defined as the durability index (I_d). Durability indices were determined both for dry samples, $(I_d)d$, and for soaked samples, $(I_d)s$.

Modified Soundness Test. This test measures the degradation of shales when subjected to five cycles of alternate wetting and drying in a sodium sulfate solution. It is more severe than the previously mentioned slaking tests, and it is more effective in distinguishing among the harder and more durable shales.

The test was modified from ASTM C 88-63, which is used to determine the resistance of aggregates to disintegration by sodium sulfate or magnesium sulfate. The standard test uses a fully saturated solution, but this is too severe for shales. After a series of trials, the saturation was reduced to 50 percent.

The Soundness Index, I_s, was defined as the percent retained by weight on the $\frac{5}{16}$-in. sieve. Durability is considered to increase with increase in I_s value.

On the basis of four simple degradation-type tests, Indiana shales can apparently be classified into four groups: (1) rocklike shales; (2) intermediate-1 shales; (3) intermediate-2 shales; and (4) soil-like shales. The flowchart for classification is shown in Figure 5.

Franklin (1981) in a recent study proposed a shale rating system based on three properties—durability, strength, and plasticity. A shale sample is assigned a rating value by first measuring its second cycle slake—durability index. Rocklike shales that have durability values greater than 80 percent for this index are further characterized by measuring their point load strength. Soil-like shales that have durability values of less than 80 percent are further characterized by measuring their plasticity index. A rating chart that has been developed based on the test results classifies shales with

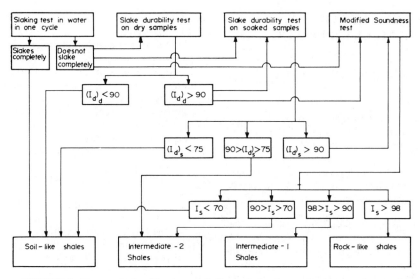

FIGURE 5. Proposed classification of Indiana shales for embankment construction (after Deo, 1972).

regard to their engineering performance such as excavating methods, foundation properties, embankment construction, and slope stabilities.

The end result of all classification systems is to aid in the establishment of construction procedures and specifications. As of now, there still exists a need for additional information that would improve the correlation between classification test values and field performance.

LEONARD E. WOOD

References

Ackenheil, A. C., 1969. Building foundations on Appalachian shales. Paper presented at the Engineering in Appalachian Shales Conference, West Virginia University, Morgantown, June, 16p.

Chenevert, M. E., 1969. Shale alteration by water adsorption, in *Fourth Conference on Drilling and Rock Mechanics,* Austin, Texas, 1141-1148.

Deo, P., 1972. Shales as embankment materials, unpublished Ph.D. thesis, School of Civil Engineering, Purdue University, Lafayette, Ind., 202p.

Eigenbrod, K. D., 1972. Progressive failure in overconsolidated clays and mudstones, unpublished Ph.D. thesis, Department of Civil Engineering, University of Alberta, Alberta, Manitoba, 355p.

Fisher, S. P.; Fanaff, A. S.; and Pickering, L. W., 1968. Landslides of southeastern Ohio, *Ohio Jour. Sci.* **68,** 65-80.

Franklin, John A., 1981. *A Shale Rating System and Tentative Applications to Shale Performance,* Transportation Research Record 790, Transportation Research Board, 2-12.

Franklin, J. A.; Broch, E.; and Walton, G., 1970. *Logging the Mechanical Character of Rock.* London: Imperial College, Rock Mechanics Research Report D-14, 10p.

Gamble, J. C., 1971. Durability—plasticity—classification of shales and other argillaceous rocks, unpublished Ph.D. thesis, University of Illinois, Urbana-Champaign, 161p.

Pettijohn, F. J., 1957. *Sedimentary Rocks.* New York: Harper & Row, 340-380.

Philbrick, S. S., 1950. Foundation problems of sedimentary rocks, in P. D. Trask (ed.), *Applied Sedimentation.* New York: Wiley, 147-167.

Reidenouer, D. R., 1970. *Shale Suitability.* Harrisburg: Pennsylvania Department of Transportation, Interim Report 1, Research Project No. 68-23, December, 160p.

Sherard, J. L.; Woodward, R. J.; Gizienski, S. F.; and Clevenger, W. A., 1963. *Earth and Rock Dams.* New York: Wiley, 724p.

Underwood, L. B., 1965. Machine tunneling on Missouri River dams, *Am. Soc. Civil Engineers Proc., Jour. Construction Div.* **91** (Co 1), 1-28.

Cross-references: *Clays, Strength of; Foundation Engineering; Rock Mechanics; Rocks, Engineering Properties.* Vol. XIV: *Dispersive Clays; Rock Weathering, Engineering Classification.*

SITE EXAMINATION—See DAMS, ENGINEERING GEOLOGY; FOUNDATION ENGINEERING; NUCLEAR PLANT SITING, OFFSHORE; PIPELINE CORRIDOR EVALUATION; PUMPING STATIONS AND PIPELINES, ENGINEERING GEOLOGY; ROCK STRUCTURE MONITORING; TUNNELS, TUNNELING.

SOIL CLASSIFICATION—See SOIL CLASSIFICATION SYSTEM, UNIFIED.

Vol. XIV: ROCK WEATHERING, ENGINEERING CLASSIFICATION.

SOIL CLASSIFICATION SYSTEM, UNIFIED

Since prehistoric times, human beings have used the earth as a construction material. Throughout history, the amount of earth used for construction purposes has progressively increased with growing populations and improved technology. The construction of giant landfills and of highways, levees, and dams reached major proportions after World War II.

Numerous forms of classification of unconsolidated sediments—"soils" or "earth"—can be found in the literature. Pedologists, geologists, sedimentologists, hydrologists, agriculturalists, and others use their own systems, adopted for their particular purposes. Unfortunately, most of these systems are not suitable for engineering requirements. A so-called Unified Soils Classification was therefore developed by a team of soil engineering specialists from the U.S. Bureau of Reclamation and Army Corps of Engineers together with Dr. A. Casagrande in 1952 through a modification of Casagrande's "Airfield Classification."

The Unified Soil Classification has become deeply rooted in engineering practices in the United States and abroad. It is now widely used in the civil engineering profession, including the two largest government construction institutions—the U.S. Army Corps of Engineers and the U.S. Bureau of Reclamation—along with practically all consulting and construction companies, the U.S. Geological Survey, and others.

According to the classification, unconsolidated, natural, or man-made "soils" are subdivided into two major groups: (1) *coarse-grained soils,* in which, by weight, more than half the particles are larger than #200 sieve size, and (2) *fine-grained soils* in which more than half the particles are smaller than #200 sieve size. (The #200 sieve size (0.074-mm-diameter openings) was selected as being the smallest particle size usually visible by the naked eye). A third, rather uncommon group of materials are highly organic peaty soils.

Fines, or particles smaller than #200 screen, are subdivided into *clays* and *silts* according to their plasticity as indicated by physical-mechanical properties such as dry strength, dilatancy (reaction to shaking) and toughness (see *Soil Mechanics*). These parameters can be determined visually in the field or under laboratory conditions using Atterberg limits (the liquid limit and plasticity index) (see *Atterberg Limits and Indices*). Clays, according to the definition, are rather plastic, while silts have no or low plasticity. Clays are further subdivided

into highly plastic "fat" clays, "lean" clays of medium to low plasticity, and "organic" clays of variable plasticity. Silts are also subdivided into slightly plastic to nonplastic, elastic, and organic.

Coarse-grained soils are divided into *sands,* in which over 50 percent of the coarse fraction is composed of "sand-size" particles and *gravels,* in which over 50 percent of the coarse fraction is composed of gravel-size particles. The break between sand- and gravel-size particles was established at ¼ "diameter" (or #4 sieve). Further subdivisions are based on "gradation" (parameter opposite to sorting) and the presence and character of fines.

To speed communication and reading, the following letter symbols were established: G, gravel; S, sand; M, silt (M stands for the Swedish *Mo,* silt); C, clay; L, lean; H, fat; O, organic; W, well-graded; P, poorly graded; and Pt, peat.

Using these subdivisions and symbols, the Unified Soil Classification recognizes the following 15 soil types:

GW—well-graded gravel: poorly sorted gravel-sand mixture with few or no fines, wide range of grain sizes.

GP—poorly graded gravel: gravel-sand mixtures with few or no fines, well-sorted, predominantly one grain size.

GM—silty gravel: poorly graded gravel-sand-silt mixtures.

GC—clayey gravel: poorly graded gravel-sand-clay mixtures.

SW—well-graded sand: poorly sorted sand or gravelly few or no fines.

SP—poorly graded sand: well-sorted or gravelly sands, few or no fines.

SM—silty sand: poorly graded (well-sorted) sand-silt mixtures.

SC—clayey sand: poorly graded sand-clay mixtures.

ML—inorganic silt: rock flour and other non- or slightly plastic fines.

CL—inorganic clay: low- to medium-plasticity fines.

OL—organic silt-clay: organic low-plasticity silt and clays.

MH—elastic silt: inorganic micaceous, diatomaceous, fine sandy-silty elastic fines.

CH—fat clay: inorganic, high-plasticity clays.

OH—organic clay: organic medium- to high-plasticity clays.

Pt—peat and other highly organic deposits.

A borderline dual classification is used, particularly in the field, to indicate material with characteristics approaching two soil types. For example, SC-CL, CH-CL.

A chart for laboratory classification of fine-grained soils is shown in Figure 1. A more detailed description of the system can be obtained from a

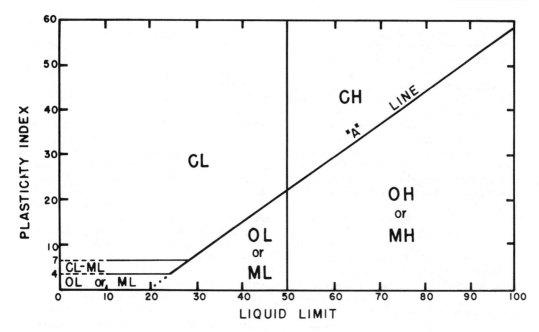

FIGURE 1. Simplified graph for laboratory classification of fine-grained "soils" according to the Unified Soil Classification System.

variety of "Earth Manuals" and "Technical Instruction Manuals" (e.g., U.S. Bureau of Reclamation, 1974).

The Uniform Soils Classification has proved to be very simple and adaptable for practical uses and is now deeply rooted in applied geology. Its most important feature is the possibility of classification using either visual field data or laboratory testing. It does not require prolonged training and complicated laboratory test procedures. The system, based exclusively on physical properties of soils, has, however, four limitations, which users should keep in mind:

(1) It uses two different principles of classification: gradation (for coarse particles) and plasticity (for fine particles). (2) It ignores geological character or origin of "earth" or "soils," which may lead to serious misinterpretations or errors. Natural materials are catalogued together with man-made excavated material. For example, the same soil symbol, SM, could be applied to alluvium and to some decomposed granitic rocks. (3) It ignores the geochemical character of "soils." (4) The measurements are not metric. For example, the breaking point between "fines" and "coarse" sediments is 0.074 mm (#200 screen), gravel particles have diameters over 4.76 mm (#4 screen), and so on. A conversion to metric units is now under way, however.

NIKOLA P. PROKOPOVICH
JOHN P. BARA

Reference

U.S. Bureau of Reclamation, 1974. *Earth Manual,* 2nd ed. Washington, D.C.: U.S. Government Printing Office.

Cross-references: *Atterberg Limits and Indices; Foundation Engineering; Geotechnical Engineering; Soil Mechanics.* Vol. XIV: *Engineering Soil Science; Soil Sampling.*

SOIL DYNAMICS—
See SOIL MECHANICS.

SOIL GEOCHEMISTRY, SURVEYS—
See Vol. XIV: PEDOGEOCHEMICAL PROSPECTING.

SOIL MECHANICS

Soil Mechanics (or *Geotechnical Engineering,* q.v.) refers to the specialist area of civil engineering requiring studies of the physical, mechanical, and chemical properties of soils, and the application of these studies to the design, analysis, and construction of civil engineering structures—principally foundations, retaining walls, sheetpile walls, cofferdams, earth and rockfill dams, natural and man-made slopes, open excavations, tunnels, and pavement structures.

It is necessary for specialists in soil mechanics to identify the soils and rocks and the structural features, then determine the relevant properties by field tests and by laboratory tests on "disturbed" and "undistributed" samples (Yong and Warkentin, 1975). These data are then used to establish the stability of the structure and to predict the settlements that inevitably will occur in the loaded soils. Every geotechnical structure requires some study, but the extent of this study is controlled by economics, by the consequences of a failure, and by other factors. A program of field and laboratory studies must be planned to take account of the practical constraints. Studies for major structures require other experts, such as engineering geologists, in the investigation team since the structural characteristics of the rocks—bedding, fissuring, dykes, faults, weathering, and so on— must be properly mapped and evaluated. It is also necessary for the geotechnical engineer to have a working knowledge of structural geology so that he can make proper use of the geological information and recognize those features of the geology of the site which could lead to the most effective design of the structure or, on the other hand, cause particular problems.

History of Soil Mechanics

It is commonly acknowledged that the "father of soil mechanics" was Karl Terzaghi (1883-1963). Not only did he make a multiplicity of basic and applied contributions to the theory and practice of soil mechanics, but his personal magnetism and technical genius attracted many outstanding engineers to the study of soil mechanics, thus identifying and establishing the uniqueness of the discipline within the civil engineering profession. It is necessary, however, to recognize and acknowledge the contributions made by Terzaghi's peers and the numerous scientists, philosophers, and engineers of the eighteenth and nineteenth centuries. Military engineers of these periods were responsible for designing structures such as earth and rock fortifications, canals, roads, and bridges. Their writings show that they had a sound qualitative understanding of construction and stability problems and some detailed understanding of earth pressures. They commonly developed empirical rules for the design of many geotechnical structures. Such an outstanding engineer was Maréchal de Vauban, who is acknowledged as the "ancestor" of soil mechanics in France. His contributions around the turn of the eighteenth century complemented those by the famous Frenchman C. A. Coulomb (1736-1806), whose name is perpetuated in the empirical shear strength law of soils and in active earth pressure-calculations. Isolated but numerous contributions can be identified through a study of the history of military and civil engineering. Table 1 is a brief list of some of the outstanding early contributors to soil mechanics.

Probably the greatest engineering advances in the understanding of strength and deformation behavior of soils occurred as a consequence of Terzaghi's development of the *effective stress law* for saturated soils. He used this law in 1925 to develop his consolidation rate theory. Before this concept was known, soils were treated as a single-phase continuum and, although the physical limitations of such a model must have been evident, earlier investigators had not realized how to isolate the effects of the water and solid phases. In present-day terminology the single-phase continuum concept is termed a *total stress model,* whereas Terzaghi's effective stress law enables us to determine the proportion of external (total) stress carried by the soil skeleton. Terzaghi was able to explain, for the first time, the role of the water phase in a stress-strain-time analysis, and it was this fundamental advance that greatly contributed

TABLE 1. Some Notable Contributors to Soil Mechanics

Contributor	Major Topics
C. A. Coulomb (1736-1806)	Earth pressures, shear strength
T. Telford (1757-1834)	Canals, earthworks, foundations
C. Berigny (1772-1842)	Grouting
A. Collin (1808-1890)	Slope stability, grouting
Sir B. Baker (1840-1907)	Earth pressures
W. J. M. Rankine (1820-1872)	Earth pressures
J. Boussinesq (1842-1929)	Stress distribution
O. Reynolds (1842-1904)	Dilatancy of sands
A. M. Atterberg (1846-1916)	Soil classification, plasticity
H. D. Krey (1866-1928)	Earth pressures
A. L. Bell (1874-1956)	Earth pressures
N. Fellenius (1876-1957)	Slope stability
K. V. Terzaghi (1883-1963)	Consolidation theory, effective stress law, foundation stability, engineering geology
O. K. Froelich (1885-1964)	Stress distribution, slope stability
D. W. Taylor (1900-1948)	Slope stability, shear strength

to the subsequent acceptance of soil mechanics as a quantitative discipline.

Terzaghi's diligent study of soil mechanics, which he developed as a discipline while at the Imperial School of Engineering, Constantinople (Istanbul) from 1616 to 1925, led to the founding of the International Society of Soil Mechanics and Foundation Engineering. The First International Conference on Soil Mechanics and Foundation Engineering was held in 1936 and the second conference in 1948. Subsequent conferences have been held every four years since 1948, and regional conferences have been organized in Europe, South East Asia, Australia, and South America. These conferences, along with other geotechnical activities, are part of the activities of national groups of the International Society. Recognizing the close relationship with engineering geology, the national groups are commonly affiliated with the International Rock Mechanics Society and the International Association of Engineering Geologists.

Several highly respected journals devoted to soil (and rock) mechanics are now published, following the success of *Géotechnique,* which was first published in 1948. Many geotechnical papers also appear in civil engineering journals. Important geotechnical journals published wholly or partly in English are *Géotechnique* (Institution of Civil Engineers, London), *Proceedings Geotechnical Division* (American Society of Civil Engineers, New York), *Canadian Geotechnical Journal* (Canadian Geotechnical Society, Toronto, Ontario), *Soils and Foundations* (Japanese Society of Soil Mechanics and Foundation Engineering, Tokyo), *Proceedings of the Norwegian Geotechnical Institute* (Norwegian Geotechnical Institute, Oslo), *Proceedings of the Swedish Geotechnical Institute* (Swedish Geotechnical Institute, Stockholm), and *Australian Geomechanics Journal* (Australian Geomechanics Society, Sydney).

Soil Properties

Phase Relationships. Soil consists of solids, water, and vapor phases (see Vol. XII, Pt. 1: *Soil*). The engineering properties are greatly influenced by the relative proportions of the phases; there are some simple volume and weight relationships that are particularly useful. Considering the soil mass of total weight W and of volume V to be composed of solids of weight W_s and volume V_s, water of weight W_w, volume V_w, and an air (vapor) phase occupying a volume V_a, then the following relationships apply:

Water content

$$w = \frac{W_w}{W_s}$$

Void ratio

$$e = \frac{V_v}{V_s} = \frac{V_w + V_a}{V_s}$$

Porosity

$$n = \frac{V_v}{V} = \frac{e}{1 + e}$$

Degree of saturation

$$S = \frac{V_a}{V_v} = \frac{G \cdot w}{e}$$

Specific gravity

$$G = \frac{W_s}{V_s \cdot \gamma_w}$$

Wet (bulk) density

$$\gamma = \frac{W}{V} = \frac{W_w + W_s}{V_v + V_s} = \frac{G + S \cdot e}{1 + e} \cdot \gamma_w = $$
$$\frac{1 + w}{1 + e} \cdot G \cdot \gamma_w$$

Dry density

$$\gamma_d = \frac{W_s}{V} = \frac{G}{1 + e} \cdot \gamma_w = \frac{G \cdot \gamma_w}{1 + \frac{w \cdot G}{S}} = \frac{\gamma}{1 + w}$$

Buoyant (submerged) density

$$\gamma' = \gamma_b = \gamma - \gamma_w = \frac{(G - 1) - e \cdot (1 - S)}{1 + e} \cdot \gamma_w$$

Table 2 gives some typical values of these properties. Table 3 is a detailed list of the specific gravity of minerals.

Solid Particles. Engineers, geologists, and soil scientists classify soils on the basis of size because the physical and mechanical properties of coarse- and fine-grained soils are dramatically different. In the past, different size classification systems have been proposed (see Vol. XII, Pt. 1: *Particle-Size Distribution*). Today, soil engineers use the following terminology: boulder (> 305 mm), cobble (15.25-305 mm), gravel (2.0-15.25 mm), sand (0.06-2.0 mm), silt (0.002-0.06 mm), and clay (< 0.002 mm, or 2 μ). Figure 1 shows the range of particle sizes in relation to measuring techniques.

The reason for the different properties of granular materials and clays is the increasing influence of electromagnetic and electrostatic forces as the

TABLE 2. Typical Properties of Soils

Soil type	w (%)	e (%)	n (%)	γ (g/cm³)	γd (g/cm³)	wLL (%)	wPL (%)	wPL (%)	L.I.	Clay content (%)	Activity
Dense, saturated, angular, well-graded quartz sand	8	20	17	2.3	2.2					0	
Loose, saturated, angular, well-graded quartz sand	34	90	47	1.9	1.4					0	
Dense, saturated, rotund, uniformly graded quartz sand	15	40	29	2.2	1.9					0	
Loose, saturated, rotund, uniformly graded quartz sand	38	100	50	1.8	1.3					0	
Normally consolidated silty clay (remolded)	64	170	63	1.6	1.0						
	90	240	71	1.5	0.8	113	70	43	0/0.5	62	0.6
Normally consolidated marine silty clay	123	325	76	1.4	0.6	137	43	94	0.9	58	1.6
Normally consolidated marine silty clay	32	85	46	1.9	1.4	36	20	16	0.8	50	0.3
Lightly overconsolidated kaolinitic clay	150	397	80	1.3	0.5	145	45	100	1.1	55	1.8
Heavily overconsolidated dated marine clay	19	50	33			44	19	25	0		
Heavily overconsolidated London clay	23	62	39	2.1	1.7	63	26	37	0	47	0.8
Norwegian quick clay	31	86	46	1.94	1.48	19	14	5	3.4	37	0.14

TABLE 3. Values of the Specific Gravity of Soil Minerals

Mineral	Specific Gravity
Augite	3.2-3.4
Aragonite	2.94
Attapulgite	2.30
Bentonite	2.13-2.18
Biotite	2.8-3.2
Calcite	2.72
Chlorite	2.6-3.0
Dolomite	2.85
Gibbsite	2.30-2.40
Gypsum	2.30
Halloysite	2.55
Hematite	4.90-5.30
Hornblende	3.0-3.5
Illite	2.60-2.85
Kaolinite	2.62-2.66
Limonite	3.8
Magnetite	5.18
Muscovite	2.7-3.0
Montmorillonite	2.75-2.80
Olivine	3.27-3.37
Orthoclase	2.54-2.57
Plagioclase	2.62-2.76
Pyrophyllite	2.85
Quartz	2.65
Serpentine	2.2-2.7
Siderite	3.83-3.88
Talc	2.8

particle size decreases. In clays, the total surface area of all the particles in a given volume (termed the *specific surface*) is several orders of magnitudes greater than the specific surface of a sand or gravel; for example, if the specific surface of a sand is 5, then the value for a clay is approximately 50,000. Because the magnitude of the interparticle forces depends on the surface area of the particles, an increase in the specific surface will increase these forces. Particles are *colloids* when the interparticle forces dominate over the gravity force. Clays are colloids, but it is sometimes important to differentiate between clays and clay sizes, since some clay-sized particles, such as rock flour, do not behave as colloids. Although the upper limit to clay size cannot be uniquely defined, it is universally agreed that the value of 2μ is quite realistic.

Figure 1 shows that dry or wet sieving can be used to establish the gradation of sizes in excess of 0.06 mm, the lower limit of size of fine sand. A sedimentation technique based on *Stokes' law*, which gives the terminal velocity of spheres falling through a fluid of infinite extent, is used to determine the quantity of silt and clay fractions. Grading curves are commonly used to present the results of the mechanical test; Figure 2 shows some typical gradings for sands and clays. It will be noted that the sand, silt, and clay sizes are further subdivided into coarse, medium, and fine fractions. The sizes shown in Figure 2 follow the system originally proposed by geologists at the Massachusetts Institute of Technology and adopted by most soil engineers. From the grading curve we can calculate the relative amounts of sand, silt, and clay sizes. Since the properties of the individual fractions are quite different, it follows that the behavior of the soil mass will depend on the proportion of sand, silt, and clay. The decription

TABLE 2. (continued)

Organic Content (%)	Silt (g/l) content	O.C.R.	C_u (kPa)	C_u/p	c' (kPa)	tan ϕ'	C_d (kPa)	tan ϕ_d	Sensitivity	S.P.T.	Dutch cone (kg/cm²)
					0		0	42/45		30/50	120/200
					0		0	30/32		4/10	20/40
					0		0	40/43		30/50	120/200
					0		0	30/32		4/10	20/40
0	0	1	140/45	0.7	0	0.60					
3.8	3.2	1	10	0.1/0.2							
		1				0.70			4/8		
3.2		1.5	10	0.6	2				1.4/3.2		
			450						1/1.5		
		> 10	430	0.27	125	26.7	200	23.2			
0.9	0.9	1	10	0.14					100		

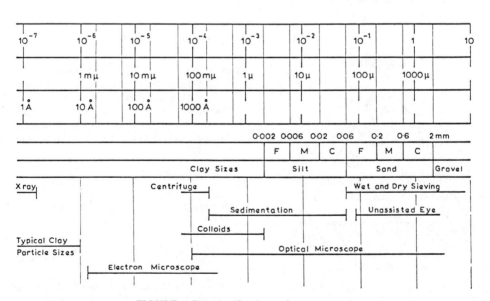

FIGURE 1. Size classification and measurement.

of the soil mass is expressed in a textural chart of the type shown in Figure 3. Such charts help the soils engineer describe the soil more precisely and possibly to anticipate likely values of properties such as permeability and plasticity.

The engineering behavior of granular soils depends on the grading of particle sizes (see *Soil Classification System, Unified*), the density of the soil, and the shape of the particles as well as the frictional properties. Several terms have been intro-duced to describe grading. A *well-graded soil* has a gradation of sizes and enables the maximum possi-ble density to be developed by the smaller parti-cles filling the voids between the larger particles. Several idealized gradings have been studied; one is shown in Figure 2. A *uniformly graded soil* has a predominance of one size and is a poor fill material, as the voids cannot be filled unless "foreign," smaller-sized particles are introduced. The slope of the grading curve is expressed by *Hazen's uni-*

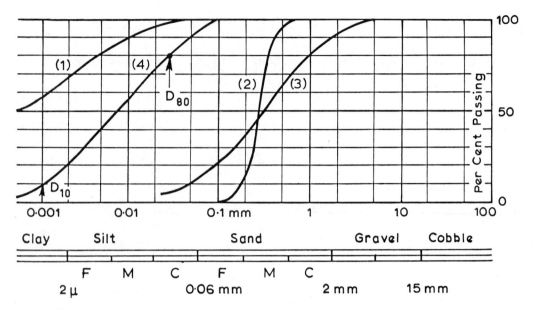

(1) Silty marine clay.
(2) Uniformly graded sand.
(3) Well graded gravel - sand.
(4) Silt.

FIGURE 2. Typical grading curves.

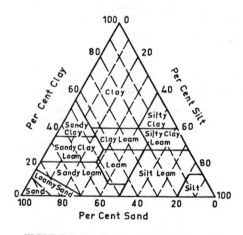

FIGURE 3. Textural chart (after Soil Survey Staff, 1951).

formity coefficient, defined as the ratio of the particle size corresponding to 80 percent finer (D_{80}) to the particle size corresponding to 10 percent finer (D_{10}). Thus a uniformly graded soil would manifest a uniformity coefficient of 1 and a well-graded material would have a coefficient of the order of 10. The term *gap-graded* refers to a deficiency of some sizes.

As well as determining the bulk and dry density,

the engineer is very interested in the relative density of a granular soil mass because the strength and compressibility properties are very sensitive to the density state. Unfortunately, two definitions of relative density are in use. In North America the relative density is defined in terms of void ratio:

$$D_{r1} = \frac{e_{max} - e}{e_{max} - e_{min}} \qquad (1a)$$

Whereas in the United Kingdom, it is defined as follows:

$$D_{r2} = \frac{\gamma_d - \gamma_{d \cdot min}}{\gamma_{d \cdot max} - \gamma_{d \cdot min}} = D_{r1} \cdot \frac{\gamma_d}{\gamma_{d \cdot max}} \qquad (1b)$$

In Table 4 the descriptions of relative density are given, together with the likely ranges of values of dry densities and the *Standard Penetration Test* value (blows per foot of a standard "split-spoon" sampler driven into the soil by a 140-lb weight dropped from a height of 30 in.), values of the Dutch cone test (static penetration expressed as the average pressure at the base of 60° cone of 10-cm² plan area to cause penetration of the cone into the soil), and the likely peak values of a drained peak-friction angle of a quartz sand for the range of density states.

Techniques for describing and measuring the

TABLE 4. Relative Densities of Cohesionless Soils

Description	γ_d (g/cm³)	Relative Density D_{r1}	D_{r2}	Standard Penetration Test (blows/ft.)	Dutch Cone (kg/cm²)	Likely Range of tan ϕ (peak)
Very loose	1.3-1.4	0-15	0-10	0-4	<20	0.58-0.62
Loose	1.4-1.5	15-35	10-30	4-10	20-40	0.62-0.70
Medium	1.5-1.7	35-65	30-60	10-30	40-120	0.70-0.84
Dense	1.7-1.8	65-85	60-80	30-50	120-200	0.84-0.93
Very dense	1.8-1.9	85-100	90-100	>50	>200	>0.93

TABLE 5. Values of Interparticle Friction (ϕ_μ) for Quartz:
Free Particles to Flat Surface, Saturated

Particle diameter (mm)	0.04	0.06	0.10	0.18	0.50	1.0
tan ϕ_μ	0.577	0.557	0.532	0.496	0.452	0.404

Source: After Rowe (1962).

shape and roundness of particles are well established (see Chap. 2 in Winterkorn and Fang, 1975). For particles of gravel size and larger, it is relatively easy to record the dimension ratios of the particles, but this task becomes rather tedious for finer particles and is rarely used in soil mechanics. The degree of roundness of particles is described by the terms *angular, subangular, subrounded, rounded,* and *well rounded,* and standard charts are used for comparative purposes.

The "basic" frictional property, designated ϕ_μ is the friction angle developed by contact of two "flat" surfaces of the soil particles. Table 5 lists some selected values for quartz sand, although these ϕ_μ values can be greatly influenced by the surface chemistry. The simple concept of friction appears to have been originated by Leonardo da Vinci, but the French scientist Amontan (1699) is credited with the basic laws: that the shearing force developed between two bodies is proportional to the normal force, and that this force is independent of gross contact area. Terzaghi (1943) explained the frictional process by showing that contact between two surfaces is confined to asperities and that the material at these points of contact is likely to yield until the normal force can be resisted. Thus, if σ is the yield stress of the material and all n contacts over a total area A are deforming plastically, the normal force is

$$N = \sum_{i=1}^{i=n} \sigma_y A_i \qquad (2a)$$

and the resultant shearing force is

$$T = \sum_{i=1}^{i=n} \tau_y A_i \qquad (2b)$$

where τ is the shearing resistance at the contacts.

Thus

$$\mu = \tan \phi_\mu = \frac{\tau_y}{\sigma_y} \qquad (3)$$

According to this "adhesion" theory, the surface roughness should not affect the value of ϕ_μ, and the effect of roughness can be determined experimentally.

In a soil mass, relative movement of the particles at a sliding contact will not be in the same direction as the overall shear movement, so that the friction angle of even a loose sand will be greater than ϕ_μ. Coulomb expressed the shear strength of a soil in 1776 by an empirical "law," still known as *Coulomb's law:* If τ is the shear strength, then

$$\tau = c + \sigma_n \tan \phi \qquad (4)$$

which is an equation to a straight line on a shear stress, normal stress σ_n (Mohr) plot with an intercept c and a slope σ. This *total* stress law was much later (1925) expressed in the more fundamental *effective* stress form by Terzaghi (1925). The modern interpretation of Coulomb's expression for shear strength is discussed in a later section of this entry.

Early work on the structure of micas and related minerals in the 1930s (e.g. Jackson and West, 1933; Hendricks and Jefferson, 1939) inspired studies on the structure of clay minerals (see Vol. XII, Pt. 1: *Clay Minerals, Silicates; Soil Components, Inorganic)* so that the general characteristics listed in Table 6 are well established. It should be noted that the two units forming the atomic structure of clay minerals are the silica tetrahedrons in which a silicon atom is equidistant from four oxygen atoms, and a hydroxyl unit in which aluminum, magnesium, or iron atoms occur in an octahedral coordination equidistant from six oxygen or hydroxyl atoms.

Commonly, aluminum atoms are found in the octahedral unit. In this only two-thirds of the possible positions of the atoms are filled (gibbsite), whereas the magnesium atoms fill all the lattice positions (brucite). The thickness of the octahedral unit is 5.05 Å, and the thickness of the silica tetrahedron is 4.93 Å.

Allophane minerals are amorphous to x-ray diffraction because the arrangement of the units is too irregular for detection by this technique. The physical properties of allophane may be similar to halloysite or kaolinite, but there is a large variability.

The clay minerals kaolinite, illite, and montmorillonite are well known to soil engineers because they are commonly occurring, and are quoted in the literature as examples of clays of small, moderate, and high activity (see Mackenzie, 1975), respectively—that is, clays that show low, moderate, and large swelling movements when an unsaturated clay is subjected to wetting. Kaolinite is composed of a single tetrahedron sheet and a single alumina octahedral sheet, 46.5 percent SiO_2, 39.5 percent Al_2O_3, and 14 percent H_2O. Illites are composed of two silica tetrahedrons covering an octahedral sheet. They have the same structure as montmorillonite except that a few of the silicon atoms are replaced by aluminum, and the resultant charge deficiency is balanced by potassium ions between the unit layers. As can be seen from the characteristics listed in Table 6, there are quite large differences between illite and montmorillonite despite this seemingly small structural difference.

The detailed structure of any of the clay minerals will vary from the idealized arrangements exemplified above due to defects in the structure (Grim, 1953). Also, because a clay will generally consist of more than one clay mineral, it is usual for soil engineers to attempt to determine the proportion of clay minerals. From a knowledge of the predominant clay mineral, it is possible to anticipate some of the properties of interest to the engineer and, in particular, the activity of the soil. Such information assists in recognizing some likely problems and the detailed information required to overcome the difficulties.

Soil-Water Relationships

Water occurs either in the combined form within the clay particles or as free water within the pores. This section discusses the influence of *free water*. A soil of considerable strength can be formed from a slurry of negligible shear strength by removal of a sufficient quantity of the free water. During the reduction of water content, there is a volume change equal to the volume of water removed provided the soil remains saturated. A stage will be reached, however, when the coarsest pores begin to drain and the soil becomes unsaturated. Specific values of water contents have been shown to be a measure of the plasticity and volume-change characteristics of a clay.

In the last 16 years of his life, A. M. Atterberg turned his attention to a study of the physical properties of soils and concluded that plasticity was an important property of a clay. He proceeded to devise the techniques, which bear his name, to measure soil plasticity. Although Atterberg first published his findings in 1908 it was not until 1925, when Terzaghi recognized the application to soil mechanics, that engineers became familiar with these tests. The specific terms associated with the Atterberg or consistency tests are the *liquid limit, the plastic limit,* the *plasticity index,* and the *liquidity index* (see *Atterberg Limits and Indices*). As the name implies, the liquid limit is the water content at which the shear strength becomes negligible and the suggested value for shear strength at the liquid limit is about 0.7 kPa (0.1 psi). Atterberg devised a standard apparatus to measure the liquid limit, and he defined the liquid limit as the water content at which 25 blows in his apparatus were just sufficient to cause the soil to flow. The plastic limit is the lowest water content at which the soil can be rolled into threads 0.3 mm (⅛ in.) without breaking. The plasticity index is the difference between the liquid and plastic limits. The liquidity index is a measure of the proximity of the natural water content to the liquid, namely:

$$\text{L.I.} = \frac{w - w_{PL}}{w_{LL} - w_{PL}} = \frac{w - w_{PL}}{w_{PI}} \qquad (5)$$

where L.I. is the liquidity index; w, the natural (*in situ*) water content; w_{PL}, the plastic limit; w_{LL}, the liquid limit; and w_{PI}, the plasticity index.

Table 2 shows some typical values of the liquid and plastic limits. There is no single value of either the liquid or plastic limit for a particular clay mineral. It is known that the value of the plastic depends on the exchangeable ions present and the plastic limit of a sodium montmorillonite is greater than a calcium, potassium, or magnesium montmorillonite. Clays with a low base-exchange capacity show less dependence of liquid and plastic limits on the ions present. It should be appreciated that the Atterberg tests are conducted on fully remolded samples; hence any property significantly dependent on the *in situ* structure of the soil cannot be logically related to the consistency limits.

Another index based on water content is the *shrinkage limit.* This value roughly corresponds to the water content at which the soil upon drying out first becomes unsaturated. It is obtained by measuring the volume change as the soil is dried out until an unsaturated state is reached. The value of the shrinkage limit lies below the plastic limit.

TABLE 6. Properties of Clay Minerals

Mineral	Octahedral Ion	Structure	Interlayer Spacing, Å	Cation Exchange Capacity (C.E.C.), mEq/100 g	Source of C.E.C. (Plus Lattice Vacancies)	Normal Particle Size	Engineering Properties
2-layer sheets							
Serpentines, $Mg_3Si_2O_5(OH)_4$	Mg^{2+}		7.3	nil	—	large	nonswelling, noncohesive
Kaolins, $Al_4Si_4O_{10}(OH)_8$	Al^{3+}		7.3	3–15	broken bonds	1–10μ	nonswelling; inactive
Halloysites, $Al_4Si_4O_{10}(OH)_8 \cdot 2H_2O$	Al^{3+}		10.2	5–10		0.1μ	very plastic and troublesome when *partially* dried, but relatively inactive otherwise
Pyrophyllite, $Al_2Si_4O_{10}(OH)_2$	Al^{3+}		9.1	nil	—	large	nonswelling, noncohesive
Montmorillonite, $(Al_{1.67}Mg_{0.33})Si_4O_{10}(OH)_2 \xrightarrow{} Na_{0.33}$	Al^{3+}		9.6 → 15.5 ∞	80–150	octahedral substitution (of every sixth cation) (Mg^{2+} for Al^{3+})	0.01–0.1μ	swelling; highly active
3-layer sheets							
Illite	Al^{3+}		ca. 10	10–40	broken bonds + some tetrahedral substitution (Al^{3+} for Si^{4+})	0.1–1μ	partial swelling, becomes active if leached
Micas e.g. $KAl_2(AlSi_3)O_{10}(OH)_2$	Al^{3+}	(K^+)	ca. 10	low ideally 0, theoretically 250	(K^+ is 12-coordinated, and the octahedral bond angles adjust to "lock on" to the cation)	large	nonswelling, noncohesive
Talc, $Mg_3Si_4O_{10}(OH)_2$	Mg^{2+}		9.3	nil	—	large	nonswelling, noncohesive
Chlorites, $Mg_3(AlSi_3)O_{10}(OH)_2 \xrightarrow{} (AlMg)_{1.5}(OH)_3$	Mg^{2+}		14.2	10–40	broken bonds + both tetrahedral and octahedral substitution (Al^{3+} for Si^{4+}) (Al^{3+} for Mg^{2+})	—	nonswelling; flows under stress, hence mining and tunneling problem
Vermiculite, $Mg_3(Si_3Al)O_{10}(OH)_2 \xrightarrow{} Mg_{0.5}$	Mg^{2+}	$O \longrightarrow (Mg^{2+})$	14.5	100–150	tetrahedral substitution	—	limited swelling; active

The *permeability* (q.v. in Vol. XII, Pt. 1) of a soil is a measure of the rate of flow of water through a soil under a specific hydraulic gradient. Except for flow through rockfill, it is usual to assume that Darcy's law is sufficiently accurate for flow predictions. This law states that the velocity of flow is proportional to the hydraulic gradient. Expressed analytically,

$$v_y = -k_y \frac{\partial h}{\partial y} = -k_y \cdot i_y \qquad (6)$$

where V_y is the "discharge velocity," defined as the ratio of the discharge Q to the total cross-sectional area at right angles to the direction of flow; k_y is the coefficient of permeability in the direction of flow y; h is the piezometric head, that is, the elevation head plus the pressure head; and i_y is the hydraulic gradient in the direction of flow.

Values of the coeeficient of permeability depend on the void ratio, particle size, structure, degree of saturation, and—for clays—the type of clay mineral. Laboratory or field measurements directly determine k_y without need for reference to the above factors. As a guide to the likely influence of compaction however, it is generally agreed that the log k_y is proportional to e. Table 7 gives some typical values of k_y.

Terzaghi and Peck (1967) were responsible for the description of permeability. The value of k_y for sands and gravel can be measured in the laboratory by a *falling-head* permeameter; a *fixed-head* permeameter is used for soils of lower permeability. Field pumping tests are the best method of establishing permeability, but care has to be taken in the method of analyzing of data.

Values of permeability are greatly influenced by the degree of saturation of the soil. There is a decrease of permeability as the water content decreases (soil suction increases). As a further illustration, the permeability of an unsaturated (moist) sand layer is low enough to prevent the passage of water to the atmosphere from an underlying clay. If the sand is wetted, however, the permeability is increased by some orders of magnitude and there is little resistance to flow.

TABLE 7. Typical Values of Coefficient of Permeability

Soil Type	k_y (cm/s)	Permeability Description
Fine gravel	10^{-2}	Medium
Coarse sand	10^{-4}	Low
Fine sand	10^{-5}	Low
Loess	10^{-4}	Low
Kaolinitic clay	10^{-5}-10^{-6}	Very low
Montmorillonitic clay	10^{-7}-10^{-8}	Practically impermeable

There are two major areas of water movement of particular interest to the geotechnical engineer:

Steady Flow State. This is either a static water-table state or a steady seepage state. In the latter case, the total head is related to position by the equation

$$k_x \frac{\partial^2 h}{\partial x^2} + k_y \frac{\partial^2 h}{\partial y^2} + k_z \frac{\partial^2 h}{\partial z^2} = 0 \qquad (7)$$

which is commonly used in the two-dimensional (Laplace) situation ($k_x = k_y$). It is then an equation to two families of curves intersecting at right angles (flow net). One family shows the flow directions, and the other family consists of lines of equal total head.

Transient Flow State. In this state, the pore pressures vary with time, and thus the effective stresses and volume vary with time. Typical applications are consolidation settlement calculations, analysis of pumping tests, and drawdown of water in earthen embankments.

Engineering Applications of Soils

As a guide to a description of the soils and the likely properties and uses, the geotechnical engineer makes use of a classification system based on the relevant physical properties (see *Soil Classification System, Unified.* Vol. XII, Pt. 1: *Clay Minerals, Silicates; Particle-Size Distribution*), and the likely properties and engineering usage are listed in Table 8.

Pore Pressures in Soils

Effective Stress Law. Soils are three-phase materials consisting of solid particles, water, and air (vapor). The strength and deformation properties are controlled by the forces between the solid particles; the magnitude of these forces is determined not only by the external stresses applied to the soil but also by the pressures developed in the water and air phases. Instead of thinking in terms of interparticle forces, the engineer finds it convenient to use stresses, that is, a quantity equal to force divided by area. This concept seems to be relatively easy to visualize when dealing with materials that can be considered to be single-phase provided we do not think of interatomic dimensions. The sciences of elasticity and plasticity use stress as a fundamental concept.

Figure 4 is a simple (mechanistic) model of a soil. It represents a section through a soil in which the forces between the particles are F_i, the pore-water pressure is u_w, and the pore-air pressure is u_a. Let us consider the simple statical relationships between the internal and external forces. For this purpose, imagine that the stress applied "externally" is σ and the section occupies an area A in the horizontal plane. Because of the domi-

TABLE 8. Engineering Applications of Soils

Typical Names of Soil Groups	Group Symbols	Important Properties				Relative Desirability for Various Uses									
		Permeability When Compacted	Shearing Strength When Compacted and Saturated	Compressibility When Compacted and Saturated	Workability as a Construction Material	Rolled Earth Dams			Canal Sections		Foundations		Roadways Fills		
						Homogeneous Embankment	Core	Shell	Erosion Resistance	Compacted Earth Lining	Seepage Important	Seepage Not Important	Frost Heave Not Possible	Frost Heave Possible	Surfacing
Well-graded gravels, gravel-sand mixtures, little or no fines	GW	pervious	excellent	negligible	excellent	—	—	1	1	—	—	1	1	1	3
Poorly graded gravels, gravel-sand mixtures, little or no fines	GP	very pervious	good	negligible	good	—	—	2	2	—	—	3	3	3	—
Silty gravels, poorly graded gravel-sand-silt mixtures	GM	semi-pervious to impervious	good	negligible	good	2	4	—	4	4	1	4	4	9	5
Clayey gravels, poorly graded gravel-sand-clay mixtures	GC	impervious	good to fair	very low	good	1	1	—	3	1	2	6	5	5	1
Well-graded sands, gravelly sands, little or no fines	SW	pervious	excellent	negligible	excellent	—	—	3 if gravelly	6	—	—	2	2	2	4
Poorly graded sands, gravelly sands, little or no fines	SP	pervious	good	very low	fair	—	—	4 if gravelly	7 if gravelly	—	—	5	6	4	—
Silty sands, poorly graded sand-silt mixtures	SM	semi-pervious to impervious	good	low	fair	4	5	—	8 if gravelly	5 erosion critical	3	7	8	10	6
Clayey sands, poorly graded sand-clay mixtures	SC	impervious	good to fair	low	good	3	2	—	5	2 erosion critical	4	8	7	6	2
Inorganic silts and very fine sands, rock flour, silty or clayey fine sands with slight plasticity	ML	semi-pervious to impervious	fair	medium	fair	6	6	—	—	6 erosion critical	6	9	10	11	—
Inorganic clays of low to medium plasticity, gravelly clays, sandy clays, silty clays, lean clays	CL	impervious	fair	medium	good to fair	5	3	—	9	3	5	10	9	7	7
Organic silts and organic silt-clays of low plasticity	OL	semi-pervious to impervious	poor	medium	fair	8	8	—	—	7 erosion critical	7	11	11	12	—
Inorganic silts, micaceous or diatomaceous fine sandy or silty soils, elastic silts	MH	semi-pervious to impervious	fair to poor	high	poor	9	9	—	—	—	8	12	12	13	—
Inorganic clays of high plasticity, fat clays	CH	impervious	poor	high	poor	7	7	—	10	8 volume change critical	9	13	13	8	—
Organic clays of medium to high plasticity	OH	impervious	poor	high	poor	10	10	—	—	—	10	14	14	14	—
Peat and other highly organic soils	Pt	—	—	—	—	—	—	—	—	—	—	—	—	—	—

Source: After Lambe and Whitman (1969).

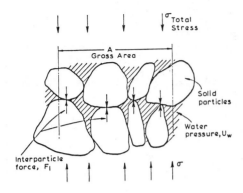

FIGURE 4. Particulate model of a granular soil.

nant importance of compressive stresses in soil mechanics, the compressive stresses are deemed positive—this is opposite to the convention used in elasticity and plasticity.) Further, let us say that the particle contact area at particle i is A_i, and the water phase is applied over a horizontal area of A_w. Then, resolving vertically,

$$\sigma A = \Sigma F_i + (u_w \cdot A_w) + u_a \cdot (A - A_w - A_i)$$

$$\sigma = \frac{\Sigma F_i}{A} + u_w \cdot \frac{A_w}{A} + u_a \cdot (1 - \frac{A_w}{A} - \frac{A_i}{A})$$

(8)

The first term on the right-handside of equation 8 is called the *effective stress, σ'* and according to the simple model, it is a measure of the forces between the particles. σ is termed the *total stress*, that is, the stress that is resisted by the combined action of the solid, liquid, and air phases.

Rearranging equation 8 gives (approximately, since A_i is small)

$$\sigma' = (\sigma - u_a) + \frac{A_w}{A}(u_w - u_a)$$

(9)

In fact, the limitations of this model require substitution of parameter χ for the area ratio of equation 9. Nevertheless, it is clear that such a parameter will depend on the space occupied in the voids by the water phase—termed the *degree of saturation, S*. There are two limiting cases of considerable importance, namely, the saturated case, $S = 1$ $(A_w \simeq A)$, and a perfectly dry soil, $S = 0$ $(A_w = 0)$.

Many soils are very close to saturation and the effective stress law, equation 9, simplifies to

$$\sigma' = \sigma - u_w$$

(10)

which applies to *any* soil, whereas the more general form of the effective stress law requires a knowledge of χ—this parameter will depend on

the type of soil and factors such as the stress history as well as the value of the degree of saturation. The difficulty of obtaining satisfactory values of X has restricted the practical application of the law.

However, the application to saturated soils simply requires a knowledge of the total stress and the pore-water pressure, (see *Hydrodynamics, Porous Media*). Similarly, the law for a dry soil is identical to equation 10 except that u_w is replaced by u_a, but since u_a is invariably zero in this case, the total effective stresses are identical. Terzaghi used the law for saturated soils in his address to the First Conference of Applied Mechanics in Delft, Holland, in 1924. In the following year his monumental book *Erdbaumechanik auf bodenphysikalischer Grundlage* was published; some consider that this event marked the "birth" of soil mechanics.

Consolidation. The fundamental and practical importance of effective stress may be applied to saturated soils by using the law to interpret soil behavior. A large area of land is initially inundated to a depth of, say, Z. Suppose the area is drained to bring the water level (water table) to ground surface. It might be anticipated that the ground surface would rise when the water loading was eliminated; in fact, the surface level remains constant. This can be readily seen to be a logical consequence of the application of the effective stress law. Before the area was drained the effective stress at a depth h below ground surface is obtained from the data shown in Figure 5a.

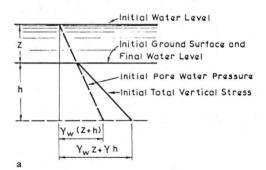

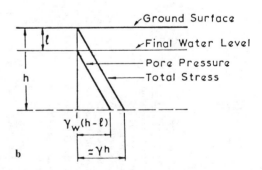

FIGURE 5. Pore pressures and total stresses in an inundated (a) and dewatered (b) soil deposit.

$$\text{Total stress} = Z \cdot \gamma_w + h \cdot \gamma \qquad (11)$$

$$\text{Pore-water pressure} = \gamma_w(Z + h) \qquad (12)$$

Then, using equation 10, the effective stress at any depth in the soil layer of bulk density γ (as defined in the preceding section, the bulk density is the total weight per unit volume), σ' is seen to be

$$\sigma' = h(\gamma - \gamma_w) \qquad (13)$$

establishing that the effective stress is independent of the height of the water level *above* the ground surface, so that the drainage will not change the effective stresses in the soil. Recalling that the effective stress is a measure of the forces between the particles, it follows that these forces are constant during drainage and that the particles will not change relative positions; hence the volume of the soil mass will remain constant. If the volume is constant, the surface level will not change.

It is of some interest to pursue this argument further. If the water table is lowered below ground level by amount l, the effective stress after the drainage is, from Figure 5b,

$$\sigma' = \gamma_1 l + (\gamma - \gamma_w)(h - l) \qquad (14)$$

where γ_1 is the bulk density of the soil above the water table (in fact, $\gamma_1 \simeq \gamma$).

The change in vertical effective stress due to the lowering of the water table is

$$\Delta\sigma' = l(\gamma_1 - \gamma + \gamma_w) \qquad (15)$$
$$\simeq l \cdot \gamma_w$$

thus establishing that the change in effective stress at any point in the soil layer below the water table ($h \geq l$) is proportional to the depth of the water table below ground surface.

An increase in effective stress means the compressive forces between the particles increase although the total stress remains approximately constant when the water table falls below the ground surface. This increase in effective stress decreases the total volume by a process of *consolidation* (q.v.), and the consequence is a reduction in surface level (settlement).

The process of consolidation can be clearly illustrated by another simple mechanistic model. Figure 6 shows a cylinder filled with water. It is sealed by a sliding piston supported on a helical spring contained within the cylinder. When the piston is loaded, sufficient water pressure is developed to support the load. There will be a small movement of the piston as the water compresses, but substantial movements will not occur until water is allowed to flow through the valve. The total volume contained within the cylinder progressively decreases at a rate determined by the flow characteristics of the valve and the water pressure within the cylinder. Now, if the spring is an analogue of the soil skeleton, the water an analogue of the soil water, the flow rate of the valve an analogue of the soil permeability, and the load an analogue of the total stress × area, then the piston movement is analogous to settlement due to consolidation, and the rate of piston movement with time is characteristic of the settlement rate in the consolidation process. It is seen that the spring (skeleton) force increases with time and that there is a reduction of volume as the consolidation proceeds. Thus, returning to the effect of reducing the water table below ground level, it is seen that the increasing effective stress is due to flow of water from the soil mass with a consequent reduction in volume, reflected by a settlement of the ground surface.

Until Terzaghi developed his analysis of the consolidation process, the settlement-time relationship of foundations on clays was not understood, so that the publication of the effective stress law and the analysis of the consolidation process had major practical as well as basic applications. To enable him to measure the compressibility and rate parameters of a particular soil, Terzaghi devised an instrument termed an *oedometer.*

The principles of the oedometer are evident through an examination of Figure 7, which shows a section through the cell containing the soil

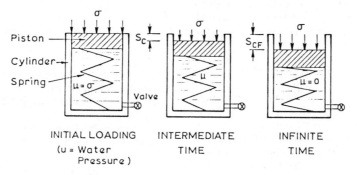

FIGURE 6. Rheological model of a saturated clay.

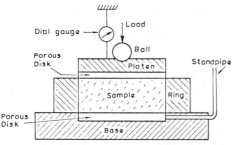

STANDARD FIXED RING OEDOMETER

FIGURE 7. Section through a standard fixed-ring oedometer.

sample. The cylinder of soil is placed in the rigid ring with as little clearance between the soil and the ring as possible. Permeable discs are placed at the top and base of the soil, and a vertical load is applied for a specified period, commonly 24 hours. The vertical compression of the sample is measured at a sequence of time intervals relative to the instant of loading. There is an immediate compression of the sample due to a variety of causes, notably, the consequence of the lateral expansion of the soil as it fills the gap between the sample and the ring, and as a result of the compression of the air in the pores. (Even if very careful procedures are adopted to saturate the sample, it is to be expected that a small amount of air will remain in the pores.) If the soil is completely saturated and the test a perfect one-dimensional test, the immediate compression is zero. At the instant of load application, the pore-water pressure theoretically increases to a value equal to the applied vertical pressure. Drainage commences immediately at a

rate depending on the permeability and compressibility of the soil and the thickness of the sample, with a consequent vertical compression of the sample.

The changes in thickness of the sample during the 24-hour loading period are recorded, Figure 8, and then the load is doubled at 24-hour intervals. From the data in Figure 8 we can obtain the thickness corresponding to each load. It is more convenient to use a nondimensional quantity, the void ratio, instead of the thickness. Thus in Figure 9 we have a relationship between the void ratio and the vertical effective stress. Assuming that the pore pressures have dissipated at the end of the loading interval, the effective stress is equal to the total stress.

Suppose a soil has been consolidated to an effective stress p_1 and an increment of pressure is applied to bring the total stress to $p + \Delta_p = p_2$. At the instant of loading, the pore pressure u is Δ_p, but as the consolidation process proceeds, u diminishes and the increase in effective stress equals the decrease in pore pressure since the total stress remains constant. This is shown in Figure 10 by the fact that cd (increase in effective stress) plus de (remaining pore pressure) equals the increment in total stress ab. The characteristics of the compression-effective stress relationship will be discussed later. The important conclusion here is that we can calculate the change in void ratio—and hence the settlement of a soil layer—if we know the initial and final effective stresses.

Let us consider the application of the analysis to the prediction of settlement of the ground surface due to the lowering of the water table (Fig. 5). Suppose the soil layer is 8 m thick, and the water table is lowered 1 m below ground level. Table 8 details the calculations.

The change in thickness of the jth sublayer is related to the change in void ratio by

$$\Delta H_j = \frac{e_i - e_f}{1 + e_i} \cdot H_j = \frac{\Delta e}{1 + e_i} \cdot H_j \qquad (16)$$

and the surface settlement at the end of consolidation is

$$S_{TF} = \sum_{j=1}^{j=n} \Delta H_j \qquad (17)$$

It is seen that the procedure adopted is to divide the soil into a number of sublayers and calculate the change in thickness of each sublayer from the change in void ratio, then summate to obtain the total change in thickness of the layer.

Equation (16) is sometimes written as

$$\Delta H_j = m_v \cdot \Delta \sigma'_1 \cdot H_j \qquad (16a)$$

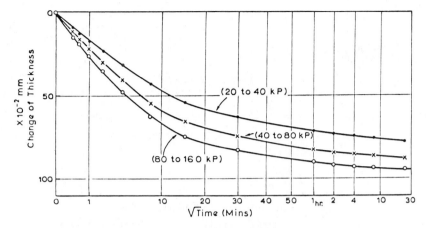

FIGURE 8. Three typical settlement-time relationships as shown by oedometer testing of saturated silty clay.

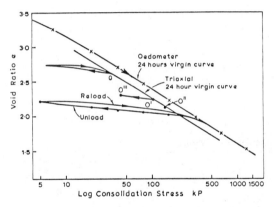

FIGURE 9. Void ratio-consolidation stress relationship in saturated silty clay following 24-hour loading cycle.

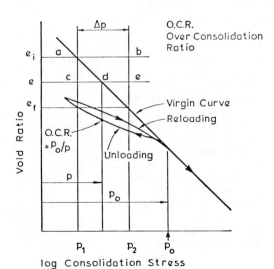

FIGURE 10. Characteristics of consolidation curves.

where $\Delta\sigma'_1$ is the increment in vertical effective stress (column 3 minus column 2 of Table 9) and m_v is termed the coefficient of volume decrease. Typical values are included in Table 10.

The *rate* of settlement can be determined by making use of the "diffusion equation" first applied to this problem by Terzaghi in 1925. The major limitation to this analysis is that the strain (settlement) must be one-dimensional.

Terzaghi's diffusion equation is

$$\frac{\partial^2 u}{\partial y^2} = c_v \cdot \frac{u}{t} \qquad (18)$$

which expresses the value of the pore pressure at a point y at time t from the instant of load application. The *coefficient of consolidation*, c_v, can be determined from the compression-time data obtained in the oedometer test (Fig. 8). The solution of this equation (Table 11) is expressed as a relationship between the degree of settlement and a time factor, T. These values are defined as

$$U_s = \frac{\text{settlement at time } t \text{ due to consolidation}}{\text{settlement at infinite time}}$$

$$= \frac{S_c}{S_{TF}} \qquad (19)$$

$$T = \frac{C_v \cdot t}{H_d^2} \qquad (20)$$

where H_d is termed the *drainage path* and is equal to the total thickness H of the layer if the pore pressure at the top of the layer is maintained at zero (permeable) and the base of the layer is impermeable (or vice versa) but H_d is equal to half the layer thickness when the top and the base are

TABLE 9. Calculation of Settlement of Ground Surface ($\gamma = 20$ kN/m³)

Depth of Sublayer (m)	Initial Effective Stress (kN/m²) (Fig. 5a)	Final Effective Stress (kn/m²) (Fig. 5b)	Initial Void Ratio, e_i	Final Void Ratio, e_f	Change in Thickness, ΔH_j (m)
0-2	20	45	2.94	2.66	0.036
2-4	60	135	2.57	2.28	0.041
4-6	100	200	2.40	2.14	0.038
6-8	140	240	2.27	2.07	0.031
					0.145 m

TABLE 10. Typical Values of the Coefficient of Volume Decrease, m_v, and Coefficient of Consolidation, c_v.

Soil Type	m_v (cm²/N)	c_v (cm²/s)
Normally consolidated silty-clay	$> 10^{-2}$	10^{-4}-10^{-3}
Compressible sandy clay	10^{-4}-10^{-3}	10^{-3}-10^{-2}
Heavily over-consolidated clay	10^{-4}	10^{-4}
Silt	10^{-2}	10^{-2}-1

TABLE 11. Degree of Settlement-Time Factor: One-Dimensional Consolidation

Degree of Settlement-Time Factor Relationships								
U_s(%)	10	20	30	40	50	60	70	80	90
T	0.008	0.031	0.071	0.126	0.197	0.287	0.403	0.567	0.848

permeable. The solution and the application will be discussed later.

Returning to the case just considered, the time required for any given degree of settlement to occur can be readily calculated from the data given in Table 11. For example, for 50 percent settlement ($U_s = 50$), the time factor is 0.197. Thus the time is

$$t = \frac{TH^2_d}{C_v} \quad (21)$$

If the bedrock underlying the layer is effectively impermeable, then $H_d = H = 8$ m, and for the silty clay $c_v = 10^{-4}$ cm²/s; thus

$$t = \frac{0.197 \times 800^2}{10^{-4} \times 60 \times 60 \times 24} = 14,600 \text{ days} \quad (22)$$

If the underlying rock was relatively permeable and was able to dissipate the excess pore processes at the base of the layer, the time would be one-quarter of the above value.

Shear Strength. Let us now examine the influence of pore pressures on soil strength and the deformations due to shear stress. It is well estab-

lished that there are changes in the relative positions of the soil particles (termed the *structure*) when there is a change in effective stress. The strength and deformation behavior is controlled by the structure and drainage conditions.

In the shear testing of soils it is common practice to subject cylindrical samples to an axially symmetric stress state, termed a *triaxial shear test*. The pore water can be allowed to drain under a constant pore pressure applied by an external pressure source, or the sample can be sealed to ensure that the water content of the sample remains constant. The latter is referred to as an *undrained state*. Similarly, the pressure in the air phase can be controlled or the air can be prevented from draining from the sample. Although the testing of saturated samples is well established, it should be mentioned that the controlled shear testing of unsaturated samples is a very difficult procedure, and it has been only in recent years that reliable tests have been carried out. Furthermore, these tests are generally conducted in research laboratories and it is incorrect to regard such tests as routine.

Most investigations in the past have been carried out on soils close to a saturated state, and thus the expertise developed over the years predominantly applies to saturated soils. Only two limiting drainage conditions can be imposed on the sample when the air content is small. The *undrained state* simulates the situation when a load is applied to a soil before appreciable dissipation of pore pressure can take place. The *drained state* is the other extreme, in which the load is applied slowly and the pore water allowed to continually drain, thus ensuring that the pore pressures developed by the stresses are dissipated.

The effective stress law has explained features of soil behavior in the undrained test, and established the relationships between the undrained and drained tests. In practice there are two types of undrained tests used, the unconsolidated-undrained test (UUD), and the consolidated-undrained test (CUD). Table 12 shows the sequence of stress application.

Pore pressures are not recorded in the UUD test. This test is applied to undisturbed samples taken from boreholes, for example, to obtain the strength data for designing isolated foundations. Investigations have established that such a test gives a close measure of the *in situ* shear strength of the soil. Since the pore pressures are not measured, the effective stresses cannot be determined. In the CUD triaxial test, the sample is consolidated by an all-round (ambient or isotropic) stress with a constant back-pressure being maintained in the pore water; then the drainage circuit is closed, and a shear stress is applied under undrained conditions.

Some details of the test apparatus are shown in Figure 11. The major features of the triaxial test include the following: (1) Constant isotropic stress is applied to the cylindrical sample. This stress is usually the minor principal total stress ("compression test"), but the stress can be made the major principal stress ("extension test"). (2) A principal stress is applied in the vertical direction—usually the major principal stress. The difference in magnitude between the major and minor principal total stresses is the deviator stress $(\sigma_1 - \sigma_3)$ and it is evident that this is also equal to the difference of the principal effective stresses $(\sigma'_1 - \sigma'_3)$. (3) Back-pressure is applied to the pore water, thus allowing flow of water to or from the sample at constant pore-water pressure or, alternatively, a sealing of the drainage line. (4) Necessary instrumentation records the cell pressure, axial load, axial compression, change in volume of pore water, or the magnitude of the pore water pressures that develop during testing.

Figure 12 shows the relationship between the deviator stress and the axial strain in a CUD test on a saturated silty clay. Before the sample was subjected to a progressively increasing deviator stress under undrained conditions, it was formed in a slurry, then consolidated with a cell pressure of 40 kPa and a back-pressure of 20 kPa; thus, by the effective stress law, the consolidation pressure was 20 kPa. Figure 9 shows the void ratio-consolidation pressure relationship for the isotropic (triaxial) consolidation stress state as well as the relationship obtained for the same soil in an oedometer test. The pore pressures developed during the undrained test are also shown in Figure 12. To emphasize the differences between the total and effective stresses, Figure 13 shows the *stress paths* during consolidation and testing. In this figure the ordinate is the axial effective stress and the abscissa is the radial effective stress. During consolidation, the effective stresses increase from O to O' along a 45° line, whereas the total stresses remain at point O'. During undrained testing the axial stress increases at constant radial total stress, but the effective radial stress must decrease since the pore pressure (Fig. 12) increases. The stress path is O'a'. If an initially identical sample is

TABLE 12. Triaxial Shear Tests (Saturated Soils)

Test	(Consolidated) Initial Stresses Stage I		(Undrained) Intermediate Stresses Stage 2		Stresses of Failure Stage 3	
	Total	Effective	Total	Effective	Total	Effective
Unconsolidated-undrained (UUD)	$\sigma_1 = 0$ $\sigma_3 = 0$	$\sigma'_1 = u_i$ $\sigma'_3 = u_i$	$\sigma_1 = p_1$ $\sigma_3 = p_1$	$\sigma'_1 = u_1$ $\sigma'_3 = u_1$	$\sigma_1 = p_1 + p_2$ $\sigma_3 = p_1$	$\sigma'_1 = p_2 (1 - A_f) + u_i$ $\sigma'_3 = A_f p_2 + u_i$
Isotropically consolidated-undrained (CUD)	$\sigma_1 = p_0$ $\sigma_3 = p_0$	$\sigma'_1 = p_0$ $\sigma'_3 = p_0$	$\sigma_1 = p_0 + p_1$ $\sigma_3 = p_0 + p_1$	$\sigma'_1 = p_0$ $\sigma'_3 = p_0$	$\sigma_1 = p_0 + p_1 + p_2$ $\sigma_3 = p_0 + p_1$	$\sigma'_1 = p_2 (1 - A_f) + p_0$ $\sigma'_3 = -A_f p_2 + p_0$
Isotropically consolidated-drained (CD)	$\sigma_1 = p_0$ $\sigma_3 = p_0$	$\sigma'_1 = p_0$ $\sigma'_3 = p_0$			$\sigma_1 = p_0 + p_2$ $\sigma_3 = p_0$	$\sigma'_1 = p_0 + p_2$ $\sigma'_3 = p_0$

Notes: σ_1 = total axial stress, σ_3 = total radial stress, σ'_1 = effective axial stress, σ'_3 = effective radial stress; u_i = initial pore pressure (suction); p_0, p_1 = increments in cell pressure; p_2 = increment in axial stress; A_f = pore pressure parameter at failure (see equation 23).

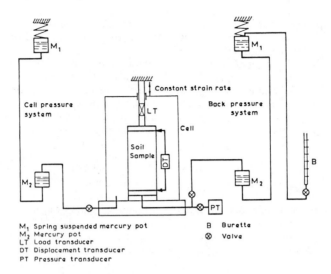

FIGURE 11. Schematic layout of triaxial shear test apparatus.

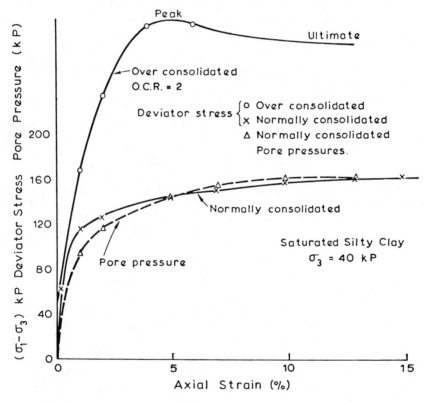

FIGURE 12. Relationship between deviator stress and axial strain in a CUD test on a saturated silty clay.

consolidated to a larger isotropic effective stress, the undrained stress path is $O''a''$ and the decreased void ratio increases the strength of the soil. In fact, the shear strength is proportional to the consolidation stress for a normally consolidated soil.

Now, examine what happens if the soil is consolidated to O' and then the consolidation stress is reduced to O'''. Referring to Figure 9, we can see that the reduction in pressure causes an increase in void ratio but the recovery does not follow $O'O$. To differentiate between the state

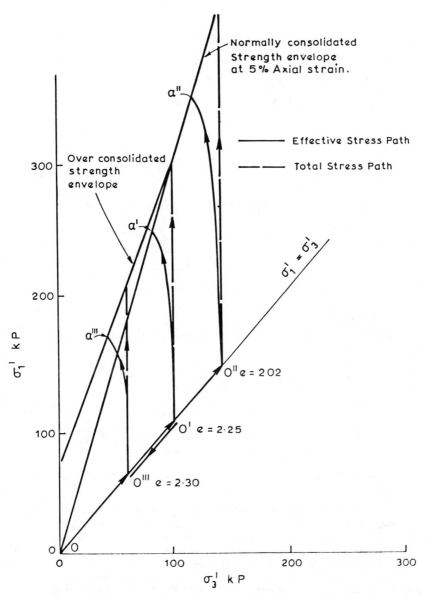

FIGURE 13. Stress paths in a CUD test on normally and overconsolidated saturated silty clay.

represented by lines OO′O″ and the state typified by O′O‴, we refer to these states as *normally consolidated* and *overconsolidated,* respectively; that is, a normally consolidated soil has never been subjected to a consolidation stress in excess of the existing stress. It will also be seen that the stress-strain curve shown in Figure 12 is no longer the "strain-hardening" curve of the normally consolidated soil and there are well-defined "peak" and "ultimate" stresses.

It will be seen in the subsequent discussions on stability that the geotechnical engineer must know the "strength" of the soil supporting the structure.

We have seen from the data plotted in Figures 12 and 13 that the strength (and the stiffness) depends on the void ratio and stress history. For a normally consolidated soil there is some difficulty in deciding what constitutes failure since there is no well-defined peak stress and it is necessary to adopt an arbitrary definition of the failure state, usually a specific axial strain, say, 5 percent. Then the effective stress Mohr's circles at failure can be plotted as shown in Figure 14. The normally consolidated strength envelope passes through the origin and, for the usual range of stresses of interest in soil mechanics, it can be regarded as a straight line

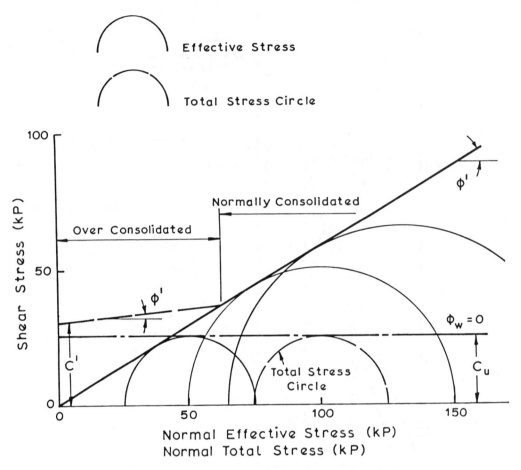

FIGURE 14. Strength envelopes in normally and overconsolidated saturated silty clay.

with $c' = 0$. Overconsolidation causes an increase in the shear strength relative to the normally consolidated state as seen in Figure 14. The cohesive intercept c' is no longer zero. It is considered that no attempt should be made to place a physical meaning on c' or ϕ', and it is best to regard them as convenient curve-fitting parameters.

Total strength envelopes are obtained by drawing an envelope to each of a series of *total stress* circles. There is a total stress circle for each effective stress circle shown in Figure 14; thus it is possible to draw the total strength envelope. In fact, this particular envelope is of no practical importance and need not be considered. However, there is one extremely important total strength envelope, termed the $\phi = 0$ envelope. Suppose one consolidates three saturated soil samples to an effective stress of 140 kPa (point O'' of Figure 13), turns off the drainage valve, and increases the cell pressure to 170 kPa for the first sample, 200 for the second, and 300 for the third. During this increase in cell pressure the increase in pore pressure is $\Delta u = B \cdot \Delta \sigma_3$ where $\Delta \sigma_3$ is the increase in cell

pressure and B is termed a pore-pressure parameter. When the soil is saturated, $B = 1$, the compressibility of the water (void) phase is $5 \times 10^{-7} \mathrm{m^2/kN}$ compared with the compressibility of the skeleton of 10^{-3} to $10^{-4} \mathrm{m^2/kN}$. Thus the increase in total stress causes an equal increase in pore pressure, that is, the effective stress in *all three samples* remains at 140 kPa. There is therefore no change in the structure of the samples, and subsequent shearing under undrained conditions gives identical shear strengths. There is only one effective stress circle at failure for the three samples, but three total stress circles of the same diameter but different mean total stresses. An envelope to these three total stress circles is a straight line at zero slope. Figure 15 shows the effective and total stress circles and $\phi_u = 0$ envelope. The importance of the $u = 0$ strength ($c = c_u$) is that this is the shear strength available under undrained conditions, and it is the same irrespective of the total stress. The undrained condition and the associated shear strength are relevant to a rapidly loaded geotechnical structure.

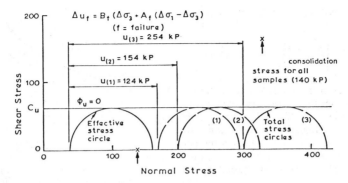

FIGURE 15. Total strength envelopes.

It is common to express the pore pressure in terms of the B parameter for an increase in σ_3, and an A parameter for an increase in deviator stress, namely:

$$\Delta u = B(\Delta\sigma_3 + A[\Delta\sigma_1 - \Delta\sigma_3]) \qquad (23)$$

and if these values are associated with failure, the subscript f is added as neither parameters are, in fact, constants.

Typical values of A_f for saturated clays are: highly sensitive clay, 1.0-1.25; normally consolidated, 0.7-1.3; lightly overconsolidated, 0.3-0.7; and heavily overconsolidated, -0.5-0.

For the total stress circle (1) of Figure 15, $\Delta u = 124$ kPa, $\Delta\sigma_3 = 30$ kPa, and $(\Delta\sigma_1 - \Delta\sigma_3) = 125$ kPa. Hence, for $B = 1$, $A_f = (124 - 30)/125 = 0.75$.

For circle (2), $\Delta u = 154$ kPa, that is, relative to circle (1) both the total stress and the pore pressure at failure had increased by 30 kPa. Hence the effective stress circle corresponding to (2) is identical to that for (1).

Stability of Geotechnical Structures

A structure is in a state of instability when a small disturbing force causes an excessive displacement. Such a state will occur when the soil cannot develop sufficient shear strength to restore a state of equilibrium. The analysis of stability of a soil mass is traditionally based on the simplified concept of a rigid-plastic stress-strain relationship so that the assessment of stability can be made without direct reference to the strain state within the soil mass. A more accurate modeling of the soil behavior can be made by use of the finite element technique; programs based on such analyses are now becoming available.

Attention will be concentrated on the fundamental earth-pressure analyses based on the rigid-plastic soil model. These analyses predict the limiting states of equilibrium of the soil mass commonly referred to as the *active* and *passive* states. A study of the pressures is fundamental to an understanding of the stability of foundations, retaining structures, and slopes. It will be seen that the technique adopted to analyze the stability of all these structures is to combine the equations of statics with a failure mechanism, and invoke sufficient further assumptions to develop a statically determinate analysis.

Coulomb (1773) can be credited with the first rational analysis of the stability of a geotechnical structure. Eighty years later Rankine published his treatise on the stability of soil masses, but it is only relatively recently that these analyses have been explained within the framework of the classical theory of plasticity.

Many perceptive accounts of slope failures have been published in the scientific literature (e.g., Collin, 1846). Early observers commonly stated that the failure surfaces were curved, but the Swedish engineer Fellenius was first to make use of this fact and consequently develop a simple statical analysis of slope stability by assuming that the failure surface was an arc of a circle. His analysis has been further developed and specifically applied to the design of earth dam embankments. The prediction of the bearing capacity of foundations is also relatively recent since Terzaghi developed the first serious practical solution in the 1930s.

It is now well established that correct solutions for the limiting stress states in many two-dimensional soil structures can be obtained by use of the theory of plasticity. When the equations of equilibrium are combined with the Mohr-Coulomb failure criterion equation, hyperbolic equations that define two families of curves are established. These curves are referred to as *stress characteristics*. For convenience, the stress characteristics are referred to as α and β lines, and the complete system of stress characteristics is a *stress field*.

The β lines are characterized by the fact that their slope is given by the first solution of the hyperbolic equations and the α lines are characterized by the second solution of the equations. It can be shown that

$$\frac{dx}{dy} = \tan\left(\theta \pm \left[\frac{\pi}{4} - \frac{\phi}{2}\right]\right) = \tan(\theta \pm \mu) \quad (24a)$$

for the β and α lines, respectively. θ is the counterclockwise angle from the vertical (y) axis to the major principal stress direction. The two families must intersect at an angle equal to 2μ and are orientated at an angle of μ to the major principal stress direction. The change in mean normal "program" stress, $d\sigma_p$, along the stress characteristics is

$$d\sigma_p \pm 2\sigma_p \cos 2\mu \, d\theta =$$

$$\frac{-\gamma \sin(\theta \pm \mu)}{+ \sin 2\mu \cdot \cos(\theta \pm \mu)} \cdot dy \quad (24b)$$

along a β line and an α line respectively. σ_p is defined as

$$\sigma_p = \sigma_N + c \cot \phi \quad (25)$$

where σ_N is mean normal stress.

Stability of Foundations. As an example of the use of the preceding approach, consider the solution for the stability of a surface-strip footing with a frictionless base. The most critical state occurs if the load is applied so quickly that there is no time for pore pressures to dissipate, that is, that soil is loaded under undrained conditions and with respect to total stresses the strength parameters are $c = c_u$, $\phi = \phi_u = 0$. For this $\phi_u = 0$ analysis integration of equation 24b gives the change in mean normal stress along a characteristic from A to B, namely:

$$\sigma_{NB} = \sigma_{NA} \pm 2c_u(\theta_B - \theta_A) \quad (26)$$

noting, Figure 16a, that $(\theta_B - \theta_A)$ is equal to the change in direction of the stress characteristic from A to B. To satisfy statical equilibrium and the failure criterion at points on the shear stress-free boundaries, the stress characteristics must intersect the boundaries at ($\pi/4$). This fact is illustrated by the Mohr-Coulomb plot shown as Figure 16b.

An approximate stress field for a $\phi_u = 0$ soil can be obtained by a combination of straight lines and arcs of circles, and Figure 17a shows one such combination. Commencing at a point a on the boundary outside the loaded area the mean stress is equal to c_u, which is the same as circle (1) of Figure 16b. Along the straight line ab the mean stress is unaltered since θ is a constant, along the arc bc of the fan XOY the change is $\pi/2$, and hence the change in mean principal stress is πc_u. The total value of the mean stress at c is $c_u(1 + \pi)$,

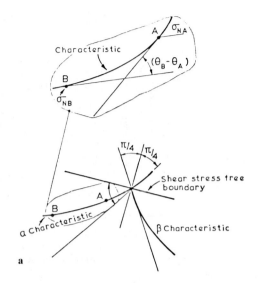

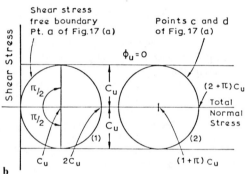

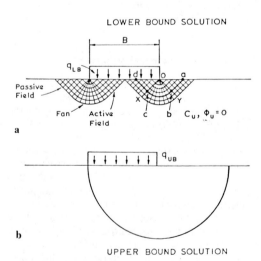

FIGURE 16. Characteristics (a) and Mohr's circles (b) for $\sigma = 0$ analysis.

FIGURE 17. A lower-bound (stress) field (a), and an upper-bound mechanism (b) of surface, smooth, strip footing.

circle (2) of Figure 15b. From c to d the value of σ_N is unchanged. At point d the major principal stress is equal to the normal contact pressure at failure q_{LB} and is seen to be equal to $c_u(2 + \pi)$.

This solution was first proposed by Prandtl in 1920 for a weightless material, but it is not a complete solution since the kinematics of the problem have not been considered. Solutions of this type—that is, solutions that satisfy the Mohr-Coulomb *failure* criterion and *equilibrium*—are termed *lower-bound* solutions and are *less than or equal to* the correct solution. A series of lower-bound solutions can be obtained, but the best solution is the one that gives the highest value of the collapse load. The most efficacious technique for establishing the correct solution with an acceptable accuracy is to determine in conjunction with the lower-bound solution a value of the collapse load based on *equilibrium* and a *mechanism of failure*. This *upper-bound solution*, q_{UB}, provides a value *equal to or greater than* the correct collapse load.

One possible mechanism is shown in Figure 17b. A rigid segment of soil rotates about a center at the edge of the footing. Equating external and internal work done as a consequence of a rotation, θ, gives

$$\tfrac{1}{2}q_{UB}\, B^2\theta = \pi B^2 c_u\theta \qquad (27)$$
$$q_{UB} = 2\pi c_u$$

Thus $2\pi c_u > q_f > (\pi + 2)c_u$, but it is possible to get a slightly better upper-bound solution by choosing a different arc of failure.

In fact, Prandtl's lower-bound solution $(5.14\, c_u)$ is also an upper-bound solution, and is therefore the correct solution. It can also be shown that this solution is independent of footing roughness and soil density.

One method of examining the kinematics of the Prandtl solution is to plot the *velocity field* from the stress field. If the ratio of principal strain rates is expressed by an equation analogous to the principal stress ratio equation 37, namely:

$$\frac{\mathring{\epsilon}_3}{\mathring{\epsilon}_1} = \tan^2\left(\frac{\pi}{4} + \frac{\Psi}{2}\right) \qquad (28)$$

Ψ is a geometric parameter analogous to ϕ and the α and β lines of the velocity field are inclined at $(\phi - \Psi)/2$ to the corresponding α and β lines of the stress field. Typical values of Ψ at failure for a dense and loose quartz sand are 0.15 and 0, respectively. Drucker (1967) established that a velocity field is kinematically admissible provided that the rate of plastic work is everywhere positive, assuming that the material behaves as an *associated flow rule material* defined by the condition that $\Psi = \phi$. The rate of plastic work $\mathring{W}$ for an *associated flow rule material*.

$$\left(\frac{\partial v_x}{\partial x} + \frac{\partial v_y}{\partial x}\right) = \frac{\mathring{W}}{c \cot \phi} \qquad (29)$$

where v_x and v_y are the velocity components in the direction of coordinate axes.

A *non-associated flow rule material* requires that $0 \le \Psi < \phi$. Of particular interest is the specific case $\Psi = 0$, since this corresponds to a state of zero *increment* in volume, and models the strain state at large strains. Such a state is termed the *critical void state*.

Granular materials are generally *assumed* to drain immediately once the load is applied, and unless partially saturated the cohesive parameter will be zero. Normally consolidated saturated clays also manifest a zero cohesion, with respect to effective stresses. Partially saturated clays and saturated silts behave as c, ϕ, γ materials with respect to total stresses, and saturated overconsolidated clays also behave as c, ϕ, γ materials but with respect to effective stresses.

Prandtl developed a lower-bound solution for the frictionless, surface, strip footing supported on a $c, \phi, \gamma = 0$ material, and an upper-bound solution for an associated flow rule material has obtained. The velocity field for a $\Psi = 0$ non-associated flow rule material was obtained by Cox (1962).

Fortunately, the plasticity solutions for the contact stress at which instability occurs under symmetrical loading of a surface-strip foundation can be expressed, to a close approximation, as

$$q_{max} = cN_c + p_oN_q + \frac{\gamma B}{2}N_\gamma \qquad (30)$$

where N_c, N_q, and N_γ are termed bearing-capacity factors, and are primarily functions of ϕ. p_0 is the total overburden pressure, γ the density, and B the width of the strip. Figure 18 shows the recent solutions obtained for the bearing-capacity factors.

Let us now apply this analysis to a practical foundation situation (see *Foundation Engineering*). Foundations of typical structures are either isolated (pad), combined, strip, raft (mat), pier, single pile, pile groups, or combinations of these basic types (see Fig. 19). The stability analysis of the particular foundation must take account of the pore pressures developed in the soil. The most expedient approach is to consider two limiting loading cases, immediate loading and long-term stability.

In immediate (but not impact) loading, the total load is applied before the excess pore pressures can dissipate. Thus the shear strength is that developed in an undrained test, and for a saturated soil $c = c_u$, $\phi = 0$, then $N_q = 1$, $N_\gamma = 0$, and thus

$$q_\gamma = (c_u \cdot N_c) + p_0 \qquad (31a)$$

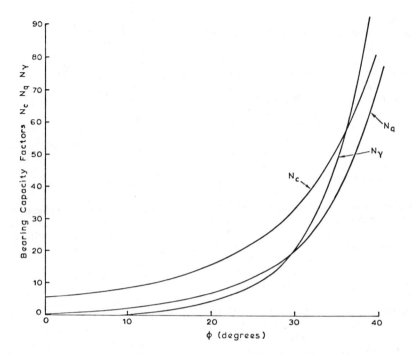

FIGURE 18. Bearing-capacity factors (after Hansen 1961, 1970).

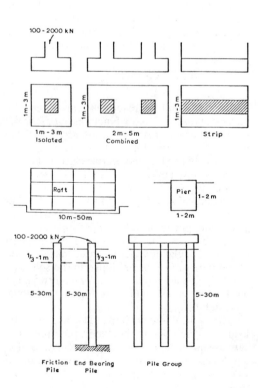

FIGURE 19. Foundation types, typical sizes, and loads.

Long-term stability corresponds to the state when dissipation of the excess pore pressures (due to the foundation load) has been completed. Then the value of c in equation 30 is c_d and the bearing-capacity factors are the values corresponding to the drained ϕ value (π_d). Both c_d and ϕ_d are obtained in a test in which the effect of a water table has to be also considered separately from the excess pore pressures. For a water table above the base level of the foundation, the appropriate modified form of equation 30 is

$$q_{max} = \left(c_d N_c + p_0'(N_q - 1) + \frac{\gamma'B}{2}N_\gamma \right) + p_0 \quad (31b)$$

where p_0' is equal to the effective vertical stress at the base of the foundation whereas p_0 is the total vertical stress at this level. γ' is $(\gamma - \gamma_w)$, the buoyant density and the values of N_c, N_q, N_γ, are taken from Figure 18.

It will be noted that the value of q_{max} (termed the *gross bearing pressure*) is the sum of the total vertical (overburden) stress of the base level of the foundation (p_0), plus the *net bearing pressure* q_{net}, the expression in the brackets on the right-hand side of equations 31a and 31b.

The *allowable bearing pressure* (q_A) is obtained by dividing the *net bearing pressure* by a factor, F, usually of the order of 3. Thus

548

$$q_A = \frac{cN_c}{F} + p_0 \text{ for immediate stability} \quad (32a)$$

and

$$q_A = \frac{cN_c + p_0'(N_q - 1) + \frac{\gamma'BN_\gamma}{2}}{F} + p_0 \quad (32b)$$

for long term stability

In many practical situations the foundation is subject to eccentric loads and inclined loads and the base is below ground level at a depth D. The maximum bearing pressure can be written in a general form to take account of these conditions:

$$q_{max} = cN_cS_cD_ci_c + p_0' (N_q - 1)S_qD_qi_q$$
$$+ \tfrac{1}{2}\gamma'B'N_\gamma S_\gamma D_\gamma i_\gamma + p_0 \quad (33)$$

in which the S, D, and i factors take account of shape, depth, and load inclination. Values of these factors are given in Table 13.

Piled foundations are formed from timber, steel, and precast and cast-in-place concrete piles; there are many patented types of pile-forming techniques. Timber piles are generally designed to carry lower loads than other types (150 to 250 kN) and are used in lengths up to 30 m. Steel H, or hollow, piles are typically designed for 500 to 1,500 kN and lengths can exceed 50 m. Common types of reinforced concrete and prestressed concrete piles are used for loads of 500 to 1,500 kN, but large-diameter piles have been used to support loads in excess of 10,000 kN.

Equations 31a and 31b give the point-bearing capacity of a single pile. The side adhesion or friction must be also added to the point-bearing pressure, and in many situations it will be found that the side friction greatly exceeds the point-bearing pressure. The stability of piles and pile groups under lateral loads requires a special analysis.

Stability of Retaining Structures. Several types of structures are designed to retain soil, notably gravity (rigid) retaining walls, crib walls, cantilever walls, counterfort or buttressed walls. sheetpile (flexible) walls, and cofferdams (see *Reinforced Earth*). To design these walls, it is necessary to determine the *in situ* pressures acting on the wall by the retained soil. Coulomb developed the first method to determine the thrust on a retaining structure, and this method—or modifications of it—is still the common design technique. It is argued that the pressure on a retaining structure is equal to the active pressure. As mentioned earlier in this section, the active pressure is the smallest lateral pressure that can be developed in a soil mass, and this active state will be developed when the retaining wall moves away from the soil. Thus if the pressure acting on the wall is in excess of the active value, movement of the wall under the action of this pressure will reduce the pressure to the active value. Very small movements on the order of 0.001 of the wall height are sufficient to develop the active state in granular soils.

A passive pressure state is developed when the structure moves into the soil mass. Consider, for example, the anchored sheetpile wall shown in Figure 20a. Under the influence of the thrust from the retained soil, the wall will move outward and an active state will develop in region A. Support beneath the dredge level is due to the movement of the wall into the soil and, as a consequence, large lateral pressures in region B can develop. The maximum possible pressure is the passive value and will occur when the soil is fully plastic; that is, it cannot develop any further shear strength. As a general rule, the active pressure can be regarded as a disturbing pressure, whereas the passive pressure is the stabilizing pressure. A considerable movement of the wall into the soil is required before the passive state is finally reached.

Coulomb's active pressure analysis is illustrated in Figure 20b, c, and d. The basic assumption is that failure results from sliding along a plane surface. The analysis shown in Figure 20b is the determination of the active thrust for a cohesionless soil. An elementary vector diagram can be drawn for each of the assumed (possible) failure planes,

TABLE 13. Values of Factors to Account for Shape, Depth, Inclination, and Eccentricity

Factor	N_c Term	N_q Term	N_γ Term
Shape, S	$S_c = 1 + 0.2\dfrac{B'}{L'}$	$S_q = 1 + 0.2\dfrac{B'}{L'}$	$S_\gamma = 1 - 0.4\dfrac{B'}{L'}$
Depth, D	$D_c = 1 + 0.35\dfrac{D}{B'}$	$D_2 = 1 + 0.35\dfrac{D}{B'}$	$D_\gamma = 1$
Inclination, i	$L_c = 1 - \dfrac{P_H}{2cB'L'}$	$L_q = 1 - 0.5\dfrac{P_H}{P_v}$	$i_\gamma = (1 - 0.7\dfrac{P_H}{P_v})^5$
Eccentricity	—	$B' = B - 2\ell_B$	$L' = L - 2\ell_L$

Note: P_H, p_v = horizontal and vertical force components, respectively.
ℓ_B, ℓ_L = eccentricity in direction of width and length, respectively.

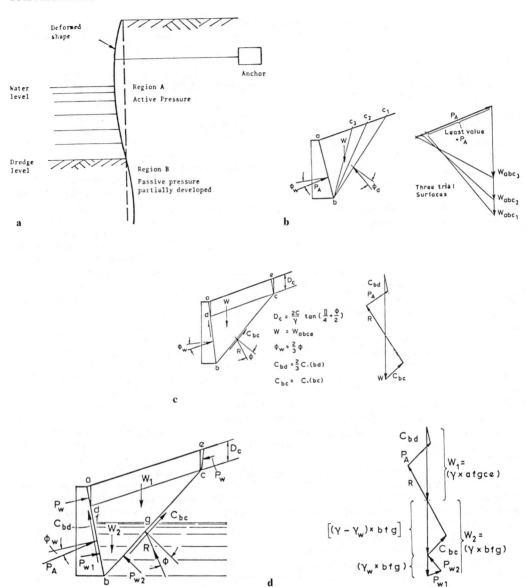

FIGURE 20. (a) Coulomb's active thrust analysis for a cohesionless soil and a $C - \sigma$ soil. (b) Basic Coulomb analysis for $C = 0$, σ, γ soil with no water table. (c) Coulomb analysis for C, σ, γ soil with tension cracks and no water table (only one vector diagram shown). (d) Coulomb analysis for C, σ, γ soil with tension cracks filled with water table of level fg (only one vector diagram shown).

provided the value of the friction angle is determined from a drained test and a value of wall friction ϕ_w is assumed. The smallest value of P is the active thrust. An extension of the analysis to a c, ϕ, γ soil is shown in Figure 20c. Allowance must be made for cracking of the soil to a depth of D_c—these cracks are termed *tension cracks*. The values of c and ϕ used in the analysis will depend on the loading conditions. If one imagines that the wall is instantaneously constructed, then the shear

parameters are the values obtained in an undrained test, whereas long-term stability requires the use of drained shear parameters. The values of wall adhesion and friction must be assumed.

Allowance should be made for water filling the tension cracks. Such thrusts do not affect the value of P_a but must be considered when examining the wall stability. If there is a water table at some level in the retained soil, the value of the active thrust can be determined by considering the buoyant density below the water table since the force components P_{w1} and P_{w2} are statically equivalent to the weight of water given by the area bfg $\times \gamma_w$. This is shown in Figure 20d.

Coulomb's analysis is an upper-bound solution, since it assumes a mechanism and fulfills force equilibrium. In fact, it gives values quite close to correct plasticity solution values provided the assumed value of ϕ_w is about $2/3\phi$ and the wall adhesion is also about $2/3c$. There is, therefore, justification for its continued use.

Rankine's solution is a lower-bound plasticity solution applicable to the specific wall-soil interfacial condition of a vertical wall with a wall friction angle equal to the slope of the backfill surface. For example, it is the lower-bound solution for a vertical wall retaining soil with a horizontal backfill surface when the friction angle developed along the interface is zero. It is not a correct lower-bound solution for any other condition of wall friction, and therefore the Coulomb analysis is a much more generally applicable analysis.

When a passive state is fully developed, the failure surfaces are almost invariably curved, so that neither the Rankine nor the Coulomb analysis is applicable. Several stability analyses have been used based on the same principles as Coulomb's analysis but using a composite failure surface. Figure 21a shows the special case when the Rankine solution is a correct lower-bound solution, for example, when a vertical wall-soil interface, horizontal backfill surface, and zero shear stress developed along the interface. Such a situation can occur only when there is no relative movement between the wall and the soil. A *positive* passive state occurs when the soil is moving upward relative to the soil, and the failure surfaces (stress field) are seen (Fig. 21b) to be composed of a curve (fan) and straight lines. In this state the angle, θ_w, between the interface (measured in a counterclockwise direction) and the direction of the major principal stress is less than $\pi/2$. Reversal of the direction of movement leads to a *negative* passive state and the stress field becomes discontinuous, and $\theta_w > \pi/2$ (Fig. 21c).

The passive (and active) thrusts can be expressed in a form very similar to the expression (30) for bearing capacity. For example, the horizontal component of passive thrust P_{PH} is

$$P_{PH} = H(cN_{cPH} + (\gamma H/2)N_{\gamma PH}) \qquad (34)$$

the subscript P referring to passive, and H to horizontal. For the vertical component H is replaced by V, and for the active state P is replaced by A. Values of the thrust factors are given in Table 14.

One important recent development enables the stress state along the wall-soil interface to be

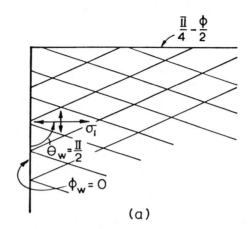

(a)

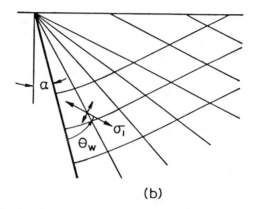

(b)

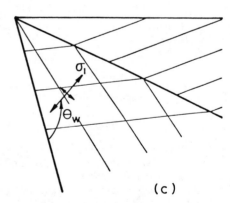

(c)

FIGURE 21. Passive pressure stress fields.

predicted rather than assumed as in the traditional analyses. When the movement of the wall or soil causes the maximum interfacial shear resistance to be developed, the cohesive and frictional components were shown to be

$$c_w = \frac{c \cos \phi \sin 2(\theta_w - \alpha)}{1 - \sin \phi \cos 2(\theta_w - \alpha)} \qquad (35a)$$

TABLE 14. Values of Passive Thrust Factors

Wall Slope α (rad)	Friction Angle (rad)	Type of Field $\psi = \sigma$	$\psi = 0$	N_{cPV} $\psi = \sigma$	$\psi = 0$	N_{cPH} $\psi = \sigma$	$\psi = 0$	$N_{\gamma PV}$ $\psi = \sigma$	$\psi = 0$	$N_{\gamma PH}$ $\psi = \sigma$	$\psi = 0$
							$i=0$ rad				
0 (0°)	0.39	C	C	3.15	2.90	5.10	4.90	1.40	1.35	3.45	2.35
	0.52	C	C	5.00	4.20	6.95	6.40	3.30	2.60	5.65	5.00
	0.70	C	C	11.1	7.10	12.2	9.65	12.0	6.60	14.2	10.2
	0.79	C	C	18.0	11.9	17.0	15.3	26.0	14.9	26.0	21.2
0.18 (10°)	0.39	C	C	2.05	1.80	4.65	4.45	0.65	0.60	3.00	2.95
	0.52	C	C	3.20	2.75	6.05	5.70	1.80	1.50	4.95	4.75
	0.70	C	C	6.85	4.50	1.01	9.30	6.55	4.50	11.0	9.10
	0.79	C	C	10.9	6.30	14.0	12.1	13.2	7.50	18.8	15.2
0.35 (20°)	0.39	C	C	1.15	1.00	4.10	4.05	0.15	0.05	2.75	2.65
	0.52	C	C	1.85	1.45	5.30	5.00	0.75	0.45	4.25	4.00
	0.70	C	C	4.05	2.50	8.60	7.40	3.20	2.40	8.80	7.40
	0.79	C	C	6.00	3.20	11.3	9.30	6.55	3.40	14.2	11.3
							$i = 0.35$ rad				
0 (0°)	0.39	C	C	2.35	1.90	3.20	3.10	0.60	0.50	1.35	1.30
	0.52	C	C	3.40	2.95	4.10	3.85	1.30	1.00	2.25	2.15
	0.70	C	C	6.15	4.50	6.10	5.75	3.95	2.65	4.70	4.15
	0.79	C	C	9.00	5.65	8.00	7.25	7.50	4.70	7.50	6.40
0.18 (10°)	0.39	C	C	1.60	1.30	2.85	2.75	0.25	0.20	1.15	1.10
	0.52	C	C	2.30	1.75	3.50	3.40	0.70	0.55	1.90	1.70
	0.70	D	C	3.95	2.60	5.10	4.55	2.10	1.35	3.60	3.25
	0.79	D	C	5.50	3.55	6.40	5.75	3.80	2.40	5.40	4.50
0.35 (20°)	0.39	D	C	1.05	0.90	2.45	2.40	0.05	0.00	1.10	1.05
	0.52	D	C	1.45	1.25	3.00	2.95	0.30	0.25	1.55	1.65
	0.70	D	C	2.35	1.50	4.15	3.60	1.00	0.60	2.80	2.45
	0.79	D	C	3.20	1.85	5.10	4.10	1.85	1.15	4.00	3.50

Note: C = continuous field and D = discontinuous field.
Source: After Lee (1974).

$$\tan \phi_w = \frac{\sin \phi \sin 2(\theta_w - \alpha)}{1 - \sin \phi \cos 2(\theta_w - \alpha)} \quad (35b)$$

where α is the slope of the interface relative to the vertical, Figure 21.

Table 15 gives calculated values of c_w and ϕ_w based on the data for a dense and a loose sand. The important conclusion is that $\phi_w < \phi'$ and is similar to the common assumption of $\phi_w = 2/3\phi$ (for example, in the British Code of Practice on Earth Retaining Structures).

From the statics at the interface, the relationships between the vertical thrust factors N_{CPV}, $N_{\gamma Pv}$, and the corresponding horizontal thrust factors N_{CPH}, $N_{\gamma PH}$, for the positive passive state can be established as

$$N_{CPV} = -N_{CPH} \cdot \tan \alpha \cdot \frac{c_w}{c}(1 + \tan^2 \alpha) \quad (36a)$$

$$N_{\gamma Pv} = N_{\gamma PH} \frac{\tan \phi_w \cos \alpha - \sin \alpha}{\tan \phi_w \sin \alpha + \cos \alpha} \quad (36b)$$

In these expressions the vertical component is the force acting on the wall in an *upward* direction. The expressions also apply to the negative active state if the passive subscript is replaced by the active subscript. The negative passive state and the positive active state are obtained by changing the sign of c_w and $\tan \phi_w$ in the relevant expressions.

The question of applicability of the rigid-plastic soil model can be partially resolved by reference to the experimental results obtained in tests recording the development of passive pressure as a function of movement. In these tests a rigid vertical wall was pushed into a dry sand mass formed with a horizontal surface. The wall was moved in a direction Z to the horizontal. Figure 22 shows the value of $N_{\gamma PH}$ as a function of Z, and the predicted frictional parameter $\tan \sigma_w$ is shown in Figure 23 for comparison with the experimental values.

It is seen that the recorded maximum thrust is less than the value predicted using $\phi = \phi_d$ at peak, but as the wall movement is increased the experimental thrusts closely agree with the thrust

TABLE 15. Values of Wall Adhesion c_w and Friction σ_w
(Dense $|\sigma_d = 0.73|$ and Loose Sand $|\sigma_d = 0.56|$

Condition		θ_w	c_w	$\tan \theta_w$	$\tan 2/3\sigma$
Positive and negative, passive	Dense	0.700	0.83	0.75	0.53
Fully rough	Loose	0.785	0.88	0.48	0.39

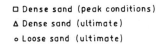

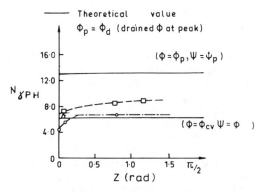

FIGURE 22. Passive pressure coefficient $K_p (= N_{\gamma PH})$ with concordance of experimental and theoretical values.

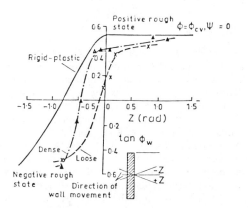

FIGURE 23. Development of wall friction as a function of direction of movement, passive state.

predicted by the plasticity theory for $\phi = \phi_{CV}$. This observed behavior would be anticipated when the effect of soil compressibility is taken into account since the peak ϕ_d would not be *simultaneously* developed in the failure region. When the wall movement is sufficient to induce large strains in the soil mass, the value in the deforming region approaches ϕ_{CV}.

It has been established that the values of thrust factors quoted in Table 1 are the correct plasticity

solutions for an associated flow rule material, provided the movement normal to the wall at the top is greater than or equal to the normal movement at the base. The solutions quoted, although satisfactory lower-bound solutions, are not the correct solutions for rotation about the top.

The correct solution for an associated flow rule material is an *upper bound* for a nonassociated flow rule material with the same ϕ value.

Stability of Slopes. Soil slopes can fail as a result of several factors, but the most common instability is due to a deep-seated, curved failure surface. Except in special situations where the shape of the failure surface is dictated by natural features such as a rock surface, the Fellenius assumption of a circular arc is accepted. As in the stability analyses of bearing pressure and lateral earth pressure, the combination of a failure mechanism and statics is sufficient to establish a factor of safety. Analysis of a homogeneous slope and a slope composed of a variety of soils is considered in the following.

When a slope is formed by excavation, there is a reduction in the total stresses as the excavation proceeds; hence the pore pressures decrease during this period. After construction has been completed, the pore pressures will usually increase as the hydraulic boundary conditions reestablish the flow pattern; hence the effective stresses decrease and the strength decreases, thus diminishing the stability. The reverse situation applies when the embankment is built up; that is, the effective stresses increase with time and the slope becomes more stable.

If the construction proceeds at a rate such that the excess pore pressures are not significantly changed, then the immediate stability can be determined by the $\phi = 0$ analysis. It is exactly the same in principle as the immediate stability analysis of a foundation. Figure 24 shows the simplest case. The shear strength is equal to the undrained cohesion so that moments about the center of the assumed failure surface give moment restraining equal to $c\,R$ and moment disturbing equal to Wx, and if we imagine that the soil strength is reduced by a factor just sufficient to cause such a factor to be considered a factor of safety, we can equate the previous expressions to give

$$F = \frac{c\,\ell R}{Wx} \qquad (37)$$

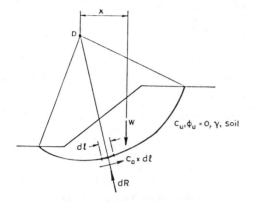

FIGURE 24. $\sigma = 0$ slope stability analysis.

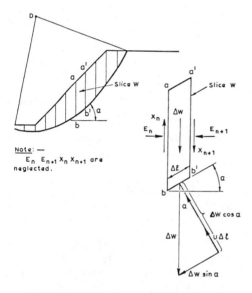

Note: —
$E_n \; E_{n+1} \; x_n \; x_{n+1}$ are
neglected.

FIGURE 25. Circular arc stability analyses.

Values of F are calculated for a series of trial surfaces, and the smallest value is deemed to be the factor of safety for the slope.

When the slope is composed of significantly different soils or if we are interested in the long-term stability (when the excess pore pressures have been dissipated), the analysis must be modified (see *Slope Stability Analysis*). Again, however, we take moments about the center of an assumed failure surface and, following a series of simplifications, we arrive at the expression for a factor of safety (Fig. 25):

$$F = \frac{\Sigma(c \cdot \Delta \ell + | \Delta W \cos \alpha - u \cdot \Delta \ell \, | \cdot \tan \phi}{\Sigma \, \Delta W \cdot \sin \alpha} \qquad (38)$$

There has been a considerable amount of examination of the validity of this approach, and a modi-

fied expression was derived by Bishop, although the principles of his analysis (circular arc failure mechanism + statics) were the same as those adopted by Fellenius.

The Bishop (1955) expression is, in terms of effective stresses,

$$F =$$

$$\frac{\Sigma(c' \cdot \Delta \ell \cdot \cos \alpha + | \Delta W - u \cdot \Delta \ell \, | \tan \phi') \cdot (1/m_a)}{\Sigma \, \Delta W \sin \alpha} \qquad (39)$$

where $m_a = \cos \alpha + \dfrac{\sin \alpha \tan \phi'}{F}$

It will be noted that the factor of safety appears on both sides of equation 39 and it is necessary to use a trial-and-error procedure to determine F for a given slip circle as well as considering a series of slip circles to establish the minimum value of F. The Fellenius expression gives a conservative value of F relative to the Bishop value, but the difference (typically 5 to 15 percent) may not be sufficient to justify the considerable increase in numerical calculations necessary to solve the Bishop equation unless, of course, a computer program is available.

When the slope is partially submerged, Figure 25b, the expression must be modified. Details are given in Terzaghi and Peck (1967).

Soil Dynamics

In many situations, soil deposits are subjected to man-made and naturally occurring dynamic loadings. Such loadings may be deliberately imposed during construction—for example, pile driving—or compaction by vibratory rolling or explosives. Shear and compression curves are generated by a source at the surface of a soil deposit to measure the elastic parameters of the layers, the thickness of layers, and the depth to bedrock. Such seismic methods find wide application in site investigations and in the analysis of road pavements.

Little need be said about the loss of life and property caused by earthquakes. Geotechnical engineers and their colleagues are expected to design structures that can withstand the most severe earthquake likely to occur during the lifetime of the structures (see *Earthquake Engineering*). This requires the determination of the spectrum of accelerations, velocities, and displacements in the soil layers, and the effects of tremors on the strength and deformation properties of the soils.

A disturbance due to an earthquake or man-made disturbance—for example, blasting or machine vibration—causes waves to be generated in the soil layers. The character of these waves can be studied analytically, provided we assume the soil behavior can be treated as linear elastic. From the equations of motion of an element of the soil, it is found that there are three types of waves: P, S, and R.

P-waves (also called primary, compressional, longitudinal, or rotational waves) have a velocity, V_D, given by

$$V_D = \sqrt{\frac{\lambda + 2G}{\rho}}$$

$$\lambda = \frac{\nu E}{(1 + \nu)(1 - 2\nu)}$$

where the relevant soil properties are defined as

E = Young's modulus
ν = Poisson's ratio
G = shear modulus
ρ = mass density ($= \gamma/g$)
γ = Density
and g = acceleration due to gravity

S-waves (secondary, or shear distortion, waves) have a velocity given by

$$V_s = \sqrt{\frac{G}{\rho}}$$

and the waves cause a distortion without volume change. Finally R-waves (Rayleigh, or surface, waves) are generated where there is a free surface and are propagated at a velocity slightly less than S-waves.

P-waves are particularly important in the immediate vicinity of the epicenter of an earthquake (particularly when the soil is a sand liable to liquefaction), whereas the predominant effects at some distance from the epicenter will be due to S-waves. R-waves are particularly significant when the disturbance is due to vibrating foundations; for example, 67 percent of the energy due to a vibrating rigid circular footing is in the form of R-waves, 26 percent in S-waves, and only 7 percent in P-waves (see Richart et al., 1970).

Earthquakes. One of the primary problems in the design of structures against earthquakes is the determination of the likely and extreme spectra of acceleration, velocity, and displacement. Table 16 lists the characteristics of the most severe earthquake in California; Newmark (1965) considers that the maximum accelerations and velocity listed for the "maximum probable earthquake" would not be exceeded anywhere in the world.

The maximum force imposed on the structure obviously depends on the maximum ground acceleration, but it must be recognized that the duration of the tremor and the frequency of the motion is also extremely significant. A series of low amplitude accelerations over a long period may lead to a progressive build-up of deflections and thus could be more disastrous than a high-intensity, short-duration motion. To study the behavior of a structure during a tremor, it is useful to plot an "acceleration response" spectrum, which is simply the relationship between the maximum acceleration in the structure as a function of the natural frequency of the structure for a series of values of the damping coefficient. This response spectrum and the associated representations for velocity and displacement can be determined for a given accelerogram of ground motion. Figure 26 shows an accelerogram for the 1940 California earthquake and the acceleration response spectrum. It is evident that the flexibility of the structure and other structural details will have a profound influence on the natural frequency and damping characteristics and hence the forces in the structure.

The use of vibrations to compact cohesionless soils is well established. It is evident that vibrations caused by earthquakes will also cause compaction of loose cohesionless soils. If the sands are saturated, a state of *liquefaction* ("quicksand") can be created; the vibrations cause a build-up of pore pressures until the total stress due to overburden is exceeded and the effective stresses, hence shear strength, are zero. One of the most outstanding examples of liquefaction following an earthquake occurred in Niigata, Japan, on June 16, 1964. The epicenter of the earthquake was approximately 50 km from the city, but the vibrations were sufficient to cause liquefaction of the fine sands in the low-lying areas, and instability (bearing-capacity) failures of many residential buildings in these areas. There are several regions where the liquefaction

TABLE 16. Probable Intensities of Maximum Motion for Major Earthquakes

Condition	Maximum Acceleration (g)	Maximum Velocity (cm/s)	Maximum Displacement (cm/s)	Duration of Major Motion (s)
Maximum recorded earthquake in California	0.32	35	30	30
Maximum probable earthquake in California	0.50	60-75	60	90-120
Extreme values considered	0.50-0.60	75-90	90-120	120-240

Source: After Newmark (1965).

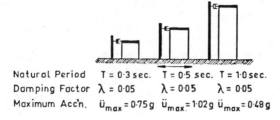

Natural Period $T = 0.3$ sec. $T = 0.5$ sec. $T = 1.0$ sec.
Damping Factor $\lambda = 0.05$ $\lambda = 0.05$ $\lambda = 0.05$
Maximum Acc'n. $\ddot{u}_{max} = 0.75$ g $\ddot{u}_{max} = 1.02$ g $\ddot{u}_{max} = 0.48$ g

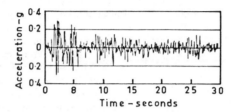

ACCELEROGRAM - EL CENTRO CALIFORNIA EARTHQUAKE
MAY 18, 1940 (N-S COMPONENT)

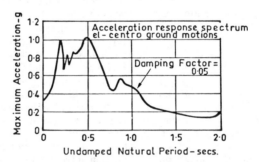

ELEVATION OF ACCELERATION
RESPONSE SPECTRUM

FIGURE 26. Accelerogram and acceleration response spectrum from El Centro, California, earthquake (after Seed, in Winterkorn and Fang, 1975).

potential is high and such deposits are heavily compacted prior to building construction (see Vol. XII, Pt.1: *Thixotropy, Thixotropisms*).

If a sand mass is subjected to vibrating or cyclic shear stresses, it tends to compact. If the water in the voids cannot drain, however, the net result is a small pore-pressure increment as a result of each cycle of loading. The pore pressures can build up to a stage at which the shear strength is too small to withstand the applied stresses. Liquefaction has also occurred in some cases when high-pore-pressure passes water from deeper to shallower deposits. In the Niigata earthquake, severe "boiling" of the sands was observed as the pore pressures were dissipated at the surface.

Laboratory techniques for measuring the pore pressures developed under similated earthquake loadings are well established. A number of cycles of a repeated load or a spectrum of loads are applied to samples compacted to the natural density state, and the pore pressures are recorded by means of transducers. It is possible, therefore, to determine whether a particular sequence of vibrations will cause liquefaction.

It is of some interest to note that the Ekofisk offshore storage tank located on the North Sea midway between Norway and England is supported on a sand of a grading similar to the Niigata sand, and it was necessary to study the liquefaction potential of the seabed, as storms cause long-term cyclic loading.

Vibrations of Foundations. When machines cause ground vibrations, it is necessary to design the foundation system to limit the amplitude of the vibrations to ensure satisfactory operation of the machines and to avoid property damage and human distress. As shown in Figure 27, the tolerable amplitude of vibration varies with frequency. In

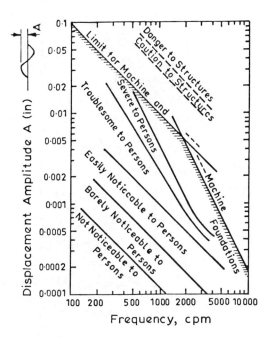

FIGURE 27. Limiting amplitudes of vibration (after Richart et al., 1970).

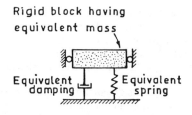

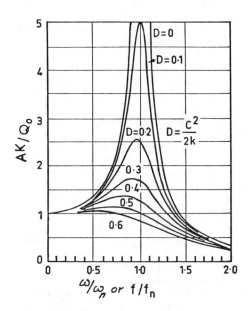

FIGURE 28. Response of mass-spring-dashpot system.

this figure a series of zones are shown to give the designer a guide to the design critera that must be satisfied. The problem is then to predict the amplitude of vibration of a machine when the mass, m, of the machine and foundation is known ($m = W/g$, where W = weight, and g = acceleration due to gravity) and the out-of-balance force can be expressed as $Q(t)$, where t = time.

The earliest attempts to find an analytical solution to the amplitude of vibration were based on the concept of a mass of soil vibrating with the machine. This concept is basically incorrect. Effective solutions have been obtained by considering the machine to be supported by a spring and dashpot system. For vertical vibrations the equation of motion is

$$m\overset{\infty}{y} + c\overset{\infty}{y} + ky = Q \cdot t \qquad (40)$$

where c is the damping characteristic and k is the (soil) spring stiffness.

The solution to this equation is shown in Figure 28 for $Q = Q_0 \sin \omega t$. If the values of c and k were known, the amplitude of vibrations could be read off this figure. It is evident that the amplitude is a function of the natural frequency ω_0 (defined as k/m) compared with the operating frequency ω, and that operation of the machine near a ω/ω_0 ratio of unity is to be avoided. This analysis is termed the *lumped-parameter method*.

A much more satisfactory representation of the soil mass is to model it by means of a linear elastic continuum. The first correct solution to the amplitude of vibration of a circular foundation was obtained by Shichter following an earlier analysis by Reissner in 1936. She showed that the form of the two solutions was very similar, and it is possible to determine from the elastic layer analysis the values of c and k for the lumped-parameter method to within a satisfactory degree of accuracy. The elastic solution for a circular foundation of radius R is expressed in terms of two parameters.

(1) the frequency factor

$$a_0 = \omega R \sqrt{\frac{\rho}{G}} = 2\pi f R \sqrt{\frac{\rho}{G}} \qquad (41a)$$

(2) the mass ratio

$$b = m/\rho R^3 = W/\gamma R^3 \qquad (41b)$$

where ω is the circular frequency of the out-of-balance force

557

Q_0 (= $2\pi f$, where f = frequency), and
ρ = mass density of the soil (= γ/g)
γ = density of the soil
G = shear modulus of the soil
m = mass of the vibrating machine plus foundation.

Figure 29 shows the analytical solution for a rigid circular footing and for the values 0, 0.25, and 0.50 for Poisson's ratio of the soil. These solutions apply to a sinusoidal disturbing force, $Q = Q_0 \sin \omega t$ and are presented as a plot of the amplitude A (the ordinate is actually A (GR/Q_0)) versus the frequency factor for a range of values of the mass ratio. Thus the amplitude can be directly determined from these plots provided the values of a_0 and b can be calculated and the elastic parameters of the soil can be satisfactorily measured.

Other parameters that may have to be taken into account in a practical situation are alternative modes of vibration, the shape of the foundation base, the depth effect, and the soil layer.

Alternative Modes of Vibration. There are six degrees of freedom in the most general case, but as far as the vibration analysis is concerned, there are only four independent modes of vibration: vertical, horizontal, torsional (rotation about the vertical axis), and rocking (rotation about a horizontal axis). There are, in fact, many cases in practice when two or more of these modes are combined (coupled)—for example, combined rocking and translatory motion.

The simplest presentation of the elastic solutions for the four modes of vibration is to use the solutions to give the mass (or inertia) ratio, the damping ratio, D, and the spring constant, which are then substituted into the plots shown as Figure 29. The values are given in Table 17. However, the elastic solutions can be applied directly without resorting to the use of the lumped-parameter method.

In this table, the values of I_ψ and I_θ are

$$I_\psi = \frac{W}{g}\left(\frac{R}{4} + \frac{H}{3}\right)$$

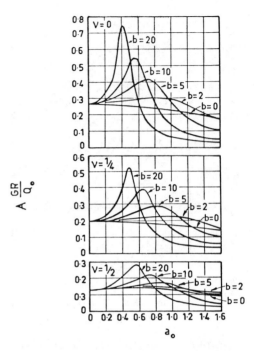

FIGURE 29. Response of linear elastic system (after Barkan, 1962).

$$I_\theta = \frac{WR}{2g}$$

assuming that all the mass can be considered to be equivalent to a circular disc of radius R and thickness H.

Shape of Foundation Base. When the foundation is square or rectangular, the simplest approximate method is to determine the radius, R, of an equivalent circular foundation by the following relationships (Richart, in Chap. 24 of Winterkorn and Fang, 1975). Vertical and translatory vibrations

$$R = \sqrt{\frac{4cd}{\pi}}$$

TABLE 17. Mass Ratio, Damping Ratio, and Spring Constant for a Rigid Circular Footing on a Semi-Infinite Homogeneous, Isotropic, Elastic Layer

Mode	Mass or Vertical Ratio	Damping Ratio, D	Spring Constant, k
Vertical	$B_z = \dfrac{(1-\nu)}{4}\dfrac{m}{\rho R^3}$	$D_z = \dfrac{0.425}{\sqrt{B_z}}$	$k_z = \dfrac{4GR}{(1-\nu)}$
Translatory	$B_x = \dfrac{(7-8\nu)}{32(1-\nu)}\dfrac{m}{\rho R^3}$	$D_x = \dfrac{0.288}{\sqrt{B_x}}$	$k_x = \dfrac{32(1-\nu)GR}{(7-8\nu)}$
Rocking	$B_\psi = \dfrac{3(1-\nu)}{8}\dfrac{I_\psi}{\rho R^5}$	$D_\psi = \dfrac{0.15}{(1+B_\psi)\sqrt{B_\psi}}$	$k_\psi = \dfrac{8GR^3}{3(1-\nu)}$
Torsional	$B_\theta = \dfrac{I_\theta}{\rho R^5}$	$D_\theta = \dfrac{0.50}{1+2B_\theta}$	$k_\theta = \dfrac{16\,GR^3}{3}$

Source: After Richart et al. (1970).

$$\text{Rocking } R = \sqrt{\frac{16cd^3}{.3\pi}}$$

$$\text{Torsion } R = \sqrt[4]{\frac{16cd(c^2 + d^2)}{6\pi}}$$

where the plan dimensions of the foundation are $2c \cdot 2d$. For rocking, d is in the plane of the rocking, and c is measured along the axis of rocking.

Depth Effect. Most machine foundations are placed at a depth D; this has the effect of increasing the resonant frequency and decreasing the amplitude of vibrations.

Soil Layer. Where the soil layer is of limited thickness and overlying rock, the rock impedes the dissipation of wave energy and the amplitude of vibrations will be greater than for a thick soil layer.

I. K. LEE

References

Barkan, D. D., 1962. *Dynamics of Bases and Foundations.* New York: McGraw-Hill.

Bishop, A. W., 1955. The use of the slip circle in the stability analysis of slopes, *Géotechnique* **5,** 7.

Collin, A., 1846. *Recherches expérimentales sur les glissements spontanés des terrains argileux accompagnées de considerations sur quelques principes de la mécanique terrestre.* Paris: Carilian-Goevry. (Collin's work is now translated by W. R. Schriever, 1956, in *Landslides in Clays.* Toronto, Ontario: University of Toronto Press, 162p.)

Coulomb, C. A., 1773. Essai sur une application des règeles de maimis et minimis à quelques problèmes de statique relatifs à l'architecture, in *Mémoires de la mathématique et de physique, présentés à l'Académie Royale des Sciences, par divers Savants, et lûs dans sés Assemblées,* vol. 7. Paris: De l'Imprimerie Royale, 343-384.

Cox, A. D., 1962. The use of non-associated flow rules in soil plasticity, *Royal Armament Research and Development Establishment Report (B), 2/63.*

Drucker, D. C., 1967. *Introduction to Mechanics of Deformable Solids.* New York: McGraw-Hill, 445p.

Grim, R. E., 1953. *Clay Mineralogy.* New York: McGraw-Hill, 384p.

Hansen, J. B., 1961. A general formula for bearing capacity, *Danish Geotech. Inst. Bull. (Copenhagen)* **11.**

Hansen, J. B., 1970. A revised and extended formula for bearing capacity, *Danish Geotech. Inst. Bull. (Cophenhagen)* **28.**

Hendricks, S. B., and Jefferson, M. E., 1939. Polymorphism in the micas, *Am. Mineral.* **24,** 729-731.

Jackson, W. W., and West, J., 1933. The crystal structure of muscovite, *Zeitschr. Krist.* **85,** 160-163.

Lambe, T. W., and Whitman, R. V., 1969. *Soil Mechanics.* New York: Wiley, 553p.

Lee, I. K. (ed.), 1974. *Soil Mechanics. New Horizons.* London: Butterworths.

Mackenzie, R. C., 1975. The classification of soil silicates and oxides, in Gieseking, J. E. (ed.), *Soil Components: Volume 2, Inorganic Components.* New York: Springer-Verlag, 1-26.

Newmark, N. M., 1965. Fifth Rankine Lecture: effects of earthquakes on dams and embankments, *Géotechnique* **15,** 2.

Reissner, E., 1936. Stationary axially symmetric vibrations of a homogeneous elastic half space, created by a vibrating mass, *Ing.-Arch.* 7, 381, 385.

Richart, Jr.; Hall, J. R.; and Woods, R. D., 1970. *Vibrations of Soils and Foundations.* Englewood Cliffs, N.J.: Prentice-Hall.

Rowe, P. W., 1962. The stress-dilatancy relation for static equilibrium of an assembly of particles in contact, *Roy. Soc. Proc.* **A269,** 500.

Soil Survey Staff, 1951. Soil survey manual, *U.S. Department of Agriculture Handbook No. 18,* 503p.

Terzaghi, K. V., 1925. *Erdbaumechanik auf bodenphysikalischer Grundlage.* Vienna: Deuticke.

Terzaghi, K., 1943. *Theoretical Soil Mechanics.* New York: Wiley, 510p.

Terzaghi, K., and Peck, R. B., 1967. *Soil Mechanics in Engineering Practice.* New York: Wiley, 729p.

Winterkorn, H. F., and Fang, H. Y. (eds.), 1975. *Foundation Engineering Handbook.* New York: Van Nostrand Reinhold.

Yong, R. N., and Warkentin, B. P., 1975. *Soil Properties and Behaviour.* Elsevier, 449p.

Cross-references: *Atterberg Limits and Indices; Clay, Engineering Geology; Consolidation, Soil; Earthquake Engineering; Foundation Engineering; Geotechnical Engineering; Grout, Grouting; Hydrodynamics, Porous Media; Lime Stabilization; Reinforced Earth; Rheology, Soil and Rock; Rock Structure Monitoring; Soil Classification System, Unified; Soil Mineralogy, Engineering Applications.* Vol. XII, Pt.1: *Electroosmosis; Flow Theory; Permeability; Soil Mechanics; Soil Stability; Soil Structure; Thixotropy, Thixotropism; Water Movement.* Vol. XIV: *Cat Clays; Dispersive Clays; Engineering Soil Science; Expansive Soils; Landslide Control; Slope Stability Analysis.*

SOIL MECHANICS, HISTORY OF

Soil mechanics, also known by the term *géotechnique* or *geotechnics,* is one of the younger basic civil engineering disciplines (see *Geotechnical Engineering).* The twentieth century—and particularly the years since 1925—has witnessed an enormous impetus in the development of soil mechanics throughout the world. Our present-day knowledge of soil mechanics (q.v.) is, however, an accumulated heritage of the past.

Soil mechanics may be defined as the discipline of engineering science that studies soils, from theoretical and practical points of view, by means that influence the way engineers build structures. Thus the soil mechanics discipline treats soil and its properties as a construction material associated with engineering (see Vol. XII, Pt. 1: *Soil).* Soil mechanics studies, both theoretically and experimentally, the effect of forces on the equilibrium and/or performance of soil-foundation systems (see *Foundation Engineering)* under static and dynamic loading conditions, as well as under the influence of water and nonfreezing and freezing temperatures.

The practice of engineering that applies the principles of soil mechanics to the design of engineering structures is called *soil engineering (*see Vol. XIV: *Engineering Soil Science).* Soil mechanics has developed from practical necessity and hence is of national importance.

In the more prehistoric civilizations, soil was used as a construction material for structures themselves: huge earth mounds for refuge during flood periods, caves to live in (e.g., loess dwellings in China), canals and ditches, and fortifications. Prehistoric man's progress in managing soils was very slow indeed.

As time progressed, soil was used in the construction of roads, canals, bridges, foundations, and fortifications, and was also used for impounding water and in building dikes and levees. Earth retaining walls supported the terraces of the famous "hanging gardens" of King Nebuchadnezzar of Babylon (from 604 to 562 B.C.).

At the peak of the Roman Empire's glory, engineers built heavy structures such as breakwaters, aqueducts, bridges, large edifices, and a great network of good roads (Vitruvius, 1914). This required considerable understanding of the performance of soil under the action of load, water, and temperature. A solid foundation and good drainage of roads were the basis of Roman engineering—principles that are still honored in modern road construction. It is known that the Romans studied soil to determine its firmness to support aqueducts and theaters.

In the Middle Ages, soil engineering extended beyond the building of roads and canals to the construction of heavy city walls with flanking towers, castles enclosed by heavy earthworks, large cathedrals, and campaniles (bell towers). The most famous example of a medieval soil mechanics problem is the Leaning Tower of Pisa, which tilts because of differential settlement. Begun in 1174 A.D., this campanile, 54.56 m (approximately 179 ft) tall, was completed in 1350. In 1910, the tower had a visible slant, and its top was 5.03 m (approximately 16.5 ft) out of plumb. From 1174 to 1933 the settlement of the center of the footing of the tower was 80 cm (approximately 2.62 ft), the lowest point of the footing settled 320 cm (approximately 10.5 ft), and the highest point settled about 160 cm (approximately 5.25 ft).

In the latter part of the seventeenth century, French military engineers contributed some empirical and analytical data pertaining to earth pressure on retaining walls for the design of revetments of fortifications. In 1661, France undertook an extensive public works program, which included the improvement of highways and the building of canals. The construction of the great fortification system along the border of France was begun in 1667 under Marquis Sebastian le Prestre de Vauban (1633-1707), who was commissary general of fortifications and Louis XIV's chief engineer, and who in 1703 became marshal of France. Vauban is regarded as one of the greatest military engineers of all times (Vauban, 1686, 1740).

It is known that at that time Vauban gave some rules for gauging the thickness of retaining walls. It is not known for sure, however, whether he based these rules on theoretical considerations or his own experience. In this respect thoughts were later expressed in France that Vauban's empirical rules appeared to be so complete that it almost seemed as if they were based on an earth-pressure theory now unknown to us.

During the times of French mercantilism (17th century), soil problems were encountered in the terrain through which the canals were dug. Also, earth retaining walls of the great fortification system along the borders of France presented earth-pressure problems in connection with their stability. France established a corps of military engineers to train, among others, experts in fortification, and in 1745 the famous École des Ponts et Chaussées was established, where engineers were educated in sound principles of physics, mechanics, and mathematics for the construction of canals, highways, and bridges.

The first theoretical contributions to soil mechanics were made by C. A. Coulomb in 1773 with his *classical earth-pressure theory*—calculations of earth pressure against a retaining wall. In this analysis, Coulomb applied the laws of friction and cohesion, and determined the earth pressure on a retaining wall from the "wedge of maximum pressure". The importance of Coulomb's theory may be recognized best by the fact that his ideas on earth pressure are recognized and still used (with a few exceptions) even today.

Later Français (1820), Navier (1821), Poncelet (1840), and Culmann (1866) further developed Coulomb's theory for practical applications by engineers. Alexandre Collin (1846) dealt with slides on slopes of canals and dams made in and of clays, and studied the forms of sliding surfaces. In 1857 W. J. M. Rankine published his theory on earth pressure and equilibrium of earth masses. Thus the impetus for the development of early soil mechanics was given by the increased activities in bridge design and construction for highways and railways. Contributions to earth-pressure theories and thus to soil mechanics were also made by Müller-Breslau (1906), Franzius, (1927), Krey (1936), Terzaghi (1925), and Fröhlich (1934).

Pioneering in practical soil mechanics must be credited to the Swedish Geotechnical Commission of the State Railways in Sweden, and the

Foundations Committee of the American Society of Civil Engineers (ASCE), both established in 1913. The year of birth of modern soil mechanics, however, is now generally recognized as 1925, when Karl Terzaghi published his book *Erdbaumechanik auf bodenphysikalischer Grundlage.* Another important step in the development of this new discipline was taken with the publication of the settlement theory of clays, coauthored by Terzaghi and Fröhlich (1936).

The basic concept of the effective stress in soil mechanics, though, is to be credited to Terzaghi (1936). Terzaghi's publications, and the work in soil mechanics done by other authorities in this field, were a great stimulus to soil mechanics studies and contributed a considerable amount of knowledge on engineering properties of soils in the United States, as well as abroad. The objectives of soil mechanics are (1) to study the physical and mechanical properties of soil, (2) to apply this knowledge for the solution of practical engineering problems, and (3) to replace by scientific methods the empirical ones of design used in foundation and soil engineering in the past.

Some typical problems encountered in soil mechanics involve the bearing capacity of soil (see Figs. 1, 2, and 3); stress distribution in soil; stability of soil foundation systems; tolerable settlement of soil and structure; effect of frost on soil and foundations; effect of vibration on soil; soil stabilization; and laying of foundations in permafrost. The combined efforts of engineers and researchers from all over the world in this discipline contributed to what may be called modern soil mechanics.

Since 1936, many conferences, congresses, short courses, and symposia on soil mechanics have been convened throughout the world to review the progress and achievements in soil mechanics. The following list summarizes the dates and locations of International Conferences on Soil Mechanics

FIGURE 2. One-sided expulsion of soil from underneath the base of a foundation by an inclined load.

FIGURE 3. Soil failure in shear: one-sided expulsion of soil from underneath the base of a foundation model by an inclined load.

and Foundation Engineering (SMFE) since 1936: First International Conference on SMFE (1936), Harvard University, Cambridge, Massachusetts; Second International Conference on SMFE (1948), Rotterdam, The Netherlands; Third International Conference on SMFE (1953), Zurich, Switzerland; Fourth International Conference on SMFE (1957), London, England; Fifth International Conference on SMFE (1961), Paris, France; Sixth International Conference on SMFE (1965), Montreal, Ontario, Canada; Seventh International Conference on SMFE (1969), Mexico City, Mexico; Eighth International Conference on SMFE (1973), Moscow, USSR; Ninth International Conference on SMFE (1977), Tokyo, Japan; and Tenth International Conference on SMFE (1981), Stockholm, Sweden.

ALFREDS RICHARDS JUMIKIS

References

Collin, Alexandre, 1846. *Recherches expérimentales sur les glissements spontanés des terrains argileux accompagnées de considérations sur quelques principes de*

FIGURE 1. Deformation of soil exhaustion of soil bearing capacity: two-sided expulsion of soil from underneath the base of a foundation by a vertical load.

la mécanique terrestre. Paris: Carillian-Goeury et van Dalmont, Editeurs, 168p. (Collin's work is now translated by W. R. Schriever under the title *Landslides in Clays,* Toronto, Ontario: University of Toronto Press, 1956, 162p.)

Coulomb, C. A., 1773. Essai sur une application des règles de maximis et minimis à quelques problèmes de statique relatifs à l'architecture, in *Mémoires de la mathématique et de physique, présentés à l'Académie Royale des Sciences, par divers Savants, et lûs dans sés Assemblées.* Paris: De l'Imprimerie Royale, 1776, vol. 7, Année 1773, pp. 343-384.

Culmann, C., 1866. *Die graphische Statik.* Zürich: Meyer and Zeller, 644p.

Français, J. F., 1820. Recherches sur la poussée de terres sur la forme et dimensions des revêtements et sur la talus d'excavation, *Mémorial de l'officier du génie* **4,** 157-206.

Franzius, O., 1927. *Der Grundbau.* Berlin: Springer-Verlag, 360p.

Fröhlich, O. K., 1934. *Druckverteilung im Baugrunde.* Vienna: Springer-Verlag, 185p.

Krey, H., 1936. *Erddruck, Erdwiderstand und Tragfähigkeit des Baugrundes.* Berlin: Wilhelm Ernst und Sohn, 347p.

Müller-Breslau, H., 1906. *Erddruck auf Stützmauern.* Stuttgart: Alfred Kroner Verlag, 160p.

Navier, C. L. M. H., 1821. Sur les lois de l'équilibre et du mourement des corps solides élastiques, *Bulletin des Sciences de la Société Philomatique de Paris,* 1823, 177-181.

Poncelet, J. V., 1840. Mémoire sur la stabilité des revêtements et de leurs fondations: note additionnelle sur les relations analytiques qui lient entre elles la poussée et la butée de la terre, *Mémorial de l'officier du génie* **13,** 261-270.

Rankine, W. J. M., 1857. On the stability of loose earth, *Royal Soc. London Philos. Trans.* **147,** pt. I, 9-27.

Statens Järnvägars Geotekniska Kommission, 1922. *Slutbetänkande 1914-1922.* Stockholm: Statens Järnvägar, 180p.

Terzaghi, K., 1925. *Erdbaumechanik auf bodenphysikalischer Grundlage.* Leipzig and Vienna: Franz Deuticke, 399p.

Terzaghi, K., 1936. The shearing resistance of saturated soils, and the angle between the planes of shear, in *Proceedings of the [1st] International Conference on Soil Mechanics and Foundation Engineering,* held June 22-26, 1936, at Harvard University, Cambridge, Mass., vol. 1, paper D-7, pp. 54-56.

Terzaghi, K., and Fröhlich, O. K., 1936. *Theorie der Setzung von Tonschichten.* Leipzig and Vienna: Franz Deuticke, 168p.

Vauban, Sebastian Le Prestre, de, 1686. *Mémoire du maréchal de Vauban sur les fortifications de Cherbourg.* Paris: V. Didron, 1851, 91p.

Vauban, S., 1740. *Mémoire, pour servir d'instruction dans la conduite des siéges et dans la défense des places, dressé par Monsieur le maréchal de Vauban, et présenté au roi Louis XIV en MDCCIV.* Leide: J. and H. Verbeek, 204p.

Vitruvius, Pollio Marcus, 1914. *The Ten Books on Architecture,* Morris Hicky Morgan (trans.). Cambridge, Mass.: Harvard University Press, 331p. (This was first published in 1649 by Ludovicum Elzevirum.)

Cross-references: *Earthquake Engineering; Foundation Engineering; Permafrost, Engineering Geology; Rheology, Soil and Rock; Rock Mechanics; Soil Mechanics.* Vol. XIV: *Expansive Soils; Geohistory, Founding Fathers USA.*

SOIL MINERALOGY, ENGINEERING APPLICATIONS

Engineering Definition of Soil

To the engineer, *soil* is any material found in the surface layer of the earth, moon, or other celestial body that is loose enough to be moved by spade and shovel (see Vol. XII, Pt. 1: *Soil*). Accordingly, the particulate components of engineering soils may range in size from small boulders to colloidally dispersed solids, and in mineral character from practically unchanged crystals and fragments of igneous, sedimentary, and metamorphic rock through a wide range of weathering products to typical products of pedogenesis (soil formation). Table 1 lists the engineering nomenclature for particle size categories. Previous subdivisions for the "soil fines" (silt, clay, colloids) have been abandoned because the most important engineering properties of the soil fines depend less on particle size than on physicochemical factors and are gauged by their interaction with water. Differentiation is made between swelling and nonswelling soil fines and the responsible minerals and mineral associations.

From the engineer's viewpoint, *soil mineralogy* (q.v. in Vol. IVB) is concerned with all minerals found in what he calls soil. This, in addition to the more common soil minerals, includes commercially important loose rock deposits such as phosphates, aluminum ores and iron ores, kaolinites, bentonites, and others used mainly as raw materials in the manufacture of ceramics, metals, and chemicals, as well as tailings from ore winning and deposits of particulate industrial and urban wastes.

Main Types of Engineering Soil Uses

Soils are used for a great many different purposes and in many different ways. In their natural or partly disturbed states, they carry the dead and live loads of terrestrial and submarine structures, including those of highways and airports (Winterkorn, 1965). Thoroughly disturbed natural soils and blends of such soils serve, usually in densified condition, as construction materials in embankments, levees, dikes, and earth dams, and in sub-bases, bases, and occasionally surfaces for transportation facilities, parking lots, and feed lots. They are also used as general or special-purpose backfill, in monolithic soil house

TABLE 1. Engineering Nomenclature for Size Categories

Name	Size Range Mineral or U.S. Sieve No.	Size Range, in mm
Boulders	> 12 in.	> 305
Cobbles [a]	3-12 in.	76.2-305
Gravel [b]	no. 4-3 in.	4.76-76.2
coarse	¾-3 in.	19.1-76.2
fine	no. 4-¾ in.	4.76-19.1
Sand	no. 200-no. 4	0.074-4.76
coarse	no. 10-no. 4	2.0-4.76
medium	no. 40-no. 10	0.42-2.0
fine	no. 200-no. 40	0.074-0.42
Soil fines or silt-clay materials [c]	< no. 200	< 0.074

[a] If rounded, otherwise field stones or rock fragments.
[b] If used as components of concretes and mortars, gravels are called "coarse aggregate" and sands "fine aggregate."
[c] Previous differentiation between silt and clay on basis of particle size (clay < 5 or 2μ) has been abandoned because clay or silt behavior for fine materials depends essentially on mineralogical character.

Source: From Winterkorn (1968), p. 39, by permission of the Soil Science Society of America.

construction, and in the making of adobe bricks. Soils stabilized by the admixture of waterproofing and/or cementing materials, or by thermal, electric, or chemical treatment, can be tailor-made to suit a wide range of construction purposes (see Vol. XII, Pt. 1: *Soil Stabilization*). Soils serve as media for surface locomotion (see *Laterite, Engineering Geology*), and are subject to such engineering operations as tunneling (q.v.), excavating (see *Pipeline Corridor Evaluation*), draining, loosening, and consolidation (see *Consolidation, Soil*). They may be comminuted (pulverized), mixed, compacted, molded, and otherwise worked for specific purposes. Separated and sized gravel and sand fractions become the volumetrically predominant constituents of mortars and concretes, and soil fines find large use as fillers in bituminous and similar compositions. Soil mineralogy plays an important role in all these uses. In combinations with soil-alien substances, such as contained in organic and inorganic cements, this role may usually be accounted for by desirable and undesirable reactions between soil and alien constituents; however, granulometric and volumetric relationships often define and limit the extent to which mineralogical factors may affect soil behavior (Winterkorn and Fang, 1975).

Soil systems can be subdivided into (a) those that in a densified state possess a continuous granular bearing skeleton of sand and/or gravel particles and (b) those without such a skeleton. The former may contain "fines" ranging from zero to a maximum value which is defined volumetrically: the combined volume of the fines and their water content under the worst expected conditions may not exceed the pore volume of the skeleton. The strength properties of such systems are governed primarily by those of the skeleton. In the absence of a bearing skeleton, soil properties are complex functions of the water content, the types, sizes, and phase volumes of the mineral constituents, the extent and physicochemical characteristics of their surfaces and interfaces, and their mutual arrangement, which reflects the previous history of the soil system (Highway Research Board, 1967).

Minerals and Mineralogy

A mineral is a naturally occurring solid of well-defined composition that can be expressed by a chemical formula. It has a characteristic internal atomic or ionic structure that produces distinctive external shapes and, together with the interatomic forces, determines the various physical and physicochemical properties of the mineral and its surfaces.

Mineralogy (q.v. in Vol. IVB) is the science of the formation, properties, and transformation of minerals. These three phases of mineralogy cover practically all our knowledge of the tangible inorganic world from information on the mechanical strength, chemical reactivity, and durability in various environments of common inorganic construction materials to the new, sophisticated science of "solid-state systems."

The interrelationships among the various physical, physicochemical, and chemical properties of a mineral permit its identification by means of tests on those that are most easily determined. Such identification renders immediately available the previously accumulated knowledge on the properties of the mineral or mineral association that qualify or disqualify it for specific engineering uses. Pertinent methods range from observations with the unaided eye of form, color, luster, opaqueness, and transparency, and simple tests for hardness, cleavage, type of fracture, and specific gravity, through inspection with the ordinary, petrographic, and pedologic microscope to evaluation of x-ray diffraction patterns and the use of ordinary and scanning electron microscopy. The latter are especially important for the visualization of the form and mutual arrangement of very small particles.

Mortars, Concretes, and Masonry

Mineral systems that possess a granular bearing skeleton and are bonded by inorganic or organic cements are called *concrete* if the largest granular fraction is of gravel size and *mortar* if it is of sand size. The specific name—for example, Portland cement concrete, bituminous concrete, soil mortar—derives from the type of binder employed. In most concretes, the volume of coarse aggregate (gravel) is about 50 percent and that of fine aggregate (sand) about 25 percent of the total volume. For the quality of the total system, that is, the concrete, the strength, toughness, and durability of the aggregate are the more important the weaker the bonding material. The pertinent properties of the aggregates derive from those of their constituent minerals. Also, the strength of bond established between cement and aggregate particles depends on the physical and chemical characteristics of the mineral surfaces and the chemistry of the bonding agent. In the case of Portland cement concrete, great problems may arise and destruction of the concrete may ensue if a highly siliceous coarse aggregate that possesses a large and accessible internal surface is used in combination with a Portland cement that has a relatively high content of alkali hydroxides. Especially susceptible to such deleterious reactions are volcanic rock and glasses of medium to high silica content, opaline and chalcedonic rocks, some phyllites, and certain zeolites. The alkali for these reactions may also be supplied by zeolites and alkali montmorillonites that are components of the aggregate. Deleterious reactions may also be caused by the presence of pyrite, marcasite, and other metal sulfides that oxidize to sulfates, which then interact with the normal cement constituents.

In bituminous concrete, the types of minerals in the gravel and sand fractions determine to a large degree the strength of bond between the bitumen and the mineral surfaces and, therefore, the ease of stripping of the bitumen from the aggregate under service conditions. This in turn governs the durability or service life of the system. Very fine sands, loess, diatomaceous earth, kaolinites, bentonites, and other natural soil fines are often used as stiffening fillers in bituminous compositions and as permeability-decreasing admixtures to Portland cement and to other mortars and concretes. In such uses, the minerals involved as well as their particle sizes are of utmost importance.

Cobbles and small boulders are used in the construction of road bases and pavements if their mineral composition is such that it provides strength, toughness, and durability sufficient for the service requirements. They are also used in masonry when they represent the most economical material available or when special architectural effects are desired. In the latter case, an additional requirement is their pigment stability, which depends on their mineralogical composition (Winkler, 1967). A special offender in the case of igneous rocks is biotite because its iron and magnesium ions are only loosely attached to the silica sheets and are given off at the beginning of the weathering process. The magnesium ion can also prove very troublesome if it is released into a chemical environment where it can form highly swelling compounds, as with certain tar acids (Winterkorn, 1968). These examples illustrate the importance of mineralogical (and therefore chemical) characterization of soil materials, especially if they are to be used in combination with other substances with which they may undergo chemical reactions. Such characterization is particularly important if one works in arid or semi-arid regions where relatively soluble minerals may have accumulated that are not normally found on the earth surface in moist temperate climates. The engineer who would try to use the white sands of New Mexico as fine aggregate in Portland cement concrete would be in for a great surprise.

Minerals Represented in Various Size Fractions with the Exception of Clays

Cobbles and Boulders. The mineral nature of cobbles and boulders is essentially that of their parent rock with a film or layer of weathering or biotic products on their surfaces, the composition of which depends on the environmental factors and the time of exposure.

Gravels and Sands. These may be, on one extreme, just mechanically comminuted parent rock with all its minerals represented or, on the other extreme, the mechanically and chemically most resistant minerals and their weathering products, with all possible gradations between the two extremes. The extent to which the mechanical and chemical breakdown has taken place depends on the environmental conditions and the time of exposure to them. Fluvial gravels and sands become more quartzitic the longer the path of transportation. In humid climates, gravels and sand tend to be siliceous and quartzitic, but they may be composed of any type of mineral in dry climates. The white sands of New Mexico are gypsum; coral and shell beach sands may consist almost exclusively of calcium carbonate.

The beach sands along the Atlantic coast change from Maine to the keys of Florida from essentially quartzitic to predominantly calcitic, and the black sands of Yellowstone Park and of some of the blue and purple beaches of the Pacific islands consist of obsidianite and similar glasses, as do some of the Alaskan and Aleutian beach sands. These sands, however, become lighter-colored and more quartzitic the closer they lie to the low-water line, that is, the more intensive the weathering and leaching process. The "sinking" beach at Los Angeles contains more unweathered feldspars than old stable beaches of similar parent material and latitude. Dolomite sands are found in the Bavarian Jura and in other locations. Sand- and gravel-sized particles used by the engineer may also be synthetic pedogenic products such as the pisoliths in laterite soils (see *Laterite, Engineering Geology*).

Silts. These may be essentially unaltered, mechanically comminuted minerals as in glacier-ground rock flower, or they may fall within a wide range of weathering products. They may also owe their origin to pedogenetic synthesis and to irreversible aggregation of more finely subdivided original weathering products. In temperate climates, silts ordinarily bear a close relationship to the minerals of their parent rock, such as feldspars, micas, and quartz.

Soil Systems Without Granular Bearing Skeletons

The engineering properties of such soil systems are complex functions of the following: (1) the phase volumes and the surface areas per unit volume of the constituent minerals; (2) the physicochemical character of the mineral surfaces and the type and amount of exchangeable ions present; (3) the water and dissolved ion content; and (4) the entire previous history of the system and of its constituents.

Obviously, prediction of engineering properties on the basis of mineral composition alone can be at best only qualitative. Required engineering information is best and most economically obtained by pertinent tests on samples that are as close as possible to the condition in which the soil is to be used. However, studies on pure clays and on physicochemically well-defined clay-water systems are important for theoretical reasons and for the practical purpose of defining property ranges and limits. They are also of direct significance for industries that use well-defined clay systems for special products and purposes.

HANS F. WINTERKORN

References

Highway Research Board, 1967. Physicochemical properties of Soils. *Highway Research Record No. 209,* 92p.

Winkler, E. M., 1967. Stone for architects, *Geotimes* **12**(4), 14-16.

Winterkorn, Hans F., 1965. The bearing of soil-water-interaction on water conduction under various energy potentials, *RILEM Bull.* (new series) **27**, 87-96.

Winterkorn, Hans F., 1968. Engineering applications of soil mineralogy, in George W. Kunze (ed.), *Mineralogy in Soil Science and Engineering.* Madison, Wisc.: Soil Science Society of America, Spec. Pub. Series No. 3, 35-51.

Winterkorn, Hans F., and Fang, H. Y., 1975. Soil Technology and Engineering Properties of Soils in Hans F. Winterkorn and H. Y. Fang (eds.), *Foundation Engineering Handbook,* New York: Van Nostrand Reinhold, 76-120.

Cross-references: *Caliche, Engineering Geology; Clay, Engineering Geology; Consolidation, Soil; Duricrust, Engineering Geology; Laterite, Engineering Geology; Soil Classification System, Unified; Soil Mechanics.* Vol. IVB: *Soil Mineralogy.* Vol. XII, Pt.1: *Clay Mineralogy; Clay Minerals, Silicates.* Vol. XIV: *Cat Clays; Dispersive Clays; Engineering Soil Science; Expansive Soils; Soil Fabric.*

SOIL SAMPLING — See Vol. XIV: PEDOGEOCHEMICAL PROSPECTING; SOIL SAMPLING.

SPARKER SURVEYS — See Vol. XIV: ACOUSTIC SURVEYS, MARINE.

SPELEOLOGY — See Vol. III.

STABILIZATION, SOIL — See CLAY, ENGINEERING GEOLOGY. Vol. XII, Pt. 1: SOIL STABLIZATION.

STREAM SEDIMENT GEOCHEMISTRY, SURVEYS — See Vol. XIV: HYDRO-GEOCHEMICAL PROSPECTING.

SUBMERSIBLES

History

One of the earliest submersible craft was built by the Dutchman Cornelius Van Drebble of Alkmaar in 1620 (Sweeney, 1970; Busby, 1976). Powered by 12 oarsmen, it sailed about 4 m below the surface of the River Thames in London from Westminster to Greenwich (a distance of 9.5 km). Most of the subsequent development of small submarines was largely directed toward military and naval objectives. It was not until the 1960s that any real attention was paid to the development of submersibles for scientific research and underwater engineering (Heirtzler and Grassle, 1976).

Definition and Specifications

A *submersible* may be defined as an underwater vehicle that relies on hull buoyancy for a substantial part of its buoyancy underwater (Flemming, 1968). Submersibles therefore differ from *sub-*

marines, which rely on large ballast tanks that are either vented by compressed air or filled with water to control buoyancy, and from *bathyscaphes,* which rely on liquid ballast (usually petrol) carried in a float to provide buoyancy. The submersible hull has a very high strength-to-weight ratio and is always fitted with portholes. Much of the equipment carried—propulsion motors, scientific sampling equipment, or underwater work tools—is fitted to the outside of the hull. Trim tanks are fitted to enable the submersible to descend and ascend and to adjust its orientation when on or near the bottom. Submersibles also have a high degree of maneuverability at slow speeds.

One of the earliest vehicles was J.-Y. Cousteau's "Diving Saucer," launched in 1959. Since 1960 over 100 different submersibles have been built with depth ranges from 100 m to 4,900 m and displacements of 5 to 10 tonnes (Busby, 1976). Most of the larger vehicles were built between 1965 and 1968. They were too sophisticated, however, and were found to be largely uneconomical to operate. Accordingly, many of these are now laid up. Specifications and operational details of submersibles in current use may be obtained from Busby (1976) and from the yearbook *Jane's Ocean Technology,* which was first published in 1974 (Trillo, 1976).

Advantages of Submersibles in Marine Geology

Manned submersibles and bathyscaphes were first used in marine geology in the mid-1960s. Until then, most knowledge of the nature of the seabed had come from samples collected by means of grabs and dredges and from photographs taken by underwater cameras operated from surface ships. There was no way in which direct scientific observations could be made *in situ* in deeper waters. In inshore shallow-water areas, Scuba diving techniques have been applied with considerable success, but scientific diving operations at depths greater than 50 m are largely impracticable. The use of submersibles therefore enables marine geologists to observe the sea floor at first hand at all depths down to 3,000 m or even deeper and to choose exactly where a rock or sediment sample should be collected in order to answer a particular question. Submersibles can also be used to deploy equipment such as time-lapse cameras, seismometers, and the like at precise locations and to recover them at a later date.

Methods

Most submersibles are fitted with a hydraulically operated telechiric manipulator. Most of these have up to six planes of movement and can grip up to 230 kg (Busby, 1976). Much experience in the

design and use of external handling tools has been gained by the "Alvin" team at the Woods Hole (Massachusetts) Oceanographic Institution (Winget, 1969), and a wide range of equipment based on these designs has been made for particular geological tasks. Sampling equipment is carried on a rack, basket, or tray attached to the front of the submersible.

In situ rock samples can be collected from well-jointed outcrops by using the manipulator, either on its own or with a fulcrum tool, to loosen and pull out a suitable joint-bounded block. Where this is not feasible, samples may be chipped off using a geological hammer carried in the claw (Winget, 1969), or a core sample can be obtained with the aid of a rock drill mounted on the front of the submersible (Fig. 1). Sediment samples can be collected by means of core tubes or scoops of varying sizes depending on the amount of sample required (Winget, 1969; Wilson, 1977).

Navigation

The problems associated with underwater navigation of submersibles have been largely solved over the last few years, and a number of systems have been developed (Busby, 1976). Most employ a long-baseline system using an array of acoustic transponders that are deployed by the submersible support ship in the area of operation prior to the dive (Fig. 2). The most common has three fixed transponders and one mobile one attached to the submersible. These are interrogated in response to signals sent at regular intervals from the support ship, the returned signals are processed by computer, and the submersible's track is plotted, giving its position to an accuracy of within 2 m. After the dive has been completed the transponders are recovered. These systems are very expensive, however, and for many geological purposes it is not essential to know precisely where one is during the dive, but the relative position of features observed and samples collected must be known after the dive. This is achieved by logging forward and athwartships current components measured by an electromagnetic log, submersible heading, depth, time, and so on throughout the dive (Wilson, 1977). These data are later processed by computer and converted into the actual track.

Geological observations made by the observer are recorded on a cassette tape recorder during the dive, and photographs and videotape records can be taken whenever required. Most dives last between three and six hours.

Investigations Using Submersibles

Geological investigations using manned submersibles have been mounted in both the Atlantic

FIGURE 1. *Pisces III* as fitted for rock drilling and sediment sampling. The rock drill is attached in place of the heavy-duty grab claw. In use, the drill is rotated into place on the outcrop. On completion, the barrel is placed in the rack attached to the left side of the drill. Four barrels are carried, enabling four rock cores to be taken during each dive. A sediment sampling scoop (labeled R) is attached to the sample basket. The manipulator claw can be seen directly above the basket. A still camera and television camera with their lights are mounted on the pan and tilt head on the port side. A second still camera, flash, and the floodlights are attached to the badge bar above the portholes.

and Pacific oceans. Some important investigations, with dates, names of submersibles used, and references to publications (where appropriate) are given in Table 1. For a bibliography of papers published up to 1969, see Ballard, and Emery (1970).

Study of part of the Mid-Atlantic Ridge sea floor spreading center was conducted in 1974 in the FAMOUS Project, which involved 47 dives and the collection of rock, sediment, and water samples from the inner-rift valley floor (*Alvin, Cyana,* and the bathyscaphe *Archimede;* Ballard and van Andel, 1977; Heirtzler and Bryan, 1975). Study of deep submarine canyons was conducted in 1975 in Bahama Banks (*Alvin*). Investigation of the Galapagos spreading center and the East Pacific Rise was conducted in (1977-1979, (*Alvin, Cyana;* Spiess et al., 1980).

Many submersibles have been built recently, and other earlier vessels modified, specifically for underwater operations in connection with the offshore oil industry, such as detailed surveying of pipeline routes and of production platform sites. These submersibles often have diver lockout facilities.

Unmanned Submersibles

Operations with manned submersibles are relatively expensive although extremely cost-effective. A number of cable-controlled unmanned vehicles (including *Telenaute 1000, Consub, Angus,* and *Snurre*) have been developed to undertake geological reconnaissance work, to take photographs and

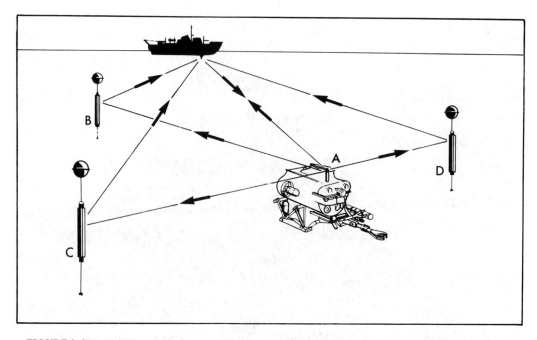

FIGURE 2. Submersible navigation system showing typical transponder array suitable for detailed survey work: **A,** mobile receiver/transmitter fitted to submersible; **B, C, D,** fixed transponders. Arrows indicate paths of interrogation and response signals between support ship, submersible, and fixed transponders.

TABLE 1. Some Important Geological Investigations Using Manned Submersibles.

Type of Investigation	Location	Date(s)	Name of Submersible	Reference Publication (if any)
Geological investigations and sampling	Okinawa Islands Great Barrier Reef of Australia	1966 1969	*Yomiuri* *Yomiuri*	
Collection of rock cores and sediments	Hudson Bay	1970	*Pisces III*	McFarlane and Trice (1972)
Study	Outer Coronado submarine canyon	1972	*Deep Quest*	Shepard and Marshall (1975)
Reconnaissance	Cobb Seamount	1972	*Sea Otter*	Schwartz and Lingbloom (1973)
Rock and sediment sampling	U.K. continental shelf	1970-1972	*Pisces III*	
	Rockall Bank	1973	*Pisces III*	
	Gulf of Mexico	1974-1976	*Diaphus*	Geyer (1977)
	Hawaii	1974-1975	*Sea Cliff*	
	Cocos Ridge	1975	*Turtle*	
	East and West North American coasts	1964-1975	*Soucoupe, Deepstar 2000, Deepstar 4000, Saucer SP300, Alvin*	
Study of sea floor spreading center	Mid-Atlantic Ridge	1974	*Alvin,* etc.	Ballard and van Andel (1977); Heirtzler and Bryan (1975)

videotapes, and to collect rock cores and sediment samples.

JOHN BRODIE WILSON

References

Ballard, R. D., and Emery, K. O., 1970. *Research Submersibles in Oceanography.* Washington, D.C.: Marine Technology Society, 70p.

Ballard, R. D., and van Andel, T. H., 1977. Project FAMOUS: Operational techniques and American submersible operations, *Geol. Soc. America Bull.,* **88,** 495-506.

Busby, R. F., 1976. *Manned Submersibles.* Washington D.C.: Office of the Oceanographer of the Navy, 764p.

Flemming, N. C., 1968. Functional Requirements for Research/Work Submersibles, *Aeronautical Jour.* **72,** 123-131.

Geyer, R. A. (ed.), 1977. *Submersibles and Their Use in Oceanography and Ocean Engineering.* Amsterdam: Elsevier, 383p.

Heirtzler, J. R., and Bryan, W. B., 1975. The floor of the Mid-Atlantic Rift, *Scientific American* **233,** 78-90.

Heirtzler, J. R., and Grassle, J. F., 1976. Deep-sea research by manned submersibles, *Science* **194,** 294-299.

McFarlane, J. R., and Trice, A. B., 1972. Core sampling in Hudson Bay, in *Fourth Annual Offshore Technology Conference,* Vol. 2, 169-174.

Schwartz, M. L., and Lingbloom, K. L., 1973. Research submersible reconnaissance of Cobb Seamount, *Geology* **1,** 31-32.

Shepard, F. P., and Marshall, N. F., 1975. Dives into Outer Coronado Canyon System, *Marine Geology* **18,** 313-323.

Spiess, F. N., et al., 1980. East Pacific Rise: Hot springs and geophysical experiments, *Science* **207,** 1421-1433.

Sweeney, J. B., 1970. *A Pictorial History of Oceanographic Submersibles.* New York: Crown, 314p.

Trillo, R. L. (ed.), 1976. *Jane's Ocean Technology 1974-1975.* London: Macdonald and Jane's, 344p.

Wilson, J. B., 1977. The role of manned submersibles in sedimentological and faunal investigations on the United Kingdom continental shelf, in R. A. Geyer (ed.), *Submersibles and Their Use in Oceanography and Ocean Engineering.* Amsterdam: Elsevier, pp. 151-167.

Winget, C. L., 1969. Hand tools and mechanical accessories for a deep submersible, *Woods Hole Oceanographic Inst. Tech. Rept. No. 69-32,* unpublished manuscript, 180p.

Cross-references: *Oceanography; Applied.* Vol. XIV: *Ocean, Oceanographic Engineering.*

SUBSURFACE EXPLORATION, SURVEYS—See PIPELINE CORRIDOR EVALUATION; ROCK STRUCTURE MONITORING: SEISMOLOGICAL METHODS; TUNNELS, TUNNELING. Vol. XIV: ACOUSTIC SURVEYS, MARINE; AUGERS, AUGERING; BOREHOLD DRILLING; EXPLORATION GEOPHYSICS; GROUNDWATER EXPLORATION; VLF ELECTRO-MAGNETIC SURVEYING; WELL LOGGING.

SURVEYS, MUSEUMS—See EARTH SCIENCE, INFORMATION AND SOURCES. Vol. XIV: GEOLOGICAL SURVEYS, STATE AND FEDERAL.

T

TERRAIN EVALUATION, MILITARY PURPOSES

Military interest in terrain goes back to the earliest times. Certain gateways have figured time and again in the history of Europe, and certain characteristics of terrain have spelled the difference between success and defeat in particular battles or campaigns. In 1356, Edward, the Black Prince, won a decisive victory at Poitiers by drawing the French attack onto soft ground where the heavily armored knights became immobilized. In 1759, General Wolfe's appreciation of the tactical significance of the landing area and ravine at l'Anse du Foulon permitted the British to achieve a complete deployment of their forces on the high ground adjacent to Quebec. Their success in the ensuing battle sealed the fate of the French regime in Canada. In 1815, Napoleon's ignorance of the sunken road at Waterloo resulted in Blucher's army achieving almost total surprise in a battle that changed the face of Europe. A century later, in the Flanders lowland, thousands of soldiers had first-hand experience of one of the most intractable types of terrain in the world (Johnson, 1921). Flanders is strategically vital in that it lies between the Artois and Ardennes uplands and provides the only corridor between France and Germany. The lowland is underlain by the gray, plastic, Ypresian clay, which is fine-grained, sensitive, and generally saturated. In this terrain, water supplies are easily contaminated, excavations and earthworks slump, trenches and gun emplacements fill with water, the explosive impact of artillery and mortar fire is diminished, and cross-country movement becomes almost impossible even after the slightest rain shower.

It was as a result of experience in Flanders that the Service Géographique of the French army was instructed to prepare the first operational terrain maps. These were simple "going" maps for tanks showing the passability of the ground in general categories. Similar going maps were produced in World War II for special operations, such as the Anzio landings in Italy and the approach to and crossing of the Rhine in 1944-1945. In general, however, they were not available and commanders relied on terrain assessments supplied through normal intelligence channels. Sometimes this information was adequate, as in the case of the Russian armored attack along the sand plain corridor

between Minsk and Baranovici, which resulted in the penetration of the German defense positions on the forest-marsh line of the Pripetz in the late summer of 1944. In other instances the lack of adequate terrain knowledge resulted in critical situations, as in Normandy in 1944, when the hedgerows of the bocage seriously delayed the Allied breakout from the beachheads, or in Iwo Jima in 1945, where the soft volcanic sands proved so costly in American lives and materiel.

It was not until after World War II that serious and sustained attention was given to the problem of evaluating and predicting terrain conditions. The stimulus was provided by the Cold War and by the realization that Central Europe could again become a battleground. As NATO forces prepared for the defense of Western Europe during the period 1950-1954, the Operational Research Section of the British Army of the Rhine (BAOR) was given responsibility for the collection of soil data and the preparation of "going" maps for the British zone in Germany, and the adjacent areas in Belgium and Holland. The basic concern in this mapping project was cross-country mobility, in particular the identification and delimitation of areas that were impassable or difficult for heavy-tracked vehicles, such as tanks (Beckett, 1955). In the preparation of these maps, considerable attention was given to the basic problems of soil trafficability and the interpretation of ground conditions from airphotos. Similar trafficability maps were prepared by the U.S. Army Engineer School, Fort Belvoir, for the American zone in Germany, and examples can be found in the army field manual on *Terrain Intelligence* (U.S. Department of the Army, 1959).

Since the early 1960s it has become apparent that terrain information has relevance to a wide range of military activities in addition to cross-country mobility (see *Military Geoscience*). Terrain evaluation has developed as a specific branch of military intelligence in response to the need for an understanding of the total complex of environmental factors occurring in a particular region. The term *terrain* can be considered as encompassing the natural features and configuration of the landscape in any given area. As such, it includes landforms, microrelief, surface and subsurface water, vegetation, soils, geology, state-of-the-ground, and climate. A listing of the significant terrain factors for military purposes as outlined by the Geographic Systems Division, U.S. Army Corps of Engineers Topographic Laboratories (Woloshin, 1968) is given in Table 1. Terrain evaluation is the process of assessing these factors individually, or in combination, in terms of their effects or anticipated effects on particular military activities. The organization and flow of information through an ideal terrain-evaluation system as conceived by the U.S. Army Corps of Engineers is presented in Figure 1. The system consists of two linked segments—the data-gathering, -processing, and -storage segment, and the operational-utilization segment. In the former, five essential components are recognized: information collection, information processing, performance-prediction modeling, data referencing, and data output. The latter has four components: operational planning, performance prediction processing, presentation and display, and reproduction-distribution. Terrain evaluation

(q.v.) is a relatively complex process, which commences with the acquisition of terrain data and terminates with the presentation of operational data indicating the anticipated effects of terrain factors on specific military activities.

Fundamental Problems of Terrain Evaluation

The terrain-evaluation process requires a careful consideration of the whole spectrum of environmental factors including their interrelationships and effects on various military activities. Certain fundamental problems arise because of the nature of terrain, and they merit consideration because of their relevance at all stages in the terrain-evaluation process.

First is the problem of *complexity*. A great many terrain factors are involved, considerable variation in the range of values is to be expected, and a

TABLE 1. Significant Terrain Factors for Military Geographic Information.

Landforms	Hydrology
Relief	*Standing Water Bodies*
Elevation	Spacing
Slope	Dimensions
Pattern	Water quantity
Spacing	Water quality
Percent area occupance	Bottom/bank material
Micro relief	Wave characteristics
Cross-section	Ice thickness
Slope	Freeze-up—breakup period
Pattern	Percent area occupance
Spacing	*Surface streams*
Percent area occupance	Areal pattern
Soil	Dimensions
Type	Current velocity
Thickness	Water quantity
Chemical composition	Water quality
Bearing strength	Bottom/bank material
Surface Friction	Ice thickness
Percent area occupance	Freeze-up-breakup period
State-of-Ground	*Springs*
Soil moisture	Spacing
Soil temperature	Pattern
Snow cover extent	Water quantity
Snow cover thickness	Water quality
Snow bearing strength	*Subsurface water*
Geology	Depth
Rock type	Water quantity
Physical properties	Water quality
Structure	Vegetation
Rock thickness	Woody/nonwoody
Climate	Stem diameter
Temperature	Stem spacing
Precipitation	Crown diameter
Wind	Canopy height
	Crown closure
	Root configuration
	Seasonal variation
	Percent area occupance

Source: After Woloshin (1968).

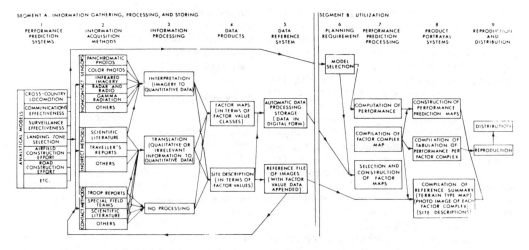

FIGURE 1. Structure and information flow in an ideal terrain evaluation system (after Grabau, 1968).

variety of intricate relationships often peculiar to a particular region have to be examined. Even within the framework of a given terrain-evaluation system, the integration of the components into a comprehensive scheme is a major task.

The second problem of terrain evaluation stems from the *lack of uniformity* in the data base. Some terrain factors, such as rock type or vegetation type, are generally defined in terms of descriptive systems; other factors, such as slope or local relief, are generally described in terms of class values, whereas factors such as soil strength, moisture content, or stream velocity can be expressed in terms of specific values. The terrain system must be sufficiently flexible to accept this kind of variation in the data input.

The third basic problem is that of *areal extent,* which arises because of the intrinsic differences between the various many terrain factors. Some, such as landform, occupy specific areas in the landscape and can be readily delimited; others, such as soil, are essentially gradational, and mapping their limits involves a certain degree of generalization. Some factors, such as slope, tend to be recorded in relation to a landscape facet, whereas with other factors, such as soil strength, the measurement refers to a point location. The terrain system must be capable of integrating the different types of data at an appropriate scale, and so inevitably different degrees of generalization are necessary for the various attributes.

The fourth problem encountered in terrain evaluation for military purposes is that of *detection* and *identification.* In many situations, the terrain analyst does not have access to the area and so the data have must be derived from airphotos or other types of imagery. For many terrain factors, the airphoto is an excellent source of precise information. Other factors, however, are not immediately identifiable by airphoto interpretation methods, with the result that the data base is often incomplete and varies in reliability.

Fifth is the problem of *scale.* Some degree of generalization is obviously necessary at many stages in the terrain-evaluation process. However, the system must be capable of handling varying amounts of detail at different levels of generalization, depending on the user's requirements. At one end of the spectrum is the strategic planner with a need for syntheses of a general nature covering large geographic areas; at the other extreme is the batallion commander with the problem of selecting a route across a particular terrain type and estimating the time required to achieve his mission objective.

The final problem is that of establishing a *basis of comparison* between different areas. The interrelationships between adjacent areas are often very significant since natural boundaries are seldom abrupt and the information relating to one terrain type is often directly relevant in the adjacent areas. Even more important is the fact that many areas have terrain analogues in other parts of the world. In a situation where access to the area of interest is not possible for political or military reasons, the analogous area can serve as a surrogate. Obviously the successful use of information derived from surrogate areas depends on the establishment of reliable procedures for determining the degree of analogy.

In view of the complexity of the terrain-evaluation process, it is hardly surprising that several different approaches to the problem have been developed. These can be subdivided into three categories depending on the methodology: landscape systems, parametric systems, and hybrid systems.

Landscape Systems for Terrain Evaluation

The concept and methodology for classifying terrain in terms of distinctive land units is well established in the geographical and ecological literature; however, the elaboration of these concepts into an operational terrain-evaluation system for military purposes is a relatively recent development. The most complete system of this type is the Oxford-MEXE system developed in the United Kingdom by researchers at the Soil Science Laboratory, Oxford University, under contract to MEXE (Military Engineering Experimental Establishment), now known as MVEE (Military Vehicles and Engineering Establishment).

In the Oxford-MEXE system the terrain unit provides the basis for the collection, processing, and storage of both terrain data and performance data resulting from military operations on a particular terrain unit. The system operates at two levels of generalization. At the lower level, landscape facets provide the fundamental terrain unit. *Facets* are small landscape subdivisions in which terrain conditions are essentially homogeneous or variable within narrow limits. At the higher level are landscape *patterns,* each characterized by a regular, repetitive sequence of facets. Both patterns and

facets can be recognized and delimited in the landscape and on airphotos.

The system was first applied in an experimental fashion by the British Army Emergency Reserve in several areas in the Middle East, East Africa, and Malaysia (MEXE Staff, 1965). This was followed by a more rigorous test of the system in the English scarplands covering an area of 5,000 km². Patterns and facets were defined and mapped at a scale of 1:63,360, and it was demonstrated that both types of units could be recognized and delimited on airphotos with a minimum of on-site inspection. The relative homogeneity of facets with regard to soil type, soil moisture conditions, and morphology was confirmed, and prediction equations were established for soil strength and soil moisture characteristics (Beckett and Webster, 1962, 1965*a*, 1965*b*).

A block diagram of one of the landscape patterns identified in the study is reproduced in Figure 2. The constituent facets are indicated, and their spatial relationships can be assessed from the diagram. It can be seen that each facet has a characteristic surface material, a typical slope or series of slope forms, and a particular position in the landscape. Many facets were found to possess remarkable uniformity both individually and

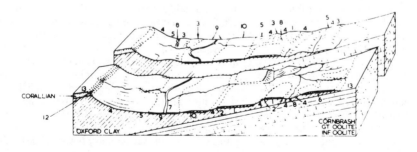

TYPE		No. 00013
OF INFO. PATTERN MASTER CARD		
CLIMATE C1 COOL TEMPERATE MARINE 25" PPTN 172 RAIN DAYS 7°F M M MAX 34°F M M MIN 1480 HRS SUN. 87 FROST DAYS		DATE 11 AUG 1961
PATTERN 1A RIVER VALLEY MATURE UPPER THAMES		COMPILER R WEBSTER
FACET 1-10	COUNTRY ROCK OXFORD CLAY (ILLITIC)	REFCE OXFORD U/61/13
LOCATION G S G S No 4620 G B, SHTS 145,157,158		

MATURE RIVER VALLEY : UPPER THAMES

DIAGRAM SHOWING MATURE RIVER VALLEY PATTERN ON OXFORD CLAY, ILLUSTRATING THE FACETS AND THEIR RELATION TO EACH OTHER IN THE LANDSCAPE

FACETS OF RIVER VALLEY AND CLAY

		FACETS OF SCARPLANDS BOUNDING THE RIVER VALLEY PATTERN
1 HIGH GRAVEL TERRACE	6 UNBEDDED GLACIAL DRIFT	12, 11 SCARP SLOPE
2 SPRING LINE	7 RIVER AND BANKS	13 DIPSLOPE
3 CLAY CREST	8 LOCAL BOTTOMLAND	
4 CLAY SLOPE	9 FLOOD PLAIN ALLUVIUM	
5 CLAY FOOTSLOPE	10 OLD ALLUVIUM, NOT FLOODED	

FIGURE 2. Block diagram of the "mature river valley pattern" — English scarplands showing the relationships of the facets (after Beckett and Webster, 1962).

collectively, whereas others showed appreciable variability even within an individual occurrence. In an attempt to assess the probable within-facet variability, field data were collected on 12 terrain parameters selected for their relevance to military engineering. Coefficients of variation were calculated and typical values of between 20 and 30 percent were obtained. In other words, for a given facet, the actual value for a given terrain factor would differ from the predicted value by more than 25 percent only once in 6 predictions, and by more than 50 percent only once in 40 predictions (Webster, 1965).

The next stage in the Oxford-MEXE program was the testing of the system on a global scale covering an entire climatic region. The subtropical arid zone was selected for study with emphasis at the initial stage on literature search and airphoto interpretation. This resulted in the identification and description of some 700 facet variants. Some of these were sufficiently similar to be grouped into what were termed *facet abstracts;* others displayed so much internal variation in soil and mesorelief that a new classifier, the *clump,* was introduced, this being analogous to the *complex* in soil mapping. A total of 56 facet abstracts and 42 clumps were defined at a mapping scale of 1:100,000, and it was demonstrated that the whole subtropical arid zone could be described in terms of 39 landscape patterns or *land systems.*

In the final stage of the study, large areas in Libya, Trucial Oman, Bahrain, Socotra, and Abdul Kuri were mapped at the facet and clump level to test the uniformity of the system. Field measurements were made in certain areas, and the data were statistically analyzed to test facet variability. Considered as a whole, the study demonstrated that a comprehensive terrain classification with a manageable number of units was feasible on a continental scale. In addition, it was shown that facets were sufficiently homogeneous and mutually exclusive to allow valid predictions within a single land system (Perrin and Mitchell, 1969-1971). Thus, there is a *prima facie* case for assuming that the land areas of the world can be subdivided into a finite number of land systems each with constituent facets. Once the terrain factors for a given land system have been assessed and the effects on particular military operations established, valid predictions can be made regarding the terrain effects in all other occurrences of the same land system or any of its constituent facets within the theater of operations. Information at the land system scale is appropriate for strategic planning at the army or corps level, whereas facet details are relevant at the scale of operations of the division, battalion, or squadron (Mitchell, 1973).

The Oxford-MEXE system has been adopted by the Indian army (Beckett, 1967), and extensive areas in various parts of India have been examined.

The procedures followed by the Indian army research groups have tended to make the facet definition more quantitative, and because of the inadequacies of the airphoto coverage of India more emphasis has been placed on field measurements.

Parametric Systems for Terrain Evaluation

The parametric approach to terrain evaluation can be broadly defined as the discrimination and classification of terrain units on the basis of selected attributes using entirely quantitative data. The advantages of the parametric approach stem from its quantitative basis, and include, first, the greater precision in stating the actual properties of the terrain compared with the subjectivity of the landscape systems, second, the greater consistency of the data, which provides a sound basis for comparison between one area and another, and third, the relative ease with which parametric data can be manipulated statistically and applied in analytic modeling using a computer.

Parametric systems have been most effectively developed by two agencies in the United States, the Environmental Protection Research Division of the Quartermaster Research and Engineering Command (QREC) and the Waterways Experiment Station of the U.S. Army Corps of Engineers. In the 1950s, QREC was responsible for producing environmental analyses and terrain handbooks for different parts of the world. As part of this program, considerable attention was given to the methods of assessing regional relief using topographic maps with contour information. The relationship between average slope and average relief was investigated for a variety of physiographic regions in the United States and Europe, and good correlations were established between relief, contour counts, and slope direction changes (Wood and Snell, 1959, 1960). The quantified terrain data system (QTDS), which was developed as a result of this work, provided a simple landform code based on the map measurement of six terrain actors: local relief, average elevation, elevation-relief ratio, average slope, slope-direction change, and grain. The terrain units derived from map analysis using this system showed good correspondence with the physiographic regions recognized by physical geographers using traditional methods; however, the QTDS units have the distinct advantage of being quantitatively defined, thus avoiding the subjectivity of nonparametric regional subdivisions.

The MEGA (Military Evaluation of Geographic Areas) Project at the Waterways Experiment Station was much broader in scope, with an overall objective of evaluating the effects of environment, particularly terrain, on military operations anywhere in the world (U.S. Army Corps of Engineers, 1963). Research effort was directed toward the development of techniques for quantifying terrain properties that were significant because of their effects on *materiel,* personnel, and operational procedures. Attention was focused on the problem of selecting key terrain factors that were readily defined, simple to map, and suitable for establishing analogues of known terrain types in other areas. These considerations led to the development of the *factor family* concept, which treats the environmental complex in terms of related attributes. The concept is based on the knowledge that terrain factors tend to associate in related groups, and that, in a general way, these groups or families tend to produce a characteristic effect on military activities. The factor families identified in the MEGA Project include surface geometry, comprising macro- and microgeometry, surface composition, hydrologic geometry, vegetation, animal life, weather, and climate. Within each factor family are a series of specific terrain attributes that are assessed quantitatively using airphoto interpretation, field survey, and literature review.

In the first stage of the MEGA program, commencing in 1953, attention was given to the problems of surface geometry with particular reference to the desert regions of the world. Procedures for determining the characteristic plan-profile (CPP) of any type of relief feature were elaborated using four indices: area occupance, peakedness, elongation, and parallelism (Van Lopik and Kolb, 1959). (In the original scheme, the basic data for the CPP system were derived from the contour maps, but later research in Canada demonstrated that the data required to calculate the indices could be obtained more efficiently using airphoto interpretation methods [Parry and Beswick, 1973].) The results of the MEGA program were presented in a series of reports (Kolb et al., 1958-1965) accompanied by factor family maps at a scale of 1:400,000, and various techniques were developed to establish the degree of analogy between terrain conditions in the western United States and in other desert regions. In addition, attention was given to the problems of quantifying terrain effects on specific military activities, such as cross-country mobility and engineering efforts in various types of construction projects.

During the period 1960-1968, research effort at the Waterways Experiment Station shifted from terrain analysis per se to the investigation of the relationship between cross-country mobility and environmental factors. The major objective of the MERS Project (Mobility Environmental Research Study) was to establish the interrelationships among the various terrain parameters as they affected trafficability, with a view to predicting vehicle performance and providing data for improved vehicle design. The first step toward this goal was to develop a unified quantitative system for describing terrain factors in a form suitable for performance prediciton analysis (Shamburger and Grabau,

1968). The MEGA factor family concept was modified to meet these requirements by retaining only factors that were relevant for mobility and critically revising the class limits. This was necessary because in prediction modeling the variation resulting from accepting a value at the upper end of a class range as opposed to one at the lower end must be reduced to a reasonable minimum level.

The procedure for compiling a factor family map for prediction modeling is accomplished by field sampling and airphoto interpretation, so that areal units can be delimited according to the class limits of particular factors. For example, the surface-geometry factor family involves four components: slope, vertical obstacle spacing, approach angle, and step height. Individual maps are compiled for each factor, as illustrated in Figure 3, and these are then synthesized by superimposition to give factor family map units. Each map unit is thus identified by a four-digit array representing the factor class values. This process is repeated for each factor family resulting in a final synthesis in the form of a terrain *factor complex* map (Figure 4a). The final step in the evaluation process is to predict vehicle performance in terms of speed, fuel consumption, or some other parameter using each array of terrain factor complex values as inputs to a computerized mathematical model. The performance data output is then substituted in the terrain factor complex map to give a performance prediction map of the type shown in Figure 4b (Benn and Grabau, 1968).

As the MERS program developed, progressive refinements were made in the procedures for acquiring all the relevant terrain data in a quantitative form and establishing relevant class limits (U.S. Army Corps of Engineers, 1963-1968). In addition, the procedures for the airphoto interpretation of terrain factors were examined in detail (Rula et al., 1963a), and sample areas were mapped using the MERS system in various parts of the world including Thailand (Rula et al., 1963b), Panama and Puerto Rico (Schreiner and Rula, 1962; McDaniel, 1966), Hawaii (Carlson, 1971), and Costa Rica (Smith, 1971). With experience gained in field testing, the analytical model for predicting vehicle performance has been modified and progressively refined (Blackmon et al., 1968-1972), and is now established as the reference model for all NATO forces—NRMM (NATO Reference Mobility Model).

Hybrid Systems for Terrain Evaluation

As the name implies, hybrid systems attempt to combine the advantages of the landscape and parametric systems. The most valuable characteristic of the former is the use of identifiable landscape units as the basis for both mapping and collating terrain data. As noted earlier, such units are generally described in naturalistic terms and

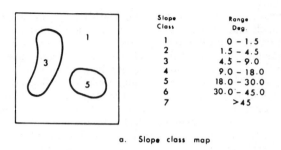

a. Slope class map

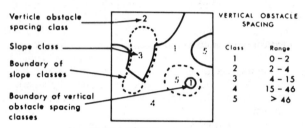

b. Vertical obstacle spacing class map superimposed on slope class map

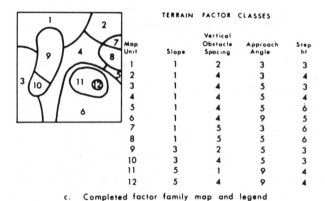

c. Completed factor family map and legend

FIGURE 3. Outline procedure for the compilation of a terrain factor family map (after Benn and Grabau, 1968).

have validity both in the field and on the airphoto. The great advantage of parametric systems lies in the quantitative data base and the use of a precise array of factor values to characterize a particular terrain type. With hybrid systems, terrain units are identified and delimited as in a landscape system and then rigorously defined in terms of fundamental terrain factor values such as surface composition and morphology. Other attribute values are added to this array as required to meet the needs of the particular survey.

Hybrid systems have been developed more or less independently by research groups in Australia and Canada. The Australian system was designed to provide a comprehensive framework for the collection and processing of civil engineering data at the preplanning, planning, and project stages,

whereas the Canadian system was developed to meet military requirements for a standard terrain-evaluation system that could provide the basic data for a full range of military operations.

The Australian system was developed by research workers in the Soil Mechanics Division of Commonwealth Scientific and Industrial Research Organization (CSIRO) (Aitchison and Grant, 1967), and susequently standardized for general application (Grant, 1968, 1973, 1974). The principal objective of the system is to identify terrain units that are essentially homogeneous with respect to certain critical factors such as natural materials (soil and rock type) and morphology (the three-dimensional form of the surface). It is recognized that other terrain factors may not be uniform within the confines of the terrain unit, and that

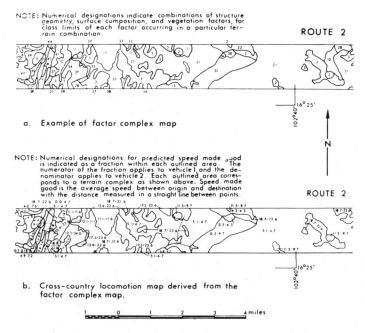

FIGURE 4. Sample strips illustrating the derivation of a cross-country mobility map from a factor complex map (after Benn and Grabau, 1968).

while some of these additional properties can be measured quantitatively, others can be conveyed only in descriptive terms.

There are four levels of evaluation in the Australian system—*province, pattern, unit,* and *component*—hence its name, the *PUCE* system of terrain evaluation. The system is hierarchical, in that each member of each class is composed of a limited number of repetitive members of the preceding class in a constant form of association. For example, a *province* consists of a constant repetitive association of terrain *patterns,* which in turn consists of a repetitive association of terrain *units,* and so on. The terrain *component* is the basic item in the system, and it is defined in terms of critical slope, rock, soil, and vegetation characteristics: the slope exhibits a constant rate of change of curvature along the major and minor axes; the lithology is constant in a uniform structural environment; the soils exhibit a consistent association, which can be expressed in terms of one class in the Unified Soil Classification system and one class in the primary profile classification for Australian soils; and the vegetation can be described in terms of a single association.

The PUCE system has been applied extensively in Australia, and the potential for military engineering applications has been assessed by the H.Q. staff of the Australian army (Lennon, 1969) with recommendations for modifications and amplifications to meet Royal Australian Engineers (RAE)

requirements. A very similar system has been developed in the Republic of South Africa by the National Institute for Road Research (Brink et al., 1968), and this has been modified for military use by the Engineering Command of the South African armed forces.

In Canada, a hybrid terrain-evaluation system has been developed by the Terrain Evaluation Group at McGill University under contract to the Defence Research Board, Department of National Defence. The McGill-DRE system was developed to meet Canadian Armed Forces requirements for a general-purpose terrain-classification and evaluation system that could provide information for cross-country mobility, engineer activity, vehicle design, and concealment and camouflage purposes. In the system design, two constraints were recognized: that access to site would be limited or impossible, and that the data output would be in the form of maps or map overlays. To meet the first requirement, it was necessary to base the system on airphoto interpretation procedures and other forms of remote sensing, with the result that much of the initial research effort was concerned with the development of classification systems of a quantitative and semi-quantitative nature that could be applied in photo and imagery interpretation. The second requirement necessitated experimentation in different types of cartographic presentation including dichrome overlays on a base map, composite factor mapping with

SURFACE COMPOSITION

☐ Consolidated rock — outcrops of granites and gneisses

⌐ Non-consolidated material

⊙ Mineral soil — poorly graded sands and silty sands,
  SP-SM Unified Soil Classification System

⊕ Organic soil — fine and coarse fibrous muskeg, types 9 and 12
  Radforth Classification System

○ Water — water bodies more than three feet deep and one acre in area

SURFACE MORPHOLOGY: MACROMORPHOLOGY

Slope steepness:

		Slope form:	
△	0 - 6°	0 - 10%	1 Convex, smooth
△2	6 - 14°	10 - 25%	2 Planar, smooth
△3	14 - 26½°	25 - 50%	3 Concave, smooth
△4	26½ - 45°	50 - 100%	4 Convex, rough
△5	Above 45°	Above 100%	5 Planar, rough
			6 Concave, rough
△	Classes I and II		7 Classes 1 and 3
△	Classes II and III		8 Classes 1,2,and 3
△	Classes III and IV		9 Classes 4,5,and 6
△	Classes I, II, and III		

SURFACE MORPHOLOGY: MICROMORPHOLOGY

◐ Positive features of mineral soil in a random, linear pattern. Slopes ⬡7 ; spacing 10 per mile.
  length 400 - 1800 ft, width–length ratio 1:4 - 1:20, amplitude 10 - 30 ft, amplitude
  non - symmetric sigmoid in section. Aeolian — fixed sand dunes

◑ Positive features of mineral soil in a random, linear pattern. Slopes ⬡7 ; spacing 18 per mile,
  length 50 - 400 ft, width–length ratio 1:2 - 1:8, amplitude less than 10 ft, spacing 18 per mile,
  irregular sigmoid in section. Aeolian — sand sheets and ripples

* ◑ Negative features in mineral soil in a random, non-linear, overlapping pattern.
  Slopes ⬡ 8 , lengths 10 - 200 ft, width–length ratio 1:1 - 1:2, amplitude 10 -
  50 ft, spacing 15 per mile, irregular cardioid in section. Glaciofluvial — kettle holes

* ● Positive features of consolidated rock in a random, non - linear pattern. Slopes
  ⬡ 9 , lengths 20 - 100 ft, width–length ratio 1:1 - 1:2, amplitude 10 - 30 ft, spacing
  calculation not possible, irregular rectilinear in section. Glacial — rock outcrops and erratics

* These symbols are not included in Fig.1

SURFACE COVER: VEGETATION STUCTURE

Height:	Stem type:	Form:
◄ More than 25ft	Woody	Trees
◁ 5 - 25ft	Woody	Young or dwarfed trees
▼ 2 - 5 ft	Woody	Tall shrubs or dwarfed trees
▶ Less than 2ft	Woody and non-woody	Low shrubs, grasses, sedges, and mosses

SURFACE COVER: VEGETATION SPACING (mean nearest neighbour distance)

⊙1 0 - 10ft	⊙4 60 - 90ft		
⊙2 10 - 15ft	⊙5 90 - 140ft		
⊙3 15 - 25ft	⊙6 140 - 220ft		
⊙ 25 - 40ft	⊙ Greater than 220 ft		
⊙ 40 - 60ft			

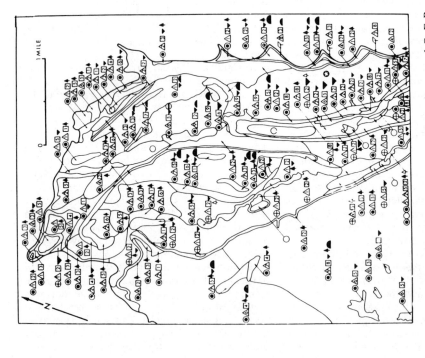

FIGURE 5. Composite terrain map and key for part of C. F. B.
Petawawa, Ontario, Canada (after Parry et al. 1968a).

symbol arrays, photo mosaic overprinting, and computer mapping.

The McGill-DRE system for terrain evaluation has been applied in six areas in Canada: the Armed Forces bases at Petawawa in Ontario (Parry et al., 1967, 1968a), Gagetown in New Brunswick (Parry et al., 1968b, 1971), the Lac Saffray area in Quebec-Labrador (Parry et al., 1974), the Mackenzie delta in the Northwest Territories (Parry et al., 1979), Churchill, Manitoba (Parry et al., 1981), and Schefferville, Quebec-Labrador (Parry et al., 1983). The studies were designed to test the effectiveness of the system at three different scales of operation—1:25,000 using airphotos at 1:5,000, 1:50,000 using airphotos at 1:40,000, 1:75,000 using airphotos at 1:36,000, and 1:250,000 using airphotos at 1:62,000.

The procedure followed in all the studies was that of separate factor evaluation—each terrain factor was examined separately and the class limits were used as mapping boundaries. The same basic factors were examined in each study—surface composition, morphology, vegetation, hydrology, and climate—but the amount of detail varied in each case. In the Petawawa study, it was possible to prepare very detailed terrain factor maps, such as the sample shown in Figure 5. This is a composite map in which the terrain units are homogeneous when considered in terms of engineering soil type, slope angle class, and slope form. Additional terrain data are serialized for each unit in a symbol array. In the Gagetown study a photo mosaic was used as a base map so that the factor class boundaries could be viewed against as detailed a terrain background as possible. In several of the analyses, the terrain data were converted from a unit-area display to a pixel display to achieve a simpler base for computer mapping. In the most recent studies emphasis has been placed on the application of the NATO Reference Mobility Model (NRMM) to selected areas in Canada and the development of a completely computerized system from airphoto analysis to final map output.

In all of these studies considerable attention has been given to the problem of manipulating terrain data to derive classification systems and mapping units that are specific for particular military uses at the tactical level. In the Petawawa study, cross-country mobility maps were produced, and in the Gagetown study various combinations of terrain factors were used to produce maps of cover and concealment, fields of fire, cross-country movement, airfield construction effort, amphibious operations, and airdrop suitability. In all of the studies of arctic and subarctic terrain, special attention has been given to winter operations (Granberg, 1974) and the prediction of snow conditions and snow trafficability has been of major concern (Parry et al., 1983).

In conclusion, it must be emphasized that the current concept of tactical and strategic deployment for combat and support demands rapid cross-country mobility, concentration of forces at designated objectives, and rapid dispersal after securing the objective. This concept of the modern mobile army requires detailed terrain information and prior knowledge of terrain effects on the full range of military operations (Cummings and White, 1981). Since the early 1950s the procedures for terrain evaluation have developed far beyond the "no-go" outline overlays of World Wars I and II, and this portion of the military art has been finally transformed into a military science (Needleman, 1969).

J. T. PARRY

References

Aitchison, G. D., and Grant, K., 1967. The PUCE program of terrain description, evaluation and interpretation for engineering purposes, in *Proceedings of the Fourth Regional Conference in Africa, Soil Mechanics and Foundation Engineering,* Section I, p. 1.

Beckett, P. H. T., 1955. *A Survey of Information Relative to the Production of Going Maps for BAOR and Its Communication Zone,* Rept. 4/55, Operational Research Section, British Army of the Rhine War Office, London, 58p.

Beckett, P. H. T., 1967. *Report on a Visit to the Terrain Evaluation Cell of the R and D Organization of the Indian Army,* Military Engineering Expt. Establishment (MEXE) Rept. 1020, 17p.

Beckett, P. H. T., and Webster, R., 1962. *The Storage and Collation of Information in Terrain,* Military Eng. Expt. Establishment (MEXE) Rept., Christchurch, England, 40p.

Beckett, P. H. T., and Webster, R., 1965a. *A Classification System for Terrain,* Military Eng. Expt. Establishment (MEXE) Rept. 872, Christchurch, England, 247p.

Beckett, P. H. T., and Webster, R., 1965b. *Field Trials of a Terrain Classification System—Organisation and Methods,* Military Eng. Expt. Establishment (MEXE) Rept. 873, Christchurch, England, 159p.

Benn, R. O., and Grabau, W. E., 1968. Terrain evaluation as a function of user requirements, in G. A. Stewart (ed.), *Land Evaluation.* Melbourne: Macmillan of Australia, 64-76.

Blackmon, C. A., Stinson, B. G.; and Stoll, J. K., 1968-1972. *An Analytical Model for Predicting Cross-Country Vehicle Performance,* U.S. Army Corps Engineers Waterways Expt. Sta. Tech. Rept. 3-783, 152p. plus appendices A-F.

Brink, A. B. A.; Partridge, T. C.; and Webster, R., 1968. Land classification and data storage for the engineering use of natural materials, *Fourth Conference Australian Road Research Board Proc., Paper 512T,* Melbourne.

Carlson, C. A., 1971. *Trafficability Prediction in Tropical Soils: Hawaii,* U. S. Army Corps Engineers Waterways Expt. Sta. Misc. Rept. 4-355:7, 48p.

Cummings, N., and White, L., 1981. *Mobility Bibliography.* Hanover: U.S. Army Corps of Engineers, Cold Regions Research and Engineering Laboratory (CREL), 313p.

Grabau, W. E., 1968. An integrated system for exploiting quantitative terrain data for military engineering purposes, in G. A. Stewart (ed.), *Land Evaluation.* Melbourne: Macmillan of Australia, 211-220.

Granberg, H. B., 1974. *Terrain analysis for winter conditions in central Quebec-Labrador,* Rapport SFM årsmöte och Konferens, med. nr 15, Samarbets organisationen för fordon-markforskning, (Swedish Society for Terrain-Vehicle Research), 111-129.

Grant, K., 1968. *A Terrain Evaluation System for Engineering,* CSIRO Div. Soil Mechanics Tech. Paper No. 2, 27p.

Grant, K., 1973. *The PUCE Program for Terrain Evaluation for Engineering Purposes. I: Principles,* CSIRO Div. Applied Geomechanics Tech. Paper 15, 32p.

Grant, K., 1974. *The PUCE Program for Terrain Evaluation for Engineering Purposes. II: Procedures for Terrain Classification,* CSIRO Div. Applied Geomechanics Tech. Paper 19, 68p.

Johnson, D. W., 1921. *Battlefields of the World War,* Am. Geog. Soc. Res. Ser. No. 3, 187p.

Kolb, C. R., et al., 1958-1965. Analogs of Yuma terrain in the desert regions of the world, U.S. Army Corps Engineers Waterways Expt. Sta. Tech. Rept. 3-630, Vol. 1-6.

Lennon, W. W., 1969. *Military Requirements for Terrain Evaluation,* Engineer in Chief, H. Q. Australian Army, Rept. 17506/69, 32p.

McDaniel, A. R., 1966. *Trafficability Predictions in Tropical Soils: Panama,* U.S. Army Corps Engineers Waterways Expt. Sta. Misc. Rept. 4-355:3, Vols. 1-3, 30p.

MEXE (Military Engineering Experimental Establishment) Staff, 1965. *The Classification of Terrain Intelligence,* Rept. 915. Oxford, England, 90p.

Mitchell, C. E., 1973. *Terrain Evaluation.* London: Longmans, 221p.

Parry, J. T., and Beswick, J. A., 1973. The application of two morphometric terrain classification systems using air photo interpretation methods, *Photogrammetria* **29**(5), 153-186.

Parry, J. T.; Heginbottom, J. A.; and Cowan, W. R., 1967. *Terrain Evaluation,* Canadian Forces Base Petawawa: Canada Defense Research Establishment Contract Rept. 2VC 5-I-69-500007, 8p.

Parry, J. T.; Heginbottom, J. A.; and Cowan, W. R., 1968a. Terrain analysis in mobility studies for military vehicles in G. A. Stewart (ed.). *Land Evaluation.* Melbourne: Macmillan of Australia, 160-170.

Parry, J. T.; Heginbottom, J. A.; and Cowan, W. R., 1968b. *Terrain Analysis,* Canadian Forces Base, Gagetown, New Brunswick: Canada Defense Research Establishment, Contract Rept. GR 700005, 18p.

Parry, J. T.; Hutchinson, I.; Dolman, A. N.; and Dredge, L., 1971. *Terrain Evaluation and Military Operations,* Canadian Forces Base, Gagetown, New Brunswick: Canada Defense Research Establishment, Contract Rept. 02GR 7090020, 10p.

Parry, J. T.; Howland, W. G.; Granberg, H. B.; Wilson, P.; and Maclean, P. A., 1974. *Terrain Evaluation—Lac Saffray,* Quebec-Labrador, Canada: Defense Research Establishment, Contract Rept. SP2 7099153, 6p.

Parry, J. T.; Zonneveld, J. M.; Howland, W. G.; Granberg, H. B.; and Barber, E., 1979. Terrain Evaluation, MacKenzie Delta, Inuvik Aklavik: Canada Defense Research Establishment, Contract Rept. 2SU77 00143, 18p.

Parry, J. T.; Howland, W. G.; Granberg, H. B.; Wilson, P.; and Maclean, P. A., 1981. *Terrain Evaluation—Churchill Transect,* Manitoba, Canada: Defense Research Establishment, Contract Rept. 8SU78-00204, 14p.

Parry, J. T.; Howland, W. G.; Granberg, H. B.; Maclean, P. A.; and Houston, L. C., 1983. *Mobility Model Developments—Terrain Characteristics,* Canada Defense Research Establishment, Contract Rept. 8SU81-00094, 120p.

Perrin, R. M. S., and Mitchell, C. W., 1969-1971. *An Appraisal of Physiographic Units for Predicting Site Conditions in Arid Areas,* Vols. 1 and 2, Military Eng. Expt. Establishment (MEXE) Rept. IIII, 313p. and 520p.

Rula, A. A.; Grabau, W. E.; and Miles, R. D., 1963a. *Forecasting Trafficability of Soils: Air Photo Approach,* Vols. 1 and 2, U.S. Army Corps Engineers Waterways Expt. Sta. Tech. Memo. 3-331:6, 218p. and 120p.

Rula, A. A., et al., 1963b. *Environmental Factors Affecting Ground Mobility in Thailand,* U.S. Army Corps Engineers Waterways Expt. Sta. Tech. Rept. 5-625, 66p. plus appendices A-H.

Schreiner, B. G., and Rula, A. A., 1962. *Operation Swamp Fox—Terrain and Soil Trafficability Observations,* U.S. Army Corps Engineers Waterways Expt. Sta. Tech. Rept. 3-609, 33p.

Shamburger, J. H., and Grabau, W. E., 1968. *Mobility Environmental Research Study: A Quantitative Method for Describing Terrain for Ground Mobility,* Vols. 1-3, U.S. Army Corps Engineers Waterways Expt. Sta. Tech. Rept. 3-726, 46p., 125p., and 207p.

Smith, M. H., 1971. *Trafficability Prediction in Tropical Soils: Costa Rica,* U.S. Army Corps Engineers Waterways Expt. Sta. Misc. Rept. 4-355:8, 37p.

U.S. Army Corps of Engineers, 1963. *Military Evaluation of Geographic Areas: Reports on Activities to 1963,* U.S. Army Corps of Engineers Waterways Expt. Sta. Misc. Paper No. 3-610, 237p.

U.S. Army Corps of Engineers, 1963-1968. *Environmental Data Collection Methods—Surface Composition, Hydrologic Geometry, Surface Geometry, Vegetation, Weather,* U.S. Army Corps Engineers Waterways Expt. Sta., Instruction Rept. Series, Vols. 1-12, 860p.

U.S. Department of the Army, 1959. *Terrain Intelligence,* Field Manual FM 30-10. Washington, D.C., 262p.

Van Lopik, J. R., and Kolb, C. R., 1959. *A Technique for Preparing Desert Terrain Analogs,* U.S. Army Corps Engineers Waterways Expt. Sta. Tech. Rept. 3-506, 90p.

Webster, R., 1965. *Minor Statistical Studies on Terrain Evaluation,* Military Eng. Expt Establishment (MEXE) Rept. 877, Christchurch, England, 61p.

Woloshin, A. J., 1968. *Comparison and Evaluation of Terrain Classification Methods,* U.S. Army Corps Engineers Topographic Labs. Contract Tech. Rept. DAAK 02-67-C0487, 27p.

Wood, W. F., and Snell, J. B., 1959. *Predictive Methods in Topographic Analysis—Relief, Slope, and Dissection on One Inch To Mile Maps in the U.S.A.,* Quartermaster Res. and Engineering Command, Environmental Protection Res. Div., Tech. Rept. EP-112, 15p.

Wood, W. F., and Snell, J. B., 1960. *A Quantitative System for Classifying Landforms,* Quartermaster Res. and Engineering Command, Environmental Protection Res. Div., Tech. Rept. EP-124, 20p.

Cross-references: *Geomorphology, Applied.* Vol. XIV: *Environmental Geology; Land Capability Analysis; Photogeology; Photo Interpretation; Terrain Evaluation Systems.*

THERMAL ANALYSIS—See Vol. IVA.

THERMAL SURVEYS—See REMOTE SENSING, ENGINEERING GEOLOGY. Vol. XIV: AERIAL SURVEYS, GENERAL; PHOTOGEOLOGY.

THERMOLUMINESCENCE—See Vol. IVB.

TOPOGRAPHIC MAPPING AND SURVEYING—See Vol. XIV.

TUNNELS, TUNNELING

A tunnel is an essentially horizontal, artificial underground opening, with a generally regular cross-section and a length that greatly exceeds its other dimensions. Tunnels are used for a wide variety of purposes. They provide essential links in many highways, railroads, and urban rapid transit systems. Urban water supply and distribution, sewage collection and disposal, hydroelectric power generation, flood control, and mining require extensive tunneling. Tunnels also have been used throughout history for both offensive and defensive military applications (Sandstrom, 1963).

The size and shape of a tunnel cross-section largely depend on the purpose of the tunnel, but to some extent are dictated by the manner in which it is driven and the geological nature of the tunnel site. Water and sewer tunnels are generally circular, rail tunnels are commonly horseshoe-shaped, and highway tunnels usually have vertical walls and an arched roof (Howard, 1967).

Construction engineers usually classify tunnels in two broad categories, depending on the character of the earth material in which they are constructed (Mayo et al., 1968). Soft-ground tunnels are driven in material that excavates easily, without drilling and blasting, such as soil, clay, hardpan gravels, and some of the softer, more incompetent rocks such as weak, poorly cemented sandstones and shales. Generally, the walls in soft-ground tunnels will not stand unsupported, and thus tightly spaced temporary shoring or primary lining must be installed concurrently with excavation and maintained close to the advancing tunnel face (see *Reinforced Earth*).

Rock tunnels are driven in harder, more competent formations where breaking the rock, usually by drilling and blasting, is the key element of the tunneling process (see *Rapid Excavation and Tunneling*). The tunnel periphery may or may not be self-supporting. Where temporary support is required, it is usually provided by rockbolts, timber sets, or steel ribs (Szechy, 1966; Helfrick et al., 1970).

Tunnel Site Investigations

Because the selection of optimum tunnel location and the choice of tunneling method depend on the characteristics and properties of the rock mass to be penetrated, thorough geological investigation of the tunnel route is of paramount importance. Geological factors that affect tunnel construction include depth and character of overburden; bedrock surface configuration; rock properties; fabric and mineralogy; structural features of the rock mass; and occurrence of ground water, abnormal rock temperatures, and gas (see *Pipeline Corridor Evaluation*).

Virtually all geological and geophysical exploration techniques are applicable to route exploration. A site investigation usually begins with study of aerial photographs and geologic maps (see Vol. XIV: *Photogeology*). The geology of the entire tunnel line is then mapped to the extent necessary to project important structural features to tunnel grade. Surface mapping (see *Field Geology*) may be supplemented by core drilling (see Vol. XIV: *Borehole Drilling*); seismic (see *Seismological Methods*), magnetometer, and sometimes gravity surveys (see Vol. XIV: *Exploration Geophysics*); and velocity and gamma-ray logging (see Vol. XIV: *Well Logging*) of boreholes (Helfrick et al., 1970). Occasionally, where the geology is particularly complex and construction is expected to be difficult, small cross-section pilot tunnels are driven to enable a detailed study of the geology of the tunnel line at tunnel grade.

The geological features critical to tunnel construction are those that determine the design and stability of the tunnel during or after construction, and those that affect the performance of equipment and methods used to drive it. It is particularly important, therefore, that the location, attitude, thickness, and character of major discontinuities such as faults and shear zones, breccia zones, contacts, and alteration zones be known in advance of penetration. Density and attitude of jointing, attitude and character of bedding planes, and degree and attitude of schistosity and foliation will influence support requirements, as will the presence of swelling clays (Szechy, 1966). Unexpected encounters with water-bearing strata have caused major difficulties in many tunnels.

Pre-construction site investigations do not always provide adequate structural information, so continuing geological study is frequently conducted during construction to check and improve the geological model of the tunnel line and guide day-to-day tunneling progress. During construction, geological investigation is usually limited to detailed mapping as the tunnel face is advanced, supplemented by occasional probing ahead of the face with short drill holes or pilot tunnels (see *Dams, Engineering Geology*).

Conventional Tunneling Methods

The conventional method of driving tunnels in rock is by a cyclical process consisting basically of *drilling, blasting* and *mucking,* and, where necessary, installing temporary support (Szechy, 1966). The choice of equipment and method is largely a function of the properties and character of the rock mass and the length and diameter of the tunnel. Tunnels may be driven full-face, in which case the entire face of the tunnel is drilled and blasted at one time. In large tunnels a top heading may be driven and then later enlarged to full size. The latter method is sometimes chosen to allow more efficient drilling and blasting procedures or it may be dictated by the instability of the ground, which requires that the tunnel crown be supported before the major portion of the face is blasted.

Drilling is commonly done with percussion drills, powered by compressed air. They may be large, heavy-duty drills mounted on booms, or light air--leg-mounted drills easily handled by one person (Fig. 1). Drill rods may be carbon or alloy steel with either integral or detachable bits, which are tipped with tungsten carbide.

Drill jumbos are used in all but the smallest tunnels. These rigs may be large, mobile work platforms or simple, movable carriages from which a number of drills can be operated simultaneously and which can be moved away from the face before blasting. For large, long tunnels jumbos are usually constructed to straddle or project over the muck pile so that drilling, mucking, and support installation may be accomplished at least partially

simultaneously (see *Rapid Excavation and Tunneling*).

Several more or less standard patterns are used for placing drill holes in the tunnel face. Most have the following features in common, with minor variations. *Rim holes* are drilled around the periphery of the designed opening to define the shape; *cut holes* are drilled at an angle toward the center to form a cone or wedge, and *relief holes* are drilled in the space between the rim and cut holes. The so-called burn cut is becoming increasingly popular, particularly for the smaller tunnel cross-sections. In this variation, the cut comprises one or more oversize holes drilled normal to the face, which are not loaded, closely surrounded by a number of parallel blast holes. In either case the cut holes are detonated first to form a free face to which the relief holes break. Rim holes are detonated last to trim the opening to the desired shape.

A *drill round* consists of all the holes that are drilled and blasted as a unit. Tunnel rounds are usually detonated electrically using either standard or millisecond-delay caps to obtain the correct sequence of detonation. Explosives used include both high-density ammonia dynamites and blasting agents such as ammonium nitrate-fuel oil mixtures (see Vol. XIV: *Blasting and Related Technology*).

A large proportion of tunneling is done in soil, sand, clay, and similar materials that are easily broken loose with hand tools such as air spades, or that may even be so unconsolidated that no breaking is necessary. The primary problem in such tunnels is stabilizing the periphery of the opening. Over the years, a wide variety of methods have been developed to cope with ground conditions ranging from stiff to fluid. Despite the range in detail, two guiding principles dominate the methods: to expose only small sections of ground at one time, and to support the excavated sections as quickly as possible (Abbott, 1956).

In some small tunnels, *forepoles* or *spilings* are forced ahead of the opening so that the face of the tunnel can be excavated under cover. In completely unconsolidated material, *lagging* is used to hold back the sides of the tunnel and *breast boards* to support the face. The breast boards are removed from the top down as muck is removed from the tunnel face.

In modern-day soft-ground tunneling, *shield* methods have largely supplanted such laborious techniques (Mayo et al., 1968). Basically, a shield is an open cylinder of steel plate that serves to support the tunnel bore immediately behind the face during excavation and installation of the primary lining. The primary lining may be sectional steel plate, precast concrete liner segments, or circular H-beams spaced at short intervals and tightly lagged with wood planking. Shield tunneling is a cyclic operation. After a ring of primary

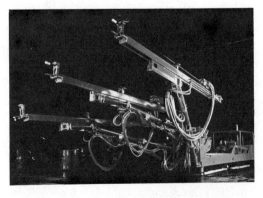

FIGURE 1. Drill jumbo for small tunnel (courtesy of Joy Manufacturing Company).

lining is erected within the tail of the shield, hydraulic jacks, anchored against the ring, begin to push the shield forward. Within the front of shield, excavation is carried out with hand tools or by mechanical methods. Bars, or plates, which form a number of working compartments in the face of the shield, perform the same function as breast boards in holding back unconsolidated ground. When the shield has been advanced sufficiently, another ring of lining is installed and the cycle repeated.

Not infrequently, tunnels must be located in water-saturated soil below the water table, under circumstances that prevent diverting the water by pumping. To penetrate such highly fluid material without inundation, air pressure equal to the hydrostatic head of the water column must be maintained within the exposed portion of the excavation in the same manner as in a diving bell. Air pressure at the face end of the tunnel is maintained by means of airlocks consisting of pairs of *airtight bulkheads* a short distance behind the face. In underwater passages, such as river crossings, a clay blanket is often placed above the tunnel to prevent "blow-outs" and losses of air pressure.

Occasionally short reaches of unstable ground such as quicksand or water-saturated muck can be penetrated by first solidifying the section by freezing, by injecting chemicals to react with the soil, or by injecting quick-setting grouts (q.v.). Tunneling then proceeds in the normal manner.

Muck Handling

Removal of broken rock or muck from a tunnel heading is usually accomplished in two steps: (1) moving it from the face, and, (2) transporting it to the point of disposal. In small rock tunnels, and some of the larger shield-driven tunnels, *hand mucking,* small overshot mechanical shovels, and scrapers are used to load small mine cars directly. In larger tunnels, rail-, crawler-, or wheel-mounted power shovels load large rail cars or trucks directly or occasionally onto a conveyor belt.

If access to the tunnel is by *portal,* the loaded trains or trucks usually travel directly to the disposal area on the surface. Tunnels serviced by *shaft* require facilities for trans-shipping muck or for hoisting muck cars. Locomotives for rail haulage may be either electric or diesel-powered. Diesel-electric locomotives and diesel-powered trucks are common in tunnels large enough to permit their use.

Tunnel Support and Lining

Where required, tunnels are supported and lined normally in two, and sometimes in three, stages. The purpose of the first, or primary, stage is to stabilize the opening and to prevent the possibility of excessive inflow of water during construction.

The primary support or lining may consist of various combinations of wood, steel, concrete, and masonry. Older, hand-driven tunnels tend toward wood and masonry lining, often meticulously constructed. Shield-driven tunnels require structurally sound linings to withstand the thrust of jacks acting against them to shove the shield forward. Such tunnels have a regular cross-section that permits the use of preformed liner sections that can be placed mechanically and bolted or welded in place. Cast iron sections, pressed steel plates, and precast concrete segments are all commonly used. Cathodic protection is often required to prevent corrosion of metal linings. The secondary, or permanent, lining—usually cast-in-place concrete that may or may not be reinforced—is installed to assure long-term structural stability and (where necessary) watertightness. For many tunnel applications, a third stage of lining may be required. For example, the final lining of modern highway tunnels is ceramic tile for improved visibility and ease of maintenance. In some water tunnels, a steel liner is installed to obtain optimum hydraulic flow characteristics or to provide the additional tensile strength necessary to withstand internal hydraulic pressures, as in penstocks and siphons. In the latter cases, final design of the total lining (primary, secondary, and tertiary) may be based on elaborate *in situ* tests to determine the mechanical moduli of the rock mass. The combined lining can then be designed to withstand that increment of internal load beyond the capacity of the rock mass (Proctor and White, 1946; Mayo et al., 1968).

Logistics of Tunnel Construction

From an operational standpoint, the limiting characteristic of tunnel construction is lack of working space. Commonly, only one working face is accessible, so that all supplies, equipment, and manpower are concentrated in a very limited area. To obtain additional work sites, shafts are often sunk at intervals along the route and new working faces opened up.

Another method is the construction of a *pioneer tunnel* parallel to the final tunnel route but a short distance to one side (20 m or more). *Cross cuts* are driven at intervals to the main tunnel line and new headings started. All service to the main tunnel faces is through the pioneer tunnel. This method, of course, has the further advantage of providing access, in advance of driving the final bore, for detailed geological study of the tunnel line at tunnel grade.

Machine Tunneling

Tunneling with mechanical boring machines, or "moles," must be considered a relatively recent development (Juergens, 1966), even though a tunneling machine was successfully used in 1881 to drive a short section of tunnel in chalk during one of the early attempts to drive a tunnel under the English Channel.

Recent tunneling machine development is a logical outgrowth of the use of cylindrical shields. In fact, most of the soft-ground tunneling machines consist of shields within which various configurations of excavating machinery have been installed. In several versions drag bits or scrapers are mounted on the spokes of a large wheel, which is rotated against the tunnel face. Muck flaps on the periphery of the wheel pick up and elevate the excavated material and discharge it onto a belt conveyor, which in turn loads into a car or another belt. Variations include multiple wheels and oscillating, rather than rotating, spokes. Another concept consists of a backhoe type of excavator. Installation of primary lining and muck handling for machine tunneling in soft ground are essentially the same as for conventional shield tunneling.

The present generation of rock tunnel borers bear a resemblance to soft-ground machines but differ substantially in significant details. The rock tunneling machines are much heavier because more power and greater thrust is required (Pattison and D'Appolonia, 1974). Anchoring to provide a base for such high thrust is obtained by gripper pads hydraulically jacked against the walls of the tunnel. The face, or cutting head, of the machine is solid except for muck and access ports. Instead of drag bits the cutting elements are disk or roller bits of the oil-well drill type. Rocks having compressive strengths as great as 2,000 kg/cm^2 have been successfully machine-tunneled.

Muck handling is essentially the same as for soft-ground machines. Ground support differs in that only a short shield, or no shield at all, is used and primary support, when necessary, can be installed close to the face. Support materials and techniques are the same as for conventional rock tunneling.

Machine tunneling is rapidly becoming widely accepted and the performance of the machines is constantly improving. Advance rates of more than 7 m/h have been attained in softer rocks under favorable conditions. These high rates of advance and the inherent lack of flexibility in machine-tunneling systems, as compared with conventional methods, make even more critical the requirement for a thorough knowledge of the geology of the tunnel route, in advance of penetration.

THOMAS E. HOWARD
JOHN. R. McWILLIAMS

References

Abbott, R. W., 1956. *American Civil Engineering Practice.* New York: Wiley.

Helfrick, H. K.; Hasselstrom, B.; and Sjogren, B., 1970. Geoscience site investigations for tunnels, the technology and potential of tunnelling, *Tuncon Proc.* **10**(1), 17-23.

Howard, Thomas E., 1967. Rapid excavation, *Sci. American* pp. 81-82.

Juergens, R., 1966. New developments in tunneling machines, *Constr. Methods and Equip.* **48**(3 and 4), 30-144 and 126-145.

Mayo, R. S.; Adair, T.; Jenny, R. J., 1968. *Tunneling, State of the Art,* U.S. Dept. Housing and Urban Devel. Doc. No. PB 176 036.

Pattison, H. C., and D'Appolonia, E., (eds.), 1974. *Rapid Excavation and Tunneling Conference Proceedings,* 2 vols. New York: American Institute of Mining Engineers, 1,842p.

Proctor, R. V., and White, T. L., 1946. *Rock Tunneling with Steel Supports.* Chicago: Commercial Shearing and Stamping Company.

Sandstrom, G., 1963. *Tunnels.* New York: Holt, Rinehart and Winston.

Szechy, K., 1966. *The Art of Tunneling.* Budapest: Academiai Kiado.

Cross-references: *Foundation Engineering; Grout, Grouting; Maps, Engineering Purposes; Pipeline Corridor Evaluation; Rapid Excavation and Tunneling; Shaft Sinking; Urban Tunnels and Subways.* Vol. XIV: *Blasting and Related Technology; Borehole Drilling; Exploration Geophysics; Mining Preplanning.*

U

UNDERGROUND CONSTRUCTION—
See CAVITY UTILIZATION: RAPID EXCAVATION AND TUNNELING; TUNNELS, TUNNELING; URBAN TUNNELS AND SUBWAYS.

UNDERGROUND ENGINEERING—
See RAPID EXCAVATION AND TUNNELING; ROCK STRUCTURE MONITORING; SHAFT SINKING; TUNNELS, TUNNELING; URBAN TUNNELS AND SUBWAYS.

UNDERGROUND SPACE—See CAVITY UTILIZATION.

UNDERGROUND STORAGE—See CAVITY UTILIZATION.

UNDERSEA TRANSMISSION LINES, ENGINEERING GEOLOGY

In siting offshore nuclear power plants (see *Nuclear Plant Siting, Offshore*), the need for an extra-high-voltage system to convey generated power ashore adds a new dimension to the state of the art of undersea transmission. The constraints of the ocean environment, coupled with a host of new technical and environmental considerations, pose problems very different from those experienced with transmission systems onshore.

The problems related to routing cables through the highly active hydraulic regime characteristics of many nearshore, estuarine, and inlet areas are typical of those that must be faced by the engineering geologist involved in such offshore works. For example, a historical data review undertaken to estimate the stability of a break in the barrier island system adjacent to the proposed site of a nuclear plant off the coast of New Jersey showed that the inlet had migrated extensively over a 132-year period. Rates of island growth of several thousand feet over five-year periods were not unusual. Scour depths in excess of 8 m were noted

over distances of greater than 700 m, with certain areas scoured to depths of up to 12 m below the present sea bed. There is no reason to suppose that these trends are highly unusual or that, in this area, these trends would not continue into the future, thus placing in jeopardy any system not designed to anticipate such changes.

Other technical considerations relate to the extreme stratigraphic variability of nearshore marine sediments, and to the fact that soils with a high thermal resistance will not permit as great a power load to be carried as soils of low resistance. The failure to design for unfavorable materials over even a short portion of the alignment may, during operation, initiate a local "hot spot" and cause a burnout of the cable.

In the discussion that follows, consideration is given to resolving environmental constraints related to routing cables through sensitive coastal fringe areas, evaluating the influence on buried cables of the thermal properties of soils surrounding the cables, and predicting the influence on the cable integrity of active coastal processes. In addition, suggestions are given for efficiently collecting meaningful field data. Also, since no American Society for Testing and Materials (ASTM) standard exists for determining the thermal properties of soils, laboratory equipment, and procedures adapted and developed for evaluating the thermal properties of soils, laboratory results and recent progress in studying the influence of heat on the thermal properties of soils by means of an analytical model are presented.

Environmental Alignment Considerations

To an engineering geologist, the ocean is an attractive location for a nuclear power plant in terms of minimized ecological effect: it offers unlimited cooling water, optimum land use, minimal environmental disruption from construction, minimal visual impact, and partial isolation from populated communities. The transmission line to shore still poses a major problem, however. The Eastern Seaboard and Gulf coastline of the United States, for example, consist almost exclusively of recreational beach or salt marsh wetlands. Such areas are extremely environmentally sensitive.

The salt marsh is a valuable nursery for aquatic life and a food producer for aquatic systems, and it serves an important function as a filter of biodegradable pollutants. It is a unique habitat, a buffer against storms, and an area of relatively unspoiled beauty on a coastline that is being rapidly developed. The wetlands are also used extensively as a recreational area. For these reasons, a number of states have enacted laws to protect the wetlands.

Marshland vegetation consists typically of smooth cord grass (*Spartina alterniflora*), which is mixed with glasswort (*Salicornia europaea*) and sea lavender (*Limonium* sp.) in some locations. *Spartina patens* and *Juncus gerardi* form pure stands in higher areas above normal high tide. If disturbed by cable installation, *Spartina alterniflora* would take a considerable time to revegetate unless elaborate restoration techniques were used.

Selection of a suitable location for the environmental alignment should include projected ecological disturbance, disruption of amenities, and the visual impact that may be imposed on local communities during investigations or installation of transmission cables. Technically viable locations, away from recreational beach areas, would offer enormous advantages in terms of environmental feasibility. The beach and the land immediately adjacent are the most critical areas. Therefore, these areas should be excluded from consideration in the selection of a switching yard. It appears that the presence of a river or inlet, which would facilitate an all-water alignment to a location inland of the critical foreshore area, would be a significant advantage.

The importance of selecting alternative routes where practical, to facilitate maximum flexibility in final selection, cannot be overemphasized. An access road on a salt marsh, for example, may offer a possible cable routing. During a single winter period, when minimal use is being made of such a secondary road, the cables can be buried beneath the road with little or no environmental disruption. Additional disruption to marine ecology may be avoided by selecting navigational channels that undergo routine dredging. Existing "clamming beds" should be identified and avoided. Shellfish or mollusks often provide a livelihood for local inhabitants and may be privately owned.

Technical Alignment Considerations

Thermal Properties of Soils. The transmission of power by means of underground cables is influenced to a very large extent by the thermal characteristics of the soils that surround these cables. A high thermal resistance will not dissipate heat rapidly away from cables, and therefore will not permit as high a power load to be carried as a low thermal resistance. The problem is compounded by the complexity of nearshore marine and estuarine sediments (Watson et al., 1974), which tend to show extreme variation in composition (Fig. 1) with corresponding variations in thermal properties.

Under these conditions, the transition from satisfactory to unsatisfactory design conditions is abrupt. Neglect of even a small lens of high-resistivity soil over only a short section of the alignment may initiate a local "hot spot" and eventually cause the cable to burn out.

In planning the alignment and providing input for the design of cables, the engineering geologist must be thoroughly familiar with the factors influencing the thermal resistivity of soils (see Vol. XII, Pt.1: *Thermal Regimes*). Some of the more important considerations include the type of soil, density, moisture content, and degree of saturation, as well as depth of burial and the operating temperature of the cables.

When one considers that the resistivity (in thermal ohms)* of quartz is 11, water 165, and air 4,000, the need for examining each of the three phases (solid material, water, and air) and their interrelationship is evident. Increasing the amount of solid material per unit volume is desirable in certain cases, but only to the point where the increased density maintains a permeability sufficient to allow for restoration of moisture, should moisture migration occur after the cable is installed.

Figure 2 shows the influence of moisture content and dry density on thermal resistivity (Salomone et al., 1974). This figure shows that granular materials generally have lower resistivity values than cohesive soils, and that soils with higher dry density values have lower resistivities than less dense soils. Granular soils showing less than 10 percent moisture content may show significant resistivity values (i.e., greater than 200 thermal ohms).

Figure 3 shows that the greatest decrease in resistivity is found from dryness to 10 percent of the volume of voids saturated and that there is little change above 30 percent (Fischer et al., 1975).

Nearshore marine and estuarine soils may consist typically of an interbedded sequence of (a) clean granular sediments (poorly graded quartzitic sands), and (b) more cohesive materials (soft to stiff clays, with variable sand, silt, and organic content; soft clayey silt; silt; and soft organic materials). The thermal properties and behavior of these two categories of soils vary widely (Table 1). Clean quartzitic sands have favorable thermal properties; quartz itself has a low thermal resistance. Sands also have a relatively high permeability, maintain saturated conditions, and tend not to dry out.

More cohesive materials show higher thermal resistivity values, but their tendency to dry (shrink and crack) over a period of time increases the thermal resistivity in the critical area surrounding the cable. Cohesive materials and any organic materi-

---

*The thermal ohm is the unit of resistivity. It is defined as the number of degrees centigrade of temperature drop that occurs when heat flows through a 1-cm cube at the rate of 1 W. The symbol used is ρ.

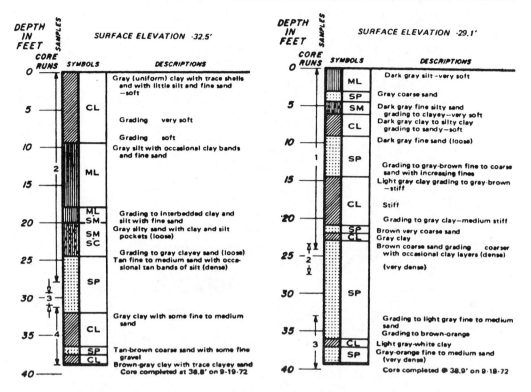

FIGURE 1. Vibracore logs showing stratigraphic complexity (after Watson et al., 1974).

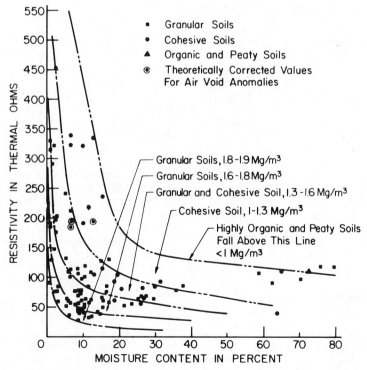

FIGURE 2. Thermal resistivity test data showing the effect of moisture
content and dry density (after Salomone et al., 1974).

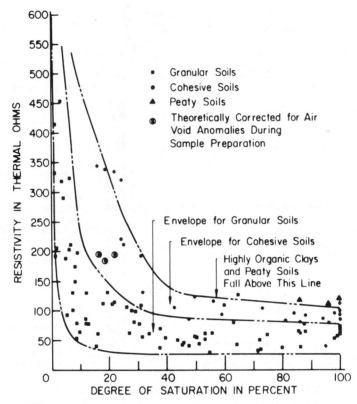

FIGURE 3. Thermal resistivity test data showing the effect of degree of saturation (after Fischer et al., 1975).

TABLE 1. Approximate Resistivity Ranges for Different Soil Types.

Soil Description	Unified Classification Symbol	Range of Thermal Resistivity, °C ·cm/W
Silty clay	CL	85-105
Silty clay with organic matter	CH/OH	120-140
Clayey silt	ML	85-105
Silt	ML	90-110
Sandy clay	CL	85-95
Sandy silt	ML	85-95
Clean uniform sand	SP	60-80
Fine to coarse sand	SP	75-95
Silty sand	SM	70-90
Silty sand and gravel	SW/SM	65-85
Clayey sand	SC	80-90
Interbedded sand and clay	SP/CL	85-95

Source: After Salomone et al. (1974).

als should therefore be avoided wherever practical.

Figure 4 shows the influence of temperature on resistivity, and Figure 5 the steady-state temperature distribution around a buried cable placed in clay. Figure 4 was derived from 250 determina-

tions of the effect of temperature on soils, together with relationships developed between thermal resistivity, moisture content, and dry density. Results suggest that it may be practically feasible to design cables for placement in clay soils; however, any design decision to place cables in clay should be based on calculations in which the rate of capillary water supply at a certain locus is compared to the rate of water removal due to thermo-osmosis.

Scour of Covering Materials. The operational integrity of the transmission system is ensured by burying cables to a depth below the maximum projected depth of scour during the life of the project. Cables uncovered by scour action would be subject to damage by strong currents or physical damage resulting from boat anchors.

The activity of the hydraulic regime and the degree of mobility and scour associated with the barrier island systems that form a substantial portion of the eastern coastline of the United States must be taken into account in designing offshore structures (see *Coastal Engineering; Coastal Inlets, Engineering Geology*). Figure 6, showing the locations of coastal landforms in 1940 superimposed on a 1972 shoreline, provides an example of the mobility of such environments. Interpretation of airphotos indicated that the once-inhabited Tuck-

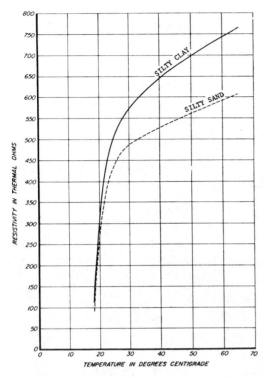

FIGURE 4. Influence of temperature on thermal resistivity (after Watson et al., 1974).

ers Island (shown in Figure 6), some 1,350 m long in 1940, had almost completely disappeared by 1951. This island was about 7,000 m long in 1920.

Therefore, installation planning should make provision for (a) routing of cables away from the most active areas, (b) burial below the historically known scour limit, or (c) stabilizing the inlet system.

Data Collection

Databank and Thermal Considerations. Sufficient detail in offshore areas is achieved by means of a coordinated approach using vibratory corings, supplemented where practical (in water depths exceeding 7 m) by high-resolution, continuous seismic profiling. The data attained by these methods require some engineering quantification obtained from undisturbed samples taken (to evaluate foundation conditions) at the proposed site, from the extension of cable route investigations on land, and from a limited program of conventional borings and undisturbed samples (see *Foundation Engineering*). A historical review of coastal stability is undertaken simultaneously with preliminary engineering investigations, and a detailed program of data collection on coastal processes initiated following preliminary approval of the engineering feasibility of the site.

A comprehensive program of coastal process study should include the following phases.

Historical Review. This review should outline coastline and bathymetric changes over the maximum period of historical record keeping. It should incorporate the maximum-use airphoto records, old maps and charts, and bathymetric data. Existing meteorological data should also be collected during this phase.

Site Data Collection. A comprehensive program of instrumentation and observation should be planned to provide sufficient input to establish the hydrodynamics of the area by subsequent numerical modeling.

Interpretation of Data. Field data should be interpreted in light of the initial historical review, engineering data, and predictive sediment transport studies (Seymour, 1983). A qualitative evaluation should be made of the limits of scour potential sediments and anticipated sediment transport trends. Flume tests should be undertaken to verify the classification of scour-resistant strata.

Analytical Modeling. Based on an adequate representative input of field data, an analytical finite element model may assist in quantitatively predicting sediment transport trends (see Vol. XIV: *Alluvial Systems Modeling*). Other approaches, where applicable, may be used to provide additional data on quantitative trends (e.g., McDougal et al., 1983; Bell et al., 1983).

Scour Prediction. The final phase of scour prediction is undertaken by interpreting the analytical model in the light of finalized engineering geological soils data and the findings of the first three study phases.

Final recommendations must include remedial plans for major structures, should hydraulic behavior not proceed exactly as predicted.

Field Investigations

Geophysics. Continuous seismic reflection profiles (see Vol. XIV: *Acoustic Surveys, Marine*) should be performed over proposed alignments prior to the vibratory coring program. A 3.5-kHz, high-resolution profiling unit would be suitable in most instances. Preliminary interpretation of collected data will assist in planning an economical sampling program. During coring operations, correlation between the stratigraphic horizons intersected by the borings and seismic reflectors found on continuous seismic profiles may be improved by obtaining a seismic profile at the location of each boring. Seismic equipment should be available during the sampling program to resolve apparent anomalies between borings. Seismic profiles are invaluable in the preparation of subsurface sections. Seismic data are most useful in areas where the water depth exceeds 8 m. Where seismic data are available, borings spaced at about 700-m intervals provide suitable correlation, but in shallow estuarine water, where stratigraphy is com-

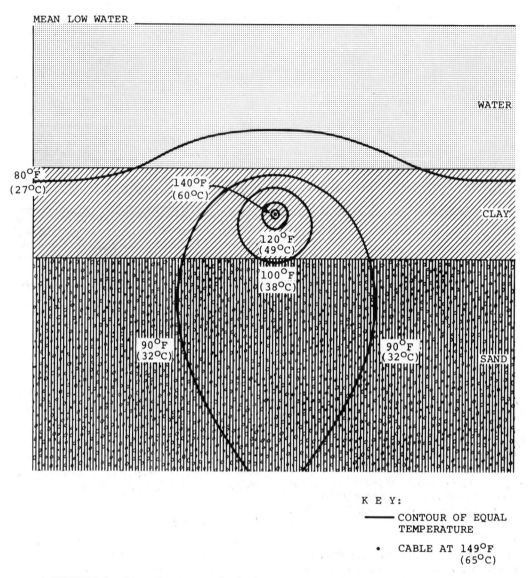

FIGURE 5. Steady-state temperature distribution around a buried cable (after Watson et al., 1974).

plicated and seismic resolution poor, spacing has to be reduced to 350 m or less.

Borings. On investigations for some extended offshore projects (such as for transmission lines) where sample disturbance may be of only secondary importance, the main data may best be provided by vibratory-cored borings (see Vol. XIV: *Borehole Drilling*). Theoretically, the core recovered represents a continuous sample. It facilitates positive identification of sediments and provides the most favorable sampling method for the construction of geological cross-sections. Sophisticated vibratory coring units are available that can penetrate to a depth of about 15 m in unconsolidated sediments. A core liner (internal diameter about 9 cm), is necessary for furnishing the comparatively large volumes (14 kg) of granular materials required for performing laboratory thermal probe tests at varying densities and moisture contents.

The technique has other advantages. It provides an extremely rapid means of sampling and it may be operated from a floating unit, enabling it to be used in a wide range of water depths (2 to 15 m) encountered along the offshore extension of transmission alignments. Thus, a consistency of sampling may be achieved for all offshore work. The principal disadvantage of the system is that the vibrating motion and long drive cause the sample to be disturbed to some extent. Dense sands or coarse granular materials and clays occasionally block

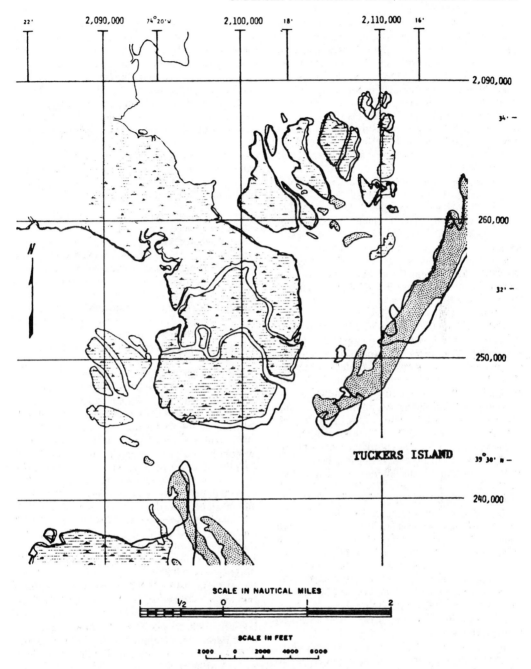

FIGURE 6. Location of coastal landforms, 1940, superimposed on 1972 shoreline (after Watson et al., 1974).

the sampler, thus impeding advance. Therefore, stringent geotechnical control is required during the coring operation to obtain satisfactory results.

Efficient core logging is provided by *penetration rates*. The number of seconds per meter of advance should be recorded during borings and noted on logs. In instances of substantial bulking or compaction of samples, changes in penetration rates assist in establishing the thicknesses of lithological horizons. Successive coring runs, made from predetermined depths, should be used. This may be achieved by hydrojetting the corer to a desired depth prior to commencing sampling. *Jetting,* in turn, gives some indication of lithology, since rates are extremely rapid in noncohesive sediments or very soft sediments and slow in cohesive materials.

591

Conversely, vibracoring rates are rapid in cohesive materials and slow in noncohesive materials.

In some instances, overlapping runs help resolve recovery anomalies positively. In other instances multiple runs are preferable to one long run since greater control is exercised. The vibrating motion of the sampler facilitates effortless penetration in cohesive materials, and short runs (2 to 3 m) in clays produce relatively undisturbed samples. In dense sands, superior recovery may, in some cases, be achieved when the core retainer is excluded from the sampler.

Core logging should be undertaken in the field, as soon as possible after the sample is recovered. Unconsolidated sediments should be described in terms of the Unified Soil Classification System (q.v.) (Casagrande, 1948). Therefore, provision should be made to perform liquid and plastic limit, moisture content, and other classification tests in the field. Hand-penetrometer and hand-torvane measurements are a useful quantitative supplement in describing cohesive materials (see *Soil Mechanics*). Cores should be logged soon after retrieval by the geologist who supervised their collection, and his log should be supplemented by geological observations. An experienced engineering geologist with a knowledge of density characteristics of the potential foundation materials should be capable of extrapolating this knowledge to estimate the relative density of noncohesive materials along the cable route. These data may be subsequently checked by a limited program of conventional borings and undisturbed sampling.

Surveying. Offshore surveying may best be carried out by one of the hydro-adapted developments of the *tellurometer.* The Cubic Autotape DM40 or the Motorola "Mini Ranger" are recommended. For seismic work, the Motorola requires a digital printout and time clock to record ranges. The power supply for the interrogator (aboard ship) is critical as the Motorola requires 28 V DC or 110 AC, which is frequently not available on small craft. During investigation the survey system should be checked at intervals for gross error against a theodolite-surveyed tower (e.g., meteorological tower) or a well-anchored bouy in the vicinity.

To obtain elevations, published tide tables may be used with ease and comparative accuracy to reduce water depths to a fixed datum. Tide tables make no provision for anomalous conditions caused by storms or strong winds of long duration, but this problem is minimized since investigations are not usually conducted during periods of very poor weather. The uncorrected water depth and time should be recorded at each boring location during the sampling operation. These data are then used to compute corrected water depths necessary for the construction of accurate cross-sections.

Sampling on Land. Both undisturbed control samples and disturbed bulk samples are required on the overland extension of the transmission route. A 10-cm hollow-stem auger is ideal for this purpose. Boring intervals of 700 m are considered to be sufficient as on-land investigations facilitate *in situ* thermal testing when "trenching" during cable installation. Details on stratigraphy and thermal properties of soils are not critical in on-land installations, since the trench may be examined for potential hot spots prior to cable burial. In areas of unfavorable conditions, a thermal sand can be placed in the critical space around cables.

Laboratory Testing and Data Reduction

Coordination. A modeling program should be established to coordinate the overall evaluation of the thermal properties of soils. This concept should provide a framework of soil types and appropriate engineering parameters to facilitate the most economical selection of samples for thermal testing. The geotechnical model may be constructed by reducing all existing vibracore and seismic data to a convenient form (subsurface sections). Qualitative data are then quantified by referral to control borings (borings from which undisturbed samples are taken) at critical locations.

Data may then be used in empirical or analytical equations that exist in the literature to develop ranges of expected thermal resistivity values.[*] The results of this study are invaluable in monitoring ongoing laboratory testing.

The soil modeling approach should be refined as more laboratory data become available. The final phase consists of using all available data to assign the most realistic ranges of thermal properties to each delineated soil stratum. For additional details regarding procedures see Salomone et al., 1974).

Laboratory Versus *in Situ* Testing. In the initial phases of thermal evaluation, laboratory procedures are considered preferable to *in situ* testing for a number of reasons. These include economics, the greater availability of laboratory-oriented expertise and equipment, the fact that greater control may be maintained in testing selected materials, and the wider range of disturbance from a natural condition that may be tested for in the laboratory (thus permitting greater flexibility in the selection of an eventual installation technique). Also, laboratory testing generally provides values more conservative than those provided by *in situ* tests (personal communication with H. F. Winterkorn).

Winterkorn, however, considers that *in situ* field

---

[*]In the U.S. study undertaken by Salomone et al. (1974), the empirically developed equations of Kersten (1949) and Van Rooyen (1959) agreed favorably with the results of laboratory tests. However, laboratory values were generally lower than those developed by Kersten in very dense sands and cohesive materials and higher in less dense, noncohesive soils.

tests are more representative than laboratory tests for assessing the thermal behavior of natural soil strata. The most opportune time for conducting *in situ* tests is considered to be after design decisions (based on the initial investigation) have been made about the mode and depth of installation. *In situ* probes may then be introduced to the design depth selected for cables by a method designed to cause disturbance to sediments similar to that anticipated by proposed installation methods.

Thermal Probe. The *thermal probe* approach is the most widely accepted laboratory approach for determining thermal values. In the offshore nuclear power plant study under discussion, an initial phase of testing was carried out at Princeton University under the supervision of H. F. Winterkorn. Subsequent testing was performed on equipment adapted from Winterkorn's design. Figures 2 and 3 and Table 1 show preliminary trends and results. The thermal probe method depends on the relationship between the thermal resistivity of a substance and the temperature rise of a line source of heat within that substance. The temperature-time characteristics resulting from a given heat input are then observed.

The representability of samples is critical. Also, since the thermal resistivity of soils depends on such fundamental factors as the type of soil, density, moisture content, and depth of burial, each sample should be tested at different densities and moisture contents. Bulk samples of granular materials should be subdivided and then tested at one of two different densities and one of three moisture contents. In addition, two runs should be made using a different heat input for each subsample to check values. Thus, tests should be performed on each granular sample, as well as on samples composed of various proportions of soils mixed to simulate backfill conditions.

Classification of samples (on the basis of particle size analyses or Atterberg limit tests) provides an independent quantitative basis for comparing soil types and thermal values. Thus, a finite limit may be established to the number of thermal determinations required to satisfy the particular soil model under consideration. Table 1 shows typical results of values developed.

Time Effects. The time effects of heat on the thermal properties of cohesive soils are significant. The results of a preliminary mathematical modeling study to simulate heat flow from a cable buried in clay are shown in Figures 4 and 5.

Conclusions

An attempt has been made to show that in a study related to the installation of offshore transmission cables, the investigator must tailor his methods to the ocean environment. Technical considerations relate not only to the thermal prop-

erties of soils, but also to the stability of these soils in an active hydrodynamic regime. In addition, in aligning the cable system through coastal fringe areas, attention to the environmental sensitivity of such areas is critical. When working offshore, adaptability, meticulous planning, and stringent coordination are, more than ever, the key to ensuring correct civil engineering site-investigation practice.

IAN WATSON

References

Bell, R. G., and Sutherland, A. J., 1983. Nonequilibrium bedload transport by steady flows, *Jour. Hydraulic Eng.* **109** (3), 351-367.

Casagrande, A., 1948. Classification and identification of soils, *Am. Soc. Civil Engineers Trans.* **113,** 901-992.

Fischer, Joseph A.; Salomone, Lawrence A.; and Watson, Ian, 1975. Influence of soils on extra high voltage offshore transmission lines, *Marine Geotechnology,* **1** (2), 141-156.

Kersten, M. S., 1949. Thermal Properties of Soils, *Eng. Expt. Sta. Bull.* **28,** 227p.

McDougal, W. G., and Hudspeth, R. T., 1983. Longshore sediment transport on non-planar beaches, *Coastal Eng.* **7** (2), 119-131.

Salomone, L. A.; Fischer, J. A.; and Watson, Ian, 1974. Procedures used to evaluate the thermal properties of soils adjacent to buried extra high voltage lines, *Am. Soc. Testing and Materials Jour.* **2** (6), 446-502.

Seymour, R. J., 1983. The nearshore sediment transport study, *Jour. Waterway, Port, Coastal, and Ocean Eng.* **109** (1), 79-85.

Van Rooyen, M., 1959. Soil thermal resistivity, *Highway Research Board Bull. 168,* 186p.

Watson, I.; Fischer, J. A.; and Salomone, L. A., 1974. Transmission lines probe a new frontier, the ocean, in *Electric Power and the Civil Engineer, Conference Papers, Power Division Specialty Conference.* New York: American Society of Civil Engineers, 557-578.

Cross-references: *Coastal Engineering; Coastal Inlets, Engineering Geology; Marine Sediments, Geotechnical Properties; Nuclear Plant Siting, Offshore; Pipeline Corridor Evaluation; Submersibles.* Vol. XIV: *Acoustic Surveys, Marine; Sea Surveys.*

URBAN ENGINEERING GEOLOGY

Public knowledge of physical environments is important to community development because properties of earth materials, landscape, climate, and local hydrology affect urban growth. Understanding the physical properties and behavior of earth materials may, for example, lead to (1) the selection of better and safer construction sites, (2) the conservation of natural resources essential for a city's development, and (3) the preservation of aesthetic resources that enhance the quality of urban life.

The purpose of an engineering geology study is

to determine the physical properties of earth materials, and to locate deposits of economic value from which the city can benefit. Costly repairs can be greatly minimized and expensive construction (over-design) may be avoided if knowledge of engineering geology (q.v.) is generally available. Savings may be achieved if natural resources are located at the planning stage and properly managed. The study of engineering geology, which provides knowledge of important geological and engineering parameters directly affecting urban areas, is essential to urban programs (see *Urban Geology*). Information obtained from engineering geology studies must be made available to urban planners, architects, realtors, building contractors, and the general public (Mathewson and Font, 1974).

Need for Engineering Geology Surveys

How necessary is an engineering geology survey? To what degree can benefits be extended from it? Consider the Waco, Texas, dam site, constructed in the early 1960s, that was built across three faults. The landform produced by the faulting was worn down prior to construction of the dam. As a consequence of the faulting, three different formations (rock types) underlay the dam (Wright and Duncan, 1972). The personnel in charge of the construction recognized the presence of one of the faults and of two of the formations, but unfortunately, they missed the third. The dam was designed with the properties of the two known formations in mind. The third—which had not been recognized—was much weaker than the other two. The dam was nearly completed when the portion resting on the undetected bedrock failed spectacularly; the crest dropped 6 m, and the downstream embankment slope moved horizontally 9 m. Bulging of the land appeared in varying degrees downstream from the dam's axis as far as 250 m, while large cracks covered a 500-m section of the embankment. Luckily, the lake behind the dam had not been filled. Corrective measures to stabilize the structure proved very expensive. And yet this failure could have been prevented through a proper *engineering geology survey*.

By way of another example, consider an exclusive Waco, Texas, residential area. The houses located at each corner of the block are intact while those in the middle of the block have large cracks running up the walls. The street pavement in front is also broken and displaced but appears undamaged near the corners. Surprisingly, a similar situation exists along a number of streets. Preparation of an *engineering geology map* (see *Maps, Engineering Purposes*) would probably show that the area is underlain by an expansive clay (see Vol. XIV: *Expansive Soils*), one that exhibits many shrink-swell characteristics. In some sections, however, the clay is overlain by deposits of sandy

gravel. Houses built on top of the swelling clay have suffered damage, while those built on the gravel have not. In short, the gravels offered a stable foundation (Font and Williamson, 1970). Had an engineering geology survey been conducted, the problem probably could have been predicted before the area was urbanized.

In another Texas suburb, rolling hills overlooking a lake are capped by a bed of hard, massive rock about 3 m thick. A clay foundation about 40 m thick is exposed on the slopes and also underlies the massive rock bed. The scenic area was urbanized without a consideration of prevailing geological conditions. Expensive houses were built on hilltops and the homeowners pleased because structural foundations rested on massive rock. In the process of building roads down to the lakeshore natural hill slopes were oversteepened. When the rainy season arrives, the soft clay exposed on the now oversteepened slopes will become saturated and loses its strength (see Vol. XIV: *Landslide Control; Slope Stability Analysis*). The weight of the houses and the massive cap rock will cause the clay to be squeezed out, just as toothpaste can be squeezed out of a tube. As the clay flows out from under the cap rock, it will remove support, causing large block failures (Font, 1977, 1978, 1979). Houses resting on the massive rock will also fail and thus will end up in the lake. This unstable situation could have been prevented.

Many case studies have shown that an engineering geology survey could have prevented tragic and regretful situations. It should be apparent that one can't ignore the physical environment; for relatively small cost, one can hire a qualified engineering geologist or geotechnical engineer to advantage.

Engineering geology studies have been generally limited to specific problems and hazards in certain limited areas. As a result, such surveys often failed to recognize all factors that affect urban expansion and community development. No established format—that is, no guideline that can be followed in the future—considers all important parameters, and each study presents the results in different ways. In most instances, these reports have been written in such a form that people untrained in engineering geology find it difficult to apply the results to adjacent property. Information from these studies is normally not available to the public because the reports are prepared only for the use of engineers, or engineering geologists (see *Engineering Geology Reports*). Consequently, the objectives of this article are (1) to outline an approach for engineering geology studies that will provide a sound basis for land use planning, (2) to determine the basic geological and engineering parameters that directly affect urban expansion, and (3) to establish a useful format for presentation of information.

Engineering Geology Decision Making

The first step in an engineering geology study is to consider the physical (geological) environment of the community. Once it has been evaluated, the second phase involves determination of specific physical and engineering properties of earth materials encountered in the area. The second phase considers parameters that may adversely or favorably affect the growing community. The third phase of the study is interpretative. The engineering geologist or geotechnical engineer is now in a position to advise the community in its development. The fourth and final step of the engineering geology survey is to integrate all available information into a recommended land use plan (Fig. 1).

The Geological Environment

Understanding of the physical environment involves consideration of six basic factors and the role that each plays. Their impact or effect on community development will be briefly discussed here.

Tectonics and Structure. Tectonics broadly involves the study of instability and deformation of the earth's crust. Earthquakes, faulting, and volcanic action are examples of major agents that affect urban and rural communities. Structural features such as faults, folded rock, structural basins, and mountain ranges are products of tectonic forces that may still be operative. It is thus imperative to recognize tectonically active areas that may pose potential hazards to community development. It is also advisable to map structures that can be responsible for costly engineering failures. An understanding of the tectonic setting and geological structures is essential to a safe urban plan.

Bedrock Geology. A knowledge of the bedrock geology is also important. The presence of different rock types in an area results in zones of widely variable physical and engineering properties. For example, the foundation design that may be ideal for one rock type can be unsatisfactory for another (see *Foundation Engineering*). Under-design (failure to provide a proper foundation), over-design (using

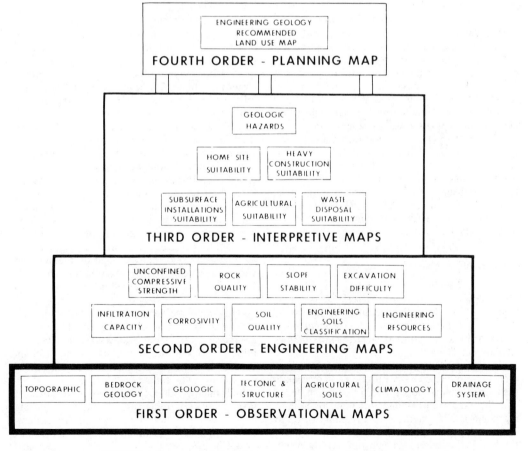

FIGURE 1. Recommended land use plan based on engineering geology.

a more complex and expensive foundation when a simpler and less expensive one could be equally safe), and slope failures, to name a few expensive problems, can be mitigated by understanding local geology (see *Reinforced Earth;* Vol. XIV: *Landslide Control*).

Surface Soils.   Surface soils, like bedrock, have different engineering properties. Not only do they affect foundations and pavements, but they often contribute to the corrosion of underground pipes and cables as well (see *Pipeline Corridor Evaluation*). Perhaps, even more important, they determine agricultural land use. Whether wishing to preserve a good agricultural soil or wanting to understand its behavior for construction purposes, one needs to consider surface soils in urban plans (Lindsay, 1974).

Climate.   Climate directly affects the behavior and response of many earth materials. Climatic factors such as rate of precipitation (rainfall) and temperature variability often contribute to periodic flooding (see *Hydrology*), topsoil erosion (see Vol. XII, Pt.1: *Water Erosion; Wind Erosion*), soil swelling or flowage (see Vol. XIV: *Expansive Soils*), and freeze-thaw damage to highways and roads (see *Geocryology*). Moreover, prevailing winds in some areas act to concentrate air pollutants.

Topography.   Topography, the shape or morphology of the land surface, also affects community planning. It may furnish either barriers or pathways for urban expansion by controlling building and transportation routes. It may even provide a natural trap for air pollutants carried by the winds.

Drainage and Water Supply.   Finally, the urban engineering geologist must consider drainage (see Vol. XIV: *Land Drainage*) and water supply (see *Hydrology;* Vol. XIV: *Groundwater Exploration*). Because communities depend on water, they must learn to manage it judiciously by protecting aquifers (water-bearing rock or soil) and avoiding their depletion or contamination by waste materials. Wise use of water does not unduly upset the natural balance.

Physical and Engineering Properties

To build effectively on soil and rock, it is important to appreciate how these materials respond to engineering works. To determine the necessary physical and engineering properties of earth materials that affect community growth, nine factors should be measured.

Unconfined Compressive Strength.   The unconfined compressive strength of a soil or rock is the maximum load per unit area that the material can withstand without failing, breaking, or collapsing (see *Rock Mechanics; Soil Mechanics*). If the load exerted on a soil by a structure is greater than that which the soil or rock can support, the structure will fail. It is, therefore, obviously desirable to

know the load capacity of the soil or bedrock prior to construction.

Rock Quality.   Rock quality indicates the extent to which a rock has been fractured or ruptured. It is a significant factor in engineering projects such as tunneling (q.v.), quarrying (see Vol. XIV), and excavating, and drainage studies.

Slope Stability.   Slope stability is a serious threat to life and property. An area should be surveyed and the maximum stable slope angle for each rock or soil type must be determined (see *Slope Stability Analysis*). It should be noted that a natural slope that has been stable for a long time may fail if overloaded, or if material is removed from its toe, decreasing its support. Gravity, the weight of the soil and the load of buildings on the top of the slope, combined with excess moisture are two common causes of slope failure (Krynine and Judd, 1957). It is therefore important that slope stability (see Vol. XIV: *Slope Stability Analysis*) be determined by a competent engineer or geologist responsible for the construction site.

Excavation Difficulty.   Excavation difficulty is in many areas an important cost factor in foundation construction. Often, the type of bedrock that offers the best foundation site is the most difficult to excavate. Blasting (see Vol. XIV), or some other expensive means of excavation is necessary in areas where hard, massive rock is exposed. On the other hand, soils and rock that are easily excavated may inadequately support heavy structures. In addition, drainage problems may arise in excavations that intersect the groundwater table.

Infiltration Capacity.   A knowledge of infiltration capacity and permeability—the ease with which water flows through the soil or bedrock—is important to the construction engineer (see Vol. XII, Pt.1: *Infiltration; Permeability*). Reservoirs, lakes, and ponds retain water if the underlying material is impervious or of very low permeability. The same material, on the other hand, would be inadequate for septic tanks because the water in the tanks would be trapped. The engineer can anticipate and correct drainage problems in advance if infiltration capacity of the soils and rocks in the construction area are known.

Corrosivity.   Underground pipes and cables frequently must be repaired or replaced because certain soils have a corrosive effect on underground structures. It is, however, possible to determine in advance the corrosivity of any soil, and corrective measures can be taken to delay, avoid, or prevent costly failures.

Soil Quality.   Certain soils tend to swell (expand) and shrink (contract) through the addition and loss of water; others tend to settle or collapse under the weight of an overlying structure (see Vol. XIV: *Dispersive Clays; Expansive Soils*). Such ground movements in developed regions may result in serious foundation damage. Soil quality can be determined in advance, thus eliminating costly failures and expensive maintenance.

Engineering Soil Classification. Engineers are often asked to design a particular structure at a specific site. If a previous engineering geology survey had been conducted, the soils of the area were probably classified under an engineering classification (see *Soil Classification System, Unified*). This information may be of great value to the engineer, who can now design with specific soil behavior in mind. Such design takes into account the properties of the particular soil; as a result, safer foundations are constructed.

Engineering Resources. Finally, communities should consider the engineering resources found in the local area. Attempts should be made to conserve commercial deposits of sand and gravel, ceramic clay, and building stones, for example, from which a growing community can always benefit. The location of these resources should be mapped and outlined early to prevent their loss as the community expands.

The Interpretative Phase

Six factors should be analyzed in the planning phase of any study.

1. *Environmental hazards.* Zones subject to flooding, active faulting, landslides, earthquakes, and other natural hazards should be located to prevent costly failures, accidents, and the loss of life.
2. *Home-site suitability:* After environmental hazards have been evaluated, the home-site suitability within the community can be determined. Sections may be classified as desirable or undesirable on the basis of foundation conditions, local resources, sewage problems, and environmental hazards.
3. *Heavy construction suitability:* Similarly, the suitability of the land for heavy construction can be determined. Areas may be classified as desirable or undesirable for heavy construction on the basis of foundation and excavation problems.
4. *Subsurface installations suitability:* Suitability for subsurface installations must be ascertained and based on soil features that might adversely affect underground structures, buried pipes, and cables (see *Cavity Utilization; Pipeline Corridor Evaluation*). Factors such as shrink-swell characteristics and potential for corrosivity dictate whether below-ground installations are practical.
5. *Agricultural suitability.* This refers to the adaptability of various garden, yard, and food plants to particular soils. This information is of value not only to the farmer who depends directly on the soil, but also to the homeowner who enhances

the aesthetic value of his property by planting a tree.
6. *Waste disposal suitability:* Certain areas may not be suitable for installation of septic tanks because of low soil permeability, which may result in the rise of septic waste to the soil surface after rainy periods, or because high permeability may lead to contamination of drinking water. Perhaps more importantly, industrial wastes must not be dumped into waterways, lakes, or coastal areas. Areas suitable for waste disposal may be delineated after careful study.

The Final Plan

The final product of an engineering geology study integrates the results of previous work into the so-called engineering geology recommended land use plan.

An effective way to transmit information is in map form (see Vol. XIV: *Maps, Logic of*). All first-order observational maps are essential to the study. Each factor must be considered, and its influence on the growing community determined. Second-order engineering maps are designed for engineers, developers, or contractors who deal with local earth materials in construction projects. Each parameter considers a basic problem that may be significant in the local area (see *Maps, Engineering Purposes*). For a particular region, it may not be necessary to compile all nine maps. For example, in zones where all earth materials exposed at the surface and encountered to considerable depths consist of soft clays and uncemented sands, the rock-quality map is not needed. It must be stressed, however, that all nine factors are usually encountered and can exert a significant influence on community growth.

Observational and engineering maps are used to compile third-order interpretative maps. Each third-order map shows the suitability of an area for a specific purpose or delineates existing and potential hazards. If an engineering geology survey lacks the interpretative portion, it has failed in its purpose. First- and second-order maps are generated for the engineering geologist or geotechnical engineer, whereas third-order interpretative maps are designed for the general public. The fourth-order map is the end result of a thorough study. It combines all previous information into one map, the *engineering geology recommended land use map,* and therefore is especially designed for the city planner. It provides a basis for outlining, from an engineering geology point of view, the best use of any region within the community. This map is the one from which the architect, the realtor, and the public will benefit most; it is the ultimate contribution that the engineering geologist can make to community decision makers.

Factors other than engineering geology also

must be considered in the final urban development plan. An equivalent to the proposed fourth-order map should be based on other disciplines such as economics, sociology, and civil engineering. The integration of all facets results in a comprehensive plan. In the final analysis, the engineering geology recommended land-use map is indispensable to ideal urban development and expansion plans.

ROBERT G. FONT

References

Font, R. G., 1977. Engineering geology of the slope instability of two overconsolidated north-central Texas shales, in D. R. Coates (ed.), *Reviews in Engineering Geology,* Vol. III. Boulder, Colo.: Geological Society of America, 205-212.

Font, R. G., 1978. Effect of anisotropies on the slope instability of heavily overconsolidated clay shales, Texas, U.S.A., in *Proceedings of the III International Congress of Engineering Geology,* Madrid, 23-26.

Font, R. G., 1979. Geotechnical properties of unstable shales in north-central Texas. Texas, *Texas Jour. Sci.* **31**(2), 119-124.

Font, R. G. and Williamson, E. F., 1970. Geologic factors affecting construction in Waco, in *Urban Geology of Greater Waco, Part IV,* Baylor Geol. Studies Bull. No. 12, 1-33.

Krynine, D. P., and Judd, W. R., 1957. *Principles of Engineering Geology and Geotechnics.* New York: McGraw-Hill, 730p.

Lindsay, J. D., 1974. Soil survey for urban development, in R. W. Simonson (ed.), *Non-Agricultural Applications of Soil Surveys.* Amsterdam: Elsevier, 35-45.

Mathewson, C. C., and Font, R. G., 1974. The geological environment—forgotten aspect in land use planning, in H. F. Ferguson (ed.), *Engineering Geology Case Histories No. 10.* Boulder, Colo.: Geological Society of America, 23-28.

Wright, S. G., and Duncan, J. M., 1972. Analysis of the Waco dam slide. *Am. Soc. Civil Engineers Proc., Jour. Soil Mechanics and Found. Div.* **SM9,** 869.

Cross-references: *Consolidation, Soil; Foundation Engineering; Geotechnical Engineering; Maps, Engineering Purposes; Permafrost, Engineering Geology; Pipeline Corridor Evaluation; Reinforced Earth; Soil Classification System, Unified; Soil Mechanics; Urban Geology; Urban Geomorphology; Urban Hydrology.* Vol. XIV: *Cities, Geologic Effects; Dispersive Clays; Engineering Soil Science; Environmental Geology; Expansive Soils; Land Capability Analysis; Land Drainage; Landslide Control; Maps, Physical Properties; Slope Stability Analysis.*

URBAN GEOLOGY

Napoleon is not usually associated with building. The lovely modern city of Paris owes much, however, to the great emperor, who by imperial decree put a stop to the mining that had been carried out beneath the streets of Paris since the time of the Romans. Much of the limestone used for the buildings of pre-nineteenth-century Paris and the gypsum necessary for interior plaster finish was obtained from quarrying the horizontal rock strata beneath the city's surface.

The stone and gypsum were simply hoisted to street level through vertical shafts. Land transportation outside the city was then so difficult as to make hauling from any great distance almost impossible.

Cave-ins at the surface had already caused trouble when underground quarrying was stopped. Ten percent of the land area of Paris was underlain by open tunnels. When close to the surface, these were clearly a hazard to further building construction, so that a special organization had to be established to make accurate surveys of the tunnels. The results were plotted on some of the very earliest maps of *urban geology,* showing the geology of the land beneath city streets. These quite beautiful maps are carefully maintained today. All excavation or foundation work in Paris must still be reported to l'Inspection Générale des Carrières de la Seine, which also supervises the maintenance of the old underground workings, some of which are the macabre catacombs of Paris (Lafay, 1958).

There is probably no other city in the world that has to control its building because of such an unusual subsurface hazard, but other great cities have other geological problems that complicate local construction. How many visitors to Mexico City realize that parts of its central area have settled more than 8 m since 1900? The Palace of Fine Arts, built in 1904, has settled more than 3 m. Its architect, warned of potential settlement, is reported to have said, "If the structure is pleasant to my eye it is structurally sound," and refused to stop the building when settlement revealed itself during early stages of construction (Thornley et al., 1955).

The ancient and beautiful city of Prague in Czechoslovakia has the reverse problem. Much of its ground surface is above the original ground level owing to the accumulation of rubble and other debris from building long since demolished. One of its most famous churches is so far below the present ground level that one has to go down a flight of steps to enter it. Today, however, there is available accurate information regarding the subsurface conditions for the whole of Prague so that modern building can proceed with certainty (Zaruba, 1948).

In North America, a paper was published as early as 1893 describing the subsurface of New York City, as revealed by a series of test borings put down along Broadway from South Ferry to 34th Street, the paper clearly showing the importance then attached to such information (Parsons, 1893). The need for maps of urban geology was recognized publicly in Canada at the turn of the century, and probably well before that by individual practitioners. In a paper presented to the Royal Society of Canada in 1900, Dr. H. M. Ami wrote "The larger cities of our Dominion, as well as those of other countries, are the centers of work and research in the pathways of science and commerce . . . . What the drill has to penetrate in any one of our larger centers of activity in Canada is a question not only of interest but also of economic value" (Ami, 1900). He proceeded to give summary accounts of the urban geology of Saint John, Montreal, Ottawa, Quebec, and Toronto—and this almost 70 years ago.

Every city and every town should have readily available accurate information about the ground upon which it is built. Despite widespread private ownership of land, this information is essential for the safety of the public and cannot properly be regarded as private property in this regard. No city or town planning worthy of the name can be carried out without a full appreciation of local geology. Not only does geology determine the natural physiographic features that may have to be modified in the framing of a master plan, but it will also influence the pattern of natural drainage, the types of materials to be excavated, and the suitability of building sites for different kinds of construction.

It is true that the design of major building foundations (see *Foundation Engineering*) will usually be the work of the specialist engineer, but an architect must have at least a general idea of local ground conditions (whether soil or solid rock, as a start) for the preliminary project planning and for later discussions with the engineers. Correspondingly, the planning of new urban developments must comprehend more than a two-dimensional plan of the area and must be based, from the outset, on at least a general appreciation of the three-dimensional character of the land under study.

The geology of any area—such as that covered by a number of city blocks or a new municipal subdivision, the planning of which is in prospect—is essentially a three-dimensional concept, extending to a depth well below that at which any surface loadings can be significant. This picture of the structural make up of a block of land (and the word *block* is singularly appropriate) is not to be thought of in isolation, for it is an integral part of the geology of the region around. The detailed study of specific building sites or planning areas will be more effectively and economically carried out against the background provided by such general knowledge of the local geology.

It may be thought necessary to put down an adequate number of test borings only at building sites that warrant such expenditure. This, however, can be a somewhat limited and possibly wasteful way of determining local ground conditions. There should be publicly available in every city a collection of all logs of test borings, correlated with the local geology. It can truly be said of such compilations of subsurface information that the whole will be of much greater value than the sum of the individual parts, important and useful as the logs of single test borings may prove to be (Legget, 1973).

Urban geology has thus a two-way importance for all those concerned with building. It provides a useful and often invaluable source of information for specific design cases. At the same time, the record can be extended if all those who gain access to further information will add it to the common store. There will be an opportunity for wider service in promoting the development of such records where they do not already exist.

There is an interesting parallel to this desirable procedure in the commendable and general practice of having centrally coordinated records of all subsurface utilities such as water mains, gas mains, and telephone and electric power cables beneath city streets. Many cities now have excellent and often quite extensive systems for recording this type of specialized subsurface information, which is usually available in the office of the city engineer for consultation by those interested.

The necessity for such records has come about because of the essential coordination of the different services now buried beneath the streets of modern cities. A committee of representatives of the several local utility organizations is usually responsible for supervising steady and accurate maintenance of these vital records. Exactly the same type of subsurface records of the urban geology of all cities should be similarly maintained and publicly available. In some, a start has been made, but it will require strong support from those concerned with building—architects, engineers, and contractors—to ensure that city engineers and their staffs are provided with the necessary facilities for the preparation and regular maintenance of these vital records.

Application to the office of the local city engineer should always be the first step in determining the general pattern of local geology. Supplementary information may often be obtained in large cities from the department of geology of the local university. Availability of geological maps of the area under study and geological reports that may be useful can be checked with the appropriate (national, state, or provincial) geological surveys. In most major cities there will be found other records of local urban geology, assembled by interested groups or individuals. To indicate just what valuable information is available in this way, following are notes on some major cities of North America:

Washington D.C. Local urban geology has been studied by the U.S. Geological Survey since 1882, and the records published in a series of notable papers, climaxed by a masterly review in 1950 by N. H. Darton.

Boston. Here the members of the Boston Society of Civil Engineers have been the interested group, publishing in their own journal a number of notable papers summarizing test borehole records from the whole urban area, with helpful comments (BSCE, 1969).

San Francisco. In more recent years the U.S. Geological Survey has made extensive surveys of the local geology, the first studies being shown on some excellent geological maps that have explanatory notes incorporated on the map sheets (Schlocker et al., 1958).

Montreal. Working cooperatively with the Geological Survey of Canada, the City of Montreal has published a comprehensive report on the local geology. The accompanying geological maps are being kept up to date as herein suggested. The City Planning Department has published in addition a guide for urban and regional planning that is a model of its kind (Anon., 1966).

Toronto. Construction of Canada's first subway, starting in 1944, led to the first assembly of local test boring records. Full geological profiles along the subway excavations, and suites of soil samples, are now available at the Royal Ontario Museum. The office of the Commission of Works of Toronto is now collecting local test boring records (Legget and Schriever, 1960).

Vancouver. The local staff of the Geological Survey of Canada has been studying the geology of the city area for many years; results have been recorded in a notable series of papers by J. E. Armstrong (1956).

Once architects or town planners have availed themselves of the information about local geology that is publicly available, they will readily appreciate the value of every contribution to this store of public knowledge. It is clearly a professional responsibility to assist in building up this accumulation of records. Accordingly, and without the legal compulsion to do so that now exists in quite a number of cities in other parts of the world, copies of all borehole logs and excavation profiles should be passed by responsible engineers and architects to the custody of the local city engineering authorities. If the office of the city engineer is not geared to accept, store, and compile such subsurface records, then professionals can help by developing civic support and the necessary financing for such an essential civic purpose.

Although the primary purpose of assembling information on local urban geology for public use is strictly utilitarian, and although the work and expense involved can be fully justified on economic grounds, the science of geology can often be an incidental beneficiary. New excavations and test borings in central city areas will often reveal geological information not previously recorded

and sometimes of great value. The excavation for the first part of Toronto's subway, (see *Urban Tunnels and Subways*), for example, revealed unique scientific information on the world-famous "Toronto interglacial beds" of soil deposited under tropical conditions between glacial periods; scientific records were enriched by papers based on the information then obtained. The department of geology at every university may therefore be expected to have a lively interest in the local urban geology and should be approached when any new project is started so that members of the staff may have the privilege of studying the new geological sections that excavation reveals (see *Pipeline Corridor Evaluation*).

A start has been made in many cities on the development of these vital civic geological records. Much more yet remains to be done, and the need increases every day as the urbanization of North America continues its rapid advance. City engineers will need all possible encouragement from architects, planners, engineers, and contractors, especially in obtaining the funds necessary for small additional staffs that may be necessary in larger cities. It may therefore be helpful to outline the ideal arrangement that can readily be shown to be economically advantageous to every town and city.

In the office of every city engineer there must be a set of topographic maps of the city, now frequently supplemented by sets of aerial photographs. Small-scale maps will show the general topography, and large-scale maps the legal controls on subdivision. In all cities, the office of the city engineer (or an office closely associated with it) should have a set of large-scale maps and associated written records, showing as accurately as possible the location of every buried utility in the city area.

Correspondingly, a set of small-scale maps should be available in the same office delineating the general pattern of the local geology in its dual aspect: the bedrock and the surficial soil deposits. Large-scale maps should contain detailed records of all test borings put down in the city area, the corresponding records from all excavations, and details of the local ground-water levels as observed in wells, test borings, and excavations. Associated with the maps should be copies of test reports on soil samples obtained in site exploration and written records of such other aspects of subsurface conditions as are available. Some cities have even arranged to have on display typical cores of the local rocks, obtained by test drilling, to permit visual inspection by those interested.

Arrangements should be made for copies of the logs of all new test borings to be deposited with the city engineer, after they have served their immediate purpose, as well as records of the geological conditions revealed by all new excavations. The procurement of such records for public use must be arranged at present by persuasion and mutual understanding. There is not yet, in North America, any legal requirement that this must be done, as is the case, for example, with the logs of holes put down in exploring for oil and natural gas. As each new record is plotted on the master-record maps, knowledge of the local geology will become more accurate and more meaningful. The maps will become the more valuable with their increasing accuracy.

The savings that can be effected in reducing the number of new boreholes (see Vol. XIV: *Borehole Drilling*) necessary to give accurate information about even one major site can easily offset a large part of the annual cost of maintaining the city's records of urban geology. In addition, such information will give a reasonable degree of certainty to the subsurface conditions suggested by new boreholes, to the immediate benefit of the owners, architects, engineers, and contractors, and to the benefit of the city itself. Savings that can be realized by the use of this information in carrying out excavation for water main and sewer trenches, and road and bridge construction, can very quickly repay the relatively small cost of such records of the local urban geology.

The necessary brevity of this entry may give the impression that the study of local geology consists merely of the collection of local subsurface records and assembling them in some convenient form. There is, however, far more to the study of urban geology than this—for every city and town. To begin with, a broad appreciation of the geology of the surrounding region is an essential starting point without which no collection of records can be fully useful. Assessment and critical study of the collected records is equally essential, not only to benefit the users of resulting urban geological maps in their more detailed studies of particular sites, but also from the purely geological point of view. Once covered up by streets and buildings, local geology cannot again be studied. It is, therefore, vitally important that, quite apart from its utilitarian uses, the study of local urban geology should be widely recognized as geologically and scientifically important.

This entry is based on Canadian Building Digest No. CBD 113, by R. F. Legget, issued in May 1969 by the Division of Building Research of the National Research Council of Canada.

ROBERT F. LEGGET

References

Ami, H. M., 1900. On the geology of the major cities of Canada, *Royal Soc. Canada Trans.,* 2nd ser. VI, sect. IV, 125.

Anon., 1966. *Physical Characteristics of the Region.* Bull. Technique No. 4, Montreal: Service d'Urbanisme, 51p.

Armstrong, J. E., 1956. *Surficial Geology of Vancouver Area, British Columbia,* Canada Geol. Survey Paper No. 55-40.

BSCE Committee on Subsoils of Boston, in cooperation with the U.S. Geological Survey, 1969. Boring data from Greater Boston, *Boston Soc. Civil Engineers Jour.* **56,** 131.

Darton, N. H., 1950. *Configuration of the Bedrock Surface of the District of Columbia and Vicinity,* U.S. Geol. Survey Prof. Paper 217.

Lafay, M., 1958. l'Inspection Générale des Carrières de la Seine, *Travaux,* p. 3.

Legget, R. F., 1973. *Cities and Geology.* New York: McGraw-Hill, 624p.

Legget, R. F., and Schriever, W. R., 1960. Site investigation for Canada's first underground railway, *Civil Eng. and Pub. Works Rev.* **55,** 73.

Parsons, W. B., 1893. Borings in Broadway, New York, *Am. Soc. Civil Engineers Trans.* **28,** 13-18.

Schlocker, J.; Bonilla, M. G.; and Radbruch, D. H., 1958. *Geology of the San Francisco North Quadrangle, California, Map 1-272.* Washington, D.C.: U. S. Geological Survey.

Thornley, J. H.; Spencer, G. B.; and Albin, P., 1955. Mexico's Palace of Fine Arts settles 10 ft., *Civil Eng.* **25,** 357.

Zaruba, Q., 1948. *Geologický Podklad a Základove Pomery Vnitrní Prahy (Geological Features and Foundation Conditions in the City of Prague),* paper 5, Prague: Geotechnica, 83p.

Cross-references: *Foundation Engineering; Maps, Engineering Purposes; Urban Engineering Geology; Urban Geomorphology; Urban Hydrology; Urban Tunnels and Subways.* Vol. XIV: *Cities, Geologic Effects.*

URBAN GEOMORPHOLOGY

Some historians have called the twentieth-century flight to the cities "the urban revolution." Schmid (1968) reported that urbanization is the dominant social, economic, and political movement in the United States. The drastic changes caused by urbanizing populations result in feedback mechanisms that affect the land-water ecosystem. These produce a variety of distortions and maladjustments whereby man changes the environment in creating a totally new "cityscape." Others, such as Gulick (1958), point out that it is in the city and its environs where greatest deterioration of the human environment is occurring. The city is an *anthropogene (human created) landscape* and constitutes a suitable field locality for study by the urban geomorphologist who must assess the type and magnitude of changes to lands and waters of the earth's surface in all physical settings. Thus *urban geomorphology* is the study of man as a physical process of change in highly populated areas where human activities transform natural systems and terrain (see Vol. XIV: Cities, Geologic Effects).

Prior to 1970 no books concentrated on the

earth science aspects of urban areas, but in the last few years several books have aided to fill this gap, such as those by Detwyler and Marcus (1972), Legget (1973), Coates (1974, 1976), and Leveson (1980). Numerous articles describing specific aspects of urban physical systems have been written, and a very helpful summary can be found in Hansen (1976). Whereas the entry on *Urban Geology* described man's need for geological data for materials, borings, excavations, and subsurface materials, the emphasis here will be placed on the description of human metamorphism of the land-water ecosystem in the urban setting. Associated themes also occur in Volume XIV under the headings of *Legal Affairs* and *Open Space.*

Hydrology

People alter the water budget of a city in many ways. They create a microclimate wherein wind, temperature, and precipitation patterns are changed by buildings and other stone, brick, and paved surfaces. The effect is to produce a "heat island." For example, Peterson (1973) has shown that night temperatures may be 6 to 12°C higher in cities than in surrounding areas. In 67 case studies yearly temperatures were about 1°C hotter in cities than in adjacent lands. Studies in several European cities indicate 10 percent more precipitation. Studies show that U.S. cities have 10 percent more cloud cover than adjacent areas, and that it is not uncommon to have 25 percent more rain during weekdays than on weekends. Such changes in the atmosphere, when coupled with pollution from the burning of hydrocarbons and other industrial contaminants, accelerates the weathering of construction materials in buildings and monuments (Winkler, 1973).

Streams in urban areas have greatly different hydrographs than those in nonurban regions (see *Urban Hydrology*). Surface-water percolation into the ground-water zone is prevented by the impervious blanket of buildings, streets, and parking lots. Sewering allows for excessively rapid flow of water through artificial conduits and abnormally quick discharge into streams. These combined effects of paving and sewering cause stream hydrographs to have higher peaks and lower troughs (Fig. 1), so that floods are accentuated and ground-water recharge is reduced. Leopold (1968) summarizes these effects to streamflow and shows how the urbanization process increases flood volumes, flood peaks, and the number of floods. For example, when 20 percent of the area is sewered and with impervious cover, the number of floods exceeding bankfull stage is doubled and stream discharge is about 1.6 times the normal value. For 50-percent sewering and paving, these values are respectively increased to four times the number of floods and 2.4 times the natural discharge. Of course, with these effects there is a concomitant depletion in availability of ground water and lowering of water tables because of reduction in recharge. In a study of the Houston metropolitan area, Johnson and Sayre (1973) concluded that an increase in impervious surfaces from 1 to 35 percent increases the magnitude of a two-year peak flood by a factor of 9 and of the 50-year peak flood by a factor of 5.

Sedimentation

Construction of new building sites and roads increases erosion and sedimentation. Deposits that reach stream channels change the character of stream beds and the flow regime of the water. Effects of sedimentation resulting from urbanization have been discussed by Wolman (1964), Wolman and Schick (1967), Guy (1976), and Fox (1976).

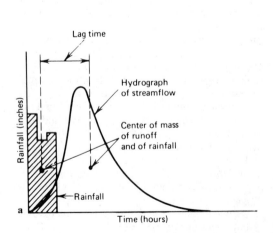

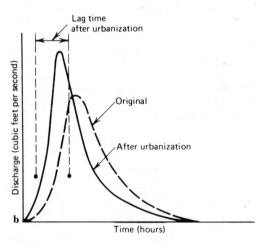

FIGURE 1. Comparison of stream hydrographs for urban and nonurban conditions: (a) elements of hydrograph for a nonurban area; (b) Change in hydrograph form after urbanization (after Leopold, 1968).

These studies in the eastern United States indicate that average sediment yield for nonurban settings is about 80 to 200 t/km² (200 to 500 t/mi²) per year, whereas sediment production of areas undergoing construction range from several thousand to 55,000 t/km² (140,000 t/mi²) per year. Fox (1976) concluded that urbanizing areas in the Patuxent River basin supplied 15 times as much sediment to river channels as was received from rural areas.

Several types of damage result from the increased erosion and sediment yields when an environment is being urbanized: (1) stream aggradation and increased flooding levels; (2) siltation in recreation and fishing sites, (3) clogging of drains, with backup problems; (4) turbid water unsuited for municipal, domestic, or industrial use; (5) damage to pumping equipment; and (6) changes in channel geometry and position. Guy (1976) provides ideas of how such damage from sedimentation can be minimized. These techniques include (1) reducing the construction and site development to areas as small as possible; (2) use of temporary vegetation and mulches on all exposed soil; (3) rapid establishment of plants and permanent types of vegetation; (5) use of shorter lengths of steep slopes; (5) use of planned engineering structures to reduce velocity and volume of water that crosses disturbed areas; (6) better use of hardened and established channels for transporting increased runoff; and (7) construction of sediment-detention basins or holding areas and settling pools.

Channel configuration changes have been discussed by Hammer (1972) and Fox (1976). Hammer reported that channel enlargement was one of the significant effects of urbanization in the Philadelphia area. Here width of channels increased 0.7 to 3.8 times in 78 small watersheds, and such changes were caused by a combination of modifications in topographic slope, implantation of impervious surfaces, and deliberate drainage alterations. Fox showed that urban channels change at rates at least three times faster than those in comparable rural settings. Bars are built and change rather unpredictably, whereas in rural areas they remain stable for many years (see *Hydrology*). Leopold (1968) documented changes in the Brandywine River basin. In a stable rural reach the stream is capable of carrying 1.5 m³/s at bankfull stage with a velocity of 0.8 m/s, and a channel 0.6 m deep and 3.3 m wide. The same reach in urbanization, however, during bankful stage must carry 4.2 m³/sec, with a width increase to 6 m and a depth of 1 m; the sediment produced by erosion would be 1,400 m³.

Water Use

Urban areas require extensive water resources, and many cities must import water since nearby sources may be inadequate. Of the 35 largest cities in the United States, 17 must import water from other areas. One out of eight Americans uses water transported more than 120 km. Several of the large New York City reservoirs are more than 160 km away, and urbanized areas of southern California now import water from the northern part of the state, more than 1,000 km away. Thus urbanization changes produce feedback to areas outside their own immediate environs, causing disturbances in other land-water ecosystems and the terrain of distant regions.

When cities or adjacent localities use too much ground water, subsidence may result, as in the areas of San Jose (California), Tokyo, Houston (Texas), and Mexico City. Here land changes range from 2.5 to more than 8 m (8 to 25 ft). Another effect occurs when withdrawal rates of coastal aquifers exceed fresh-water recharge. This causes aquifer salinization from seawater as in Brooklyn, New York (especially during the 1930-1940 period), Florida, and cities in southern California.

Geomorphic Hazards

Urban areas can be subjected to the two principal geomorphic hazards of floods and landslides. These damaging twins can be considered as part of a double-jeopardy syndrome. There is an increasing recognition and awareness that development on flood plains can be very risky. In the United States, legislation, zoning ordinances, and insurance programs alert—and in many cases, prohibit—construction on flood plains. Until recently, however (with the exception of the California grading ordinances), the general public had little knowledge of the need for constraint in developing hillslopes. Thus, when denied building permits in the flood plains, many builders develop on hillsides, thus facing the danger of landslides.

The term *flooding* is used to describe the inundation of any low-lying terrain, regardless of whether the water is from rivers, lakes, or oceans. Much of the annual toll of more than $3 billion in flood damage in the United States occurs in urban areas. Losses attributed to Hurricane Agnes in 1972 amounted to more than $3.5 billion, and those attributed to the Mississippi flood of 1973 totaled more than $1 billion. Various strategies can be used to protect or soften flood losses in urban areas, but except for complete abandonment of the area, there is little protection from such coastal flooding as hit Bangladesh on November 12, 1970, drowning several hundred thousand inhabitants, or as resulted from the tsunami that killed more than 6,000 in the Philippines in 1976. The three approaches that have been used to deal with flood hazards are (1) structural and corrective—geomorphic engineering projects that seek to control water flow by dams, levees, floodwalls, channel modifications, vegetation, and conservation structures; (2) land use management such as zoning prohibitions, building codes, tax adjustments, and open-space corridors; and (3) absorption of loss by subsidy, relief program, and insurance.

Another hazard is created when urban areas are situated downstream from a dam, for a dam failure can result in great loss of life and property. The famous Johnstown (Pennsylvania) flood of 1889 killed 2,100; the Vaiont, Italy, dam catastrophe of 1973 killed about 2,600 people; the St. Francis dam failure in California in 1928 killed 350; the Buffalo Creek dam collapse in West Virginia killed 118; and a dam break in Morvi, India, killed thousands. Each of the disasters also destroyed much property.

Landsliding is the other hazard that may afflict urban areas. Slope stability failures create losses of hundreds of millions of dollars annually in the United States. The three principal factors that trigger landslides are earthquakes, excessive precipitation, and human activities (see Vol. XIV: *Slope Stability Analysis*). The most devastating disaster occurred in Kansu Province, China, during the 1920 earthquake that jarred the loess-covered hills, sending materials cascading into adjacent populated valleys and killing about 200,000. The 1970 earthquake in Peru killed 21,000 in the cities of Yungay and Ranrahirca when rock avalanches and debris flows overwhelmed the inhabitants. The 1964 Alaska earthquake killed more than 100 people in Anchorage when the Bootlegger Cove Clay failed and produced liquefaction flow landsliding. Exceptionally heavy rainfall in 1966 and 1967 in mid-southern Brazil caused tens of thousands of landslides, killing 2,700 people; property losses were so high that they were never completely calculated. In Aberfan, Wales, 144 people were killed in 1966 when rains soaked a spoil bank from coal-mining operations that landslided and smothered the residents.

Urban areas in California have long suffered from flooding and landsliding. The grading ordinances in southern California (see Vol. XIV: *Legal Affairs*) have greatly helped reduce losses in that part of the state. Winter rainstorms in 1968-1969, however, caused $25.4 million worth of slope failure losses in the San Francisco Bay region (Nilson et al., 1976), and $6.5 billion (Slosson, 1969) in the Los Angeles area. The Portuguese Bend landslide accelerated its motion after housing developments were constructed on the hillslopes. In a classic lawsuit of property owners versus the County of Los Angeles, the government was judged to have aided in causing the landslide because of road building and had to pay $5.36 million damages to homeowners.

Studies of slope stability and landsliding in urban areas formed an important part of the massive U.S. Geological Survey and Department of Housing and Urban Development investigation of the San Francisco Bay region (Nilson and Brabb, 1977).

West Virginia has completed landslide studies of the seven major urban areas of that state (Lessing and Erwin, 1977). The first step for reduction of landslide damages is careful mapping and recognition of landslide-prone terrain. The surface symptoms may include (1) abnormal contour changes and breaks in slope, (2) anomalous vegetation changes, (3) frontal bulge, (4) lateral tears, (5) remnant scars at head or within the terrain and (6) unusual moisture conditions.

Landslide prevention and control (except for legal aspects or avoidance of the area) constitute geomorphic engineering—the use of man-made structures to change a physical process (see Vol. XIV: *Landslide Control*). Alfors et al (1973) showed the importance for use of the full range of landslide remedy methods. Unless such measures are employed, the landsliding losses in California from 1970-2000 were predicted to be $9.85 billion; such potential losses can be reduced 99 percent if the appropriate management and engineering control methods are used.

The following methods are in use for minimizing landslide damages: (1) surface-water control, to prevent such waters from entering the risk area by diversions; (2) removal of subsurface water from critical soils and rocks by means of drains, tiles, interceptor trenches, and wells (see Vol. XIV: *Land Drainage;*) (3) cut-and-fill grading, including excavation and removal of material at the slide head, regrading of uneven topography, and hillside benching; (4) building of restraining structures, the most common being retaining walls, buttresses and shear keys, and rockbolts (see *Reinforced Earth*).

Land Use

An understanding of geomorphic principles is vital to the environmental planner in the design and use of urban areas (see Vol. XIV: *Environmental Geology; Environmental Management; Legal Affairs; Open Space*). Resource inventories of surface and subsurface materials should be made before they are paved over and become unobtainable. If stream channels must be changed or rerouted, the geometry of the man-made channels should resemble as closely as possible the planimetric and cross-sectional shapes of natural channels prior to modification (Keller, 1976). To prevent the evils of urban sprawl, which consume needlessly large areas of prime farmland and cause excessive waste of service facilities and resources, the new developments should conserve space and harmonize with nature. Highways should be planned to restrict the amount of natural interference (Parizek, 1971).

Although most goods, services, and resources that are imported to urban areas are necessary for human activities, enormous amounts of waste products result, which must be exported from the city.

The burial method of the sanitary landfill is used by many cities (see Vol. XIV: *Artificial Fill*). These by-products are commonly placed in the outer environs, and unless designed and managed properly, they can create deleterious impacts of the adjacent land-water ecosystems. Therefore cities affect not only the area within the city limits, but also cause influences in the surrounding countryside— its land, its water, and its air.

DONALD R. COATES

References

Alfors, J. T.; Burnett, J. L.; and Gay, T. E., Jr., 1973. Urban geology master plan for California, *California Div. Mines and Geology Bull. 198,* 112p.

Coates, D. R. (ed.), 1974. *Environmental Geomorphology and Landscape Conservation. Vol. II, Urban Areas.* Stroudsburg, Pa.: Dowden, Hutchinson & Ross, 454p.

Coates, D. R. (ed.), 1976. Urban geomorphology, *Geol. Soc. America Spec. Paper 174,* 454p.

Detwyler, T. R., and Marcus, M. G. (eds.), 1972. *Urbanization and Environment.* North Scituate, Mass.; Duxbury Press, 287p.

Fox, H. L., 1976. The urbanizing river: a case study in the Maryland Piedmont, in D. R. Coates (ed.), *Geomorphology and Engineering.* Stroudsburg, Pa.: Dowden, Hutchinson & Ross, 245-271.

Gulick, L., 1958. The city's challenge in resource use, in H. Jarrett (ed.), *Perspectives in Conservation.* Baltimore: Johns Hopkins University Press, 115-137.

Guy, H. P., 1976. Sediment-control methods in urban development: some examples and implications, in D. R. Coates (ed.), *Urban Geomorphology,* Geol. Soc. America Spec. Paper 174, 21-35.

Hammer, T. R., 1972. Stream channel enlargement due to urbanization, *Water Resources Research* **8,** 1530-1540.

Hansen, W. R., et al. 1976. Geologic constraints in the urban environment, *Geol. Soc. America Rept. of Comm. on Environmental and Public Policy,* 12p.

Johnson, S. L., and Sayre, D. M., 1973. Effects of urbanization on floods in the Houston metropolitan area, *U.S. Geol. Survey Water Resources Inv. 3-73,* 50p.

Keller, E. A., 1976. Channelization: environmental, geomorphic, and engineering aspects, in D. R. Coates (ed.), *Geomorphology and Engineering.* Stroudsburg, Pa.: Dowden, Hutchinson & Ross, 115-140.

Legget, R. F., 1973. *Cities and Geology,* New York: McGraw-Hill, 624p.

Leopold, L. B., 1968. Hydrology for urban land planning—a guidebook on the hydrologic effects of urban land use, *U.S. Geol. Survey Circ. 554,* 18p.

Lessing, P., and Erwin, R. B., 1977. Landslides in West Virginia, in D. R. Coates (ed.), *Landslides,* Geol. Soc. America Eng. Geology Revs. 3, 245-254.

Leveson, D. L., 1980. *Geology and the Urban Environment.* New York: Oxford University Press, 386p.

Nilson, R. H., and Brabb, E. E., 1977. Slope-stability studies in the San Francisco Bay region, California, in D. R. Coates (ed.), *Landslides,* Geol. Soc. America Eng. Geology Revs. 3, 235-243.

Nilson, T. H.; Taylor, F. A.; and Brabb, E. E., 1976. Recent landslides in Alameda County, California

(1940-1971): an estimate of economic losses and correlations with slope, rainfall, and ancient landslide deposits, *U.S. Geol. Survey Bull. 1398,* 21p.

Parizek, R. R., 1971. Impact of highways on the hydrogeologic environment, in D. R. Coates (ed.), *Environmental Geomorphology.* Binghamton State University of New York, Publications in Geomorphology, 151-199.

Peterson, J. T., 1973. The climate of cities: a survey of recent literature, in G. McBoyle (ed.), *Climate in Review.* Boston: Houghton Mifflin, 264-285.

Schmid, A. A., 1968. *Converting Land from Rural to Urban Uses.* Baltimore: Johns Hopkins University Press for Resources for the Future, 103p.

Slosson, J. E., 1969. The role of engineering geology in urban planning, in *The Governor's Conference on Environmental Geology,* Colorado Geol. Survey Spec. Pub. No. 1, 8-15.

Winkler, E. M., 1973. *Stone: Properties, Durability in Man's Environment.* New York: Springer-Verlag, 230p.

Wolman, M. G., 1964. *Problems Posed by Sediment Derived from Construction Activities in Maryland.* Report to Maryland Water Pollution Control Commission (now Department of Water Resources), 125p.

Wolman, M. G., and Schick, A. P., 1967. Effects of construction on fluvial sediment, urban and suburban areas of Maryland, *Water Resources Research* **3,** 451-464.

Cross-references: *Channelization and Bank Stabilization; Earthquake Engineering; Foundation Engineering; Geomorphology, Applied; River Engineering; Urban Engineering Geology; Urban Hydrology. Vol. XIV: Canals and Waterways; Cities, Geologic Effects; Expansive Soils; Land Drainage; Landslide Control; Slope Stability Analysis.*

URBAN HYDROLOGY

Although obviously not a new field, urban hydrology has recently been the subject of considerable interest and concern (Leopold, 1968; Johnson, 1971). Urbanization results in radical changes in land use and in the interaction between land and water. Hence the hydrology of urbanized areas differs notably from that of the same land in its preceding rural condition. *Urban hydrology* is the scientific application of hydrologic principles and knowledge to the planning and management of urban areas and their surroundings. It embraces all aspects of the interactions of man and water in occupancy of land. It includes the special hydrologic studies needed to accomplish these ends and deals with minimizing the adverse effects of man's use of land and water and with maximizing the effective use of the available water resources (see *Hydrogeology and Geohydrology*).

More than two-thirds of the U.S. population is urban, occupying about 7 percent of the land. This clear majority may well control the destiny not only of urban areas, but also of the remaining 93 percent of the land. Population trends in the

United States suggest that the urban population in the year 2000 may be three-fourths of the total population of the nation, and more than the total population of today. By that time, the urbanized area is expected to increase to perhaps 10 percent of the total land area of the nation. Sound management of total resources, and especially of urban areas, will then be increasingly urgent.

As urbanization proceeds, an increasing proportion of the total land area is covered with impermeable surfaces such as roofs and pavement. Rainfall, which formerly trickled slowly through vegetated areas or soaked into the ground, now runs quickly over the surface to streams. This creates one of the major problems of urban hydrology—the safe transmission and rapid disposal of greatly increased surface runoff. One role of the urban hydrologist is to estimate the quantities of maximum runoff and, on this basis, to design storm sewers, discharge channels, and disposal and treatment facilities. Impervious surfaces also reduce infiltration, soil moisture, and ground-water recharge. When combined with the increased ground-water withdrawal that often accompanies urbanization, this often results in lowered water tables and decreased ground-water yields (Rantz, 1970; Schneider, 1970).

Human activity in an urban environment produces large quantities of wastes that can find their way into and degrade the quality of the natural waters of the area. Surface streams receive both solid particles (sediment) and dissolved matter. Ground water normally receives only dissolved substances. The control of these contaminants and/or the correction of pollution is a major function of urban hydrology (Thomas and Schneider, 1970). Another important facet of urban hydrology is the provision of adequate quantities of good water for a concentrated population. If sources of water are not available in the immediate vicinity, a search in more distant areas becomes necessary. Thus, urban hydrology involves the planning and development of public water supplies and of the works for the disposal of liquid wastes (Schneider et al., 1973).

Urban hydrology is concerned with the management of flood plains under urban conditions. Many urban areas have been built on flood plains because these areas are level, easily built on, and usually dry. A flood plain is a strip of relatively smooth land adjacent to a river channel, constructed by the present river in its existing regimen, and covered with water when the river overflows its banks at times of high water. Buildings and other objects on the flood plain tend to be submerged or swept away at times of high water. Measures to prevent or reduce the loss of life and property on urbanized parts of flood plains thus become an important part of urban hydrology. The choice of measures to accomplish this includes structural measures, such as levees, dikes, channel improvement, and upstream storage, and nonstructural measures such as zoning and building regulations. The best solution in any given situation depends on local circumstances, and a combination of structural and nonstructural measures often produces the best results (Bue, 1967; McHarg, 1969).

The use and control of water for the aesthetic enhancement of the urban environment—for example, watering of desirable vegetation, judicious draining of urban swamps to increase their beauty and utility, and providing parkland lakes and water courses—also fall within the purview of urban hydrology. Closely allied to this function is the provision of water and the adaptation of natural bodies of water in urban localities to best serve the recreational needs of a crowded populace. This may range from the provision of a sprinkler on a city fire hydrant on a hot summer day to the construction of lakes in city parks that are used for rowing, fishing, skiing, or swimming. It includes the provision of public swimming pools, and the adaptation of the shores of major bodies of water to varied recreational uses.

Thus we see that urban hydrology covers a wide range of water management and control within and around cities. Where water is a threat or inconvenience, it must be controlled. Where it can serve a useful or satisfying purpose, it must flow in appropriate channels. Where human ignorance or greed results in structures that impinge on the natural channels or reduce the utility or quality of water, restrictions must be imposed so that we can live safely and harmoniously with the water that is an essential and inescapable part of our environment and that serves many of our needs.

Urban hydrology is related to numerous other sciences and disciplines, and when applied in conjunction with them, affords for wise planning and management of the entire urban environment (see *Hydrology*). The goal of such management should be to permit the efficient use of the hydrologic regime, and to preserve this vital resource for the benefit of future generations.

PHILIP E. LaMOREAUX
PAUL H. MOSER
HENRY C. BARKSDALE

References

Bue, Conrad D., 1967. Flood information for flood plain planning, *U.S. Geol. Survey Circ. 539,* 10p.

Johnson, James H., 1971. *Urban Geology—An Introductory Analysis.* Oxford, England: Pergamon Press, 188p.

Leopold, L. B., 1968. Hydrology for urban land planning—a guidebook on the hydrologic effects of urban land use, *U.S. Geol. Survey Circ. 554,* 18p.

McHarg, Ian, 1969. *Design with Nature.* Garden City, N.Y.: American Museum of Natural History, Natural History Press, 197p.

Rantz, S. E., 1970. Water in the urban environment: urban sprawl and flooding in southern California, *U.S. Geol. Survey Circ. 601-B,* 11p.

Schneider, William J., 1970. Water in the urban environment: hydrologic implications of solidwaste disposal, *U.S. Geol. Survey Circ. 601-F,* 9p.

Schneider, William J.; Rickert, David A.; and Spieker, Andrew M., 1973. Water in the urban environment: role of water in urban planning and management, *U.S. Geol. Survey Circ. 601-H,* 10p.

Thomas, Harold E., and Schneider, William J., 1970. Water in the urban environment: water as an urban resource and nuisance, *U.S. Geol. Survey Circ. 601-D,* 9p.

Cross-references: *Alluvial Plains, Engineering Geology; Hydrogeology and Geohydrology; Hydrology; River Engineering; Urban Geology.* Vol. XIV: *Canals and Waterways; Land Drainage.*

URBAN TUNNELS AND SUBWAYS

City planners and engineers are charged with the responsibility of deciding which construction method is best in the central business district. When dealing with the necessity of routing a large linear transportation system—be it for water, trunk sewers, or mass transit—the economic and social impact of their decision is most important in congested urban environments.

The three main modes of linear urban construction are (1) tunnels, (2) near-surface "cut-and cover," such as for pipelines and some subway excavations, and (3) surface and elevated conveyance, essentially following street patterns. Major urban subways of the world and transit distances are listed in Table 1.

Urban Geology

The subsurface geological condition is one of the most important factors that must be considered before construction methods are adopted. For example, an urban area underlain by sedimentary rock and a low water table, in which a modern tunnel boring machine can make good excavation progress (see *Rapid Excavation and Tunneling*), would be ideal for tunneling beneath all the utilities and surface structures, be independent of street patterns, and should prove several times more economical than a surface or near-surface conveyance.

Conversely, an urban area underlain by loose sediments, such as recent alluvium or estuarine deposits, and with a high water table, may dictate surface or cut-and-cover construction as the only feasible method, in spite of resultant utility relocation, community disruption, and loss of business along the route during the construction period

(two to five years at a given locality). The marriage of construction method with geological conditions is important because construction costs are commonly 70 percent of the total cost of large linear projects, such as a rapid transit system.

Tunneling Methods

Soft-Ground Tunnels. The term *soft-ground* includes sedimentary formations such as sandstone, shale, marl, limestone, alluvium, terrace deposits, and glacial till. If these formations are dry, or can be dewatered easily, then use of a tunnel boring machine (TBM) is the most economical method of excavation. Rates of progress are commonly 700 m per month, including primary lining. With improvements in design and cutters, TBMs have actually lowered the cost of tunneling in the past several years on some jobs in the United States. For example, a recent, long water tunnel in southern California, 7 m in diameter, was bid at $3.5 million per 2.6 km, including concrete lining; the tunnel access points were located 5 km apart (Proctor and Hoffman, 1974).

Cities known to be underlain largely by soft-ground conditions especially suitable for high-speed TBMs are Los Angeles, Denver, Minneapolis-St. Paul, Dallas, Kansas City, Chicago, St. Louis, Cincinnati, Indianapolis, and others (see Table 2).

Hard Rock Tunnels. Rock tunneling is generally in igneous rocks and metamorphic rocks. In very hard rocks tunneling usually progresses most rapidly by the "conventional" method of drilling and blasting. Shallow rock in urban areas (e.g., New York City and Washington, D.C.) imposes a severe constraint on the excavation progress by restricting blasting to daylight hours and limiting the amount of explosives used in each blast.

Deep tunnels mitigate the surface vibrations due to blasting, but require deeper subway stations. Four cities in the world (Leningrad, London, New York, and Washington, D. C.) have portions of their subways deeper than 60 m. Deeper tunnels usually encounter less weathered rock, and the joints and fractures in the rock are tighter and less frequent, thus requiring fewer tunnel supports. The effects of ground water in hard-rock tunnels is minimal; the water is simply pumped out without causing much delay in tunneling progress. Most TBMs have not proved economically successful in very hard rock, mainly because of high cutter wear and maintenance.

Compressed Air Tunnels. This type of tunneling is the most expensive because subsurface conditions are usually soft, saturated soils such as the San Francisco Bay mud with which BART had to contend (Kuesel, 1969) (Fig. 1).

TABLE 1. Urban Subways of the World

City	First Year of Operation	Total Rapid Transit Miles (km)	Total Subway Miles (km)
Athens	?	16(26)	1(2)
Atlanta (MARTA)	1977	50(80)	9(14)
Baku Metro, USSR	1967	6(10)	6(10)
Barcelona Metro	1924	28(45)	28(45)
Berlin U Bahn (East & West)	1902	61(98)	52(83)
Bonn, W. Germany	1974	7(11)	7(11)
Boston—Mass. Bay Trans. Authority	1897	70(112)	17(27)
Budapest Tramway Organization	1896	6(10)	6(10)
Buenos Aires Metro	1913	49(78)	30(48)
Chicago Transit Authority	1943	88(141)	9(14)
Cologne, W. Germany	1974	120(192)	18(29)
Copenhagen S-Toge	?	41(66)	?
Fort Worth M&O	1964	1(2)	0.4(0.6)
Frankfurt Stadtwerke	1968	83(133)	20(32)
Glasgow Corp. Transport	1897	7(11)	7(11)
Haifa Carmelit Subway	1959	1(2)	1(2)
Hamburg U Bahn (HHA)	1912	55(88)	16(26)
Helsinki	1976	45(72)	?
Kiev Metro	1960	15(24)	10(16)
Leningrad Metro	1955	30(48)	30(48)
Lisbon Metro	1959	8(13)	8(13)
London Transport	1863	254(409)	100(162)
Madrid Metro	1919	36(58)	36(58)?
Mexico City Metro	1967	26(42)	26(42)
Milan Metro	1964	14(22)	12(19)
Montreal Metro	1967	16(26)	16(26)
Moscow Metro	1935	100(162)	57(92)
Munich U Bahn	1972	22(35)	10(16)
Nagoya City Transit Bureau, Japan	1957	14(22)	14(22)
New York City Transit Authority	1904	240(388)	137(221)
New York Port Authority Trans-Hudson (PATH)	1908	14(22)	7(11)
Newark City Subway (TNJ)	1903	13(21)	4(6)
Osaka Municipal Trans. Bureau	1933	42(67)	42(67)
Oslo Kommune Tunnelbane	1928	25(40)	4(6)
Paris Metro (RATP)	1900	144(230)	105(168)
Peking	1969	?	?
Philadelphia Trans. Co. & PATCO	1907	45(72)	18(29)
Rome (STEFER)	1955	19(30)	9(14)
Rotterdam Electric Tram	1968	6(10)	3(5)
San Francisco (BART)	1972	75(120)	21(34)
Santiago, Chile	1976	65(104)	12(19)
São Paulo, Brazil	1974	42(67)	42(67)?
Sapporo, Japan	1972	8(13)	8(13)
Seoul, Korea	1975	6(10)	6(10)
Stockholm T Bana	1950	40(64)	16(26)
Stuttgart, Germany	1974	?	11(18)
Sydney Metro, Australia	1926	7(11)	7(11)
Tbilisi Metro, USSR	1966	6(10)	5(8)
Tokyo Metro (TRTA)	1927	95(153)	95(152)
Toronto Transit Commission	1954	21(34)	17(27)
Vienna Municipal Stadtbahn	1898	17(27)	4(6)
Warsaw Metro	?	?	3(5)
Washington, D.C. (WMTA)	1975	98(158)	45(72)
Yokohama, Japan	1974	46(74)	?

Source: After Howson (1972), Walton and Proctor (1976), and U.S. Geological Survey (1974).

TABLE 2. Urban Geology for Underground Construction

1970 SMSA Metropolitan Area	Rock Types[a]	Rapid Transit Status[b]
New York, NY	HR—granite schist, clay	1, 2
Los Angeles, CA	M—sandstone, shale	3
Chicago, IL	M—clay, limestone	1, 2
Philadelphia, PA	HR—granite schist	1, 2
Detroit, MI	M—clay, sandstone, limestone	3
San Francisco, CA	M—clay, sandstone, shale	1, 2
Washington, DC	HR & M—clay, sandstone, granite	2
Boston, MA	HR & M—clay, argillite, volcanic, granite	1, 2
Pittsburgh, PA	M—sandstone, shale, limestone	3
St. Louis, MO	M—limestone	3
Baltimore, MD	HR—clay, granite schist	2
Cleveland, OH	M—sandstone, clay	1
Houston, TX	M—sand, clay (saturated)	3
Newark, NJ	M—clay, sandstone	1
Minneapolis, St. Paul, MN	M—sandstone, till	3
Dallas, TX	M—sandstone, limestone	?
Seattle, WA	M—clay, till	3
Anaheim, CA	M—sandstone, sand	
Milwaukee, WI	M—limestone, clay, till	3
Atlanta, GA	HR & M—granite, clay	2
Cincinnati, OH	M—limestone, till	
Paterson, NJ	M—sandstone, basalt	
San Diego, CA	HR & M—clay, sandstone, granite	3
Buffalo, NY	M—limestone, sandstone	3
Miami, FL	—sand (saturated)	3
Kansas City, KS	M—limestone, shale	3
Denver, CO	M—sandstone, shale	3
Riverside-San Bernardino, CA	HR & M—soft sandstone, granite	
Indianapolis, IN	M—limestone, till	
San Jose, CA	—sand (saturated)	3
New Orleans, LA	—clay (saturated)	3
Tampa, FL	—sand (saturated)	
Portland, OR	HR & M—basalt, soft sandstone	3

[a]M = Typical "mole" ground
HR = Hard rock

[b]1 = Has rapid transit
  2 = Rapid transit funded or under construction
  3 = Rapid transit proposed

Source: After Walton and Proctor (1976).

Excavation by TBM or shield must include bulkheading or breast boarding to prevent the working face from coming into the tunnel. This method is used when a high water table cannot be lowered prior to construction operations, due either to impermeable sediments or to fear of causing surface settlement with dewatering.

Cut-and-Cover Method

A typical subway cut-and-cover construction process involves the following steps: (1) designing and installing underpinning for all the buildings along the route with foundations close to the cut; (2) relocating or avoiding utility lines and other subsurface installations along the route; (3) drilling in soldier piles or slurry walls on both sides of the projected cut down the streets along the route, wherever there is any depth of unconsolidated

overburden requiring support; (4) taking up the pavement and installing a deck in sections, while maintaining one or more lanes of traffic; (5) while the decking is completed, beginning excavation by lifting muck with cranes at intervals of a few blocks for haulage by trucks on the city streets; (6) after excavation is completed, beginning construction with steel and ready-mixed concrete brought by trucks through city streets to crane locations; and (7) on completion of structures, placing backfill and restoring the pavement. The process requires several years of surface disruption at any given locality along the route.

Advantages and Disadvantages of Tunnel and Cut-and-Cover

The benefits and costs listed here do not all have the same weight. Consideration must be given to

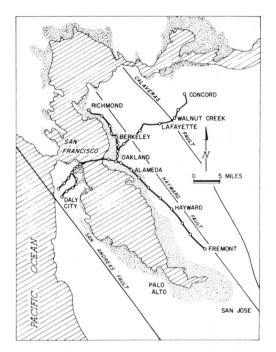

FIGURE 1. The 75-mile Bay Area Rapid Transit system. Subways comprise 20 miles including 10 miles tunneled, 6 miles cut-and-cover, and almost 4 miles of sunken tube. Stippled areas are Bay Mud (from Proctor and Hoffman, 1974).

Advantages of Subway Construction by Tunneling. The main advantages of tunnels include minimum surface disturbance (only at construction shaft sites and at station sites); shorter construction period; few street closures and detours with minimal traffic congestion; no visual clutter along the route; no surface noise, dust, or fumes along the route; construction is done only beneath utilities or substructures; no underpinning requirements; no loss of tax revenue to city due to loss of businesses adjacent to route; tunnels are not confined to street patterns; lower right-of-way costs when route must cross private property because the subsurface is in easement, not fee purchase; and, finally, the fact that tunneling is safer than surface and near-surface structures during an earthquake.

Some comments on seismic hazards may be helpful relative to subways. The greatest damage in any earthquake is due to ground shaking, not surface fault rupture. When a strong earthquake occurs, a large area is subjected to strong ground shaking, and past experience has shown that tunnels (and deep caves) are less damaged than surface or elevated facilities. This is logical because a tunnel is surrounded by the medium that is moving, and moves with it. At the surface, where a ground-air interface exists, and other seismic waves come into play, the shaking and the resultant damage are more significant. Thus, earthquake hazard considerations strongly favor deep tunnels.

Advantages of Subway Construction by Cut-and-Cover. Advantages of cut-and-cover methods focus on: less expensive subway stations; ready access along entire route during construction; less ground-water disposal in near-surface excavation; no concentrated disposal of ventilation air as with

the weight (or ranking) each advantage or benefit may accrue in each particular community situation (Fig. 2). Similarly, the disadvantages, costs, and negative impacts are not of equal weight (Proctor and Hoffman, 1974).

CONSTRUCTION METHOD	NOISE	TRAFFIC DISRUPTION	SURFACE WATER DISPOSAL	VIBRATION	UTILITIES RELOCATION	RIGHT-OF-WAY COSTS	COMMUNITY DISRUPTION	CONSTRUCTION COSTS
CUT-AND-COVER	●	●	●	●	●	○	●	●₁
MOLE TUNNELING	○	○	●					○₃
HARD-ROCK TUNNELING	○	○	●	●				●₂
ELEVATED GUIDEWAY	●	●				●	●	○₄

●   HIGH COST OR SERIOUS IMPACT

○   LOW COST OR IMPACT

BLANK   NEGLIGIBLE COST OR IMPACT

FIGURE 2. Assessment of impacts of transit construction in the urban environment (after Perazich and Fischman, 1966).

a tunnel; and no blasting vibrations, as in hard-rock tunnels within about 70 m of the surface.

Advantages of Subways Versus Elevated Guideways. Subways offer advantages over elevated guideways because they can be used for civil defense shelters after construction; they provide auxiliary conduit space overhead or under the rails for telephone, water, power; and there is no withdrawal of open space, or possibility of architecturally dated guideway support columns (e.g., the New York "el" and the Chicago "loop").

Conclusion

Geological advice early in the planning stage is a prerequisite to deciding modes of mass transit construction in urban areas (Legget, 1973). Unfortunately, the most current subway construction in the United States has been in cities with geological conditions unfavorable to high-speed TBM excavation (e.g., San Francisco, Washington, D.C., Baltimore, and New York City). As a result, planners in other cities have assumed the tunneling option is the most expensive alternative. This is simply not true in many cities. Furthermore, when the real costs of social and psychological impacts attendant with surface disruption are included with the total rapid transit costs, tunneled subways often emerge as the least expensive alternative.

RICHARD J. PROCTOR

References

Howson, H. F., 1972. *The Rapid Transit Railways of the World.* London: George Allen and Unwin, 183p.

Kuesel, T. R., 1969. "BART" subway construction: planning and costs, *Civil Engineering* (March), 60-65.

Legget, R. F., 1973. *Cities and Geology.* New York: McGraw-Hill, 624p.

Perazich, G., and Fischman, L. L., 1966. Methodology for evaluating cost and benefits of alternative urban transportation systems, *Highway Research Rec.* **148,** 59-71.

Proctor, R. J., and Hoffman, G. A., 1974. Planning subways by tunnel or cut-and-cover—some cost-benefit comparisons, in *Proceedings of the 2nd Rapid Excavation and Tunneling Conference,* Vol. 1. New York: American Institute of Mining Engineers, 51-63.

U.S. Geological Survey, 1974. *Summary of Geologic and Hydrologic Information Pertinent to Tunneling in Selected Urban Areas.* Prepared for U.S. Department of Transportation, No. DOT-TST-75-49, available at National Tech. Information Service, Springfield, VA., 1974.

Walton, Matt, and Proctor, R. J., 1976. Urban tunnels—an option for the transit crisis, *Am. Soc. Civil Engineers Transp. Eng. Jour.* **102** (November), 715-726.

Cross-references: *Earthquake Engineering; Pipeline Corridor Evaluation; Rapid Excavation and Tunneling; Tunnels, Tunneling; Urban Engineering Geology; Urban Geology.*

V

VECTOR ANALYSIS — See Vol. I.

W

WATER ENGINEERING — See CHANNELIZATION AND BANK STABILIZATION; COASTAL INLETS, ENGINEERING GEOLOGY; RIVER ENGINEERING; WELLS, AERIAL; WELLS, WATER. Vol. XIV: ALLUVIAL SYSTEMS MODELING; CANALS AND WATERWAYS.

WATER GEOCHEMISTRY, SURVEYS — See Vol. XIV: HYDROCHEMICAL PROSPECTING.

WATER SAMPLING — See Vol. XIV: HYDROCHEMICAL PROSPECTING.

WELL DATA SYSTEMS

The term *well data systems* is applied to various information files that include the factual information on historical drilling operations. These files are prepared by the cooperative effort of segments of the oil and gas industry and other organizations. Information from current drilling activity is added to the historical files periodically as an updating procedure.

History

Efforts to develop a means of storing the very large quantity of drilling information were initiated in the 1950s by several study projects in the state of California and in the province of Alberta,

Canada (Walsh, 1973). These efforts were not successful, due primarily to limitations in computer equipment and programming knowledge at the time. In the early 1960s, efforts to convert well information to a machine form were reinitiated by the oil and gas industry, and in the fall of 1961 a group known as the Permian Basin Well Data System was informally organized in Midland, Texas. Eighteen months later, on March 18, 1963, the Permian Basin Well Data System was officially incorporated as the first operational well data system. At approximately the same time that the Permian Basin system was coming into existence, a similar project to encode well data from the Rocky Mountain region was being created in Denver, Colorado, by Petroleum Information Corporation, a commercial scouting agency that had been closely associated with the earlier efforts in California and Canada. Shortly after the Permian Basin and Rocky Mountain efforts materialized, study groups were also formed in the Gulf Coast and mid-continent areas as well as in the state of Michigan and in eastern Canada.

Coverage and File Size

The areas covered by current well data systems are shown in Figure 1. A total of about 2.2 million wells have been drilled in the United States and Canada through 1967, and more than half of these wells rest within the area covered by the various active well data systems. At the end of 1967 the systems had encoded the drilling information from some 450,000 of these wells. The information of interest to the geologists and engineers concerning drilling operations consists of items such as the name and location of the boring, the drill stem tests and production tests run during its drilling,

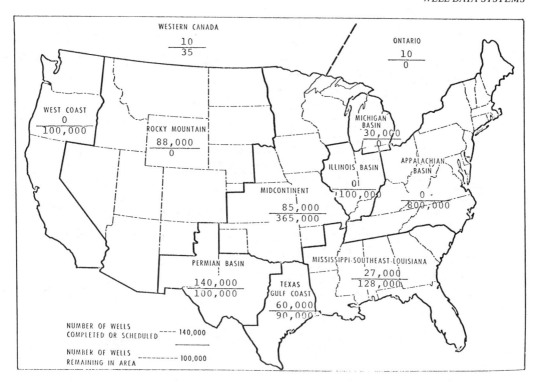

FIGURE 1. Coverage of cooperative well data systems and their status.

the completion procedures used in the well, and any unusual difficulties encountered in the drilling operation. The average well report contains about 2,000 characters of information. Thus, to date some 900 million characters have been reduced to magnetic tape form for machine manipulation (Dillon, 1967).

Data Flow Procedure

The general flow of information from its origin into a well data system file is shown in Figure 2. The basic source of all drilling information is the well itself. Information on the well is, by custom, initially available to the operators of wells and their partners, if any. A limited quantity of information is also available to service companies who provide support or special services in the well's drilling or completion. The basic decision as to what information can be released on the well rests with the operator, although in some states in the United States and in most of the Canadian provinces the operator is required to release much of the information within a certain period of time after the well's completion.

The information that is available to anyone at the time of drilling or completion of the well, or shortly thereafter, is said to be "unrestricted." All information withheld by the operator is referred to as "restricted" on the diagram. As a general indus-

try practice in most countries, all information acquired by a service company is withheld until the operator specifically authorizes its release, or until a regulatory agency requires that it be released.

A certain amount of information on each well is required to be reported on various official forms to the regulatory bodies, such as the Texas Railroad Commission, the California Division of Oil and Gas, the New Mexico Conservation Commission, and so on. By law, these items of information become public data and are available to anyone through the regulatory files. Much of the remainder of the unrestricted data is gathered by oil scouts during the drilling of the well through personal visits to the well site or by some contact with the operator. These scouts either represent individual companies or belong to a commercial scouting organization that distributes the gathered data to its members. The unrestricted data cease to be available, except through a direct request for the information from the operator, once the particular operation is finished or the well is completed. If the well is completed as a producer, additional information on the well's production history may be available.

The well data systems concern themselves with the unrestricted data as well as with restricted information that has been released subsequent to the well's completion. Merged in with this information are special items of data calculated by the

613

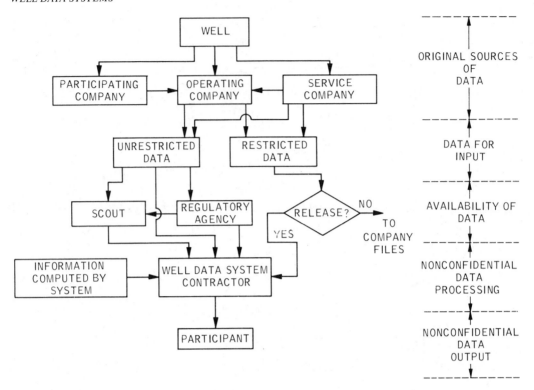

FIGURE 2. Flow of information from the well to the participant in a typical well data system.

system. Ultimate distribution of the data is in punchcard, disc, or magnetic tape form to the members of the system or to the subscribers, depending on the organizational structure of the system's management.

Information on North American drilling activity is, for example, reported by members of the Committee on Statistics of Drilling (CSD) of the American Association of Petroleum Geologists (AAPG). Basic data is reported for all wells, but CSD district chairmen report additional interpretive data on exploratory wells only. Data are gathered in the field and transmitted to the American Petroleum Institute (API) in Washington, D.C., where they are then computer processed. The API publishes monthly, quarterly, and annual statistical tables of these data; year-end tables may also be obtained from this computer file.

Annual well data for wells drilled in the United States are available on magnetic tape in essentially the same form as the API-AAPG (CSD) individual well ticket (Johnston, 1980). Tapes containing both exploratory and development data may be ordered through the American Petroleum Institute (Washington, D.C.) for a moderate fee plus computer costs for processing and copying. Annual exploratory well tapes only can be ordered from the American Association of Petroleum Geologists (Tulsa, Oklahoma) for a fee plus processing costs. Special requests for parts of tapes are per-

mitted where the data are provided on a proportionate cost formula. Most tapes are available in the current format from 1969 to the present.

The non-confidential items to be processed by the contractor for a well data system include (1) information computed by the system for the benefit of all of its participants (well coordinates, corrected well classifications, and other special information introduced by an individual member); (2) information from the files of the participating companies that is common to all of the companies, generally information that was initially acquired by a scouting group or organization; and (3) useful items of restricted data that have been cleared for release by the well's operator.

Formats

The organization of the information as it is distributed to the member companies is generally that of a series of 80-column punchcards. This very basic format is used as it is easily convertible to their own internal files by all member companies regardless of the particular type of computer hardware that they might have. This "card-image" approach is bulky, but it does produce the information in the most easily digested form for all clients (Brandon, 1965). The number of cards of data for any given well may number in the hundreds, although the average well numbers between 30 and 50 cards.

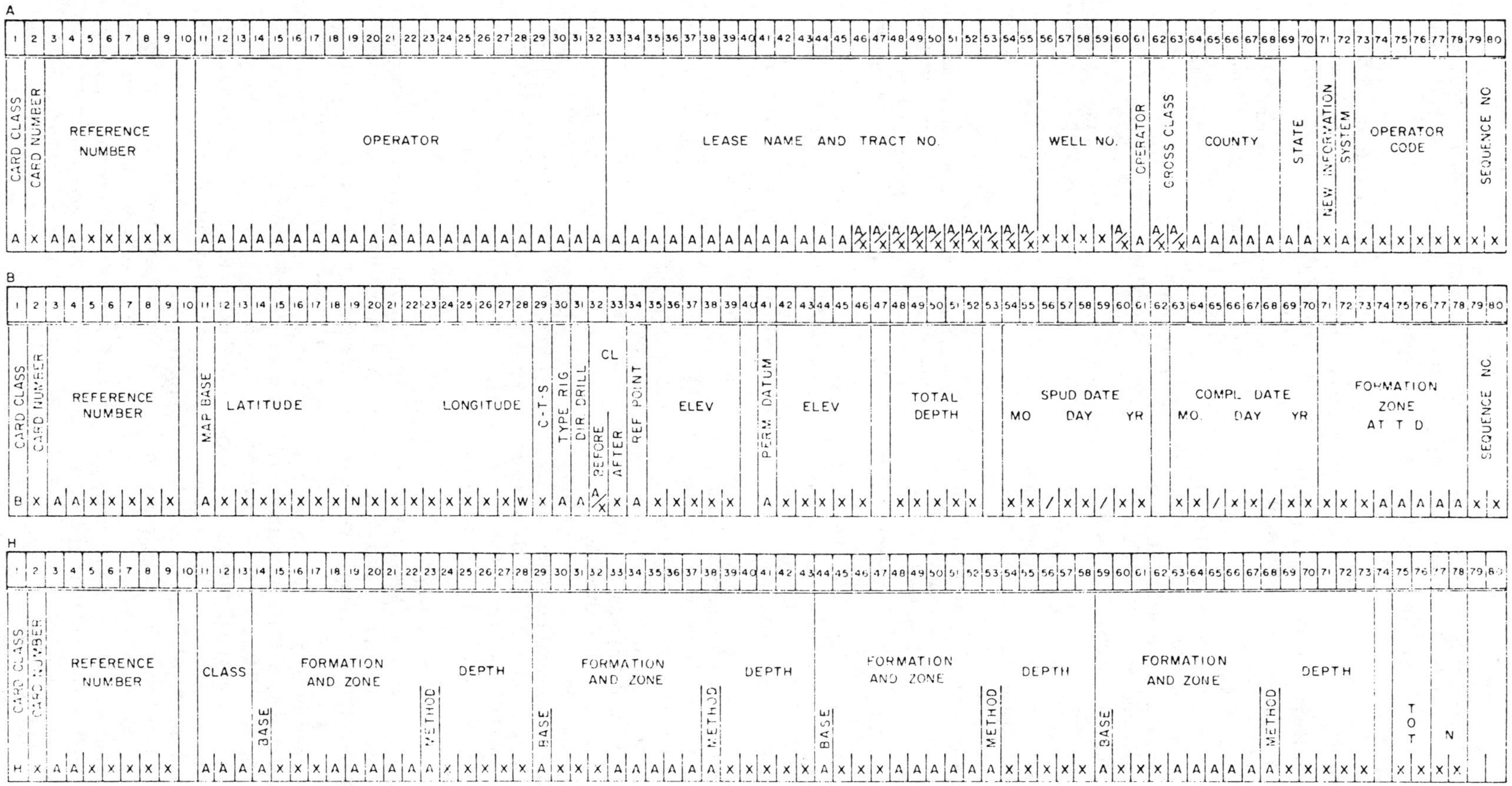

FIGURE 3. Permian Basin Well Data System formats.

The format of the blank cards from the Permian Basin Well Data System that deal with well identification, location, and formation tops are shown in Figure 3.

Individual Systems

The Permian Basin Well Data System covers much of western Texas and southeastern New Mexico. Included in this system is information from some 140,000 wells in 68 counties. Surveys of an additional 96 counties, containing some 100,000 wells, are yet to be completed. These counties lie mostly to the east and immediately to the north of the ones that have been finished. In early 1968 a total of 14 companies belonged to the Permian Basin Well Data System.

The Rocky Mountain Well History Control System covers most of the western and north central United States except for the Pacific Coast area. This system includes a total of some 88,000 wells. Approximately 14 companies have participated in this acquisition of the historical data and 15 companies are subscribing to current well information.

The Michigan Basin Well Data System includes the state of Michigan and a few counties in Ohio and Indiana. Approximately 30,000 wells have been drilled in this area, and information from these wells has been placed in computer form. Seven companies and two universities supported this system.

The Northeastern United States Well Data System is essentially inactive. Only a handfull of deep wells in this area have been encoded, and almost 800,000 wells remain to be done. The shallowness of many of the tests and the paucity of major company activity in this area have slowed the encoding of information from this portion of the country.

The Mississippi-Southeast Louisiana Well Data System embraces about 150,000 wells in seven states. Approximately one-fifth of this information has now been encoded. Eight oil and gas companies support this effort.

The Texas Gulf Coast System also includes approximately 150,000 wells. About 60,000 are estimated to have been completed up to the beginning of 1968, with the work being supported by six companies.

The Mid-Continent Well History Control System has processed some 85,000 wells, with 365,000 remaining to be done. There is a possibility that the completion of the Mid-Continent coverage may lag for the same reason as the encoding of well data from the Appalachian Basin is being retarded. Seven companies are actively involved in this project.

A large quantity of data on drilling activity in the Illinois Basin, which embraces the state of Illinois, most of Indiana, and western Kentucky, is available. No organized effort to encode these data has developed, however, except for a project in the planning stage by the Illinois State Geological Survey.

The West Coast of the United States is estimated to contain about 100,000 wells, and the general lack of a cooperative data exchange on the West Coast has slowed the creation of a well data system for that area.

In Canada, two well data systems are active. In western Canada the system is known as Petrodata, and is operated by a commercial scouting and engineering service. Approximately 10,000 of the 40,000 wells drilled in western Canada have been encoded. In eastern Canada the University of Western Ontario has led the development of a well data system that has encoded all of the approximately 10,000 wells that have been drilled in that area.

ED L. DILLON

References

Brandon, B. H., 1965. Electronic storage, retrieval, and processing of well data, in *Computer Techniques for the Petroleum Geologists.* Ann Arbor, Michigan: The University of Michigan Engineering Summer Conferences.

Dillon, E. L., 1967. Expanding role of the computer in geology, *Am. Assoc. Petroleum Geologists Bull.* **1,** 1185-1201.

Johnston, R. R., 1980. North American drilling activity in 1979, *Am Assoc. Petroleum Geologists Bull.* **64** (9), 1295-1330.

Walsh, J. D., 1973. Oil field automation, in G. V. Chilingar and C. M. Beeson (eds.), *Surface Operations in Petroleum Production.* New York: American Elsevier, 339-348.

Cross-references: *Computerized Resources Information Bank; Earth Science, Information and Sources.* Vol. XIV: *Computers in Geology; Geostatistics; Punch Cards, Geologic Referencing; Well Logging.*

WELLS, AERIAL

Artificial condensers (Hitier, 1925) found in the ruins of the ancient Greek colony of Theodosia, Crimea, were used as a secondary source of water supply in an arid climate. A network of condensers, conduits, and cisterns provided dependable amounts of water for the town and its gardens more than 2,000 years ago. This structure for condensing atmospheric water vapor in the air was built above the ground surface instead of below it. Hence the term *aerial well.*

F. I. Siebold's discovery, in about 1883, of the ancient aerial wells revealed 13 enormous pyramids of crushed calcareous rock fragments, 5 to 10 cm in size. The wells were located on the crests of mountains surrounding Theodosia, 300 to 320 m above sea level. The size of the pyramidal rock

FIGURE 1. Knapen's experimental aerial well at Trans, Vars, France.

piles spread over a distance of about 3 km and measured 30 m in length, 25 m in width, and 10 m in height.

Upon entering the cool interior of the pile of crushed rock through voids, atmospheric vapor becomes chilled and is transformed into water. Siebold estimated that each of these 13 condensers would yield about 55,400 liters of water per day, supplying Theodosia with approximately 721,000 liters of water per day.

Investigations by Chaptal (1932) and Knapen (1928) showed that there is an important source of (soil) moisture in addition to rain and dew, and confirmed experimentally that it is possible to obtain rather large quantities of water, by the process of condensation of atmospheric water vapor on rocky surfaces, under appropriate conditions. Figure 1 shows Knapen's experimental aerial well at Trans, Vars, France.

ALFREDS R. JUMIKIS

References

Chaptal, L., 1932. La captation de la vapeur d'eau atmosphérique, *Annales Agronomiques* **4,** 540-555.
Hitier, H., 1925. On Condensers of atmospheric vapor in antiquity, *Acad. Agriculture C. R.* 679-683.
Knapen, A., 1928. Mémoires sur le puits aérien, *Soc. Ingénieurs Civils de France Bull.* 139-140.

Cross-reference: *Wells, Water.*

WELLS, WATER

A *water well* is a pit, shaft, or hole sunk into the earth to exploit a supply of water. Wells also may be used to inject waste water into deep aquifers or to recharge ground-water supplies artificially.

Although most water wells are the traditional vertical ones, horizontal wells are common in some areas. As an example, the *kanats* of Persia are horizontal infiltration galleries connected by shafts (see Geomorphology, Applied); similar galleries are used in the chalk aquifers of southeastern England. Horizontal wells may be used to "skim off" less dense fresh water floating above sea or other salt water. Collector wells, often used to include infiltration from a river, are large-diameter shafts with radiating, horizontal collector pipes.

History

The use of wells to obtain water began, by necessity, in arid regions. The Bible contains numerous references to wells and well construction. Roman cities, before the building of the famous aqueducts, depended entirely on wells or cisterns for water. In the Middle East, as early as 800 B.C., infiltration galleries connected by shafts (*kanats*) were used to obtain large supplies of water; *kanat* systems are often several kilometers long and may be as much as 150 m below the ground surface. Shafts are spaced along the tunnels at an average of 40 per km. Centuries ago in China, wells were drilled manually by percussion methods and cased with bamboo. Some of these wells are almost 1,700 m deep and took three generations to complete.

In the Western world, flowing wells, first drilled near Artois, France, created interest in improved drilling methods (see Vol. XIV: *Borehole Drilling*). Early wells in the United States were primarily drilled to obtain salt; in the late 1800s, deep wells helped develop much of the western part of the nation.

The oil industry has innovated most modern drilling techniques, and many recent water well-drilling methods are adaptations from the industry. Modern pumps allow economical lifts of large volumes of water from depth, and today drilled wells have become one of the standard methods of supplying clear, good-quality water throughout the world (Anderson, 1969).

Use

About one-fifth of all water used in the United States is obtained from wells or springs. More than 75 percent of individual (mostly rural) water systems depend on ground water (see Vol. XIV: *Groundwater Exploration*). Private wells are common throughout the country, although wells for community supplies are more common in the southern and western states than in the eastern ones.

Construction

In unconsolidated materials (soft rock), wells may be dug, bored, jetted, driven, or drilled. In the past, many wells were hand dug, but power equipment has almost completely replaced hand digging. Dug wells may be a meter or so in diameter. They generally range from 3 to 15 m in depth, although the city of Spokane, Washington, obtains water from dug wells more than 85 m deep. Dug wells are cribbed with wood, brick, rock, concrete, or metal (see *Reinforced Earth*).

Wells can be bored by hand or with power-driven augers (see Vol. XIV: *Augers, Augering*). Jetted wells are constructed using a stream of water (often in conjunction with driving) to wash open a hole ahead of a special well point. The method is best adapted to holes 10 cm or less in diameter. Driven wells utilize special points, are small in diameter, and are commonly completed at shallow depths. Often a series of well points will be jetted or driven for a special purpose, such as lowering the water table during a construction project.

Wells in solid rock formations (hard rock) must be drilled; most methods are adaptations (or combinations) of the cable tool or rotary method. *Cable tool drilling* (also called percussion or churn drilling) consists of lifting and dropping a heavy metal bit. The resultant broken rock (cuttings) must be removed from the bottom of the hole with a special bailer. In soft formations, casing must be driven closely behind the bit to keep the hole from collapsing.

The *hydraulic rotary method* uses a rotating bit. Continuously circulating fluid, generally a specially prepared mud containing bentonite or other clay, removes cuttings and helps stabilize the walls of the hole in unconsolidated formations. Cone-shaped roller bits are used in hard rock, whereas "drag bits" with short blades cut soft rock. In the *normal rotary method*, the fluid is injected through a hollow drill stem and the cuttings are washed upward to the surface around the drill stem.

In the *reverse rotary method*, the circulation of drilling fluid is such that cuttings are pulled upward inside the hollow drillpipe. Recently, water wells have been drilled using air, or a mixture of air and water, as the drilling fluid; either roller-type bits or special pneumatic-hammer bits are used. A further refinement of the air rotary method uses a special "hammer" that drives the casing directly behind the drill bit.

Completion

Drilled wells are completed to provide minimum resistance to water entrance. In hard rock, water enters the uncased hole and no completion is needed; in unconsolidated materials, however, a perforated casing or screen allows water to enter

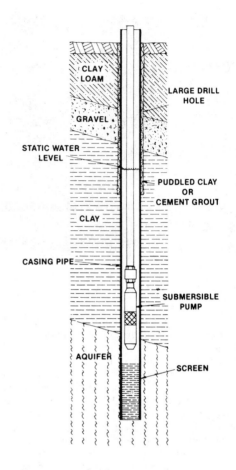

FIGURE 1. Typical drilled well, showing how contaminated water can be sealed off using proper construction techniques (from *Water System and Treatment Handbook*, 1983, with permission of the Water Systems Council).

while protecting the hole from collapse. The casing can be perforated before installation by machine or with an acetylene torch, or perforated in place with a casing knife (Fig. 1).

Screens, used instead of perforated casings, are available in various materials, diameters, and slot sizes. Openings in both screens and casings should be large enough to allow 50 to 60 percent of the sand grains to enter the well and be removed. Additional well completion consists of placing an envelope of sorted gravel or coarse sand around the casing. The gravel pack acts as a strainer, cuts down on frictional losses, and protects the screen or casing.

Development

After completion, water wells are developed for maximum production by removing the fine material through the casing or screen. The simplest method of development is by pumping at a rate higher than when the well will be in service. The more common method is surging, either with a plunger or with compressed air. High-velocity jetting, backwashing, addition of solid CO_2 (dry ice), and explosives have also been used to develop wells.

Pumps

Water is obtained from wells by natural flow or, more likely, by pumping. Pumps used in water wells include suction pumps, rotary pumps, jet pumps, airlift pumps, piston pumps, and centrifugal (turbine) pumps. Turbine-type pumps with completely submersible motors have recently come into extensive use (Gibson and Singer, 1971).

Ground-Water Protection

In the last several decades it has become apparent that ground water becomes contaminated by careless water well drilling and completion techniques. Once contaminated, it may take tens of years for a ground water supply to renew and purify itself. Almost all state governments recognize this problem and have set minimum construction standards (*Water System and Treatment Handbook*) for water wells or have a contractor licensing program. Many state (or county) governments also require water well permits or completion reports and have a continuing water well inspection program.

SYLVIA H. ROSS

References

Anderson, Keith E. (ed.), 1969. *Water Well Handbook.* Rolla, Mo.: Missouri Water Well and Pump Contractors Association, 281p.

Gibson, U. P., and Singer, R. D., 1971. *Water Well Manual.* Berkeley, Calif.: Premier Press, 156p.

Water System and Treatment Handbook, 1983. 8th ed. Chicago: Water Systems Council.

Cross-references: *Pumping Stations and Pipelines, Engineering Geology.* Vol. XIV: *Augers, Augering; Borehole Drilling; Groundwater Exploration.*

AUTHOR CITATION INDEX

SUBJECT INDEX

Abandoned channel, 9
  environment, 9
Abandoned course environment,
  10
Abandoned formation pressures,
  causes of, 175
Abandoned river channel, 9, 10,
  131
Absolute orientation photography,
  400
Absorption terrace. *See* Terraces
Accelerogram, earthquake, 143,
  145, 556
Acoustics, underwater operations,
  381
Active layer, frozen ground, 386
Active stress, 493
Activity ratio, 26
Adit, 514
Adjacent settlement, 420
Aerial photographs, 167, 403, 445
  classification of, 392
  geometry of, 393-394
Aerial photography, 454
Aerial survey, 393
Aerial triangulation, 401
Aerial well, 616
Aeroembolism, caisson disease,
  196
Age discordance, geochronology,
  218
Agricultural suitability, 597
"Airfield Classification," 524
Airphoto interpretation, 451
Air pollution mapping, 455
Airtight bulkhead, 583
Albedo, snow and ice, 225
Alluvial apron, 6
Alluvial channels, 473
Alluvial fan, flood hazard, 246
Alluvial inlet, 69
Alluvial plain, 1, 2, 18
ALLUVIAL PLAINS,
  ENGINEERING GEOLOGY,
  1-2
Alluvial valley, 1, 2, 123
ALLUVIAL VALLEY ENGI-
  NEERING, **2-12**
Alluvial valley fill, 2
Alsman blocks, coastal
  engineering structure, 65, 66
Alternate bars, 477
Amphiboles, boxwork, 366
Amplification, earthquake wave
  train, 149
Analytical techniques,
  geochemistry, 210

Anchor blocks, pipe foundations,
  446
Anelastic behavior, minerals, 486
Anode, electrode, 159
Anomaly, geochemical, 208
Antidune, 475
Apparatus designs, rock
  mechanics, 485
Apparent age, geochronology, 216
Apparent angle of friction, 60
Apparent cohesion, 60
Applied geochemistry. *See*
  GEOCHEMISTRY,
  APPLIED
Applied geology. *See* GEOLOGY,
  APPLIED
Applied geomorphology. *See*
  GEOMORPHOLOGY,
  APPLIED
Applied oceanography. *See*
  OCEANOGRAPHY,
  APPLIED
Applied research, 238
Aquifer, definition of, 286
Aquifer exploration, 284
Aquifer storage, Mississippi
  Valley, 6
Argillaceous rock, classification
  of, 522
Aridisol, U.S. soil classification, 32
Aridity index, 13
ARID LANDS, ENGINEERING
  GEOLOGY, **12-25**
Arid zone, 12
Artesian conditions, ground water,
  286
Artesian pressure, 60
Articulated concrete mattresses,
  continuous bank protection,
  43
Artificial bypassing, 71
Artificial condensers, 616
Aseismic design, 150-152
Associated flow rule material, 547
Associations, societies, and
  institutes. *See*
  CONFERENCES, CON-
  GRESSES, AND SYMPOSIA
ASTM Method D-2167, 325
ASTM Test D-1556, 325
Atchafalaya River, 10
Atmospheric quality, 454
Atterberg, or consistency tests,
  532
Atterberg limits, 346, 519
ATTERBERG LIMITS AND
  INDICES, **25-27**

Attitudes, fractures in rocks, 491
Australian land system concept,
  262
Average, or mean photo scale, 395
Avulsion, 9
Axis of tilt, aerial photographs,
  394

Backfill, trucking, 425
Background, natural elemental
  concentrations, 208
Background research. *See*
  Oriented research
Backswamp clays, 11
Backswamp deposits, 11
Backswamp environment, 11
Badland drainage trench, longi-
  tudinal profile, 257
Baldwin Hills Storage Basin,
  U.S., 122
Bank protection
  channel modification, 4
  river-training structures, 43
Barbela Dam, Pakistan, 113
Bar fingers, 132
Bars, in alluvial channels, 477
Base-level plains, 18
Basin hydrological cycle, 298
Basin hydrology, 300
Bathyscaphes, 566
Bauxite, 140
Bay-sound deposits, Mississippi
  deltaic plain, 135
Beaches
  fill, 29
  gradient, 29
  sand, 138, 564
  shell, 38
BEACH REPLENISHMENT,
  ARTIFICIAL, **28-31**
Bearing capacity
  factors, 548
  marine soils, 349
  pile, 197-198
Bearing pressure, allowable, 548
Bed configuration, river, 474
Bedforms, river, 473
Bedrock geology, 595-596
Bed roughness
  sand channels, 475
  streams, 474
"Beef" roads, Australia, 141
Belled-out caisson, 193
Bell hole protection, 424
Belt apparatus, rock mechanics,
  483
Bendora Dam, Australia, 116-117

631